Theoretical and Mathematical Physics

The series founded in 1975 and formerly (until 2005) entitled *Texts and Monographs in Physics* (TMP) publishes high-level monographs in theoretical and mathematical physics. The change of title to *Theoretical and Mathematical Physics* (TMP) signals that the series is a suitable publication platform for both the mathematical and the theoretical physicist. The wider scope of the series is reflected by the composition of the editorial board, comprising both physicists and mathematicians.

The books, written in a didactic style and containing a certain amount of elementary background materials, bridge the gap between advanced textbooks and research monographs. They can thus serve as basis for advanced studies, not only for lectures and seminars at graduate level, but also for scientists entering a field of research.

Jiří Blank · Pavel Exner · Miloslav Havlíček

Hilbert Space Operators in Quantum Physics

Second Edition

AIP PRESS

Springer

Jiří Blank†
Charles University
Prague
Czechia

Pavel Exner
Doppler Institute
Břehová 7
11519 Prague
and Nuclear Physics Institute
Czech Academy of Sciences
25068 Řež near Prague
Czech Republic
exner@ujf.cas.cz

Miloslav Havlíček
Doppler Institute
and Faculty of Nuclear Sciences
and Physical Engineering
Czech Technical University
Trojanova 13
12000 Prague
Czech Republic
havlicek@fjfi.cvut.cz

ISBN 978-90-481-8012-7 e-ISBN 978-1-4020-8870-4

Printed on acid-free paper

9 8 7 6 5 4 3 2 1

springer.com

To our wives and daughters

Preface to the second edition

Almost fifteen years later, and there is little change in our motivation. Mathematical physics of quantum systems remains a lively subject of intrinsic interest with numerous applications, both actual and potential.

In the preface to the first edition we have described the origin of this book rooted at the beginning in a course of lectures. With this fact in mind, we were naturally pleased to learn that the volume was used as a course text in many points of the world and we gladly accepted the offer of *Springer Verlag* which inherited the rights from our original publisher, to consider preparation of a second edition.

It was our ambition to bring the reader close to the places where real life dwells, and therefore this edition had to be more than a corrected printing. The field is developing rapidly and since the first edition various new subjects have appeared; as a couple of examples let us mention quantum computing or the major progress in the investigation of random Schrödinger operators. There are, however, good sources in the literature where the reader can learn about these and other new developments.

We decided instead to amend the book with results about new topics which are less well covered, and the same time, closer to the research interests of one of us. The main change here are two new chapters devoted to quantum waveguides and quantum graphs. Following the spirit of this book we have not aspired to full coverage — each of these subjects would deserve a separate monograph — but we have given a detailed enough exposition to allow the interested reader to follow (and enjoy) fresh research results in this area. In connection with this we have updated the list of references, not only in the added chapters but also in other parts of the text in the second part of the book where we found it appropriate.

Naturally we have corrected misprints and minor inconsistencies spotted in the first edition. We thank the colleagues who brought them to our attention, in particular to Jana Stará, who indicated numerous improvements. As with the first edition, we have asked a native speaker to try to remove the foreign "accent" from our writing; we are grateful to Mark Harmer for accepting this role.

Prague, December 2007

Pavel Exner
Miloslav Havlíček

Preface

Relations between mathematics and physics have a long and entangled tradition. In spite of repeated clashes resulting from the different aims and methods of the two disciplines, both sides have always benefitted. The place where contacts are most intensive is usually called mathematical physics, or if you prefer, physical mathematics. These terms express the fact that mathematical methods are needed here more to understand the essence of problems than as a computational tool, and conversely, the investigated properties of physical systems are also inspiring from the mathematical point of view.

In fact, this field does not need any advocacy. When A. Wightman remarked a few years ago that it had become "socially acceptable", it was a pleasant understatement; by all accounts, mathematical physics is flourishing. It has long left the adolescent stage when it cherished only oscillating strings and membranes; nowadays it has built synapses to almost every part of physics. Evidence that the discipline is developing actively is provided by the fruitful oscillation between the investigation of particular systems and synthetizing generalizations, as well as by discoveries of new connections between different branches.

The drawback of this rapid development is that it has become virtually impossible to write a textbook on mathematical physics as a single topic. There are, of course, books which cover a wide range of problems, some of them indeed monumental, but even they are like cities which govern the territory while watching the frontier slowly moving towards the gray distance. This is simply the price we have to pay for the flood of ideas, concepts, tools, and results that our science is producing.

It was not our aim to write a poor man's version of some of the big textbooks. What we want is to give students basic information about the field, by which we mean an amount of knowledge that could constitute the basis of an intensive one–year course for those who already have the necessary training in algebra and analysis, as well as in classical and quantum mechanics. If our exposition should kindle interest in the subject, the student will be able, after taking such a course, to read specialized monographs and research papers, and to discover a research topic to his or her taste. We have mentioned that the span of the contemporary mathematical physics is vast; nevertheless the cornerstone remains where it was laid by J. von Neumann, H. Weyl, and the other founding fathers, namely in regions connected with quantum theory. Apart from its importance for fundamental problems such as the constitution of matter, this claim is supported by the fact that quantum theory is gradually becoming a basis for most branches of *applied* physics, and has in this way entered our everyday life.

The mathematical backbone of quantum physics is provided by the theory of linear operators on Hilbert spaces, which we discuss in the first half of this book. Here we follow a well–trodden path; this is why references in this part aim mostly at standard book sources, even for the few problems which maybe go beyond the standard curriculum. To make the exposition self–contained without burdening the main text, we have collected the necessary information about measure theory, integration, and some algebraic notions in the appendices.

The physical chapters in the second half are not intended to provide a self–contained exposition of quantum theory. As we have remarked, we suppose that the reader has background knowledge up to the level of a standard quantum mechanics course; the present text should rather provide new insights and help to reach a deeper understanding. However, we attempt to describe the mathematical foundations of quantum theory in a sufficiently complete way, so that a student coming from mathematics can start his or her way into this part of physics through our book.

In connection with the intended purpose of the text, the character of referencing changes in the second part. Though the material discussed here is with a few exceptions again standard, we try in the notes to each chapter to explain extensions of the discussed results and their relations to other problems; occasionally we have set traps for the reader's curiosity. The notes are accompanied by a selective but quite broad list of references, which map ways to the areas where real life dwells.

Each chapter is accompanied by a list of problems. Solving at least some of them in full detail is the safest way for the reader to check that he or she has indeed mastered the topic. The problem level ranges from elementary exercises to fairly complicated proofs and computations. We have refrained from marking the more difficult ones with asterisks because such a classification is always subjective, and after all, in real life you also often do not know in advance whether it will take you an hour or half a year to deal with a given problem.

Let us add a few words about the history of the book. It originates from courses of lectures we have given in different forms during the past two decades at Charles University and the Czech Technical University in Prague. In the 1970s we prepared several volumes of lecture notes; ten years later we returned to them and rewrote the material into a textbook, again in Czech. It was prepared for publication in 1989, but the economic turmoil which inevitably accompanied the welcome changes delayed its publication, so that it appeared only recently.

In the meantime we suffered a heavy blow. Our friend and coauthor, Jiří Blank, died in February 1990 at the age of 50. His departure reminded us of the bitter truth that we usually are able to appreciate the real value of our relationships with fellow humans only after we have lost them. He was always a stabilizing element of our triumvirate of authors, and his spirit as a devoted and precise teacher is felt throughout this book; we hope that in this indirect way his classes will continue.

Preparing the English edition was therefore left to the remaining two authors. It has been modified in many places. First of all, we have included two chapters and some other material which was prepared for the Czech version but then left out due to editorial restrictions. Though the aim of the book is not to report on the present state of research, as we have already remarked, the original manuscript was finished four years ago and we felt it was necessary to update the text and references in some places. On the other hand, since the audience addressed by the English text is different — and is equipped with different libraries — we decided to rewrite certain parts from the first half of the book in a more condensed form.

One consequence of these alterations was that we chose to do the translation ourselves. This decision contained an obvious danger. If you write in a language which you did not master during your childhood, the result will necessarily contain some unwanted comical twists reminiscent of the famous character of Leo Rosten. We are indebted to P. Moylan and, in particular, to R. Healey, who have read the text and counteracted our numerous petty attacks against the English language; those clumsy expressions that remain are, of course, our own.

There are many more people who deserve our thanks: coauthors of our research papers, colleagues with whom we have had the pleasure of exchanging ideas, and simply friends who have supported us during difficult times. We should not forget about students in our courses who have helped just by asking questions; some of them have now become our colleagues. In view of the book complex history, the list should be very long. We prefer to thank all of them anonymously. However, since every rule should have an exception, let us name J. Dittrich, who read the manuscript and corrected numerous mistakes. Last but not least we want to thank our wives, whose patience and understanding made the writing of this book possible.

Prague, July 1993

Pavel Exner
Miloslav Havlíček

Contents

Chapter 1

Some notions from functional analysis

1.1 Vector and normed spaces

The notion of a vector space is obtained by axiomatization of the properties of the three–dimensional space of Euclidean geometry, or of configuration spaces of classical mechanics. A **vector** (or *linear*) **space** V is a set $\{x, y, \dots\}$ equipped with the operations of summation, $[x, y] \mapsto x + y \in V$, and multiplication by a complex or real number α, $[\alpha, x] \mapsto \alpha x \in V$, such that

(i) The summation is commutative, $x + y = y + x$, and associative, $(x + y) + z = x + (y + z)$. There exist a zero element $0 \in V$, and an inverse element $-x \in V$, to any $x \in V$ so that $x + 0 = x$ and $x + (-x) = 0$ holds for all $x \in V$.

(ii) $\alpha(\beta x) = (\alpha\beta)x$ and $1x = x$.

(iii) The summation and multiplication are distributive, $\alpha(x + y) = \alpha x + \alpha y$ and $(\alpha + \beta)x = \alpha x + \beta x$.

The elements of V are called *vectors*. The set of numbers (or *scalars*) in the definition can be replaced by any algebraic field F. Then we speak about a vector space over F, and in particular, about a *complex* and *real* vector space for $F = \mathbb{C}, \mathbb{R}$, respectively. A vector space without further specification in this book always means a complex vector space.

1.1.1 Examples: (a) The space $\mathbb{C}^n$ consists of n–tuples of complex numbers with the summation and scalar multiplication defined componentwise. In the same way, we define the real space $\mathbb{R}^n$.

(b) The space ℓ^p, $1 \le p \le \infty$, of all complex sequences $X := \{\xi_j\}_{j=1}^{\infty}$ such that $\sum_{j=1}^{\infty} |\xi_j|^p < \infty$ for $p < \infty$ and $\sup_j |\xi_j| < \infty$ if $p = \infty$, with the summation and scalar multiplication defined as above; the *Minkowski inequality* implies $X + Y \in \ell^p$ for $X, Y \in \ell^p$ (Problem 2).

(c) The space $C(J)$ of continuous complex functions on a closed interval $J \subset \mathbb{R}$ with $(\alpha f+g)(x) := \alpha f(x)+g(x)$. In a similar way, we define the space $C(X)$ of continuous functions on a compact X and spaces of *bounded* continuous functions on more general topological spaces (see the next two sections).

A *subspace* $L \subset V$ is a subset, which is itself a vector space with the same operations. A minimal subspace containing a given subset $M \subset V$ is called the *linear hull (envelope)* of M and denoted as M_{lin} or $\mathrm{lin}(M)$. Vectors $x_1, \dots, x_n \in V$ are **linearly independent** if $\alpha_1 x_1 + \dots + \alpha_n x_n = 0$ implies that all the numbers $\alpha_1, \dots, \alpha_n$ are zero; otherwise they are *linearly dependent*, which means some of them can be expressed as a linear combination of the others. A set $M \subset V$ is *linearly independent* if each of its finite subsets consists of linearly independent vectors.

This allows us to introduce the *dimension* of a vector space V as a maximum number of linearly independent vectors in V. Among the spaces mentioned in Example 1.1.1, $\mathbb{C}^n$ and $\mathbb{R}^n$ are n–dimensional ($\mathbb{C}^n$ is $2n$–dimensional as a real vector space) while the others are infinite–dimensional. A *basis* of a finite–dimensional V is any linearly independent set $B \subset V$ such that $B_{lin} = V$; it is clear that $\dim V = n$ *iff* V has a basis of n elements. Vector spaces V, V' are said to be (algebraically) *isomorphic* if there is a bijection $f : V \to V'$, which is linear, $f(\alpha x + y) = \alpha f(x) + f(y)$. Isomorphic spaces have the same dimension; for finite–dimensional spaces the converse is also true (Problem 3).

There are various ways to construct new vector spaces from given ones. Let us mention two of them:

(i) If $V_1, \dots, V_N$ are vector spaces over the same field; then we can equip the Cartesian product $V := V_1 \times \dots \times V_N$ with a summation and scalar multiplication defined by $\alpha[x_1, \dots, x_N] + [y_1, \dots, y_N] := [\alpha x_1 + y_1, \dots, \alpha x_N + y_N]$. The axioms are obviously satisfied; the resulting vector space is called the *direct sum* of $V_1, \dots, V_N$ and denoted as $V_1 \oplus \dots \oplus V_N$ or $\sum_j^{\oplus} V_j$. The same term and symbols are used if $V_1, \dots, V_N$ are subspaces of a given space V such that each $x \in V$ has a unique decomposition $x = x_1 + \dots + x_N$, $x_j \in V_j$.

(ii) If W is a subspace of a vector space V, we can introduce an equivalence relation on V by $x \sim y$ if $x - y \in W$. Defining the vector–space operations on the set $\tilde{V}$ of equivalence classes by $\alpha\tilde{x} + \tilde{y} := (\alpha x + y)^{\tilde{}}$ for some $x \in \tilde{x}$, $y \in \tilde{y}$, we get a vector space, which is called the *factor space* of V with respect to W and denoted as V/W.

1.1.2 Example: The space $\mathcal{L}^p(M, d\mu)$, $p \geq 1$, where μ is a non–negative measure, consists of all measurable functions $f : M \to \mathbb{C}$ satisfying $\int_M |f|^p d\mu < \infty$ with pointwise summation and scalar multiplication — *cf.* Appendix A.3. The subset $\mathcal{L}_0 \subset \mathcal{L}^p$ of the functions such that $f(x) = 0$ for μ–almost all $x \in M$ is easily seen to be a subspace; the corresponding factor space $L^p(M, d\mu) := \mathcal{L}^p(M, d\mu)/\mathcal{L}_0$ is then formed by the classes of μ–equivalent functions.

A map $f : V \to \mathbb{C}$ on a vector space V is called a *functional*; if it maps into the reals we speak about a *real* functional. A functional f is *additive* if $f(x+y) = f(x) + f(y)$ holds for all $x, y \in V$, and *homogeneous* if $f(\alpha x) = \alpha f(x)$ or *antihomogeneous* if $f(\alpha x) = \bar{\alpha} f(x)$ for $x \in V$, $\alpha \in \mathbb{C}$. An additive (anti)homogeneous functional is called *(anti)linear.* A real functional p is called a *seminorm* if $p(x+y) \le p(x) + p(y)$ and $p(\alpha x) = |\alpha| p(x)$ holds for any $x, y \in V$, $\alpha \in \mathbb{C}$; this definition implies that p maps V into $\mathbb{R}^+$ and $|p(x) - p(y)| \le p(x-y)$. The following important result is valid (see the notes to this chapter).

1.1.3 Theorem (Hahn–Banach): Let p be a seminorm on a vector space V. Any linear functional f_0 defined on a subspace $V_0 \subset V$ and fulfilling $|f_0(y)| \le p(y)$ for all $y \in V_0$ can be extended to a linear functional f on V such that $|f(x)| \le p(x)$ holds for any $x \in V$.

A map $F := V \times \cdots \times V \to \mathbb{C}$ is called a *form*, in particular, a *real* form if its range is contained in $\mathbb{R}$. A form $F : V \times V \to \mathbb{C}$ is *bilinear* if it is linear in both arguments, and *sesquilinear* if it is linear in one of them and antilinear in the other. Most frequently we shall drop the adjective when speaking about sesquilinear forms; we shall use the "physical" convention assuming that such a form is antilinear in the *left* argument. For a given F we define the *quadratic form* (generated by F) by $q_F : q_F(x) = F(x,x)$; the correspondence is one–to–one as the *polarization formula*

$$F(x,y) = \frac{1}{4}\Big(q_F(x+y) - q_F(x-y)\Big) - \frac{i}{4}\Big(q_F(x+iy) - q_F(x-iy)\Big)$$

shows. A form F is *symmetric* if $F(x,y) = \overline{F(y,x)}$ for all $x, y \in V$; it is *positive* if $q_F(x) \ge 0$ for any $x \in V$ and *strictly positive* if, in addition, $F(x) = 0$ holds for $x = 0$ only. A positive form is symmetric (Problem 6) and fulfils the **Schwarz inequality**,

$$|F(x,y)|^2 \le q_F(x) q_F(y) .$$

A *norm* on a vector space V is a seminorm $\|\cdot\|$ such that $\|x\| = 0$ holds *iff* $x = 0$. A pair $(V, \|\cdot\|)$ is called a **normed space**; if there is no danger of misunderstanding we shall speak simply about a normed space V.

1.1.4 Examples: (a) In the spaces $\mathbb{C}^n$ and $\mathbb{R}^n$, we introduce

$$\|x\|_\infty := \max_{1 \le j \le n} |\xi_j| \quad \text{and} \quad \|x\|_p := \left(\sum_{j=1}^n |\xi_j|^p\right)^{1/p}, \; p \ge 1,$$

for $x = \{\xi_1, \dots, \xi_n\}$; the norm $\|\cdot\|_2$ on $\mathbb{R}^n$ is often also denoted as $|\cdot|$. Analogous norms are used in ℓ^p (see also Problem 8).

(b) In $L^p(M, d\mu)$, we introduce

$$\|f\|_p := \left(\int_M |f|^p d\mu\right)^{1/p} .$$

The relation $\|f\|_p=0$ implies $f(x)=0$ μ–a.e. in M, so f is the zero element of $L^p(M,d\mu)$. If we speak about $L^p(M,d\mu)$ as a normed space, we always have in mind this natural norm though it is not, of course, the only possibility. If the measure μ is discrete with countable support, $L^p(M,d\mu)$ is isomorphic to ℓ^p and we recover the norm $\|\cdot\|_p$ of the previous example.

(c) By $L^\infty(M,d\mu)$ we denote the set of classes of μ–equivalent functions $f : M\to\mathbb{C}$, which are bounded a.e., *i.e.*, there is $c>0$ such that $|f(x)|\le c$ for μ–almost all $x\in M$. The infimum of all such numbers is denoted as $\sup\text{ess}\,|f(x)|$. We can easily check that $L^\infty(M,d\mu)$ is a vector space and $f\mapsto\|f\|_\infty := \sup\text{ess}_{\,x\in M}|f(x)|$ is a norm on it.

(d) The space $C(X)$ can be equipped with the norm $\|f\|_\infty := \sup_{x\in X}|f(x)|$.

A strictly positive sesquilinear form on a vector space V is called an **inner** (or **scalar**) **product**. In other words, it is a map $(\cdot,\cdot)$ from $V\times V$ to $\mathbb{C}$ such that the following conditions hold for any $x,y,z\in V$ and $\alpha\in\mathbb{C}$:

(i) $(x,\alpha y+z)=\alpha(x,y)+(x,z)$

(ii) $(x,y)=\overline{(y,x)}$

(iii) $(x,x)\ge 0$ and $(x,x)=0$ *iff* $x=0$

A vector space with an inner product is called a **pre–Hilbert space**. Any such space is at the same time a normed space with the norm $\|x\| := \sqrt{(x,x)}$; the *Schwarz inequality* then assumes the form

$$|(x,y)|\;\le\;\|x\|\,\|y\|\,.$$

The above norm is said to be *induced by the inner product.* Due to conditions (i) and (ii) it fulfils the *parallelogram identity,*

$$\|x+y\|^2+\|x-y\|^2\;=\;2\|x\|^2+2\|y\|^2\,;$$

on the other hand, it allows us to express the inner product by polarization,

$$(x,y)\;=\;\frac{1}{4}\Big(\|x+y\|^2-\|x-y\|^2\Big)\;-\;\frac{i}{4}\Big(\|x+iy\|^2-\|x-iy\|^2\Big)\,.$$

These properties are typical for a norm induced by an inner product (Problem 11).

Vectors x,y of a pre–Hilbert space V are called *orthogonal* if $(x,y)=0$. A vector x is *orthogonal to a set* M if $(x,y)=0$ holds for all $y\in M$; the set of all such vectors is denoted as $M^\perp$ and called the *orthogonal complement* to M. Inner–product linearity implies that it is a subspace, $(M^\perp)_{lin}=M^\perp$, with the following simple properties

$$(M_{lin})^\perp=M^\perp\,,\quad M_{lin}\subset(M^\perp)^\perp\,,\quad M\subset N\Rightarrow M^\perp\supset N^\perp\,.$$

A set M of nonzero vectors whose every two elements are orthogonal is called an *orthogonal set*; in particular, M is *orthonormal* if $\|x\| = 1$ for each $x \in M$. Any orthonormal set is obviously linearly independent, and in the opposite direction we have the following assertion, the proof of which is left to the reader.

1.1.5 Theorem (Gram-Schmidt): Let N be an at most countable linearly independent set in a pre–Hilbert space V, then there is an orthonormal set $M \subset V$ of the same cardinality such that $M_{lin} = N_{lin}$.

1.2 Metric and topological spaces

A *metric* on a set X is a map $\varrho : X \times X \to [0, \infty)$, which is symmetric, $\varrho(x,y) = \varrho(y,x)$, $\varrho(x,y) = 0$ *iff* $x = y$, and fulfils the *triangle inequality,*

$$\varrho(x,z) \leq \varrho(x,y) + \varrho(y,z)\,,$$

for any $x, y, z \in X$; the pair (X, ϱ) is called a **metric space** (we shall again for simplicity often use the symbol X only). If X is a normed space, one can define a metric on X by $\varrho(x,y) := \|x-y\|$; we say it is *induced by the norm* (see also Problems 15 and 16).

Let us first recall some basic notions and properties of metric spaces. An ε–*neighborhood* of a point $x \in X$ is the open ball $U_\varepsilon(x) := \{\, y \in X \,:\, \varrho(y,x) < \varepsilon \}$. A point x is an *interior point* of a set M if there is a $U_\varepsilon(x) \subset M$. A set is *open* if all its points are interior points, in particular, any neighborhood of a given point is open. A union of an arbitrary family of open sets is again an open set; the same is true for *finite* intersections of open sets.

The *closure* $\overline{M}$ of a set M is the family of all points $x \in X$ such that the intersection $U_\varepsilon(x) \cap M \neq \emptyset$ for any $\varepsilon > 0$. A point $x \in \overline{M}$ is called *isolated* if there is $U_\varepsilon(x)$ such that $U_\varepsilon(x) \cap M = \{x\}$, otherwise x is a *limit* (or *accumulation*) point of M. The closure points of M which are not interior form the *boundary* $\operatorname{bd} M$ of M. A set is *closed* if it coincides with its closure, and $\overline{M}$ is *the smallest closed set* containing M (*cf.* Problem 17). In particular, the whole X and the empty set $\emptyset$ are closed and open at the same time.

A set M is said to be *dense in a set* $N \subset X$ if $\overline{M} \supset N$; it is *everywhere dense* if $\overline{M} = X$ and *nowhere dense* if $X \setminus \overline{M}$ is everywhere dense. A metric space which contains a countable everywhere dense set is called *separable.* An example is the space $\mathbb{C}^n$ with any of the norms of Example 1.1.4a where a dense set is formed, *e.g.*, by n–tuples of complex numbers with rational real and imaginary parts; other examples will be given in the next chapter (see also Problem 18).

A sequence $\{x_n\} \subset X$ *converges to a point* $x \in X$ if to any $U_\varepsilon(x)$ there is n_0 such that $x_n \in U_\varepsilon(x)$ holds for all $n > n_0$. Since any two mutually different points $x, y \in X$ have disjoint neighborhoods, each sequence has at most one limit. Sequences can also be used to characterize closure of a set (Problem 17).

Next we recall a few notions related to maps $f : X \to X'$ of metric spaces. The map f is *continuous at a point* $x \in X$ if to any $U'_\varepsilon(f(x))$ there is a $U_\delta(x)$ such that $f(U_\delta(x)) \subset U'_\varepsilon(f(x))$; alternatively we can characterize the local continuity using sequences (Problem 19). On the other hand, f is (*globally*) *continuous* if the pull–back $f^{(-1)}(G')$ of any open set $G' \subset X'$ is open in X.

An important class of continuous maps is represented by *homeomorphisms, i.e.*, bijections $f : X \to X'$ such that both f and f^{-1} are continuous. It is clear that in this way any family of metric spaces can be divided into equivalence classes. A homeomorphism maps, in particular, the family τ of open sets in X bijectively onto the family τ' of open sets in X'; we say that homeomorphic metric spaces are *topologically equivalent*. Such spaces can still differ in metric properties. As an example, consider the spaces $\mathbb{R}$ and $(-\frac{\pi}{2}, \frac{\pi}{2})$ with the same metric $\varrho(x,y) := |x-y|$; they are homeomorphic by $x \mapsto \arctan x$ but only the first of them contains unbounded sets. A bijection $f := X \to X'$ which preserves the metric properties, $\varrho'(f(x), f(y)) = \varrho(x,y)$, is called *isometry*; this last named property implies continuity, so any isometry is a homeomorphism.

A homeomorphism $f : V \to V'$ of normed spaces is called *linear homeomorphism* if it is simultaneously an isomorphism. Linearly homeomorphic spaces therefore also have the same algebraic structure; this is particularly simplifying in the case of finite dimension (Problem 21). In addition, if the identity $\|f(x)\|_{V'} = \|x\|_V$ holds for any $x \in V$ we speak about a *linear isometry.*

A sequence $\{x_n\}$ in a metric space X is called *Cauchy* if to any $\varepsilon > 0$ there is n_ε such that $\varrho(x_n, x_m) < \varepsilon$ for all $n, m > n_\varepsilon$. In particular, any convergent sequence is Cauchy; a metric space in which the converse is also true is called *complete.* Completeness is one of the basic "nontopological" properties of metric spaces: recall the spaces $\mathbb{R}$ and $(-\frac{\pi}{2}, \frac{\pi}{2})$ mentioned above; they are homeomorphic but only the first of them is complete.

1.2.1 Example: Let us check the *completeness of* $L^p(M, d\mu)\,, p \ge 1$, *with a σ–finite measure* μ. Suppose first $\mu(M) < \infty$ and consider a Cauchy sequence $\{f_n\} \subset L^p$. By the Hölder inequality, it is Cauchy also with respect to $\|\cdot\|_1$, so for any $\varepsilon > 0$ there is $N(\varepsilon)$ such that $\|f_n - f_m\|_1 < \varepsilon$ for $n, m > N(\varepsilon)$. We pick a subsequence, $g_n := f_{k_n}$, by choosing $k_1 := N(2^{-1})$ and $k_{n+1} := \max\{k_n + 1,\, N(2^{-n-1})\}$, so $\|g_{n+1} - g_n\|_1 < 2^{-n}$, and the functions $\varphi_n := |g_1| + \sum_{\ell=1}^{n-1} |g_{\ell+1} - g_\ell|$ obey

$$\int_M \varphi_n d\mu \le \|g_1\|_1 + \sum_{\ell=1}^{n-1} 2^{-\ell} < 1 + \|g_1\|_1 \,.$$

Since they are measurable and form a nondecreasing sequence, the monotone–convergence theorem implies existence of a finite $\lim_{n\to\infty} \varphi_n(x)$ for μ–a.a. $x \in M$. Furthermore, $|g_{n+p} - g_n| \le \varphi_{n+p} - \varphi_n$, so there is a function f which is finite μ-a.e. in M and fulfils $f(x) = \lim_{n\to\infty} g_n(x)$. The sequence $\{g_n\}$ has been picked from a Cauchy sequence and it is therefore Cauchy also, $\|g_n - g_m\|_p < \varepsilon$ for all $n, m > \tilde{N}(\varepsilon)$ for a suitable $\tilde{N}(\varepsilon)$. On the other hand, $\lim_{m\to\infty} |g_n(x) - g_m(x)|^p = |g_n(x) - f(x)|^p$ for μ–a.a. $x \in M$, so Fatou's lemma implies $\|g_n - f\|_p \le \varepsilon$ for all $n > \tilde{N}(\varepsilon)$; hence $f \in L^p$ and $\lim_{n\to\infty} \|f_n - f\|_p = 0$ (Problem 24).

If μ is σ–finite and $\mu(M)=\infty$, there is a disjoint decomposition $\bigcup_{j=1}^{\infty} M_j = M$ with $\mu(M_j)<\infty$. The already proven completeness of $L^p(M_j,d\mu)$ implies the existence of functions $f^{(j)}\in L^p(M_j,d\mu)$ which fulfil $\|f_n^{(j)}-f^{(j)}\|_p\to 0$ as $n\to\infty$; then we can proceed as in the proof of completeness of ℓ^p (*cf.* Problem 23).

Other examples of complete metric spaces are given in Problem 23. Any metric space can be extended to become complete: a complete space (X',ϱ') is called the *completion* of (X,ϱ) if (i) $X\subset X'$ and $\varrho'(x,y)=\varrho(x,y)$ for all $x,y\in X$, and (ii) the set X is everywhere dense in X' (this requirement ensures minimality — *cf.* Problem 25).

1.2.2 Theorem: Any metric space (X,ϱ) has a completion. If $(\tilde X,\tilde\varrho)$ is another completion of (X,ϱ), there is an isometry $f: X'\to\tilde X$ which preserves X, *i.e.*, $f(x)=x$ for all $x\in X$.

Sketch of the proof: Uniqueness follows directly from the definition. Existence is proved constructively by the so–called *standard completion procedure* which generalizes the Cantor construction of the reals. We start from the set of all Cauchy sequences in (X,ϱ). This can be factorized if we set $\{x_j\}\sim\{y_j\}$ for the sequences with $\lim_{j\to\infty}\varrho(x_j,y_j)=0$. The set of equivalence classes we denote as X^* and define $\varrho^*([x],[y]) := \lim_{j\to\infty}\varrho(x_j,y_j)$ to any $[x],[y]\in X^*$. Finally, one has to check that this definition makes sense, *i.e.*, that ϱ^* does not depend on the choice of sequences representing the classes $[x],[y]$, ϱ^* is a metric on X^*, and (X^*,ϱ^*) satisfies the requirements of the definition. ∎

The notion of topology is obtained by axiomatization of some properties of metric spaces. Let X be a set and τ a family of its subsets which fulfils the following conditions *(topology axioms):*

(t1) $X\in\tau$ and $\emptyset\in\tau$.

(t2) If I is any index set and $G_\alpha\in\tau$ for all $\alpha\in I$; then $\bigcup_{\alpha\in I} G_\alpha\in\tau$.

(t3) $\bigcap_{j=1}^{n} G_j\in\tau$ for any finite subsystem $\{G_1,\dots,G_n\}\subset\tau$.

The family τ is called a *topology*, its elements *open sets* and the set X equipped with a topology is a **topological space**; when it is suitable we write (X,τ).

A family of open sets in a metric space (X,ϱ) is a topology by definition; we speak about the *metric–induced topology* τ_ϱ, in particular, the *norm–induced topology* if X is a vector space and ϱ is induced by a norm. On the other hand, finding the conditions under which a given topology is induced by a metric is a nontrivial problem (see the notes). Two extreme topologies can be defined on any set X: the *discrete topology* $\tau_d := 2^X$, *i.e.*, the family of all subsets in X, and the *trivial topology* $\tau_0 := \{\emptyset, X\}$. The first of them is induced by the discrete metric, $\varrho_d(x,y):=1$ for $x\neq y$, while (X,τ_0) is not metrizable unless X is a one–point set.

An open set in a topological space X containing a point x or a set $M\subset X$ is called a *neighborhood* of the point X or the set M, respectively. Using this concept, we can adapt to topological spaces most of the "metric" definitions presented above, as well as some simple results such as those of Problems 17a, c, 19b, topological equivalence of homeomorphic spaces, *etc.* On the other hand, equally elementary metric–space properties may not be valid in a general topological space.

1.2.3 Example: Consider the topologies τ_{fin} and τ_{count} on $X = [0,1]$ in which the closed sets are all finite and almost countable subsets of X, respectively. If $\{x_n\} \subset X$ is a simple sequence, $x_n \neq x_m$ for $n \neq m$; then any neighborhood $U(x)$ contains all elements of the sequence with the exception of a finite number; hence *the limit is not unique* in (X, τ_{fin}). This is not the case in (X, τ_{count}) but there only very few sequences converge, namely those with $x_n = x_{n_0}$ for all $n \geq n_0$, which means, in particular, that we cannot use sequences to characterize local continuity or points of the closure.

Some of these difficulties can be solved by introducing a more general notion of convergence. A partially ordered set I is called *directed* if for any $\alpha, \beta \in I$ there is $\gamma \in I$ such that $\alpha \prec \gamma$ and $\beta \prec \gamma$. A map of a directed index set I into a topological space X, $\alpha \mapsto x_\alpha$, is called a *net* in X. A net $\{x_\alpha\}$ is said to *converge* to a point $x \in X$ if to any neighborhood $U(x)$ there is an $\alpha_0 \in I$ such that $x_\alpha \in U(x)$ for all $\alpha \succ \alpha_0$. To illustrate that nets in a sense play the role that sequences played in metric spaces, let us mention two simple results the proofs of which we leave to the reader (Problem 29).

1.2.4 Proposition: Let (X, τ), (X', τ') be topological spaces; then

(a) A point $x \in X$ belongs to the closure of a set $M \subset X$ *iff* there is a net $\{x_\alpha\} \subset M$ such that $x_\alpha \to x$.

(b) A map $f : X \to X'$ is continuous at a point $x \in X$ *iff* the net $\{f(x_\alpha)\}$ converges to $f(x)$ for any net $\{x_\alpha\}$ converging to x.

Two topologies can be compared if there is an inclusion between them, $\tau_1 \subset \tau_2$, in which case we say that τ_1 is *weaker* (*coarser*) than τ_2; while the latter is *stronger* (*finer*) than τ_1. Such a relation between topologies has some simple consequences — see, *e.g.*, Problem 32. In particular, continuity of a map $f : X \to Y$ is preserved when we make the topology in Y weaker or in X stronger. In other cases it may not be preserved; for instance, Problem 3.9 gives an example of three topologies, $\tau_w \subset \tau_s \subset \tau_u$, on a set $X := \mathcal{B}(\mathcal{H})$ and a map $f : X \to X$ which is continuous with respect to τ_w and τ_u but not τ_s.

1.2.5 Example: A frequently used way to construct a topology on a given X employs a family $\mathcal{F}$ of maps from X to a topological space $(\tilde{X}, \tilde{\tau})$. Among all topologies such that each $f \in \mathcal{F}$ is continuous there is one which is the weakest; its existence follows from Problem 30, where the system $\mathcal{S}$ consists of the sets $f^{(-1)}G)$ for each $G \subset \tilde{\tau}$, $f \in \mathcal{F}$. We call this the *$\mathcal{F}$–weak topology.*

For any set M in a topological space (X, τ) we define the *relative topology* τ_M as the family of intersections $M \cap G$ with $G \subset \tau$; the space (M, τ_M) is called a *subspace* of (X, τ). Other important notions are obtained by axiomatization of properties of open balls in metric spaces. A family $\mathcal{B} \subset \tau$ is called a *basis* of a topological space (X, τ) if any nonempty open set can be expressed as a union of elements of $\mathcal{B}$. A family $\mathcal{B}_x$ of neighborhoods of a given point $x \in X$ is called a *local basis* at x if any neighborhood $U(x)$ contains some $B \in \mathcal{B}_x$. A trivial example of both a basis and a local basis is the topology itself; however, we are naturally more interested in cases where bases are rather a "small part" of it. It is easy to see that local bases can be used to compare topologies.

1.2.6 Proposition: Let a set X be equipped with topologies τ, τ' with local bases $\mathcal{B}_x$, $\mathcal{B}'_x$ at each $x \in X$. The inclusion $\tau \subset \tau'$ holds *iff* for any $B \in \mathcal{B}_x$ there is $B' \in \mathcal{B}'_x$ such that $B' \subset B$.

To be a basis of a topology or a local basis, a family of sets must meet certain consistency requirements (*cf.* Problem 30c, d); this is often useful when we define a particular topology by specifying its basis.

1.2.7 Example: Let (X_j, τ_j), $j = 1, 2$, be topological spaces. On the Cartesian product $X_1 \times X_2$ we define the standard topology $\tau_{X_1 \times X_2}$ determined by τ_j, $j = 1, 2$, as the weakest topology which contains all sets $G_1 \times G_2$ with $G_j \in \tau_j$, *i.e.*, $\tau_{X_1 \times X_2} := \tau(\tau_1 \times \tau_2)$ in the notation of Problem 30b. Since $(A_1 \times A_2) \cap (B_1 \times B_2) = (A_1 \cap B_1) \times (A_2 \cap B_2)$, the family $\tau_1 \times \tau_2$ itself is a basis of $\tau_{X_1 \times X_2}$; a local basis at $[x_1, x_2]$ consists of the sets $U(x_1) \times V(x_2)$, where $U(x_1) \in \tau_1$, $V(x_2) \in \tau_2$ are neighborhoods of the points x_1, x_2, respectively. The space $(X_1 \times X_2,\, \tau_{X_1 \times X_2})$ is called the *topological product* of the spaces (X_j, τ_j), $j = 1, 2$.

Bases can also be used to classify topological spaces by the so–called countability axioms. A space (X, τ) is called *first countable* if it has a countable local basis at any point; it is *second countable* if the whole topology τ has a countable basis. The second requirement is actually stronger; for instance, a nonseparable metric space is first but not second countable (*cf.* Problem 18; some related results are collected in Problem 31). The most important consequence of the existence of a countable local basis, $\{U_n(x) : n = 1, 2, \dots\} \subset \tau$, is that one can pass to another local basis $\{V_n(x) : n = 1, 2, \dots\}$, which is ordered by inclusion, $V_{n+1} \subset V_n$, setting $V_1 := U_1$ and $V_{n+1} := V_n \cap U_{n+1}$. This helps to partially rehabilitate *sequences* as a tool in checking topological properties (Problem 33a).

The other problem mentioned in Example 1.2.3, namely the possible nonuniqueness of a sequence limit, is not related to the cardinality of the basis but rather to the degree to which a given topology separates points. It provides another classification of topological spaces through *separability axioms:*

T_1 To any $x, y \in X$, $x \neq y$, there is a neighborhood $U(x)$ such that $y \notin U(x)$.

T_2 To any $x, y \in X$, $x \neq y$, there are disjoint neighborhoods $U(x)$ and $U(y)$.

T_3 To any closed set $F \subset X$ and a point $x \notin F$, there are disjoint neighborhoods $U(x)$ and $U(F)$.

T_4 To any pair of disjoint closed sets F, F', there are disjoint neighborhoods $U(F)$ and $U(F')$.

A space (X, τ) which fulfils the axioms T_1 and T_j is called *T_j–space*, T_2–spaces are also called *Hausdorff*, T_3–spaces are *regular,* and T_4–spaces are *normal*. For instance, the spaces of Example 3 are T_1 but not Hausdorff; one can find examples showing that the whole hierarchy is nontrivial (see the notes). In particular, any metric space is normal. The question of limit uniqueness that we started with is answered affirmatively in Hausdorff spaces (see Problem 29).

1.3 Compactness

One of the central points in an introductory course of analysis is the Heine–Borel theorem, which claims that given a family of open intervals covering a closed bounded set $F \subset \mathbb{R}$, we can select a finite subsystem which also covers F. The notion of compactness comes from axiomatization of this result. Let M be a set in a topological space (X, τ). A family $\mathcal{P} := \{M_\alpha : \alpha \in I\} \subset 2^X$ is a *covering* of M if $\bigcup_{\alpha \in I} M_\alpha \supset M$; in dependence on the cardinality of the index set I the covering is called finite, countable, *etc.*We speak about an *open* covering if $\mathcal{P} \subset \tau$. The set M is **compact** if an arbitrary open covering of M has a *finite* subsystem that still covers M; if this is true for the whole of X we say that the topological space (X, τ) is compact. It is easy to see that compactness of M is equivalent to compactness of the space (M, τ_M) with the induced topology, so it is often sufficient to formulate theorems for compact spaces only.

1.3.1 Proposition: Let (X, τ) be a compact space, then

(a) Any infinite set $M \subset X$ has at least one accumulation point.

(b) Any closed set $F \subset X$ is compact.

(c) If a map $f : (X, \tau) \to (X', \tau')$ is continuous, then $f(X)$ is compact in (X', τ').

Proof: To check (a) it is obviously sufficient to consider countable sets. Suppose $M = \{x_n : n = 1, 2, \dots\}$ has no accumulation points; then the same is true for the sets $M_N := \{x_n : n \geq N\}$. They are therefore closed and their complements form an open covering of X with no finite subcovering. Further, let $\{G_\alpha\}$ be an open covering of F; adding the set $G := X \setminus F$ we get an open covering of X. Any finite subcovering $\mathcal{G}$ of X is either contained in $\{G_\alpha\}$ or it contains the set G; in the latter case $\mathcal{G} \setminus G$ is a finite covering of the set F. Finally, the last assertion follows from the appropriate definitions. ▮

Part (a) of the proposition represents a particular case of a more general result (see the notes) which can be used to define compactness; another alternative definition is given in Problem 36. Compactness has an important implication for the way in which the topology separates points.

1.3.2 Theorem: A compact Hausdorff space is normal.

Proof: Let F, R be disjoint closed sets and $y \in R$. By assumption, to any $x \in F$ one can find disjoint neighborhoods $U_y(x)$ and $U_x(y)$. The family $\{U_y(x) : x \in F\}$ covers the set F, which is compact in view of the previous proposition; hence there is a finite subsystem $\{U_y(x_j) : j = 1, \dots, n\}$ such that $U_y(F) := \bigcup_{j=1}^n U_y(x_j)$ is a neighborhood of F. Moreover, $U(y) := \bigcap_{j=1}^n U_{x_j}(y)$ is a neighborhood of the point y and $U(y) \cap U_y(F) = \emptyset$. This can be done for any point $y \in R$ giving an open covering $\{U(y) : y \in R\}$ of the set R; from it we select again a finite subsystem $\{U(y_k) : k = 1, \dots, m\}$ such that $U(R) := \bigcup_{k=1}^m U(y_k)$ is a neighborhood of R which has an empty intersection with $U(F) := \bigcap_{k=1}^m U_{y_k}(F)$. ∎

1.3.3 Theorem: Let X be a Hausdorff space, then

(a) Any compact set $F \subset X$ is closed.

(b) If the space X is compact, then any continuous bijection $f : X \to X'$ for X' Hausdorff is a homeomorphism.

Proof: If $y \notin F$, the neighborhood $U(y)$ from the preceding proof has an empty intersection with F, so $y \notin \overline{F}$. To prove (b) we have to check that $f(F)$ is closed in X' for any closed $F \subset X$; this follows easily from (a) and Proposition 1.3.1c. ∎

A set M in a topological space is called *precompact* (or *relatively compact*) if $\overline{M}$ is compact. A space X is *locally compact* if any point $x \in X$ has a precompact neighborhood; it is *σ–compact* if any *countable* covering has a finite subcovering.

Let us now turn to compactness in metric spaces. There, any compact set is closed by Theorem 1.3.3 and bounded — from an unbounded set we can always select an infinite subset which has no accumulation point. However, these conditions are not sufficient. For instance, the closed ball $S_1(0)$ in ℓ^2 is bounded but not compact: its subset consisting of the points $X_j := \{\delta_{jk}\}_{k=1}^{\infty}$, $j = 1, 2, \dots$, has no accumulation point because $\|X_j - X_k\| = \sqrt{2}$ holds for all $j \neq k$.

To be able to characterize compactness by metric properties we need a stronger condition. Given a set M in a metric space (X, ϱ) and $\varepsilon > 0$, we call a set N_ε an *ε–lattice* for M if to any $x \in M$ there is a $y \in N_\varepsilon$ such that $\varrho(x, y) \le \varepsilon$ (N_ε may not be a subset of M but by using it one is able to construct a 2ε–lattice for M which is contained in M). A set M is *completely bounded* if it has a *finite* ε–lattice for any $\varepsilon > 0$; if the set X itself is completely bounded we speak about a completely bounded metric space. If M is completely bounded, the same is obviously true for $\overline{M}$. Any completely bounded set is bounded; on the other hand, any infinite orthonormal set in a pre–Hilbert space represents an example of a set which is bounded but not completely bounded.

1.3.4 Proposition: A σ–compact metric space is completely bounded. A completely bounded metric space is separable.

Proof: Suppose that for some $\varepsilon > 0$ there is no finite ε–lattice. Then $X \setminus S_\varepsilon(x_1) \neq \emptyset$ for an arbitrarily chosen $x_1 \in X$, otherwise $\{x_1\}$ would be an ε–lattice for X. Hence there is $x_2 \in X$ such that $\varrho(x_1, x_2) > \varepsilon$ and we have $X \setminus (S_\varepsilon(x_1) \cup S_\varepsilon(x_2)) \neq \emptyset$ *etc.*; in this way we construct an infinite set $\{x_j : j = 1, 2, \dots\}$ which fulfils $\varrho(x_j, x_k) > \varepsilon$ for all $j \neq k$, and therefore it has no accumulation points. As for the second part, if N_n is a $(1/n)$–lattice for X, then $\bigcup_{n=1}^\infty N_n$ is a countable everywhere dense set. ∎

1.3.5 Corollary: Let X be a metric space; then the following conditions are equivalent:

(i) X is compact.

(ii) X is σ–compact.

(iii) Any infinite set in X has an accumulation point.

1.3.6 Theorem: A metric space is compact *iff* it is complete and completely bounded.

Proof: Let X be compact; in view of Proposition 1.3.4 it is sufficient to show that it is complete. If $\{x_n\}$ is Cauchy, the compactness implies existence of a convergent subsequence so $\{x_n\}$ is also convergent (Problem 24). On the other hand, to prove the opposite implication we have to check that any $M := \{x_n : n = 1, 2, \dots\} \subset X$ has an accumulation point. By assumption, there is a finite 1–lattice N_1 for X, hence there is $y_1 \in N_1$ such that the closed ball $S_1(y_1)$ contains an infinite subset of M. The ball $S_1(y_1)$ is completely bounded, so we can find a finite $(1/2)$–lattice $N_2 \subset S_1(y_1)$ and a point $y_2 \in N_2$ such that the set $S_{1/2}(y_2) \cap M$ is infinite. In this way we get a sequence of closed balls $S_n := S_{2^{1-n}}(y_n)$ such that each of them contains infinitely many points of M and their centers fulfil $y_{n+1} \in S_n$. The closed balls of doubled radii then satisfy $S_{2^{1-n}}(y_{n+1}) \subset S_{2^{2-n}}(y_n)$ and M has an accumulation point in view of Problem 26. ∎

1.3.7 Corollary: (a) A set M in a complete metric space X is precompact *iff* it is completely bounded. In particular, if X is a finite–dimensional normed space, then M is precompact *iff* it is bounded.

(b) A continuous real–valued function f on a compact topological space X is bounded and assumes its maximum and minimum values in X.

Proof: The first assertion follows from Problem 25. If $\overline{M}$ is compact, it is bounded so M is also bounded. To prove the opposite implication in a finite–dimensional normed space, we can use the fact that such a space is topologically isomorphic to $\mathbb{C}^n$ (or $\mathbb{R}^n$ in the case of a real normed space — see Problem 21). As for part (b), the set $f(X) \subset \mathbb{R}$ is compact by Proposition 1.3.1c, and therefore bounded. Denote

$\alpha := \sup_{x\in X} f(x)$ and let $\{x_n\} \subset X$ be a sequence such that $f(x_n) \to \alpha$. Since X is compact there is a subsequence $\{x_{k_n}\}$ converging to some x_s and the continuity implies $f(x_s) = \alpha$. In the same way we can check that f assumes a minimum value. ∎

1.4 Topological vector spaces

We can easily check that the operations of summation and scalar multiplication in a normed space are continuous. Let us now see what would follow from such a requirement when we combine algebraic and topological properties. A vector space V equipped with a topology τ is called a **topological vector space** if

(tv1) The summation maps continuously $(V\times V,\, \tau_{V\times V})$ to (V,τ).

(tv2) The scalar multiplication maps continuously $(\mathbb{C}\times V,\, \tau_{\mathbb{C}\times V})$ to (V,τ).

(tv3) (V,τ) is Hausdorff.

In the same way, we define a topological vector space over any field. Instead of (tv3), we may demand T_1–separability only because the first two requirements imply that T_3 is valid (Problem 39).

A useful tool in topological vector spaces is the family of translations, $t_x : V \to V$, defined for any $x \in V$ by $t_x(y) := x+y$. Since $t_x^{-1} = t_{-x}$, the continuity of summation implies that any translation is a homeomorphism; hence if G is an open set, then $x + G := t_x(G)$ is open for all $x \in V$; in particular, U is a neighborhood of a point x *iff* $U = x + U(0)$, where $U(0)$ is a neighborhood of zero. This allows us to define a topology through its local basis at a single point (Problem 40).

Suppose a map between topological vector spaces (V,τ) and (V',τ') is simultaneously an algebraic isomorphism of V, V' and a homeomorphism of the corresponding topological spaces, then we call it a *linear homeomorphism* (or *topological isomorphism*). As in the case of normed spaces (*cf.* Problem 21), the structure of a finite–dimensional topological vector space is fully specified by its dimension.

1.4.1 Theorem: Two finite–dimensional topological vector spaces, (V,τ) and (V',τ'), are linearly homeomorphic *iff* $\dim V = \dim V'$. Any finite–dimensional topological vector space is locally compact.

Proof: It is sufficient to construct a linear homeomorphism of a given n–dimensional (V,τ) to $\mathbb{C}^n$. We take a basis $\{e_1,\dots,e_n\} \subset V$ and construct $f : V \to \mathbb{C}^n$ by $f\left(\sum_{j=1}^n \xi_j e_j\right) := [\xi_1,\dots,\xi_n]$; in view of the continuity of translations we have to show that f and f^{-1} are continuous at zero. According to (tv1), for any $U(0) \in \tau$ we can find neighborhoods $U_j(0)$ such that $\sum_{j=1}^n x_j \in U(0)$ for $x_j \in U_j(0)\,,\ j = 1,\dots,n$ and f^{-1} is continuous by Problem 42a. To prove that f is continuous we use the fact that V is Hausdorff: Proposition 1.3.1 and Theorem 1.3.3 together with the already proven continuity of f^{-1} ensure that $S_\varepsilon := \{x \in V : \|f(x)\| = \varepsilon\} = f^{(-1)}(K_\varepsilon)$ is closed for any $\varepsilon > 0$; we have denoted here by K_ε the ε–sphere in $\mathbb{C}^n$. Since $0 \notin S_\varepsilon$, the set $G := V \setminus S_\varepsilon$ is a neighborhood of zero, and by Problem 42b there is a balanced neighborhood $U \subset G$ of zero; this is possible only if $\|f(x)\| < \varepsilon$ for all $x \in U$. ∎

Next we want to discuss a class of topological vector spaces whose properties are closer to those of normed spaces. In distinction to the latter the topology in them is not specified generally by a single (semi)norm but rather by a family of them. Let $\mathcal{P} := \{p_\alpha : \alpha \in I\}$ be a family of seminorms on a vector space V where I is an arbitrary index set. We say that $\mathcal{P}$ *separates points* if to any nonzero $x \in V$ there is a $p_\alpha \in \mathcal{P}$ such that $p_\alpha(x) \neq 0$. It is clear that if $\mathcal{P}$ consists of a single seminorm p it separates points *iff* p is a norm. Given a family $\mathcal{P}$ we set

$$B_\varepsilon(p_1, \dots, p_n) := \{\, x \in V : p_j(x) < \varepsilon \,,\, j = 1, \dots, n \,\} ;$$

the collection of these sets for any $\varepsilon > 0$ and all finite subsystems of $\mathcal{P}$ will be denoted as $\mathcal{B}_0^{\mathcal{P}}$. In view of Problem 40, $\mathcal{B}_0^{\mathcal{P}}$ defines a topology on V which we denote as $\tau^{\mathcal{P}}$.

1.4.2 Theorem: If a family $\mathcal{P}$ of seminorms on a vector space V separates points, then $(V, \tau^{\mathcal{P}})$ is a topological vector space.

Proof: By assumption, to a pair x, y of different points there is a $p \in \mathcal{P}$ such that $\varepsilon := \frac{1}{2}p(x-y) > 0$. Then $U(x) := x + B_\varepsilon(p)$ and $U(y) := y + B_\varepsilon(p)$ are disjoint neighborhoods, so the axiom T_2 is valid. The continuity of summation at the point $[0,0]$ follows from the inequality $p(x+y) \le p(x) + p(y)$; for the scalar multiplication we use $p(\alpha x - \alpha_0 x_0) \le |\alpha - \alpha_0|\, p(x_0) + |\alpha|\, p(x - x_0)$. ∎

A topological vector space with a topology induced by a family $\mathcal{P}$ separating points is called *locally convex*. This name has an obvious motivation: if $x, y \in B_\varepsilon(p_1, \dots, p_n)$, then $p_j(tx + (1-t)x) \le t p_j(x) + (1-t) p_j(x)$ holds for any $t \in [0,1]$ so the sets $B_\varepsilon(p_1, \dots, p_n)$ are convex. The convexity is preserved at translations, so the local basis of $\tau^{\mathcal{P}}$ at each $x \in V$ consists of convex sets (see also the notes).

1.4.3 Example: The family $\mathcal{P} := \{p_x := |(x, \cdot)| : x \in V\}$ in a pre-Hilbert space V generates a locally convex topology which is called the *weak topology* and is denoted as τ_w; it is easy to see that it is weaker than the "natural" topology induced by the norm.

1.4.4 Theorem: A locally convex space (V, τ) is metrizable *iff* there is a countable family $\mathcal{P}$ of seminorms which generates the topology τ.

Proof: If V is metrizable it is first countable. Let $\{U_j : j = 1, 2, \dots\}$ be a local basis of τ at the point 0. By definition, to any U_j we can find $\varepsilon > 0$ and a finite subsystem $\mathcal{P}_j \subset \mathcal{P}$ such that $\bigcap_{p \in \mathcal{P}_j} B_\varepsilon(p) \subset U_j$. The family $\mathcal{P}' := \bigcup_{j=1}^\infty \mathcal{P}_j$ is countable and generates a topology $\tau^{\mathcal{P}'}$ which is not stronger than $\tau := \tau^{\mathcal{P}}$; the above inclusion shows that $\tau^{\mathcal{P}'} = \tau$. On the other hand, suppose that τ is generated by a family $\{p_n : n = 1, 2, \dots\}$ separating points; then we can define a metric ϱ as in Problem 16 and show that the corresponding topology satisfies $\tau_\varrho = \tau$ (Problem 43). ∎

A locally convex space which is complete with respect to the metric used in the proof is called a *Fréchet space* (see also the notes).

1.4.5 Example: The set $\mathcal{S}(\mathbb{R}^n)$ consists of all infinitely differentiable functions $f:\mathbb{R}^n \to \mathbb{C}$ such that

$$\|f\|_{J,K} := \sup_{x\in\mathbb{R}^n} |x^J (D^K f)(x)| < \infty$$

holds for any multi–indices $J := [j_1,\dots,j_n]$, $K := [k_1,\dots,k_n]$ with j_r, k_r non–negative integers, where $x^J := \xi_1^{j_1}\dots\xi_n^{j_n}$, $D^K := \partial^{|K|}/\partial\xi_1^{k_1}\dots\partial\xi_n^{k_n}$ and $|K| := k_1+\dots+k_n$. It is easy to see that any such f and any derivative $D^K f$ (as well as polynomial combinations of them) tend to zero faster than $|x^J|^{-1}$ for each J; we speak about *rapidly decreasing* functions. It is also clear that any $\|\cdot\|_{J,K}$ is a seminorm, with $\|f\|_{0,0} = \|f\|_\infty$, and the family $\mathcal{P} := \{\|\cdot\|_{J,K}\}$ separates points. The corresponding locally convex space $\mathcal{S}(\mathbb{R}^n)$ is called the *Schwartz space*; one can show that it is complete, *i.e.*, a Fréchet space (see the notes).

An important subspace in $\mathcal{S}(\mathbb{R}^n)$ consists of infinitely differentiable functions with a compact support; we denote it as $C_0^\infty(\mathbb{R}^n)$. It is dense,

$$\overline{C_0^\infty(\mathbb{R}^n)} = \mathcal{S}(\mathbb{R}^n), \tag{1.1}$$

with respect to the topology of $\mathcal{S}(\mathbb{R}^n)$ (Problem 44).

1.5 Banach spaces and operators on them

A normed space which is complete with respect to the norm–induced metrics is called a **Banach space**. We have already met some frequently used Banach spaces — see Example 1.2.1 and Problem 23. In view of Problem 21, any finite–dimensional normed space is complete; in the general case we have the following completeness criterion, the proof of which is left to the reader (see also Example 1.5.3b below).

1.5.1 Theorem: A normed space V is complete *iff* to any sequence $\{x_n\}\subset V$ such that $\sum_{n=1}^\infty \|x_n\| < \infty$ there is an $x\in V$ such that $x = \lim_{n\to\infty}\sum_{k=1}^n x_k$ (or in short, *iff* any absolutely summable sequence is summable).

Given a noncomplete norm space, we can always extend it to a Banach space by the standard completion procedure (Problem 46). A set M in a Banach space $\mathcal{X}$ is called *total* if $\overline{M_{lin}} = \mathcal{X}$. Such a set is a basis if M is linearly independent and $\dim\mathcal{X} < \infty$, while an infinite–dimensional space can contain linearly independent total sets, which are not Hamel bases of $\mathcal{X}$ (*cf.* the notes to Section 1.1).

1.5.2 Lemma: (a) If M is total in a Banach space $\mathcal{X}$, then any set $N\subset\mathcal{X}$ dense in M is total in $\mathcal{X}$.

(b) A Banach space which contains a countable total set is separable.

Proof: Part (a) follows from the appropriate definitions. Suppose that $M = \{x_1, x_2, \dots\}$ is total in $\mathcal{X}$ and $\mathbb{C}_{rat}$ is the countable set of complex numbers with rational real and imaginary parts; then the set $L := \{\sum_{j=1}^{n} \gamma_j x_j \,:\, \gamma_j \in \mathbb{C}_{rat},\, n < \infty\}$ is countable. Since $\mathbb{C}_{rat}$ is dense in $\mathbb{C}$, we get $\overline{L} = \mathcal{X}$. ▮

1.5.3 Examples: (a) The set $\mathcal{P}(a, b)$ of all complex polynomials on (a, b) is an infinite–dimensional subspace in $C[a, b] := C([a, b])$. By the Weierstrass theorem, any $f \in C[a, b]$ can be approximated by a uniformly convergent sequence of polynomials; hence $C[a, b]$ is a complete envelope of $(\mathcal{P}(a, b), \|\cdot\|_\infty)$. The set $\{x^k : k = 0, 1, \dots\}$ is total in $C[a, b]$, which is therefore separable.

(b) Consider the sequences $E_k := \{\delta_{jk}\}_{j=1}^{\infty}$ in ℓ^p, $p \ge 1$. For a given $X := \{\xi_j\} \in \ell^p$, the sums $X_n := \sum_{j=1}^{n} \xi_j E_j$ are nothing else than truncated sequences, so $\lim_{n\to\infty} \|X - X_n\|_p = 0$. Hence $\{E_k : k = 1, 2 \dots\}$ is a countable total set and ℓ^p is separable. Notice also that the sequence $\{\xi_j E_j\}_{j=1}^{\infty}$ is summable but it may not be absolutely summable for $p > 1$.

(c) Consider next the space $L^p(\mathbb{R}^n, d\mu)$ with an arbitrary Borel measure μ on $\mathbb{R}^n$. We use the notation of Appendix A. In particular, $\mathcal{J}^n$ is the family of all bounded intervals in $\mathbb{R}^n$; then we define $S^{(n)} := \{\chi_J : J \in \mathcal{J}^n\}$. It is a subset in L^p and the elements of its linear envelope are called *step functions*; we can check that $S^{(n)}$ is total in $L^p(\mathbb{R}^n, d\mu)$ (Problem 47). Combining this result with Lemma 1.5.2 we see that the subspace $C_0^\infty(\mathbb{R}^n)$ is dense in $L^p(\mathbb{R}^n, d\mu)$; in particular, for the Lebesgue measure on $\mathbb{R}^n$ the inclusions $C_0^\infty(\mathbb{R}^n) \subset \mathcal{S}(\mathbb{R}^n) \subset L^p(\mathbb{R}^n)$ yield

$$\overline{(C_0^\infty(\mathbb{R}^n))_p} = \overline{(\mathcal{S}(\mathbb{R}^n))_p} = L^p(\mathbb{R}^n)\,. \tag{1.2}$$

(d) Given a topological space (X, τ) we call $C_\infty(X)$ the set of all continuous functions on X with the following property: for any $\varepsilon > 0$ there is a compact set $K \subset X$ such that $|f(x)| < \varepsilon$ outside K. It is not difficult to check that $C_\infty(X)$ is a closed subspace in $C(X)$ and $\overline{C_0(X)} = C_\infty(X)$, where $C_0(X)$ is the set of continuous functions with compact support (Problem 48). In the particular case $X = \mathbb{R}^n$, $C_0^\infty(\mathbb{R}^n)$ is dense in $C_\infty(\mathbb{R}^n)$ (see the notes), so

$$\overline{(C_0^\infty(\mathbb{R}^n))_\infty} = \overline{(\mathcal{S}(\mathbb{R}^n))_\infty} = C_\infty(\mathbb{R}^n)\,. \tag{1.3}$$

There are various ways in, which it is possible to construct new Banach spaces from given ones. We mention two of them (see also Problem 49):

(i) Let $\{\mathcal{X}_j : j = 1, 2, \dots\}$ be a countable family of Banach spaces. We denote by $\mathcal{X}$ the set of all sequences $x := \{x_j\}$, $x_j \in \mathcal{X}_j$, such that $\sum_j \|x_j\|_j < \infty$, and equip it with the "componentwise" defined summation and scalar multiplication. The norm $\|X\|_\oplus := \sum_j \|x_j\|_j$ turns it into a Banach space; the completeness can be checked as for ℓ^p (Problem 23). The space $(\mathcal{X}, \|\cdot\|_\oplus)$ is called the *direct sum* of the spaces $\mathcal{X}_j$, $j = 1, 2, \dots$, and denoted as $\sum_j^\oplus \mathcal{X}_j$.

(ii) Starting from the same family $\{\mathcal{X}_j : j = 1, 2, \dots\}$, one can define another Banach space (which is sometimes also referred to as a direct sum) if we change the above norm to $\|X\|_\infty := \sup_j \|x_j\|_j$ replacing, of course, $\mathcal{X}$ by the set of sequences for which $\|X\|_\infty < \infty$. The two Banach spaces are different unless the family $\{\mathcal{X}_j\}$ is finite; the present construction can easily be adapted to families of any cardinality.

A map $B : V_1 \to V_2$ between two normed spaces is called an *operator*; in particular, it is called a **linear operator** if it is linear. In this case we conventionally do not use parentheses and write the image of a vector $x \in V_1$ as Bx. In this book we shall deal almost exclusively with linear operators, and therefore the adjective will usually be dropped. A linear operator $B : V_1 \to V_2$ is said to be **bounded** if there is a positive c such that $\|Bx\|_2 \le c\|x\|_1$ for all $x \in V_1$; the set of all such operators is denoted as $\mathcal{B}(V_1, V_2)$ or simply $\mathcal{B}(V)$ if $V_1 = V_2 := V$. One of the elementary properties of linear operators is the equivalence between continuity and boundedness (Problem 50).

The set $\mathcal{B}(V_1, V_2)$ becomes a vector space if we define on it summation and scalar multiplication by $(\alpha B + C)x := \alpha Bx + Cx$. Furthermore, we can associate with every $B \in \mathcal{B}(V_1, V_2)$ the non–negative number

$$\|B\| := \sup_{S_1} \|Bx\|_2 \,,$$

where $S_1 := \{x \in V_1 : \|x\|_1 = 1\}$ is the unit sphere in V_1 (see also Problem 51).

1.5.4 Proposition: The map $B \mapsto \|B\|$ is a norm on $\mathcal{B}(V_1, V_2)$. If V_2 is complete, the same is true for $\mathcal{B}(V_1, V_2)$, *i.e.*, it is a Banach space.

Proof: The first assertion is elementary. Let $\{B_n\}$ be a Cauchy sequence in $\mathcal{B}(V_1, V_2)$; then for all n, m large enough we have $\|B_n - B_m\| < \varepsilon$, and therefore $\|B_n x - B_m x\|_2 \le \varepsilon\|x\|_1$ for any $x \in V_1$. As a Cauchy sequence in V_2, $\{B_n x\}$ converges to some $B(x) \in V_2$. The linearity of the operators B_n implies that $x \mapsto Bx$ is linear, $B(x) = Bx$. The limit $m \to \infty$ in the last inequality gives $\|Bx - B_n x\|_2 \le \varepsilon\|x\|_1$, so $B \in \mathcal{B}(V_1, V_2)$ by the triangle inequality, and $\|B - B_n\| \le \varepsilon$ for all n large enough. ∎

The norm on $\mathcal{B}(V_1, V_2)$ introduced above is called the *operator norm*. It has an additional property: if $C : V_1 \to V_2$ and $B : V_2 \to V_3$ are bounded operators, and BC is the *operator product* understood as the composite mapping $V_1 \to V_3$, we have $\|B(Cx)\|_3 \le \|B\|\,\|Cx\|_2 \le \|B\|\,\|C\|\,\|x\|_1$ for all $x \in V_1$, so BC is also bounded and

$$\|BC\| \le \|B\|\,\|C\| \,. \tag{1.4}$$

Let V_1 be a subspace of a normed space $\tilde{V}_1$. An operator $B : V_1 \to V_2$ is called a *restriction* of $\tilde{B} : \tilde{V}_1 \to V_2$ to the subspace V_1 if $Bx = \tilde{B}x$ holds for all $x \in V_1$, and on the other hand, $\tilde{B}$ is said to be an *extension* of B; we write $B = \tilde{B} \upharpoonright V_1$ or $B \subset \tilde{B}$. Another simple property of bounded operators is that they can be extended uniquely by continuity.

1.5.5 Theorem: Assume that $\mathcal{X}_1$, $\mathcal{X}_2$ are Banach spaces and V_1 is a dense subspace in $\mathcal{X}_1$; then any $B \in \mathcal{B}(V_1, \mathcal{X}_2)$ has just one extension $\tilde{B} \in \mathcal{B}(\mathcal{X}_1, \mathcal{X}_2)$, and moreover, $\|\tilde{B}\| = \|B\|$.

Proof: For any $x \in \mathcal{X}$ we can find a sequence $\{x_n\} \subset V_1$ that converges to x. Since B is bounded, the sequence $\{Bx_n\}$ is Cauchy, so there is a $y \in \mathcal{X}_2$ such that $\|Bx_n - y\|_2 \to 0$. We can readily check that y does not depend on the choice of the approximating sequence and the map $x \mapsto y$ is linear; we denote it as $\tilde{B}$. If $x \in V_1$ one can choose $x_n = x$ for all n, which means $\tilde{B} \restriction V_1 = B$. Passing to the limit in the relation $\|Bx_n\|_2 \le \|B\|\,\|x_n\|_1$ we get $\tilde{B} \in \mathcal{B}(\mathcal{X}_1, \mathcal{X}_2)$ and $\|\tilde{B}\| \le \|B\|$; since $\tilde{B}$ is an extension of B the two norms must be equal. Suppose finally that $B = C \restriction V_1$ for some $C \in \mathcal{B}(\mathcal{X}_1, \mathcal{X}_2)$. We have $Cx_n = Bx_n$, $n = 1, 2, \dots$, for any approximating sequence, and therefore $C = \tilde{B}$. ∎

Notice that in view of Proposition 1.5.4, $\mathcal{B}(V_1, \mathcal{X})$ is complete even if V_1 is not. The approximation procedure used in the proof to define $\tilde{B}$ is often of practical importance, namely if we study an operator whose action on some dense subspace is given by a simple formula.

1.5.6 Example (Fourier transformation): Following the usual convention we denote the scalar product of the vectors $x, y \in \mathbb{R}^n$ by $x \cdot y$ and set

$$\hat{f}(y) := (2\pi)^{-n/2} \int_{\mathbb{R}^n} e^{-ix\cdot y} f(x)\, dx \quad \text{and} \quad \check{f}(y) := \hat{f}(-y) \tag{1.5}$$

for any $f \in \mathcal{S}(\mathbb{R}^n)$ and $y \in \mathbb{R}^n$. The function $\hat{f}$ is well–defined and one can check that it belongs to $\mathcal{S}(\mathbb{R}^n)$ (Problem 52), *i.e.*, that $\mathcal{F}_0 : \mathcal{F}_0 f = \hat{f}$ is a linear map of $\mathcal{S}(\mathbb{R}^n)$ onto itself. We want to prove that $\tilde{\mathcal{F}}_0 : f \mapsto \check{f}$ is its inverse. To this end, we use the relation from Problem 52; choosing $g_\varepsilon := e^{-\varepsilon^2|x|^2/2}$ we get

$$\int_{\mathbb{R}^n} e^{ix\cdot y - \varepsilon^2|x|^2/2} \hat{f}(y)\, dy = \int_{\mathbb{R}^n} e^{-|z|^2/2} f(x+\varepsilon z)\, dz$$

for any $\varepsilon > 0$. The two integrated functions can be majorized independently of ε; the limit $\varepsilon \to 0+$ then yields $\tilde{\mathcal{F}}_0 \mathcal{F}_0 f = f$. Since $\check{f}(x) = \hat{f}(-x)$ we also get $\mathcal{F}_0 \tilde{\mathcal{F}}_0 f = f$ for all $f \in \mathcal{S}(\mathbb{R}^n)$, *i.e.*, the relation

$$(\mathcal{F}_0^{-1} f)(x) = (\mathcal{F}_0 f)(-x)\,.$$

Using Theorem 1.5.5, we shall now construct two important extensions of the operator $\mathcal{F}_0$. For the moment, we denote by $\mathcal{S}_p(\mathbb{R}^n)$ the normed space $(\mathcal{S}(\mathbb{R}^n), \|\cdot\|_p)$; we know from (1.2) that it is dense in $L^p(\mathbb{R}^n)$, $p \ge 1$.

(i) Since $\mathcal{S}(\mathbb{R}^n)$ is a subset of $C_\infty(\mathbb{R}^n)$ and $\|\hat{f}\|_\infty \le (2\pi)^{-n/2}\|f\|_1$, the operator $\mathcal{F}_0$ can also be understood as an element of $\mathcal{B}(\mathcal{S}_1(\mathbb{R}^n), C_\infty(\mathbb{R}^n))$. As such it extends uniquely to the operator $\mathcal{F} \in \mathcal{B}(L^1(\mathbb{R}^n), C_\infty(\mathbb{R}^n))$; it is easy to check that its action on any $f \in L^1(\mathbb{R}^n)$ can be expressed again by the first one of the relations (1.5),

$$(\mathcal{F}f)(y) = (2\pi)^{-n/2} \int_{\mathbb{R}^n} e^{-ix\cdot y} f(x)\, dx\,.$$

The function $\mathcal{F}f$ is called the *Fourier transform* of f. We have $\mathcal{F}f \in C_\infty(\mathbb{R}^n)$, and therefore

$$\lim_{|x|\to\infty} (\mathcal{F}f)(x) = 0;$$

this relation is often referred to as the *Riemann-Lebesgue lemma.*

(ii) Using once more the relation from Problem 52, now with $g := \overline{\hat{f}} = \check{\bar{f}}$, we find

$$\int_{\mathbb{R}^n} |\hat{f}(y)|^2 dy = \int_{\mathbb{R}^n} |f(y)|^2 dy$$

for any $f \in \mathcal{S}(\mathbb{R}^n)$. This suggests another possible interpretation of the operator $\mathcal{F}_0$ as an element of $\mathcal{B}(\mathcal{S}_2(\mathbb{R}^n),\, L^2(\mathbb{R}^n))$. Extending it by continuity, we get the operator $F \in \mathcal{B}(L^2(\mathbb{R}^n))$, which is called the *Fourier–Plancherel operator* or briefly *FP–operator*; if it is suitable to specify the dimension of $\mathbb{R}^n$ we denote it as F_n. The above relation shows that $\|Ff_k\|_2 = \|f_k\|_2$ holds for the elements of any sequence $\{f_k\} \subset \mathcal{S}_2$ approximating a given $f \in L^2$, and therefore

$$\|Ff\|_2 = \|f\|_2$$

for all $f \in L^2(\mathbb{R}^n)$; this implies that F is surjective (Problem 53). Hence the Fourier–Plancherel operator is *a linear isometry of* $L^2(\mathbb{R}^n)$ *onto itself.*

We are naturally interested in how F acts on the vectors from $L^2 \setminus \mathcal{S}$. There is a simple functional realization for $n = 1$ (see Example 3.1.6). In the general case, the right sides of the relations (1.5) express $(Ff)(y)$ and $(F^{-1}f)(y)$ as long as $f \in L^2 \cap L^1$. To check this assume first that $\operatorname{supp} f \subset J$, where J is a bounded interval in $\mathbb{R}^n$ and consider a sequence $\{f_k\}$ approximating f according to Problem 47d. By the Hölder inequality $\|f-f_k\|_1 \le \mu(J)^{1/2}\|f-f_k\|_2$, so $\|f-f_k\|_1 \to 0$, and consequently, the functions $\mathcal{F}_0 f_k$ converge uniformly to $\mathcal{F}f$. On the other hand $\mathcal{F}_0 f_k = Ff_k$, so $\|\mathcal{F}_0 f_k - Ff\|_2 = \|f_k - f\|_2 \to 0$; the sought expression of Ff then follows from the result mentioned in the notes to Section 1.2. A similar procedure can be used for a general $f \in L^2 \cap L^1$: one approximates it, *e.g.*, by the functions $f_j := f\chi_j$, where χ_j are characteristic functions of the balls $\{\, x \in \mathbb{R}^n : |x| \le j \,\}$.

For the remaining vectors, $f \in L^2 \setminus L^1$, the right side of (1.5) no longer makes sense and Ff must be defined as a limit, *e.g.*, $\|Ff - Ff_j\|_2 \to 0$, where f_j are the truncated functions defined above. The last relation is often written as

$$(Ff)(y) = \text{l.i.m.}_{\,j\to\infty}(2\pi)^{-n/2} \int_{|x|\le j} e^{-ix\cdot y} f(x)\, dx\,,$$

where the symbol l.i.m. (*limes in medio*) means convergence with respect to the norm of $L^2(\mathbb{R}^n)$.

A particular role is played by operators that map a given normed space into $\mathbb{C}$. We call $\mathcal{B}(V,\mathbb{C})$ the *dual space* to V and denote it as V^*; its elements are *bounded linear functionals.* Comparing it with the algebraic dual space defined in the notes, we see that V^* is a subspace of V^f, and moreover, a Banach space with respect to the operator norm; the two dual spaces do not coincide unless $\dim V < \infty$ (Problem 54). The Hahn–Banach theorem has some simple implications.

1.5.7 Proposition: Let V_0 be a subspace of a normed space V; then

(a) For any $f_0 \in V_0^*$ there is an $f \in V^*$ such that $f \upharpoonright V_0 = f_0$ and $\|f_0\| = \|f\|$

(b) If $\overline{V_0} \neq V$, then to any $z \notin \overline{V_0}$ one can find $f_z \in V^*$ with $\|f_z\| = 1$ such that $f_z \upharpoonright V_0 = 0$ and $f_z(z) = d(z) := \inf_{y\in V_0} \|z-y\|$

Proof: The first assertion follows from Theorem 1.1.3 with $p = \|f_0\| \, \|\cdot\|$. To prove (b), we have to check first that $x \mapsto d(x)$ is a seminorm on V and $d(x) = 0$ holds *iff* $x \in \overline{V_0}$. Then we take the subspace $V_1 := \{\, x = y+\alpha z \,:\, y \in V_0,\, \alpha \in \mathbb{C} \,\}$ and set $f_1(x) := \alpha d(z)$, in particular, $f_1(z) = d(z)$. This is obviously a linear functional and $|f_1(y+\alpha z)| = d(\alpha z) \le \|y+\alpha z\|$, so $f_1 \in V_1^*$ and $\|f_1\| \le 1$. On the other hand, we have $d(z) = |f_1(y-z)| \le \|f_1\| \, \|y-z\|$; hence $\|f_1\| = 1$ and the sought functional f_z is obtained by extending f_1 to the whole V in accordance with the already proven part (a). ∎

1.5.8 Corollary: (a) To any nonzero $x \in V$ there is a functional $f_x \in V^*$ such that $f_x(x) = \|x\|$ and $\|f_x\| = 1$.

(b) The family V^* separates points of V.

(c) If the dual space $\mathcal{X}^*$ to a Banach space $\mathcal{X}$ is separable, then $\mathcal{X}$ is also separable.

Proof: The first two assertions follow immediately from Proposition 1.5.7. Let further $\{f_n : n = 1,2,\dots\}$ be a dense set in $\mathcal{X}^*$. To any nonzero f_n we can find a unit vector $x_n \in \mathcal{X}$ such that $|f_n(x_n)| > \frac{1}{2}\,\|f_n\|$. In view of Lemma 1.5.2 it is sufficient to check that $M := \{x_n : n = 1,2,\dots\}$ is total in $\mathcal{X}$. Let us assume that $V_0 := \overline{M_{lin}} \neq \mathcal{X}$; then Proposition 1.5.7 implies the existence of a functional $f \in \mathcal{X}^*$ such that $\|f\| = 1$ and $f(x) = 0$ for $x \in V_0$. For any $\varepsilon > 0$ we can find a nonzero f_n such that $\|f_n - f\| < \varepsilon$, *i.e.*, $\|f_n\| > 1-\varepsilon$; hence we arrive at the contradictory conclusion $\varepsilon > \|f_n - f\| \ge |f_n(x_n) - f(x_n)| = |f_n(x_n)| > \frac{1}{2}\,\|f_n\| > \frac{1}{2}(1-\varepsilon)$. ∎

One of the basic problems in the Banach–space theory is to describe fully $\mathcal{X}^*$ for a given $\mathcal{X}$. We limit ourselves to one example; more information can be found in the notes and in the next chapter, where we shall show how the problem simplifies when $\mathcal{X}$ is a Hilbert space.

1.5.9 Example: The dual $(\ell^p)^*$, $p \ge 1$, is linearly isomorphic to ℓ^q, where $q := p/(p-1)$ for $p > 1$ and $q := \infty$ for $p = 1$. To demonstrate this, we define

$$f_Y(X) := \sum_{k=1}^{\infty} \xi_k \eta_k$$

for any sequences $X := \{\xi_k\} \in \ell^p$ and $Y := \{\eta_k\} \in \ell^q$; then the Hölder inequality implies that f_Y is a bounded linear functional on ℓ^p and $\|f_Y\| \le \|Y\|_q$. The map $Y \mapsto f_Y$ of ℓ^q to $(\ell^p)^*$ is obviously linear; we claim that this is the sought isometry. We have to check its invertibility. We take an arbitrary $f \in (\ell^p)^*$ and set $\eta_k := f(E_k)$, where E_k are the sequences introduced in Example 1.5.3b; then it follows from the continuity of f that the sequence $Y_f := \{\eta_k\}_{k=1}^{\infty}$ fulfils $f = f_{Y_f}$. To show that $Y_f \in \ell^q$, consider first the case $p > 1$. The vectors $X_n := \sum_{k=1}^{n} \operatorname{sgn}(\overline{\eta}_k)\,|\eta_k|^{q-1}$, where $\operatorname{sgn} z := z/|z|$ if $z \ne 0$ and zero otherwise, fulfil $\|X_n\|_p = \left(\sum_{k=1}^{n} |\eta_k|^q\right)^{1/p}$ and $f(X_n) = \sum_{k=1}^{n} |\eta_k|^q$, so the inequality $|f(X_n)| \le \|f\|\,\|X_n\|_p$ yields

$$\left(\sum_{k=1}^{\infty} |\eta_k|^q\right)^{1/q} \le \|f\|\,, \qquad n = 1, 2, \dots.$$

If $p = 1$ we have $|f(E_n)| = |\eta_n|$ and $\|E_n\|_1 = 1$, so $\sup_n |\eta_n| \le \|f\|$. Hence in both cases the sequence $Y_f \in \ell^q$, and the obtained bounds to its norm in combination with the inequality $\|f_Y\| \le \|Y\|_q$, which we proved above, yield $\|f\| = \|Y_f\|_q$.

The dual space of a given V is normed, so we can define the *second dual* $V^{**} := (V^*)^*$ as well as higher dual spaces. For any $x \in V$ we can define $J_x \in V^*$ by $J_x(f) := f(x)$. The map $x \mapsto J_x$ is a linear isometry of V to a subspace of V^{**} (Problem 55); if its range is the whole V^{**} the space V is called *reflexive*. It follows from the definition that any reflexive space is automatically Banach. In view of Example 1.5.9, the spaces ℓ^p are reflexive for $p > 1$, and the same is true for $L^p(M, d\mu)$ (see the notes). On the other hand, ℓ^1 and $C(K)$ are not reflexive, and similarly, $L^1(M, d\mu)$ is not reflexive unless the measure μ has a finite support. Below we shall need the following general property of reflexive spaces, which we present without proof.

1.5.10 Theorem: Any closed subspace of a reflexive space is reflexive.

The notion of dual space extends naturally to topological vector spaces: the dual to (V, τ) consists of all continuous linear functionals on V; in this case we often denote it alternatively as V'. It allows us to define the *weak topology* τ_w on V as the weakest topology with respect to which any $f \in V'$ is continuous, or the V'–weak topology in the terminology introduced in Example 1.2.5 ; in the next chapter we shall see that the definition is consistent with that of Example 1.4.3. We have $\tau_w \subset \tau$ because each $f \in V'$ is by definition continuous with respect to τ, and it is easy to check that τ_w coincides with the topology generated by the family $\mathcal{P}_w := \{p_f : f \in V'\}$, where $p_f(x) := |f(x)|$.

1.5.11 Proposition: If (V, τ) is a locally convex space, then $\mathcal{P}_w$ is separating points of V and the space (V, τ_w) is also locally convex.

Proof: If V is a normed space, the assertion follows from Corollary 1.5.8; for the general case see the notes. ∎

If a sequence $\{x_n\} \subset V$ converges to some $x \in V$ with respect to τ_w, we write

$x = \text{w-lim}_{n\to\infty} x_n$ or $x_n \overset{w}{\to} x$. Let us list some properties of weakly convergent sequences (only the case $\dim V = \infty$ is nontrivial — see Problem 56).

1.5.12 Theorem: (a) $x_n \overset{w}{\to} x$ *iff* $f(x_n) \to f(x)$ holds for any $f \in V^*$.

(b) Any weakly convergent sequence in a normed space is bounded.

(c) If $\{x_n\}$ is a bounded sequence in a normed space V and $g(x_n) \to g(x)$ for all g of some total set in $F \subset V^*$, then $x_n \overset{w}{\to} x$.

Proof: The first assertion follows directly from the definition. Further, we use the uniform boundedness principle which will be proven in the next section: if $x_n \overset{w}{\to} x$, then the family $\{\psi_n\} \subset V^{**}$ with $\psi_n(f) := f(x_n)$ fulfils the assumptions of Theorem 1.6.1, which yields $\|\psi_n\| = \|x_n\| < c$ for some $c > 0$. As for part (c), we have $g(x_n) \to g(x)$ for all $g \in F_{lin}$. Since F is total by assumption, for any $f \in V^*$, $\varepsilon > 0$ there is a $g \in F_{lin}$ and a positive integer $n(\varepsilon)$ such that $\|f-g\| < \varepsilon$ and $|g(x_n)-g(x)| < \varepsilon$ for $n > n(\varepsilon)$; this easily yields $|f(x_n)-f(x)| \le (1+\|x_n\|+\|x\|)\varepsilon$. However, the sequence $\{x_n\}$ is bounded, so $f(x_n) \to f(x)$. ∎

1.5.13 Example (weak convergence in ℓ^p, $p > 1$): In view of Examples 3b and 9, the family of the functionals $f_k(\{\xi_j\}_{j=1}^{\infty}) := \xi_k$, $k = 1, 2, \dots$, is total in $(\ell^p)^*$. This means that a sequence $\{X_n\} \subset \ell^p$, $X_n := \{\xi_j^{(n)}\}_{j=1}^{\infty}$, converges weakly to $X := \{\xi_j\}_{j=1}^{\infty} \in \ell^p$ *iff* it is bounded and $\xi_j^{(n)} \to \xi_j$ for $j = 1, 2, \dots$. For instance, the sequence $\{E_n\}$ of Example 3b converges weakly to zero; this illustrates that the two topologies are different because $\{E_n\}$ is not norm–convergent.

A topological vector space (V, τ) is called *weakly complete* if any sequence $\{x_n\} \subset V$ such that $\{f(x_n)\}$ is convergent for each $f \in V'$ converges weakly to some $x \in V$. A set $M \subset V$ is *weakly compact* if any sequence $\{x_n\} \subset M$ contains a weakly convergent subsequence.

1.5.14 Theorem: Let $\mathcal{X}$ be a reflexive Banach space, then

(a) $\mathcal{X}$ is weakly complete.

(b) A set $M \subset \mathcal{X}$ is weakly compact *iff* it is bounded.

Proof: (a) Let $\{x_n\} \subset \mathcal{X}$ be such that $\{f(x_n)\}$ is convergent for each $f \in V^*$. The same argument as in the proof of Theorem 12 implies existence of a positive c such that the sequence $\{\psi_n\} \subset \mathcal{X}^{**}$, $\psi_n(f) := f(x_n)$, fulfils $|\psi_n(f)| \le c\|f\|$, $n = 1, 2, \dots$, for all $f \in \mathcal{X}^*$. The limit $\psi(f) := \lim_{n\to\infty} \psi_n(x)$ exists by assumption, the map $f \mapsto \psi(f)$ is linear, and the last inequality implies $f \in \mathcal{X}^{**}$. Since $\mathcal{X}$ is reflexive, there is an $x \in \mathcal{X}$ such that $\psi(f) = f(x)$ for all $f \in \mathcal{X}^*$, *i.e.*, $x_n \overset{w}{\to} x$.

(b) If M is not bounded there is a sequence $\{x_n\} \subset M$ such that $\|x_n\| > n$; then no subsequence of it can be weakly convergent. Suppose on the contrary that M is bounded and consider a sequence $X := \{x_n\} \subset M$; it is clearly sufficient to assume that X is simple, $x_n \neq x_m$ for $n \neq m$. In view of Theorem 10, $\mathcal{Y} := \overline{\{x_n\}_{lin}}$ is a separable and reflexive Banach space, so $\mathcal{Y}^{**}$ is also separable, and Corollary 8 implies that $\mathcal{Y}^*$ is separable too. Let $\{g_j : j = 1, 2, \dots\}$ be a dense set in $\mathcal{Y}^*$. Since X is bounded the same is true for $\{g(x_n)\}$; hence there is a subsequence $X_1 := \{x_n^{(1)}\}$ such that $\{g_1(x_n^{(1)})\}$ converges. In a similar way, $\{g_2(x_n^{(1)})\}$ is bounded, so we can pick a subsequence $X_2 := \{x_n^{(2)}\} \subset X_1$ such that $\{g_2(x_n^{(2)})\}$ converges, *etc.*This procedure yields a chain of sequences, $X \supset \cdots \supset X_j \supset X_{j+1} \supset \cdots$, such that $\{g_j(x_n^{(j)})\}_{n=1}^{\infty}$, $j = 1, 2, \dots$, are convergent. Now we set $y_n := x_n^{(n)}$, so $y_n \in X_j$ for $n \geq j$ and $\{g_j(y_n)\}_{n=1}^{\infty}$ converges for any j; then $\{g(y_n)\}_{n=1}^{\infty}$ is convergent for all $g \in \mathcal{Y}^*$ due to Theorem 12c. The already proven part (a) implies the existence of $y \in \mathcal{Y}$ such that $g(y_n) \to g(y)$ for any $g \in \mathcal{Y}^*$. Finally, we take an arbitrary $f \in \mathcal{X}^*$ and denote $g_f := f \upharpoonright \mathcal{Y}$. Since $g_f \in \mathcal{Y}^*$ we have $f(y_n) = g_f(y_n) \to g_f(y) = f(y)$, and therefore $y_n \xrightarrow{w} y$. ∎

1.6 The principle of uniform boundedness

Any Banach space $\mathcal{X}$ is a complete metric space so the Baire category theorem is valid in it (*cf.* Problem 27). Now we are going to use this fact to derive some important consequences for bounded operators on $\mathcal{X}$.

1.6.1 Theorem (uniform boundedness principle): Let $\mathcal{F} \subset \mathcal{B}(\mathcal{X}, V_1)$, where $\mathcal{X}$ is a Banach space and $(V_1, \|\cdot\|_1)$ is a normed space. If $\sup_{B\in\mathcal{F}} \|Bx\|_1 < \infty$ for any $x \in \mathcal{X}$, then there is a positive c such that $\sup_{B\in\mathcal{F}} \|B\| < c$.

Proof: Since any operator $B \in \mathcal{F}$ is continuous, the sets $M_n := \{x \in \mathcal{X} : \|Bx\|_1 \leq n \text{ for all } B \in \mathcal{F}\}$ are closed. Due to the assumption, we have $\mathcal{X} = \bigcup_{n=1}^{\infty} M_n$ and by the Baire theorem, at least one of the sets M_n has an interior point, *i.e.*, there is a natural number $\tilde{n}$, an $\tilde{x} \in M_{\tilde{n}}$, and an $\varepsilon > 0$ such that all x fulfilling $\|x-\tilde{x}\| < \varepsilon$ belong to $M_{\tilde{n}}$, and therefore $\sup_{B\in\mathcal{F}} \|Bx\|_1 \leq \tilde{n}$. Let $y \in \mathcal{X}$ be a unit vector. We set $x_y := \frac{\varepsilon}{2} y$; then $x_y + \tilde{x} \in M_{\tilde{n}}$ and

$$\|By\|_1 = \frac{2}{\varepsilon} \|Bx_y\|_1 \leq \frac{2}{\varepsilon} \left(\|B(x_y+\tilde{x})\|_1 + \|B\tilde{x}\|_1 \right) \leq \frac{4\tilde{n}}{\varepsilon} \, ;$$

this implies $\|B\| \leq \frac{4\tilde{n}}{\varepsilon}$ for all $B \in \mathcal{F}$. ∎

In what follows, $\mathcal{X}$, $\mathcal{Y}$ are Banach spaces, U_ε and V_ε are open balls in $\mathcal{X}$ and $\mathcal{Y}$, respectively, of the radius $\varepsilon > 0$ centered at zero. By N^o we denote the *interior* of a set $N \subset \mathcal{Y}$, *i.e.*, the set of all its interior points. Any operator $B \in \mathcal{B}(\mathcal{X}, \mathcal{Y})$ is continuous, so the pull–back $B^{(-1)}(G)$ of an open set $G \subset \mathcal{Y}$ is open in $\mathcal{X}$. If B is surjective, the converse is also true.

1.6.2 Theorem (open–mapping theorem): If an operator $B \in \mathcal{B}(\mathcal{X}, \mathcal{Y})$ is surjective and $G \subset \mathcal{X}$ is an open set, then the set BG is open in $\mathcal{Y}$.

We shall first prove a technical result.

1.6.3 Lemma: Let $B \in \mathcal{B}(\mathcal{X},\mathcal{Y})$ and $\varepsilon > 0$. If $(\overline{BU_\varepsilon})^o \neq \emptyset$ or $(BU_\varepsilon)^o \neq \emptyset$; then $0 \in (\overline{BU_\eta})^0$ or $0 \in (BU_\eta)^0$, respectively, holds for any $\eta > 0$.

Proof: Let y_0 be an interior point of $\overline{BU_\varepsilon}$; then there is $\delta > 0$ such that $y_0 + V_\delta \subset \overline{BU_\varepsilon}$, *i.e.*, to any $y \in V_\delta$ there exists a sequence $\{x_n^{(y)}\} \subset U_\varepsilon$ such that $z_n^{(y)} := Bx_n^{(y)} \to y + y_0$. In particular, $z_n^{(0)} \to y_0$, so $z_n^{(y)} - z_n^{(0)} \to y$, and since $\|x_n^{(y)} - x_n^{(o)}\|_{\mathcal{X}} < 2\varepsilon$ we get $V_\delta \subset \overline{BU_{2\varepsilon}}$. In view of Problem 33, $c\overline{BU_\varepsilon} = \overline{cBU_\varepsilon}$ holds for any $c > 0$. This implies that $V_{\eta'} \subset \overline{BU_\eta}$, $\eta' := \frac{\eta\delta}{2\varepsilon}$, so 0 is an interior point of $\overline{BU_\eta}$. A similar argument applies to $(BU_\varepsilon)^o \neq \emptyset$. ∎

Proof of Theorem 1.6.2: To any $x \in G$ we can find U_η such that $x + U_\eta \subset G$, *i.e.*, $Bx + BU_\eta \subset BG$. If there is $V_\delta \subset BU_\eta$ the set BG is open; hence it is sufficient to check $0 \in (BU_\eta)^o$ for any $\eta > 0$. We write $\mathcal{X} = \bigcup_{n=1}^\infty U_n$; since B is surjective, we have $\mathcal{Y} = \bigcup_{n=1}^\infty BU_n$ and the Baire category theorem implies $(\overline{BU_{\tilde n}})^o \neq \emptyset$ for some positive integer $\tilde n$. We shall prove that $\overline{BU_{\tilde n}} \subset BU_{2\tilde n}$.

Due to the lemma, $\overline{BU_{\tilde n}}$ contains a ball V_δ, and this further implies $V_{\delta_j} \subset \overline{BU_{n_j}}$ for $j = 1, 2, \dots$, where $\delta_j := \delta/2^j$ and $n_j := \tilde n/2^j$. Let $y \in \overline{BU_{\tilde n}}$, so any neighborhood of y contains elements of $BU_{\tilde n}$; in particular, for the neighborhood $y + V_{\delta_1}$ we can find $x_1 \in U_{\tilde n}$ such that $Bx_1 \in y + V_{\delta_1}$, and therefore also $y - Bx_1 \in V_{\delta_1} \subset \overline{BU_{n_1}}$. Repeating the argument we see that there is an $x_2 \in U_{n_1}$ such that $y - Bx_1 - Bx_2 \in V_{\delta_2} \subset \overline{BU_{n_2}}$ *etc*; in this way we construct a sequence $\{x_j\} \subset \mathcal{X}$ such that $\|x_j\|_{\mathcal{X}} < 2\tilde n/2^j$ and

$$\left\| y - \sum_{k=1}^{j} Bx_k \right\|_{\mathcal{Y}} < \delta_j \,.$$

Then Theorem 1.6.1 implies the existence of $\lim_{j\to\infty} \sum_{k=1}^{j} x_j =: x \in \mathcal{X}$; we have $y = Bx$ because B is continuous. Now $\|x\|_{\mathcal{X}} \le \sum_{k=1}^\infty \|x_k\|_{\mathcal{X}} < 2\tilde n$, so $y \in BU_{2\tilde n}$; this proves $\overline{BU_{\tilde n}} \subset BU_{2\tilde n}$. Since $(\overline{BU_{\tilde n}})^o \neq \emptyset$ the set $BU_{2\tilde n}$ has an interior point; using the lemma again we find $0 \in (BU_\eta)^o$ for any $\eta > 0$. ∎

Theorem 1.6.2 further implies the following often used result, the proof of which is left to the reader (Problem 58).

1.6.4 Corollary (inverse–mapping theorem): If $B \in \mathcal{B}(\mathcal{X},\mathcal{Y})$ is a bijection, then B^{-1} is a continuous linear operator from $\mathcal{Y}$ to $\mathcal{X}$.

In the rest of this section, T means a linear operator defined on a subspace D_T of a Banach space $\mathcal{X}$ and mapping D_T into a Banach space $\mathcal{Y}$; we do *not* assume T to be continuous. The subspace D_T is called the **domain** of the operator T and is alternatively denoted as $D(T)$.

The set $\Gamma(T) := \{\, [x, Tx] \,:\, x \in D_T \,\}$ is called the **graph** of the operator T; it is a subspace in the Banach space $\mathcal{X} \oplus \mathcal{Y}$. In general, a subspace $\Gamma \subset \mathcal{X} \oplus \mathcal{Y}$ is said to be a *graph* if each element $[x, y] \in \Gamma$ is determined by its first component. Any graph Γ determines the linear operator T_Γ from $\mathcal{X}$ to $\mathcal{Y}$ with the domain $D(T_\Gamma) := \{\, x \in \mathcal{X} : [x, y] \in \Gamma \,\}$ by $T_\Gamma x := y$ for each $[x, y] \in \Gamma$. It is clear that

$\Gamma(T_\Gamma) = \Gamma$ and conversely, a linear operator is uniquely determined by its graph, $T_{\Gamma(T)} = T$. An operator T is called **closed** if $\Gamma(T)$ is a closed set in $\mathcal{X} \oplus \mathcal{Y}$; the definition of the direct product gives the following equivalent expression.

1.6.5 Proposition: An operator T is closed *iff* for any sequence $\{x_n\} \subset D_T$ such that $x_n \to x$ and $Tx_n \to y$, we have $x \in D_T$ and $y = Tx$.

Any bounded operator is obviously closed. On the other hand, there are closed operators which are not bounded (Problem 59). If $\Gamma(T)$ is not closed but its closure in $\mathcal{X} \oplus \mathcal{Y}$ is a graph (which may not be true) the operator T is said to be *closable* and the closed operator

$$\overline{T} := T_{\overline{\Gamma(T)}}$$

is called the **closure** of the operator T; we have $\Gamma(\overline{T}) = \overline{\Gamma(T)}$. Since $\overline{\Gamma(T)}$ is the smallest closed set containing $\Gamma(T)$, the closure is *the smallest closed extension* of the operator T. Moreover, $\overline{\Gamma(T)}$ is a subspace in $\mathcal{X} \oplus \mathcal{Y}$, so it is a graph *iff* $[0, y] \in \overline{\Gamma(T)}$ implies $y = 0$. This property makes it possible to describe the closure sequentially.

1.6.6 Proposition: (a) An operator T is closable *iff* any sequence $\{x_n\} \subset D_T$ such that $x_n \to 0$ and $Tx_n \to y$ fulfils $y = 0$.

(b) A vector x belongs to $D(\overline{T})$ *iff* T is closable and there is a sequence $\{x_n\} \subset D_T$ such that $x_n \to x$ and $\{Tx_n\}$ is convergent; if this is true, we have $Tx_n \to \overline{T}x$.

We have mentioned that a closed operator may not be continuous. However, this can happen only if the operator is *not* defined on the whole space $\mathcal{X}$.

1.6.7 Theorem (closed–graph theorem): A closed linear operator $T : \mathcal{X} \to \mathcal{Y}$ with $D_T = \mathcal{X}$ is continuous.

Proof: $\Gamma(T)$ is by assumption a closed subspace in $\mathcal{X} \oplus \mathcal{Y}$, so it is a Banach space with the direct–sum norm, $\|[x, y]\|_\oplus = \|x\|_\mathcal{X} + \|y\|_\mathcal{Y}$. The map $S_1 : S_1([x, Tx]) = x$ is a continuous linear bijection from $\Gamma(T)$ to $\mathcal{X}$, and therefore S_1^{-1} is continuous due to Corollary 1.6.4. The map $S_2 : S_2([x, Tx]) = Tx$ is again continuous, and the same is true for the composite map $S_2 \circ S_1^{-1} = T$. ∎

1.7 Spectra of closed linear operators

We denote by $\mathcal{C}(\mathcal{X})$ the set of all closed linear operators from a Banach space $\mathcal{X}$ to itself; since such operators are not necessarily bounded, one has to pay attention to their domains. A complex number λ is called an **eigenvalue** of an operator $T \in \mathcal{C}(\mathcal{X})$ if $T - \lambda I$ is not injective, *i.e.*, if there is a nonzero vector $x \in D_T$ such that $Tx = \lambda x$. Any vector with this property is called an *eigenvector* of T (corresponding to the eigenvalue λ). The subspace $\mathrm{Ker}\,(T - \lambda)$ is the respective *eigenspace* of T and its dimension is the *multiplicity* of the eigenvalue λ; in particular, the latter is *simple* if $\dim \mathrm{Ker}\,(T-\lambda) = 1$. Here and in the following we use $T-\lambda$ as a shorthand for $T-\lambda I$ where I is the unit operator on $\mathcal{X}$.

A subspace $L \subset \mathcal{X}$ is called *T–invariant* if $Tx \in L$ holds for all $x \in L \cap D_T$; we see that any eigenspace of T is T–invariant. Furthermore, Proposition 1.6.5 gives the following simple result.

1.7.1 Proposition: Any eigenspace of an operator $T \in \mathcal{C}(\mathcal{X})$ is closed.

Let us now ask under which conditions the equation $(T-\lambda)x = y$ can be solved for a given $y \in \mathcal{X}$, $\lambda \in \mathbb{C}$. Recall first the situation when $\mathcal{X} := V$ is a finite–dimensional vector space. We know that the equation with $y = 0$ has then a nontrivial solution if λ belongs to the spectrum of T which is defined as the set of all eigenvalues,

$$\sigma(T) := \{\lambda \in \mathbb{C} : \text{there is a nonzero } x \in V\,,\ Tx = \lambda x\,\}\,.$$

Since $\lambda \in \sigma(T)$ holds *iff* $\det(T-\lambda) = 0$, to find the spectrum is a purely algebraic problem; $\sigma(T)$ is a nonempty set having at most $\dim V$ elements. On the other hand, the equation has a unique solution if $\lambda \notin \sigma(T)$ and this solution depends continuously on y. In that case therefore $T-\lambda$ is a bijection of V to itself, and we get an alternative way to describe the spectrum,

$$\sigma(T) = \mathbb{C} \setminus \rho(T)\,, \quad \rho(T) := \{\lambda \in \mathbb{C} : (T-\lambda)^{-1} \in \mathcal{B}(V)\,\}\,.$$

In an infinite–dimensional Banach space the two expressions are no longer equivalent: the inverse $(T-\lambda)^{-1}$ exists if λ is not an eigenvalue of T, but it may be neither bounded nor defined on the whole $\mathcal{X}$. The definition can be then formulated with the help of the second expression. Taking into account that, for $T \in \mathcal{C}(\mathcal{X})$ such that $T-\lambda$ is invertible, the conditions $(T-\lambda)^{-1} \in \mathcal{B}(\mathcal{X})$ and $\mathrm{Ran}\,(T-\lambda) = \mathcal{X}$ are equivalent by the closed–graph theorem and Problem 60b, we can define the **spectrum** of $T \in \mathcal{C}(\mathcal{X})$ as the set of all complex numbers λ for which $T-\lambda$ is *not* a bijection of D_T onto $\mathcal{X}$. This may happen in two (mutually exclusive) cases:

(i) $T-\lambda$ is not injective, *i.e.*, λ is an eigenvalue of T.

(ii) $T-\lambda$ is injective but $\mathrm{Ran}\,(T-\lambda) \ne \mathcal{X}$.

The set of all eigenvalues of T forms its *point spectrum* $\sigma_p(T)$. The remaining part of the spectrum which corresponds to case (ii) is divided as follows: the *continuous spectrum* $\sigma_c(T)$ consists of those λ such that $\mathrm{Ran}\,(T-\lambda)$ is dense in $\mathcal{X}$, while the points where $\overline{\mathrm{Ran}\,(T-\lambda)} \ne \mathcal{X}$ form the *residual spectrum* $\sigma_r(T)$ of T. In this way, the spectrum decomposes into three disjoint sets,

$$\sigma(T) = \sigma_p(T) \cup \sigma_c(T) \cup \sigma_r(T)\,. \tag{1.6}$$

The set

$$\rho(T) := \mathbb{C} \setminus \sigma(T) = \{\lambda \in \mathbb{C} : (T-\lambda)^{-1} \in \mathcal{B}(\mathcal{X})\,\}$$

is called the *resolvent set* of the operator T and its elements are *regular values.* The map $R_T : \rho(T) \to \mathcal{B}(\mathcal{X})$ defined by $R_T(\lambda) := (T-\lambda)^{-1}$ is called the **resolvent** of the operator T; the same name is also often used for its values, *i.e.*, the operators $R_T(\lambda)$. The starting point for derivation of basic properties of the spectrum is the following result, which is an operator analogy of the geometric–series sum.

1.7.2 Lemma: Let $B \in \mathcal{B}(\mathcal{X})$ and $\|I-B\| < 1$; then B is invertible, $B^{-1} \in \mathcal{B}(\mathcal{X})$, and

$$B^{-1} = \lim_{n\to\infty} \sum_{j=0}^{n} (I-B)^j =: \sum_{j=0}^{\infty} (I-B)^j .$$

Proof: Denote $S_n := \sum_{j=0}^{n} (I-B)^j$. The inequality (1.4) gives $\|S_n-S_m\| \le \sum_{j=m+1}^{n} \|I-B\|^j$ for any positive integers m and $n > m$, so $\{S_n\}$ is Cauchy, and since $\mathcal{B}(\mathcal{X})$ is complete, it converges to some $S \in \mathcal{B}(\mathcal{X})$. We have $BS_n = S_nB = I+S_n-S_{n+1}$; passing to the limit we get $BS = SB = I$, *i.e.*, $S = B^{-1}$. ∎

1.7.3 Theorem: The resolvent set of an operator $T \in \mathcal{C}(\mathcal{X})$ is open, containing together with any λ_0 also its neighborhood $U(\lambda_0) := \{\, \lambda \in \mathbb{C} : |\lambda-\lambda_0| < \|R_T(\lambda_0)\|^{-1}\}$, and the resolvent is given by

$$R_T(\lambda) = \sum_{j=0}^{\infty} R_T(\lambda_0)^{j+1}(\lambda-\lambda_0)^j$$

for each $\lambda \in U(\lambda_0)$.

Proof: The operator $B_\lambda := I - R_T(\lambda_0)(\lambda-\lambda_0)$ obeys the assumptions of the lemma if $\lambda \in U(\lambda_0)$, then B_λ^{-1} exists and belongs to $\mathcal{B}(\mathcal{X})$. Using the obvious identities $(T-\lambda)R_T(\lambda_0) = B_\lambda$ and $R_T(\lambda_0)(T-\lambda)x = B_\lambda x$ for all $x \in D_T$, we can check that $(T-\lambda)^{-1}$ exists and $(T-\lambda)^{-1} = R_T(\lambda_0)B_\lambda^{-1} \in \mathcal{B}(\mathcal{X})$; hence $\lambda \in \rho(T)$ and $R_T(\lambda_0)B_\lambda^{-1} = R_T(\lambda)$. To finish the proof, we have to substitute for B_λ^{-1} the expansion from the above lemma. ∎

1.7.4 Corollary: $\sigma(T)$ is a closed set.

To formulate the next assertion, we need one more notion. A vector–valued function $f : \mathbb{C} \to V$, where V is a locally convex space, is said to be *analytic* in a region $G \subset \mathbb{C}$ if the derivative

$$f'(\lambda_0) := \lim_{\lambda\to\lambda_0} \frac{f(\lambda)-f(\lambda_0)}{\lambda-\lambda_0}$$

exists for any $\lambda_0 \in G$. One can extend to such functions some standard results of the theory of analytic functions such as the Liouville theorem; if V is, in addition, a Banach space the generalized Cauchy theorem and Cauchy integral formula are also valid — see the notes for more information.

1.7.5 Theorem: Let B be a bounded operator on a Banach space $\mathcal{X}$; then the resolvent $R_B : \rho(B) \to \mathcal{B}(\mathcal{X})$ is an analytic function, the spectrum $\sigma(B)$ is a nonempty compact set, and $r(B) := \sup\{\, |\lambda| : \lambda \in \sigma(T)\,\} = \lim_{n\to\infty} \|B^n\|^{1/n}$.

Proof: By Problem 62, $\lim_{\lambda\to\lambda_0}(R_B(\lambda)-R_B(\lambda_0))(\lambda-\lambda_0)^{-1} = R_B(\lambda_0)^2$ holds for any $\lambda_0 \in \rho(B)$, so the resolvent is analytic in $\rho(B)$. It is analytic also at the point $\lambda = \infty$. To check this we set $\mu := \lambda^{-1}$; then $(I-\mu B)^{-1}$ exists due to Lemma 1.7.2 provided $|\mu| < \|B\|^{-1}$. The set $\{\lambda : |\lambda| > \|B\|\}$ is thus contained in $\rho(B)$, so the spectrum is bounded and therefore compact in view of Corollary 1.7.4. The obvious identity $R_B(\lambda) = -\mu(I-\mu B)^{-1}$ yields

$$\frac{d}{d\mu} R_B(\mu^{-1})\Big|_{\mu=0} = -\lim_{\mu\to 0}(I-\mu B)^{-1} = -I;$$

now we may use the generalized Liouville theorem mentioned above according to which a vector-valued function is analytic in the extended complex plane *iff* it is constant. Hence the assumption $\sigma(B) = \emptyset$ implies $R_B(\lambda) = C$, *i.e.*, $(B-\lambda)C = I$ for some $C \in \mathcal{B}(\mathcal{X})$ and all $\lambda \in \mathbb{C}$, which is impossible, however.

The above argument shows at the same time that $r(B) \le \|B\|$. Applying Lemma 1.7.2 to the operator $I-\mu B$ we get an alternative expression for the resolvent,

$$R_B(\lambda) = -\lambda^{-1}I - \sum_{k=1}^{\infty} \lambda^{-(k+1)} B^k, \tag{1.7}$$

where the series converges with respect to the norm in $\mathcal{B}(\mathcal{X})$ if the numerical series $\sum_k \|B^k\|\,|\lambda|^{-k}$ is convergent; by the Cauchy–Hadamard criterion, its radius of convergence is $r_0(B) := \limsup_{n\to\infty}\|B^n\|^{1/n}$. Now we set $n = jk+m$, where $0 \le m < j$, then $\|B^n\|^{1/n} \le \|B^j\|^{k/n}\|B\|^{m/n}$. Choosing an arbitrary fixed j and performing the limit $k\to\infty$, we get $\limsup_{n\to\infty}\|B^n\|^{1/n} \le \|B^j\|^{1/j}$, and therefore $\limsup_{n\to\infty}\|B^n\|^{1/n} \le \inf_n \|B^n\|^{1/n} \le \liminf_{n\to\infty}\|B^n\|^{1/n}$; this means that the sequence $\{\|B^n\|^{1/n}\}$ converges to $r_0(B)$. Since the resolvent exists for $|\lambda| > r_0(B)$, we have $r(B) \le r_0(B) \le \|B\|$; the second inequality follows from the above estimate with $j = 1$. To finish the proof, we use the Cauchy integral formula according to which the coefficients of the series (1.7) equal

$$B^k = -\frac{1}{2\pi i}\int_{C_r} z^k R_B(z)\,dz = -\frac{r^{k+1}}{2\pi}\int_0^{2\pi} e^{i(k+1)\theta} R_B(r\,e^{i\theta})\,d\theta,$$

where $r > r(B)$. The circle C_r is a compact subset of $\rho(B)$, so the function $R_B(\cdot)$ is continuous and bounded on it by the already proved analyticity and Corollary 1.3.7b, *i.e.*, $M_r := \max_{z\in C_r}\|R_B(z)\| < \infty$. This yields the estimate $\|B^k\| \le M_r\, r^{k+1}$, which shows that $r_0(B) \le r$ for any $r > r_0(B)$, and therefore $r_0(B) \le r(B)$. ∎

The number $r(B)$ is called the *spectral radius* of the operator B; we have proven that it does not exceed the operator norm, $r(B) \le \|B\|$. Boundedness of the spectrum means that the resolvent set $\rho(B)$ is nonempty for a bounded B. On the other hand, it may happen that $\rho(T) = \emptyset$ or $\sigma(T) = \emptyset$ if $T \in \mathcal{C}(\mathcal{X}) \setminus \mathcal{B}(\mathcal{X})$.

1.7.6 Examples: (a) Consider the operators T and $T^{(0)} := T\restriction D^{(0)}$ of Problem 59. The function $f_\lambda : f_\lambda(x) = e^{\lambda x}$ belongs to D_T and $Tf_\lambda = \lambda f_\lambda$, so $\sigma(T) = \sigma_p(T) = \mathbb{C}$ and $\rho(T) = \emptyset$. On the other hand, for a given $\lambda \in \mathbb{C}$ we define the operator $S_\lambda : (S_\lambda g)(x) = \int_0^x e^{\lambda(x-t)} g(t)\, dt$ which maps $C[0,1]$ to $D^{(0)}$. We have $(S_\lambda g)' - \lambda(S_\lambda g) = g$ for any $g \in C[0,1]$ and $S_\lambda(f' - \lambda f) = f$ for all $f \in D^{(0)}$, and therefore $S_\lambda = (T^{(0)} - \lambda)^{-1}$; in view of the closed–graph theorem and Problem 60b, S_λ is bounded, so we get $\rho(T^{(0)}) = \mathbb{C}$ and $\sigma(T^{(0)}) = \emptyset$.

(b) The same conclusions are valid for the operator $\tilde{P} : \tilde{P}f := -if'$ on $L^2(0,1)$ whose domain $D(\tilde{P}) := AC[0,1]$ consists of all functions that are absolutely continuous on the interval $[0,1]$ with the derivatives in $L^2(0,1)$, and its restriction $P^{(0)} := P\restriction\{f \in AC[0,1] : f(0) = 0\}$. However, one has to use a different method to check that $\tilde{P}$ and $P^{(0)}$ are closed; we postpone the proof to Example 4.2.5.

Notes to Chapter 1

Section 1.1 A linearly independent set $B \subset V$ such that $B_{lin} = V$ is called a *Hamel basis* of the space V and its cardinality is the *algebraic* dimension of V. Such a basis exists in any vector space (Problem 4). We are more interested, however, in other bases which allow us to express elements of a space as "infinite linear combinations" of the basis vectors. This requires a topology; we shall return to that problem in the next chapter.

A set $C \subset V$ is *convex* if together with any two points it contains the line segment connecting them. Any subspace $L \subset V$ is convex, and the intersection of any family of convex sets is again convex. If a point $x \in C$ does not belong to the line segment connecting some $y, z \in C$ different from x, it is called *extremal*. Equivalently, $x \in C$ is an extremal point of C if $x = ty + (1-t)z$ with $t \in (0,1)$, $y, z \in C$ implies $x = y = z$.

The Hahn–Banach theorem is proven in most functional–analysis textbooks — see, *e.g.*, ⟦KF⟧, ⟦RS 1⟧, ⟦Tay⟧. The set of all linear functionals on a given V becomes a vector space if we equip it with the operations defined by $(\alpha f + g)(x) := \alpha f(x) + g(x)$; we call it the *algebraic dual space* of V and denote as V^f. If V is finite–dimensional, one can check easily that V and V^f are isomorphic.

We often need to know whether a given family $\mathcal{F}$ of maps $F : X \to Y$ is "large enough" to contain for any pair of different elements $x, y \in X$ a map f such that $f(x) \neq f(y)$; if this is true we say that $\mathcal{F}$ *separates points.* If X, Y are vector spaces and the maps in $\mathcal{F}$ are linear, the family separates points if for any non–zero $x \in X$ there is $f_x \in \mathcal{F}$ such that $f_x(x) \neq 0$. This is true, in particular, for $\mathcal{F} = X^f$ (Problem 7).

Section 1.2 As a by–product of the argument of Example 1.2.1 we have obtained the following result: if a sequence $\{f_n\} \subset L^p(M, d\mu)$ fulfils $\|f_n - f\|_p \to 0$, then there is a subsequence $\{f_{n_k}\}$ such that $f_{n_k}(x) \to f(x)$ for μ–a.a. $x \in M$. In fact, this is true for any positive measure and the space $L^p(M, d\mu)$ is still complete in this case — see, *e.g.*, ⟦Jar 2⟧, Thm.68; ⟦DS⟧, Sec.III.6.

An alternative way to introduce the notion of a topology is through axiomatization of properties of the closure — see Problem 28 and ⟦Kel⟧ for more details. The described construction of the topological product extends easily to any finite number of topological spaces; for further generalizations see, *e.g.*, ⟦Nai 1⟧, Sec.I.2.12. A discussion of the countability and separability axioms and their relations can be found in topology monographs such as ⟦Al⟧ or ⟦Kel⟧, and also in most functional–analysis textbooks. For instance, any second countable regular space is normal; the fundamental result of Uryson claims that a second countable space is normal *iff* its topology is induced by a metric, thus giving a partial answer to the metrizability problem.

A topological space (X,τ) is *connected* if it cannot be written as a union of two non–empty disjoint open sets; this is equivalent to the requirement that X and $\emptyset$ are the only two sets which are simultaneously closed and open. A set $M\subset X$ is *connected* if the space (M,τ_M) with the induced topology is connected. A continuous map $\varphi:\,[0,1]\to X$ is called a *curve* connecting the points $\varphi(0)$ and $\varphi(1)$. A topological space (or its subset) whose any two points can be connected by a curve is said to be *arcwise* (or *linearly*) *connected.* Such spaces are connected but the converse is not true.

Section 1.3 There is a standard procedure called *one–point compactification* which allows us to construct for a given noncompact (X,τ) a compact space (X',τ') such that (X,τ) is its subspace and $X'\setminus X\equiv\{x_0\}$ is a one–point set; the topology τ' consists of all sets $\{x_0\}\cup(X\setminus F)$, where F is a closed compact set in X — for more details see, *e.g.*, ⟦Tay⟧, Sec.2.31. A simple example is the compactification of $\mathbb{C}$ by adding to it the point ∞.

Any compact space is by definition σ–compact but the converse is not true — *cf.* ⟦KF⟧, Sec.IV.6.4. An infinite set in a σ–compact space has again at least one accumulation point; this is clear from the proof of Proposition 1.3.1. In some cases the notions of compactness and σ–compactness coincide (*cf.* Corollary 1.3.5 and Problem 37c).

A net $\{y_\beta\}_{\beta\in J}$ is called a *subnet* of a net $\{x_\alpha\}_{\alpha\in I}$ if there is a map $\varphi:J\to I$ such that (i) $y_\beta=x_{\varphi(\beta)}$ for all $\beta\in J$, (ii) for any $\alpha'\in J$ there is a $\beta'\in J$ such that $\beta\succ\beta'$ implies $\varphi(\beta)\succ\alpha'$. Using this definition, we can state the *Bolzano–Weierstrass theorem:* A topological space is compact *iff* any net in X has a convergent subnet. A proof can be found, *e.g.*, in ⟦RS 1⟧, Sec.IV.3.

A map f of a metric space (X,ϱ) to (X',ϱ') is *uniformly continuous* if to any $\varepsilon>0$ there is $\delta>0$ such that $\varrho'(f(x_1),f(x_2))<\varepsilon$ holds for any pair of points $x_1,x_2\in X$ fulfilling $\varrho(x_1,x_2)<\varepsilon$. One can easily prove the following useful result: a continuous map of a *compact* space (X,ϱ) to (X',ϱ') is uniformly continuous.

Section 1.4 A converse to Theorem 1.4.1 was proved by F. Riesz: *any locally compact topological vector space is finite–dimensional* — see, *e.g.*, ⟦Tay⟧, Sec.3.3. Some metric–space notions do not extend directly to topological spaces but can be used after a suitable generalization. For instance, a set M in a topological vector space is *bounded* if to any neighborhood U of zero there is an $\alpha\in\mathbb{C}$ such that $M\subset\alpha U$. In view of Problem 42b, there is a positive b such that $M\subset\beta U$ holds for all $|\beta|\geq b$. It is easy to see that in a normed space this definition is equivalent to the requirement of existence of a $c_M>0$ such that $\|x\|<c_M$ for all $x\in M$.

A locally convex space is often defined as a topological vector space in which any neighborhood of the point 0 contains a *convex* neighborhood of zero. We have seen that this is true for $(V,\tau^{\mathcal{P}})$. On the other hand, if a topological vector space (V,τ) has the stated property, there is a family of seminorms on V which separate points and generate the topology τ — see, *e.g.*, ⟦Tay⟧, Sec.3.8 — so the two definitions are really equivalent.

A net $\{x_\alpha\}_{\alpha\in I}$ in a topological vector space V is *Cauchy* if for any neighborhood U of 0 there is a $\gamma\in I$ such that $x_\alpha-x_\beta\in U$ holds for all $\alpha\succ\gamma$, $\beta\succ\gamma$. In particular, in a locally convex space with a topology generated by a family $\mathcal{P}$ the condition reads: for any $\varepsilon > 0$ and $p \in \mathcal{P}$, there is a $\gamma \in I$ such that $p(x_\alpha - x_\beta) < \varepsilon$ for all $\alpha \succ \gamma$, $\beta \succ \gamma$. The space V is *complete* if any Cauchy net in it converges to some $x \in V$. In a similar way, we define a Cauchy net in a metric space. Since such a net in a complete metric space is convergent — see, *e.g.*, ⟦DS 1⟧, Lemma I.7.5 — a locally convex space with a topology generated by a countable family $\mathcal{P}$ is Fréchet *iff* it is complete in the sense of the above definition. The completeness of $\mathcal{S}(\mathbb{R}^n)$ is proved, *e.g.*, in ⟦RS 1⟧, Sec.V.3.

Section 1.5 Combining the results of Example 1.5.3c with the inclusion $C_0^\infty(\mathbb{R}^n) \subset C(\mathbb{R}^n) \cap L^p(\mathbb{R}^n, d\mu)$ we find that the set $C(\mathbb{R}^n) \cap L^p(\mathbb{R}^n, d\mu)$ is also dense in $L^p(\mathbb{R}^n, d\mu)$. This can be proved directly for a wider class of L^p spaces — *cf.* ⟦KF⟧, Sec.VII.1.2. The proof of density of $C_0^\infty(\mathbb{R}^n)$ in $C_\infty(\mathbb{R}^n)$ can be found, *e.g.*, in ⟦Yo⟧, Sec.1.1. In a similar way, the set $C_0^\infty(\Omega)$ for an open connected set $\Omega \subset \mathbb{R}^n$, which we shall mention below, is dense in $L^2(\Omega)$ — *cf.* ⟦RS 4⟧, Sec.XIII.14.

The map $\mathcal{F} : L^1(\mathbb{R}^n) \to C_\infty(\mathbb{R}^n)$ is *injective* as can checked easily: the relation $\mathcal{F}g = 0$ for $g \in L^1$ implies $\int_{\mathbb{R}^n} f(x)g(x)\,dx = 0$ for all $f \in \mathcal{S}(\mathbb{R}^n)$; then $\int_J g(x)\,dx = 0$ holds by Problem 47 for any bounded interval $J \subset \mathbb{R}^n$ and therefore also for any Borel set in $\mathbb{R}^n$; this in turn means $g = 0$. On the other hand, $\mathcal{F}$ is *not* surjective — see, *e.g.*, ⟦KF⟧, Sec.VIII.4.7; ⟦Jar 2⟧, Sec.XIII.11.

The product of functions $f, g \in \mathcal{S}(\mathbb{R}^n)$ belongs to $\mathcal{S}(\mathbb{R}^n)$. The relation from Problem 52 implies that its Fourier transform is $\widehat{fg} = (2\pi)^{-n/2}\hat{f} * \hat{g}$, where

$$(f * g)(x) := \int_{\mathbb{R}^n} f(x-y)g(y)\,dy = \int_{\mathbb{R}^n} g(x-y)f(y)\,dy\,.$$

The map $[f, g] \mapsto f * g$ is called the *convolution*. It is a binary operation on $\mathcal{S}(\mathbb{R}^n)$ which is obviously bilinear and commutative; the relation $f * g = (2\pi)^{n/2}\widehat{\check{f}\check{g}}$ shows that it is also associative. Some important extensions of the convolution are discussed in ⟦Yo⟧, Sec.VI.3; ⟦RS 2⟧, Sec.IX.1.

While $(\ell^1)^*$ is isomorphic to ℓ^∞, the dual $(\ell^\infty)^*$ is *not* isomorphic to ℓ^1. This follows from Corollary 1.5.8c because ℓ^1 is separable but ℓ^∞ is nonseparable due to Problem 18. The situation with the spaces $L^p(X, d\mu)$ is similar: $(L^p)^*$, $p \geq 1$, is linearly isomorphic to L^q by $f_\varphi : f_\varphi(\psi) = \int_X \varphi\psi\,d\mu$ — see ⟦Ru 1⟧, Sec.6.16; ⟦Tay⟧, Sec.7.4. The spaces $(L^\infty)^*$ and L^1 are again nonisomorphic with the exception of the trivial case when the measure μ has a finite support; the expression of $(L^\infty)^*$ is given in ⟦DS 1⟧, Sec.IV.8. These results have implications for the reflexivity of the considered spaces as mentioned in the text. The proof of Theorem 1.5.10 can be found in ⟦DS 1⟧, Sec.II.3.23.

Another example of Banach spaces in which one can find a general form of a bounded linear functional is represented by the spaces $C(K)$ and $C_{\mathbb{R}}(K)$ of continuous functions (complex or real–valued, respectively) on a compact Hausdorff space K. The *Riesz–Markov theorem* associates the functionals with Borel measures on K. In the particular case of $\mathcal{X} \equiv C_{\mathbb{R}}[a, b]$ any such measure corresponds to a function F of a bounded variation; the theorem then claims that for any $f \in \mathcal{X}^*$ there is F of a bounded variation such that $f(\varphi) = \int_{[a,b]} \varphi\,dF$ holds for all $\varphi \in \mathcal{X}$, and moreover, $\|f\|$ is equal to the total variation of F — for details see, *e.g.*, ⟦KF⟧, Sec.VI.6.6 or ⟦RN⟧, Sec.50. The general formulation and proof of the Riesz–Markov theorem can be found in ⟦DS 1⟧, Sec.IV.6.

One of the most important examples of duals to topological vector spaces is the space $\mathcal{S}(\mathbb{R}^n)'$ whose elements are called *tempered distributions*, and the dual of the complete locally convex space $\mathcal{D}(\Omega) \equiv (C_0^\infty(\Omega), \tau)$ for a given open connected set $\Omega \subset \mathbb{R}^n$. The space $\mathcal{D}(\Omega)$ is the so–called inductive limit of the locally convex spaces $(C_0^\infty(K_n), \tau_n)$, where $\{K_n\}$ is a nondecreasing sequence of compact sets such that $\bigcup_n K_n = \Omega$ and the topologies τ_n are determined by families of seminorms analogous to that of Example 1.4.5. The elements of $\mathcal{D}(R^n)'$ are called *distributions* on Ω; we have, in particular, $\mathcal{S}(\mathbb{R}^n)' \subset \mathcal{D}(\mathbb{R}^n)'$. There is an extensive literature on the theory of distributions; in addition to special monographs such ⟦ Schw 1 ⟧ or ⟦ GŠ ⟧ the reader may consult, *e.g.*, ⟦ RS 1–2 ⟧, Chaps.V and IX; ⟦ Yo ⟧, Sec.I.8 and Chap.VI; ⟦ KF ⟧, Sec.XIV.4, *etc.*

The proof of Proposition 1.5.11 for a general locally convex space can be found in ⟦ DS 1 ⟧, Sec.V.2. A set M in a Banach space $\mathcal{X}$ is weakly compact *iff* its weak closure is a compact set in the topological space $(\mathcal{X}, \tau_w)$ — *cf.* ⟦ DS 1 ⟧, Thm.V.6.1. If a set $F \subset V'$ on a topological vector space (V, τ) is separating points, then $\{p_f := |f(\cdot)| \,:\, f \in F\}$ determines a locally convex topology which is called the F–topology on V; it is the weakest topology in which all the functionals $f \in F$ are continuous. It is clear that the smaller the set F is, the weaker is the corresponding F–topology.

Section 1.6 In addition to the norm $\|\cdot\|_\oplus$ on $\mathcal{X} \oplus \mathcal{Y}$ one sometimes introduces $|\!|\!|\cdot|\!|\!|$ by $|\!|\!|[x_1, x_2]|\!|\!| := \left(\|x_1\|_1^2 + \|x_2\|_2^2\right)^{1/2}$; we can easily check that the two norms induce the same topology. It is sometimes useful to regard the graph of a given T as a subset in the Banach space $(\mathcal{X} \oplus \mathcal{Y}, |\!|\!|\cdot|\!|\!|)$. Since the properties of the graph discussed here are characterized by topological means only, they are not sensitive to such a modification.

Section 1.7 The notions of eigenvalue, eigenvector *etc.*make sense for any linear operator; however, if T is not closed it can have an eigenspace which is also not closed (Problem 61). On the other hand, the spectrum has been defined for the closed operators only because otherwise the notion is trivial: if there is a $\lambda \in \mathbb{C}$ such that $(T-\lambda)^{-1} \in \mathcal{B}(\mathcal{X})$, then $T \in \mathcal{C}(\mathcal{X})$ in view of Problem 60b, so $\sigma(T) = \mathbb{C}$ holds if T is not closed.

The theory of analytic vector–valued functions is discussed, *e.g.*, in ⟦ DS 1 ⟧, Sec.III.4; ⟦ HP ⟧, Secs.3.10–3.15; ⟦ Nai 1 ⟧, Secs.3.12 and 4.7; ⟦ Ru 2 ⟧, Chap.3; ⟦ Schw 2 ⟧, Chap.VII. The case when the values of such a function are bounded operators on a Hilbert space is of particular importance; we can then introduce functions which are analytic weakly (strongly, in the operator norm) with respect to different topologies which we shall define in Section 3.1; however, the adjective may be dropped because the three notions coincide — see ⟦ HP ⟧, Sec.3.10; ⟦ RS 1 ⟧, Thm.VI.4. The resolvent expansion (1.7) is sometimes called the *Neumann series* after C. Neumann, who introduced it in the theory of integral equations. An alternative proof of Theorem 1.7.5 which avoids use of the Cauchy formula is given in ⟦ BR 1 ⟧, Prop.2.2.2.

Problems

1. Let $p > 1$ and $q := p/(p-1)$, then the *Hölder inequality,*

$$\sum_j |\xi_j \eta_j| \le \left(\sum_j |\xi_j|^p\right)^{1/p} \left(\sum_j |\eta_j|^q\right)^{1/q},$$

is valid for any n–tuples $\{\xi_j\}, \{\eta_j\} \in \mathbb{C}^n$ or sequences $\{\xi_j\} \in \ell^p$ and $\{\eta_j\} \in \ell^q$.

Analogously, if $|f|^p, |g|^q \in L^1(M,d\mu)$, then $fg \in L^1(M,d\mu)$ and

$$\int_M |fg|d\mu \le \left(\int_M |f|^p d\mu\right)^{1/p} \left(\int_M |g|^q d\mu\right)^{1/q} .$$

Hint: We have $p^{-1}a^p + q^{-1}b^q \ge ab$ for any $a,b \ge 0$.

2. Let $p \ge 1$. The *Minkowski inequality,*

$$\left(\sum_j |\xi_j + \eta_j|^p\right)^{1/p} \le \left(\sum_j |\xi_j|^p\right)^{1/p} + \left(\sum_j |\eta_j|^p\right)^{1/p},$$

is valid for any n–tuples $\{\xi_j\}, \{\eta_j\} \in \mathbb{C}^n$ or sequences from ℓ^p, as well as the analogous integral relation for functions $f,g \in L^p(M,d\mu)$.

3. Algebraic isomorphism preserves linear independence. An n–dimensional complex or real vector space is isomorphic to $\mathbb{C}^n$ and $\mathbb{R}^n$, respectively.

4. Using the Zorn lemma, prove that a Hamel basis exists in any vector space V, and moreover, that for any linearly independent set $M \subset V$ there is a Hamel basis containing M.

5. Let V_1, V_2 be subspaces of a vector space V. Prove

(a) $\dim V_1 + \dim V_2 = \dim(V_1+V_2) + \dim(V_1 \cap V_2)$, where $V_1+V_2 := (V_1 \cup V_2)_{lin}$ is the *algebraic sum* of the subspaces V_1, V_2.

(b) $(V_1 \oplus V_2)/V_1$ is algebraically isomorphic to V_2.

(c) $\dim V/V_1 = \dim V - \dim V_1$ provided $\dim V_1 < \infty$.

6. Let q_F be the quadratic form generated by a sesquilinear form F. Prove

(a) F is symmetric *iff* $q_F(x) \in \mathbb{R}$ for all $x \in \mathbb{R}$.

(b) If F is positive, $p_F : p_F(x) = \sqrt{q_F(x)}$ is a seminorm and the Schwarz inequality is valid; it turns into equality *iff* there are $\alpha, \beta \in \mathbb{C}$, $|\alpha| + |\beta| \ne 0$, such that $q_F(\alpha x + \beta y) = 0$.

7. The algebraic dual V^f is separating points in a vector space V.

8. Let $1 \le p \le q$; then $\|x\|_\infty \le \|x\|_q \le \|x\|_p \le n^{1/p}\|x\|_\infty$ holds for any $x \in \mathbb{C}^n$. The first two inequalities are also valid in ℓ^p, where $\|\cdot\|_q$ makes sense for any $q \ge p$. In this case, however, there are no positive k,k' such that $\|X\|_p \le k\|X\|_\infty$ and $\|X\|_p \le k'\|X\|_q$ with $p<q$ holds for all $X \in \ell^p$.

9. Let $1 \le p \le q$ and $\mu(M) < \infty$; then $\|\cdot\|_q$ is a norm on $L^p(M,d\mu)$ and the inequality $\mu(M)^{-1/q}\|f\|_q \le \mu(M)^{-1/p}\|f\|_p$ holds for any $f \in L^p(M,d\mu)$. Furthermore, $L^\infty(M,d\mu) \subset L^p(M,d\mu)$ and $\|f\|_p \le \mu(M)^{1/p}\|f\|_\infty$ for any $f \in L^\infty(M,d\mu)$.

10. Suppose that a sequence $\{f_n\}_{n=1}^\infty \subset L^\infty(M,d\mu)$ is uniformly bounded, *i.e.*, there is a positive c such that $\|f_n\|_\infty \le c$ for all n, and that $f_n(x) \to f(x)$ μ–a.e. in M; then $f \in L^\infty(M,d\mu)$ and $\|f\|_\infty \le c$.

11. A norm on a vector space is induced by an inner product if and only if it fulfils the parallelogram identity.
Hint: Using the identity show that $r(x,y) := \left(\|x+y\|^2 - \|x-y\|^2\right)/4$ is additive. This implies further that it is homogeneous for rational α, and hence for all $\alpha \in \mathbb{R}$ by continuity.

12. Among the norms of Examples 1.1.4, only $\|\cdot\|_2$ is induced by an inner product.

13. Prove Theorem 1.1.5. Using it, show that any finite–dimensional pre–Hilbert space contains an orthonormal basis.
Hint: For $N = \{y_1, y_2, \dots\}$ consider the vectors $x_n := z_n/\|z_n\|$, where $z_n := y_n - \sum_{k=1}^{n-1} (x_k, y_n) x_k$.

14. To any finite subset $\{x_1, \dots, x_n\} \subset V$ its *Gram determinant* $\Gamma(x_1, \dots, x_n)$ corresponds to the matrix whose entries are the inner products (x_j, x_k), $j,k = 1, \dots, n$. It is non–negative and equals zero *iff* the vectors $x_1, \dots, x_n$ are linearly dependent.

15. A metric ϱ on a vector space V is induced by a norm *iff* it is translation–invariant, $\varrho(x+z, y+z) = \varrho(x,y)$, and homogeneous at dilatations, $\varrho(\alpha x, \alpha y) = |\alpha|\varrho(x,y)$ for all $x,y,z \in V$, $\alpha \in \mathbb{C}$.

16. Let $\mathcal{P} = \{p_n : n = 1, 2, \dots\}$ be a countable family of seminorms on a vector space V which is separating points; then $\varrho : \varrho(x,y) = \sum_{n=1}^{\infty} 2^{-n} \frac{p_n(x-y)}{1+p_n(x-y)}$ is a metric on V which is induced by no norm.

17. Let (X, ϱ) be a metric space. Prove

(a) A set M is closed *iff* $X \setminus M$ is open. The intersection of an arbitrary family of closed sets is a closed set; the same is true for finite unions of closed sets.

(b) A point x belongs to $\overline{M}$ *iff* there is a sequence $\{x_n\} \subset M$ which converges to x.

(c) $M \subset \overline{M} = \overline{\overline{M}}$ for any $M \subset X$. Furthermore, $M \subset N \Rightarrow \overline{M} \subset \overline{N}$ and $\overline{M \cup N} = \overline{M} \cup \overline{N}$, while for intersections we have only $\overline{M \cap N} \subset \overline{M} \cap \overline{N}$.

(d) For a set $M \subset X$ the following conditions are equivalent: (i) M is nowhere dense, (ii) $\overline{M}$ has no interior points, (iii) M is dense in no open ball.

18. Let M be everywhere dense in a metric space X, and let $\mathcal{B}$ be a family of disjoint open balls in X; then $\operatorname{card} \mathcal{B} \le \operatorname{card} M$. In particular, X is nonseparable if there exists $\mathcal{B}$ which is not countable. Using this result, prove that the spaces ℓ^∞ and $L^\infty(\mathbb{R}, dx)$ are nonseparable.

19. Let $f : X \to X'$ be a mapping between two metric spaces. Prove

(a) f is continuous at a point $x \in X$ *iff* $f(x_n) \to f(x)$ holds for any sequence $\{x_n\}$ which converges to x.

(b) f is continuous *iff* it is continuous at each point $x \in X$.

20. Consider a pair of normed spaces $V_j := (V, \|\cdot\|_j)$, $j = 1, 2$, with the same V.

(a) The norm $\|\cdot\|_2$ is said to be *dominated by* $\|\cdot\|_1$ if there is $c > 0$ such that $\|x\|_2 \leq c\|x\|_1$ holds for all $x \in V$. Prove that this is equivalent to continuity of the identical map $V_1 \to V_2$, and this in turn is equivalent to the relation $\tau_2 \subset \tau_1$ between the corresponding norm–induced topologies.

(b) The two norms are *equivalent* if they are mutually dominated; this is true *iff* the identical map $V_1 \to V_2$ is a homeomorphism. If this is the case, V_1 is complete *iff* V_2 is complete, and the two norms induce the same topology.

(c) If V_j are Banach spaces and $\|\cdot\|_2$ is dominated by $\|\cdot\|_1$, then the two norms are mutually equivalent.

Hint: (c) Use the inverse–mapping theorem.

21. Finite–dimensional normed spaces V, V' are linearly homeomorphic *iff* they have the same dimension, $\dim V = \dim V'$.
Hint: Any two norms on a finite–dimensional space are mutually equivalent.

22. Let M be a set in a separable metric space (X, ϱ); then (M, ϱ) is separable.
Hint: If $\{x_j : j = 1, 2, \dots\}$ is everywhere dense, for any natural number k we can find $y_{jk} \in M$ such that $\varrho(x_j, y_{jk}) < \varrho(x_j, M) + \frac{1}{k}$.

23. Check completeness of the following metric spaces: $\mathbb{C}^n$ and $\mathbb{R}^n$ with the norms of Example 1.1.4a, ℓ^∞ and ℓ^p with $p \geq 1$, $C(X)$ and $L^\infty(M, d\mu)$.
Hint: If $\left\{\{\xi_k^{(n)}\}_{k=1}^\infty : n = 1, 2, \dots\right\} \subset \ell^p$ is Cauchy, we have $\sum_{k=1}^N |\xi_k^{(n)} - \xi_k^{(m)}|^p < \varepsilon$ for any N. As for $C(X)$, a uniformly convergent sequence of continuous functions has a continuous limit.

24. If a Cauchy sequence $\{x_n\}_{n=1}^\infty$ contains a convergent subsequence, $\lim_{n\to\infty} x_{n_k} = x$, then $\lim_{n\to\infty} x_n = x$.

25. Let Y be a subset of a complete metric space (X, ϱ). The space (Y, ϱ) is complete *iff* Y is closed.

26. A metric space X is complete *iff* any sequence of closed balls $S_n \equiv S_{\varepsilon_n}(x_n)$ such that $S_{n+1} \subset S_n$, $n = 1, 2, \dots$, and $\lim_{n\to\infty} \varepsilon_n = 0$ has a nonempty intersection, $\bigcap_{n=1}^\infty S_n \neq \emptyset$.

27. Using the previous result, prove the *Baire category theorem:* If a metric space X is complete and $\{M_n : n = 1, 2, \dots\}$ is a countable family of its nowhere dense subsets, then $\bigcup_{n=1}^\infty M_n \neq X$.
Hint: Since $\overline{M_1} \neq X$, there is an open ball U_1 disjoint with M_1 and a closed ball $S_1 \subset U_1$. Furthermore, M_2 is nowhere dense, so there is an open ball $U_2 \subset S_1$ disjoint with M_2 *etc.*; in this way we construct a sequence of embedded balls whose intersection is not contained in any of the sets M_n.

28. Suppose a map $M \mapsto [M]$ on 2^X satisfies the *Kuratowski axioms:* $M \subset [M]$, $[M \cup N] = [M] \cup [N]$, $[\emptyset] = \emptyset$ and $[[M]] = [M]$; then the set family $\{G \in 2^X : [X \setminus G] = X \setminus G\}$ is a topology on X and the corresponding closure $\overline{M}$ is equal to $[M]$ for any $M \subset X$.

29. Prove Proposition 1.2.4. Show that a topological space (X,τ) is Haussdorf *iff* any net $\{x_\alpha\} \subset X$ has at most one limit.

30. Prove: (a) If $\{\tau_\alpha : \alpha \in I\}$ is a family of topologies on X, then $\bigcap_{\alpha\in I} \tau_\alpha$ is a topology which is weaker than each τ_α.

(b) To any $\mathcal{S} \subset 2^X$ there is a topology $\tau(\mathcal{S})$ such that $\mathcal{S} \subset \tau(\mathcal{S})$ and any other topology τ containing $\mathcal{S}$ is stronger than $\tau(\mathcal{S})$. A basis of $\tau(\mathcal{S})$ consists of the set X together with the intersections of all finite subsystems of $\mathcal{S}$.

(c) A system $\mathcal{B} \subset 2^X$ is a basis of the topology $\tau(\mathcal{B})$ on X *iff* (i) for any $x \in X$ there is a $B \in \mathcal{B}$ such that $x \in B$ and (ii) for any $B, C \in \mathcal{B}$ and $x \in B \cap C$ there is a $D \in \mathcal{B}$ such that $x \in D \subset B \cap C$.

(d) Suppose that a nonempty system $\mathcal{B}_x \subset 2^X$ corresponds to any $x \in X$ and the following conditions are valid: (i) $x \in B$ for all $B \in \mathcal{B}_x$, (ii) for any pair $C, D \in \mathcal{B}_x$ there is a set $B \in \mathcal{B}_x$ such that $B \subset C \cap D$, (iii) if $B \in \mathcal{B}_x$ and $y \in B$, then there is a $C \in \mathcal{B}_y$ such that $C \subset B$. Then the systems $\mathcal{B}_x$, $x \in X$, are local bases of the topology $\tau(\mathcal{B})$, where $\mathcal{B} := \bigcup_{x\in X} \mathcal{B}_x$, and $\mathcal{B}$ is a basis of this topology.

31. Prove: (a) If a topological space (X,τ) is separable, then (X,τ') with a weaker topology, $\tau' \subset \tau$, is also separable.

(b) A second countable topological space is separable.

(c) A metric space is separable *iff* it is second countable.

(d) The topological spaces (X,τ_{fin}) and (X,τ_{count}) of Example 1.2.3 are not first countable; the former is separable while the latter is not.

32. Let $\tau_1 \subset \tau_2$ be two topologies on a given X. Prove

(a) $(\overline{M})_{\tau_2} \subset (\overline{M})_{\tau_1}$ for any $M \subset X$.

(b) A net $\{x_\alpha\}$ which converges in (X,τ_2) converges in (X,τ_1) too.

33. Let (X,τ) and (X',τ') be topological spaces. Prove

(a) If a map $f : X \to X'$ is continuous at a point $x \in X$ and a sequence $\{x_n\} \subset X$ converges to x in (X,τ), then $\{f(x_n)\}$ converges to $f(x)$ in (X',τ'). If (X,τ) is first countable, also the converse is true (compare to Problem 19a and Proposition 1.2.4b), and the closure points of a set $M \subset X$ can be characterized by sequences as in Problem 17b

(b) Let $\mathcal{B}_x \subset \tau$, $\mathcal{B}'_{x'} \subset \tau'$ be local bases at the points x and $x' := f(x)$; then f is continuous at x *iff* for any $B' \in \mathcal{B}'_{x'}$ the pull–back $f^{(-1)}(B')$ contains some $B \in \mathcal{B}_x$.

(c) If f is continuous, we have $f(\overline{M}) \subset \overline{f(M)}$ for any $M \subset X$.

34. Let (X_j,τ_j), $j = 1,2$, be topological spaces. Prove

(a) $\tau_{X_1\times X_2}$ is the weakest among the topologies, with respect to which the maps $f_j : f_j(x_1,x_2) = x_j$, $j = 1,2$, are continuous.

(b) In the topological product, we have $\overline{M_1 \times M_2} = \overline{M_1} \times \overline{M_2}$ for any $M_j \subset X_j$.

35. A topological space (X, τ) fulfils the axiom T_3 *iff* to any neighborhood $U(x)$ of a point $x \in X$ there is a neighborhood $V(x)$ such that $\overline{V(x)} \subset U(x)$.

36. A family $\mathcal{C} \subset 2^X$ is called *centered* if any *finite* subsystem of it has a nonempty intersection. A topological space (X, τ) is compact (σ–compact) *iff* any (countable) centered family of closed sets in it has a nonempty intersection.

37. Let X be a topological space. Prove

(A) if X is a T_1–space, $x \in X$ is an accumulation point of a set $M \subset X$ *iff* its arbitrary neighborhood contains infinitely many points of the set M.

(b) If X is a T_1–space and any infinite set $M \subset X$ has an accumulation point, then X is σ–compact.

(c) A second countable space is compact *iff* it is σ–compact.

Hint: (b) Given a centered family $\{F_n : n = 1, 2, \dots\}$, construct the sets $R_N := \bigcap_{n=1}^{N} F_n$, pick a point x_N from any R_N and show that we can select an infinite set $\{x_{N_k} : k = 1, 2, \dots\}$ whose accumulation point belongs to all R_N. (c) Any open covering of X has a countable subcovering.

38. If X is a connected topological space and a map $f : X \to X'$ is continuous, then $f(X)$ is a connected set.

39. A vector space V equipped with a topology τ such that the summation and scalar multiplication are continuous satisfies the axiom T_3; this means that (tv3) may be replaced by the T_1–separability requirement.
Hint: For any $U(0)$ there is $U'(0)$ such that $x, y \in U'(0)$ implies $x - y \in U(0)$. From here we get $\overline{U'(0)} \subset U(0)$, which allows us to use Problem 35.

40. Let V be a vector space and $\mathcal{B}_0 \subset 2^X$ a family with the following properties:

(i) $0 \in B$ for all $B \in \mathcal{B}_0$.

(ii) For any pair $C, D \in \mathcal{B}_0$ there is a $B \in \mathcal{B}_0$ such that $B \subset C \cap D$.

(iii) If $B \in \mathcal{B}_0$ and $y \in B$, then there is $C \in \mathcal{B}_0$ such that $y + C \subset B$.

Then the family $\mathcal{B}_x := x + \mathcal{B}_0$ for any $x \in V$ is a local basis of the weakest topology containing $\mathcal{B} := \bigcup_{x \in V} \mathcal{B}_x$ and $\mathcal{B}$ is a basis of this topology. Under additional conditions (see Theorem 1.4.2), $(V, \tau(\mathcal{B}))$ is a topological vector space.

41. Let V be a topological vector space. The closure of a subspace in V is a subspace, and the intersection of any family of closed subspaces is a closed subspace.
Hint: Use Problems 33c and 34.

42. Let (V, τ) be a topological vector space; then

(a) If $U(0)$ is a neighborhood of zero, then for any $x \in V$ there is an $a_x > 0$ such that $\alpha x \in U(0)$ holds for $0 < |\alpha| \le a_x$.

(b) A set $M \subset V$ is called *balanced* if $M = \bigcup_{|\alpha|\le 1} \alpha M$. Prove that any neighborhood $U(0)$ contains a balanced neighborhood $V(0)$ of zero.

Hint: Use continuity of scalar multiplication.

43. Let ϱ be the metric of Problem 16, then the corresponding topology coincides with the locally convex topology generated by the family $\mathcal{P}$. A sequence $\{x_n\} \subset V$ is Cauchy in the metric space (V,ϱ) *iff* to any n and $\varepsilon > 0$ there is a $j(\varepsilon,n)$ such that $p_n(x_k - x_\ell) < \varepsilon$ holds for all $k,\ell > j(\varepsilon,n)$.

44. Prove the relation (1.1).
Hint: Any $f \in \mathcal{S}(\mathbb{R}^n)$ can be approximated with respect to all seminorms $\|\cdot\|_{J,K}$ by the functions $f_\varepsilon : f_\varepsilon(x) = f(x)h(\varepsilon x)$, where h is a function of $C_0^\infty(\mathbb{R}^n)$ such that $h(x) = 1$ for $|x| \le 1$.

45. Prove Theorem 1.5.1.
Hint: A Cauchy sequence $\{x_k\}$ contains $\{x_{k_n}\}$ such that $\|x_{k_{n+1}} - x_{k_n}\| < 2^{-n}$.

46. A complete envelope $(\tilde V,\varrho)$ of a normed space V is a Banach space which is unique up to linear isometries preserving V. In particular, if V is a pre–Hilbert space we can introduce on $\tilde V$ the inner product $(\cdot,\cdot)_*$ such that $\varrho(x,y) = \sqrt{(x-y,x-y)_*}$.

47. Prove: (a) If μ is a σ–finite measure on a σ–algebra $\mathcal{A} \subset 2^X$, then the set $\{\chi_M : M \in \mathcal{A},\ \mu(M) < \infty\}$ is total in $L^p(X,d\mu)$.

(b) The set $S^{(n)}$ of Example 1.5.3c is total in $L^p(\mathbb{R}^n,d\mu)$.

(c) If J is a bounded interval in $\mathbb{R}^n$, then there is a sequence $\{f_k\} \subset C_0^\infty(\mathbb{R}^n)$ of functions with supports in J, $|f_k(x)| \le 1$, which converges to χ_J pointwise in the interior of J, as well as in the $\|\cdot\|_p$ norm.

(d) Let $f \in L^p(J,d\mu)$, where J is a bounded interval in $\mathbb{R}^n$ and μ a Borel measure; then there is a sequence $\{f_k\} \subset C_0^\infty(\mathbb{R}^n)$ such that $\operatorname{supp} f_k \subset J$ and $\|f_k - f\|_p \to 0$.

Hint: (a) If $\mu(X) < \infty$, one has $L^p \subset L^1$ and to any $f \in L^1$ there is a σ–simple function g_ε such that $|f(x) - g_\varepsilon(x)| < \varepsilon$ a.e. in M, (b) for any $M \in \mathcal{B}^n$ and $\varepsilon > 0$ there is $R \in \mathcal{R}^n$ such that $\mu(R\Delta M)^{1/p} < \varepsilon$ for all $x \in X$.

48. The space $(C_\infty(X), \|\cdot\|_\infty)$ is complete and $C_0(X)$ is dense in it.

49. Let $\mathcal{X}$ be a Banach space. If $\mathcal{Z}$ is a closed subspace in $\mathcal{X}$, prove that $\|\tilde x\|_{\mathcal{Z}} := \inf_{z\in\mathcal{Z}} \|x - z\|$ is a norm on the factor space $\tilde{\mathcal{X}} := \mathcal{X}/\mathcal{Z}$ and the corresponding normed space is complete.
Hint: Use Theorem 1.5.1.

50. Let $B : V_1 \to V_2$ be a linear operator; then the following conditions are equivalent:

(i) $B \in \mathcal{B}(V_1,V_2)$.

(ii) B is continuous.

(iii) B is continuous at a point $x \in V_1$.

51. The operator norm can be expressed equivalently as

$$\|B\| = \sup_{x\in V_1\setminus\{0\}} \frac{\|Bx\|_2}{\|x\|_1} = \sup_{x\in B_1(0)} \|Bx\|_2 = \inf \mathcal{C}_B\,,$$

where $B_1(0)$ is the closed unit ball in the normed space V_1 and $\mathcal{C}_B$ is the set of all c such that $\|Bx\|_2 \le c\|x\|_1$ for all $x \in V_1$.

52. Show that the Fourier transformation maps $\mathcal{S}(\mathbb{R}^n)$ onto itself, and check that

$$\int_{\mathbb{R}^n} e^{ix\cdot y} g(y)\hat f(y)\,dy = \int_{\mathbb{R}^n} \hat g(y) f(x+y)\,dy$$

holds for any $f,g \in \mathcal{S}(\mathbb{R}^n)$ and $x \in \mathbb{R}^n$.
Hint: To estimate $y^J(D^K\hat f)(y)$, use an integration by parts together with the fact that $x \mapsto (1+|x|^{2n})^{-1}$ belongs to $L^1(\mathbb{R}^n)$.

53. The Fourier–Plancherel operator is surjective, $\mathrm{Ran}\,F = L^2(\mathbb{R}^n)$, and its inverse F^{-1} is the continuous extension of $\mathcal{F}_0^{-1}$.

54. Let V be a normed space. Prove that

(a) The dual spaces V^* and V^f coincide *iff* $\dim V < \infty$.

(b) If $\dim V = \infty$, there are unbounded linear functionals on V.

Hint: (b) Choose a countable subset $\{e_j\}$ in a suitable Hamel basis of V and define $f : V \to \mathbb{C}$ by $f(e_j) := j$ and $f(x) = 0$ outside $\{e_j\}_{lin}$.

55. Let V be a normed space. If the functionals J_x on V^* are defined by $J_x(f) := f(x)$, then the map $x \mapsto J_x$ is a linear isometry of V to a subspace in V^{**}.

56. Let (V,τ) be a finite–dimensional topological vector space, then the weak topology τ_w coincides with τ.

57. Given a Banach space $\mathcal{X}$, operators $B, C \in \mathcal{B}(\mathcal{X})$, and sequences $\{B_n\}$, $\{C_n\} \subset \mathcal{B}(\mathcal{X})$, prove

(a) If $B_n x \to Bx$ and $C_n x \to Cx$, then $B_n C_n x \to BCx$.

(b) If $B_n y \to By$ for all y from a total set in $\mathcal{X}$ and the sequence $\{\|B_n\|\}$ is bounded, then $B_n x \to Bx$ for all $x \in \mathcal{X}$.

58. Prove the inverse–mapping theorem.

59. Define the operator T from $\mathcal{X} \equiv C[0,1]$ to itself with the domain $D_T := \{f \in \mathcal{X} : f' \in \mathcal{X}\}$ by $Tf := f'$. Show that T is unbounded and closed, and that the same is true for its restriction $T^{(0)}$ to $D^{(0)} := \{f \in D_T :\ f(0) = 0\}$.

60. Let $\mathcal{X}$, $\mathcal{Y}$, $\mathcal{Z}$ be Banach spaces. Prove

(a) A linear operator $T : \mathcal{X} \to \mathcal{Y}$ is closable if there is a closed $S : \mathcal{X} \to \mathcal{Y}$ such that $\Gamma(T) \subset \Gamma(S)$; in that case $\overline{T} \subset S$.

(b) If $T : \mathcal{X} \to \mathcal{Y}$ is closed and invertible, then T^{-1} is a closed linear operator from $\mathcal{Y}$ to $\mathcal{X}$. The operator T is unbounded if there is a sequence $\{y_n\} \subset \operatorname{Ran} T$ such that $T^{-1}y_n \to 0$ and $\{y_n\}$ does not converge to zero; on the other hand, if T is unbounded, then there is a sequence of unit vectors $y_n \in \operatorname{Ran} T$ which has no accumulation points and fulfils $T^{-1}y_n \to 0$.

(c) Let $S : \mathcal{X} \to \mathcal{Y}$ be a closed linear operator; if $T : \mathcal{Y} \to \mathcal{Z}$ is linear, invertible, and $T^{-1} \in \mathcal{B}(\mathcal{Z}, \mathcal{Y})$; then $TS : \mathcal{X} \to \mathcal{Z}$ is closed.

61. Define $T : L^p(0,1) \to L^p(0,1)$ with the domain $D_T := C[0,1]$ by $(Tf)(x) = f(0)x$. Prove that T is neither closed nor closable, and find an eigenspace of T which is not closed.

62. Let $T \in \mathcal{C}(\mathcal{X})$, then the *first resolvent identity* (or *Hilbert identity*),

$$R_T(\lambda) - R_T(\mu) = (\lambda - \mu) R_T(\lambda) R_T(\mu) ,$$

holds for any $\lambda, \mu \in \rho(T)$ and the operators $R_T(\lambda)$, $R_T(\mu)$ commute mutually.

63. Let $T, S \in \mathcal{C}(\mathcal{X})$ with $D_T \subset D_S$ and $\lambda \in \rho(T) \cap \rho(S)$; then the *second resolvent identity,*

$$R_T(\lambda) = R_S(\lambda) - R_S(\lambda)(T - S) R_T(\lambda) ,$$

is valid. Moreover, let $T - S = UV$, where $D_V = D_S$, $D_U = V D_T$, $\operatorname{Ran} V = \mathcal{X}$, and V is invertible, then the operator $I + V R_S(\lambda) U$ is also invertible and

$$R_T(\lambda) = R_S(\lambda) - R_S(\lambda) U (I + V R_S(\lambda) U)^{-1} V R_S(\lambda) .$$

64. Given an operator $T \in \mathcal{C}(\mathcal{X})$, we define its *regularity domain* $\pi(T)$ as the set of all $\lambda \in \mathbb{C}$ for which there is $c(\lambda) > 0$ such that $\|(T - \lambda)x\| \ge c(\lambda)\|x\|$ holds for all $x \in D_T$. Prove that

(a) $\pi(T)$ is an open set.

(b) $\operatorname{Ran}(T - \lambda)$ is a closed subspace in $\mathcal{X}$ for each $\lambda \in \pi(T)$.

(c) $\rho(T) \subset \pi(T)$.

65. The operator Q on $C[0,1]$ defined by $(Qf)(x) := xf(x)$ is bounded, $\|Q\| = 1$, and we have $\sigma_p(Q) = \sigma_c(Q) = \emptyset$ while $\sigma_r(Q) = [0,1]$.

66. Given a bounded operator B on a Banach space $\mathcal{X}$, we can define $e^B \in \mathcal{B}(\mathcal{X})$ as the norm limit $e^B := \lim_{N\to\infty} \sum_{n=1}^{N} \frac{1}{n!} B^n$. Prove that

(a) $\|e^B\| \le e^{\|B\|}$.

(b) $BC = CB$ implies $e^{B+C} = e^B e^C = e^C e^B$.

(c) The map $z \mapsto e^{Bz}$ is continuous with respect to the operator norm.

Chapter 2

Hilbert spaces

2.1 The geometry of Hilbert spaces

In Section 1.1 we introduced pre–Hilbert spaces and derived some of their simple properties, in particular, the Schwarz inequality, parallelogram identity, and polarization formula. A pre–Hilbert space which is complete with respect to the metric induced by the inner product is called a **Hilbert space**. Any such space is at the same time a Banach space, so it has all of the properties discussed above.

If two Hilbert spaces are linearly isometric they have the same algebraic and metric properties; we usually say briefly that they are *isomorphic*. By polarization, an isomorphism of Hilbert spaces preserves the inner product. Furthermore, the standard procedure of Theorem 1.2.2 allows us to complete any pre–Hilbert space, uniquely up to an isomorphism, to a Hilbert space (*cf.* Problem 1.46). On the other hand, not every notion is adapted from Banach–space theory without a modification. For instance, given Hilbert spaces $\mathcal{H}_j$, $j = 1, 2$, we can equip their vector–space direct sum with the inner product defined by

$$([x_1, x_2], [y_1, y_2]) := (x_1, y_1)_1 + (x_2, y_2)_2;$$

it is easy to check that the pre–Hilbert space obtained in this way is complete, *i.e.*, a Hilbert space which is called the *direct sum of the Hilbert spaces* $\mathcal{H}_1, \mathcal{H}_2$ and is denoted as $\mathcal{H}_1 \oplus \mathcal{H}_2$. It does not coincide with the Banach–space direct sum of $\mathcal{H}_1$ and $\mathcal{H}_2$ though, of course, they are topologically equivalent (see the notes to Section 1.6). The construction extends to any finite number of Hilbert spaces; for infinite families the direct sum will be introduced in Section 2.3 below.

Using the Schwarz inequality, we can check easily that the inner product in $\mathcal{H}$ is jointly continuous, *i.e.*, it is continuous as a map $\mathcal{H} \oplus \mathcal{H} \to \mathbb{C}$. Combining this with the properties of the orthogonal complement mentioned in Section 1.1 we get

$$M^\perp = \overline{(M^\perp)_{lin}} = \left(\overline{M_{lin}}\right)^\perp .$$

In particular, the orthogonal complement to any subset of a Hilbert space is a closed subspace. The existence of an inner product together with the completeness requirement allow us to derive for any Hilbert space several results which have a simple geometric meaning.

2.1.1 Proposition (orthogonal–projection lemma): Let $\mathcal{G}$ be a closed subspace in $\mathcal{H}$; then for any $x \in \mathcal{H}$ there is just one vector $y_x \in \mathcal{G}$ such that the distance $\varrho(x,\mathcal{G}) = \|x-y_x\|$.

Proof: By definition, $\varrho(x,\mathcal{G}) := d := \inf_{y\in\mathcal{G}} \|x-y\|$, so there is a sequence $\{y_n\} \subset \mathcal{G}$ such that $\lim_{n\to\infty} \|x-y_n\| = d$. The parallelogram identity implies that $\{y_n\}$ is Cauchy, $\|y_n-y_m\|^2 = 2\|x-y_n\|^2 + 2\|x-y_m\|^2 - 4\|x-(y_n+y_m)/2\|^2 \le 2\|x-y_n\|^2 + 2\|x-y_m\|^2 - 4d^2 \to 0$ for $n,m\to\infty$. Since $\mathcal{G}$ is closed, $\{y_n\}$ converges to some $y_x \in \mathcal{G}$ and $\lim_{n\to\infty} \|x-y_n\| = \|x-y_x\|$. The uniqueness is checked again by the parallelogram identity. ∎

The vector y_x is called the *orthogonal projection* of x onto the subspace $\mathcal{G}$.

2.1.2 Theorem (orthogonal–decomposition theorem): Let $\mathcal{G}$ be a closed subspace in $\mathcal{H}$; then to each vector $x \in \mathcal{H}$ there is just one $y \in \mathcal{G}$ and $z \in \mathcal{G}^\perp$ such that $x = y+z$.

Proof: The uniqueness follows from $\mathcal{G}\cap\mathcal{G}^\perp = \{0\}$ (see Problem 2). To prove the existence, we choose for y the orthogonal projection of x onto $\mathcal{G}$, so it is sufficient to check that $z := x-y \in \mathcal{G}^\perp$, or equivalently, $(e,z)=0$ for any unit vector $e\in\mathcal{G}$. We have $\|z\| = \inf_{y'\in\mathcal{G}} \|x-y'\|$ by the orthogonal–projection lemma; hence choosing $y' = y + (e,z)e$ we get $\|z\|^2 \le \|z-(e,z)e\|^2 = \|z\|^2 - |(e,z)|^2$, which is possible only if $(e,z)=0$. ∎

The theorem thus ensures existence of the orthogonal decomposition $\mathcal{H} = \mathcal{G}\oplus\mathcal{G}^\perp$ for any closed subspace $\mathcal{G}\subset\mathcal{H}$. Combining this with the properties of the orthogonal complement (Problem 2) we get the following useful criterion.

2.1.3 Proposition: A set $M\subset\mathcal{H}$ is total *iff* $M^\perp = \{0\}$.

We saw in the previous chapter that the problem of finding the dual of a given Banach space has no universal solution and may be difficult in some cases. In Hilbert spaces the situation is much simpler.

2.1.4 Theorem (Riesz lemma): To any $f\in\mathcal{H}^*$ there is just one vector $y_f\in\mathcal{H}$ such that $f(x) = (y_f,x)$ holds for all $x\in\mathcal{H}$. The map $f\mapsto y_f$ is an antilinear isometry of the spaces $\mathcal{H}^*$ and $\mathcal{H}$.

Proof: Since f is linear and continuous, $\operatorname{Ker} f$ is a closed subspace in $\mathcal{H}$. If $\operatorname{Ker} f = \mathcal{H}$ we have $f=0$ and $y_f=0$. In the opposite case there is a unit vector $y\in(\operatorname{Ker} f)^\perp$; then $z_x := f(x)y - f(y)x$ belongs to $\operatorname{Ker} f$ for any $x\in\mathcal{H}$. Hence $0 = (y,z_x) = f(x) - (y,x)f(y)$ and we can choose $y_f = \overline{f(y)}y$. The uniqueness of y_f and the injectivity of the map $f\mapsto y_f$ follow from the fact that $\mathcal{H}^\perp = \{0\}$; the map is at the same time surjective because $(y,\cdot)$ is a bounded linear functional for any $y\in\mathcal{H}$. The relation $f(x) = (y_f,x)$ implies $\|f\| \le \|y_f\|$ by the Schwarz inequality; on the other hand, setting $x=y_f$ we get $\|y_f\|^2 = f(y_f) \le \|f\|\,\|y_f\|$, so together we have $\|f\| = \|y_f\|$. ∎

In view of Theorem 1.5.12a, a sequence $\{x_n\}$ in a Hilbert space $\mathcal{H}$ converges weakly to a vector x *iff* $(y, x_n) \to (y, x)$ for all $y \in \mathcal{H}$. This yields the following useful result.

2.1.5 Proposition: If a sequence $\{x_n\} \subset \mathcal{H}$ converges weakly to $x \in \mathcal{H}$ and $\|x_n\| \to \|x\|$, then it also converges to x in the norm topology.

We have mentioned the role of bases in a finite–dimensional vector space. Using the inner product, we can extend this notion to infinite–dimensional Hilbert spaces. Let $\mathcal{E} := \{\, e_\alpha : \alpha \in I \,\}$ be an orthonormal set in $\mathcal{H}$, not necessarily total. The numbers $\xi_\alpha := (e_\alpha, x)$ for a given $x \in \mathcal{H}$ are called **Fourier coefficients** of the vector x with respect to $\mathcal{E}$. Let $\{\, \alpha_j : j = 1, 2, \dots \}$ be an at most countable subset of the index set I; then the *Bessel inequality,*

$$\sum_j \left|(e_{\alpha_j}, x)\right|^2 \le \|x\|^2 ,$$

is valid (Problem 1). Using it, we can easily derive the following properties of orthonormal sets.

2.1.6 Proposition: (a) The set X of nonzero Fourier coefficients of a vector x with respect to a given $\mathcal{E}$ is at most countable.

(b) Any orthonormal set $\{e_j\}$ fulfils $\text{w-lim}_{j\to\infty} e_j = 0$.

Proof: The sets $X_n := \{\, \xi_\alpha : \alpha \in I,\ |\xi_\alpha| \ge n^{-1} \,\}$, $n = 1, 2, \dots$, are finite, so $X = \bigcup_{n=1}^\infty X_n$ is at most countable. The series $\sum_j |(e_j, x)|^2$ converges, which implies that $\lim_{j\to\infty}(e_j, x) = 0$ for any $x \in \mathcal{H}$. ∎

Since the sequence $\{\xi_{\alpha_j}\}$ of nonzero Fourier coefficients fulfils $\sum_{j=1}^\infty |\xi_{\alpha_j}|^2 \le \|x\|^2$, the sequence of the vectors $y_n := \sum_{j=1}^n \xi_{\alpha_j} e_{\alpha_j}$ is Cauchy. Its limit, which we denote as y, belongs to the closed subspace $\overline{\mathcal{E}_{lin}}$, and we can easily check that $(e_\alpha, x - y) = 0$ holds for any $\alpha \in I$, *i.e.*, $x - y \in \mathcal{E}^\perp$. This means that y is the orthogonal projection of x onto $\overline{\mathcal{E}_{lin}}$, and therefore

$$\|x\|^2 = \|y\|^2 + \|x-y\|^2 = \sum_{j=1}^\infty \left|\xi_{\alpha_j}\right|^2 + \|x-y\|^2 .$$

In particular, if $\mathcal{E}$ is a total orthonormal set in $\mathcal{H}$, we have

$$x = y = \lim_{n\to\infty} y_n =: \sum_{j=1}^\infty \xi_{\alpha_j} e_{\alpha_j} \qquad \text{and} \qquad \|x\|^2 = \sum_{j=1}^\infty \left|\xi_{\alpha_j}\right|^2 .$$

The first of these relations shows that the sum of the series on the right side is independent of the order of the terms. This is true even in the case when $\mathcal{E}$ is not total: if $\{\xi_{\beta_k}\}$ is another sequence formed of the set of nonzero Fourier coefficients of the vector x, then $z := \lim_{n\to\infty} \sum_{k=1}^{\infty} \xi_{\beta_k} e_{\beta_k}$ is again the orthogonal projection of x onto $\overline{\mathcal{E}_{lin}}$, and therefore $z = y$ by Proposition 2.1.1. In this way, total orthonormal sets allow us to characterize any vector $x \in \mathcal{H}$ by means of its Fourier coefficients. The set $\{\, e_{\alpha_j} : j = 1, 2, \dots\}$ in the above formulas depends in general on x but this fact is not important. If $\mathcal{H}$ is separable then any orthonormal set in it is at most countable (see Proposition 2.2.1 below) and we get

$$x = \sum_{j=1}^{\infty} (e_j, x)e_j \,, \qquad \|x\|^2 = \sum_{j=1}^{\infty} |(e_j, x)|^2 \,;$$

the first relation is called the **Fourier expansion** of the vector x, the other is the **Parseval identity**. In a nonseparable $\mathcal{H}$ the sets of Fourier coefficients are not countable; then it is sufficient to replace the sequences by the corresponding nets,

$$\begin{aligned} x &= \lim_{\mathcal{S}} \sum_{\alpha\in K} (e_\alpha, x)e_\alpha =: \sum_{\alpha\in\mathcal{S}} (e_\alpha, x)e_\alpha \,, \\ \|x\|^2 &= \lim_{\mathcal{S}} \sum_{\alpha\in K} |(e_\alpha, x)|^2 = \sup_{K\in\mathcal{S}} \sum_{\alpha\in K} |(e_\alpha, x)|^2 =: \sum_{\alpha\in\mathcal{S}} |(e_\alpha, x)|^2 \,, \end{aligned}$$

where $\mathcal{S}$ is the family of all finite subsets in I ordered by inclusion; the relations are now checked as in Proposition 3.2.13 below.

Total orthonormal sets may serve therefore as the sought generalization of the (orthonormal) bases in finite–dimensional vector spaces; any such set will be called an **orthonormal basis**. In view of Proposition 2.1.3, an orthonormal set $\mathcal{E}$ is an orthonormal basis *iff* $\mathcal{E}^\perp = \{0\}$, or equivalently, if it is maximal, *i.e.*, it is not a proper subset of another orthonormal set.

2.1.7 Theorem: Any Hilbert space $\mathcal{H}$ has an orthonormal basis. Any two bases in a given $\mathcal{H}$ have the same cardinality.

Proof: The family $\mathcal{S}$ of all orthonormal sets in $\mathcal{H}$ is partially ordered by set inclusion; then Zorn's lemma implies existence of its maximal element $\mathcal{E}$. If there were a nonzero $x \in \mathcal{E}^\perp$, then $\mathcal{E}$ would be a proper subset of the orthonormal set $\mathcal{E} \cup \{x/\|x\|\}$, but this contradicts its maximality. The other assertion can be proven using Proposition 2.1.6 (see the notes). ∎

Hence we may introduce the *Hilbertian dimension* of $\mathcal{H}$ as the cardinality of any orthonormal basis of it; note that it need not coincide with the algebraic dimension of $\mathcal{H}$ unless both of them are finite — *cf.* Problem 4. Unless stated otherwise, the term dimension in the following always means the Hilbertian dimension.

2.1.8 Theorem: Hilbert spaces $\mathcal{H}$, $\mathcal{H}'$ are isomorphic *iff* $\dim\mathcal{H} = \dim\mathcal{H}'$.

Proof: If $\dim\mathcal{H} = \dim\mathcal{H}'$ there are orthonormal bases $\mathcal{E}$, $\mathcal{E}'$ of the same cardinality, so we can construct a bijection $V_0 : \mathcal{E} \to \mathcal{E}'$. Its linear extension is a bijection from $\mathcal{E}_{lin}$ to $\mathcal{E}'_{lin}$ which preserves the inner product. Hence it is bounded, and since $\overline{\mathcal{E}_{lin}} = \mathcal{H}$, it has a unique continuous extension $V \in \mathcal{B}(\mathcal{H}, \mathcal{H}')$; it is easy to check that it is isometric and is therefore the sought isomorphism. On the other hand, if there exists an isomorphism $V : \mathcal{H} \to \mathcal{H}'$ and $\mathcal{E}$ is an orthonomal basis in $\mathcal{H}$ then $V\mathcal{E}$ is an orthonormal set of the same cardinality as $\mathcal{E}$, which is moreover total in $\mathcal{H}'$ because $(V\mathcal{E})^\perp = V\mathcal{E}^\perp = \{0\}$. ∎

2.2 Examples

Since Hilbert spaces represent the raw material for most of the problems discussed in this book, it is appropriate to list now those which will appear frequently. Since most of them are separable, the following result is useful.

2.2.1 Proposition: Any orthonormal set in a separable $\mathcal{H}$ is at most countable. A Hilbert space $\mathcal{H}$ is separable *iff* it has at most a countable orthonormal basis.

Proof: If $\mathcal{E} := \{\, e_\alpha : \alpha \in I \,\}$ is an orthonormal set then $\{\, U_\varepsilon(e_\alpha) : \alpha \in I \,\}$ with $\varepsilon \in (0, 2^{-1/2}]$ is a disjoint set of open balls in $\mathcal{H}$ of the same cardinality as $\mathcal{E}$, so the first assertion follows from Problem 1.18. Together with Theorem 2.1.7, this implies in turn the existence of an at most countable orthonormal basis in a separable $\mathcal{H}$, while the opposite implication follows from Lemma 1.5.2. ∎

This means, for instance, that the space ℓ^2 represents a canonical realization of an infinite–dimensional separable Hilbert space. It would be a mistake, however, to conclude that it is sufficient to know one Hilbert space of each dimension only. The problem is that we are usually interested in an operator or a family of operators on a given $\mathcal{H}$ whose properties may be derived easily in one realization while being disguised completely in another one. With a certain overstatement, we can say that if Dirac had thought of operators only as infinite–dimensional matrices he would never been able to formulate the equation of motion of a relativistic electron.

In the following examples, we shall construct countable orthonormal bases in some frequently used Hilbert spaces, thereby proving their separability.

2.2.2 Example (the space $L^2(a,b)$, $b{-}a < \infty$): In view of the obvious isomorphism we may consider, *e.g.*, the space $L^2(0,2\pi)$. The set $\mathcal{E}_T := \{\, e_k : k = 0, \pm1, \pm2, \dots \}$ with $e_k(x) := (2\pi)^{-1/2} e^{ikx}$ is orthonormal; we shall show that it is total. Let $f \in \mathcal{E}_T^\perp$. The function $g : g(x) = \int_0^x f(s)\, ds + C$ is absolutely continuous in $[0, 2\pi]$ and an integration by parts yields $0 = (e_k, f) = (2\pi)^{-1/2}[g(2\pi) - g(0)] - ik(e_k, g)$ for $k = 0, \pm1, \pm2, \dots$. This implies, in particular, $g(2\pi) = g(0) = C$, and choosing C so that $\int_0^{2\pi} g(x)\, dx = 0$, we get $(e_k, g) = 0$ for $k = 0, \pm1, \pm2, \dots$. By Fejér's theorem, g can be approximated uniformly by trigonometric polynomials: for any $\varepsilon > 0$ there is $T_\varepsilon = \sum_{k=-n}^{n} \alpha_k e_k$ such that $\|g - T_\varepsilon\|_\infty < \varepsilon$. Since $(T_\varepsilon, g) = 0$, we get

$$\|g\|^2 = |(g - T_\varepsilon, g)| \le \|g - T_\varepsilon\|\, \|g\| < \varepsilon \|g\| ,$$

so $g = 0$; in view of its continuity we have $g(x) = 0$ for all $x \in [0, 2\pi]$, and therefore $f(x) = 0$ a.e. in $[0, 2\pi]$, which is what we set out to prove. The set $\mathcal{E}_T$ (or a sequence arranged of it) is called the *trigonometric basis* in $L^2(0, 2\pi)$.

In a similar way, one checks that the set $\{ x^l : l = 0,1,\dots \}$ is total in $L^2(a,b)$; since it is linearly independent it can be orthogonalized by Theorem 1.1.5. This yields, in particular, the orthonormal basis $\mathcal{E}_P := \{ (l+\frac{1}{2})^{1/2} P_l : l = 0,1,\dots \}$ in $L^2(-1,1)$, where

$$P_l : P_l(x) = (2^l\, l!)^{-1} \frac{d^l}{dx^l} (x^2-1)^l$$

are *Legendre polynomials.*

2.2.3 Example (the spaces $L^2(\mathbb{R})$ and $L^2(\mathbb{R}^+)$): Let us start with $L^2(\mathbb{R})$. We again want to construct orthonormal bases by orthogonalization of a suitable total linearly independent set; this time we use $\mathcal{T}^{(\alpha)} := \{ f_k^{(\alpha)} : k = 0,1,\dots \}$, where $f_k^{(\alpha)}(x) := |x|^{\alpha+1/2} e^{-x^2/2} x^k$. We assume $\alpha > -1$, so $\mathcal{T}^{(\alpha)}$ is contained in $L^2(\mathbb{R})$; we shall show that it is total. If $g \in \left(\mathcal{T}^{(\alpha)}\right)^{\perp}$ we have $\sum_{k=0}^n \frac{(i\lambda)^k}{k!} \left(g, f_k^{(\alpha)}\right) = 0$ for any real λ and a natural number n, and therefore

$$\int_{\mathbb{R}} e^{-i\lambda x} g(x) f_0^{(\alpha)}(x)\, dx = 0\,, \quad \lambda \in \mathbb{R}\,,$$

by the dominated–convergence theorem, or $\mathcal{F}(g f_0^{(\alpha)}) = 0$. Since $\mathcal{F}$ is injective (see the notes to Section 1.5) and $f_0^{(\alpha)}(x) > 0$ for all $x \in \mathbb{R}$, we conclude that $g(x) = 0$ a.e. in $\mathbb{R}$. Gram–Schmidt procedure applied to $\mathcal{T}^{(\alpha)}$ then yields the orthonormal basis $\mathcal{E}_L^{(\alpha)} := \{ g_n^{(\alpha)} : n = 0,1,\dots \}$ with

$$\begin{aligned} g_{2k}^{(\alpha)}(x) &:= \left(\frac{k!}{\Gamma(k+\alpha+1)}\right)^{1/2} |x|^{\alpha+1/2} e^{-x^2/2}\, L_k^{(\alpha)}(x^2) \\ g_{2k+1}^{(\alpha)}(x) &:= \left(\frac{k!}{\Gamma(k+\alpha+2)}\right)^{1/2} |x|^{\alpha+1/2} e^{-x^2/2}\, x\, L_k^{(\alpha+1)}(x^2)\,, \end{aligned}$$

where $L_k^{(\alpha)} : L_k^{(\alpha)}(r) = \frac{e^r}{k!} r^{-\alpha} \frac{d^k}{dr^k} \left(e^{-r} r^{k+\alpha}\right)$ are *Laguerre polynomials.* In the case $\alpha = -1/2$ we conventionally use the functions $h_n := (-1)^{[n/2]} g_n^{(-1/2)}$, where $[\cdot]$ denotes the integer part, or

$$h_n(x) = (2^n\, n!)^{-1/2}\, \pi^{-1/4}\, e^{-x^2/2}\, H_n(x)\,, \tag{2.1}$$

which express in terms of *Hermite polynomials* $H_n : H_n(x) = (-1)^n e^{x^2} \frac{d^n}{dx^n} e^{-x^2}$ (see also Problem 6). As for the space $L^2(\mathbb{R}^+)$, one can use the results derived above to check that $\mathcal{F}_L^{(\alpha)} := \{ \psi_n^{(\alpha)} : n = 0,1,\dots \}$ with

$$\psi_n^{(\alpha)}(x) := \left(\frac{n!}{\Gamma(n+\alpha+1)}\right)^{1/2} x^{\alpha/2}\, e^{-x/2}\, L_n^{(\alpha)}(x)$$

and $\alpha > -1$ is an orthonormal basis (Problem 7).

Given an orthonormal basis in $L^2(\mathbb{R})$ we can construct orthonormal bases in the spaces $L^2(\mathbb{R}^n)$, $n = 2,3,\dots$, and at the same time check their separability, using the following more general result.

2.2.4 Proposition: Let μ, ν be σ–finite non–negative measures on sets M, N, respectively, such that the spaces $L^2(M, d\mu)$ and $L^2(N, d\nu)$ are separable (see the notes), and let further $\mathcal{E} := \{ e_k : k = 1, 2, \dots \}$ and $\mathcal{F} := \{ f_l : l = 1, 2, \dots \}$ be orthonormal bases in these spaces. Then $L^2(M \times N, d(\mu \otimes \nu))$ is separable and the set $\{ g_{kl} : k, l = 1, 2, \dots \}$ with $g_{kl} : g_{kl}(x, y) = e_k(x) f_l(y)$ is an orthonormal basis in it.

Proof: By the Fubini theorem, $\{g_{kl}\}$ is an orthonormal set in $L^2(M \times N, d(\mu \otimes \nu)) =: L^2_\otimes$, so it remains for us to check that it is total. Let $(h, g_{kl}) = 0$ for $h \in L^2_\otimes$ and $k, l = 1, 2, \dots$. The y–cut of the function h, $h^y : h^y(x) = h(x, y)$, then belongs to $L^2(M, d\mu)$ for ν–a.a. $y \in N$, and $H : H(y) = \|h^y\|^2$ belongs to $L^1(N, d\nu)$. The assumption can be thus written as

$$\int_N H_k(y) \overline{f_l}(y)\, d\nu(y) = 0 \,, \qquad k, l = 1, 2, \dots,$$

where $H_k(y) := (e_k, h^y)$. Due to the σ–finiteness of ν, the functions H_k are measurable, and the Schwarz inequality yields $|H_k(y)|^2 \le \|h^y\|^2 = H(y)$ for ν–a.a. $y \in N$, so $H_k \in L^2(N, d\nu)$. Since $\mathcal{E}$ and $\mathcal{F}$ are orthonormal bases, the relations $(H_k, f_l) = 0$ imply $H_k(y) = 0$ for ν–a.a. $y \in N$, and this in turn means $h^y = 0$ for ν–a.a. $y \in N$. Hence we get $\|h\|^2_\otimes = \int_N \|h^y\|^2 d\nu(y) = 0$, *i.e.*, $h = 0$. ∎

A frequently used generalization of the spaces $L^2(X, d\mu)$ is obtained if we replace complex functions by functions with values in some auxiliary Hilbert space.

2.2.5 Example (vector–valued functions): Let $\mathcal{G}$ be a separable Hilbert space and μ a non–negative σ–finite measure defined on a σ–algebra $\mathcal{A} \subset 2^X$. It is easy to check that the functions $f : X \to \mathcal{G}$ with the pointwise defined summation and scalar multiplication form a vector space; we denote it as $\mathcal{V}(X, \mathcal{G})$ or simply $\mathcal{V}$. A function F is measurable if the complex–valued functions $(g, F(\cdot))_\mathcal{G}$ are measurable for all $g \in \mathcal{G}$. The set of all measurable vector–valued functions in $\mathcal{V}$ is a subspace $\mathcal{V}_\mu \subset \mathcal{V}$; using the Parseval identity for some orthogonal basis in $\mathcal{G}$ we find that $\|F(\cdot)\|^2_\mathcal{G}$ is measurable for any $F \in \mathcal{V}_\mu$. Consider next the set

$$\mathcal{L}^2(X, d\mu; \mathcal{G}) := \left\{ F \in \mathcal{V}_\mu : \int_X \|F(x)\|^2_\mathcal{G}\, d\mu(x) < \infty \right\},$$

which is a subspace in $\mathcal{V}_\mu$ as can be seen from the inequality $\|F(x) + G(x)\|^2_\mathcal{G} \le 2(\|F(x)\|^2_\mathcal{G} + \|G(x)\|^2_\mathcal{G})$, and its subspace $\mathcal{L}_0 := \{ F \in \mathcal{V}_\mu \colon F(x) = 0 \;\; \mu\text{–a.e. in } X \}$. The factor space

$$L^2(X, d\mu; \mathcal{G}) := \mathcal{L}^2(X, d\mu; \mathcal{G}) / \mathcal{L}_0$$

thus consists of the equivalence classes of functions that differ mutually on a μ–zero set; with the usual license we shall refer to them again as vector–valued functions. $L^2(X, d\mu; \mathcal{G})$ is obviously a pre–Hilbert space with respect to the inner product

$$(F, G) := \int_X (F(x), G(x))_\mathcal{G}\, d\mu(x) \,.$$

It is moreover complete as we shall see in Example 2.3.4 below; we call it a *Hilbert space of vector–valued functions.* In the case $\mathcal{G} = \mathbb{C}$ this space, of course, coincides with $L^2(X, d\mu)$.

In the rest of this section we are going to discuss another important Hilbert space. Consider the measure $\mu_g : \mu_g(M) = \frac{1}{\pi}\int_M e^{-|z|^2} dz$ on $\mathbb{C}$, where $dz = d(\mathrm{Re}\, z)\, d(\mathrm{Im}\, z)$ refers to the Lebesgue measure. By $A^2(\mathbb{C})$ we denote the pre–Hilbert space of all functions which are analytic in $\mathbb{C}$ and square–integrable with respect to μ_g, with the inner product defined as in $L^2(\mathbb{C}, d\mu_g)$, *i.e.*,

$$(f, g) := \frac{1}{\pi}\int_{\mathbb{C}} \overline{f(z)} g(z)\, e^{-|z|^2} dz\,.$$

In distinction to L^2, the elements of $A^2(\mathbb{C})$ are continuous functions; hence $f = g$ means $f(z) = g(z)$ for *all* $z \in \mathbb{C}$ and $(f, f) = 0$ implies $f = 0$.

We want to show that $A^2(\mathbb{C})$ is complete with respect to the metric induced by the inner product. To this end, we consider the subspace $\mathcal{P} \subset A^2(\mathbb{C})$ of all polynomials on $\mathbb{C}$ and the Hilbert space $\tilde{\mathcal{P}}$ obtained by the standard completion of $\mathcal{P}$. We can check directly that the functions $u_n : u_n(z) := z^n (n!)^{-1/2}$, $n = 0, 1, \dots$, form an orthonormal set. Since this is a Hamel basis in $\mathcal{P}$, and therefore total in $\tilde{\mathcal{P}}$, the space $\tilde{\mathcal{P}}$ is separable.

2.2.6 Lemma: Let $P \in \mathcal{P}$; then the relations $P(w) = (e_w, P)$ and $|P(w)| \le \|P\|\, e^{|w|^2/2}$ hold for $w \in \mathbb{C}$, where $e_w(z) := e^{\overline{w}z}$.

Proof: We can compute directly that $\|e_w\| = e^{|w|^2/2}$, so the second relation follows from the first; it is sufficient to check the latter for $P = u_n$. The inequality $\left|\sum_{k=0}^N \frac{1}{k!}(\overline{w}z)^k\right| \le e^{|wz|}$ valid for any N and $z \in \mathbb{C}$ allows us to use the dominated–convergence theorem, which yields

$$\int_{\mathbb{C}} u_n(z)\overline{e_w(z)}\, d\mu_g(z) = \lim_{N\to\infty} \sum_{k=0}^N \frac{w^k}{\sqrt{k!}} \int_{\mathbb{C}} u_n(z)\overline{u_k(z)}\, d\mu_g(z) = u_n(w)\,. \quad ∎$$

Now we are able to prove the basic properties of the space $A^2(\mathbb{C})$.

2.2.7 Theorem: (a) $A^2(\mathbb{C})$ is a separable Hilbert space.

(b) Let $f \in A^2(\mathbb{C})$ and $w \in \mathbb{C}$; then

$$f(w) = (e_w, f) = \frac{1}{\pi}\int_{\mathbb{C}} f(z)\, \overline{e_w(z)}\, e^{-|z|^2/2} dz\,.$$

(c) The norm convergence in $A^2(\mathbb{C})$, $\|f_n - f\| \to 0$, implies the pointwise convergence, $f_n(z) \to f(z)$ for all $z \in \mathbb{C}$.

(d) $A^2(\mathbb{C})$ contains just those analytic functions $f = \sum_{k=0}^\infty c_k u_k$ for which $\sum_{k=0}^\infty |c_k|^2 < \infty$.

Proof: To prove (a) we have to check that $A^2(\mathbb{C}) = \tilde{\mathcal{P}}$. Let $\{P_n\} \subset \mathcal{P}$ be a Cauchy sequence; then the corresponding sequence of μ_g–equivalent classes is Cauchy in the Hilbert space $L^2(\mathbb{C}, d\mu)$, so there is a class $[F] \in L^2$ such that $\|F - P_n\| \to 0$ and a subsequence $\{P_{n_k}\}$ converges to F pointwise μ_g–almost everywhere (*cf.* Example 1.2.1). The above lemma then implies that $\lim_{n\to\infty} P_n(w) \equiv f(w)$ exists for all $w \in \mathbb{C}$ and the convergence is uniform in any compact $K \subset \mathbb{C}$. Hence f is holomorphic (see the notes), $f \in [F]$ and $f \in A^2$, *i.e.*, $\tilde{\mathcal{P}} \subset A^2(\mathbb{C})$. To prove the opposite inclusion, we first note that $\tilde{\mathcal{P}}^\perp = \{u_n : n = 0, 1, \dots\}^\perp$, so a function $f \in A^2(\mathbb{C})$, $f(z) = \sum_{k=0}^\infty a_k z^k$, belongs to $\tilde{\mathcal{P}}^\perp$ *iff* $\int_{\mathbb{C}} f(z)\overline{u_n(z)}\, d\mu_g(z) = 0$ for all n. Since $f\overline{u}_n \in \mathcal{L}(\mathbb{C}, d\mu)$ and the series converges uniformly on compact sets, we get the relation

$$0 = (u_n, f) = \lim_{R\to\infty} \sum_{k=0}^\infty \frac{a_k}{\sqrt{n!}} \int_{|z|\le R} z^k \bar{z}^n\, d\mu_g(z) = a_n \sqrt{n!}$$

for any n, *i.e.*, $A^2(\mathbb{C}) \cap \tilde{\mathcal{P}}^\perp = \{0\}$. Combining this with the inclusion $\tilde{\mathcal{P}} \subset A^2(\mathbb{C})$, we arrive at the sought result.

The reproducing property of part (b) is valid for each polynomial. For any $f \in A^2(\mathbb{C})$ there is a sequence $\{P_n\} \subset \mathcal{P}$ such that $\|f - P_n\| \to 0$. Repeating the above argument we find that $P_n(w) \to f(w)$, so (b) follows from inner–product continuity; using the Schwarz inequality, we get assertion (c). Finally, we have shown that $\{u_n : n = 0, 1, \dots\}$ is an orthonormal basis in $A^2(\mathbb{C})$, so any $f \in A^2(\mathbb{C})$ can be written in the form of the Fourier expansion, $f = \sum_{n=0}^\infty c_n u_n$, with $\|f\|^2 = \sum_{n=0}^\infty |c_n|^2$. On the other hand, if a function f is holomorphic and fulfils $\sum_{n=0}^\infty |c_n|^2 < \infty$ then the non–negative polynomials $Q_n(z) := |\sum_{k=0}^n c_k u_k(z)^2|$ fulfil $Q_n(z) \to |f(z)|^2$ for all $z \in \mathbb{C}$ and

$$\int_{\mathbb{C}} Q_n(z)\, d\mu_g(z) = \sum_{k=0}^n |c_k|^2 \le \sum_{k=0}^\infty |c_k|^2 < \infty,$$

so $f \in L^2(\mathbb{C}, d\mu_g)$ by Fatou's lemma. ∎

The space $A^2(\mathbb{C})$ is called *Hilbert space of analytic functions*; the same name is used for the spaces of those functions which are analytic and square integrable in some region $G \subset \mathbb{C}$ (see the notes to this section and to Section 8.1). The space $A^2(\mathbb{C})$ is naturally embedded into $L^2(\mathbb{C}, d\mu_g)$, *i.e.*, it is linearly isometric to its subspace. Since it is a proper subspace it has some special properties, *e.g.*, that expressed by assertion (c). The most important among them is the reproducing property of assertion (b). It implies, in particular, that the set $\mathcal{C} := \{ e_w : w \in \mathbb{C} \}$ is total in $A^2(\mathbb{C})$ because $\mathcal{C}^\perp = \{0\}$. However, $\mathcal{C}$ is not orthonormal and it cannot be orthogonalized as a uncountable set in a separable Hilbert space. The elements of $\mathcal{C}$, which will later be called *coherent states*, have the following remarkable property: substituting $f(z) = (e_z, f)$ to the formula of Theorem 2.2.7b and using $\overline{e_w(z)} = e_z(w)$, we get the relation

$$f(w) = \frac{1}{\pi} \int_{\mathbb{C}} (e_z, f)\, e_z(w)\, e^{-|z|^2/2} dz, \tag{2.2}$$

which can be regarded as a continuous analogy of the Fourier expansion.

2.2.8 Remark: Coherent states represent a more general concept; we return to this topic in Section 8.1. In some Hilbert spaces, however, families of coherent spaces are obtained as an isomorphic image of the set $\mathcal{C}$. As an example, consider the operator $V : L^2(\mathbb{R}) \to A^2(\mathbb{C})$ obtained by linear and continuous extension of the map $V_0 : V_0 h_n = u_n$, $n = 0, 1, \dots$, where the h_n are the vectors (2.1). As in the proof of Theorem 2.1.8, we can check that V is an isomorphism of the two spaces. The vectors $\psi_w := e^{-|w|^2/2} V^{-1} e_w$ thus have unit norm for any $w \in \mathbb{C}$ and

$$\psi_w(x) = \frac{1}{\sqrt[4]{\pi}} e^{[\bar{w}^2 - |w|^2 - (x - \sqrt{2}\bar{w})^2]/2}\,; \tag{2.3}$$

they are again called coherent states.

2.3 Direct sums of Hilbert spaces

In Section 2.1 we introduced the notion of direct sum for a finite system of spaces. If all of them happen to be mutually orthogonal subspaces of a given Hilbert space, then there is a simple way to characterize the direct sum topologically.

2.3.1 Proposition: Let $\mathcal{G}_1, \mathcal{G}_2, \dots, \mathcal{G}_N$ be closed mutually orthogonal subspaces of a given $\mathcal{H}$; then $\bigoplus_{j=1}^N \mathcal{G}_j$ is the minimal closed subspace containing all $\mathcal{G}_j$.

Proof: By definition, $\mathcal{H}_N := \bigoplus_{j=1}^N \mathcal{G}_j = \left\{ x = \sum_{j=1}^N x_j \,:\, x_j \in \mathcal{G}_j \right\}$, so it is contained in any subspace that contains all $\mathcal{G}_j$. It remains for us to check the closedness, *i.e.*, that $\lim_{n\to\infty} x_n \in \mathcal{H}_N$ holds for each Cauchy sequence $\{x^{(n)}\} \subset \mathcal{H}_N$. The decomposition $x^{(n)} = \sum_{j=1}^N x_j^{(n)}$, $x_j \in \mathcal{G}_j$, together with the orthogonality of the subspaces, implies $\|x^{(n)} - x^{(m)}\|^2 = \sum_{j=1}^N \|x_j^{(n)} - x_j^{(m)}\|^2$ so the sequences $\{x_j^{(n)}\}_{n=1}^\infty$ are also Cauchy, and due to the closedness of the subspaces each of them converges to some $x_j \in \mathcal{G}_j$; the vector $x := \sum_{j=1}^N x_j \in \mathcal{H}_N$ then fulfils $\lim_{n\to\infty} \|x^{(n)} - x\|^2 = \lim_{n\to\infty} \sum_{j=1}^N \|x_j^{(n)} - x_j\|^2 = 0$. ■

The space $\bigoplus_{j=1}^N \mathcal{G}_j$ is called the *orthogonal sum* of the subspaces $\mathcal{G}_1, \dots, \mathcal{G}_N$. Proposition 2.3.1 allows us to extend this notion to any family $\{\mathcal{G}_\alpha : \alpha \in I,\, \mathcal{G}_\alpha \neq \{0\}\}$ of subspaces in a given $\mathcal{H}$: we define the **orthogonal sum** $\sum_{\alpha\in I}^{\oplus} \mathcal{G}_\alpha$ as the minimal closed subspace containing all $\mathcal{G}_\alpha$.

For a finite system of subspaces, the notions of direct and orthogonal sum are thus equivalent, and any element of $\bigoplus_{j=1}^N \mathcal{G}_j$ can be expressed as the sum of its projections to the subspaces $\mathcal{G}_j$. This last property also remains essentially true in the general case. Consider the subspace $\left(\bigcup_{\alpha\in I} \mathcal{G}_\alpha\right)_{lin}$, which is called the *algebraic sum* of the subspaces $\mathcal{G}_\alpha$ and is denoted as $\sum_{\alpha\in I} \mathcal{G}_\alpha$ (compare with Problem 1.5). If I is finite it coincides with $\sum_{\alpha\in I}^{\oplus} \mathcal{G}_\alpha$, while in the general case we have

$$\sum_{\alpha\in I}^{\oplus} \mathcal{G}_\alpha = \overline{\sum_{\alpha\in I} \mathcal{G}_\alpha} \tag{2.4}$$

since the subspace on the right side is closed and contains all $\mathcal{G}_\alpha$ and hence also $\sum_{\alpha\in I}^{\oplus}\mathcal{G}_\alpha$; at the same time the relations $\mathcal{G}_\alpha \subset \sum_{\alpha\in I}^{\oplus}\mathcal{G}_\alpha$ together with the closedness of $\sum_{\alpha\in I}^{\oplus}\mathcal{G}_\alpha$ yield the opposite inclusion. Combining the relation (2.4) with some simple consequences of the orthogonal–decomposition theorem (Problem 13) we get the sought expression of the elements of $\sum_{\alpha\in I}^{\oplus}\mathcal{G}_\alpha$ in terms of the projections to the subspaces $\mathcal{G}_\alpha$. We formulate this for a countable I; the generalization to an arbitrary index set is left to the reader (Problem 14).

2.3.2 Theorem: Let $\{\mathcal{G}_j : j = 1,2,\dots\}$ be a family of mutually orthogonal closed subspaces of a given $\mathcal{H}$, then

$$\sum_{j=1}^{\infty}{}^{\oplus}\mathcal{G}_j \equiv \bigoplus_{j=1}^{\infty}\mathcal{G}_j = \left\{ x\in\mathcal{H} : x = \lim_{n\to\infty}\sum_{j=1}^{n} x_j\,,\ x_j\in\mathcal{G}_j \right\}.$$

Proof: Denote the right side of the last relation as $\mathcal{G}$. Obviously $\mathcal{G}\subset\overline{\sum_{j=1}^{\infty}\mathcal{G}_j}$, so $\mathcal{G}\subset\bigoplus_{j=1}^{\infty}\mathcal{G}_j$ holds in view of the relation (2.4). To check the opposite inclusion, take a vector $x\in\bigoplus_{j=1}^{\infty}\mathcal{G}_j$ and denote by x_j its projections to $\mathcal{G}_j$. Due to Problem 13b, $s=\sum_{j=1}^{n}x_j$ is the projection of x to $\bigoplus_{j=1}^{n}\mathcal{G}_j$, thus $\|s_n\|^2 = \sum_{j=1}^{n}\|x_j\|^2 \le \|x\|^2$, $n=1,2,\dots$. Problem 13c together with the already proven inclusion $\mathcal{G}\subset\bigoplus_{j=1}^{\infty}\mathcal{G}_j$ implies that $s_n\to s\in\bigoplus_{j=1}^{\infty}\mathcal{G}_j$; hence $x-s\in\bigoplus_{j=1}^{\infty}\mathcal{G}_j$. We have

$$(x-s,y) = \lim_{n\to\infty}\left(x-\sum_{j=1}^{n}x_j,\,y\right) = (x-x_k,y) = 0$$

for any $y\in\mathcal{G}_k$, $k=1,2,\dots$, because x_k is the projection of x to $\mathcal{G}_k$. Problem 13a now gives $x-s\in(\bigoplus_{j=1}^{\infty}\mathcal{G}_j)^{\perp}$, so $x=s$, *i.e.*, $x\in\mathcal{G}$. ∎

2.3.3 Example: Consider the space $\mathcal{H} := L^2(X,d\mu)$, where μ is a non–negative σ–finite measure defined on $\mathcal{A}\subset 2^X$, and the family of the subspaces $\mathcal{G}_j$ corresponding to a decomposition $X=\bigcup_{j=1}^{\infty}X_j$ into a system of disjoint finite–measure subsets, $\mathcal{G}_j := \{\chi_{X_j}f : f\in L^2\}$. The measures $\mu_j : d\mu_j = \chi_{X_j}d\mu$ are finite and the spaces $\mathcal{G}_j$ are obviously isomorphic with $L^2(X_j,d\mu_j)$. This implies their closedness, $\overline{\mathcal{G}_j} = \mathcal{G}_j$, while $\mathcal{G}_j\perp\mathcal{G}_k$ follows from $X_j\cap X_k=\emptyset$ for $j\ne k$. For any $f\in\mathcal{H}$ we construct the sequence of vectors $f_n := \sum_{j=1}^{n}\chi_{X_j}f = \chi_{X^{(n)}}f$, where $X^{(n)} := \bigcup_{j=1}^{n}X_j$. It converges pointwise to f and $\|f_n\|\le\|f\|$, so $\|f_n-f\|^2\le 4\|f\|^2$; the dominated–convergence theorem then implies $\|f_n-f\|\to 0$. Hence we get the decomposition $L^2(X,d\mu) = \bigoplus_{j=1}^{\infty}\mathcal{G}_j$ corresponding to $X=\bigcup_{j=1}^{\infty}X_j$, where the spaces $\mathcal{G}_j$ are isomorphic to $L^2(X_j,d\mu_j)$.

In Section 2.1 we have described how one constructs the direct sum $\mathcal{H}_1\oplus\mathcal{H}_2$ to a given pair of Hilbert spaces $\mathcal{H}_1$, $\mathcal{H}_2$. The subspaces $\mathcal{H}^{(1)} := \{[x_1,0] : x_1\in\mathcal{H}_1\}$ and $\mathcal{H}^{(2)} := \{[0,x_2] : x_2\in\mathcal{H}_2\}$ in $\mathcal{H}_1\oplus\mathcal{H}_2$ are obviously orthogonal and closed, naturally isomorphic to $\mathcal{H}_1$ and $\mathcal{H}_2$, respectively, and their orthogonal sum is $\mathcal{H}_1\oplus\mathcal{H}_2$. The above results thus give us an idea of how to extend the notion of direct sum to an arbitrary family of Hilbert spaces.

Let $\mathcal{M} := \{ \mathcal{H}_\alpha : \alpha \in I \}$ be such a family. We do not here require injectivity of the map $\alpha \mapsto \mathcal{H}_\alpha$; the set $\mathcal{M}$ can even consist of copies of the same Hilbert space. The quantities referring to $\mathcal{H}_\alpha$ will be indexed by α. By $\mathcal{H}$ we denote the subset in the Cartesian product $\mathsf{X}_{\alpha\in I}\mathcal{H}_\alpha$ consisting of the maps $\alpha \mapsto X(\alpha) \in \mathcal{H}_\alpha$ which satisfy the following conditions:

(i) $I(X) := \{ \alpha : x(\alpha) \neq 0_\alpha \}$ is at most countable.

(ii) $\sum_{\alpha\in I(X)} \|X(\alpha)\|_\alpha^2 < \infty$.

The set $\mathcal{H}$ becomes a vector space if we set $(aX+Y)(\alpha) := aX(\alpha)+Y(\alpha)$, $a \in \mathbb{C}$, and linear combinations satisfy condition (ii) in view of the Minkowski inequality; in the same way as for ℓ^2 we can check that

$$X \longmapsto |||X||| := \left(\sum_{\alpha\in I(X)} \|X(\alpha)\|_\alpha^2 \right)^{1/2}$$

is a norm on $\mathcal{H}$. Each of the norms $\|\cdot\|$ fulfils the parallelogram identity; hence the norm $|||\cdot|||$ is induced by the inner product

$$(X,Y) := \sum_{\alpha\in I(X)\cap I(Y)} (X(\alpha),Y(\alpha))_\alpha \,;$$

to check the correctness of the definition we have to employ the Schwarz and Hölder inequalities to prove that the series is absolutely convergent, and the right side is therefore independent of the order of summation. Let us finally show that the space $(\mathcal{H}, |||\cdot|||)$ is complete. Let $\{X_n\} \subset \mathcal{H}$ be a Cauchy sequence; the set $I_{\{X_n\}} := \bigcup_{n=1}^\infty I(X_n)$ is at most countable, so we can write it as $\{\alpha_j : j = 1,2,\dots\}$. The condition $|||X_n-X_m||| < \varepsilon$ for $n,m > n(\varepsilon)$ implies that there is $X(\alpha_j) \in \mathcal{H}_{\alpha_j}$ to any j such that $X_n(\alpha_j) \to X(\alpha_j)$, and at the same time, the inequalities $\sum_{j=1}^N \|X_n(\alpha_j) - X_m(\alpha_j)\|_{\alpha_j}^2 < \varepsilon^2$ for $N = 1,2,\dots$. Taking first the limit $m \to \infty$ and then $N \to \infty$, we get $\sum_{j=1}^\infty \|X_n(\alpha_j) - X(\alpha_j)\|_{\alpha_j}^2 \le \varepsilon^2$ for $n > n(\varepsilon)$, so if we define $X(\alpha) := 0$ for all indices $\alpha \in I \setminus I_{\{X_n\}}$, the vector X belongs to $\mathcal{H}$ and the sequence $\{X_n\}$ converges to it.

The Hilbert space constructed in this way is called the *direct sum* of the spaces $\mathcal{H}_\alpha$ and is denoted as $\sum_{\alpha\in I}^{\oplus} \mathcal{H}_\alpha$. If I is a finite set, the conditions (i) and (ii) are fulfilled automatically and we return to the construction of Section 2.1. In the countable case, condition (i) can be dropped and the definition simplifies to the following form: $\bigoplus_{j=1}^\infty \mathcal{H}_j$ is the set of all sequences $\{x_j\}_{j=1}^\infty$ with $x_j \in \mathcal{H}_j$ such that $\sum_{j=1}^\infty \|x_j\|_j^2 < \infty$, with the componentwise defined operations of summation and scalar multiplication, and the inner product

$$(X,Y) := \sum_{j=1}^\infty (x_j, y_j)_j \,.$$

2.3.4 Example (vector–valued function revisited): We use the notation of Example 2.2.5. Let $\mathcal{G}$ be a separable Hilbert space of dimension N; we shall show that the corresponding Hilbert space $L^2(X, d\mu; \mathcal{G})$ of vector–valued functions is isomorphic to $\bigoplus_{j=1}^{N} L^2(X, d\mu)$, thus proving its completeness. We shall consider only the case $N = \infty$ since most part of the following argument is otherwise trivial. We choose a fixed orthonormal basis $\{e_j\} \subset \mathcal{G}$ and define the complex–valued functions $x \mapsto F_j(x) := (e_j, F(x))_{\mathcal{G}}$ to any measurable vector–valued function F; these are obviously measurable and fulfil $\|F(x)\|_{\mathcal{G}}^2 = \sum_j |F_j(x)|^2 < \infty$ for all $x \in X$. Conversely, the last condition together with the measurability of the functions imply that $x \mapsto F(x) := \sum_j F_j(x) e_j$ defines a measurable vector–valued function.

We want to show that $F \mapsto V(F) := \{F_j\}_{j=1}^{\infty}$ is the sought isomorphism. Let $F \in L^2(X, d\mu; \mathcal{G})$; then the relation $\|F\|^2 = \int_X \sum_j |F_j(x)|^2 d\mu(x)$ gives $F_j \in L^2(X, d\mu)$ for $j = 1, 2, \dots$ and $\int_X \Phi_n(x) d\mu(x) \le \|F\|^2$ for $n = 1, 2, \dots$, where we set $\Phi_n := \sum_{j=1}^{n} |F_j|^2$. The monotone–convergence theorem implies the identity $\|F\|^2 = \sum_j \int_X |F_j(x)|^2 d\mu(x)$, so $\operatorname{Ran} V \subset \sum_j^{\oplus} L^2(X, d\mu)$ and the map V is norm preserving. It remains to check its surjectivity. We have

$$\int_X \Phi_n d\mu \;=\; \sum_{j=1}^{n} \int_X |F_j|^2 d\mu \;\le\; \|\{F_j\}\|_{\oplus}^2$$

for any sequence $\{F_j\} \in \sum_j^{\oplus} L^2(X, d\mu)$, so using the monotone–convergence theorem once again, we get $\lim_{n\to\infty} \Phi_n(x) = \sum_j |F_j(x)|^2 < \infty$ for μ–a.a. $x \in X$. It follows that $x \mapsto F(x) := \sum_j F_j(x) e_j$ is a measurable vector–valued function and

$$\int_X \|F(x)\|_{\mathcal{G}}^2 d\mu(x) \;=\; \int_X \lim_{n\to\infty} \Phi_n \, d\mu \;=\; \lim_{n\to\infty} \int_X \Phi_n \, d\mu \;=\; \|\{F_j\}\|_{\oplus}^2 \,,$$

i.e., $F \in L^2(X, d\mu; \mathcal{G})$ and $\{F_j\} = V(F)$.

The spaces of vector–valued functions can be further generalized if we replace the auxiliary space $\mathcal{G}$ by a family of Hilbert spaces $\mathcal{H}(x)$ labelled by a variable x. Suppose we have a measure space $(X, \mathcal{A}, \mu)$ with a non–negative σ–finite measure, a nonempty family $\mathcal{R}$ of separable Hilbert spaces and a map $x \mapsto \mathcal{H}(x)$ from X to $\mathcal{R}$. We define the "dimension function" by $d:\, d(x) = \dim \mathcal{H}(x)$ and assume that it is measurable, *i.e.*, $X_n := d^{(-1)}(n) \in \mathcal{A}$ for $n = 1, 2, \dots$; then $X_0 := X \setminus \bigcup_{n=1}^{\infty} X_n$ also belongs to $\mathcal{A}$. The elements of the set $\mathcal{V} := \times_{x\in X} \mathcal{H}(x)$ are again called *vector-valued functions* and $\mathcal{V}$ becomes a vector space when equipped with the pointwise defined algebraic operations.

There are several ways to define measurability for these vector–valued functions (see the notes); we choose the simplest one. A set $\{E_k : k = 1, 2, \dots\} \subset \mathcal{V}$ is called a *measurability basis* if $\{E_k(x)\}_{k=1}^{\infty}$ is an orthonormal basis in $\mathcal{H}(x)$ for any $x \in X$ and $E_k(x) = 0$ holds if $x \in X_n$ and $k = n+1, n+2, \dots$. To a given measurability basis we define measurable vector–valued functions as those $F \in \mathcal{V}$ such that $x \mapsto (E_k(x), F(x))_x$ is measurable for all $k = 1, 2, \dots$; they form a subspace $\mathcal{V}_\mu(\{E_k\})$ in $\mathcal{V}$. In the particular case when $\mathcal{H}(x) = \mathcal{G}$ for all $x \in X$ the measurability basis is formed by the constant vector–valued functions $E_k(\cdot) = e_k$, where $\{e_k\}$ is an orthonormal basis in $\mathcal{G}$, and we arrive back at the definition of measurability from Problem 11a. Furthermore, $\mathcal{V}_\mu(\{E_k\})$ is in fact independent of the measurability basis used (Problem 17), so we denote it as $\mathcal{V}_\mu$ in the following.

The map $x \mapsto \mathcal{H}(x)$ together with the subspace $\mathcal{V}_\mu \subset \mathcal{V}$ is called a *measurable Hilbert–space field* on $(X, \mathcal{A}, \mu)$. If two such fields have the same dimension function d they are essentially identical (Problem 17). Now we are able to introduce the notion of the **direct integral** of a measurable field $(x \mapsto \mathcal{H}(x), \mathcal{V}_\mu)$ on $(X, \mathcal{A}, \mu)$ in analogy with Example 2.2.5: we set

$$\int_X^{\oplus} \mathcal{H}(x)\, d\mu(x) := \mathcal{L}^2/\mathcal{L}_0\,,$$

where $\mathcal{L}^2 := \left\{ F \in \mathcal{V}_\mu : \int_X \|F(x)\|_x^2 d\mu(x) < \infty \right\}$ is factorized with respect to the subspace $\mathcal{L}_0 := \{ F \in \mathcal{V}_\mu : F(x) = 0 \;\; \mu\text{–a.e.} \}$; the inner product at the factor space is introduced by

$$(F, G) := \int_X (F(x), G(x))_x d\mu(x)\,.$$

We leave to the reader to check the completeness and other properties of the direct integral (Problem 18). Note further that the direct sum considered above turns out to be a particular case of the direct integral referring to a discrete measure.

2.3.5 Example: Let $I := \{x_j : j = 1, 2, \dots\}$ be a countable subset in X. Let μ_I be a measure on X supported by I and such that $\mu_I(M) = \mathrm{card}\,(M \cap I)$ for any $M \subset X$. All the vector–valued functions are obviously μ_I–measurable, $\mathcal{V}_{\mu_I} = \mathcal{V}$, for any map $x \mapsto \mathcal{H}(x)$. Since $\mu_I(X \backslash I) = 0$, each $F \in \int_X^{\oplus} \mathcal{H}(x)\, d\mu_I(x)$ is fully determined by its values at the points x_j, $j = 1, 2, \dots$, and moreover, we have $\|F\|^2 = \sum_{j=1}^{\infty} \|F(x_j)\|_j^2$; it follows that $\int_X^{\oplus} \mathcal{H}(x)\, d\mu_I(x) = \bigoplus_{j=1}^{\infty} \mathcal{H}(x_j)$.

2.4 Tensor products

Let $\mathcal{H}_1, \mathcal{H}_2$ be arbitrary Hilbert spaces; then a pair $(\mathcal{H}, \otimes)$, where $\mathcal{H}$ is a Hilbert space and $\otimes : \mathcal{H}_1 \times \mathcal{H}_2 \to \mathcal{H}$ is a bilinear map, is called a *realization* of the *tensor product* of $\mathcal{H}_1$ and $\mathcal{H}_2$ if the following conditions are valid:

(t1) $(x \otimes y, x' \otimes y') = (x, x')_1 (y, y')_2$ holds for all $x, x' \in \mathcal{H}_1$ and $y, y' \in \mathcal{H}_2$.

(t2) The set $\{ x \otimes y : x \in \mathcal{H}_1,\, y \in \mathcal{H}_2 \}$ is total in $\mathcal{H}$.

After formulating such a definition we must first check the question of existence.

2.4.1 Theorem: For any pair $\mathcal{H}_1, \mathcal{H}_2$ there is at least one realization of their tensor product.

Proof: Consider the vector space $\mathcal{H}_1 \diamond \mathcal{H}_2$, whose elements are linear combinations of the "objects" $x \diamond y$ with $x \in \mathcal{H}_1$ and $y \in \mathcal{H}_2$ where the equality $x \diamond y = x' \diamond y'$

means $x = x'$ and $y = y'$, and any set of mutually different objects is linearly independent, so the set $\{ x \diamond y : x \in \mathcal{H}_1, y \in \mathcal{H}_2 \}$ is a Hamel basis of $\mathcal{H}_1 \diamond \mathcal{H}_2$. To be concrete, we may take for $\mathcal{H}_1 \diamond \mathcal{H}_2$ the set of all functions $f : \mathcal{H}_1 \times \mathcal{H}_2 \to \mathbb{C}$ with $f^{(-1)}(\mathbb{C} \setminus \{0\})$ finite; the object $x \diamond y$ is realized by the function $e_{x,y}$ such that $e_{x,y}(u,v) = 1$ for $[u,v] = [x,y]$ and $e_{x,y}(u,v) = 0$ otherwise. Now let $L \subset \mathcal{H}_1 \diamond \mathcal{H}_2$ be the subspace spanned by the elements of the form

$$l = \sum_{jk} \alpha_j \beta_k x_j \diamond y_k - \left(\sum_j \alpha_j x_j \right) \diamond \left(\sum_k \beta_k y_k \right) .$$

For a given $[x,y] \in \mathcal{H}_1 \times \mathcal{H}_2$ we denote by $x \diamond y$ the set of all $f \in \mathcal{H}_1 \diamond \mathcal{H}_2$ such that $f = x \diamond y \pmod{L}$; this defines the map $\otimes : \mathcal{H}_1 \times \mathcal{H}_2 \to (\mathcal{H}_1 \diamond \mathcal{H}_2)/L$, which is obviously bilinear.

Next we shall define on $(\mathcal{H}_1 \diamond \mathcal{H}_2)/L$ an inner product which would fulfil the axiom (t1). We associate the number $s(x \diamond y, x' \diamond y') := (x,x')_1 (y,y')_2$ with any pair $x \diamond y$, $x' \diamond y'$; since $\{x \diamond y\}$ is a Hamel basis the map defined in this way extends uniquely to a sesquilinear form s on $\mathcal{H}_1 \diamond \mathcal{H}_2$ (Problem 20). Moreover, $s(f,g) = 0$ if at least one of f, g belongs to L; from here it follows that $s(f_u, g_v)$ for given $u, v \in (\mathcal{H}_1 \diamond \mathcal{H}_2)/L$ does not depend on the choice of the elements $f_u \in u$ and $g_v \in v$. Hence the relation $\langle u, v \rangle := s(f_u, g_v)$ defines a form on $(\mathcal{H}_1 \diamond \mathcal{H}_2)/L$ which is easily seen to be sesquilinear. We shall show that it is strictly positive. We take an arbitrary $u := \sum_{j=1}^n x_j \otimes y_j$ and choose orthonormal bases $\{e_1, \dots, e_{n_1}\}$, $\{f_1, \dots, f_{n_2}\}$ in the subspaces $\{x_1, \dots, x_n\}_{lin} \subset \mathcal{H}_1$ and $\{y_1, \dots, y_n\}_{lin} \subset \mathcal{H}_2$, respectively. The expansions $x_j = \sum_r c_{jr} e_r$ and $y_k = \sum_s d_{ks} f_s$ give $u = \sum_{rs} a_{rs} e_r \otimes f_s$ with $a_{rs} := \sum_j c_{jr} d_{js}$. Then $\langle u, u \rangle = \sum_{rs} |a_{rs}|^2 \ge 0$, where the left side is zero only if $u = 0$. By the standard completion of the inner product space $(\mathcal{H}_1 \diamond \mathcal{H}_2)/L$, we finally arrive at a Hilbert space $\mathcal{H}$ which satisfies the definition requirements (t1) and (t2). ∎

The next question concerns relations between different tensor–product realizations. They appear to be quite simple.

2.4.2 Proposition: Let $(\mathcal{H}, \otimes)$ and $(\mathcal{H}, \otimes')$ be realizations of the tensor product of $\mathcal{H}_1$ and $\mathcal{H}_2$; then there is a unique isomorphism $V : \mathcal{H} \to \mathcal{H}'$ such that $x \otimes' y = V(x \otimes y)$ holds for all $[x,y] \in \mathcal{H}_1 \times \mathcal{H}_2$.

Proof: By Problem 20, $x \otimes y = \tilde{x} \otimes y$ implies $x \otimes' y = \tilde{x} \otimes' y$ for any $[x,y]$, $[\tilde{x}, \tilde{y}] \in \mathcal{H}_1 \times \mathcal{H}_2$; hence the relation $x \otimes' y = V(x \otimes y)$ defines a bijection from $\{ x \otimes y : [x,y] \in \mathcal{H}_1 \times \mathcal{H}_2 \}$ to $\{ x \otimes' y : [x,y] \in \mathcal{H}_1 \times \mathcal{H}_2 \}$ which preserves the inner product. The sought isomorphism is then obtained by its linear and continuous extension; the argument is the same as in Theorem 2.1.8. ∎

These results demonstrate that the tensor–product realizations for a given pair $\mathcal{H}_1$ and $\mathcal{H}_2$ exist and are unique up to an isomorphism. This is the starting point for an abstract definition of the tensor product (see the notes). For our purposes, however, it is not needed: the **tensor product** of Hilbert spaces $\mathcal{H}_1$, $\mathcal{H}_2$ will simply be understood as any of its realizations; in concrete situations there is usually some standard map $\otimes$ which we shall specify when introducing the particular tensor product. For simplicity we shall often write $\mathcal{H} = \mathcal{H}_1 \otimes \mathcal{H}_2$ remembering that the formula is just a shorthand expression for the fact that the tensor product of $\mathcal{H}_1$ and $\mathcal{H}_2$ is realized in $\mathcal{H}$ through a map $\otimes$.

2.4.3 Remark: One defines the *algebraic tensor product* of arbitrary subspaces $L_1 \subset \mathcal{H}_1$ and $L_2 \subset \mathcal{H}_2$ by $L_1 \underline{\times} L_2 := \{ x \otimes y \,:\, x \in L_1,\, y \in L_2 \}_{lin}$. We can easily check the relation $\overline{L_1 \underline{\times} L_2} = \overline{L_1} \otimes \overline{L_2}$ (Problem 23); in particular $\mathcal{G}_1 \otimes \mathcal{G}_2 = \overline{\mathcal{G}_1 \underline{\times} \mathcal{G}_2}$ holds for closed subspaces $\mathcal{G}_j \subset \mathcal{H}_j$.

2.4.4 Proposition: (a) If M_r is total in $\mathcal{H}_r$, $r = 1, 2$, then the set $M_{12} := \{ x \otimes y \,:\, x \in M_1,\, y \in M_2 \}$ is total in $\mathcal{H}_1 \otimes \mathcal{H}_2$.

(b) If $\{ e_{\alpha_r}^{(r)} \,:\, \alpha_r \in I_r \}$ is an orthonormal basis in $\mathcal{H}_r$, $r = 1, 2$, then the vectors $e_{\alpha_1}^{(1)} \otimes e_{\alpha_2}^{(2)}$, $\alpha_r \in I_r$, form an orthonormal basis in $\mathcal{H}_1 \otimes \mathcal{H}_2$. In particular, $\mathcal{H}_1 \otimes \mathcal{H}_2$ is separable if the same is true for $\mathcal{H}_1$ and $\mathcal{H}_2$.

Proof: Denote $L_r := (M_r)_{lin}$; in view of the above remark it is enough to prove that $x \otimes y \in \overline{L_1 \underline{\times} L_2}$ for all $x \in \mathcal{H}_1$ and $y \in \mathcal{H}_2$. Since the M_r are total by assumption, it follows from the estimate that

$$\|x \otimes y - x_n \otimes y_n\| \le \|(x - x_n) \otimes y_n\| + \|x \otimes (y_n - y)\| = \|x - x_n\|_1 \|y_n\|_2 + \|x\|_1 \|y_n - y\|_2 \,.$$

The remaining statement is an easy consequence of part (a). ∎

2.4.5 Example: Under the assumptions of Proposition 2.2.4 we define a map $\otimes :$ $L^2(M, d\mu) \times L^2(N, d\nu) \to L^2(M \times N, d(\mu \otimes \nu))$ by $(f \otimes g)(x, y) := f(x) g(y)$. It is obviously bilinear, $(f \otimes g, \tilde{f} \otimes \tilde{g}) = (f, \tilde{f})_\mu (g, \tilde{g})_\nu$ by the Fubini theorem, and $\{ f \otimes g : f \in L^2(M, d\mu),\, g \in L^2(N, d\nu) \}$ is total in view of Proposition 2.2.4. Hence the requirements (t1), (t2) are valid and we may write

$$L^2(M \times N, d(\mu \otimes \nu)) \;=\; L^2(M, d\mu) \otimes L^2(N, d\nu) \,. \tag{2.5}$$

In a similar way, we can check that the Hilbert space $L^2(X, d\mu; \mathcal{G})$ of Example 2.2.5 can be expressed in the tensor–product form

$$L^2(X, d\mu; \mathcal{G}) \;=\; L^2(X, d\mu) \otimes \mathcal{G}; \tag{2.6}$$

the corresponding bilinear map is $(f \otimes \phi)(x) := f(x) \phi$ (*cf.* Problem 11b).

Notes to Chapter 2

Section 2.1 The concept of an abstract (separable) Hilbert space belongs to J. von Neumann, who formulated it at the end of the 1920s [vN 1,2]; it was extended a few years later to nonseparable spaces by H. Löwig and F. Rellich. The name was chosen in honor of D. Hilbert, who investigated the ℓ^2 and L^2 spaces at the beginning of the twentieth century.

The orthogonal–projection lemma admits a generalization: it is clear from the proof that $\mathcal{G}$ may be replaced by any closed convex set in $\mathcal{H}$. The existence of the orthogonal decomposition $\mathcal{H} = \mathcal{G} \oplus \mathcal{G}^{\perp}$ for *any* closed subspace $\mathcal{G} \subset \mathcal{H}$ is a characteristic feature of Hilbert spaces, distinguishing them among Banach spaces — *cf.* ⟦DS 1⟧, Sec.VI.12. Proof of the second part of Theorem 2.2.7 is somewhat lengthy — see, *e.g.*, ⟦DS 1⟧, Sec.IV.4.

The dual $\mathcal{H}^*$ is a Hilbert space with the inner product $(\cdot,\cdot)_*$ defined by $(f,g)_* = (y_g, y_f)$; using this result, one can check that any Hilbert space is reflexive. It is also easy to see that the kernel of any functional $f \in \mathcal{H}^*$ has codimension one, since $\operatorname{Ker} f = \{y_f\}^{\perp}$. To check that $\mathcal{H}$ and $\mathcal{H}^*$ are *linearly* isometric, we can combine the Riesz lemma with the isometric involution J defined for a given orthonormal basis $\{e_\alpha : \alpha \in I\}$ by $Jx := \sum_{\alpha\in I} (x, e_\alpha) e_\alpha$.

Section 2.2 For Fejér's theorem see, *e.g.*, ⟦Jar 2⟧, Thm.191. Separability of the spaces $L^2(M, d\mu)$ depends on the properties of the measure μ. Denote the corresponding σ–algebra as $\mathcal{A}$; then a system $\mathcal{N} \subset \mathcal{A}$ is called a *basis of the measure* μ if $\inf\{\mu(M\Delta N) : N \in \mathcal{A}\} = 0$ holds for each $M \in \mathcal{A}$ with $\mu(M) < \infty$. Assume that μ is σ–finite; then a space $L^2(M, d\mu)$ is separable *iff* μ has a countable basis (Problem 8) and the same is true for the L^p spaces, $p \geq 1$.

The functions from the closed linear envelope of the set $\{e_\lambda : e_\lambda(x) = e^{i\lambda x},\ \lambda \in \mathbb{R}\} \subset C(\mathbb{R})$ are called *almost periodic;* they form a subspace in the Banach space $C(\mathbb{R})$ with the norm $\|\cdot\|_\infty$, which we denote as AP. H. Bohr has proven the following criterion: a function $f : \mathbb{R} \to \mathbb{C}$ is almost periodic *iff* it is continuous and to any $\varepsilon > 0$ one can find $l > 0$ such that any interval of the length l contains at least one t such that $|f(x+t) - f(x)| < \varepsilon$ holds for all $x \in \mathbb{R}$. In addition to $\|\cdot\|_\infty$, the space AP can be equipped with another norm,

$$p(f)^2 := \lim_{T\to\infty} \frac{1}{2T} \int_{-T}^{T} |f(x)|^2 dx$$

(Problem 9), which is obviously induced by an inner product. The completion of AP with respect to $p(\cdot)$ is thus a Hilbert space, which is not separable because $\{e_\lambda : \lambda \in \mathbb{R}\}$ is an orthonormal set in it, which is not countable. The basic properties of almost periodic functions, are discussed in ⟦DS 1⟧, Sec.IV.7; ⟦RN⟧, Secs.101, 102.

In the proof of Theorem 2.2.7 we have used the fact that the limit of a sequence of holomorphic functions which converges uniformly on compact sets is holomorphic — *cf.* ⟦Ru 1⟧, Thm.10.28. If $G \subset \mathbb{C}$ is an open connected set we can in the same way as above construct the Hilbert space $A^2(G)$ of functions, which are analytic and square integrable in G. However, the measure used to define the inner product is not necessarily μ_g; if G is bounded, for instance, we can replace it by the Lebesgue measure. In particular, if G is the unit disc $D_1 := \{z \in \mathbb{C} : |z| < 1\}$ then the functions $e_n : e_n(z) = \sqrt{\frac{n+1}{\pi}}\, z^n$, $n = 0, 1, \dots$, form an orthonormal basis in $A^2(D_1)$ — see ⟦Hal 2⟧, Secs.24, 25.

Section 2.3 The orthogonal sum $\sum_{\alpha\in I}^{\oplus} \mathcal{G}_\alpha$ is sometimes called also the *Hilbertian sum* of the subspaces. If $\mathcal{E} \equiv \{e_j\}$ is an orthonormal set in $\mathcal{H}$, and $\mathcal{G}_j$ are the corresponding one–dimensional subspaces, then Theorem 2 reduces to the Fourier expansion of x with respect to $\mathcal{E}$. In this sense the relation of Problem 13c represents a generalization to the Parseval identity; in the same way one can generalize the Bessel inequality.

As in the case of a finite family, the direct sum $\sum_{\alpha\in I} \mathcal{H}_\alpha$ can be written as the orthogonal sum of the subspaces $\mathcal{H}^{(\alpha)} := \{X \in \sum_{\alpha\in I} \mathcal{H}_\alpha : X(\beta) = 0_\beta \text{ for } \beta \neq \alpha\}$, which are naturally isomorphic to $\mathcal{H}_\alpha$; this is checked easily using Problem 14.

Direct integrals of Hilbert spaces were introduced by J. von Neumann. They have numerous applications, *e.g.*, in spectral theory of self–adjoint operators ([[BS]], Chap.7; [[Mau]], Chap.IX), diagonalization of commutative W^*–algebras ([[Di 1]], Sec.II.1), decomposition of C^*–algebras or representations of locally compact topological groups to irreducible components ([[Nai 1]], Chap.VIII; [[BR 1]], Sec.4.4), *etc.* Some authors like [[Nai 1]], Sec.41, define the direct integral only for the particular class of maps $x \mapsto \mathcal{H}(x)$ described in Problem 18. Our definition of measurability is adopted from [[BS]], Sec.7.1; a more general approach can be found, *e.g.*, in [[Di 1]], Sec.II.1; [[BR 1]], Sec.4.4.1.

Section 2.4 The abstract definition of the tensor product is formulated in terms of the theory of categories — see, *e.g.*, [[KGv]], Sec.I.3 and Problem 61. The construction used to prove Theorem 1 is adopted essentially from [[We]], Sec.3.4. There are alternative ways to construct the space $\mathcal{H}$: in [[RS 1]], Sec.II.4, one finds a construction based on bilinear forms on $\mathcal{H}_1 \times \mathcal{H}_2$; another method uses continuous antilinear operators $T_{xy} : \mathcal{H}_2 \to \mathcal{H}_1$ defined for any $x \in \mathcal{H}_1$, $y \in \mathcal{H}_2$ by $T_{xy} := (\cdot, y)_2 x$ — see Problem 3.39b.

All the definitions and results formulated here for a pair of Hilbert spaces extend easily to any finite family; the map $[x_1, \dots, x_n] \mapsto x_1 \otimes \cdots \otimes x_n$ is then multilinear, fulfils the condition $(x_1 \otimes \cdots \otimes x_n, x_1' \otimes \cdots \otimes x_n') = \prod_{j=1}^n (x_j, x_j')_j$, and the set $\{ x_1 \otimes \cdots \otimes x_n : x_j \in \mathcal{H}_j \}$ is total in $\mathcal{H}_1 \otimes \cdots \otimes \mathcal{H}_n$. The notion of a tensor product also extends to infinite families of Hilbert spaces [vN 3]; however, we shall not need it in the present book.

We have introduced the algebraic tensor product for subspaces of a Hilbert space. If V_1, V_2 are arbitrary vector spaces, a *realization* of their *algebraic tensor product* is a vector space V together with a bilinear map $\otimes : V_1 \times V_2 \to V$ fulfilling the following conditions:

(i) $\{ x \otimes y : x \in V_1, y \in V_2 \}_{lin} = V$.

(ii) For any other pair $(V', \otimes')$ which fulfil requirement (i) there is a linear operator $T : V \to V'$ such that $x \otimes' y = T(x \otimes y)$ holds for all $[x, y] \in V_1 \times V_2$.

From here one can again arrive at an abstract definition of the tensor product — see, *e.g.*, [[MB]], Sec.IX.8. It is possible, however, to identify the algebraic tensor product with one of its realizations, in which case we write symbolically $V = V_1 \underline{\otimes} V_2$. To fulfil condition (ii), it is sufficient that the map $[x, y] \mapsto x \otimes y$ has properties (a), (b) of Problem 20; this ensures consistency of the present definition with that of Remark 3. The existence of an algebraic–tensor–product realization can be checked again constructively. We can start, *e.g.*, as in the proof of Theorem 2.4.1; however, the procedure is now different since there is no inner product (Problem 22).

The notion of a tensor product is also introduced for Banach and locally convex spaces. The definition starts from the algebraic tensor product of the appropriate vector spaces. Given Banach spaces $\mathcal{X}_r$ with the norms $\|\cdot\|_r$, $r = 1, 2$, we call a call a norm $\|\cdot\|$ on $\mathcal{X}_1 \underline{\otimes} \mathcal{X}_2$ a *cross–norm* if $\|x \otimes y\| = \|x\|_1 \|y\|_2$ for all $x \in \mathcal{X}_1$, $y \in \mathcal{X}_2$. In distinction to the Hilbert–space case, the last condition does not determine the norm on $\mathcal{X}_1 \underline{\otimes} \mathcal{X}_2$ uniquely; there is, however, a maximum cross–norm defined by $\| \sum_k x_k \otimes y_k \|_\pi := \inf \left\{ \sum_j \|x_j'\|_1 \|y_j'\|_2 \right\}$, where the infimum is taken over the set of vectors $\sum_j x_j' \otimes y_j' \in \mathcal{X}_1 \underline{\otimes} \mathcal{X}_2$ fulfilling $\sum_j x_j' \otimes y_j' = \sum_k x_k \otimes y_k$. If $\|\cdot\|_\alpha$ is a cross–norm on $\mathcal{X}_1 \underline{\otimes} \mathcal{X}_2$, then the Banach space obtained by the standard completion is denoted $\mathcal{X}_1 \otimes_\alpha \mathcal{X}_2$; for $\mathcal{X}_1 \otimes_\pi \mathcal{X}_2$ we also use the symbol $\mathcal{X}_1\hat{\otimes}\mathcal{X}_2$. The cross–norms on $\mathcal{X}_1\underset{\smile}{\times}\mathcal{X}_2$ and their relations to the cross–norms on $\mathcal{X}_1^*\underset{\smile}{\times}\mathcal{X}_2^*$ are discussed in ⟦Sch⟧, see also ⟦KGv⟧, Sec.III.1.4; for an extension of the construction of $\mathcal{X}_1\hat{\otimes}\mathcal{X}_2$ to locally convex spaces see ⟦Gr⟧.

Problems

1. Prove (a) Let $\mathcal{G}$ be a finite–dimensional subspace in a Hilbert space $\mathcal{H}$; then if $\{e_1,\dots,e_n\}$ is an orthonormal basis in $\mathcal{G}$, the orthogonal projection y_x of a vector $x\in\mathcal{H}$ to $\mathcal{G}$ is given by $y_x=\sum_{j=1}^n(e_j,x)e_j$.

(b) The Bessel inequality.

Hint: (a) The distance of x from $y\in\mathcal{G}$ is a function of n complex parameters; find its minimum.

2. Let M and $\mathcal{G}$ be a subset and a closed subspace, respectively, in a Hilbert space $\mathcal{H}$. Prove

(a) $M\cap M^\perp\subset\{0\}$, in particular, $M^\perp=\{0\}$ if M is total

(b) $\mathcal{G}^{\perp\perp}=\mathcal{G}$, and more generally, $M^{\perp\perp}=\overline{M_{lin}}$, which implies $\mathcal{H}=M^\perp\oplus\overline{M_{lin}}$

3. Let L_1, L_2 be subspaces in a Hilbert space $\mathcal{H}$ and $\dim L_2<\infty$. Prove

(a) If L_1 is closed, the algebraic sum L_1+L_2 is a closed subspace.

(b) If T is a linear operator on $\mathcal{H}$ such that $L_1+L_2\subset D_T$ and the subspace TL_1 is closed, then there is a subspace $M\subset(TL_1)^\perp$ such that $\dim M\le\dim L_2$ and $T(L_1+L_2)=TL_1\oplus M$.

4. There is no Hilbert space with a countably infinite Hamel basis.
Hint: Use the Gram–Schmidt theorem.

5. Let $\mathcal{H}$ be a separable Hilbert space.

(a) Prove the existence of a countable orthonormal basis in $\mathcal{H}$ without using the Zorn lemma.

(b) If $\mathcal{G}$ is a dense subspace in $\mathcal{H}$, there is an orthonormal basis $\mathcal{E}$ of $\mathcal{H}$ contained in $\mathcal{G}$.

Hint: (a) From a countable dense set M one can pick a linearly independent subset N such that $M\subset N_{lin}$. (b) If $\{x_n:\,n=1,2,\dots\}$ is dense in $\mathcal{H}$, there is $\{y_{nm}:\,n,m=1,2,\dots\}\subset\mathcal{G}$ such that $\|x_n-y_{nm}\|<m^{-1}$.

6. The basis $\mathcal{E}_H\equiv\{h_n:\,n=0,1,\dots\}$ in $L^2(\mathbb{R})$ defined by (2.1) is contained in $\mathcal{S}(\mathbb{R})$ and its elements are eigenvectors of the Fourier–Plancherel operator, $Fh_n=\hat{h}_n=(-i)^nh_n\,,\ n=0,1,\dots$.

7. The set $\mathcal{F}_L^{(\alpha)}$, $\alpha>-1$, of Example 2.2.3 is an orthonormal basis in $L^2(\mathbb{R}^+)$.

8. Let $(X, \mathcal{A}, \mu)$ be a measure space with a σ–finite μ; then the space $L^2(X, d\mu)$ is separable *iff* μ has a countable basis.
Hint: If $\{f_n : n = 1, 2, \dots\}$ is dense in $L^2(X, d\mu)$, we can find f_n such that $\|\chi_M - f_n\| < m^{-1}$ to any m and $M \in \mathcal{A}$ with $\mu(M) < \infty$. Denote $M_{nm} := \{x : |f_n(x) - 1| < 2^{-m}\}$ and show that $\mu(M \Delta M_{nm}) < m^{-1}$.

9. Let f be an almost periodic function.

(a) Suppose that a sequence $\{f_n\} \subset \{e_\lambda : \lambda \in \mathbb{R}\}_{lin}$ fulfils $\|f_n - f\|_\infty \to 0$. Define $F_n(T) := \frac{1}{2T} \int_{-T}^{T} |f_n(x)|^2 dx$ and $F(T)$ corresponding in the same way to f. Check that $\{F_n\}$ converges to F uniformly in $\mathbb{R}^+$ and use this result to prove the existence of the limit $p(f)^2 \equiv \lim_{T\to\infty} F(T)$ and the relations

$$p(f) = \lim_{n\to\infty} p(f_n)\,, \quad p(f_n)^2 = \lim_{T\to\infty} \frac{1}{2T} \int_{-T}^{T} |f_n(x+t)|^2 dx$$

for any $t \in \mathbb{R}$.

(b) $p(f) > 0$ unless f is zero.

Hint: (b) Use the Bohr criterion (see the notes to Section 2.2).

10. The map $f \longmapsto F_w(f) := f(w)$ is for any $w \in \mathbb{C}$ a bounded linear functional on $A^2(\mathbb{C})$ whose norm is $\|F_w\| = e^{|w|^2/2}$.

11. Prove: (a) A vector–valued function $F : X \to \mathcal{G}$ is measurable *iff* the functions $(e_j, F(\cdot))$ are measurable for all elements of some orthonormal basis $\{e_j\}$ of the space $\mathcal{G}$.

(b) If $\{\, f_\alpha : \alpha \in I \,\}$ is a total set in $L^2(X, d\mu)$ and $\{e_j\}$ is an orthonormal basis in $\mathcal{G}$, then the set $\{\, f_\alpha(\cdot) e_j : \alpha \in I\,,\ j = 1, 2, \dots \}$ is total in $L^2(X, d\mu; \mathcal{G})$.

12. Check the relation (2.3) for coherent states in $L^2(\mathbb{R})$ and prove that

$$\int_{\mathbb{R}} x|\psi_w(x)|^2\, dx \;=\; \sqrt{2}\,\mathrm{Re}\, w\,, \qquad \int_{\mathbb{R}} \overline{\psi_w(x)}\, \frac{d\psi_w}{dx}(x)\, dx \;=\; -i\sqrt{2}\,\mathrm{Im}\, w\,.$$

Hint: Use part (b) of Theorem 2.2.7 and the generating function of Hermite polynomials, $\sum_{k=0}^{\infty} \frac{z^k}{k!} H_k(x) = e^{-z^2+2zx}$ for $z \in \mathbb{C}$ and $x \in \mathbb{R}$.

13. Let $\{\, \mathcal{G}_\alpha : \alpha \in I \,\}$ be a family of mutually orthogonal closed subspaces of a given Hilbert space $\mathcal{H}$; then

(a) $\left(\sum_{\alpha\in I}^{\oplus} \mathcal{G}_\alpha\right)^\perp = \bigcap_{\alpha\in I} \mathcal{G}_\alpha^\perp\,,$

(b) If x_α is the projection of a vector $x \in \mathcal{H}$ to $\mathcal{G}_\alpha$ and $\{\alpha_1, \dots, \alpha_n\}$ is a finite subset in I, then the projection of x to the subspace $\bigoplus_{j=1}^{n} \mathcal{G}_{\alpha_j}$ equals $\sum_{j=1}^{n} x_{\alpha_j}$.

(c) Given a countable subset $\{\alpha_j : j = 1, 2, \dots \} \subset I$ and a sequence $\{x_j\}$ with $x_j \in \mathcal{G}_{\alpha_j}$, set $s_n := \sum_{j=1}^{n} x_j$. The sequence $\{s_n\}$ converges *iff* $\sum_{j=1}^{\infty} \|x_j\|^2 < \infty$; if this condition is valid and $x := \lim_{n\to\infty} s_n$, then x_j is the projection of x to $\mathcal{G}_{\alpha_j}$ and $\|x\|^2 = \sum_{j=1}^{\infty} \|x_j\|^2$.

14. Let $\{\mathcal{G}_\alpha : \alpha \in I\}$ be a family of mutually orthogonal closed subspaces of a given $\mathcal{H}$, where I is an arbitrary index set. The subspace $\sum_{\alpha\in I}^{\oplus} \mathcal{G}_\alpha \subset \mathcal{H}$ consists just of those vectors for which there is an at most countable $\{\alpha_k : k = 1, 2, \dots\} \subset I$ such that $x = \lim_{k\to\infty} \sum_{j=1}^{k} x_j$ with $x_j \in \mathcal{G}_{\alpha_j}$.

15. Let $\mathcal{E}_j$ be an orthonormal basis in a Hilbert space $\mathcal{H}_j$, $j = 1, 2, \dots$; then $\mathcal{E} := \bigcup_{j=1}^{\infty}\{0, \dots 0, \mathcal{E}_j, 0, \dots\}$ is an orthonormal basis in $\bigoplus_{j=1}^{\infty} \mathcal{H}_j$. In particular, $\mathcal{H}$ is separable provided all the spaces $\mathcal{H}_j$ are separable.

16. Any measurability basis $\{E_k\}$ belongs to $\mathcal{V}_\mu$. A vector-valued function F belongs to $\mathcal{V}_\mu$ *iff* $x \mapsto (F(x), G(x))_x$ is measurable for all $G \in \mathcal{V}_\mu$.

17. Let $(x \mapsto \mathcal{H}(x), \mathcal{V}_\mu)$ and $(x \mapsto \mathcal{H}'(x), \mathcal{V}'_\mu)$ be measurable Hilbert-space fields such that $\dim \mathcal{H}(x) = \dim \mathcal{H}'(x)$ holds for all $x \in X$; then there is $x \mapsto U(x)$ such that

(a) $U(x)$ is an isomorphism of the spaces $\mathcal{H}(x)$ and $\mathcal{H}'(x)$ for any $x \in X$.

(b) A vector-valued function F' belongs to $\mathcal{V}'_\mu$ *iff* there is $F \in \mathcal{V}_\mu$ such that $F'(x) = U(x)F(x)$ for all $x \in X$.

18. Let $(x \mapsto \mathcal{H}(x), \mathcal{V}_\mu)$ be a measurable Hilbert-space field with a measurability basis $\{E_k : k = 1, 2, \dots\}$, and denote $\mathcal{H} \equiv \int_X^{\oplus} \mathcal{H}(x)\, d\mu(x)$. Prove

(a) If $f_k(x) := (E_k(x), F(x))_x$ for a given $F \in \mathcal{H}$, then the sequence $\{f_k\}$ belongs to $L^2(X, d\mu)$ and

$$(F, G) = \int_X \sum_{k=1}^{\infty} \bar{f}_k g_k \, d\mu = \sum_{k=1}^{\infty} (f_k, g_k)_{L^2};$$

in particular, $F = 0$ *iff* $f_k = 0$ for $k = 1, 2, \dots$.

(b) If a sequence $\{f_k\} \subset L^2(X, d\mu)$ fulfils $\sum_{k=1}^{\infty} \|f_k\|_{L^2}^2 < \infty$, then there is just one $F \in \mathcal{H}$ such that $(E_k(x), F(x))_x = f_k(x)$ holds for μ-a.a. $x \in X$.

(c) The space $\mathcal{H}$ is complete, *i.e.*, a Hilbert space.

(d) If $\{f_\alpha : \alpha \in I\}$ is a total set in $L^2(X, d\mu)$, then the vector-valued functions $x \mapsto F_{\alpha k}(x) := f_\alpha E_k(x)$ form a total set in $\mathcal{H}$.

(e) If $L^2(X, d\mu)$ is separable, the same is true for $\mathcal{H}$.

19. Let $X = \bigcup_{n=0}^{\infty} X_n$ be the decomposition of a set X corresponding to a given dimension function d. Choose a sequence $\{\mathcal{G}_n\}_{n=0}^{\infty}$ of separable Hilbert spaces such that $\dim \mathcal{G}_n = n$, $n = 1, 2, \dots$, and $\dim \mathcal{G}_0 = \infty$ and define $\mathcal{H}_d(x) := \mathcal{G}_n$ for $x \in X_n$, $n = 0, 1, \dots$. Using orthonormal bases in the spaces $\mathcal{G}_n$ construct a measurability basis in $\times_{x\in X}\mathcal{H}_d(x)$. Prove that the map $V : F \mapsto \{F_n\}_{n=0}^{\infty}$ with $F_n := F \restriction X_n$ is an isomorphism of the spaces $\int_X^{\oplus} \mathcal{H}(x)\, d\mu(x)$ and $\bigoplus_{n=0}^{\infty} L^2(X_n, d\mu; \mathcal{G}_n)$.

20. Let $x \otimes y \in \mathcal{H}_1 \otimes \mathcal{H}_2$; then

(a) $x \otimes y = 0$ *iff* at least one of the vectors x, y is zero.

(b) If $x \otimes y$ is nonzero and equal to $x' \otimes y'$, then there is a nonzero $\alpha \in \mathbb{C}$ such that $x' = \alpha x$ and $y' = \alpha^{-1} y$.

21. Prove: (a) If $\dim\mathcal{H}_1 = n < \infty$ then $\mathcal{H}_1 \otimes \mathcal{H}_2$ is isomorphic to $\bigoplus_{k=1}^{n} \mathcal{H}_2$.

(b) Let $\mathcal{H}_1 = \sum_{\alpha\in I}^{\oplus} \mathcal{G}_\alpha$, where the $\mathcal{G}_\alpha$ are mutually orthogonal closed subspaces, then the subspaces $\mathcal{G}_\alpha \otimes \mathcal{H}_2 \subset \mathcal{H}_1 \otimes \mathcal{H}_2$ are mutually orthogonal and $\mathcal{H}_1 \otimes \mathcal{H}_2 = \sum_{\alpha\in I}^{\oplus} \mathcal{G}_\alpha \otimes \mathcal{H}_2$.

22. Let V_1, V_2 be arbitrary vector spaces. Show that a bilinear map $[x,y] \mapsto x \otimes y \in V_1 \diamond V_2/L$ satisfies implications (a), (b) of Problem 20.
Hint: Check under which conditions $x \diamond y \in L$ or $x \diamond y - x' \diamond y' \in L$, respectively.

23. $\overline{L_1 \underline{\times} L_2} = \overline{L_1} \otimes \overline{L_2}$ holds for any subspaces $L_j \subset \mathcal{H}_j$.

24. Let $\mathcal{E}^{(k)} = \{e_j^{(k)}\}$ and $\mathcal{F} = \{f_k\}$ be orthonormal bases in the Hilbert spaces $\mathcal{H}_1$ and $\mathcal{H}_2$, respectively; then $\{\, e_j^{(k)} \otimes f_k : j = 1, \dots, \dim\mathcal{H}_1,\ k = 1, \dots, \dim\mathcal{H}_2 \,\}$ is an orthonormal basis in $\mathcal{H}_1 \otimes \mathcal{H}_2$.

Chapter 3

Bounded operators

3.1 Basic notions

The space $\mathcal{B}(\mathcal{H})$ of all bounded linear operators on a given Hilbert space $\mathcal{H}$ represents a particular case of $\mathcal{B}(V,V')$ discussed in Section 1.5, so all of the results derived there apply. At the same time, the existence of an inner product in $\mathcal{H}$ brings new information. For instance, we know that $\mathcal{B}(\mathcal{H})$ is a Banach space with the norm of Problem 1.51; now we are able to express it alternatively as

$$\|B\| = \sup_{x,y\in S_1} |(x,By)| = \sup_{x,y\in B_1} |(x,By)|,$$

where S_1, B_1 are the unit sphere and the unit ball in $\mathcal{H}$, respectively. Another simple consequence concerns bounded sesquilinear forms on $\mathcal{H}$, *i.e.*, the forms f to which a positive c exists such that $|f(x,y)| \le c\|x\|\,\|y\|$ for all $x,y \in \mathcal{H}$. Using elementary properties of the norm together with Riesz's lemma, we get

3.1.1 Proposition: There is a one–to–one correspondence between bounded operators and bounded sesquilinear forms; it relates $B \in \mathcal{B}(\mathcal{H})$ to $f_B : f_B(x,y) = (x,By)$.

For any $B \in \mathcal{B}(\mathcal{H})$, we can define the form f_B^* by $f_B^*(x,y) := \overline{f_B(y,x)}$. Since it is again sesquilinear and bounded, there is just one operator $B^* \in \mathcal{B}(\mathcal{H})$ such that $f_B^*(x,y) = (x,B^*y)$, in other words,

$$(y,Bx) = (B^*y,x) \tag{3.1}$$

for all $x,y \in \mathcal{H}$; we call it the **adjoint** to B.

3.1.2 Theorem: (a) The map $B \mapsto B^*$ is an antilinear isometry of $\mathcal{B}(\mathcal{H})$ onto itself, which fulfils $(BC)^* = C^*B^*$ and $B^{**} = B$ for any $B,C \in \mathcal{B}(\mathcal{H})$.

(b) If B is invertible the same is true for B^* and $(B^*)^{-1} = (B^{-1})^*$.

Proof: Antilinearity and the last two relations in (a) follow from (3.1) and Problem 3. Inserting $x = B^*y$ into (3.1), we get $\|B^*y\|^2 \le \|y\|\,\|B\|\,\|B^*y\|$, so $\|B^*\| \le \|B\|$;

replacing B by B^* we prove $B \mapsto B^*$ is an isometry. To check (b), notice that B^{-1} is bounded and use $B^{-1}B = BB^{-1} = I$ together with part (a). ∎

3.1.3 Example: To a given orthonormal basis $\mathcal{E} := \{e_j\}_{j=1}^{\infty}$ in a separable $\mathcal{H}$ define S_0 on $\mathcal{E}_{lin}$ as a linear extension of $S_0 e_j := e_{j+1}$, $j = 1, 2, \dots$. It is bounded, $\|S_0 x\| = \|x\|$ for any $x \in \mathcal{E}_{lin}$, so its continuous extension S called the *right–shift operator* has $\|S\| = 1$. It follows from (3.1) that

$$S^* e_j = \sum_{k=1}^{\infty} (e_k, S^* e_j) e_k = \begin{cases} 0 & \dots \quad j = 1 \\ e_{j-1} & \dots \quad \text{otherwise} \end{cases}$$

and due to continuity, S^* is fully determined by these relations; we call it the *left–shift operator.*

3.1.4 Proposition: $\operatorname{Ker} B^* = (\operatorname{Ran} B)^{\perp}$ for any $B \in \mathcal{B}(\mathcal{H})$.

Proof: Since B^* is continuous, $\operatorname{Ker} B^*$ is a closed subspace. If $x \in \operatorname{Ker} B^*$ and $z = By$, then $(x, z) = (B^* x, y) = 0$ so $\operatorname{Ker} B^* \subset (\operatorname{Ran} B)^{\perp}$. On the other hand, $z \in (\operatorname{Ran} B)^{\perp}$ implies $0 = (z, By) = (B^* z, y)$ for all $y \in \mathcal{H}$, *i.e.*, $z \in \operatorname{Ker} B^*$. ∎

As an illustration of how the adjoint operator may be used, let us prove the following representation of bounded operators in $L^2(\mathbb{R})$.

3.1.5 Proposition: Let $B \in \mathcal{B}(L^2(\mathbb{R}))$; then there is a function $G : \mathbb{R}^2 \to \mathbb{C}$ such that

(a) The function $G_x := \overline{G(x, \cdot)}$ belongs to $L^2(\mathbb{R})$ and $G_0 = 0$.

(b) $x \mapsto (G_x, f)$ is absolutely continuous in any finite interval $[a, b] \subset \mathbb{R}$ for all $f \in L^2(\mathbb{R})$.

(c) $(Bf)(x) = \frac{d}{dx} \int_{\mathbb{R}} G(x, y) f(y)\, dy$ for all $f \in L^2(\mathbb{R})$ and almost all $x \in \mathbb{R}$.

(d) G is essentially unique: suppose a function $\tilde{G} := \mathbb{R}^2 \to \mathbb{C}$ fulfils (a) and there is a total set $D \subset L^2(\mathbb{R})$ such that (b),(c) are valid with $\tilde{G}$ for all $f \in D$, then $G(x, y) = \tilde{G}(x, y)$ holds for all $x \in \mathbb{R}$ and almost all $y \in \mathbb{R}$.

Proof: Let J_x be the interval with the endpoints $0, x$. Setting $e_x := \operatorname{sgn}(x) \chi_{J_x}$, we have $e_0 = 0$, $e_x \in L^2(\mathbb{R})$, so $G_x := B^* e_x$ satisfies (a) and

$$(G_x, f) = (e_x, Bf) = \operatorname{sgn}(x) \int_{J_x} (Bf)(y)\, dy$$

for any $f \in L^2(\mathbb{R})$. The function $x \mapsto F(x) := \int_a^x (Bf)(y)\, dy$ is absolutely continuous on any finite interval (a, b) and fulfils there $(G_x, f) = F(x) + \text{const}$; hence (b) is valid and

$$(Bf)(x) = F'(x) = \frac{d}{dx} (G_x, f) = \frac{d}{dx} \int_{\mathbb{R}} G(x, y) f(y)\, dy$$

for almost all $x \in (a,b)$, and therefore a.e. in $\mathbb{R}$. Finally, consider a function $\tilde{G}$ specified in part (d). Then $\frac{d}{dx}(G_x, f) = \frac{d}{dx}(\tilde{G}_x, f)$ for almost all $x \in \mathbb{R}$ and any $f \in D$; in view of the absolute continuity and the identity $\tilde{G}_0 = G_0 = 0$, we have $(G_x, f) = (\tilde{G}_x, f)$ for all $x \in \mathbb{R}$, $f \in D$, so G_x and $\tilde{G}_x$ represent the same vector in $L^2(\mathbb{R})$. ∎

3.1.6 Example (a functional realization of the FP–operator on $L^2(\mathbb{R})$): In the same way as in Example 1.5.6 we can check that

$$\left(F^{-1}h\right)(x) = (2\pi)^{-1/2} \int e^{ixy} h(y)\, dy \tag{3.2}$$

for any $h \in L^2 \cap L^1$. Moreover, we have $(f, Fg) = (FF^{-1}f, Fg) = (F^{-1}f, g)$ for any $f, g \in L^2$, so $F^* = F^{-1}$. Notice that this relation and the formula (3.2) hold in $L^2(\mathbb{R}^n)$ as well; we have only to replace $(2\pi)^{-1/2}$ by $(2\pi)^{-n/2}$ and e^{ixy} by $e^{ix\cdot y}$ where $x \cdot y$ means the inner product in $\mathbb{R}^n$.

Since $e_x \in L^2 \cap L^1$, the functions G and G^* referring to $F^* = F^{-1}$ are easily found: the operators F, F^{-1} act at any $f \in L^2(\mathbb{R})$ as

$$(Ff)(x) = (2\pi)^{-1/2} \frac{d}{dx} \int_{\mathbb{R}} \frac{e^{-ixy} - 1}{-iy} f(y)\, dy\,,$$

$$\left(F^{-1}f\right)(x) = (2\pi)^{-1/2} \frac{d}{dx} \int_{\mathbb{R}} \frac{e^{ixy} - 1}{iy} f(y)\, dy\,.$$

The functions G, G^* here are complex conjugated, which is not true in general for the representation of a pair B, B^* — *cf.* Problem 6.

Next we show that the notion of a matrix representation known from linear algebra can be extended to bounded operators. We shall assume that $\mathcal{H}$ is separable and shall choose an orthonormal basis $\mathcal{E} = \{e_j\}_{j=1}^{\infty} \subset \mathcal{H}$. To any $B \in \mathcal{B}(\mathcal{H})$, we define an infinite matrix by

$$B_{jk} := (e_j, Be_k)\ , \quad j,k = 1,2,\dots; \tag{3.3}$$

its k–th column is formed by the Fourier coefficients of the vector Be_k, so Parseval's identity gives $\sum_{j=1}^{\infty} |B_{jk}|^2 < \infty$, $k = 1,2,\dots$. The same argument applied to B^*e_k yields $\sum_{j=1}^{\infty} |B_{kj}|^2 < \infty$, $k = 1,2,\dots$. Furthermore, we can express the relation between the Fourier coefficients of x and Bx by

$$(e_j, Bx) = \sum_{k=1}^{\infty} B_{jk}(e_k, x)\,,$$

where the absolute convergence of the series is checked easily using the Hölder inequality. In this way, we have defined a mapping of $\mathcal{B}(\mathcal{H})$ onto the set of infinite matrices with square summable rows and columns, which is obviously linear and satisfies the standard matrix rules

$$(BC)_{jk} = \sum_{l=1}^{\infty} B_{jl}C_{lk}\ , \quad (B^*)_{jk} = \overline{B_{kj}}\ ;$$

we speak about the *matrix representation* of the operator B with respect to the basis $\mathcal{E}$.

On the other hand, a matrix (B_{jk}) is said to represent an operator B if there is a separable Hilbert space $\mathcal{H}$ with an orthonormal basis $\mathcal{E}$ such that the relations (3.3) hold. A finite matrix always represents a bounded operator on an appropriate finite–dimensional space, while in the general case the problem is slightly more complicated.

3.1.7 Theorem: An infinite matrix $(b_{jk})_{j,k=1}^{\infty}$ represents a bounded operator *iff* the mapping

$$(X,Y) \longmapsto F(X,Y) := \sum_{j=1}^{\infty} \left(\bar{\xi}_j \sum_{k=1}^{\infty} b_{jk}\eta_k \right),$$

where $X = \{\xi_k\}$, $Y = \{\eta_k\}$ are any vectors of ℓ^2, is a bounded sesquilinear form on ℓ^2.

Proof: The necessary condition is simple. On the other hand, choosing $Y_l = \{\delta_{lj}\}_{j=1}^{\infty}$ we find from the boundedness of F that $\sum_{k=1}^{\infty} b_{kl}\bar{\xi}e_k$ converges for any $\{\xi_k\} \in \ell^2$, and therefore $\sum_{k=1}^{\infty} |b_{kl}|^2 < \infty$. Hence we can pick a separable $\mathcal{H}$ with an orthonormal basis $\mathcal{E} = \{e_l\}$ and define $e_l \mapsto f_l := \sum_{k=1}^{\infty} b_{kl}e_k$. Denote by B_0 the linear extension of this map. We can easily check that it is bounded, and its continuous extension to $\mathcal{H}$ is the sought operator B, which is by definition represented by the matrix (b_{jk}). ∎

The space $\mathcal{B}(\mathcal{H})$ can be equipped with topologies other than the topology τ_u generated by the operator norm. Let us now mention two of them. Consider first the system of seminorms $\mathcal{P}_s := \{p_x : x \in \mathcal{H},\, p_x(B) := \|Bx\|\}$, which is obviously separating points. The topology generated by $\mathcal{P}_s$ is called the **strong operator topology** (we use the symbol τ_s) and the corresponding locally convex space will be denoted as $\mathcal{B}_s(\mathcal{H})$; its local basis is formed by the sets

$$V_\varepsilon(M) := \{\, B \in \mathcal{B}(\mathcal{H}) : p_x(B) < \varepsilon\,,\ x \in M \,\}$$

for all $\varepsilon > 0$ and any finite set $M \subset \mathcal{H}$. A sequence $\{B_n\}$ converges in $\mathcal{B}_s(\mathcal{H})$ to an operator B *iff* $B_n x \to Bx$ for all $x \in \mathcal{H}$; we use the notation $B_n \xrightarrow{s} B$ or $B = \text{s-lim}_{n\to\infty} B_n$. The following results are easy consequences of the uniform boundedness principle.

3.1.8 Proposition: (a) *Strong completeness of* $\mathcal{B}(\mathcal{H})$ *:* Let $\{B_n\}$ be a sequence such that $\{B_n x\}$ is Cauchy for any $x \in \mathcal{H}$; then there is $B \in \mathcal{B}(\mathcal{H})$ such that $B_n \xrightarrow{s} B$.

(b) *Sequential continuity of multiplication:* If $B_n \xrightarrow{s} B$ and $C_n \xrightarrow{s} C$; then the product sequence $B_nC_n \xrightarrow{s} BC$.

In distinction to the operator–norm topology, however, assertion (b) does not imply continuity of the multiplication (understood as a map $\mathcal{B}_s(\mathcal{H}) \times \mathcal{B}_s(\mathcal{H}) \to \mathcal{B}_s(\mathcal{H})$).

The reason is that the topological space $\mathcal{B}_s(\mathcal{H})$ is not first countable with the exception of the trivial case of a finite–dimensional $\mathcal{H}$.

3.1.9 Theorem: If $\dim\mathcal{H} = \infty$, the operator multiplication is not continuous on $\mathcal{B}_s(\mathcal{H})$.

Proof: We may assume that $\mathcal{H}$ is separable and $\mathcal{E} = \{e_j\}_{j=1}^{\infty}$ is an orthonormal basis; it is sufficient to find for any neighborhoods $V_r := V_{\varepsilon_r}(M_r)$, $r = 1,2$, a pair of operators $B_r \in V_r$ such that $B_1B_2 \notin V_{1/2}(\{x\})$, where $x \in \mathcal{H}$ is a unit vector. Consider the shift operators of Example 3.1.3: to any $\delta > 0$, $n = 1,2,\dots$, we define $B_{n,\delta} := \frac{1}{\delta}(S^*)^n$, $C_{n,\delta} := \delta S^n$. The operator $B_{n,\delta}C_{n,\delta} = I$ does not belong to $V_{1/2}(\{x\})$ for any n,δ. Choosing $\delta = \delta_2 := \varepsilon_2\left(2\max_{y\in M_2}\|y\|\right)^{-1}$, we have $C_{n,\delta_2} \in V_2$ for any natural number n. Moreover, $(S^*)^n \xrightarrow{s} 0$, so we can find a natural number n_1 (depending on $\varepsilon_1, \delta_2, M_1$) such that $\|(S^*)^{n_1}z\| < \varepsilon_1\delta_2$ for all $z \in M_1$, and therefore $B_{n_1,\delta_2} \in V_1$. ∎

It is easy to check that the two topologies on $\mathcal{B}(\mathcal{H})$ are related by $\tau_u \supset \tau_s$. The proved theorem tells us, in particular, that the inclusion is nontrivial. This can be illustrated in an even easier way (Problem 8).

Let us now pass to the **weak operator topology** τ_w on $\mathcal{B}(\mathcal{H})$. It is generated by the system of seminorms $\mathcal{P}_w := \{p_{xy} : x,y \in \mathcal{H},\, p_{xy}(B) := |(y,Bx)|\}$, which again separates points. The corresponding local basis is formed by

$$W_\varepsilon(M,N) := \{B \in \mathcal{B}(\mathcal{H}) : p_{xy}(B) < \varepsilon,\, x \in M,\, y \in N\}$$

and the locally convex space $(\mathcal{B}(\mathcal{H}), \tau_w)$ will be denoted as $\mathcal{B}_w(\mathcal{H})$. The weak operator convergence of a sequence $\{B_n\}$ to an operator B is therefore equivalent to $(y,B_nx) \to (y,Bx)$ for all $x,y \in \mathcal{H}$; we use the notation $B_n \xrightarrow{w} B$ or $B = \text{w-lim}_{n\to\infty}B_n$. The space $\mathcal{B}(\mathcal{H})$ is also weakly complete.

3.1.10 Proposition: Let $\{B_n\} \subset \mathcal{B}(\mathcal{H})$ be a sequence such that $\{(y,B_nx)\}$ is Cauchy for any $x,y \in \mathcal{H}$; then there is a $B \in \mathcal{B}(\mathcal{H})$ such that $B_n \xrightarrow{w} B$.

Proof: The weak completeness of $\mathcal{H}$ implies that B_nx converges for any $x \in \mathcal{H}$ to some $z(x) = Bx$. Since the B_n are linear operators, B is linear also. The sequence $\{B_nx\}$ is weakly convergent and therefore bounded; the uniform boundedness principle then implies $\|B_n\| \le c$ for some c and $n = 1,2,\dots$. We have $|(B_nx,Bx)| \le c\|x\|\,\|Bx\|$ and $|(B_nx,Bx)| \to \|Bx\|^2$, so $B \in \mathcal{B}(\mathcal{H})$. ∎

On the other hand, the multiplication in $\mathcal{B}_w(\mathcal{H})$ is not even sequentially continuous (Problem 9). In general, the nontrivial inclusion $\tau_w \subset \tau_s$ is valid (Problems 9, 10); more about topologies on $\mathcal{B}(\mathcal{H})$ will be said in Section 6.3.

3.2 Hermitean operators

An operator $A \in \mathcal{B}(\mathcal{H})$ is said to be **Hermitean** if $A = A^*$; by properties of the inner product this is equivalent to the condition $(x,Ax) \in \mathbb{R}$ for all $x \in \mathcal{H}$. The set of all Hermitean operators forms a *real* subspace in $\mathcal{B}(\mathcal{H})$; however, it is not closed with respect to multiplication. The analogy between Hermitean operators and real numbers inspires the decomposition $B = \operatorname{Re} B + i \operatorname{Im} B$ with

$$\operatorname{Re} B := \frac{1}{2}(B + B^*) , \quad \operatorname{Im} B := \frac{1}{2i}(B - B^*)$$

valid for any $B \in \mathcal{B}(\mathcal{H})$.

The *numerical range* of a linear operator $T : \mathcal{H} \to \mathcal{H}$ with the domain D_T is defined as $\Theta(T) := \{ (x, Tx) : x \in D_T , \|x\| = 1 \}$. A bounded operator A is therefore Hermitean *iff* $\Theta(A) \subset \mathbb{R}$; the numbers

$$m_A := \inf \Theta(A) , \quad M_A = \sup \Theta(A)$$

are its *lower* and *upper bound*, respectively. They provide us with an alternative definition for the operator norm.

3.2.1 Proposition: The relations $\|A\| = \max(|m_A|, |M_A|) = \sup_{\|x\|=1} |(x, Ax)|$ hold for any Hermitean operator A.

Proof: Call the right side c_A. The last identity as well as the inequality $c_A \le \|A\|$ verify trivially. For a Hermitean A, the polarization formula gives $\operatorname{Re}(y, Ax) = \frac{1}{4}(q_A(x+y) - q_A(x-y))$, where $q_A(z) := (z, Az)$, and therefore

$$|\operatorname{Re}(y, Ax)| \le \frac{1}{4} c_A \left(\|x+y\|^2 + \|x-y\|^2 \right) = \frac{1}{2} c_A \left(\|x\|^2 + \|y\|^2 \right)$$

for all $x, y \in \mathcal{H}$. In particular, if x is a unit vector with $Ax \ne 0$, then one has to choose $y = Ax/\|Ax\|$ to get $\|Ax\| \le c_A$. ∎

3.2.2 Example: Consider $L^2(a, b)$ with $b - a < \infty$ and define the operator Q by $(Qf)(x) := xf(x)$ for any $f \in L^2(a, b)$. It is Hermitean with $m_Q = a$, $M_Q = b$, and $\|Q\| = \max(|a|, |b|)$.

An operator $A \in \mathcal{B}(\mathcal{H})$ whose numerical range is such that $m_A \ge 0$ ($m_A > 0$) is called **positive** (**strictly positive**); we use the symbols $A \ge 0$ and $A > 0$, respectively. Positive operators satisfy the Schwarz inequality,

$$|(y, Ax)|^2 \le (x, Ax)(y, Ay) . \tag{3.4}$$

3.2.3 Example: For an arbitrary $B \in \mathcal{B}(\mathcal{H})$ we have $(x, B^*Bx) = \|Bx\|^2 \ge 0$, so $B^*B \ge 0$. Then $\|B^*B\| = M_{B^*B}$ and Proposition 3.2.1 implies

$$\|B^*B\| = \|B\|^2 ; \tag{3.5}$$

this equality plays an important role in the theory of operator algebras — *cf.* Section 6.1. In particular, for a Hermitean A we get by induction $\|A^{2^n}\| = \|A\|^{2^n}$, so its spectral radius equals

$$r(A) = \lim_{n\to\infty} \|A^{2^n}\|^{2^{-n}} = \|A\| .$$

The notion of positivity allows us to introduce inequalities between Hermitean operators: we write $A \geq B$ if the operator $A - B$ is positive. Operator inequalities can be manipulated using simple rules analogous to those of numerical inequalities; we leave this to the reader. It is also clear that the norm is nondecreasing on the set of positive operators: $A \geq B \geq 0$ implies $\|A\| \geq \|B\|$.

Operator inequalities can be also used to define the supremum and infimum for an arbitrary set $\{ A_\alpha : \alpha \in I \}$ of Hermitean operators: $A^{(s)} := \sup_{\alpha \in I} A_\alpha$ is a Hermitean operator such that (i) $A_\alpha \leq A^{(s)}$ for any $\alpha \in I$, (ii) if B is a Hermitean operator such that $A_\alpha \leq B$ for all $\alpha \in I$, then $A^{(s)} \leq B$. In a similar way, we define the infimum. The supremum and infimum may not exist, in general, because $\leq$ defines only a partial ordering on the set of Hermitean operators. Existence is guaranteed, however, for fully ordered subsets.

3.2.4 Theorem: Let $\{A_n\}$ be a nondecreasing sequence of Hermitean operators, and B be a Hermitean operator such that $A_n \leq B$, $n = 1, 2, \dots$. Then there is a Hermitean $A := \text{s-lim}_{n\to\infty} A_n$ and $\sup_n A_n = A$.

Proof: Denote $f_n(x) := (x, A_n x)$ for any $x \in \mathcal{H}$. The operator $A_{mn} := A_m - A_n$ is positive for $m > n$, and inserting $A = A_{mn}$, $y = A_{mn}x$ into (3.4) we get

$$\|A_{mn}x\|^4 \leq |f_m(x) - f_n(x)| \left(A_{mn}x, A_{mn}^2 x\right) .$$

Norm monotonicity yields $(A_{mn}x, A_{mn}^2 x) \leq \|A_{mn}\|^3 \|x\|^2 \leq \|B - A_1\|^3 \|x\|^2$ and since $\{f_n(x)\}$ is a nondecreasing sequence bounded by (x, Bx), the sequence $\{A_n x\}$ is Cauchy, and by strong completeness, $\{A_n\}$ converges strongly to some $A \in \mathcal{B}(\mathcal{H})$. Then $f_n(x) \to (x, Ax)$ so $(x, Ax) = \sup_n f_n(x)$ for all $x \in \mathcal{H}$; this means that A is Hermitean and equal to $\sup_n A_n$. ∎

3.2.5 Remark: In the same way, $\text{s-lim}_{n\to\infty} A_n = \inf_n A_n$ holds for any nonincreasing below bounded sequence $\{A_n\}$. Let us stress that the limit of $\{A_n\}$ with respect to operator–norm topology may not exist — *cf.* Problem 8.

As an illustration to the above theorem, let us show how it can be used to construct for a positive A a positive operator B satisfying $B^2 = A$; it is natural to call the latter the *square root* of A and denote it as $\sqrt{A}$.

3.2.6 Proposition: To any positive operator A there is a unique square root. Moreover, the conditions $AB = BA$ and $\sqrt{A}B = B\sqrt{A}$ are equivalent for any $B \in \mathcal{B}(\mathcal{H})$.

Proof: Since the case $A = 0$ is trivial we may consider positive operators A with $\|A\| = 1$ only. It follows from (3.5) that $\|\sqrt{A}\| = 1$, so $0 \leq A \leq I$ and $0 \leq \sqrt{A} \leq I$. Defining $X := I - A$, $Y := I - \sqrt{A}$ we rewrite the condition $\sqrt{A}^2 = A$ as the equation $Y = \frac{1}{2}(X + Y^2)$, which can be solved by iteration,

$$Y_1 := \frac{1}{2}X\,, \quad Y_{n+1} := \frac{1}{2}\left(X + Y_n^2\right) \quad \text{for} \quad n = 1, 2, \dots . \tag{3.6}$$

Each Y_n is then a polynomial in X with positive coefficients so, in particular, $\{Y_n : n = 1, 2, \dots\}$ is a commutative set. We have $Y_{n+1} - Y_n = \frac{1}{2}(Y_n - Y_{n-1})$

$(Y_n + Y_{n-1})$ and another induction argument shows that the differences $Y_{n+1} - Y_n$ are again polynomials in X with positive coefficients; hence $\{Y_n\}$ is nondecreasing. It also follows from (3.6) that $Y_n \le I$ for all n. Theorem 3.2.4 then implies the existence of a positive $\text{s-lim}_{n\to\infty} Y_n =: Y \le I$, and performing the limit in (3.6) we get the sought solution, $\sqrt{A} = I - Y$.

If B commutes with A, it commutes with X too, so $BY_n = Y_nB$. Taking the limit we get $B\sqrt{A} = \sqrt{A}B$. The opposite implication is trivial, so it remains to check the uniqueness. Suppose there are positive B, C such that $B^2 = C^2 = A$; we already know that this means $BY_n = Y_nB$. Set $D := B - C$ and $y = Dx$ for any $x \in \mathcal{H}$. Then $(y, By) + (y, Cy) = (y, (B^2 - C^2)x) = 0$, and since B, C are positive, we get $(y, By) = (y, Cy) = 0$. It follows further that $\sqrt{B}y = \sqrt{C}y = 0$, so $Dy = 0$ and $\|Dx\|^2 = (Dy, x) = 0$, *i.e.*, $D = 0$. ∎

The notion of a square root allows us to associate with any $B \in \mathcal{B}(\mathcal{H})$ a positive operator defined by

$$|B| := \sqrt{B^*B}$$

which has some properties similar to the complex–number modulus. However, the analogy should be used with a caution (Problems 14, 15).

One of most important classes of positive operators consists of **projections**. Let $\mathcal{G}$ be a closed subspace in $\mathcal{H}$; for any $x \in \mathcal{H}$ we denote by y_x its orthogonal projection to $\mathcal{G}$. It follows from Theorem 2.1.2 that the map $x \mapsto y_x$ is a linear operator; we denote it as $E_\mathcal{G}$. By definition, $E_\mathcal{G}$ is bounded, $\operatorname{Ran} E_\mathcal{G} = (\operatorname{Ker} E_\mathcal{G})^\perp = \mathcal{G}$ and $\|E_\mathcal{G}\| = 1$ provided $\mathcal{G} \ne \{0\}$. Moreover $I - E_\mathcal{G}$ is the projection to the subspace $\mathcal{G}^\perp$, so $E_\mathcal{G} + E_{\mathcal{G}^\perp} = I$.

3.2.7 Examples: (a) Let $\mathcal{E} = \{e_1, \dots, e_n\}$ be an orthonormal set in $\mathcal{H}$. The projection $E^{(n)}$ to $\mathcal{G} := \mathcal{E}_{lin}$ acts as $E^{(n)}x = \sum_{j=1}^n (e_j, x)e_j$.

(b) The subspace $\mathcal{G}_t := \{ f : f(x) = 0 \text{ for a.a. } x > t\} \subset L^2(\mathbb{R})$ is closed and the corresponding projection E_t acts as

$$(E_t h) = \begin{cases} h(x) & \dots \quad x \le t \\ 0 & \dots \quad x > t \end{cases}$$

3.2.8 Theorem: An operator E defined on the whole $\mathcal{H}$ is a projection *iff* it is Hermitean and $E^2 = E$. Any projection E is positive, $\operatorname{Ran} E$ is closed and consists exactly of the vectors which satisfy $Ex = x$.

Proof: If E is a projection, the identity $E^2 = E$ follows from the definition, and $((I-E)x, Ex) = 0$ gives $(x, Ex) = \|Ex\|^2 \in \mathbb{R}$, *i.e.*, Hermiticity. On the other hand, let E be Hermitean and fulfil $E^2 = E$. For any $y \in \overline{\operatorname{Ran} E}$, $y = \lim_{n\to\infty} Ex_n$, we have $Ey = \lim_{n\to\infty} E^2x_n = \lim_{n\to\infty} Ex_n = y$ due to continuity, so $\operatorname{Ran} E$ is closed and its elements fulfil $Ey = y$. The relation $x = Ex + (I-E)x$ defines its orthogonal decomposition for any $x \in \mathcal{H}$; hence E is a projection. ∎

Projections E, F are said to be mutually orthogonal if $\mathrm{Ran}\, E \perp \mathrm{Ran}\, F$; this is equivalent to the condition $EF = FE = 0$. The following simple relations between projections are often useful.

3.2.9 Proposition: Let E, F be projections on $\mathcal{H}$.

(a) $E+F$ is a projection *iff* E, F are mutually orthogonal; in that case we have $\mathrm{Ran}\,(E+F) = \mathrm{Ran}\, E \oplus \mathrm{Ran}\, F$.

(b) The following conditions are equivalent:

(i) $E - F$ is a projection.

(ii) $E \geq F$.

(iii) $\mathrm{Ran}\, E \supset \mathrm{Ran}\, F$.

(iv) $EF = FE = F$.

if these are satisfied, $\mathrm{Ran}\,(E-F)$ is the orthogonal complement to $\mathrm{Ran}\, F$ in $\mathrm{Ran}\, E$.

(c) EF is a projection *iff* $EF = FE$; then $\mathrm{Ran}\, EF = \mathrm{Ran}\, E \cap \mathrm{Ran}\, F$.

3.2.10 Remark: Let us denote $\mathcal{E} := \mathrm{Ran}\, E$, $\mathcal{F} := \mathrm{Ran}\, F$. If the projections commute, their product projects to the subspace $\mathcal{E} \cap \mathcal{F}$. In the general case, $\mathcal{E} \cap \mathcal{F}$ is still a closed subspace, and we can ask whether the corresponding projection G can be expressed with the help of E and F. The sought formula reads

$$G = \operatorname*{s-lim}_{n\to\infty} (EFE)^n = \operatorname*{s-lim}_{n\to\infty} (EF)^n \tag{3.7}$$

or the same with E and F interchanged. To prove this, we first show that the operators $G_n := (EFE)^n$ fulfil $0 \leq G_{n+1} \leq G_n \leq I$ so $G := \text{s-lim}_{n\to\infty} G_n$ exists by Theorem 3.2.4 and $0 \leq G \leq I$; we have $G_n^2 = G_{2n}$, so $G^2 = G$ and G is a projection. Performing the limit in $EG_n = G_nE = G_n$, we get $\mathrm{Ran}\, G \subset \mathcal{E}$; similarly $G_nFG_n = G_{2n+1}$ gives the relation $GFG = G$ which implies $\|FG - G\|^2 = \|(FG-G)^*(FG-G)\| = \|G - GFG\| = 0$, and therefore $\mathrm{Ran}\, G \subset \mathcal{F}$. On the other hand, $G_n x = x$ for any $x \in \mathcal{E} \cap \mathcal{F}$, which yields $Gx = x$; together we get $\mathrm{Ran}\, G = \mathcal{E} \cap \mathcal{F}$. The second identity in (3.7) follows from $FG = GF = G$.

3.2.11 Proposition: Any projections E, F on a given $\mathcal{H}$ satisfy $\|E - F\| \leq 1$. If $\|E - F\| < 1$, then $\dim \mathrm{Ran}\, E = \dim \mathrm{Ran}\, F$ and $\dim \mathrm{Ker}\, E = \dim \mathrm{Ker}\, F$.

Proof: Since the vectors $E(I-F)x$ and $(E-I)Fx$ are orthogonal for any $x \in \mathcal{H}$, we have $\|(E-F)x\|^2 = \|E(I-F)x\|^2 + \|(E-I)Fx\|^2 \leq \|(I-F)x\|^2 + \|Fx\|^2 = \|x\|^2$. The remaining assertion follows from Problem 18. ∎

Proposition 3.2.9 generalizes easily to any finite system of projections; in combination with Theorem 3.2.4 it gives the following result for sequences.

3.2.12 Theorem: To any monotonic sequence $\{E_n\}_{n=1}^{\infty}$ of projections there is a projection $E := \text{s-lim}_{n\to\infty} E_n$. Its range satisfies $\operatorname{Ran} E = \bigcap_{n=1}^{\infty} \operatorname{Ran} E_n$ if the sequence is nonincreasing, and $\operatorname{Ran} E = \overline{\bigcup_{n=1}^{\infty} \operatorname{Ran} E_n}$ if it is nondecreasing.

Let $\{E^{(n)}\}_{n=1}^{\infty}$ be a nondecreasing sequence of projections; we set $E_1 := E^{(1)}$ and $E_{n+1} := E^{(n+1)} - E^{(n)}$, $n = 1, 2, \dots$; then $\{ E_j : j = 1, 2, \dots \}$ is a set of mutually orthogonal projections due to Proposition 3.2.9. Since $\operatorname{Ran} E^{(n)} = \bigoplus_{j=1}^{n} \operatorname{Ran} E_j$ and the algebraic sum of subspaces $\sum_{j=1}^{\infty} \operatorname{Ran} E_j$ equals $\bigcup_{j=1}^{\infty} \operatorname{Ran} E^{(j)}$, we see that

$$E := \underset{n\to\infty}{\text{s-lim}} \sum_{j=1}^{n} E_j =: \sum_{j=1}^{\infty} E_j \tag{3.8}$$

projects to the orthogonal sum $\bigoplus_{j=1}^{\infty} \operatorname{Ran} E_j$. This result generalizes to any system $\{ E_\alpha : \alpha \in I \}$ of mutually orthogonal projections. We denote by E the projection to the subspace $\sum_{\alpha\in I}^{\oplus} \operatorname{Ran} E_\alpha$ (Problem 2.14) and define $\sum_{\alpha\in I} E_\alpha := E$. In particular, $\{ E_\alpha : \alpha \in I \}$ is said to be a *complete system of projections* if $E = I$. The relation (3.8) and Theorem 3.2.12 have the following generalization.

3.2.13 Proposition: Let $\{ E_\alpha : \alpha \in I \}$ be a system of mutually orthogonal projections; then

$$\sum_{\alpha\in I} E_\alpha = \underset{\mathcal{S}}{\text{s-lim}}\, E_K = \sup_{\mathcal{S}} E_K \,,$$

i.e., $Ex = \lim_{\mathcal{S}} E_K x$ for any $x \in \mathcal{H}$, where $\mathcal{S} := \{ K \in I : K \text{ finite} \}$ and $E_K := \sum_{\alpha\in K} E_\alpha$.

Proof: To any $x \in \mathcal{H}$ and $\varepsilon > 0$ there is an at most countable $\{\alpha_k\}$ such that $E_\alpha x = 0$ for $\alpha \notin \{\alpha_k\}$ and $\|Ex - \sum_{k=1}^{n} E_{\alpha_k} x\| < \varepsilon$ for all $n > n(\varepsilon, x)$. The family $\mathcal{S}$ is obviously directed, so there is a set $K \in \mathcal{S}$ such that $K \supset \{\alpha_1, \dots, \alpha_{n(\varepsilon,x)}\}$. This means that $K \cap \{\alpha_k\} = \{\alpha_1, \dots, \alpha_{n_K}\}$, where $n_K \ge n(\varepsilon, x)$, and therefore $\|Ex - E_K x\| < \varepsilon$. Furthermore, $E_K \le E$ and $(x, Ex) = \lim_{\mathcal{S}} (x, E_K x)$, so the second identity holds too. ▮

3.3 Unitary and isometric operators

An isomorphism $U : \mathcal{H} \to \mathcal{G}$ between two Hilbert spaces is often called a **unitary operator**; this means that it maps surjectively $\mathcal{H}$ onto $\mathcal{G}$ and $\|Ux\|_{\mathcal{G}} = \|x\|_{\mathcal{H}}$ for all $x \in \mathcal{H}$. Using the appropriate definitions, we easily check

3.3.1 Proposition: An operator $U \in \mathcal{B}(\mathcal{H}, \mathcal{G})$ is unitary *iff* it is surjective and $(Ux, Uy)_{\mathcal{G}} = (x, y)_{\mathcal{H}}$ for all $x, y \in \mathcal{H}$. If $\mathcal{G} = \mathcal{H}$, this is further equivalent to the following: U is linear, everywhere defined, and $U^{-1} = U^*$.

3.3.2 Example (substitution operators): Consider a continuous bijective mapping $\varphi : \mathbb{R}^n \to \mathbb{R}^n$ which satisfies the following conditions:

(i) Each component map $\varphi_j : \mathbb{R}^n \to \mathbb{R}$ of φ has continuous partial derivatives $\partial_k \varphi_j$, $k = 1, \dots, n$.

(ii) There are positive numbers a, b such that the Jacobian $D_\varphi := \det(\partial_k \varphi_j)$ fulfils $b^{-1} \le |D_\varphi(x)| \le a$ for all $x \in \mathbb{R}^n$.

The set of all such bijections will be denoted as Φ. The substitution theorem implies

$$\int_{\mathbb{R}^n} |f(x)|^2\, dx = \int_{\mathbb{R}^n} |f(\varphi(x))|^2 |D_\varphi(x)|\, dx$$

for any $f \in L^2(\mathbb{R}^n)$, and we readily check that φ^{-1} belongs again to Φ. Define now the operator U_φ by

$$(U_\varphi f)(x) := |D_\varphi(x)|^{1/2} (f \circ \varphi)(x);$$

it is a product of the operator of multiplication by $|D_\varphi(\cdot)|^{1/2}$, which is bounded and invertible by assumption, and of the "substitution" operator $S_\varphi : S_\varphi f = f \circ \varphi$. The latter is also bounded since $\|f\|^2 \ge \frac{1}{b} \|S_\varphi f\|^2$, and moreover, $S_\varphi S_{\varphi^{-1}} = I$, so $\operatorname{Ran} S_\varphi = L^2(\mathbb{R}^n)$ and the same is true for $\operatorname{Ran} U_\varphi$. Finally, the substitution relation shows that U_φ preserves the norm, $\|U_\varphi f\|^2 = \|f\|^2$, and therefore it is unitary. As a particular example, let us mention the *reflection* (or *parity*) *operator* $R : (Rf)(x) = f(-x)$ for all $x \in \mathbb{R}^n$. The definition relation implies $R^2 = I$, *i.e.*, $R^{-1} = R$, which means the operator R is simultaneously unitary and Hermitean (see Problem 20).

To check that a given operator U is unitary it is often enough to verify the definition properties on a subset of $\mathcal{H}$ only.

3.3.3 Proposition: (a) Let U_0 be a densely defined linear operator on $\mathcal{H}$ with $\overline{\operatorname{Ran} U_0} = \mathcal{H}$ such that $\|U_0 x\| = \|x\|$ for all $x \in D(U_0)$; then it has just one unitary extension.

(b) Let $M \subset \mathcal{H}$ be a total linearly independent subset. If $V : M \to \mathcal{H}$ is such that VM is again total and $(Vx, Vy) = (x, y)$ for all $x, y \in M$, then there is just one unitary U fulfilling $U \restriction M = V$.

Proof: The existence of a unique extension $U \in \mathcal{B}(\mathcal{H})$, $\|U\| = 1$, follows from Theorem 1.5.5. By assumption, for any $y \in \mathcal{H}$ we can find a sequence $\{x_n\} \subset \mathcal{H}$ such that $U_0 x_n \to y$. Since U_0 is norm preserving, $\{x_n\}$ is Cauchy, so it converges to some x and $y = \lim_{n\to\infty} U_0 x_n = Ux$, *i.e.*, $\operatorname{Ran} U = \mathcal{H}$. Furthermore, $\|Ux\| = \lim_{n\to\infty} \|U_0 x_n\| = \lim_{n\to\infty} \|x_n\| = \|x\|$, which proves (a). Let U_0 be the linear extension of V. The subspaces M_{lin} and $(VM)_{lin}$ are dense by assumption, and inner-product linearity yields $\|U_0 z\|^2 = \|z\|^2$, so part (a) applies. ∎

A linear operator V on $\mathcal{H}$ is called **isometric** if $D(V)$, $\operatorname{Ran} V$ are closed subspaces and V is unitary as a map $D(V) \to \operatorname{Ran} V$; it is clear that V is a unitary operator on $\mathcal{H}$ *iff* $D(V) = \operatorname{Ran} V = \mathcal{H}$. If $D(V) \ne \mathcal{H}$ we can extend V to an operator $W \in \mathcal{B}(\mathcal{H})$ called a **partial isometry** such that $V_W := W \restriction (\operatorname{Ker} W)^\perp$ is isometric. The extension is clearly unique and obtained by setting $Wx := 0$ for

$x \in D(V)^\perp$. The ranges of W and V_W coincide, so $\mathrm{Ran}\, W$ is closed; we call it the *final subspace* of W. On the other hand, $(\mathrm{Ker}\, W)^\perp$ is the *initial subspace* of W.

Any unitary operator is a partial isometry. Other examples are projections, shift operators S, S^*, or the operators P_{xy} of Problem 6 ; the initial and final subspaces in all these cases are easily found. An inverse operator to W does not exist unless $\mathrm{Ker}\, W = \{0\}$; however, we can replace it by W^* in the following sense.

3.3.4 Proposition: The adjoint to a partial isometry W is also a partial isometry, with the initial and final subspaces interchanged, and the relation $V_{W^*} = V_W^{-1}$ holds between the corresponding isometric operators. The operators W^*W and WW^* are projections to the initial and final subspace of W, respectively.

Proof: Denote by E_i, E_f the projections to $(\mathrm{Ker}\, W)^\perp$, $\mathrm{Ran}\, W$, respectively. We have $(Wx, y) = (V_W E_i x, E_f y) = (E_i x, V_W^{-1} E_f y) = (x, V_W^{-1} E_f y)$ for any $x, y \in \mathcal{H}$, so $W^* = V_W^{-1} E_f$. Furthermore, $(\mathrm{Ker}\, W^*)^\perp = \mathrm{Ran}\, W = E_f \mathcal{H}$, so $W^* \upharpoonright (\mathrm{Ker}\, W^*)^\perp = V_W^{-1}$. Hence the first assertion is valid, and it is sufficient to check the second one for W^*W. We have $W^*Wx = W^*V_W E_i x = V_W^{-1} V_W E_i x = E_i x$ for any $x \in \mathcal{H}$, so $W^*W = E_i$. ∎

In conclusion we shall prove one more theorem inspired by the analogy between bounded operators and complex numbers (see also Problem 24).

3.3.5 Theorem (polar decomposition for bounded operators): To any $B \in \mathcal{B}(\mathcal{H})$ there is just one partial isometry W_B such that $B = W_B|B|$ and $\mathrm{Ker}\, W_B = \mathrm{Ker}\, B$. Furthermore, the identity $\mathrm{Ran}\, W_B = \overline{\mathrm{Ran}\, B}$ holds.

Proof: Define $W_0 : \mathrm{Ran}\,|B| \to \mathrm{Ran}\, B$ by $W_0|B|x := Bx$. Since $|B|$ need not be invertible, we have to check whether W_0 is well defined (which would imply its linearity). This follows from the relation $\|Bx\| = \|\,|B|x\|$ (Problem 14), which further implies that W_0 is norm preserving. Its continuous extension is an isometric operator that maps $\overline{\mathrm{Ran}\,|B|} = (\mathrm{Ker}\,|B|)^\perp = (\mathrm{Ker}\, B)^\perp$ to $\overline{\mathrm{Ran}\, B}$, and the corresponding partial isometry W_B satisfies $W_B|B|x = Bx$ for all $x \in \mathcal{H}$. To prove uniqueness, suppose there is a partial isometry W with the needed properties so that $(W_B - W)|B| = 0$. Using the orthogonal decomposition $x = y + z$ with $y \in \mathrm{Ker}\, B$ and $z \in (\mathrm{Ker}\, B)^\perp = \overline{\mathrm{Ran}\,|B|}$ we find $(W_B - W)x = (W_B - W)z = \lim_{n\to\infty}(W_B - W)|B|v_n$, where $|B|v_n \to z$. By assumption, $(W_B - W)|B|v_n = 0$, so $(W_B - W)x = 0$ for any $x \in \mathcal{H}$. ∎

3.4 Spectra of bounded normal operators

We know from linear algebra that a linear operator B on a finite–dimensional $\mathcal{H}$ has an orthonormal basis of eigenvectors *iff* B commutes with B^*. Now we are going to generalize this result to bounded linear operators on an infinite–dimensional Hilbert space. An operator $B \in \mathcal{B}(\mathcal{H})$ is dubbed **normal** if $BB^* = B^*B$. The set of all bounded normal operators on $\mathcal{H}$ will be denoted $\mathcal{N}(\mathcal{H})$; it obviously contains the real subspace of Hermitean operators and the group $\mathcal{U}(\mathcal{H})$ of unitary operators but the whole set $\mathcal{N}(\mathcal{H})$ is neither a subspace nor a group. We start from simple modifications of the definition.

3.4.1 Proposition: The following conditions are equivalent:

(a) $B \in \mathcal{N}(\mathcal{H})$

(b) $B^* \in \mathcal{N}(\mathcal{H})$

(c) The operators $\mathrm{Re}\, B$, $\mathrm{Im}\, B$ commute

(d) $\|(B-\lambda)x\| = \|(B^*-\overline{\lambda})x\|$ for all $x \in \mathcal{H}$, $\lambda \in \mathbb{C}$

The last condition implies that $\lambda \in \sigma_p(B)$ *iff* $\overline{\lambda} \in \sigma_p(B^*)$, and moreover, the corresponding eigenspaces $\mathrm{Ker}\,(B-\lambda) =: N_B(\lambda)$ satisfy $N_{B^*}(\overline{\lambda}) = N_B(\lambda)$. In particular, $\sigma_p(B^*) = \emptyset$ *iff* $\sigma_p(B) = \emptyset$.

3.4.2 Theorem: Let $B \in \mathcal{N}(\mathcal{H})$; then

(a) If λ, μ are different eigenvalues of B, one has $N_B(\lambda) \perp N_B(\mu)$.

(b) If a subspace $L \subset \mathcal{H}$ is B–invariant, then $L^\perp$ is B^*–invariant. In particular, $N_B(\lambda)^\perp$ is B–invariant for any $\lambda \in \sigma_p(B)$.

(c) The residual spectrum of B is empty.

Proof: Consider nonzero vectors $x \in N_B(\lambda), y \in N_B(\mu)$; we have $B^*y = \overline{\mu}y$, which together with $Bx = \lambda x$ and $(y, Bx) = (B^*y, x)$ gives $(\lambda-\mu)(y,x) = 0$ proving (a). Consider further $x \in L^\perp$; by assumption $By \in L$ for any $y \in L$, so $(y, B^*x) = (By, x) = 0$, *i.e.*, $B^*x \in L^\perp$. In particular, $N_{B^*}(\overline{\lambda}) = N_B(\lambda)$ is B^*–invariant, so $N_B(\lambda)^\perp$ is B^{**}–invariant. Finally, if $\lambda \in \sigma_r(B)$, *i.e.*, $\mathrm{Ran}\,(B-\lambda)^\perp \neq \{0\}$, then $\overline{\lambda}$ is an eigenvalue of B^* by Proposition 3.1.4; hence $\lambda \in \sigma_p(B)$ in contradiction with the assumption. ∎

3.4.3 Theorem: The resolvent set $\rho(B)$ of any $B \in \mathcal{N}(\mathcal{H})$ coincides with its *regularity domain;* in other words, $\lambda \in \rho(B)$ *iff* there is a positive $c \equiv c(\lambda)$ such that $\|(B-\lambda)x\| \geq c\|x\|$ for all $x \in \mathcal{H}$.

Proof: If $\lambda \in \rho(B)$, the above condition is satisfied for $c = \|R_B(\lambda)\|^{-1}$ since $\|y\| \geq \|R_B(\lambda)\|^{-1}\|R_B(\lambda)y\|$ and any $y \in \mathcal{H}$ can be written as $y = (B-\lambda)x$. On the other hand, the condition implies $\mathrm{Ker}\,(B-\lambda) = \{0\}$, so $(B-\lambda)^{-1}$ exists. Proposition 3.4.1 gives $\mathrm{Ker}\,(B^*-\overline{\lambda}) = \{0\}$, so $\overline{\mathrm{Ran}\,(B-\lambda)} = \mathcal{H}$; it remains to prove that $\mathrm{Ran}\,(B-\lambda)$ is closed. Let a sequence $\{y_n\} \subset \mathrm{Ran}\,(B-\lambda)$ converge to y; then there is $\{x_n\} \subset \mathcal{H}$ fulfilling $y_n = (B-\lambda)x_n$, and the condition shows that $\{x_n\}$ is Cauchy. Setting $x := \lim_{n\to\infty} x_n$, we have $(B-\lambda)x = y$ by continuity. ∎

3.4.4 Corollary: (a) Suppose that $B \in \mathcal{N}(\mathcal{H})$, then $\lambda \in \sigma(B)$ *iff* $\inf\{\|(B-\lambda)x\| : x \in S_1\} = 0$, where S_1 is the unit sphere in $\mathcal{H}$, *i.e.*, *iff* there is a sequence $\{x_n\}$ of unit vectors such that $(B-\lambda)x_n \to 0$.

(b) The spectra of unitary and Hermitean operators are subsets of the unit circle and the real axis, respectively.

Proof: The first part is obvious. A unitary operator U fulfils

$$\|(U-\lambda)x\|^2 = \|Ux\|^2 + |\lambda|^2\|x\|^2 - 2\,\mathrm{Re}\,(Ux,\lambda x) \geq (1-|\lambda|)^2\|x\|^2,$$

so $\lambda \in \rho(U)$ if $|\lambda| \neq 1$ in view of the above theorem. In the same way, we obtain the relation $\|(A-\lambda)x\|^2 = \|(A-\lambda_1)x\|^2 + |\lambda_2|^2\|x\|^2$ for a Hermitean A and $\lambda = \lambda_1 + i\lambda_2$, which shows that $\sigma(A) \subset \mathbb{R}$. ∎

The spectrum of a Hermitean operator can be further localized as follows (see also the notes to Section 3.2).

3.4.5 Proposition: The spectrum of a Hermitean A is a subset of the interval $[m_A, M_A] \subset \mathbb{R}$ whose endpoints, *i.e.*, the lower and upper bounds to A, belong to $\sigma(A)$.
Proof: In view of the corollary, we have to check $(-\infty, m_A) \cup (M_A, \infty) \subset \rho(A)$. If, for instance, $\lambda \in (M_A, \infty)$ then the definition of M_A gives $(x,(\lambda-A)x) \geq \lambda - M_A$ for any unit vector x, so $\|(A-\lambda)x\| \geq |((A-\lambda)x,x)| \geq \lambda - M_A$. Let us check further that $m_A \in \sigma(A)$. To this end, we apply the inequality (3.4) with $\|x\| = 1$ and $y = (A-m_A)x$ to the positive operator $A - m_A$. This yields $\|(A-m_A)x\|^4 \leq (x,(A-m_A)x)\|A-m_A\|\,\|(A-m_A)x\|^2$, so by definition of the lower bound we have

$$\inf_{x\in S_1}\|(A-m_A)x\|^2 \leq \|A-m_A\|\left(\inf_{x\in S_1}(x,Ax) - m_A\right) = 0.$$

In the same way, we can check that $M_A \in \sigma(A)$. ∎

3.4.6 Examples: (a) If E is a nontrivial projection, then $m_E = 0$ and $M_E = 1$, and furthermore, we easily find $\|(E-\lambda)x\|^2 \geq \min\{(1-\lambda)^2, \lambda^2\}\|x\|^2$ for any $\lambda \in (0,1)$, so $\sigma(E) = \{0,1\}$.

(b) Let Q be the multiplication operator of Example 3.2.2 and $\lambda \in (a,b)$. We set $f_n := \sqrt{n}\,\chi_{I_n}$, where $I_n = (a,b) \cap \left(\lambda, \lambda + \frac{1}{n}\right)$; if n is large enough we have $\|f_n\| = 1$ and $\|(Q-\lambda)f_n\|^2 = (3n^2)^{-1} \to 0$. Since Q has no eigenvalues and $\sigma_r(Q) = \emptyset$ by Theorem 3.4.2, we obtain $\sigma(Q) = \sigma_c(Q) = [a,b]$.

We have mentioned that an operator on a finite-dimensional $\mathcal{H}$ is normal *iff* it has an orthonormal basis of eigenvectors. If $\dim\mathcal{H} = \infty$ this condition is no longer necessary; however, it is still sufficient. An operator $B \in \mathcal{B}(\mathcal{H})$ will be said to have a **pure point spectrum** if its eigenvalues form an orthonormal basis; this definition is justified by the following result.

3.4.7 Theorem: Any operator $B \in \mathcal{B}(\mathcal{H})$ with a pure point spectrum is normal and $\sigma(B) = \overline{\sigma_p(B)}$.

Proof: Let $\mathcal{E} = \{e_\alpha : \alpha \in I\}$ be an orthonormal basis such that $Be_\alpha = \lambda_\alpha e_\alpha$ for any $\alpha \in I$. It follows easily that $B^* e_\alpha = \overline{\lambda} e_\alpha$, so $[B^*, B]\mathcal{E} = \{0\}$ and by linear and continuous extension, B is normal. Since $\sigma(B)$ is closed by definition and contains $\sigma_p(B)$, it is sufficient to show that $\sigma(B) \subset \overline{\sigma_p(B)}$. Suppose $\lambda \notin \overline{\sigma_p(B)}$ so $|\lambda - \lambda_\alpha| \geq \delta$ for some $\delta > 0$ and all $\alpha \in I$. For any $x \in \mathcal{H}$, there is at most countable $I_x := \{\alpha_j\} \subset I$ such that $(e_\alpha, x) = (e_\alpha, Bx) = 0$ for $\alpha \notin I_x$; then

$$\|(B-\lambda)x\|^2 = \sum_{j=1}^{\infty} |(e_{\alpha_j}, (B-\lambda)x)|^2 = \sum_{j=1}^{\infty} |\lambda_{\alpha_j} - \lambda|^2 |(e_{\alpha_j}, x)|^2 \geq \delta^2 \|x\|^2$$

and $\lambda \notin \sigma(B)$ in view of Theorem 3.4.3. ∎

3.5 Compact operators

A linear everywhere defined operator C on $\mathcal{H}$ is said to be **compact** if it maps any bounded subset of $\mathcal{H}$ to a precompact set. Since $\mathcal{H}$ is a metric space, this is equivalent to the requirement that any bounded $\{x_n\} \subset \mathcal{H}$ contains a subsequence $\{x_{n_k}\}$ such that $\{Cx_{n_k}\}$ converges. The set of all compact operators is denoted as $\mathcal{K}(\mathcal{H})$; alternatively we use the symbol $\mathcal{J}_\infty(\mathcal{H})$ or $\mathcal{J}_\infty$. A compact operator cannot be unbounded since any precompact set is completely bounded by Corollary 1.3.7a, and therefore bounded, so we have

$$\mathcal{K}(\mathcal{H}) \subset \mathcal{B}(\mathcal{H}).$$

If $\dim \mathcal{H} = \infty$ the inclusion is nontrivial, as the unit operator or a projection to an infinite–dimensional subspace of $\mathcal{H}$ illustrate. On the other hand, a bounded operator is always compact if it is *finite–dimensional*, *i.e.*, $\dim \operatorname{Ran} B < \infty$ since BM is a bounded subset of $\operatorname{Ran} B$ for any bounded $M \subset \mathcal{H}$, and is therefore precompact (Corollary 1.3.7a). In particular, the above inclusion turns into identity on a finite–dimensional $\mathcal{H}$.

3.5.1 Theorem: An operator is compact *iff* it maps any weakly convergent sequence to a convergent one.

Proof: Suppose that $C \in \mathcal{K}(\mathcal{H})$ and $x_n \xrightarrow{w} x$ does *not* imply $Cx_n \to Cx$. Then there is a positive ε and a growing sequence $\{n_k\}$ of natural numbers such that $\|Cx - y_k\| \geq \varepsilon$ holds for $y_k := Cx_{n_k}$, $k = 1, 2, \ldots$. In view of Theorem 1.5.12b, $\{x_{n_k}\}$ is bounded, so one can select from $\{y_k\}$ a subsequence $\{y_{k_j}\}$ which converges to some y. This means, in particular, $y_{k_j} \xrightarrow{w} y$. On the other hand, $Cx_n \xrightarrow{w} Cx$ since C is bounded, so together we get $y = Cx$ in contradiction with the assumption. The opposite implication follows from the reflexivity of $\mathcal{H}$ (*cf.* Theorem 1.5.14b): one can select a weakly convergent subsequence $\{x_{n_k}\}$ from any bounded $\{x_n\} \subset \mathcal{H}$, and $\{Cx_{n_k}\}$ converges by assumption. ∎

3.5.2 Theorem: $\mathcal{K}(\mathcal{H})$ is a closed subspace in $\mathcal{B}(\mathcal{H})$ with respect to the operator norm topology. If $C \in \mathcal{K}(\mathcal{H})$, then its adjoint C^* is compact, as are the operators BC and CB for any bounded B.

Proof: The vector structure of $\mathcal{K}(\mathcal{H})$ and compactness of BC, CB follow from the previous theorem. If a sequence $\{x_n\}$ converges weakly to some x, we have $\|C^*x_n - C^*x\|^2 \le \|x_n - x\|\,\|CC^*(x_n - x)\|$; however, $\{x_n\}$ is bounded and CC^* is compact in view of the just proved result, so $CC^*x_n \to CC^*x$. This implies that $\|C^*x_n - C^*x\| \to 0$, which means C^* is compact too.

It remains to prove that $\mathcal{K}(\mathcal{H})$ is closed. Suppose that $\|C_n - B\| \to 0$ for $\{C_n\} \subset \mathcal{K}(\mathcal{H})$ and $B \in \mathcal{B}(\mathcal{H})$, and let $M \subset \mathcal{H}$ be bounded, $\|x\| \le K$ for all $x \in M$. We need to find a finite ε–lattice to the set BM for any $\varepsilon > 0$. There is $n(\varepsilon)$ such that $\|C_{n(\varepsilon)} - B\| < \frac{\varepsilon}{2K}$, and since $C_{n(\varepsilon)}$ is compact, there is a finite $\frac{\varepsilon}{2}$–lattice N for $C_{n(\varepsilon)}M$, *i.e.*, we can find $y \in N$ such that $\|C_{n(\varepsilon)}x - y\| < \frac{\varepsilon}{2}$ for any $x \in M$. Then $\|Bx - y\| < \|B - C_{n(\varepsilon)}\|\,\|x\| + \|C_{n(\varepsilon)}x - y\| < \varepsilon$, so N is the sought ε–lattice. ∎

3.5.3 Remark: Referring to Appendix B.1 for the appropriate definitions, we can state the above theorem concisely: $\mathcal{K}(\mathcal{H})$ is a closed two–sided $*$–ideal in $\mathcal{B}(\mathcal{H})$. We shall use this algebraic terminology also in the following section.

3.5.4 Corollary: If C is a compact operator on an infinite–dimensional Hilbert space $\mathcal{H}$, then $0 \in \sigma(C)$.

Proof: A bounded inverse C^{-1} does not exist, since otherwise $C^{-1}C = I$ would be compact. ∎

Finite–dimensional operators are more than a simple example of compact operators. They allow us to approximate them, which is extremely important, *e.g.*, in practical solutions of operator equations.

3.5.5 Theorem: An operator $B \in \mathcal{B}(\mathcal{H})$ is compact *iff* there is a sequence $\{B_n\}$ of finite–dimensional operators converging to it with respect to the operator norm.

Proof: Let $C \in \mathcal{K}(\mathcal{H})$ with $\dim \operatorname{Ran} C = \infty$ and denote by $\{e_j\}_{j=1}^{\infty}$ an orthonormal basis in $E\mathcal{H} := \overline{\operatorname{Ran} C}$ (*cf.* Problem 33). Let E_n be the projection to $\{e_1, \dots, e_n\}_{lin}$, then $E_n \xrightarrow{s} E$ and $E_nC \to EC$ (Problem 32), so $\{E_nC\}$ is the sought sequence. The opposite implication follows from the operator–norm closedness of $\mathcal{K}(\mathcal{H})$. ∎

Next we are going to show what the spectrum of a compact operator looks like. We start from an auxiliary result.

3.5.6 Lemma: $\operatorname{Ran}(C - \lambda)$ is closed for a compact C and any nonzero $\lambda \in \mathbb{C}$.

Proof: Let $\{x_n\} \subset (\operatorname{Ker}(C-\lambda))^{\perp}$ be a sequence such that $(C-\lambda)x_n =: y_n \to y$. If $\{x_n\}$ is bounded, then due to compactness there is a subsequence $\{x_{n_k}\}$ such that $\{Cx_{n_k}\}$ is convergent, and therefore $x_{n_k} = \frac{1}{\lambda}(Cx_{n_k} - y_{n_k})$ converges to some $x \in \mathcal{H}$. Then $y = \lim_{k\to\infty}(C - \lambda)x_{n_k} = (C - \lambda)x$, so $y \in \operatorname{Ran}(C - \lambda)$. It remains to check that $\{x_n\}$ is bounded. Suppose that there is $\{x_{n_k}\}$ such that $\lim_{k\to\infty} \|x_{n_k}\| = \infty$. We set $z_k := x_{n_k}/\|x_{n_k}\|$. There is again a subsequence $\{z_{k(l)}\}$ such that $\{Cz_{k(l)}\}$ converges, and in the same way as above, we find that $z_{k(l)} \to z \in (\operatorname{Ker}(C - \lambda))^{\perp}$

and $\|z\| = 1$. However, this is in contradiction with

$$(C-\lambda)z = \lim_{l\to\infty}(C-\lambda)z_{k(l)} = \lim_{l\to\infty}\|x_{n_{k(l)}}\|^{-1}y_{n_{k(l)}} = 0\,. \quad \blacksquare$$

3.5.7 Theorem (Riesz–Schauder): If C is a compact operator on $\mathcal{H}$, then

(a) Any nonzero point of the spectrum is an eigenvalue, $\sigma_p(C)\setminus\{0\} = \sigma(C)\setminus\{0\}$.

(b) Any nonzero eigenvalue has a finite multiplicity.

(c) The spectrum accumulates at most at zero.

(d) The set of eigenvalues is at most countable, $\sigma_p(C) = \{\lambda_j : j = 1,\dots,N\}$ with $N \le \infty$. The eigenvalues can be ordered, $|\lambda_j| \ge |\lambda_{j+1}|$, and $\lim_{j\to\infty}\lambda_j = 0$ if $N = \infty$.

Proof: Assume that $\sigma(C)$ contains a nonzero λ which is not an eigenvalue so the inverse $(C-\lambda)^{-1}$ exists and $\mathrm{Ran}\,(C-\lambda) \ne \mathcal{H}$. We define $\mathcal{G}_n := (C-\lambda)^n\mathcal{H}$, $n = 0,1,\dots$; then $(C-\lambda)^n - (-\lambda)^n$ is for $n \ge 1$ a polynomial in C without an absolute term, so it is compact and $\mathcal{G}_n$ is closed by Lemma 3.5.6. Furthermore, $\mathcal{G}_{n+1} \subset \mathcal{G}_n$ and $\mathcal{G}_{n+1} \ne \mathcal{G}_n$ for any n, which follows from $\mathcal{G}_1 = \mathrm{Ran}\,(C-\lambda) \ne \mathcal{H}$ and the existence of $(C-\lambda)^{-n}$. Hence there are unit vectors $x_n \in \mathcal{G}_{n-1}\cap\mathcal{G}_n^{\perp}$, $n = 1,2,\dots$, and $\{x_n\}$ is an orthonormal set, so $x_n \stackrel{w}{\to} 0$ and $Cx_n \to 0$. Now we take the decomposition $Cx_n = \lambda x_n + (C-\lambda)x_n$, where $(C-\lambda)x_n \in \mathcal{G}_n$ is orthogonal to x_n; it means $\|Cx_n\| \ge |\lambda|\,\|x_n\| = |\lambda|$. This is in contradiction with $Cx_n \to 0$ thus proving (a).

Let $\lambda \in \sigma_p(C)\setminus\{0\}$. Since C acts as a λ–multiple of the unit operator on the corresponding subspace, we obtain (b). To prove (c), we have in view of (a) to check that the eigenvalues cannot accumulate at a nonzero λ. Suppose that there is $\{\lambda_n\} \subset \sigma_p(C)$ such that $\lambda_n \ne \lambda_m$ for $n \ne m$ and $\lambda_n \to \lambda$. The corresponding set of eigenvectors $\{x_n : n = 1,2,\dots\}$ is therefore linearly independent; applying the orthogonalization procedure we get an orthonormal set $\{y_n : n = 1,2,\dots\}$ such that $(C-\lambda)y_n \in \{y_1,\dots,y_n\}_{lin}$. Then $Cy_n = \lambda_n y_n + (C-\lambda_n)y_n$ is an orthonormal decomposition, so $\|Cy_n\| \ge |\lambda_n| \to |\lambda|$. However, this contradicts the condition $Cy_n \to 0$, which follows from the orthonormality of $\{y_n\}$ and the compactness of the operator C.

Consider finally the sets $M_n := \{\lambda \in \sigma_p(C) : |\lambda| \ge n^{-1}\}$. Due to (c), they have no accumulation points, and since they are bounded, they must be finite. Hence $\sigma_p(C)\setminus\{0\} = \bigcup_{n=1}^{\infty} M_n$ is at most countable; if it is infinite, we have $\lim_{j\to\infty}\lambda_j = 0$ for any numbering of the eigenvalues. Passing to the disjoint sets $N_1 := M_1$, $N_{m+1} := M_{m+1}\setminus M_n$, $n = 1,2,\dots$, we see that the eigenvalues may be numbered so that $|\lambda_j| \ge |\lambda_{j+1}|$. $\blacksquare$

3.5.8 Corollary (Fredholm alternative): Consider the equation

$$x - \lambda Cx = y$$

for a compact operator C, a complex λ, and $y \in \mathcal{H}$. One and only one of the following possibilities is valid:

(i) The equation has a unique solution x_y for any $y \in \mathcal{H}$, in particular, $x_0 = 0$.

(ii) The equation with a zero right side has a nontrivial solution.

Proof: Everything follows from part (a) of the theorem: if $\lambda^{-1} \notin \sigma(C)$, then $x_y = -\lambda^{-1}(C - \lambda^{-1})y$; in the opposite case λ^{-1} is an eigenvalue and any of the corresponding eigenvectors solves the equation with $y = 0$. ∎

For compact operators, a converse to Theorem 3.4.7 is true.

3.5.9 Theorem (Hilbert–Schmidt): Any normal compact operator has a pure point spectrum.

Proof: Let $\{\lambda_n : n = 1, \dots, N\}$ be the set of nonzero eigenvalues of a normal operator B, L_n the corresponding eigenspaces, and $m_n = \dim L_n < \infty$ the multiplicity of λ_n. The subspaces L_n are mutually orthogonal and contained in $(\mathrm{Ker}\, B)^\perp$. We have to check that $\mathcal{G} := \bigoplus_{n=1}^\infty L_n$ equals $(\mathrm{Ker}\, B)^\perp$, since the sought orthonormal basis will be then obtained as a union of bases in L_n and a basis in $\mathrm{Ker}\, B$.

By definition, $\mathcal{G} \subset (\mathrm{Ker}\, B)^\perp$. Each of the subspaces L_n is simultaneously an eigenspace of B^* corresponding to $\overline{\lambda}_n$; since the set $\{\lambda_n\}$ is bounded, the subspace $\mathcal{G}$ is B^*–invariant. Using Theorem 3.4.2 again we see that $\mathcal{G}^\perp$ is B–invariant. Hence $B \upharpoonright \mathcal{G}^\perp$ is a normal compact operator with no nonzero eigenvalues, and therefore $B \upharpoonright \mathcal{G}^\perp = 0$ (Problem 34). In other words, $\mathcal{G}^\perp \subset \mathrm{Ker}\, B$, so $\mathcal{G} \supset (\mathrm{Ker}\, B)^\perp$. ∎

With Theorem 3.5.7 in mind, we shall always assume that the eigenvalues of a normal compact operator B are ordered, $|\lambda_n| \geq |\lambda_{n+1}|$. Denoting $J_B := \dim(\mathrm{Ker}\, B)^\perp$, we can construct the sequence $\{\lambda(j)\}_{j=1}^{J_B}$ in which any nonzero eigenvalue is repeated as many times as its multiplicity is, and $|\lambda(j)| \geq |\lambda(j+1)|$, $j = 1, \dots, J_B$. The corresponding eigenvectors which form an orthonormal basis in $(\mathrm{Ker}\, B)^\perp$ can be numbered so that

$$Be_j = \lambda(j) e_j \,, \quad j = 1, 2, \dots, J_B \,.$$

Using this notation, we can derive a useful expression of the eigenvalues representing a modification of the *minimax principle,* which we shall discuss in Section 14.2.

3.5.10 Corollary: We have

$$|\lambda(j)| = \min_{\mathcal{G}_j} \ell_B(\mathcal{G}_j) \,, \quad j = 1, 2, \dots, J_B$$

where $\mathcal{G}_j \subset \mathcal{H}$ is any subspace of dimension $\leq j - 1$, and

$$\ell_B(\mathcal{G}_j) := \sup \{ \, \|Bx\| \, : \, x \in \mathcal{G}_j^\perp \,, \, \|x\| = 1 \, \}$$

(the supremum can be replaced by maximum — *cf.* Problem 37). In particular, $|\lambda(1)| = \|B\|$.

Proof: Denote $\mathcal{F}_j := \{e_1, \dots, e_j\}_{lin}$. In view of Problem 18, there is a unit vector $y \in \mathcal{F}_j \cap \mathcal{G}_j^\perp$ for which we get

$$\ell_B(\mathcal{G}_j)^2 \geq \|By\|^2 = \left\| \sum_{k=1}^{j} (e_k, y)\, Be_k \right\|^2 = \sum_{k=1}^{j} |\lambda(k)\, (e_k, y)|^2 \geq |\lambda(j)|^2 \,.$$

On the other hand, any unit vector $x \in \mathcal{F}_{j-1}^{\perp}$ satisfies

$$\|Bx\|^2 = \sum_{k=1}^{\infty} |(B^*e_k, x)|^2 = \sum_{k=j}^{\infty} |\lambda(k)|^2 \, |(e_k, x)|^2 \leq |\lambda(j)|^2 ,$$

so $\min_{\mathcal{G}_j} \ell_B(\mathcal{G}_j) = \ell_B(\mathcal{F}_{j-1}) = |\lambda(j)|$. ∎

The existence of a basis of eigenvectors allows us easily to express the action of a normal compact operator. It appears, moreover, that a similar formula can be written for any compact C. To derive it we use the fact that $|C|$ is compact too (Problem 35), so

$$|C|x = \sum_{j=1}^{J_C} \mu(j)\, (e_j, x)\, e_j$$

holds for any $x \in \mathcal{H}$, where $J_C = \dim(\operatorname{Ker}|C|)^{\perp} = \dim(\operatorname{Ker} C)^{\perp}$. The positive numbers $\mu(j) \equiv \mu_C(j)$, *i.e.*, the nonzero eigenvalues of $|C|$ are called *singular values* of the compact operator C. We assume they are ordered, $\mu(j) \geq \mu(j+1)$, and $|C|e_j = \mu(j)e_j$ for $j = 1, \dots, J_C$. Furthermore, using Problem 14 we get the following minimax–type expression,

$$\mu_C(j) = \min_{\mathcal{G}_j} \left\{ \sup\{ \|Cx\| \,:\, x \in \mathcal{G}_j^{\perp} ,\, \|x\| = 1 \} \right\} . \tag{3.9}$$

The vectors e_j belong to $(\operatorname{Ker} C)^{\perp}$, which is the initial space of the partial isometry in the polar decomposition $C = W|C|$. Then the vectors $f_j := We_j$, $j = 1, \dots, J_C$, form an orthonormal basis in $\overline{\operatorname{Ran} C}$; we can also write them as

$$f_j = \mu(j)^{-1} C e_j .$$

Now applying W to the above expression of $|C|x$, we get the sought formula,

$$C = \sum_{j=1}^{J_C} \mu(j)\, (e_j, \cdot)\, f_j , \tag{3.10}$$

which is called the **canonical form** of the compact operator C. Notice that it has a practical importance. If $J_C = \infty$, the partial sums of this series yield a sequence of finite–dimensional operators which is easily seen to converge to C in the operator norm. Hence we have a concrete form of the approximating sequence whose existence is ensured by Theorem 3.5.5.

3.6 Hilbert–Schmidt and trace–class operators

Now we are going to discuss two important classes of compact operators; to make the formulation easier, $\mathcal{H}$ in this section will always mean an infinite–dimensional separable Hilbert space. An operator $B \in \mathcal{B}(\mathcal{H})$ is called **Hilbert–Schmidt** if there is an orthonormal basis $\mathcal{E}_B = \{f_j\}_{j=1}^{\infty} \subset \mathcal{H}$ such that $\sum_{j=1}^{\infty} \|Bf_j\|^2 < \infty$. The set of all Hilbert–Schmidt operators on a given $\mathcal{H}$ is denoted as $\mathcal{J}_2(\mathcal{H})$ or simply $\mathcal{J}_2$.

3.6.1 Proposition: Let $B \in \mathcal{J}_2$. Then $B^* \in \mathcal{J}_2$, and furthermore

$$\sum_{j=1}^{\infty} \|Be_j\|^2 < \infty$$

for any orthonormal basis $\mathcal{E} = \{e_j\} \subset \mathcal{H}$, with the sum independent of $\mathcal{E}$.

Proof: We denote the sum of the series in question by $N_{\mathcal{E}}(B)$. The Parseval relation gives $\|Be_j\|^2 = \sum_k |(f_k, Be_j)|^2 = \sum_k |(B^*f_k, e_j)|^2$, and since a series with non-negative terms may be rearranged, we get

$$N_{\mathcal{E}}(B) = \sum_k \sum_j |(e_j, B^*f_k)|^2 = \sum_k \|B^*f_k\|^2 = N_{\mathcal{E}_B}(B^*)\,.$$

Choosing first $\mathcal{E} = \mathcal{E}_B$ we find $N_{\mathcal{E}_B}(B^*) = N_{\mathcal{E}_B}(B) < \infty$, so $B^* \in \mathcal{J}_2$, and using the last identity, we get $N_{\mathcal{E}}(B) = N_{\mathcal{E}_B}(B)$ for any $\mathcal{E}$. ∎

Since $\sqrt{N_{\mathcal{E}}(B)}$ does not depend on the choice of $\mathcal{E}$, we may use it to define the *Hilbert–Schmidt norm* of an operator B by

$$\|B\|_2 := \left(\sum_{j=1}^{\infty} \|Be_j\|^2 \right)^{1/2} . \tag{3.11}$$

3.6.2 Theorem: $\mathcal{J}_2(\mathcal{H})$ is a two–sided $*$–ideal in $\mathcal{B}(\mathcal{H})$ and $\|\cdot\|_2$ is a norm on it which fulfils

$$\|B\|_2 \geq \|B\|$$

for all $B \in \mathcal{J}_2$. Any Hilbert–Schmidt operator is compact, $\mathcal{J}_2(\mathcal{H}) \subset \mathcal{K}(\mathcal{H})$.

Proof: If $C \in \mathcal{J}_2$, the relation $\|(B+C)e_j\|^2 \leq \|Be_j\|^2 + 2\|Be_j\|\,\|Ce_j\| + \|Ce_j\|^2$ in combination with the Hölder inequality yields $\|B+C\|_2 \leq \|B\|_2 + \|C\|_2$. We have $\|\alpha B\|_2 = |\alpha|\,\|B\|_2$, and $\|B\|_2 = 0$ implies $Be_j = 0\,,\ j = 1,2,\dots$, so $B = 0$ by continuity; thus $(\mathcal{J}_2,\, \|\cdot\|_2)$ is a normed space. For any unit $x \in \mathcal{H}$ we can find an orthonormal basis $\{f_j\}$ such that $x = f_1$, then $\|Bx\| \leq \left(\sum_k \|Bf_k\|^2\right)^{1/2} = \|B\|_2$.

The inequalities $\|DBe_j\| \leq \|D\|\,\|Be_j\|$ yield $DB \in \mathcal{J}_2$ for any $D \in \mathcal{B}(\mathcal{H})$. We know already that $B \in \mathcal{J}_2$ implies $B^* \in \mathcal{J}_2$; it follows that $D^*B^* \in \mathcal{J}_2$ and $BD = (D^*B^*)^* \in \mathcal{J}_2$. Finally, let E_n be the projection to $\{e_1, \dots, e_n\}_{lin}$; then $B_n := E_nB$ is a finite–dimensional operator and

$$\|(B-B_n)x\|^2 = \sum_{j=n+1}^{\infty} |(e_j, Bx)|^2 \leq \|x\|^2 \sum_{j=n+1}^{\infty} \|B^*e_j\|^2 .$$

Since $B^* \in \mathcal{J}_2$, we get $B_n \to B$, so B is compact by Theorem 3.5.5. ∎

3.6.3 Remark: Since any $\mathcal{B} \in \mathcal{J}_2$ is compact, there is an orthonormal basis formed by the eigenvectors of $|B|$; using it we obtain the following useful expression,

$$\|B\|_2 = \left(\sum_{j=1}^{\infty} \mu_B(j)^2 \right)^{1/2} .$$

This provides us with a criterion for a compact operator to belong to $\mathcal{J}_2$ (Problem 38) showing, in particular, that $\mathcal{J}_2(\mathcal{H}) \subset \mathcal{K}(\mathcal{H})$ is a proper inclusion and $\mathcal{J}_2$ is not complete with respect to the norm $\|\cdot\|$ (see also Theorem 6.4.8). It is clear from the definition that the norm $\|\cdot\|_2$ satisfies the parallelogram law, and therefore it is generated by the scalar product

$$(B,C)_2 := \sum_{j=1}^{\infty} (Be_j, Ce_j) = \sum_{j=1}^{\infty} (e_j, B^*Ce_j) \, ; \tag{3.12}$$

we can readily check that the series converges absolutely and its sum is independent of the orthonormal basis used.

3.6.4 Proposition: $\mathcal{J}_2(\mathcal{H})$ equipped with the inner product $(\cdot,\cdot)_2$ is a Hilbert space.

Proof: Consider a sequence $\{B_n\} \subset \mathcal{J}_2$ which is Cauchy with respect to $\|\cdot\|_2$. In view of Theorems 3.6.2 and 3.5.2, it is Cauchy with respect to the operator norm too and converges to some $B \in \mathcal{K}(\mathcal{H})$. Choose $\varepsilon > 0$ and an orthonormal basis $\{e_j\} \subset \mathcal{H}$, then

$$\sum_{j=1}^{N} \|(B_n - B_m)e_j\|^2 \le \|B_n - B_m\|_2^2 < \varepsilon^2$$

holds for all m, n large enough and any $N = 1, 2, \dots$. Performing first the limit $n \to \infty$, then $N \to \infty$, we get $B \in \mathcal{J}_2$ and $\|B - B_m\|_2 < \varepsilon$ for a sufficiently large m, *i.e.*, $\lim_{m\to\infty} \|B - B_m\|_2 = 0$. ∎

Next we are going to derive a functional realization of Hilbert–Schmidt operators on a space $L^2(M, d\mu)$ with a σ–finite measure μ such that $L^2(M, d\mu)$ is separable (*cf.* Problem 5).

3.6.5 Theorem: An operator $B \in \mathcal{B}(L^2(M, d\mu))$ is Hilbert–Schmidt *iff* there is $k_B \in L^2(M \times M, d(\mu \otimes \mu))$ such that

$$(Bf)(x) = \int_M k_B(x,y)\, f(y)\, d\mu(y)$$

for all $f \in L^2(M, d\mu)$. In that case,

$$\|B\|_2^2 = \int_{M\times M} |k_B(x,y)|^2 d(\mu \otimes \mu)(x,y) \, .$$

Proof: Choose an orthonormal basis $\{g_n\} \subset L^2(M, d\mu) =: L^2$; then the vectors h_{mn} : $h_{mn}(x, y) = g_m(x)\overline{g_n(y)}$ form an orthonormal basis in $L^2(M \times M, d(\mu \otimes \mu)) =: L^2_\otimes$. Given $B \in \mathcal{J}_2(L^2)$, we define $b_{mn} := (g_m, Bg_n)$. Since $\|B\|_2^2 = \sum_{m,n} |b_{mn}|^2 < \infty$, the relation

$$k_B := \sum_{m,n} b_{mn} h_{mn}$$

defines an element of $L^2_\otimes$, and this in turn corresponds to an integral operator of Problem 5 which we denote as K_B. Using the Fubini theorem, we get

$$\begin{aligned}(g_m, K_B g_n) &= \int_{M\times M} \overline{h_{mn}(x, y)}\, k_B(x, y)\, d(\mu \otimes \mu)(x, y) \\ &= (h_{mn}, k_B)_\otimes = b_{mn} = (g_m, Bg_n),\end{aligned}$$

and since the two operators are bounded, one has $B = K_B$. On the other hand, using Problem 5 we associate an operator $B_k := K$ with any $k \in L^2_\otimes$. One more application of the Fubini theorem gives

$$\begin{aligned}\|k\|_\otimes^2 &= \sum_{m,n} |(h_{mn}, k)_\otimes|^2 \\ &= \sum_{m,n} \left| \int_M \overline{g_m(x)} \left(\int_M k(x, y)\, g_n(y)\, d\mu(y) \right) d\mu(x) \right|^2 \\ &= \sum_n \|B_k g_n\|^2 = \|B_k\|_2^2 .\end{aligned}$$

This finishes the proof, and shows that the Hilbert spaces $\mathcal{J}_2(L^2(M, d\mu))$ and $L^2_\otimes$ are isomorphic by the map $B \mapsto k_B$. ∎

Another important class of compact operators is obtained if we generalize the notion of the trace known from linear algebra. Consider first a *positive* operator $A \in \mathcal{B}(\mathcal{H})$ and set

$$\operatorname{Tr} A := \sum_j (e_j, Ae_j),$$

where $\{e_j\}$ is an orthonormal basis in $\mathcal{H}$. The sum (either finite or infinite) makes sense and is independent of the choice of the basis; it follows from Proposition 3.6.1 applied to $B = \sqrt{A}$.

An operator $B \in \mathcal{B}(\mathcal{H})$ is said to be of the **trace class** if $\operatorname{Tr}|B| < \infty$; the set of all such operators on a given $\mathcal{H}$ is denoted as $\mathcal{J}_1(\mathcal{H})$ or $\mathcal{J}_1$.

3.6.6 Proposition: (a) Any trace–class operator B is Hilbert–Schmidt, and therefore compact, $\mathcal{J}_1(\mathcal{H}) \subset \mathcal{J}_2(\mathcal{H}) \subset \mathcal{K}(\mathcal{H})$, and

$$\operatorname{Tr}|B| = \left\| \sqrt{|B|} \right\|_2^2 = \sum_{j=1}^\infty \mu_B(j), \qquad \operatorname{Tr}|B| \geq \|B\|_2 \geq \|B\| .$$

(b) An operator $B \in \mathcal{B}(\mathcal{H})$ belongs to the trace class *iff* it is a product of two Hilbert–Schmidt operators.

Proof: Part (a) is checked easily. It implies $\sqrt{|B|} \in \mathcal{J}_2$ for any $B \in \mathcal{J}_1$; hence by polar decomposition, B is the product of $W\sqrt{|B|}$ and $\sqrt{|B|}$. Conversely, assume $C, D \in \mathcal{J}_2$. Using the polar decomposition once more, $CD = W|CD|$, in combination with the Schwarz and Hölder inequalities, we get

$$\begin{aligned}\operatorname{Tr}|CD| &= \sum_j |(e_j, W^*CDe_j)| \\ &\le \left(\sum_j \|C^*We_j\|^2 \sum_k \|De_k\|^2\right)^{1/2} = \|D\|_2\, \|C^*W\|_2 < \infty\end{aligned}$$

because $C^*W \in \mathcal{J}_2$. ∎

3.6.7 Theorem: $\mathcal{J}_1(\mathcal{H})$ is a two–sided $*$–ideal in $\mathcal{B}(\mathcal{H})$. It is a complete space with respect to the norm $\|\cdot\|_1 := \operatorname{Tr}|\cdot|$.

Proof: The algebraic property follows from part (b) of the previous proposition. Using part (a) together with (3.9) we find $\mu_{B+C}(j) \le \mu_B(j) + \mu_C(j)$, which implies $\operatorname{Tr}|B+C| \le \operatorname{Tr}|B| + \operatorname{Tr}|C|$; the other properties of the norm are easily verified. To prove completeness, we can use continuity of the singular values with respect to the operator norm (Problem 36), and repeat the argument from Proposition 3.6.4. ∎

Now we are able to extend the notion of *trace* to nonpositive operators, thereby justifying the name given to the set $\mathcal{J}_1$.

3.6.8 Theorem: (a) Let B be a trace–class operator, then

$$\operatorname{Tr} B := \sum_{j=1}^{\infty} (e_j, Be_j)$$

is well–defined in the sense that the series on the right side converges absolutely and its sum is independent of the choice of the orthonormal basis $\{e_j\}$.

(b) The map $B \mapsto \operatorname{Tr} B$ is a bounded linear functional on $(\mathcal{J}_1, \|\cdot\|_1)$ with the unit norm. Moreover, the relations

$$\operatorname{Tr} B^* = \overline{\operatorname{Tr} B}\,, \qquad \operatorname{Tr}(BC) = \operatorname{Tr}(CB)$$

hold for any $B \in \mathcal{J}_1$, $C \in \mathcal{B}(\mathcal{H})$.

Proof: Consider the orthonormal bases $\{e_j\}$ and $\{f_j\}$ where the former is arbitrary and the latter is formed by the eigenvectors of $|B|$. First we check the absolute convergence,

$$\sum_j |(e_j, Be_j)| \le \sum_j \sum_k |(B^*e_j, f_k)(f_k, e_j)| = \sum_j \sum_k |(e_j, Bf_k)(f_k, e_j)|\,.$$

Rearranging the last series, we get the estimate

$$\sum_j |(e_j, Be_j)| \leq \sum_k \left(\sum_j |(e_j, Bf_k)|^2 \sum_l |(e_l, f_k)|^2 \right)^{1/2} = \sum_k \|Bf_k\| = \mathrm{Tr}\,|B|$$

which shows that the series on the left side converges, and at the same time the double series $\sum_{j,k} (e_j, Bf_k)(f_k, e_j)$ converges absolutely, so its sum is independent of the order of summation; it follows that

$$\sum_j (e_j, Be_j) = \sum_j \sum_k (e_j, Bf_k)(f_k, e_j) = \sum_k \sum_j (f_k, e_j)(e_j, Bf_k) = \sum_k (f_k, Bf_k)\,,$$

which proves part (a).

The linearity of $\mathrm{Tr}\,(\cdot)$ is obvious. Using the polar decomposition $B = W_B|B|$, we obtain $|\mathrm{Tr}\,B| \leq \sum_j \mu_B(j)\,|(f_j, W_B f_j)| \leq \sum_j \mu_B(j)$, which shows that

$$|\mathrm{Tr}\,B| \leq \mathrm{Tr}\,|B|\,;$$

On the other hand, $\mathrm{Tr}\,E = 1$ for any one–dimensional projection. The first one of the remaining relations follows directly from the definition. If U is a unitary operator, we have $\mathrm{Tr}\,(BU) = \sum_j (e_j, BUe_j) = \sum_j (Ue_j, UBUe_j) = \mathrm{Tr}\,(UB)$, and since any bounded operator can be expressed as a linear combination of four unitary operators (Problem 22), the theorem is proved. ∎

Let us finally mention a particular family of trace–class operators which will be important in quantum–mechanical applications: $W \in \mathcal{J}_1(\mathcal{H})$ is called a **statistical operator** if it is positive and fulfils the normalization condition $\mathrm{Tr}\,W = 1$. Its eigenvalues therefore coincide with its singular values, $w_j \equiv \mu_W(j)$, with a possible repetition according to their multiplicity, and correspond to a complete set of eigenvectors, $We_j = w_j e_j$. The normalization condition then reads

$$\sum_j w_j = 1\,.$$

It follows trivially that $0 \leq w_j \leq 1$, $j = 1, 2, \ldots$, and moreover, $w_j = 1$ can hold only if the other eigenvalues are zero, *i.e.*, if W is a one–dimensional projection; it is further equivalent to the condition $\mathrm{Tr}\,W^2 = \mathrm{Tr}\,W$. There is an alternative geometrical formulation.

3.6.9 Proposition: The set $\mathcal{W}$ of all statistical operators on a given $\mathcal{H}$ is convex. An operator $W \in \mathcal{W}$ is one–dimensional projection *iff* it is an extremal point of $\mathcal{W}$, *i.e.*, *iff* the condition $W = \alpha W_1 + (1-\alpha)W_2$ with $0 < \alpha < 1$ and $W_1, W_2 \in \mathcal{W}$ implies $W_1 = W_2 = W$.

Proof: If $W^2 \neq W$, then W has an eigenvalue $\lambda \in (0,1)$ corresponding to a one–dimensional projection E. We can write $W = \lambda E + (1-\lambda)W'$, where $W' := (1-\lambda)^{-1}(W - \lambda E)$ is a statistical operator, so W is not an extremal point. On the other hand, suppose that $W^2 = W$ and $W = \alpha W_1 + (1-\alpha)W_2$ for some $\alpha \in (0,1)$. We have

$$\operatorname{Tr} W^2 = \alpha^2 \operatorname{Tr} W_1^2 + 2\alpha(1-\alpha) \operatorname{Tr} W_1 W_2 + (1-\alpha)^2 \operatorname{Tr} W_2^2 = 1,$$

and since $\operatorname{Tr} W_1 W_2 \le 1$ for any statistical operators W_1, W_2 (*cf.* Problem 40), this is possible only if $\operatorname{Tr} W_1^2 = \operatorname{Tr} W_2^2 = 1$. Then W_j is a projection to a one–dimensional subspace spanned by a unit vector e_j and the condition $\operatorname{Tr} W_1 W_2 = 1$ implies $|(e_1, e_2)| = 1$, so the two vectors differ at most by a phase factor. ∎

Notes to Chapter 3

Section 3.1 The adjoint operator can be defined even without the inner–product structure. Let $\mathcal{X}, \mathcal{Y}$ be Banach spaces, then the *dual* (or *Banach adjoint*) operator to a given $B \in \mathcal{B}(\mathcal{X}, \mathcal{Y})$ is defined by $(B^*g)(x) := g(Bx)$ for all $x \in \mathcal{X}$, $g \in \mathcal{Y}^*$. By definition, B^* is linear and $\|B^*\| \le \|B\|$, so it belongs to $\mathcal{B}(\mathcal{Y}^*, \mathcal{X}^*)$; using Corollary 1.5.8 we can check that $\|B^*\| = \|B\|$. Notice that the map $B \mapsto B^*$ is linear, in distinction to the Hilbert–space adjoint discussed in this section. The latter belongs to $\mathcal{B}(\mathcal{H})$, however, rather than to $\mathcal{B}(\mathcal{H}^*)$, and the correspondence between the two adjoints is in this case given by the antilinear isometry between $\mathcal{H}$ and $\mathcal{H}^*$.

The other notions discussed here also admit generalizations. The integral representation of Proposition 3.1.5 extends to any bounded mapping from $L^1(X, d\mu)$ with a σ–finite μ to $L^1(Y, d\nu)$, where Y is a metric space and ν a Borel measure on Y — *cf.* ⟦DS 1⟧, Sec.VI.8. The matrix representation can be used for the operators $B \in \mathcal{B}(\mathcal{X}, \mathcal{Y})$, where $\mathcal{X}, \mathcal{Y}$ are Banach spaces having the so–called Schauder basis — *cf.* ⟦Tay⟧, Sec.4.51. Finally, strong and weak operator topologies can be defined on $\mathcal{B}(\mathcal{X}, \mathcal{Y})$ for any Banach $\mathcal{X}, \mathcal{Y}$; we have to replace p_x of the above definition by $p_x : p_x(B) = \|Bx\|_{\mathcal{Y}}$ and p_{xy} by $p_{f,x} : p_{f,x}(B) = |f(Bx)|$ with $x \in \mathcal{X}$, $f \in \mathcal{Y}^*$. The corresponding topological spaces are discussed in ⟦DS 1⟧, Sec.VI.1.

Section 3.2 Basic properties of the numerical range can be derived easily (Problem 27); they extend mostly to unbounded operators — *cf.* ⟦Sto⟧, Sec.IV.3. Using the numerical range, we can define some classes of operators. For instance, $B \in \mathcal{B}(\mathcal{H})$ is *antihermitean* if $\Theta(B)$ is a subset of the imaginary axis; this is equivalent to the condition $B^* = -B$. Other definitions of this type will be mentioned in the notes to Section 4.1. Projections can generally be associated with any decomposition $\mathcal{H} = L + L'$, not necessarily orthogonal; the projections introduced in this section are then called *orthogonal projections*. Since we shall use almost exclusively the latter, we drop the adjective.

Section 3.3 An antilinear isometry $U : \mathcal{H} \to \mathcal{G}$ is called an *antiunitary operator*. Examples of such operators are, for instance, the complex conjugation on $L^2(M, d\mu)$ or the map $\mathcal{H}^* \to \mathcal{H}$ given by the Riesz lemma. As in Problem 20, we can check that $U : \mathcal{H} \to \mathcal{G}$ is antiunitary *iff* it is surjective and $(Ux, Uy)_{\mathcal{G}} = (y, x)_{\mathcal{H}}$ for any $x, y \in \mathcal{H}$.

Section 3.5 The definition of compactness extends to Banach space operators $C \in \mathcal{B}(\mathcal{X}, \mathcal{Y})$. The present proofs show that the set of compact operators is closed again and the sufficient condition of Theorem 1 remains valid. Compactness is also preserved when passing to the dual operator; this is known as the Schauder theorem — *cf.* ⟦Ka⟧, Sec.III.4;

⟦Yo⟧, Sec.X.5. If the space is reflexive, Theorem 3.3.1 holds in both directions. The Riesz–Schauder theorem extends to operators from $\mathcal{B}(\mathcal{X})$ — *cf.* ⟦Yo⟧, Sec.X.5. Corollary 3.5.10 also follows from the minimax principle: since B is normal, $|\lambda_j|$ are eigenvalues of $|B|$, so one has to apply Theorem 14.2.1 to the operator B^*B.

Section 3.6 An operator $C \in \mathcal{B}(\mathcal{X},\mathcal{Y})$, where $\mathcal{X}$, $\mathcal{Y}$ are arbitrary Banach spaces, belongs to the trace class if there are bounded sequences $\{f_n\} \subset \mathcal{X}^*$, $\{y_n\} \subset \mathcal{Y}$ and a sequence of complex numbers fulfilling $\sum_j |c_n| < \infty$ such that $Cx = \sum_n c_n f_n(x) y_n$ for any $x \in \mathcal{X}$. Such an operator is approximated by finite–dimensional operators and is therefore compact. In a similar way, we define a trace–class map from a locally convex space to a Banach space; using this A. Grothendieck introduced the important notion of nuclear space — *cf.* ⟦Yo⟧, App. to Chap.X.

The definition of the classes $\mathcal{J}_1$ and $\mathcal{J}_2$ with the help of singular values suggests a natural generalization: we define the class $\mathcal{J}_p$ for any real $p \geq 1$ as consisting of the compact operators C whose singular values fulfil $\{\mu_C(j)\} \in \ell^p$. Since the set of nonzero singular values of a compact C in a nonseparable $\mathcal{H}$ is at most countable, the definition makes sense in any Hilbert space. Investigation of the classes $\mathcal{J}_p$ was initiated by J.W. Calkin in the 1940s — *cf.* [Ca 1]. The recent state of knowledge in this field including numerous applications is summarized in the monograph ⟦Si 3⟧. The reader can easily check that any $\mathcal{J}_p$ is again a two–sided $*$–ideal in $\mathcal{B}(\mathcal{H})$, which is complete w.r.t. the norm

$$\|C\|_p := \left(\sum_j \mu_C(j)^p \right)^{1/p} .$$

These norms fulfil $\|B\| \leq \|B\|_q \leq \|B\|_p$ for any $q \geq p \geq 1$ to which the inclusions $\mathcal{K}(\mathcal{H}) \supset \mathcal{J}_q \supset \mathcal{J}_p$ correspond. Compact operators are associated with ℓ^∞ sequences of singular values; this justifies the alternative notation $\mathcal{K}(\mathcal{H}) =: \mathcal{J}_\infty(\mathcal{H})$. The factorization assertion of Proposition 3.6.6 generalizes as follows (see ⟦We⟧, Sec.7.1): to any $p,q \geq 1$ fulfilling $pq \geq p+q$ set $r := pq/(p+q)$, then $B \in \mathcal{J}_r$ *iff* there are $C \in \mathcal{J}_p$, $D \in \mathcal{J}_q$ such that $B = CD$.

Since $\mathcal{J}_2$ has a Hilbert–space structure, any bounded linear functional f on it is of the form $f = \mathrm{Tr}\,(C_f\,\cdot)$ for some $C_f \in \mathcal{J}_2$, and $\|f\| = \|C_f\|_2$. Similarly we can prove that $\mathcal{J}_1^*$ is linearly isometric to $\mathcal{B}(\mathcal{H})$ (Problem 42), while $\mathcal{J}_p^*$ is isometric to $\mathcal{J}_q$ with $q := p/(p-1)$, and $\mathcal{J}_1^*$ to $\mathcal{J}_\infty := (\mathcal{K}(\mathcal{H}), \|\cdot\|)$. For more details see ⟦Si 3⟧, Sec.3.

Problems

1. Multiplication of bounded operators (defined as a composite mapping) is jointly continuous with respect to the operator norm: $B_n \to B$ and $C_n \to C$ imply $B_nC_n \to BC$.

2. Let D be a proper subspace in $\mathcal{H}$. Then any $B_0 \in \mathcal{B}(D,\mathcal{H})$ has just one extension $B \in \mathcal{B}(\mathcal{H})$ such that $B \upharpoonright D^\perp = 0$; this operator satisfies $\|B\| = \|B_0\|$.

3. Let L, L' be dense subspaces in $\mathcal{H}$. The operators $B, C \in \mathcal{B}(\mathcal{H})$ are equal provided $(x, By) = (x, Cy)$ for all $(x,y) \in L \times L'$, or $(x, Bx) = (x, Cx)$ for all $x \in L$.

4. We have $B\overline{M} \subset \overline{BM}$ for any $B \in \mathcal{B}(\mathcal{H})$ and $M \subset \mathcal{H}$; the inclusion turns into identity if B is invertible.

5. Consider $\mathcal{H} = L^2(M, d\mu)$ with a σ–finite μ. To any $k \in L^2(M \times M, d(\mu \otimes \mu))$ we define the *Hilbert–Schmidt integral operator* by $(Kf)(x) := \int_M k(x,y) f(y)\, d\mu(y)$. It satisfies $\|K\|^2 \le \int_{M\times M} |k|^2\, d(\mu\otimes\mu)$, and the adjoint operator to K is given by the formula $(K^*f)(x) = \int_M \overline{k(y,x)} f(y)\, d\mu(y)$.

6. The operator $P_{xy} := (y, \cdot)x$ is bounded for any $x, y \in \mathcal{H}$, $\|P_{xy}\| = \|x\|\,\|y\|$. It satisfies $P_{xy}^* = P_{yx}$ and $P_{xy}P_{uv} = (y,u)\, P_{xv}$.

7. The shift operators of Example 3.1.3 have the following properties:

(a) $\|S^n\| = \|(S^*)^n\| = 1$ and $(S^*)^n S^n = I$. Furthermore, $S^n(S^*)^n$ is a partial isometry from $\mathcal{H}$ to $\{e_1, \dots e_n\}^\perp$.

(b) $\text{s-lim}_{n\to\infty}(S^*)^n = 0$, while $\{S^n\}$ does not converge strongly.

8. Using the shift operators of the previous problem, define $Q_n := S^n(S^*)^n$. The sequence $\{Q_n\}$ of projections converges strongly to zero, but it does not converge with respect to the operator norm.

9. Let $\dim \mathcal{H} = \infty$. Using Problem 7 show that

(a) The map $B \mapsto B^*$ is continuous with respect to the operator norm and weak operator topologies while it is not continuous with respect to the strong operator topology.

(b) $B_n \xrightarrow{w} B$ and $C_n \xrightarrow{w} C$ does *not* imply $B_nC_n \xrightarrow{w} BC$.

10. The topologies τ_w, τ_s, and τ_u on $\mathcal{B}(\mathcal{H})$ coincide for $\dim \mathcal{H} < \infty$.

11. A bounded operator B on a separable $\mathcal{H}$ is Hermitean *iff* its matrix representation with respect to any basis fulfils $B_{jk} = \overline{B_{kj}}$, $j,k = 1,2,\dots$.

12. The product of Hermitean operators A, B is Hermitean *iff* they commute, $AB = BA$. The product of commuting positive operators is positive.

13. Let λ be an eigenvalue of a positive A; then $\sqrt{\lambda}$ is an eigenvalue of $\sqrt{A}$.

14. We have $\|Bx\| = \|\,|B|x\|$ for any $x \in \mathcal{H}$ and $\|B\| = \|\,|B|\,\|$. The relation $|B| = B$ holds *iff* B is positive.

15. Find bounded operators B, C such that

(a) $|BC| \neq |B||C|$

(b) $|B^*| \neq |B|$

(c) $|B+C| \le |B| + |C|$ is not valid

Hint: In the last case use $B = \sigma_3 + I$, $C = \sigma_1 - I$, where the σ_j are Pauli matrices.

16. If L is a subspace in $\mathcal{H}$ and E is the projection to $\overline{L}$, then $\|Ex\| = \sup\{\,|(x,y)| : y \in L,\ \|y\| = 1\,\}$ holds for any $x \in \mathcal{H}$.

17. Prove Proposition 3.2.9.

18. Let E, F be arbitrary projections on $\mathcal{H}$. Prove:

(a) $\mathrm{Ran}\, E \cap \mathrm{Ker}\, F = \{0\}$ implies $\dim E \le \dim F$.

(b) $\mathrm{Ran}\, E \cap \mathrm{Ker}\, F \neq \{0\}$ implies $\|E - F\| = 1$.

(c) $\|E - F\| = \max\{\sigma_{EF}, \sigma_{FE}\}$, where we have denoted $\sigma_{EF} := \sup\{\, \|(I-F)x\| : x \in E\mathcal{H}\,,\ \|x\| = 1\,\} = \sup\{\, d(x, E\mathcal{H}) : x \in E\mathcal{H}\,,\ \|x\| = 1\,\}$.

Hint: To prove (a) for a non–separable $\mathcal{H}$ use the following assertion: if I is an index set of infinite cardinality m and $\{M_\alpha : \alpha \in I\}$ with $\mathrm{card}\, M_\alpha \le m$ for each α, then $\mathrm{card} \bigcup_{\alpha \in I} M_\alpha \le m$.

19. Suppose that a sequence $\{E_n\}$ of projections, not necessarily monotonic, converges weakly to a projection E, then $E = \text{s-lim}_{n\to\infty} E_n$.

20. Prove: (a) If the map $U : \mathcal{H} \to \mathcal{H}$ is surjective and preserves the inner product, then it is linear.

(b) A unitary U which satisfies $U^2 = I$ is Hermitean; on the other hand, any $U \in \mathcal{B}(\mathcal{H})$ which is simultaneously unitary and Hermitean satisfies $U^2 = I$.

(c) If U is unitary, $UL^\perp = (UL)^\perp$ holds for any subspace $L \subset \mathcal{H}$.

21. The set $\mathcal{U}(\mathcal{H})$ of all unitary operators on a given $\mathcal{H}$ is a group with respect to operator multiplication. If $\mathcal{H} = L^2(\mathbb{R})$, the set $\{\, U_\Phi : \varphi \in \Phi \,\}$ of Example 3.3.2 is a subgroup in $\mathcal{U}(L^2(\mathbb{R}))$. Which topology makes $\mathcal{U}(\mathcal{H})$ a topological group?

22. Any Hermitean operator is a linear combination of two unitary operators.
Hint: If $\|A\| = 1$, consider $A \pm i\sqrt{I - A^2}$.

23. The Fourier–Plancherel operator F on $L^2(\mathbb{R}^n)$ satisfies $F^2 = R$, where R is the reflection operator, and $F^4 = I$.
Hint: Show that $Ff = F^{-1}Rf$ for $f \in L^2 \cap L^1$.

24. The polar decomposition of Theorem 3.3.5 cannot be generally replaced by $B = |B|W$. For a Hermitean A, however, one has $W^* = W$ and $W|A| = |A|W$.
Hint: Try a shift operator, and for the second part use Proposition 3.2.6.

25. Let V_r, $r = 1, 2$, be isometric operators with the domains $D_r = E_r\mathcal{H}$. If $D_1 \perp D_2$ and $\mathrm{Ran}\, V_1 \perp \mathrm{Ran}\, V_2$, then $W := V_1E_1 + V_2E_2$ is a partial isometry with the initial subspace $D_1 \oplus D_2$ and final subspace $\mathrm{Ran}\, V_1 \oplus \mathrm{Ran}\, V_2$.

26. Given a sequence $s := \{\lambda_j\} \in \ell^\infty$, one can construct an operator B_s with a pure point spectrum, $\sigma(B_s) = \overline{s}$, and $\|B_s\| = \sup_j |\lambda_j|$, on any infinite–dimensional Hilbert space $\mathcal{H}$.

27. Let $B \in \mathcal{B}(\mathcal{H})$. Its numerical range satisfies

(a) $\Theta(B)$ is a convex set which is generally neither open nor closed.

(b) The spectrum of B is contained in $\overline{\Theta(B)}$.

(c) If B is not normal, $\overline{\Theta(B)}$ need not equal be to the convex hull of $\sigma(B)$.

Hint: To prove (b), use Theorem 4.7.1; for part (c) consider $B = \begin{pmatrix} 1 & -2a \\ 0 & 2 \end{pmatrix}$ and $f = 3^{-1/2} \begin{pmatrix} \sqrt{2} \\ \mathrm{sgn}\, a \end{pmatrix}$.

28. If a sequence $\{B_n\} \subset \mathcal{N}(\mathcal{H})$ converges strongly to $B \in \mathcal{N}(\mathcal{H})$; then $B_n^* \xrightarrow{s} B^*$ (compare to Problem 9).

29. Prove: (a) Suppose $0 \in \sigma(B) \setminus \sigma_p(B)$ for an operator $B \in \mathcal{N}(\mathcal{H})$; then $\mathrm{Ran}\, B$ is not closed. Find examples of such operators with (i) a pure point spectrum, (ii) a purely continuous spectrum.

(b) The spectral radius of an operator $B \in \mathcal{N}(\mathcal{H})$ equals its norm, $r(B) = \|B\|$.

Hint: (b) Use the relation $r(B) = r(B^*)$ and (3.5).

30. A bounded operator B is m–dimensional *iff* there are linearly independent sets $\{f_1, \dots, f_m\}$, $\{g_1, \dots, g_m\} \subset \mathcal{H}$ such that $B = \sum_{j=1}^m (f_j, \cdot) g_j$. In that case, we also have $\dim \mathrm{Ran}\, B^* = m$.

31. $\mathcal{K}(\mathcal{H})$ is closed with respect to neither the strong nor the weak operator topology unless $\dim \mathcal{H} < \infty$.

32. Let $B_n \xrightarrow{s} B$ in $\mathcal{B}(\mathcal{H})$ and $C \in \mathcal{K}(\mathcal{H})$; then $\|B_n C - BC\| \to 0$.

33. $(\mathrm{Ker}\, C)^\perp$ is separable for any compact C.
Hint: Let $\{e_\alpha\}$ be an orthonormal basis in $(\mathrm{Ker}\, C)^\perp$; then $\{\, e_\alpha :\ \|Ce_\alpha\| \ge n^{-1} \,\}$ must be finite for any n.

34. A normal compact operator B has at least one non–zero eigenvalue unless $B = 0$.
Hint: Consider first a Hermitean B; in the general case use the commutativity of $\mathrm{Re}\, B$, $\mathrm{Im}\, B$.

35. Prove: (a) $C \in \mathcal{B}(\mathcal{H})$ is compact *iff* $Ce_j \to 0$ holds for any orthogonal set $\{e_j\} \subset (\mathrm{Ker}\, C)^\perp$.

(b) $C \in \mathcal{B}(\mathcal{H})$ is compact *iff* $|C|$ is compact.

(c) A positive operator A is compact *iff* $\sqrt{A}$ is compact.

(d) If $B \ge A \ge 0$ and B is compact, then A is compact.

36. The singular values of a compact operator C satisfy

(a) $\mu_C(j) = \mu_{C^*}(j)$

(b) $\mu_{BC}(j) \le \|B\| \mu_C(j)$ and $\mu_{CB}(j) \le \|B\| \mu_C(j)$ for any $B \in \mathcal{B}(\mathcal{H})$

(c) $|\mu_C(j) - \mu_D(j)| \le \mu_{C-D}(j) \le \|C - D\|$

37. Set $\lambda := \sup\{\, \|Cx\| :\ x \in S_1 \,\}$, where C is a compact operator and S_1 is the unit sphere in a closed subspace $\mathcal{G} \subset \mathcal{H}$. Then there is a unit vector $y \in \mathcal{G}$ such that $\lambda = \|Cy\|$.
Hint: Use the weak compactness of bounded sets in $\mathcal{H}$.

38. Prove: (a) A compact operator B is Hilbert–Schmidt or of the trace class *iff* the corresponding sequence of singular values $\{\mu_B(j)\}$ belongs to ℓ^2 or ℓ^1, respectively.

(b) The set of finite–dimensional operators is dense in the Hilbert space $\mathcal{J}_2$.

(c) The same set is dense in $(\mathcal{J}_1, \|\cdot\|_1)$.

39. Prove: (a) The operator P_{xy} of Problem 6 is of the trace class for any $x, y \in \mathcal{H}$ and $\mathrm{Tr}\, P_{xy} = (y, x)$. If $\{e_j\}_{j=1}^{\infty}$ is an orthonormal basis in $\mathcal{H}$, then $\{ P_{e_j e_k} : j, k = 1, 2, \dots \}$ is an orthonormal basis in the Hilbert space $\mathcal{J}_2(\mathcal{H})$, which is therefore separable.

(b) Let $x \mapsto \overline{x}$ be an antilinear isomorphism of the spaces $\mathcal{H}$ and $\mathcal{H}^*$, and define $\overline{x} \otimes y := P_{xy}$; then one has $\mathcal{J}_2(\mathcal{H}) = \mathcal{H}^* \otimes \mathcal{H}$. Extend this result to a pair of Hilbert spaces $\mathcal{H}_1$, $\mathcal{H}_2$.

40. Let $B \in \mathcal{B}(\mathcal{H})$ and $C \in \mathcal{J}_1(\mathcal{H})$; then

(a) $\mathrm{Tr}\,|C| = \mathrm{Tr}\,|C^*|$.

(b) $\mathrm{Tr}\,|BC| \le \|B\|\,\mathrm{Tr}\,|C|$.

(c) Due to the previous assertion, the maps $f_C := \mathrm{Tr}\,(\cdot\, C)$ and $g_B := \mathrm{Tr}\,(B\, \cdot)$ are bounded linear functionals on $\mathcal{B}(\mathcal{H})$ and $\mathcal{J}_1(\mathcal{H})$, respectively. Their norms are $\|f_C\| = \mathrm{Tr}\,|C|$ and $\|g_B\| = \|B\|$.

41. If C is a trace class operator and a sequence $\{B_n\}$ converges weakly to some B, then $\mathrm{Tr}\,(B_n C) \to \mathrm{Tr}\,(BC)$.
Hint: Use boundedness of $\{\|B_n\|\}$ to check that $\sum_j |(e_j, (B_n - B)Ce_j)|$ converges uniformly with respect to n.

42. The spaces $\mathcal{J}_1^*$ and $\mathcal{B}(\mathcal{H})$ are linearly isometric: for any bounded linear functional g on the Banach space $(\mathcal{J}_1, \|\cdot\|_1)$, there is a bounded B_g such that $g = \mathrm{Tr}\,(B_g \cdot)$.
Hint: Use the fact that $\{ P_{xy} : x, y \in \mathcal{H} \}$ is total in $\mathcal{J}_1$ — see Problem 39.

Chapter 4

Unbounded operators

4.1 The adjoint

A linear operator T on $\mathcal{H}$ is said to be *densely defined* if its domain is dense, $\overline{D(T)} = \mathcal{H}$; we denote the set of all such operators as $\mathcal{L}(\mathcal{H})$. Any bounded operator is densely defined. However, Example 4.1.3 below shows that $\mathcal{L}(\mathcal{H})$ contains unbounded operators as well; in such a case there is no standard way to extend T to the whole space space $\mathcal{H}$.

The main importance of the fact that an operator is densely defined is that it allows us to generalize the concept of the adjoint. If $T \in \mathcal{L}(\mathcal{H})$ then to any $y \in \mathcal{H}$ there is obviously at most one vector y^* such that $(y, Tx) = (y^*, x)$ holds for all $x \in D(T)$; the difference from the bounded case is that for some y the vector y^* may not exist. Using this we define the **adjoint operator** T^* of T by $T^*y := y^*$ with the domain $D(T^*)$ consisting of those y for which y^* exist. Let us list first its elementary properties.

4.1.1 Proposition: (a) The operator T^* is linear and satisfies $(y, Tx) = (T^*y, x)$ for any $x \in D(T)$ and $y \in D(T^*)$.

(b) $\operatorname{Ker} T^* = (\operatorname{Ran} T)^\perp$, in particular, $\operatorname{Ker} T^*$ is a closed subspace.

(c) $S \supset T$ implies $S^* \subset T^*$.

Proof: Assertions (a),(c) are evident. If $y \in \operatorname{Ker} T^*$, then (a) gives $(y, Tx) = 0$ for all $x \in D(T)$, so $y \in (\operatorname{Ran} T)^\perp$; this condition in turn implies $y \in D(T^*)$ and $T^*y = 0$. ∎

The sum or product of operators $T, S \in \mathcal{L}(\mathcal{H})$ may not be densely defined, and therefore Theorem 3.1.2 is replaced by the following collection of weaker statements, the proof of which is left to the reader (Problem 2).

4.1.2 Proposition: (a) $(\alpha T)^* = \overline{\alpha} T^*$.

(b) $T^{**} \supset T$ provided $T^* \in \mathcal{L}(\mathcal{H})$.

(c) If T is invertible and $T^{-1} \in \mathcal{L}(\mathcal{H})$, then T^* is also invertible and its inverse $(T^*)^{-1} = (T^{-1})^*$.

(d) Let $S+T$ be densely defined; then $(S+T)^* \supset S^* + T^*$ and the inclusion turns to identity if at least one of the operators T, S is bounded.

(e) If TS is densely defined then $(TS)^* \supset S^*T^*$; the two operators are equal if $T \in \mathcal{B}(\mathcal{H})$.

The concept of an adjoint allows us to define several important operator classes. A densely defined operator A is called **symmetric** if $A \subset A^*$; this is equivalent to the condition $(y, Ax) = (Ay, x)$ for all $x, y \in D_A$. If $A \in \mathcal{L}(\mathcal{H})$ is equal to its adjoint, $A = A^*$, it is called **self–adjoint**. The corresponding subsets in $\mathcal{L}(\mathcal{H})$ are denoted as $\mathcal{L}_s(\mathcal{H})$ and $\mathcal{L}_{sa}(\mathcal{H})$, respectively. Every self–adjoint operator is symmetric, $\mathcal{L}_{sa}(\mathcal{H}) \subset \mathcal{L}_s(\mathcal{H})$; for bounded operators the notions symmetric, self–adjoint, and Hermitean coincide.

4.1.3 Example (the operator Q on $L^2(\mathbb{R})$): We put $(Q\psi)(x) := x\psi(x)$ on the domain $D(Q) := \left\{ \psi \in L^2(\mathbb{R}) : \int_{\mathbb{R}} x^2 |\psi(x)|^2 dx < \infty \right\}$. It is densely defined since $D(Q)$ contains, *e.g.*, the Schwartz space $\mathcal{S}(\mathcal{R})$, and is symmetric because $(\phi, Q\psi) = (Q\phi, \psi)$ holds by definition for all $\phi, \psi \in D(Q)$. It is also easy to see that Q is unbounded: we have $\|Q\phi_n\|^2 = (2n+1)(2n+2)$ for unit vectors $\phi_n : \phi_n(x) = (2(2n!))^{-1/2} x^n e^{-|x|/2}$, $n = 1, 2, \dots$.

We shall show that $D(Q^*) = D(Q)$. To any $\psi \in D(Q^*)$ there is $\psi^* \in L^2$ such that $\int_{\mathbb{R}} \phi(x) \overline{[x\psi(x) - \psi^*(x)]}\, dx = 0$ holds for each $\phi \in D(Q)$. Since any L^2 function with a bounded support belongs to $D(Q)$ we can replace ϕ by $\phi\chi_n$ in the above condition, where χ_n is the characteristic function of the interval $(-n, n)$. Using the fact that $D(Q)$ is dense in $L^2(\mathbb{R})$ we find $x\psi(x) = \psi^*(x)$ for a.a. $x \in (-n, n)$, $n = 1, 2, \dots$, *i.e.*, $x\psi(x) = \psi^*(x)$ a.e. in $\mathbb{R}$. We have $\psi^* \in L^2$ by assumption, so ψ belongs to $D(Q)$ and the operator Q is self–adjoint.

Let A be a symmetric operator. Any symmetric A' such that $A \subset A'$ is called a *symmetric extension* of the operator A. A symmetric operator is *maximal* if it has no proper symmetric extensions, *i.e.*, if the relation $A \subset A'$ for a symmetric A' implies $A = A'$. By Proposition 4.1.1, we have $A \subset A' \subset A'^* \subset A^*$, so *any self–adjoint operator is maximal.* On the other hand, there are maximal symmetric operators which are not self–adjoint (*cf.* Example 4.2.5 below).

4.1.4 Example: Let $\mathcal{E} := \{e_j\}_{j=1}^{\infty}$ be an orthonormal basis in a separable $\mathcal{H}$. To any sequence $s := \{s_j\}_{j=1}^{\infty}$ of complex numbers we can define the linear operator $\dot{T}_s$ with the domain $D(\dot{T}_s) = \mathcal{E}_{lin}$ by $\dot{T}_s e_j := s_j e_j$, $j = 1, 2, \dots$. It obviously belongs to $\mathcal{L}(\mathcal{H})$; it is bounded *iff* the sequence s is bounded.

Now we want to find the adjoint $\dot{T}_s^*$. Let $y := \sum_{j=1}^{\infty} \eta_j e_j$ belong to $D(\dot{T}_s^*)$ so there is a vector $z := \sum_{j=1}^{\infty} \zeta_j e_j$ such that $(z, x) = (y, \dot{T}_s x)$ holds for all $x \in \mathcal{E}_{lin}$. Choosing in particular $x = e_j$, $j = 1, 2, \dots$, we get $\overline{s}_j \eta_j = \zeta_j$ and the Parseval identity gives $D(\dot{T}_s^*) \subset D_s$, where we have denoted

$$D_s := \left\{ x = \sum_{j=1}^{\infty} \xi_j e_j \in \mathcal{H} : \sum_{j=1}^{\infty} |\xi_j s_j|^2 < \infty \right\};$$

the parallelogram identity together with the inclusion $\mathcal{E} \subset D_s$ implies that it is a dense subspace in $\mathcal{H}$. On the other hand, we have $(y, \dot{T}_s x) = \sum_{j=1}^{\infty} \overline{\eta}_j s_j \xi_j$ for any $y \in D_s$ and $x \in \mathcal{E}_{lin}$. The sum $\sum_j \eta_j \overline{s}_j e_j$ is convergent by assumption and determines a vector $y_s \in \mathcal{H}$, so the last relation can be written as $(y, \dot{T}_s x) = (y_s, x)$ for all $x \in \mathcal{E}_{lin}$; this means that $y \in D_s$. In this way, we have found the operator $\dot{T}_s^*$: it has the domain $D(\dot{T}_s^*) = D_s$ and acts on any $y := \sum_j \eta_j e_j \in D_s$ as $\dot{T}_s^* y = \sum_{j=1}^{\infty} \eta_j \overline{s}_j e_j$.

Since $D_s \neq \mathcal{E}_{lin}$, the operator $\dot{T}_s$ is not self-adjoint; it is symmetric *iff* $\{s_j\} \subset \mathbb{R}$. It is also obvious from the above argument that $\dot{T}_s$ extends naturally to the operator T_s defined by $T_s x := \sum_{j=1}^{\infty} \xi_j s_j e_j$ for any $x := \sum_j \xi_j e_j \in D_s =: D(T_s)$. We see that $\dot{T}_s^* = T_{\overline{s}}$ where $\overline{s} := \{\overline{s}_j\}$. Repeating the above argument we find $T_s^* = \dot{T}_s^*$ so $T_s^* = T_{\overline{s}}$; this means that T_s is self-adjoint *iff* $\{s_j\}$ is a real sequence. Other properties of the operators T_s can be found in Problem 6.

The operator T_s in the previous example is defined on the whole $\mathcal{H}$ only if it is bounded. Similarly the operator Q of Example 4.1.3 is not defined on the whole $L^2(\mathbb{R})$; for instance, the vector $\psi : \psi(x) = (1+x^2)^{-1/2}$ does not belong to $D(Q)$. This appears to be a common property of all unbounded symmetric operators (see also Problem 7).

4.1.5 Theorem (Hellinger–Toeplitz): Let A be a symmetric operator on $\mathcal{H}$ with $D_A = \mathcal{H}$; then A is bounded.

Proof: A symmetric operator defined on the whole $\mathcal{H}$ is necessarily self-adjoint and therefore closed (see Example 4.2.1b below); the assertion then follows from the closed-graph theorem. ∎

4.2 Closed operators

Now we are going to discuss how the general results on closed operators dealt with in Sections 1.6 and 1.7 can be specified for the set $\mathcal{C}(\mathcal{H})$ of closed operators on a given Hilbert space. Let us start with some simple examples.

4.2.1 Examples: (a) Suppose T is densely defined and $\{y_n\} \subset D(T^*)$ is a sequence such that $y_n \to y$ and $T^* y_n \to z$. Since $(y_n, Tx) = (T^* y_n, x)$ holds for any $x \in D_T$, the limit $n \to \infty$ yields $(y, Tx) = (z, x)$, so $y \in D(T^*)$ and $z = T^* y$. This means that the adjoint T^* is closed for any $T \in \mathcal{L}(\mathcal{H})$.

(b) Consider further a symmetric operator A. Since A^* is closed, A is closable, $A \subset \overline{A} \subset A^*$. Let $x, y \in D(\overline{A})$, so there is a sequence $\{x_n\} \subset D_A$ such that $x_n \to x$, $Ax_n \to \overline{A}x$, and an analogous sequence for the vector y. The relations $(y_n, Ax_n) = (Ay_n, x_n)$ then yield $(y, \overline{A}x) = (\overline{A}y, x)$; hence any symmetric operator has a symmetric closure. In particular, any self–adjoint operator is closed, $A = A^* = \overline{A}$.

Passing to the closure is a standard way of extending an operator. Of course, it is by far less universal than the continuous extension of bounded operators. First of all, the closure may not exist for a densely defined operator, as Problem 1.61 illustrates. Moreover, even if it exists its domain is in general still different from $\mathcal{H}$. For instance, the closure $\overline{A}$ of an unbounded symmetric operator A is again unbounded and symmetric, so $D(\overline{A}) \neq \mathcal{H}$ by Theorem 4.1.5.

Nevertheless, the closure represents a unique extension for any closable operator. This motivates the following definition: a symmetric operator A is called **essentially self–adjoint** (we use the abbreviation *e.s.a.*) if $\overline{A}$ is self–adjoint. Since any self–adjoint extension A' of a symmetric A is closed, we have $\overline{A} \subset A'$; if A is *e.s.a.* we get $A' = \overline{A}$. In other words, an essentially self–adjoint operator has a unique self–adjoint extension, namely its closure (see also Problem 8).

The converse is also true: a symmetric operator which has just one self–adjoint extension is *e.s.a.* (see Section 4.7 below). If an operator T is closed, any subspace $D \subset D(T)$ such that $\overline{T \restriction D} = T$ is called its *core* . In particular, D is a core for a self–adjoint A *iff* $A \restriction D$ is *e.s.a.*; we usually say briefly that A is *e.s.a.* on D.

4.2.2 Example (the operators T_s revisited): The relation $T_s = T_{\bar{s}}^*$ implies that T_s is closed for any sequence s. The operator $\dot{T}_s$, from which we started, fulfils $\dot{T}_s \subset T_s$, and therefore $\overline{\dot{T}_s} \subset T_s$. On the other hand, any $x \in D_s$ can be approximated by the vectors $x_n := \sum_{j=1}^n (e_j, x)e_j$; then $T_s x_n = \sum_{j=1}^n s_j(e_j, x)e_j \to T_s x$, which means that x belongs to the domain of $\overline{\dot{T}_s}$. Together we find that $\mathcal{E}_{lin}$ is a core for T_s; in particular, that $\dot{T}_s$ is *e.s.a.* if s is a real sequence.

Properties of the closure can be investigated using the notion of the operator graph introduced in Section 1.6. We define the unitary operator U on $\mathcal{H} \oplus \mathcal{H}$ by $U[x, y] := [-y, x]$; then the graph of T^* can be expressed as $\Gamma(T^*) = (U\Gamma(T))^\perp$ (Problem 9). Since the orthogonal complement is closed, we again arrive at the conclusion of Example 1a.

4.2.3 Theorem: Let $T \in \mathcal{L}(\mathcal{H})$; then $\overline{T}$ exists *iff* T^* is densely defined. If this condition is valid, we have $T^{**} = \overline{T}$ and $(\overline{T})^* = T^*$.

Proof: If $T^* \in \mathcal{L}(\mathcal{H})$ then T^{**} exists, and $T^{**} \supset T$, and since T^{**} is closed T is closable. On the contrary, let $\overline{T}$ exist, so $\Gamma(\overline{T}) = \overline{\Gamma(T)}$. Using the operator U introduced above, we have $U^2\Gamma = \Gamma$ and $U(\Gamma)^\perp = (U\Gamma)^\perp$ for any subspace $\Gamma \subset \mathcal{H} \oplus \mathcal{H}$, and therefore $\Gamma(\overline{T}) = \overline{\Gamma(T)} = \overline{U^2\Gamma(T)} = (U^2\Gamma(T))^{\perp\perp} = (U(U\Gamma(T))^\perp)^\perp$. Substituting $\Gamma(T^*)$ for $(U\Gamma(T))^\perp$ we get $\Gamma(\overline{T}) = (U\Gamma(T^*))^\perp$, and since $\Gamma(\overline{T})$ is a graph, T^* is densely defined and $\Gamma(\overline{T}) = \Gamma(T^{**})$ by Problem 9. The second relation follows from the first, and from the closedness of the operator T, because $(\overline{T})^* = T^{***} = (T^*)^{**} = \overline{T^*} = T^*$. ∎

The subset of all closed operators in $\mathcal{L}(\mathcal{H})$, *i.e.*, the intersection $\mathcal{L}(\mathcal{H}) \cap \mathcal{C}(\mathcal{H})$ is denoted as $\mathcal{L}_c(\mathcal{H})$. Theorem 4.2.3 implies the following simple relations between the spectral properties of T and T^*.

4.2.4 Corollary: Let $T \in \mathcal{L}_c(\mathcal{H})$; then

(a) $\lambda \in \sigma(T)$ *iff* $\overline{\lambda} \in \sigma(T^*)$,

(b) $((T-\mu)^{-1})^* = (T^* - \overline{\mu})^{-1}$ holds for any regular value μ of T.

Proof: If $\mu \in \rho(T)$ then Proposition 4.1.2c implies $\overline{\mu} \in \rho(T^*)$ and the identity of part (b). This also means that $\overline{\lambda} \in \sigma(T^*)$ implies $\lambda \in \sigma(T)$. By Theorem 4.2.3, $T^* \in \mathcal{L}(\mathcal{H})$ and $T^{**} = T$, so the argument can be repeated with T replaced by T^* thus giving assertion (a). ∎

4.2.5 Example (the operator P): Let $J \subset \mathbb{R}$ be an open interval with the endpoints a, b, $-\infty \le a < b \le \infty$; we want to construct a closed symmetric operator corresponding to the formal differential expression $-i\,\frac{d}{dx}$. The problem clearly consists of finding a suitable domain. It can contain only those $\psi \in L^2$ for which the derivative ψ' is finite a.e. and square integrable in J. The first requirement is fulfilled if ψ is supposed to be absolutely continuous in J (see Remark A.3.9). We denote

$$AC(J) := \left\{ \psi \in L^2(J) \,:\, \psi \text{ absolutely continuous in } J,\ \psi' \in L^2(J) \right\} ;$$

this subspace is dense in $L^2(J)$ containing, *e.g.*, $C_0^\infty(J)$ (*cf.* Example 1.5.3c). If ϕ, ψ belong to $AC(J)$, an easy integration by parts yields

$$(\phi, -i\psi') = -i[\phi,\psi] - (-i\phi', \psi)\,,$$

where we have denoted $[\phi,\psi] := \lim_{x\to b-} \overline{\phi(x)}\psi(x) - \lim_{x\to a+} \overline{\phi(x)}\psi(x)$. An analogous argument shows that $\alpha\,\frac{d}{dx}$ can correspond to a symmetric operator only if $\mathrm{Re}\,\alpha = 0$; with later applications of this operator in quantum mechanics in mind we choose $\alpha = -i$.

Introducing the operator $\tilde{P} : \tilde{P}\psi := -i\psi'$ with the domain $D(\tilde{P}) = AC(J)$, we can rewrite the above relation as $(\phi, \tilde{P}\psi) = -i[\phi,\psi] + (\tilde{P}\phi, \psi)$; this shows that $\tilde{P}$ is not symmetric unless $J = \mathbb{R}$ (Problem 12a); it is also clear that a densely defined operator $\dot{P} \subset \tilde{P}$ is symmetric if $[\phi,\psi] = 0$ holds for all ϕ, ψ of its domain. We shall show that $\dot{P}$ can be chosen in such a way that $\dot{P}^* = \tilde{P}$.

Let $\dot{D}$ be the set of all $\psi \in AC(J)$ with a compact support contained in J. The operator $\dot{P} := \tilde{P} \restriction \dot{D}$ is densely defined, because $C_0^\infty(J) \subset \dot{D}$. We have $[\phi,\psi] = 0$ for all $\phi \in AC(J)$, $\psi \in \dot{D}$, and therefore $(\phi, \dot{P}\psi) = (\tilde{P}\phi, \psi)$, which means that $\tilde{P} \subset \dot{P}^*$. To prove the opposite inclusion, one has to check that if $(\phi, \dot{P}\psi) = (\eta, \psi)$ holds for given $\phi, \eta \in L^2(J)$ and any $\psi \in \dot{D}$, then $\phi \in AC(J)$. We take a compact interval $K := [\alpha, \beta] \subset J$ and define on it $\eta_K^*(x) := \int_\alpha^x \eta(t)\,dt + c$, where the number c will be specified a little later. The function η_K^* is absolutely continuous in K and $(\eta_K^*)' = \chi_K \eta$ holds in the L^2 sense. Now let ψ_K be an arbitrary function from $\dot{D}$ with the support contained in K; then integrating the right side of $(\phi, \dot{P}\psi_K) = (\eta, \psi_K)$ by parts, we get $(\eta_K^* + i\phi, \psi_K') = 0$. Consider next the function η_K^{**}, which is defined by $\eta_K^{**}(x) := \int_\alpha^x (\eta_K^* + i\phi)(t)\, dt$ for $x \in K$ and equals zero outside this interval. We now choose the number c in such a way that $\eta_K^{**}(\beta) = 0$; then η_K^{**} belongs to $\dot{D}$ and we may set $\psi_K = \eta_K^{**}$ in which case the relation $(\eta_K^* + i\phi, \psi_K') = 0$ yields $\eta_K^* = -i\chi_K\phi$. Hence ϕ is absolutely continuous on K and $\phi'(x) = i\eta(x)$ for a.a. $x \in K$. Since K was an arbitrary compact interval in J, we finally arrive at the desired result, $\phi \in AC(J)$.

Thus the operator $\dot{P}$ defined above satisfies $\dot{P}^* = \tilde{P}$; it implies in particular that $\tilde{P}$ is closed. In view of Theorem 4.2.3, the closure $P := \overline{\dot{P}}$ has the same adjoint, $P^* = \tilde{P}$, and $\tilde{P}^* = P$. Now we want to show how the domain of P can be characterized by means of boundary conditions.

We have to distinguish the cases when J is a finite interval (a, b), or $J = (0, \infty)$ as a representative of semifinite intervals, or finally $J = \mathbb{R}$. In view of Problem 12, any $\psi \in AC(J)$ has one–sided limits at the endpoints of the interval; we denote them as $\psi(a), \psi(b)$. Let us define

$$D := \begin{cases} \{\psi \in AC(a,b) : \psi(a) = \psi(b) = 0\} \\ \{\psi \in AC(0,\infty) : \psi(0) = 0\} \\ AC(\mathbb{R}) \end{cases}$$

for the three cases, respectively. Theorem 4.2.3 readily implies that the operator P is unique in the following sense: if P_D is a closed symmetric operator such that $P_D^* = \tilde{P}$, then $P_D = P$. Using this fact we shall check that $D(P) = D$.

We set $P_D := \tilde{P} \upharpoonright D$. By Problem 12, this operator is symmetric and $P_D^* \supset \tilde{P}$; on the other hand, $\dot{P} \subset P_D$ implies $P_D^* \subset \tilde{P}$, so it remains for us to prove that P_D is closed. This is obvious for $J = \mathbb{R}$ when $P_D = \tilde{P}$ is self–adjoint. In the remaining two cases it is sufficient to check the inclusion $D(P_D^{**}) \subset D$. Let $\phi \in D(P_D^{**})$, *i.e.*, $(\phi, P_D^*\psi) = (P_D^{**}\phi, \psi)$ for all $\psi \in D(P_D^*) = AC(J)$. Since $P_D^* = \tilde{P}$ and the inclusion $P_D \subset \tilde{P}$ together with the closedness of $\tilde{P}$ imply $P_D^{**} \subset \tilde{P}$ we can rewrite this condition as $(\phi, \tilde{P}\psi) = (\tilde{P}\phi, \psi)$; from this we see that $[\phi, \psi] = 0$ must hold for all $\psi \in AC(J)$. The functions ψ may assume any values at the finite endpoints of J; hence $\phi(a) = \phi(b) = 0$ must be valid for the finite interval and $\phi(0) = 0$ for $J = (0, \infty)$, *i.e.*, $\phi \in D$.

Concluding the above argument, we have shown that to any interval $J \subset \mathbb{R}$ the formal expression $-i\frac{d}{dx}$ determines just one unbounded closed symmetric operator $P \subset \tilde{P}$ which satisfies the condition $P^* = \tilde{P}$; its domain is $D(P) = D$.

In the case of a finite or semi–infinite J the operator P is not self–adjoint, so we have to ask whether it has nontrivial symmetric extensions. If $J = (0, \infty)$ the answer is *negative:* the domain of such an extension should contain a function $\psi \in AC(J)$ with $\psi(0) \neq 0$; however, then $[\psi, \psi] \neq 0$, so the operator cannot be symmetric. The operator P on $L^2(0, \infty)$ thus represents an example of a maximal symmetric operator which is *not* self–adjoint.

On the other hand, in the case of a finite $J = (a, b)$ the condition $[\psi, \psi] = |\psi(b)|^2 - |\psi(a)|^2 = 0$ can be fulfilled provided $\psi(b) = \theta\psi(a)$ where $|\theta| = 1$. Moreover, if we define $D_\theta := \{\psi \in AC(J) : \psi(b) = \theta\psi(a)\}$ then $[\phi,\psi] = 0$ holds for any $\phi,\psi \in D_\theta$. Hence there is a bijective correspondence between symmetric extensions of the operator P on $L^2(a,b)$ and complex numbers with $|\theta| = 1$; each of these operators $P_\theta := \tilde{P} \restriction D_\theta$ is self–adjoint.

In Section 1.7 we described how the spectrum of a closed operator T can be classified. If T is a Hilbert–space operator it is useful to introduce another subset of $\sigma(T)$. It is called the *essential spectrum* and denoted as $\sigma_{ess}(T)$; it consists of all $\lambda \in \mathbb{C}$ to which there is a sequence of unit vectors $x_n \in D_T$ which has no convergent subsequence and satisfies $(T-\lambda)x_n \to 0$.

4.2.6 Proposition: The spectrum of an operator $T \in \mathcal{C}(\mathcal{H})$ decomposes as $\sigma(T) = \sigma_p(T) \cup \sigma_r(T) \cup \sigma_{ess}(T)$. Moreover, $\sigma_c(T) = \sigma_{ess}(T) \setminus (\sigma_p(T) \cup \sigma_r(T))$.

Proof: If $\lambda \in \rho(T)$ the resolvent $R_T(\lambda)$ is bounded and $\|(T-\lambda)x\| \geq \|R_T(\lambda)\|^{-1}\|x\|$ holds for any $x \in D_T$; from here it follows that $\lambda \neq \sigma_{ess}(T)$, and therefore $\sigma_{ess}(T) \subset \sigma(T)$. Now let λ be an arbitrary point of the spectrum; if there is $c_\lambda > 0$, such that $\|(T-\lambda)x\| \geq c_\lambda\|x\|$ for any $x \in D_T$, then $\mathrm{Ran}\,(T-\lambda)$ is a closed subspace by Problem 1.64 and the condition $\lambda \in \sigma(T)$ implies $\mathrm{Ran}\,(T-\lambda) \neq \mathcal{H}$, *i.e.*, $\lambda \in \sigma_r(T)$. In the opposite case $\inf\{\,\|(T-\lambda)x\| \,:\, x \in D_T,\ \|x\| = 1\} = 0$, so there is a sequence of unit vectors $x_n \in D_T$ such that $(T-\lambda)x_n \to 0$. If some subsequence fulfils $x_{n_k} \to x$, then $\|x\| = 1$ and $Tx_{n_k} \to \lambda x$, and since the operator T is closed, we get $x \in D_T$ and $Tx = \lambda x$, *i.e.*, $\lambda \in \sigma_p(T)$. On the other hand, if there is no convergent subsequence, $\lambda \in \sigma_{ess}(T)$ by definition. This proves the first decomposition; combining it with the fact that the decomposition (1.6) is disjoint we get the remaining assertion. ∎

In distinction to (1.6), the decomposition of $\sigma(T)$ derived here is in general not disjoint. For instance, any eigenvalue of T of an infinite multiplicity belongs to $\sigma_{ess}(T)$ because the corresponding normalized eigenvectors form an orthonormal set from which it is impossible to select a convergent subsequence. Notice also that the last mentioned condition can be replaced in the definition of $\sigma_{ess}(T)$ by the requirement that the set $\{x_n : n = 1,2,\dots\}$ is noncompact. The nonexistence of a convergent subsequence easily implies noncompactness; on the other hand, if the set is noncompact it contains an infinite subset $\{x_{n_k} : k = 1,2,\dots\}$ without an accumulation point; hence the sequence $\{x_{n_k}\}$ has no convergent subsequence. Let us finally mention a remarkable property of densely defined closed operators.

4.2.7 Theorem: If $T \in \mathcal{L}_c(\mathcal{H})$, then T^*T is a positive self–adjoint operator and $D(T^*T)$ is a core for T; similarly TT^* is positive self–adjoint and $D(TT^*)$ is a core for T^*.

Proof: We have $(x, T^*Tx) = \|Tx\|^2 \geq 0$ for any $x \in D(T^*T)$, so T^*T is positive. The relation $T^{**} = T$ in combination with Problem 9 gives $\Gamma(T)\oplus U\Gamma(T^*) = \mathcal{H}\oplus\mathcal{H}$, so to any $x \in \mathcal{H}$, there are vectors $u \in D(T)$ and $v \in D(T^*)$ such that $[x,0] = [u,Tu] + [-T^*v, v]$. It follows that $u \in D(T^*T)$ and $x = (I+T^*T)u$. The vector x is arbitrary, which means that the range of the operator $S := I+T^*T$ is the whole

$\mathcal{H}$. Since T^*T is positive, S is invertible; using the fact that $(y, Sx) = (Sy, x)$ holds for all $x, y \in D(T^*T)$ we find that S^{-1} is symmetric. In addition, it is bounded by Theorem 4.1.5, because $D(S^{-1}) = \operatorname{Ran} S = \mathcal{H}$; thus S is self-adjoint by Problem 3 and the same is true for $T^*T = S - I$.

Next we denote $T_D := T \upharpoonright D(T^*T)$; then the second assertion is valid *iff* $\Gamma(T) = \overline{\Gamma(T_D)}$, and in view of the inclusion $T_D \subset T$ and the closedness of T, this is further equivalent to the requirement that the orthogonal complement of $\Gamma(T_D)$ in $\Gamma(T)$ consists of the zero vector only. Suppose that $([x, Tx], [y, Ty]) = 0$ holds for some $x \in D(T)$ and all $y \in D(T^*T)$. It follows that $0 = (x, (I+T^*T)y) = (x, Sy)$, and since $\operatorname{Ran} S = \mathcal{H}$, we have $x = 0$. Finally, $T \in \mathcal{L}_c(\mathcal{H})$ implies $T^* \in \mathcal{L}_c(\mathcal{H})$ and $T^{**} = T$, so the claim remains valid after the interchange of T and T^*. ∎

4.3 Normal operators. Self-adjointness

The notion of a normal operator introduced in Section 3.4 also extends to unbounded operators; we say that an operator T on $\mathcal{H}$ is **normal** if it is densely defined, closed, and satisfies $T^*T = TT^*$. The set of all normal operators on a given $\mathcal{H}$ will be denoted as $\mathcal{L}_n(\mathcal{H})$. The roles of T and T^* in the definition can be switched, so if T is normal the same is true for T^* and vice versa. Any self-adjoint operator is obviously normal, *i.e.*, $\mathcal{L}_{sa}(\mathcal{H}) \subset \mathcal{L}_n(\mathcal{H}) \subset \mathcal{L}_c(\mathcal{H})$. The definition includes the requirement of equality of the domains, $D(T^*T) = D(TT^*)$, which is not always easy to check. For this reason, the following criterion is useful.

4.3.1 Theorem: A densely defined closed operator T is normal *iff* $D(T) = D(T^*)$ and $\|Tx\| = \|T^*x\|$ holds for all $x \in D(T)$.

Proof: Suppose that $T^*T = TT^*$; then the common domain $D := D(T^*T) = D(TT^*)$ is by Theorem 4.2.7 a core for both T and T^*, and since $\|Tx\| = \|T^*x\|$ holds for all $x \in D$, the same is true on $D(T) = D(T^*)$ (Problem 11c). Conversely, let $\|Tx\| = \|T^*x\|$ hold for any $x \in \tilde{D} := D(T) = D(T^*)$, so $(Tx, Ty) = (T^*x, T^*y)$ for all $x, y \in \tilde{D}$ by the polarization formula. If $x \in D(T^*T)$, *i.e.*, $x \in \tilde{D}$ and $Tx \in \tilde{D}$, then $(Tx, Ty) = (T^*Tx, y)$ so we have $(T^*Tx, y) = (T^*x, T^*y)$ for any $y \in \tilde{D} = D(T^*)$. Hence $T^*x \in D(T^{**}) = D(T) = \tilde{D}$ and $TT^*x = T^*Tx$; in other words, $T^*T \subset TT^*$. Finally, interchanging the roles of T and T^*, we find that T is normal. ∎

4.3.2 Remark: In fact, the closedness of T need not be assumed (*cf.* Problem 11b). Any normal operator T is maximal in the sense that it has no nontrivial normal extension: if $S \in \mathcal{L}_n(\mathcal{H})$ fulfils $S \supset T$, then $D(S) = D(S^*) \subset D(T^*) = D(T)$.

4.3.3 Example (the operators T_f): Let $(X, \mathcal{A}, \mu)$ be a measure space with a σ–finite measure. Given a measurable function $f : X \to \mathbb{C}$, we set

$$D_f := \left\{ \psi \in L^2(X, d\mu) \,:\, \int_X |f|^2 |\psi|^2 d\mu < \infty \right\}$$

and define T_f with the domain $D(T_f) := D_f$ by $(T_f\psi)(x) := f(x)\psi(x)$; we call it the *operator of multiplication* by f. Such operators play an important role in spectral theory as we shall see in Section 5.3. Let us list their basic properties, which follow from the general results derived up to now:

(a) T_f *is densely defined.* To check this, we use the nondecreasing sequence of sets $M_n^f := \{x \in X : |f(x)| \le n\}$, $n = 1, 2, \dots$, with $\bigcup_{n=1}^{\infty} M_n^f = X$. Let χ_n be the characteristic function of M_n^f; then for any vector $\psi \in L^2(X, d\mu)$ the function $\chi_n \psi$ belongs to D_f and $\|\chi_n\psi - \psi\| \to 0$ by the dominated–convergence theorem, *i.e.*, $\overline{D}_f = L^2(X, d\mu)$.

(b) $T_f = T_g$ *iff* $f(x) = g(x)$ *for* μ*–a.a.* $x \in X$. The sufficient condition is obvious. To check the necessary condition, we use the sets $N_n := M_n^f \cap M_n^g$. Their characteristic functions again belong to $D_f = D_g$, and since $\bigcup_n N_n = X$, it is enough to show that $f(x) = g(x)$ for μ–a.a. $x \in N_n$; this follows easily from the relation $\|(T_f - T_g)\chi_{N_n}\| = 0$.

(c) Assertion (a) implies existence of T_f^*. Mimicking the argument of Example 4.1.3 we check that $T_f^* = T_{\bar{f}}$; the two operators have the same domain D_f and $\|T_f\psi\| = \|T_{\bar{f}}\|$ for any $\psi \in D_f$, so

   (i) T_f is normal.

   (ii) T_f is self–adjoint *iff* $f(x) \in \mathbb{R}$ a.e. in X.

(d) If $f \in L^\infty(X, d\mu)$, we have $D_f = L^2(X, d\mu)$ and $\|T_f\psi\|^2 = \int_X |f|^2|\psi|^2 d\mu \le \|f\|_\infty^2 \|\psi\|^2$ for any $\psi \in L^2$, so T_f is bounded and $\|T_f\| \le \|f\|_\infty^2$. To show that the converse is also true we have to check that for a given $T_f \in \mathcal{B}(L^2(X, d\mu))$, the set $N := \{x \in X : |f(x)| > \|T_f\|\}$ is μ–zero. Let $R \in \mathcal{A}$ be any set with $\mu(R) < \infty$, then $\|T_f \chi_{R\cap N}\|^2 \le \|T_f\|^2 \mu(R \cap N)$; in other words, $\int_X (|f|^2 - \|T_f\|^2)\chi_{R\cap N} d\mu \le 0$. But $|f(x)| > \|T_f\|$ on N, so the inequality can hold only if $\mu(R \cap N) = 0$; since μ is σ–finite we arrive at the desired result. We see that the condition $f \in L^\infty$ is necessary and sufficient for T_f to be bounded; if it is valid we have $\|T_f\| = \|f\|_\infty$.

Let us now turn to spectral properties of normal operators. In the same way as in Section 3.4, we can check that $\lambda \in \sigma_p(T)$ *iff* $\overline{\lambda} \in \sigma_p(T^*)$ and the corresponding eigenspaces coincide,

$$\operatorname{Ker}(T - \lambda) = \operatorname{Ker}(T^* - \overline{\lambda}). \tag{4.1}$$

The eigenspaces corresponding to different eigenvalues, $\lambda \ne \mu$, are mutually orthogonal, $\operatorname{Ker}(T - \lambda) \perp \operatorname{Ker}(T - \mu)$. Furthermore, Proposition 4.1.1b implies that the residual spectrum is empty, $\sigma_r(T) = \emptyset$, and $\lambda \in \sigma_p(T)$ *iff* $\overline{\operatorname{Ran}(T - \lambda)} \ne \mathcal{H}$. Theorem 3.4.3 also extends to the unbounded case.

4.3.4 Theorem: The resolvent set of a normal operator T coincides with its regularity domain, *i.e.*, $\lambda \in \rho(T)$ *iff* there is $c(\lambda)$ such that $\|(T - \lambda)x\| \ge c(\lambda)\|x\|$ holds for all $x \in D(T)$; this is further equivalent to $\operatorname{Ran}(T - \lambda) = \mathcal{H}$.

Proof: The argument of Theorem 3.4.3 can be followed up to the point where we have to check that $\mathrm{Ran}\,(T-\lambda)$ is closed. Here we must use the fact that T is closed (see Problem 1.64); due to this $\mathrm{Ran}\,(T-\lambda)=\mathcal{H}$ implies $(T-\lambda)^{-1}\in\mathcal{B}(\mathcal{H})$. ∎

4.3.5 Corollary: (a) Let $T\in\mathcal{L}_n(\mathcal{H})$; then the following conditions are equivalent:

(i) $\lambda\in\sigma(T)$.

(ii) $\inf\{\,\|(T-\lambda)x\|\,:\,x\in D(T),\ \|x\|=1\,\}=0$.

(iii) There is a sequence of unit vectors $x_n\in D(T)$ such that $(T-\lambda)x_n\to 0$.

(iv) $\mathrm{Ran}\,(T-\lambda)\neq\mathcal{H}$.

(b) The spectrum of a self–adjoint operator A lies on the real axis. In particular, if A is below bounded then $\inf\sigma(A)\geq\inf\{\,(x,Ax)\,:\,x\in D(A),\ \|x\|=1\}$ (in fact, an identity is valid — *cf.* Proposition 5.4.1a below).

4.3.6 Example (spectrum of the resolvent for $T\in\mathcal{L}_n(\mathcal{H})$): Let μ be a regular value of a normal operator T. Since $R_T(\mu)$ is normal (Problem 18a) we can use the above results to determine its spectrum. First of all, $0\in\rho(R_T(\mu))$ *iff* $\mathcal{H}=\mathrm{Ran}\,(R_T(\mu))=D(T)$, and this is in turn equivalent to the boundedness of T; hence the spectrum of $R_T(\mu)$ contains zero *iff* T is unbounded. Further, let $\lambda\neq 0$ and $y\in\mathrm{Ran}\,(R_T(\mu)-\lambda)$, *i.e.*, $y=(R_T(\mu)-\lambda)x$ for some $x\in\mathcal{H}$. The vector $z:=-\lambda R_T(\mu)x$ belongs to D_T and satisfies the identity $y=(T-\mu-\lambda^{-1})z$, which means that $y\in\mathrm{Ran}\,(T-\mu-\lambda^{-1})$. The same argument in the reverse order then yields the identity $\mathrm{Ran}\,(R_T(\mu)-\lambda)=\mathrm{Ran}\,(T-\mu-\lambda^{-1})$, and by Corollary 4.3.5a, $\lambda\in\sigma(R_T(\mu))$ *iff* $\mu+\lambda^{-1}\in\sigma(T)$. This gives the sought expression for $\sigma(R_T(\mu))$ of a normal operator T,

$$\sigma(R_T(\mu))=\begin{cases}\{\,\lambda:\,\lambda=(\nu-\mu)^{-1},\ \nu\in\sigma(T)\,\} & \ldots\ T\text{ bounded}\\ \{\,\lambda:\,\lambda=(\nu-\mu)^{-1},\ \nu\in\sigma(T)\,\}\cup\{0\} & \ldots\ T\text{ unbounded}\end{cases}$$

Using further the fact that the spectral radius of a bounded normal operator equals its norm (see Problem 3.29) we get

$$\|R_T(\mu)\|=\sup\left\{\frac{1}{|\nu-\mu|}\,:\,\nu\in\sigma(T)\right\}. \tag{4.2}$$

In the same way we can find the spectrum of the operator $TR_T(\mu)$, which is again normal and bounded; we have $\sigma(TR_T(\mu))=\{\,\lambda:\,\lambda=\nu(\nu-\mu)^{-1},\ \nu\in\sigma(T)\,\}$, in particular $\|TR_T(\mu)\|=\sup_{\nu\in\sigma(T)}|\nu(\nu-\mu)^{-1}|$.

4.3.7 Example (the operators T_f revisited): Now we want to find the spectrum of the multiplication operators T_f of Example 4.3.3. Given a measurable function $f:X\to\mathbb{C}$ we put $R^{(\mu)}_{ess}(f):=\{\,\lambda\in\mathbb{C}:\,\mu(f^{(-1)}((\lambda-\varepsilon,\lambda+\varepsilon)))\neq 0\ \text{ for any }\varepsilon>0\,\}$;

this set is called the *essential range* of the function f (with respect to μ; the index μ will be omitted when it is clear from the context). We shall prove that

$$\sigma(T_f) = R_{ess}(f)\,.$$

Let $\lambda \in R_{ess}(f)$. We denote $M_\varepsilon := f^{(-1)}((\lambda-\varepsilon, \lambda+\varepsilon))$; since μ is σ–finite we may assume $\mu(M_\varepsilon) < \infty$ without loss of generality. The function f is bounded on M_ε, so $\psi_\varepsilon := \mu(M_\varepsilon)^{-1}\chi_{M_\varepsilon}$ belongs to D_f. It has a unit norm and $\|(T_f - \lambda)\psi_\varepsilon\|^2 = (\mu(M_\varepsilon))^{-1}\int_{M_\varepsilon}|f-\lambda|^2 d\mu < \varepsilon^2$, and since ε was arbitrary, $\lambda \in \sigma(T_f)$ by Corollary 4.3.5a. On the other hand, if $\lambda \notin R_{ess}(f)$ then there is $\varepsilon > 0$ such that $\mu(M_\varepsilon) = 0$; hence we have $\int_X |\psi|^2 d\mu = \int_{X\setminus M_\varepsilon}|\psi|^2 d\mu$ for any $\psi \in L^2(X, d\mu)$, and consequently

$$\|(T_f-\lambda)\psi\|^2 = \int_{X\setminus M_\varepsilon}|f(t)-\lambda|^2|\psi(t)|^2 d\mu(t) \geq \varepsilon^2\int_{X\setminus M_\varepsilon}|\psi|^2 d\mu = \varepsilon^2\|\psi\|^2$$

for all $\psi \in D_f$, *i.e.*, $\lambda \notin \sigma(T_f)$ by Theorem 4.3.4. In particular, if μ is the Lebesgue measure on $\mathbb{R}^n$ and $f : \mathbb{R}^n \to \mathbb{C}$ is continuous, then $\sigma(T_f) = \overline{\mathrm{Ran}\, T_f}$.

In Section 3.4 we introduced the notion of pure point spectrum for $B \in \mathcal{B}(\mathcal{H})$. When the operators in question are unbounded, we have to pay attention to their domains; for instance, the operators $\dot{T}_s$ and T_s from Example 4.1.4 have the same orthonormal basis of eigenvectors but, of course, $\dot{T}_s \neq T_s$. In the general case, we therefore introduce the notion for normal operators only: we say that an operator $T \in \mathcal{L}_n(\mathcal{H})$ has a *pure point spectrum* if there is an orthonormal basis $\mathcal{E} := \{e_\alpha\}_{\alpha\in I} \subset D(T)$ consisting of eigenvectors of T, *i.e.*, $Te_\alpha = \lambda_\alpha e_\alpha$.

4.3.8 Proposition: If T is a normal operator with a pure point spectrum, then $\sigma(T) = \overline{\sigma_p(T)}$ and $D(T) = \left\{ x \in \mathcal{H} : \sum_{\alpha\in I}|(e_\alpha, x)|^2|\lambda_\alpha|^2 < \infty \right\}$.

Proof: The assertion about the spectrum is proved in the same way as in Theorem 3.4.7. Denote the right side of the latter relation as $D_\mathcal{E}$. If $x \in D(T)$ then $\|Tx\|^2 = \sum_\alpha|(e_\alpha, Tx)|^2 = \sum_\alpha|(T^*e_\alpha, x)|^2 = \sum_\alpha|\lambda_\alpha|^2|(e_\alpha, x)|^2$, so $D(T) \subset D_\mathcal{E}$. In view of Proposition 2.1.6a and the Hölder inequality, $D_\mathcal{E}$ is a subspace in $\mathcal{H}$; the relation $\tilde{T}x := \sum_\alpha \lambda_\alpha(e_\alpha, x)e_\alpha$ defines on it an operator which satisfies $\tilde{T} \supset T$. Mimicking the argument of Example 4.1.4 we can check that $D(\tilde{T}^*) = D_\mathcal{E}$ and $\tilde{T}^*x = \sum_\alpha \overline{\lambda}_\alpha(e_\alpha, x)e_\alpha$, so $\|\tilde{T}x\| = \|\tilde{T}^*x\|$ holds for all $x \in D_\mathcal{E}$. Hence $\tilde{T}$ is normal, and therefore $\tilde{T} = T$ by Remark 4.3.2, *i.e.*, $D(T) = D_\mathcal{E}$. ∎

Let us next mention an important application of Theorem 4.3.4.

4.3.9 Theorem (self–adjointness criterion): Let A be a symmetric operator; then the following conditions are equivalent:

(a) A is self–adjoint.

(b) A is closed and $\mathrm{Ker}\,(A^* \mp i) = \{0\}$.

(c) $\mathrm{Ran}\,(A \pm i) = \mathcal{H}$;

the numbers $\pm i$ can be replaced by the pair $z, \overline{z}$ for any $z \in \mathbb{C} \setminus \mathbb{R}$.

Proof: A self-adjoint operator A is closed and satisfies $\mathrm{Ker}\,(A^* \mp i) = \{0\}$ because $A^* \mp i = A \mp i$ and $\sigma_p(A) \subset \sigma(A) \subset \mathbb{R}$; thus (a) implies (b). If (b) is valid, then $\overline{\mathrm{Ran}\,(A \pm i)} = \mathcal{H}$ by Proposition 4.1.1b and $\|(A \pm i)x\| \geq \|x\|$ holds for any $x \in D(A)$ (Problem 4); in combination with the closedness of A this implies that $\mathrm{Ran}\,(A \pm i)$ is closed (*cf.* Problem 1.64). Finally, let (c) be valid, and denote $z := (A^* + i)y$ for any $y \in D(A^*)$. Since $\mathrm{Ran}\,(A + i) = \mathcal{H}$ and $A \subset A^*$, there is a vector $y_0 \in D(A)$ such that $z = (A^* + i)y_0$. Then $y - y_0 \in \mathrm{Ker}\,(A^* + i) = (\mathrm{Ran}\,(A - i))^\perp = \{0\}$ so $y = y_0$; this implies $D(A) = D(A^*)$. ∎

4.3.10 Corollary (essential self-adjointness criterion): Let A be symmetric; then the following conditions are equivalent:

(a) A is *e.s.a.*.

(b) $\mathrm{Ker}\,(A^* \mp i) = \{0\}$.

(c) $\mathrm{Ran}\,(A \pm i)$ is dense in $\mathcal{H}$;

the numbers $\pm i$ can again be replaced by any nonreal $z, \overline{z}$.

Proof is left to the reader (Problem 20).

4.3.11 Example (operator Q revisited): Denote $Q_0 := Q \upharpoonright C_0^\infty(\mathbb{R})$. In view of (1.2), to any $\psi \in L^2(\mathbb{R})$ and $\varepsilon > 0$ there is $\phi \in C_0^\infty(\mathbb{R})$ such that $\|\psi - \phi\| < \varepsilon$. The functions $\eta_\pm : \eta_\pm(x) = \phi(x)(x \pm i)^{-1}$ then also belong to $C_0^\infty(\mathbb{R})$, so $\|\psi - (Q_0 \pm i)\eta_\pm\| = \|\psi - \phi\| < \varepsilon$, *i.e.*, $\overline{\mathrm{Ran}\,(Q_0 \pm i)} = L^2(\mathbb{R})$. Hence the operator Q is *e.s.a.* on $C_0^\infty(\mathbb{R})$; this is also true for other operators (Problem 17).

It is often difficult to check the self-adjointness of a given operator using either the definition or the above criterion. One way to counter this problem is based on a perturbative approach: we have a self-adjoint operator A and look for a class of symmetric S such that $A + S$ is self-adjoint. A simple example is given in Problem 3d; however, self-adjointness is stable with respect to a much wider class of symmetric perturbations. Let us mention one frequently used theorem at this point; other results of this type will be discussed later.

Let A, S be linear operators on $\mathcal{H}$; we say that S is *relatively bounded with respect to* A (or briefly, *A–bounded*) if $D(S) \supset D(A)$ and there are non-negative a, b such that the relation

$$\|Sx\| \leq a\|Ax\| + b\|x\| \tag{4.3}$$

holds for all $x \in D(A)$. The infimum of all $a \geq 0$ for which there is $b \geq 0$ such that the above inequality is valid is called the *relative bound* of S with respect to A, or the *A–bound* of the operator S; notice that the inequality may not hold if we substitute for a the A–bound itself (Problem 21). It is useful sometimes to work with squared norms. It is clear that the condition

$$\|Sx\|^2 \leq \alpha^2\|Ax\|^2 + \beta^2\|x\|^2$$

with $\alpha = a$ and $\beta = b$ implies (4.3), and this in turn implies the above inequality with $\alpha^2 = (1+\varepsilon)a^2$ and $\beta^2 = (1+\varepsilon^{-1})b^2$ for any $\varepsilon > 0$, so the infimum of all $\alpha \geq 0$ for which it is valid with some β coincides with the A–bound of S.

4.3.12 Theorem (Kato–Rellich): Suppose that A is self–adjoint, and S is symmetric and A–bounded with the A–bound less than one; then

(a) $A+S$ is self–adjoint.

(b) If $D \subset D(A)$ is a core for A then $A+S$ is also *e.s.a.* on D.

Proof: By assumption, $A+S$ is symmetric, $D(A+S) = D(A)$, and the "quadratic" inequality holds for some $\alpha < 1$. In view of Problem 3d, we may assume $\alpha \neq 0$; we set $\gamma := \beta/\alpha$ and rewrite the condition in the form $\|Sx\|^2 \leq \alpha^2 \|(A \mp i\gamma)x\|^2$ (*cf.* Problem 4). The numbers $\pm i\gamma$ are thus regular values of the operator A, so $D(A) = R_A(\pm i\gamma)\mathcal{H}$. Substituting $x = R_A(\pm i\gamma)y$ in the condition, we get for any $y \in \mathcal{H}$ the relation $\|SR_A(\pm i\gamma)y\| \leq \alpha\|y\|$. Hence the operators $B_\pm := SR_A(\pm i\gamma)$ are bounded with $\|B_\pm\| \leq \alpha < 1$, and the $B_\pm + I$ are regular by Lemma 1.7.2. This means that for any $z \in \mathcal{H}$ there are $y_\pm \in \mathcal{H}$ such that $z = (B_\pm + I)y_\pm$, and to them in turn we can find $x_\pm \in D(A)$ such that $(A \mp i\gamma)x_\pm = y_\pm$. Then $z = (SR_A(\pm i\gamma) + I)(A \mp i\gamma)x_\pm = (S + A \mp i\gamma)x_\pm$, *i.e.*, $\mathrm{Ran}\,(S + A \mp i\gamma) = \mathcal{H}$ and $A+S$ is self–adjoint due to Theorem 4.3.9.

Now let D be a core for A; then for any $x \in D(A)$ there is a sequence $\{x_n\} \subset D$ such that $x_n \to x$ and $Ax_n \to Ax$. Due to the assumption, we get $\|(A+S)(x_n - x)\|^2 \leq 2\|A(x_n-x)\|^2 + 2\|S(x_n-x)\|^2 \leq 2(1+\alpha^2)\|A(x_n-x)\|^2 + 2\beta^2\|x_n - x\|^2$, and since $D(A+S) = D(A)$ we obtain $\overline{(A+S) \upharpoonright D} \supset A+S$. On the other hand, $(A+S) \upharpoonright D \subset A+S$ and $A+S$ is closed as we have already proven, so the opposite inclusion is valid too. ∎

4.4 Reducibility. Unitary equivalence

Let T be a linear operator on $\mathcal{H}$ with the domain D_T. If $\mathcal{G}$ is a closed T–invariant subspace, then $T \upharpoonright \mathcal{G}$ is an operator on $\mathcal{G}$; we call it the *part of* T *in* $\mathcal{G}$. Suppose further that subspaces $\mathcal{G}_1$ and $\mathcal{G}_2 := \mathcal{G}_1^\perp$ are both T–invariant, and denote by T_j the part of T in $\mathcal{G}_j$, $j = 1,2$, then we define the operator $T_1 \oplus T_2$ with the domain $D(T_1 \oplus T_2) := (D_T \cap \mathcal{G}_1) \oplus (D_T \cap \mathcal{G}_2)$ by $(T_1 \oplus T_2)(x_1 + x_2) := T_1x_1 + T_2x_2$ for $x_j \in D_T \cap \mathcal{G}_j$; we call it the *orthogonal sum* of T_1 and T_2. We obviously have $T_1 \oplus T_2 = T$ if $D_T = \mathcal{H}$, while in the general case only the inclusion $T_1 \oplus T_2 \subset T$ is valid unless the projections E_j to $\mathcal{G}_j$ map the domain D_T to itself.

4.4.1 Example: Consider the operator P from Example 4.2.5 on $L^2(-1,1)$ and let E be the operator of multiplication by $\chi_{(0,1)}$. The subspaces $\mathrm{Ran}\,E$ and $(\mathrm{Ran}\,E)^\perp$ are obviously P–invariant, but $E\psi$ belongs to D_P for $\psi \in D_P$ only if $\psi(0) = 0$.

This motivates the following definition: a linear operator T on $\mathcal{H}$ is **reducible** if there is a nontrivial projection E such that (i) $ED_T \subset D_T$, and (ii) the subspaces

$E\mathcal{H}$ and $(I-E)\mathcal{H}$ are T–invariant; in the opposite case T is **irreducible**. If T satisfies the above conditions, we also say that it is *reduced* by the projection E or by the subspace $E\mathcal{H}$. It is clear that T is reduced by $E\mathcal{H}$ *iff* it is reduced by $(E\mathcal{H})^\perp$. Moreover, in the notation used above $T = T_1 \oplus T_2$ *iff* the subspace $\mathcal{G}_1 =: E_1\mathcal{H}$ reduces T; in that case $D(T_j) := D_T \cap \mathcal{G}_j = E_j D_T$, where $E_2 := I - E_1$. This results easily extends to any finite number of orthogonal subspaces; for an infinite orthogonal sum an extra condition is needed (Problem 23).

4.4.2 Theorem (reducibility criterion): A linear operator T is reduced by a projection E *iff* $ET \subset TE$; this is further equivalent to the conditions $ED_T \subset D_T$ and $ETx = TEx$ for all $x \in D_T$.

Proof: Let T be reduced by E. We take any $x \in D_T$; by assumption the vectors Ex, $(I-E)x$ belong to D_T and $TEx \in E\mathcal{H}$, $T(I-E)x \in (E\mathcal{H})^\perp$, so $ETx = ETEx + ET(I-E)x = ETEx = TEx$. On the other hand, the condition $ET \subset TE$ implies $ED_T \subset D_T$ and $Tx = TEx = ETx \in E\mathcal{H}$ for any $x \in E\mathcal{H} \cap D_T$; in the same way we check T–invariance of the subspace $(I-E)\mathcal{H}$. ∎

4.4.3 Example (reduction of a normal operator): Let λ be an eigenvalue of a normal operator T. Since T is closed the corresponding eigenspace $\mathrm{Ker}\,(T-\lambda)$ is also closed, and furthermore, $\mathrm{Ker}\,(T-\lambda) \subset D_T$ implies that the projection $E(\lambda)$ to the eigenspace satisfies $E(\lambda)D_T \subset D_T$. The subspace $\mathrm{Ker}\,(T-\lambda)$ is obviously T–invariant, and since $(Ty, x) = (y, T^*x) = \overline{\lambda}(y, x) = 0$ is valid for any $y \in (\mathrm{Ker}\,(T-\lambda))^\perp \cap D_T$ and $x \in \mathrm{Ker}\,(T-\lambda)$, the orthogonal complement $(\mathrm{Ker}\,(T-\lambda))^\perp$ is also T–invariant. Hence a normal operator is reduced by any of its eigenspaces.

If μ, λ are different eigenvalues of T, the corresponding eigenspaces are orthogonal, $E(\mu)E(\lambda) = 0$. If $\sigma_p(T) := \{\lambda_1, \dots, \lambda_n\}$ is finite, then $E_n := \sum_{j=1}^n E(\lambda_j)$ is the projection to $\bigoplus_{j=1}^n \mathrm{Ker}\,(T-\lambda_j)$, and each of the projections $E(\lambda_j)$ reduces T; the same is true for E_n. Further, let $\sigma_p(T) := \{\lambda_j : j = 1, 2, \dots\}$ be countably infinite and denote by E the projection to $\mathcal{H}_p := \bigoplus_{j=1}^\infty \mathrm{Ker}\,(T-\lambda_j)$, *i.e.*, $E = \sum_{j=1}^\infty E(\lambda_j) := \text{s-lim}_{n\to\infty} E_n$. Since $EE_n = E_nE = E_n$ for any n, we have $y_n := E_nEx = E_nx \in D_T$ for each $x \in D_T$ and $y_n \to y$. At the same time, E_n reduces T, so $Ty_n = TE_nx = E_nTx \to ETx$; due to the closedness of T we get $Ex \in D_T$, *i.e.*, $ED_T \subset D_T$ and $TEx = ETx$. Hence the subspace $\mathcal{H}_p$ reduces the operator T, and its parts $T_p := T \restriction \mathcal{H}_p$ and $T_c := T \restriction \mathcal{H}_p^\perp$ are normal operators on Hilbert spaces $\mathcal{H}_p$ and $\mathcal{H}^\perp$, respectively (Problem 24b). Notice that the assumption that $\sigma_p(T)$ is countable was used just for convenience; in the same way we can check using Proposition 3.2.13 that $\mathcal{H}_p := \sum_{\lambda\in\sigma_p(T)}^{\oplus} N(\lambda)$ reduces the operator T for $\sigma_p(T)$ of any cardinality.

It is clear from the construction that $\mathcal{H}_p$ contains an orthonormal basis consisting of eigenvectors of T; on the other hand, T_c has no eigenvalues. Hence we have found that any normal operator decomposes into the orthogonal sum

$$T = T_p \oplus T_c\,, \qquad \sigma(T) = \sigma(T_p) \cup \sigma(T_c) = \overline{\sigma_p(T)} \cup \sigma(T_c)\,,$$

where T_p is a normal operator with a pure point spectrum while $\sigma_p(T_c) = \emptyset$; the decomposition of the spectrum follows from Problem 24b and Proposition 4.3.8. The formulas hold, in particular, for any self-adjoint A, in which case the operators A_p and A_c are also self-adjoint.

If T is bounded, the condition of Theorem 4.4.2 means that the operators E, T commute, $ET = TE$. Motivated by this we extend the notion of commutativity in the following way: suppose that T is any linear operator on $\mathcal{H}$ and $B \in \mathcal{B}(\mathcal{H})$; then we say that T and B *commute* if $BT \subset TB$, or more explicitly, if $BD_T \subset D_T$ and $BTx = TBx$ holds for any $x \in D_T$. According to this definition, for instance, any bounded invertible B commutes with its inverse while, of course, $BB^{-1} \neq B^{-1}B$ if $\operatorname{Ran} B \neq \mathcal{H}$. Commutativity can be checked using the resolvent of T.

4.4.4 Theorem: A closed linear operator T with a nonempty resolvent set commutes with a bounded B *iff* $R_T(\mu)B = BR_T(\mu)$ for at least one $\mu \in \rho(T)$; in that case the relation holds for all $\mu \in \rho(T)$.

Proof: To check the sufficient condition, notice that any $x \in D_T$ can be written as $x = R_T(\mu)y$; then $Bx = BR_T(\mu)y = R_T(\mu)By$ so $Bx \in \operatorname{Ran} R_T(\mu) = D_T$. Moreover, $BTx = B(T-\mu+\mu)R_T(\mu)y = By + \mu Bx$ and $TBx = (T-\mu)BR_T(\mu)y + \mu Bx = By + \mu Bx$, so $BTx = TBx$ for all $x \in D_T$. Conversely, the inclusion $BT \subset TB$ means that $B(T-\mu) \subset (T-\mu)B$ for any $\mu \in \mathbb{C}$, which in turn gives $R_T(\mu)B = BR_T(\mu)$ provided $\mu \in \rho(T)$ (Problem 25a). ∎

4.4.5 Corollary: Let T, S be closed operators with nonempty resolvent sets. If there are $\lambda_0 \in \rho(T)$ and $\mu_0 \in \rho(S)$ such that $R_T(\lambda_0)$ commutes with $R_S(\mu_0)$, then $R_T(\lambda)$ and $R_S(\mu)$ commute for all $\lambda \in \rho(T)$, $\mu \in \rho(S)$.

This property can serve as an alternative definition of commutativity for self-adjoint operators (*cf.* Problem 5.23b); it is essentially the only situation when we deal with commuting operators, which are both of unbounded.

Linear operators T on $\mathcal{H}$ and S on $\mathcal{G}$ are said to be **unitarily equivalent** if there is an unitary operator $U : \mathcal{G} \to \mathcal{H}$ such that $T = USU^{-1}$. This condition means, in particular, that $D_T = UD_S$ and $\operatorname{Ran} T = U \operatorname{Ran} S$. The unitary operators in $\mathcal{B}(\mathcal{H})$ form a group with respect to multiplication, so the unitary equivalence between operators on a given $\mathcal{H}$ is reflexive, symmetric, and transitive — hence the name — and we can decompose a family of operators into equivalence classes. Unitary equivalence preserves many operator properties; we shall list a few of them (see also Problem 27).

4.4.6 Proposition: Let T, S be unitarily equivalent, $T = USU^{-1}$; then

(a) If S is densely defined, the same holds true for T and $T^* = US^*U^{-1}$. In particular, if S is symmetric or self-adjoint, then T is also symmetric or self-adjoint, respectively.

(b) If S is invertible, so is T and $T^{-1} = US^{-1}U^{-1}$.

(c) If S is closed the same is true for T; we have $\sigma(T) = \sigma(S)$ and $\Gamma(T) = U_\oplus\Gamma(S)$, where $U_\oplus[x, y] := [Ux, Uy]$.

Proof: In view of Problem 3.4, $\overline{D_S} = \mathcal{G}$ implies $\overline{UD_S} = \mathcal{H}$. Using Proposition 4.1.2e and Problem 2c, we can easily check $T^* = US^*U^{-1}$, which means, in particular, that $D(T^*) = UD(S^*)$; from here we get the remaining part of assertion (a). Next we denote $T' := US^{-1}U^{-1}$. We have $D(T') = U \operatorname{Ran} S = \operatorname{Ran} T$ and $T'Tx = x$ for all $x \in D_T$ so $T' = T^{-1}$; this proves (b). The operator $U_\oplus : \mathcal{G} \oplus \mathcal{G} \to \mathcal{H} \oplus \mathcal{H}$ is unitary and $U_\oplus \Gamma(S) = \Gamma(T)$ holds by definition. If $S \in \mathcal{C}(\mathcal{H})$ we have $U_\oplus \Gamma(S) = \underline{U_\oplus \Gamma(S)} = \overline{U_\oplus \Gamma(S)}$, so $T \in \mathcal{C}(\mathcal{H})$. The operators $T-\lambda$ and $S-\lambda$ are unitarily equivalent for any $\lambda \in \mathbb{C}$; if $\lambda \in \rho(S)$ then $\operatorname{Ran}(T-\lambda) = U \operatorname{Ran}(S-\lambda) = \mathcal{H}$ and $\operatorname{Ker}(T-\lambda) = U \operatorname{Ker}(S-\lambda) = \{0\}$, so $\lambda \in \rho(T)$. Interchanging the roles of T, S we get $\rho(T) = \rho(S)$, and therefore $\sigma(T) = \sigma(S)$. ∎

4.4.7 Example: We shall show that the operators P and Q on $L^2(\mathbb{R})$ are unitarily equivalent. The Fourier–Plancherel operator F is unitary by definition; hence $T := F^{-1}QF$ is self–adjoint and its domain consists of the vectors $\psi = F^{-1}\phi$ with $\phi \in D(Q)$. Since $D(Q) \subset L^1(\mathbb{R})$ (Problem 5), we can write the identity $(2\pi)^{1/2}\psi(x) = i\int_{\mathbb{R}} \frac{e^{ixy}-1}{iy}\, y\phi(y)\, dy + \int_{\mathbb{R}} \phi(y)\, dy$. It follows from Proposition 3.1.5 and Example 3.1.6 that ψ is absolutely continuous in $\mathbb{R}$ and

$$-i\psi'(x) \;=\; (2\pi)^{-1/2}\, \frac{d}{dx} \int_{\mathbb{R}} \frac{e^{ixy}-1}{iy}\, (Q\phi)(y)\, dy \;=\; (F^{-1}Q\phi)(x);$$

hence $\psi' \in L^2(\mathbb{R})$ and $-i\psi' = F^{-1}QF\psi$, which we may write as $T \subset P$. Finally, both operators are self–adjoint and therefore equal each other, *i.e.*,

$$P \;=\; F^{-1}QF\,. \tag{4.4}$$

This unitary equivalence allows us to study the properties of P using the simpler operator Q. For instance, we immediately get $\sigma(P) = \sigma(Q) = \mathbb{R}$.

4.5 Tensor products

Now we want to show how one can construct to given operators T_r on Hilbert spaces $\mathcal{H}_r$, $r = 1, \dots, n$, an operator on the tensor product $\mathcal{H}_1 \otimes \cdots \otimes \mathcal{H}_n$. For simplicity we shall assume a pair of Hilbert spaces; the extension to any finite n is straightforward.

First, let us have a pair of bounded operators $B_r \in \mathcal{B}(\mathcal{H}_r)$. We define the operator $B_1 \underline{\times} B_2$ called the **tensor product** of B_1, B_2 by $(B_1 \underline{\times} B_2)(x \otimes y) := B_1 x \otimes B_2 y$ for any $x \in \mathcal{H}_1$, $y \in \mathcal{H}_2$ and extend it linearly to $\mathcal{H}_1 \underline{\times} \mathcal{H}_2$; the correctness of this definition can be checked using Problem 2.20.

4.5.1 Lemma: The operator $B_1 \underline{\times} B_2$ is bounded with $\|B_1 \underline{\times} B_2\| \le \|B_1\|\, \|B_2\|$.

Proof: Since we can write $B_1 \underline{\times} B_2 = \underline{B}_1^0 \underline{B}_2^0$, where $\underline{B}_1^0 := B_1 \underline{\times} I_2$ and $\underline{B}_2^0 := I_1 \underline{\times} B_2$, it is sufficient to prove that $\|\underline{B}_r^0 u\| \le \|B_r\|\, \|u\|$ holds for any $u \in \mathcal{H}_1 \underline{\times} \mathcal{H}_2$. Let $u = \sum_{j=1}^N x_j \otimes y_j$ and choose orthonormal bases $\{e_1, \dots, e_n\}$, $\{f_1, \dots, f_m\}$ in the subspaces spanned by $\{x_1, \dots, x_N\}$ and $\{y_1, \dots, y_N\}$, respectively; then $u =$

$\sum_{r,s} a_{rs} e_r \otimes f_s$ and $\|\underline{B}_1^0 u\|^2 = \|\sum_s g_s \otimes f_s\|^2$, where we denote $g_s := \sum_r a_{rs} B_1 e_r$. However, the set $\{g_s \otimes f_s : s = 1, \dots, m\}$ is orthonormal, so $\|\underline{B}_1^0 u\|^2 = \sum_s \|g_s\|_1^2 \le \|B_1\|^2 \|u\|^2$; the relation for $r = 2$ is checked in the same way. ∎

As a bounded densely defined operator, $B_1 \underline{\times} B_2$ has a unique extension to $\mathcal{H}_1 \otimes \mathcal{H}_2$ which we call the **tensor product** of the operators B_1, B_2 and denote as $B_1 \otimes B_2$. We have $\|B_1 \otimes B_2\| \ge \|B_1 x \otimes B_2 y\| = \|B_1 x\|_1 \|B_2 y\|_2$ for unit vectors $x \in \mathcal{H}_1$, $y \in \mathcal{H}_2$, and the inequality remains valid if we take the supremum over x, y; in combination with Lemma 1 this gives

$$\|B_1 \otimes B_2\| = \|B_1\| \, \|B_2\| \,. \tag{4.5}$$

The map $[B_1, B_2] \mapsto B_1 \otimes B_2$ defined in this way is obviously bilinear and has the following simple properties, the proof of which is left to the reader (Problem 30, see also Problems 31–34).

4.5.2 Theorem: Let $B_r, C_r \in \mathcal{B}(\mathcal{H}_r)$, $r = 1, 2$; then

(a) $B_1 C_1 \otimes B_2 C_2 = (B_1 \otimes B_2)(C_1 \otimes C_2)$ and $(B_1 \otimes B_2)^* = B_1^* \otimes B_2^*$.

(b) If the operators B_r are invertible, the same is true for $B_1 \otimes B_2$ and the inverse $(B_1 \otimes B_2)^{-1} = B_1^{-1} \otimes B_2^{-1}$.

(c) If the operators B_r are normal (unitary, Hermitean, projections), then $B_1 \otimes B_2$ is respectively normal (unitary, Hermitean, a projection).

Let us return to the general case of operators T_r with the domains $D(T_r) \subset \mathcal{H}_r$, $r = 1, 2$. We define the operator $T_1 \otimes T_2$ with the domain $D(T_1) \underline{\times} D(T_2)$ by

$$(T_1 \otimes T_2) \left(\sum_j x_j \otimes y_j \right) := \sum_j T_1 x_j \otimes T_2 y_j$$

for any n–tuples $\{x_j\} \subset D(T_1)$ and $\{y_j\} \subset D(T_2)$; we call it the **tensor product** of T_1 and T_2.

4.5.3 Remarks: (a) Strictly speaking, $T_1 \otimes T_2$ corresponds rather to $T_1 \underline{\times} T_2$ according to the previous definition. The problem is that each definition is natural in its context. For bounded operators $B_1 \underline{\times} B_2$ played an auxiliary role only due to the existence of its unique extension; this is no longer true in the unbounded–operator case, where $D(T_1) \underline{\times} D(T_2)$ represents the suitable "smallest" domain for the tensor product.

(b) The tensor product $\mathcal{H}_1 \otimes \mathcal{H}_2$ can have different realizations to which, of course, different realizations of $T_1 \otimes T_2$ correspond. Using Proposition 2.4.2, however, it is easy to check that any two of them are unitarily equivalent.

4.5.4 Theorem: Let operators T_r, $r = 1, 2$, be densely defined; then

(a) $T_1 \otimes T_2$ is densely defined and $(T_1 \otimes T_2)^* \supset T_1^* \otimes T_2^*$.

(b) If the operators T_r are closable, so is $T_1 \otimes T_2$ and $\overline{T_1 \otimes T_2} \supset \overline{T}_1 \otimes \overline{T}_2$.

Proof: Since $\overline{D(T_r)} = \mathcal{H}_r$, Proposition 2.4.4a implies that $T_1 \otimes T_2$ is densely defined; the inclusion then follows from the appropriate definitions. Using Theorem 4.2.3 we find that $T_1 \otimes T_2$ is closable if T_r are closable; furthermore, $x_n \to x$ and $y_n \to y$ imply $x_n \otimes y_n \to x \otimes y$, so Proposition 1.6.6 yields the remaining inclusion. ∎

4.5.5 Example: In general, the above inclusions are not identities even if one of the operators is bounded. To illustrate this, consider an unbounded self-adjoint operator A on $\mathcal{H}_1$ and a one-dimensional projection E on a two-dimensional $\mathcal{H}_2$, in which we choose an orthonormal basis $\{e, f\}$ such that $e \in \operatorname{Ran} E$. Using Problem 2.21a we find $D(A \otimes E) = \{x \otimes e + y \otimes f : x, y \in D_A\}$; on the other hand, a simple argument shows that the adjoint has the domain $D((A \otimes E)^*) = \{x \otimes e + y \otimes f : x \in D_A,\ y \in \mathcal{H}_1\}$ and acts on it as $(A \otimes E)^*(x \otimes e + y \otimes f) = Ax \otimes e$. Since A is unbounded, we have $D(A \otimes E) \neq D((A \otimes E)^*)$. In particular, $A \otimes E$ is not self-adjoint, but it is *e.s.a.* because $(A \otimes E)^*$ is symmetric.

Tensor products of unbounded operators again fulfil natural algebraic rules; however, we should be careful now about the domains.

4.5.6 Proposition: Let T_r, S_r be linear operators on $\mathcal{H}_r$, $r = 1, 2$; then

$$(T_1 + S_1) \otimes T_2 = T_1 \otimes T_2 + S_1 \otimes T_2 \quad \text{and} \quad (T_1 S_1) \otimes (T_2 S_2) \subset (T_1 \otimes T_2)(S_1 \otimes S_2)\,.$$

In particular, if S_r is invertible and $\operatorname{Ran} S_r \subset D(T_r)$ for at least one r, then the last relation turns to identity.

Proof: The inclusion $(T_1 + S_1) \otimes T_2 \subset T_1 \otimes T_2 + S_1 \otimes T_2$ is obvious. Further, let $x \in (D(T_1) \underline{\times} D(T_2)) \cap (D(S_1) \underline{\times} D(T_2))$, *i.e.*, $x = \sum_{j=1}^n x_j \otimes y_j = \sum_{k=1}^m x'_k \otimes y'_k$ with $x_j \in D(T_1)$, $x'_k \in D(S_1)$ and $y_j, y'_k \in D(T_2)$. Let $\{e_l\}_{l=1}^N$ be an orthonormal basis in $\{y_1, \dots, y_n, y'_1, \dots, y'_m\}_{lin} \subset D(T_2)$, so $y_j = \sum_l c_{jl} e_l$ and $y'_k = \sum_l d_{kl} e_l$; then the vectors $\tilde{x}_l := \sum_{j=1}^n c_{jl} x_j \in D(T_1)$ and $\tilde{x}'_l := \sum_{k=1}^n d_{kl} x'_k \in D(S_1)$ satisfy $\sum_l (\tilde{x}_l - \tilde{x}'_l) \otimes e_l = 0$. Since the e_l are orthogonal, we get $\tilde{x}_l = \tilde{x}'_l$, $l = 1, \dots, N$, *i.e.*, $\tilde{x}_l \in D(T_1) \cap D(S_1)$. Finally $e_l \in D(T_2)$, so $x = \sum_l \tilde{x}_l \otimes e_l$ belongs to $D(T_1 + S_1) \otimes D(T_2)$, which proves the first relation. The remaining inclusion follows directly from the definition; we leave to the reader to check the sufficient conditions for it to turn to identity (Problem 35). ∎

4.6 Quadratic forms

The simple correspondence between bounded operators and bounded sesquilinear forms expressed by Proposition 3.1.1 poses the question whether one can extend it to the unbounded-operator case. We shall see that domain considerations make the problem much more complicated and an affirmative answer can be obtained only for certain classes of operators.

First we must extend the definitions of Section 1.1 to include unbounded forms which may not be everywhere defined. Let D be a subspace in $\mathcal{H}$; a sesquilinear form $f : D \times D \to \mathbb{C}$ will be briefly referred to as a **form** on $\mathcal{H}$ with the *domain* $D(f) := D$. If $\overline{D(f)} = \mathcal{H}$ the form f is said to be *densely defined.* In view of the polarization formula, there is a bijective correspondence between f and the quadratic form $q_f : q_f(x) = f(x,x)$. It is thus reasonable to simplify the notation and use the same symbol f for both, setting $f[x] := f(x,x)$ for any $x \in D(f)$.

We can define naturally the operation of summation and scalar multiplication for forms, with the domains obeying the same rules as for summation of operators; restrictions and extensions also have the usual meaning. The form associated with the unit operator is sometimes denoted as $e : e(x,y) = (x,y)$. The set $\Theta(f) := \{ f[x] : x \in D(f),\, \|x\| = 1 \}$ is called the *numerical range* of f; in the same way as in the operator case we find that f is symmetric *iff* $\Theta(f) \subset \mathbb{R}$. A symmetric form s is *below bounded* if $m_s := \inf \Theta(s) > -\infty$ and *positive* if $m_s \ge 0$. Notice that if s is below bounded there is always a "shifted" form which is positive, *e.g.*, $s_0 := s - m_s e$. If s is positive, then the *Schwarz inequality,*

$$|s(x,y)| \le (s[x]s[y])^{1/2},$$

is valid for arbitrary vectors $x, y \in D(s)$; this implies, in particular, the relations $s[x+y]^{1/2} \le s[x]^{1/2} + s[y]^{1/2}$ and $|s[x]^{1/2} - s[y]^{1/2}| \le s[x-y]^{1/2}$.

4.6.1 Examples: (a) Let T be an operator on $\mathcal{H}$ with the domain D_T. The form $f_T : f_T(x,y) = (x, Ty)$ with the domain $D(f_T) := D_T$ is said to be *generated* by T. Its numerical range clearly coincides with $\Theta(T)$; in particular if T is symmetric and below bounded the same is true for f_T.

(b) A linear operator $S : \mathcal{H} \to \mathcal{H}_1$ defines a positive form p with $D(p) := D_S$ by $p(x,y) := (Sx, Sy)_1$.

(c) Let $\mathcal{H} = L^2(a,b)$ with $b-a < \infty$. Given real numbers c_a, c_b and a real–valued function $V \in L^2(a,b)$ such that $V(x) \ge m$ a.e. for some m, we set $D(t) := AC[a,b]$ and

$$t(\phi,\psi) := \int_a^b \left(\overline{\phi'(x)}\psi'(x) + V(x)\overline{\phi(x)}\psi(x) \right) dx + c_b\overline{\phi(b)}\psi(b) + c_a\overline{\phi(a)}\psi(a).$$

The form t is densely defined and symmetric. The integral part is obviously below bounded; we can choose $m := \operatorname{inf\,ess}_{x\in(a,b)} V(x)$. The other two parts are also below bounded. To check this, we express $\phi(b) = (g_n, \phi') + (g_n', \phi)$ integrating by parts with $g_n(x) := \left(\frac{x-a}{b-a}\right)^n$, and $\phi(a)$ similarly using the function $h_n(x) := -g_n(b+a-x)$. It yields

$$t[\phi] \ge \|\phi'\|^2 + m\|\phi\|^2 - |c_b|\,|\phi(b)|^2 - |c_a|\,|\phi(a)|^2 \ge A\|\phi'\|^2 - B\|\phi\|^2,$$

where $A := 1 - (|c_a| + |c_b|)\|g_n\|^2$ and $B := 2(|c_a| + |c_b|)\|g_n'\|^2 - m$. Since $\|g_n\| = \left(\frac{b-a}{2n+1}\right)^{1/2}$, we can choose n so that $A > 0$; hence the form t is below bounded with $m_t \ge -B$ (corresponding to the chosen n).

Given a symmetric below bounded form s we can equip its domain $D(s)$ with the inner product $(x,y)_s := s(x,y) + (1-m_s)(x,y)$; this can be also expressed as $(x,y)_s = (x,y)_{s_0} = s_0(x,y) + (x,y)$, where s_0 is the positive form introduced above. In this way $D(s)$ becomes a pre–Hilbert space which we denote as $\mathcal{H}_s$; if it is complete, *i.e.*, a Hilbert space, the form s is called *closed.*

A sequence $\{x_n\} \subset D(s)$ is Cauchy with respect to $\|\cdot\|_s$ *iff* it converges to some $x \in \mathcal{H}$ and for any $\varepsilon > 0$ there is an n_ε such that $|s[x_n - x_m]| < \varepsilon$ holds for any $n, m > n_\varepsilon$; for the sake of brevity, we shall use the symbol $x_n \stackrel{(s)}{\longrightarrow} x$ as a shorthand for these conditions.

4.6.2 Proposition: The following conditions are equivalent:

(a) s is a closed form.

(b) $s + ae$ is closed for any $a \in \mathbb{R}$.

(c) if a sequence $\{x_n\} \subset D(s)$ fulfils $x_n \stackrel{(s)}{\longrightarrow} x$, then $x \in D(s)$ and $s[x-x_n] \to 0$.

Proof is left to the reader (Problem 38).

4.6.3 Example: Let p be the form of Example 4.6.1b; then a sequence fulfils $x_n \stackrel{(s)}{\longrightarrow} x$ *iff* $x_n \to x$ and $\{Sx_n\}$ is convergent in $\mathcal{H}_1$; hence p is closed *iff* the operator S is closed. For instance, let $\mathcal{H} = \mathcal{H}_1 = L^2(a,b)$ and $S := -i\frac{d}{dx}$ with the domain $D(S) := AC[a,b]$, *i.e.*, $S = P^*$ in the notation of Example 4.2.5; then $p : p(\psi,\phi) = \int_a^b \overline{\phi'(x)}\psi'(x)\,dx$ is a closed form.

4.6.4 Lemma: If $x_n \stackrel{(s)}{\longrightarrow} x$ and $y_n \stackrel{(s)}{\longrightarrow} y$, then the sequence $\{s(x_n,y_n)\}$ is Cauchy; in particular, $s(x_n,y_n) \to s(x,y)$ if the form s is closed.

Proof: The conditions $x_n \stackrel{(s)}{\longrightarrow} x$, $y_n \stackrel{(s)}{\longrightarrow} y$ imply $(x_n,y_n) \to (x,y)$, so we can assume without loss of generality that s is positive. The assertion then follows from the estimate

$$\begin{aligned} |s(x_n,y_n) - s(x_m,y_m)| &\le |s(x_n - x_m, y_n)| + |s(x_m, y_n - y_m)| \\ &\le s[x_n - x_m]^{1/2} s[y_n]^{1/2} + s[y_n - y_m]^{1/2} s[x_n]^{1/2} \end{aligned}$$

combined with the fact that the sequences $\{s[x_n]\}$, $\{s[y_n]\}$ are Cauchy by the Schwarz inequality and therefore bounded. If the form s is closed, then the conditions $x_n \stackrel{(s)}{\longrightarrow} x$, $y_n \stackrel{(s)}{\longrightarrow} y$ imply $s[x_n - x] \to 0$ and $s[y_n - y] \to 0$, so it is sufficient to put $x_m = x$ and $y_m = y$ in the above estimate. ∎

Not every symmetric form is closed, so we have to ask about existence of closed extensions. The situation is less transparent than in the operator case, where we could lean on the concept of a graph, and the results are generally weaker; we shall see below that there are symmetric below bounded forms with no closed extension.

A form s is called *closable* if there is a closed form $t \supset s$. The following theorem shows that a closable form always has a smallest closed extension, which we call the

closure of s and denote as $\overline{s}$. If s is a closed form and $D \subset D(s)$ is a subspace such that $\overline{s \restriction D} = s$ it is called a *core* for s.

4.6.5 Theorem: (a) A symmetric below bounded form is closable *iff* $x_n \xrightarrow{(s)} 0$ implies $s[x_n] \to 0$.

(b) Let s be closable and denote by $D(\overline{s})$ the subspace consisting of those $x \in \mathcal{H}$ for which there is a sequence $\{x_n\} \subset D(s)$ such that $x_n \xrightarrow{(s)} x$. Then the relation

$$\overline{s}(x,y) := \lim_{n\to\infty} s(x_n, y_n)$$

with $x_n \xrightarrow{(s)} x$ and $y_n \xrightarrow{(s)} y$ defines on $D(\overline{s})$ the form $\overline{s}$ which is the smallest closed extension of s, and moreover, $m_{\overline{s}} = m_s$.

Proof: If $t \supset s$ is a closed form then $x_n \xrightarrow{(s)} 0$ implies $x_n \xrightarrow{(t)} 0$, and $0 = \lim_{n\to\infty} t[x_n] = \lim_{n\to\infty} s[x_n]$ due to the closedness of t. Suppose on the contrary that $x_n \xrightarrow{(s)} 0$ implies $s[x_n] \to 0$; we shall show that the relation $\overline{s}(x,y) := \lim_{n\to\infty} s(x_n, y_n)$ defines on $D(\overline{s})$ a closed form such that $t \supset \overline{s}$ holds for any closed extension t of s. Existence of the limit follows from Lemma 4.6.4; its independence of the chosen sequences follows from the estimate used in the proof of the lemma. In view of Problem 38b, $D(\overline{s})$ is a subspace and $\overline{s}$ is a form which extends s; it is furthermore clear from the definition relation that $\overline{s}$ is symmetric, below bounded, and $m_{\overline{s}} \ge m_s$, which in combination with the inclusion $\overline{s} \supset s$ gives $m_{\overline{s}} = m_s$.

It remains to check that $\overline{s}$ is the smallest closed extension. Let $x \in D(\overline{s})$ and $x_n \xrightarrow{(s)} x$, so to any $\varepsilon > 0$ there is n_ε such that $|s[x_n - x_m]| < \varepsilon$ for $n, m > n_\varepsilon$. Passing to the limit $m \to \infty$ we get $|\overline{s}[x_n - x]| \le \varepsilon$ if $n > n_\varepsilon$; hence (i) $D(s)$ is dense in the pre-Hilbert space $\mathcal{H}_{\overline{s}} = D(\overline{s})$ and (ii) if a sequence $\{x_n\} \subset D(s)$ is Cauchy with respect to $\|\cdot\|_s$, then there is $x \in \mathcal{H}_{\overline{s}}$ such that $\|x_n - x\|_s \to 0$. This allows us to check that $\mathcal{H}_{\overline{s}}$ is complete. Let a sequence $\{y_n\} \subset \mathcal{H}_{\overline{s}}$ be Cauchy with respect to $\|\cdot\|_{\overline{s}}$; then due to (i) there is a sequence $\{x_n\} \subset D(s)$ such that $\|x_n - y_n\|_{\overline{s}} < n^{-1}$ for $n = 1, 2, \dots$. This implies that the sequence $\{x_n\}$ is Cauchy with respect to $\|\cdot\|_{\overline{s}}$, and therefore also with respect to $\|\cdot\|_s$; then $\|x_n - x\|_{\overline{s}} \to 0$ by (ii). Finally, $\|y_n - x\|_{\overline{s}} \le \|x_n - x\|_{\overline{s}} + n^{-1} \to 0$, so $\mathcal{H}_{\overline{s}}$ is complete and $\overline{s}$ is a closed form.

Finally, consider a closed form $t \supset s$ and any $x \in D(\overline{s})$. By assumption, there is a sequence $\{x_n\} \subset D(s)$ such that $x_n \xrightarrow{(s)} x$; then $x_n \xrightarrow{(t)} x$ and the closedness of t implies $x \in D(t)$. Using the definition of $\overline{s}$ we get $\overline{s}[x] = \lim_{n\to\infty} s[x_n] = \lim_{n\to\infty} t[x_n] = t[x]$, so $t \supset s$. ∎

4.6.6 Example: Consider again the form p of Example 4.6.1b. In view of Proposition 1.6.6, $x_n \xrightarrow{(p)} 0$ implies $p[x_n] \to 0$ *iff* operator S is closable. Hence a nonclosable *positive* form corresponds to any operator which is not closable. For instance, choose $p : p(\phi, \psi) = \overline{\phi(0)}\psi(0)$ on $L^2(\mathbb{R})$ with $D(p) := AC(\mathbb{R})$. Then $\phi_n \xrightarrow{(p)} 0$ for a sequence $\{\phi_n\} \subset D(p)$ means $\phi_n \to 0$ and $\phi_n(0) \to \alpha$; however, α need not be zero, as we can easily check.

Fortunately, this cannot happen with a form which is generated by a symmetric below bounded operator.

4.6.7 Proposition: A form s which is generated by a symmetric below bounded operator A is closable, and its closure fulfils $\overline{s}(x,y) = (x,Ay)$ for any $x \in D(\overline{s})$ and $y \in D_A = D(s)$.

Proof: In view of Theorem 4.6.5, it is sufficient to check that $x_n \xrightarrow{(s)} 0$ implies $s[x_n] \to 0$. Suppose first that A is positive, so s is also positive; then using the Schwarz inequality we find that $\{s[x_n]\}$ is Cauchy, and therefore bounded, $s[x_n] < M$ for all n. Another application of the Schwarz inequality gives

$$\begin{aligned} s[x_n] &= s(x_n, x_n - x_m) + s(x_n, x_m) \\ &\le s[x_n]^{1/2} s[x_n - x_m]^{1/2} + (Ax_n, x_m) < (M\varepsilon)^{1/2} + \|Ax_n\|\,\|x_m\| \end{aligned}$$

for all $n, m > n_\varepsilon$. Since $x_n \xrightarrow{(s)} 0$ includes the condition $x_n \to 0$, the limit $m \to \infty$ leads to $s[x_n] \le (M\varepsilon)^{1/2}$ for all $n > n_\varepsilon$, *i.e.*, $s[x_n] \to 0$. If $\inf \Theta(A) = m_A < 0$, then the form $s_0 := s - m_A e$ is positive, and by Proposition 4.6.2, $x_n \xrightarrow{(s)} 0$ implies $s_0[x_n] \to 0$ so $|s[x_n]| \le s_0[x_n] + |m_A|\,\|x_n\|^2 \to 0$; this proves the existence of the closure. For any $x \in D(\overline{s})$, there is by Theorem 4.6.5 a sequence $\{x_n\} \subset D_A$ such that $x_n \xrightarrow{(s)} x$; then $\overline{s}(x,y) = \lim_{n\to\infty} s(x_n, y) = \lim_{n\to\infty} (x_n, Ay) = (x, Ay)$ holds for any $y \in D_A$. ∎

The following theorem plays a key role in the theory of unbounded forms.

4.6.8 Theorem (representation theorem): Let s be a densely defined form which is closed, symmetric, and below bounded; then there is a self–adjoint operator A such that

(a) $D_A \subset D(s)$ and $s(x,y) = (x,Ay)$ for any $x \in D(s)$, $y \in D_A$.

(b) The domain D_A is a core for s.

(c) If there are vectors $x \in D(s)$ and $z \in \mathcal{H}$ such that $s(x,y) = (z,y)$ holds for all $y \in D(s)$, then $x \in D_A$ and $z = Ax$.

(d) Let T be a linear operator such that $D_T \subset D(s)$ and $(Tx,y) = s(x,y)$ for all $x \in D_T$, $y \in D(s)$; then $T \subset A$. In particular, $T = A$ if T is self–adjoint; hence the operator A is determined uniquely by condition (a).

(e) A is below bounded with $\inf \Theta(A) = m_s$.

Proof: First suppose that $m_s = 0$. Since s is closed, $\mathcal{H}_s$ is a Hilbert space; for any $y \in \mathcal{H}$ we define the antilinear functional $l_y : l_y(x) := (x,y)$ on it. We have $\|x\| \le \|x\|_s$, so l_y is continuous on $\mathcal{H}_s$, and by an easy modification of the Riesz lemma proof, there is just one $y' \in \mathcal{H}_s$ such that $(x,y) = (x,y')_s$, and the map $C : y \mapsto y'$ is linear. In particular, setting $x = y'$ we get $\|y'\|_s^2 \le \|y'\|\,\|y\| \le \|y'\|_s\,\|y\|$, *i.e.*,

$\|Cy\| \le \|Cy\|_s \le \|y\|$ so $C \in \mathcal{B}(\mathcal{H})$. The relation $(x,y) = (x,y')_s$ can be written as $(x,y) = (x,Cy)_s = s(x,Cy) + (x,Cy)$, so $s(x,Cy) = (x, y - Cy)$ holds for any $x \in D(s)$ and $y \in \mathcal{H}$. Hence if $Cy = 0$ we have $(x,y) = 0$ for all $x \in D(s)$, and since this set is dense in $\mathcal{H}_s$, this further implies $y = 0$, so C is invertible. We also have $(Cu,v) = (Cu,Cv)_s = \overline{(Cv,u)} = \overline{(Cv,Cu)_s} = (u,Cv)$ for any pair $u,v \in \mathcal{H}$, which means that C^{-1} is self-adjoint by Problem 3c. Now denoting $z = Cy$ in $s(x,Cy) = (x,y-Cy)$ we find that the self-adjoint operator $A := C^{-1} - I$ satisfies $s(x,z) = (x,Az)$ for all $x \in D(s)$ and $z \in D_A = \operatorname{Ran} C \subset D(s)$. Finally, in the general case when $m_s \ne 0$ the sought operator A equals $A_0 + m_s I$, where A_0 corresponds to the positive form $s_0 := s - m_s e$; this proves assertion (a).

To check (b) it is sufficient to show that the subspace $D_A = \operatorname{Ran} C$ is dense in $\mathcal{H}_s$ (Problem 38d). Since $\|x\|_s^2 = (x,Cx)_s$, any vector from the orthogonal complement of $\operatorname{Ran} C$ in $\mathcal{H}_s$ is zero, and therefore $(\overline{D}_A)_s = \mathcal{H}_s$. Assertion (c) is obtained if we write the condition $s(x,y) = (z,y)$ for $y \in D_A$ in the form $(x,Ay) = (z,y)$; this implies $x \in D(A^*) = D_A$ and $z = A^*x = Ax$. Finally, (d) follows from (c), and (e) from (b) in combination with Theorem 4.6.5. ∎

4.6.9 Corollary: The representation theorem defines the bijection $s \mapsto A_s$ from the set of all densely defined, closed symmetric below bounded forms to the set of all self-adjoint below bounded operators; A_s is bounded *iff* s is bounded.

Proof: Let $A_s = A_t$ and denote $D := D(A_s) = D(A_t)$, then $s \restriction D = t \restriction D$ and injectivity of $s \mapsto A_s$ follows from Theorem 4.6.8b. To check the surjectivity consider the form s_A generated by a below bounded self-adjoint operator A; by Proposition 4.6.7 it has a closure $u := \overline{s}_A$ and $u(x,y) = (x,Ay)$ for all $x \in D(s_A)$, $y \in D_A$, so $A = A_u$ by Theorem 4.6.8d. ∎

Operator A_s is said to be *associated* with form s. If A is a below bounded self-adjoint operator and s is the corresponding form, $A_s = A$, then $D(s)$ is called the *form domain* of operator A and is alternatively denoted as $Q(A)$.

4.6.10 Example: The positive form p of Example 4.6.1b is densely defined and closed if $S \in \mathcal{L}_c(\mathcal{H})$. We have $(Sx,Sy) = (x,A_p y)$ for all $x \in D_S$ and $y \in D(A_p) \subset D_S$, which means $Sy \in D(S^*)$ and $S^*Sy = A_p y$, or $A_p \subset S^*S$. However, S^*S is self-adjoint by Theorem 4.2.7, so we have $A_p = S^*S$, and $D(S)$ is the form domain of operator A_p.

The representation theorem provides a way to construct a self-adjoint extension to a given symmetric operator.

4.6.11 Theorem (Friedrichs' extension): Suppose that A_0 is a symmetric below bounded operator, s is the closure of the form generated by A_0, and A_s is the self-adjoint operator associated with s; then $A_s \supset A_0$ and $\inf \Theta(A_s) = \inf \Theta(A_0)$. Moreover, A_s is the only self-adjoint extension of A_0 such that $D_{A_s} \subset D(s)$.

Proof: By Proposition 4.6.7, the form s_0 generated by A_0 has a closure, $s := \overline{s}_0$, and $s(x,y) = (x,Ay)$ holds for all $x \in D(s)$ and $y \in D(A_0) = D(s_0)$, so $A_0 \subset A_s$ and $\inf \Theta(A_s) = m_s = m_{s_0} = \inf \Theta(A_0)$ due to Theorems 5 and 8e. Now let A be a self–adjoint extension of A_0 fulfilling the stated conditions, and denote by t the form generated by A. Since $t \supset s$, we have $\bar{t} \supset s$ and $A = A_{\bar{t}}$ by Proposition 4.6.7 and Theorem 4.6.8d. Then $s(x,y) = t(x,y) = (x, Ay)$ holds for any $x \in D(s)$, $y \in D_A$; using Theorem 4.6.8d again, we get $A \subset A_s$, *i.e.*, $A = A_s$ because the two operators are self–adjoint. ∎

4.6.12 Remark: Given two below bounded self–adjoint operators A_1 and A_2, we denote by s_1 and s_2 the corresponding forms, $A_r := A_{s_r}$ for $r = 1, 2$. They are densely defined, symmetric, below bounded, and closed; the same is true for the form $s := s_1 + s_2$ with the exception that it may not be densely defined (Problem 38c). If s is densely defined, then by the representation theorem there is a unique self–adjoint operator A_s which we call the *form sum* of A_1 and A_2; sometimes it is denoted as $A_1 \dotplus A_2$. Since $D(A_1 + A_2) \subset D(S_1) \cap D(s_2) = D(s)$ and $s(x,y) = s_1(x,y) + s_2(x,y) = (x, (A_1 + A_2)y)$, we have $A_1 + A_2 \subset A_s$ by Theorem 4.6.8d, *i.e.*, the form sum extends the usual operator sum; notice that $A_1 + A_2$ may even not be densely defined. Moreover, if $A_1 + A_2 \in \mathcal{L}(\mathcal{H})$ it is symmetric and therefore has the Friedrichs extension A_F; however, it differs in general from the form sum, $A_F \neq A_s$ (Problem 40).

The result about stability of self–adjointness proved at the end of Section 4.3 has its form counterpart. To formulate it, we have to reintroduce the notion of relative boundedness. Suppose that s, t are symmetric forms on $\mathcal{H}$ and s is below bounded, then we say that t is *s–bounded* if $D(s) \subset D(t)$ and there are non–negative a, b such that

$$|t[x]| \leq a|s[x]| + b\|x\|^2 \tag{4.6}$$

holds for any $x \in D(s)$; the infimum of all a for which the condition is valid with some b is called the *s–bound* of the form t. Since an s–bounded form is at the same time bounded with respect to $s + ce$ for any $c \in \mathbb{R}$ and the relative bound is the same, we may assume without loss of generality that s is positive.

4.6.13 Lemma: Let s, t be symmetric forms. If s is positive, $D(s) \subset D(t)$, and condition (4.6) holds for some $a < 1$, then

(a) $s + t$ is below bounded with $m_{s+t} \geq -b$.

(b) $s + t$ is closed *iff* s is closed.

(c) $s + t$ is closable *iff* the same is true for s and $D(s + t) = D(s)$.

Proof: It follows from the assumption that $(s+t)[x] \geq (1-a)s[x] - b\|x\|^2$ holds for any $x \in D(s+t)$, so $m_{s+t} \geq -b$. In a similar way, we find $|(s+t)[x]| \leq (a+1)s[x] + b\|x\|^2$ and $s[x] \leq (1-a)^{-1}(|(s+t)[x]| + b\|x\|^2)$. These inequalities show that conditions $x_n \xrightarrow{(s)} x$ and $x_n \xrightarrow{(s+t)} x$ are equivalent; assertions (b) and (c) then follow from Proposition 4.6.2 and Theorem 4.6.5. ∎

Combining this result with the representation theorem, Corollary 4.6.9 and Theorem 4.6.5, we obtain the mentioned counterpart of Theorem 4.3.12.

4.6.14 Theorem (KLMN–theorem): Let A be a self–adjoint below bounded operator and s_A the form generated by it. If a symmetric form t fulfils condition (4.6) for $s = s_A$, some $a < 1$ and any $x \in D_A$, then there is a unique self–adjoint operator $\tilde{A}$ associated with the form $\overline{s_A} + t$. The form domains of the operators A, $\tilde{A}$ coincide, and

$$(x, \tilde{A}y) = (x, Ay) + t(x, y)$$

holds for all $x \in Q(A)$, $y \in D_A$. Moreover, the operator $\tilde{A}$ is below bounded, $\inf \Theta(\tilde{A}) \geq (1-a)m_A - b$, and any core of A is at the same time a core for $\tilde{A}$.

In particular, if the form t is generated by a symmetric B, then $\tilde{A}$ is the form sum of the two operators, and by Theorem 4.6.8c, it coincides with their operator sum, $\tilde{A} := A \dot{+} B = A + B$. Such a result follows from the Kato–Rellich theorem without the requirement that A is below bounded; however, the relative–boundedness assumption in Theorem 4.6.14 is weaker than in the operator case.

4.6.15 Proposition: Suppose that A is a self–adjoint below bounded operator, and B is symmetric and A–bounded with the relative bound $a < 1$. Let s, t be the forms generated by A and B, respectively; then t is s–bounded with the relative bound $\leq a$; if t is closable, the same is true for the closures $\bar{t}$ and $\bar{s}$.

Proof: To any $\alpha > a$, there is a β such that $\|Bx\| \leq \alpha\|Ax\| + \beta\|x\|$ holds for all $x \in D_A$. By Theorem 4.3.12 and Problem 22, the operator $A - \kappa B$ is self–adjoint for any $\kappa \in (-\alpha^{-1}, \alpha^{-1})$, and its lower bound $m(\kappa) := \inf \Theta(A - \kappa B)$ can be estimated from below by $m_A - |\kappa| \max\{\beta(1-\alpha|\kappa|)^{-1}, \beta + \alpha|m_A|\}$. Then $\kappa t[x] = \kappa(x, Bx) \leq -m(\kappa)\|x\|^2 + (x, Ax)$ for any $x \in D_A$, and changing κ to $|\kappa| \operatorname{sgn} t[x]$ we get the estimate

$$|t[x]| \leq |\kappa|^{-1}|s[x]| + \left(-\frac{m_A}{|\kappa|} + \max\left\{\frac{\beta}{1-\alpha|\kappa|}, \alpha|m_A| + \beta\right\}\right)\|x\|^2$$

for all $x \in D(s)$, which shows that t is s–bounded and its relative bound does not exceed $\inf |\kappa|^{-1} = \alpha$. Since α can be chosen arbitrarily close to a, this proves the first assertion; the second follows from the above estimate, Proposition 4.6.7, and Theorem 4.6.5. ∎

4.7 Self–adjoint extensions

We have already on several occasions encountered the problem of finding self–adjoint extensions to a given symmetric operator, *e.g.*, in Example 4.2.5 where we solved it by elementary means for the operator P on $L^2(a, b)$. In this section we are going to present a general method which allows us to construct and classify all self–adjoint extensions.

We start with some properties of the regularity domain $\pi(T)$ of a linear operator T on $\mathcal{H}$; recall that we defined it as a set of those $\lambda \in \mathbb{C}$ to which there is a $c_\lambda > 0$ such that $\|(T-\lambda)x\| \geq c_\lambda\|x\|$ holds for any $x \in D_T$ (see Problem 1.64).

In particular, if A is a symmetric operator then $\pi(A) \supset \mathbb{C}\setminus\mathbb{R}$ by Problem 4, so it has at most two connected components (*i.e.*, open arcwise connected subsets); it is connected if A is below bounded. Similarly an isometric operator V satisfies $\|(V-\lambda)x\| \ge |1-|\lambda||\,\|x\|$, so $\pi(V) \supset \mathbb{C}\setminus\{\lambda\in\mathbb{C} : |\lambda|=1\}$ again consists of at most two connected components.

The orthogonal complement to $\mathrm{Ran}\,(T-\lambda)$ will be called the *deficiency subspace* of T with respect to λ, and its dimension will be denoted by $\mathrm{def}\,(T-\lambda)$. If T is densely defined, then $\mathrm{def}\,(T-\lambda) = \dim\mathrm{Ker}\,(T^*-\overline{\lambda})$ by Proposition 4.1.1b.

4.7.1 Theorem: The map $\lambda \mapsto \mathrm{def}\,(T-\lambda)$ is constant on any connected component of the regularity domain $\pi(T)$.

Proof: Let G be a connected component of $\pi(T)$, and for any $\lambda\in G$ denote by E_λ the projection onto $\mathrm{Ran}\,(T-\lambda)^\perp$. By assumption, any two points $\lambda_1,\lambda_2\in G$ can be connected by a curve which is a compact set in $\mathbb{C}$ (as a continuous image of the segment $[0,1]$); hence it is sufficient to check that the map $\lambda\mapsto\mathrm{def}\,(T-\lambda)$ is locally constant. This is true if $\sup\{\,\|(I-E_\mu)E_\lambda x\| : x\in\mathcal{H},\ \|E_\lambda x\|=1\,\}<1$, and the same inequality with λ,μ interchanged are valid for all μ in some neighborhood of the point λ (*cf.* Proposition 3.2.11 and Problem 3.18c). Since G is a part of the regularity domain, we have

$$\|(T-\mu)x\| \ge \|(T-\lambda)x\| - \|(\mu-\lambda)x\| \ge \frac{2}{3}c_\lambda\|x\|$$

for $|\mu-\lambda| < \frac{1}{3}c_\lambda$. The norm in question can be expressed by Problem 3.16 as $\|(I-E_\mu)E_\lambda x\| = \sup\{\,|(E_\lambda x,y)| : y\in\mathrm{Ran}\,(T-\mu),\ \|y\|=1\,\}$, and using the fact that $E_\lambda x\in\mathrm{Ran}\,(T-\lambda)^\perp$, we find

$$\begin{aligned}\|(I-E_\mu)E_\lambda x\| &= \sup\left\{\frac{|(E_\lambda x,(T-\mu)z)|}{\|(T-\mu)z\|} : z\in D_T,\ z\ne 0\right\}\\ &\le \sup\left\{\frac{|\mu-\lambda|\,\|z\|}{\|(T-\mu)z\|} : z\in D_T,\ z\ne 0\right\} < \frac{1}{2}\end{aligned}$$

for any $x\in\mathcal{H}$ with $\|E_\lambda x\|=1$. The same argument, with the above estimate replaced by $\|(T-\lambda)x\|\ge c_\lambda\|x\|$, yields $\sup\{\,\|(I-E_\lambda)E_\mu x\| : x\in\mathcal{H},\ \|E_\mu x\|=1\,\}<\frac{1}{3}$, which concludes the proof. ∎

The open upper and lower complex halfplanes which we denote as $\mathbb{C}^\pm$ are connected components of the regularity domain for any symmetric operator A, so the function $\lambda\mapsto\mathrm{def}\,(A-\lambda)$ assumes a constant value at each of them; we call these numbers **deficiency indices** of the operator A and denote them by $n_\pm$,

$$n_\pm(A) := \mathrm{def}\,(A\mp i) = \dim\mathrm{Ker}\,(A^*\pm i)\,.$$

Since $\overline{A}^* = A^*$, one has $n_\pm(\overline{A}) = n_\pm(A)$. By Theorem 4.3.9, a symmetric operator A is self-adjoint *iff* it is closed and $n_\pm(A)=0$; the condition $n_\pm(A)=0$ is necessary and sufficient for A to be *e.s.a.* The deficiency indices are usually written as the ordered pair $(n_+(A),n_-(A))$.

4.7.2 Corollary: If the regularity domain of a symmetric operator A contains at least one real number λ, then $n_+(A) = n_-(A) = \operatorname{def}(A-\lambda)$; this is true, in particular, for any below bounded symmetric A.

4.7.3 Example: Let us ask about the deficiency indices of operator P in Example 4.2.5. Since $D(P^*) = AC(J)$ and $P^*\psi = -i\psi'$, the problem reduces to finding the solutions to the equation $i\psi' \pm i\psi = 0$ which belong to $AC(J)$. Since the general solution is of the form $\psi_\pm(x) = c_\pm e^{\pm x}$ we have $n_+(P) = n_-(P) = 1$ if J is a finite interval and $n_+(P) = n_-(P) = 0$ for $J = \mathbb{R}$. On the other hand, in the case of $J = (0,\infty)$ only ψ_- belongs to $AC(J)$, so the deficiency indices are not equal, $n_+(P) = 0$ while $n_-(P) = 1$.

If A is a closed symmetric operator, then $\lambda \in \pi(A)$ is a regular value of A *iff* $\operatorname{Ran}(A-\lambda) = \mathcal{H}$, *i.e.*, $\operatorname{Ker}(A^* - \bar{\lambda}) = \{0\}$. Hence $\lambda \in \mathbb{C}^+$ belongs to $\sigma(A)$ *iff* $n_+(A) \neq 0$, and the analogous statement is valid for $\mathbb{C}^-$. Taking into account the closedness of the spectrum, we arrive at the following conclusion.

4.7.4 Proposition: (a) Let A be a closed symmetric operator; then just one of the following alternatives is valid:

(i) $\sigma(A) = \mathbb{C}$ which happens *iff* both $n_\pm(A)$ are nonzero.

(ii) $\sigma(A) = \mathbb{C} \setminus \mathbb{C}^-$ *iff* $n_+(A) \neq 0$ and $n_-(A) = 0$.

(iii) $\sigma(A) = \mathbb{C} \setminus \mathbb{C}^+$ *iff* $n_+(A) = 0$ and $n_-(A) \neq 0$.

(iv) $\sigma(A) \subset \mathbb{R}$ *iff* $n_+(A) = n_-(A) = 0$.

(b) A closed symmetric operator is self–adjoint if it has a real regular value.

Any symmetric operator obeys the inclusion $A \subset A^*$. The following theorem allows us to describe the domain of A^* using $D(A)$ and the deficiency subspaces.

4.7.5 Theorem (the first von Neumann formula): Let A be a closed symmetric operator; then to any $x \in D(A^*)$ there is a unique decomposition

$$x = x_0 + x_+ + x_- , \qquad A^*x = Ax_0 - i(x_+ - x_-),$$

with $x_0 \in D(A)$ and $x_\pm \in \operatorname{Ker}(A^* \pm i)$.

Proof: We denote $\mathcal{K}_\pm := \operatorname{Ker}(A^* \pm i)$. Let z be the projection of $A^*x - ix$ to $\mathcal{K}_+$, so $A^*z + iz = 0$. Since the operator A is closed, $\operatorname{Ran}(A-i) = \overline{\operatorname{Ran}(A-i)} = \mathcal{K}_+^\perp$; hence to $y := A^*x - ix - z$ there is an $x_0 \in D(A)$ such that $y = (A-i)x_0 = (A^* - i)x_0$. The decomposition $A^*x - ix = y + z$ then implies $A^*\left(x - x_0 - \frac{i}{2}z\right) = i\left(x - x_0 - \frac{i}{2}z\right)$, so the vector $x_- := x - x_0 - \frac{i}{2}z$ belongs to $\mathcal{K}_-$. At the same time $x_+ := \frac{i}{2}z \in \mathcal{K}_+$ which means that $x_0, x_\pm$ are the sought vectors. It remains for us to check the uniqueness; this requires to show that $y_0 + y_+ + y_- = 0$ with $y_0 \in D(A)$, $y_\pm \in \mathcal{K}_\pm$ implies $y_0 = y_+ = y_- = 0$. Applying the operator $(A^* - i)$ to $y_0 + y_+ + y_-$ we get $(A-i)y_0 = 2iy_+$, but $y_+ \in \mathcal{K}_+$ and $(A-i)y_0 \in \operatorname{Ran}(A-i) = \mathcal{K}_+^\perp$; hence $y_+ = 0$

and $y_0 \in \mathrm{Ker}\,(A-i)$, which finally gives $y_0 = 0$. The second formula follows easily from the first. ∎

The obtained decomposition represents a starting point to construct self–adjoint extensions to a given symmetric operator. There are different ways to do that; we shall employ a simple trick which allows us to reduce the problem to constructing unitary extensions to an isometric operator. The inspiration comes from the function $w : w(x) = \frac{x-i}{x+i}$, which maps the real axis bijectively to the unit circle in $\mathbb{C}$ with the point $z = 1$ removed; the trick is based on an operator analogy of this relation. For brevity, we denote by $\mathcal{L}_{cs}(\mathcal{H})$ the set of all closed symmetric operators on $\mathcal{H}$.

4.7.6 Lemma: To any $A \in \mathcal{L}_{cs}(\mathcal{H})$, the relation $V_A := (A-i)(A+i)^{-1}$ defines an isometric operator, and $\mathrm{Ran}\,(I - V_A) = D_A$.

Proof: The operators $(A \pm i)^{-1}$ exist, so V_A maps $\mathrm{Ran}\,(A+i)$ to $\mathrm{Ran}\,(A-i)$; the two subspaces are closed since A is closed. If $y \in \mathrm{Ran}\,(A+i)$ we have $y = (A+i)x$ and $V_A y = (A-i)x$ for some $x \in D_A$, and $\|(A+i)x\| = \|(A-i)x\|$, which proves the isometry. Furthermore, the relation $\mathrm{Ran}\,(I - V_A) = (I - V_A)\mathrm{Ran}\,(A+i)$ implies that $y \in \mathcal{H}$ belongs to $\mathrm{Ran}\,(I - V_A)$ *iff* there is $x \in D_A$ such that $y = (I - V_A)(A+i)x = (A+i)x - (A-i)x = 2ix$, so $\mathrm{Ran}\,(I - V_A) = D_A$. ∎

We call $C : A \mapsto V_A$ defined in this way the *Cayley transformation*; it maps $\mathcal{L}_{cs}(\mathcal{H})$ to the set $\mathcal{V}(\mathcal{H})$ of all isometric operators on $\mathcal{H}$ with $\overline{\mathrm{Ran}\,(I - V)} = \mathcal{H}$.

4.7.7 Theorem: (a) The Cayley transformation is a bijection between $\mathcal{L}_{cs}(\mathcal{H})$ and $\mathcal{V}(\mathcal{H})$ with $C^{-1}(V) = i(I + V)(I - V)^{-1}$.

(b) The set $\mathcal{L}_{sa}(\mathcal{H})$ of all self–adjoint operators is mapped by C to the set $\mathcal{U}_1(\mathcal{H})$ of all unitary operators on $\mathcal{H}$ which do not have $\lambda = 1$ as an eigenvalue.

Proof: Let $V \in \mathcal{V}(\mathcal{H})$ and suppose that $Vy = y$ for some $y \in D_V$. Since V is isometric, we have $(x, y) = (z - Vz, y) = 0$ for any $x = (I - V)z$ with $z \in D_V$; then the condition $\overline{\mathrm{Ran}\,(I - V)} = \mathcal{H}$ gives $y = 0$, which means that $I - V$ is invertible and $A_V := i(I+V)(I-V)^{-1}$ is a densely defined operator with the domain $D(A_V) = \mathrm{Ran}\,(I - V)$. We shall check that it is symmetric and closed. The relation $(x, A_V y) = (A_V x, y)$ for all $x, y \in D(A_V) = \mathrm{Ran}\,(I - V)$ follows easily from the fact that V preserves the inner product. For any $[x, y] \in \overline{\Gamma(A_V)}$, there is a sequence $\{z_n\} \subset D_V$ such that $(I - V)z_n \to x$ and $i(I + V)z_n \to y$, so $z_n \to \frac{1}{2}(x - iy)$ and $Vz_n \to -\frac{1}{2}(x+iy)$. Since V is isometric, the subspace D_V is closed and the vector $z := \frac{1}{2}(x - iy)$ belongs to it. Furthermore, V is continuous, so $Vz = -\frac{1}{2}(x+iy)$; this yields $x = (I - V)z$ and $y = i(I + V)z$, *i.e.*, $[x, y] \in \Gamma(A_V)$. To prove (a), it remains to check that $A_V = C^{-1}(V)$. Let $y \in D(C(A_V))$, so $y = (A_V+i)x$ for some $x \in D(A_V)$, then there is $z \in D_V$ such that $x = (I - V)z$ and $A_V x = i(I + V)z$. This implies $C(A_V)y = (A_V-i)x = i(I + V)z - i(I - V)z = 2iVz = Vy$; the same argument in the reversed order yields the inclusion $D_V \subset D(C(A_V))$, so together we have $C(A_V) = V$. In the same way, we can prove the identity $A_{C(A)} = A$.

If A is self–adjoint, then $\mathrm{Ran}\,(A \pm i) = \mathcal{H}$, so $U := C(A)$ is unitary, and $1 \notin \sigma_p(U)$ due to the condition $\overline{\mathrm{Ran}\,(I - U)} = \mathcal{H}$. It remains for us to prove that $A_U := C^{-1}(U)$ is self–adjoint for any $U \in \mathcal{U}_1(\mathcal{H})$, *i.e.*, that $A_U^* \subset A_U$. Let $y \in D(A_U^*)$ so $(y, A_U x) = (y^*, x)$ holds for some $y^* \in \mathcal{H}$ and all $x \in D(A_U) = (I-U)\mathcal{H}$; this is equivalent to the relation $i(y,(I+U)z) = (y^*,(I-U)z)$ for any $z \in \mathcal{H}$, which implies $(I+U^*)y = i(I-U^*)y^*$. Now we have to apply operator U and add $(I-U)y$ to both sides to get the expression $y = \frac{1}{2}(I-U)(y - iy^*)$ which shows that $y \in \mathrm{Ran}\,(I-U) = D(A_U)$. ∎

With these prerequisites we are now able to describe the above mentioned reformulation of the original problem in terms of unitary extensions of a given isometric operator.

4.7.8 Proposition: Let A be a closed symmetric operator and $V := C(A)$ its Cayley transform; then

(a) An operator $V' \supset V$ is the Cayley transform of some nontrivial closed symmetric extension A' of A *iff* the following conditions hold simultaneously:

(i) There are closed subspaces $\mathcal{G}_\pm \subset \mathrm{Ran}\,(A \pm i)^\perp$ with $\dim \mathcal{G}_+ = \dim \mathcal{G}_- > 0$.

(ii) $D(V') = \mathrm{Ran}\,(A+i) \oplus \mathcal{G}_+$ and $\mathrm{Ran}\, V' = \mathrm{Ran}\,(A-i) \oplus \mathcal{G}_-$.

(iii) If $x = y+z$ with $y \in \mathrm{Ran}\,(A+i)$ and $z \in \mathcal{G}_+$, then $V'x = Vy + \tilde{V}z$, where $\tilde{V}$ is an isometric operator from $\mathcal{G}_+$ to $\mathcal{G}_-$.

(b) If the above conditions are valid and, in addition, $\dim \mathcal{G}_+ = \dim \mathcal{G}_- = d < \infty$, then $n_\pm(A') = n_\pm(A) - d$.

(c) Operator A is maximal *iff* at least one of its deficiency indices is zero.

Proof: By definition, $D(V') = \mathrm{Ran}\,(A'+i)$ and $\mathrm{Ran}\, V' = \mathrm{Ran}\,(A'-i)$, and the subspaces $\mathcal{G}_\pm := \mathrm{Ran}\,(A' \pm i) \cap \mathrm{Ran}\,(A \pm i)^\perp$ satisfy condition (ii). Furthermore, the isometric operator $\tilde{V} := V' \restriction \mathcal{G}_+$ satisfies (iii); we shall show that $\mathrm{Ran}\, \tilde{V} = \mathcal{G}_-$. We have $(Vy, \tilde{V}z) = (V'y, V'z) = (y,z) = 0$ for any $z \in \mathcal{G}_+$, $y \in \mathrm{Ran}\,(A+i)$, so $\tilde{V}z \in (V\,\mathrm{Ran}\,(A+i))^\perp = \mathrm{Ran}\,(A-i)^\perp$, and since $\tilde{V}z \in \mathrm{Ran}\, V' = \mathrm{Ran}\,(A'-i)$ holds at the same time, we get $\tilde{V}z \in \mathcal{G}_-$, *i.e.*, $\mathrm{Ran}\,\tilde{V} \subset \mathcal{G}_-$. On the other hand, any vector $z' \in \mathcal{G}_- \subset \mathrm{Ran}\,(A'-i) = \mathrm{Ran}\, V'$ can, due to (ii), be written as $z' = V'(y_0+z_0)$ with $y_0 \in \mathrm{Ran}\,(A+i)$ and $z_0 \in \mathcal{G}_+$. We have $V'y_0 = Vy_0 \in \mathrm{Ran}\,(A-i) \subset \mathcal{G}_-^\perp$, so $0 = (Vy_0, z') = (y_0, y_0+z_0) = \|y_0\|^2$, and therefore $z' = V'z_0 = \tilde{V}z \in \mathrm{Ran}\,\tilde{V}$. Hence the isometric operator $\tilde{V}$ maps $\mathcal{G}_+$ on $\mathcal{G}_-$, which implies condition (a). On the other hand, conditions (ii) and (iii) in view of Problem 3.25 determine an isometric operator $V' \supset V = C(A)$; then $V' = C(A')$, where $A' := C^{-1}(V') \in \mathcal{L}_{cs}(\mathcal{H})$ is an extension of A.

To prove (b), it suffices to notice that $\mathcal{H} = \mathrm{Ran}\,(A' \pm i)^\perp \oplus \mathrm{Ran}\,(A \pm i) \oplus \mathcal{G}_\pm$, so $\mathrm{Ran}\,(A \pm i)^\perp = \mathrm{Ran}\,(A' \pm i)^\perp \oplus \mathcal{G}_\pm$. Suppose finally, *e.g.*, that $n_-(A) = 0$; then $\mathrm{Ran}\,(A+i) = D_V = \mathcal{H}$, so V has no nontrivial isometric extensions, and therefore A has no nontrivial self–adjoint extensions. On the other hand, if both deficiency indices are nonzero then the above described construction yields an isometric extension of V, so A cannot be maximal. ∎

These results allow us to classify all symmetric extensions of a given symmetric operator A; without loss of generality we can assume that A is closed and discuss its closed symmetric extensions only. The following situations are possible:

(i) At least one of the deficiency indices is zero; then A is a maximal symmetric operator which is self-adjoint *iff* $n_+(A) = n_-(A) = 0$.

(ii) Both deficiency indices are nonzero and $n_+(A) \neq n_-(A)$. Suppose for definiteness that $n_+(A) > n_-(A)$; then for any $\mathcal{G}_+ \subset \mathrm{Ran}\,(A+i)^\perp$ with $\dim \mathcal{G}_+ \leq n_-(A)$ there is a closed symmetric extension $A' \supset A$. In particular, in the case $\mathcal{G}_+ = \mathrm{Ran}\,(A+i)^\perp$ the above construction yields a maximal isometric extension V' of $V := C(A)$ because $D(V') = \mathcal{H}$; the operator $A' := C^{-1}(V')$ then has $\mathrm{Ran}\,(A'+i) = D(V') = \mathcal{H}$, *i.e.*, $n_-(A') = 0$. It is therefore a maximal operator which is not self-adjoint because $n_+(A') = n_+(A) - n_-(A) > 0$; the operator A has no self-adjoint extensions.

(iii) The deficiency indices are nonzero and equal each other, $n_+(A) = n_-(A) =: d$. Choosing $\mathcal{G}_\pm = \mathrm{Ran}\,(A \pm i)^\perp$, we obtain a unitary extension V' of $C(A)$, so $C^{-1}(V')$ is self-adjoint. If $d < \infty$, any maximal extension of A is self-adjoint by Proposition 4.7.8b. The situation is more complicated if the deficiency indices are infinite; then some maximal extensions of A are self-adjoint and some not. The last named case occurs, *e.g.*, if $\mathcal{G}_+ = \mathrm{Ran}\,(A+i)^\perp$ and $\mathcal{G}_-$ is a *proper* infinite-dimensional subspace in $\mathrm{Ran}\,(A-i)^\perp$; the construction then yields an operator V' which is maximal isometric but not unitary, so $C^{-1}(V')$ is maximal but not self-adjoint.

Let us summarize this discussion.

4.7.9 Theorem: (a) A closed symmetric operator A has a nontrivial closed symmetric extension *iff* both the deficiency indices $n_\pm(A)$ are nonzero.

(b) The operator A has self-adjoint extensions *iff* its deficiency indices equal each other.

(c) Let $n_+(A) = n_-(A) =: d$. If $d < \infty$ then any maximal extension of A is self-adjoint; in the opposite case, there are non-selfadjoint maximal extensions.

With a little effort, the construction of Proposition 4.7.8a provides an explicit expression for symmetric extensions of a given operator $A \in \mathcal{L}_{cs}(\mathcal{H})$.

4.7.10 Theorem (the second von Neumann formula): We adopt the notation of Proposition 4.7.8, then for any $y' \in D(A')$ there are unique $y \in D(A)$ and $x_0 \in \mathcal{G}_+$ such that

$$y' = y + (I - \tilde{V})x_0\,, \quad A'y' = Ay + i(I + \tilde{V})x_0\,;$$

the first formula allows us to write the domain of A' in a direct-sum form, $D(A') = D(A) \oplus (I - \tilde{V})\mathcal{G}_+$.

Proof: Since $D(A') = (I-V')D(V')$, condition (ii) of Proposition 4.7.8a implies that for any $y' \in D(A')$ there are $x \in D_V = \mathrm{Ran}\,(A+i)$ and $x_0 \in \mathcal{G}_+$ such that $y' = (I-V')(x+x_0)$; condition (iii) then gives $(I-V')x = (I-V)x \in D_A$ and $(I-V')x_0 = (I-\tilde{V})x_0$. Therefore denoting $y := (I-V)x$, we get the first relation, and the inverse Cayley transformation, $A' = C^{-1}(V')$, yields the second,

$$A'y' = i(I+V')(x+x_0) = i(I+V)x + i(I+\tilde{V})x_0 = Ay + i(I+\tilde{V})x_0 .$$

The uniqueness of the decomposition follows from the invertibility of $(I-V')$: if $y+(I-\tilde{V})x_0 = 0$ for some $y \in D_A$ and $x_0 \in \mathcal{G}_+$, then $0 = (I-V)x+(I-\tilde{V})x_0 = (I-V')(x+x_0)$ holds for $x := (I-V)^{-1}y \in D_V$, so $x+x_0 = 0$; however, the vectors x, x_0 belong to mutually orthogonal subspaces; hence $x = x_0 = 0$ and also $y = (I-V)x = 0$. ∎

If $d := \dim \mathcal{G}_+ < \infty$, the formulas of the theorem can be given an even more explicit form. We choose orthonormal bases $\{e_j\}_{j=1}^d$ and $\{f_k\}_{k=1}^d$ in the subspaces $\mathcal{G}_+$, $\mathcal{G}_-$, respectively; then there is a bijective correspondence between isometric operators $\tilde{V} : \mathcal{G}_+ \to \mathcal{G}_-$ and $d\times d$ unitary matrices (u_{jk}) given by $\tilde{V}e_k = \sum_{j=1}^d u_{jk} f_k$. Writing $x_0 \in \mathcal{G}_+$ as $x_0 = \sum_{k=1}^d \xi_k e_k$, we get

$$y' = y+\sum_{k=1}^d \xi_k \left(e_k - \sum_{j=1}^d u_{jk} f_j \right) , \quad A'y' = Ay+i\sum_{k=1}^d \xi_k \left(e_k + \sum_{j=1}^d u_{jk} f_j \right) . \tag{4.7}$$

In particular, these relations define, for an operator $A \in \mathcal{L}_{cs}(\mathcal{H})$ with finite deficiency indices (d,d), a map from the set of all $d \times d$ unitary matrices on the set of all self-adjoint extensions of A; using the injectivity of the Cayley transformation together with Proposition 4.7.8a, we see that this map is bijective.

4.7.11 Example (operator P revisited): We have found the deficiency indices in Example 4.7.3; according to it and Theorem 4.7.9, P is self-adjoint on $J = \mathbb{R}$ and maximal symmetric with no self-adjoint extensions for $J = (0,\infty)$. If $J = (a,b)$ is finite, the deficiency subspaces are spanned by the unit vectors $\eta_+ : \eta_+(x) = c\,e^{a-x}$ and $\eta_- : \eta_-(x) = c\,e^{x-b}$, where $c := \left(\frac{1}{2}(1-e^{2(a-b)})\right)^{-1/2}$; then the isometric operators from $\mathcal{G}_+$ to $\mathcal{G}_-$ correspond to complex numbers μ of the unit circle, $|\mu| = 1$. By Theorem 4.7.10, the domain of operator $P(\mu)$ consists of the vectors $\psi = \phi + \xi(\eta_+ - \mu\eta_-)$ with $\phi \in D(P)$ and $\xi \in \mathbb{C}$, on which it acts as

$$P(\mu)\psi = -i\phi' + i\xi(\eta_+ + \mu\eta_-) = -i\phi' + i\xi(-\eta_+' + \mu\eta_-') = -i\psi' .$$

It is easy to find the relation with the parametrization used in Example 4.2.5. Since $\phi(a) = \phi(b) = 0$, we have $P(\mu) = P_\theta$ with $\theta := (e^{a-b} - \mu)(1 - \mu\, e^{a-b})^{-1}$, where the last relation obviously describes a bijective map of the unit circle onto itself.

Next we shall derive some simple spectral properties of self-adjoint extensions of a given closed symmetric operator A with *finite* deficiency indices (n,n). Let us start with the eigenvalues. Given $\lambda \in \mathbb{C}$, we denote $N_A(\lambda) := \mathrm{Ker}\,(A-\lambda)$ and

$m_A(\lambda) := \dim N_A(\lambda)$, so if $\lambda \in \sigma_p(A)$, then $m_A(\lambda)$ is its multiplicity and $N_A(\lambda)$ is the corresponding eigenspace; otherwise $m_A(\lambda) = 0$.

We shall again consider closed extensions of A only. Let $A' \supset A$ be a symmetric extension, not necessarily self–adjoint. Obviously $N_{A'}(\lambda) \supset N_A(\lambda)$ for any $\lambda \in \mathbb{C}$, and due to the closedness of A, A', both subspaces are closed. We denote by $\Delta(\lambda)$ the orthogonal complement to $N_A(\lambda)$ in $N_{A'}(\lambda)$, *i.e.*, $N_{A'}(\lambda) = N_A(\lambda) \oplus \Delta(\lambda)$. Since $N_A(\lambda) = N_{A'}(\lambda) \cap D(A)$, we have $\Delta(\lambda) \cap D(A) = \{0\}$; on the other hand, $\Delta(\lambda) \oplus D(A) \subset D(A')$, so

$$\dim \Delta(\lambda) = \dim(\Delta(\lambda) \oplus D(A))/D(A) \le \dim D(A')/D(A) \le n$$

due to Problems 1.5 and 47; this shows how much the extension can increase the eigenvalue multiplicity:

4.7.12 Proposition: Let A' be a symmetric extension of an operator $A \in \mathcal{L}_{cs}(\mathcal{H})$ with finite deficiency indices (n, n). If λ is an eigenvalue of A', then $m_{A'}(\lambda) \le m_A(\lambda) + n$.

We shall show that if the deficiency indices are finite, the extension does not affect the essential spectrum. First, we prove an auxiliary result. Given an operator $A \in \mathcal{L}_{cs}(\mathcal{H})$, and $\lambda \in \mathbb{C}$, we define A_λ as the part of A in the subspace $N_A(\lambda)^\perp$. Obviously $A_\lambda = A$ if $\lambda \notin \sigma_p(A)$; it is also easy to check that A_λ is a closed symmetric operator on the Hilbert space $N_A(\lambda)^\perp$, the resolvent $(A_\lambda - \lambda)^{-1}$ exists, and $\mathrm{Ran}\,(A_\lambda - \lambda) = \mathrm{Ran}\,(A - \lambda)$.

4.7.13 Proposition: Let $A \in \mathcal{L}_{cs}(\mathcal{H})$. A complex number λ belongs to $\sigma_{ess}(A)$ *iff* at least one of the following conditions is valid:

(i) λ is an eigenvalue of infinite multiplicity.

(ii) The operator $(A - \lambda)^{-1}$ is unbounded.

Proof: The sufficient condition is easy in both cases. Suppose on the contrary that λ is not an eigenvalue of infinite multiplicity, and at the same time $(A - \lambda)^{-1}$ is bounded. Let $\{x_k\} \subset D_A$ be an arbitrary sequence of unit vectors such that $(A - \lambda)x_k \to 0$; to prove $\lambda \notin \sigma_{ess}(A)$, we have to show that it contains a convergent subsequence. We denote by y_k and z_k the projections of x_k to the subspaces $N_A(\lambda)$ and $N_A(\lambda)^\perp$, respectively; then $(A_\lambda - \lambda)z_k = (A - \lambda)x_k \to 0$, and therefore also $z_k \to 0$ because $(A - \lambda)^{-1}$ is bounded. On the other hand, $\{y_k\}$ is a bounded set in the subspace $N_A(\lambda)$ which is finite–dimensional by assumption; hence there is a convergent subsequence $\{y_{k_l}\}$, and the corresponding sequence $\{x_{k_l}\}$ is then also convergent. ∎

4.7.14 Theorem: Suppose that A, $A' \in \mathcal{L}_{cs}(\mathcal{H})$ and operator A has deficiency indices (n, n). If $A \subset A'$, then $\sigma_p(A) \subset \sigma_p(A')$ and $\sigma_{ess}(A) \subset \sigma_{ess}(A')$; in particular, the essential spectra coincide, $\sigma_{ess}(A) = \sigma_{ess}(A')$, if $n < \infty$.

Proof: The inclusions follow directly from the appropriate definitions; it remains to check that $\lambda \notin \sigma_{ess}(A)$ implies $\lambda \notin \sigma_{ess}(A')$ for $n < \infty$. In view of the two previous propositions, λ is not an eigenvalue of A' with infinite multiplicity; hence we have to show that if $(A-\lambda)^{-1}$ is bounded the same is true for $(A'-\lambda)^{-1}$. We denote $L := (I-\tilde{V})\mathcal{G}_+$, then $\dim L = \dim D(A')/D(A) \le n$ by Problem 47, so $\mathrm{Ran}\,(A'-\lambda) = (A'-\lambda)(D(A)\oplus L) = \mathrm{Ran}\,(A-\lambda)\oplus M$, where $\dim M \le n$ and $M \perp \mathrm{Ran}\,(A-\lambda)$. Since the subspace $\mathrm{Ran}\,(A-\lambda)$ is closed the same is true for $\mathrm{Ran}\,(A'-\lambda) = \mathrm{Ran}\,(A'_\lambda-\lambda)$, *i.e.*, the domain of the closed operator $(A'_\lambda-\lambda)^{-1}$; the result then follows from the closed–graph theorem. ∎

We shall conclude this section by proving a useful result which allows us to compare the extension resolvents directly. Let A_1, A_2 be self–adjoint extensions of a closed symmetric operator A. We call A the *maximal common part* of A_1, A_2 if it is an extension to any $\dot{A}$ such that $\dot{A}\subset A_1$ and $\dot{A}\subset A_2$; it is clear that for any pair A_1, A_2 the maximal common part always exists.

4.7.15 Theorem (Krein formula): (a) Suppose that the maximal common part of self–adjoint operators A_1, A_2 has finite deficiency indices (n,n); then the relation

$$(A_1-z)^{-1}-(A_2-z)^{-1} = \sum_{j,k=1}^{n} \lambda_{jk}(z)\,(y_k(\bar{z}),\cdot)\,y_j(z)$$

holds for any $z\in\rho(A_1)\cap\rho(A_2)$, where the matrix $(\lambda_{jk}(z))$ is nonsingular and $y_j(z)$, $j=1,\dots,n$, are linearly independent vectors from $\mathrm{Ker}\,(A^*-z)$.

(b) The functions $\lambda_{jk}(\cdot)$ and $y_j(\cdot)$ can be chosen to be analytic in $\rho(A_1)\cap\rho(A_2)$. In fact, we can define

$$y_j(z) := y_j(z_0)+(z-z_0)(A_2-z)^{-1}y_j(z_0)$$

for $j=1,\dots,n$ and $z\in\rho(A_2)$, where $z_0\in\mathbb{C}\backslash\mathbb{R}$ and $y_j(z_0)$ are fixed linearly independent vectors in $\mathrm{Ker}\,(A^*-z_0)$; then

$$(\lambda(z_1))^{-1}_{jk} = (\lambda(z_2))^{-1}_{jk} - (z_1-z_2)(y_k(\bar{z}_1),y_j(z_2))$$

holds for $z_1,z_2\in\rho(A_1)\cap\rho(A_2)$ and $j,k=1,\dots,n$.

Proof: We denote the maximal common part as A_0 and $D_z := (A_1-z)^{-1}-(A_2-z)^{-1}$. If $x=(A_0-z)y$ with $y\in D(A_0)$, the relation $D_z x = (R_{A_1}(z)-R_{A_2}(z))(A_0-z)y = y-y=0$ gives $D_z x=0$ for an arbitrary $x\in\mathrm{Ker}\,(A_0^*-\bar{z}) = \mathrm{Ran}\,(A_0-z)$. On the other hand, we have $(y,D_z x) = ((R_{A_1}(\bar{z})-R_{A_2}(\bar{z}))y,x)=0$ for $y\in\mathrm{Ran}\,(A_0-\bar{z})$ and $x\in\mathrm{Ker}\,(A_0^*-z)$, which means that $D_z x\in\mathrm{Ker}\,(A_0^*-z)$ for $x\in\mathrm{Ker}\,(A_0^*-\bar{z})$. These subspaces are by assumption n–dimensional, so it is possible to choose orthonormal bases $\{e_j(z)\}_{j=1}^n$ and $\{e_j(\bar{z})\}_{j=1}^n$ in $\mathrm{Ker}\,(A_0^*-z)$ and $\mathrm{Ker}\,(A_0^*-\bar{z})$, respectively, and to express the action of D_z as

$$D_z x = \sum_{j=1}^{n} c_j(x)e_j(z)$$

for any $x \in \mathcal{H}$. The functions $c_j(\cdot)$ are linear functionals which fulfil $|c_j(x)| \le \sum_{k=1}^n |c_k(x)| \le \|D_z x\|$, and since D_z is bounded for $z \in \rho(A_1) \cap \rho(A_2)$, they are bounded. Then the Riesz lemma implies the existence of vectors $f_j(z)$, $j = 1, \dots, n$, such that $c_j(x) = (f_j(z), x)$. Now the vectors $e_j(z)$ are linearly independent, so $D_z x = 0$ for $x \in (\mathrm{Ker}\,(A_0^* - \overline{z}))^\perp$ implies $c_j(x) = 0$, $j = 1, \dots$; hence $f_j(z) \in \mathrm{Ker}\,(A_0^* - \overline{z})$ for $j = 1, \dots, n$ and we can express them using the basis $\{e_j(\overline{z})\}_{j=1}^n$ as $f_j(z) = \sum_{j,k=1}^n \overline{\lambda_{jk}(z)} e_k(\overline{z})$, which gives

$$D_z x = \sum_{j,k=1}^n \lambda_{jk}(z)(e_k(\overline{z}), x) e_j(z)\,.$$

Suppose that the matrix–valued function $(\lambda_{jk}(\cdot))$ is singular at some $z_0 \in \rho(A_1) \cap \rho(A_2)$; then the vectors $f_j(z_0)$, $j = 1, \dots, n$, are linearly dependent so there is $y \in \mathrm{Ker}\,(A_0^* - z)$ perpendicular to all of them. This would mean $D_{z_0} y = 0$; in other words, $(A_1 - z_0)^{-1} y = (A_2 - z_0)^{-1} y$, which further implies $y \in D(A_1) \cap D(A_2)$, and $A_1 y - A_2 y = (A_1 - z_0) D_{z_0} (A_2 - z_0) y = 0$ by the second resolvent identity in contradiction with the assumption that A_0 is the maximal common part of A_1, A_2. The proof of part (b) is left to the reader (Problem 52). ∎

4.8 Ordinary differential operators

In fact, the title of this section is too bold. We shall be concerned with a particular class of symmetric second–order operators corresponding to the formal expression

$$\ell := -\frac{d^2}{dx^2} + V(x) \tag{4.8}$$

on an interval $J := (a, b) \subset \mathbb{R}$; however, most of the results and methods discussed here can be extended to a much wider class of ordinary differential operators (see the notes). The real–valued function V is assumed to belong to the class $L_{loc}(a, b)$ of *locally integrable* functions, *i.e.*, integrable on any compact interval $K \subset (a, b)$. We shall present a universal procedure which generalizes the argument of Example 4.2.5 and associates with the formal expression (4.8) a pair of operators $\tilde{H}$ and H on $L^2(a, b)$ such that $\tilde{H}$ is the maximal operator corresponding to ℓ and H is a closed symmetric operator with $H^* = \tilde{H}$. In cases when the two operators do not coincide, we shall discuss self–adjoint extensions of H.

The formal operator (4.8) is *regular* if $b - a < \infty$ and V is integrable on $L(a, b)$; otherwise it is *singular*. In the same way we can classify the endpoints, *e.g.*, a is *regular* if $a > -\infty$ and V is integrable on $L(a, c)$ for some $c > a$ and is otherwise *singular*; ℓ is clearly regular *iff* both the endpoints a, b are regular. By J_ℓ we denote the interval (a, b) amended by the regular endpoints, for instance, $J_\ell = [a, b)$ if the left endpoint is regular, *etc.*; the function V is by definition locally integrable in J_ℓ.

Given an interval $I \subset \mathbb{R}$ we denote by $ac(I)$ the set of all functions $f : I \to \mathbb{C}$ such that f and f' are absolutely continuous in I. It follows from the definition that any function $f \in ac(I)$ has a second derivative a.e. in I which is locally integrable. Returning to the formal operator (4.8) we see that if $f \in ac(J_\ell)$, then the function $x \mapsto (\ell[f])(x)$ is defined a.e. in J_ℓ and is locally integrable there. The absolute continuity allows us to perform integrations by parts which play an important role in the theory (Problem 53); in this sense $ac(J_\ell)$ is the natural (maximal) domain of the operator (4.8).

Let $g \in L_{loc}(J_\ell)$; then a function $f \in ac(J_\ell)$ is said to solve the differential equation $\ell[f] = g$ if $(\ell[f])(x) = g(x)$ holds a.e. in J. Using standard methods (see the notes) one can prove the following existence and uniqueness result.

4.8.1 Theorem: Let ℓ be the formal operator (4.8) and $g \in L_{loc}(J_\ell)$; then for any $c \in J_\ell$ and arbitrary complex γ_0, γ_1 there is just one solution of the equation $\ell[f] = g$ which satisfies the boundary conditions $f(c) = \gamma_0$ and $f'(c) = \gamma_1$.

To a given formal expression ℓ, we define the operator $\tilde{H}$ on $L^2(a,b)$ by $\tilde{H}\psi := \ell[\psi]$; its domain is the subspace

$$\tilde{D} := \{\, \psi : \psi \in ac(J_\ell),\ \ell[\psi] \in L^2(a,b) \,\}.$$

The Lagrange formula of Problem 53 can be used for $[c,d] \subset J_\ell$, and moreover, the limits $[\phi,\psi]_a := \lim_{x\to a+}[\phi,\psi]_x$ and $[\phi,\psi]_b := \lim_{x\to b-}[\phi,\psi]_x$ exist for any $\phi,\psi \in \tilde{D}$, even in the cases when the endpoints are singular or the one–sided limits of the functions ϕ,ϕ',ψ,ψ' in them make no sense; this follows from the fact that $\ell[\overline{\phi}]\psi - \overline{\phi}\ell[\psi] \in L^1(a,b)$. This yields the relation

$$(\tilde{H}\phi,\psi) - (\phi,\tilde{H}\psi) = [\phi,\psi]_b - [\phi,\psi]_a \tag{4.9}$$

for any $\phi,\psi \in \tilde{D}$, which shows that $\tilde{H}$ is not symmetric.

Following the program mentioned in the introduction, we are now going to construct operator the H. We begin with the regular case where we define it as the restriction $H := \tilde{H} \restriction D$ to the set

$$D := \{\, \psi \in \tilde{D} : \psi(a) = \psi'(a) = \psi(b) = \psi'(b) = 0 \,\}.$$

To check that H has the required properties we need an auxiliary result.

4.8.2 Lemma: (a) If ϕ_0, ϕ_1 are linearly independent solutions to the equation $\ell[\phi] = 0$, then $\operatorname{Ker}\tilde{H} = (\operatorname{Ran} H)^\perp = \{\phi_0,\phi_1\}_{lin}$.

(b) For any $\gamma_0,\gamma_1,\delta_0,\delta_1 \in \mathbb{C}$, there is a function $\phi \in \tilde{D}$ such that $\phi^{(r)}(a) = \gamma_r$ and $\phi^{(r)}(b) = \delta_r$, $r = 0,1$, where we have denoted $\phi^{(0)} := \phi$ and $\phi^{(1)} := \phi'$.

(c) If the relation $(\phi, H\psi) = (\eta,\psi)$ is valid for some $\phi,\eta \in L^2(a,b)$ and all $\psi \in D$, then $\phi \in \tilde{D}$ and $\eta = \tilde{H}\phi = \ell[\phi]$.

Proof: The relation $\operatorname{Ker}\tilde{H} = \{\phi_0,\phi_1\}_{lin}$ is evident. Suppose that $\eta \in \operatorname{Ran} H$, *i.e.*, $\eta = \ell[\psi]$ for some $\psi \in D$; then the Lagrange formula together with the conditions $\psi^{(r)}(a) = \psi^{(r)}(b) = 0$, $r = 0,1$, gives $\int_a^b \overline{\eta}(x)\phi(x)\,dx = 0$ for any $\phi \in \operatorname{Ker}\tilde{H}$, so

$\operatorname{Ran} H \subset (\operatorname{Ker} \tilde{H})^{\perp}$. On the other hand, let $\eta \in (\operatorname{Ker} \tilde{H})^{\perp}$; then Theorem 4.8.1 implies the existence of $\psi \in ac[a,b]$ such that $\ell[\psi] = \eta$ and $\psi(a) = \psi'(a) = 0$. Using the last conditions in combination with the Lagrange formula, we get $0 = (\eta, \phi_r) = (\ell[\psi], \phi_r) = [\psi, \phi_r]_b$, $r = 0, 1$. Hence it is sufficient to choose the functions ϕ_0, ϕ_1 so that $\phi_s^{(r)}(b) = \delta_{rs}$; then $\psi(b) = \psi'(b) = 0$, so $\psi \in D$ and $\eta = \ell[\psi] = H\psi \in \operatorname{Ran} H$.

To prove (b), first consider the case $\gamma_0 = \gamma_1 = 0$. Using the pair of linearly independent solutions which satisfy the conditions $\phi_s^{(r)}(b) = \delta_{rs}$, we construct the function $\eta := k_0\phi_0 + k_1\phi_1$. The constants k_0, k_1 can be chosen (uniquely) so that $(\phi_0, \eta) = -\delta_1$ and $(\phi_1, \eta) = \delta_0$. Now $\eta \in ac[a,b] \subset L^1(a,b)$, so by Theorem 4.8.1 we can find $\phi_b \in ac[a,b]$ such that $\ell[\phi_b] = \eta$ and $\phi_b(a) = \phi_b'(a) = 0$; at the same time $\ell[\phi_b] \in L^2(a,b)$ so $\phi_b \in \tilde{D}$, and the conditions $\phi_s^{(r)}(b) = \delta_{rs}$ imply finally $\delta_1 = -(\phi_0, \ell[\phi_b]) = \phi_b'(b)$ and $\delta_0 = (\phi_1, \ell[\phi_b]) = \phi_b(b)$. In the same way we construct the function $\phi_a \in \tilde{D}$ which satisfies the conditions $\phi_a^{(r)}(a) = \gamma_r$ and $\phi_a^{(r)}(b) = 0$; then $\phi := \phi_a + \phi_b$ is the sought function.

As for (c), since $\eta \in L^1(a,b)$ there is $\zeta \in \tilde{D}$ such that $\ell[\zeta] = \eta$. We have $\psi \in D$, so the Lagrange formula implies $(\eta, \psi) = (\zeta, H\psi)$. Since the left side is equal to $(\phi, H\psi)$ with the help of the already proven assertion (a) we get $\phi - \zeta \in (\operatorname{Ran} H)^{\perp} = (\operatorname{Ker} \tilde{H})^{\perp\perp} = \operatorname{Ker} \tilde{H}$; the last identity follows from the fact that $\operatorname{Ker} \tilde{H}$ is finite-dimensional. Hence there are $\alpha, \beta \in \mathbb{C}$ such that $\phi = \zeta + \alpha\phi_0 + \beta\phi_1$; this implies $\phi \in \tilde{D}$ and $\ell[\phi] = \ell[\zeta] = \eta$. ∎

4.8.3 Proposition: If the formal operator (4.8) is regular, then $H := \tilde{H} \upharpoonright D$ is a closed symmetric operator such that $H^* = \tilde{H}$; it has deficiency indices $(2,2)$.

Proof: Let $\eta \in D^{\perp}$; then Lemma 4.8.2c with $\phi = 0$ gives $\eta = \ell[\phi] = 0$, so H is densely defined. The same assertion implies $H^* \subset \tilde{H}$, while the opposite inclusion follows from (4.9); together we get $H^* = \tilde{H}$. To check the closedness, we have to prove $H^{**} \subset H$. Since H is symmetric, it satisfies $H^{**} \subset H^*$, and therefore $H^{**}\psi = \ell[\psi]$ on its domain. We also have $(\psi, H^*\phi) = (H^{**}\psi, \phi)$ for any $\phi \in D(H^*) = D(\tilde{H})$, *i.e.*, $(\psi, \ell[\phi]) - (\ell[\psi], \phi) = 0$ so the Lagrange formula implies $[\psi, \phi]_b = [\psi, \phi]_a$. By Lemma 4.8.2b, the boundary values of $\psi^{(r)}$ are arbitrary; hence the last identity gives $\psi^{(r)}(a) = \psi^{(r)}(b) = 0$ for $r = 0, 1$, *i.e.*, $\psi \in D$. Finally, the relation $H^* = \tilde{H}$ means that the deficiency indices equal the number of linearly independent solutions to the equation $\ell[\phi] = \mp i\phi$; by Theorem 4.8.1, there are just two such solutions in the regular case. ∎

The singular case is more complicated; operator H is now constructed in two steps. In the first step, $\tilde{H}$ has to be restricted to a sufficiently small domain; it is natural to use the "minimal" restriction $\dot{H} := \tilde{H} \upharpoonright \dot{D}$ with

$$\dot{D} := \{\, \psi \in \tilde{D} \,:\, \operatorname{supp} \psi \subset (a,b) \text{ is compact} \,\}.$$

4.8.4 Proposition: Operator $\dot{H}$ is symmetric and $\dot{H}^* = \tilde{H}$; its deficiency indices are (n,n), where $0 \le n \le 2$.

Proof: Let $(\phi, \dot{H}\psi) = (\eta, \psi)$ for some $\phi, \eta \in L^2(a,b)$ and all $\psi \in \dot{D}$; we have to show that $\phi \in \tilde{D}$ and $\eta = \tilde{H}\phi$. On a compact interval $K := [c,d] \subset (a,b)$, the formal expression ℓ is regular, and due to the previous proposition it defines a closed symmetric operator on $L^2(K)$ which we denote as H_K. We take any $\psi \in D(H_K)$ and extend it to (a,b) putting $\psi(x) := 0$ for $x \in (a,b)\backslash K$; then $(\dot{H}\psi)(x) = (H_K\psi)(x)$ a.e. in K and $(\dot{H}\psi)(x) = 0$ outside this interval, and therefore $(\phi, H_K\psi) = (\eta,\psi)_K$. By Lemma 4.8.2c, the functions ϕ and ϕ' are absolutely continuous in K and $\eta(x) = (\ell[\phi])(x)$ a.e. in K, and since K was arbitrary, we get $\phi \in \tilde{D}$ and $\eta = \tilde{H}\phi$; it allows us to mimic the argument of the previous proof. The deficiency indices are again given by the number of linearly independent solutions to $\ell[\phi] = \mp i\phi$ which belong to $L^2(a,b)$. In distinction to the regular case, the last condition is not fulfilled automatically; we only know that there are at most two such solutions and the deficiency indices equal each other by Problem 49. ∎

The sought closed symmetric operator generated by ℓ is now defined as $H := \overline{\dot{H}}$. It follows that $H^* = \dot{H}^* = \tilde{H}$ and $\tilde{H}^* = H$, and these conditions in turn show that in the regular case both definitions of the operator H coincide. Its domain is $D(H) := D$; it can be equivalently characterized by means of the boundary form (4.9) — see Problem 55.

4.8.5 Example: Consider the simplest case when $V = 0$, *i.e.*, $\ell = -d^2/dx^2$; we write T, D_T, *etc.*, instead of H, D, respectively. By Problem 13, the subspace $\tilde{D}_T$ coincides with the domain of the operator $\tilde{P}^2$ so $\tilde{T} = \tilde{P}^2$. In the case $J = \mathbb{R}$, the operator P is self-adjoint, so $T = \tilde{T}^* = (\tilde{P}^2)^* \subset (\tilde{P}^*)^2 = \tilde{P}^2 = \tilde{T}$; in combination with $T \subset \tilde{T}$ this relation shows that the operator T on $L^2(\mathbb{R})$ is self-adjoint.

If $J = (0,\infty)$, the formal operator ℓ is singular with one regular endpoint at zero. In view of Problem 55, the domain D_T now consists of all $\psi \in \tilde{D}_T = AC^2(0,\infty)$ which fulfil the condition $[\phi,\psi]_0 = \lim_{x\to\infty}[\phi,\psi]_x$ for any $\phi \in \tilde{D}_T$; however, $\phi' \in L^2(0,\infty)$, so the function $x \mapsto [\phi,\psi]_x$ belongs to $L^1(0,\infty)$ for all $\phi,\psi \in \tilde{D}_T$, which means that $\lim_{x\to\infty}[\phi,\psi]_x = 0$. Hence ψ belongs to D_T *iff* $\overline{\phi(0)}\psi'(0) = \overline{\phi'(0)}\psi(0)$ for all $\phi \in \tilde{D}_T$, and using Problem 54a, we finally get

$$D_T = \{\psi \in AC^2(0,\infty) : \psi(0) = \psi'(0) = 0\}.$$

The equation $-f'' = if$ has two linearly independent solutions, $f_\pm(x) := e^{\pm\epsilon x}$, where $\epsilon := e^{-\pi i/4}$, but only f_- belongs to $L^2(0,\infty)$; hence the deficiency indices are $(1,1)$ in this case.

If ℓ is singular with one regular endpoint, the conclusions of Proposition 4 can be strengthened in the following way.

4.8.6 Proposition: Suppose that the formal operator (4.8) has just one regular endpoint; for definiteness, let it be a. Then

(a) The domain $D(H)$ consists just of the vectors $\psi \in \tilde{D}$ which satisfy the conditions $\psi(a) = \psi'(a) = 0$ and $[\phi,\psi]_b = 0$ for any $\phi \in \tilde{D}$.

(b) The deficiency indices of H are $(1,1)$ or $(2,2)$.

(c) Operator H has no eigenvalues.

Proof: To prove (a), we have to check that the stated conditions are equivalent to those of Problem 55. Since a is a regular endpoint, $\psi(a) = \psi'(a) = 0$ implies $[\phi,\psi]_b = 0$ for any $\phi \in \tilde{D}$, and therefore $[\phi,\psi]_b - [\phi,\psi]_a = 0$. The opposite implication requires us to check that the last condition gives $\psi(a) = \psi'(a) = 0$. By Problem 54a, we can choose for any $c \in (a,b)$ functions $\phi_0, \phi_1 \in \tilde{D}$ such that $\phi_r^{(s)}(a) = \delta_{rs}$ for $r,s = 0,1$ and $\phi_r(x) = 0$ for $x \in (c,b)$, so $[\phi_r,\psi]_b := \lim_{x\to b-}[\phi_r,\psi]_x = 0$, *i.e.*, $[\phi_r,\psi]_a = 0$, which implies $\psi(a) = \psi'(a) = 0$ in view of the conditions $\phi_r^{(s)}(a) = \delta_{rs}$.

Now denote the deficiency indices of H by (n,n). The first von Neumann formula implies that the dimension of the subspace $\mathrm{Ker}\,(H^*+i) \oplus \mathrm{Ker}\,(H^*-i)$ is $\dim \tilde{D}/D = 2n$, so it is sufficient to check that the factor space $\tilde{D}/D$ contains at least two linearly independent vectors. Consider the equivalence classes $\tilde{\phi}_r$ corresponding to the functions $\phi_r \in \tilde{D}$ which satisfy the conditions $\phi_r^{(s)}(a) = \delta_{rs}$, and suppose that $\beta_0\tilde{\phi}_0 + \beta_1\tilde{\phi}_1 = 0$, *i.e.*, that there is $\psi \in D$ such that $\beta_0\phi_0 + \beta_1\phi_1 = \psi$. Assertion (a) then implies $\psi(a) = \psi'(a) = 0$, which together with the conditions $\phi_r^{(s)}(a) = \delta_{rs}$ yields $\beta_0 = \beta_1 = 0$ thus proving assertion (b). Finally, the relation $H\psi = \lambda\psi$ means that ψ solves the equation $\ell[\psi] = \lambda\psi$ with the boundary conditions $\psi(a) = \psi'(a) = 0$, so $\psi = 0$ by Theorem 4.8.1. ∎

This result allows us to treat the general singular case using the *splitting trick* which we are now going to describe. Given a singular expression ℓ on (a,b), we choose a point $c \in (a,b)$ and consider ℓ on the intervals $J_1 := (a,c)$ and $J_2 := (c,b)$. Let $\dot{H}_r$ be the corresponding symmetric operators on $L^2(J_r)$ as described above; it is clear that the direct sum $\dot{H}_1 \oplus \dot{H}_2$ is (up to the natural isomorphism) a restriction of $\dot{H}$. The operator $\dot{H}_1 \oplus \dot{H}_2$ is symmetric and its closure fulfils $\overline{\dot{H}_1 \oplus \dot{H}_2} = H_1 \oplus H_2$, where $H_r := \overline{\dot{H}}_r$ (Problem 29c); moreover, the inclusion $\dot{H}_1 \oplus \dot{H}_2 \subset \dot{H}$ implies $H_1 \oplus H_2 \subset H$. The direct sum $H_1 \oplus H_2$ is called a *splitting* of operator H.

The splitting $H_1 \oplus H_2$ provides a way to determine the deficiency indices of H from $n_\pm(H_r)$, $r = 1,2$. To demonstrate this, let us first prove the relation

$$\dim(D(H)/D(H_1 \oplus H_2)) = 2\,.$$

It follows from Proposition 4.8.6a that $\psi(c) = \psi'(c) = 0$ holds for any $\psi \in D(H_1 \oplus H_2)$. Furthermore, there are functions $\phi_0, \phi_1 \in D_H$ such that $\phi_s^{(r)}(c) = \delta_{rs}$ for $r,s = 0,1$; to see this, it is sufficient to take a closed interval $[c_1,c_2] \subset (a,b)$ containing the point c and to construct by Problem 54a a square integrable function which fulfils the boundary conditions at c and equals zero outside of $[c_1,c_2]$. The subspace $\mathcal{G} := \{\phi_0,\phi_1\}_{lin}$ is two–dimensional because the Wronskian $W(\phi_0,\phi_1) = 1$ at the point $x = c$, $\mathcal{G} \cap D(H_1 \oplus H_2) = \{0\}$ and $\mathcal{G} \oplus D(H_1 \oplus H_2) \subset D_H$; it remains for us to prove that the last inclusion is in fact an identity.

Let $\phi \in D_H$ and $\psi := \phi - \phi(c)\phi_0 - \phi'(c)\phi_1$, so $\psi(c) = \psi'(c) = 0$; in view of Proposition 6a we have to check that $[\psi,\eta_1]_a = 0$ for all $\eta_1 \in \tilde{D}_1$ and $[\psi,\eta_2]_b = 0$

for all $\eta_2 \in \tilde{D}_2$. We use Problem 54a again; due to it, we can extend a given $\eta_1 \in \tilde{D}_1$ to a function $\eta \in \tilde{D}$ which is zero in the vicinity of the point b. Then $[\phi, \eta]_b = 0$ and by Problem 55, $0 = [\phi, \eta]_a = [\phi, \eta_1]_a$. Combining this with the fact that ϕ_0, ϕ_1 are zero in the vicinity of a, we get $[\psi, \eta_1]_a = 0$. In the same way we can check $[\psi, \eta_2]_b = 0$, so together we get $\psi \in D(H_1 \oplus H_2)$.

We know that H and $H_1 \oplus H_2$ are closed symmetric operators; hence the just proved relation, in combination with Proposition 4.7.8b, the second von Neumann formula, and Problems 46 and 55, yields the identity

$$n_\pm(H) \;=\; n_\pm(H_1 \oplus H_2) - 2 \;=\; n_\pm(H_1) + n_\pm(H_2) - 2\,,$$

which represents the desired expression of the deficiency indices of operator H by means of those of H_1 and H_2. Since c is a regular endpoint for both the last named operators, we have $1 \le n_\pm(H_r) \le 2$ so, for instance, H is self–adjoint *iff* $n_\pm(H_r) = 1$ for $r = 1, 2$.

It is often not easy to find the deficiency indices for a singular ℓ directly; however, there is a variety of alternative methods using the behavior of function V. One of them will be mentioned below; we shall return to this question in a more general context in Section 14.1. The splitting trick allows us to reduce the problem to investigation of the formal operators with one regular endpoint, which have the following remarkable property.

4.8.7 Theorem: Let the formal operator (4.8) on (a,b) have one regular endpoint; then the following conditions are equivalent:

(a) The deficiency indices of the operator H are $(2,2)$.

(b) To any $\lambda \in \mathbb{C}$, all solutions of the equation $\ell[\psi] = \lambda\psi$ belong to $L^2(a,b)$.

(c) There is $\lambda \in \mathbb{R}$ such that all solutions of $\ell[\psi] = \lambda\psi$ belong to $L^2(a,b)$.

Proof: The most difficult part is to show that (b) follows from (a). The implication is evident for $\lambda \in \mathbb{C} \setminus \mathbb{R}$; hence we consider $\lambda \in \mathbb{R}$. Let the functions $\phi_1, \phi_2 \in \tilde{D}$ form a basis in Ker$\,(H^*{-}i)$; their Wronskian is constant (Problem 54b) and with a proper normalization we can set $W := \phi_1\phi_2' - \phi_1'\phi_2 = 1$. Let ψ be any solution to $\ell[\psi] = \lambda\psi$ and define $g := (\lambda{-}i)\psi$, so the equation can be rewritten as $\ell[\psi] - i\psi = g$. Assume for definiteness that a is the regular endpoint, $\psi \in ac[a,b)$, and the function g is locally integrable in $[a,b)$. In view of Theorem 1 and Problem 54b, ψ obeys, for any $\alpha_1, \alpha_2 \in \mathbb{C}$, the equation

$$\psi(x) \;=\; \alpha_1\phi_1(x) + \alpha_2\phi_2(x) + (\lambda - i)\int_a^x [\phi_1(x)\phi_2(y) - \phi_2(x)\phi_1(y)]\psi(y)\,dy$$

a.e. in (a,b). Denoting $\phi := |\phi_1| + |\phi_2|$ and $m := \max\{|\alpha_1|, |\alpha_2|\}$ we get the inequality

$$|\psi(x)|^2 \;\le\; 2(m\phi(x))^2 + 2\left[\,|\lambda{-}i|\,\phi(x)\int_a^x \phi(y)|\psi(y)|\,dy\right]^2 .$$

Next, we use the conditions $\phi \in L^2(a,b)$ and $\psi \in ac[a,b)$; the second of these implies $\psi \in L^2(a,x)$, so using the Hölder inequality we get

$$|\psi(x)|^2 \le 2(m\phi(x))^2 + M\phi(x)^2 \int_a^x |\psi(y)|^2\, dy\,,$$

where we have denoted $M := 2|\lambda - i|^2 \int_a^b \phi(x)^2 dx$. Due to the absolute continuity of the integral, there is a $c \in (a,b)$ such that $\int_c^b \phi(x)^2 dx < \frac{1}{2M}$; then

$$\int_c^z \left[\phi(x)^2 \int_a^x |\psi(y)|^2 dy\right] dx \le \int_c^z \phi(x)^2 dx \int_a^z |\psi(y)|^2 dy \le \frac{1}{2M} \int_a^z |\psi(y)|^2 dy\,.$$

In combination with the above estimate, this gives $\int_c^z |\psi(x)|^2 dx \le 2m^2 \int_c^z \phi(x)^2 dx + \frac{1}{2}\int_a^z |\psi(x)|^2 dx$, so

$$\int_c^z |\psi(x)|^2 dx \le 4m^2 \int_c^z \phi(x)^2 dx + \int_a^c |\psi(x)|^2 dx\,,$$

which means that $\psi \in L^2(c,b)$ due to Fatou's lemma, and since $\psi \in ac[a,b)$, we finally get $\psi \in L^2(a,b)$.

Now (c) is a trivial consequence of (b), so it remains to prove that it implies (a). By Proposition 4.8.6c, H has no eigenvalues, so we can use Problem 47c and Proposition 4.8.6b which give $\mathrm{def}\,(H-\lambda) \le n_\pm(H) \le 2$, and since $\mathrm{def}\,(H-\lambda) = 2$ by assumption, the proof is completed. ∎

4.8.8 Remark: Combining Theorem 4.8.7 with Proposition 4.8.6 and the splitting trick, we get the so–called *Weyl alternative* for solutions of the equation $\ell[\psi] = \lambda\psi$ corresponding to the formal operator (4.8). At each of the endpoints *just one* of the following possibilities is valid:

(i) For any $\lambda \in \mathbb{C}$, all solutions are square integrable in the vicinity of the endpoint.

(ii) For any $\lambda \in \mathbb{C}$, there is at least one solution which is *not* square integrable at the vicinity of the endpoint.

Moreover, in case (ii) for any $\lambda \in \mathbb{C}\backslash\mathbb{R}$ there is just one square integrable solution which is unique up to a multiplicative constant.

Notice that the alternative includes those cases with regular endpoints for which (i) is trivially valid. Possibility (i) is called the *limit–circle case* and corresponds to the deficiency indices $(2,2)$, while (ii) is the *limit–point case* with deficiency indices $(1,1)$; this terminology was introduced by H. Weyl.

To conclude the section, let us present a simple result indicated above, which illustrates one way to determine the deficiency indices from the behavior of the function V.

4.8.9 Proposition: Let ℓ be the formal operator (4.8) and suppose that the function V is below bounded on J; then

(a) If $J = (a, \infty)$ or $J = (-\infty, b)$ and a or b, respectively, is a regular endpoint, then the deficiency indices of the corresponding operator H are $(1, 1)$.

(b) If $J = \mathbb{R}$, the operator H is self–adjoint.

Proof: In case (a) it is sufficient to show that the equation $\ell[\psi] = 0$ has at least one solution which does not belong to $L^2(J)$. Such a solution exists if $\psi' \in L^2(J)$ follows from the conditions $\ell[\psi] = 0$ and $\psi \in L^2(J)$, since the assumption of existence of linearly independent solutions ψ_1, ψ_2 to $\ell[\psi] = 0$ implies then that the function $W := \psi_1\psi_2' - \psi_1'\psi_2$ equals a nonzero constant on J, and at the same time $W \in L^1(J)$, which is impossible because the interval J is unbounded. Suppose therefore $\ell[\psi] = 0$ for some nonzero $\psi \in L^2(J)$; the function V is real–valued, so we can assume without loss of generality that the same is true for ψ. By assumption, $V\psi^2 = \psi''\psi$, which yields for any $x \in (a, \infty)$ the estimate

$$\psi'(x)\psi(x) \geq \psi'(a)\psi(a) + \int_a^x \psi'(y)^2 dy + V_0 \int_a^x \psi(y)^2 dy \,,$$

where $V_0 := \inf_{x \in J} V(x)$. Hence $\psi' \notin L^2(a, \infty)$ implies $\lim_{x\to\infty} \psi'(x)\psi(x) = +\infty$, so the function $\psi(\cdot)^2$ is growing as $x \to \infty$; however, this contradicts the condition $\psi \in L^2(a, \infty)$; the proof for $J = (-\infty, b)$ is the same. Assertion (b) now follows from (a), by which the corresponding operators $H_\pm$ on the half–axes have the deficiency indices $(1, 1)$, and the splitting–trick formula which yields $n_\pm(H) = n_\pm(H_+) + n_\pm(H_-) - 2 = 0$. ∎

4.9 Self–adjoint extensions of differential operators

Now we are going to continue the discussion started in the previous section by investigating self–adjoint extensions of the closed symmetric operators H corresponding to formal expressions (4.8). It is clear that any such extension H' is a restriction of the operator $\tilde{H}$, which means that it acts as a differential operator, $H'\phi = \ell[\phi]$ for all $\phi \in D(H')$.

We shall consider only the nontrivial case when at least one of the endpoints corresponds to the limit–circle case, so H is not self–adjoint. Since the deficiency indices are finite, $n = 1$ or 2, we can use formulas (4.7). Suppose that vectors $\eta_j \in \tilde{D}$, $1 \leq j \leq n$, form an orthonormal basis in $\mathrm{Ker}\,(H^* - i)$; then $\ell[\overline{\eta}_j] = -i\overline{\eta}_j$, so $\{\overline{\eta}_j\}_{j=1}^n$ is an orthonormal basis in $\mathrm{Ker}\,(H^* + i)$. The self–adjoint extensions are parametrized by $n \times n$ unitary matrices u; we have $H_u\phi = \ell[\phi]$ for $\phi \in D(H_u)$, where the domain is defined by

$$D_u := D(H_u) = \left\{ \phi \in \tilde{D} \,:\, \phi = \psi + \sum_{k=1}^n \gamma_k \left(\eta_k - \sum_{j=1}^n u_{jk}\overline{\eta}_j \right) ,\ \psi \in D,\ \gamma_k \in \mathbb{C} \right\} .$$

4.9.1 Example: Consider again the operator T corresponding to the formal expression $-d^2/dx^2$, now on an interval $(0,b)$; for simplicity we set $b := \sqrt{2}\pi$. We shall construct a class of self–adjoint extensions; the general analysis is left to the reader (Problem 57). The deficiency indices are $(2,2)$ and the functions η_j in the above formula can be chosen as $\eta_1(x) := c\, e^{\epsilon x}$ and $\eta_2(x) := c\, e^{\pi-\epsilon x}$, where $c := 2^{1/4}(e^{2\pi}-1)^{-1/2}$ and $\epsilon := e^{-\pi i/4}$.

The class in question will correspond to the matrices $u_\vartheta := \frac{\bar\epsilon}{\sqrt{2}} \begin{pmatrix} -i & \bar\vartheta \\ \vartheta & -i \end{pmatrix}$, with ϑ from the unit circle in $\mathbb{C}$. Each of these matrices has the eigenvalues 1, $-i$, so they are all unitarily equivalent. The domain D_ϑ of the operator $T_\vartheta := T_{u_\vartheta}$ consists of the functions

$$\phi = \psi + \gamma_1\left(\eta_1 - \frac{\epsilon}{\sqrt{2}}\bar\eta_1 - \frac{\bar\epsilon}{\sqrt{2}}\vartheta\bar\eta_2\right) + \gamma_2\left(\eta_2 - \frac{\bar\epsilon}{\sqrt{2}}\bar\vartheta\bar\eta_1 - \frac{\epsilon}{\sqrt{2}}\bar\eta_2\right)$$

with $\psi \in D$ and $\gamma_1,\gamma_2 \in \mathbb{C}$, but this formula is clearly not very transparent. An alternative is to use boundary conditions; computing $\phi(0)$, $\phi'(0)$, $\phi(b)$, $\phi'(b)$ from the last formula we can directly check the relations

$$\phi(b) = \theta\phi(0)\,, \quad \phi'(b) = \theta\phi'(0)\,,$$

where we have denoted $\theta := (e^\pi - \vartheta)(\vartheta e^\pi - 1)^{-1}$ (compared to Example 4.7.11, this represents yet another bijection of the unit circle onto itself). Using the first von Neumann formula, we can show that the converse is also true, *i.e.*, if $\phi \in \tilde D$ satisfies the stated boundary conditions, then it can be expressed as the above mentioned linear combination with some $\psi \in D_T$ and complex γ_k. Hence $D_\vartheta = \{\, \phi \in \tilde D_T :\ \phi(b) = \theta\phi(0),\ \phi'(b) = \theta\phi'(0)\,\}$; this means that $T_\vartheta = P_\theta^2$, where P_θ are the self–adjoint extensions of the operator P from Example 4.2.5.

This result allows us to find the spectrum of T_ϑ. In view of to the relation $T_\vartheta = P_\theta^2$, it has no negative eigenvalues; zero is an eigenvalue only if $\theta = 1$, *i.e.*, $\vartheta = 1$. If $\lambda > 0$ the general solution of the equation $-\phi'' = \lambda\phi$ is $x \mapsto \gamma\, e^{i\sqrt{\lambda}x} + \delta\, e^{-i\sqrt{\lambda}x}$; substituting into the boundary conditions and introducing the real parameter $d := \frac{1}{2\pi}\arg\vartheta$ (so that $d \mapsto \vartheta$ maps the interval $[0,1)$ bijectively onto the unit circle) we find that a solution contained in D_ϑ exists *iff* $\lambda = \lambda_k := 2(k+d)^2$, $k = 0, \pm1, \pm2, \dots$. The corresponding normalized eigenvalues are $\phi_k :\ \phi_k(x) = b^{-1/2}e^{i\sqrt{2}(k+d)x}$; in view of Example 2.2.2 these functions form a basis in $L^2(0,b)$, so the operator T_ϑ has a pure point spectrum, $\sigma(T_\vartheta) = \{\, (2(k+d)^2 :\ k = 0, \pm1, \dots\,\}$. Moreover, the spectrum is simple if $d \in (0,\frac12)\cup(\frac12,1)$, *i.e.*, $\mathrm{Im}\,\vartheta \neq 0$. In the case $d = 0$ the eigenvalue $\lambda_0 = 0$ is simple while the other eigenvalues have multiplicity two; the same is true for every eigenvalue if $d = \frac12$.

This implies, in particular, that unitarily equivalent matrices u need not give rise to unitarily equivalent operators. In the present case all the matrices u_ϑ are unitarily equivalent while the extensions corresponding to a pair of different $d_1, d_2 \in (0,\frac12)$ have different spectra, so they cannot be unitarily equivalent. On the other hand, the operators $T_{\vartheta(d)}$ and $T_{\vartheta(1-d)}$ for $\vartheta \in (0,1)$ have the same simple spectra, so they are unitarily equivalent (see the notes to Section 4.4).

The moral of this example is that boundary conditions were much better suited to handle the extensions than the general formulas of Section 4.7. This suggests that we should see whether the same is true for a general H corresponding to the formal expression (4.8); a suitable starting point is the relation (4.9) — *cf.* Problem 58.

Given a subspace L in a vector space V, we call vectors $x_1,\dots,x_n \in V$ *linearly independent modulo* L if no nontrivial linear combination of them belongs to L, *i.e.*, if the condition $\sum_{j=1}^n \alpha_j x_j \in L$ implies $\alpha_1 = \dots = \alpha_n = 0$.

4.9.2 Theorem: (a) Let (n,n) be the deficiency indices of the operator H. If the functions $\omega_k \in \tilde D$, $1 \le k \le n$, are linearly independent modulo D and $[\omega_k,\omega_j]_b - [\omega_k,\omega_j]_a = 0$ holds for $1 \le j,k \le n$, then the functions $\phi \in \tilde D$ which satisfy the conditions

$$[\omega_k,\phi]_b - [\omega_k,\phi]_a = 0\,, \quad 1 \le k \le n\,,$$

form the domain of a self–adjoint extension of the operator H.

(b) In this way we get all self–adjoint extensions; in other words, if a subspace D' fulfilling $D \subset D' \subset \tilde D$ is such that $\tilde H \restriction D'$ is self–adjoint, then there are functions ω_k with the above described properties such that $D' = \{\phi \in \tilde D : [\omega_k,\phi]_b - [\omega_k,\phi]_a = 0,\ 1 \le k \le n\}$.

Proof: By Theorem 4.7.5, $\tilde D = D \oplus \mathcal{G}$, where $\mathcal{G} := \mathrm{Ker}\,(H^*-i) \oplus \mathrm{Ker}\,(H^*+i)$; hence any ω_k can be expressed as $\omega_k = \psi_k + \eta_k$ with $\psi_k \in D$ and $\eta_k \in D$. Using Problem 55 we find that ω_k can be replaced by η_k in the assumptions; we shall check that the subspace

$$D_\omega := \{\phi \in \tilde D : [\omega_k,\phi]_b - [\omega_k,\phi]_a = 0,\ 1 \le k \le n\}$$

coincides with $D' := D \oplus \{\eta_1,\dots,\eta_n\}_{lin}$. Problem 55 gives $D' \subset D_\omega$; to prove the opposite inclusion, we complete $\{\eta_k\}$ to a basis $\{\eta_r\}_{r=1}^{2n} \subset \mathcal{G}$, so for any $\phi \in D_\omega$, there is a $\psi \in D$ and complex numbers $\beta_1,\dots,\beta_{2n}$ such that $\phi = \psi + \sum_{r=1}^{2n} \beta_r \eta_r$. In view of the assumptions, we get then the system of equations

$$\sum_{r=n+1}^{2n} \beta_r \left([\eta_k,\eta_r]_b - [\eta_k,\eta_r]_a\right) = 0\,, \quad 1 \le k \le n\,,$$

which is solved by $\beta_{n+1} = \dots = \beta_{2n} = 0$ only; otherwise a nontrivial solution to the system $\left[\eta_r, \sum_{k=1}^n c_k\eta_k\right]_b - \left[\eta_r, \sum_{k=1}^n c_k\eta_k\right]_a = 0$ would also exist for $n+1 \le r \le 2n$ but this would mean $\sum_{k=1}^n c_k\eta_k \in D$. Hence we get $\phi \in D'$, *i.e.*, $D' = D_\omega$ and using Problem 55 again we conclude the proof of assertion (a).

The domain D' of any extension is of the form $D' = D \oplus L$, where L is an n–dimensional subspace in $\mathcal{G}$. Let $\{\omega_k\}_{k=1}^n$ be a basis in L, so the functions are linearly independent modulo D. Since $\omega_k \in D'$, the assumptions of part (a) are valid by Problem 55; it remains for us to prove that any $\phi \in \tilde D$ which satisfies the condition $[\omega_k,\phi]_b - [\omega_k,\phi]_a = 0$, $1 \le k \le n$, belongs to D'. To this end, we have to complete the set $\{\omega_k\}$ to a basis in $\mathcal{G}$ and to repeat the argument from the first part of the proof. ▮

Now we shall discuss in more detail some situations in which this result can be further simplified, so that we need not check that given functions ω_k are linearly independent modulo D. This happens, *e.g.*, when formal operator ℓ is regular.

4.9.3 Corollary: Let ℓ be a regular expression (4.8) on (a,b); then

(a) Any pair f_1, f_2 of linearly independent functionals on $\tilde{D}$ of the form

$$f_k(\phi) := \alpha_{k1}\phi(a) - \alpha_{k2}\phi'(a) - \beta_{k1}\phi(b) + \beta_{k2}\phi'(b)\,,$$

where the complex coefficients fulfil the conditions $\alpha_{k1}\overline{\alpha}_{j2} - \alpha_{k2}\overline{\alpha}_{j1} = \beta_{k1}\overline{\beta}_{j2} - \beta_{k2}\overline{\beta}_{j1}$ for $j,k = 1,2$, determines a self–adjoint extension $H(f_1,f_2)$ of H with the domain

$$D(f_1,f_2) := \{\,\phi\in\tilde{D}:\ f_k(\phi)=0,\ k=1,2\,\}\,.$$

(b) To any self–adjoint extension $H'\supset H$, there are linearly independent functionals f_1, f_2 on $\tilde{D}$ satisfying the above conditions and such that $D(H') = D(f_1,f_2)$.

Proof: By Lemma 4.8.2b, there are functions $\omega_1,\omega_2\in\tilde{D}$ such that $\omega_k^{(2-j)}(a) = \overline{\alpha}_{kj}$ and $\omega_k^{(2-j)}(b) = \overline{\beta}_{kj}$, so the functionals f_1, f_2 can be expressed as

$$f_k(\phi) = [\omega_k,\phi]_b - [\omega_k,\phi]_a$$

and ω_1,ω_2 fulfil the assumptions of Theorem 4.9.2a. On the other hand, for given $\omega_1,\omega_2\in\tilde{D}$ which obey these assumptions, the last relation defines functionals f_1, f_2, and we can readily check that they are linearly independent *iff* ω_1,ω_2 are linearly independent modulo D. ▮

The functionals f_k are linearly independent *iff* the same is true for the vectors $(\alpha_{k1}, -\alpha_{k2}, -\beta_{k1}, \beta_{k2})$, $k=1,2$, in $\mathbb{C}^4$. In view of the conditions that the coefficients have to satisfy, four real parameters are left, as must be the case because the extensions correspond bijectively to 2×2 unitary matrices. Among these, the following two–parameter class is important.

4.9.4 Example (separated boundary conditions:) Suppose that each of the functionals f_k contains quantities referring to one endpoint only, *e.g.*, $f_1(\phi) = \alpha_{11}\phi(a) - \alpha_{12}\phi'(a)$ and $f_2(\phi) = \beta_{21}\phi(b) - \beta_{22}\phi'(b)$. The assumptions of the previous corollary together with the fact that $H(f_1,f_2) = H(\gamma f_1,\delta f_2)$ for any nonzero $\gamma,\delta\in\mathbb{C}$ show that without loss of generality the coefficients can be chosen real and in the following normalized form,

$$f_1(\phi) \;=\; \phi(a)\,\cos\xi - \phi'(a)\,\sin\xi\,,\quad f_2(\phi) \;=\; \phi(b)\,\cos\eta + \phi'(b)\,\sin\eta\,,$$

for a pair of parameters $\xi, \eta \in [0, \pi)$; the sign convention reflects the natural symmetry of the problem. Some properties of the operators $H_{[\xi,\eta]} := H(f_1, f_2)$ are collected in Problems 59 and 60. Among the conditions $f_k(\phi) = 0$ a distinguished role is played by those corresponding to values 0 and $\frac{\pi}{2}$ of the parameter. The operator $H_{[0,\eta]}$ is said to satisfy the *Dirichlet boundary condition*, $\phi(a) = 0$ at the endpoint a, while $H_{[\frac{\pi}{2},\eta]}$ satisfies the *Neumann boundary condition*, $\phi'(a) = 0$, and similarly for the other endpoint.

Another situation in which Theorem 2 can be simplified occurs when ℓ has one singular endpoint, and there it corresponds to the limit–point case, *i.e.*, the operator H has deficiency indices $(1, 1)$. Suppose for definiteness that the endpoint a is regular; then we have the following result.

4.9.5 Corollary: For any $\xi \in [0, \pi)$, there is a self–adjoint extension H_ξ of H with the domain

$$D_\xi := \{ \phi \in \tilde{D} : \phi(a) \cos\xi - \phi'(a) \sin\xi = 0 \},$$

and on the other hand, to any self–adjoint extension $H' \supset H$ there is just one $\xi \in [0, \pi)$ such that $H' = \tilde{H} \upharpoonright D_\xi$.

Proof: First we shall prove that $[\eta, \phi]_b := \lim_{x \to b-} [\eta, \phi]_x = 0$ holds for all $\phi, \eta \in \tilde{D}$. We use the functions $\phi_0, \phi_1 \in \tilde{D}$ constructed in the proof of Proposition 4.8.6 which can be expressed as $\phi_r = \psi_r + \eta_r$, where $\psi_r \in D$ and $\eta_r \in \mathcal{G}$ with $\eta_r^{(s)}(a) = \delta_{rs}$. Hence η_0, η_1 are linearly independent and form a basis in $\mathcal{G}$, so any $\phi \in \tilde{D}$ can be written as $\phi = \psi + \gamma_0\eta_0 + \gamma_1\eta_1$ with $\psi \in D$. By Proposition 4.8.6, $[\eta, \psi]_b = 0$ for any $\eta \in \tilde{D}$, and since the ϕ_r are zero in the vicinity of b, we have $[\eta, \eta_r]_b = [\eta, \phi_r]_b = 0$, *i.e.*, $[\eta, \phi]_b = 0$. The assertion now follows from Theorem 2. To any $\xi \in [0, \pi)$, we can choose a function $\omega \in \tilde{D}$ such that $\omega'(a) = \cos\xi$ and $\omega(a) = \sin\xi$ which in view of the relation $[\omega, \phi]_b = 0$ fulfils the assumptions of the theorem. On the other hand, for a given H' there is a nonzero function $\omega \in \tilde{D} \setminus D$ such that $\overline{\omega(a)}\omega'(a) = \omega(a)\overline{\omega'(a)}$; the domain of H' then equals $D' = \{ \phi \in \tilde{D} : \overline{\omega'(a)}\phi(a) - \overline{\omega(a)}\phi'(a) = 0 \}$. Hence it is sufficient to choose $\xi = 0$ for $\omega(a) = 0$ and $\xi = \arctan(\omega'(a)/\omega(a))$ otherwise. The injectivity of the map $\xi \mapsto H_\xi$ is easily checked using Problem 54a. ∎

Notice that there are alternative parametrizations of the above set of self–adjoint extensions; for instance, we can use $c := \cot\xi$ setting $H(c) := \tilde{H} \upharpoonright D(c)$, where $D(c) := \{ \phi \in \tilde{D} : \phi'(a) - c\phi(a) = 0 \}$ for $c \in \mathbb{R}$ and $D(\infty) := \{ \phi \in \tilde{D} : \phi(a) = 0 \}$ corresponding to the Dirichlet boundary condition.

4.9.6 Example (the operator T on $(0, \infty)$): By the corollary, the operator T of Example 4.8.5 has a one–parameter family of self–adjoint extensions whose domains are the subspaces in $\tilde{D}_T = AC^2(0, \infty)$ specified by the above boundary conditions at $a = 0$; we denote them as $T(c)$. Let us determine their spectra. The equation $-\phi'' = \lambda\phi$ has a solution in $AC^2(0, \infty)$ only if $\lambda < 0$; it is $\phi(x) = \alpha\, e^{-kx}$ where $k := \sqrt{-\lambda}$. It follows that $T(\infty)$ as well as $T(c)$ for $c \ge 0$ have no eigenvalues, while $T(c)$ with $c < 0$ has just one eigenvalue $\lambda = -c^2$ of unit multiplicity.

Next we shall show that $\sigma(T(\infty)) = [0,\infty)$ holds for the operator with Dirichlet boundary conditions. The spectrum is contained in $\mathbb{R}^+$ because $T(\infty)$ is positive; indeed, $\lim_{x\to\infty} \overline{\phi(x)}\phi'(x) = 0$ due to Problem 12a, so $(\phi, T(\infty)\phi) = \|\phi'\|^2 \ge 0$, and therefore $\inf \sigma(T(\infty)) \ge 0$. To prove the opposite inclusion it is sufficient to choose a suitable sequence of unit vectors from $D(T(\infty))$, say

$$\phi_n \,:\, \phi_n(x) = \sqrt{\frac{6}{n}}\,\left(e^{-x/n} - e^{-x/2n}\right) e^{ikx}\,,$$

for any $k \ge 0$; by a straightforward computation one shows that $\|\phi_n'' + k^2\phi_n\|^2 \to 0$, so Corollary 4.3.5a applies giving the sought result.

This allows us to find the spectra of all the other extensions $T(c)$. Each of them is self-adjoint, so its residual spectrum is empty, as well as $\sigma_{res}(T(\infty))$. On the other hand, the deficiency indices of H are $(1,1)$, so $\sigma_{ess}(T(c)) = \sigma_{ess}(T(\infty)) = [0,\infty)$ in view of Theorem 4.7.14. Combining these results, we get

$$\sigma(T(c)) = \left\{ \begin{array}{lll} [0,\infty) & \ldots & c \ge 0 \\ \\ \{-c^2\} \cup [0,\infty) & \ldots & c < 0 \end{array} \right.$$

In conclusion, let us mention the explicit form of the resolvent for a regular ℓ with separated boundary conditions described in Example 4.9.4 (references to the singular case are given in the notes).

4.9.7 Theorem: Let ℓ be a regular expression (4.8) on (a,b) and H the corresponding closed symmetric operator. Further, let $H_{[\xi,\eta]}$ be a self-adjoint extension of H; then

(a) The resolvent of $H_{[\xi,\eta]}$ is for any $\lambda \in \rho(H_{[\xi,\eta]})$ a Hilbert–Schmidt integral operator with the kernel

$$g_\lambda(x,y) := -W_{ab}^{-1} \left\{ \begin{array}{lll} \psi_b(x)\psi_a(y) & \ldots & x \ge y \\ \psi_a(x)\psi_b(y) & \ldots & x < y \end{array} \right.$$

where ψ_a, ψ_b are the solutions to $\ell[\psi] = \lambda\psi$ fulfilling the boundary conditions $\psi_a(a) = \sin\xi,\ \psi_a'(a) = \cos\xi$ and $\psi_b(b) = -\sin\eta,\ \psi_b'(b) = \cos\eta$, and where $W_{ab} := W(\psi_a, \psi_b)$ is their Wronskian.

(b) The spectrum of $H_{[\xi,\eta]}$ is pure point without accumulation points and simple, *i.e.*, each eigenvalue has a unit multiplicity.

Proof: The functions ψ_a and ψ_b exist due to Theorem 4.8.1 and they are linearly independent, since otherwise they would belong to $D(H_{[\xi,\eta]})$ and λ would be an eigenvalue; hence W_{ab} equals a nonzero constant on $[a,b]$ due to Problem 54b. As elements of $ac[a,b]$, the two functions are square integrable, and therefore the kernel g_λ belongs to $L^2((a,b)\times(a,b))$. We denote the corresponding integral operator as

G_λ; then by straightforward computation we check that $(\ell-\lambda)[G_\lambda\phi]=\phi$ holds for any $\phi\in L^2(a,b)$, and that the function $G_\lambda\phi$ satisfies the corresponding boundary conditions, *i.e.*, that $(H_{[\xi,\eta]}-\lambda)\phi=\phi$ for all $\phi\in L^2(a,b)$, which proves (a).

In view of Problem 18a and the Hilbert–Schmidt property, G_λ is a normal compact operator; since it is injective, zero is its regular value. By Theorems 3.5.7 and 3.5.9, there is now an orthonormal basis $\{\phi_j\}_{j=1}^{\infty}\subset L^2(a,b)$ and a sequence of nonzero numbers $\{\mu_j\}_{j=1}^{\infty}$ such that $G_\lambda\phi_j=\mu_j\phi_j$ and $\lim_{j\to\infty}\mu_j=0$. This implies $\phi_j\in D(H_{[\xi,\eta]})$ and $H_{[\xi,\eta]}=(\mu_j^{-1}+\lambda)\phi_j$, so $H_{[\xi,\eta]})$ has a pure point spectrum. The existence of an accumulation point would mean that a subsequence fulfilled $\mu_{j_k}^{-1}\to\mu-\lambda$; however, that would contradict the condition $\mu_j\to 0$. Finally, the spectrum is simple due to the fact that the solution of $\ell[\phi]=\mu_j\phi$ is determined by the boundary conditions uniquely up to a multiplicative constant. ∎

Notes to Chapter 4

Section 4.1 As in the bounded–operator case, some classes of densely defined operators can be specified using their numerical range. If A is symmetric, *e.g.*, we have $\Theta(A)\subset\mathbb{R}$ in view of Problem 3a. A symmetric operator is called *below* or *above bounded* if $\inf\Theta(A)>-\infty$ or $\sup\Theta(A)<\infty$, respectively. In particular, A is said to be *positive* if $\Theta(A)\subset[0,\infty)$. Likewise, we define the *dissipative* and *accretive* operators as those whose numerical range is contained in the lower complex halfplane $\{\,z:\ \mathrm{Im}\,z\le 0\,\}$ and in $\{\,z:\ \mathrm{Re}\,z\ge 0\,\}$, respectively. More generally, an operator T is *sectorial* if $\Theta(T)$ is a subset of a sector $\{\,z:\ |\arg(z-z_0)|\le\theta\,\}$ of the complex plane with an angle of $2\theta<\pi$.

The adjoint to a densely defined Banach–space operator $T:\mathcal{X}\to\mathcal{Y}$ is defined as follows: its domain $D(T^*)$ is the subspace of all $g\in\mathcal{Y}^*$ for which there is an $f\in\mathcal{X}^*$ such that $g(Tx)=f(x)$ holds for all $x\in D_T$, and $T^*g:=f$ for these g. Hence T^* is a linear operator from $\mathcal{Y}^*$ to $\mathcal{X}^*$. If $\mathcal{X},\mathcal{Y}$ are reflexive and T is closable, then T^* is densely defined and $T^{**}=\overline{T}$ — *cf.* [[Ka]], Sec.III.5.5.

Using the coherent states introduced in Section 2.2, we can derive a useful integral representation for a class of unbounded operators on $A^2(\mathbb{C})$. Consider the set $\mathcal{T}$ consisting of operators which are densely defined and such that $\mathcal{C}\subset D(T)\cap D(T^*)$; then Theorem 2.2.7b yields $(Tf)(w)=(e_w,Tf)=\int_{\mathbb{C}}\overline{(T^*e_w)(z)}f(z)\,e^{-|z|^2/2}dz$ for any $f\in D(T)$, $w\in\mathbb{C}$. Using the theorem once more, we get

$$(Tf)(w)=\int_{\mathbb{C}}(e_w,Te_z)\,f(z)\,e^{-|z|^2/2}dz\quad\text{and}\quad(T^*g)(w)=\int_{\mathbb{C}}\overline{(e_z,Te_w)}\,g(z)\,e^{-|z|^2/2}dz$$

for any $g\in D(T^*)$; in the second relation we have taken into account that $\mathcal{C}$ is total in $A^2(\mathbb{C})$, so the adjoint is densely defined and belongs to $\mathcal{T}$. In this way, we get a sort of matrix representation with respect to the "overcomplete basis" $\mathcal{C}$.

Section 4.2 The condition $\|x_n\|=1$ in the definition of the essential spectrum can be replaced by the requirement that sequence $\{x_n\}$ is bounded; we can easily check that the two definitions are equivalent. We have to stress that there are alternative definitions of $\sigma_{ess}(T)$ in the literature; however, all of them coincide when T is self–adjoint — *cf.* [[Ka]], Sec.X.1.2. The notion of the essential spectrum can also be introduced for closed operators on a Banach space — see [[Ka]], Sec.IV.5.6. Theorem 4.2.7 implicitly contains the following non–trivial statement: the subspaces $D(T)$ and $D(T^*)$ are dense in $\mathcal{H}$ for any $T \in \mathcal{L}_c(\mathcal{H})$.

Section 4.3 If S is a symmetric perturbation to a self–adjoint A which is A–bounded with the relative bound one, then Theorem 4.3.12 is replaced by a weaker assertion known as the *Wüst theorem:* $A+S$ is in this case *essentially* self–adjoint on any core of A. More about stability of self–adjointness with respect to symmetric perturbations can be found, *e.g.*, in ⟦Ka⟧, Sec.V.4; ⟦RS 2⟧, Sec.X.2; ⟦We⟧, Sec.5.3.

Section 4.4 Given a family of operators $\{T_\alpha : \alpha \in I\}$, on Hilbert spaces $\mathcal{H}_\alpha$ with the domains $D_\alpha \subset \mathcal{H}_\alpha$, we can construct the operator $T := \sum_{\alpha\in I}^{\oplus} T_\alpha$ on $\mathcal{H} := \sum_{\alpha\in I}^{\oplus} \mathcal{H}_\alpha$ called the *direct sum* of the operator family $\{T_\alpha\}$ in the following way: we set

$$D := \left\{ X \in \mathcal{H} : X(\alpha) \in D_\alpha, \sum_\alpha \|T_\alpha X(\alpha)\|_\alpha^2 < \infty \right\}$$

and define $(TX)(\alpha) := T_\alpha X(\alpha)$ for any $X \in D$. It is easy to see that any subspace $\mathcal{H}^{(\alpha)} := \{X : X(\beta) = 0 \text{ for } \beta \neq \alpha\}$ reduces T, and if we denote $T^{(\alpha)} := T \upharpoonright \mathcal{H}^{(\alpha)}$ there is a natural unitary equivalence between the direct sum $\sum_{\alpha\in I}^{\oplus} T_\alpha$ and the orthogonal sum $\sum_{\alpha\in I}^{\oplus} T^{(\alpha)}$. Some properties of direct sums are given in Problem 29. In a similar way, one can introduce the *direct integral* of an operator–valued function $T : X \to \int_X^{\oplus} \mathcal{H}(\xi)\, d\mu(\xi)$. The modifications are obvious: the sum in the definition in the domain is replaced by $\int_X \|T(\xi)X(\xi)\|_\xi^2 d\mu(\xi)$, *etc.*

Proposition 4.4.6c shows that the unitary transformations $T \mapsto UTU^{-1}$ do not change $\sigma(T)$; we say that the spectrum is a *unitary invariant.* Other unitary invariants are the eigenvalues and their multiplicities. We could ask which are the quantities $\{p_\alpha\}$ such that $p_\alpha(T) = p_\alpha(S)$ implies that the operators T, S are unitarily equivalent. If, *e.g.*, T and S are normal operators with a pure point spectrum which have same eigenvalues including their multiplicities, then to any eigenvalue λ there is an isometric operator V_λ that maps $\text{Ker}\,(S-\lambda)$ to $\text{Ker}\,(T-\lambda)$, and we can readily check that $U := \sum_{\lambda\in\sigma_p}^{\oplus} V_\lambda$ is unitary and $T = USU^{-1}$. We shall see in the next chapter that in a sense this result extends to any normal operator.

Section 4.6 A form is said to be *sectorial* if its numerical range is contained in a sector $\{z \in \mathbb{C} : |\arg(z-\gamma)| \leq \theta\}$ of the complex plane for some $\gamma \in \mathbb{C}$ and $0 \leq \theta < \frac{\pi}{2}$. Most of the results discussed here have a straightforward generalization to sectorial forms — we refer, *e.g.*, to ⟦ Ka ⟧, Chap.VI, ⟦ RS 1 ⟧, Sec.VIII.6. The Friedrichs extension is in a sense privileged among all self–adjoint extensions of a symmetric operator A; by Theorem 4.6.11 it is the only one with the property that its operator domain is contained in $D(s)$. Moreover, the Friedrichs extension has the smallest form domain of all self–adjoint extensions of A — see Problems 38, 39, and the notes to Section 4.9. Theorem 4.6.14 is named after T. Kato, J. Lions, P. Lax, A. Milgram and E. Nelson.

Section 4.7 We sometimes want to know only that $n_+(A) = n_-(A)$ holds for a given symmetric operator A, so its self–adjoint extensions exist. A sufficient condition is given in Corollary 4.7.2. Another way to check that the deficiency indices equal each other uses a conjugation operator commuting with A (see Problems 47, 48); this idea comes from J. von Neumann as does the most of the theory presented in this section. On the spaces $L^2(X, d\mu)$ the complex conjugation, $\psi \mapsto \overline{\psi}$, is usually used: the deficiency indices of a symmetric A on $L^2(X,d\mu)$ are equal if $\psi \in D_A$ implies $\overline{\psi} \in D_A$ and $(A\overline{\psi})(x) = (\overline{A\psi})(x)$ holds for any $\psi \in D_A$ a.e. in X. Various physical applications of self–adjoint extensions will be discussed in Sections 14.6 and 15.4.

Section 4.8 The proof of Theorem 1 is based on construction of a convergent iteration sequence; it can be found, *e.g.*, in ⟦Nai 2⟧, Sec.16.1 or ⟦DS 2⟧, Sec.XIII.1. We can extend to the equations considered here many results of the classical theory of differential equations (with smooth coefficients), *e.g.*, the existence of a fundamental system of solutions to the homogeneous equation and the variation–of–constants method of solving a nonhomogeneous equation (Problem 54). In addition, most proofs of the classical theory — see, *e.g.*, ⟦Kam⟧ — can also be used with appropriate modifications.

In the general theory of ordinary symmetric differential operators one considers the so–called formally self–adjoint expressions

$$\ell_n \,:\, \ell_n[f] = \sum_{j=0}^{n}(-1)^j \frac{d^j}{dx^j}\left(p_{n-j}\frac{d^j f}{dx^j}\right), \quad n = 1,2,\dots\,,$$

where $p_0,\dots,p_n$ are real–valued functions on an interval (a,b); in particular, for $n=1$ we get the *Sturm–Liouville operator,* which for $p_0 = 1$ yields the expression (4.8). If the functions $p_0^{-1},p_1,\dots,p_n$ are locally integrable in (a,b), then ℓ_n again determines a pair of operators $\tilde H$ and H on $L^2(a,b)$, the first of which has the maximal domain consisting of all $\phi \in L^2(a,b)$ for which $\ell_n[\phi]$ makes sense, and $\ell_n[\phi] \in L^2(a,b)$. The operator H is uniquely determined by the requirement that it is symmetric, closed, and satisfies $H^* = \tilde H$; its deficiency indices are (m,m), where $0 \le m \le n$. The origin of the limit–point/limit–circle terminology is explained, *e.g.*, in ⟦RS 2⟧, notes to Sec.X.2.

Proposition 4.8.9a actually holds under much weaker assumptions about the function V, *e.g.*, it is sufficient that $V(x) > -M(x)$ holds for all x large enough, where M is a positive nondecreasing differentiable function on (a,∞) such that the function $M'M^{-3/2}$ is bounded from above and $\int_a^\infty M^{-1/2}(x)\,dx = \infty$; the proof can be found in ⟦Nai 2⟧, Sec.23, or with slightly modified assumptions in ⟦DS 2⟧, Sec.VIII.6. An illustrative discussion of the halfline case is given in ⟦RS 2⟧, App. to Sec.X.1. From another point of view, Proposition 4.8.9 represents a particular case of the self–adjointness results we shall discuss in Section 14.1.

Section 4.9 By Problem 40, the forms of Example 4.6.1c with $V = 0$ are associated with the self–adjoint extensions $T_{[\xi,\eta]}$ of Example 4, if we set $c_a := \cot\xi$ and $c_b := -\cot\eta$, while the one with the coefficients $c_a = c_b = 0$ and the restricted domain $\{\phi \in AC[a,b] \,:\, \phi(a) = \phi(b) = 0\}$ corresponds to the Dirichlet operator $T_{[0,0]}$, which is the Friedrichs extension of $T = P^2$. Similar conclusions can be made for the operator T on a halfline (Problem 62). These examples well illustrate some differences between the form and operator formalisms; while all the self–adjoint extensions $T_{[\xi,\eta]}$ act as the same differential expression on *different* domains, the corresponding forms are different but have the *same* domain $AC[a,b]$, with the exception of a single form whose domain is smaller and which is associated with the Friedrichs extension of the minimal operator T.

The explicit integral–operator representations for the resolvent in the singular case, as well as those for the regular case with nonseparated boundary conditions, are similar to that of Theorem 4.9.7; they can be found, *e.g.*, in ⟦Nai 2⟧, Sec.19; ⟦DS 2⟧, Sec.XIII.3; see also ⟦We⟧, Sec.8.4.

Problems

1. If $\mathcal{H}$ is separable, then for any $T \in \mathcal{L}(\mathcal{H})$ we can find a countable set $\{x_n\} \subset D_T$ which is dense in $\mathcal{H}$.

2. Prove: (a) The properties of an adjoint collected in Proposition 4.1.2. Show that the boundedness conditions in assertions (d),(e) are in general not necessary.

   (b) The conditions $T \in \mathcal{L}(\mathcal{H})$, $S \in \mathcal{B}(\mathcal{H})$ and $TS \in \mathcal{L}(\mathcal{H})$ do *not* imply $(TS)^* = S^*T^*$.

   (c) If $T, S \in \mathcal{L}(\mathcal{H})$ and S has a bounded inverse, then TS belongs to $\mathcal{L}(\mathcal{H})$ and $(TS)^* = S^*T^*$.

3. Prove: (a) An operator $A \in \mathcal{L}(\mathcal{H})$ is symmetric *iff* its numerical range $\Theta(A)$ is contained in $\mathbb{R}$.

   (b) A symmetric operator A is self-adjoint if $\operatorname{Ran} A = \mathcal{H}$.

   (c) If A is self-adjoint and invertible, then A^{-1} is self-adjoint.

   (d) The sum of a self-adjoint and a Hermitean operator is self-adjoint.

4. If A is symmetric then $\|(A-\lambda)x\|^2 = \|(A - \operatorname{Re}\lambda)x\|^2 + |\operatorname{Im}\lambda|^2\|x\|^2$ holds for any $\lambda \in \mathbb{C}$ and $x \in D_A$.

5. The domain of the operator Q on $L^2(\mathbb{R})$ satisfies $D(Q) \subset L^1(\mathbb{R})$.
   Hint: Apply the Hölder inequality to $x^{-1}x\psi(x)$ on a suitable interval.

6. Let T_s be the operator from Example 4.1.4. Prove

   (a) $D_s \neq \mathcal{H}$ provided the sequence $\{s_j\}$ is unbounded.

   (b) $T_{s+t} \supset T_s + T_t$, where the inclusion turns to equality *iff* $D_s \supset D_{s+t}$ or $D_t \supset D_{s+t}$.

   (c) $T_{st} \supset T_sT_t$, where the equality holds *iff* $D_t \supset D_{st}$.

   (d) T_s^{-1} exists *iff* $s_j \neq 0$ for all j; in that case we have $T_s^{-1} = T_{s^{-1}}$, where $s^{-1} := \{s_j^{-1}\}$.

   Hint: (a) If s is unbounded, there is an increasing sequence of positive integers $\{n_j\}$ such that $|s_{n_j}| > j$.

7. Prove: (a) Theorem 4.1.5 without reference to the closed-graph theorem.

   (b) A closed operator T on $\mathcal{H}$ with $D_T = \mathcal{H}$ is bounded.

   Hint: (a) Apply the uniform boundedness principle to $\{\, f_y : y \in \mathcal{H},\ \|y\| = 1\,\} \subset \mathcal{H}^*$, where $f_y(x) := (y, Ax)$. (b) Show that T^* is bounded and use Theorem 4.2.3.

8. Prove: (a) If A is a symmetric operator then the following conditions are equivalent: (i) A^* is symmetric, (ii) A^* is self-adjoint, (iii) A is essentially self-adjoint.

   (b) If an operator A is *e.s.a.* the same is true for any symmetric extension A' of it, and $\overline{A'} = \overline{A}$.

9. Let U be the operator on $\mathcal{H} \oplus \mathcal{H}$ given by $U[x,y] := [-y,x]$. A linear operator T on $\mathcal{H}$ is densely defined if $(U\Gamma(T))^{\perp}$ is a graph; in that case $\Gamma(T^*) = (U\Gamma(T))^{\perp}$.

10. Using operator V on $\mathcal{H} \oplus \mathcal{H}$ defined by $V[x,y] := [y,x]$, prove

(a) T is invertible *iff* $V\Gamma(T)$ is a graph; in that case, $\Gamma(T^{-1}) = V\Gamma(T)$.

(b) If T is invertible and closed, T^{-1} is also closed.

(c) If T is invertible and $T^{-1} \in \mathcal{L}(\mathcal{H})$, then $\Gamma((T^{-1})^*) = V\Gamma(T^*)$.

11. Prove: (a) Let T be a linear operator on $\mathcal{H}$. Its domain D_T can be equipped with the inner product $(x,y)_T := (x,y)+(Tx,Ty)$; then T is closed *iff* $(D_T, (\cdot,\cdot)_T)$ is a Hilbert space.

(b) Using the previous result show that T is closed if there is a closed operator S such that $D_S = D_T$ and $\|Tx\| = \|Sx\|$ for all $x \in D_T$.

(c) If the operators $T, S \in \mathcal{C}(\mathcal{H})$ have a common core D and $\|Tx\| = \|Sx\|$ holds for all $x \in D$, then $D_T = D_S$ and $\|Tx\| = \|Sx\|$ is valid for any $x \in D_T$.

12. Using the notation of Example 4.2.5, prove

(a) If $\phi \in AC(a,\infty)$ then $\lim_{x\to\infty} \phi(x) = 0$; an analogous assertion is valid for $\phi \in AC(-\infty, b)$.

(b) $AC(\overline{J}) = AC(J)$ for any interval $J \subset \mathbb{R}$.

(c) The operator P is unbounded; construct sequences $\{\psi_n\} \subset D(P)$ of unit vectors such that $\lim_{n\to\infty} \|P\psi_n\| = \infty$ for J finite, semifinite, and infinite.

(d) The operator $P^{(0)}$ on $L^2(0,1)$ defined as the restriction of $\tilde{P}$ to $D^{(0)} := \{\, \psi \in AC(0,1) : \psi(0) = 0 \,\}$ is closed but not symmetric; its adjoint acts as $P^{(0)*} := -i\psi'$ on $D(P^{(0)*}) = \{\, \psi \in AC(0,1) : \psi(1) = 0 \,\}$.

Hint: (a) Use the identity $\int_a^x (\overline{\phi}\phi' + \phi\overline{\phi'})(t)\, dt = |\phi(x)|^2 - |\phi(a)|^2$.

13. Given an interval $J \subset \mathbb{R}$ denote $AC^2(J) := \{\, \psi \in L^2(J) : \psi' \in AC(J) \,\}$. Prove that $AC^2(J) = D(\tilde{P}^2) = D_2$, where $D(\tilde{P}^2) := \{\, \psi \in AC(J) : \psi' \in AC(J) \,\}$ and $D_2 := \{\, \psi \in L^2 : \psi, \psi'$ absolutely continuous in J, $\psi'' \in L^2(J) \,\}$.

14. If T is a closed operator, then its regularity domain $\pi(T)$ defined in Problem 1.64 satisfies $\mathbb{C}\backslash\pi(T) = \sigma(T)\backslash\pi(T) = \sigma_p(T) \cup \sigma_{ess}(T)$.

15. (a) Let μ_d be a discrete measure concentrated on a countable subset $\{x_j\} \subset X$. Then any function $f : X \to \mathbb{C}$ is measurable and the corresponding operator T_f on $L^2(X, d\mu_d)$ is unitarily equivalent to T_{s_f}, where $s_f := \{f(x_j)\}_{j=1}^{\infty}$. Find the corresponding unitary operator.

(b) The conclusions of Problem 6b–d extend to the operators T_f of Example 4.3.3 with the following modification: the inverse T_f^{-1} exists *iff* $f(x) \neq 0$ μ–a.e. in X; in that case $T_f^{-1} = T_g$, where g is defined by $g(x) := f(x)^{-1}$ for $x \in f^{(-1)}(\mathbb{C} \setminus \{0\})$ and arbitrarily in the remaining points.

16. Let T_f be the operator from Example 4.3.3 on $L^2(X, d\mu)$ with a finite measure, $\mu(X) < \infty$. Prove

(a) If the measure ν on X is defined by $\nu(M) := \int_M e^{-|f(x)|^2} d\mu(x)$, then f belongs to $L^p(X, d\nu)$ for any $p \geq 1$, the operator $V := T_g$ with $g(x) := e^{|f(x)|^2/2}$ is unitary from $L^2(X, d\mu)$ to $L^2(X, d\nu)$, and VT_fV^{-1} is the operator of multiplication by f on $L^2(X, d\nu)$.

(b) Suppose that $f \in L^p(X, d\mu)$ for some $p > 2$ and set $r := \frac{2p}{p-2}$; then $L^r(X, d\mu) \subset D(T_f)$ and any D dense in $L^r(X, d\mu)$ is a core for T_f.

17. Let μ be a Borel measure on $\mathbb{R}^d$ and $f \in L^2_{loc}(\mathbb{R}^d, d\mu)$; then $C_0^\infty(\mathbb{R}^d)$ is a core for the operator T_f on $L^2(\mathbb{R}^d, d\mu)$.

18. Prove: (a) The resolvent of a normal operator T is normal for any $\mu \in \rho(T)$.

(b) If A is self–adjoint and $\lambda \in \sigma(A)$, then $\text{s-lim}_{\eta \to 0} R_A(\lambda + i\eta)$ does not exist.

Hint: (b) Since A is closed this would mean $\text{Ran}\,(A - \lambda) = \mathcal{H}$.

19. Find the spectrum of an arbitrary self–adjoint extension P_θ of the operator P on $L^2(0,1)$ and show that it is pure point, while $\sigma_{ess}(P_\theta) = \sigma_{ess}(P) = \emptyset$.
Hint: To prove $\sigma_{ess}(P) = \emptyset$ use Example 1.7.6.

20. Prove Corollary 4.3.10.

21. Find an example of operators A, S such that S is A–bounded with the A–bound equal to one but for any $b \geq 0$ there is $\psi_b \in D_A$ such that $\|S\psi_b\| > \|A\psi_b\| + b\|\psi_b\|$.
Hint: Consider the operators of multiplication by x^2 and $x^2 + x$ on $L^2(\mathbb{R})$, respectively.

22. Suppose that A, S satisfy the assumptions of Theorem 4.3.12, and in addition, A is below bounded. Then $A + S$ is also below bounded and its lower bound can be estimated by

$$m_{A+S} \geq m_A - \max\left\{ \frac{b}{1-a}\,,\; a|m_A| + b \right\},$$

where a, b are the numbers appearing in (4.3).
Hint: $\|SR_A(\mu)\| < 1$ if μ is less than the expression on the right side .

23. Let $\{E_j\}_{j=1}^\infty$ be a complete family of projections in $\mathcal{H}$, *i.e.*, $E_jE_k = \delta_{jk}E_j$ and $\sum_{j=1}^\infty E_j = I$ (strong operator convergence). Further, let T be a closed linear operator which is reduced by all the projections E_j; its part in the subspace $E_j\mathcal{H}$ is denoted as T_j. Then $x \in D_T$ *iff* $E_jx \in D_T$, $j = 1, 2, \ldots$, and $\sum_{j=1}^\infty \|T_jx\|^2 < \infty$; for any such vector x we have $Tx = \sum_{j=1}^\infty T_jx$. The analogous assertion holds for any complete family $\{E_\alpha\}$.

24. Prove: (a) Suppose that T is densely defined and E is a projection such that $ED_T \subset D_T$ and the subspace $E\mathcal{H}$ is T–invariant; then its complement $(E\mathcal{H})^\perp$ is T^*–invariant. In particular, if T is symmetric then the above conditions are sufficient for it to be reduced by E.

(b) If a projection E reduces a normal operator T then it also reduces T^*, and $T_1 := T \upharpoonright ED_T$, $T_2 := T \upharpoonright (I-E)D_T$ are normal operators on Hilbert spaces $E\mathcal{H}$ and $(E\mathcal{H})^\perp$, respectively; moreover, we have $\sigma(T) = \sigma(T_1) \cup \sigma(T_2)$. In particular, if T is self-adjoint (*e.s.a.*, unitary), the same is true for operators T_1 and T_2.

25. Prove: (a) Let T be an invertible operator on $\mathcal{H}$, in general unbounded, which commutes with $B \in \mathcal{B}(\mathcal{H})$; then T^{-1} commutes with B.

(b) Let $\{B_n\} \subset \mathcal{B}(\mathcal{H})$ be a sequence of operators each of which commutes with a closed T and $B_n \xrightarrow{w} B$; then $BT \subset TB$.

26. Let T, S be linear operators on $\mathcal{H}$, $B, C \in \mathcal{B}(\mathcal{H})$ and $BT \subset TB$. Prove

(a) If $BS \subset SB$, then $B(T+S) \subset (T+S)B$ and $BTS \subset TBS$.

(b) If $CT \subset TC$, then $(B+C)T \subset T(B+C)$ and $BCT \subset TBC$.

(c) If T is densely defined, then $B^*T^* \subset T^*B^*$.

(d) If T is closable, then $B\overline{T} \subset \overline{T}B$.

27. Let operators T and S be unitarily invariant, $T = USU^{-1}$; then

(a) The operators T and S have the same eigenvalues and their multiplicities.

(b) If S is bounded the same is true for T and $\|T\| = \|S\|$.

(c) If S is normal, then T is normal; in addition, if S has a pure point spectrum the same is true for T.

(d) If S is closable, then T is also closable and $\overline{T} = U\overline{S}U^{-1}$.

(e) If D is a core for S, then UD is a core for T; in particular, T is *e.s.a.* on UD provided S is *e.s.a.* on D.

(f) If a projection E reduces S, then UEU^{-1} is a projection and reduces T.

28. Projections E, F which satisfy the condition $\|E-F\| < 1$ are unitarily equivalent. *Hint:* Use Proposition 3.2.11.

29. Let $T := \sum_{\alpha\in I}^{\oplus} T_\alpha$ be the direct sum of an operator family. Prove

(a) If all the operators T_α are bounded and $\sup_{\alpha\in I} \|T_\alpha\|_\alpha < \infty$, then T is bounded and $\|T\| := \sup_{\alpha\in I} \|T_\alpha\|_\alpha$.

(b) If all T_α are densely defined, so is T and $T^* = \sum_{\alpha\in I}^{\oplus} T_\alpha^*$; in particular, if all the operators T_α are normal (symmetric, *e.s.a.*, self-adjoint), then T is respectively normal (symmetric, *e.s.a.*, self-adjoint).

(c) If all T_α are closed the same is true for T and $\overline{T} = \sum_{\alpha\in I}^{\oplus} \overline{T}_\alpha$.

30. Prove Theorem 4.5.2.

31. Let $B_r, C_r \in \mathcal{B}(\mathcal{H}_r)$, $r = 1, 2$. Prove

(a) $B_1 \otimes B_2 = 0$ *iff* at least one of the operators B_1, B_2 is zero.

(b) If $B_1 \otimes B_2 = C_1 \otimes C_2 \neq 0$, then $C_1 = \alpha B_1$ and $C_2 = \alpha^{-1} B_2$ holds for some nonzero $\alpha \in \mathbb{C}$.

32. Prove the following properties of the bounded–operator tensor product:

(a) $B_1 \otimes B_2 = \underline{B}_1 \underline{B}_2 = \underline{B}_2 \underline{B}_1$, where $\underline{B}_1 := B_1 \otimes I_2$ and $\underline{B}_2 := I_1 \otimes B_2$.

(b) $|B_1 \otimes B_2| = |B_1| \otimes |B_2|$.

(c) If A_1, A_2 are positive so is $A_1 \otimes A_2$ and $\sqrt{A_1 \otimes A_2} = \sqrt{A_1} \otimes \sqrt{A_2}$.

(d) If $A_r \geq A_r' \geq 0$, $r = 1, 2$, then $A_1 \otimes A_2 \geq A_1' \otimes A_2' \geq 0$.

(e) If E_r, $r = 1, 2$, is a projection, then $\mathrm{Ran}\,(E_1 \otimes E_2) = \mathrm{Ran}\, E_1 \otimes \mathrm{Ran}\, E_2$.

33. Suppose that E_r, F_r are nonzero projections on $\mathcal{H}_r$, $r = 1, 2$, then the projections $\underline{E} := E_1 \otimes E_2$ and $\underline{F} := F_1 \otimes F_2$ are also nonzero and the following analogue of Proposition 3.2.9 is valid:

(a) $\underline{E} + \underline{F}$ is a projection *iff* $E_1 + F_1$ *or* $E_2 + F_2$ is a projection.

(b) $\underline{E} - \underline{F}$ is a projection *iff* $E_1 - F_1$ *and* $E_2 - F_2$ are projections.

(c) $\underline{EF}$ is a nonzero projection *iff* $E_1 F_1$ and $E_2 F_2$ are nonzero projections; then we have $(\mathrm{Ran}\, E_1 \otimes \mathrm{Ran}\, E_2) \cap (\mathrm{Ran}\, F_1 \otimes \mathrm{Ran}\, F_2) = (\mathrm{Ran}\, E_1 \cap \mathrm{Ran}\, F_1) \otimes (\mathrm{Ran}\, E_2 \cap \mathrm{Ran}\, F_2)$.

34. The Fourier–Plancherel operator on $L^2(\mathbb{R}^n)$ can be expressed as $F_n = F \otimes \ldots \otimes F$.

35. (a) Finish the proof of Proposition 4.5.6.

(b) Find an example of operators for which $(T_1 S_1) \otimes (T_2 S_2) \neq (T_1 \otimes T_2)(S_1 \otimes S_2)$.

Hint: (b) Choose T_1 unbounded symmetric and $\mathrm{Ker}\, T_2 = \{0\}$.

36. Given a tensor product $\mathcal{H}_1 \otimes \mathcal{H}_2$ with $\dim \mathcal{H}_2 < \infty$, prove

(a) If $T_1 \in \mathcal{C}(\mathcal{H}_1)$ and T_2 is an invertible operator on $\mathcal{H}_2$, then $T_1 \otimes T_2$ is closed.

(b) If A_1 is self–adjoint the same is true for $A_1 \otimes I_2$.

37. Let $f : \mathbb{R}^n \to \mathbb{R}$ be a measurable function with $\inf\mathrm{ess}_{x \in \mathbb{R}^n} f(x) > -\infty$; then the form s given by $s(\phi, \psi) := \int_{\mathbb{R}^n} f \overline{\phi} \psi \, dx$ with the domain $D(s) := \{\, \phi \in L^2(\mathbb{R}^n) : \int_{\mathbb{R}^n} |f|\, |\phi|^2 dx < \infty \,\}$ is densely defined, symmetric, below bounded, and closed.

38. (a) Prove Proposition 4.6.2.

(b) Let s be a symmetric below bounded form and $\alpha \in \mathbb{C}$; then the conditions $x_n \xrightarrow{(s)} x$ and $y_n \xrightarrow{(s)} y$ imply $\alpha x_n + y_n \xrightarrow{(s)} \alpha x + y$.

(c) If s_r, $r = 1, 2$, are symmetric below bounded forms, the same is true for $s := s_1 + s_2$. If s_r are closed so is s; if they are closable, s is also closable and $\overline{s} \subset \overline{s}_1 + \overline{s}_2$.

(d) Let s be a closed form, then a subspace $D \subset D(s)$ is a core for it *iff* it is dense in $\mathcal{H}_s$.

39. Suppose that A is a symmetric below bounded operator, s is the closure of the form generated by A, and A_s is the corresponding Friedrichs extension. Let $\tilde{A}$ be a self-adjoint extension of A and denote by $\tilde{s}$ the closure of the corresponding form; then $D(\tilde{s}) \supset D(s)$.

40. Prove: (a) Form t of Example 4.6.1c with $V=0$ is closed and the operator A_t associated with it is given by $A_t\phi := -\phi''$ with $D(A_t) := \{\, \phi \in AC^2(\mathbb{R}) : \phi'(a) - c_a\phi(a) = 0,\ \phi'(b) + c_b\phi(b) = 0 \,\}$ (compare with Example 4.9.4).

(b) Let t_0 be the restriction of t to $D(t_0) := \{\, \phi \in AC[a,b] : \phi(a) = \phi(b) = 0 \,\}$. Show that this form is closed and the associated self-adjoint operator is A_{t_0} : $A_{t_0}\phi = -\phi''$ with $D(A_{t_0}) := \{\, \phi \in AC^2(\mathbb{R}) : \phi(a) = \phi(b) = 0 \,\}$.

(c) Are these results preserved in the case of $V \neq 0$?

(d) Find an example of self-adjoint A_1, A_2 such that their form sum differs from Friedrichs extension of their operator sum, $A_1 \dotplus A_2 \neq (A_1 + A_2)_F$.

Hint: Use a relative-boundedness argument. As for (d), try $A_r := \frac{1}{2} A_t^{(r)}$, where $A_t^{(1)}$, $A_t^{(2)}$ are the operators of part (a) corresponding to different pairs $[c_a, c_b]$.

41. Let V be an isometric operator; then $\operatorname{def}(V-\lambda) = \dim(\operatorname{Ran} V)^\perp$ if $|\lambda| < 1$ and $\operatorname{def}(V-\lambda) = \dim(D_V)^\perp$ if $|\lambda| > 1$.

42. If A is symmetric and B Hermitean, then $n_\pm(A) = n_\pm(A+B)$.
Hint: Mimicking the proof of Theorem 4.7.1 show that $t \mapsto \operatorname{def}(A + tB \mp i)$ is constant on $\mathbb{R}$.

43. Let A be a closed symmetric operator. Prove

(a) The decomposition $D(A^*) = D(A) \oplus \operatorname{Ker}(A^* + i) \oplus \operatorname{Ker}(A^* - i)$ of Theorem 4.7.5 is orthogonal with respect to the inner product $(x,y)_A := (A^*x, A^*y) + (x,y)$ on $D(A^*)$.

(b) $\operatorname{Im}(x, A^*x) = \|x_-\|^2 - \|x_+\|^2$ holds for any $x \in D(A^*)$.

44. Prove: (a) A real number λ is an eigenvalue of an operator $A \in \mathcal{L}_{cs}(\mathcal{H})$ *iff* $\mu := \frac{\lambda - i}{\lambda + i}$ is an eigenvalue of its Cayley transform $C(A)$.

(b) A self-adjoint operator A commutes with $B \in \mathcal{B}(\mathcal{H})$ *iff* $C(A)B = BC(A)$.

(c) Let $U : \mathcal{H} \to \mathcal{H}_1$ be a unitary operator; then UAU^{-1} belongs to $\mathcal{L}_{cs}(\mathcal{H}_1)$ for any $A \in \mathcal{L}_{cs}(\mathcal{H})$ and $UC(A)U^{-1} = C(UAU^{-1})$; similarly, UVU^{-1} belongs to $\mathcal{V}(\mathcal{H}_1)$ for any $V \in \mathcal{V}(\mathcal{H})$ and $UC^{-1}(V)U^{-1} = C^{-1}(UVU^{-1})$.

45. Let S be the right-shift operator from Example 3.1.3. Show that the domain of $A_S := C^{-1}(S)$ is $D(A_S) = \left\{ x = \sum_{j=1}^\infty \xi_j e_j : \sum_{n=1}^\infty \left| \sum_{j=1}^n \xi_j \right|^2 < \infty \right\}$ and find the Fourier coefficients $(e_k, A_S x)$ for any $x \in D(A_S)$ and $k = 1, 2, \dots$. What are the deficiency indices of A_S?

46. The Cayley transform of operator P on $L^2(0,\infty)$ from Example 4.2.5 is $C(P) = S^* \restriction \{\psi_0\}^\perp$ where S^* is the left-shift operator with respect to the orthonormal basis $\{\psi_n\}_{n=0}^\infty \subset L^2(0,\infty)$, where $\psi_n(x) := \sqrt{2}\, e^{-x} L_n(2x)$ — *cf.* Example 2.2.3.

47. Let A be a closed symmetric operator with finite deficiency indices (n,n). Using the notation of Theorem 4.7.10, prove

(a) $\dim(I-\tilde{V})\mathcal{G}_+ = \dim\mathcal{G}_+ \le n$.

(b) $\dim(D(A')/D(A)) \le n$; the relation turns to equality *iff* A' is self-adjoint.

(c) $\mathrm{def}\,(A-\lambda) \le n$ holds for any $\lambda\in\mathbb{R}$ which is not an eigenvalue of A.

Hint: (c) $A^*\upharpoonright(D(A)\oplus\mathrm{Ker}\,(A^*-\lambda))$ is a symmetric extension of A.

48. Let J be a *conjugation operator* on $\mathcal{H}$, *i.e.*, an antiunitary operator such that $J^2=I$. Prove

(a) The relations $JL^\perp = (JL)^\perp$, $J\overline{L} = \overline{JL}$ and $\dim J\overline{L} = \dim\overline{L}$ hold for any subspace $L\subset\mathcal{H}$.

(b) If subspaces $L,L'\subset\mathcal{H}$ fulfil $JL\subset L'$ and $JL'\subset L$, then $JL=L'$ and $JL'=L$.

49. Using the results of the previous problem, prove the following *von Neumann theorem:* Let A be a symmetric operator. If there is a conjugation operator J such that $JA\subset AJ$, then $n_+(A)=n_-(A)$.

50. Let A be a closed symmetric operator, then

(a) If $\lambda\notin\sigma_{ess}(A)$, then $\mathrm{Ran}\,(A-\lambda)$ is a closed subspace.

(b) If $\lambda\in\sigma_{ess}(A)$ is not an eigenvalue of infinite multiplicity, then $\mathrm{Ran}\,(A-\lambda)$ is not closed.

51. Let A be a closed symmetric operator with finite deficiency indices (n,n). If $\mathrm{def}\,(A-\lambda)<n$ holds for some real number λ, then the latter belongs to the spectrum of any self-adjoint extension A' of A. In addition, if $\lambda\notin\sigma_p(A)$ then it belongs to $\sigma_{ess}(A')$.

52. Prove Theorem 4.7.15b.
Hint: Use the operator $U(z_1,z_2) := (A_0-z_2)(A_0-z_1)^{-1}$ and the first resolvent identity.

53. Prove *Lagrange's formula* for the formal operator (4.8), *i.e.*, the identity

$$\int_c^d \left(\ell[\overline{f}]g - \overline{f}\ell[g]\right)(x)\,dx = [f,g]_d - [f,g]_c$$

with $[f,g]_x := (\overline{f}g' - \overline{f}'g)(x)$ for all $f,g\in ac(J_\ell)$ and any compact $[c,d]\subset J_\ell$.

54. Let ℓ be the formal operator (4.8) on $J\equiv(a,b)$. Prove

(a) If the endpoint a is regular, then for any $c\in(a,b)$ and arbitrary $\gamma_0,\gamma_1\in\mathbb{C}$ there is $\phi\in\tilde{D}$ such that $\phi(a)=\gamma_0$, $\phi'(a)=\gamma_1$, and $\phi(x)=0$ on (c,b).

(b) Let $g \in L_{loc}(J_\ell)$; if f_1, f_2 are linearly independent solutions to the equation $\ell[f] = g$, then their *Wronskian* $W := f_1 f_2' - f_1' f_2$ is a nonzero constant function on J, and to any $c \in J_\ell$ the function

$$f := W^{-1} \int_c^x [f_1(\cdot) f_2(y) - f_2(\cdot) f_1(y)] g(y)\, dy$$

is the solution corresponding to boundary conditions $f(c) = f'(c) = 0$.

55. Let H be the closed symmetric operator corresponding to the expression (4.8); then $\psi \in \tilde{D}$ belongs to $D \equiv D(H)$ *iff* $[\phi, \psi]_b - [\phi, \psi]_a = 0$ is valid for any $\phi \in \tilde{D}$. If ℓ is regular, it is further equivalent to the conditions $\psi^{(r)}(a) = \psi^{(r)}(b) = 0$, $r = 0, 1$.

56. Let A_r, $r = 1, 2$, be a closed symmetric operator on a Hilbert space $\mathcal{H}_r$, and $A := A_1 \oplus A_2$; then

(a) $n_\pm(A) = n_\pm(A_1) + n_\pm(A_2)$

(b) $\sigma_{ess}(A) = \sigma_{ess}(A_1) \cup \sigma_{ess}(A_2)$.

Hint: (b) Use Proposition 4.7.13.

57. (a) Perform the calculations in Example 4.9.1.

(b) Using the same method, construct all self–adjoint extensions of T.

58. With the notation of Section 4.8, let D' be a subspace such that $D \subset D' \subset \tilde{D}$; then the operator $H' := H \restriction D'$ is self–adjoint *iff* the following conditions are valid simultaneously:

(i) $[\phi, \psi]_b - [\phi, \psi]_a = 0$ for any $\phi, \psi \in D'$

(ii) If $\phi \in \tilde{D}$ satisfies $[\phi, \psi]_b - [\phi, \psi]_a = 0$ for all $\psi \in D'$, then $\phi \in D'$

59. Let T_1 be the operator of Example 4.9.1 corresponding to the boundary conditions $\phi(b) = \phi(0)$ and $\phi'(b) = \phi'(0)$. Prove that the resolvent $(T_1 - z)^{-1}$ is an integral operator with the kernel

$$g_z(x,y) := -\frac{2}{k \sin kb} \left[\sin k(x_< + b) \sin kx_> \; - \; \cos k\left(x + \frac{b}{2}\right) \cos k\left(y + \frac{b}{2}\right) \right],$$

where $k := \sqrt{z}$, $x_< := \min\{x, y\}$ and $x_> := \max\{x, y\}$, and show that the function $z \mapsto g_z(x, y)$ has poles only at the points of $\sigma(T_1)$. Find the resolvent kernels for the other extensions of Example 4.9.1.
Hint: Use Theorem 4.9.7 and Krein's formula.

60. Let $H_{[\xi,\eta]}$ be the operator of Example 4.9.4. Prove

(a) The map $[\xi, \eta] \mapsto H_{[\xi,\eta]}$ is injective on $[0, \pi) \times [0, \pi)$.

(b) If $V(b+a-x) = V(x)$ holds for a.a. $x \in (a, b)$, then $H_{[\xi,\eta]}$ and $H_{[\eta,\xi]}$ are unitarily equivalent by $U :\ (Uf)(x) = f(b+a-x)$.

61. Let $T_{[\xi,\eta]}$ be the operator corresponding to the expression $-d^2/dx^2$ on (a, b), $l := b-a < \infty$, with separated boundary conditions, and $T_{[\xi]} := T_{[\xi,\xi]}$.

(a) Find the spectrum of $T_{[\xi,0]}$.

(b) Find the spectrum of $T_{[\xi,-\xi]}$.

(c) With the exception of the lowest eigenvalue, the spectra of $T_D := T_{[0]}$ and $T_N := T_{[\frac{\pi}{2}]}$ coincide: we have $\lambda_0^N = 0$ and $\lambda_n^N = \lambda_n^D = \left(\frac{\pi n}{l}\right)^2$, $n = 1, 2, \dots$.

(d) To any $\xi \in \left(0, \frac{\pi}{2}\right) \cup \left(\frac{\pi}{2}, \pi\right)$, there is just one eigenvalue of $T_{[\xi]}$ in each interval $\left(\left(\frac{\pi n}{l}\right)^2, \left(\frac{\pi(n+1)}{l}\right)^2\right)$, $n = 0, 1, \dots$, and it is a monotonic function of ξ.

(e) How many negative eigenvalues has the operator $T_{[\xi]}$ and for which values of the parameter?

62. Let T be the operator of Example 4.8.5 on $L^2(0, \infty)$. Show that its Friedrichs extension is the operator $T(\infty)$ with the Dirichlet boundary condition from Example 4.9.6. Find the forms associated with the other self–adjoint extensions $T(c)$.

Chapter 5

Spectral theory

5.1 Projection–valued measures

The central point of this chapter is the spectral theorem which we shall prove in Section 5.3; before doing that we must generalize the standard integration theory to the case of measures whose values are projections in a given Hilbert space rather than real or complex numbers. Given a pair $(X, \mathcal{A})$ where X is a set and $\mathcal{A}$ a σ–field of its subsets, we define a **projection–valued** (or **spectral**) **measure** as a map $E : \mathcal{A} \to \mathcal{B}(\mathcal{H})$ which fulfils the following conditions:

(pm1) $E(M)$ is a projection for any $M \in \mathcal{A}$.

(pm2) $E(\mathbb{R}^d) = I$.

(pm3) $E\left(\bigcup_n M_n\right) = \sum_n E(M_n)$ holds for any at most countable disjoint system $\{M_n\} \subset \mathcal{A}$; if it is infinite, then the right side is understood as the strong limit of the corresponding sequence of partial sums.

The additivity requirement implies, in particular, $E(\emptyset) = 0$. If sets $M, N \in \mathcal{A}$ are disjoint, then $E(M \cup N) = E(M) + E(N)$ is a projection, and therefore $E(M)E(N) = 0$; this means that the projection $E\left(\bigcup_n M_n\right)$ in (pm3) corresponds to the orthogonal sum $\bigoplus_n \operatorname{Ran} E(M_n)$ of the corresponding subspaces. As in the case of numerical measure, we introduce the notions of an E–zero set, and of a proposition–valued function which is valid E–a.e. in X.

In view of the intended application, we restrict our attention to projection–valued measures on $\mathbb{R}^d$ defined on the σ–field $\mathcal{A} := \mathcal{B}^d$ of Borel sets. Let us first list some of their elementary properties (Problem 1).

5.1.1 Proposition: (a) Let M, N be any Borel sets in $\mathbb{R}^d$. Then

$$\begin{aligned} E(M \cap N) &= E(M)E(N) = E(N)E(M)\,, \\ E(M \cup N) &= E(M) + E(N) - E(M \cap N)\,; \end{aligned}$$

the inclusion $M \subset N$ implies $E(M) \le E(N)$, and $E(M) = E(N)$ holds *iff* $E(M \Delta N) = 0$.

(b) If $\{M_n\}$ is a nondecreasing and $\{N_n\}$ a nonincreasing sequence in $\mathcal{B}^d$, then

$$E\left(\bigcup_{n=1}^{\infty} M_n\right) = \operatorname*{s-lim}_{n\to\infty} E(M_n)\,, \quad E\left(\bigcap_{n=1}^{\infty} N_n\right) = \operatorname*{s-lim}_{n\to\infty} E(N_n)\,.$$

(c) The map $\mu_x : \mathcal{B}^d \to \mathbb{R}^+$ defined to any $x \in \mathcal{H}$ by $\mu_x(M) := (x, E(M)x)$ is a finite Borel measure, and similarly $\nu_{xy} : \nu_{xy}(M) := (x, E(M)y)$ is for any pair of vectors $x, y \in \mathcal{H}$ a complex measure with $\operatorname{Re}\nu_{xy} = \frac{1}{4}(\mu_{x+y} - \mu_{x-y})$ and $\operatorname{Im}\nu_{xy} = \frac{1}{4}(\mu_{x-iy} - \mu_{x+iy})$.

5.1.2 Examples: (a) Suppose that $\{P_j\} \in \mathcal{B}(\mathcal{H})$ is an at most countable complete system of mutually orthogonal nonzero projections and $\Lambda := \{\lambda_j\} \subset \mathbb{R}^d$ has the same cardinality; then we define $E_D(M) := \sum_j \chi_M(\lambda_j)P_j$ for any $M \in \mathcal{B}^d$. It is clear that E_D fulfils the conditions (pm1) and (pm2). Let $\{M_n\}_{n=1}^{\infty} \subset \mathcal{B}^d$ be a disjoint system and $M := \bigcup_{n=1}^{\infty} M_n$, then $(x, E_D(M)x) = \sum_j \sum_n \chi_{M_n}(\lambda_j)(x, P_j x)$ holds for any $x \in \mathcal{H}$. The double series has non–negative terms so it can be rearranged; this yields the identity $(x, E_D(M)x) = \sum_{n=1}^{\infty}(x, E_D(M_n)x)$. Since the family $\{P_j\}$ is orthogonal the same is true for $\{E_D(M_n)\}$ and the last relation implies $E_D(M) = \sum_{n=1}^{\infty} E_D(M_n)$, where the series converges with respect to the strong operator topology, *i.e.*, the condition (pm3). The measure $E_D(\cdot)$ is said to be *discrete* with the support $\overline{\Lambda}$.

(b) Given $M \in \mathcal{B}^d$ we define $E(M)$ as the operator of multiplication by the characteristic function of this set, $E(M) := T_{\chi_M}$ in the notation of Example 4.3.3. If M and N are disjoint, we have $\chi_{M\cup N} = \chi_M + \chi_N$, and therefore $E(M \cup N) = E(M) + E(N)$; the σ–additivity can be checked easily using the dominated–convergence theorem. Finally, $E(\mathbb{R}^d) = I$, so E is a projection–valued measure.

In the particular case $X = \mathbb{R}$ it is useful also to introduce another class of maps whose values are projections on a given $\mathcal{H}$. We say that a nondecreasing map $t \mapsto E_t$ is a *spectral decomposition* (or *decomposition of unity*) if it is right–continuous, $E_{u+0} := \operatorname{s-lim}_{t\to u+} E_t = E_u$, and satisfies the conditions

$$\operatorname*{s-lim}_{t\to-\infty} E_t = 0\,, \quad \operatorname*{s-lim}_{t\to+\infty} E_t = I\,. \tag{5.1}$$

This definition makes sense due to the fact that the assertion of Theorem 3.2.12 extends to one–parameter families of projections (Problem 2a). A spectral decomposition will often be denoted as $\{E_t\}$.

5.1.3 Proposition: Any projection–valued measure E on $\mathbb{R}$ determines a spectral decomposition by $E_t := E(-\infty, t]$.

Proof: Monotonicity follows from the additivity of E. The relation $(-\infty, t] = \bigcap_{n=1}^{\infty}(-\infty, t + n^{-1}]$ together with Proposition 1b gives $E_{t+0} = E_t$, and in the same way we check the relations (5.1). ∎

Moreover, a little later we shall see that this natural correspondence between the two classes of projection–valued maps in fact holds in the other direction too.

5.1.4 Example *(spectral decompositions for* $\dim\mathcal{H} < \infty$*):* In this case any point $u \in \mathbb{R}$ has a right neighborhood $(u, u+\delta)$ on which a given $\{E_t\}$ is constant, $E_t = E_u$. Suppose that this is not true; then there is a decreasing sequence $\{t_n\}$ converging to u such that $E_{t_n} \neq E_u$ for all n, and the right continuity of $\{E_t\}$ implies $\text{s-lim}_{n\to\infty} E_{t_n} = E_u$. We may assume without loss of generality that all E_{t_n} are mutually different, but then $\{(E_{t_{n+1}} - E_{t_n})\mathcal{H}\}$ is an infinite family of orthogonal subspaces, and this contradicts the assumption $\dim\mathcal{H} < \infty$. The same argument yields the existence of a left neighborhood $(u-\delta, u)$ on which $E_t = E_{u-0}$, and real numbers $\alpha < \beta$ such that $E_\alpha = 0$ and $E_\beta = I$; in particular, any point at which $t \mapsto E_t$ is continuous has a neighborhood where the map is constant.

This allows us to describe the general form of a spectral decomposition on a finite–dimensional $\mathcal{H}$. We denote $u_1 := \sup\{\, t : E_t = 0 \,\}$. It is clear that $E_{u_1-0} = 0$ and u_1 cannot be a continuity point, so $E_{u_1} \neq 0$. In the case $E_{u_1} \neq I$ we set $u_2 := \sup\{\, t \geq u_1 : E_t = E_{u_1} \,\}$, *etc.*; the process must end at last at the N–th step, where $N := \dim\mathcal{H}$. Concluding the argument, we can say that for a given $\{E_t\}$ there are real numbers $u_1 < u_2 < \cdots < u_n$ and a complete system of orthogonal projections $\{P_j\}_{j=1}^n$, where $1 \leq n \leq \dim\mathcal{H}$, such that $E_t = \sum_{j=1}^n \chi_{(-\infty,t]}(u_j) P_j$ holds for any $t \in \mathbb{R}$.

Next we shall show how projection–valued measures can be constructed starting from a suitable projection–valued set function; this is a direct extension of the standard procedure for numerical–valued measures — see Appendix A.2. Suppose we have a function $\dot{E}$ which associates a projection $\dot{E}(J)$ with any bounded interval $J \subset \mathbb{R}^d$ and satisfies the following conditions:

(i) $\dot{E}(\cdot)$ is additive.

(ii) $\dot{E}(J)\dot{E}(K) = 0$ if $J \cap K = \emptyset$.

(iii) The function $\tilde{\mu}_x := (x, \dot{E}(\cdot)x)$ is regular for each $x \in \mathcal{H}$ and the relation $\sup\{\, \tilde{\mu}_x(J) : J \in \mathcal{J}^d \,\} = \|x\|^2$ holds for any interval $J \in \mathcal{J}^d$.

The condition (ii) implies, in particular, the identity $\dot{E}(J \cap K) = \dot{E}(J)\dot{E}(K) = \dot{E}(K)\dot{E}(J)$ for any $J, K \in \mathcal{J}^d$.

By Theorem A.2.3, there is a unique Borel measure μ_x on $\mathbb{R}^d$ such that $\mu_x(J) = \tilde{\mu}_x(J)$ holds for any $J \in \mathcal{J}^d$. The condition (iii) then gives

$$\mu_{\alpha x} = |\alpha|^2 \mu_x \,, \quad \mu_x(\mathbb{R}^d) = \|x\|^2;$$

to check the last relation, it is sufficient to realize that there is a sequence $\{J_n\}$ of bounded intervals such that $\mu_x(M_n) > \|x\|^2 - n^{-1}$. Given $\{\mu_x\}$, we can associate with any pair $x, y \in \mathcal{H}$ the signed measures $\varrho_{xy} := \frac{1}{4}(\mu_{x+y} - \mu_{x-y})$ and $\tau_{xy} := \frac{1}{4}(\mu_{x+iy} - \mu_{x-iy})$, and the complex measure ν_{xy} defined by $\nu_{xy} := \varrho_{xy} - i\tau_{xy}$. In particular, if $x = y$ we get $\nu_{xx} = \varrho_{xx} = \mu_x$ from the first of the above relations, and the polarization formula implies

$$\nu_{xy}(J) = (x, \dot{E}(J)y), \quad J \in \mathcal{J}^d.$$

To demonstrate that the set function $\dot{E}$ can be extended to a projection–valued measure, we first prove some auxiliary assertions.

5.1.5 Lemma: Let M, N be Borel sets in R^d; then

(a) There is a positive operator $E(M) \in \mathcal{B}(\mathcal{H})$ such that $\nu_{xy}(M) = (x, E(M)y)$ holds for any $x, y \in \mathcal{H}$.

(b) An operator $B \in \mathcal{B}(\mathcal{H})$ commutes with $E(M)$ if it commutes with $\dot{E}(J)$ for all $J \in \mathcal{J}^d$.

(c) $E(M)E(N) = E(N)E(M) = E(M \cap N)$.

Proof: To prove (a), it is sufficient to check that the form $[x, y] \mapsto f_M(x, y) := \nu_{xy}(M)$ is sesquilinear, bounded, and positive. We have $\nu_{x,\alpha y+z}(J) = \alpha\nu_{xy}(J) + \nu_{xz}(J)$ for any $x, y, z \in \mathcal{H}$, $\alpha \in \mathbb{C}$ and $J \in \mathcal{J}^d$; then by Proposition A.4.2 the same relation holds for all $M \in \mathcal{B}^d$, which means that the form f_M is linear in the right argument. In the same way we get $\nu_{xy}(M) = \overline{\nu_{yx}(M)}$, so it is also sesquilinear and symmetric. The relation $f_M(x, x) = \mu_x(M)$ shows it is positive, and finally the Schwartz inequality yields $|f_M(x, y)|^2 \le \mu_x(M)\mu_y(M) \le \mu_x(\mathbb{R}^d)\mu_y(\mathbb{R}^d) \le \|x\|^2\|y\|^2$. The condition $B\dot{E}(J) = \dot{E}(J)B$ implies $\nu_{B^*x,y}(J) = \nu_{x,By}(J)$ for all $x, y \in \mathcal{H}$ and $J \in \mathcal{J}^d$; using Proposition A.4.2 again we get the same identity for all $M \in \mathcal{B}^d$, so (b) follows from the already proven assertion (a).

Let $\mu_x^{(N)}$ be the measure generated by μ_x and the function χ_N, so $\mu_x^{(N)}(M) := \mu_x(M \cap N) = (x, E(M \cap N)x)$, where the last relation follows from assertion (a). Hence $\mu_x^{(J)}(K) = (x, E(J)E(K)x) = \mu_y(K)$ holds for any pair $J, K \in \mathcal{J}^d$, where we have denoted $y := E(J)x$. By Theorem A.2.3, $\mu_x^{(J)}(M) = \mu_y(M)$ for any $M \in \mathcal{B}^d$ and since the vector x was arbitrary, assertion (a) gives $E(M \cap J) = E(J)E(M)E(J)$. The operators $E(M), E(J)$ commute by assertion (b) so the right side can be rewritten as $E(M)E(J)$, *i.e.*, we get $\mu_x^{(N)}(J) = (x, E(J)E(N)x)$. The commutativity of $E(J), E(M)$ implies $E(J) = \sqrt{E(N)}E(J)\sqrt{E(N)}$ by Proposition 3.2.6, so $\mu_x^{(N)}(J) = \mu_z(J)$ with $z := \sqrt{E(N)}x$; repeating the above argument, we arrive at $\mu_x^{(N)}(M) = (x, \sqrt{E(N)}E(M)\sqrt{E(N)}x) = (x, E(M)E(N)x)$, which concludes the proof. ∎

5.1.6 Theorem: Any set function $\dot{E}$ on $\mathcal{J}^d$ which fulfils conditions (i)–(iii) extends uniquely to a projection–valued measure E on $\mathbb{R}^d$. Furthermore, if an operator $B \in \mathcal{B}(\mathcal{H})$ commutes with $E(J)$ for any $J \in \mathcal{J}^d$, then $BE(M) = E(M)B$ holds for all Borel $M \subset \mathbb{R}^d$.

Proof: Assertion (a) of the lemma tells us that $E(\cdot)$ is an extension of $\dot{E}$ to the σ–field $\mathcal{B}^d$ and $E(\mathbb{R}^d) = I$, and putting $M = N$ in (c) we find that $E(M)$ is a projection. The σ–additivity of E follows from the σ–additivity of the measure μ_x and Theorem 3.2.12, and the uniqueness is a consequence of Theorem A.2.3. The last assertion is implied by assertion (b) of the lemma. ∎

Now we want to prove the above mentioned converse to Proposition 3.

5.1.7 Corollary: There is a bijective correspondence between the spectral decompositions $\{E_t\}$ on $\mathcal{H}$ and the projection–valued measures $E : \mathbb{R} \to \mathcal{B}(\mathcal{H})$.

Proof: A spectral decomposition $\{E_t\}$ determines a set function $\dot{E}$ on $\mathcal{J}$ which fulfils conditions (i) and (ii) — see Problem 3b — and the corresponding function $\tilde{\mu}_x$ is regular due to the right continuity of $t \mapsto E_t$. Indeed, in view of Problem 3a to any $\varepsilon > 0$ we can find a neighborhood $(a, a+\delta)$ in which $(x, E_t x) < (x, E_a x) + \frac{1}{2}\varepsilon$, so $(x, E_{a+\delta-0}x) < (x, E_a x) + \varepsilon$. By the same argument, there is $\eta > 0$ such that $(x, E_{b-\eta}x) > (x, E_{b-0}x) - \varepsilon$, and therefore $\tilde{\mu}_x(J_F) > \tilde{\mu}_x(J) - 2\varepsilon$ holds for $J := (a,b)$ and $J_F := [a+\delta, b-\eta]$; similarly we check the regularity for the other types of intervals. Finally, the condition $\sup\{\, \tilde{\mu}_x(J) : J \in \mathcal{J}^d \,\} = \|x\|^2$ is implied by (5.1), so the result follows from Theorem 5.1.6 in combination with Proposition 5.1.3. ∎

5.1.8 Example: The spectral decomposition from Example 5.1.4 is associated with the discrete projection valued–measure $E_D : E_D(M) = \sum_{j=1}^n \chi_m(u_j) P_j$ supported by the set $\{u_j : j = 1, \dots, n\}$ — *cf.* Example 5.1.2a.

Let us mention another application of Theorem 5.1.6. Suppose we have projection–valued measures E on $\mathbb{R}^r$ and F on $\mathbb{R}^s$, both with values in $\mathcal{B}(\mathcal{H})$, which commute mutually, *i.e.*, $E(M)F(N) = F(N)E(M)$ for any $M \in \mathcal{B}^r$ and $N \in \mathcal{B}^s$. By Lemma 5.1.5b, this property is equivalent to $E(J)F(K) = F(K)E(J)$ for all $J \in \mathcal{J}^r$, $K \in \mathcal{J}^s$. A projection–valued measure P on $\mathbb{R}^{r+s}$ is called the *direct product* of E and F if $P(M \times N) = E(M)F(N)$ holds for any $M \in \mathcal{B}^r$ and $N \in \mathcal{B}^s$.

5.1.9 Proposition: Under the stated assumptions, there is a unique direct product $P : \mathcal{B}^{r+s} \to \mathcal{B}(\mathcal{H})$ of E and F.

Proof: Any nonempty interval $\tilde{J} \in \mathcal{J}^{r+s}$ can be expressed uniquely as $\tilde{J} = J \times K$ with $J \in \mathcal{J}^r$ and $K \in \mathcal{J}^s$, and by the commutativity assumption $E(J)F(K)$ is a projection, so $\dot{P}$ defined by $\dot{P}(\tilde{J}) = E(J)F(K)$ for $\tilde{J} \neq \emptyset$ and $\dot{P}(\emptyset) = 0$ maps $\tilde{\mathcal{J}}^{r+s}$ to the set of projections on $\mathcal{H}$. Using the Cartesian product properties, we easily check that $\dot{P}$ fulfils conditions (i) and (ii). Consider now the function $\tilde{\mu}_x := (x, \dot{P}(\cdot)x)$ and a pair of intervals $\tilde{J}_\alpha = J_\alpha \times K_\alpha \in \mathcal{J}^{r+s}$, $\alpha = 1, 2$. We shall use the identity

$$\tilde{\mu}_x(\tilde{J}_1) - \tilde{\mu}_x(\tilde{J}_2) = (x, [E(J_1) - E(J_2)]F(K_1)x) + (x, E(J_2)[F(K_1) - F(K_2)]x);$$

if $\tilde{J}_1 \supset \tilde{J}_2$, then $J_1 \supset J_2$, so $E(J_1) - E(J_2)$ is a projection and the first term on the right side equals $\|F(K_1)[E(J_1) - E(J_2)]x\|^2 \le \mu_x^{(E)}(J_1) - \mu_x^{(E)}(J_2)$. The analogous estimate for the second term gives

$$\tilde{\mu}_x(\tilde{J}_1) - \tilde{\mu}_x(\tilde{J}_2) \le \mu_x^{(E)}(J_1) - \mu_x^{(E)}(J_2) + \mu_x^{(F)}(K_1) - \mu_x^{(F)}(K_2);$$

hence the regularity of $\tilde{\mu}_x$ follows from the same property of $\mu_x^{(E)}$ and $\mu_x^{(F)}$. A similar argument with $\tilde{J}_1$ replaced by $\mathbb{R}^{r+s}$ shows that $\sup\{\tilde{\mu}_x(\tilde{J}) : \tilde{J} \in \mathcal{J}^{r+s}\} = \|x\|^2$, so $\dot{P}$ extends to a projection–valued measure on $\mathbb{R}^{r+s}$.

It remains for us to prove the relation $P(M\times N) = E(M)F(N)$ for any $M \in \mathcal{B}^r$ and $N \in \mathcal{B}^s$; recall that the left side makes sense due to Example A.1.5. Using the identity $M\times N = (M\times\mathbb{R}^s) \cap (\mathbb{R}^r\times N)$ and Proposition 5.1.1a, we see that it is sufficient to check the relations $P(M\times\mathbb{R}^s) = E(M)$ and $P(\mathbb{R}^r\times N) = F(N)$. The first of these can be rewritten as $\mu_x(M\times\mathbb{R}^s) = \mu_x^{(E)}(M)$ for all $x \in \mathcal{H}$, where the numerical measures μ_x, $\mu_x^{(E)}$ correspond to P and E, respectively. We know this holds for intervals, and by σ–additivity, for unions of any at most countable disjoint system $\{J_k\} \subset \mathcal{J}^r$, in particular, for any open set $G \subset \mathbb{R}^r$. The measure $\mu_x^{(E)}$ is regular so there is a nonincreasing sequence of open sets $\{G_n^{(x)}\}$ such that each of them contains M and

$$\mu_x^{(E)}(M) \,=\, \lim_{n\to\infty} \mu_x^{(E)}(G_n^{(x)}) = \mu_x^{(E)}(M_x)\,,$$

where $M_x := \bigcap_{n=1}^{\infty} G_n^{(x)}$. Then $\{G_n^{(x)}\times\mathbb{R}^s\}$ is a nondecreasing sequence of sets containing $M\times\mathbb{R}^s$, and we get $\mu_x(M\times\mathbb{R}^s) = \lim_{n\to\infty} \mu_x^{(E)}(G_n^{(x)}) = \mu_x^{(E)}(M)$, so $\mu_x^{(E)}(M) = 0$ implies $\mu_x(M\times\mathbb{R}^s) = 0$. In the same way, we obtain the relation $\mu_x(M_x\times\mathbb{R}^s) = \mu_x^{(E)}(M)$; combining these two results, we arrive at

$$\mu_x^{(E)}(M) \,=\, \mu_x(M_x\times\mathbb{R}^s) \,=\, \mu_x((M_x\setminus M)\times\mathbb{R}^s) + \mu_x(M\times\mathbb{R}^s) \,=\, \mu_x(M\times\mathbb{R}^s)\,.$$

This is the desired result; similarly we can check the identity $P(\mathbb{R}^r\times N) = F(N)$ for any $N \in \mathcal{B}^s$. ∎

5.2 Functional calculus

Let us turn now to integration with respect to projection–valued measures. We shall proceed in two steps; first we restrict our attention to bounded functions. Let E be a projection–valued measure on $\mathbb{R}^d$. The set of functions $\varphi : \mathbb{R}^d \to \mathbb{C}$ which are E–a.e. defined and Borel will be denoted as $L^\infty(\mathbb{R}^d, dE)$; as usual, we identify the functions which differ at most on an E–zero set. As in the case of numerical–valued measures, we can demonstrate that $L^\infty(\mathbb{R}^d, dE)$ is a Banach space with the norm $\|\cdot\|_\infty$, where $\|\varphi\|_\infty$ is defined as the smallest $c \ge 0$ such that $|\varphi(x)| \le c$ holds for E–a.a. $x \in \mathbb{R}^d$ (alternatively we can use the essential range of $|\varphi|$ — *cf.* Problem 5).

The integral will now be a bounded linear mapping of the space $L^\infty(\mathbb{R}^d, dE)$ into $\mathcal{B}(\mathcal{H})$. We shall define it first on the set of simple Borel functions $\sigma : \mathbb{R}^d \to \mathbb{C}$ which we denote as S_d. If $\sigma = \sum_j \alpha_j \chi_{M_j}$, where $\{M_j\} \subset \mathcal{B}^d$ is a finite disjoint system with $\bigcup_j M_j = \mathbb{R}^d$, we set

$$\mathcal{T}_b(\sigma) \equiv \int_{\mathbb{R}^d} \sigma(t)\, dE(t) := \sum_j \alpha_j E(M_j)\,, \tag{5.2}$$

where t in the integration variable stands for $(t_1,\dots,t_d)$. The function σ can have different representations unless all the numbers α_j are mutually different; however, this does not affect the right side in view of the additivity of the measure E. We shall also write $\mathcal{T}_b^{(E)}(\varphi)$ if the dependence on the projection–valued measure should be stressed; the integration domain and variables will be often dropped.

5.2.1 Proposition: The map $\mathcal{T}_b : S_d \to \mathcal{B}(\mathcal{H})$ is linear and $\|\mathcal{T}_b(\sigma)\| = \|\sigma\|_\infty$. Moreover, the relations $\mathcal{T}_b(\overline{\sigma}) = \mathcal{T}_b(\sigma)^*$ and $\mathcal{T}_b(\sigma\tau) = \mathcal{T}_b(\sigma)\mathcal{T}_b(\tau) = \mathcal{T}_b(\tau)\mathcal{T}_b(\sigma)$ hold for any $\sigma,\tau \in S_d$.
Proof is left to the reader (Problem 6a).

The last property is called *multiplicativity*. Notice that when equipped with the pointwise multiplication, $(\varphi\psi)(x) := \varphi(x)\psi(x)$, the space $L^\infty(\mathbb{R}^d, dE)$ becomes a commutative algebra, and S_d is a subalgebra in it; using the terminology of Appendix B, we can state the assertion concisely as follows: $\mathcal{T}_b$ is an isometric $*$–morphism of the subalgebra $S_d \subset L^\infty(\mathbb{R}^d, dE)$ into the Banach algebra $\mathcal{B}(\mathcal{H})$.

In the next step we extend the mapping $\mathcal{T}_b$ to the whole space $L^\infty(\mathbb{R}^d, dE)$. This is possible in view of Theorem 1.5.5, because $\mathcal{T}_b$ is continuous and S_d is dense in $L^\infty(\mathbb{R}^d, dE)$ by Proposition A.2.2; hence we may define

$$\mathcal{T}_b(\varphi) \equiv \int_{\mathbb{R}^d} \varphi(t)\, dE(t) := \operatorname*{u-lim}_{n\to\infty} \mathcal{T}_b(\sigma_n)\,, \tag{5.3}$$

where $\{\sigma_n\} \subset S_d$ is any sequence such that $\|\varphi - \sigma_n\|_\infty \to 0$.

The "matrix elements" of this operator, $(x, \mathcal{T}_b(\varphi)y) = \lim_{n\to\infty}(x, \mathcal{T}_b(\sigma_n)y)$, can be expressed in an integral form; using $\left|\int \varphi\, d\mu_x - \int \sigma_n\, d\mu_x\right| \le \|\varphi - \sigma_n\|_\infty \|x\|^2$, we get $(x, \mathcal{T}_b(\varphi)x) = \int \varphi\, d\mu_x$, and by polarization this result extends to the nondiagonal case,

$$\left(x, \int_{\mathbb{R}^d} \varphi(t)\, dE(t)\, y\right) = \int_{\mathbb{R}^d} \varphi(t)\, d\nu_{xy}(t) \tag{5.4}$$

for any $x, y \in \mathcal{H}$. Roughly speaking, inner product and integration can be interchanged when matrix elements of $\int \varphi\, dE$ for a bounded φ are computed.

5.2.2 Theorem (functional calculus for bounded functions): The map $\varphi \mapsto \mathcal{T}_b(\varphi)$ defined by (5.3) has the following properties:

(a) It is linear and multiplicative.

(b) $\|\mathcal{T}_b(\varphi)\| = \|\varphi\|_\infty$ and $\mathcal{T}_b(\overline{\varphi}) = \mathcal{T}_b(\varphi)^*$ for any $\varphi \in L^\infty(\mathbb{R}^d, dE)$.

(c) The operator $\mathcal{T}_b(\varphi)$ is normal for any $\varphi \in L^\infty(\mathbb{R}^d, dE)$ and $\sigma(\mathcal{T}_b(\varphi)) = R_{ess}^{(E)}(\varphi)$.

(d) If $\{\varphi_n\} \subset L^\infty(\mathbb{R}^d, dE)$ is a sequence such that $\varphi(t) := \lim_{n\to\infty} \varphi_n(t)$ exists for E–a.a. $t \in \mathbb{R}^d$ and the set $\{\, \|\varphi_n\|_\infty : n = 1, 2, \dots \}$ is bounded, then the function φ belongs to $L^\infty(\mathbb{R}^d, dE)$ and $\mathcal{T}_b(\varphi) = \operatorname{s-lim}_{n\to\infty} \mathcal{T}_b(\varphi_n)$.

Proof: Linearity is due to Proposition 5.2.1 and Theorem 1.5.5. Let $\varphi, \psi \in L^\infty(\mathbb{R}^d, dE)$ be uniformly approximated by sequences $\{\sigma_n\}$, $\{\tau_n\} \subset S_d$. Since the multiplication in $L^\infty(\mathbb{R}^d, dE)$ is continuous, we have $\lim_{n\to\infty} \|\varphi\psi - \sigma_n\tau_n\|_\infty = 0$, and therefore $\mathcal{T}_b(\varphi\psi) = \text{u-}\lim_{n\to\infty} \mathcal{T}_b(\sigma_n\tau_n)$, so Proposition 5.2.1 together with Problem 3.1 yields the multiplicativity. The definition relation (5.3) gives $\|\mathcal{T}_b(\varphi)\| = \lim_{n\to\infty} \|\mathcal{T}_b(\sigma_n)\| = \lim_{n\to\infty} \|\sigma_n\|_\infty = \|\varphi\|_\infty$, and similarly $\mathcal{T}_b(\overline{\varphi}) = \mathcal{T}_b(\varphi)^*$ follows from Proposition 5.2.1 and the continuity of the map $B \mapsto B^*$ in the operator–norm topology.

Assertions (a) and (b) further imply $\mathcal{T}_b(\varphi)^*\mathcal{T}_b(\varphi) = \mathcal{T}_b(|\varphi|^2) = \mathcal{T}_b(\varphi)\mathcal{T}_b(\varphi)^*$, so the operator $\mathcal{T}_b(\varphi)$ is normal. In a similar way,

$$\|(\mathcal{T}_b(\varphi) - \lambda)x\|^2 = (x, \mathcal{T}_b(|\varphi - \lambda|^2)x) = \int_{\mathbb{R}^d} |\varphi(t) - \lambda|^2 d\mu_x(t);$$

by Corollary 3.4.4 a number λ belongs to $\sigma(\mathcal{T}_b(\varphi))$ if the infimum of this expression over all unit vectors x is zero. Suppose that $\lambda \in R^{(E)}_{ess}(\varphi)$ and denote $M_{\varepsilon,\lambda} := \varphi^{(-1)}((\lambda-\varepsilon, \lambda+\varepsilon))$. The projection $E(M_{\varepsilon,\lambda})$ is by assumption nonzero for any $\varepsilon > 0$; at the same time it equals $\mathcal{T}_b(\chi_{M_{\varepsilon,\lambda}})$, so we get

$$\|(\mathcal{T}_b(\varphi) - \lambda)x\|^2 = \int_{\mathbb{R}^d} |\varphi(t) - \lambda|^2 \chi_{M_{\varepsilon,\lambda}}(t)\, d\mu_x(t) < \varepsilon^2$$

for any unit vector $x \in \operatorname{Ran} E(M_{\varepsilon,\lambda})$, *i.e.*, $\lambda \in \sigma(\mathcal{T}_b(\varphi))$. Conversely, let $\lambda \notin R^{(E)}_{ess}(\varphi)$; then $M_{\varepsilon,\lambda}$ is an E–zero set for some $\varepsilon > 0$ and $\mu_x(\mathbb{R}^d \setminus M_{\varepsilon,\lambda}) = \mu_x(\mathbb{R}^d) = \|x\|^2$ holds for all $x \in \mathcal{H}$. This implies $\|(\mathcal{T}_b(\varphi) - \lambda)x\|^2 \ge \varepsilon^2\|x\|^2$, so $\lambda \notin \sigma(\mathcal{T}_b(\varphi))$, thus proving assertion (c). Using Problem 1.10 we readily find that φ belongs to $L^\infty(\mathbb{R}^d, dE)$. We have

$$\|(\mathcal{T}_b(\varphi) - \mathcal{T}_b(\varphi_n))x\|^2 = \int_{\mathbb{R}^n} |\varphi(t) - \varphi_n(t)|^2 d\mu_x(t),$$

and since $\|\varphi_n\|_\infty$ are bounded and μ_x is a finite measure, assertion (d) follows from the dominated–convergence theorem. ∎

Some simple consequences of this theorem are collected in Problems 6–8.

5.2.3 Example: If E_D is a discrete projection–valued measure on $\mathbb{R}$ with a finite support $\{\lambda_1, \dots, \lambda_N\}$, then the operators $\int \varphi(t)\, dE(t)$ have a simple form. Any function $\varphi \in L^\infty(\mathbb{R}^d, dE)$ is in this case E–a.e. equal to the simple function with the values $\varphi(\lambda_j)$ at $t = \lambda_j$ and zero otherwise; hence $\mathcal{T}_b(\varphi) = \sum_{j=1}^N \varphi(\lambda_j)P_j$, where $P_j := E_D(\{\lambda_j\})$.

Moreover, for any $\varphi \in L^\infty(\mathbb{R}, dE)$ we can find a polynomial Q_φ of a degree $\le N-1$ such that $\mathcal{T}_b(\varphi) = \mathcal{T}_b(Q_\varphi)$. Indeed, by Problem 6c this identity is equivalent to the conditions $Q(\lambda_j) = \varphi_j$, $j = 1, \dots, N$, which yield a system of linear equations for the coefficients of Q_φ. Its determinant is $\prod_{j=2}^N \prod_{k=1}^{j-1} (\lambda_j - \lambda_k)$, so it has just one solution, provided the numbers λ_j are mutually different. Using linearity and multiplicativity, we can further rewrite this result as follows,

$$\int_{\mathbb{R}} \varphi(t)\, dE(t) = \sum_{k=0}^{N-1} a_k A_D^k, \quad A_D := \int_{\mathbb{R}} t\, dE_D(t),$$

where the numbers $a_0, \dots, a_{N-1}$ solve the system $\sum_{k=0}^{N_1} a_k \lambda_j^k = \varphi(\lambda_j)$, $j = 1, \dots, N$. In particular, this result holds for any projection–valued measure with values in a finite–dimensional $\mathcal{H}$.

5.2.4 Example: Consider the projection–valued measure P on $\mathbb{R}^2$ which is determined by a pair of commuting spectral decompositions $\{E_t\}$ and $\{F_t\}$ (*cf.* Problem 3c); the corresponding projection–valued measures on $\mathbb{R}$ will be denoted as E and F, respectively. To a given Borel $\varphi : \mathbb{R} \to \mathbb{C}$ we define the function $\varphi_1 : \mathbb{R}^2 \to \mathbb{C}$ by $\varphi_1(t,u) := \varphi(t)$; in the notation of Appendix A.1 we can write this as $\varphi_1 = \varphi \times e$. We shall show that $\int \varphi_1 \, dP = \int \varphi \, dE$.

Since $\varphi_1^{(-1)}(K) = \varphi^{(-1)}(K) \times \mathbb{R}$ holds for any $K \subset \mathbb{C}$, the function φ_1 is Borel, and by definition of the measure P, the conditions $\varphi \in L^\infty(\mathbb{R}, dE)$ and $\varphi_1 \in L^\infty(\mathbb{R}^2, dP)$ are equivalent, and $\|\varphi\|_\infty = \|\varphi_1\|_\infty$. If $\varphi \in L^\infty(\mathbb{R}, dE)$, there is a sequence $\{\sigma_n\}$ of simple functions, $\|\sigma_n - \varphi\|_\infty \to 0$; then the functions $\sigma_n \times e$ are also simple, $\int (\sigma_n \times e) dP = \int \sigma_n \, dE$, and $\|\sigma_n \times e - \varphi_1\|_\infty \to 0$, so we get the sought relation,

$$\int_{\mathbb{R}^2} \varphi(t) \, dP(t,u) = \int_{\mathbb{R}} \varphi(t) \, dE(t) \, .$$

The same is true for the integration over the other variable (see also Problem 9).

The following proposition shows how the functional–calculus rules combine with the Bochner integral of Appendix A.5. For simplicity, we present it with rather strong assumptions which are, however, sufficient for our future purposes.

5.2.5 Proposition: Let the function $\psi : \mathbb{R} \times J \to \mathbb{C}$, where $J \subset \mathbb{R}$ is any interval, fulfil the following conditions:

(i) $\psi(\cdot, u)$ is a bounded Borel function for any $u \in J$.

(ii) $\psi(t, \cdot)$ belongs to $\mathcal{L}(J)$ for any $t \in \mathbb{R}$.

(iii) the functions $\varphi(\cdot) := \int_J \psi(\cdot, u) \, du$ and $\eta(\cdot) := \int_{\mathbb{R}} \psi(t, \cdot) \, dt$ are Borel and bounded.

(iv) the operator–valued function $u \mapsto B(u) := \mathcal{T}_b(\psi(\cdot, u))$ is continuous on J in the operator–norm topology, and if J is noncompact the limits at its endpoints exist; moreover, $\|B(\cdot)\| \in \mathcal{L}(J)$.

Then $B(\cdot) \in \mathcal{B}(J; \mathcal{B}(\mathcal{H}))$ and

$$\int_J B(u) \, du = \int_J \left(\int_{\mathbb{R}} \psi(t,u) \, dE(t) \right) du = \int_{\mathbb{R}} \left(\int_J \psi(t,u) \, du \right) dE(t) = \mathcal{T}_b(\varphi) \, .$$

Proof: The vector–valued function $B(\cdot)$ is integrable due to Proposition A.5.1. We denote $C := \int_J B(u) \, du$; then Proposition A.5.2, together with (5.4), yields

$$(x, Cx) = \int_J (x, B(u)x) \, du = \int_J \left(\int_{\mathbb{R}} \psi(t,u) \, d\mu_x(t) \right) du$$

for any $x \in \mathcal{H}$. Furthermore, $\eta \in \mathcal{L}(\mathbb{R}, d\mu_x)$ by assumption (iii), so ψ is integrable on $\mathbb{R}\times J$ with respect to the product of μ_x and the Lebesgue measure, and

$$(x, Cx) \,=\, \int_{\mathbb{R}} \left(\int_J \psi(t,u)\, du \right) d\mu_x(t) \,=\, \int_{\mathbb{R}} \varphi(t)\, d\mu_x(t) \,=\, (x, \mathcal{T}_b(\varphi)x)$$

by the Fubini theorem; since both operators are bounded, we get $C = \mathcal{T}_b(\varphi)$. ∎

Let us now pass to unbounded functions. Given a projection–valued measure E on $\mathbb{R}^d$ with values in $\mathcal{B}(\mathcal{H})$, we denote by $\Phi_E(\mathbb{R}^d)$ the set of all E–a.e. defined complex Borel functions. If the equality of two elements and the algebraic operations have the same meaning as in the bounded case, $\Phi_E(\mathbb{R}^d)$ becomes a commutative algebra containing $L^\infty(\mathbb{R}^d, dE)$ as a subalgebra. Our aim is to extend $\mathcal{T}_b$ to a mapping which associates with any function $\varphi \in \Phi_E(\mathbb{R}^d)$ an operator, in general unbounded, with the domain

$$D_\varphi \,:=\, \left\{ x \in \mathcal{H} \,:\, \int_{\mathbb{R}^d} |\varphi(t)|^2 d\mu_x(t) < \infty \right\}.$$

5.2.6 Proposition: D_φ is dense in $\mathcal{H}$ for any $\varphi \in \Phi_E(\mathbb{R}^d)$.

Proof: We have $\|E(M)(\alpha x+y)\|^2 \le 2|\alpha|^2\mu_x(M) + 2\mu_y(M)$ for any $M \in \mathcal{B}^d$ and all $x, y \in \mathcal{H}$, $\alpha \in \mathbb{C}$, so D_φ is a subspace. The sets $M_n := \{\, t \in \mathbb{R}^d : |\varphi(t)| \le n \,\}$, are Borel and form a nondecreasing system such that $E\left(\bigcup_{n=1}^\infty M_n\right) = E(\mathbb{R}^d) = I$ because the φ is by assumption E–a.e. defined. Then $\|E(M_n)x - x\| \to 0$ follows from Proposition 5.1.1b, so it is sufficient to check that $x_n := E(M_n)x$ belongs to D_φ for $n = 1, 2, \dots$. We have $\mu_{x_n}(N) = \|E(N)x_n\|^2 = \|E(N \cap M_n)x\|^2 = \int_N \chi_{M_n} d\mu_x$ for any $N \in \mathcal{B}^d$; hence Proposition A.3.4 gives

$$\int_{\mathbb{R}^d} |\varphi(t)|^2 d\mu_{x_n}(t) \,=\, \int_{\mathbb{R}^d} |\varphi(t)|^2 \chi_{M_n}\, d\mu_x(t) \,\le\, n^2\|x\|^2,$$

and therefore $x_n \in D_\varphi$. ∎

To construct the operator $\int \varphi(t)\, dE(t)$ for a given $\varphi \in \Phi_E(\mathbb{R}^d)$, we take a sequence $\{\varphi_n\} \subset L^\infty(\mathbb{R}^d, dE)$ such that $\varphi_n(t)$ converges to $\varphi(t)$ and $|\varphi_n(t)| \le |\varphi(t)|$ holds E–a.e. in $\mathbb{R}^d$; these requirements are fulfilled, *e.g.*, for $\varphi_n^{(0)} := \varphi\chi_{M_n}$, where the M_n are the sets introduced in the above proof. Let $x \in D_\varphi$; then

$$\|\mathcal{T}_b(\varphi_n)x - \mathcal{T}_b(\varphi_m)x\|^2 \le 2\int_{\mathbb{R}^d} |\varphi_n(t) - \varphi(t)|^2 d\mu_x(t) + 2\int_{\mathbb{R}^d} |\varphi_m(t) - \varphi(t)|^2 d\mu_x(t),$$

so the sequence $\{\mathcal{T}_b(\varphi_n)x\}$ is Cauchy by the dominated–convergence theorem. Thus it has a limit in $\mathcal{H}$, and a similar argument shows that it is independent of the choice of sequence $\{\varphi_n\}$, *i.e.*, that the relation

$$\mathcal{T}(\varphi)x \,:=\, \lim_{n\to\infty} \mathcal{T}_b(\varphi_n)x\,, \quad x \in D_\varphi\,,$$

determines a densely defined operator on $\mathcal{H}$ which is linear because $\mathcal{T}_b(\varphi_n)$ are linear; we call it the *integral of* φ *with respect to the measure* E and denote it as

$$\int_{\mathbb{R}^d} \varphi(t)\, dE(t) \;\equiv\; \mathcal{T}(\varphi)\,. \tag{5.5}$$

Since $\|\mathcal{T}_b(\varphi_n)x\|^2 = \int |\varphi_n|^2 d\mu_x$ and the right side has a limit due to the dominated–convergence theorem, the definition implies

$$\|\mathcal{T}(\varphi)x\|^2 = \int_{\mathbb{R}^d} |\varphi(t)|^2 d\mu_x(t)\,; \tag{5.6}$$

this shows that D_φ is the natural choice for the domain of $\mathcal{T}(\varphi)$. In particular, $D_\varphi = \mathcal{H}$ if $\varphi \in L^\infty(\mathbb{R}^d, dE)$ and $\mathcal{T}(\varphi) = \mathcal{T}_b(\varphi)$ as we check, choosing $\varphi_n = \varphi$ for all n, *i.e.*, the mapping $\mathcal{T}$ represents an extension to $\mathcal{T}_b$ (see also Problem 10a).

Next we shall derive functional–calculus rules for unbounded functions. Domain considerations make the proofs more difficult in comparison with the bounded case, as it already illustrates the following assertion, which extends the identity (5.4).

5.2.7 Proposition: Let $y \in D_\varphi$ for a given $\varphi \in \Phi_E(\mathbb{R}^d)$ and $x \in \mathcal{H}$; then φ is integrable with respect to the complex measure $\nu_{xy} := (x, E(\cdot)y)$ and

$$(x, \mathcal{T}(\varphi)y) \;=\; \int_{\mathbb{R}^d} \varphi(t)\, d\nu_{xy}(t)\,.$$

Moreover, $\nu_{x,\mathcal{T}(\varphi)y}(M) = \int_M \varphi(t)\, d\nu_{xy}(t)$ holds for any $M \in \mathcal{B}^d$.

Proof: The integrability is equivalent to the condition $\int |\varphi(t)|\, d|\nu_{xy}|(t) < \infty$, where $|\nu_{xy}|$ is the total variation of ν_{xy}. Let $\{N_k\} \subset \mathcal{B}^d$ be a disjoint decomposition of a Borel set M. By the Schwarz inequality, $|\nu_{xy}(N_k)| \le (\mu_x(N_k)\mu_y(N_k))^{1/2}$, so using the definition of $|\nu_{xy}|$ from Appendix A.4 we get the estimate

$$|\nu_{xy}|(M) \;\le\; (\mu_x(M)\mu_y(M))^{1/2}\,.$$

It follows from the monotone–convergence theorem that there is a nondecreasing sequence of non–negative simple functions $\sigma_n := \sum_j c_{nj}\chi_{M_{nj}}$ which converges to φ everywhere in $\mathbb{R}^d$, fulfils $\sigma_n(t) \le |\varphi(t)|$, and $\lim_{n\to\infty} \int \sigma_n\, d|\nu_{xy}| \;=\; \int |\varphi|\, d|\nu_{xy}|$. At the same time $\sum_j c_{nj}^2 \mu_y(M_{nj}) \;\le\; \int |\varphi(t)|^2\, d\mu_y(t)$; then Hölder inequality together with the above estimate imply

$$\begin{aligned}\int_{\mathbb{R}^d} |\varphi(t)|\, d|\nu_{xy}|(t) \;&=\; \lim_{n\to\infty} \sum_j c_{nj}|\nu_{xy}|(M_{nj})\\ &\le\; \mu_x(\mathbb{R}^d) \lim_{n\to\infty} \Big(\sum_j c_{nj}^2 \mu_y(M_{nj})\Big)^{1/2} \;\le\; \mu_x(\mathbb{R}^d) \left(\int_{\mathbb{R}^d} |\varphi(t)|^2 d\mu_y(t)\right)^{1/2},\end{aligned}$$

where the right side is finite provided $y \in D_\varphi$. To prove the first identity, we take a sequence $\{\varphi_n\} \subset L^\infty(\mathbb{R}^d, dE)$ which satisfies the requirements of the definition, and use the relation (5.4),

$$(x, \mathcal{T}(\varphi)y) \;=\; \lim_{n\to\infty} (x, \mathcal{T}_b(\varphi_n)y) \;=\; \lim_{n\to\infty} \int_{\mathbb{R}^d} \varphi_n(t)\, d\nu_{xy}(t) \;=\; \int_{\mathbb{R}^d} \varphi(t)\, d\nu_{xy}(t)\,,$$

where in the final step we have employed the dominated–convergence theorem. This implies in turn $\nu_{x,\mathcal{T}(\varphi)y}(M) = (x, E(M)\mathcal{T}(\varphi)y) = \int \varphi(t)\, d\nu_{E(M)x,y}(t)$, so the remaining assertion follows from Problem 8. ∎

5.2.8 Example: Consider the projection–valued measure $E^{(\varrho)} : E^{(\varrho)}(M) = T_{\chi_M}$ from Example 5.1.2b on $L^2(\mathbb{R}^d, d\varrho)$, where ϱ is a given Borel measure; we shall show that $\mathcal{T}^{(\varrho)}(f) := \int f(t)\, dE^{(\varrho)}(t)$ coincides for any Borel function $f : \mathbb{R}^d \to \mathbb{C}$ with the operator T_f of Example 4.3.3. We have $(\varphi, E^{(\varrho)}(M)\psi) = \int_M \overline{\varphi(t)}\psi(t)\, d\varrho(t)$ for any $\varphi, \psi \in L^2(\mathbb{R}^d, d\varrho)$ and $M \in \mathcal{B}^d$, so the condition $\int |f(t)|^2 |\psi(t)|^2 d\varrho(t) < \infty$ is equivalent to $\int |f(t)|^2 d\mu_\psi^{(\varrho)}(t) < \infty$ by Proposition A.3.4, which means that $\mathcal{T}^{(\varrho)}(f)$ and T_f have the same domain. Using the above proposition, we get

$$(\varphi, \mathcal{T}^{(\varrho)}(f)\psi) = \int_{\mathbb{R}^d} f(t)\, d\nu_{\varphi,\psi}^{(\varrho)} = \int_{\mathbb{R}^d} f(t)\overline{\varphi(t)}\psi(t)\, d\varrho(t) = (\varphi, T_f \psi)$$

for any $\psi \in D(T_f)$ and $\varphi \in L^2(\mathbb{R}^d, d\varrho)$, *i.e.*, $\mathcal{T}^{(\varrho)}(f)\psi = T_f\psi$ for all $\psi \in D(T_f)$.

This example suggests why the general functional–calculus rules listed below, which generalize Theorem 5.2.2, are closely analogous to the properties of the operators T_f from Example 4.3.3, and the proofs are also almost identical.

5.2.9 Theorem (functional calculus — the general case): Let E be a measure on $\mathbb{R}^d$ with values in $\mathcal{B}(\mathcal{H})$; then the mapping $\mathcal{T} : \Phi_E(\mathbb{R}^d) \to \mathcal{L}(\mathcal{H})$ defined above has the following properties:

(a) Homogeneity, $\mathcal{T}(\alpha\varphi) = \alpha\mathcal{T}(\varphi)$ for any $\alpha \in \mathbb{C}$.

(b) $\mathcal{T}(\varphi+\psi) \supset \mathcal{T}(\varphi) + \mathcal{T}(\psi)$, with the inclusion turning into identity *iff* at least one of the relations $D_{\varphi+\psi} \subset D_\varphi$ or $D_{\varphi+\psi} \subset D_\psi$ is valid.

(c) $\mathcal{T}(\varphi\psi) \supset \mathcal{T}(\varphi)\mathcal{T}(\psi)$, where the domain of the right side is $D_{\varphi\psi} \cap D_\psi$; an identity is valid *iff* $D_{\varphi\psi} \subset D_\psi$.

(d) Injectivity, $\mathcal{T}(\varphi) = \mathcal{T}(\psi)$ implies $\varphi(t) = \psi(t)$ for E–a.a. $t \in \mathbb{R}^d$.

(e) $\mathcal{T}(\varphi)^* = \mathcal{T}(\overline{\varphi})$.

(f) $\mathcal{T}(\varphi)$ is invertible *iff* the set $\operatorname{Ker}\varphi := \varphi^{(-1)}(\{0\})$ is E–zero, in which case $\mathcal{T}(\varphi)^{-1} = \mathcal{T}(\varphi^{-1})$.

Proof: The homogeneity is obvious. If the conditions $x \in D_\varphi$ and $x \in D_\psi$ hold simultaneously, we also have $\int |\varphi(t)+\psi(t)|^2 d\mu_x(t) < \infty$, *i.e.*, $x \in D_{\varphi+\psi}$; then $(y, \mathcal{T}(\varphi+\psi)x) = \int (\varphi(t)+\psi(t))\, d\nu_{xy}(t) = (y, (\mathcal{T}(\varphi)+\mathcal{T}(\psi))x)$ holds for any $y \in \mathcal{H}$ by Proposition 5.2.7. The remaining part of (b) follows from the relations $D_\varphi \cap D_\psi = D_{\varphi+\psi} \cap D_\varphi$ and $D_\varphi \cap D_\psi = D_{\varphi+\psi} \cap D_\psi$ which are easy consequences of the inclusion $D_\varphi \cap D_\psi = D_{\varphi+\psi}$.

A vector x belongs to the domain D of the product $\mathcal{T}(\varphi)\mathcal{T}(\psi)$ *iff* the conditions $\int |\psi(t)|^2 d\mu_x(t) < \infty$ and $\int |\varphi(t)|^2 d\mu_{\mathcal{T}(\psi)x}(t) < \infty$ are valid simultaneously;

the second of these can be rewritten as $\int |\varphi(t)|^2|\psi(t)|^2 d\mu_x(t) < \infty$ (Problem 11a), *i.e.*, $x \in D_{\varphi\psi}$. Together we get $D = D_\psi \cap D_{\varphi\psi}$, and the two sets are equal if $D_{\varphi\psi} \subset D_\psi$. By Proposition 5.2.7,

$$(y, \mathcal{T}(\varphi)\mathcal{T}(\psi)x) = \int_{\mathbb{R}^d} \varphi(t)\, d\nu_{y,\mathcal{T}(\psi)x}(t) = \int_{\mathbb{R}^d} \varphi(t)\psi(t)\, d\nu_{yx}(t) = (y, \mathcal{T}(\varphi\psi)x)$$

holds for all $x \in D$ and $y \in \mathcal{H}$, which proves (c).

Let $M_{\varphi\psi}$ be the subset of $\mathbb{R}^d$ on which both φ, ψ are defined; to prove (d), it is sufficient to show that the set $N := \{t \in M_{\varphi\psi} : \varphi(t) \neq \psi(t)\}$ is E–zero. This can be expressed as the union of the nondecreasing sequence of its subsets $N_n := \{t \in M_{\varphi\psi} : |\varphi(t) - \psi(t)| \geq n^{-1}, |\varphi(t)| \leq n\}$, so $E(N) = \text{s-lim}_{n\to\infty} E(N_n)$ and the relation $E(N) = 0$ holds if $x_n := E(N_n)x = 0$ for all n and $x \in \mathcal{H}$. Clearly $x_n \in D_\varphi$ for all n, and since $D_\varphi = D_\psi$ by assumption, assertions (a) and (b) give $\mathcal{T}(\varphi - \psi)x_n = 0$. Now the relation (5.6) implies

$$0 = \|\mathcal{T}(\varphi - \psi)x_n\|^2 = \int_{N_n} |\varphi(t) - \varphi(t)|^2 d\mu_x(t) \geq n^{-2}\|E(N_n)x\|^2$$

for any n, *i.e.*, $E(N_n)x = 0$.

The adjoint $\mathcal{T}(\varphi)^*$ exists because $D_\varphi = D_{\overline{\varphi}}$ is dense in $\mathcal{H}$; using Theorem 5.2.2b we easily find that $\mathcal{T}(\overline{\varphi}) \subset \mathcal{T}(\varphi)^*$. It remains for us to prove that if $(y, \mathcal{T}(\varphi)x) = (z, x)$ holds for some $y, z \in \mathcal{H}$ and all $x \in D_\varphi$, then $y \in D_\varphi$. We set $y_n := \mathcal{T}_b(\overline{\eta_n})y$, where $\eta_n := \varphi\chi_{M_n}$ and M_n are the sets from the proof of Proposition 5.2.6; then $y_n \in D_\varphi$ since $\int |\varphi(t)|^2 d\mu_{y_n}(t) = \int_{M_n} |\varphi(t)|^4 d\mu_y(t) < \infty$. Next we set $x = y_n$; then the Schwarz inequality gives $|(y, \mathcal{T}(\varphi)y_n)| \leq \|z\|\,\|y_n\|$. The left side can be rewritten by Proposition 5.2.7 as

$$(y, \mathcal{T}(\varphi)y_n) = \int_{\mathbb{R}^d} \varphi(t)\overline{\eta_n(t)}\, d\mu_y(t) = \|y_n\|^2.$$

Thus we get $\int |\eta_n(t)|^2 d\mu_y(t) \leq \|z\|^2$, $n = 1, 2, \ldots$, which means that $y \in D_\varphi$ by the Fatou lemma.

Finally, if $E(\operatorname{Ker}\varphi) = 0$ then the relation $\psi(t) := \varphi(t)^{-1}$ for $t \in \mathbb{R}^d \backslash \operatorname{Ker}\varphi$ defines a function $\psi \in \Phi_E(\mathbb{R}^d)$. Since $\varphi(t)\psi(t) = 1$ holds E–a.e., assertion (c) implies $D(\mathcal{T}(\psi)\mathcal{T}(\varphi)) = D_\varphi$ and $\mathcal{T}(\psi)\mathcal{T}(\varphi)x = x$ for all $x \in D_\varphi$, and the same relations with the roles of φ, ψ interchanged; hence the operator $\mathcal{T}(\varphi)^{-1}$ exists and $\mathcal{T}(\varphi)^{-1} = \mathcal{T}(\psi)$. To check the opposite implication, we employ the identity $\varphi\chi_{\operatorname{Ker}\varphi} = 0$, which yields by (c) the relation $\mathcal{T}(\varphi)E(\operatorname{Ker}\varphi) = 0$ from which $E(\operatorname{Ker}\varphi) = 0$ follows if $\mathcal{T}(\varphi)$ is invertible. ∎

5.2.10 Example: Let Q_N be a polynomial of N–th order, $Q_N(z) := \sum_{n=0}^N \alpha_n z^n$, with complex coefficients such that $\alpha_N \neq 0$. If $\varphi \in L^\infty(\mathbb{R}^d, dE)$, the assertions (a)–(c) of the theorem yield

$$\mathcal{T}(Q_N \circ \varphi) = \mathcal{T}\left(\sum_{n=0}^N \alpha_n \varphi^n\right) = \sum_{n=0}^N \alpha_n \mathcal{T}(\varphi)^n;$$

with a little more effort one can show that the same is true for any $\varphi \in \Phi_E(\mathbb{R}^d)$. Indeed, let $x \in D(\varphi^n)$ (we use the alternative notation $D(\varphi) := D_\varphi$ whenever it is convenient), then $x \in D(\varphi^m)$ for $m < n$ by the Hölder inequality because the measure μ_x is finite, *i.e.*, $D(\varphi^n) \subset D(\varphi^m)$ if $n > m$, and therefore

$$\mathcal{T}(\varphi^n) = \mathcal{T}(\varphi)^n$$

follows by induction from Theorem 5.2.9c. Furthermore, due to the assumption there is a positive c such that $\frac{1}{2} < |Q_N(z)\alpha_N^{-1}z^{-N}| < \frac{3}{2}$ holds for all $|z| > c$, so Problem 10c gives $D((Q_N\circ\varphi)\chi_{M^c}) = D(\varphi^N\chi_{M^c})$, where we have denoted $M^c := \{t \in \mathbb{R}^d : |\varphi(t)| > c\}$; using the fact that both functions are bounded on $\mathbb{R}^d\backslash M^c$ we conclude that $D(Q_N\circ\varphi) = D(\varphi^N)$. Theorem 5.2.9b now implies $\mathcal{T}(Q_N\circ\varphi) = \alpha_N\mathcal{T}(\varphi)^N + \mathcal{T}(Q_M\circ\varphi)$, where Q_M is a polynomial of a degree $\le N-1$; the desired formula is then obtained by induction.

If $f \in \Phi_E(\mathbb{R}^d)$ is a real–valued function and the polynomial coefficients are real, this result can be combined with Theorem 5.2.9e showing that the operator $\sum_{n=0}^N c_n\mathcal{T}(f)^n$ is self–adjoint for any real $c_0, c_1, \dots, c_N$; in particular, any power of $\mathcal{T}(f)$ is self–adjoint.

Theorem 5.2.9 implies other important properties of the operators $\mathcal{T}(\varphi)$.

5.2.11 Theorem: (a) $\mathcal{T}(\varphi)$ is normal for any $\varphi \in \Phi_E(\mathbb{R}^d, dE)$ and the spectrum $\sigma(\mathcal{T}(\varphi)) = R_{ess}^{(E)}(\varphi)$.

(b) A complex number λ is an eigenvalue of $\mathcal{T}(\varphi)$ *iff* $E(\varphi^{(-1)}(\{\lambda\})) \ne 0$, in which case the corresponding eigenspace is $N_{\mathcal{T}(\varphi)}(\lambda) = \operatorname{Ran} E(\varphi^{(-1)}(\{\lambda\}))$.

(c) $\mathcal{T}(\varphi)$ is self–adjoint *iff* $\varphi(t) \in \mathbb{R}$ for E–a.a. $t \in \mathbb{R}^d$.

(d) If $B \in \mathcal{B}(\mathcal{H})$ commutes with the measure E, then $B\mathcal{T}(\varphi) \subset \mathcal{T}(\varphi)B$ for any function $\varphi \in \Phi_E(\mathbb{R}^d)$.

Proof: We have $\mathcal{T}(\varphi) = \mathcal{T}(\overline{\varphi})^*$, so $\mathcal{T}(\varphi)$ is closed. Furthermore, $D_\varphi = D_{\overline{\varphi}}$ and (5.6) implies $\|\mathcal{T}(\varphi)x\| = \|\mathcal{T}(\varphi)^*x\|$ for all $x \in D_\varphi$; hence $\mathcal{T}(\varphi)$ is normal by Theorem 4.3.1. The remaining part of (a) is obtained in the same way as Theorem 5.2.2c. To prove (b), we denote $M_\lambda := \varphi^{(-1)}(\{\lambda\})$. If a nonzero x fulfils $E(M_\lambda)x = x$, then $\int_{\mathbb{R}^d} |\varphi(t)|^2 d\mu_x(t) = \int_{M_\lambda} |\varphi(t)|^2 d\mu_x(t) = |\lambda|^2\|x\|^2$, so $x \in D_\varphi$ and the relation (5.6) together with Theorem 9b gives

$$\|(\mathcal{T}(\varphi)-\lambda)x\|^2 = \int_{M_\lambda} |\varphi(t)-\lambda|^2 d\mu_x(t) = 0\,,$$

i.e., $\lambda \in \sigma(\mathcal{T}(\varphi))$ and $N_{\mathcal{T}(\varphi)}(\lambda) \supset \operatorname{Ran} E(M_\lambda)$. On the other hand, let $x \in D_\varphi$ be a unit vector such that $\mathcal{T}(\varphi)x = \lambda x$. This implies $\int |\varphi(t)-\lambda|^2 d\mu_x(t) = 0$, which means that $\mu_x(\mathbb{R}^d \setminus M_\lambda) = 0$, and therefore $\mu_x(M_\lambda) = \|x\|^2$, *i.e.*, $E(M_\lambda)x = x$. Assertion (c) follows from (d) and (e) of Theorem 5.2.9. Finally, the commutativity of operator B with $E(M)$ for any $M \in \mathcal{B}^d$ implies $\nu_{y,Bx} = \nu_{B^*y,x}$, and in a similar way, we get $\mu_{Bx}(M) \le \|B\|^2\mu_x(M)$. The last relation shows that D_φ is B–invariant; then

$$(y, \mathcal{T}(\varphi)Bx) = \int_{\mathbb{R}^d} \varphi(t)\, d\nu_{y,Bx}(t) = (B^*y, \mathcal{T}(\varphi)x) = (y, B\mathcal{T}(\varphi)x)$$

holds for any $x \in D_\varphi$ and all $y \in \mathcal{H}$. ∎

In conclusion, let us mention how composite functions are integrated with respect to a projection–valued measure (see also Problem 15).

5.2.12 Proposition: Suppose that E is a projection–valued measure on $\mathbb{R}^d$ and $w : \mathbb{R}^d \to \mathbb{R}^n$ is such that its "component" functions $w_j : \mathbb{R}^d \to \mathbb{R}$, $j = 1, \dots, n$, are Borel. Define $E_w : E_w(M) = E(w^{(-1)}(M))$; then the composite function $\varphi \circ w$ belongs to $\Phi_E(\mathbb{R}^d)$ for any $\varphi \in \Phi_{E_w}(\mathbb{R}^n)$, and

$$\int_{\mathbb{R}^n} \varphi(t)\, dE_w(t) = \int_{\mathbb{R}^d} (\varphi \circ w)(t)\, dE(t)\,.$$

Proof: Let φ be defined on $\mathbb{R}^n \backslash N$, where $E_w(N) = 0$; then $\psi := \varphi \circ w$ is defined on $w^{(-1)}(\mathbb{R}^n \backslash N)$, whose complement is an E–zero set; using the properties of Borel sets, we can check that $\psi^{(-1)}(M) = w^{(-1)}(\varphi^{(-1)}(M))$ belongs to $\mathcal{B}^d$ for $M \in \mathcal{B}$, so ψ is Borel, *i.e.*, $\psi \in \Phi_E(\mathbb{R}^d)$. Theorem A.3.10 easily implies $D_\varphi^{(E_w)} \subset D_{\varphi \circ w}^{(E)}$ and the relation $(x, \mathcal{T}^{(E_w)}(\varphi)x) = (x, \mathcal{T}^{(E)}(\varphi \circ w)x)$ for all $x \in D_\varphi^{(E_w)}$, which extends by polarization to any pair $x, y \in D_\varphi^{(E_w)}$, and since this domain is dense in $\mathcal{H}$, we get $\mathcal{T}^{(E_w)}(\varphi) \subset \mathcal{T}^{(E)}(\varphi \circ w)$; however, the two operators are normal, and therefore equal each other (*cf.* Remark 4.3.2). ∎

5.3 The spectral theorem

The main goal of this section is to prove the spectral theorem for self–adjoint operators, which in a sense represents a cornerstone of the theory. Let us first formulate it.

5.3.1 Theorem: (a) To any self–adjoint A, there is just one projection–valued measure E_A such that

$$A = \int_{\mathbb{R}} t\, dE_A(t)\,. \tag{5.7}$$

(b) A bounded operator B commutes with A *iff* it commutes with $E_t^{(A)} := E_A(-\infty, t]$ for all $t \in \mathbb{R}$.

The formula (5.7) is usually called the *spectral decomposition* of the operator A. The proof will proceed in several steps. We start by checking the assertion for Hermitean operators; from them we pass to bounded normal ones, in particular, to unitary operators, and in the last step we shall employ the Cayley transformation.

Therefore let A be a Hermitean operator on $\mathcal{H}$; we denote the interval between its lower and upper bound by $J_A := [m_A, M_A]$. To any polynomial P with real coefficients, $P(t) := \sum_{j=0}^{n} c_j t^j$, we can define the Hermitean operator $P(A) := \sum_{j=0}^{n} c_j A^j$, where $A^0 := I$. We can easily check that the map $P \mapsto P(A)$ is linear and multiplicative, and also monotonic, which means that $P(A) \geq 0$ if $P(t) \geq 0$ on J_A (Problem 16b). Our aim is extend this to a wider class of real-valued functions on J_A in such a way that the mentioned properties will be preserved.

Consider first the family $\mathcal{K}_0$ of functions $f : J_A \to [0, \infty)$ such that there is a nondecreasing sequence $\{f_n\} \subset C(J_A)$ which converges pointwise to f; it is obvious that f is closed with respect to sums, products, and multiplication by positive numbers. Instead of continuous functions, the elements of $\mathcal{K}_0$ can be approximated from above by polynomials.

5.3.2 Proposition: If $f \in \mathcal{K}_0$ and $\{f_n\} \subset C(J_A)$ is a sequence with the stated properties, then there is a sequence $\{P_n\}$ of polynomials such that $f_n(t) \leq P_n(t)$ and $P_n(t) \leq P_{n+1}(t)$ for any $t \in J_A$ and $n = 1, 2, \ldots$, and $\lim_{n\to\infty} P_n(t) = f(t)$. If $\{Q_n\}$ is another sequence of polynomials with these properties, then for any n we can find m_n so that $Q_m(t) < P_n(t) + n^{-1}$ and $P_m(t) < Q_n(t) + n^{-1}$ holds for any $t \in J_A$ and all $m > m_n$.

Proof: By the Weierstrass theorem, the function $g_n := f_n + 3.2^{-n-2}$ can be approximated by a polynomial P_n so that $|g_n(t) - P_n(t)| < 2^{-n-2}$ holds for any $t \in J_A$, *i.e.*, $2^{-n-1} < P_n(t) - f_n(t) < 2^{-n}$; this proves the first assertion. In the remaining part the polynomials appear symmetrically, so it is enough to prove one of the inequalities. Since $Q_m(t) - P_m(t) \to 0$, for any n and $t \in J_A$ we can find a positive integer $m(n,t)$ such that $Q_m(t) - P_m(t) < n^{-1}$ for all $m > \max\{m(n,t), n\}$. Since $\{P_m(t)\}$ is nonincreasing, we have $Q_m(t) < P_n(t) + n^{-1}$; the inequality then follows from Problem 17. ∎

The monotonicity condition here is important, because due to it the sequence $\{P_n(A)\}$ is nonincreasing, and since it consists of positive operators, it converges to a positive $P := \text{s-lim}_{n\to\infty} P_n(A)$ (*cf.* Remark 3.2.5). Successively performing the limits $m \to \infty$ and $n \to \infty$ in the inequalities $Q_m(A) \leq P_n(A) + n^{-1}$ and $P_m(A) \leq Q_n(A) + n^{-1}$ which follow from the proposition, we see that the limiting operator is independent of the choice of the sequence $\{P_n\}$ approximating the function f; in this way, we get the map $\dot{T}_A : \dot{T}_A f = \text{s-lim}_{n\to\infty} P_n(A)$ from $\mathcal{K}_0$ to the set of positive operators on $\mathcal{H}$. It has the following simple properties, the proof of which is left to the reader (Problem 16a).

5.3.3 Proposition: The map $\dot{T}_A$ is additive, multiplicative, and $\dot{T}_A(cf) = c\dot{T}_A(f)$ for any $f \in \mathcal{K}_0$ and $c \geq 0$. If functions $f, g \in \mathcal{K}_0$ satisfy $f(t) \geq g(t)$ for all $t \in J_A$, then $\dot{T}_A(f) \geq \dot{T}_A(g)$.

The map $\dot{T}_A$ does not extend $P \mapsto P(A)$ because the polynomials assuming negative values on J_A are not contained in $\mathcal{K}_0$, but there is a common extension to these two maps. We introduce the set $\mathcal{K}$ consisting of the functions $\varphi : J_A \to \mathbb{R}$

which can be expressed as $\varphi = \varphi_+ - \varphi_-$ for some $\varphi_\pm \in \mathcal{K}_0$. We can easily check that $\mathcal{K}$ is a real algebra with pointwise multiplication, which contains all real functions continuous on J_A; in particular, a restriction to J_A of any real polynomial belongs to $\mathcal{K}$. Given a function $\varphi \in \mathcal{K}$, we ascribe to it the Hermitean operator

$$T_A(\varphi) := \dot{T}_A(\varphi_+) - \dot{T}_A(\varphi_-);$$

the definition makes sense, because if $\varphi = \tilde{\varphi}_+ - \tilde{\varphi}_-$ is valid at the same time, then the additivity of $\dot{T}_A$ implies $\dot{T}_A(\varphi_+) + \dot{T}_A(\tilde{\varphi}_-) = \dot{T}_A(\tilde{\varphi}_+) + \dot{T}_A(\varphi_-)$, so $T_A(\varphi) = \dot{T}_A(\tilde{\varphi}_+) - \dot{T}_A(\tilde{\varphi}_-)$.

5.3.4 Proposition: The map T_A is linear, multiplicative, and monotonic. For any $\varphi \in \mathcal{K}$, there is a sequence $\{P_n\}$ of real polynomials such that $\varphi(t) = \lim_{n\to\infty} P_n(t)$ for any $t \in J_A$ and $T_A(\varphi) = \text{s-lim}_{n\to\infty} P_n(A)$.

Proof: The additivity is obvious; to prove the linearity, we have to decompose $\alpha\varphi$ into the difference of $\epsilon\alpha\varphi_\epsilon$ and $\epsilon\alpha\varphi_{-\epsilon}$, where $\epsilon := \operatorname{sgn}\alpha$, and in a similar way we can check the multiplicativity. If $\varphi \le \psi$ on J_A, then $\varphi_+ + \psi_- \le \psi_+ + \varphi_-$, which implies $T_A(\varphi) \le T_A(\psi)$ by Proposition 5.3.3. Finally, if $\{P_n^{(\pm)}\}$ are sequences approximating the functions $\varphi_\pm$, we set $P_n := P_n^{(+)} - P_n^{(-)}$; it is easy to check that $T_A(\varphi) = \text{s-lim}_{n\to\infty} P_n(A)$. ∎

The map T_A extends both $\dot{T}_A$ and $P \mapsto P(A)$; to check the last assertion it is sufficient to realize that any polynomial can be written as a sum of a polynomial positive on J_A and a constant.

5.3.5 Proposition: For any $u \in \mathbb{R}$, define $e_u := \chi_{(-\infty,u]\cap J_A} \restriction J_A$; then the map $u \mapsto E_u := T_A(e_u)$ is a spectral decomposition on $\mathcal{H}$.

Proof: Each of the operators E_u is positive and $E_u^2 = E_u$ follows from the multiplicativity of T_A, so E_u is a projection; using further the monotonicity of T_A, we get $E_u \le E_v$ for $u < v$. The relations (5.1) are satisfied trivially; hence it remains to check that $u \mapsto E_u$ is right continuous at any $u \in [m_A, M_A)$. Consider the continuous functions $g_n := \max\{0, \min\{1, 1 - n(n-1)(\cdot - u - n^{-1})\}\}$ for $n > (M_A - u)^{-1} + 1$ which fulfil the inequalities $g_{n+1}(t) \le e_{u+1/n}(t) \le g_n(t)$ for each $t \in J_A$, and $\lim_{n\to\infty} g_n(t) = e_u(t)$. Let P_n be polynomials approximating g_n according to Proposition 5.3.2; then $E_u = \text{s-lim}_{n\to\infty} P_n(A)$ and $P_n(A) \ge E_{u+1/n} \ge E_u$; however, the operators $E_{u+1/n} - E_u$ are projections, so $(x, (P_n - E_u)x) \ge \|(E_{u+1/n} - E_u)x\|^2$ for all $x \in \mathcal{H}$, and therefore $\text{s-lim}_{n\to\infty} E_{u+1/n} = E_u$. ∎

With these prerequisites, we are able to finish the first step of the proof.

5.3.6 Proposition: The spectral theorem holds for any Hermitean A, and in this case the measure E_A is supported by J_A.

Proof: Let E_A be the projection–valued measure generated by $\{T_A(e_u)\}$. Obviously $E_A(-\infty, m_A) = E_A(M_A, \infty) = 0$ and the function $id : id(x) = x$ belongs to $L^\infty(\mathbb{R}, dE_A)$; we shall show that $\mathcal{T}_b^{(E_A)}(id) = A$. Let D be any division $m_A = t_0 < t_1 < \ldots < t_n = M_A$ of the interval J_A. We denote $L_0 := [t_0, t_1]$ and $L_k :=$

$(t_{k-1},t_k]$, $k=2,\dots,n$; then $\chi_{L_1}\upharpoonright J_A = e_{t_1}$ and $\chi_{L_k}\upharpoonright J_A = e_{t_k}-e_{t_{k-1}}$ for $k=2,\dots,n$, so the characteristic functions χ_{L_k}, $k=1,\dots,n$ belong to $\mathcal{K}$ in view of Problem 16f and $T_A(\chi_{L_k}) = E_A(L_k)$. Furthermore, the simple function $s_D := \sum_{k=1}^n t_k\chi_{L_k}$ also belongs to $\mathcal{K}$ and $T_A(s_D) = \sum_{k=1}^n t_k T_A(\chi_{L_k}) = \mathcal{T}_b^{(E_A)}(s_D)$ by (5.2). Given $\varepsilon>0$, we can choose D such that $\max_{1\le k\le n}(t_k-t_{k-1}) < \varepsilon$; then $\|\mathcal{T}_b^{(E_A)}(id) - \mathcal{T}_b^{(E_A)}(s_D)\| < \varepsilon$ by Theorem 5.2.2b, and at the same time, $\|T_A(id\upharpoonright J_A) - T_A(s_D)\| < \varepsilon$ by Problem 16d. Since $T_A(id\upharpoonright J_A) = A$ due to Problem 16e, the last two inequalities give $\|\mathcal{T}_b^{(E_A)}(id) - A\| < 2\varepsilon$, and therefore $\mathcal{T}_b^{(E_A)}(id) = A$.

Suppose that F is another spectral measure corresponding to the operator A, then $\mathcal{T}_b^{(F)}(f) = T_A(f\upharpoonright J_A)$ holds by Problem 18a for any real function f whose restriction to J_A belongs to $\mathcal{K}_0$. In particular, choosing $f := \chi_{(-\infty,u]}$ for $u \ge m_A$ we get $F(-\infty,u] = E_A(-\infty,u]$. The same is obviously true for $u < m_A$, so the two measures are identical, $F = E_A$, in view of Problem 3d.

The condition $BE_t^{(A)} = E_t^{(A)}B$, $t\in\mathbb{R}$, implies by Problem 3c $BE_A(M) = E_A(M)B$ for any Borel set M, so $BA = AB$ follows from the functional–calculus rules. To prove the opposite implication, we employ Proposition 5.3.4. To any $t\in\mathbb{R}$, there is a sequence $\{P_n^{(t)}\}$ of polynomials such that $E_t^{(A)} = \text{s-lim}_{n\to\infty} P_n^{(t)}(A)$; the condition $BA = AB$ then implies $BP_n^{(t)}(A) = P_n^{(t)}(A)B$ and the limit $n\to\infty$ concludes the proof. ∎

Now let B be a bounded normal operator, $B = A_1 + iA_2$. Its real and imaginary part commute, $A_1A_2 = A_2A_1$, so the identity , $E_{A_1}(M_1)E_{A_2}(M_2) = E_{A_2}(M_2)E_{A_1}(M_1)$, M_1, $M_2\in\mathcal{B}$, is valid for the corresponding projection–valued measures. We denote their direct product as F_B and use the natural isometry which allows us to identify $\mathbb{R}^2$ and $\mathbb{C}$. Since A_1, A_2 are bounded, the function $id_{\mathbb{C}}$: $id_{\mathbb{C}}(z) = z$ belongs to $L^\infty(\mathbb{C}, dF_B)$, and the relation

$$\int_{\mathbb{C}} z\, dF_B(z) = A_1 + iA_2 = B \tag{5.8}$$

follows from Problem 9. This yields another modification of the spectral theorem.

5.3.7 Proposition: Let B be a bounded normal operator; then

(a) There is a projection–valued measure F_B on $\mathbb{C}$ supported by $\sigma(B)$ for which the formula (5.8) is valid. A projection–valued measure F on $\mathbb{C}$ such that $\mathcal{T}_b^{(F)}(id_{\mathbb{C}}) = B$ and $F(M_1\times M_2) = E_{A_1}(M_1)E_{A_2}(M_2)$ for all $M_1, M_2 \in \mathcal{B}$ coincides with F_B.

(b) A number $\lambda\in\mathbb{C}$ belongs to the spectrum of B *iff* $F_B(U_\varepsilon)\ne 0$ holds for any ε–neighborhood of it; furthermore, λ is an eigenvalue *iff* $F_B(\{\lambda\})\ne 0$, and in that case the corresponding eigenspace is $N_B(\lambda) = \operatorname{Ran} F_B(\{\lambda\})$.

Proof: We have already checked part (a) with the exception of the support claim; assertion (b) follows from Theorem 5.2.11. The resolvent set $\rho(B)$ is open and due to (a) any point of it has a neighborhood U such that $F_B(U) = 0$. Hence $(x, F_B(\rho(B))x) = 0$ holds for all $x\in\mathcal{H}$ (see Appendix A.2), so $F_B(\rho(B)) = 0$. ∎

In the next step we pass to unitary operators. We know that a unitary operator U is normal and its spectrum is a subset of the unit circle $C := \{ z \in \mathbb{C} : |z| = 1 \}$, which means that $F_U(\mathbb{C}\backslash C) = 0$. The circle can be parametrized by $\mathrm{Re}\, z = \cos t$ and $\mathrm{Im}\, z = \sin t$ with $t \in [0, 2\pi)$; using this we define the function $w : \mathbb{C} \to \mathbb{R}$ which maps $\mathbb{C}\backslash C$ to zero and $w(z) = t$ at the circle points. It is easy to check that w is Borel; in view of Problem 15a, this determines a projection–valued measure on $\mathbb{R}$ by $E_U(M) := F_U(w^{(-1)}(M))$ for all $M \in \mathcal{B}$. It is obvious that $E_U(-\infty, 0) = E_U[2\pi, \infty) = 0$, and Propositions 5.3.7a and 5.2.12 imply

$$U = \int_{\mathbb{R}} e^{it}\, dE_U(t)\,; \tag{5.9}$$

hence we get the spectral theorem for unitary operators (see also Problem 20).

5.3.8 Proposition: For any unitary operator U, there is a projection–valued measure E_U on $\mathbb{R}$ such that (5.9) is valid. Moreover, the spectral decomposition is unique, *i.e.*, if E is a projection–valued measure fulfilling $E(-\infty, 0) = E[2\pi, \infty) = 0$ and $U = \int_{\mathbb{R}} e^{it}\, dE(t)$, then $E = E_U$.

Proof: It remains for us to check the uniqueness of the spectral measure which is equivalent to the relation $\int \chi_{[0,t]}(u)\, dE(u) = \int \chi_{[0,t]}(u)\, dE_U(u)$ for all $t \in [0, 2\pi)$. If there is a measure E with the stated properties, then Theorem 5.2.2 implies $\int T(u)\, dE(u) = \int T(u)\, dE_U(u)$ for any trigonometric polynomial, and furthermore, $\int \psi(u)\, dE(u) = \int \psi(u)\, dE_U(u)$ for any function $\psi : \mathbb{R} \to \mathbb{C}$ which is continuous and 2π–periodic due to Problem 6a and the Fejér theorem. In particular, we can choose for ψ the piecewise linear function which is 2π–periodic and is equal to 1 and 0 in the intervals $[0, t]$ and $[t+1/n, 2\pi - 1/n]$, respectively; taking the limit $n \to \infty$ and using Problem 6a once more we get the sought identity. ∎

5.3.9 Example (spectral decomposition of the FP–operator): Consider the Fourier–Plancherel operator F_n on $L^2(\mathbb{R}^n)$. If $n = 1$, it has a pure point spectrum due to Problem 2.6, $\sigma_p(F) = \{1, -i, -1, i\}$, and by induction we can check that the same is true for any n. The corresponding spectral measure is then

$$E_{F_n} :\; E_{F_n}(M) = \sum_{k=0}^{3} \chi_{w^{(-1)}(M)}(e^{i\pi k/2})\, P_k\,,$$

where the projections $P_k := E_{F_n}(\{\frac{\pi k}{2}\})$ refer to the eigenvalues $e^{\pi i k/2}$ and w is the function used above; it is discrete and supported by $\{0, \frac{\pi}{2}, \pi, \frac{3\pi}{2}\}$. By Problems 20b and 22, the eigenprojections may be expressed as $P_k = \sum_{j=0}^{3} c_j^{(k)} F_n^j$, and using the argument of Example 5.2.3, we get a system of equations for the coefficients which is solved readily by $c_j^{(k)} = \frac{1}{4}(-i)^{jk}$; this yields the formula

$$E_{F_n}(M) = \frac{1}{4} \sum_{j,k=0}^{3} \chi_M\left(\frac{\pi k}{2}\right) (-i)^{jk}\, F_n^j$$

for any Borel set $M \subset \mathbb{R}$.

In the final step, we combine Proposition 5.3.8 with the Cayley transform of a self–adjoint operator A. We start with an auxiliary result.

5.3.10 Proposition: Suppose that A is self–adjoint and E is a projection–valued measure on $\mathbb{R}$ such that the relation (5.7) is valid. Then $E(w^{(-1)}(\cdot))$, where $w(t) := \pi + 2\arctan t$, is the spectral measure of the unitary operator $C(A)$.

Proof: Since w is continuous, $F := E(w^{(-1)}(\cdot))$ is a projection–valued measure due to Problem 15a, and moreover, $F(-\infty,0] = F[2\pi,\infty) = 0$. The function $\eta : \eta(s) = e^{is}$ belongs to $L^\infty(\mathbb{R},dF)$ and $e^{is} = (w^{-1}(s)-i)(w^{-1}(s)+i)^{-1}$. Using Proposition 5.2.12, we get

$$\int_{\mathbb{R}} e^{is}\, dF(s) = \int_{\mathbb{R}} \frac{t-i}{t+i}\, dE(t);$$

it remains to prove that the right side, which we can write as $\mathcal{T}(\varphi)$ with $\varphi(t) := \frac{t-i}{t+i}$ equals $C(A)$. The operator $\mathcal{T}(\varphi)$ is unitary by Problem 6c. Since A is self–adjoint, for any $y \in \mathcal{H}$ there is $z \in D_A$ such that $y = (A+i)z = \mathcal{T}(id+i)z$, and $\mathcal{T}(\varphi)\mathcal{T}(id+i) = \mathcal{T}(id-i)$ due to Theorem 5.2.9c because the two operators have the same domain D_A. This yields $(x,\mathcal{T}(\varphi)y) = (x,(A-i)z) = (x,(A-i)(A+i)^{-1}y)$ for any $x \in \mathcal{H}$, *i.e.*, $\mathcal{T}(\varphi) = C(A)$. ∎

Now we are ready for the *proof of the spectral theorem*: Let $U := C(A)$ be the Cayley transform of A and F the spectral measure of U. Due to Theorem 4.7.7b, $\lambda = 1$ is not an eigenvalue of U, so $F(\{0\}) = 0$ by Problem 21, and therefore $F(-\infty,0] = F[2\pi,\infty) = 0$. Consider the function $v : \mathbb{R} \to \mathbb{R}$ defined by $v(s) := \tan(\frac{s-\pi}{2})$ for $s \in (0,2\pi)$ and $v(s) := 0$ otherwise; it is easy to see that it is Borel and $v(s) = \tan(\frac{s-\pi}{2})$ F–a.e. in $\mathbb{R}$. Using the functional–calculus rules in combination with (5.9) and the inverse Cayley transformation, $A = (I+U)(I-U)^{-1}$, we get the relation $A = i\,\mathcal{T}^{(F)}(1+\eta)\mathcal{T}^{(F)}((1-\eta)^{-1}) \subset \mathcal{T}^{(F)}\left(i\,\frac{1+\eta}{1-\eta}\right)$, where we have denoted again $\eta(s) := e^{is}$. This further implies $A \subset \mathcal{T}^{(F)}(v)$, because $i\,\frac{1+\eta(s)}{1-\eta(s)} = \tan(\frac{s-\pi}{2})$ provided s is not an integer multiple of 2π; however, both operators are self–adjoint, and therefore equal each other. The relation (5.7) now follows from Proposition 5.2.12 if we define $E_A := F(v^{(-1)}(\cdot))$.

The uniqueness of the spectral decomposition is a consequence of Propositions 5.3.8 and 5.3.10. The condition $BA \subset AB$ implies $BU = UB$ by Problem 4.44b, and since $E_t^{(A)} = F(0,w(t)] = F(-\infty,w(t)]$ we get $E_t^{(A)}B = BE_t^{(A)}$ due to Problem 20c. The opposite implication is checked in the same way as in Proposition 5.3.6. ∎

Part (b) of Theorem 5.3.1 allows us to extend the notion of commutativity to any pair A, A' of self–adjoint operators, in general unbounded: we say that they *commute* if the corresponding spectral measures E and E' commute. By Problem 3c, this is further equivalent to the condition $E_tE'_s = E'_sE_t$ for all $t,s \in \mathbb{R}$, which is due to the spectral theorem valid *iff* A commutes with $\{E'_s\}$ and A' commutes with $\{E_t\}$. Taking Proposition 5.3.6 into account, we see that if one of the operators, say A, is bounded the commutativity is equivalent to $AA' \subset A'A$, *i.e.*, the new definition is consistent with the old one (see also Problem 23b).

Theorem 1 has a purely existential character; it tells us nothing about how the spectral measure can be found for a particular self–adjoint operator A. To solve this problem, we either have to guess the spectral measure and prove afterwards that it satisfies the relation (5.7), or find it using another self–adjoint operator whose spectral decomposition is known.

5.3.11 Examples: (a) Let A be a self–adjoint operator on a separable $\mathcal{H}$ with a pure point spectrum, $\sigma_p(A) = \{\lambda_j : j = 1, 2\dots\}$, where the eigenvalues λ_j correspond to a complete system of eigenprojections P_j. Let $E : E(M) = \sum_j \chi_M(\lambda_j)P_j$ be the discrete projection–valued measure from Example 5.1.2a; we shall show that $\mathcal{T}^{(E)}(id) = A$. The set $\mathbb{R}\backslash\sigma_p(A)$ is μ_x–zero for any $x \in \mathcal{H}$; hence the domain of $\mathcal{T}^{(E)}(id)$ consists of the vectors which satisfy

$$\int_{\mathbb{R}} t^2\, d\mu_x(t) = \sum_j \lambda_j^2\, (x, P_j x) < \infty\,,$$

and $(x, \mathcal{T}^{(E)}(id)x) = \sum_j \lambda_j\, (x, P_j x)$ holds for any D_{id} in view of Proposition 5.2.7. The projections P_j reduce the operator A by Example 4.4.3, $P_j A \subset A P_j$, so $\|Ax\|^2 = \sum_j \lambda_j^2\, \|P_j x\|^2 < \infty$ holds for any $x \in D_A$, *i.e.*, $D_A \subset D_{id}$. In a similar way, we check the inclusion $A \subset \mathcal{T}^{(E)}(id)$, and since the operators are self–adjoint, they equal each other. The action of the operator A given by the right side of (5.7) can in this case be expressed explicitly as $Ax = \sum_j P_j Ax = \sum_j AP_j x = \sum_j \lambda_j P_j x$ for any $x \in D_A$.

(b) Let E be a projection–valued measure on $\mathbb{R}^d$ and $w : \mathbb{R}^d \to \mathbb{R}$ a Borel function which is defined E–a.e. The operator $\int_{\mathbb{R}} w(t)\, dE(t)$ is self–adjoint and $E_w := E(w^{(-1)}(\cdot))$ is its spectral measure due to Proposition 5.2.12.

(c) Suppose that A is self–adjoint and U is unitary, then $A' := UAU^{-1}$ is also self–adjoint and the spectral measures of A, A' are related by $E'(\cdot) = UE(\cdot)U^{-1}$ in view of Problems 4 and 12.

(d) Similarly, let A be reduced by a projection P; then its part $A' := A \upharpoonright PD_A$ is again self–adjoint and its spectral measure is given by $E'(\cdot) := E(\cdot) \upharpoonright P\mathcal{H}$ due to Problems 3e and 13.

5.4 Spectra of self–adjoint operators

The existence of the spectral decomposition for a self–adjoint operator A provides a tool for characterizing and classifying the spectrum. To begin with, let us collect some results which follow immediately from Theorem 5.2.11.

5.4.1 Proposition: Let A be a self–adjoint operator and λ a real number; then

(a) $\lambda \in \sigma(A)$ *iff* $E_A(\lambda-\varepsilon, \lambda+\varepsilon) \neq 0$ for any $\varepsilon > 0$.

(b) $\lambda \in \sigma_p(A)$ *iff* $E_A(\{\lambda\}) \neq 0$, and $N_A(\lambda) = \mathrm{Ran}\,(E_A(\{\lambda\}))$ is the corresponding eigenspace. Any isolated point of the spectrum is an eigenvalue.

(c) The set $\sigma(A)$ is nonempty and $\mathbb{R}\backslash\sigma(A)$ is E_A–zero; the operator A is bounded *iff* its spectrum is bounded.

5.4.2 Remarks: (a) The fact that $\sigma(A)$ is nonempty represents an important result of the spectral theory; recall that in distinction to the bounded case covered by Theorem 1.7.5, an unbounded closed operator can have an empty spectrum — *cf.* Examples 1.7.6. Assertion (c) also allows us to write the formula (5.7) as

$$A = \int_{\sigma(A)} t\, dE_A(t);$$

this shows that the numerical range of A satisfies the relations $\inf\Theta(A) = \inf\sigma(A)$ and $\sup\Theta(A) = \sup\sigma(A)$. Moreover, the spectrum is the minimal closed set with an E_A–zero complement, *i.e.*, the spectral measure E_A is supported by $\sigma(A)$.

(b) The spectral measure of a single point can be written using the spectral decomposition $\{E_t^{(A)}\}$ as $E_A(\{\lambda\}) = E_\lambda^{(A)} - E_{\lambda-0}^{(A)}$; then it follows from assertion (b) that $\sigma_p(A)$ consists just of the discontinuity points of the map $t \mapsto E_t^{(A)}$. Furthermore, the residual spectrum of a self–adjoint operator is empty, so (a) in combination with the decomposition (1.6) shows that $\sigma_c(A)$ consists of the points in which $t \mapsto E_t^{(A)}$ is nonconstant but continuous.

The essential spectrum of a self–adjoint operator A is related to the dimensionality of its spectral measure.

5.4.3 Theorem: (a) A real number λ belongs to the essential spectrum of A *iff* $\dim \mathrm{Ran}\, E_A(\lambda-\varepsilon, \lambda+\varepsilon) = \infty$ holds for any $\varepsilon > 0$.

(b) $\sigma_{ess}(A)$ is a closed set.

Proof: Let $\dim \mathrm{Ran}\, E_A(\lambda-\varepsilon, \lambda+\varepsilon) = \infty$ for any $\varepsilon > 0$, and consider the sequence of neighborhoods $U_n := (\lambda-1/n, \lambda+1/n)$. Proposition 5.1.1b implies $\text{s-lim}_{n\to\infty} E_A(U_n) = E_A(\{\lambda\})$; if there is an $\tilde n$ such that $E_A(U_n) = E_A(U_{\tilde n})$ for all $n > \tilde n$, then $\dim E_A(\{\lambda\}) = \dim E_A(U_{\tilde n}) = \infty$, so λ is an eigenvalue of infinite multiplicity. In the opposite case we can choose a subsequence $\{U_{n_k}\}$ such that $E_A(U_{n_{k+1}}) \neq E_A(U_{n_k})$ for $k = 1, 2, \dots$; since the sets $\Delta_k := U_{n_{k+1}} \backslash U_{n_k}$ are disjoint, the projections $E_k := E_A(\Delta_k)$ form an orthogonal family. We pick a unit vector x_k in each $\mathrm{Ran}\, E_k$; then $\|(A-\lambda)x_k\|^2 = \int_{\Delta_k} |t-\lambda|^2 d\mu_{x_k}(t) < n_k^{-2}$. In view of the orthonormality, $\{x_k\}$ has no convergent subsequence, so again $\lambda \in \sigma_{ess}(A)$.

Assume on the contrary that the condition is not valid; then λ is not an eigenvalue of infinite multiplicity; in view of Proposition 4.7.13 we have to show that λ is a regular value of the operator $A_\lambda := A \restriction (N_A(\lambda))^\perp$. By assumption,

$\dim \operatorname{Ran} E_A(U_\varepsilon(\lambda)) < \infty$ for some $\varepsilon > 0$; hence there is a positive $\delta < \varepsilon$ such that $E_A(\lambda-\delta,\lambda) = E_A(\lambda,\lambda+\delta) = 0$ (*cf.* Example 5.1.4), and therefore $(N_A(\lambda))^\perp = \operatorname{Ran}(I - E_A(U_\delta(\lambda)))$. The operator A_λ is self–adjoint due to Problem 4.24b, so $\lambda \in \rho(A_\lambda)$ *iff* there is a $c > 0$ such that $\|(A_\lambda-\lambda)x\|^2 \ge c\|x\|^2$ for all $x \in D(A_\lambda)$, *i.e.*, all $x \in D_A$ fulfilling the condition $E_A(U_\delta(\lambda)) = 0$; the functional–calculus rules show that the inequality is valid with $c = \delta$. This proves assertion (a), which in turn easily implies (b). ∎

Another useful criterion combines a modification of the definition from Section 4.2 with topological properties of the essential spectrum.

5.4.4 Theorem: Let $\lambda \in \mathbb{R}$; then the following conditions are equivalent:

(a) $\lambda \in \sigma_{ess}(A)$.

(b) there is a sequence $\{x_n\} \subset D_A$ of unit vectors which converges weakly to zero and $\lim_{n\to\infty} \|(A-\lambda)x_n\| = 0$.

(c) λ is an accumulation point of $\sigma(A)$ or an eigenvalue of infinite multiplicity.

Proof: In the previous proof we have shown that (b) follows from (a); recall that any orthonormal sequence converges weakly to zero. On the other hand, a sequence $\{x_n\}$ which satisfies condition (b) contains no convergent subsequence, since $x_{n_k} \to 0$ contradicts the requirement $\|x_{n_k}\| = 1$ for all k. It remains for us to show that (a) and (c) are equivalent. If condition (c) is not valid, then either $\lambda \notin \sigma(A)$ or it is an isolated point of the spectrum with $\dim \operatorname{Ran} E_A(\{\lambda\}) < \infty$, so $\lambda \notin \sigma_{ess}(A)$ due to Theorem 5.4.3. On the contrary, $\lambda \notin \sigma_{ess}(A)$ implies $\dim \operatorname{Ran} E_A(U_\varepsilon(\lambda)) =: n < \infty$ for some $\varepsilon > 0$. In the case $n = 0$ we have $\lambda \notin \sigma(A)$; otherwise it is an isolated point of the spectrum with a multiplicity not exceeding n. ∎

The complement $\sigma_d(A) := \sigma(A)\backslash\sigma_{ess}(A)$ is called the **discrete spectrum** of the operator A; it consists of all isolated eigenvalues of a finite multiplicity. If the essential spectrum is empty, $\sigma(A) = \sigma_p(A)$, we say that operator A has a *purely discrete* spectrum. Any such operator has a pure point spectrum but the converse is not true (Problem 24). The fact that $\sigma_{ess}(A) = \emptyset$ can be expressed equivalently by means of the operators $(A-\lambda)^{-1}$ (see Problem 25b); this is why instead of a purely discrete spectrum we often speak about operators *with a compact resolvent.*

Theorem 5.4.4 easily implies that the essential spectrum does not change if we add a compact Hermitean operator to A (Problem 26). It appears that this stability result remains valid for a much wider class of perturbations. An operator T, in general unbounded, is said to be *relatively compact* with respect to a self–adjoint A, or briefly A*–compact*, if $D_T \supset D_A$ and the operator $T(A-i)^{-1}$ is compact; in such a case the first resolvent formula together with Theorem 3.5.2 shows that $TR_A(\lambda)$ is compact for any $\lambda \in \rho(A)$.

5.4.5 Proposition: Suppose that A is self–adjoint and T is A–compact, then

(a) A sequence $\{x_n\} \subset D(A)$ which satisfies $\|x_n\|^2 + \|Ax_n\|^2 < c$ for a positive c and all $n = 1, 2, \dots$ contains a subsequence $\{x_{n_k}\}$ such that $\{Tx_{n_k}\}$ is convergent.

(b) If $A+T$ is self-adjoint and the operator T is $(A+T)$–bounded, then it is also $(A+T)$–compact.

(c) If the operator T is symmetric and A–compact, then $A+T$ is self-adjoint and T is $(A+T)$–compact.

Proof: We set $y_n := (A-i)x_m$. The parallelogram identity gives $\|y_n\|^2 \le 2\|x_n\|^2 + 2\|Ax_n\|^2$, so $\{y_n\}$ is bounded and by Theorems 1.5.14 and 3.5.2 there is a convergent subsequence $\{TR_A(i)y_{n_k}\} = \{Tx_{n_k}\}$. To prove (b), we take a bounded sequence $\{y_n\}$. The vectors $z_n := R_{A+T}(i)y_n$ belong to $D(A+T) = D(A)$ and there are real α, β such that $\|Tz_n\|^2 \le \alpha^2\|(A+T)z_n\|^2 + \beta^2\|z_n\|^2$ holds for all n. Furthermore, $\|(A+T)z_n\| = \|y_n + iz_n\|^2 \le 2\|y_n\|^2 + 2\|R_{A+T}(i)\|^2\|y_n\|^2$; these two inequalities show that the sequence $\{\|z_n\|^2 + \|Az_n\|^2\}$ is bounded, so $TR_{A+T}(i)$ is compact by (a).

Next we employ the inequality $\|Tx\| \le a_n(\|Ax\| + n\|x\|)$ with $a_n := \|TR_A(in)\|$, which is valid for any $x \in D(A)$ and $n = 1, 2, \dots$; we shall show that $a_n \to 0$. We define $B_n := I - i(n-1)R_A(-in) = (A+i)R_A(-in)$; then the functional-calculus rules in combination with Problems 8 and 23a yield

$$\|B_n y\|^2 = \int_{\mathbb{R}} |t+i|^2 d\mu_{R_A(-in)y}(t) = \int_{\mathbb{R}} \frac{t^2+1}{t^2+n^2}\, d\mu_y(t)$$

for any $y \in \mathcal{H}$, and therefore $\|B_n y\| \to 0$ by the dominated-convergence theorem, *i.e.*, $\text{s-lim}_{n\to\infty} B_n = 0$. The compactness of $TR_A(i)$ implies that its adjoint is also compact; then $\|TR_A(i)B_n^*\| = \|B_n(TR_A(i))^*\| \to 0$ by Problem 3.32, and since $B_n^* = I + i(n-1)R_A(in) = (A-i)R_A(in)$, we get $a_n = \|TR_A(i)B_n^*\| \to 0$. Hence T is A–bounded with zero relative bound; in particular, it is self-adjoint by the Kato–Rellich theorem. Choosing n large enough, we have $a_n < \frac{1}{2}$, and therefore $\|Tx\| \le \|Ax\| - \|Tx\| + n\|x\| \le \|(A+T)x\| + n\|x\|$ for any $x \in D(A)$, *i.e.*, T is $(A+T)$–bounded; to conclude the proof, we have to apply assertion (b). ∎

5.4.6 Theorem: The essential spectrum of a self-adjoint A is stable with respect to a symmetric A–compact perturbation T, *i.e.*, $\sigma_{ess}(A+T) = \sigma_{ess}(A)$.

Proof: It is sufficient to check that $Tx_n \to 0$ for any sequence $\{x_n\}$ which satisfies condition (b) of Theorem 5.4.4; in view of the compactness of $TR_A(i)$, this is equivalent to $\text{w-lim}_{n\to\infty}(A-i)x_n = 0$. Since $\text{w-lim}_{n\to\infty} x_n = 0$ by assumption, we have $((A+i)y, x_n) = (y, (A-i)x_n) \to 0$ for any $y \in D_A$. Let $\lambda \in \sigma_{ess}(A)$; then the sequence $\{(A-i)x_n\}$ is bounded because $\limsup_{n\to\infty}\|(A-i)x_n\| \le |\lambda - i|$, and therefore $\text{w-lim}_{n\to\infty}(A-i)x_n = 0$ due to Theorem 1.5.12c. In this way, we get $\sigma_{ess}(A) \subset \sigma_{ess}(A+T)$, and since Proposition 5.4.5c allows us to switch the roles of A and $A+T$, the two sets are equal. ∎

Spectral decomposition provides another way to decompose the spectrum of a self-adjoint operator A which is based on the relations of the measures $\mu_x :=$

$(x, E_A(\cdot)x)$ to the Lebesgue measure on $\mathbb{R}$. Let $\mathcal{B}_0$ be the family of all Borel sets of zero Lebesgue measure; then we introduce

$$\begin{aligned} \mathcal{H}_{ac} &:= \{ x \in \mathcal{H} : \mu_x(N) = 0 \text{ for all } N \in \mathcal{B}_0 \}, \\ \mathcal{H}_s &:= \{ x \in \mathcal{H} : \text{there is } N_x \in \mathcal{B}_0 \text{ such that } \mu_x(\mathbb{R}\setminus N_x) = 0 \}; \end{aligned}$$

using the terminology of Appendix A.3, we may say that $\mathcal{H}_{ac}$ and $\mathcal{H}_s$ consists just of the vectors for which the measure μ_x is respectively absolutely continuous and singular with respect to the Lebesgue measure. Alternatively, we can use the function $\sigma_x : \sigma_x(t) = (x, E_t^{(A)}x)$. In the case of $x \in \mathcal{H}_{ac}$ we have $\mu_x(J) = \sigma_x(b) - \sigma_x(a)$ for any interval J with the endpoints $a < b$, and therefore $\mathcal{H}_{ac}$ consists of the vectors $x \in \mathcal{H}$ for which the function σ_x is absolutely continuous.

5.4.7 Proposition: The subspaces $\mathcal{H}_{ac}$ and $\mathcal{H}_s$ are closed and mutually orthogonal, $\mathcal{H} = \mathcal{H}_{ac} \oplus \mathcal{H}_s$, and moreover, they reduce the operator A.

Proof: Clearly $x \in \mathcal{H}_{ac}$ *iff* $E_A(N)x = 0$ for all $N \in \mathcal{B}_0$, and $y \in \mathcal{H}_s$ *iff* $E_A(N_y)y = y$ for some $N_y \in \mathcal{B}_0$. Then $(x, y) = (x, E_A(N_y)y) = (E_A(N_y)x, y) = 0$, so $\mathcal{H}_{ac} \subset \mathcal{H}_s^\perp$. To check that this is in fact an identity, we take any $N \in \mathcal{B}_0$ and $x \in \mathcal{H}_s^\perp$. Since $E_A(N)x \in \mathcal{H}_s$, we get $\mathcal{H}_{ac} = \mathcal{H}_s^\perp$. It remains for us to prove that $\mathcal{H}_s$ is a closed subspace reducing the operator A. Let a sequence $\{y_n\} \subset \mathcal{H}_s$ converge to some y; then there are sets $N_n \in \mathcal{B}_0$ such that $E_A(N_n)y_n = y_n$. Their union $N := \bigcup_{n=1}^\infty N_n$ also belongs to $\mathcal{B}_0$, so $E_A(N)y_n = E_A(N_n)y_n = y_n$ and the limit gives $E_A(N)y = y$. Hence the set $\mathcal{H}_s$ is closed; in a similar way we check that it is a closed subspace. It reduces A *iff* $E_t^{(A)}x \in \mathcal{H}_s$ holds for any $x \in \mathcal{H}_s$ and $t \in \mathbb{R}$; this is true because $E_A(N_x)x = x$, and the operators $E_A(N_x)$ and $E_t^{(A)}$ commute. ∎

Due to Problem 4.24b, the operators $A_{ac} := A \upharpoonright \mathcal{H}_{ac}$ and $A_s := A \upharpoonright \mathcal{H}_s$, which we call respectively the *absolutely continuous* and *singular part* of the operator A, are self-adjoint. Their spectra are respectively called the *absolutely continuous* and *singular spectrum* of A, and we have

$$A = A_{ac} \oplus A_s , \quad \sigma(A) = \sigma_{ac}(A) \cup \sigma_s(A) . \tag{5.10}$$

The operator A also has other orthogonal-sum decompositions; we know from Example 4.4.3 that $A = A_p \oplus A_c$, where $A := A \upharpoonright \mathcal{H}_p$ and $\mathcal{H}_p := \sum_{\lambda \in \sigma_p(A)}^{\oplus} N(\lambda)$. It is not difficult to check the inclusion $\mathcal{H}_p \subset \mathcal{H}_s$ (Problem 27); if $\mathcal{H}_p$ is a proper subspace we denote its orthogonal complement in $\mathcal{H}_s$ by $\mathcal{H}_{sc}$, and set $A_{sc} := A \upharpoonright \mathcal{H}_{sc}$ and $\sigma_{sc}(A) := \sigma(A_{sc})$. By definition, A_{sc} has no eigenvalues. This means, in particular, that $\mathcal{H}_{sc}$ is either infinite-dimensional or trivial, and furthermore, that the spectral decomposition of A_{sc} is continuous. The operator A_{sc} is called the *singularly continuous part* of A and $\sigma_{sc}(A)$ is its *singularly continuous spectrum.* Problem 4.24b now gives $A_s = A_p \oplus A_{sc}$ and $\sigma_s(A) = \overline{\sigma_p(A)} \cup \sigma_{sc}(A)$; combining these relations with (5.10) we get

$$A = A_p \oplus A_{ac} \oplus A_{sc} , \quad \sigma(A) = \overline{\sigma_p(A)} \cup \sigma_{ac}(A) \cup \sigma_{sc}(A) . \tag{5.11}$$

5.4.8 Remark: The operator $A_c := A \upharpoonright \mathcal{H}_p^\perp$ has no eigenvalues, *i.e.*, a purely discrete spectrum; this is why the continuous spectrum of a self–adjoint operator A is often defined as the spectrum of A_c. This definition differs from that of Section 1.7 because the sets $\sigma_p(A)$ and $\sigma(A_c)$ are in general not disjoint (Problem 24c). An isolated eigenvalue which is not an isolated point of $\sigma(A)$ belongs to $\sigma(A_c)$; such eigenvalues are said to be *embedded* in the continuous spectrum.

5.4.9 Examples: (a) Consider the space $L^2(\mathbb{R}, d\mu)$ with an arbitrary Borel measure μ and the operator $Q_\mu : (Q_\mu\psi)(x) = x\psi(x)$. Let $\mu = \mu_{ac} + \mu_s$ be the Lebesgue decomposition of μ with respect to the Lebesgue measure on $\mathbb{R}$; then there is a set $N_\mu \in \mathcal{B}_0$ such that $\mu_s(\mathbb{R}\backslash N_\mu) = 0$ and $\mu_{ac}(N_\mu) = 0$. Denote $\mathcal{H}_j := L^2(\mathbb{R}, d\mu_j)$; it is easy to check that

$$(U[\psi_{ac}, \psi_s])(x) := \begin{cases} \psi_{ac}(x) & \dots \quad x \in \mathbb{R}\backslash N_\mu \\ \psi_s(x) & \dots \quad x \in N_\mu \end{cases}$$

is a unitary operator from $\mathcal{H}_{ac} \oplus \mathcal{H}_s$ to $L^2(\mathbb{R}, d\mu)$ and a vector $\psi \in \mathcal{H}_s$ *iff* $\psi = \chi_{N_\mu}\psi$. Since $(E_Q(M)\psi)(x) = \chi_M(x)\psi(x)$ by Example 5.3.11b, we conclude that the absolutely continuous and singular subspaces are in this case isomorphic to $L^2(\mathbb{R}, d\mu_{ac})$ and $L^2(\mathbb{R}, d\mu_s)$, respectively. In particular, the spectrum of the operator Q on $L^2(\mathbb{R})$ is purely absolutely continuous, $\mathcal{H}_{ac}(Q) = L^2(\mathbb{R})$ and $\sigma(Q) = \sigma_{ac}(Q) = \mathbb{R}$.

(b) Let us ask under which condition the operator $T_f \in \mathcal{L}_{sa}(L^2(\mathbb{R}^n))$ corresponding to a real Borel f has a purely absolutely continuous spectrum, *i.e.*, $\mathcal{H}_{ac}(T_f) = L^2(\mathbb{R}^n)$ and $\sigma_{ac}(T_f) = \sigma(T_f)$. Due to Examples 5.2.8 and 5.3.11b, the spectral measure of T_f is $E_f := E(f^{(-1)}(\cdot))$, so

$$\|E_f(N)\psi\|^2 = \int_{f^{(-1)}(N)} |\psi(x)|^2 dx$$

holds for any $\psi \in L^2(\mathbb{R}^n)$ and $N \in \mathcal{B}$. It is sufficient therefore to require that $f^{(-1)}(N) \subset \mathbb{R}^n$ has Lebesgue measure zero for all $N \in \mathcal{B}$ of Lebesgue measure zero — see, *e.g.*, Problem 28.

5.5 Functions of self–adjoint operators

The spectral theorem together with functional–calculus rules gives us a tool to associate a class of operators with a given self–adjoint A. As usual, E_A is the spectral measure of A and $\Phi^{(A)} := \Phi_{E_A}$ will denote the set of all E_A–a.e. defined Borel functions $\varphi : \mathbb{R} \to \mathbb{C}$. In particular, any function continuous on the spectrum belongs to this set, $C(\sigma(A)) \subset \Phi^{(A)}$.

The operator $\mathcal{T}^{(E_A)}(\varphi) \equiv \int_{\mathbb{R}} \varphi(t)\, dE_A(t)$ is for any $\varphi \in \Phi^{(A)}$ fully determined by the function φ and the operator A. It is customary to denote these operators as

$\varphi(A)$ and call them *functions* of A. This notation is less cumbersome, particularly when we deal with elementary functions; one should keep in mind, of course, that φ rather than A is the "variable" in such a "function".

5.5.1 Examples: (a) The multiplication operator Q_μ on $L^2(\mathbb{R}, d\mu)$ with a Borel measure μ can be written as $Q_\mu = T_{id}$. Moreover, Example 5.2.8 implies

$$T_f = \int_{\mathbb{R}} f(t)\, dE^{(\mu)}(t)$$

for any Borel function $f : \mathbb{R} \to \mathbb{C}$ where the projection–valued measure is determined by $E^{(\mu)}(M) := T_{\chi_M}$, $M \in \mathcal{B}$; in the particular case $f = id$ the above formula represents the spectral decomposition of the operator Q_μ. Given a function $\varphi \in \Phi^{(Q_\mu)}$ we can extend it arbitrarily to a function defined on the whole $\mathbb{R}$ denoted by the same symbol; since the extension concerns an $E^{(\mu)}$–zero set, Problem 6c implies $\varphi(Q_\mu) = T_\varphi$.

(b) *Functions of the operator P on $L^2(\mathbb{R})$*: By Example 4.4.7, P is unitarily equivalent to the operator Q; in view of Problem 12 and Example 5.3.11c, this implies $\Phi^{(P)} = \Phi^{(Q)}$ and $\varphi(P) = F^{-1}\varphi(Q)F$ for any $\varphi \in \Phi^{(P)}$. Since the Fourier–Plancherel operator has a simple functional realization, we are in some cases able to express the operator $\varphi(P)$ explicitly. For instance, e^{iaP} is for any $a \in \mathbb{R}$ a unitary "substitution" operator which acts at an arbitrary $\psi \in L^2(\mathbb{R})$ as

$$\left(e^{iaP}\psi\right)(x) = \psi(x+a)\,.$$

The translation operator $U_a : U_a\psi = \psi(\cdot + a)$ is unitary by Example 3.3.2, so it is sufficient to check the identity $e^{iaP}\psi = U_a\psi$ for all ψ of some dense set in $L^2(\mathbb{R})$. Using Example 1.5.6 we easily find that $e^{iaQ}F\psi = FU_a\psi$ holds for $\psi \in L^2(\mathbb{R}) \cap L^1(\mathbb{R})$; hence the result follows from the above unitary equivalence.

As another illustration, consider a function $\varphi \in L^2(\mathbb{R})$; then $\varphi(P)$ has the following integral–operator representation,

$$(\varphi(P)\psi)(x) = (2\pi)^{-1/2} \int_{\mathbb{R}} (F\varphi)(y-x)\psi(y)\, dy\,. \tag{5.12}$$

To check this formula, notice that the product of φ and $\eta := F\psi$ belongs to $L^1(\mathbb{R})$, so the relation (3.2) yields $\varphi(P)\psi(x) = (F^{-1}\varphi\eta)(x) = (2\pi)^{-1/2}(\theta_x, F\psi)$ for a.a. $x \in \mathbb{R}$, where $\theta_x(y) := \overline{e^{-ixy}\varphi(y)}$. Let T_n denote the operator of multiplication by $\chi_{(-n,n)}$; then $T_n\theta_x \in L^2 \cap L^1$ and since F is unitary, we have $(\theta_x, F\psi) = \lim_{n\to\infty}(F^{-1}T_n\theta_x, \psi)$. Next we have to write the expression on the right side explicitly, and to make a simple substitution in the appropriate integral; using then the identity $U_x\psi = e^{ixP}\psi$ proven above, we get

$$\sqrt{2\pi}\,(\varphi(P)\psi)(x) = \lim_{n\to\infty}(F^{-1}T_n\theta_0, e^{ixP}\psi) = \int_{\mathbb{R}} \overline{(F^{-1}\bar\varphi)(y)}\psi(x+y)\, dy\,.$$

However, $\overline{(F^{-1}\bar{\varphi})(y)} = (F\varphi)(y)$, so the relation (5.12) is valid for an arbitrary $\psi \in D(\varphi(P)) = F^{-1}D(\varphi(Q))$ and a.a. $x \in \mathbb{R}$.

The spectral measure of a self–adjoint A is itself a function of A and as such it can be expressed by means of other functions of the same operator.

5.5.2 Example (Stone formula): Given $\varepsilon > 0$, a finite interval $J := [a,b]$ with $b > a$, and a number $u \in J$, we define the function $\psi_\varepsilon^{(u)} : \mathbb{R} \to \mathbb{R}^+$ by $\psi_\varepsilon^{(u)}(t) := \mathrm{Im}\,(t-u-i\varepsilon)^{-1} = \varepsilon[(t-u)^2+\varepsilon^2]^{-1}$, and furthermore,

$$\varphi_\varepsilon(t) := \int_a^b \psi_\varepsilon^{(u)}(t)\,du = \arctan\left(\frac{b-t}{\varepsilon}\right) - \arctan\left(\frac{a-t}{\varepsilon}\right);$$

by Theorem 5.2.2d we then have $\text{s-lim}_{\varepsilon\to0+}\varphi_\varepsilon(A) = \frac{\pi}{2}\,[E_A[a,b]+E_A(a,b)]$. To express the left side of the last relation, notice that the function $(t,u) \mapsto \psi_\varepsilon^{(u)}(t)$ fulfils the assumptions (i)–(iii) of Proposition 5.2.5. At the same time, we have $\psi_\varepsilon^{(u)} = \frac{1}{2i}(R_A(u+i\varepsilon) - R_A(u-i\varepsilon))$, so the first resolvent formula in combination with the inequality $\|R_A(u \pm i\varepsilon)\| < \varepsilon^{-1}$ gives the estimate $\|\psi_\varepsilon^{(u)}(A) - \psi_\varepsilon^{(v)}(A)\| \le \varepsilon^{-2}|u-v|$, which shows that assumption (iv) is valid too. Hence the proposition may be applied: we get $\varphi_\varepsilon^{(A)} = \frac{1}{2i}\int_a^b [R_A(u+i\varepsilon) - R_A(u-i\varepsilon)]\,du$, which together with the above limiting relation provides an expression of the spectral measure in terms of the resolvent,

$$E_A[a,b] + E_A(a,b) = \frac{1}{\pi i}\,\underset{\varepsilon\to0+}{\text{s-lim}} \int_a^b [R_A(u+i\varepsilon) - R_A(u-i\varepsilon)]\,du\,, \tag{5.13}$$

which is known as the **Stone formula**.

Since the resolvent set of a self–adjoint A is E_A–zero, the operator $\varphi(A)$ is by Theorem 5.2.9d fully determined by the restriction $\varphi \restriction \sigma(A)$. If this function is continuous, we can strengthen the assertion of Theorem 5.2.11a.

5.5.3 Proposition: Let $\varphi \in C(\sigma(A))$, then $\sigma(\varphi(A)) = \overline{\varphi(\sigma(A))}$; in particular, $\sigma(\varphi(A)) = \varphi(\sigma(A))$ if A is bounded.

Proof: If $y = \varphi(x)$ for some $x \in \sigma(A)$, then due to the assumed continuity we can find for any $\delta > 0$ a neighborhood $U(y)$ such that $(x-\delta, x+\delta) \subset \varphi^{(-1)}(U(y))$. Since $x \in \sigma(A)$, we have $E_A(x-\delta, x+\delta) \ne 0$, and therefore $E_A(\varphi^{(-1)}(U(y))) \ne 0$, *i.e.*, $y \in R_{ess}^{(E_A)}(\varphi) = \sigma(\varphi(A))$. This yields the inclusion $\varphi(\sigma(A)) \subset \sigma(\varphi(A))$, which implies $\overline{\varphi(\sigma(A))} \subset \sigma(\varphi(A))$ because the spectrum is closed. On the other hand, any point $z \notin \overline{\varphi(\sigma(A))}$ has a neighborhood $U(z)$ disjoint with $\varphi(\sigma(A))$; hence $\varphi^{(-1)}(U(z)) \cap \sigma(A) = \emptyset$ holds for the pull–backs, which means that $z \notin \sigma(\varphi(A))$. This proves the first assertion; the rest follows from the fact that the spectrum of a Hermitean A is compact (*cf.* Proposition 1.3.1c and Theorem 1.3.3b). ∎

The functions of a self–adjoint A obey the usual composition rules. If $w \in \Phi^{(A)}$ is a real–valued function, then the spectral measure of $w(A)$ is $E_w := E(w^{(-1)}(\cdot))$ by Example 5.3.11b, and Proposition 5.2.12 gives $(\varphi\circ w)(A) = \varphi(w(A))$ for any Borel function $\varphi : \mathbb{R} \to \mathbb{C}$ which is E_w–a.e. defined.

5.5.4 Example: The function $|id| : x \mapsto |x|$ is Borel and positive, so we can associate with any $A \in \mathcal{L}_{sa}(\mathcal{H})$ the positive operator $|A| := |id|(A)$ with the domain $D(|A|) = D(A)$; in particular, $|A| = A$ if A is positive. The composition rule yields the identity $|A| = (A^2)^{1/2}$, which shows that for Hermitean operators the present definition of $|A|$ coincides with that of Section 3.2.

Next, we want to show that the set of all functions of a given self–adjoint A can be characterized solely by its commutativity properties. Given a set $\mathcal{S}$ of linear operators, not necessarily bounded, on a Hilbert space $\mathcal{H}$ we define its *commutant* as $\mathcal{S}' := \{ B \in \mathcal{B}(\mathcal{H}) : BT \subset TB \text{ for all } T \in \mathcal{S} \}$, and its *extended bicommutant* as

$$\mathcal{S}''_{ex} := \{ T \in \mathcal{L}_c(\mathcal{H}) : BT \subset TB \text{ for all } B \in \mathcal{S}' \};$$

it is clear how these definitions modify the general algebraic notions of commutant and bicommutant — *cf.* Appendix B.1. The assertion of Theorem 5.3.1b can now be written as $\{A\}' = \{E_t^{(A)} : t \in \mathbb{R}\}' = \{E_A(M) : M \in \mathcal{B}\}'$, where the last identity follows from Problem 3. In combination with Proposition 5.2.11b, these relations yield the inclusion

$$\{ \varphi(A) : \varphi \in \Phi^{(A)} \} \subset \{A\}''_{ex};$$

we are going to show that in the case of a *separable* $\mathcal{H}$ it turns into identity. We need an auxiliary assertion.

5.5.5 Lemma: Let A be a self–adjoint operator on a separable $\mathcal{H}$ and denote by $\mathcal{G}(x)$ the closed linear hull of the set $\{E_A(J)x : J \in \mathcal{J}\}$, where $\mathcal{J}$ is the family of all bounded intervals in R. Then there is an orthonormal set $\{x_n\}_{n=1}^N$, $N \le \infty$, such that $\mathcal{H} = \bigoplus_{n=1}^N \mathcal{G}(x_n)$.

Proof: By the functional–calculus rules, a vector y belongs to $\mathcal{G}(x)$ *iff* there is a sequence $\{s_n\}$ of step functions such that $y = \text{s-lim}_{n\to\infty} s_n(A)$. Since the product of a step function with χ_J for each $J \in \mathcal{J}$ is again a step function, the last relation implies $E_A(J)$–invariance of the subspace $\mathcal{G}(x)$ for any $x \in \mathcal{H}$. Denoting the corresponding projection as $P(x)$ we get $P(x)E_A(J) = E_A(J)P(x)$ for an arbitrary $J \in \mathcal{J}$, and furthermore, $P(x)x = x$, where we have used the relation $x = \lim_{n\to\infty} E_A(-n,n)x$.

Since $\mathcal{H}$ is separable there is a countable set $M := \{y_j\} \subset \mathcal{H}$ dense in $\mathcal{H}$; without loss of generality we may assume that all its elements are nonzero. We set $x_1 := y_1/\|y_1\|$, so $P(x_1)y_1 = y_1$. If $P(x_1)y_j = y_j$ holds for all $j \ge 2$ the condition $\overline{M} = \mathcal{H}$ implies $P(x_1) = I$ and the assertion is valid with $N = 1$. In the opposite case we set $j_2 = \min\{ j : P(x_1)y_j \ne y_j \}$ and define $x_2 := x_2'/\|x_2'\|$ where $x_2' := (I - P(x_1))y_{j_2}$; we shall show that the projections $P(x_1)$ and $P(x_2)$ are orthogonal. The above derived properties of the projections $P(x)$ show that

$$P(x_1)P(x_2)z = \lim_{n\to\infty} P(x_1)s_n(A)z = \frac{1}{\|x_2'\|} \lim_{n\to\infty} s_n(A)P(x_1)(I - P(x_1))y_{j_2} = 0$$

for any $z \in \mathcal{H}$, *i.e.*, $P(x_1)P(x_2) = 0$. Moreover, using the relations $P(x_2)x_2' = x_2'$ and $y_{j_2} = x_2' + P(x_1)y_{j_2}$ we can check that $(P(x_1)+P(x_2))y_j = y_j$ holds for $j = j_2$, and thus for any $1 \le j \le j_2$. In the next step we choose j_3 as the smallest j such that $(P(x_1)+P(x_2))y_j \ne y_j$, *etc.*; by induction we prove that after the n–th step we have an orthonormal set $\{x_1,\dots,x_n\}$ such that the corresponding projections $P(x_1),\dots,P(x_n)$ form an orthogonal family and $\sum_{k=1}^n P(x_k)y_j = y_j$ holds for $1 \le j \le j_n$. The process is either terminated after a finite number of steps, so $Py_j = y_j$ for all j where $P := \sum_n P(x_n)$, or the last relation is valid with $P :=$ s-$\lim_{n\to\infty} \sum_{k=1}^n P(x_k)$. In both cases P is the projection onto $\bigoplus_n \mathcal{G}(x_n)$, and the condition $\overline{M} = \mathcal{H}$ implies $P = I$, concluding the proof. ∎

5.5.6 Theorem: Let A be a self–adjoint operator on a separable $\mathcal{H}$; then the condition $T \in \{A\}''_{ex}$ is valid *iff* $T = \varphi(A)$ for some Borel function $\varphi : \mathbb{R} \to \mathbb{C}$.

Proof: We have to check the necessary condition. First we shall find for any $y \in D_T$ a Borel function $\varphi_y : \mathbb{R} \to \mathbb{C}$ such that $y \in D(\varphi_y(A))$ and $\varphi_y(A)y = Ty$. By the proof of the previous lemma, $P(x) \in \{E_t^{(A)}\}' = \{A\}'$ for any $x \in \mathcal{H}$, and since $T \in \{A\}''_{ex}$ by assumption, we get the inclusion $P(x)T \subset TP(x)$. This in turn implies $P(y)Ty = TP(y)y = Ty$, *i.e.*, $Ty \in \mathcal{G}(y)$ for any $y \in D_T$; hence there is a sequence $\{s_n\}$ of step functions such that $Ty =$ s-$\lim_{n\to\infty} s_n(A)y$. This means, in particular, that the sequence $\{s_n(A)y\} \subset \mathcal{H}$ is Cauchy, and by the identity (5.6), $\{s_n\}$ is Cauchy in $L^2(\mathbb{R}, d\mu_y)$. The measure $\mu_y = (y, E_A(\cdot)y)$ is Borel, so the sequence converges to a Borel function $\varphi_y \in L^2(\mathbb{R}, d\mu_y)$; it follows that $y \in D(\varphi_y(A))$ and $\varphi_y(A) = Ty$.

In the next step we shall construct a vector $\tilde{y} \in D_T$ such that $TB\tilde{y} = \varphi(A)B\tilde{y}$ holds for $\varphi := \varphi_{\tilde{y}}$ and all $B \in \{A\}'$; we shall use the orthonormal set $\{x_n\}$ of Lemma 5.5.5. Since T is reduced by $P(x)$, so $P(x)D_T \subset D_T$, and there is a dense set $\{y_j\} \subset D_T$ (*cf.* Problem 4.1), the construction used in the previous proof shows that we may assume $\{x_n\} \subset D_T$. We set $\tilde{y} := \lim_{m\to\infty} \sum_{n=1}^m c_n x_n$, where $\{c_n\} \subset \ell^2$ is a sequence of nonzero complex numbers such that $\sum_{n=1}^\infty |c_n|^2 \|Tx_n\|^2 < \infty$; we can choose, *e.g.*, $c_n := \left(2^n(1+\|Tx_n\|^2)\right)^{-1/2}$. These requirements in combination with the relation $Tx_n \in \mathcal{G}(x_n)$ and the orthogonality of the subspaces ensure existence of the limit $\lim_{m\to\infty} \sum_{n=1}^m c_n Tx_n$; the closedness of T then implies $\tilde{y} \in D_T$. Due to the previous part of the proof, $\tilde{y} \in D(\varphi(A))$ and $\varphi(A)\tilde{y} = T\tilde{y}$, and since the operators T and $\varphi(A)$ belong to $\{A\}''_{ex}$, we have $B\tilde{y} \in D_T \cap D(\varphi(A))$ for any $B \in \{A\}'$ and $TB\tilde{y} = BT\tilde{y} = B\varphi(A)\tilde{y} = \varphi(A)B\tilde{y}$.

To prove that φ is the sought function we introduce the projections $E_l := E_A(M_l)$, $l = 1, 2, \dots$, where $M_l := \{t \in \mathbb{R} : |\varphi(t)| \le l\}$. These belong to $\{A\}'$ for any l, the operators $\varphi(A)E_l$ are bounded (see Proposition 5.2.6), and s-$\lim_{l\to\infty} E_l = I$. Moreover, the commutant $\{A\}'$ contains all the projections $P(x_n)$ as well as any step function $s(A)$, so we may use the above derived relation with $B = E_l s(A)P(x_n)$. We know that $P(x_n)\tilde{y} = c_n x_n$, and since the coefficients c_n have been chosen nonzero, we get $TE_l s(A)x_n = \varphi(A)E_l s(A)x_n$ for all $l, n = 1, 2, \dots$ and any step function s.

Consider an arbitrary vector $z \in \mathcal{H}$. By Lemma 5.5.5, it can be expressed as $z = \lim_{m\to\infty} \sum_{n=1}^m z_n$ with $z_n \in \mathcal{G}(x_n)$, and for any $m, n = 1, 2, \dots$ there is a step function s_{nm} such that $\|z_n - s_{nm}(A)x_n\| < 2^{-n}m^{-1}$. This implies $z = \lim_{m\to\infty} z^{(m)}$, where $z^{(m)} := \sum_{n=1}^{m} s_{nm}(A)x_n \in D_T$. The operator $\varphi(A)E_l$ is bounded, so $\varphi(A)E_l z = \lim_{m\to\infty}\sum_{n=1}^{m}\varphi(A)E_l s_{nm}(A)x_n = \lim_{m\to\infty} TE_l z^{(m)}$; since T is closed, we have $E_l z \in D_T$ and $\varphi(A)E_l z = TE_l z$ for $l = 1,2,\dots$. The rest of the argument is simple: if $z \in D_T$, then $TE_l z = E_l Tz \to Tz$ and the closedness of $\varphi(A)$ yields $z = \lim_{l\to\infty} E_l z \in D(\varphi(A))$ and $\varphi(A)z = Tz$, *i.e.*, $T \subset \varphi(A)$; interchanging the roles of T and $\varphi(A)$, we get the opposite inclusion. ∎

In the remaining part of this section we are going to show how the above results generalize to "functions" of a finite set of commuting self–adjoint operators $A_1,\dots,A_N$. For the sake of simplicity we shall consider the case $N = 2$; the extension to other finite sets is straightforward.

We shall therefore consider a pair of commuting self–adjoint operators A, A' with the spectral measures $E := E_A$ and $F := E_{A'}$, respectively; recall that the commutativity is by definition equivalent to $E(M)F(N) = F(N)E(M)$ for any $M, N \in \mathcal{B}$. By Proposition 5.1.9, the pair E, F determines a unique product measure P. We denote the set of all complex Borel functions on $\mathbb{R}^2$ which are P–a.e. defined as Φ_P; then the operators $\varphi(A, A') := \mathcal{T}^{(P)}(\varphi)$ with $\varphi \in \Phi_P$ will again be called *functions of the commuting self–adjoint operators* A, A'. Using properties of the Cartesian product together with Proposition 5.4.1c we easily find that the set $\mathbb{R}^2\backslash(\sigma(A)\times\sigma(A'))$ is P–zero, so Φ_P contains all Borel functions defined on $\sigma(A)\times\sigma(A')$; in particular, we have $C(\sigma(A)\times\sigma(A')) \subset \Phi_P$.

In some cases, $\varphi(A, A')$ can be expressed in terms of functions of the operators A and A'. As an illustration, we shall extend the conclusions of Example 5.2.4 to unbounded functions. Consider $\varphi \in \Phi_E$ and a Borel function φ_1 on $\mathbb{R}^2$ which fulfils F–a.e. the relation $\varphi_1 = \varphi\times e$, *i.e.*, $\varphi_1(x, y) = \varphi(x)$, and therefore belongs to Φ_P. If s is a σ–simple function then $s_1 := s\times e$ is again σ–simple, and since $\mu_x^{(E)}(M) = \mu_x^{(P)}(M\times\mathbb{R})$ for all $x \in \mathcal{H}$ and $M \in \mathcal{B}(\mathcal{H})$, we see that $s \in L^p(\mathbb{R}, d\mu_x^{(E)})$ implies $s_1 \in L^p(\mathbb{R}^2, d\mu_x^{(P)})$ and the corresponding integrals equal each other; we shall use this fact to prove the identity $\varphi_1(A, A') = \varphi(A)$.

Without loss of generality we may assume that the functions φ and φ_1 are real–valued; then the operators $\varphi_1(A, A')$ and $\varphi(A)$ are self–adjoint and it is sufficient, *e.g.*, to check the inclusion $\varphi(A) \subset \varphi_1(A, A')$. Let $x \in D(\varphi(A))$ so there is a sequence of σ–simple functions $s_n \in L^2(\mathbb{R}, d\mu_x^{(E)})$ such that $\|\varphi^2 - s_n\|_\infty \to 0$. The sequence $\{\varphi_1^2 - s_n\times e\} \subset L(\mathbb{R}^2, d\mu_x^{(P)})$ then converges to zero uniformly in $\mathbb{R}^2$; hence $\varphi_1^2 \in L(\mathbb{R}^2, d\mu_x^{(P)})$, *i.e.*, $x \in D(\varphi_1(A, A'))$. In a similar way, we check the identity $(x, \varphi(A)x) = (x, \varphi_1(A, A')x)$, which yields the sought inclusion by the standard polarization and density arguments. The same reasoning applies to functions which depend on the second argument only; see also Problem 33b.

5.5.7 Proposition: With the above notation, the relations

$$\begin{aligned} (\varphi\times e)(A, A') &= \varphi(A)\,, \quad (e\times\psi)(A, A') = \psi(A')\,, \\ ((\varphi\times e) + (e\times\psi))(A, A') &\supset \varphi(A) + \psi(A')\,, \\ (\varphi\psi)(A, A') &\supset \varphi(A)\psi(A') \end{aligned}$$

are valid for any $\varphi \in \Phi_E$ and $\psi \in \Phi_F$; the necessary and sufficient conditions under which the inclusions turn into identities are given by Theorem 5.2.9b,c.

5.5.8 Example: Consider the space $L^2(R^2, d(\mu_1 \otimes \mu_2))$, where μ_1, μ_2 are Borel measures on $\mathbb{R}$, and the operators Q_r of the multiplication by the r–th coordinate defined as $Q_r := T_{id_r}$ with $id_r(x_1, x_2) := x_r$. The product measure $\mu_1 \otimes \mu_2$ is again Borel; then due to Example 5.2.8 $E_\otimes : E_\otimes(M) = T_{\chi_M}$ is a projection–valued measure and $Q_r = \mathcal{T}^{(E_\otimes)}(id_r)$. The operators Q_r are self–adjoint because the functions id_r are real–valued, and their spectral measures are given by $E_{Q_r} = E_\otimes(id_r^{(-1)}(\cdot))$, so $E_{Q_1}(M) = E_\otimes(M \times \mathbb{R})$ and $E_{Q_2}(N) = E_\otimes(\mathbb{R} \times N)$. Multiplying the last two expressions we get the relation $P(M \times N) = E_\otimes(M \times N)$ for any $M, N \in \mathcal{B}$, so the two projection–valued measures coincide, $P = E_\otimes$. Hence the operator $\varphi(Q_1, Q_2) := \mathcal{T}^{(P)}(\varphi)$ is for any $\varphi \in \Phi_P$ nothing else than the multiplication by this function, *i.e.*, $\varphi(Q_1, Q_2) = T_\varphi$.

Notice that in view of Example 2.4.5, the operators Q_r can be expressed alternatively in the tensor–product form, $Q_1 = \overline{Q_{\mu_1} \otimes I_2}$ and $Q_2 = \overline{I_1 \otimes Q_{\mu_2}}$; we have to use the fact that the operators whose closures are taken on the right side are *e.s.a.* (see Theorem 5.7.2 below).

5.5.9 Example: Let $p_j(t) := t^j$; then Example 5.2.10 gives $(p_j \times p_k)(A, A') \subset p_j(A)p_k(A') = A^j(A')^k$. Interchanging the operators A, A' we see that $A^j(A')^k x = (A')^k A^j x$ holds for all $x \in D(A^j(A')^k) \cap D((A')^k A^j)$ and any $j, k = 1, 2, \ldots$. In particular, commuting self–adjoint operators obey the relation

$$(AA' - A'A)x = 0\,, \quad x \in D(AA') \cap D(A'A)\,.$$

Let us stress, however, that in distinction to the bounded–operator case this condition is necessary but in general not sufficient for commutativity of A and A' (*cf.* Example 8.2.1).

Some other properties of commuting self–adjoint operators are collected in Problem 33. In conclusion, let us mention how Theorem 6 generalizes to the present situation. Using Problem 33f we get the inclusion

$$\{\,\varphi(A, A') \,:\, \varphi \in \Phi_P\,\} \subset \{A, A'\}''_{ex};$$

it turns into identity if the underlying Hilbert space is separable.

5.5.10 Theorem: Let A, A' be commuting self–adjoint operators on a separable $\mathcal{H}$; then $T \in \{A, A'\}''_{ex}$ *iff* $T = \varphi(A, A')$ for a Borel $\varphi : \mathbb{R}^2 \to \mathbb{C}$.

Proof: Mimicking the proof of Lemma 5.5.5 we get the identity $\mathcal{H} = \bigoplus_{n=1}^N \mathcal{G}(x_n)$, where $\{x_n\}_{n=1}^N$ is an orthonormal set and $\mathcal{G}(x_n) := \{\, P(J)x_n \,:\, J \in \mathcal{J}^2\,\}$. In view of the above inclusion, we have to find to a given $T \in \mathcal{L}_c(\mathcal{H})$, which satisfies $BT \subset TB$ for all $B \in \{P(\cdot)\}'$, a Borel function $\varphi : \mathbb{R}^2 \to \mathbb{C}$ such that $\varphi(A, A') = T$; to this end the proof of Theorem 5.5.6 can be used with a few evident modifications. ∎

5.6 Analytic vectors

For any $B \in \mathcal{B}(\mathcal{H})$ we can define e^B as the appropriate power series convergent in the operator–norm limit — see Problem 1.66. This definition makes no sense if T is unbounded; however, it is possible that the series $\sum_{k=1}^{\infty} \frac{1}{k!} T^k x$ converges for some $x \in C^\infty(T) := \bigcap_{k=1}^{\infty} D(T^k)$. Notice that $C^\infty(T)$ is dense for a self-adjoint T; otherwise it can consist of the zero vector only, even if T is symmetric (Problem 35). A vector $x \in C^\infty(T)$ is called an **analytic vector** of the operator T if the power series $\sum_{n=0}^{\infty} \|T^n x\| \frac{z^n}{n!}$ has a non–zero convergence radius; we denote the latter as $r_T(x)$ or simply $r(x)$. It follows from the definition that the analytic vectors of a given T form a subspace which contains, in particular, all eigenvectors of T.

5.6.1 Proposition: Let x be an analytic vector of a self–adjoint operator A; then

$$x \in D\left(e^{izA}\right) , \quad e^{izA}x = \lim_{n\to\infty} \sum_{k=0}^{n} \frac{(iz)^k}{k!} A^k x$$

holds for any complex z with $|z| < r(x)$.

Proof: We denote $\eta_z(t) := e^{izt}$; then Theorems 5.2.2d and 5.2.9c yield

$$\int_{\mathbb{R}} \eta_z \chi_{(-j,j)} dE_A = \eta_z(A) E_A(-j,j) = \operatorname*{s-lim}_{n\to\infty} \sum_{k=0}^{n} \frac{(iz)^k}{k!} A^k E_A(-j,j)$$

for any positive integer j. From here we get the estimate

$$\left(\int_{\mathbb{R}} |\eta_z|^2 \chi_{(-j,j)} d\mu_x \right)^{1/2} \le \sum_{k=0}^{\infty} \frac{|z|^k}{k!} \|A^k E_A(-j,j)x\| \le \sum_{k=0}^{\infty} \frac{|z|^k}{k!} \|A^k x\| < \infty;$$

together with Fatou's lemma, this implies $\int_{\mathbb{R}} |\eta_z|^2 d\mu_x < \infty$, *i.e.*, $x \in D(\eta_z(A))$. Next we use the estimate $\left| e^{izt} - \sum_{k=0}^{n} \frac{(izt)^k}{k!} \right| \le 2\, e^{r|t|}$; since $x \in C^\infty(A) \cap D(e^{izA})$ and $\int_{\mathbb{R}} e^{2r|t|} d\mu_x(t) < \infty$ holds for any $r \in (0, r(x))$, the assertion follows from the identity (5.6) and the dominated–convergence theorem. ∎

The concept of an analytic vector provides us with a tool for checking essential self–adjointness as the following theorem shows.

5.6.2 Theorem (Nelson): A symmetric operator whose analytic vectors form a total set is essentially self–adjoint.

Proof: Consider first the case when a symmetric A has equal deficiency indices. This ensures the existence of self–adjoint extensions; let S be one of them. If x is an analytic vector of A, it is at the same time an analytic vector of S and $r_A(x) = r_S(x) =: r(x)$; we claim that the function $F : z \mapsto (y, e^{izS}x)$ is for any $y \in \mathcal{H}$ analytic in the strip $G := \{ z \in \mathbb{C} : |\mathrm{Im}\, z| < \frac{1}{2} r(x) \}$. Setting $z = u + iv$ we can write it as $F(z) = f(u,v) + ig(u,v)$, where

$$f(u,v) := \int_{\mathbb{R}} e^{-vt} \cos(ut)\, d\nu_{yx}(t) , \quad g(u,v) := \int_{\mathbb{R}} e^{-vt} \sin(ut)\, d\nu_{yx}(t)$$

and $\nu_{yx} := (y, E_S(\cdot)x)$; we have to check that the functions f and g have in $\mathbb{R}\times(-\frac{1}{2}r(x), \frac{1}{2}r(x))$ continuous partial derivatives which fulfil the Cauchy–Riemann conditions $\partial_u f = \partial_v g$ and $\partial_v f = -\partial_u g$. This is obviously true for the integrated functions, and moreover, their partial derivatives are in $\mathbb{R}\times[-r,r]$, where $r \in (0, \frac{1}{2}r(x))$, majorized by the function $h:\ h(t) = |t|e^{r|t|}$. The Hölder inequality gives $\left(\int |h(t)|^2 d\mu_x(t)\right)^2 \le \int t^4 d\mu_x(t) \int e^{4r|t|} d\mu_x(t)$, so the estimate of the previous proof in combination with the assumption $x \in C^\infty(S)$ shows that $h \in \mathcal{L}^2(\mathbb{R}, d\mu_x)$. Then h is ν_{yx}–integrable by Proposition 5.2.7 and the analyticity of F follows from the dominated–convergence theorem; Proposition 1 yields the Taylor expansion of the function F around the origin,

$$F(z) = \sum_{k=0}^{\infty} (y, S^k x)\,\frac{(iz)^k}{k!} = \sum_{k=0}^{\infty} (y, A^k x)\,\frac{(iz)^k}{k!}$$

for $|z| < r(x)$. In view of Corollary 4.3.10, we have to check that $\operatorname{Ker}(A^* \pm i) = \{0\}$. We take $w \in \operatorname{Ker}(A^*-i)$ and set $y = w$ in the Taylor expansion. Since $(A^*)^k w = i^k w$, it gives $F(z) = e^z(w,x)$ in the disc $\{\, z : |z| < \frac{1}{2}r(x) \,\}$. However, the two functions are analytic in G, and therefore the identity holds for all $z \in G$ (see the notes), in particular, $F(t) = e^t(w,x)$ for any $t \in \mathbb{R}$. At the same time, e^{itS} is unitary, so F is bounded on $\mathbb{R}$; this is possible only if $(w,x) = 0$. The vector x has been chosen from the set of analytic vectors which is dense by assumption; hence $w = 0$ and $\operatorname{Ker}(A^*-i) = \{0\}$. The relation $\operatorname{Ker}(A^*+i) = \{0\}$ can be checked in the same way.

Consider finally the general case where the deficiency indices need not equal each other. The operator $A_\oplus := A \oplus (-A)$ on $\mathcal{H}\oplus\mathcal{H}$ is again symmetric, and by Problem 4.56a we have $n_+(A_\oplus) = n_-(A_\oplus) = n_+(A) + n_-(A)$; in particular, $A_\oplus$ is *e.s.a.* *iff* the same is true for A. If x, y is a pair of analytic vectors of A, then $[x,y]$ is an analytic vector of $A_\oplus$. This shows that $A_\oplus$ has a dense set of analytic vectors in $\mathcal{H}\oplus\mathcal{H}$ and by the first part of the proof it is *e.s.a.* ∎

There is a useful modification of Theorem 5.6.2 based on the notion of a *semianalytic vector* of a given T. This is what we call any vector $x \in C^\infty(T)$ such that the power series $\sum_{n=0}^\infty \frac{z^{2n}}{(2n)!}\|T^n x\|$ has a nonzero convergence radius (which we denote by $\varrho(T)$).

5.6.3 Theorem (Nussbaum): A positive operator whose semianalytic vectors form a total set is *e.s.a.*.

Proof: Let A be positive; then by Theorem 4.6.11 it has a positive self–adjoint extension $S \supset A$. If x is a semianalytic vector of A it is at the same time a semianalytic vector of S; mimicking the preceding proof we can check that it belongs for any $|z| < \varrho(x)$ to the domain of the operator $\cos(z\sqrt{S})$. In particular, we have $\int_0^\infty \cosh(2r\sqrt{t})\, d\mu_x(t) < \infty$ for any $r \in (0, \varrho(x))$, and therefore also $\int_0^\infty e^{2r\sqrt{t}}\, d\mu_x(t) < \infty$. In the same way as above we check that the function $F:\ F(z) = (y, \cos(z\sqrt{S})x)$ is for any $y \in \mathcal{H}$ analytic in $G := \{\, z \in \mathbb{C} : |\mathrm{Im}\, z| < \frac{1}{2}\varrho(x) \,\}$; the relevant majorizing function which allows us to differentiate under the integral sign is in this case $h(t) := \frac{1}{2}\sqrt{t}\,(e^{r\sqrt{t}}+1)$. The Taylor expansion now reads

$$\cos(z\sqrt{S})x \;=\; \lim_{n\to\infty}\sum_{k=0}^{n}(-1)^k\,\frac{z^{2k}}{(2k)!}\,A^k x$$

for $|z|<\varrho(x)$. Choosing an arbitrary vector $w\in \mathrm{Ker}\,(A^*\mp i)$, we get from here

$$F(z) \;=\; (w,x)\sum_{k=0}^{\infty}(\pm i)^k\,\frac{z^{2k}}{(2k)!}$$

for $|z|<\varrho(x)$; however, the right side is an entire function so the identity extends to the whole strip G. In particular, we have $|F(u)| = 2^{-1/2}|(w,x)|[\cosh(\sqrt{2}u)+\cos(\sqrt{2}u)]^{1/2}$ for all $u\in\mathbb{R}$, and since $\cos(z\sqrt{S})$ is a bounded operator, we finally find $(w,x)=0$ for any semianalytic vector of A, *i.e.*, $\mathrm{Ker}\,(A^*\pm i)=\{0\}$. ∎

5.6.4 Example: Consider the operators $A_\pm := 2^{-1/2}(Q\mp iP)\restriction\mathcal{S}(\mathbb{R})$ on $L^2(\mathbb{R})$ which obviously map $\mathcal{S}(\mathbb{R})$ into itself. The operator H_λ with the domain $D(H_\lambda):=\mathcal{S}(\mathbb{R})$ defined for any $\lambda\in\mathbb{R}$ by

$$H_\lambda := A_+A_- + \frac{1}{2} + \frac{\lambda}{4}(A_+ + A_-)^4 = \left[\frac{1}{2}(Q^2+P^2)+\lambda Q^4\right]\restriction\mathcal{S}(\mathbb{R}) \qquad (5.14)$$

is symmetric; it is positive provided $\lambda\ge 0$. We shall show that the h_n, $n=1,2,\dots$, defined by (2.1) are semianalytic vectors of the operator H_λ. It follows from (5.14) that H_λ^k can for any positive integer k be expressed as a sum of 18^k terms of the form $c_\lambda^k A_1A_2\dots A_{4k}$, where $|c_\lambda|\le\max\{1,|\lambda|/4\}$ and each A_j, $j=1,\dots,4k$, is equal to one of the operators I, A_+, A_-.

Using the well-known functional relation between Hermite polynomials we find $A_+h_n=\sqrt{n+1}h_{n+1}$ and $A_-h_n=\sqrt{n}h_{n-1}$, where in the case $n=0$ the last relation reads $A_-h_0=0$. By induction, we get the estimate

$$\|A_1\dots A_m h_n\| \le \prod_{j=1}^{m}\sqrt{n+j}, \qquad m=1,2,\dots$$

which shows that the series $\sum_{k=0}^{\infty}\frac{z^{2k}}{(2k)!}\,\|H_\lambda^k h_n\|$ converges if $|z|<(72\,|c_\lambda|)^{-1/2}$. Thus by Nussbaum's theorem, H_λ is *e.s.a.* for any $\lambda\ge 0$. Notice that the above estimate does not imply *analyticity* of the vectors h_n (see also Problem 36).

5.7 Tensor products

Given a pair of self-adjoint operators A_1 and A_2, it is natural to ask about the relations between their spectral properties and those of their tensor product, or more generally, polynomial combinations of tensor products. We shall consider operators of the type

$$P[A_1,A_2] := \sum_{k=0}^{n_1}\sum_{l=0}^{n_2} a_{kl}(A_1^k\otimes A_2^l) \qquad (5.15)$$

with the domain $D_P \equiv D(P[A_1,A_2]) := D(A_1^{n_1})\underline{\times}D(A_2^{n_2})$, where n_1, n_2 are non-negative integers and a_{kl} are real coefficients. We shall use the notation introduced in Section 4.6, $\underline{A}_1 := A_1 \otimes I_2$ and $\underline{A}_2 := I_1 \otimes A_2$.

5.7.1 Remark: Without loss of generality, we may assume that a_{kn_2} and $a_{n_1 l}$ are nonzero for some k,l. Then D_P is the natural domain for the operator (5.15); recall that $D(A_r^n) \supset D(A_r^m)$ holds for $n<m$. With Remark 4.5.3a in mind, we shall also assume that at least one of the operators A_1, A_2 is unbounded; a reformulation of the results for the case when both of them are Hermitean is easy. Finally, the results extend in a straightforward way to polynomial combinations of tensor products $A_1^{k_1} \otimes \cdots \otimes A_N^{k_N}$ for any finite N.

5.7.2 Theorem: The operator $P[A_1,A_2]$ given by (5.15) is essentially self-adjoint, and the same is true for its restriction $\dot P[A_1,A_2] := P[A_1,A_2] \restriction D_1\underline{\times}D_2$, where D_r is any core of the operator $A_r^{n_r}$, $r=1,2$.

Proof: The operators $A_r^{n_r}$ are self-adjoint, so $P[A_1,A_2]$ is densely defined and symmetric by Theorem 4.5.4 and Proposition 4.1.2. Let E_r denote the spectral measure of A_r; the set $M_r := \bigcup_{j=1}^\infty \mathrm{Ran}\, E_r(-j,j)$ is obviously dense in $\mathcal{H}_r$. Hence $M := \{\, x_1 \otimes x_2 :\ x_r \in M_r \,\}$ is total in $\mathcal{H}_1 \otimes \mathcal{H}_2$ by Proposition 2.4.4a; to prove the first assertion it is sufficient, due to Nelson's theorem, to show that any element of M is an analytic vector of $P[A_1,A_2]$. Let $x_r \in \mathrm{Ran}\, E_r(-j_r,j_r)$; then the functional-calculus rules yield the estimate $\|A_r^k x_r\|_r^2 \le j_r^{2k}\|x_r\|^2$, which implies $M_r \subset \bigcap_{k=1}^\infty D(A_r^k) =: C^\infty(A_r)$, and therefore $M \subset C^\infty(P[A_1,A_2])$. The same estimate gives

$$\begin{aligned} \|P[A_1,A_2]^N(x_1\otimes x_2)\| &= \Big\| \sum_{k_N l_N} a_{k_N l_N} \cdots \sum_{k_1 l_1} a_{k_1 l_1} A_1^{k_1+\cdots+k_N} x_1 \otimes A_2^{l_1+\cdots+l_N} x_2 \Big\| \\ &\le (|P|(j_1,j_2))^N \|x_1 \otimes x_2\|\,, \end{aligned}$$

where we have denoted $|P|(j_1,j_2) := \sum_{kl} |a_{kl}| j_1^k j_2^l$; thus the appropriate power series converges with the radius $r(x_1\otimes x_2)=\infty$.

The operator $\dot P[A_1,A_2]$ is again densely defined, and therefore symmetric as a restriction of a symmetric operator. Its closure is contained in $\overline{P[A_1,A_2]}$, and since we already know that this operator is self-adjoint, it remains to check the inclusion $P[A_1,A_2] \subset \overline{\dot P[A_1,A_2]}$, *i.e.*, to show that to any $z \in D_P$ there is a sequence $\{z_j\} \subset D_1\underline{\times}D_2$ fulfilling $z_j \to z$ and $\dot P[A_1,A_2]z_j \to P[A_1,A_2]z$. Let $z = x_1 \otimes x_2$ with $x_r \in D(A_r^{n_r})$. Since D_r is a core for $A_r^{n_r}$ by assumption, there are sequences $\{x_r^j\}_{j=1}^\infty \subset D_r$ such that $\lim_{j\to\infty} x_1^j \otimes x_2^j = x_1 \otimes x_2$ and $\lim_{j\to\infty} A_r^{n_r} x_r^j = A_r^{n_r} x_r$. It follows that $\lim_{j\to\infty}(A_1^k \otimes A_2^l)(x_1 \otimes x_2) = (A_1^k \otimes A_2^l)(x_1\otimes x_2)$ for $0 \le k \le n_1$ and $0 \le l \le n_2$ (see Problem 37), and therefore also $\lim_{j\to\infty} \dot P[A_1,A_2](x_1^j \otimes x_2^j) = P[A_1,A_2](x_1\otimes x_2)$; a linear extension then yields the sought result. ∎

Before we start discussing spectral properties of the operator $\overline{P[A_1,A_2]}$ let us mention some simple properties of the operators $\underline{A}_r$ the proof of which is left to the reader (*cf.* Problem 38).

5.7.3 Proposition: Let $A_r \in \mathcal{L}_{sa}(\mathcal{H}_r)$, $r = 1, 2$, and denote by $E_r(\cdot)$ the corresponding spectral measures; then

(a) $\underline{E}_r(\cdot)$ is the spectral measure of the self-adjoint operator $\overline{\underline{A}}_r$.

(b) The operators $\overline{\underline{A}}_1$ and $\overline{\underline{A}}_2$ commute.

(c) The spectra of A_r and $\overline{\underline{A}}_r$ coincide.

By definition, there is a bijective correspondence between the operator $P[A_1, A_2]$ and the real polynomial $P: P(s,t) = \sum_{k=0}^{n_1} \sum_{l=0}^{n_2} a_{kl} s^k t^l$. On the other hand, since the operators $\overline{\underline{A}}_1$ and $\overline{\underline{A}}_2$ commute, there is a unique self-adjoint operator $P(\overline{\underline{A}}_1, \overline{\underline{A}}_2)$ associated with the function $P : \mathbb{R}^2 \to \mathbb{R}$. We shall show that

$$P(\overline{\underline{A}}_1, \overline{\underline{A}}_2) = \overline{P[A_1, A_2]}. \tag{5.16}$$

Indeed, Proposition 4.5.6 implies $A_1^k \otimes A_2^l \subset (\underline{A}_1)^k (\underline{A}_2)^l \subset (\overline{\underline{A}}_1)^k (\overline{\underline{A}}_2)^l$, and by functional calculus, $P(\overline{\underline{A}}_1, \overline{\underline{A}}_2) \supset \sum_{kl} a_{kl} p_{kl}(\overline{\underline{A}}_1, \overline{\underline{A}}_2)$, where $p_{kl}(s,t) := s^k t^l$. Finally, $p_{kl}(\overline{\underline{A}}_1, \overline{\underline{A}}_2) \supset (\overline{\underline{A}}_1)^k (\overline{\underline{A}}_2)^l$ follows from Proposition 5.5.7, so together we have $P(\overline{\underline{A}}_1, \overline{\underline{A}}_2) \supset P[A_1, A_2]$; since the left side operator is self-adjoint and its right side counterpart is *e.s.a.*, we arrive at the stated relation.

5.7.4 Theorem: The closure of $P[A_1, A_2]$ is given by (5.16). The corresponding spectral measure is $E_P : E_P(M) = \chi_{P_M}(\overline{\underline{A}}_1, \overline{\underline{A}}_2)$, $M \in \mathcal{B}$, where $P_M := P^{(-1)}(M)$, and the spectra of the operators in question are related by

$$\sigma\left(\overline{P[A_1, A_2]}\right) = \overline{P(\sigma(A_1) \times \sigma(A_2))}.$$

Proof: The remaining assertions follow from Example 5.3.11b and Problem 33c in combination with Proposition 5.7.3c. ∎

5.7.5 Example: Define the operators $A_\Pi := \overline{A_1 \otimes A_2}$ and $A_\Sigma := \overline{\underline{A}_1 + \underline{A}_2}$. By the theorem, their spectra are $\sigma(A_\Pi) = \overline{M_\Pi}$ and $\sigma(A_\Sigma) = \overline{M_\Sigma}$, respectively, where $M_\Pi := \{ uv : u \in \sigma(A_1),\, v \in \sigma(A_2) \}$ and $M_\Sigma := \{ u{+}v : u \in \sigma(A_1),\, v \in \sigma(A_2) \}$. Notice that the closure is essential here because the sets M_Π and M_Σ need not be closed. Consider, for instance, self-adjoint operators with pure point spectra $\sigma(A_1) := \{ j + \frac{1}{2j} : j = 1, 2, \dots \}$ and $\sigma(A_2) := \{ -j : j = 1, 2, \dots \}$ (*e.g.*, the operators T_{s_r} of Example 4.1.4 for the appropriate sequences s_r). The spectra have no accumulation points; however, any integer obviously belongs to $\sigma(A_\Sigma) \backslash M_\Sigma$.

Similar relations are valid between the point spectra of the operators under consideration — see Problem 39.

5.8 Spectral representation

To motivate the problem we are going to discuss here, consider first a self-adjoint operator A with a pure point spectrum and suppose that each eigenvalue λ_j of A is of multiplicity one. The corresponding eigenvectors e_j by assumption form an orthonormal basis. We define the vector $y := \sum_{j=1}^{\infty} j^{-1}e_j$ and set $\mu(M) := (y, E_A(M)y)$ for any Borel M, where $E(\cdot)$ is the spectral measure of A — see Example 5.3.11a. The set $\mathbb{R}\backslash\sigma_p(A)$ is E_A–zero, so any element $\psi \in L^2(\mathbb{R}, d\mu)$ is uniquely determined by the sequence $\{\psi(\lambda_j)\}$. Since there is a bijective correspondence between the vectors e_j and the eigenvalues we can define the map $V : \mathcal{H} \to L^2(\mathbb{R}, d\mu)$ which associates $Vx : (Vx)(\lambda_j) = j\xi_j$ with any $x := \sum_j \xi_j e_j$. It is obviously an isomorphism. A vector $\psi \in L^2(\mathbb{R}, d\mu)$ belongs to the subspace VD_A *iff* $\int_{\mathbb{R}} |t\psi(t)|^2 d\mu(t) = \sum_j |\lambda_j \psi(\lambda_j)\, j^{-1}|^2 < \infty$ and $(VAV^{-1}\psi)(\lambda_j) = \lambda_j \psi(\lambda_j)$ holds for any $j = 1, 2, , \dots$, *i.e.*, $(VAV^{-1}\psi)(x) = x\psi(x)$ for μ–a.a. $x \in \mathbb{R}$.

Hence a self–adjoint operator with a pure point nondegenerate spectrum is unitarily equivalent to the multiplication operator, $VAV^{-1} = Q_\mu$, on $L^2(\mathbb{R}, d\mu)$ with a finite Borel measure μ. This considerably simplifies the analysis of such operators because multiplication operators are easy to handle, and the spectral properties of a self–adjoint operator are either unitary invariants or they transform in a transparent way. It is therefore natural to ask whether the conclusions of the above example extend to an arbitrary self–adjoint operator. We introduce the notion of a **spectral representation** (sometimes also called a *canonical form*) of a given self–adjoint A on a Hilbert space $\mathcal{H}$: by definition it is an operator T_f of multiplication by a real function f on $L^2(X, d\mu)$ which is unitarily equivalent to A; *i.e.*, there is a unitary operator $V : \mathcal{H} \to L^2(X, d\mu)$ such that $VAV^{-1} = T_f$.

It was important in the introductory example that all eigenvalues were assumed to be simple; owing to this we were able to construct the isomorphism V using the vector y. To introduce an analogy of this vector in the general case, note that $jP_jy = e_j$, where P_j are the appropriate eigenprojections so the set $\{ P_jy : j = 1, 2, \dots \}$ is total. Given an operator $A \in \mathcal{L}_{sa}(\mathcal{H})$ we call $y \in \mathcal{H}$ a *generating vector* of A if the set $\{ E_A(J)y : J \in \mathcal{J} \}$ is total in $\mathcal{H}$; here, as usual, $\mathcal{J}$ means the family of bounded intervals in $\mathbb{R}$. If an operator A has a generating vector, we say it has a **simple spectrum**.

5.8.1 Proposition: If $A \in \mathcal{L}_{sa}(\mathcal{H})$ has a simple spectrum, then $\mathcal{H}$ is separable and any eigenvalue of A is simple.

Proof: By assumption, A has a generating vector y. Let $\mathcal{J}_r$ be a subsystem in $\mathcal{J}$ consisting of intervals with rational endpoints; the set $\{ E_A(J)y : J \in \mathcal{J}_r \}$ is clearly countable. We denote it as $M_r(y)$ while $M(y)$ will be the analogous set corresponding to the whole family $\mathcal{J}$. Using the right continuity of the spectral decomposition and Proposition 5.1.3 we find that $M(y) \subset \overline{M_r(y)}$; the separability of $\mathcal{H}$ then follows from Lemma 1.5.2b. Let further λ be an eigenvalue of A and denote the corresponding eigenspace $\operatorname{Ran} E_A(\lambda)$ as $N(\lambda)$. If $z \in N(\lambda) \cap \{y\}_{lin}^{\perp}$, we have

$$(E_A(J)y, z) \;=\; (y, E_A(J \cap \{\lambda\})z) \;=\; \chi_J(\lambda)(y, z) \;=\; 0$$

for any interval $J \subset \mathbb{R}$, so $z \in \{ E(J)y : J \subset \mathbb{R} \}^{\perp} = \{0\}$. Problem 3.18a then implies $\dim N(\lambda) \le 1$, and since the opposite inequality is valid for any eigenspace, we get $\dim N(\lambda) = 1$. ∎

5.8.2 Example: The multiplication operator Q_μ on $L^2(\mathbb{R}, d\mu)$ where μ is an arbitrary Borel measure has a simple spectrum. To prove this claim we employ the fact that $\{\chi_J : J \in \mathcal{J}\}$ is total in $L^2(\mathbb{R}, d\mu)$ by Example 1.5.3c in combination with the relation $E^\mu(M) := E_{Q_\mu}(M) = T_{\chi_M}$ — *cf.* Example 5.2.8. If the measure μ is finite, then the function $\psi : \psi(x) = 1$ can be chosen as the generating vector. In the case $\mu(\mathbb{R}) = \infty$ the situation is slightly more complicated. Since μ is supposed to be Borel, there is a disjoint decomposition $\{J_k\} \subset \mathcal{J}$ such that $\mathbb{R} = \bigcup_k J_k$ and $\mu(J_k) < \infty$ for all k. The set $\{\chi_{J_k}\} \subset L^2(\mathbb{R}, d\mu)$ is orthogonal and the vectors $\psi_n := \sum_{k=1}^n (2^k \mu_k)^{-1/2} \chi_{J_k}$ with $\mu_k := \max\{\mu(J_k), 1\}$ form a Cauchy sequence. We shall check that $\psi := \lim_{n\to\infty} \psi_n$ is a generating vector of Q_μ. Indeed, any $J \subset \mathcal{J}$ has a finite disjoint decomposition $J = \bigcup_{k=k_1}^{k_2} (J_k \cap J)$, so $\chi_J = \sum_{k=k_1}^{k_2} (2^k \mu_k)^{1/2} E^\mu(J_k \cap J)\psi$. It means that the set $\{E^{\mu)}(J)\psi : J \in \mathcal{J}\}_{lin}$ contains $\{\chi_J : J \in \mathcal{J}\}$, and since the latter is total, the assertion is proved.

An important property of the class of operators with a simple spectrum is that we can easily find their spectral representation.

5.8.3 Theorem: An operator $A \in \mathcal{L}_{sa}(\mathcal{H})$ with a simple spectrum is unitarily equivalent to the operator $\mathbb{Q}_\mu$ on $L^2(\mathbb{R}, d\mu)$, where μ is the finite Borel measure which corresponds to the spectral measure E_A and a generating vector y of A by $\mu := (y, E_A(\cdot)y)$.

Proof: Consider the subspace $S \subset L^2(\mathbb{R}, d\mu)$ consisting of simple Borel functions; by Problem 1.47b we have $\overline{S} = L^2(\mathbb{R}, d\mu)$. The operator $\sigma(A)$ is bounded for any $\sigma \in S$ and the functional–calculus rules imply $G := \{x \in \mathcal{H} : x = \sigma(A)y,\ \sigma \in S\}$ is a subspace; since it contains the set $\{E_A(J)y : J \in \mathcal{J}\}$ we have $\overline{G} = \mathcal{H}$. Now we put $V_0\sigma(A)y := \sigma$. The map $V_0 : G \to L^2(\mathbb{R}, d\mu)$ is well defined because $\sigma(A)y = \tilde{\sigma}(A)y$ implies $\|\sigma - \tilde{\sigma}\|_\mu = 0$, where $\|\cdot\|_\mu$ is the norm in $L^2(\mathbb{R}, d\mu)$, as can be seen from the relation (5.4). Moreover, we have $\|\sigma(A)y\|^2 = \|\sigma\|_\mu^2 = \|V_0\sigma(A)y\|_\mu^2$, and since $\operatorname{Ran} V_0 = S$, the continuous extension of V_0 is a unitary operator from $\mathcal{H}$ to $L^2(\mathbb{R}, d\mu)$; we denote it as V.

We shall show that $VAV^{-1} = Q_\mu$. The operator on the left side is self–adjoint with the spectral measure $F := VE_A(\cdot)V^{-1}$ — *cf.* Example 5.3.11c. We have $\chi_M\sigma \in S$ for any $M \in \mathcal{B}$ and $\sigma \in S$, so $F(M)\sigma = V\chi_M(A)\sigma(A)y = \chi_M\sigma$. Since the operators F and T_{χ_M} are bounded and S is dense in $L^2(\mathbb{R}, d\mu)$, the last relation implies $F(M) = T_{\chi_M}$ for all $M \in \mathcal{B}$. Using the notation of Example 5.2.8, we end up with $F = E^{(\mu)}$ and $VAV^{-1} = \int_{\mathbb{R}} t\, dE^{(\mu)}(t) = T_{id} = Q_\mu$. ∎

If the spectrum of A is not simple the spectral representation does also exist but the construction is more complicated; the generating vector has to be replaced by a certain generating set. We limit ourselves to formulating the result; references containing the proof are given in the notes.

5.8.4 Theorem: Let A be a self–adjoint operator on a separable $\mathcal{H}$; then there is a space $L^2(X, d\mu)$ with a finite measure, a measurable function $f : X \to \mathbb{R}$, and a unitary operator $W : \mathcal{H} \to L^2(X, d\mu)$ such that $WAW^{-1} = T_f$.

In the rest of this section, we shall mention an algebraic criterion for an operator to have a simple spectrum; in view of Proposition 5.8.1 we may restrict our attention to the case of a separable $\mathcal{H}$. We know from Theorem 5.5.6 that the bicommutant $\{A\}''$ consists of the operators $\varphi(A)$ with $\varphi \in L^\infty(\mathbb{R}, dE_A)$. On the other hand, any such operator commutes with A, so $\{A\}'' \subset \{A\}'$. If A is an operator with a pure point spectrum considered in the opening of this section, we can check by elementary means that the validity of the opposite inclusion, *i.e.*, the relation $\{A\}'' = \{A\}'$ is necessary and sufficient for A to have a simple spectrum (Problem 42). To extend this result to operators with a nonempty continuous spectrum, we shall need the following equivalent definition of the generating vector.

5.8.5 Proposition: The condition $\overline{\{\, By : B \in \{A\}'' \}} = \mathcal{H}$ is necessary and sufficient for y to be a generating vector of an operator $A \in \mathcal{L}_{sa}(\mathcal{H})$.

Proof: If y is a generating vector the condition is valid because the operators $E_A(J)$ belong to $\{A\}''$. To show that it is at the same time necessary, consider a vector $x \in \{\, E_A(J)y \,:\, J \in \mathcal{J} \}^\perp$. The complex measure $\nu_{xy} := (x, E_A(\cdot)y)$ then fulfils $\nu_{xy}(J) = 0$ for all $J \in \mathcal{J}$, and therefore also $\nu_{xy}(M) = 0$ for any $M \in \mathcal{B}$ (Proposition A.4.2). For any $B \in \{A\}''$ there is a function $\varphi \in L^\infty(\mathbb{R}, dE_A)$ such that $B = \varphi(A)$; then $(x, By) = \int_{\mathbb{R}} \varphi(t)\, d\nu_{xy}(t) = 0$, and since the vectors By by assumption form a total set, we obtain $x = 0$, *i.e.*, y is a generating vector. ∎

5.8.6 Theorem: Let A be a self–adjoint operator on a separable $\mathcal{H}$; then the following conditions are equivalent:

(a) A has a simple spectrum.

(b) There is a vector $y \in \mathcal{H}$ such that $\overline{\{\, By : B \in \{A\}'' \}} = \mathcal{H}$.

(c) $\{A\}' = \{A\}''$, *i.e.*, any bounded operator B which commutes with A equals $\varphi(A)$, where φ is a bounded Borel function.

Proof: The equivalence of (a) and (b) follows from the preceding proposition. Assume further that (c) is valid and consider arbitrary $z \in \mathcal{H}$ and $\varepsilon > 0$. By Lemma 5.5.5 we can find an integer n_ε so that $\|z - \sum_{n=1}^{n_\varepsilon} P(x_n)z\| < \varepsilon$, where the $P(x_n)$ are projections onto the mutually orthogonal subspaces $\{\, E_A(J)x_n \,:\, J \subset \mathcal{J} \}$ and the vectors x_n form an orthonormal set of a cardinality $N \le \infty$. Denote $y := \sum_{n=1}^{N} n^{-1} x_n$, so $x_n = nP(x_n)y$. By the proof of Lemma 5.5.5, we can find step functions s_n , $n = 1, \dots, n_\varepsilon$, such that $\|P(x_n)z - s_n(A)x_n\| < \varepsilon n_\varepsilon^{-1}$; substituting for x_n, we get $\|z - By\| < 2\varepsilon$, where $B := \sum_{n=1}^{n_\varepsilon} n s_n(A)P(x_n)y$. The above mentioned proof in combination with the assumption gives $P(x_n) \in \{A\}' = \{A\}''$ for all n; hence the relation $\|z - By\| < 2\varepsilon$ yields (b).

To prove that (b) implies (c) it is sufficient to check that $\{A\}'$ is a commutative set — *cf.* Proposition B.1.2. In view of the decomposition $B = \mathrm{Re}\, B + i \mathrm{Im}\, B$ we have to show only that $C_1C_2 = C_2C_1$ holds for any Hermitean C_1, C_2 belonging to $\{A\}'$. By assumption, for any $x \in \mathcal{H}$ one can find a sequence $\{B_n\} \subset \{A\}''$ such that $B_n y \to x$. Since each of the operators C_1, C_2 commutes with all B_n

we see that the sought relation is valid *iff* $C_1C_2y = C_2C_1y$, and using assumption (b) once more we can rewrite it equivalently as $(C_1C_2y, By) = (C_2C_1y, By)$ for all $B \in \{A\}''$. By Problem 43c, there are sequences of Hermitean $B_n^{(r)} \in \{A\}''$, $r = 1, 2$, such that $B_n^{(r)}y \to C_ry$. Then $(C_1C_2y, By) = (C_2y, BC_1y) = \lim_{n\to\infty}(y, B_n^{(2)}BB_n^{(1)}y)$ and the analogous expression is valid for (C_2C_1y, By); since $\{A\}''$ is a commutative set the proof is finished. ∎

5.9 Groups of unitary operators

A family $\{\, U(s) \,:\, s \in \mathbb{R} \,\}$ of unitary operators on a given $\mathcal{H}$ is called a *strongly continuous one–parameter unitary group* (for brevity, one or both of the first two adjectives are often dropped) if the map $s \mapsto U(s)$ is continuous in the strong operator topology and $U(t+s) = U(t)U(s)$ holds for all $t, s \in \mathbb{R}$. It follows from the definition that the set $\{\, U(s) \,:\, s \in \mathbb{R} \,\}$ is commutative, and furthermore, the unitarity implies $U(0) = I$ and $U(-s) = U(s)^{-1} = U(s)^*$.

Given a unitary group, we denote $D := \{\, x \in \mathcal{H} : \lim_{s\to 0} s^{-1}(U(s)-I)x \text{ exists} \,\}$ and define the operator T with the domain $D_T := D$ by

$$Tx := \lim_{s\to 0} \frac{U(s) - I}{is}\, x;$$

it is called the **generator** of the group $\{\, U(s) \,:\, s \in \mathbb{R} \,\}$. The group determines the domain D uniquely. Moreover, the group condition $U(t+s) = U(t)U(s)$ implies $U(s)D \subset D$ for any $s \in \mathbb{R}$ and

$$TU(s)x = U(s)Tx = \lim_{h\to 0} \frac{U(s+h) - U(s)}{ih}\, x = -i\,\frac{dU}{ds}(s)\, x\,. \tag{5.17}$$

Note first that a unitary group can be associated with any self–adjoint operator.

5.9.1 Proposition: Let A be a self–adjoint operator; then $\{\, e^{isA} \,:\, s \in \mathbb{R} \,\}$ is a strongly continuous unitary group and A is its generator.

Proof: The first assertion follows easily from functional calculus; it remains to check that the generator T coincides with A. For any $s \neq 0$ the function $\varphi_s : \varphi_s(t) = (e^{ist} - 1)/is$ is continuous and bounded, *i.e.*, it belongs to $L^\infty(\mathbb{R}, dE_A)$. Let $x \in D_A$; then the relation (5.6) gives $\|(\varphi_s(A) - A)x\|^2 = \int_{\mathbb{R}} |\varphi_s(t) - t|^2 d\mu_x(t)$. The estimate $|\varphi_s(t) - t| \le 2|t|$ in combination with the dominated–convergence theorem shows that $\lim_{s\to 0} \varphi_s(A)x$ exists and equals Ax; hence $A \subset T$. On the other hand, let $x \in D_T$ so there is $\lim_{s\to 0} \varphi_s(A)x =: y$. Then $\int_{\mathbb{R}} |\varphi_s(t)|^2 d\mu_x(t) < (\|y\| + 1)^2$ holds for all sufficiently small nonzero s, and since $\lim_{s\to 0} |\varphi_s(t)|^2 = t^2$, we get from Fatou's lemma the condition $id \in L^2(\mathbb{R}, d\mu_x)$, *i.e.*, $x \in D_A$. ∎

It appears that this example exhausts all possible strongly continuous unitary groups on $\mathcal{H}$. This is the content of the following theorem, which has a theoretical and practical importance comparable to that of the spectral theorem.

5.9.2 Theorem (Stone): For any strongly continuous one–parameter unitary group $\{U(s) : s \in \mathbb{R}\}$ there is just one self–adjoint operator A such that $U(s) = e^{isA}$ holds for all $s \in \mathbb{R}$.

Proof: If the operator A with the stated properties exists, it is unique in view of Proposition 5.9.1. To construct it we shall first apply the definition relation (5.17) to the vectors of a particular dense subset in $\mathcal{H}$. We define $x_f := \int_{\mathbb{R}} f(t)U(t)x\,dt$ for any $f \in C_0^\infty(\mathbb{R})$ and $x \in \mathcal{H}$; the integral exists because the function $f(\cdot)U(\cdot)x$ is continuous and $\|f(t)U(t)x\| \le \|f\|_\infty \chi_K(t)\|x\|$, where K is the compact support of f — *cf.* Appendix A.5. In particular, choosing $f := j_\varepsilon$ with $j_\varepsilon(t) := \varepsilon^{-1}j(t/\varepsilon)$, where the support of j is contained in $[-1,1]$ and $\int_{-1}^{1} j(t)\,dt = 1$, we get

$$\|x - x_{j_\varepsilon}\| \le \int_{\mathbb{R}} j_\varepsilon(t)\|(U(t)-I)x\|\,dt \le \sup_{t\in[-\varepsilon,\varepsilon]} \|(U(t)-I)x\|$$

for any $x \in \mathcal{H}$; the strong continuity of $U(\cdot)x$ then implies that the subspace $D_0 := \{\, x_f : x \in \mathcal{H},\ f \in C_0^\infty(\mathbb{R})\,\}_{lin}$ is dense in $\mathcal{H}$.

Let T be the generator of $\{U(s) : s \in \mathbb{R}\}$; we shall show that $D_0 \subset D_T$ and $Tx_f = ix_{f'}$ for all $x \in \mathcal{H}$ and $f \in C_0^\infty$. We have

$$\frac{U(s)-I}{is}\, x_f = \frac{1}{is}\int_{\mathbb{R}} f(t)[U(s+t) - U(t)]x\,dt = \int_{\mathbb{R}} \frac{f(t-s) - f(t)}{is}\, U(t)x\,dt$$

for any $s \ne 0$. The derivative f' again belongs to C_0^∞, and it is therefore bounded, $|f'(t)| \le C_f$ for all $t \in \mathbb{R}$. It follows that the norm of the integrated function on the right side is majorized by $C_f\chi_K(t)\|x\|$, where K is the compact support of f. Hence Theorem A.5.3 may be applied; performing the limit $s \to 0$ in the last relation we get $Tx_f = i\int_{\mathbb{R}} f'(t)U(t)x\,dt = ix_{f'}$ for any $x \in D_0$; in other words, we have found an explicit expression for the operator $A_0 := T \upharpoonright D_0$.

The subspace D_0 is obviously A_0–invariant. Moreover, $U(s)x_f = x_{f_{-s}}$, where $f_{-s} := f(\cdot - s)$, and since $C_0^\infty(\mathbb{R})$ is invariant with respect to translations, D_0 is also $U(s)$–invariant for all $s \in \mathbb{R}$ and $A_0U(s)z = U(s)A_0z = -i\frac{dU}{ds}(s)\,z$ for any $z \in D_0$. The operator A_0 is densely defined, and using the unitarity of $U(s)$ we can readily check the identity $(A_0x_f, y_g) = (x_f, A_0y_g)$ for any $x, y \in \mathcal{H}$ and $f, g \in C_0^\infty(\mathbb{R})$ which means that A_0 is symmetric. Suppose now that there is a vector $y \in D(A_0^*)$ such that $A_0^*y = iy$; then

$$\frac{d}{ds}(y, U(s)z) = \left(y, \frac{dU}{ds}(s)z\right) = i(A_0^*y, U(s)z) = (y, U(s)z)$$

holds for all $z \in D_0$. This differential equation with the initial condition $U(0) = I$ is solved by $(y, U(s)z) = (y, z)\,e^s$; to avoid contradiction with the boundedness of the function $(y, U(\cdot)z)$, the vector y must belong to $D_0^\perp$. However, D_0 is dense, so $\mathrm{Ker}\,(A_0^*-i) = 0$. The relation $\mathrm{Ker}\,(A_0^*+i) = 0$ can be checked in the same way, *i.e.*, A_0 is essentially self–adjoint.

It remains for us to prove that the operator $A := \overline{A_0}$ fulfils $V(s) := e^{isA} = U(s)$ for all $s \in \mathbb{R}$; it is obviously sufficient to check that the function $y : y(s) =$

$U(s)z - V(s)z$ assumes zero value only for any $z \in D_0$. We know that $\frac{dU}{ds}(s)z$ exists and equals $iA_0U(s)z$. The condition $z \in D_0 \subset D_A$ in combination with the identity $\|E_A(M)V(s)z\|^2 = \|E_A(M)z\|^2$, $M \in \mathcal{B}$, implies $V(s)z \in D_A$; on the other hand, $U(s)D_0 \subset D_0 \subset D_A$, so $y(s) \in D_A$ for all $s \in \mathbb{R}$. Using Proposition 5.9.1, we get $y'(s) = iAU(s)z - iAV(s)z = iAy(s)$; hence $\frac{d}{ds}\|y(s)\|^2 = 2\mathrm{Re}\,(Ay(s), y(s)) = 0$. Finally, $y(0) = 0$ so $y(s) = 0$ holds for all $s \in \mathbb{R}$ and $z \in D_0$. ∎

Stone's theorem implies an expression for the commutant of a unitary group.

5.9.3 Proposition: $\{U(s) : s \in \mathbb{R}\}' = \{E_A(\cdot)\}' = \{E_t^{(A)} : t \in \mathbb{R}\}' = \{A\}'$.

Proof: An operator $B \in \{E_A(\cdot)\}'$ belongs to $\{U(s) : s \in \mathbb{R}\}'$ by functional calculus. Since the second and the third relation follow from Theorem 5.1.6, Problem 3c and the spectral theorem, it remains for us to check the inclusion $\{U(s) : s \in \mathbb{R}\}' \subset \{E_t^{(A)} : t \in \mathbb{R}\}'$. For any $t \in \mathbb{R}$ and $n \geq 2$, consider the periodic extension to $f_n^{(t)} : f_n^{(t)}(u) = \max\{0, \min\{1, \frac{1}{2}(n^2+1) - n|t-u - \frac{1}{2n}(n^2+1)|\}\}$ on $[t-n, t+n]$; it is not difficult to check that $\lim_{n\to\infty} f_n^{(t)}(u) = \chi_{(-\infty,t]}(u)$ for any $u \in \mathbb{R}$. By Fejér's theorem there are trigonometric polynomials P_n such that $\|f_n^{(t)} - P_n\|_\infty < n^{-1}$ for $n = 2, 3, \dots$; thus $\lim_{n\to\infty} P_n(u) = \chi_{(-\infty,t]}(u)$ and $|P_n(u)| \leq |f_n^{(t)}(u)| + n^{-1} \leq 2$, so $P_n(A)x \to E_t^{(A)}x$ for all $x \in \mathcal{H}$ by Theorem 5.2.2d. This means that the relation $[B, E_t^{(A)}] = 0$ follows from $[B, P_n(A)] = 0$, which is certainly true if $B \in \{e^{isA} : s \in \mathbb{R}\}'$. ∎

Using functional–calculus rules, this gives us a commutativity criterion for self–adjoint operators (compare to Problem 23b).

5.9.4 Corollary: Self–adjoint operators A, A' commute *iff* $[e^{isA}, e^{itA'}] = 0$ holds for all $s, t \in \mathbb{R}$.

5.9.5 Examples: (a) Given a Borel function $f : \mathbb{R}^d \to \mathbb{R}$ and a Borel measure ϱ on $\mathbb{R}^d$, consider the operators of multiplication by $e^{isf(\cdot)}$ on $L^2(\mathbb{R}^d, d\varrho)$ for $s \in \mathbb{R}$. By Example 5.2.8, they can be expressed as $\int_{\mathbb{R}} e^{isf(x)} dE^{(\varrho)}(x)$, where $E^{(\varrho)}$ is the appropriate spectral measure; the same argument as in Proposition 5.9.1 shows that they form a strongly continuous unitary group. Its generator A is equal to the operator T_f of multiplication by f. This follows from the estimate $|\frac{1}{s}(e^{isf(x)} - 1)| \leq |f(x)|$ for any $s \neq 0$ and $x \in \mathbb{R}^d$, which allows us to apply the dominated–convergence theorem to the relation (5.17); we find that any $\psi \in D(T_f)$ belongs to D_A and $A\psi = T_f\psi$, *i.e.*, $T_f \subset A$, and since the two operators are self–adjoint they equal each other.

(b) The dilation $x \mapsto e^s x$ of the real axis defines for any $s \in \mathbb{R}$ the unitary operator $U_d(s) : (U_d(s)\psi)(x) = e^{s/2}\psi(e^s x)$ on $L^2(\mathbb{R})$ (see Example 3.3.2); we can easily check that the family $\{U_d(s) : s \in \mathbb{R}\}$ fulfils the group condition. Using the dominated–convergence theorem we can prove the relation $\lim_{s\to 0}(\phi, U_d(s)\phi) = \|\phi\|^2$ for any $\phi \in C_0^\infty(\mathbb{R})$, and since this set is dense, it extends to all $\phi \in L^2(\mathbb{R})$; moreover, the polarization identity in combination with the group condition give w-$\lim_{s\to t} U_d(s) = U_d(t)$ for any $t \in \mathbb{R}$, so $U_d(\cdot)$ is strongly continuous by Problem 44.

To find the generator A_d of $\{ U_d(s) : s \in \mathbb{R} \}$, consider a vector $\phi \in C_0^\infty(\mathbb{R})$; its support is contained in a certain compact interval $[a,b]$. If $0 < |s| < 1$ the function $f_s : f_s(x) = \frac{1}{is}(e^{s/2}\phi(e^s x) - \phi(x)) + \frac{i}{2}\phi(x) + ix\phi'(x)$ is continuous and its support is again contained in a certain compact (s–independent) interval J_{ab}, for instance, $J_{ab} := [a/e, be]$ for $a \ge 0$. We have $\lim_{s\to 0} f_s(x) = 0$ for any $x \in \mathbb{R}$ and $\frac{1}{s}(e^{s/2}\phi(e^s x) - \phi(x)) = \frac{1}{2}e^{\tilde s/2}\phi(e^{\tilde s}x) + x\, e^{3\tilde s/2}\phi'(e^{\tilde s}x)$, where $\tilde s$ belongs to the open interval with the endpoints 0 and s. Hence f_s is for any $s \in (-1,1)\backslash\{0\}$ majorized by $(e^{1/2}\|\phi\|_\infty + 2|x|\, e^{3/2}\|\phi'\|_\infty)\chi_{J_{ab}}$; the dominated–convergence theorem then yields $A_d\phi = (QP - \frac{i}{2})\phi$, where Q, P are the operators introduced in Examples 4.1.3 and 4.2.5, respectively. Finally, $C_0^\infty(\mathbb{R})$ is obviously $U_d(s)$–invariant for any $s \in \mathbb{R}$, and it is therefore a core for A_d by Problem 46, *i.e.*,

$$A_d = \overline{\left(QP - \frac{i}{2}\right) \upharpoonright C_0^\infty(\mathbb{R})} = \overline{\frac{1}{2}(QP + PQ) \upharpoonright C_0^\infty(\mathbb{R})}\,.$$

The task of finding the generator to a given unitary group is usually easier if the group can be expressed by means of "simpler" unitary groups; for instance, Proposition 5.9.3 shows that $\{ U(s) : s \in \mathbb{R} \}$ is reduced by a subspace *iff* the same is true for its generator which is then the orthogonal sum of the "component" generators. As another example, consider unitary groups $\{ U_r(s) : s \in \mathbb{R} \}$ on $\mathcal{H}_r$, $r = 1,2$, with the generators A_r. The operators $U_\otimes(s) := U_1(s) \otimes U_2(s)$ on $\mathcal{H}_1 \otimes \mathcal{H}_2$ are unitary for any $s \in \mathbb{R}$ and fulfil the group condition; using the sequential continuity of multiplication together with Problem 45, we can check that $\{ U_\otimes(s) : s \in \mathbb{R} \}$ is a unitary group. Denote its generator as $A_\otimes$. The relation (5.17) yields $A_\otimes(x \otimes y) = A_1 x \otimes y + x \otimes A_2 y$ for all $x \in D(A_1)$ and $y \in D(A_2)$, *i.e.*, $\underline{A_1} + \underline{A_2} \subset A_\otimes$. However, the operator on the left side is *e.s.a.* by Theorem 5.7.2, so we have proven the following result (see also Problem 49).

5.9.6 Proposition: The group $\{ U_\otimes(s) : s \in \mathbb{R} \}$ is generated by $A_\otimes = \overline{\underline{A_1} + \underline{A_2}}$.

If Hermitean operators A, B on the *same* Hilbert space commute, then the products $e^{isA}e^{isB}$ form a unitary group which is generated by A+B. This need not be true if the operators are unbounded but the assertion still makes sense in the functional–calculus sense (*cf.* Problem 33e). No such result is valid in the noncommutative case; instead, we have the following limiting relation.

5.9.7 Theorem (Trotter formula): Suppose that A, B are self–adjoint operators and $C := A+B$ is *e.s.a.*; then the corresponding unitary groups are related by

$$e^{it\overline{C}} = \operatorname*{s-lim}_{n\to\infty} \left(e^{itA/n}e^{itB/n}\right)^n, \quad t \in \mathbb{R}\,.$$

Proof: We shall consider only the particular case when C is self–adjoint; references to the complete proof are given in the notes. In view of Theorem 5.2.2 we have to check the identity $\lim_{n\to\infty}\left[\left(e^{itA/n}e^{itB/n}\right)^n - \left(e^{it/n}\right)^n\right]x = 0$; since the norm of the operator in the square bracket is ≤ 2 for any $t \in \mathbb{R}$, it is sufficient to do so for all x of some dense set, say, all $x \in D_C$. We denote $K(u) := \frac{1}{u}(e^{iuA}e^{iuB} - e^{iuC})$ for $u \ne 0$ and $K(0) := 0$; then the estimate

$$\begin{aligned}\Big\| \Big[\left(e^{itA/n}e^{itB/n}\right)^n - \left(e^{it/n}\right)^n\Big] x \Big\| \\ = \left\| \sum_{k=0}^{n-1} \left(e^{itA/n}e^{itB/n}\right)^k \left(e^{itA/n}e^{itB/n} - e^{itC/n}\right) \left(e^{itC/n}\right)^{n-k-1} x \right\| \\ \le \sum_{k=0}^{n-1} \frac{|t|}{n} \left\| K\left(\frac{t}{n}\right) e^{i(n-k-1)tC/n} x \right\| \le |t| \sup_{|s|\le|t|} \left\| K\left(\frac{t}{n}\right) e^{isC} x \right\|\end{aligned}$$

shows that we have to check $\lim_{u\to 0}\sup_{|s|\le|t|} \left\|K(u)e^{isC}x\right\| = 0$ for all $x \in D_C$. Since $D_C = D_A \cap D_B$ the identity

$$K(u)x = \frac{e^{iuA} - I}{u}x + e^{iuA}\,\frac{e^{iuB} - I}{u}x - \frac{e^{iuC} - I}{u}x$$

easily implies $\lim_{u\to 0} K(u)x = iAx + iBx - iCx = 0$; hence there is $\delta(x) > 0$ such that $\|K(u)x\| < 1$ holds for $|u| < \delta(x)$. On the other hand, the above identity gives $\|K(u)x\| \le 6\|x\|/\delta(x)$; together we find that the function $\|K(\cdot)x\|$ is bounded, $\|K(u)x\| \le C_x$ for some $C_x > 0$ and all $u \in \mathbb{R}$.

We equip D_C with the inner product $[x, y] \mapsto (x, y)_C := (x, y) + (Cx, Cy)$. The operator C is self–adjoint and therefore closed, so we get a Hilbert space (Problem 4.11a), which will be denoted $\mathcal{H}_C$. If $K \in \mathcal{B}(\mathcal{H})$ the restriction $K \upharpoonright D_C$ belongs to $\mathcal{B}(\mathcal{H}_C, \mathcal{H})$ in view of the inequality $\|x\| \le \|x\|_C$; the uniform boundedness principle then implies the existence of $k > 0$ such that $\|K(u)x\| \le k\|x\|_C$ for all $u \in \mathbb{R}$ and $x \in D_C$. Let $M \subset D_C$ be a completely bounded set in $\mathcal{H}_C$, so we can find for any $\varepsilon > 0$ a finite ε–lattice N_ε, and the relation $\lim_{u\to 0} K(u)x = 0$ yields $\|K(u)y\| < \varepsilon$ for all $y \in N_\varepsilon$. Using the uniform boundedness proved above we get $\lim_{u\to 0}\sup_{z\in M}\|K(u)z\| = 0$; hence it is sufficient to check that the set $M_x := \{\, e^{isC}x \,:\, |s| \le |t| \,\}$ is completely bounded in $\mathcal{H}_C$ for any $x \in D_C$. We have $M_x \subset D_C$ because D_C is invariant under e^{isC}, and moreover it is not difficult to verify that $s \mapsto e^{isC}x$ is continuous for any $x \in D_C$ also as a map from $\mathbb{R}$ to $\mathcal{H}_C$. The set M_x as the image of the compact interval $[-|t|, |t|]$ is $\mathcal{H}_C$–compact and thus, *a fortiori*, completely bounded in $\mathcal{H}_C$. ∎

Notes to Chapter 5

Section 5.1 The right–continuity condition in the definition of spectral decomposition is a matter of convention; some authors, for instance, [[AG]] or [[BS]], require $\{E_t\}$ to be left–continuous. The direct–product construction easily extends to any finite commuting family of projection–valued measures on Euclidean spaces; moreover, the result holds also under much more general assumptions — see, *e.g.*, [[BS]], Sec.5.2. It should be stressed that while Proposition 5.1.9 is analogous to Theorem A.3.12 for numerical–valued measures, the measure $\mu_x := (x, P(\cdot)x)$ is *not* the product of the corresponding measures for E and F, as the reader can easily check.

Section 5.3 Our proof of the spectral theorem using the Cayley transformation follows J. von Neumann's original argument — see [vN 2] or ⟦vN⟧, Sec.II.9. There are alternative proofs. For instance, Riesz and Lorch used the fact that any self–adjoint operator A can be expressed as an orthogonal sum of Hermitean operators — see ⟦RN⟧, Sec.120. Another way is to construct the spectral measure via the resolvent of A — *cf.* ⟦We⟧, Sec.7.3. The Fejér theorem claims that a continuous periodic function can be approximated uniformly by trigonometric polynomials — see, *e.g.*, ⟦Jar 2⟧, Thm.191. Proposition 5.3.8 can be proved without reference to Hermitean and bounded normal operators. The argument is similar to that of Proposition 5.3.6; one extends the map which associates the operator $\dot{T}(U) := \sum_{k=-n}^{n} c_k U^k$ with a trigonometric polynomial $T : T(s) = \sum_{k=-n}^{n} c_k\, e^{iks}$ — *cf.* ⟦AG⟧, Secs.77–78.

Section 5.4 More information about the stability of the essential spectrum can be found, *e.g.*, in ⟦RS 4⟧, Sec.XIII.4 or ⟦We⟧, Sec.9.2. Many authors, *e.g.*, ⟦RN⟧or ⟦RS 1⟧, prefer the continuous–spectrum definition of Remark 8; the difference between the two definitions usually causes no problems.

Section 5.5 The Stone formula has various modifications. For instance, using the fact that the spectral measure is right continuous, we get the expression

$$E_t^{(A)} - E_s^{(A)} = \frac{1}{2\pi} \underset{\delta\to 0+}{\text{s-lim}} \left(\underset{\varepsilon\to 0+}{\text{s-lim}} \int_{s+\delta}^{t+\delta} [R_A(u+i\varepsilon) - R_A(u-i\varepsilon)]\, du \right)$$

for $E_t^{(A)} := E_A(-\infty, t]$; this relation can serve as a basis for a proof of the spectral theorem — *cf.* ⟦We⟧, Sec.7.3. Theorem 5.5.6 was formulated by F. Riesz (who derived inspiration from an earlier result of J. von Neumann) for the case when both A and T are bounded, and extended by I. Mimura to the unbounded case — see ⟦RN⟧, Sec.129. The separability assumption is substantial; the result may not be valid in a nonseparable Hilbert space. Notice also that the investigation of the operators $\varphi(A)$ can always be reduced to the case when A is bounded — *cf.* Problem 32b.

Section 5.6 In the proof of Theorem 5.6.2 we have used the fact that if two analytic functions coincide on a disc, they coincide on the whole common domain of analyticity — see, *e.g.*, ⟦Mar⟧, Sec.VI.6. In combination with Problem 35c, Nelson's theorem shows that a closed symmetric operator is self–adjoint *iff* it has a dense set of analytic vectors. Another consequence of Theorem 5.6.2 concerns symmetric (not necessarily closed) operators whose eigenvectors form an orthonormal basis in $\mathcal{H}$. Since each eigenvector is an analytic vector, any such operator is *e.s.a.*.

Section 5.8 The proof of Theorem 5.8.4 can be found, *e.g.*, in ⟦AG⟧, Sec.86; ⟦Nai 2⟧, Sec.18.6; ⟦DS 2⟧, Sec.XIII.5. The theorem also extends to the case when $\mathcal{H}$ is not separable — see ⟦KGv⟧, Sec.V.2. Let us stress that the spectral representation of a given $A \in \mathcal{L}_{sa}(\mathcal{H})$ is far from unique, because if A has a generating vector it will have plenty of them — *cf.* Problem 40c.

Section 5.9 An alternative proof of the Stone theorem uses the fact that for any $x \in \mathcal{H}$ the function $(x, U(\cdot)x)$ is of *positive type*, which means that the matrix with the elements

$(x, U(s_j - s_k)x)$, $j,k = 1,\dots,n$, is positive for any natural n and arbitrary $s_1,\dots,s_n$. By *Bochner's theorem* (see, *e.g.*, ⟦AG⟧, Sec.70) any function $f : \mathbb{R} \to \mathbb{C}$ of positive type is the Fourier image of a non-negative Borel measure on R. Hence we can associate with a continuous unitary group a map $x \mapsto \mu_x$ such that $(x, U(t)x) = \int_{\mathbb{R}} e^{-its} d\mu_x(t)$. By polarization, it corresponds to a complex measure ν_{xy}. Finally, one has to show that there is a projection-valued measure E such that $(x, E(J)y) = \nu_{xy}(J)$ holds for any interval J and all $x, y \in \mathcal{H}$; see ⟦RN⟧, Sec.138 for details.

By definition, the generator of a unitary group $\{U(s) : s \in \mathbb{R}\}$ is determined by the function $U(\cdot)$ in a neighborhood of the point $s = 0$. Hence if $\{V(s) : s \in \mathbb{R}\}$ is a unitary group such that $U(s) = V(s)$, $|s| < \delta$ for some $\delta > 0$, then the generators of the two groups coincide and $U(s) = V(s)$ by Stone's theorem holds for any $s \in \mathbb{R}$. Notice also that if $U(s)$ is a unitary operator, Proposition 5.3.8 implies the existence of a Hermitean $A(s)$ such that $U(s) = e^{iA(s)}$. This operator is obviously not unique. If $U(s)$ belongs to a strongly continuous unitary group, the same relation holds with $\tilde{A}(s) := sA$, where A is the appropriate generator; a Hermitean $A(s)$ can be obtained by "folding" the operator $\tilde{A}(s)$ to the interval $[0, 2\pi)$.

By Problem 3.21, the set $\mathcal{U}(\mathcal{H})$ is a group. Any one-parameter unitary group is an Abelian subgroup in it; at the same time it is a unitary representation of the one-parameter translation group. Stone's theorem can be reformulated as the assertion about existence of a spectral decomposition corresponding to such a unitary representation. This result extends to strongly continuous unitary representations of any locally compact commutative group; this is the contents of the so-called *SNAG theorem* (M.H. Stone, M.A. Naimark, W. Ambrose and R. Godement) — see, *e.g.*, ⟦RN⟧, Sec.140. We shall need only the generalization of Theorem 5.9.2 to the group of translations of the Euclidean space $\mathbb{R}^n$, which is given in Problem 48.

Theorem 5.9.7 represents a generalization to the classical Lie formula for matrix semigroups. It was originally proved in [Tro 1]; another proof can be found in [Cher 1], see also ⟦Cher⟧. Some results of these section extend to the case of strongly continuous contractive semigroups, *i.e.*, families $\{C(t) : t \ge 0\}$ such that $C(\cdot)$ is strongly continuous, $\|C(t)\| \le 1$ and $C(t+s) = C(t)C(s)$ holds for all $t, s \ge 0$. By a formula that differs from (5.17) by the imaginary unit only, one can associate with such a semigroup a unique accretive operator A which is called its generator. There is no direct analogy to Stone's theorem; the right side of the expression $C(t) = e^{-At}$ remains formal. On the other hand, the resolvent of A has an integral representation analogous to that of Problem 31b, and the Trotter formula also extends to operator semigroups. More information about these problems can be found in the monograph ⟦Da 2⟧.

Problems

1. Prove Proposition 5.1.1.

2. Prove: (a) Let $\{E_t\}_{t\in\mathbb{R}}$ be a nondecreasing family of projections on a given $\mathcal{H}$, *i.e.*, $E_t \le E_u$ if $t < u$, then the one-sided limits $E_{u-0} := \text{s-lim}_{t\to u-} E_t = \sup_{t<u} E_t$ and $E_{u+0} := \text{s-lim}_{t\to u+} E_t = \inf_{t>u} E_t$ exist at any $u \in \mathbb{R}$, as well as at $u = \pm\infty$ in the first and the second case, respectively, and the operators $E_{u\pm 0}$ are projections.

(b) Let E be a spectral measure on R, then s-$\lim_{t\to u} E(\{t\}) = 0$ for any $u \in \mathbb{R}$ and analogous relations are valid at $u = \pm\infty$.

Hint: (b) Use the generalized Bessel inequality — see the notes to Section 2.3.

3. Let $\{E_t\}$ be a spectral decomposition on $\mathcal{H}$. Prove

(a) To any $x \in \mathcal{H}$, the relation $\sigma_x(t) := (x, E_t x)$ defines a nondecreasing right–continuous function such that $\lim_{t\to-\infty} \sigma_x(t) = 0$ and $\lim_{t\to+\infty} \sigma_x(t) = \|x\|^2$.

(b) The map $\dot{E}$ defined on the set $\mathcal{J}$ of all bounded intervals in $\mathbb{R}$ by $\dot{E}(J) = E_{b-0} - E_a,\ E_b - E_a,\ E_{b-0} - E_{a-0},\ E_b - E_{a-0}$ for $J = (a,b),\ (a,b],\ [a,b),\ [a,b]$, respectively, is additive, and $\dot{E}(J)\dot{E}(K) = \dot{E}(K)\dot{E}(J) = \dot{E}(J\cap K)$ holds for any $J, K \in \mathcal{J}$.

(c) If a bounded operator commutes with E_t for any $t \in \mathbb{R}$, it also commutes with $\dot{E}(J)$ for all $J \in \mathcal{J}$.

(d) A projection–valued measure E is generated by $\{E_t\}$, *i.e.*, $E(J) = \dot{E}(J)$ for all $J \in \mathcal{J}$ *iff* $E(-\infty, t] = E_t$ holds for any $t \in \mathbb{R}$.

(e) Let a projection P reduce the set $\{E_t\}$, *i.e.*, $PE_t = E_t P$ for all $t \in \mathbb{R}$; then $t \mapsto PE_tP$ is a spectral decomposition on the subspace $\operatorname{Ran} P$, and the corresponding spectral measures fulfil $E_P(M) = PE(M)P$ for any Borel M.

4. Let $V : \mathcal{H} \to \mathcal{H}'$ be a unitary operator, then

(a) If $E : \mathcal{B}^d \to \mathcal{B}(\mathcal{H})$ is a projection–valued measure, then $E'(M) := VE(M)V^{-1}$ defines a measure on $\mathbb{R}^d$ whose values are projections in $\mathcal{B}(\mathcal{H}')$.

(b) If $\{E_t\}$ is a spectral decomposition in $\mathcal{H}$, then $\{E'_t\}$, $E'_t := VE_tV^{-1}$, is a spectral decomposition in $\mathcal{H}'$ and the corresponding spectral measures fulfil $E'(M) = VE(M)V^{-1}$ for any $M \in \mathcal{B}$.

5. If $\varphi : \mathbb{R}^d \to \mathbb{C}$ is a Borel function and $R^{(E)}_{ess} := \{\, \lambda \in \mathbb{C} : E(\varphi^{(-1)}((\lambda-\varepsilon, \lambda+\varepsilon))) \neq 0$ for any $\varepsilon > 0 \,\}$ its *essential range* with respect to a projection–valued measure E (compare to Example 4.3.7), then

(a) $R^{(E)}_{ess}(\varphi)$ is a closed set which is contained in $\overline{\operatorname{Ran}\varphi}$.

(b) $\varphi \in L^\infty(\mathbb{R}^d, dE)$ *iff* $\max R^{(E)}_{ess}(|\varphi|) < \infty$, and then $\|\varphi\|_\infty = \max R^{(E)}_{ess}(|\varphi|)$.

6. Prove: (a) Proposition 5.2.1.

(b) If the functions φ_n, $n = 1, 2, \dots$, and φ from $L^\infty(\mathbb{R}^d, dE)$ satisfy the relation $\|\varphi_n - \varphi\|_\infty \to 0$, then $\mathcal{T}_b(\varphi) = \text{u-}\lim_{n\to\infty} \mathcal{T}_b(\varphi_n)$.

(c) The operator $\mathcal{T}_b(\varphi)$ is Hermitean (zero, unitary) if the values of the function φ are respectively real, zero, belong to the unit circle E–a.e. in $\mathbb{R}^d$.

7. Let E be the projection–valued measure on $\mathbb{R}$ generated by a spectral decomposition $\{E_t\}$ and denote $A_E := \int_{\mathbb{R}} t\, dE(t)$. If $\alpha := \sup\{\, t \in \mathbb{R} : E_t = 0 \,\} > -\infty$ and $\beta := \inf\{\, t \in \mathbb{R} : E_t = I \,\} < \infty$, then the operator A_E is Hermitean and the numbers α, β equal its lower and upper bound, respectively.

8. The identity $\overline{\int_{\mathbb{R}^d} \varphi(t)\, d\nu_{xy}(t)} = \int_{\mathbb{R}^n} \overline{\varphi(t)}\, d\nu_{yx}(t)$ is valid for all $x, y \in \mathcal{H}$ and $\varphi \in L^\infty(\mathbb{R}^d, dE)$, and moreover, $\nu_{x,\mathcal{T}_b(\varphi)y}(M) = \int_M \varphi(t)\, d\nu_{xy}(t)$ for any Borel M.

9. With the notation of Example 5.2.4, prove that the relations

$$\int_{\mathbb{R}^2} (\varphi(t)+\psi(u))\, dP(t,u) = \int_{\mathbb{R}} \varphi(t)\, dE(t) + \int_{\mathbb{R}} \psi(u)\, dF(u)\,,$$

$$\int_{\mathbb{R}^2} \varphi(t)\psi(u))\, dP(t,u) = \int_{\mathbb{R}} \varphi(t)\, dE(t) \int_{\mathbb{R}} \psi(u)\, dF(u) = \int_{\mathbb{R}} \psi(u)\, dF(u) \int_{\mathbb{R}} \varphi(t)\, dE(t)$$

are valid for any $\varphi \in L^\infty(\mathbb{R}, dE)$ and $\psi \in L^\infty(\mathbb{R}, dF)$ (see Proposition 5.5.7 for the generalization of these results to unbounded functions).

10. Prove: (a) The operator $\mathcal{T}(\varphi) := \int_{\mathbb{R}^d} \varphi(t)\, dE(t)$ is bounded *iff* $\varphi \in L^\infty(\mathbb{R}^d, dE)$.

(b) Suppose that a projection–valued measure on $\mathbb{R}$ has an unbounded support and $\varphi \in \Phi_E(\mathbb{R})$ satisfies the condition $\lim_{t\to\pm\infty} |\varphi(t)| = \infty$; then the operator $\mathcal{T}(\varphi)$ is unbounded. Could the condition be weakened?

(c) If $\varphi, \psi \in \Phi_E(\mathbb{R}^d)$ and there are $c, d > 0$ such that $d^{-1}|\psi(t)| \le |\varphi(t)| \le c|\psi(t)|$ holds for E–a.a. $t \in \mathbb{R}^d$, then the corresponding integrals with respect to E have the same domains, $D_\varphi = D_\psi$.

Hint: (a) Show that the set $\{\, t \in \mathbb{R}^d : |\varphi(t)| > \|\mathcal{T}(\varphi)\| \,\}$ is E–zero.

11. Let $\varphi, \psi \in \Phi_E(\mathbb{R}^d)$ and $M \in \mathcal{B}^d$. Prove

(a) $\mu_{\mathcal{T}(\varphi)x}(M) = \int_M |\varphi(t)|^2 d\mu_x(t)$ holds for any $x \in D_\varphi$.

(b) $\mathcal{T}(\varphi)$ is reduced by the projection $E(M)$.

(c) If the functions φ, ψ are real–valued, then $\mathcal{T}(\varphi + i\psi) = \mathcal{T}(\varphi) + i\mathcal{T}(\psi)$.

12. Suppose that $V : \mathcal{H} \to \mathcal{H}'$ is unitary and E is a measure on R^d whose values are projections in $\mathcal{B}(\mathcal{H})$; then $\Phi_E(\mathbb{R}^d) = \Phi_{E'}(\mathbb{R}^d)$, where E' is the projection–valued measure of Problem 4a, and $\mathcal{T}^{(E')}(\varphi) = V\mathcal{T}^{(E)}(\varphi)V^{-1}$ for any $\varphi \in \Phi_E(\mathbb{R}^d)$.

13. Let E be a projection–valued measure on $\mathbb{R}^d$. If a projection P commutes with $E(M)$ for any $M \in \mathcal{B}^d$, then $PE(\cdot)P$ is a projection–valued measure on R^d with values in $\mathcal{B}(P\mathcal{H})$ and the relation $\mathcal{T}^{(E)}(\varphi)x = P\mathcal{T}^{(E)}(\varphi)Px = \mathcal{T}^{(PEP)}(\varphi)x$ holds for any $\varphi \in \Phi_E(\mathbb{R}^d)$ and all $x \in PD_\varphi^{(E)}$.

14. The identities $\mathcal{T}(\varphi+\psi) = \overline{\mathcal{T}(\varphi)+\mathcal{T}(\psi)}$ and $\mathcal{T}(\varphi\psi) = \overline{\mathcal{T}(\varphi)\mathcal{T}(\psi)} = \overline{\mathcal{T}(\psi)\mathcal{T}(\varphi)}$ hold for arbitrary functions $\varphi, \psi \in \Phi_E(\mathbb{R}^d)$.

15. Suppose that E is a projection–valued measure on $\mathbb{R}^d$ and $w : \mathbb{R}^d \to \mathbb{R}^n$ is a map whose component functions $w_j : \mathbb{R}^d \to \mathbb{R}$, $j = 1, \dots, n$, are Borel; then

(a) The relation $E_w(M) := E(w^{(-1)}(M))$, $M \in \mathcal{B}^n$, defines a projection–valued measure on $\mathbb{R}^n$.

(b) $\varphi \in L^\infty(\mathbb{R}^n, dE_w)$ implies $\varphi \circ w \in L^\infty(\mathbb{R}^d, dE)$ and $\|\varphi\|_\infty = \|\varphi \circ w\|_\infty$.

16. Prove: (a) Proposition 5.3.3.

(b) If a polynomial P fulfils $P(t) \geq 0$ for all $t \in J_A$, then $P(A) \geq 0$.

(c) If a real–valued function is continuous on J_A, it belongs to $\mathcal{K}$.

(d) If a function $\varphi \in \mathcal{K}$ satisfies $|\varphi(t)| < c$ for all $t \in J_A$, then $\|T_A(\varphi)\| < c$.

(e) $T_A(P \upharpoonright J_A) = P(A)$ holds any real polynomial P.

(f) The function $e_u := \chi_{(-\infty,u] \cap J_A} \upharpoonright J_A$ belongs to $\mathcal{K}_0$ for any $u \in \mathbb{R}$.

Hint: (b) Any root of P in J_A has an even multiplicity.

17. Suppose that M is a compact set in a topological space and $\{f_n\}$ is a sequence of real functions continuous on M such that $\{f_n(x)\}$ is nondecreasing for all $x \in M$. If there is a c such that $f_{n_x}(x) < c$ holds for any $x \in M$ and some n_x, then $f_n(x) < c$ for each $x \in M$ and all n large enough.
Hint: M is covered by open sets $U(x)$ such that $f_{n_x}(x) < c$ for any $y \in U(x)$.

18. Let A be a Hermitean operator and E_A its spectral measure; then

(a) If a function $f : \mathbb{R} \to \mathbb{R}$ is such that $f \upharpoonright J_A \in \mathcal{K}_0$, then f belongs to $L^\infty(\mathbb{R}, dE_A)$ and $\mathcal{T}_b^{(E_A)}(f) = T_A(f \upharpoonright J_A)$.

(b) For any interval $J \subset \mathbb{R}$, there is a sequence $\{P_n\}$ of real polynomials such that $E_A(J) = \text{s-lim}_{n\to\infty} P_n(A)$.

19. Let B be a bounded normal operator on a separable $\mathcal{H}$ with a pure point spectrum, $\sigma_p(B) = \{\, \lambda_k : k = 1, 2, \dots \}$, and denote by $\{P_k\}$ the corresponding family of projections to the eigenspaces; then $F_B : F_B(M) = \sum_k \chi_M(\lambda_k) P_k$ (strong operator convergence) is the spectral measure of B.

20. Prove: (a) Let B be a bounded normal operator and F_B the corresponding spectral measure. To any pair of intervals $J, K \subset \mathbb{R}$, there is a sequence of Hermitean operators of the form $S_n(B, B^*) := \sum_{r=0}^{N} \sum_{s=0}^{r} a_{rs}^{(n)} B^{r-s} (B^*)^s$, $n = 1, 2, \dots$, such that $F_B(J \times K) = \text{s-lim}_{n\to\infty} S_n(B, B^*)$.

(b) If U is unitary, then to any $t \in \mathbb{R}$ there is a sequence of trigonometric polynomials, $T_n(u) := \sum_k c_k^{(n)} e^{iku}$, such that $E_t^{(U)} = \text{s-lim}_{n\to\infty} \sum_k c_k^{(n)} U^k$.

(c) A bounded operator B commutes with a unitary U *iff* it commutes with $E_t^{(U)} := E_U(-\infty, t]$ for any $t \in [0, 2\pi)$.

Hint: (a) Use Proposition 5.3.2 for $\mathrm{Re}\, B$ and $\mathrm{Im}\, B$. (c) Modify the corresponding argument from Proposition 5.3.6.

21. Let U be a unitary operator and E_U the corresponding spectral measure; then a complex number $e^{i\alpha}$, $0 \leq \alpha < 2\pi$, belongs to $\sigma_p(U)$ *iff* $E_U(\{\alpha\}) \neq 0$, and in that case the corresponding eigenspace is $N_U(e^{i\alpha}) = \mathrm{Ran}\, E_U(\{\alpha\})$.

22. Let $P_n(F_d) := \sum_{k=-M_n}^{N_n} a_k^{(n)} F_d^k$ with $a_k^{(n)} \in \mathbb{C}$ be quasipolynomial functions of the Fourier–Plancherel operator F_d on $L^2(\mathbb{R}^d)$. If the sequence $\{P_n(F_d)\}$ is strongly convergent, then there are complex c_j such that $\text{s-lim}_{n\to\infty} P_n(F_d) = \sum_{j=0}^{3} c_j F_d^j$.

23. Prove: (a) The resolvent of a self–adjoint operator A can for any $\mu \in \rho(A)$ be expressed as $R_A(\mu) = \int_{\mathbb{R}} \frac{1}{t-\mu}\, dE_A(t)$.

(b) Self–adjoint operators A_1, A_2, in general unbounded, commute *iff* their resolvents commute, $R_{A_1}(\lambda)R_{A_2}(\mu) = R_{A_2}(\mu)R_{A_1}(\lambda)$ for any $\lambda, \mu \in \mathbb{C}\backslash\mathbb{R}$.

24. Find examples of self–adjoint operators with

(a) A pure point spectrum which is not purely discrete.

(b) A pure point spectrum such that the discrete spectrum is empty.

(c) Eigenvalues embedded in the continuous spectrum.

25. Prove: (a) If A is a Hermitean operator on $\mathcal{H}$, $\dim \mathcal{H} = \infty$, then $\sigma_{ess}(A) \neq \emptyset$.

(b) A self–adjoint operator A has an empty essential spectrum *iff* $(A-\lambda)^{-1}$ is compact for some $\lambda \in \rho(A)$, and in such a case, the resolvent is compact for any $\lambda \in \rho(A)$.

Hint: (a) $\sigma(A)$ has an accumulation point. (b) Use formula (3.10).

26. Prove *Weyl's theorem:* Suppose that A is a self–adjoint operator and C is Hermitean and compact; then $\sigma_{ess}(A+C) = \sigma_{ess}(A)$.

27. Let A be a self–adjoint operator; then $\mathcal{H}_p \subset \mathcal{H}_s$ and $\sigma(A \restriction \mathcal{H}_p^{\perp}) = \sigma_{ac}(A) \cup \sigma_{sc}(A)$. Moreover, both $\sigma_{ac}(A)$ and $\sigma_{sc}(A)$ are contained in $\sigma_{ess}(A)$.

28. Let $h : \mathbb{R}^n \to \mathbb{R}$ be defined by $h(x_1, \dots, x_n) := \sum_{j=1}^n x_j^2$; then the corresponding multiplication operator T_h has a purely absolutely continuous spectrum, $\sigma(T_h) = \sigma_{ac}(T_h) = \sigma_{ess}(T_h) = \mathbb{R}^+$.

29. Let A be a self–adjoint operator with a pure point spectrum. Using the notation of Example 5.3.11a, prove that any function φ defined on $\sigma_p(A)$ belongs to $\Phi^{(A)}$, the domain of $\varphi(A)$ consists of all $x \in \mathcal{H}$ such that $\sum_j |\varphi(\lambda_j)|^2 (x, P_j x) < \infty$, and $\varphi(A)x = \sum_j \varphi(\lambda_j) P_j x$ holds for any $x \in D(\varphi(A))$. In particular, $\varphi \in L^\infty(\mathbb{R}, dE_A)$ *iff* the sequence $\{\varphi(\lambda_j)\}$ is bounded; then $\varphi(A) = \sum_j \varphi(\lambda_j) P_j$ in the strong operator topology, while in general the series does not converge with respect to the operator norm.

30. Consider the operators P on $L^2(\mathbb{R})$ and $H := P^2$.

(a) Given a bounded interval $J \subset \mathbb{R}$ find a functional realization of $E_P(J)$.

(b) Find a functional realization of the spectral decomposition $\{E_t^{(H)}\}$.

31. Prove: (a) A unitary operator U can be written as $U = e^{iA_U}$, where $A_U := \int_0^{2\pi} t\, dE_U(t)$. A number λ belongs to $\sigma(U)$ *iff* $\lambda = e^{is}$ for some $s \in \sigma(A_U)$; in particular, $1 \in \sigma(U)$ *iff* $E_\varepsilon^{(U)} \neq 0$ or $E_{2\pi-\varepsilon}^{(U)} \neq 0$ holds for any $\varepsilon > 0$.

(b) Let A be a self–adjoint operator and $U(s) := e^{-iAs}$, $s \in \mathbb{R}$; then the resolvent can for any $x \in \mathcal{H}$ and $z \in \mathbb{C}\backslash\mathbb{R}$ be expressed as

$$R_A(z)x = \pm i \int_0^\infty e^{\pm izs} U(\pm s)x\, ds\,,$$

with the upper and lower sign for $\operatorname{Im} z > 0$ and $\operatorname{Im} z < 0$, respectively.

32. Prove: (a) Let A be a self–adjoint operator and $w : \mathbb{R} \to \mathbb{R}$ a Borel function; then $\lambda \in \sigma_p(A)$ implies $w(\lambda) \in \sigma_p(w(A))$, while the converse need not be true.

(b) To any self–adjoint A there is a Hermitean operator $\tilde{A}$ and a continuous real strictly monotone function f such that $\mathrm{Ran}\, f = \mathbb{R}$ and $A = f(\tilde{A})$.

33. Let A, A' be commuting self–adjoint operators. Using the notation of Section 5.5, prove:

(a) If $f_T : \mathbb{R}^2 \to \mathbb{R}^2$ is the transposition, $f_T(t,s) = (s,t)$, and $\tilde{P}$ denotes the product measure of F and E, then the function $\varphi_T := \varphi \circ f_T$ belongs to $\Phi_{\tilde{P}}$ *iff* $\varphi \in \Phi_P$, and in that case, $\varphi(A, A') = \varphi_T(A', A)$.

(b) Complete the proof of Proposition 5.5.7.

(c) Let $\varphi \in C(\sigma(A) \times \sigma(A'))$; then $\sigma(\varphi(A \times A')) = \overline{\varphi(\sigma(A) \times \sigma(A'))}$. In particular, the closure is not required if the operators A and A' are Hermitean.

(d) If $w : \mathbb{R}^2 \to \mathbb{R}^2$ fulfils the assumption of Problem 15, then $(\varphi \circ w)(A, A') = \varphi(A_1, A_2)$ holds for any Borel $\varphi : \mathbb{R}^2 \to \mathbb{C}$, where $A_j := w_j(A, A')$.

(e) $e^{i\lambda A} e^{i\lambda A'} = e^{i\lambda\sigma(A,A')}$ holds for any $\lambda \in \mathbb{R}$, where $\sigma(s,t) := s+t$. Why can the right side not in general be written as $e^{i\lambda(A+A')}$?

(f) $\{A, A'\}' = \{E(\cdot)\}' \cap \{F(\cdot)\}' = \{P(\cdot)\}'$.

(g) Extend these results to the case of commuting self–adjoint $A_1, \dots, A_n$.

34. Let A be a positive self–adjoint operator; then its domain D_A is a core for $A^{1/2}$.

35. Prove: (a) Any self–adjoint operator has a dense set of analytic vectors.

(b) Suppose that the operators T and S are unitarily equivalent, $T = USU^{-1}$, and x is an analytic vector of S; then Ux is an analytic vector of T and $r_S(x) = r_T(Ux)$.

(c) Let A be symmetric. Suppose that a subspace $D \subset D_A$ is A–invariant and the analytic vectors of A are dense in D; then A is *e.s.a.* on D.

(d) Find a symmetric operator A with $C^\infty(A) = \{0\}$.

Hint: (a) Consider the vectors $E_A(-n,n)x$ for an arbitrary $x \in \mathcal{H}$. (d) Restrict the operator P of Example 4.2.5 on $L^2(0,\pi)$ to $\{s_k : k = 1, 2, \dots\}_{lin}$, where $s_k(x) := (2/\pi)^{1/2} \sin kx$.

36. Each of the vectors h_n given by (2.1) is an analytic vector of the operators Q and P on $L^2(\mathbb{R})$.

37. Suppose that $A \in \mathcal{L}_{sa}(\mathcal{H})$ and $n \geq 2$; then $\|A^k x\| \leq \|A^n x\|^{k/n} \|x\|^{(n-k)/n}$ holds for all $x \in D(A^n)$ and any positive integer $k \leq n$.

38. Let $A_r \in \mathcal{L}_{sa}(\mathcal{H}_r)$, $r = 1, 2$.

(a) Prove Proposition 5.7.3.

(b) $\sigma_p(\overline{A_r}) = \sigma_p(A_r)$; if A_r has a pure point spectrum the same is true for $\overline{A_r}$.

(c) $f(\overline{A}_1) = \overline{f(A_1) \otimes I_2}$ and $f(\underline{A}_2) = \overline{I_1 \otimes f(A_2)}$ holds for an arbitrary real Borel function f; the closure is not needed if f is bounded.

39. Consider the operator (5.15) corresponding to given A_1, A_2. Prove

(a) $P(\sigma_p(A_1) \times \sigma_p(A_2)) \subset \sigma(\overline{P[A_1, A_2]})$; the relation turns into identity if the spaces $\mathcal{H}_r$ are separable and the spectra of the operators A_r are pure point,

(b) In the general case, the two sets need not be equal.

Hint: (b) Consider the operator $\overline{A_1 \otimes A_2}$, where A_1 is a nontrivial projection of finite codimension and $A_2 := Q$ on $\mathcal{H}_2 := L^2(\mathbb{R})$.

40. Prove: (a) If $V : \mathcal{H} \to \mathcal{H}'$ is a unitary operator and $y \in \mathcal{H}$ is a generating vector of some $A \in \mathcal{L}_{sa}(\mathcal{H})$, then $y' := Vy$ is a generating vector of VAV^{-1}.

(b) Suppose that $A \in \mathcal{L}_{sa}(\mathcal{H})$ has a simple spectrum and the operator Q_μ on $L^2(\mathbb{R}, d\mu)$ is its spectral representation; then $\varphi(A)$ for any complex Borel function φ is unitarily equivalent to the operator T_φ on $L^2(\mathbb{R}, d\mu)$.

(c) If y is a generating vector of $A \in \mathcal{L}_{sa}(\mathcal{H})$ and $B \in \mathcal{B}(\mathcal{H})$ is such that $BA \subset AB$ and $\operatorname{Ran} B = \mathcal{H}$, then the set $\{ E_A(J)By : J \in \mathcal{J} \}$ is total in $\mathcal{H}$.

41. Using Theorem 5.8.3, prove

(a) The measure $\mu_x := (x, E_A(\cdot)x)$ for any $x \in \mathcal{H}$ is generated by the function $\psi_x := Vx$ and the measure μ, *i.e.*, $\mu_x(M) = \int_M |\psi_x(t)|^2 d\mu(t)$ holds for all $M \in \mathcal{B}$. In particular, any μ–zero set is E_A–zero, so $L^2(\mathbb{R}, d\mu) \subset \Phi^{(A)}$.

(b) The generating vector y, for any $\psi \in L^2(\mathbb{R}, d\mu)$, belongs to the domain of $\psi(A)$ and $V^{-1}\psi = \psi(A)y$. This implies, in particular, that $\{ \psi(A)y : \psi \in \Psi \}$ is total in $\mathcal{H}$ for any total set $\Psi \subset L^2(\mathbb{R}, d\mu)$.

42. Let $A \in \mathcal{L}_{sa}(\mathcal{H})$ on a separable $\mathcal{H}$ have a pure point spectrum. Without reference to Theorem 5.8.6, prove that $\sigma(A)$ is simple *iff* $\{A\}' = \{A\}''$.

43. Let $A \in \mathcal{L}_{sa}(\mathcal{H})$. Prove:

(a) If there are $y \in \mathcal{H}$ and and a set $\Phi \subset \Phi^{(A)}$ such that $y \in D(\varphi(A))$ for all $\varphi \in \Phi$ and $\{ \varphi(A)y : \varphi \in \Phi \}$ is total in $\mathcal{H}$, then the spectrum of A is simple.

(b) If A is *bounded*, then its spectrum is simple *iff* the set $\{ A^n y : n = 0, 1, \dots \}$ is total for some $y \in \mathcal{H}$.

(c) Suppose that there is $y \in \mathcal{H}$ such that $\overline{\{ By : B \in \{A\}'' \}} = \mathcal{H}$; then for any Hermitean $C \in \{A\}''$ and $\varepsilon > 0$ we can find a Hermitean operator $B \in \{A\}''$ such that $\|(C - B)y\| < \varepsilon$.

Hint: (b) Use (a) and Problem 41 for the set of all polynomials. (c) We have $\tilde{B}^* \in \{A\}''$ for any $\tilde{B} \in \{A\}''$ and $C - \tilde{B}$ is normal since $\{A\}''$ is commutative.

44. Let $\{ U(s) : s \in \mathbb{R} \}$ be a family of unitary operators on $\mathcal{H}$ fulfilling the group condition $U(t+s) = U(t)U(s)$ for all $t, s \in \mathbb{R}$. The map $s \mapsto U(s)$ is strongly continuous *iff* it is weakly continuous. If these conditions are valid, the group is operator–norm continuous *iff* its generator is a bounded operator.

45. Let D be a dense subspace in $\mathcal{H}$ and suppose that a linear operator $U_0(s)$ on D corresponds to any $s \in \mathbb{R}$ in such a way that $U_0(s)D \subset D$ and $\overline{U_0(s)D} = \mathcal{H}$; moreover, $\|U_0(s)x\| = \|x\|$ holds for all $x \in D$, the map $s \mapsto U_0(s)x$ is continuous in $\mathbb{R}$, and $U_0(t+s)x = U_0(t)U_0(s)x$ holds for any $t, s \in \mathbb{R}$. Denote by $U(s)$ the continuous extension of $U_0(s)$; then $\{ U(s) : s \in \mathbb{R} \}$ is a strongly continuous unitary group.

46. Let A be a self-adjoint operator on $\mathcal{H}$. If a subspace $D \subset D_A$ is dense in $\mathcal{H}$ and $e^{itA}D \subset D$ for all $t \in \mathbb{R}$, then D is a core for A.
Hint: Use the argument from the proof of Theorem 5.9.2.

47. Consider strongly continuous unitary groups $\{ U(s) = e^{isA} : s \in \mathbb{R} \}$ and $\{ \tilde{U}(s) = e^{is\tilde{A}} : s \in \mathbb{R} \}$. They are unitarily equivalent, *i.e.*, there is a unitary operator V such that $\tilde{U}(\cdot) = VU(\cdot)V^{-1}$ *iff* $\tilde{A} = VAV^{-1}$.

48. Prove the following generalization to Theorem 5.9.2: Suppose that a unitary-valued map $U : \mathbb{R}^n \to \mathcal{B}(\mathcal{H})$ is strongly continuous and $U(t+s) = U(t)U(s)$ holds for all $t := [t_1, \dots, t_n]$ and $s := [s_1, \dots, s_n] \in \mathbb{R}^n$; then there are commuting self-adjoint operators $A_1, \dots, A_n$ such that $U(t) := e^{i(t_1A_1 + \cdots + t_nA_n)}$ holds for all $t \in \mathbb{R}^n$.
Hint: Use Corollary 5.9.4 and Proposition 5.5.7.

49. It follows from Proposition 5.9.6 that $\underline{U}_r(s) = e^{is\underline{A}_r}$, $r = 1, 2$, holds for all $s \in \mathbb{R}$. Prove that these relations in turn imply $U_\otimes(s) = e^{is(\underline{A}_1 + \underline{A}_2)}$.
Hint: Use (5.16) and Problem 33e.

50. Without reference to Example 5.5.1b, show that the translation operators $U_\tau(s)\psi := \psi(\cdot + s)$ form a strongly continuous group which is generated by $A_\tau := P$.

51. Consider rotations of the plane, $[x, y] \mapsto [x_\theta, y_\theta]$ with $x_\theta := x\cos\theta - y\sin\theta$ and $y_\theta := x\sin\theta + y\cos\theta$. Define the operators $U_\varrho(\theta)$ on $L^2(\mathbb{R}^2)$ by $(U_\varrho(\theta)\psi)(x, y) := \psi(x_\theta, y_\theta)$ and show that $\{ U_\varrho(\theta) : \theta \in \mathbb{R} \}$ is a unitary group, its generator A_ϱ is *e.s.a.* on $C_0^{(\infty)}(\mathbb{R}^2)$, and $(A_\varrho\phi)(x, y) = -i(x\partial_y - y\partial_x)\phi(x, y)$ for all $\phi \in C_0^{(\infty)}(\mathbb{R}^2)$.
Hint: Modify the argument of Example 5.9.5b.

Chapter 6

Operator sets and algebras

Up to now we have been discussing properties of single operators or certain commutative operator families. Now we turn our attention to more complicated operator sets. An efficient way to study their structure is to use algebraic methods; this is the main topic of this chapter. A drawback of this approach is that the sets under consideration must be closed under algebraic operations, which is ensured automatically only if the involved operators are bounded; however, we already know that in many cases properties of unbounded operators can be studied through their bounded functions. Algebras of unbounded operators can also be introduced; we shall mention this topic briefly at the end of the chapter.

6.1 C^*–algebras

Consider a topological algebra $\mathcal{A}$, which is equipped with an involution — we refer to Appendices B.1 and B.2 for the necessary prerequisites. If the involution is continuous with respect to the given topology, $\mathcal{A}$ is called a *topological $*$–algebra.* If $\mathcal{A}$ is a normed algebra and $\|a^*\| = \|a\|$ holds for any $a \in \mathcal{A}$ we speak about a *normed $*$–algebra;* the involution in $\mathcal{A}$ is obviously continuous. In a similar way we define a *Banach $*$–algebra.*

6.1.1 Example: The algebra $\mathcal{B}(\mathcal{H})$ provides an illustration that the first of the above definitions is nontrivial. We know that $\mathcal{B}(\mathcal{H})$ is a $*$–algebra and at the same time a topological algebra with respect to τ_s (*cf.* Example B.2.2). However, by Problem 3.9a the involution is not continuous in the strong operator topology unless $\dim \mathcal{H} < \infty$. On the other hand, the weak operator topology makes $\mathcal{B}(\mathcal{H})$, as well as any $*$–subalgebra of it, a topological $*$–algebra; algebras of this type will be discussed in Section 6.3 below. Furthermore, in view of Theorem 3.1.2a, $\mathcal{B}(\mathcal{H})$ is a topological $*$–algebra with respect to the operator–norm topology. Recall that by Example 3.2.3 it has an additional property, namely that $\|B^*B\| = \|B\|^2$ holds for any operator $B \in \mathcal{B}(\mathcal{H})$.

Inspired by the last mentioned relation, we can formulate an important definition.

A Banach $*$–algebra is called a C^*–**algebra** if

$$\|a^*a\| = \|a\|^2 \tag{6.1}$$

holds for all $a \in \mathcal{A}$. The example then says that $\mathcal{B}(\mathcal{H})$ with the operator–norm topology is a C^*–algebra and the same is, of course, true for its closed $*$–subalgebras. Notice that the involution must be isometric if condition (6.1) is valid: we have $\|a\|^2 = \|a^*a\| \le \|a^*\|\,\|a\|$, *i.e.*, $\|a\| \le \|a^*\|$, and similarly $\|a^*\| \le \|a\|$.

Let us review the basic properties of C^*–algebras; without loss of generality we may consider only those algebras with the unit element (see the notes). First we want to show that morphisms of C^*–algebras are automatically continuous; this is a consequence of Proposition B.2.5a and of the following assertion.

6.1.2 Proposition: Let φ be a $*$–morphism of a Banach $*$–algebra $\mathcal{A}$ into a C^*–algebra $\mathcal{B}$; then $\|\varphi(a)\|_{\mathcal{B}} \le \|a\|_{\mathcal{A}}$ holds for any $a \in \mathcal{A}$.

Proof: Suppose first that $\mathcal{A}$ has the unit element; then $\varphi(\mathcal{A})$ is a subalgebra in $\mathcal{B}$ with the unit element $\varphi(e)$, so $\sigma_{\mathcal{B}}(\varphi(a)) \subset \sigma_{\varphi(\mathcal{A})}(\varphi(a))$ holds for any $a \in \mathcal{A}$. Moreover, $(\varphi(a) - \lambda\varphi(e))^{-1} = \varphi((a-\lambda e)^{-1})$ exists provided $a - \lambda e$ is invertible, which means that $\sigma_{\varphi(\mathcal{A})}(\varphi(a)) \subset \sigma_{\mathcal{A}}(a)$. In combination with Theorem B.2.4d, it implies the inequalities $r(\varphi(a)) \le r(a) \le \|a\|_{\mathcal{A}}$. Let b be an arbitrary Hermitean element of $\mathcal{B}$. By induction we get $\|b^m\|_{\mathcal{B}}^{1/m} = \|b\|_{\mathcal{B}}$ where $m = 2^n$; hence the limit $n \to \infty$ yields $r(b) = \|b\|_{\mathcal{B}}$.

By assumption φ is a $*$–morphism, so the element $\varphi(a^*a)$ is Hermitean and it is sufficient to combine the above results,

$$\|\varphi(a)\|_{\mathcal{B}}^2 = \|\varphi(a^*a)\|_{\mathcal{B}} = r(\varphi(a^*a)) \le \|a^*a\|_{\mathcal{A}} \le \|a^*\|_{\mathcal{A}}\|a\|_{\mathcal{A}} = \|a\|_{\mathcal{A}}^2 .$$

If $\mathcal{A}$ or both algebras have no unit element, we can extend them to the Banach $*$–algebra $\tilde{\mathcal{A}}$ and the C^*–algebra $\tilde{\mathcal{B}}$, which have the unit elements. The assertion is valid for the $*$–morphism $\tilde\varphi : \tilde\varphi(\alpha e_{\mathcal{A}} + a) = \alpha e_{\mathcal{B}} + \varphi(a)$, and thus, *a fortiori*, for its restriction $\varphi = \tilde\varphi \restriction \mathcal{A}$. ∎

By Theorem B.2.4d, the spectral radius in a Banach $*$–algebra depends on the element itself. In C^*–algebras this result can be substantially strengthened.

6.1.3 Theorem: Suppose that $\mathcal{A}$ is a C^*–algebra and $\mathcal{B}$ its subalgebra containing the unit element of $\mathcal{A}$; then $\sigma_{\mathcal{B}}(b) = \sigma_{\mathcal{A}}(b)$ holds for any element $b \in \mathcal{B}$.

The relation $\sigma_{\mathcal{A}}(b) \subset \sigma_{\mathcal{B}}(b)$ is obvious. To prove the opposite inclusion we need an auxiliary result.

6.1.4 Lemma: Let $\mathcal{A}$ be a Banach algebra with the unit element.

(a) If $\{a_n\}$ is a Cauchy sequence in the set $\mathcal{R}$ of all invertible elements of $\mathcal{A}$ and $a := \lim_{n\to\infty} a_n \in \mathrm{bd}\,\mathcal{R} := \overline{\mathcal{R}} \setminus \mathcal{R}$; then $\lim_{n\to\infty} \|a_n^{-1}\| = \infty$.

(b) Let $\mathcal{B} \subset \mathcal{A}$ be a closed subalgebra containing the unit element of $\mathcal{A}$; then $\mathrm{bd}\,(\sigma_{\mathcal{B}}(b)) \subset \sigma_{\mathcal{A}}(b)$ holds for all $b \in \mathcal{B}$.

Proof: If (a) is not valid, there is K such that $\|a_n^{-1}\| < K$ for infinitely many values of the index n. At the same time, $\|a_n - a\| < K^{-1}$ holds for all sufficiently large n, so we can choose n such that $\|e - a_n^{-1}a_n\| = \|a_n^{-1}(a_n - a)\| < 1$. By Theorem B.2.4a and Proposition B.1.1a, elements $a_n^{-1}a$ and a are then invertible; however, this contradicts the assumption $a \in \mathrm{bd}\,\mathcal{R}$ because the set $\mathcal{R}$ is open.

Denote the sets of all invertible elements in $\mathcal{A}$ and $\mathcal{B}$ by $\mathcal{R}_\mathcal{A}$ and $\mathcal{R}_\mathcal{B}$, respectively. They are open and $\mathcal{R}_\mathcal{B} \subset \mathcal{R}_\mathcal{A}$. For any $b \in \mathrm{bd}\,\mathcal{R}_\mathcal{B}$ we can find a sequence $\{b_n\} \subset \mathcal{R}_\mathcal{B}$ such that $b_n \to b$; if b belonged simultaneously to $\mathcal{R}_\mathcal{A}$, then $b_n^{-1} \to b^{-1}$ by continuity of the inversion, *i.e.*, the sequence $\{\|b_n^{-1}\|\}$ would be bounded. Since this contradicts assertion (a), we have $\mathcal{R}_\mathcal{A} \cap \mathrm{bd}\,\mathcal{R}_\mathcal{B} = \emptyset$. Furthermore, we can easily check that $\lambda \in \mathrm{bd}(\rho_\mathcal{B}(b))$ implies $b - \lambda e \in \mathrm{bd}\,\mathcal{R}_\mathcal{A}$, so $b - \lambda e \notin \mathcal{R}_\mathcal{A}$; in other words, $\lambda \in \sigma_\mathcal{A}(b)$. This means that $\sigma_\mathcal{A}(b)$ contains the set $\mathrm{bd}(\rho_\mathcal{B}(b))$; however, the latter coincides with the boundary of the complement, $\mathrm{bd}(\rho_\mathcal{B}(b)) = \mathrm{bd}(\sigma_\mathcal{B}(b))$. ∎

Proof of Theorem 6.1.3: The element a^*a is Hermitean for $a \in \mathcal{A}$, so $\sigma_\mathcal{A}(a^*a) \subset \mathbb{R}$ by Problem 2b. Hence each point of the spectrum belongs to its boundary (with respect to the complex plane) and Lemma 4b gives

$$\sigma_\mathcal{B}(a^*a) \subset \mathrm{bd}(\sigma_\mathcal{B}(a^*a)) \subset \sigma_\mathcal{A}(a^*a) \subset \sigma_\mathcal{B}(a^*a)\,,$$

i.e., $\sigma_\mathcal{B}(a^*a) = \sigma_\mathcal{A}(a^*a)$. Next we want to check the same relation for an arbitrary element $b \in \mathcal{B}$. We choose $\lambda \notin \sigma_\mathcal{A}(b)$ and denote $a := b - \lambda e$; we have to show that $\lambda \notin \sigma_\mathcal{B}(b)$, *i.e.*, $a^{-1} \in \mathcal{B}$. This is true for $c := a^*a$; since it is invertible, c^{-1} belongs to $\mathcal{B}$. The element $a_l := (a^*a)^{-1}a^*$ is contained in $\mathcal{B}$ and $a_l a = e$, and similarly, $a_r := a^*(aa^*)^{-1}$ belongs to $\mathcal{B}$ and $aa_r = e$. However, the inverse in $\mathcal{B}$ is unique so $a_l = a_r = a^{-1} \in \mathcal{B}$. ∎

The structure of a C^*–algebra simplifies considerably in the commutative case (see the notes; compare also with Theorem B.2.6).

6.1.5 Theorem (Gel'fand–Naimark): A commutative C^*–algebra with the unit element is isometrically $*$–isomorphic to the algebra $C(\Delta)$ of continuous functions on a certain compact Hausdorff space Δ.

This structural result can also be employed to solve various problems in noncommutative C^*–algebras; we have to select a commutative subalgebra and represent it functionally. The simplest case corresponds to the situation when this subalgebra is generated by a single element $a \in \mathcal{A}$; in this way we can construct the functional calculus on C^*–algebras in analogy with the results of Section 5.2. The same idea provides an alternative proof of the spectral theorem for Hermitean operators; references are given in the notes.

Let us finally describe briefly how a new C^*–algebra can be constructed from a given family of C^*–algebras. There are several ways, namely

(i) Restriction to a closed $*$–subalgebra in a C^*–algebra.

(ii) Factorization of a C^*–algebra with respect to a closed $*$–ideal.

(iii) Let $\{\mathcal{A}_\alpha : \alpha \in I\}$ be a family of C^*–algebras. We choose the subset $\mathcal{A} := \{a = [a_\alpha] : a_\alpha \in \mathcal{A}_\alpha,\, \|a\|_\infty := \sup_{\alpha\in I} \|a_\alpha\|_\alpha < \infty\}$ in their Cartesian product; since any $\mathcal{A}_\alpha$ is a Banach space, $\mathcal{A}$ may also be regarded as a Banach space (see Section 1.5). In addition, if we introduce the operations of multiplication and involution componentwise, $[a_\alpha][b_\alpha] := [a_\alpha b_\alpha]$ and $[a_\alpha]^* := [a_\alpha^*]$; then $\mathcal{A}$ becomes a C^*–algebra (*cf.* Problem 4), which is called the *direct sum* of the C^*–algebras $\mathcal{A}_\alpha$, $\alpha \in I$.

(iv) Consider finally C^*–algebras $\mathcal{A}_1$, $\mathcal{A}_2$ with the norms $\|\cdot\|_j$, $j = 1,2$. A bilinear map $\otimes : \mathcal{A}_1 \times \mathcal{A}_2 \to \mathcal{A}$, where $\mathcal{A}$ is a C^*–algebra with the norm $\|\cdot\|$, which satisfies the requirements $(a_1\otimes a_2)(b_1\otimes b_2) = a_1b_1\otimes a_2b_2$, $(a_1\otimes a_2)^* = a_1^*\otimes a_2^*$ and $\|a_1\otimes a_2\| = \|a_1\|_1\|a_2\|_2$ for any $a_j \in \mathcal{A}_j$ and such that the set $\{a_1\otimes a_2 : a_j \in \mathcal{A}_j\}$ is total in $\mathcal{A}$ is called a *realization* of the *tensor product* of $\mathcal{A}_1$ and $\mathcal{A}_2$. One can show that such a realization always exists and it is essentially unique — *cf.* Problem 5 and the notes.

6.2 GNS construction

The properties of C^*–algebras derived above do not exhaust the consequences of condition (6.1). Using it, for instance, we can derive various spectral properties for elements of C^*–algebras analogous to those of bounded operators (Problem 2). This similarity is not accidental; we are going to show now that any C^*–algebra can be represented faithfully in some $\mathcal{B}(\mathcal{H})$.

First we have to introduce the notion of positivity. Let $\mathcal{A}$ be a C^*–algebra with the unit element. An element $a \in \mathcal{A}$ is said to be *positive* if it is Hermitean and $\sigma(a) \subset [0,\infty)$. This is written symbolically as $a \ge 0$, and moreover, $a \ge b$ means $a-b \ge 0$. A set $\mathcal{P}$ in a vector space is called a *cone* if the elements αa, $a+b$ belong to $\mathcal{P}$ for all $a,b \in \mathcal{P}$, $\alpha \ge 0$. A cone is always a convex set: if α, $1-\alpha$ are nonzero numbers, then $\alpha a+(1-\alpha)a \in \mathcal{P}$ holds for any $a,b \in \mathcal{P}$. Positive elements have the following properties (see the notes).

6.2.1 Theorem: Let $\mathcal{A}$ be a C^*–algebra with the unit element; then $a^*a \ge 0$ holds for all $a \in \mathcal{A}$, and conversely, any positive element $b \in \mathcal{A}$ can be expressed as $b = a^*a$ for some $a \in \mathcal{A}$. The sets $\mathcal{A}_\pm := \{a \in \mathcal{A} : \pm a \ge 0\}$ are closed cones in $\mathcal{A}$ such that $\mathcal{A}_+ \cap \mathcal{A}_- = \{0\}$.

A linear functional f on a $*$–algebra $\mathcal{A}$ is *positive* if $f(a^*a) \ge 0$ for all $a \in \mathcal{A}$. Let $\mathcal{A}$ be a Banach $*$–algebra; then a positive functional f on $\mathcal{A}$ normalized by $\|f\| = 1$ is called a *state* (on the algebra $\mathcal{A}$). Positivity of a functional is denoted as $f \ge 0$ while $f \ge g$ again means $f-g \ge 0$. If $\mathcal{A}$ is a C^*–algebra, a functional f is positive, by Theorem 6.2.1, *iff* $f(b) \ge 0$ holds for all positive elements $b \in \mathcal{A}$.

6.2.2 Proposition: Suppose that f is a positive functional on a C^*–algebra with the unit element, and a,b are arbitrary elements of $\mathcal{A}$; then

(a) $f = 0$ if $f \le 0$ holds at the same time.

(b) $f(a^*) = \overline{f(a)}$.

(c) The generalized Schwarz inequality is valid, $|f(a^*b)|^2 \le f(a^*a)f(b^*b)$.

(d) $|f(a)|^2 \le f(e)f(a^*a) \le f(e)^2 r(a^*a)$.

(e) f is continuous and $\|f\| = f(e)$.

Proof: By assumption, $f((a^* + \overline{\alpha}e)(a + \alpha e)) \ge 0$; choosing $\alpha = 1, i$, we find that $f(a)+f(a^*)$ and $i(f(a^*)-f(a))$ are real, *i.e.*, assertion (b). Furthermore, (c) follows from the fact that $[a, b] \mapsto f(a^*b)$ is a positive symmetric sesquilinear form; substituting $b = e$ we get the first inequality in (d). The element $c := r(a^*a)e - a^*a$ is positive, so $f(c) \ge 0$, and since f is linear we have $f(a^*a) \le f(e)r(a^*a)$. Next we use (6.1) in combination with Theorem B.2.4d; it yields

$$|f(a)|^2 \le f(e)f(a^*a) \le f(e)^2 r(a^*a) \le f(e)^2\|a^*a\| \le f(e)^2\|a\|^2 .$$

This proves (d), the continuity of f, and $\|f\| \le f(e)$; the opposite inequality follows from $\|e\| = 1$. It remains for us to prove (a). By assumption $f(a^*a) = 0$ holds for any $a \in \mathcal{A}$; hence (d) implies $f(a) = 0$. ∎

After these preliminaries, let us turn to the mentioned result concerning the existence of a faithful representation for any C^*–algebra. The proof is based on a constructive method devised by I. Gel'fand, M. Naimark and I. Segal.

6.2.3 Theorem (GNS construction): Let $\mathcal{A}$ be a Banach $*$–algebra with the unit element. For any positive functional f on $\mathcal{A}$ there is a Hilbert space $\mathcal{H}$ and a representation $\pi : \mathcal{A} \to \mathcal{B}(\mathcal{H})$ with a cyclic vector ψ_0 such that $f(a) = (\psi_0, \pi(a)\psi_0)$ holds for any $a \in \mathcal{A}$. If $\{\mathcal{H}', \pi', \psi_0'\}$ is another triplet with the same properties, then there is a unitary operator $U \in \mathcal{B}(\mathcal{H}, \mathcal{H}')$ such that $\psi_0' = U\psi_0$ and $U\pi(a) = \pi'(a)U$ for all $a \in \mathcal{A}$.

Proof: The starting point for the construction of $\mathcal{H}$ is the algebra $\mathcal{A}$ itself. The form $\varphi : \mathcal{A} \times \mathcal{A} \to \mathbb{C}$ defined by $\varphi(a, b) := f(a^*b)$ is sesquilinear, symmetric, and positive. Hence we can use it to construct the inner product; however, since $f(a^*a) = 0$ does not imply $a = 0$ a factorization is needed. By Proposition 2c, $\mathcal{J}_f := \{\, a : f(a^*a) = 0 \,\}$ is a subspace in $\mathcal{A}$, and moreover,

$$|f((ab)^*ab)|^2 \le f((a^*ab)^*a^*ab)f(b^*b) = 0$$

holds for all $a \in \mathcal{A}$, $b \in \mathcal{J}_f$, so $\mathcal{J}_f$ is also a left ideal. On the factor space $\mathcal{A}/\mathcal{J}_f$ we define an inner product by $(\tilde{a}, \tilde{b}) := f(a^*b)$, where a, b are some elements representing the classes $\tilde{a}, \tilde{b} \in \mathcal{A}/\mathcal{J}_f$. Of course, $(\tilde{a}, \tilde{b})$ must not depend on the chosen representatives, *i.e.*, $f(a^*b)$ must be zero if at least one of the elements a, b belongs to $\mathcal{J}_f$; this follows from Proposition 6.2.2c and the relation $f(a^*b) = \overline{f(b^*a)}$. The form $(\cdot, \cdot)$ is obviously sesqilinear, symmetric, and positive; $(\tilde{a}, \tilde{a}) = 0$ holds *iff* $\tilde{a} = \mathcal{J}_f$, which is the zero element of $\mathcal{A}/\mathcal{J}_f$. In this way we have defined the inner product on $\mathcal{A}/\mathcal{J}_f$; the Hilbert space $\mathcal{H}$ is then obtained by the standard completion of the factor space $\mathcal{A}/\mathcal{J}_f$.

Let us turn to the representation π. We define the operator $\pi_0(a)$ on $\mathcal{A}/\mathcal{J}_f$ for any $a \in \mathcal{A}$ by $\pi_0(a)\tilde{b} := \widetilde{ab}$, where b is an element representing the class $\tilde{b}$; this makes sense because $\mathcal{J}_f$ is a left ideal in $\mathcal{A}$. The operator $\pi_0(a)$ is obviously linear; let us check that it is bounded. We have

$$\|\pi_0(a)\tilde{b}\|^2 = (\widetilde{ab}, \widetilde{ab}) = f(b^*a^*ab) = \varphi_b(a^*a)$$

for any $\tilde{b} \in \mathcal{A}/\mathcal{J}_f$, where $\varphi_b(c) := f(b^*cb)$. The identity implies, in particular, that the functional φ_b is positive, so Proposition 6.2.2e yields

$$\|\pi_0(a)\tilde{b}\|^2 \le \varphi_b(e)\|a^*a\|_{\mathcal{A}} \le f(b^*b)\|a\|^2_{\mathcal{A}} = \|\tilde{b}\|^2\|\tilde{a}\|^2_{\mathcal{A}}\,.$$

By Theorem 1.5.5, $\pi_0(a)$ has then a unique continuous extension to the space $\mathcal{H}$, which we denote as $\pi(a)$. The norm remains preserved, $\|\pi(a)\|_{\mathcal{B}(\mathcal{H})} \le \|a\|_{\mathcal{A}}$. Since $\mathcal{A}/\mathcal{J}_f$ is dense in $\mathcal{H}$, to prove that $\pi(\cdot)$ is a representation of $\mathcal{A}$ it is sufficient to check the relations

$$\pi(\alpha a+b)\tilde{c} = \alpha\pi(a)\tilde{c} + \pi(b)\tilde{c}\,, \quad \pi(ab)\tilde{c} = \pi(a)\pi(b)\tilde{c}\,, \quad (\tilde{c}, \pi(a^*)\tilde{d}) = (\tilde{c}, \pi(a)^*\tilde{d})$$

for any $a, b \in \mathcal{A}$, $\tilde{c}, \tilde{d} \in \mathcal{A}/\mathcal{J}_f$ and $\alpha \in \mathbb{C}$. The first two of these follow directly from the definition, and $(\tilde{c}, \pi(a^*)\tilde{d}) = (\tilde{c}, \widetilde{a^*d}) = f(c^*a^*d) = (\widetilde{ac}, \tilde{d}) = (\pi(a)\tilde{c}, \tilde{d})$ yields the third. Furthermore, $\pi(\mathcal{A})\tilde{e} = \mathcal{A}/\mathcal{J}_f$ and $\mathcal{A}/\mathcal{J}_f$ is dense in $\mathcal{H}$, so $\psi_0 := \tilde{e}$ is a cyclic vector of the representation π and

$$(\psi_0, \pi(a)\psi_0) = (\tilde{e}, \tilde{a}) = f(e^*a) = f(a)$$

holds for all $a \in \mathcal{A}$. Hence we have proven that there is at least one triplet $\{\mathcal{H}, \pi, \psi_0\}$ with the required properties; for the sake of brevity we speak about a *GNS triplet* (corresponding to algebra $\mathcal{A}$ and functional f) and π is called a *GNS representation.*

Assume finally that $\{\mathcal{H}', \pi', \psi_0'\}$ is another GNS triplet. We denote $\mathcal{H}_0 := \pi(\mathcal{A})\psi_0 = \mathcal{A}/\mathcal{J}_f$ and $\mathcal{H}_0' := \pi'(\mathcal{A})\psi_0'$, and define the operator $U_0 : \mathcal{H}_0 \to \mathcal{H}_0'$ by $U_0\pi(a)\psi_0 := \pi'(a)\psi_0'$ for any $a \in \mathcal{A}$; the relations

$$\|U_0\pi(a)\psi_0\|^2_{\mathcal{H}'} = (\psi_0', \pi'(a^*a)\psi_0')_{\mathcal{H}'} = f(a^*a) = \|\pi(a)\psi_0\|^2$$

show that U_0 is bijective and norm–preserving. At the same time it is linear, and since the subspaces $\mathcal{H}_0$, $\mathcal{H}_0'$ are by assumption dense in $\mathcal{H}$ and $\mathcal{H}'$, respectively, it can, due to Proposition 3.3.3a, be extended continuously to a unitary operator $U : \mathcal{H} \to \mathcal{H}'$. Substituting $a := e$ into the last relation we get $U\psi_0 = \psi_0'$, and consequently, $U\pi(a)\psi_0 = \pi'(a)U\psi_0$. We apply the operator $\pi'(b)$ to both sides of this identity,

$$\pi'(b)U\pi(a)\psi_0 = \pi'(ba)U\psi_0 = U\pi(ba)\psi_0 = U\pi(b)\pi(a)\psi_0,$$

and since $\pi(\mathcal{A})\psi_0$ is dense in $\mathcal{H}$, it follows that $\pi'(b)U = U\pi(b)$ holds for any element $b \in \mathcal{A}$. ∎

6.2.4 Examples: (a) Let $\mathcal{A}$ be a closed subalgebra in $\mathcal{B}(\mathcal{H})$ containing the unit operator. Given an arbitrary vector $\psi \in \mathcal{H}$, we define the functional

$$f_\psi : f_\psi(B) = (\psi, B\psi) \quad \text{for all} \quad B \in \mathcal{A}.$$

It is obviously positive. Moreover, the subspace $\mathcal{A}\psi \subset \mathcal{H}$ is invariant with respect to the operators of $\mathcal{A}$, so $\mathcal{H}_\psi := \overline{\mathcal{A}\psi}$ is also invariant (Problem 1.33c). This allows us to define the map $\pi_\psi : \mathcal{A} \to \mathcal{B}(\mathcal{H}_\psi)$ by $\pi_\psi(B) := B_\psi := B \restriction \mathcal{H}_\psi$. It is not difficult to check that $\{\mathcal{H}_\psi, \pi_\psi, \psi\}$ is a GNS triplet corresponding to the functional f_ψ. A particularly simple situation occurs if $\mathcal{H}_\psi = \mathcal{H}$; then π_ψ is the identical mapping and ψ is simultaneously a cyclic vector of the operator algebra $\mathcal{A}$.

(b) Not every positive functional on a C^*–algebra $\mathcal{A} \subset \mathcal{B}(\mathcal{H})$ can be expressed in the above form. To describe a more general case, consider a statistical operator W on $\mathcal{H}$ (*cf.* Section 3.6) with the spectral decomposition $W = \sum_{k=1}^N w_k E_k$, where $N := \dim(\operatorname{Ker} W)^\perp$ and the E_k are one–dimensional projections corresponding to normalized eigenvectors ψ_k with the nonzero eigenvalues w_k, $k = 1, \dots, N$; the latter fulfil the condition $\operatorname{Tr} W = \sum_{k=1}^N w_k = 1$. Using operator W we define the positive functional

$$f_W : f_W(B) = \operatorname{Tr}(WB) = \sum_{k=1}^N w_k(\psi_k, B\psi_k)$$

on $\mathcal{A}$. The subspaces $\mathcal{H}_k := \overline{\mathcal{A}\psi_k}$ are again invariant with respect to the algebra $\mathcal{A}$; we construct the Hilbert space $\mathcal{H}_W := \bigoplus_{k=1}^N \mathcal{H}_k^W$, where the spaces $\mathcal{H}_k^W$ are obtained from $\mathcal{H}_k$ by replacing the inner product with $(\cdot,\cdot)_k := w_k(\cdot,\cdot)$. In other words, the elements of $\mathcal{H}_W$ are sequences $\Phi = \{\phi_k\}_{k=1}^N$ with $\phi_k \in \mathcal{H}_k$ such that $\|\Phi\|_W < \infty$; here the norm $\|\cdot\|_W$ is induced by the inner product $(\Phi, \Psi)_W := \sum_{k=1}^N w_k(\phi_k, \psi_k)$. Given an operator $B \in \mathcal{A}$ we define $\pi_W(B) := B_W$ on $\mathcal{H}_W$ by

$$B_W\{\phi_k\}_{k=1}^N := \{B\phi\}_{k=1}^N;$$

it is easy to see that $\|B_W\| \le \|B\|$. If $\mathcal{A}$ contains the unit operator then the normalized vector $\Psi_W := \{\psi_k\}_{k=1}^N$ and $(\Psi_W, \pi_W(B)\Psi_W)_W = f_W(B)$ holds for any $B \in \mathcal{A}$.

The map $\pi_W : \mathcal{A} \to \mathcal{B}(\mathcal{H}_W)$ is a representation of algebra $\mathcal{A}$. However, this does not mean that $\{\mathcal{H}_W, \pi_W, \Psi_W\}$ is a GNS triplet corresponding to f_W, because Ψ_W need not be a cyclic vector of π_W. The algebra $\mathbb{C}(\mathcal{H}) := \{\alpha I : \alpha \in \mathbb{C}\}$ of scalar operators can be taken as an example; the set $\pi_W(\mathbb{C}(\mathcal{H}))\Psi_W = (\Psi_W)_{lin}$ is clearly not dense in $\mathcal{H}_W$ unless $N = 1$. On the other hand, we shall show that Ψ_W is cyclic for π_W if the projections E_k, $k = 1, \dots, N$, belong to algebra $\mathcal{A}$. To check this claim, suppose that $(\Phi, \pi_W(B)\Psi_W)_W = \sum_{k=1}^N w_k(\phi_k, B\psi_k) = 0$ for $\Phi \in \mathcal{H}_W$ and all $B \in \mathcal{A}$.

Choosing $B = CE_n$ with $C \in \mathcal{A}$ we get $(\phi_n, C\psi_n) = 0$ for $n = 1, \dots, N$; however, the set $\mathcal{A}\psi_n$ is dense in $\mathcal{H}_n$, so $\phi_n = 0$ for all n, and also $\Phi = 0$. In this case $\{\mathcal{H}_W, \pi_W, \Psi_W\}$ is the sought GNS triplet corresponding to the functional f_W. Notice also that the representation π_W is reducible unless $N = 1$; it follows from the definition that $\pi_W(B) = \bigoplus_{k=1}^N \pi_k(B)$, where $\pi_k(B) := B \upharpoonright \mathcal{H}_k$ for any $B \in \mathcal{A}$.

(c) A GNS representation need not be faithful. To illustrate this claim, consider a nontrivial orthogonal–sum decomposition $\mathcal{H} := \mathcal{H}_1 \oplus \mathcal{H}_2$ and the algebra $\mathcal{A} := \mathcal{B}(\mathcal{H}_1) \oplus \mathcal{B}(\mathcal{H}_2)$. If the functional f_ψ, used above, corresponds to a nonzero vector $\psi \in \mathcal{H}_1$, the GNS representation is given by $\pi_\psi(B) = B_1$ for any $B := B_1 \oplus B_2$ because $\mathcal{H}_\psi = \mathcal{H}_1$. Its kernel $\mathcal{A}_2 := \{B \in \mathcal{A} :\ B_1 = 0\}$ is obviously nontrivial and forms a closed ideal in algebra $\mathcal{A}$ (not in $\mathcal{B}(\mathcal{H})$, of course!).

Example 6.2.4c shows that Theorem 6.2.3 itself does not solve the mentioned problem. In fact, we want to show that any C^*–algebra $\mathcal{A}$ has a representation $\pi : \mathcal{A} \to \mathcal{B}(\mathcal{H})$, which is not only faithful but at the same time reproduces the metric properties. This requires the map π to be simultaneously an isometry; in that case we speak of an *isometric representation.* To construct such a representation we need a rich enough family of positive functionals on $\mathcal{A}$. Its existence is ensured by the following result.

6.2.5 Proposition: Let $\mathcal{A}$ be a C^*–algebra with the unit element. For any nonzero element $a \in \mathcal{A}$ there is a positive functional f_a such that $f_a(e) = 1$ (so f_a is a state on $\mathcal{A}$) and at the same time, $f_a(a^*a) = \|a\|^2$.

Proof is a modification of the argument used to prove the Hahn–Banach theorem (see the notes). First one constructs a functional f with the required properties on the set $\mathcal{A}_R$ of all Hermitean elements in $\mathcal{A}$, which forms a real Banach space. We set $f(\alpha e+\beta a^*a) := \alpha+\beta\|a^*a\|$ on the subspace $\{e, a^*a\}_{lin}$ and extend it by adding vectors. The most difficult part is to check that the positivity is preserved at each step; the procedure is completed by Zorn's lemma. Having constructed f we define $f_a :\ f_a(b) = f((b+b^*)/2)+if((b-b^*)/2i)$. ∎

6.2.6 Theorem: An arbitrary C^*–algebra $\mathcal{A}$ has an isometric representation $\pi : \mathcal{A} \to \mathcal{B}(\mathcal{H})$ on some Hilbert space $\mathcal{H}$.

Proof: Without loss of generality we may assume that $\mathcal{A}$ has the unit element. By Proposition 6.2.5, we can associate a positive functional f_a with any nonzero element $a \in \mathcal{A}$, and furthermore a GNS triplet $\{\mathcal{H}_a, \pi_a, \psi_a\}$ corresponds to f_a. We define the Hilbert space in question as the direct sum, $\mathcal{H} := \bigoplus_{0 \ne a \in \mathcal{A}} \mathcal{H}_a$; recall that any vector $\{\phi_a :\ 0 \ne a \in \mathcal{A}\} \in \mathcal{H}$ has at most countably many nonzero components. The representation π is constructed as the direct sum of representations π_a: we set $\pi(b)\Phi := \{\pi_a(b)\phi_a :\ 0 \ne a \in \mathcal{A}\}$ for any $b \in \mathcal{A}$. Proposition 6.1.2 implies $\|\pi_a(b)\| \le \|b\|_{\mathcal{A}}$, and therefore $\|\pi(b)\| = \sup_{0 \ne a \in \mathcal{A}} \|\pi_a(b)\| \le \|b\|_{\mathcal{A}}$, *i.e.*, the operators $\pi(b)$ are bounded. It is easy to check that the map $\pi : \mathcal{A} \to \mathcal{B}(\mathcal{H})$ is a representation. It is also isometric: cyclic vectors ψ_a are normalized by Theorem 6.2.3 and Proposition 6.2.5, $\|\psi_a\|_a^2 = f_a(e) = 1$; hence

$$\|\pi_b(b)\|^2 \ge \|\pi_b(b)\psi_b\|_b^2 = (\psi_b, \pi_b(b)^*\pi_b(b)\psi_b)_b = f_b(b^*b) = \|b\|_{\mathcal{A}}^2 .$$

In combination with the above estimate, this yields $\|b\|_{\mathcal{A}} \le \|\pi_b(b)\| \le \|\pi(b)\| \le \|b\|_{\mathcal{A}}$ for any $b \in \mathcal{A}$, which concludes the proof. ∎

The representation π constructed in the proof is, of course, faithful but it does serve practical purposes; the space $\mathcal{H}$ is extremely large (recall that any vector has infinitely many components corresponding to multiples of a single element of $\mathcal{A}$) and the representation π is "excessively reducible". We usually employ other methods to construct isometric representations of C^*–algebras; the importance of Theorem 6.2.6 is that it guarantees their existence.

In conclusion let us mention another application of the GNS representation. Let $\mathcal{A}$ again be a C^*–algebra with the unit element and let us denote by $S_{\mathcal{A}}$ the family of all states on $\mathcal{A}$. It is easy to see that $S_{\mathcal{A}}$ is convex: the functional $f := \alpha f_1 + (1-\alpha) f_2$ is obviously positive for any $f_1, f_2 \in S_{\mathcal{A}}$ and non–negative numbers α, $1-\alpha$, and $\|f\| = f(e) = \alpha f_1(e) + (1-\alpha) f_2(e) = 1$, *i.e.*, f is a state on $\mathcal{A}$. In the notes to Section 1.1 we have defined an extremal point of a convex set; the extremal points of $S_{\mathcal{A}}$ are called *pure states.* The following important result is valid (see the notes).

6.2.7 Theorem (Segal): Let $\mathcal{A}$ be a C^*–algebra with the unit element. A state $f \in S_{\mathcal{A}}$ is pure *iff* the corresponding GNS representation π_f is irreducible.

6.3 W^*–algebras

Our next topic deals with algebras of bounded operators on some fixed Hilbert space $\mathcal{H}$. In addition to the operator C^*–algebras considered in the preceding sections, *i.e.*, norm–closed $*$–subalgebras in $\mathcal{B}(\mathcal{H})$, we can in this case also define other classes of algebras due to the fact that $\mathcal{B}(\mathcal{H})$ can be equipped with different topologies.

6.3.1 Remark: The topologies τ_u, τ_s, τ_w of Section 3.1 are not the only ones which can be defined on $\mathcal{B}(\mathcal{H})$. We shall describe two more topologies. Let $\tilde{\mathcal{H}}$ be the set of all sequences $\Phi = \{\phi_k\} \subset \mathcal{H}$ such that $\sum_k \|\phi_k\|^2 < \infty$. We set

$$p_\Phi(B) := \left(\sum_{k=1}^{\infty} \|B\phi_k\|^2 \right)^{1/2}$$

for all $\Phi \in \tilde{\mathcal{H}}$ and $B \in \mathcal{B}(\mathcal{H})$. It follows from the Minkowski inequality that $p_\Phi : \mathcal{B}(\mathcal{H}) \to [0,\infty)$ is a seminorm, and it is easy to check that the family $\{p_\Phi\}_{\sigma s} := \{ p_\Phi : \Phi \in \tilde{\mathcal{H}} \}$ separates points. The corresponding locally convex topology on $\mathcal{B}(\mathcal{H})$ is called *σ–strong*. In a similar way, we define the seminorms

$$p_{\phi\Psi}(B) := \left| \sum_{k=1}^{\infty} (\phi_k, B\psi_k) \right|$$

and the system $\{p_{\Phi\Psi}\}_{\sigma w} := \{\, p_{\Phi\Psi} : \Phi, \Psi \in \tilde{\mathcal{H}} \,\}$ determines the *σ–weak topology* on $\mathcal{B}(\mathcal{H})$. Both definitions are conveniently reformulated if we regard $\tilde{\mathcal{H}}$ as the direct sum $\tilde{\mathcal{H}} = \bigoplus_{k=1}^{\infty} \mathcal{H}_k$, with $\mathcal{H}_1 = \mathcal{H}_2 = \ldots = \mathcal{H}$ equipped with the inner product

$$(\Phi, \Psi)_{\tilde{\mathcal{H}}} := \sum_{k=1}^{\infty} (\phi_k, \psi_k)\,.$$

If we use $\tilde{B}$ to denote the direct sum of identical copies of operator B, $\tilde{B}\Phi := \{B\phi_k\}$, then the seminorms are expressed as $p_\Phi(B) = \|\tilde{B}\Phi\|_{\tilde{\mathcal{H}}}$ and $p_{\Phi\Psi}(B) = |(\Phi, \tilde{B}\Psi)_{\tilde{\mathcal{H}}}|$. The five topologies we have introduced on $\mathcal{B}(\mathcal{H})$ are related by

$$\begin{array}{ccccc} \tau_u & \supset & \tau_{\sigma s} & \supset & \tau_{\sigma w} \\ & & \cup & & \cup \\ & & \tau_s & \supset & \tau_w \end{array}$$

where all the inclusions are nontrivial unless $\dim \mathcal{H} < \infty$ (see Problem 6 and the notes). The indices appearing here will be used in the following to distinguish these topologies, *e.g.*, $(\overline{\mathcal{S}})_{\sigma w}$ will mean the σ–weak closure of set $\mathcal{S}$, *etc.*

After this preliminary, let us introduce a new class of operator algebras. For a set $\mathcal{S} \subset \mathcal{B}(\mathcal{H})$, the definition of the commutant from Section 5.5 simplifies to $\mathcal{S}' := \{\, B \in \mathcal{B}(\mathcal{H}) : BC = CB,\ C \in \mathcal{S} \,\}$, and we introduce the *bicommutant* by $\mathcal{S}'' := (\mathcal{S}')'$. One should keep in mind that these definitions are related to $\mathcal{B}(\mathcal{H})$; the commutant of a subset $\mathcal{S}$ in an operator algebra $\mathcal{A}$ introduced in Appendix B.1 is in the present notation equal to $\mathcal{S}' \cap \mathcal{A}$. If $\mathcal{S}$ is a symmetric subset in $\mathcal{B}(\mathcal{H})$, then $\mathcal{S}'$ and $\mathcal{S}''$ are by Proposition B.1.3f ∗–subalgebras in $\mathcal{B}(\mathcal{H})$. Hence the following definition makes sense: a ∗–algebra $\mathcal{A} \subset \mathcal{B}(\mathcal{H})$ is called a **W^*–algebra** (or a *von Neumann algebra*) if it coincides with its bicommutant, $\mathcal{A} = \mathcal{A}''$.

Though the notion is introduced in a purely algebraic way, W^*–algebras have an equivalent topological characterization. By Proposition B.2.1c and Example B.2.2, a W^*–algebra $\mathcal{A}$ is closed with respect to weak operator topology. Moreover, taking into account the relations between the topologies on $\mathcal{B}(\mathcal{H})$ and Example 6.1.1, we arrive at the following result.

6.3.2 Proposition: Any W^*–algebra $\mathcal{A} \subset \mathcal{B}(\mathcal{H})$ is weakly closed. As a consequence, it is closed in all the other topologies of Remark 6.3.1; in particular, it is a C^*–algebra.

It is natural to ask whether an arbitrary weakly closed ∗–subalgebra in $\mathcal{B}(\mathcal{H})$ is at the same time a W^*–algebra, or whether this is true for any operator C^*–algebra. In both cases the answer is negative.

6.3.3 Examples: (a) Let $E \in \mathcal{B}(\mathcal{H})$ be a nontrivial projection. The algebra $\mathcal{A} := \mathcal{A}_0(\{E\})$, which consists of all multiples of E, is weakly closed, but it does not contain the scalar operators which belong to $\mathcal{A}_W(\{E\})$ (Problem 8). Here $\mathcal{A}_W(\mathcal{S})$ means the smallest W^*–algebra containing $\mathcal{S}''$; it is easy to see that $\mathcal{A}$ has to be completed with multiples of the unit operator.

(b) Let Q be the operator of Example 3.2.2 on a finite interval $J := [a,b]$. In view of Theorem 5.2.2, $\mathcal{A}_C := \{ f(Q) : f \in C(J) \}$ is isometrically $*$–isomorphic to the functional C^*–algebra $C(J)$; hence it is a C^*–subalgebra in $\mathcal{B}(L^2(J))$. To check that it is not weakly closed, we have to choose a suitable sequence $\{g_n\} \subset C(J)$, *e.g.*, $g_n(x) := \max\{1, ((x-a)/(c-a))^n\}$ for some $c \in (a,b)$. It follows from the already quoted Theorem 5.2.2 that $\text{w-lim}_{n\to\infty} g_n(Q) = \chi_{[c,b]}(Q)$, so this operator belongs to $(\overline{\mathcal{A}_C})_w$ but not to $\mathcal{A}_C$.

We therefore want to find a necessary and sufficient condition under which a weakly closed $*$–algebra is a W^*–algebra. We notice first that we can associate with any set $\mathcal{S} \subset \mathcal{B}(\mathcal{H})$ a projection $E_\mathcal{S}$ such that $BE_\mathcal{S} = B$ holds for all $B \in \mathcal{S}$ (Problem 8c). Set $\mathcal{S}$ is called *nondegenerate* if $E_\mathcal{S} = I$; this is true, in particular, if $\mathcal{S}$ contains the unit operator.

6.3.4 Theorem: Let $\mathcal{S}$ be a symmetric subset in $\mathcal{B}(\mathcal{H})$; then

(a) $\mathcal{A}_\mathcal{S} := \{ B \in \mathcal{S}'' : B = E_\mathcal{S} B E_\mathcal{S} \}$ is a weakly closed $*$–subalgebra in $\mathcal{B}(\mathcal{H})$.

(b) The algebra $\mathcal{A}_w(\mathcal{S}) := (\overline{\mathcal{A}_0(\mathcal{S})})_w$ coincides with the σ–strong closure $(\overline{\mathcal{A}_0(\mathcal{S})})_{\sigma s}$.

(c) $\mathcal{A}_w(\mathcal{S}) = \mathcal{A}_\mathcal{S}$.

Proof: The set $\mathcal{A}_\mathcal{S}$ is obviously a $*$–subalgebra in $\mathcal{S}''$. If $B \in (\overline{\mathcal{A}_\mathcal{S}})_w$; then for any $\varepsilon > 0$ and $\phi, \psi \in \mathcal{H}$ there is $C \in \mathcal{A}_\mathcal{S}$ such that $B - C$ is contained in the weak neighborhood $W_\varepsilon(\{\phi\}, \{\psi, E_\mathcal{S}, \psi\})$ of zero (*cf.* Section 3.1). The relation $CE_\mathcal{S} = C$ then implies

$$|(\phi, B(I-E_\mathcal{S})\psi)| \le |(\phi, (B-C)\psi)| + |(\phi, (B-C)E_\mathcal{S}\psi)| < 2\varepsilon\,,$$

so $(\phi, B(I-E_\mathcal{S})\psi) = 0$ for all $\phi, \psi \in \mathcal{H}$, *i.e.*, $B = BE_\mathcal{S}$. In the same way we prove $B = E_\mathcal{S} B$, which together gives assertion (a).

The most difficult part of the proof is the inclusion $\mathcal{A}_\mathcal{S} \subset (\overline{\mathcal{A}_0(\mathcal{S})})_{\sigma s}$. In view of Problem 7, it is sufficient to show that any σ–strong neighborhood $U_\varepsilon(C; \Phi)$ of an operator $C \in \mathcal{A}_\mathcal{S}$ contains some operator $B \in \mathcal{A}_0(\mathcal{S})$. We use the notation of Remark 1 and define $\tilde{E}$ as the projection onto the closure of the subspace $M_\Phi := \{ \tilde{B}\Phi : B \in \mathcal{A}_0(\mathcal{S}) \} \subset \tilde{\mathcal{H}}$. To prove the stated inclusion we have to check that $\tilde{E}\tilde{C}\Phi = \tilde{C}\Phi$, *i.e.*, $\tilde{C}\Phi \in \overline{M_\Phi}$, since then for any $\varepsilon > 0$ there is an operator $B(\varepsilon) \in \mathcal{A}_0(\mathcal{S})$ such that $p_\Phi(B(\varepsilon)-C) = \|\tilde{B}(\varepsilon)\Phi - \tilde{C}\Phi\|_{\tilde{\mathcal{H}}} < \varepsilon$.

The set M_Φ can be written as $\tilde{\mathcal{A}}_0(\mathcal{S})\Phi$, where $\tilde{\mathcal{A}}_0(\mathcal{S}) := \{ \tilde{B} : B \in \mathcal{A}_0(\mathcal{S}) \}$. Thus $\tilde{\mathcal{A}}_0(\mathcal{S}) M_\Phi \subset M_\Phi$ and since the operators in question are bounded, we also have $\tilde{\mathcal{A}}_0(\mathcal{S}) \overline{M_\Phi} \subset \overline{M_\Phi}$ by Problem 1.33c, *i.e.*, $\tilde{E}\tilde{B}\tilde{E} = \tilde{E}\tilde{B}$ for all $\tilde{B} \in \tilde{\mathcal{A}}_0(\mathcal{S})$. We can easily check that set $\tilde{\mathcal{A}}_0(\mathcal{S})$ is symmetric; hence the same identity also holds for $\tilde{B}^*$, which implies $\tilde{B}\tilde{E} = \tilde{E}\tilde{B}$. In addition, $\tilde{E}\tilde{B}\Phi = \tilde{B}\Phi$ follows from the definition of the projection $\tilde{E}$, so $\tilde{B}\tilde{E}\Phi = \tilde{E}\tilde{B}\Phi = \tilde{B}\Phi$ for any $\tilde{B} \in \tilde{\mathcal{A}}_0(\mathcal{S})$. In other words, all operators of set $\tilde{\mathcal{A}}_0(\mathcal{S})$ are reduced by the projection $\tilde{E}$.

Next we introduce the projections $\tilde{E}_k : \tilde{E}_k\Phi = \{0, \dots, 0, \phi_k, 0, \dots\}$; they satisfy the condition $\sum_{k=1}^\infty \tilde{E}_k = \tilde{I}$, where the convergence of this series and similar ones is understood with respect to the topology τ_s. Given an operator $\tilde{A} \in \tilde{\mathcal{B}}(\mathcal{H})$ we define $\tilde{A}_{ij} := \tilde{E}_i \tilde{A} \tilde{E}_j$; using the natural isomorphisms $V_j : \tilde{E}_j \tilde{\mathcal{H}} \to \mathcal{H}$ we can associate the operators $A_{ij} := V_i \tilde{A}_{ij} V_j^{-1} \in \mathcal{B}(\mathcal{H})$ with these "matrix elements". The relation $\tilde{E}\tilde{B}\Phi = \tilde{B}\Phi$ implies $\tilde{E}_k \tilde{B}(I-\tilde{E})\Phi = 0$ for all $k = 1, 2, \dots$, and therefore

$$0 = \sum_{i,j=1}^{\infty} \tilde{E}_k \tilde{B} \tilde{E}_i (I-\tilde{E}) \tilde{E}_j \Phi = \sum_{i,j=1}^{\infty} \delta_{ki} \tilde{B}_{kk} (\delta_{ij} \tilde{E}_j - \tilde{E}_{ij}) \Phi$$

because $\tilde{B}$ is by definition reduced by all the projections $\tilde{E}_k$. Summing over i and using operator V_k, we put the last relation in the form $B(\phi_k - \sum_{j=1}^{\infty} E_{kj}\phi_j) = 0$. Since it is true for any $B \in \mathcal{A}_0(\mathcal{S})$, we get

$$\phi_k - \sum_j E_{kj}\phi_j \in \bigcap_{B \in \mathcal{A}_0(\mathcal{S})} \operatorname{Ker} B \subset \bigcap_{B \in \mathcal{S}} \operatorname{Ker} B = \operatorname{Ker} E_{\mathcal{S}} .$$

Operator C belongs by assumption to $\mathcal{A}_{\mathcal{S}}$, so $C = CE_{\mathcal{S}}$ and the above relation implies $C(\phi_k - \sum_{j=1}^{\infty} E_{kj}\phi_j) = 0$. Moreover, the operators $\tilde{B} \in \tilde{\mathcal{A}}_0(\mathcal{S})$ commute with $\tilde{E}$, and at the same time, with all $\tilde{E}_k$. Hence

$$\tilde{B}_{ii} \tilde{E}_{ij} = \tilde{E}_i \tilde{B} \tilde{E} \tilde{E}_j = \tilde{E}_i \tilde{E} \tilde{B} \tilde{E}_j = \tilde{E}_{ij} \tilde{B}_{jj} ,$$

and using unitary operator V_i again we arrive at $BE_{ij} = E_{ij}B$ for all $i, j = 1, 2, \dots$. It follows that $E_{ij} \in \mathcal{S}'$, and since operator C belongs to $\mathcal{S}''$ we have $C\phi_k = \sum_{j=1}^{\infty} E_{kj} C\phi_j$. This is equivalent to $\tilde{E}_k \tilde{C}\Phi = \tilde{E}_k \tilde{E} \tilde{C}\phi$, and summing over k we arrive at the sought relation $\tilde{E}\tilde{C}\Phi = \tilde{C}\Phi$.

The rest of the proof is easy. By Problem 8c, $B = E_{\mathcal{S}} B E_{\mathcal{S}}$ holds for any $B \in \mathcal{S}$, and therefore also for all $B \in \mathcal{A}_0(\mathcal{S})$. At the same time, $\mathcal{A}_0(\mathcal{S}) \subset \mathcal{S}''$ so $\mathcal{A}_0(\mathcal{S}) \subset \mathcal{A}_{\mathcal{S}}$. Due to (a), the algebra $\mathcal{A}_{\mathcal{S}}$ is weakly closed; then the above result in combination with Remark 6.3.1 gives $(\overline{\mathcal{A}_0(\mathcal{S})})_w \subset \mathcal{A}_{\mathcal{S}} \subset (\overline{\mathcal{A}_0(\mathcal{S})})_{\sigma s} \subset (\overline{\mathcal{A}_0(\mathcal{S})})_w$. ∎

The answer to our question can be derived easily from the proved theorem. If a weakly closed $*$–algebra $\mathcal{A}$ contains the unit operator, we have $\mathcal{A}_{\mathcal{A}} = \mathcal{A}''$, so $(\overline{\mathcal{A}_0(\mathcal{A})})_w = (\overline{\mathcal{A}})_w = \mathcal{A}$. On the other hand, $\mathcal{A} = \mathcal{A}''$ implies $I \in \mathcal{A}$ because the unit operator belongs to the commutant of any subset in $\mathcal{B}(\mathcal{H})$. Hence we have

6.3.5 Corollary: A weakly closed $*$–algebra $\mathcal{A} \subset \mathcal{B}(\mathcal{H})$ is a W^*–algebra *iff* it contains the unit operator.

The second question posed above also has a simple answer: in view if the relations between the topologies an operator C^*–algebra is a W^*–algebra if it is weakly (which is the same as σ–strongly) closed and contains the unit operator. A related problem, namely under which conditions an abstract C^*–algebra can be represented isometrically by a W^*–algebra, is much more complicated (see the notes). The difference between the two classes is illustrated by the fact that W^*–algebras contain "sufficiently many" projections; recall Example 6.3.3b which shows that an operator C^*–algebra may contain trivial projections only. We denote by $\mathcal{A}^E$ the set of all projections in an algebra $\mathcal{A} \subset \mathcal{B}(\mathcal{H})$. It contains clearly the unit operator, so $\mathcal{A}_w(\mathcal{A}^E)$ is a W^*–algebra.

6.3.6 Proposition: $\mathcal{A}_w(\mathcal{A}^E) = \mathcal{A}$.

Proof: By Problem 8a the left side is equal to $(\mathcal{A}^E)''$. The inclusion $\mathcal{A}^E \subset \mathcal{A}$ implies $\mathcal{A}^E \subset (\mathcal{A}^E)'' \subset \mathcal{A}'' = \mathcal{A}$, so an arbitrary projection belongs to $\mathcal{A}$ *iff* it is contained in $(\mathcal{A}^E)''$. Next we use Lemma 6.5.3 which will be proven below; it claims that a weakly closed algebra with the unit element contains a Hermitean operator A *iff* it contains all projections of its spectral decomposition $\{E_t\}$. Hence $\{E_t\} \subset \mathcal{A}^E \subset (\mathcal{A}^E)''$, and using the lemma once again we get $A \in (\mathcal{A}^E)''$. Since any $B \in \mathcal{A}$ is a linear combination of two Hermitean operators, the assertion is proved. ∎

In the rest of this section we are going to describe how W^*–algebras can be classified. To this end, we have to introduce several new notions. An algebra $\mathcal{A}$ is called a *factor* if its *center* $\mathcal{Z}_\mathcal{A} := \mathcal{A} \cap \mathcal{A}'$ consists of scalar operators only; simple examples are $\mathcal{B}(\mathcal{H})$ itself and the algebra $\mathbb{C}(\mathcal{H})$ of scalar operators. A W^*–algebra satisfies the identity $\mathcal{A} = \mathcal{A}''$ by definition, so if $\mathcal{A}$ is a factor the same is true for $\mathcal{A}'$. Notice that a factor may be reducible as an operator set (*cf.* Section 6.7 below); it is important that it is not reduced by a projection belonging to $\mathcal{A}$. The opposite extremum is represented in a sense by a commutative algebra, which is equal to its center and reduced by any of its projections.

Let E be a projection in $\mathcal{B}(\mathcal{H})$. For any $B \in \mathcal{B}(\mathcal{H})$ the operator EBE is fully determined by its part in the subspace $E\mathcal{H}$, which we denote as B_E; this allows us to associate the operator set $\mathcal{S}_E := \{\, B_E : B \in \mathcal{S}\,\}$ with any $\mathcal{S} \subset \mathcal{B}(\mathcal{H})$. In particular, we have $(\mathcal{A}_E)' = (\mathcal{A}')_E$ provided $\mathcal{A}$ is a W^*–algebra and $E \in \mathcal{A}$ (*cf.* Problem 10). It follows that $\mathcal{A}_E$ is a W^*–algebra; it is called the *reduced W^*–algebra* (corresponding to projection E). A projection $E \in \mathcal{A}$ is said to be *Abelian* (with respect to algebra $\mathcal{A}$) if the reduced algebra $\mathcal{A}_E$ is Abelian; in particular, any minimal projection in $\mathcal{A}$ is Abelian (Problem 11). Projections E, F from a W^*–algebra $\mathcal{A} \subset \mathcal{B}(\mathcal{H})$ are called *equivalent* if $\mathcal{A}$ contains an operator $U \in \mathcal{A}$ such that $U^*U = E$ and $UU^* = F$, *i.e.*, a partial isometry with the initial space $E\mathcal{H}$ and the final space $F\mathcal{H}$. Projections E, F are equivalent with respect to $\mathcal{B}(\mathcal{H})$ *iff* they have the same dimension; in general the identity of the dimensions is a necessary but not sufficient condition for equivalence of the projections.

Simple properties of W^*–algebra $*$–morphisms are listed in Problem 12. An important class of maps between W^*–algebras $\mathcal{A} \subset \mathcal{B}(\mathcal{H})$ and $\mathcal{B} \subset \mathcal{B}(\mathcal{G})$ consists of spatial isomorphisms: a $*$–isomorphism $\varphi : \mathcal{A} \to \mathcal{B}$ is called *spatial* if there is a unitary operator $U \in \mathcal{B}(\mathcal{H}, \mathcal{G})$ such that $\varphi(B) = UBU^{-1}$ holds for all $B \in \mathcal{A}$. In such a case φ is at the same time a spatial isomorphism of algebras $\mathcal{B}(\mathcal{H})$ and $\mathcal{B}(\mathcal{G})$, and it represents a bijective correspondence between the centers of algebras $\mathcal{A}, \mathcal{B}$ as well as between their commutants, $\varphi(\mathcal{A}') = \mathcal{B}'$; it is also obvious that $\dim \mathcal{H} = \dim \mathcal{G}$. This last relation means, in particular, that not every isomorphism of W^*–algebras is spatial; a simple example is provided by the natural isomorphism

$\varphi : \mathbb{C}(\mathcal{H}) \to \mathbb{C}$, $\varphi(\alpha I) = \alpha$, for $\dim \mathcal{H} > 1$.

Next, consider a family of W^*–algebras $\mathcal{A}_\alpha \in \mathcal{B}(\mathcal{H}_\alpha)$, where α runs over an index set J. Their direct product introduced in Section 6.1 can be identified with an $*$–operator algebra on $\mathcal{H} := \bigoplus_{\alpha\in J} \mathcal{H}_\alpha$ if we associate the operator $B : B[\phi_\alpha] = [B_\alpha \phi_\alpha]$ with any $[B_\alpha]$. This algebra has the following properties (Problem 13).

6.3.7 Proposition: The norm of B equals $\|B\| = \|[B_\alpha]\| = \sup_{\alpha\in J} \|B_\alpha\|_\alpha$. If we denote $\mathcal{A} := \bigoplus_{\alpha\in J} \mathcal{A}_\alpha$ and $\mathcal{B} := \bigoplus_{\alpha\in J} \mathcal{A}'_\alpha$, then $\mathcal{A}' = \mathcal{B}$ and $\mathcal{B}' = \mathcal{A}$. The projections $E_\alpha := [\delta_{\alpha\beta} I_\beta]$, where I_β is the unit operator on $\mathcal{H}_\beta$, are contained in $\mathcal{A}$ for all $\beta \in J$; an operator $C \in \mathcal{B}(\mathcal{H})$ is contained in $\bigoplus_{\alpha\in J} \mathcal{B}(\mathcal{H}_\alpha)$ *iff* it commutes with all E_α.

This implies, in particular, that $\mathcal{A} := \bigoplus_{\alpha\in J} \mathcal{A}_\alpha$ is a W^*–algebra; we call it the *direct sum* of the W^*–algebras $\mathcal{A}_\alpha$. By Problem 12, the center of $\mathcal{A}$ is $\mathcal{Z}_\mathcal{A} = \bigoplus_{\alpha\in J} \mathcal{Z}_{\mathcal{A}_\alpha}$. On the other hand, if the center of a W^*–algebra $\mathcal{A}$ contains a complete system of projections, we can express $\mathcal{A}$ as the direct product of the corresponding reduced algebras (Problem 14).

Finally, consider W^*–algebras $\mathcal{A}_j \subset \mathcal{B}(\mathcal{H}_j)$, $j = 1, 2$. The W^*–algebra generated by the set $\mathcal{A}_1 \times \mathcal{A}_2 := \{ B_1 \otimes B_2 : B_j \in \mathcal{A}_j \}$ is called the *tensor product* of the two algebras and is denoted as $\mathcal{A}_1 \otimes \mathcal{A}_2$. The set $\mathcal{A}_1 \times \mathcal{A}_2$ contains the unit operator, so $\mathcal{A}_1 \otimes \mathcal{A}_2 = (\mathcal{A}_1 \times \mathcal{A}_2)''$ follows from Theorem 6.3.4. Notice also that since $\mathcal{A}_1$, $\mathcal{A}_2$ are at the same time C^*–algebras, we have to compare the present definition with that in Section 6.1. There, the realization of the tensor product in $\mathcal{B}(\mathcal{H}_1 \otimes \mathcal{H}_2)$ is also constructed as a closure of the algebra $\mathcal{A} := \mathcal{A}_0(\mathcal{A}_1 \times \mathcal{A}_2)$. However, in general only the inclusion $(\overline{\mathcal{A}})_w \supset (\overline{\mathcal{A}})_u$ is valid; hence the two tensor products are mutually different; some authors use the terms *tensor C^*–product* and *tensor W^*–product* to distinguish them.

After these preliminaries, let us describe how W^*–algebras can be classified. Since the corresponding proofs are not needed in the following, and are for the most part complicated, we limit ourselves to formulating the results; the appropriate references are given in the notes. In the rest of the section the symbol $\mathcal{A}$ always means a W^*–algebra in some $\mathcal{B}(\mathcal{H})$.

We call $\mathcal{A}$ a *type I* (or *discrete*) algebra if it is $*$–isomorphic to a W^*–algebra with an Abelian commutant. For instance, $\mathcal{A} := \mathbb{C}(\mathcal{H})$ is type I because it is $*$–isomorphic to the algebra $\mathbb{C}$, though $\mathcal{A}' = \mathcal{B}(\mathcal{H})$ is noncommutative unless $\dim \mathcal{H} = 1$. In the particular case when $\mathcal{A}$ is a discrete factor, a more detailed classification is possible.

6.3.8 Proposition: Let $\mathcal{A}$ be a factor; then the following conditions are equivalent:

(a) $\mathcal{A}$ is discrete.

(b) $\mathcal{A}$ contains minimal projections (*cf.* Problem 11).

(c) There are Hilbert spaces $\mathcal{H}_1, \mathcal{H}_2$ such that $\mathcal{A}$ can be identified with the tensor product $\mathcal{B}(\mathcal{H}_1) \otimes \mathbb{C}(\mathcal{H}_2)$.

(d) There exist a Hilbert space $\mathcal{G}$ and an index set J such that $\mathcal{A}$ is spatially isomorphic to the algebra $\tilde{\mathcal{A}} := \{\, \tilde{B} : \tilde{B}[\phi_\alpha] = [B\phi_\alpha]\,,\, B \in \mathcal{B}(\mathcal{G}) \,\}$ on the Hilbert space $\bigoplus_{\alpha\in J} \mathcal{G}$.

Hence any discrete factor is $*$–isomorphic to algebra $\mathcal{B}(\mathcal{G})$ for some Hilbert space $\mathcal{G}$; we classify these factors according to the dimension of $\mathcal{G}$ to algebras of *type* I_n, $n = 1, 2, \dots$, and *type* I_∞.

This classification extends to a wider class of discrete algebras. A W^*–algebra is called *homogeneous* if it contains a family $\{\, E_\alpha : \alpha \in J \,\}$ of mutually orthogonal and equivalent Abelian projections such that $\sum_{\alpha\in J} E_\alpha = I$; such an algebra belongs respectively to type I_n or I_∞ according to the cardinality of index set J. Correctness of the definition follows from the fact that any two such families of projections have the same cardinality. Any homogeneous algebra is discrete. On the other hand, any discrete factor $\mathcal{A}$ is homogeneous (Problem 15), so its type is defined in two different ways. It is easy to see, however, that the two definitions are equivalent: $\mathcal{A}$ is $*$–isomorphic to algebra $\mathcal{B}(\mathcal{G})$ in which the corresponding family consists of the one–dimensional projections referring to vectors of some orthonormal basis; thus its cardinality is $\dim \mathcal{G}$.

Commutative algebras are particularly simple. In such a case the unit operator is an Abelian projection with respect to $\mathcal{A}$; hence any Abelian W^*–algebra is discrete, homogeneous and of type I_1. On the other hand, for general discrete algebras we have the structural result according to which any such algebra is spatially isomorphic to a direct sum of homogeneous algebras belonging to mutually different types:

6.3.9 Proposition: In an arbitrary discrete W^*–algebra $\mathcal{A}$ we can find a system $\{\, E_j : j \in J \,\}$ of mutually orthogonal projections such that $\mathcal{A}$ can be identified with the direct sum $\bigoplus_{j\in J} \mathcal{A}_{E_j}$. Moreover, every reduced algebra $\mathcal{A}_{E_j}$ is homogeneous and of type I_{n_j}, $n_j \le \infty$, and $n_j \ne n_k$ for any pair $j, k \in J$.

Continuous algebras are the complement to the class of discrete W^*–algebras. A W^*–algebra is said to be *continuous* if its center contains no nonzero projection E such that the reduced algebra $\mathcal{A}_E$ is discrete. It is clear from the definition that any factor is either discrete or continuous. An important feature of continuous algebras is that they contain no minimal projections.

6.3.10 Proposition: An algebra $\mathcal{A}$ is continuous *iff* any projection $E \in \mathcal{A}$ can be expressed as a sum of two orthogonal equivalent projections.

It also can be proved that any algebra reduced from a discrete (continuous) algebra is respectively discrete or continuous; hence a W^*–algebra cannot be discrete and continuous at the same time. This does not, however, mean that a W^*–algebra should be either discrete or continuous; this is true only for factors. In the general case we have the following structural theorem.

6.3.11 Proposition: Any W^*–algebra $\mathcal{A}$ contains orthogonal projections $E, F \in \mathcal{Z}_\mathcal{A}$, $E+F = I$, such that $\mathcal{A}_E$ is discrete and $\mathcal{A}_F$ continuous. In other words, $\mathcal{A}$ is spatially isomorphic to a direct sum of a discrete and a continuous algebra.

To get a more detailed classification we introduce traces. Let $\mathcal{A}_+$ be the cone of positive operators in $\mathcal{A}$; then a **trace** is a map $\tau : \mathcal{A}_\tau \to \mathbb{R}^+$, $\mathcal{A}_\tau \subset \mathcal{A}_+$, with the following properties:

(i) $\tau(\alpha B+C) = \alpha\tau(B)+\tau(C)$ for all $B, C \in \mathcal{A}_\tau$, $\alpha \in \mathbb{C}$.

(ii) $\tau(UBU^{-1}) = \tau(B)$ holds for all $B \in \mathcal{A}_\tau$ and all unitary operators $U \in \mathcal{A}$.

In addition, we set $\tau(B) = \infty$ for $B \in \mathcal{A} \setminus \mathcal{A}_\tau$. Condition (ii) is equivalent to the requirement $\tau(B^*B) = \tau(BB^*)$ for all $B \in \mathcal{A}$. A trace τ is called *finite* if $\mathcal{A}_\tau = \mathcal{A}_+$, and *semifinite* if for any nonzero $B \in \mathcal{A}_+$ there is a nonzero positive operator $C \in \mathcal{A}_\tau$ such that $C \le B$. A trace τ is said to be *faithful* if $\tau(B) = 0$ implies $B = 0$, and *normal* if for any sequence $\{A_j\} \subset \mathcal{A}_+$, such that $A :=$ w-$\lim_{n\to\infty} \sum_{j=1}^n A_j \in \mathcal{A}_+$ exists, the relation $\tau(A) = \sum_{j=1}^\infty \tau(A_j)$ is valid; an alternative definition will be given in the next section. If $\mathcal{H}$ is separable, for instance, then the map $\tau : \tau(B) = \mathrm{Tr}\, B$ is a faithful normal trace on $\mathcal{B}(\mathcal{H})$ (Problem 16). Its domain is $\mathcal{B}(\mathcal{H})_\tau = \mathcal{J}_1$, so τ is finite if $\dim \mathcal{H} < \infty$ and semifinite otherwise.

An algebra $\mathcal{A}$ is said to be *finite* (*semifinite*) if for any nonzero $B \in \mathcal{A}_+$ there is a finite (respectively, semifinite) normal trace τ_B on $\mathcal{A}$ such that $\tau(B) \ne 0$. On the other hand, if there is no finite (semifinite) normal trace on $\mathcal{A}$, the algebra is called *properly infinite* (respectively, *purely infinite*). It follows from Proposition 6.3.8 and Problem 16 that any discrete factor is semifinite. More generally, discrete algebras are semifinite; in particular, any algebra of type I_n, where n is a positive integer, is finite. Continuous W^*–algebras are classified in the following way. If $\mathcal{A}$ is semifinite it is of *type* II. If, in addition, it is finite we speak about about *type* II_1, while a continuous $\mathcal{A}$ which is semifinite and properly infinite belongs to *type* II_∞. A purely infinite algebra is always infinite; we call it a *type* III algebra. One can prove in a constructive way that all the listed types exist.

Each continuous factor belongs to just one of the types II_1, II_∞, III, and since any factor is either discrete or continuous, we have the following result.

6.3.12 Theorem: Any factor belongs to just one of the types I_n , $n = 1, 2, \dots$, I_∞, II_1, II_∞, or III.

A general W^*–algebra can be divided, from the viewpoint of finiteness, in a way similar to Proposition 6.3.11.

6.3.13 Proposition: In the center of any W^*–algebra $\mathcal{A}$ there are pairs of mutually orthogonal projections E_1, F_1 and E_2, F_2 such that $E_1 \le E_2$ and $E_j + F_j = I$, and the reduced algebra $\mathcal{A}_{E_1}$ (respectively, $\mathcal{A}_{E_2}$, $\mathcal{A}_{F_1}$, $\mathcal{A}_{F_2}$) is finite (respectively, semifinite, properly infinite, purely infinite).

Combining these results we arrive at a complete classification of W^*–algebras.

6.3.14 Theorem: Any W^*–algebra is spatially isomorphic to a direct sum of algebras of types I_n , $n = 1, 2, \dots$, I_∞, II_1, II_∞, and III; each type is contained at most once in the direct sum.

6.4 Normal states on W^*–algebras

Any state f on a W^*–algebra $\mathcal{A} \subset \mathcal{B}(\mathcal{H})$ is additive by definition; in particular, $f(E_1+\cdots+E_n) = f(E_1)+\cdots+f(E_n)$ holds for orthogonal projections $E_1, \dots, E_n$. On the other hand, a similar relation may not be valid for an infinite family of projections. Strengthening the additivity requirement we obtain an important subclass in the set of all states: a state f on $\mathcal{A}$ is called **normal** provided

$$f\left(\sum_{\alpha\in J} E_\alpha\right) = \sum_{\alpha\in J} f(E_\alpha) \tag{6.2}$$

holds for any family $\{ E_\alpha : \alpha \in J \}$ of mutually orthogonal projections in $\mathcal{A}$, and furthermore, a nonzero positive functional g on $\mathcal{A}$ is said to be *normal* if the state $\|f\|^{-1}f$ is normal.

Relation (6.2) needs a comment. If index set J is at most countable its meaning is clear: we arrange projections E_α into a sequence and define $\sum_{\alpha\in J} E_\alpha$ as the limit of partial sums and the right side as the sum of the corresponding series; the numbers $f(E_\alpha)$ are non–negative, so the order of summation is not important. In the case of a nonseparable $\mathcal{H}$ the set J may be uncountable; then $\sum_{\alpha\in J} E_\alpha$ is given by Proposition 3.2.13. On the other hand, motivated by the countable case, we *define* the right side of (6.2) as $\sup_{K\in\mathcal{S}} \sum_{\alpha\in K} f(E_\alpha)$, where $\mathcal{S} \subset 2^J$ is the family of all finite subsets in J, *i.e.*, as a limit of the net of finite partial sums (see also the notes).

Our main aim in this section is to derive conditions which specify the subset of normal states in the set of all states, and to find a general expression for a normal state. The result, the proof of which will proceed through a series of lemmas, can be formulated as follows.

6.4.1 Theorem: Let f be a state on a W^*–algebra $\mathcal{A} \subset \mathcal{B}(\mathcal{H})$; then the following conditions are equivalent:

(a) f is normal.

(b) f is σ–weakly continuous.

(c) There is a statistical operator $W \in \mathcal{J}_1$ such that $f(B) = \mathrm{Tr}\,(WB)$ holds for all $B \in \mathcal{A}$.

6.4.2 Remark: A statistical operator on a separable Hilbert space $\mathcal{H}$ was defined in Section 3.6 as a positive trace–class operator W such that $\mathrm{Tr}\, W = 1$. This definition naturally extends to the case of a nonseparable $\mathcal{H}$, where the operator W is required to be positive with $(\mathrm{Ker}\, W)^\perp = \mathrm{Ran}\, W$ separable and such that the nontrivial part of W, *i.e.*, the restriction $W \restriction (\mathrm{Ker}\, W)^\perp$ satisfies the stated conditions.

6.4.3 Lemma: A functional f which is the limit of a sequence of σ–strongly continuous linear functionals on a W^*–algebra $\mathcal{A}$ is σ–strongly continuous.

Proof: In view of Proposition 6.3.2, it is sufficient to check that f is σ–strongly continuous on the unit ball $\mathcal{A}_1 := \{ B \in \mathcal{A} : \|B\| \le 1 \}$ (see the notes); this follows from the estimate $|f(B)| \le |f_n(B)| + \|f-f_n\| \, \|B\|$ in combination with the assumption $\|f-f_n\| \to 0$ as $n \to \infty$. ∎

6.4.4 Lemma: Let f be a linear functional on a W^*–algebra $\mathcal{A} \subset \mathcal{B}(\mathcal{H})$; then the following assertions are equivalent:

(a) There is a sequence $\Phi = \{\phi_k\} \in \tilde{\mathcal{H}}$ (in the notation of Remark 6.3.1) such that $|f(B)| \le p_\Phi(B)$ holds for any $B \in \mathcal{A}$.

(b) There are sequences $\Phi, \Psi \in \tilde{\mathcal{H}}$ such that the relation $f(B) = \sum_{k=1}^{\infty} (\phi_k, B\psi_k)$ holds for all $B \in \mathcal{A}$.

(c) f is σ–weakly continuous.

(d) f is σ–strongly continuous.

Proof: It is sufficient to check the chain of implications (a) ⇒ (b) ⇒ (c) ⇒ (d) ⇒ (a). The intermediate two are simple. To prove the last one, assume that (a) is not valid; then for any $\Phi \in \tilde{\mathcal{H}}$ there is an operator $B_\Phi \in \mathcal{A}$ such that $|f(B_\Phi)| > \|\tilde{B}_\Phi \Phi\|_{\tilde{\mathcal{H}}}$. By Problem 7, the σ–strong continuity implies, in particular, that we can find $\delta > 0$ and a nonzero vector $\Psi \in \tilde{\mathcal{H}}$ such that $|f(B)| < 1$ holds for all $B \in \mathcal{A}$ fulfilling the condition $\|\tilde{B}\Psi\|_{\tilde{\mathcal{H}}} < \delta$. The last inequality is valid, *e.g.*, for $C := \|\tilde{B}_\Phi \Phi\|_{\tilde{\mathcal{H}}}^{-1} B_\Phi$ with $\Phi := \frac{2}{\delta}\Psi$. This is, however, impossible because then $f(C) = \|\tilde{B}_\Phi \Phi\|_{\tilde{\mathcal{H}}}^{-1} f(B_\Phi) < 1$ in contradiction to the assumption.

It remains for us to check that (b) follows from (a). We define the linear functional $g_0 : g_0(\tilde{B}\Phi) = f(B)$ on the subspace $\{ \tilde{B}\Phi : B \in \mathcal{A} \} \subset \tilde{\mathcal{H}}$; the definition is correct since $\tilde{B} \ne \tilde{C}$ implies $B \ne C$. By assumption, $|f(B)| \le \|\tilde{B}\Phi\|_{\tilde{\mathcal{H}}}$ holds for all $B \in \mathcal{A}$, so the functional g_0 is bounded; in view of Proposition 1.5.7a we can extend it to a bounded functional $g : \tilde{\mathcal{H}} \to \mathbb{C}$. Finally the Riesz lemma gives $g(\Phi') = (\Psi, \Phi')_{\tilde{\mathcal{H}}}$ for some $\Psi \in \tilde{\mathcal{H}}$ and all $\Phi' \in \tilde{\mathcal{H}}$; in particular, $f(B) = g(\tilde{B}\Phi) = (\Psi, \tilde{B}\Phi)_{\tilde{\mathcal{H}}}$ for any $B \in \mathcal{A}$, which is just assertion (b) up to the interchange of Φ and Ψ. ∎

6.4.5 Lemma: Let f be a positive functional on a W^*–algebra $\mathcal{A} \subset \mathcal{B}(\mathcal{H})$; then

(a) If $f(B) \le (\psi, B\psi)$ holds for some vector $\psi \in \mathcal{H}$ and all $B \in \mathcal{A}_+$, then there is an operator $C \in \mathcal{A}'$ such that $f(B) = (C\psi, BC\psi)$ for all $B \in \mathcal{A}$.

(b) If f expresses in the form $f(B) = (\phi, B\psi)$ for some $\phi, \psi \in \mathcal{H}$, then there is a vector $\chi \in \mathcal{H}$ such that $f(B) = (\chi, B\chi)$ for all $B \in \mathcal{A}$.

Proof: We define the form g on the subspace $\mathcal{A}\psi$ by $g(B_1\psi, B_2\psi) := f(B_1^* B_2)$ for all $B_1, B_2 \in \mathcal{A}$. This makes sense: it follows from the assumption and Proposition 6.2.2c that $|f(C^*B)|^2 \le f(B^*B)(\psi, C^*C\psi)$, so $f(C^*B) = 0$ for any $B, C \in \mathcal{A}$ if $C\psi = 0$; the same is true for $f(B^*C)$. The form g is obviously symmetric and positive, and the Schwarz–inequality argument shows that it is bounded,

$|g(B_1\psi, B_2\psi)| \le \|B_1\psi\|^2\|B_2\psi\|^2$. Hence it can be extended to a bounded form on $\overline{\mathcal{A}\psi} \subset \mathcal{H}$ and by Proposition 3.1.1 there is a positive operator $A_g \in \mathcal{B}(\overline{\mathcal{A}\psi})$ such that $f(B_1^*B_2) = (B_1\psi, A_g B_2\psi)$. Then we also have $(B_1\psi, A_g B_3 B_2\psi) = f(B_1^* B_3 B_2) = f((B_3^*B_1)^*B_2) = (B_3^*B_1\psi, A_g B_2\psi)$, *i.e.*,

$$(A_g B_3 - B_3 A_g) \upharpoonright \overline{\mathcal{A}\psi} = 0 \quad \text{for all } B_3 \in \mathcal{A}.$$

Thus $A := A_g \oplus 0$, where 0 is the zero operator on $(\overline{\mathcal{A}\psi})^\perp$, is positive and belongs to $\mathcal{A}'$. By Proposition 3.2.6 the same is true for $C := A^{1/2}$ and $f(B) = (\psi, AB\psi) = (\psi, C^2B\psi) = (C\psi, BC\psi)$ for any $B \in \mathcal{A}$.

To prove part (b), we use the parallelogram identity which yields $4f(A) = 2(\phi, A\psi) + 2(\psi, A\phi) = (\phi+\psi, A(\phi+\psi)) - (\phi-\psi, A(\phi-\psi))$ for any $A \in \mathcal{A}$. Hence $4f(B) \le (\phi+\psi, B(\phi+\psi))$ holds if B is positive, and by (a) there is an operator $C \in \mathcal{A}'$ such that $f(B) = (\chi, B\chi)$, where $\chi := \frac{1}{2}C(\phi+\psi)$. ∎

6.4.6 Lemma: Let f, g be normal positive functionals on a W^*–algebra $\mathcal{A} \subset \mathcal{B}(\mathcal{H})$. If $f(E) < g(E)$ holds for some projection $E \in \mathcal{A}$, then $\mathcal{A}$ contains a nonzero projection $F \le E$ such that $f(B) \le g(B)$ for all operators $B \in \mathcal{A}$ which satisfy the condition $0 \le B \le F$.

Proof: Consider the family of projections $M := \{\, P \le E : f(P) \ge g(P) \,\}$, which is partially ordered by operator inequalities. Let N be a completely ordered subset in M. If it is countable we can write it as a nondecreasing sequence $\{P_n\}$; by Theorem 3.2.12 it converges strongly to some projection P. The functionals f, g are supposed to be normal, so $f(P) = \lim_{n\to\infty} f(P_n) \ge \lim_{n\to\infty} g(P_n) = g(P)$, *i.e.*, the set N has an upper bound. The same is true for an uncountable N when the sequence has to be replaced by a net of projections. Zorn's lemma then implies existence of at least one maximal element E_0.

The projection $F := E - E_0$ is nonzero since otherwise $f(E) = f(E_0) \ge g(E_0) = g(E)$. Assume that $\mathcal{A}$ contains an operator B such that $0 \le B \le F$ and $f(B) > g(B)$. We denote its spectral measure by $E_B(\cdot)$; due to Proposition 6.3.2 and Lemma 6.5.3 which we shall prove below, the algebra $\mathcal{A}$ then also contains the projections $E_B(J)$ for any interval $J \subset \mathbb{R}$. We shall check that for some of these projections the inequality $f(E_B(J)) \le g(E_B(J))$ cannot be valid. In the opposite case, we could choose a suitable sequence $\{B_N\}$, say

$$B_N := \sum_{k=1}^{N} \frac{k}{N} E_B(J_k^N), \quad J_k^N := \left(\frac{k-1}{N}, \frac{k}{N}\right],$$

which approximates B in the operator norm in view of Theorem 5.2.2; we use the fact that the spectrum of B is contained in the interval $[0,1]$. The functionals f, g are bounded by Proposition 6.2.2e, so $f(B) = \lim_{N\to\infty} f(B_N) \le \lim_{N\to\infty} g(B_N) = g(B)$ in contradiction with the assumption. This means that there is at least one *nonzero* projection $E_1 := E_B(J_1)$ corresponding to some $J_1 \subset [0,1]$ such that $f(E_1) > g(E_1)$.

The inequalities $0 \le B \le F$ imply $0 \le \|B^{1/2}\psi\|^2 \le \|F\psi\|^2$ for any $\psi \in \mathcal{H}$, *i.e.*, $\operatorname{Ker} F \subset \operatorname{Ker} B$ and $E_B(\mathbb{R}\setminus\{0\}) \le F$. Then $E_1 := E_B(J_1) \le E_B(\mathbb{R}\setminus\{0\}) \le F$

so $E_1 \le F \le E$, and furthermore, $E_0E_1 = (E-F)E_1 = 0$. This means that E_0+E_1 is a projection which should belong to M; however, this is impossible since E_0 is a maximal element in M. ∎

Proof of Theorem 6.4.1: The most complicated part is to check that (a) implies (b). Given a projection $E \in \mathcal{A}$ we define $f_E(B) := f(BE)$ and denote $M := \{\, E \in \mathcal{A} : f_E \text{ is } \sigma\text{–strongly continuous} \,\}$. Let $\{E_j\} \subset M$ be a sequence of mutually orthogonal projections; we denote $P_k := \sum_{j=1}^{k} E_j$ and $P := \text{s-lim}_{k\to\infty} P_k$. The inequality

$$|f(BE)|^2 \le f(B^*B)f(E) \le f(E)\|B\|^2$$

holds for any projection $E \in \mathcal{A}$ and for all $B \in \mathcal{A}$; choosing $E := P-P_k$ we get $\|f_P - f_{P_k}\| \le f(P-P_k)$. The state f is normal by assumption, so we have $\lim_{k\to\infty} f(P-P_k) = 0$, and therefore f_P is σ–strongly continuous, *i.e.*, P belongs to M. This argument easily extends to the situation when $\{P_k\}$ is replaced by a nondecreasing net of projections $\{P_\alpha\}$: the projection $P := \sup_\alpha P_\alpha$ again belongs to M. It now follows from Zorn's lemma that the set M contains at least one maximal element E_0.

Suppose that the projection $E_1 := I - E_0$ is nonzero. We choose a vector $\phi \in E_1\mathcal{H}$, $\|\phi\| > 1$, and define the functional $g_\phi : g_\phi(B) = (\phi, B\phi)$ which is obviously positive and normal. We have $f(E_1) \le \|E_1\| < \|\varphi\|^2 = g_\phi(E_1)$; hence by Lemma 6.4.6, $\mathcal{A}$ contains a nonzero projection $E_2 \le E_1$ such that $f(B) \le (\phi, B\phi)$ holds for all $B \in \mathcal{A}$ with $0 \le B \le E_2$. The operator $B := \|C\|^{-2}E_2C^*CE_2$ for any nonzero $C \in \mathcal{A}$ obeys this condition, so $\|C\|^{-2}f(E_2C^*CE_2) \le \|C\|^{-2}\|CE_2\phi\|^2$, and furthermore,

$$|f(CE_2)|^2 \le f(I)f(E_2C^*CE_2) \le \|CE_2\phi\|^2 .$$

Thus the functional f_{E_2} is strongly continuous, and therefore also σ–strongly continuous. By assumption, $E_2E_0 = E_2(I-E_1) = 0$, *i.e.*, E_2+E_0 is a projection which should belong to M, but this contradicts the fact that the projection E_0 is maximal. In this way we get $E_0 = I$, which means that the functional f itself is σ–strongly continuous, and by Lemma 6.4.4 it is at the same time σ–weakly continuous.

Next we are going to show that (b) implies the existence of a vector $\Phi \in \tilde{\mathcal{H}}$ (in the notation of Remark 6.3.1) such that $\|\Phi\|^2_{\tilde{\mathcal{H}}} = \sum_{k=1}^\infty \|\phi_k\|^2 = 1$ and

$$f(B) = \sum_{k=1}^{\infty} (\phi_k, B\phi_k) \tag{6.3}$$

for an arbitrary $B \in \mathcal{A}$. First we define the functional $\tilde{f}$ on the W^*–algebra $\tilde{\mathcal{A}} := \{\, \tilde{B} : B \in \mathcal{A} \,\} \subset \mathcal{B}(\tilde{\mathcal{H}})$ by $\tilde{f}(\tilde{B}) := f(B)$. In view of Lemma 6.4.4, $\tilde{f}(\tilde{B}) = (\Psi', \tilde{B}\Psi)_{\tilde{\mathcal{H}}}$ for some $\Psi, \Psi' \in \tilde{\mathcal{H}}$, and since $\tilde{f}$ is positive, Lemma 6.4.5 implies existence of a vector $\Phi \in \tilde{\mathcal{H}}$ such that $\tilde{f}(\tilde{B}) = (\Phi, \tilde{B}\Phi)_{\tilde{\mathcal{H}}}$ for all $B \in \mathcal{A}$, *i.e.*, expression (6.3). Moreover, f is by assumption a state, so $f(I) = \|\Phi\|^2_{\tilde{\mathcal{H}}} = 1$.

Relation (6.3) in turn implies the condition (a). Let $\{\, E_\alpha : \alpha \in J \,\}$ be a family of mutually orthogonal projections in $\mathcal{A}$ and $E := \sum_{\alpha\in J} E_\alpha$. For any k there is an at most countable subset $J_k \subset J$ such that $E_\alpha \phi_k = 0$ for $\alpha \in J \setminus J_k$. The set $K := \bigcup_{k=1}^\infty J_k$ is also at most countable; hence the corresponding subfamily of projections can be ordered into a sequence $\{E_j\}$ and $\sum_{j=1}^\infty E_j = E$. Then $f(E) = \sum_{k=1}^\infty (\phi_k, E\phi_k) = \sum_{k=1}^\infty \sum_{j=1}^\infty \|E_j E\phi_k\|^2$, and moreover, $f(E) \le \|E\| = 1$, so the last series converges absolutely and we may rearrange it,

$$f(E) = \sum_{j=1}^\infty \sum_{k=1}^\infty (\phi_k, E_j\phi_k) = \sum_{j=1}^\infty f(E_j) = \sum_{\alpha\in J} f(E_\alpha),$$

i.e., the state f is normal. It remains for us to check that expression (6.3) is equivalent to condition (c). If the former is valid, we can define $W\psi := \sum_{k=1}^\infty (\phi_k, \psi)\phi_k$, where the series converges with respect to the norm in $\mathcal{H}$. The estimate

$$\left\| \sum_{k=m+1}^{n} (\phi_k, \psi)\phi_k \right\| \le \sum_{k=m+1}^{n} |(\phi_k, \psi)|\, \|\phi_k\| \le \|\psi\| \sum_{k=m+1}^{n} \|\phi_k\|^2$$

shows that operator W is defined on the whole $\mathcal{H}$; it is easy to check that W is positive and $\|W\| \le 1$. Let $\{\psi_\alpha\}$ be a total orthonormal set in $\mathcal{H}$. Each of the vectors ϕ_k has at most countably many nonzero Fourier coefficients; hence there is an at most countable subset in $\{\psi_\alpha\}$ such that its complement belongs to $\operatorname{Ker} W$. We arrange it into a sequence $\{\psi_j\}$; then $\operatorname{Tr} W = \sum_j (\psi_j, W\psi_j) = \sum_j \sum_k |(\phi_k, \psi_j)|^2 = \sum_k \|\phi_k\|^2 = 1$. This means that W is a statistical operator; in view of Theorem 3.6.8a the series expressing $\operatorname{Tr}(WB)$ converges absolutely and

$$\begin{aligned}\operatorname{Tr}(WB) &= \sum_j (\psi_j, WB\psi_j) = \sum_j \sum_k (\phi_k, B\psi_j)(\psi_j, \phi_k) \\ &= \sum_j \sum_k (B^*\phi_k, \psi_j)(\psi_j, \phi_k) = \sum_k (B^*\phi_k, \phi_k) = f(B),\end{aligned}$$

i.e., assertion (c). On the other hand, if W is a statistical operator with the required properties and $\sum_k w_k(\psi_k, \cdot)\psi_k$ is its spectral decomposition, we set $\phi_k := \sqrt{w_k}\,\psi_k$; in the case $\dim \operatorname{Ran} W < \infty$ we append zeros to the obtained finite sequence. It is straightforward to check that the functional f is expressed in the form (6.3) by means of the vector Φ constructed in this way. ∎

The statistical operator W, which corresponds to the state f by the proved theorem, is in general not unique, and it need not belong to the algebra $\mathcal{A}$. A simple illustration is provided by the algebra $\mathbb{C}(\mathcal{H})$ of scalar operators on which there is just one state $f: f(\alpha I) = \alpha$ which is normal and pure. Any statistical operator from $\mathcal{B}(\mathcal{H})$ can in this case be taken for W, and moreover no such operator belongs to $\mathbb{C}(\mathcal{H})$ if $\dim \mathcal{H} = \infty$ (see also Problem 17). We shall show that the nonuniqueness can be removed provided we restrict our attention to a particular class of W^*–algebras. Assume that

(ao1) $\mathcal{A}$ is a direct sum of type I factors.

(ao2) Any minimal projection in $\mathcal{A}$ is one–dimensional.

in view of Proposition 6.3.8 any algebra which obeys these requirements can be identified with

$$\mathcal{A} = \bigoplus_{\alpha\in J} \mathcal{B}(\mathcal{H}_\alpha)\,, \tag{6.4}$$

where $\{\mathcal{H}_\alpha\}$ is some system of Hilbert spaces, $\sum_{\alpha\in J}^{\oplus} \mathcal{H}_\alpha = \mathcal{H}$. Below we shall encounter algebras of this type as algebras of observables of quantum mechanical systems. Condition (ao2) is then a natural simplicity requirement; it is useless to associate an n–tuple of *identical* operators with a given observable.

6.4.7 Theorem: Suppose that a W^*–algebra $\mathcal{A} \subset \mathcal{B}(\mathcal{H})$ satisfies the conditions (ao1) and (ao2); then

(a) There is a bijective correspondence between the normal states f on $\mathcal{A}$ and the statistical operators $W \in \mathcal{J}_1(\mathcal{H})$ which are reduced by all the subspaces $\mathcal{H}_\alpha$, such that

$$f(B) = \mathrm{Tr}\,(WB) \tag{6.5}$$

holds for all $B \in \mathcal{A}$.

(b) A normal state f on $\mathcal{A}$ is pure *iff* the corresponding statistical operator W is a one–dimensional projection.

Proof: Due to Theorem 1 a normal state f can be expressed in the form (6.3) and $\mathcal{A}$ can be identified with algebra (6.4). Let E_α denote the projections onto the subspaces $\mathcal{H}_\alpha$. Each of the vectors ϕ_k has at most countably many nonzero components $\phi_{k\alpha} := E_\alpha\phi_k$ corresponding to a subset $J_k \subset J$, so the system $\{\,\phi_{k\alpha} : \alpha \in J_k\,,\ k = 1,2,\dots\,\}$ can be ordered into a sequence $\{\phi^{(j)}\}$. Any operator $B \in \mathcal{A}$ commutes with all E_α, so we get $f(B) = \sum_{k=1}^\infty \sum_{\alpha\in J_k} (\phi_{k\alpha}, B\phi_{k\alpha})$; it follows from $\sum_{k=1}^\infty \sum_{\alpha\in J_k} \|\phi_{k\alpha}\|^2 = \sum_{k=1}^\infty \|\phi_k\|^2 = 1$ that the series converges absolutely and we can rearrange it,

$$f(B) = \sum_{j=1}^\infty (\phi^{(j)}, B\phi^{(j)})\,.$$

Thus we have found for f a new expression of type (6.3), where now each of the vectors $\phi^{(j)}$ belongs to some subspace $\mathcal{H}_\alpha$. We associate a statistical operator W with the sequence $\{\phi^{(j)}\}$ in the same way as in the proof of Theorem 1; then

$$WE_\alpha\psi = \sum_{j=1}^\infty (\phi^{(j)}, E_\alpha\psi)\phi^{(j)} = \sum_{k=1}^\infty (\phi_{k\alpha}, \psi)\phi_{k\alpha} = \sum_{j=1}^\infty (\phi^{(j)}, \psi)E_\alpha\phi^{(j)} = E_\alpha W\psi$$

holds for any $\alpha \in J\,,\ \psi \in \mathcal{H}$, *i.e.*, W commutes with all the projections E_α and therefore it belongs to algebra $\mathcal{A}$.

Suppose now that statistical operators W_1, W_2 correspond to a given f, so that both of them satisfy relation (6.5). The operator $W := W_1 - W_2$ is Hermitean and of the trace class, and $\mathrm{Tr}\,(WB) = 0$ holds for all $B \in \mathcal{A}$. In particular, $\mathrm{Tr}\,W^2 = \|W\|_2 = 0$, hence $W = 0$. The map $f \mapsto W$ is thus injective. It is also surjective: for any statistical operator W the relation (6.5) defines a state on $\mathcal{A}$, and mimicking the argument from the previous proof we can show that it is normal.

To prove (b) we employ Theorem 6.2.7. The GNS representation π_W corresponding to state (6.5) was constructed in Example 6.2.4b; we have seen that it is reducible if $\dim(\operatorname{Ker} W)^{\perp} > 1$. In the opposite case W is a one–dimensional projection corresponding to a vector $\psi \in \mathcal{H}$. Operator W commutes with all E_α, so ψ must belong to the subspace $\mathcal{H}_\beta$ for some $\beta \in J$. Due to (6.4) the GNS representation π_ψ of Example 6.4.2a fulfils $\mathcal{H}_\psi = \mathcal{H}_\beta$ and $\pi_\psi(\mathcal{A}) = \mathcal{B}(\mathcal{H}_\beta)$, which means that it is irreducible. ∎

In conclusion, let us briefly mention the states which are not normal; for simplicity we shall consider only the algebra $\mathcal{A} = \mathcal{B}(\mathcal{H})$. We have shown that a normal state on $\mathcal{A}$ corresponds to a unique statistical operator W_f which in turn defines the GNS representation $\pi_f := \pi_{W_f}$ (*cf.* Example 6.2.4b); in this case any $B \in \mathcal{A}$ is represented by the direct sum of its N identical copies, so π_f is clearly faithful. By negation, f is not normal if the corresponding GNS representation is not faithful. It is easy to see that the kernel of π_f forms in that case a nontrivial ideal in $\mathcal{B}(\mathcal{H})$ which, in view of Proposition 6.1.2, is closed with respect to the operator norm. However, the following result is valid (see Problems 19–21).

6.4.8 Theorem (Calkin): If $\mathcal{H}$ is separable, then the only nontrivial ideal in $\mathcal{B}(\mathcal{H})$, which is closed with respect to the operator norm, is the ideal $\mathcal{K}(\mathcal{H})$ of all compact operators in $\mathcal{B}(\mathcal{H})$.

Hence $f(B) = 0$ must hold for any $B \in \mathcal{K}(\mathcal{H})$ if π_f is not faithful. In view of the role played by the quantities $f(A)$ in quantum theory, which will be discussed below, this suggests that non–normal states can have rather pathological properties.

6.5 Commutative symmetric operator sets

From now on we are going to consider more general operator sets including those which contain unbounded operators. With the needs of the following chapters in mind, however, we shall be interested primarily in the case when the unbounded operators are self–adjoint.

An operator set is called *symmetric* if, together with each $T \in \mathcal{S}$, it also contains its adjoint; this notion obviously makes sense only if all the operators contained in $\mathcal{S}$ are densely defined. The symmetry is preserved at some set operations (Problem 22). Examples of symmetric operator sets are easily found. The set $\mathcal{L}_{sa}(\mathcal{H})$ of all self–adjoint operators on a given $\mathcal{H}$ is symmetric. On the other hand, the set $\mathcal{L}_s(\mathcal{H})$ of all symmetric operators is *not* symmetric because the adjoint to a symmetric operator need not be symmetric (*cf.* Example 4.2.5). If a set $\mathcal{S}$ of bounded operators is symmetric, then its linear envelope $\mathcal{S}_{lin}$ is also symmetric; this is not true in the general case since the operators of $\mathcal{S}_{lin}$ need not be densely defined.

The notion of commutativity has been introduced in situations when either at least one of the two operators is bounded (Section 4.4) or both of them belong to a particular class, *e.g.*, they are self–adjoint. Motivated by this, we define $\mathcal{L}_{b,sa}(\mathcal{H}) := \mathcal{B}(\mathcal{H}) \cup \mathcal{L}_{sa}(\mathcal{H})$. A set $\mathcal{S} \subset \mathcal{L}_{b,sa}(\mathcal{H})$ is called *commutative* if arbitrary operators $T, S \in \mathcal{S}$ commute mutually. Any subset of a commutative set is again commutative; in particular, an intersection of commutative sets is commutative. Operator sets $\mathcal{S}_1$, $\mathcal{S}_2 \in \mathcal{L}_{b,sa}(\mathcal{H})$, in general noncommutative, are said to be *commuting* if any operator $T_1 \in \mathcal{S}_1$ commutes with all $T_2 \in \mathcal{S}_2$. The union of commuting commutative sets is a commutative set. Instead of studying a given $\mathcal{S} \subset \mathcal{L}_{b,sa}(\mathcal{H})$ it is sometimes useful to treat some other set constructed from it. For instance, we can define

$$\begin{aligned} \mathcal{S}_R &:= \{\mathcal{S} \cap \mathcal{L}_{sa}(\mathcal{H})\} \cup \{\mathrm{Re}\, B,\, \mathrm{Im}\, B :\ B \in \mathcal{S} \cap \mathcal{B}(\mathcal{H})\}, \\ \mathcal{S}_p &:= \{\mathcal{S} \cap \mathcal{B}(\mathcal{H})\} \cup \{E_A(M) :\ A \in \mathcal{L}_{sa}(\mathcal{H}) \cap \mathcal{S},\ M \text{ Borel}\}, \\ \mathcal{S}_f &:= \{\mathcal{S} \cap \mathcal{B}(\mathcal{H})\} \cup \{f(A) :\ A \in \mathcal{L}_{sa}(\mathcal{H}) \cap \mathcal{S},\ f \text{ bounded Borel}\}; \end{aligned}$$

the first set consists of self–adjoint operators only, while the other two are subsets in $\mathcal{B}(\mathcal{H})$. Some of their simple properties are listed in Problems 24, 25.

The task of describing the structure of an operator set is considerably simplified if there are functional relations between its elements, which allow us to choose a subset of "independent" operators and to use these to express the remaining ones.

6.5.1 Examples: (a) Consider operators $A, B \in \mathcal{B}(\mathbb{C}^n)$ such that A is Hermitean with a simple spectrum, $\sigma(A) = \{\lambda_1, \dots, \lambda_n\}$, so the corresponding eigenvectors form an orthonormal basis $\{e_j\}_{j=1}^n$ in $\mathbb{C}^n$. If the operators A, B commute, e_j are at the same time eigenvectors of B; mimicking the argument of Example 5.2.3 we readily check that there is a polynomial Q of degree $\le n-1$ such that $B = Q(A)$.

(b) Let E_1, E_2, E_3 be mutually orthogonal one–dimensional projections on $\mathbb{C}^3$. The operators $A_1 := E_1 + E_2$ and $A_2 := E_2 + E_3$ are Hermitean and commute mutually; however, neither of them is a function of the other. We can, of course, find a Hermitean A commuting with all E_j such that A_1, A_2 are its functions: it is sufficient to put $A := \sum_{j=1}^3 \lambda_j E_j$, where λ_j are mutually different real numbers.

(c) If an operator set $\mathcal{S}$ is noncommutative or nonsymmetric, there may be no Hermitean operator such that all elements of $\mathcal{S}$ would be its functions. Consider, *e.g.*, the operator $B \in \mathcal{B}(\mathbb{C}^2)$ represented by the matrix $\begin{pmatrix} 0 & 1 \\ 0 & 0 \end{pmatrix}$. The set $\{B\}$ is not symmetric while $\{B, B^*\}$ is not commutative; the operator B cannot be a function of a Hermitean operator since otherwise it would have to be normal.

A Hermitean $A \in \mathcal{B}(\mathcal{H})$ is called a *generating operator* of a set $\mathcal{S} \subset \mathcal{L}_{b,sa}(\mathcal{H})$ if for any $T \in \mathcal{S}$ there is a Borel function f_T such that $T = f_T(A)$. It is clear that the generating operator is not unique: another Hermitean operator $\tilde{A}$ such that $A = h(\tilde{A})$ for some Borel function h is also a generating operator of the set $\mathcal{S}$ and $\tilde{f}_T := f_T \circ h$. The above examples show that the generating operator may not belong to $\mathcal{S}$ and it may not exist if $\mathcal{S}$ is not symmetric or commutative. A sufficient condition for existence of a generating operator is based on the following result.

6.5.2 Theorem (von Neumann): Suppose that $\mathcal{H}$ is separable and $\mathcal{A}$ is a weakly closed commutative $*$–subalgebra in $\mathcal{B}(\mathcal{H})$; then there exists a Hermitean operator $A \in \mathcal{A}$ which generates $\mathcal{A}$, *i.e.*, $\mathcal{A} = \mathcal{A}_w(\{A\})$.

6.5.3 Lemma: Let $\mathcal{A} \subset \mathcal{B}(\mathcal{H})$ be a weakly closed algebra. A Hermitean operator A with the spectral decomposition $\{E_t\}$ belongs to $\mathcal{A}$ *iff* the algebra contains the projections E_t for all $t < 0$ and $I - E_t$ for $t \geq 0$. If this family of projections is denoted as $\{P_t^{(A)}\}$, then $\mathcal{A}_w(\{A\}) = \mathcal{A}_w(\{P_t^{(A)}\})$.

Proof: Let us begin with the sufficient condition. The spectrum of A is contained in some interval $(-b, b)$. Given a positive integer n, we can write it as the union of the subintervals $J_k^n := (b(k-1)/n, bk/n)$, $k = -n+1, \dots, n$. By Problem 5.3b, the corresponding spectral projections are $E_A(J_k^n) = E_{bk/n} - E_{b(k-1)/n}$. They belong to $\mathcal{A}$ for $k < 0$ due to the assumption, and furthermore, the identity $E_A(J_k^n) = (I - E_{bk/n}) - (I - E_{b(k-1)/n})$ shows that the same is true for $k \geq 0$. Then the algebra also contains the operators $A_n := \sum_{k=-n+1}^{n} \frac{bk}{n} E_A(J_k^n)$, $n = 1, 2, \dots$, so combining the weak closedness of $\mathcal{A}$ with Theorem 5.2.2d, we find $A \in \mathcal{A}$.

The functional–calculus rules also yield the necessary condition. By Proposition 5.3.4 a sequence of polynomials $\{P_n\}$ can be found to a given $s \in (-b, 0)$ such that $\lim_{n\to\infty} P_n(t) = \chi_{[-b,s]}(t)$. The same is true for the sequence of homogeneous polynomials $\tilde{P}_n(t) := P_n(t) - P_n(0)$; we have $\tilde{P}_n(A) \in \mathcal{A}$ even if $I \notin \mathcal{A}$. Using the weak closedness together with Theorem 5.2.2d once again, we get $E_s \in \mathcal{A}$ for $s < 0$; in the same way we can check that $I - E_s \in \mathcal{A}$ holds for $s \geq 0$. The last assertion follows easily from the proved equivalence. ∎

6.5.4 Lemma: Let $\mathcal{H}$ be separable; then we can select from any infinite set $\mathcal{S} \subset \mathcal{B}(\mathcal{H})$ a countable subset $\mathcal{S}_0$ such that for any $B \in \mathcal{S}$ there is a sequence $\{B_n\} \subset \mathcal{S}_0$ fulfilling the conditions $\text{s-lim}_{n\to\infty} B_n = B$ and $\text{s-lim}_{n\to\infty} B_n^* = B^*$.

Proof: Since the set $\mathcal{S}$ can be expressed as the countable union $\bigcup_{n=1}^{\infty} \{ B \in \mathcal{S} : n-1 \leq \|B\| < n \}$, it is sufficient to assume that there is a K such that $\|B\| \leq K$ holds for all $B \in \mathcal{S}$. We construct the Hilbert space $\tilde{\mathcal{H}}$ as the countable direct sum of identical copies of the space $\mathcal{H}$ (as in Remark 6.3.1); the latter is separable by assumption, so the same is true for $\tilde{\mathcal{H}}$. We choose a countable set $\{\phi_j\}$ of nonzero vectors, which is dense in $\mathcal{H}$ and associate with each operator $B \in \mathcal{S}$ the vector

$$\Phi_B := \{c_1 B\phi_1, c_1 B^*\phi_1, c_2 B\phi_2, c_2 B^*\phi_2, \dots\} \in \tilde{\mathcal{H}}$$

with $c_j := 2^{-j}\|\phi_k\|^{-1}$. The subset $\Phi_{\mathcal{S}} := \{ \Phi_B : B \in \mathcal{S} \}$ in the separable metric space $\tilde{\mathcal{H}}$ is separable itself (Problem 1.22), so we can choose a countable set $\mathcal{S}_0 \subset \mathcal{S}$ such that the corresponding $\Phi_{\mathcal{S}_0}$ is dense in $\Phi_{\mathcal{S}}$; in other words, for any $B \in \mathcal{S}$

there is a sequence $\{B_n\}\subset\mathcal{S}_0$ such that

$$\|\Phi_B-\Phi_{B_n}\|^2_{\tilde{\mathcal{H}}}=\sum_{j=1}^{\infty}c_j^2\|B\phi_j-B_n\phi_j\|^2+\sum_{j=1}^{\infty}c_j^2\|B^*\phi_j-B_n^*\phi_j\|^2\longrightarrow 0$$

as $n\to\infty$; hence $B_n\phi_j\to B\phi_j$ and $B_n^*\phi_j\to B^*\phi_j$ for all j. Using the estimate $\|B\phi-B_n\phi\|\le\|B\phi-B\phi_j\|+\|B\phi_j-B_n\phi_j\|+\|B_n\phi_j-B_n\phi\|$, we get the inequality $\limsup_{n\to\infty}\|B\phi-B_n\phi\|\le 2K\|\phi-\phi_j\|$ for an arbitrary $\phi\in\mathcal{H}$; however, the set $\{\phi_j\}$ is supposed to be dense in $\mathcal{H}$, so the right side can be made smaller than any positive number. This follows that $\text{s-lim}_{n\to\infty}B_n=B$; the other relation can be proved in the same way. ∎

6.5.5 Lemma: A weakly closed commutative $*$–subalgebra $\mathcal{A}\subset\mathcal{B}(\mathcal{H})$ on a separable $\mathcal{H}$ contains a sequence $\{E_k\}$ of mutually commuting projections which generates it, $\mathcal{A}=\mathcal{A}_w(\{E_k\})$.

Proof: Since $\mathcal{A}$ is a $*$–algebra by assumption, we have $\mathcal{A}_R\subset\mathcal{A}\subset\mathcal{A}_0(\mathcal{A}_R)$, and therefore $\mathcal{A}=\mathcal{A}_w(\mathcal{A})=\mathcal{A}_w(\mathcal{A}_R)$. Using the notation of Lemma 6.4.3, we introduce $\mathcal{S}_P:=\{\,P_t^{(A)}:\,A\in\mathcal{A}_R,\ t\in\mathbb{R}\}$. This set is commutative by Problem 24 because $\mathcal{S}_P\subset(\mathcal{A}_R)_f$. Moreover, $\mathcal{S}_P\subset\mathcal{A}_R$ follows from Lemma 6.4.3 so $\mathcal{A}_w(\mathcal{S}_P)\subset\mathcal{A}_w(\mathcal{A}_R)=\mathcal{A}$, and on the other hand by the same lemma, $\mathcal{S}_P\subset\mathcal{A}_w(\mathcal{S}_P)$ yields $\mathcal{A}_R\subset\mathcal{A}_w(\mathcal{S}_P)$, *i.e.*, together we get $\mathcal{A}=\mathcal{A}_w(\mathcal{S}_P)$. Finally, Lemma 6.4.4 allows us to choose from $\mathcal{S}_P$ a countable subset $\{E_k\}$ of mutually commuting projections. Clearly $\mathcal{A}_w(\{E_k\})\subset\mathcal{A}$; at the same time, for any $E\in\mathcal{S}_P$ there is a subsequence $\{E_{k_j}\}$ which converges to E strongly, and therefore also weakly. Hence $\mathcal{S}_P\subset\overline{(\{E_k\})}_w\subset\mathcal{A}_w(\{E_k\})$, so $\mathcal{A}_w(\{E_k\})\supset\mathcal{A}_w(\mathcal{S}_P)=\mathcal{A}$. ∎

Proof of Theorem 6.5.2: The just proved lemma tells us that algebra $\mathcal{A}$ is generated by a sequence $\{E_k\}$ of mutually commuting projections. To construct a spectral decomposition from it, we must replace $\{E_k\}$ by an ordered family of projections. We set $F_1:=E_1$, and furthermore, $F_2:=E_2F_1,\ F_1+E_2(I-F_1)$. It follows from the commutativity of E_1,E_2 that F_1,F_2,F_3 are projections which commute mutually and belong to $\mathcal{A}$; moreover, they satisfy the inequalities $F_2\le F_1\le F_3$. In the next step we add the operator E_3 and construct the commuting projections

$$F_4:=E_3F_2\,,\quad F_5:=F_2{+}E_3(F_1{-}F_2)\,,\quad F_6:=F_1{+}E_3(F_3{-}F_1)\,,\quad F_7:=F_3{+}E_3(I{-}F_3)\,,$$

which belong to $\mathcal{A}$ and satisfy the inequalities $F_4\le F_2\le F_5\le F_1\le F_6\le F_3\le F_7$, *etc.*; in the k–th step we add E_k to the already constructed $2^{k-1}{-}1$ projections and "insert" the projection $F'':=F{+}E_k(F'{-}F)$ between each neighboring $F\le F'$; the left (right) neighbor of the minimal (respectively, maximal) projection is supposed to be zero (respectively, the unit operator). We have $F\le F''\le F'$; the operators E_k are obtained from the relation

$$E_k=\sum_{j=2^{k-1}}^{2^k-1}F_j-\sum_{j=1}^{2^{k-1}-1}F_j\,,$$

which shows that any homogeneous polynomial composed of the operators E_k can be expressed at the same time as a polynomial in the operators F_j, so the corresponding weakly closed algebras coincide, $\mathcal{A}_w(\{E_k\}) = \mathcal{A}_w(\{F_j\})$, in view of Example B.1.4.

The sought spectral decomposition is now constructed as follows: we divide the interval $[-1,0)$ into three equal parts and label the middle open interval $\left(-\frac{2}{3},-\frac{1}{3}\right)$ as J_1. Then we divide the remaining two intervals again into three parts and set $J_2 := \left(-\frac{8}{9},-\frac{7}{9}\right)$ and $J_3 := \left(-\frac{2}{9},-\frac{1}{9}\right)$; repeating this procedure we get a sequence $\{J_j\}$ of disjoint open intervals. We denote $M_C := [-1,0) \setminus \bigcup_{j=1}^{\infty} J_j$ and associate with any $t \in \mathbb{R} \setminus M_C$ the projection

$$E_t := \begin{cases} 0 & \dots \quad t < -1 \\ F_j & \dots \quad t \in J_j \\ I & \dots \quad t \geq 0 \end{cases}$$

The map $t \mapsto E_t$ defined in this way on $\mathbb{R} \setminus M_C$ is obviously continuous and nondecreasing; to get a spectral decomposition we have to extend it to the complement M_C. The latter is a Cantor–type set, and therefore nowhere dense in $\mathbb{R}$. Hence for any $t \in \mathbb{R}$ there is a nondecreasing sequence $\{t_n\} \subset \mathbb{R} \setminus M_C$ such that $t_n \to t$. The corresponding sequence of projections $\{E_{t_n}\}$ is also nondecreasing and converges strongly to a projection; we set $E_t :=$ s-$\lim_{n\to\infty} E_{t_n}$. The definition makes sense: if $\{t'_n\}$ is any other sequence converging to t, we have s-$\lim_{n\to\infty} E_{t'_n} = E_t$. Indeed, one can choose subsequences $\{t_{k_n}\}$ and $\{t'_{l_n}\}$ in such a way that $\{t_{k_1}, t'_{l_1}, t_{k_2}, t'_{l_2}, \dots\}$ is again nondecreasing, so the corresponding sequence of projections is nondecreasing and any subsequence of it converges to the same limit. It is now straightforward to check that the map $t \mapsto E_t$ is also right–continuous and nondecreasing at the points of the set M_C, *i.e.*, that it is a spectral decomposition.

The inclusion $\{F_j\} \subset \{E_t : t < 0\}$ in combination with Lemma 6.4.5 yields $\mathcal{A} = \mathcal{A}_w(\{E_k\}) = \mathcal{A}_w(\{F_j\}) \subset \mathcal{A}(\{E_t : t < 0\})$; however, the set $\{E_t : t < 0\}$ belongs by construction to the weakly closed algebra $\mathcal{A}$, so the opposite inclusion is also valid; together we get $\mathcal{A} = \mathcal{A}(\{E_t : t < 0\})$. We have $I - E_t = 0$ for $t \geq 0$, and therefore $\{E_t : t < 0\} = \{P_t^{(A)}\}$, where A is the Hermitean operator corresponding to the spectral decomposition $\{E_t\}$ (its spectrum is contained in the set M_C); it follows from Lemma 6.4.3 that $A \in \mathcal{A}$ and $\mathcal{A}_w(\{A\}) = \mathcal{A}_w(\{P_t^{(A)}\}) = \mathcal{A}$. In view of Theorem 6.3.4, this means that $\mathcal{A} \subset \{A\}''$, and Theorem 5.5.6 tells us that any element of the set $\{A\}'' \subset \{A\}''_{ex}$ can be expressed in the form $\varphi(A)$ for some Borel function φ. ∎

An analogous structural result is valid for more general operator families, in particular those containing unbounded self–adjoint operators.

6.5.6 Corollary: Any commutative symmetric set $\mathcal{S} \subset \mathcal{L}_{b,sa}(\mathcal{H})$ on a separable $\mathcal{H}$ has a generating operator A, which is contained in $\mathcal{S}''$.

Proof: If $\mathcal{S}$ contains unbounded operators, we construct to it the set $\mathcal{S}_a \subset \mathcal{B}(\mathcal{H})$ (*cf.* Problem 25). A function f_T^a is Borel *iff* the same is true for $\tan f_T^a(\cdot)$, and moreover, $\mathcal{S}'' = \mathcal{S}_a''$; hence it is sufficient to consider the case $\mathcal{S} \subset \mathcal{B}(\mathcal{H})$. Due to the assumption, $\mathcal{A}_w(\mathcal{S}) = \overline{\mathcal{A}_0(\mathcal{S})}_w$ is an Abelian $*$–subalgebra in $\mathcal{B}(\mathcal{H})$, so there is a Hermitean operator $A \in \mathcal{A}_w(\mathcal{S})$ which generates it, $\mathcal{A}_w(\mathcal{S}) \subset \{A\}''$. This means that $\mathcal{S} \subset \{A\}''$, and since $\mathcal{S}'' \supset \mathcal{A}_w(\mathcal{S})$ by Theorem 6.3.4, the operator A belongs to $\mathcal{S}''$. ∎

This result also extends to other operator sets, for which it still makes sense to speak of commutativity of unbounded operators (Problem 26).

6.6 Complete sets of commuting operators

Let $\mathcal{S}$ be a commutative subset in $\mathcal{L}_{sa}(\mathcal{H})$. The preceding section was devoted to internal properties of such sets; now we want to ask which conditions would ensure that $\mathcal{S}$ cannot be completed by "independent" commuting operators. Without loss of generality, the latter may be supposed to be bounded; then we have to require that the commutant $\mathcal{S}'$ consists solely of functions of the operators from $\mathcal{S}$, or in other words, $\mathcal{S}' \subset \mathcal{S}''$. A set $\mathcal{S}$ with this property is called a **complete set of commuting** (self–adjoint) **operators**; for the sake of brevity we shall use the shorthand CSCO.

6.6.1 Theorem: Let $\mathcal{S}$ be a commutative subset in $\mathcal{L}_{sa}(\mathcal{H})$ where $\mathcal{H}$ is separable; then the following conditions are equivalent:

(a) $\mathcal{S}$ is a CSCO.

(b) The algebra $\mathcal{S}''$ is maximal Abelian, $\mathcal{S}' = \mathcal{S}''$.

(c) The set $\mathcal{S}$ has a cyclic vector, *i.e.*, there is $\varphi \in \mathcal{H}$ such that $\overline{\mathcal{S}''\varphi} = \mathcal{H}$.

Proof: In view of Corollary 6.5.6 and Problem 26b, there is a generating operator A of $\mathcal{S}$ such that $\mathcal{S}' = \{A\}'$ and $\mathcal{S}'' = \{A\}''$. The implications (b) ⇒ (c) ⇒ (a) then follow from Theorem 5.8.6; it remains for us to check (a) ⇒ (b). By Problem 24, the set $\mathcal{S}_f$ is commutative and $\mathcal{S}' = (\mathcal{S}_f)'$. Hence we have $\mathcal{S}_f \subset (\mathcal{S}_f)'$ which implies $(\mathcal{S}_f)' \supset (\mathcal{S}_f)''$, *i.e.*, $\mathcal{S}' \supset \mathcal{S}''$. However, the set $\mathcal{S}$ is a CSCO by assumption, so $\mathcal{S}' = \mathcal{S}''$. ∎

In view of Theorem 5.8.6, $\mathcal{S}$ is a CSCO provided it is generated by an operator $A \in \mathcal{S}''$ having a simple spectrum. This is particularly illustrative if $\mathcal{S} = \{A_j\}_{j=1}^N$ is finite and the operators A_j have pure point spectra. We shall again assume that the Hilbert space $\mathcal{H}$ is separable; the eigenvalues and the corresponding spectral projections of A_j will be denoted as $\lambda_k^{(j)}$ and $P_k^{(j)}$, respectively. Since $\mathcal{S}$ is supposed to be commutative, all the eigenprojections commute mutually and $P_{\{k\}} := \prod_{j=1}^N P_{k_j}^{(j)}$ is a projection for any N–tuple $\{k\} := \{k_1, \dots, k_N\}$ of positive integers; it is easy to check that the projections referring to different N–tuples are mutually orthogonal. We are interested, of course, only in those $\{k\}$ to which a nonzero projection

$P_{\{k\}}$ corresponds. We denote this set as K_N; it is at most countable by assumption, so one can arrange it into a sequence, $K_N = \{\{k\}_1, \dots, \{k\}_n, \dots\}$, and set $P^{(n)} := P_{\{k\}_n}$.

6.6.2 Proposition: A set $\mathcal{S} = \{A_j\}_{j=1}^N$ of commuting self–adjoint operators on a separable $\mathcal{H}$ with pure point spectra is a CSCO *iff* $\dim P^{(n)} = 1$ holds for all $\{k\}_n \in K_N$.

Proof: To check the sufficient condition, we pick a unit vector ϕ_n from each subspace $P^{(n)}\mathcal{H}$. In view of Problem 27, $\{\phi_n\}$ is an orthonormal basis in $\mathcal{H}$. We define $\psi := \sum_n 2^{-n}\phi_n$; the set $M := \{\, P^{(n)}\psi : \{k\}_n \in K_N \,\}$ is therefore total in $\mathcal{H}$ and its linear envelope is contained in $\mathcal{S}''\psi$. It follows that $\mathcal{H} = \overline{M_{lin}} \subset \overline{\mathcal{S}''\psi} \subset \mathcal{H}$, *i.e.*, ψ is cyclic for $\mathcal{S}$.

Assume on the contrary that $\mathcal{S}$ is a CSCO with a cyclic vector ψ; then for any nonzero $\phi \in P^{(m)}\mathcal{H}$ there is a sequence $\{B_n\} \subset \mathcal{S}''$ such that $B_n\psi \to \phi$. Since $\mathcal{S}$ is commutative, Problem 27 yields $\phi = P^{(m)}\phi = \lim_{n\to\infty} B_n P^{(m)}\psi$; the vector ϕ is nonzero so the same is true for $P^{(m)}\psi$. Suppose that the subspace $P^{(m)}\mathcal{H}$ contains a vector $\chi \perp P^{(m)}\psi$; the corresponding projection is denoted as E_χ. By assumption, χ is a common eigenvector of the operators A_j, so $E_\chi \in \mathcal{S}' = \mathcal{S}''$. Then $B\chi = BE_\chi\chi = E_\chi B\chi$ holds for any $B \in \mathcal{S}''$, *i.e.*, $B\chi = \alpha(B)\chi$. It follows that $(B\psi, \chi) = (B\psi, P^{(m)}\chi) = \alpha(B^*)(P^{(m)}\psi, \chi) = 0$ or $\chi \in \{\mathcal{S}''\psi\}^\perp$; however, ψ is a cyclic vector for $\mathcal{S}$, and therefore $\chi = 0$. ∎

Complete sets can have different cardinalities, but the most interesting are those which consist of a few operators only. The extremal case is a CSCO containing a single self–adjoint operator; in view of Theorem 5.8.6 this happens *iff* the operator has a simple spectrum. The preceding proposition provides an illustration for operators with pure point spectra; recall some other important cases:

6.6.3 Examples: (a) Each of the sets $\{Q\}$ and $\{P\}$ is a CSCO in $L^2(\mathbb{R})$ — see Examples 5.8.2, 4.4.7, Theorem 5.5.6, and Problem 5.12.

(b) More generally, due to Example 5.8.2, $\{Q_\mu\}$ is a CSCO on $L^2(\mathbb{R}, d\mu)$ for any Borel measure μ on $\mathbb{R}$.

In other situations a "natural" CSCO consists of several operators. An example is given in Problem 28; in the same way we can prove the following more general result.

6.6.4 Proposition: Suppose that self–adjoint operators A_j on separable Hilbert spaces $\mathcal{H}_j$ have cyclic vectors ϕ_j, $j = 1, 2$; then the set $\{\underline{A}_1, \underline{A}_2\}$ is a CSCO on $\mathcal{H}_1 \otimes \mathcal{H}_2$ and $\phi_1 \otimes \phi_2$ is its cyclic vector.

Our aim is now to extend this result to the situation, where the CSCO in each of the spaces $\mathcal{H}_j$ need not consist of a single operator. First, for symmetric sets $\mathcal{S}_j \subset \mathcal{B}(\mathcal{H}_j)$, $j = 1, 2$, we introduce the following notation: $\underline{\mathcal{S}}_j := \{\underline{B}_j : B_j \in \mathcal{S}_j\}$ and $\mathcal{S}_\Sigma := \underline{\mathcal{S}}_1 \cup \underline{\mathcal{S}}_2$; alternatively we shall write $\mathcal{S}_1 \otimes I_2 := \underline{\mathcal{S}}_1$ and $I_1 \otimes \mathcal{S}_2 := \underline{\mathcal{S}}_2$.

6.6.5 Theorem: The W^*–algebras generated by symmetric sets $\mathcal{S}_j \subset \mathcal{B}(\mathcal{H}_j)$, $j = 1, 2$, and by the set $\mathcal{S}_\Sigma \subset \mathcal{B}(\mathcal{H}_1 \otimes \mathcal{H}_2)$ are related by

$$\mathcal{A}_W(\underline{\mathcal{S}}_1 \cup \underline{\mathcal{S}}_j) \;=\; \mathcal{A}_W(\mathcal{S}_1) \otimes \mathcal{A}_W(\mathcal{S}_2)\,.$$

Proof: As the first step, we shall derive the identities

$$(\underline{\mathcal{S}}_1)'' = \mathcal{S}_1'' \otimes I_2\,, \quad (\underline{\mathcal{S}}_2)'' = I_1 \otimes \mathcal{S}_2''\,. \tag{6.6}$$

Let $\{\psi_\alpha\}$ be an orthonormal basis in $\mathcal{H}_2$, and denote by $E^{(\alpha)}$ the corresponding one–dimensional projections. We define the unitary operators $V_\alpha : \mathcal{H}_1 \to \mathcal{H}_1 \otimes E^{(\alpha)}\mathcal{H}_2 = \underline{E}_2^{(\alpha)}\mathcal{H}$ by $V_\alpha\phi := \phi \otimes \psi_\alpha$, and extend the inverse operators to the whole $\mathcal{H} := \mathcal{H}_1 \otimes \mathcal{H}_2$ setting $V_\alpha^+\chi := V_\alpha^{-1}\underline{E}_2^{(\alpha)}\chi$. It is easy to check that V_α and V_α^+ as well as $V_\alpha V_\beta^+$ for any α, β are partial isometries (*cf.* Problem 30). We shall show that an operator $B \in \mathcal{B}(\mathcal{H})$, which commutes with all $V_\alpha V_\beta^+$, is of the form $B_1 \otimes I_2$ for some $B_1 \in \mathcal{B}(\mathcal{H}_1)$. Due to the assumption, there is γ such that

$$V_\beta^+ B V_\alpha \;=\; V_\gamma^+ V_\gamma V_\beta^+ B V_\alpha \;=\; V_\gamma^+ B V_\gamma V_\beta^+ V_\alpha \;=\; \delta_{\alpha\beta} V_\gamma^+ B V_\gamma;$$

putting $\alpha = \beta$ we get $V_\alpha^+ B V_\alpha = V_\gamma^+ B V_\gamma := B_1$ for all α. Since V_α, V_α^+ are partial isometries, B_1 belongs to $\mathcal{B}(\mathcal{H}_1)$. Furthermore, the relation $V_\beta^+ B V_\alpha = \delta_{\alpha\beta} B_1$ implies

$$B(\phi\otimes\psi_\alpha) \;=\; \sum_\beta V_\beta V_\beta^+ B(\phi\otimes\psi_\alpha) \;=\; \sum_\beta V_\beta \delta_{\alpha\beta} B_1\phi \;=\; B_1\phi\otimes\psi_\alpha \;=\; (B_1\otimes I_2)(\phi\otimes\psi_\alpha)\,;$$

however, the two operators are linear and bounded, so $B = B_1 \otimes I_2$ follows from their coincidence on the total set $\mathcal{H}_1 \underline{\times} \{\psi_\alpha\}$.

To prove the relations (6.6) we define the set $V(\mathcal{S}_1) := \{\, B \in \mathcal{B}(\mathcal{H}) : V_\beta^+ B V_\alpha \in \mathcal{S}_1 \cup \{0\}$ for all $\alpha, \beta \,\}$. Using Problem 30, we find that

$$\underline{C}_1 B \;=\; \sum_{\alpha,\beta,\gamma} V_\alpha V_\alpha^+ (C \otimes I_2) V_\beta V_\beta^+ B V_\gamma V_\gamma^+ \;=\; \sum_{\beta,\gamma} V_\beta C (V_\beta^+ B V_\gamma) V_\gamma^+$$

holds for any $B \in \mathcal{B}(\mathcal{H})$ and $C \in \mathcal{B}(\mathcal{H}_1)$. In the same way we can express the product $B\underline{C}_1$; it shows that the operators B, $\underline{C}_1$ commute *iff* the same is true for C and $V_\beta^+ B V_\alpha$ with any α, β. Choosing $B \in V(\mathcal{S}_1)$, $C \in \mathcal{S}_1'$, and $B \in (\underline{\mathcal{S}}_1)'$, $C \in \mathcal{S}_1$, we get from here respectively the inclusions

$$\mathcal{S}_1' \otimes I_2 \subset V(\mathcal{S}_1)' \quad \text{and} \quad (\underline{\mathcal{S}}_1)' \subset V(\mathcal{S}_1')\,.$$

Suppose for a while that $I_1 \in \mathcal{S}_1$; then the operators $V_\alpha V_\beta^+$ belong to $V(\mathcal{S}_1)$ for all α, β and each element of the commutant $V(\mathcal{S}_1)'$ is of the form $C \otimes I_2$. However, such an operator commutes with all $B \in V(\mathcal{S}_1)$ *iff* $[C, V_\beta^+ B V_\alpha] = 0$ holds for all α, β, *i.e.*, *iff* $C \in \mathcal{S}_1'$. This fact, together with the first of the above inclusions, yields the identity $\mathcal{S}_1' \otimes I_2 = V(\mathcal{S}_1)'$ which is valid under the condition $I_1 \in \mathcal{S}_1$.

Hence we may apply it to the commutant $\mathcal{S}_1'$, which contains the unit operator, obtaining $V(\mathcal{S}_1')' = \mathcal{S}_1'' \otimes I_2 \supset \underline{\mathcal{S}}_1$. This in turn implies $V(\mathcal{S}_1') \subset V(\mathcal{S}_1')'' \subset (\underline{\mathcal{S}}_1)'$ which, in combination with the second of the above inclusions, gives $V(\mathcal{S}_1') = (\underline{\mathcal{S}}_1)'$. Then the commutants of these sets also coincide, so we finally arrive at the first of the relations (6.6), $(\underline{\mathcal{S}}_1)'' = V(\mathcal{S}_1')' = \mathcal{S}_1'' \otimes I_2$; the other identity is checked in the same way.

The rest of the proof is easy. In view of (6.6), any operators $B_j \in \mathcal{S}_j'' = \mathcal{A}_W(\mathcal{S}_j)$, $j = 1,2$, belong to $(\underline{\mathcal{S}}_j)''$, so they commute with any $C \in (\mathcal{S}_\Sigma)' = (\underline{\mathcal{S}}_1)' \cap (\underline{\mathcal{S}}_2)'$ and $[B_1 \otimes B_2, C] = [\underline{B}_1 \underline{B}_2, C] = 0$. Hence $B_1 \otimes B_2 \in (\mathcal{S}_\Sigma)''$, which implies the inclusion $\mathcal{A}_W(\mathcal{S}_1) \otimes \mathcal{A}_W(\mathcal{S}_2) \subset \mathcal{A}_W(\mathcal{S}_\Sigma)$. At the same time, $\mathcal{S}_\Sigma = \underline{\mathcal{S}}_1 \cup \underline{\mathcal{S}}_2 \subset \{ B_1 \otimes B_2 : B_j \in \mathcal{A}_W(\mathcal{S}_j) \}$; combining these inclusions with the definition of the W^*–algebra tensor product, we conclude the proof. ∎

The proved theorem provides the sought extension of Proposition 6.6.4.

6.6.6 Corollary: Let $\mathcal{S}_j$, $j = 1,2$, be a CSCO on a separable Hilbert space $\mathcal{H}_j$; then the set $\mathcal{S} := \{ \overline{\underline{A}_1} : A_1 \in \mathcal{S}_1 \} \cup \{ \overline{\underline{A}_2} : A_2 \in \mathcal{S}_2 \}$ is a CSCO on $\mathcal{H}_1 \otimes \mathcal{H}_2$. If ϕ_j is a cyclic vector of the set $\mathcal{S}_j$, $j = 1,2$, then $\phi_1 \otimes \phi_2$ is cyclic for $\mathcal{S}$.

Proof: Assume first that $\mathcal{S}_j \subset \mathcal{B}(\mathcal{H}_j)$; then the operators A_j are closed and $\mathcal{S} = \mathcal{S}_\Sigma$. As a CSCO, each of the sets $\mathcal{S}_j$ has a cyclic vector which we denote as ϕ_j. Since $\{ B_j\phi_j : B_j \in \mathcal{S}_j'' \}$ is total in $\mathcal{H}_j$, the set $\{ B_1\phi_1 \otimes B_2\phi_2 : B_j \in \mathcal{S}_j'' \}$ is total in $\mathcal{H} := \mathcal{H}_1 \otimes \mathcal{H}_2$. Moreover, $\{ B_1 \otimes B_2 : B_j \in \mathcal{S}_j'' \}_{lin} \subset \{ B_1 \otimes B_2 : B_j \in \mathcal{S}_j'' \}'' = \mathcal{S}''$ where the last relation follows from the theorem, so

$$\mathcal{H} = \overline{\{ B_1\phi_1 \otimes B_2\phi_2 : B_j \in \mathcal{S}_j'' \}_{lin}} \subset \overline{\mathcal{S}''(\phi_1 \otimes \phi_2)} \subset \mathcal{H}\,.$$

If the sets $\mathcal{S}_j$ contain unbounded operators, we construct to them the sets $\mathcal{S}_{j,a} \subset \mathcal{B}(\mathcal{H}_j)$ as in Problem 25. We have $\mathcal{S}_j'' = \mathcal{S}_{j,a}''$; hence $\mathcal{S}_{j,a}$ is a CSCO with the same cyclic vector as ϕ_j, and by the already proven result, $\underline{\mathcal{S}}_{1,a} \cup \underline{\mathcal{S}}_{2,a}$ is a CSCO with the cyclic vector $\phi_1 \otimes \phi_2$. Finally, we use Problem 5.28c due to which $\arctan \overline{\underline{A}_j} = \underline{\arctan A_j}$, *i.e.*, $\underline{\mathcal{S}}_{1,a} \cup \underline{\mathcal{S}}_{2,a} = \mathcal{S}_a$, and since $\mathcal{S}_a'' = \mathcal{S}''$ the proof is completed. ∎

6.7 Irreducibility. Functions of non-commuting operators

In Section 4.4 we have introduced the notions of invariant subspace, irreducibility, *etc.*, for a given operator; now we are going to do the same for operator sets. We say that a subspace $L \subset \mathcal{H}$ is an *invariant subspace* of an operator set $\mathcal{S}$ if it is an invariant subspace of any operator $T \in \mathcal{S}$. A closed subspace $\mathcal{G} \subset \mathcal{H}$ *reduces* the set $\mathcal{S}$ if it reduces each operator of this set; we shall again say alternatively that $\mathcal{S}$ is reduced by the corresponding projection $E_\mathcal{S}$. An operator set $\mathcal{S}$ is called **irreducible** if there is no nontrivial closed subspace which reduces it; in the opposite case it is called *reducible*.

Irreducibility is equivalent to nonexistence of a nontrivial projection, which would commute with all the operators of $\mathcal{S}$. In the case when $\mathcal{S}$ is symmetric we have a simpler criterion analogous to Problem 4.24a (see Problem 33a). If all the involved operators are bounded we need not worry about their domains: a symmetric set $\mathcal{S} \subset \mathcal{B}(\mathcal{H})$ is reducible *iff* it has a nontrivial closed invariant subspace. It is also obvious that for the operator sets $\mathcal{S}_1 \subset \mathcal{S}_2$ reducibility of $\mathcal{S}_2$ implies reducibility of $\mathcal{S}_1$, and conversely, irreducibility of $\mathcal{S}_2$ follows from irreducibility of the set $\mathcal{S}_1$.

If a closed subspace $\mathcal{G} \subset \mathcal{H}$ reduces an operator set $\mathcal{S}$, we can construct the sets $\mathcal{S}_\mathcal{G} := \{T \upharpoonright \mathcal{G} : T \in \mathcal{S}\}$ and $\mathcal{S}_{\mathcal{G}^\perp}$; any operator $T \in \mathcal{S}$ is the orthogonal sum of its parts in the subspaces $\mathcal{G}$, $\mathcal{G}^\perp$. The sets $\mathcal{S}_\mathcal{G}$ and $\mathcal{S}_{\mathcal{G}^\perp}$ have the same cardinality as $\mathcal{S}$; however, their structure is in general less complicated since they act on "smaller" spaces. If at least one of them is reducible we can continue and take the next step; however, it may happen that we never end up with irreducible sets (even if $\mathcal{S}$ consists of a single operator — see Example 6.7.2b below). A good illustration of the reduction process is provided by the classification of W^*–algebras described in Section 6.3.

It is not always convenient to check irreducibility of a given operator set directly from the definition, so we want to find an equivalent expression to be used instead. Two simple criteria can serve this purpose.

6.7.1 Theorem: Let $\mathcal{S} \subset \mathcal{L}_{b,sa}(\mathcal{H})$, $\dim \mathcal{H} > 1$, be a symmetric set; then the following conditions are equivalent:

(a) $\mathcal{S}$ is irreducible.

(b) The commutant of $\mathcal{S}$ consists of scalar operators only, $\mathcal{S}' = \mathbb{C}(\mathcal{H})$.

(c) Any nonzero vector $\phi \in \mathcal{H}$ is cyclic for $\mathcal{S}''$.

Proof: We shall check the chain of implications (a) $\Rightarrow$ (b) $\Rightarrow$ (c) $\Rightarrow$ (a). Suppose that $\mathcal{S}$ is irreducible and $C \in \mathcal{S}'$. If C is a projection it must equal either 0 or I. If C is a Hermitean operator with the spectral decomposition $\{E_\lambda\}$, then Theorem 5.3.1 implies $\{C\}' = \{E_\lambda\}'$ so $\{E_\lambda\} \subset \{E_\lambda\}'' = \{C\}'' \subset \mathcal{S}''$. Since the spectral decomposition is monotonic and cannot contain nontrivial projections, there is a $\lambda_0 \in \mathbb{R}$ such that $E_\lambda = 0$ for $\lambda < \lambda_0$ and $E_\lambda = I$ for $\lambda \geq \lambda_0$, *i.e.*, $C = \lambda_0 I$. Finally, let C be any bounded operator from $\mathcal{S}'$. The set $\mathcal{S}$ is symmetric by assumption, so $\mathcal{S}_p$ is also symmetric and the same is true for $\mathcal{S}' = \mathcal{S}'_p$. The Hermitean operators $\mathrm{Re}\, B$ and $\mathrm{Im}\, B$ belong to $\mathcal{S}'$; hence they are scalar and C is also scalar; in other words, we get (b).

The relation $\mathcal{S}' = \mathbb{C}(\mathcal{H})$ implies $\mathcal{S}'' = \mathcal{B}(\mathcal{H})$, so $\mathcal{S}''\phi = \mathcal{H}$ holds for any nonzero vector $\phi \in \mathcal{H}$, and the latter is therefore cyclic. Finally, suppose that the set $\mathcal{S}$ is not irreducible; then there is a nontrivial projection $E \in \mathcal{S}' = (\mathcal{S}'')'$. Any nonzero vector $\phi \in E\mathcal{H}$ then satisfies $\mathcal{S}''\phi \subset E\mathcal{H}$, which means that it cannot be cyclic for the set $\mathcal{S}''$. ∎

6.7.2 Examples: (a) A symmetric *commutative* set $\mathcal{S} \subset \mathcal{L}_{b,sa}(\mathcal{H})$ is always reducible provided $\dim \mathcal{H} > 1$. It is sufficient to assume that $\mathcal{S}$ contains at least one nonscalar operator T since otherwise we would have $\mathcal{S} \subset \mathbb{C}(\mathcal{H})$ and $\mathcal{S}' = \mathcal{B}(\mathcal{H})$. If T is bounded it belongs to $\mathcal{S}'$. If it is unbounded self-adjoint, $\mathcal{S}'$ contains its spectral decomposition $\{E_\lambda\}$ and as T is not scalar by assumption, we can find at least one nontrivial projection in $\mathcal{S}'$. In all cases therefore $\mathcal{S}' \neq \mathbb{C}(\mathcal{H})$, so the claim is proved.

(b) By the previous example, a set $\{A\}$ consisting of a single self-adjoint operator is reducible unless $\dim \mathcal{H} = 1$, and the same is true for $\{\, A \restriction E_A(M)\mathcal{H}\}$, where M is any Borel set in $\mathbb{R}$. Hence the complete decomposition of $\{A\}$ to irreducible (one-point) sets is possible only if A possesses a total set of eigenvectors, *i.e.*, a pure point spectrum.

(c) Consider the right-shift operator S on an infinite-dimensional separable Hilbert space $\mathcal{H}$ with an orthonormal basis $\{\phi_j\}_{j=1}^{\infty}$ introduced in Example 3.1.3. We want to prove that the pair $\{S, S^*\}$ is an irreducible family. All operators $C \in \{S, S^*\}'$ satisfy $S^*C\phi_1 = CS^*\phi_1 = 0$, and therefore $C\phi_1 = \alpha\phi_1$ because $\operatorname{Ker} S^* = \{\phi_1\}_{lin}$. Then any basis element satisfies $C\phi_j = CS^{j-1}\phi_1 = S^{j-1}C\phi_1 = \alpha\phi_j$, and since $C \in \mathcal{S}'$ is bounded by definition, we get $C = \alpha I$.

(d) It follows trivially from the previous example that the set $\mathcal{B}(\mathcal{H})$ is also irreducible. This is true, of course, for any dimension of $\mathcal{H}$: the operators $P_{\psi\phi}$ of Problem 3.6 are bounded for any $\phi, \psi \in \mathcal{H}$ and therefore any nonzero $\phi \in \mathcal{H}$ is cyclic for $\mathcal{B}(\mathcal{H})$.

(e) We shall prove finally that the operators $\{Q, P\}$ form an irreducible set on $L^2(\mathbb{R})$. We know from Example 6.6.3a that $\{Q\}$ is a CSCO, which means that for any operator $B \in \{Q, P\}'$ there is a bounded Borel function f_B such that $B = f_B(Q)$. Furthermore, $BP \subset PB$ holds by assumption, so the function $B\psi = f_B(\cdot)\psi(\cdot)$ belongs to $D(P)$ for all $\psi \in D(P)$. To any bounded interval $J \subset \mathbb{R}$ we can choose $\psi_J \in D(P)$ such that $\psi_J(x) = 1$ for $x \in J$; it follows that f_B is absolutely continuous in $\mathbb{R}$. The inclusion $BP \subset PB$ implies, in addition,

$$-if_B(x)\psi'(x) \;=\; (BP\psi)(x) \;=\; (PB\psi)(x) \;=\; -i(f_B(x)\psi(x))',$$

i.e., $f_B'(x)\psi(x) = 0$ for all $x \in D(P)$ and a.a. $x \in \mathbb{R}$. In view of the absolute continuity of the function f_B, we end up with $f_B(x) = c$ for all $x \in \mathbb{R}$, which means that B is scalar.

Now we want to find a way to construct an irreducible set on the tensor product $\mathcal{H}_1 \otimes \mathcal{H}_2$ from irreducible sets $\mathcal{S}_j$ on the "component" Hilbert spaces $\mathcal{H}_j$, $j = 1, 2$. It appears that we can use the same construction for that purpose as in the case of commutative sets discussed in the preceding section.

6.7.3 Theorem: Let $\mathcal{S}_j \subset \mathcal{L}_{b,sa}(\mathcal{H}_j)$ be symmetric irreducible operator sets on Hilbert spaces $\mathcal{H}_j$, $j = 1, 2$; then $\mathcal{S} := \{\, \underline{A_1} : A_1 \in \mathcal{S}_1 \,\} \cup \{\, \underline{A_2} : A_2 \in \mathcal{S}_2 \,\}$ is an irreducible set on $\mathcal{H}_1 \otimes \mathcal{H}_2$.

Proof: Due to the assumption, $\mathcal{S}_j' = \mathbb{C}(\mathcal{H}_j)$ so $\mathcal{S}_j'' = \mathcal{B}(\mathcal{H}_j)$. First, let $\mathcal{S}_j \subset \mathcal{B}(\mathcal{H}_j)$; then $\mathcal{S} = \mathcal{S}_\Sigma$ and Theorem 6.6.5 implies $\mathcal{A}_W(\mathcal{S}_\Sigma) = (\mathcal{S}_\Sigma)'' = \mathcal{S}_1'' \otimes \mathcal{S}_2'' = \mathcal{B}(\mathcal{H}_1) \otimes \mathcal{B}(\mathcal{H}_2)$, and therefore $(\mathcal{S}_\Sigma)' = (\mathcal{B}(\mathcal{H}_1) \otimes \mathcal{B}(\mathcal{H}_2))' = \mathbb{C}(\mathcal{H}_1 \otimes \mathcal{H}_2)$ (see the notes to Section 6.6). Moreover, it is obvious that $\mathcal{S}_\Sigma$ is symmetric; hence it is irreducible by Schur's lemma. If some of the sets $\mathcal{S}_j$ contain unbounded operators we use the sets $\mathcal{S}_{j,a}$ as in the proof of Corollary 6.6.6: we have $\mathcal{S}'' = \mathcal{S}_a'' = (\mathcal{S}_{1,a} \otimes \mathcal{S}_{2,a})'' = \mathcal{S}_{1,a}'' \otimes \mathcal{S}_{2,a}''$, and since $\mathcal{S}_{j,a}'' = \mathcal{S}_j'' = \mathcal{B}(\mathcal{H}_j)$, the set $\mathcal{S}$ is again irreducible. ∎

6.7.4 Remark: It is not difficult to see that the proved assertion is nontrivial: other sets consisting of simple combinations of the operators from $\mathcal{S}_1$, $\mathcal{S}_2$ need not be irreducible. As an example, consider the sets $\mathcal{S}_1 = \mathcal{S}_2 = \{\, \pi(L_j) : j = 1,2,3 \,\} \subset \mathcal{B}(\mathbb{C}^2)$ with $\pi(L_j) := \frac{1}{2}\sigma_j$, where σ_j are the Pauli matrices, and define

$$\pi_\Sigma(L_j) \;:=\; \pi(L_j) \otimes I + I \otimes \pi(L_j)$$

and $\pi_\Sigma(L^2) := \sum_{j=1}^3 \pi(L_j)^2$ on the tensor product $\mathbb{C}^2 \otimes \mathbb{C}^2 = \mathbb{C}^4$. The last named operator is nonscalar (we leave the proofs to the reader — Problem 34), and commutes with all the $\pi_\Sigma(L_j)$, so the set $\mathcal{S}_{12} := \{\, \pi_\Sigma(L_j) : j = 1,2,3 \,\}$ is reducible. On the other hand, $\pi_\Sigma(L^2)$ does not commute with the elements of $\mathcal{S}_\Sigma := \{\, \pi(L_j) \otimes I,\, I \otimes \pi(L_j) : j = 1,2,3 \,\}$, as should be the case, because each $\pi(L_j)$ has a simple spectrum, so the sets $\mathcal{S}_j$ are irreducible and by Theorem 6.7.3 the same is true for $\mathcal{S}_\Sigma$.

There is also another viewpoint from which irreducibility can be discussed. We know that if $\mathcal{S} = \{A_1, \dots, A_n\}$ is a finite family of commuting self–adjoint operators, its bicommutant consists exclusively of functions of the operators $A_1, \dots, A_n$. The situation is considerably more complicated if the operators do not commute. The only naturally defined functions are then polynomials, and even in that case caution is needed because such a polynomial need not be densely defined if some of the operators A_j are unbounded.

This difficulty can, however, be overcome by replacing the A_j by their bounded functions and taking polynomials of these operators, *i.e.*, the algebra $\mathcal{A}_0(\mathcal{S}_f)$ for the set $\mathcal{S}_f$ defined in Section 6.5; recall that the unit operator as a bounded function of any A_j is contained in $\mathcal{A}_0(\mathcal{S}_f)$. Now by Theorem 6.3.4 and Problem 24, this set generates the W^*–algebra $(\mathcal{S}_f)'' = \mathcal{S}''$, which means that any operator of $\mathcal{S}''$ can be approximated by the polynomials from $\mathcal{A}_0(\mathcal{S}_f)$ in the weak operator topology (and also in the strong, σ–weak, and σ–strong topologies; it should be remembered that none of them is first countable, so the approximation can be realized by a net, but in general not by a sequence).

In this sense therefore the elements of $\mathcal{S}''$ can be regarded as functionally dependent on the operators $A_1, \dots, A_n$. The importance of irreducible sets now stems from the fact that if $\mathcal{S}$ is irreducible, then this conclusion extends, due to Theorem 1, to any bounded operator on $\mathcal{H}$. For instance, Example 6.7.2e tells us that every $B \in \mathcal{B}(L^2(\mathbb{R}))$ can be approximated by polynomials in bounded functions of operators Q and P.

On the other hand, there is no general way to introduce a functional calculus for noncommuting operators. To illustrate the difficulties which would arise, consider a pair of noncommuting Hermitean operators A_1, A_2. The function $F: f(x,y) = xy$ is a polynomial, so it should be associated with an analogous operator polynomial; however, both A_1A_2 and A_2A_1 are candidates for the role of $f(A_1, A_2)$ as is more generally $\alpha A_1A_2 + (1-\alpha)A_2A_1$ for any $\alpha \in \mathbb{R}$. The function f is real-valued so the corresponding operator should be Hermitean; this requires $f(A_1, A_2) = \frac{1}{2}(A_1A_2 + A_2A_1)$ but such a map is not multiplicative, because

$$(f^2)(A_1, A_2) = \frac{1}{2}(A_1^2A_2^2 + A_2^2A_1^2) \neq \frac{1}{4}(A_1A_2 + A_2A_1)^2 = (f(A_1, A_2))^2 .$$

Hence the considerations of Sections 5.2 and 5.5 cannot be extended to the case of noncommuting operators; if the symbol $f(A_1, \dots, A_2)$ is nevertheless used in such a situation, it is always a shorthand for a particular operator.

6.8 Algebras of unbounded operators

We have seen how useful algebraic methods are for the analysis of operator sets. It is natural to ask whether some of these notions and results can be extended to the case of unbounded operators. Since the problem is, of course, technically more complicated we limit ourselves to a few remarks; in the notes we give a guide for further reading.

If a set $\mathcal{S} \subset \mathcal{L}(\mathcal{H})$ has to be equipped with an algebraic structure we must ensure that the algebraic operations make sense for its elements. A way to do that is to assume existence of a common dense invariant subspace. Let D be a dense subspace in a Hilbert space $\mathcal{H}$; then the symbol $\mathcal{L}^+(D)$ will denote the set of all operators in $\mathcal{L}(\mathcal{H})$ such that

(i) $D(T) = D$ and D is invariant with respect to T, $TD \subset D$.

(ii) The adjoint T^* exists, $D(T^*) \supset D$, and $T^*D \subset D$.

The set $\mathcal{L}^+(D)$ has the structure of a $*$-algebra: the sums and products of the operators from $\mathcal{L}^+(D)$ belong to $\mathcal{L}^+(D)$, and the involution can be defined by $T \mapsto T^+$, where $T^+ := T^* \upharpoonright D$. Any subalgebra $\mathcal{A} \subset \mathcal{L}^+(D)$ which contains the unit operator is called an *Op*-algebra.* Algebra $\mathcal{A}$ is called *closed* if $D = \bigcap_{T\in\mathcal{A}} D(\overline{T})$ and *self-adjoint* if $D = \bigcap_{T\in\mathcal{A}} D(T^*)$. We also say that $\mathcal{A}$ is *standard* if any symmetric operator $A \in \mathcal{A}$ is *e.s.a.*

6.8.1 Example: Consider again the operators Q, P on $\mathcal{H} = L^2(\mathbb{R})$. By Examples 1.5.6 and 4.4.7, the subspace $\mathcal{S} := \mathcal{S}(\mathbb{R})$ is dense in $\mathcal{H}$ and invariant with respect to Q, P; hence $Q_\mathcal{S} := Q \upharpoonright \mathcal{S}$ and $P_\mathcal{S} := P \upharpoonright \mathcal{S}$ belong to $\mathcal{L}^+(\mathcal{S}(\mathbb{R}))$. Now consider the algebra $\mathcal{A}$ consisting of all nonhomogeneous polynomials in $Q_\mathcal{S}$, $P_\mathcal{S}$; using the relation

$$QP\psi - PQ\psi = -i\psi \tag{6.7}$$

we can rewrite it as $\mathcal{A} = \{\, T = \sum_{j,k=1}^{n} \alpha_{jk} Q^j P^k : \alpha_{jk} \in \mathbb{C},\ n \text{ positive integer} \,\}$. It follows from Proposition 4.1.2 that $T^+ = \sum_{j,k=1}^{n} \overline{\alpha}_{jk} P^k Q^j$. Using the commutation relation (6.7) we can express this operator as $T^+ = \sum_{j,k=1}^{n} \tilde{\alpha}_{jk} Q^j P^k$, which means that $\mathcal{A}$ is an *Op**–algebra; it can be proven that it is closed and self–adjoint (see the notes).

The algebra $\mathcal{A}$ defines a topological structure on D. It is easy to check that the family of seminorms $\{\, p_T = \|T\cdot\| : T \in \mathcal{A} \,\}$ separates points, and the corresponding topology $\tau_{\mathcal{A}}$ is the weakest locally convex topology in which any operator $T \in \mathcal{A}$ is continuous as a map $(D, \tau_{\mathcal{A}}) \mapsto \mathcal{H}$. Since $I \in \mathcal{A}$ by assumption, $\tau_{\mathcal{A}}$ is stronger than the topology induced by the Hilbert–space norm of $\mathcal{H}$ in D. Further, let $\mathcal{M}$ be the family of all bounded sets in $(D, \tau_{\mathcal{A}})$ (see the notes to Section 1.4); then we associate with any $M \in \mathcal{M}$ the seminorm $s_M : s_M(T) = \sup_{\phi,\psi \in \mathcal{M}} |(\phi, T\psi)|$. The system $\{\, s_M : M \in \mathcal{M} \,\}$ separates points; hence it defines on $\mathcal{A}$ a locally convex topology τ_D which is called *uniform* (see Problem 35 and the notes).

6.8.2 Theorem: An *Op**–algebra $\mathcal{A} \subset \mathcal{L}^+(D)$ with the topology τ_D is a topological $*$–algebra.

The algebra $(\mathcal{A}, \tau_D)$ is called an *O*–algebra.* Some results which are valid in algebras of bounded operators can be extended to *Op**–algebras and *O**–algebras. We mention several of them in the notes; here we limit ourselves to a few words about states on *Op**–algebras.

By Proposition 6.2.2e, any positive functional on a C^*–algebra is continuous. This suggests how to formulate the definition in the case of unbounded operators: a *state* on an *Op**–algebra $\mathcal{A} \subset \mathcal{L}^+(D)$ is a positive functional $F : \mathcal{A} \to \mathbb{C}$ such that $f(I) = 1$; the positivity here means $f(T^+T) \ge 0$ for all $T \in \mathcal{A}$. A linear functional $g : \mathcal{A} \to \mathbb{C}$ is said to be *strongly positive* if $g(T) \ge 0$ holds on the set $\mathcal{A}_+ := \{\, T \in \mathcal{A} : (\phi, T\phi) \ge 0,\ \phi \in D \,\}$. In the case of bounded operators the two notions coincide in view of Theorem 6.2.1; generally one has only $\mathcal{P}(\mathcal{A}) := \{\, T^+T : T \in \mathcal{A} \,\} \subset \mathcal{A}_+$, so any strongly positive functional is positive but the converse is not true.

We want to know whether a state on an *Op**–algebra $\mathcal{A} \subset \mathcal{L}^+(D)$ can again be expressed by means of a statistical operator, as in Theorems 6.4.1 and 6.4.7. We shall mention one assertion; for its proof and related results we refer to literature quoted in the notes. We denote

$$\mathcal{J}_1(\mathcal{A}) := \{\, W \in \mathcal{B}(\mathcal{H}) : \overline{WT},\ \overline{W^*T} \in \mathcal{J}_1(\mathcal{A}),\ T \in \mathcal{A}, \text{ and } W\mathcal{H} \subset D,\ W^*\mathcal{H} \subset D \,\}$$

and $\mathcal{J}_1(\mathcal{A})_+ := \{\, W \in \mathcal{J}_1(\mathcal{A}) : W \ge 0 \,\}$. We have $I \in \mathcal{A}$, so $\mathcal{J}_1(\mathcal{A})$ is a subset of $\mathcal{J}_1(\mathcal{H})$ and $\mathcal{J}_1(\mathcal{A})_+ \subset \mathcal{T}(\mathcal{H})_+$; we also denote $\tau_+ := \tau_{\mathcal{L}^+(D)}$.

6.8.3 Theorem: Let (D, τ_+) be a Fréchet space; then the following conditions are equivalent:

(a) Any bounded set in (D, τ_+) is precompact.

(b) If f is a strongly positive functional on $\mathcal{A}$ and $\tau_{\mathcal{A}} = \tau_+$, then there is an operator $W \in \mathcal{J}_1(\mathcal{L}^+(D))_+$ such that $f(T) = \mathrm{Tr}\,(\overline{WT}) = \mathrm{Tr}\,(TW)$ holds for all $T \in \mathcal{A}$.

(c) Any strongly positive functional f on $\mathcal{L}^+(D)$ is of the form $f(B) = \mathrm{Tr}\,(TW)$ for some $W \in \mathcal{J}_1(\mathcal{L}^+(D))_+$.

Notes to Chapter 6

Section 6.1 The notion of C^*-algebra was introduced by I.M. Gel'fand and M.A. Naimark in 1943. Some mathematicians use the term B^*–algebra in the same sense (see ⟦Ru 2⟧, Chap.11, ⟦Ti⟧); a normed $*$–algebra fulfilling condition (6.1) is called a completely regular algebra — *cf.* ⟦Nai 1⟧, Secs.16 and 24. The original definition also included the requirement that $e{+}a^*a$ must be invertible for all $a \in \mathcal{A}$; only later it became clear that this is fulfilled automatically — see Theorem 6.2.1 and ⟦BR 1⟧, notes to Chap.2. Let $\tilde{\mathcal{A}}$ be the extension obtained by addition of the unit element to a C^*–algebra $\mathcal{A}$; then there is a unique extension of the norm of $\mathcal{A}$, which makes $\tilde{\mathcal{A}}$ a C^*–algebra. This extension is given by $\|[\alpha, a]\|_{\tilde{\mathcal{A}}} := \sup_{\|b\|\le 1} \|\alpha b{+}ab\|$ (*cf.* Problems 1 and 3). It is slightly more complicated to check that such a norm has the C^*–property — see, *e.g.*, ⟦Di 2⟧, Sec.1.3.8.

Theorem 5 is the main result of Gel'fand's theory of commutative Banach algebras, which is discussed, *e.g.*, in ⟦BR 1⟧, Sec.2.3.5; ⟦Mau⟧, Chap.VIII; ⟦Ru 2⟧, Chap.11; ⟦Sa⟧, Sec.1.2.; [Si 1], Sec.4. The space Δ consists of all multiplicative functionals on $\mathcal{A}$, *i.e.*, such that $f(a)f(b) = f(ab)$, and the isomorphism $\mathcal{A} \to C(\Delta)$ is given by the so–called Gel'fand transformation, $\hat{a}(f) := f(a)$ for all $f \in \Delta$. The application of Gel'fand's theory to proof of the spectral theorem is discussed in ⟦Mau⟧, Sec.IX.2; ⟦Nai 1⟧, Sec.17.4; ⟦Ru 2⟧, Chap.12.

If we define a C^*–algebra by factorization with respect to an ideal, the only nontrivial thing is to check that the factor norm introduced in Appendix B.2 has the property (6.1) — *cf.* ⟦BR 1⟧, Prop.2.2.19 or ⟦Di 2⟧, Sec.1.8.2. Instead of "direct sum", the term *C^*-product* of the algebras $\mathcal{A}_\alpha$, $\alpha \in I$, is sometimes used — see ⟦Di 2⟧, Sec.1.3.3. The direct sum of C^*–algebras does not coincide with the direct sum of the Banach spaces $\mathcal{A}_\alpha$ — *cf.* Section 1.5. The definition of the tensor product is adopted from ⟦Di 2⟧, Sec.2.12.15; ⟦BR 1⟧, Sec.2.7.2. In general, there are more candidates for the role of the tensor product (see the notes to Section 2.4); a discussion can be found in [Lan 1].

Section 6.2 The definition of a cone appears in the literature in various modifications — see, *e.g.*, ⟦Nai 1⟧, Sec.3.10; ⟦RS 1⟧, Sec.IV.4. The proof of Theorem 6.2.1 uses the functional calculus on C^*–algebras; it can be found in ⟦BR 1⟧, Sec.2.2.2; ⟦Di 2⟧, Sec.1.6; ⟦Ru 2⟧, Sec.11.28; ⟦Sa⟧, Sec.1.4. Proposition 6.2.2 remains valid in Banach $*$–algebras with the unit element but the proof is more complicated — *cf.* ⟦Ru 2⟧, Sec.11.31. Other properties of positive functionals are discussed in ⟦BR 1⟧, Sec.2.3.2.

The GNS construction was formulated in the papers [GN 1], [Seg 1]. Due to Theorem 6.2.3 all GNS representations are unitarily equivalent, so it is possible to speak about properties of such a representation without a more detailed specification. The sketched proof of Proposition 6.2.5 is given with all details in ⟦Ru 2⟧, Thm.12.39; see also ⟦BR 1⟧, Lemma 2.3.2; ⟦Di 2⟧, Thm.2.6.1.

A pure state f on $\mathcal{A}$ can be defined equivalently as such that the inequality $f \geq g$, where g is a positive functional, implies $g = \lambda f$, $\lambda \in [0,1]$ — see [[BR 1]], Sec.2.3.2. Nonextremal points of $S_{\mathcal{A}}$ are called *mixed states.* Theorem 6.2.7 was proved for the first time in [Seg 1]; see also [[BR 1]], Thm.2.3.19; [[Di 2]], Sec.2.5 or [Si 1], Sec.5. For a necessary and sufficient condition under which a pure state can be extended from a subalgebra $\mathcal{B} \subset \mathcal{A}$ to the algebra $\mathcal{A}$ — *cf.* [And 1].

Section 6.3 The topologies introduced in Remark 6.3.1 are sometimes also called *ultrastrong* and *ultraweak;* however, these notions may be misleading because $\tau_{\sigma w}$ is stronger than τ_w. The fact that all the five topologies are mutually different in an infinite-dimensional Hilbert space is proven in [[Di 2]], Sec.I.3. This still does not exhaust all nonequivalent locally convex topologies on $\mathcal{B}(\mathcal{H})$; another pair of them can be found in [[BR 1]], Sec.2.4.1.

The foundations of the theory of W^*-algebras were formulated in a series of papers by J. von Neumann starting from 1929. There are numerous monographs devoted to this subject, *e.g.*, [[BR 1]], Chap.2; [[Di 1]] and Appendix A to [[Di 2]]; [[Em]], Chap.2; [[Nai 1]], Chap.VII; [[Sa]], *etc.,* where additional information can be found. Corollary 6.3.5 provides us with an alternative way to define a W^*-algebra; in that case the condition $I \in \mathcal{A}$ is sometimes dropped — see [[Nai 1]], Sec.34.1. Another easy consequence of Theorem 4 is the so-called *bicommutant theorem:* let $\mathcal{A}$ be a nondegenerate $*$-subalgebra in $\mathcal{B}(\mathcal{H})$; then the following conditions are equivalent:

(a) $\mathcal{A} = \mathcal{A}''$.

(b) $\mathcal{A}$ is closed in any of the topologies τ_w, τ_s, $\tau_{\sigma w}$, $\tau_{\sigma s}$.

(c) $\mathcal{A}$ is closed in all of the topologies τ_w, τ_s, $\tau_{\sigma w}$, $\tau_{\sigma s}$.

This result can be extended to some other topologies on $\mathcal{B}(\mathcal{H})$. Moreover, the conditions (a)–(c) are equivalent to closedness of the unit ball, $\mathcal{A}_1 := \{ B \in \mathcal{A} : \|B\| \leq 1 \}$; see [[BR 1]], Thm.2.4.11; [[Di 1]], Sec.I.4. Still another consequence of Theorem 4 is the von Neumann density theorem (Problem 9); a stronger result is the Kaplansky density theorem — *cf.* [[BR 1]], Thm.2.4.16; [[Di 1]], Sec.I.5.

There is also an abstract way to define W^*-algebras as a particular class among all C^*-algebras with the unit element. Given a W^*-algebra $\mathcal{A} \subset \mathcal{B}(\mathcal{H})$ we define its *predual* $\mathcal{A}_*$ as the set of all σ-weakly continuous linear functionals on $\mathcal{A}$; since any such functional is norm continuous we have $\mathcal{A}_* \subset \mathcal{A}^*$. It can be shown that $\mathcal{A}_*$ is a Banach space with respect to $\|\cdot\|_{\mathcal{A}^*}$ and $\mathcal{A}$ is isometrically $*$-isomorphic to the dual space of $\mathcal{A}_*$ by $\varphi : (\varphi(a))(f) = f(a)$ — see [[BR 1]], Prop.2.4.18; [Si 1], Thm.6.4. We usually say briefly that $\mathcal{A}$ is dual to $\mathcal{A}_*$; hence the name. The converse to this result is the so-called *Sakai theorem:* A C^*-algebra with the unit element can be represented isometrically by means of some W^*-algebra on a Hilbert space $\mathcal{H}$ *iff* it can be identified with a dual of some Banach space (*cf.* [[Sa]], Thm.1.16.17). This can be used to formulate the mentioned abstract definition.

The described classification of W^*-algebras was formulated in a series of papers by F.J. Murray and J. von Neumann, published in 1936–43. The proofs of Proposition 6.3.8 to Theorem 6.3.14 can be found, *e.g.*, in [[Di 1]], Secs.I.6–9, III.1–3; [[Nai 1]], Secs.36–38, or [[Sa]], Chap.2 together with many details and related results. The type of a given factor can be determined equivalently using the *relative dimension,* $d_{\mathcal{A}}$, which is defined as the restriction of a normal faithful trace on $\mathcal{A}$ to the set $\mathcal{A}^E$ of all projections in $\mathcal{A}$; this makes sense because any two normal faithful semifinite traces on $\mathcal{A}$ are proportional to each other. The function $d_{\mathcal{A}}$ has the following properties: (i) $d_{\mathcal{A}}(E) = 0$ *iff* $E = 0$, (ii) $d_{\mathcal{A}}(E_1) = d_{\mathcal{A}}(E_2)$ if E_1, E_2 are equivalent with respect to , $\mathcal{A}$, (iii) $d_{\mathcal{A}}(E_1) \le d_{\mathcal{A}}(E_2)$ if $E_1 \le E_2$, (iv) $d_{\mathcal{A}}(E_1{+}E_2) = d_{\mathcal{A}}(E_1){+}d_{\mathcal{A}}(E_2)$ if $E_1E_2 = 0$, and finally, (v) $d_{\mathcal{A}}(E) < \infty$ for all $E \in \mathcal{A}^E$ *iff* the reduced algebra $\mathcal{A}_E$ is finite. The range of the relative dimension determines the type of a factor; for different types, with a suitable normalization, the set $d_{\mathcal{A}}(\mathcal{A}^E)$ equals

$$\begin{array}{lll} \{1,2,\dots,n\} & \dots & I_n \\ \{1,2,\dots,\infty\} & \dots & I_\infty \\ [0,1] & \dots & II_1 \\ [0,\infty] & \dots & II_\infty \\ \{0,\infty\} & \dots & III \end{array}$$

Type I algebras usually appear in quantum mechanical applications (see Sections 7.4 and 7.6 below) while in quantum field theory and statistical physics we often meet algebras of type III — see, *e.g.*, [Ara 1] or ⟦BR 1⟧, Sec.5.2 and the notes to it. This has stimulated a search for a finer classification of these algebras; it appears that there are infinitely many nonisomorphic factors of type III, the so–called type III_λ for $\lambda \in [0,1]$. More details are given in [Con 1] or ⟦BR 1⟧, Sec.2.7.3, where some results concerning a classification of C^*–algebras are also mentioned. Numerous examples of W^*–algebras of types II and III are given in ⟦Sa⟧, Chap.4.

Section 6.4 The direct sum $\sum_{\alpha\in J} E_\alpha$ in the definition of a normal state belongs to $\mathcal{A}$ due to ⟦Di 1⟧, Appendix II. A state f on $\mathcal{A}$ is normal *iff* $f(\sup_{\alpha\in J} A_\alpha) = \sup_{\alpha\in J} f(A_\alpha)$ holds for any nondecreasing net $\{A_\alpha\} \subset \mathcal{A}_+$ — see, *e.g.*, [Kad 1] or ⟦BR 1⟧, Sec.2.3.4; ⟦Di 1⟧, Sec.I.4. This property is often used as a definition of normal states, while the states which satisfy relation (6.2) are called *completely additive.* Another equivalent formulation is the following (⟦BR 1⟧, Thm.2.7.11): f is normal *iff* $f(A) = \sum_{j=1}^{\infty} f(A_j)$ holds for any sequence $\{A_j\} \subset \mathcal{A}_+$ such that $A := \text{w-lim}_{n\to\infty} \sum_{j=1}^{n} A_j \in \mathcal{A}_+$. This is not surprising since the right side of relation (6.2) is finite only if there is at most countable number of nonzero values $f(E_\alpha)$.

The presented proof of Theorem 1 is adapted from ⟦Di 1⟧; a somewhat different proof can be found in ⟦BR 1⟧, Thm.2.4.21. The result easily extends to σ–weakly continuous functionals on $\mathcal{A}$, which are neither positive nor normalized (Problem 17). Notice also that any state is normal if $\dim \mathcal{H} < \infty$; then only assertion (c) provides nontrivial information. The proof of Lemma 6.4.3 leans on the following non-trivial result: a linear functional f on a weakly closed subspace $\mathcal{M} \in \mathcal{B}(\mathcal{H})$ is σ–weakly (σ–strongly) continuous *iff* its restriction $f \upharpoonright \mathcal{M}_1$ to the unit ball is σ–weakly (respectively, σ–strongly) continuous — *cf.* ⟦Di 1⟧, Sec.I.3 for more details.

The GNS representation corresponding to a normal state need not in general be faithful as Example 6.2.4c illustrates. Theorem 6.4.8 comes from the paper [Ca 1], where an example can be also found of a faithful representation of the factor algebra $\mathcal{B}(\mathcal{H})/\mathcal{K}(\mathcal{H})$, which proves the existence of noninjective representations of $\mathcal{B}(\mathcal{H})$.

Section 6.5 The term "symmetric set" is often replaced by $*$–*invariant set,* which is more appropriate in a sense when the set contains unbounded operators which are not closed. The main result of this section is Theorem 6.5.2, which is usually called the *theorem on a generating operator;* it was proved first by J. von Neumann — see ⟦Di 1⟧, Sec.I.7; ⟦RN⟧, Sec.IX.1 or ⟦AG⟧, Secs.90, 92. In the case of a separable $\mathcal{H}$, Theorem 5.5.6 allows us to define the generating operator equivalently by the requirement $\mathcal{S} \subset \{A\}''_{ex}$.

Section 6.6 Given a subset $\mathcal{S}$ of $\mathcal{B}(\mathcal{H})$ we call $\phi \in \mathcal{H}$ a *cyclic vector* if $\overline{\mathcal{A}_0(\mathcal{S})\phi} = \mathcal{H}$. In particular, if $\mathcal{S}$ is a $*$–algebra containing the unit operator, then ϕ is cyclic for $\mathcal{S}$ *iff* it is cyclic for $\mathcal{S}''$ (Problem 32). The generating vector of a Hermitean operator A defined in Section 5.8 is at the same time a cyclic vector of algebra $\{A\}''$ — *cf.* Problem 5.43.

Problem 28 has an interesting consequence. If $A \in \mathcal{S}''$ is a generating operator of the set $\mathcal{S} = \{Q_1, Q_2\}$, then there are real–valued Borel functions g and f_j such that $A = g(Q_1, Q_2)$ and $Q_j = f_j(A)$, $j = 1, 2$. Changing the functions $f_j \circ g$ on a $(\mu_1 \otimes \mu_2)$–zero set if necessary we can achieve that the relations $f_j(g(x_1, x_2)) = x_j$ hold for all $[x_1, x_2]$ from any fixed bounded interval $J \subset \mathbb{R}^2$; in this way we get a bijection between J and a certain subset of $\mathbb{R}$. A similar map, the so–called *Peano curve*, is used in set theory to prove that $\mathbb{R}^2$ and $\mathbb{R}$ have the same cardinality — see, *e.g.*, ⟦ vN ⟧, Sec.II.10, or ⟦ Al ⟧, Sec.5.2.

If $B \in \mathcal{B}(\mathcal{H}_1 \otimes \mathcal{H}_2)$ commutes with all operators of the form $I_1 \otimes C$ with $C \in \mathcal{B}(\mathcal{H}_2)$, it commutes particularly with $V_\alpha V_\beta^+$ defined in the proof of Theorem 6.6.5, and therefore $B = B_1 \otimes I_2$ for some $B_1 \in \mathcal{B}(\mathcal{H}_1)$, *i.e.*, we get the relation $(\mathbb{C}(\mathcal{H}_1) \otimes \mathcal{B}(\mathcal{H}_2))' = \mathcal{B}(\mathcal{H}_1) \otimes \mathbb{C}(\mathcal{H}_2)$. In a similar way, we find $(\mathcal{B}(\mathcal{H}_1) \otimes \mathcal{B}(\mathcal{H}_2))' = \mathbb{C}(\mathcal{H}_1) \otimes \mathbb{C}(\mathcal{H}_2)$. These two identities represent particular cases of the relation $(\mathcal{A}_1 \otimes \mathcal{A}_2)' = \mathcal{A}_1' \otimes \mathcal{A}_2'$, which is valid for any semi–infinite W^*–algebras $\mathcal{A}_1$, $\mathcal{A}_2$ — see ⟦ Di 1 ⟧, Sec.I.6.

Section 6.7 The definition of irreducibility of operator sets corresponds to the one–operator case discussed in Section 4.4. It is worth mentioning that this terminology does not fully correspond to that used in the representation theory of groups and algebras (*cf.* Appendix B.1). In view of Problem 33a, a representation is irreducible *iff* its image is an irreducible operator set, provided the latter is symmetric; this is true, for instance, if we deal with representations of $*$–algebras or unitary representations of groups. In the general case those representations whose image is a reducible set are called *completely reducible* — see, *e.g.*, ⟦ Žel ⟧, Sec.16.

There is another viewpoint from which representations can be classified. A representation π of an algebra $\mathcal{A}$ by means of linear operators on some vector space V is called *algebraically irreducible* if $\pi(\mathcal{A})$ has no nontrivial invariant subspace. If V is a topological vector space and $\pi(\mathcal{A})$ has no nontrivial closed subspace, π is said to be *topologically irreducible* — *cf.* ⟦ Di 2 ⟧, Sec.2.3; ⟦ Kir ⟧, Sec.7.1; ⟦ Nai 3 ⟧, Sec.III.4. It is clear that the algebraic irreducibility of a given representation implies its topological irreducibility, while the converse is not true even if $\pi(\mathcal{A}) \subset \mathcal{B}(\mathcal{H})$ in spite of Problem 33d, because it is not ensured that $\overline{L} \neq \mathcal{H}$ unless $\dim \mathcal{H} < \infty$. Irreducibility of representations in the following always means the topological irreducibility; in the case of representations by unbounded operators the definition contains additional requirements on the domains of operators from $\pi(\mathcal{A})$ — *cf.* ⟦ BaR ⟧, Secs.11.1 and 16.5.

The irreducibility criteria of Theorem 1 are standard — see, *e.g.*, ⟦ BR 1 ⟧, Prop.2.3.8 or ⟦ Di 2 ⟧, Prop.2.3.1; we exclude the trivial case $\dim \mathcal{H} = 1$. The equivalence of conditions (a) and (b) is usually referred to as *Schur's lemma*. On the other hand, in group representation theory this term often means the following implication (⟦ Ham ⟧, Sec.3.14; ⟦ Boe ⟧, Sec.I.7): if a set $\mathcal{S}$ of operators on a finite–dimensional vector space V has no nontrivial invariant subspace, then $\mathcal{S}' = \mathbb{C}(V)$ (see also Problem 33d); the same is true even if V has a countable algebraic dimension — *cf.* ⟦ Kir ⟧, Sec.8.2. The name is sometimes also used for a more general result about intertwining operators between a pair of representations of the same group or algebra — see ⟦Ham⟧, Sec.3.14; ⟦Želⅿ⟧, Sec.20; ⟦Kir⟧, Sec.8.2. An analogy of the Schur lemma for real vector spaces of a countable dimension is given in ⟦Kir⟧, Sec.8.2; a generalization to operator sets in a real Hilbert space can be found in [CM 1].

The conclusion about expressing each bounded operator by means of an irreducible set can be strengthened substantially in a finite–dimensional vector space V. The symmetry assumption is not required (after all, it makes sense only if V is equipped with an inner product) and the irreducibility condition can also be modified: if $\mathcal{S} \subset \mathcal{L}(V)$ has no nontrivial invariant subspace, then any (linear) operator on V is a polynomial in the elements of $\mathcal{S}$, *i.e.*, $\mathcal{A}_0(\mathcal{S}) = \mathcal{L}(V)$; this result is usually called the *Burnside theorem* — see, *e.g.*, ⟦Žel⟧, Sec.21.

The fact that functional calculus cannot, in general, be introduced, for noncommuting operators does not mean that there are no reasonable definitions of functions for particular classes of noncommuting self–adjoint operators. A prime example is provided by operators P and Q for which one can use the Fourier transformation to introduce a family of their functions called *pseudo–differential operators* — see, *e.g.*, ⟦Hör⟧; they have recently been applied to semiclassical analysis of Schrödinger operators — *cf.* ⟦Hel⟧.

Section 6.8 The theory of unbounded operator algebras has two main sources of inspiration: the representation theory of Lie algebras and algebraic quantum field theory. We might add that it has not yet reached its maturity; the results were found only in journal literature until recently the monograph ⟦Schm⟧ appeared covering the subject. The reader can find a much deeper treatment there; we also refer to the review papers [Bor 1], [Las 3] and [Vas]. The notion of *Op**–algebra was introduced by G. Lassner (1972); properties of these algebras are discussed, *e.g.*, in [Epi 1], [Las 1–3], [Pow 1], [VSH 1], and in the other papers quoted below. The set $\mathcal{L}^+(D)$ is alternatively denoted as $\mathcal{C}_D$ — see [Epi 1], [ET 1–3] — and a separate term is sometimes introduced for *Op**–algebras $\mathcal{A}$ in which $(I+T^+T)^{-1}$ belongs to $\mathcal{A} \cap \mathcal{B}(\mathcal{H})$ (compare the notes to Section 6.1). Example 6.8.1 represents a particular case of the algebra $\mathcal{S} \subset \mathcal{L}^+(\mathcal{S}(\mathbb{R}^n))$ generated by the restriction of the operators Q_j, P_k, $j,k = 1,\dots,n$, to the subspace $\mathcal{S}(\mathbb{R}^n) \subset L^2(\mathbb{R}^n)$; the proof that $\mathcal{S}$ is closed and self–adjoint can be found in [Pow 1], Sec.5.

Proof of Theorem 6.8.2 is given in [Las 2]. In addition to τ_D, there are other ways to equip $\mathcal{A}$ with a topological structure [Las 1–3]; all these topologies generalize the operator–norm topology on $\mathcal{B}(\mathcal{H})$; in general, however, they do not yield a continuous involution. Other possible topological structures are discussed, *e.g.*, in [ArJ 1], [Schm 2]. In the cases when (D,τ_A) is a Fréchet space the domains D for closed *Op**–algebras can be classified — see [LT 2].

While *Op**–algebras in a sense represent an analogy to C^*–algebras, some special classes of them have been studied with the aim of constructing an unbounded counterpart to the theory of W^*–algebras — *cf.* [AJ 1], [ET 3]. On the other hand, there are structures which generalize the concept of an *Op**–algebra; let us mention, for instance, the *partial* $*$*–algebras* discussed in [AK 1], [AIT 1], where the product is defined only for some pairs of operators. We encounter such objects in quantum mechanics, *e.g.*, when, together with the operators Q, P of Example 1, we also consider $H := (2m)^{-1}P^2 + V(Q)$ which need not preserve $\mathcal{S}(\mathbb{R})$.

In *Op**–algebras we may use matrix representations: an operator $T \in \mathcal{L}^+(D)$ on a separable $\mathcal{H}$ is uniquely determined by the numbers $\{(\phi_j, T\phi_k) : j,k = 1,2,\dots\}$ for any orthonormal basis $\{\phi\}_{j=1}^{\infty} \subset D$ — see [Epi 1], [ET 1]. Some results concerning ideals in $\mathcal{B}(\mathcal{H})$ can also be extended to $\mathcal{L}^+(D)$. For instance, $\mathcal{L}^+(D)$ contains several ideals which turn into $\mathcal{K}(\mathcal{H})$ in the case $D = \mathcal{H}$; two of them are such that the set $\mathcal{F}(D)$ of all finite–dimensional operators from $\mathcal{L}^+(D)$ is dense in them with respect to τ_D — *cf.* [Tim 1]. In some cases one can prove that $\mathcal{L}^+(D)$ contains just one τ_D–closed ideal $\mathcal{C}$; it is also possible to construct a faithful representation of the factor algebra $\mathcal{L}^+(D)/\mathcal{C}$, which generalizes the Calkin result mentioned in the notes to Sec.6.4 — see [LöT 1]. The paper [Tim 2] describes a way of using the ideals in $\mathcal{B}(\mathcal{H})$, *e.g.*, $\mathcal{J}_p$, $p \geq 1$, to construct ideals in $\mathcal{L}^+(D)$.

A more complete version of Theorem 6.8.3 and related results can be found in [Schm 1]. It is worth mentioning that this assertion is considerably weaker than Theorem 6.4.1 since in the case $D = \mathcal{H}$ condition (a) holds only if $\dim \mathcal{H} < \infty$. Similar restrictions apply to other results on continuity of the functionals which admit a trace representation — *cf.* [LT 1]. On the other hand, an assertion analogous to Theorem 6.4.1 can be proven for a class of generalized W^*–algebras mentioned above — see [AJ 1].

Problems

1. The relations $\|a\| = \sup_{\|b\|\leq 1} \|ab\| = \sup_{\|b\|=1} \|ab\| = \max_{\|b\|=1} \|ab\|$ are valid in any C^*–algebra.

2. Let $\mathcal{A}$ be a C^*–algebra with the unit element. Prove

(a) The relation $a^*a = e$ implies $r(a) = 1$. In addition, if the element a is unitary, then $\sigma_{\mathcal{A}}(a) \subset \{\lambda \in \mathbb{C} : |\lambda| = 1\}$.

(b) $r(a) = \|a\|$ holds for any normal $a \in \mathcal{A}$, and moreover $\sigma_{\mathcal{A}}(a) \subset [-\|a\|, \|a\|]$ provided a is Hermitean.

(c) If p is a complex polynomial, then $\sigma_{\mathcal{A}}(p(a)) = p(\sigma_{\mathcal{A}}(a))$ holds for any $a \in \mathcal{A}$.

Hint: (a) Use Theorem B.2.4d. (b) If $a = a^*$, $|\lambda| > \|a\|$, consider the element $(a+i|\lambda|e)(a-i|\lambda|e)^{-1}$. (c) Use the root decomposition of $p(a)-\lambda e$.

3. Let $\mathcal{A}$ be a $*$–algebra with the unit element; then there is at most one norm which makes it a C^*–algebra.
Hint: Use the relation $\|a\| = r(a^*a)^{1/2}$.

4. The direct sum of C^*–algebras $\mathcal{A}_\alpha$, $\alpha \in I$, with the norm $\|[a_\alpha]\|_\infty := \sup_{\alpha\in I} \|a_\alpha\|_\alpha$ is a C^*–algebra.

5. To given C^*–algebras $\mathcal{A}_1$, $\mathcal{A}_2$ there is at least one realization of their tensor product. If $(\mathcal{A}, \otimes)$ and $(\tilde{\mathcal{A}}, \tilde{\otimes})$ are two such realizations, then there is a isometric $*$–isomorphism $\varphi : \mathcal{A} \to \tilde{\mathcal{A}}$ such that $\varphi(a_1)\tilde{\otimes}\varphi(a_2) = \varphi(a_1 \otimes a_2)$.
Hint: Use Theorem 6.2.6.

6. Prove the relations between the five topologies in Remark 6.3.1.

7. The neighborhoods $U_\varepsilon(B; \Phi) := \{C : p_\Phi(C-B) < \varepsilon\}$ for all $\Phi \in \tilde{\mathcal{H}}$ form a basis of the σ–strong topology in $\mathcal{B}(\mathcal{H})$. Is the analogous assertion valid for the σ–weak

topology?
Hint: Components of Φ_j, $j = 1, \dots, n$, can be arranged into a single vector $\Phi \in \tilde{\mathcal{H}}$.

8. Prove: (a) Let $\mathcal{S}$ be a symmetric subset in $\mathcal{B}(\mathcal{H})$; then $\mathcal{S}'$ is a W^*–algebra. Moreover, the W^*–algebra $\mathcal{A}_W(\mathcal{S})$ generated by $\mathcal{S}$ equals $\mathcal{S}''$, and $\mathcal{A}_W(\mathcal{S}) = \mathcal{A}_W((\overline{\mathcal{S}})_w)$.

(b) Let E be a projection different from $0, I$. The algebra $\mathcal{A} := \{\alpha E : \alpha \in \mathbb{C}\}$ is weakly closed and $\mathcal{A}'' = \{\alpha E + \beta(I - E) : \alpha, \beta \in \mathbb{C}\}$.

(c) Given a set $\mathcal{S} \subset \mathcal{B}(\mathcal{H})$ we define $\mathcal{G}_{\mathcal{S}} := (\bigcap_{B \in \mathcal{S}} \operatorname{Ker} B)^{\perp}$ and denote the corresponding projection as $E_{\mathcal{S}}$. Then $BE_{\mathcal{S}} = B$ holds for all $B \in \mathcal{S}$; if $\mathcal{S}$ is symmetric we also have $E_{\mathcal{S}} B = B$.

Hint: (b) Any $B \in (\overline{\mathcal{A}})_w$ is of the form $B = \alpha E$, $\alpha \in \mathbb{C}$.

9. Prove the *von Neumann density theorem:* Any nondegenerate $*$–subalgebra $\mathcal{A} \subset \mathcal{B}(\mathcal{H})$ is dense in $\mathcal{A}''$ with respect to any of the topologies $\tau_w, \tau_s, \tau_{\sigma w}, \tau_{\sigma s}$.

10. Let $\mathcal{A}$ be a W^*–algebra; then the relations $(\mathcal{A}_E)' = (\mathcal{A}')_E$ and $\mathcal{Z}_{\mathcal{A}_E} = (\mathcal{Z}_{\mathcal{A}})_E$ hold for any projection $E \in \mathcal{A}$.

11. Let $\mathcal{A} \subset \mathcal{B}(\mathcal{H})$ be a W^*–algebra. A projection $E \in \mathcal{A}$ is said to be *minimal* if there are no nonzero projections E_1, E_2 in $\mathcal{A}$ such that $E = E_1 + E_2$. In that case, $\|B\phi\| = \|BE\|\,\|\phi\|$ holds for all $\phi \in E\mathcal{H}$, $B \in \mathcal{A}$. Any minimal projection in $\mathcal{A}$ is Abelian.
Hint: Use Lemma 6.5.3.

12. Let $\varphi : \mathcal{A} \to \mathcal{B}$ be a $*$–morphism of W^*–algebras; then

(a) If $B \in \mathcal{A}$ is Hermitean (respectively, positive, a projection, unitary), the same is true for $\varphi(B)$.

(b) $\|\varphi(B)\| \le \|B\|$ holds for any $B \in \mathcal{A}$; it turns into identity if φ is injective.

(c) If $A \in \mathcal{A}$ is Hermitean and g is a continuous real–valued function, then $\varphi(g(A)) = g(\varphi(A))$.

(d) $\varphi(\mathcal{Z}_{\mathcal{A}}) = \mathcal{Z}_{\mathcal{B}}$ holds provided φ is bijective.

Hint: (b) Use Proposition 6.1.2. (c) Prove the assertion first for polynomials, then use Theorem 5.2.2.

13. Prove Proposition 6.3.7.

14. Suppose that $\mathcal{A} \subset \mathcal{B}(\mathcal{H})$ is a W^*–algebra whose center contains a complete system of projections $\{E_\alpha : \alpha \in J\}$ (*cf.* Section 3.2); then there is a spatial isomorphism between $\mathcal{A}$ and $\bigoplus_{\alpha \in \mathcal{J}} \mathcal{A}_{E_\alpha}$.

15. Using Proposition 6.3.8 show that any discrete factor is homogeneous.

16. The map $\tau : \tau(B) = \operatorname{Tr} B$ is a faithful normal trace on $\mathcal{B}(\mathcal{H})$.

17. The assumption that W commutes with all E_α in Theorem 6.4.7 is essential: with the exception of the trivial case, $\operatorname{card} J = 1$, there are other statistical operators without this property which satisfy relation (6.5).
Hint: Consider the algebra $\mathcal{A} := \mathbb{C} \oplus \mathbb{C}$.

18. For any σ–weakly continuous linear functional on a W^*–algebra $\mathcal{A} \subset \mathcal{B}(\mathcal{H})$ there is a trace class operator W such that $f(B) = \operatorname{Tr}(WB)$ holds for all $B \in \mathcal{A}$. In a similar way, generalize Theorem 6.4.7a.
Hint: Modify the proof of Theorem 6.4.1.

19. Suppose that $\mathcal{J}$ is a two–sided ideal in algebra $\mathcal{B}(\mathcal{H})$ on a separable $\mathcal{H}$, and E, F are projections; then

(a) $\mathcal{J}^E = \{0\}$ implies $\mathcal{J} = \{0\}$.

(b) If E, F have the same dimension and $E \in \mathcal{J}$, then also $F \in \mathcal{J}$.

(c) Any projection in $\mathcal{J}$ is finite–dimensional.

(d) If $E \geq F$ and $E \in \mathcal{J}$, then also $F \in \mathcal{J}$.

Hint: (a) Let $B \in \mathcal{J}$; then the spectral decomposition of $A := B^*B$ contains a nonzero projection $E_A(J)$ such that $A \upharpoonright E_A(J)\mathcal{H}$ is invertible. (b) Use a suitable partial isometry.

20. Prove Theorem 6.3.8.
Hint: If $\mathcal{J}$ is such an ideal, the set $\mathcal{J}^E$ consists of all finite–dimensional projections. To prove $\mathcal{K}(\mathcal{H}) \subset \mathcal{J} \subset \mathcal{K}(\mathcal{H})$, one can consider Hermitean operators only; use the argument from the preceding problem to show that $E_A(J) \in \mathcal{J}$ provided $0 \notin J$.

21. Let $\mathcal{H}$ be a separable Hilbert space; then

(a) Algebra $\mathcal{B}(\mathcal{H})$ has no nontrivial two–sided ideal which is weakly (strongly, σ–weakly, σ–strongly) continuous.

(b) Algebra $\mathcal{K}(\mathcal{H})$ has no nontrivial two–sided norm–closed ideal.

22. Let $\mathcal{S}$, $\mathcal{S}_1$, $\mathcal{S}_2$ be symmetric operator sets; then

(a) Any $T \in \mathcal{S}$ is closable and $\overline{T} \in \mathcal{S}$.

(b) If the set $\mathcal{S}_1 \setminus \mathcal{S}_2$ consists of closed operators only, it is symmetric; find an example showing that the closedness assumption is essential.

(c) The union (intersection) of any family of symmetric sets is symmetric.

23. A real subalgebra in the set of all Hermitean operators on a given $\mathcal{H}$ is commutative.

24. Let $\mathcal{S}_R$, $\mathcal{S}_p$, $\mathcal{S}_f$ be the sets introduced in Section 6.5 to a given $\mathcal{S} \subset \mathcal{L}_{sa}(\mathcal{H})$; then

(a) $\mathcal{S}_p$ is symmetric *iff* $\mathcal{S}$ is symmetric; in that case $\mathcal{S}_f$ is also symmetric. The set $\mathcal{S}_f$ can be symmetric even if $\mathcal{S}$ and $\mathcal{S}_p$ are not symmetric.

(b) If any of the sets $\mathcal{S}$, $\mathcal{S}_p$, $\mathcal{S}_f$ is commutative, the other two are also commutative.

(c) If $\mathcal{S}$ is symmetric, it is commutative *iff* the same is true for $\mathcal{S}_R$.

(d) $\mathcal{S}' = (\mathcal{S}_p)' = (\mathcal{S}_f)' = (\mathcal{S}_R)'$; if $\mathcal{S}$ is symmetric we also have $\mathcal{S}' = (\mathcal{S}_R)'$.

25. Given a set $\mathcal{S} \subset \mathcal{L}_{b,sa}(\mathcal{H})$ we construct $\mathcal{S}_a$ in the following way: we keep all bounded operators and replace each unbounded self–adjoint $A \in \mathcal{S}$ by $A_a := \arctan A$. Show that set $\mathcal{S}_a$ has the same properties as $\mathcal{S}_p$ in the preceding problem.

26. Prove: (a) Let $\mathcal{H}$ be separable. Corollary 6.5.6 extends to the situation when $\mathcal{S}$ is a commutative symmetric subset in $\mathcal{L}_{b,n}(\mathcal{H})$, where the last named set is obtained by adding to $\mathcal{B}(\mathcal{H})$ all normal operators on $\mathcal{H}$.

(b) Suppose that A is a generating operator of a set $\mathcal{S} \subset \mathcal{L}_{b,sa}(\mathcal{H})$; then $\{A\}' = \mathcal{S}'$ and $\{A\}'' = \mathcal{S}''$.

(c) $\mathcal{S}'' \supset \bigcup_{B\in\mathcal{S}}\{B\}''$ holds for any set $\mathcal{S} \subset \mathcal{L}_{b,sa}(\mathcal{H})$ while the opposite inclusion need not be valid.

27. Let $\mathcal{S} = \{A\}_{j=1}^N$ be a commutative family of self–adjoint operators on a separable Hilbert space $\mathcal{H}$. Using the notation introduced in Section 6.6, show that $P^{(n)} \in \mathcal{S}''$ holds for all $\{k\}_n \in K_N$ and $\sum_n P^{(n)} = I$.

28. The result of Example 6.6.3a can be extended: the sets $\{f(Q)\}$ and $\{f(P)\}$ are CSCO for any continuous strictly monotonic function $f : \mathbb{R} \to \mathbb{R}$. What does the commutant $\{f(Q)\}'$ look like if the function f is continuous but only *piecewise* strictly monotonic?
Hint: Cf. [Nas 1].

29. Define the operators Q_j, $j = 1,2$ on $L^2(\mathbb{R}^2, d(\mu_1 \otimes \mu_2))$, where the measures μ_j are supposed to be Borel, not necessarily finite, by $(Q_j\psi)(x_1,x_2) := x_j\psi(x_1,x_2)$. Prove:

(a) The set $\{Q_1, Q_2\}$ is a CSCO.

(b) $Q_1 = \overline{Q_{\mu_1} \otimes I_2}$ and $Q_2 = \overline{I_1 \otimes Q_{\mu_2}}$.

Hint: (a) Let ϕ_j be a cyclic vector for Q_{μ_j} on $L^2(\mathbb{R}, d\mu_j)$; show that $\phi_{12} : \phi_{12}(x_1,x_2) = \phi_1(x_1)\phi_2(x_2)$ is cyclic for $\{Q_1, Q_2\}$.

30. Show that the operators V_α defined in the proof of Theorem 6.6.5 obey the relations $V_\alpha^+(B \otimes I_2)V_\beta = \delta_{\alpha\beta}B$ for any $B \in \mathcal{B}(\mathcal{H}_1)$, and moreover, $V_\alpha V_\beta^+ = (\psi_\beta, \cdot)\psi_\alpha$.

31. The W^*–algebras generated by symmetric sets $\mathcal{S}_j \subset \mathcal{B}(\mathcal{H}_j)$, $j = 1,2$, satisfy the relation $\mathcal{A}_W(\underline{\mathcal{S}}_1) = \mathcal{A}_W(\mathcal{S}_1) \otimes \mathbb{C}(\mathcal{H}_2)$ and the analogous identity is valid for $\mathcal{A}_W(\underline{\mathcal{S}}_2)$. In addition, if each of the sets $\mathcal{S}_j$ contains the unit operator, then also $\mathcal{A}_W(\{ A_1 \otimes A_2 : A_j \in \mathcal{S}_j \}) = \mathcal{A}_W(\mathcal{S}_1) \otimes \mathcal{A}_W(\mathcal{S}_2)$.

32. Let $\mathcal{A}$ be a $*$–subalgebra in $\mathcal{B}(\mathcal{H})$ containing the unit operator. If $\mathcal{A}''\phi$ is dense in $\mathcal{H}$ for some vector $\phi \in \mathcal{H}$, then also $\overline{\mathcal{A}\phi} = \mathcal{H}$.

33. Prove: (a) A symmetric operator set $\mathcal{S}$ is reducible *iff* it has a nontrivial closed invariant subspace $\mathcal{G}$ and $E_\mathcal{G}D(T) \subset D(T)$ holds for any $T \in \mathcal{S}$.

(b) Let $\mathcal{S} \subset \mathcal{L}_{b,sa}(\mathcal{H})$ be a symmetric set. If any of the sets $\mathcal{S}$, $\mathcal{S}''$, $\mathcal{S}_R$, $\mathcal{S}_p$, $\mathcal{S}_f$ and $\mathcal{S}_a$ (*cf.* Section 6.5 and Problem 25) is irreducible, then the others are also irreducible.

(c) If $L \subset \mathcal{H}$ is an invariant for $\mathcal{S} \subset \mathcal{B}(\mathcal{H})$, the same is true for $\overline{L}$.

(d) A nonsymmetric $\mathcal{S} \subset \mathcal{B}(\mathcal{H})$ can have an invariant subspace even if $\mathcal{S}' = \mathbb{C}(\mathcal{H})$.

Hint: (d) Consider the set of upper triangle matrices in $\mathcal{B}(\mathbb{C}^2)$.

34. Check the claims made in Remark 6.7.4.
Hint: Choose the basis in which L_3 is diagonal and use the commutation relations between Pauli matrices.

35. Let $\mathcal{A} \subset \mathcal{L}^+(D)$ be an Op^*–algebra. If $D = \mathcal{H}$; then $\mathcal{A} \subset \mathcal{B}(\mathcal{H})$ and the uniform topology coincides with the operator–norm topology in $\mathcal{B}(\mathcal{H})$.

Chapter 7

States and observables

From now on we shall discuss how the theory of Hilbert–space operators explained in the previous chapters is used in the treatment of quantum systems. To begin with, we have to say something about states and observables; these notions play a fundamental role in any branch of physics but for quantum theory a thorough analysis of them is exceptionally important.

7.1 Basic postulates

A good manner in physics is to start from classical mechanics. A classical system of point particles is completely described at a given instant if we know the values of all generalized coordinates and momenta; this information serves at the same time as the initial condition for the solution of the equations of motion, which determine the time evolution of a state as a trajectory in phase space. The simplest example is a point particle whose motion is constrained to a line; the phase space is the plane $\mathbb{R}^2$. A family of allowed trajectories is associated with any Hamiltonian, for instance, concentric circles correspond to the harmonic oscillator, $H(q,p) := (2m)^{-1}(p^2+m^2\omega^2q^2)$, with a proper choice of units.

In classical physics we usually disregard the way in which a given system has achieved its moving state. Generally speaking, this is the result of its previous history, which may include the evolution governed by the appropriate dynamical laws as well as a sequence of operations performed intentionally by an experimentalist, or possibly a combination of both. In some physical disciplines the "spontaneous" way is the only one, or at least it is strongly preferred, say, in astrophysics or in geophysics where human–made states (for example, artificial quakes of the Earth's crust) are rare. On the other hand, in the case of microscopic objects (elementary particles, atomic nuclei, *etc.*) we mostly deal with the *preparation of a state* in the true sense.

The reason why the difference between the two ways is more than a technical matter is hidden in the *measurements* we use to determine the state. Classical physics supposes — quite rightly in its own domain — that the influence of the measuring process on the investigated object can be made arbitrarily small. This is no longer true when we treat microscopic objects; the tools we use for measuring are macroscopic and experience tells us that their influence cannot be neglected. This fact itself need not represent an obstacle if we were able to formulate the dynamical laws governing the process of measurement. The trouble is that we do not know how to describe in a deterministic way the system consisting of a microscopic object and a measuring apparatus. Quantum theory provides, for instance, information about the behavior of an atom interacting with an electromagnetic field, but we are not able to predict *which* emulsion grain will blacken or *which one* of the Geiger counters placed in line with the Stern–Gerlach apparatus will click. We can compute and verify experimentally only the *probabilities* with which these events occur.

The conditions under which a state is prepared have different impacts on the final result; according to that we can regard them as *substantial* or *nonsubstantial.* There is no general criterion, which would allow us to decide about the substantiality of a given condition; an analysis of a particular physical situation is always required. Another general feature of the state preparation is its *replicability.* This is not only easily realized when we deal with microscopic objects, but it also represents a fundamental requirement, which must be fulfilled to verify the probabilistic predictions of the theory. To replicate a state means to prepare it according to the same prescription, with all the substantial conditions preserved.

Hence we arrive at the formulation which is usually presented as the *definition of a state*: a state is a result of a sequence of physical manipulations with the system, which together form a preparation of this state. Two states coincide if all the substantial conditions of their preparation are identical. In a similar way we can formulate the "operational" definition of an *observable* (*dynamical variable*): we associate with it a suitable instrument (measuring apparatus), which displays (records) a measured value when we let it interact with the system.

Of course, such a definition does not take into account the fact that the same physical quantity can be measured in substantially different ways. If we want to measure, for instance, the longitudinal momentum in a beam of charged particles produced by an accelerator (or more precisely, to find whether its value belongs to a chosen interval), we can use a magnetic separator, a pair of detectors with a delayed–coincidence link, or a differential Cherenkov counter provided the particles are relativistic. Hence using the above mentioned "operational" definitions one must keep in mind that there is no bijective relation between the observables and measuring instruments. Since measurement is an unavoidable part of the state preparation, the reproduction "under the identical substantial conditions" requires that the results of all the performed measurements should coincide irrespective of the experimental procedures used. On the other hand, in Section 7.5 below we shall discuss which collections of measurements have to give the same results in order to ensure that the states coincide.

7.1.1 Remarks: (a) There is also no clear distinction between systems and states or families of states. For example, two photons and an electron–positron pair can be regarded as two states of the same system since (under the appropriate kinematic conditions) one of these entities can evolve into the other. Similarly the pions π^+, π^0, π^- can be understood as three isotopic states of a single particle; the reader will easily find other examples.

(b) The substantial conditions of state preparation do not include — not only in the microscopic world — the time and place when and where the event occurred. E. Wigner has stressed that this observation represented historically the first invariance law, which enabled physics to be established as an exact science.

(c) In addition to dynamical variables, which are always related to a particular system, there are other measurable quantities. The most prominent example is time, which is a universal parameter for all nonrelativistic systems. Some characteristics of external fields acting on the system may also appear in the theory as parameters; recall the Stark or Zeeman effect. Other examples are provided by various universal constants: the electron charge e, Planck's constant $\hbar$, *etc.*

After this introduction we are going to discuss the ways, in which the states and observables are described in quantum theory. We begin with the following postulates:

(Q1a) A complex Hilbert space $\mathcal{H}$ called the **state (Hilbert) space** is associated with any quantum system.

(Q1b) A **ray**, *i.e.*, a one-dimensional subspace in $\mathcal{H}$ corresponds to any state of the system under consideration.

The states described by rays are called *pure;* a more general concept of a state together with the appropriate modification of the postulate (Q1b) will be given in Section 7.3 below. Each ray Ψ is spanned by some unit vector $\psi \in \Psi$; in common parlance the difference between the two objects is often disregarded. Any two unit vectors $\psi, \psi' \in \Psi$ differ at most by a phase factor, $\psi' = e^{i\gamma}\psi$ with $\gamma \in \mathbb{R}$.

Before we proceed, let us recall the considerations used to motivate the above postulates. They are based on an analysis of simple experiments, such as observation of the interference effect in polarized light understood as a beam of corpuscular photons, spin measurements by a pair of Stern–Gerlach devices of differing mutual orientation, *etc.* To explain the results, one has to adopt the following assumptions:

(i) If we perform a measurement of a certain observable on a system, which is in a state Ψ, then as a consequence the state changes into one of a set $\{\Psi_j\}_{j=1}^N$ which is determined by the observable. In this sense therefore any state of the system can be regarded as composed of the states $\{\Psi_j\}_{j=1}^N$.

(ii) The outcome of the measurement has a *probabilistic character*, *i.e.*, one is generally able to determine only the probability $P(\Psi, \Psi_j)$ with which the mentioned transition occurs. This probability becomes certainty if Ψ coincides with some Ψ_j; then we have $P(\Psi_k, \Psi_j) = \delta_{jk}$.

A more detailed analysis of particular experiments, where the number N is finite, shows that the set of states, which are composed in the above mentioned sense of N states, depends on $2(N-1)$ real parameters. A simple way to realize such a composition mathematically is to suppose that there is an injective map $\mathcal{R}$ from the set of states to the set of rays in some complex vector space. A state Ψ is then called the *superposition* of states Ψ_1, Ψ_2 provided there are vectors $\psi_j \in \mathcal{R}(\Psi_j)$ such that $\psi_1 + \psi_2$ belongs to $\mathcal{R}(\Psi)$. It is easy to check that the definition makes sense. If the map $\mathcal{R}$ is surjective at the same time we say that the *superposition principle* is valid. In general this need not be the case; we shall return to this problem in Section 7.4. For practical reasons in the following we shall employ the same symbols for states and for the corresponding rays.

Let us return now to assumptions (i) and (ii). Using the just introduced terminology we can say that Ψ is a superposition of the states $\{\Psi_j\}_{j=1}^N$; when the measurement is performed, it passes to the state Ψ_j with the probability $P(\Psi, \Psi_j)$. Since it must with certainty pass to some state, one has $\sum_{j=1}^N P(\Psi, \Psi_j) = 1$. If we assume that the state space is equipped with an inner product $(\cdot,\cdot)$ such that the states Ψ_j are mutually orthogonal with respect to it, we may set $P(\Psi, \Psi_j) := |(\psi, \psi_j)|^2 \|\psi\|^{-2} \|\psi_j\|^{-2}$, where ψ, ψ_j are any nonzero vectors from the respective rays. The function P defined in this way satisfies the relation $P(\Psi_j, \Psi_k) = \delta_{jk}$ and the normalization condition is valid due to the Parseval identity.

We are usually able to prepare the considered system not only in some of the states of the set $\{\Psi_j\}_{j=1}^N$ but also in other which are their superpositions; for that it is sufficient to measure on the system another observable. It is therefore useful to extend the above definition to any pair of states and to introduce the transition probability between them by $P(\Phi, \Psi) := |(\phi, \psi)|^2 \|\phi\|^{-2} \|\psi\|^{-2}$. It is straightforward to check that the right side is independent of the choice of vectors representing the rays Φ, Ψ. In particular, we have

$$P(\Phi, \Psi) = |(\phi, \psi)|^2 \tag{7.1}$$

for any unit vectors $\phi \in \Phi$ and $\psi \in \Psi$.

The simple experiments mentioned above correspond to the situations when the number N of independent states is finite. One can also proceed in the same way if the set $\{\Psi_j\}_{j=1}^N$ is infinite; however, then we are faced with the problem whether the state space should include also all "infinite superpositions" of the states Ψ_j. The answer is rather a matter of convention because in an actual measurement we are never able to distinguish such a "superposition" from a suitable superposition of a large enough number of states; we postpone a more detailed discussion of this problem to Section 7.4. The requirement of mathematical simplicity leads us to the assumption that the state space is complete, *i.e.*, a Hilbert space.

The presented considerations should not be regarded as an attempt to "derive" the Hilbert space structure of the state space; they merely illustrate the heuristic way which leads to the mentioned postulates. They served the founding fathers of quantum mechanics in this way more than half a century ago, and the theory that they based on them was afterwards confirmed by its correct predictions. We may nevertheless ask whether quantum theory could not with the same success use a state space of another structure. To answer this question, we must first axiomatize the properties of the measurements and then discuss whether the postulates (Q1a) and (Q1b) represent the only possible way to satisfy these conditions. We shall treat this problem in Chapter 13 where we shall see that, for example, the choice of a *complex* Hilbert space is to certain extent arbitrary.

In quantum *mechanics* we usually add another requirement of a mathematical nature: we suppose that

(Q1a) The state space of a quantum mechanical system has the structure of a complex *separable* Hilbert space.

In fact, the class of separable Hilbert spaces is rich enough not only for quantum mechanics but also for most problems of quantum field theory and statistical physics. The convenience of the assumption is that it allows us to use countable orthonormal bases, and moreover, we know that some of the results of the previous chapters were derived under the separability assumption. There is a heuristic argument supporting the strengthened version of the postulate; we shall mention it in Remark 11.1.1b.

Let us turn now to a description of the observables. We add a new postulate to the list, namely

(Q1c) A self–adjoint operator on the state space is associated with any observable of the system.

As in the case of the state space, we postpone the discussion of the ways in which this correspondence is realized. To motivate the postulate, we shall again use a simple system with N independent states Ψ_j, which can arise as a result of measuring an observable A. We know that the rays $\{\Psi_j\}_{j=1}^N$ are mutually orthogonal, and we also have the sequence $\{a_j\}_{j=1}^N$ of real numbers representing possible outcomes of the measurement; this uniquely determines a self–adjoint operator A such that Ψ_j is for any j the eigenspace corresponding to the eigenvalue a_j. The postulate (Q1c) extends this correspondence to all observables including those with a nonempty continuous spectrum. We have to say more, however, about the results we can obtain when a particular observable is measured.

7.1.2 Example: The Hilbert space $\mathbb{C}^2$ describes, *e.g.*, electron spin states. The spin projections at the j–th axis correspond to the operators $S_j := \frac{1}{2}\sigma_j$, where σ_j are the Pauli matrices (in fact, $S_j := \frac{1}{2}\hbar\sigma_j$; for the sake of simplicity we put $\hbar = 1$ as we shall almost everywhere in the following). Each of them has a simple spectrum, $\sigma(S_j) = \{\frac{1}{2}, -\frac{1}{2}\}$ and its spectral decomposition is $S_j = \frac{1}{2}E_j^{(+)} - \frac{1}{2}E_j^{(-)}$, where $E_j^{(\pm)} := \frac{1}{2}(I \pm \sigma_j)$ (Problem 1). Suppose now we are measuring the quantity S_j on an electron whose state is represented by a unit vector $\chi \in \mathbb{C}^2$. Experience tells us that

(a) The eigenvalues $\pm\frac{1}{2}$ of the operator S_j are the only possible outcome of the measurement.

(b) The probability of finding any particular one of the spin component values is $w_\pm \equiv w(\{\pm\frac{1}{2}\}, S_j; \chi) = (\chi, E_j^{(\pm)}\chi)$.

(c) For the mean value $\langle S_j\rangle_\chi := \frac{1}{2}w_+ - \frac{1}{2}w_-$, we then easily get $\langle S_j\rangle_\chi = (\chi, S_j\chi)$.

Similar conclusions can be made for other simple systems; motivated by them we can extend our list of postulates. To simplify the notation, we shall use mostly the same symbol for an observable and for the operator associated with it. Suppose therefore we have a system with the state space $\mathcal{H}$ and an observable described by a self-adjoint operator A, we denote the corresponding projection–valued measure and spectral decomposition as E_A and $\{E_\lambda^{(A)}\}$, respectively. If the system is in a state Ψ determined by a unit vector $\psi \in \mathcal{H}$, then

(Q2a) The possible outcome of measuring the observable A will be the points of the spectrum $\sigma(A)$ of the operator A.

(Q2b) The probability of finding the measured value in a Borel set $\Delta \subset \mathbb{R}$ is

$$w(\Delta, A; \psi) = \int_\Delta d(\psi, E_\lambda^{(A)}\psi) = \|E_A(\Delta)\psi\|^2.$$

The assertion (c) in the example has followed from (b). In the same sense the postulate (Q2b) implies that the mean value of the measurement results is given by the relation

$$\langle A\rangle_\psi = (\psi, A\psi) \tag{7.2}$$

provided the right side of this relation makes sense; we shall prove this in a more general context in Section 7.3 below. It is clear that the quantity $w(\Delta, A; \psi)$ in (Q2b) does not depend on the choice of the vector ψ representing the state Ψ. We must also check that it can indeed be interpreted as a probability. In the Kolmogorov axiomatic approach, a probability measure is a (non–negative, σ–additive) measure on a σ–algebra of events which is normalized, *i.e.*, the measure of the whole σ–algebra is equal to one. These requirements are satisfied because $w(\cdot, A; \psi)$ is a Borel measure according to Proposition 5.1.1c and $w(\sigma(A), A; \psi) = 1$; an event Δ here means that measurement result is contained in the set Δ.

The measurement of spin components discussed in Example 7.1.2 represents the simplest nontrivial (dichotomic) case of a measurement: there are only two mutually exclusive results. Measurements of that type are often called a *yes–no experiment.* Almost all actual measurements, of course, have a more complicated structure; the importance of *yes–no* experiments stems from the fact that the measurement of any observable can, at least in principle, be regarded as a collection of *yes–no* experiments. One can always divide the scale of the apparatus into intervals and ask: will the measured value be found in a given interval — yes or no? The most illustrative example of this equivalence is provided by various multichannel–analyzer devices frequently used in experimental physics.

The self–adjoint operator corresponding to a *yes–no* experiment has by postulate (Q2a) just two different eigenvalues $\alpha, \beta \in \mathbb{R}$. Without loss of generality we may suppose that these values are $\alpha = 1$ (yes, the positive result) and $\beta = 0$ (no, the negative result); hence each *yes–no* experiment is associated with a projection onto some subspace in $\mathcal{H}$. This makes the identification of an observable with a collection of *yes–no* experiments more transparent. If a projection P_Δ corresponds to the *yes–no* experiment "Is the measured value of the observable A contained in the set Δ?", then we have $w(\{1\}, P_\Delta; \psi) = w(\Delta, A; \psi)$. Since $E_{P_\Delta}(\{1\}) = P_\Delta$ the postulate (Q2b) gives

$$(\psi, P_\Delta \psi) = (\psi, E_A(\Delta)\psi)$$

for any vector ψ to which some state of the system corresponds. If this is true for all $\psi \in \mathcal{H}$ we obtain $P_\Delta = E_A(\Delta)$ for any Borel set $\Delta \subset \mathbb{R}$. We shall see in Section 7.4 that, in general, there are vectors which are associated with no state, but the conclusion will remain valid because we shall at the same time adopt a restriction on the set of admissible observables. Projections corresponding to the mentioned *yes–no* experiments are therefore nothing else than the values of the spectral measure E_A.

We already know what kinds of mathematical objects are ascribed to the states and observables, and what the predictions of measurement results can look like; it remains to answer the question of what happens to the state of the system, which has suffered a measurement. In the heuristic consideration presented above we have stated the answer for the case of an observable whose spectrum is pure point and simple. In the general situation we use equivalence with the collection of *yes–no* experiments and formulate the answer for the observable associated with the spectral projection $E_A(\Delta)$ as an additional postulate:

(Q3) If the result of the experiment is positive (the measured value is contained in the set Δ), the system after the measurement will be in the state described by the unit vector $E_A(\Delta)\psi/\|E_A(\Delta)\psi\|$, in the opposite case we have to replace $E_A(\Delta)$ on $E_A(\mathbb{R} \setminus \Delta) = I - E_A(\Delta)$.

This makes sense, of course, only if the norm appearing in the denominator is nonzero; but this condition causes no trouble. For instance, $\|E_A(\Delta)\psi\| = 0$ holds *iff* the probability of finding the value in Δ is zero (provided the system is in the state Ψ); in this case therefore we get a negative result with certainty, $\|E_A(\mathbb{R} \setminus \Delta)\psi\| = 1$, and the state of the system after the measurement will be described again by the vector ψ. The *yes–no* experiment under consideration can also, however, be regarded as a *filter*, *i.e.*, one can take into account only those cases when the measurement has given a positive result; the assumption $\|E_A(\Delta)\psi\| = 0$ then means that the filter is closed for the system in the state Ψ and it naturally makes no sense to speak about what will happen after the passage through it.

The postulate (Q3) has the following important consequence: if a *yes–no* experiment yields, *e.g.*, a positive result, then *the same* result will be found if the experiment is repeated immediately. Indeed, the vector $\psi' = E_A(\Delta)\psi/\|E_A(\Delta)\psi\|$ is reproduced by $E_A(\Delta)$, so the probability $w(\{1\}, E_A(\Delta); \psi') = 1$ by the postulate (Q2b). This is also in agreement with experience; one should be aware, however, of the indefiniteness hidden in the expression "immediately repeated". In an actual experiment we have to pay attention whether the change of state between the two measurements may be neglected; this question can be answered for a particular system only from a knowledge of its dynamics.

We have motivated the postulates by simple arguments about observables with pure point spectra; the conclusions are easily seen to be valid for any observable of this type.

7.1.3 Example: Let an observable A be described by a self–adjoint operator on a separable $\mathcal{H}$ whose spectrum is pure point, $\sigma(A) = \overline{\sigma_p(A)}$ with $\sigma_p(A) = \{\lambda_j : j = 1, 2, \dots\}$. Denote the projections onto the eigenspaces by $P_j := E_A(\{\lambda_j\})$. If the state before the measurement is described by a unit vector $\psi \in \mathcal{H}$, then

(a) The possible outcomes of the measurement are the eigenvalues λ_j , $j = 1, 2, \dots$.

(b) The probability of finding the value λ_j is $w(\{\lambda_j\}, A; \psi) = (\psi, P_j\psi) = \|P_j\psi\|^2$.

(c) If we find the value λ_j, the state after the measurement is described by the vector $P_j\psi/\|P_j\psi\|$.

We assume implicitly that the device can distinguish any two eigenvalues of the operator A; it is not difficult to modify the result for the case of an apparatus with limited resolution. It is important to keep in mind that this differs substantially from the situation, where the device does measure a particular value from a set $\Delta := \{\lambda_1, \dots, \lambda_n\}$ but we register only that the result lies within Δ; a more detailed discussion will be given in Section 7.5.

An attentive reader may object to the assertion (a) of the above example that in view of the postulate (Q2a) the measurement can *in addition* yield values, which belong to the closure of $\sigma_p(A)$ but not to the set itself. The answer is that the difference has no measurable consequences: due to Proposition 5.4.1b $E_A(\{\lambda\}) = 0$ holds for any $\lambda \in \sigma(A)\backslash\sigma_p(A)$, and therefore $w(\{\lambda\}, A; \psi) = 0$. A different conclusion would be strange anyhow, because no measurement can distinguish "arbitrarily close" quantities (see also the notes). Let us finally mention two other simple consequences of the postulates.

7.1.4 Example: Given a state Φ described by a unit vector ϕ, consider the *yes–no* experiment "Is the system in the state Φ?". Let E_Φ denote the corresponding projection. The quantity $w(\{1\}, E_\Phi; \psi)$ is nothing else than the transition probability (7.1); hence we have $\|E_\Phi\psi\|^2 = |(\phi, \psi)|^2$ for any state Ψ. If a state of the system corresponds to any nonzero $\psi \in \mathcal{H}$, the last relation implies $E_\Phi = (\phi, \cdot)\phi$, *i.e.*, E_Ψ is the projection onto the ray Φ in $\mathcal{H}$. The same is true even if the assumption is not valid, as we shall see in Section 7.4.

7.1.5 Example: Consider a pair of Borel sets $\Delta_1, \Delta_2 \in \mathbb{R}$. The probability $w(\Delta_1 \cap \Delta_2, A; \psi)$ can be expressed through a pair of consecutive measurements. Suppose, for example, that we have first performed the *yes–no* experiment $E_A(\Delta_1)$: with the probability $\|E_A(\Delta_1)\psi\|^2$ we find the positive result and the state after the measurement will be described by $\psi' = E_A(\Delta_1)\psi/\|E_A(\Delta_1)\psi\|$. Next, we perform the *yes–no* experiment $E_A(\Delta_2)$, where the probability of the positive outcome is

$$\|E_A(\Delta_2)\psi'\|^2 = \frac{\|E_A(\Delta_2)E_A(\Delta_1)\psi\|^2}{\|E_A(\Delta_1)\psi\|^2}.$$

The two described measurements clearly represent independent events; hence the joint probability of finding both positive results can be expressed in the product form, $w(\Delta_2,\Delta_1,A;\psi) = w(\Delta_2,A;\psi')w(\Delta_1,A;\psi) = \|E_A(\Delta_2)E_A(\Delta_1)\psi\|^2$, and it follows from Proposition 5.1.1 that it equals $w(\Delta_1\cap\Delta_2,A;\psi)$. Interchanging the order of the *yes–no* experiments, we get

$$w(\Delta_1\cap\Delta_2,A;\psi) = w(\Delta_2,\Delta_1,A;\psi) = w(\Delta_1,\Delta_2,A;\psi)$$

and also the state after the complete measurement is in both cases given by the same vector, namely $E_A(\Delta_1\cap\Delta_2)\psi/\|E_A(\Delta_1\cap\Delta_2)\psi\|$. Hence the *yes–no* experiment $E_A(\Delta_1\cap\Delta_2)$ is equivalent to the experiments $E_A(\Delta_j)$, $j=1,2$, performed in any order if the positive result of $E_A(\Delta_1\cap\Delta_2)$ means a positive result in the two measurements; this conclusion easily extends to any finite number of measurements (*cf.* Problem 12).

7.2 Simple examples

Before we proceed let us mention several elementary systems known from the introductory chapters of quantum mechanics textbooks. Apart from the cases when the state Hilbert space is finite–dimensional, the simplest situation occurs if $\mathcal{H} = L^2(\mathbb{R})$ or a subspace of it corresponding to an interval in $\mathbb{R}$. In reality, state Hilbert spaces have a more complicated structure, but we can often get a useful one–dimensional model by separation of variables, as we shall see in Section 11.5.

7.2.1 Example (a spinless particle on line): The state Hilbert space is $\mathcal{H} = L^2(\mathbb{R})$ and the fundamental dynamical variables of the "one–dimensional" particle are its **position** and **momentum** represented by operators Q and P from Examples 4.1.3 and 4.2.5, respectively; a motivation for this choice will be discussed in Section 8.2.

Consider first the position operator Q. We know that its spectrum is purely continuous, $\sigma(Q) = \mathbb{R}$, so any real number can be a result of position measurement. Suppose that the state of the particle is described by a unit vector $\psi \in L^2(\mathbb{R})$ (it is customary to speak about a *wave function* if $\mathcal{H}$ is a functional space); then the postulate (Q2b) in combination with Example 5.5.1a gives

$$w(\Delta,Q;\psi) = \int_\Delta |\psi(x)|^2 dx \tag{7.3}$$

for any Lebesgue measurable $\Delta\subset\mathbb{R}$, in particular, for an arbitrary Borel subset of the real line. Since the probability measure $w(\cdot,Q;\psi)$ is generated by the function

$|\psi(\cdot)|^2$ and the Lebesgue measure, we usually say that $|\psi(x)|^2$ is the *probability density* of finding the particle at a point x. If the particle is found in a set Δ, its state after measurement is given by the vector

$$\psi_\Delta : \psi_\Delta(x) = \frac{(E_Q(\Delta)\psi)(x)}{\|E_Q(\Delta)\psi\|} = \frac{\chi_\Delta(x)\psi(x)}{\sqrt{w(\Delta, Q; \psi)}},$$

where χ_Δ is the characteristic function of the set Δ. The operator P can be treated using unitary equivalence (4.4): we have $P = F^{-1}QF$ where F is the Fourier–Plancherel operator. By Examples 5.3.11c and 5.5.1b, the probability of finding the particle momentum in a Borel set Δ is equal to $w(\Delta, P; \psi) = \|E_Q(\Delta)F\psi\|^2$. This expression can be rewritten in the form $w(\Delta, P; \psi) = \int_\Delta |\phi(k)|^2 dk$, where $\phi := F\psi$ can be computed as in Example 3.1.6. If the wave function *in addition* belongs to $L^1(\mathbb{R})$, the formula simplifies to the Fourier transformation, $\phi(k) = (2\pi)^{-1/2} \int_{\mathbb{R}} e^{-ikx}\psi(x)\, dx$, and we arrive at the textbook result.

Another important observable for a one–dimensional particle (of a mass m) is its *kinetic energy* described by the operator $H_0 := \frac{1}{2m}P^2$. By Proposition 5.5.3, it has a purely continuous spectrum, $\sigma(H_0) = \sigma_c(H_0) = \mathbb{R}^+$; the probability of finding the kinetic–energy value in a set Δ is easily seen to be

$$w(\Delta, H_0; \psi) = \int_{h^{(-1)}(\Delta)} |\phi(k)|^2 dk,$$

where $h(k) := k^2/2m$. If the particle is free (noninteracting) then operator H_0 also, of course, describes its total energy.

7.2.2 Example (a particle on halfline): The state space is in this case $\mathcal{H} := L^2(\mathbb{R}^+)$. The position operator $Q : (Q\psi)(x) = x\psi(x)$ with the domain $D(Q) := \{\psi : \int_0^\infty x^2|\psi(x)|^2 dx < \infty\}$ is self–adjoint and the conclusions of the previous example modify easily for it. In distinction to it (and to the situation in classical mechanics), there is *no momentum observable* for a particle on halfline: we know from Example 4.2.5 that there is no self–adjoint operator corresponding to the formal expression $-i\, d/dx$.

The most interesting case is the operator of kinetic energy, which can be defined but in a nonunique way. For simplicity we neglect the numerical factor, *i.e.*, we put $m = 1/2$. In Example 4.9.6 we have shown that there is a one–parameter family of operators corresponding to the expression $-d^2/dx^2$, namely $H_{0,c} := T(c)$ with the domain $D(c) := \{\psi \in AC^2(0,\infty) : \psi'(0) - c\psi(0) = 0\}$ for $c \in \mathbb{R}$ and $D(\infty) := \{\psi \in AC^2(0,\infty) : \psi(0) = 0\}$. The spectrum $\sigma(H_{0,c}) \supset [0,\infty)$, where inclusion turns into identity for $c \geq 0$, while in the case $c < 0$ we have an additional eigenvalue $-c^2$ corresponding to the eigenvector $\phi_c : \phi_c(x) = \sqrt{-2c}\, e^{cx}$.

How are we to understand these facts ? We shall present a heuristic argument (which can nevertheless be made rigorous — see the notes). For any $k \in \mathbb{R}$ and $\varepsilon > 0$ we define the vector $\phi_{k,\varepsilon} : \phi_{k,\varepsilon}(x) = (e^{-ikx} + R_\varepsilon e^{ikx})\, e^{-\varepsilon x^2}$, where the coefficient R_ε is chosen in such a way that $\phi_{k,\varepsilon} \in D(c)$, *i.e.*,

$$R_\varepsilon := \frac{ik + \varepsilon + c}{ik - \varepsilon - c}.$$

Since s-$\lim_{\varepsilon\to 0}(H_{0,c}-k^2)\phi_{k,\varepsilon}=0$ we may regard the function $\phi_{k,0}$ as a generalized eigenvector of $H_{0,c}$ corresponding to eigenvalue k^2 belonging to the continuous spectrum (compare to Corollary 4.3.5). It is not normalized, of course, but it shows how the wave functions behave in the vicinity of the point $x=0$. The expression $\phi_k(x)=e^{-ikx}+R\,e^{ikx}$ with $R:=R_0$ can be understood as the superposition of the incident wave and the reflected wave whose phase is changed on $\arg R$, for each c this quantity is a different function of k (or of the energy variable k^2). In addition, in the case $c<0$ the particle can exist in a state bound to the "barrier", which is described by the vector ϕ_c. Hence we see that different self–adjoint extensions of the formal Hamiltonian $-d^2/dx^2$ describe *different physical situations.*

7.2.3 Example: Next, we shall mention the operators describing the position and momentum of a real free spinless particle whose state space is $\mathcal{H}:=L^2(\mathbb{R}^3)$. To the three Cartesian components of the position vector $x=(x_1,x_2,x_3)$ we ascribe the operators $Q_j:\ (Q_j\psi)(x)=x_j\psi(x)\,,\ j=1,2,3,$ with the domains $D(Q_j)=\{\psi\in L^2(\mathbb{R}^3):\int_{\mathbb{R}^3}x_j^2|\psi(x)|^2dx<\infty\}$. By Examples 4.3.3 and 4.3.7, these are self–adjoint with the purely continuous spectra, $\sigma(Q_j)=\mathbb{R}$, and

$$(E_{Q_j}(\Delta)\psi)(x)\;=\;\chi_\Delta(x_j)\psi(x)$$

holds for any Borel set $\Delta\subset\mathbb{R}$, from here it is easy to express the probabilities $w(\Delta,Q_j;\psi)$ and the state of the particle after such a measurement by analogy with Example 7.2.1. Recall also that the Q_j are related to the operator Q of this example by the relations $Q_1=\overline{Q\otimes I\otimes I}$, *etc.*, which can be checked as in Problem 6.29.

In a similar way, we associate with the Cartesian components of the momentum the operators P_j, $j=1,2,3,$ defined by $P_1=\overline{P\otimes I\otimes I}$, *etc.*; they are self–adjoint by Theorem 5.7.2. We shall check that the operators Q_j and P_j are again unitarily equivalent,

$$P_j\;=\;F_3^{-1}Q_jF_3\,,\quad j=1,2,3\,, \tag{7.4}$$

by means of the Fourier–Plancherel operator F_3. Using Proposition 4.5.6 and Problem 4.34, we find the inclusion $P_1\subset F_3^{-1}Q_1F_3$, and the relation $Q=FPF^{-1}$ implies in the same way $Q_1\subset F_3P_1F_3^{-1}$. Since inclusions between operators are preserved by unitary equivalence, we get the relation (7.4); the proof for $j=2,3$ is analogous.

Momentum–component operators have the domains $D(P_j)=F_3^{-1}D(Q_j)$. However, the way in which we have defined them specifies their action directly only for some vectors, namely those of the form $\psi(x)=\sum_{k=1}^n\prod_{l=1}^3\psi_l^{(k)}(x_l)$, where $\psi_j^{(k)}\in D(P_j)$ and $\psi_l^{(k)}\in L^2(\mathbb{R})$ for the other two values of the index l different from j; the definition of the unbounded–operator tensor product then yields

$$(P_j\psi)(x)\;=\;-i\left(\frac{\partial\psi}{\partial x_j}\right)(x)\,. \tag{7.5}$$

This formula is, of course, also valid for other vectors; we shall check that it holds for all $\psi\in\mathcal{S}(\mathbb{R}^3)$. By Example 1.5.6, F_3^{-1} maps $\mathcal{S}(\mathbb{R}^3)$ onto itself, so $\mathcal{S}(\mathbb{R}^3) \subset D(Q_j)$ implies $\mathcal{S}(\mathbb{R}^3) \subset D(P_j)$. Furthermore, we have $\psi(x) = (2\pi)^{-3/2} \int_{\mathbb{R}^3} e^{ik\cdot x}(F_3\psi)(k)\,dk$ for any $\psi \in \mathcal{S}(\mathbb{R}^3)$. The integrated function is majorized by $|(Q_j F_3\psi)(\cdot)|$ independently of x, hence the integration and differentiation may be interchanged,

$$-i\left(\frac{\partial\psi}{\partial x_j}\right)(x) = (2\pi)^{-3/2}\int_{\mathbb{R}^3} e^{ik\cdot x} k_j (F_3\psi)(k)\,dk = (F_3^{-1}Q_jF_3\psi)(x) = (P_j\psi)(x),$$

which is what we set out to prove. Since $\mathcal{S}(\mathbb{R}^3)$ is a core for P_j, the relation (7.5) can be used as an equivalent definition of the momentum–component operators (*cf.* Problem 5 and the notes). Notice finally that the same conclusions can be drawn for the operators Q_j, P_j on $L^2(\mathbb{R}^n)$ for any positive integer n. If $n = 3N$, for instance, then such operators describe the positions and momenta coordinates in a system of N spinless particles.

One of the most important observables is the **total energy** of the system; the corresponding operator is conventionally called **Hamiltonian**. We have already encountered free–particle Hamiltonians in the first two examples; now we want to mention two other simple systems.

7.2.4 Example (harmonic oscillator): Consider again the state space $\mathcal{H} := L^2(\mathbb{R})$. The operator

$$H:\ (H\psi)(y) = -\psi''(y) + y^2\psi(y)$$

with the domain $D(H) := \{\psi \in \mathcal{H}:\ \psi, \psi'$ absolutely continuous, $\int_{\mathbb{R}} |-\psi''(y) + y^2\psi(y)|^2 dy < \infty\}$ is up to the dimensional factor $\frac{1}{2}\hbar\omega$ identical with the harmonic–oscillator Hamiltonian

$$H_{m,\omega} := -\frac{\hbar^2}{2m}\frac{d^2}{dx^2} + \frac{1}{2}m\omega^2x^2$$

if we set $y := \sqrt{\frac{m\omega}{\hbar}}\,x$. It is well known that H has a simple spectrum consisting of the eigenvalues $\lambda = 2n+1$, $n = 0, 1, 2, \ldots$, and the corresponding eigenvectors are the Hermite functions (2.1) which form the orthonormal basis $\mathcal{E}_H$ in $L^2(\mathbb{R})$. It follows from Example 4.2.2 that H is *e.s.a.* on the linear envelope of $\mathcal{E}_H$ as well as, *e.g.*, on $\mathcal{S}(\mathbb{R})$.

In addition, H is self–adjoint due to Proposition 4.8.9b; we shall show that it can be expressed as $H = P^2 + Q^2$. We easily get $P^2 + Q^2 \subset H$; to check the opposite inclusion we use the fact that $\mathcal{S}(\mathbb{R})$ is invariant with respect to both P^2 and Q^2, and the relation $[P, Q^n]\psi = -inQ^{n-1}\psi$ holds for any $\psi \in \mathcal{S}(\mathbb{R})$. It yields

$$\begin{aligned}\|(P^2+Q^2)\psi\|^2 &= ((P^2+Q^2)^2\psi, \psi)\\ &= ((P^4+Q^4+2PQ^2P + [P,[P,Q^2]])\psi, \psi)\\ &= \|P^2\psi\|^2 + \|Q^2\psi\|^2 + 2\|QP\psi\|^2 - 2\|\psi\|^2,\end{aligned}$$

and therefore the inequality

$$\|(P^2+Q^2)\psi\|^2 + 2\|\psi\|^2 \ge \|P^2\psi\|^2 + \|Q^2\psi\|^2. \tag{7.6}$$

Since $\mathcal{S}(\mathbb{R})$ is a core for H, we can find to an arbitrary $\phi \in D(H)$ a sequence $\{\phi_n\} \subset \mathcal{S}(\mathbb{R})$ such that $\phi_n \to \phi$ and $H\phi_n \to H\phi$. Furthermore, one has $H\phi_n = (P^2 + Q^2)\phi_n$ so the inequality (7.6) can be rewritten as

$$\|H(\phi_n - \phi_m)\|^2 + 2\|(\phi_n - \phi_m)\|^2 \geq \|P^2(\phi_n - \phi_m)\|^2 + \|Q^2(\phi_n - \phi_m)\|^2 .$$

It follows that the sequences $\{P^2\phi_n\}$ and $\{Q^2\phi_n\}$ are Cauchy, and since both operators are closed we find that ϕ belongs to $D(P^2) \cap D(Q^2)$, *i.e.*, the inclusion $D(H) \subset D(P^2) \cap D(Q^2)$ which concludes the proof.

7.2.5 Example (rectangular potential well): This time we consider the operator

$$H : \ (H\psi)(x) \ = \ -\psi''(x) + V(x)\psi(x)$$

with the domain $D(H) := D(P^2)$, where $V := -V_0\chi_{[-a,a]}$ for some positive V_0, a. The operator V of multiplication by the function V coincides by Example 5.2.8 with $V(Q)$; in the present case it is bounded, $\|V\| = V_0$. Hence we can write $H = P^2 + V$. Up to the factor $\hbar^2/2m$, this is the Hamiltonian of a particle interacting with the rectangular–well shaped potential; the operators $H_0 := P^2$ and V correspond to its kinetic and potential energy, respectively. The boundedness of the potential V implies further that H is below bounded, $(\psi, H\psi) \geq -V_0\|\psi\|^2$, so $\sigma(H) \subset [-V_0, \infty)$.

Let us investigate the spectrum in more detail. The point part of it is simple and well known: there is a finite number of eigenvalues, all of them contained in the interval $(-V_0, 0)$. If we introduce $k := \sqrt{E+V_0}$ and $\kappa := \sqrt{-E}$ for $E \in (-V_0, 0)$, then the eigenvalues $E_n := -\kappa_n^2$ are given by the solutions to equations $ka\tan(ka) = \kappa a$ and $ka\cot(ka) = -\kappa a$, these are conventionally numbered starting from the smallest, $E_n < E_{n+1}$, $n = 1, 2, \dots$. The respective eigenvectors are

$$\psi_n : \ \psi_n(x) \ = \ \begin{cases} C_n \sin\left(k_n x - \frac{n\pi}{2}\right) & \dots \quad |x| \leq a \\ \\ C_n (\operatorname{sgn} x)^{n+1} \sin\left(k_n a - \frac{n\pi}{2}\right) e^{\kappa_n (a-|x|)} & \dots \quad |x| \geq a \end{cases}$$

where the normalization factor is $C_n := \sqrt{\kappa_n/(1+a\kappa_n)}$. There are just N eigenvalues, where N is given by the inequalities $(N-1)\pi < 2\sqrt{V_0 a^2} \leq N\pi$. Recall that the eigenvalues are found in such a way that we match the solutions of the appropriate differential equations in the three intervals so that the resulting function is square integrable and continuous together with its first derivative at the points $x = \pm a$. In other words, we choose among the candidates for the role of eigenvector those functions, which *belong to the domain* of the Hamiltonian. Similarly one can check that H has no eigenvalue outside the interval $(-V_0, 0)$.

Since the eigenvalues are simple and have no accumulation point, the remaining points of the spectrum belong to $\sigma_{ess}(H)$. It is easy to ascertain that the spectrum contains any point situated above the well, *i.e.*, $\sigma(H) \supset \mathbb{R}^+$ (Problem 8). To check that the points of the interval $[-V_0, 0)$ with the exception of the eigenvalues do not belong to the spectrum, we employ Theorem 5.4.6. We know from Example 7.2.1 that $\sigma_{ess}(P^2) = \mathbb{R}^+$, so

$$\sigma_c(H) = \sigma_{ess}(H) = \mathbb{R}^+$$

provided the operator V is P^2–compact. The resolvent $(P^2+\kappa^2)^{-1}$ is bounded for any $\kappa > 0$, and by Problem 9, $V(P^2+\kappa^2)^{-1}$ is an integral operator with the kernel $\frac{1}{2\kappa}V(x)\,e^{-\kappa|x-y|}$. This function is square integrable so $V(P^2+\kappa^2)^{-1}$ is Hilbert–Schmidt according to Theorem 3.6.5, and therefore compact.

7.3 Mixed states

Let us now return to analysis of the formalism. The next thing we have to discuss is the postulate (Q1b) because the correspondence it describes is not general enough in its present form.

7.3.1 Example: Suppose that an electron whose spin state is determined by a unit vector $\chi \in \mathbb{C}^2$ passes though the Stern–Gerlach apparatus (*i.e.*, a slit with a nonhomogeneous magnetic field followed by a pair of detectors $D_\pm$) adjusted along the j–th axis, and that after the passage the electrons are again brought to a single beam by means of a magnetic collimator. We disregard the other dynamical variables of the electron as well as the fact that hydrogen atoms rather than single electrons, in fact, pass through the device. If one of the detectors $D_\pm$ clicks, it happens by Example 7.1.2 with the probability $w_\pm := \|E_j^{(\pm)}\chi\|^2$ and the electrons leave the collimator in one of the states $\phi_\pm := E_j^{(\pm)}\chi/\|E_j^{(\pm)}\chi\|$. Assume that next we measure on them, for instance, the k–th spin component. This can be done in different ways:

(a) We choose, *e.g.*, the electrons whose j–th spin component has been determined to be $+\frac{1}{2}$, while those in the state ϕ_- are allowed to go unregistered. The probability of finding the value $\frac{\alpha}{2}$, where $\alpha = \pm 1$, is then

$$w_\alpha^{(+)} := w\left(\left\{\frac{\alpha}{2}\right\}, S_k; \phi_+\right) = (\phi_+, E_k^{(\alpha)}\phi_+)\,.$$

On the other hand, if we decide not to register the electrons in the state ϕ_+, then we get the value $\frac{\alpha}{2}$ with the probability $w_\alpha^{(-)} = (\phi_-, E_k^{(\alpha)}\phi_-)$.

(b) Another possibility is to cease distinguishing between the electrons and to measure S_k on each electron that has passed through the first device. In that case the probability of finding the value $\frac{\alpha}{2}$ is equal to $\tilde{w}_\alpha := w_+w_\alpha^{(+)} + w_-w_\alpha^{(-)}$. The vectors ϕ_+, ϕ_- form an orthonormal basis in $\mathbb{C}^2$, thus substituting for $w_\alpha^{(\pm)}$ we can rewrite the last expression as

$$\tilde{w}_\alpha = \mathrm{Tr}\,(E_k^{(\alpha)}W)\,,$$

where $W := w_+E_j^{(+)} + w_-E_j^{(-)}$, notice that the operator W is positive and its trace equals one, *i.e.*, it is a statistical operator. This allows us to compute the probability $\tilde{w}_\alpha$ without knowing in which particular state the electron had been after the first measurement; it is sufficient to know that any electron was in some of these states, together with the probability of this event.

Let us stress that the measurement described in part (b) differs, of course, from the situation when the detectors are not switched on; in that case the state of the electron after the passage through the first apparatus is described again by the vector χ. This can easily be checked experimentally: choosing, for example, $j=3,\ k=1$, and the electron state χ before the first measurement such that $S_1\chi = \frac{1}{2}\chi$, we find $\tilde{w}_\alpha = \frac{1}{2}$ while $w\left(\left\{\frac{\alpha}{2}\right\}, S_1; \chi\right) = \delta_{\alpha,1}$

Measurements of type (b) occur frequently in actual experiments. This motivates us to extend the notion of a state. The states described by rays in the state space will be called **pure**. More generally, we shall associate with a state a statistical operator $W := \sum_j w_j E_j$ and interpret it in such a way that the system is with probability w_j in the pure state $\Phi_j := E_j\mathcal{H}$. Pure states obviously represent a particular case in which some w_j is equal to one. The other states for which all $w_j < 1$ are called **mixed**. The respective postulate can be now formulated as follows:

(Q1b) A statistical operator (density matrix) on the state space is associated with any state of the system.

In combination with the definition of a mixed state, this implies

(Q2b) The probability that measuring an observable A on the system in a state W we find a value contained in a Borel set $\Delta \subset \mathbb{R}$ equals

$$w(\Delta, A; W) \;=\; \mathrm{Tr}\,(E_A(\Delta)W)\,,$$

and therefore we replace the postulate (Q2b) of Section 7.1 by this condition; it is clear that if W is a pure state, $W = E_\psi$, we return back to the previous formulation.

Mixed states are important from the practical point of view. We know that the probabilistic prediction of quantum theory can be verified only by performing the same measurement on a large number of identical copies of the system. In an actual experiment, however, it is usually technically impossible to achieve the situation in which all elements of such a family would be in a given pure state; we know mostly that particular copies are in some states from a subset in the state space together with the corresponding probabilities. In that case it is useful to assume that they are all in the same mixed state described by the appropriate statistical operator.

7.3.2 Example (polarization density matrix): We want to find the general form of a statistical operator on $\mathbb{C}^2$. Any Hermitean operator is there described by a real linear combination of the matrices $I,\ \sigma_1,\ \sigma_2,\ \sigma_3$. The condition $\mathrm{Tr}\,W = 1$ implies

$$W \;=\; \frac{1}{2}(I + \xi\cdot\sigma)\,,$$

where $\xi \cdot \sigma := \sum_{j=1}^{3} \xi_j \sigma_j$, $\xi_j \in \mathbb{R}$, because $\mathrm{Tr}\, \sigma_j = 0$. Moreover, the operator W must be positive. Its eigenvalues are easily found, $w^{(\pm)} = \frac{1}{2}(1 \pm |\xi|)$ where $|\xi| := (\sum_{j=1}^{3} \xi_j^2)^{1/2}$; this means that W is a statistical operator *iff* $|\xi| \le 1$. If we measure the j–th component of the spin, the mean value of the results is

$$\langle S_j \rangle_W = \mathrm{Tr}\,(W S_j) = \frac{1}{2}\, \xi_j \,.$$

The vector ξ, usually called *polarization*, is thus the doubled mean value of the spin. The density matrix W can describe, for instance, the spin state of an electron beam produced by an accelerator. If $|\xi| = 1$, we have $W^2 = W$ so the state W is pure (Problem 10). The opposite extreme occurs when $\xi = 0$; then $W = \frac{1}{2} I$ and all outcomes of spin measurements are equally probable. Such a density matrix corresponds to an unpolarized beam.

To accept the postulates formulated above, we have to check that $w(\cdot, A; W)$ of (Q2b) is indeed a probability measure on $\mathcal{B}$. At the same time we would like to generalize to the present situation the expression (7.2) for mean values. This is the content of the following assertion.

7.3.3 Theorem: Suppose that A is self–adjoint and W is a statistical operator on $\mathcal{H}$, then

(a) The map $w(\cdot, A; W)$ of (Q2b) is a probability measure on the σ–algebra $\mathcal{B}$; it is identical with the Lebesgue–Stieltjes measure $\mu_W^{(A)}$ generated by the function $f_W^{(A)} : f_W^{(A)}(\lambda) = \mathrm{Tr}\,(E_\lambda^{(A)} W)$.

(b) The mean value of (the results of measuring) the observable A in the state W is given by

$$\langle A \rangle_W := \int_{\mathbb{R}} \lambda \, d\mu_W^{(A)}(\lambda) = \mathrm{Tr}\,(AW) = \mathrm{Tr}\,(\overline{WA}) \qquad (7.7)$$

provided the right side makes sense, *i.e.*, $AW \in \mathcal{J}_1$.

7.3.4 Remark: For the sake of illustrativeness, the integrals with respect to the mentioned measure are often written in a more explicit form, *e.g.*,

$$\langle A \rangle_W := \int_{\mathbb{R}} \lambda \, d\mathrm{Tr}\,(E_\lambda^{(A)} W) \,.$$

Notice that the right side can also exist in some cases when $\mathrm{Tr}\,(AW)$ makes no sense; for example, if the state is pure, $W = E_\psi$, then $\mathrm{Tr}\,(AW) = (\psi, A\psi)$ is defined for $\psi \in D(A)$ while for the existence of the integral $\psi \in D(|A|^{1/2})$ is sufficient.

Proof of Theorem 7.3.3: By Theorem 3.6.7 $E_A(\Delta) W \in \mathcal{J}_1$, so the trace of this operator makes sense. The spectral decomposition of W is of the form $W = \sum_j w_j E_j$, where E_j are the one–dimensional projections corresponding to normalized eigenfunctions

$\phi_j : W\phi_j = w_j\phi_j$, $j = 1, 2, \dots$. The set function $w(\cdot, A; W)$ defined in (Q2b) then satisfies the relation

$$w(\Delta) := w(\Delta, A; W) = \sum_j w_j \mu_j(\Delta) \le 1$$

for any $\Delta \in \mathcal{B}$, where $\mu_j = (\phi_j, E_A(\cdot)\phi_j)$ are the measures referring to the pure states ϕ_j. We obviously have $w(\emptyset) = 0$ and $w(\mathbb{R}) = 1$, it remains to check the σ–additivity. Let $\Delta = \bigcup_{k=1}^{\infty} \Delta_k$ be a disjoint union of Borel sets; then the last formula gives $w(\Delta) = \sum_j w_j \mu_j \left(\bigcup_k \Delta_k\right) = \sum_j w_j \left(\sum_k \mu_j(\Delta_k)\right)$. The elements of the series on the right side are non–negative so we may rearrange it,

$$w(\Delta) = \sum_k \left(\sum_j w_j \mu_j(\Delta_k) \right) = \sum_k w(\Delta_k).$$

Hence $w(\cdot, A; W)$ is a probability measure on $\mathcal{B}$. The function $f_W^{(A)}$ is clearly nondecreasing and bounded; we shall check that it is right–continuous. Using the spectral decomposition of W, we find

$$f_W^{(A)}(\eta) - f_W^{(A)}(\lambda) = \sum_j w_j (\phi_j, (E_\eta^{(A)} - E_\lambda^{(A)})\phi_j)$$

for any pair of real numbers $\eta > \lambda$. The series on the right side is majorized by $\sum_j w_j$ independently of μ, so the summation may be interchanged with the limit $\eta \to \lambda+$; the properties of the spectral measure then imply $\lim_{\eta \to \lambda+} f_W^{(A)}(\eta) = f_W^{(A)}(\lambda)$. Using the definition of $f_W^{(A)}$, we easily get $\mu_W^{(A)}(J) = w(J)$ for any interval $J \subset \mathbb{R}$, and therefore $\mu_W^{(A)}(\Delta) = w(\Delta)$ for any $\Delta \in \mathcal{B}$ according to Theorem A.2.3; this concludes the proof of part (a).

The operator A is self–adjoint and W is Hermitean, so $\overline{(WA)^*} = A^*W^* = AW$. Since $\mathcal{J}_1$ is a $*$–ideal in $\mathcal{B}(\mathcal{H})$, it contains the operator $\overline{WA} = (WA)^{**}$ and $\mathrm{Tr}\,(\overline{WA}) = \overline{\mathrm{Tr}\,((WA)^*)} = \overline{\mathrm{Tr}\,(AW)}$. Next, we use the decomposition $W = \sum_j w_j E_j$. The operator $AW \in \mathcal{J}_1$ is everywhere defined; hence $\phi_j \in D(A)$ if $w_j > 0$. The quantity

$$\mathrm{Tr}\,(AW) = \sum_j (\phi_j, AW\phi_j) = \sum_{\{j:\, w_j > 0\}} w_j (\phi_j, A\phi_j)$$

is real; this proves the second identity in (b). To check the first one, it is enough to show that $AW \in \mathcal{J}_1$ implies the existence of the integrals $\int_{\mathbb{R}^\pm} |\lambda|\, d\mu_W^{(A)}(\lambda)$ and the relations

$$\int_{\mathbb{R}^\pm} |\lambda|\, d\mu_W^{(A)}(\lambda) = \pm \mathrm{Tr}\,(E_A(\mathbb{R}^\pm)AW).$$

Let $\{s_n\}$ be a nondecreasing sequence of non–negative simple functions on $\mathbb{R}^+$ such that $\lim_{n\to\infty} s_n(x) = x$ for all $x \ge 0$, and furthermore, $\int_0^\infty \lambda\, d\mu_W^{(A)}(\lambda) =$

$\lim_{n\to\infty}\int_0^\infty s_n(\lambda)\,d\mu_W^{(A)}(\lambda)$ (*cf.* the remark following Proposition A.2.2 and Theorem A.3.2). For any simple function $s := \sum_j c_j\chi_{\Delta_j}$ we get from the already proven assertion (a)

$$\int_0^\infty s(\lambda)\,d\mu_W^{(A)}(\lambda) \,=\, \sum_j \mathrm{Tr}\,(E_A(\Delta_j)W) \,=\, \mathrm{Tr}\,(s(A)W) \,=\, \sum_j w_j(\phi_j, s(A)\phi_j)\,,$$

so the integral equals $\lim_{n\to\infty}\sum_j w_j(\phi_j, s_n(A)\phi_j)$. We know that $\phi_j \in D(A)$ if $w_j > 0$; this implies the existence of the integral $\int_0^\infty \lambda\,d\mu_j(\lambda) = (\phi_j, E_A(\mathbb{R}^+)A\phi_j)$. At the same time, the inequality $s_n(\lambda) \le \lambda$ gives

$$(\phi_j, E_A(\mathbb{R}^+)A\phi_j) \,\ge\, (\phi_j, s_n(A)\phi_j) \,=\, \int_0^\infty s_n(\lambda)\,d\mu_j(\lambda)$$

for all n and j, so $\lim_{n\to\infty}(\phi_j, s_n(A)\phi_j) = (\phi_j, E_A(\mathbb{R}^+)A\phi_j)$ by the monotone convergence theorem. The above estimate also shows that the sum $\sum_j w_j(\phi_j, s_n(A)\phi_j)$ can be majorized independently of n, so the summation may be interchanged with the limit. We have therefore obtained the sought relation for the interval $\mathbb{R}^+$; the proof of the other one is analogous. ∎

The postulate (Q3) readily implies in which state the system will be after the measurement when originally it was in a mixed state. We again formulate the result for the *yes–no* experiment $E_A(\Delta)$, a more general situation will be discussed in Section 7.5:

(Q3) Suppose that the system before measurement is in a state W. If the result is positive (the measured value is contained in the set Δ), then the state after measurement is described by the statistical operator

$$W' \,:=\, \frac{E_A(\Delta)WE_A(\Delta)}{\mathrm{Tr}\,(E_A(\Delta)W)}\,.$$

We can speak about the state W' only when the probability $w(\Delta, A; W)$ is nonzero; in that case the right side makes sense and W' is again a statistical operator (Problem 11). Also the consequence of the postulate (Q3) mentioned in Section 7.1 remains valid: if the *yes–no* experiment $E_A(\Delta)$ is repeated immediately, we get the same result with certainty.

7.4 Superselection rules

The postulate (Q1b) does not tell us whether any ray in the state Hilbert space $\mathcal{H}$ of a given system corresponds to some state. Similarly we can ask whether any self–adjoint operator on $\mathcal{H}$ describes an observable. The answer to both questions is, in general, negative.

7.4.1 Example: Proton and neutron can be regarded as two isotopic states of a single particle — a nucleon; the corresponding state Hilbert space is $\mathbb{C}^2$. If we put the proton charge $e = 1$, then the charge operator is represented by the matrix $Q := \frac{1}{2}(\sigma_3 + I)$, and its eigenstates $\psi_p : Q\psi_p = \psi_p$ and $\psi_n : Q\psi_n = 0$ describe the proton and the neutron state, respectively. Up to this point, the situation is the same as in Example 7.1.2. There is, however, a substantial difference between the two cases: experimental experience tells us that

(a) No realizable state of the nucleon corresponds to a nontrivial superposition of the proton and the neutron state, *i.e.*, to a vector $\psi = \alpha\psi_p + \beta\psi_n$ with nonzero coefficients α, β.

(b) Any observable A commutes with the charge operator $Q = \begin{pmatrix} 1 & 0 \\ 0 & 0 \end{pmatrix}$; hence it can be expressed as $A = \lambda_+ E_3^{(+)} + \lambda_- E_3^{(-)}$, where $E_3^{(+)} = I - E_3^{(-)} := Q$.

The second claim follows from the first. Had an observable $B := \mu_1 E_1 + \mu_2 E_2$ noncommuting with Q existed, then its one–dimensional spectral projections could not commute with Q either. A measurement of B would produce the nucleon in a state described by of the eigenvectors of B; however, the latter are not contained in the rays $E_3^{(\pm)}\mathbb{C}^2$.

Consider now a general quantum system with the state Hilbert space $\mathcal{H}$. The example shows that $\mathcal{H}$ can contain unit vectors to which no state of the system corresponds. Denote by F the set of all vectors to which some state does correspond. Without loss of generality, we may assume that F is total in $\mathcal{H}$. This follows from the postulate (Q3) according to which we have $E_A(\Delta)F \subset F$ for any observable A and a Borel set Δ, and therefore also $E_A(\Delta)\mathcal{H}_F \subset \mathcal{H}_F$ where $\mathcal{H}_F := \overline{F_{lin}}$. Let E_F be the projection to the subspace $\mathcal{H}_F$, then the vector $E_A(\Delta)E_F\psi$ belongs to $\mathcal{H}_F$ for any $\psi \in \mathcal{H}$, *i.e.*, $E_A(\Delta)E_F\psi = E_F E_A(\Delta)E_F\psi$. Hence the projections $E_A(\Delta)$ and E_F commute for all $\Delta \in \mathcal{B}$, and by the spectral theorem, the observable A is reduced by the subspace $\mathcal{H}_F$. It is clear from the postulate (Q2b) that no information can be obtained experimentally about the part of A in $\mathcal{H}_F^\perp$, because all vectors corresponding to realizable states of the system are contained in $\mathcal{H}_F$, and therefore we may put $\mathcal{H} = \mathcal{H}_F$.

A set $M \subset F$ is said to be *coherent* if there are no nonempty orthogonal sets M_1, M_2 such that $M = M_1 \cup M_2$; if, in addition, M is not a proper subset of another coherent set we call it a *maximal coherent set*. The closed subspace $\overline{M_{lin}}$ spanned by a maximal coherent set M is called a **coherent subspace**.

7.4.2 Proposition: Any state Hilbert space $\mathcal{H}$ contains a family $\{\mathcal{H}_\alpha : \alpha \in J\}$ of mutually orthogonal coherent subspaces such that $\mathcal{H} = \sum_{\alpha\in J}^{\oplus} \mathcal{H}_\alpha$.

Proof: Since the set F is total by assumption, it is sufficient to find the decomposition $F = \bigcup_{\alpha\in J} F_\alpha$, where $\{F_\alpha\}$ is a family of mutually orthogonal maximal coherent sets. The sought orthogonal–sum decomposition then holds with $\mathcal{H}_\alpha := \overline{(F_\alpha)_{lin}}$, this follows from the inclusions $F \subset \sum_{\alpha\in J}^{\oplus} \mathcal{H}_\alpha \subset \mathcal{H}$ and the condition $F^\perp = \{0\}$.

To get the sets F_α we introduce on F the following equivalence relation: $\phi \sim \psi$ if there is a coherent set $M \subset F$, which contains both of them. The reflexivity and symmetry of the relation are obvious, so we have to check its transitivity. If

$\phi \sim \psi$ and $\psi \sim \chi$, there are coherent sets M_1, M_2 such that $\phi, \psi \in M_1$ and $\psi, \chi \in M_2$; if $M := M_1 \cup M_2$ was not coherent we could express it as a union of nonempty orthogonal sets N_1, N_2. Suppose, for instance, that $\psi \in N_1$ so $M_1 \cap N_1$ is nonempty. At the same time it is orthogonal to $M_1 \cap N_2$, and since $M_1 = (M_1 \cap N_1) \cup (M_1 \cap N_2)$ is coherent, we have $M_1 \cap N_2 = \emptyset$. The coherence of M_2 implies in the same way $M_2 \cap N_2 = \emptyset$, so together we get $N_2 = (M_1 \cap N_2) \cup (M_2 \cap N_2) = \emptyset$ in contradiction to the assumption. Hence the set M is coherent and $\phi \sim \chi$.

We now use the described relation and identify the sets F_α with the equivalence classes. We must first check that they are mutually orthogonal. If the vectors $\phi, \psi \in F$ are nonorthogonal, the point set $\{\phi, \psi\}$ is coherent so $\phi \sim \psi$. By negation, if ϕ, ψ belong to different equivalence classes they must be orthogonal. The sets F_α are coherent by definition; we have to show that they are maximal. Let $N \supset F_\alpha$ be a coherent set; then $N = N_\alpha \cup F_\alpha$ where $N_\alpha := \bigcup_{\beta \neq \alpha}(N \cap F_\beta)$. The orthogonality of the family $\{F_\alpha : \alpha \in J\}$ implies $N_\alpha \perp F_\alpha$, and since N is coherent we conclude that $N_\alpha = \emptyset$, *i.e.*, F_α is maximal coherent. ∎

Hence the state Hilbert space $\mathcal{H}$ of any quantum system decomposes into an orthogonal sum of coherent subspaces. In particular, if $\mathcal{H}$ is separable the index set J is at most countable. Any vector $\psi \in F$ belongs to just one of the subspaces $\mathcal{H}_\alpha$; in other words if $\phi \in \mathcal{H}$ has nonzero projections to at least two coherent subspaces, it does not describe a realizable state of the system.

Proposition 7.4.2 says nothing about which vectors within a particular coherent subspace correspond to realizable states. It follows from experimental experience that F_α are subspaces in $\mathcal{H}$, for the sake of simplicity we assume, in addition, that they are *closed*, *i.e.*,

$$\mathcal{H}_\alpha = \mathcal{H}_\alpha \cap F = F_\alpha\,, \quad \alpha \in J; \tag{7.8}$$

we shall return to discussion of this assumption a little later. A system is called **coherent** if its state space is coherent, $\mathcal{H} = F$.

7.4.3 Remark: Up to now we have been discussing pure realizable states. Consider a mixed state described by the statistical operator $W := \sum_j w_j E_j$. The projections E_j correspond to pure states; hence if the state W has to be realizable the corresponding rays must be contained in some of the coherent subspaces. The assumption (7.8) yields the opposite implication: if W is reduced by all coherent subspaces, then any pure state contained in the mixture is realizable. Together we find that a statistical operator W describes a realizable state of the system *iff* it is reduced by all the coherent subspaces.

The symbol $\mathcal{O}$ will denote the family of all observables of a given system. An important subset in it consists of *bounded observables* corresponding to Hermitean operators on $\mathcal{B}(\mathcal{H})$; we denote this by $\mathcal{O}_b$. We shall also use the W^*–algebra $\mathcal{A} := \mathcal{A}_W(\mathcal{O}_b)$ generated by the bounded observables. For simplicity we refer to this as to the *algebra of observables* of the system; we have to keep in mind, of course, that not every element of $\mathcal{A}$ is associated with an observable.

7.4.4 Theorem: Suppose the state space decomposes as in Proposition 7.4.2. Let assumption (7.8) be valid; then

(a) The algebra of observables is reduced by all the coherent subspaces; if the index set J is at most countable we have

$$\mathcal{A} \subset \sum_{\alpha\in J}^{\oplus} \mathcal{B}(\mathcal{H}_\alpha). \tag{7.9}$$

(b) If one of the sets $\mathcal{O}$, $\mathcal{O}_b$, and $\mathcal{A}$ is irreducible, the same is true for the other two; in that case the system is coherent.

(c) Any self–adjoint operator associated with some observable is reduced by all the coherent subspaces.

Proof: In order to prove the (a), it is sufficient due to Proposition 6.3.6 to check that each projection $E \in \mathcal{A}$ is reduced by all the projections E_α corresponding to the coherent subspaces. Suppose that there is a projection $E \in \mathcal{A}$ and $\alpha \in J$ such that E does not commute with E_α. Then we can choose a unit vector $\psi \in \mathcal{H}_\alpha$ such that both $E_\alpha E\psi$ and $(I-E_\alpha)E\psi$ are nonzero. To prove this claim assume first that $(I-E_\alpha)E\phi = 0$ for all $\phi \in \mathcal{H}_\alpha$. Then we have $(I-E_\alpha)EE_\alpha = 0$ and at the same time, $E_\alpha E(I-E_\alpha) = ((I-E_\alpha)EE_\alpha)^* = 0$, *i.e.*, $EE_\alpha = E_\alpha EE_\alpha = E_\alpha E$, but the projections E, E_α do not commute by assumption. Hence there is a unit vector $\psi \in \mathcal{H}_\alpha$ such that $(I-E_\alpha)E_\alpha\psi \neq 0$. If $E_\alpha E\psi = 0$, then we would have $(I-E_\alpha)E\psi = E\psi$, *i.e.*, $E\psi \in \mathcal{H}_\alpha^\perp$. In that case, however, $\|E\psi\|^2 = (\psi, E\psi) = 0$ so $E\psi = 0$, which is impossible.

According to the assumption (7.8) the vector ψ corresponds to a realizable state of the system. The projection E belongs to the algebra of observables; the probability of a positive result in the *yes–no* experiment E is $\|E\psi\|^2 \neq 0$ and the state after the measurement is then described by the vector $\psi' := E\psi/\|E\psi\|$. However, the latter has nonzero orthogonal components in both $\mathcal{H}_\alpha$ and $\mathcal{H}_\alpha^\perp$, and therefore no realizable state of the system can be ascribed to it; this concludes the proof of assertion (a).

Next we shall check that if any of the sets $\mathcal{O}$, $\mathcal{O}_b$, $\mathcal{A}$ is irreducible, so also are the other two. It follows from Problem 6.8a that $\mathcal{A} = \mathcal{O}_b''$ and so $\mathcal{A}' = \mathcal{O}_b'$. In Section 7.1 we identified the measurement of any observable A with the family of *yes–no* experiments $\mathcal{E}_A := \{E_A(\Delta) : \Delta \in \mathcal{B}\}$, so $A \in \mathcal{O}$ means $\mathcal{E}_A \subset \mathcal{O}_b$. Using the notation of Section 6.5 we can write this as the identity $\mathcal{O}_b = \mathcal{O}_p$; then Problem 6.24 gives $\mathcal{A}' = \mathcal{O}_b' = \mathcal{O}'$ and the sought result follows from Schur's lemma. If the sets under consideration are irreducible, we have $\mathcal{A}' = \mathbb{C}(\mathcal{H})$ and therefore $\mathcal{A} = \mathcal{A}'' = \mathcal{B}(\mathcal{H})$, which means in view of the part (a) that the system is coherent. Finally, using once again the equivalence of the observable A with the family $\mathcal{E}_A$ together with the spectral theorem we obtain assertion (c). ∎

The assertion (c) of the theorem can be used to check whether a given system is coherent. If we find among the operators representing the observables of the system an irreducible subset, the set $\mathcal{O}$ is itself irreducible and the system is coherent.

7.4.5 Examples: (a) *A particle on line:* It is sufficient to take the position and momentum operators; the set $\{Q, P\}$ is by Example 6.7.2e irreducible, so a one–dimensional particle represents a coherent system and $\mathcal{A} = \mathcal{B}(L^2(\mathbb{R}))$.

(b) *A system of spinless particles:* Consider now the operators $Q_j,\ P_j,\ j = 1,\dots,n$, on $L^2(\mathbb{R}^n)$ — see Example 7.2.3. The preceding example in combination with Theorem 6.7.3 shows that the set $\{Q_1,\dots,Q_n,P_1,\dots,P_n\}$ is irreducible; hence the corresponding system (for instance, a system of N spinless particles for $n = 3N$) is coherent and $\mathcal{A} = \mathcal{B}(L^2(\mathbb{R}^n))$.

We use the term **superselection rules** for the restriction to the set of admissible states, which is represented by the decomposition of the state space into the orthogonal sum of coherent subspaces. These rules are usually determined by a particular family of observables; the latter are sometimes called *superselection operators.* We have mentioned in Example 7.4.1 that the electric charge belongs to this family; other examples are the baryon number or the integrity/half-integrity of the spin. The superselection operators are usually of the form $A := \sum_{\alpha\in J} \lambda_\alpha E_\alpha$, *i.e.*, they have pure point spectra and the corresponding eigenspaces are the coherent subspaces $\mathcal{H}_\alpha$. All such operators commute mutually; this fact is usually referred to as *commutativity of the superselection rules.*

Let us now return to assumption (7.8) about the closedness of the set of states. If we want to use it, we have to accept that the mean values of some important observable — for example, energy — may not be defined for some states. Unfortunately, this sometimes happens to states which are physically interesting.

7.4.6 Example (Breit–Wigner formula): In Section 9.6 we shall discuss how unstable systems can be treated in quantum theory. In the simplest approximation we associate with such a system the state ψ_u such that the projection of the spectral measure of the Hamiltonian onto the corresponding one–dimensional subspace has the following form

$$d(\psi_u, E_\lambda^{(H)}\psi_u) \;=\; \frac{\Gamma}{2\pi}\,\frac{d\lambda}{(\lambda-\lambda_0)^2+\frac{1}{4}\Gamma^2}\,,$$

where $\lambda_0,\ \Gamma$ are constants characteristic of the considered decay process. The mean value $\langle H\rangle_{\psi_u}$ makes no sense because the integral $\int_{\mathbb{R}} |\lambda|\, d(\psi_u, E_\lambda^{(H)}\psi_u)$ does not converge.

The mean value of energy, of course, does exist for all pure states which are represented by vectors from $D(H)$ (and obey the superselection rules). More generally, one can introduce the notion of a *finite–energy state* W as such that the integral $\int_{\mathbb{R}} |\lambda|\, d\,\mathrm{Tr}\,(E_\lambda^{(H)}W)$ converges; the mean value $\langle H\rangle_W$ for these states is given by the relation (7.7). In particular, the set of pure finite–energy states is nothing else than the form domain $Q(H)$ of the Hamiltonian.

One might therefore attempt to replace assumption (7.8) by the requirement that the admissible (pure) states correspond to the dense subset $F_\alpha := \mathcal{H}_\alpha \cap Q(H)$

in each one of the coherent subspaces. However, in that case a question arises whether the "true" physical states should not also exhibit finite mean values of other important observables such as positions and momenta coordinates, *etc.* Fortunately, the difference between such conjectures and assumption (7.8) has *no measurable consequences.* The reason is that experimentally we determine only the probabilities of the postulate (Q2b) and not the mean values directly; thus the convergence of integrals like (7.7) is a matter of our extrapolation.

To illustrate this claim in more detail, let us introduce the set $B(H)$ of *bounded-energy states:* it includes all W such that the measure $w(\cdot,H;W)$ has a compact support, in other words, $w(\mathbb{R}\backslash\Delta_b,H;W)=0$ for some interval $\Delta_b:=(-b,b)$. Any $W\in B(H)$ is obviously a finite-energy state; pure states of $B(H)$ are described by analytic vectors of the Hamiltonian. We shall show that $B(H)$ is dense in the set of all states in the trace-norm topology.

7.4.7 Proposition: For any state W there is a one-parameter family $\{W_b\}\subset B(H)$ such that $\lim_{b\to\infty}\mathrm{Tr}\,|W-W_b|=0$.

Proof: Given a statistical operator W we set $W_b:=N_bE_bWE_b$, where $E_b:=E_H(\Delta_b)$ and $N_b^{-1}:=\mathrm{Tr}\,(E_bW)$; the definition makes sense for all b larger than some $b_0\ge 0$ since $\lim_{b\to\infty}N_b^{-1}=1$ due to Problem 3.41. We employ the estimate

$$\begin{aligned}\mathrm{Tr}\,|W-W_b| &\le \mathrm{Tr}\,|W-E_bW|+\mathrm{Tr}\,|E_b(W-WE_b)|+|1-N_b|\,\mathrm{Tr}\,|E_bWE_b|\\ &\le 2\,\mathrm{Tr}\,|W-E_bW|+N_b-1\,.\end{aligned}$$

By the polar-decomposition theorem, there exists a partial isometry U such that $|W-E_bW|=U^*(I-E_b)W$. We express the trace by means of the basis $\{\phi_j\}$ consisting of the eigenvectors of the operator W, this yields

$$\mathrm{Tr}\,|W-E_bW| = \sum_j w_j(U\phi_j,(I-E_b)\phi_j) \le \sum_j w_j\|(I-E_b)\phi_j\|\,.$$

The series on the right side can be majorized independently of b, the sought result is then obtained using the relation $\text{s-lim}_{b\to\infty}(I-E_b)=0$. ∎

Experimentally we cannot decide (even in principle) whether a given state has bounded energy, because any actual energy measurement tells us that the system is in a state W such that for the *yes–no* experiments $E_H(\Delta_k)$, positive numbers ε_k, $k=1,\dots,n$, and some $W^{(0)}\in\mathcal{J}_1(\mathcal{H})$ the inequalities

$$|\mathrm{Tr}\,(E_H(\Delta_k)(W-W^{(0)}))|<\varepsilon_k$$

are valid. Proposition 7.4.7 and Problem 3.40b show that $\lim_{b\to\infty}\mathrm{Tr}\,(A(W-W_b))=0$ holds for any bounded observable A; hence for all b large enough the states W and W_b cannot be experimentally distinguished.

7.5 Compatibility

We know that a measurement changes the state of the system. If several measurements are performed successively, the result may depend on the order in which they are done; hence it generally makes no sense to speak about a simultaneous observation of the corresponding dynamical variables. In some cases, however, the order is irrelevant. Observables A_1, A_2 are said to be **compatible** if

$$w(\Delta_2, A_2; \Delta_1, A_1; W) = w(\Delta_1, A_1; \Delta_2, A_2; W)$$

holds for any state W of the system and arbitrary Borel sets $\Delta_1, \Delta_2 \subset \mathbb{R}$; more generally, the observables of a family $\{A_\beta : \beta \in I\}$, where I is any index set, are compatible if any two of them are mutually compatible. We again implicitly assume that the two measurements follow immediately one after the other, so the change of state in the meantime can be neglected.

7.5.1 Proposition: The observables $\{A_\beta : \beta \in I\}$ are compatible if the corresponding self–adjoint operators form a commutative set. For each finite subset $\{A_{\beta_1}, \dots, A_{\beta_n}\}$ we have

$$w(\Delta_n, A_{\beta_n}; \dots; \Delta_1, A_{\beta_1}; W) = w(\Delta_{\pi(n)}, A_{\beta_{\pi(n)}}; \dots; \Delta_{\pi(1)}, A_{\beta_{\pi(1)}}; W),$$

where π is any permutation of the set $\{1, \dots, n\}$. The compatibility is reflexive and symmetric but not transitive.

Proof: It is evident from the definition of a commutative operator set that it is sufficient to check the equivalence for a pair of operators A_1, A_2. If they commute, the projections $E_{A_1}(\Delta_1)$, $E_{A_2}(\Delta_2)$ also commute for any Borel sets $\Delta_1, \Delta_2 \subset \mathbb{R}$, the corresponding probabilities then coincide by Problem 12. To check the necessary condition it is enough to consider pure states $W := E_\psi$ only. The relation between the probabilities then becomes $(\psi, EE'E\psi) = (\psi, E'EE'\psi)$, where we write for brevity $E := E_{A_1}(\Delta_1)$ and $E' := E_{A_2}(\Delta_2)$. This is valid by assumption for all realizable states, *i.e.*, for any vector ψ contained in some of the coherent subspaces $E_\alpha\mathcal{H}$. The projections E, E' commute by Theorem 7.4.4 with all E_α, and the same is true for the operator $C := EE'E - E'EE'$. The condition can be rewritten as $(E_\alpha\phi, CE_\alpha\phi) = (\phi, CE_\alpha\phi) = 0$ for any $\phi \in \mathcal{H}$, which readily implies $C = 0$. Denote further $B := EE' - E'E$; then $C = 0$ gives $B^*B = 0$, and therefore $B = 0$ according to identity (3.5). Hence it follows from the compatibility assumption that the projections $E_{A_1}(\Delta_1)$ and $E_{A_2}(\Delta_2)$ commute for any Borel $\Delta_1, \Delta_2 \in \mathbb{R}$, *i.e.*, the commutativity of A_1, A_2.

The reflexivity and symmetry are obvious from the definition. Any observable is compatible with the trivial one represented by the unit operator. Hence if the compatibility were transitive, the observable algebra would have to be Abelian; it is easy to check that this is not the case with the exception of the trivial case when each coherent subspace is one–dimensional. ∎

7.5.2 Example: Consider the operators Q_j of Example 7.2.3. By Proposition 5.7.3 they commute, which means that the Cartesian coordinates are compatible. The same is true for operators P_j: it follows from (7.4) that the Cartesian coordinates of the momentum are also compatible. Both conclusions are confirmed by experience; they allow us, in particular, to speak about measuring positions or momenta in a system of N particles. On the other hand, the Cartesian coordinates are *incompatible* with the corresponding momentum coordinates because the operators Q_j and P_j do not commute. This has important consequences which we shall discuss in the next chapter.

According to Proposition 5.1.9 we are able to associate with commuting self–adjoint operators $A_1, \dots, A_n$ a unique projection–valued measure E on $\mathbb{R}^n$ such that $E(\Delta_1 \times \cdots \times \Delta_n) = E_{A_1}(\Delta_1) \dots E_{A_n}(\Delta_n)$ for any Borel $\Delta_1, \dots, \Delta_n \subset \mathbb{R}$. Given a state W we can then define the map $w : \mathcal{B}^n \to [0,1]$ by

$$w(\Delta, \{A_1, \dots, A_n\}; W) := \operatorname{Tr}(E(\Delta)W) \tag{7.10}$$

for any $\Delta \in \mathcal{B}^n$. We can extend to it a part of the assertion of Theorem 7.3.3; it is sufficient to realize that the corresponding proof does not employ the fact that $E_A(\cdot)$ is a projection–valued measure on $\mathbb{R}$.

7.5.3 Proposition: The map $w(\cdot, \{A_1, \dots, A_n\}; W)$ defined by (7.10) is a probability measure on $\mathbb{R}^n$ for any compatible observables $A_1, \dots, A_n$.

Now we want to know how this measure is related to the probability of finding the result of measuring $A_1, \dots, A_n$ regarded as the n–tuple of real numbers $\{\lambda_1, \dots, \lambda_n\}$ in a set $\Delta \subset \mathbb{R}^n$. In the case when $\Delta := \Delta_1 \times \cdots \times \Delta_n$ is an interval in $\mathbb{R}^n$ the sought probability is $\operatorname{Tr}(E(\Delta)W)$ according to Problem 12. At the same time we know that there is no other Borel measure with this property; if two Borel measures coincide on intervals in $\mathbb{R}^n$ they are identical by Theorem A.2.3. This argument is the motivation behind the natural generalization of the postulates, which is again confirmed by experimental evidence:

(Q2b) The probability of finding the result of a simultaneous measurement of compatible observables $A_1, \dots, A_n$ in a set $\Delta \in \mathcal{B}^n$ when the system is in a state W equals $\operatorname{Tr}(E(\Delta)W)$, *i.e.*, it is expressed by the relation (7.10) regarded as an identity.

(Q3) The state after such a measurement is described (in case of the positive result) by the statistical operator $(\operatorname{Tr}(E(\Delta)WE(\Delta)))^{-1}E(\Delta)WE(\Delta)$.

The extension covers situations when the measurement cannot be reduced to a finite sequence of elementary acts, for instance, ascertaining the presence of a particle in a spherical volume.

7.5.4 Examples: (a) Consider the operators Q_j of the preceding example. The measure E is easily found: we have $E(\Delta) = \chi_\Delta(Q)$, where Q stands for $\{Q_1, \dots, Q_n\}$, so Example 5.5.8 gives $(E(\Delta)\psi)(x) = \chi_\Delta(x)\psi(x)$. In particular, if the system is in a pure state Ψ we get from here

$$w(\Delta, Q; \psi) = \int_\Delta |\psi(x)|^2 dx$$

for any set $\Delta \in \mathcal{B}^n$. This represents a natural extension to (7.3); recall that this formula historically played an important role in constituting quantum mechanics (the statistical interpretation of the wave function postulated by M. Born, P. Dirac and P. Jordan in 1926).

(b) *A particle with nonzero spin:* Spin has a double meaning: firstly, the triplet of observables S_j, $j = 1,2,3$, and secondly, a number $s = 0, \frac{1}{2}, 1, \frac{3}{2}, \dots$. The state space of such a particle is $\mathcal{H} := L^2(\mathbb{R}^3;\mathbb{C}^{2s+1})$, due to (2.6) it can be expressed as $\mathcal{H} = L^2(\mathbb{R}^3) \otimes \mathbb{C}^{2s+1}$, where the spaces $L^2(\mathbb{R}^3)$ and $\mathbb{C}^{2s+1}$ correspond to the configuration and spin states of the particle, respectively. The operators S_j are usually defined by means of the orthonormal basis $\{\chi_m\}_{m=-s}^{s}$ of the eigenvectors of S_3,

$$(S_1 \pm iS_2)\chi_m := \sqrt{(s\mp m)(s\pm m+1)}\,\chi_{m\pm 1}\,, \quad S_3\chi_m := m\chi_m; \tag{7.11}$$

we can easily check (Problem 15) that they are Hermitean and satisfy the relations

$$[S_j, S_k] = i\epsilon_{jkl}S_l\,, \quad S^2 := S_1^2 + S_2^2 + S_3^2 = s(s+1)I_s\,, \tag{7.12}$$

where I_s is the unit operator on $\mathbb{C}^{2s+1}$ and we have adopted the standard convention according to which one sums over the repeated indices in the commutation relations. The spectra of these operators coincide, $\sigma(S_j) = \{-s, -s+1, \dots, s\}$.

Dynamical variables such as position, momentum, spin, *etc.*, are, however, ascribed to the particle as a single entity, and therefore operators on the total state space $\mathcal{H}$ should correspond to them. They are of the form

$$\underline{Q}_j := \overline{Q_j \otimes I_s}\,, \quad \underline{P}_j := \overline{P_j \otimes I_s}\,, \quad \underline{S}_j := \overline{I_c \otimes S_j}\,,$$

where $j = 1,2,3$ and I_c is the unit operator on $L^2(\mathbb{R}^3)$ (we shall use the underlined symbols only if it is necessary to stress the relations to the operators on the "component" spaces). According to Proposition 5.7.3, they are self–adjoint and $\sigma(\underline{Q}_j) = \sigma(\underline{P}_j) = \mathbb{R}$ while $\sigma(\underline{S}_j) = \{-s, -s+1, \dots, s\}$. In view of the above mentioned relations no two of the operators $\underline{S}_j$ commute, but all of them commute with $\underline{Q}_k$ and $\underline{P}_k$ for $k = 1,2,3$. Moreover, the commutativity of Q_j, Q_k implies that the operators $\underline{Q}_k$, $k = 1,2,3$, commute mutually, and the same conclusion can be drawn for $\underline{P}_k$, $k = 1,2,3$. Hence we can choose for a set of compatible observables one of the spin components (or a real linear combination of them) together with the components of position or momentum; most frequently one uses the sets

$$\mathcal{S}_{qs} := \{\underline{Q}_1, \underline{Q}_2, \underline{Q}_3, \underline{S}_3\}, \quad \mathcal{S}_{ps} := \{\underline{P}_1, \underline{P}_2, \underline{P}_3, \underline{S}_3\}.$$

Notice that $\underline{S}^2 := I_c \otimes S^2 = s(s+1)I$; this provides an example of a particular observable represented by a multiple of the unit operator. The measurement of spin squared can yield only the value $s(s+1)$; this gives us right to speak about a particle with spin s.

Let us now return to the problem of how the state of the system is changed after measurement. In the above formulation of the postulate (Q3) we have assumed that we register the results of all performed measurements. However, this is not the most general case.

7.5.5 Example: Suppose that we perform a sequence of compatible *yes–no* experiments $E := \{E_1,\dots,E_n\}$ on the system in a state W; the result is an ordered n–tuple $r := \{r_1,\dots,r_n\}$, where the numbers r_j assume the values $0,1$. To be able to express the probabilities $w(r) := w(r,E;W)$ we introduce the projections

$$E(r) := \prod_{j=1}^{n}(E_j\delta_{1,r_j} + (I-E_j)\delta_{0,r_j})\,.$$

Obviously, $E(r)E(r') = 0$ for $r \ne r'$, and furthermore, $\sum_{r\in M} E(r) = I$ where the sum runs over the set M of all the 2^n different n–tuples.

Up to now we have just a particular case of the measure E which is supported here by the set M; the probability $w(r)$ is then given by $\mathrm{Tr}\,(E(r)W)$ and the state after the measurement is described by the operator $w(r)^{-1}E(r)WE(r)$. Suppose now that we have registered only a part of the results, *i.e.*, that we do not distinguish the states of a subset $M_{reg} \subset M$. Examples:

(i) We register only the result $\tilde r_1$ of the first experiment E_1; then $M_{reg} := \{r : r_1 = \tilde r_1\}$.

(ii) If we remember only that the result contained the number one k–times, then $M_{reg} := \{r : r_1 + \cdots + r_n = k\}$.

(iii) If we have registered the result completely, then M_{reg} consists of a single r–tuple. Conversely, if we have registered nothing one has $M_{reg} = M$.

The state of the system after such a measurement is the mixture of the states $w(r)^{-1}E(r)WE(r)$ for all $r \in M_{reg}$ with the weights $w(r)$; in other words

$$W' = N \sum_{r\in M_{reg}} E(r)WE(r)\,,$$

where the normalization factor is given by $N^{-1} := \sum_{r\in M_{reg}} w(r) = \mathrm{Tr}\,(E(M_{reg})W)$, the orthogonality of the projections $E(r)$ implies that $E(M_{reg}) := \sum_{r\in M_{reg}} E(r)$ is a projection. In the particular case when we register nothing the state is described by $W'_M := \sum_{r\in M} E(r)WE(r)$. It is easy to check that

$$W' = \frac{E(M_{reg})W'_M E(M_{reg})}{\mathrm{Tr}\,(E(M_{reg})W'_M)}\,.$$

This means that the considered measurement may be regarded as consisting of two operations: first we let the system pass through the device without registering the results and afterwards we perform the *yes–no* experiment $E(M_{reg})$. Notice that these results simplify if the *yes–no* experiments $E_1, \dots, E_n$ in question are disjoint (Problem 14b).

Functional relations between observables are not a self–evident matter in quantum theory. In classical physics the relation $B = f(A_1, \dots, A_n)$ between dynamical variables has a clear meaning: if the A_j assume the values a_j, then the quantity B assumes the value $f(a_1, \dots, a_n)$. Compatible observables are important, in particular, because they make it possible to define similar relations. In the quantum case, however, we are allowed to speak only about probabilities of the measurement outcomes. We shall thus formulate the definition as follows: let $\{A_1, \dots, A_n\}$ be compatible observables and $f : \mathbb{R}^n \to \mathbb{R}$ a Borel function; then the relation $B = f(A_1, \dots, A_n)$ means that

$$w(\Delta, B; W) = w(f^{(-1)}(\Delta), \{A_1, \dots, A_n\}; W) \tag{7.13}$$

holds for any state W and an arbitrary Borel set $\Delta \subset \mathbb{R}$, where $B, A_1, \dots A_n$ are the self–adjoint operators corresponding to the observables under consideration. The operator $A := f(A_1, \dots, A_n)$ is self–adjoint by the definition given in Section 5.5; we shall check that $B = A$.

According to assumption (7.13), $(\psi, E_B(\Delta)\psi) = (\psi, E(f^{(-1)}(\Delta))\psi)$ holds for any pure realizable state Ψ, where E is the projection–valued measure corresponding to the operators $A_1, \dots, A_n$. Both projections are reduced by all coherent subspaces; it readily implies $E_B(\Delta) = E(f^{(-1)}(\Delta))$ for any $\Delta \in \mathcal{B}$. On the other hand, we have $E_A(\Delta) = E(f^{(-1)}(\Delta))$ due to Proposition 5.2.12, *i.e.*, $E_A(J) = E_B(J)$ for any interval $J \subset \mathbb{R}$, and therefore the two operators coincide. We have proven in this way that $B = f(A_1, \dots, A_n)$ as defined above implies the same functional relation between the corresponding self–adjoint operators. Hence we shall employ the same symbols again in the following for both the observables and the respective operators. Using in addition Theorem 5.5.10 and Problem 5.33c, we arrive at the following conclusion.

7.5.6 Proposition: Suppose that $A_1, \dots, A_n$ are compatible observables and $f : \mathbb{R}^n \to \mathbb{R}$ a Borel function; then the observable $A := f(A_1, \dots, A_n)$ is compatible with $A_1, \dots, A_n$. If the function is continuous, we have $\sigma(A) = \overline{f(A_1, \dots, A_n)}$.

7.5.7 Remarks: (a) The above argument can be reversed. If we have observables represented by the commuting operators $A_1, \dots, A_n$, then the relation $B = f(A_1, \dots, A_n)$ defines a new observable which satisfies (7.13). Its interpretation is clear: we measure it by the same device as the compatible family $A_1, \dots, A_n$ but change the scale: the point λ on it corresponds to the points of $f^{(-1)}(\lambda)$ on the original scale.

(b) Postulate (Q2a) means implicitly that the measured values are real numbers. In practice we often meet measurable quantities assuming complex values — let us mention just the example of the scattering operator which we shall discuss in Chapter 15. They are usually represented by complex functions of commuting self-adjoint operators so no problems arise: we proceed as above using complex numbers to scale the apparatus. It is therefore useful to regard complex Borel functions of "real" observables as *observables in a broader sense;* by Theorem 5.2.11 they are described by normal operators.

7.5.8 Example: Cartesian coordinates of momentum are compatible by Example 7.5.2. The kinetic energy of a spinless particle of mass m is described by a function of them, $H_0 := h(P)$, where $h(p) := (2m)^{-1}(p_1^2+p_2^2+p_3^2)$; according to Example 5.2.10 and the functional-calculus rules, this is equivalent to

$$H_0 = \frac{1}{2m}P^2 := \frac{1}{2m}(P_1^2+P_2^2+P_3^2).$$

Since the function h is real-valued, H_0 is self-adjoint and Proposition 7.5.6 gives $\sigma(H_0) = \mathbb{R}^+$. Moreover, it follows from the equivalence (7.4) that $H_0 = F_3^{-1}T_hF_3$, where T_h is the operator of multiplication by h. This yields the expression for the domain, $D(H_0) = F_3^{-1}D_h$, where $D_h := \{\psi \in L^2(\mathbb{R}^3) : \int_{\mathbb{R}^3} x^4|\psi(x)|^2dx < \infty\}$. In addition, using Examples 5.3.11c and 5.4.9b we get

$$\mathcal{H}_{ac}(H_0) = L^2(\mathbb{R}^3), \quad \sigma_{ac}(H_0) = \sigma_{ess}(H_0) = \sigma(H_0) = \mathbb{R}^+. \tag{7.14}$$

Another way to write the operator H_0 is with the Laplacian: the relation (7.5) together with $P_j\mathcal{S}(\mathbb{R}^3) \subset \mathcal{S}(\mathbb{R}^3)$ imply

$$(H_0\psi)(x) = -\frac{1}{2m}(\Delta\psi)(x) := -\frac{1}{2m}\left(\frac{\partial^2\psi}{\partial x_1^2} + \frac{\partial^2\psi}{\partial x_2^2} + \frac{\partial^2\psi}{\partial x_3^2}\right)(x) \tag{7.15}$$

for all $\psi \in \mathcal{S}(\mathbb{R}^3)$. This is true also for other vectors from $D(H_0)$, what is important is that the operator $-(2m)^{-1}\Delta$ is *e.s.a.* on $\mathcal{S}(\mathbb{R}^3)$ (Problem 16).

Let us mention one more property of the operator H_0, namely that its domain $D(H_0) \subset L^\infty(\mathbb{R}^3)$, and for any $a > 0$ one can find a b such that

$$\|\psi\|_\infty \le a\|H_0\psi\| + b\|\psi\| \tag{7.16}$$

holds for all $\psi \in D(H_0)$, where $\|\cdot\| = \|\cdot\|_2$ is the usual norm in $L^2(\mathbb{R}^3)$. To check this inequality, consider an arbitrary vector $\phi := F_3\psi \in D_h$. The identity $\phi = (1+h)^{-1}(1+h)\phi$ together with the fact that the functions $(1+h)^{-1}$ and $(1+h)\phi$ belong to $L^2(\mathbb{R}^3)$ imply $\phi \in L^1(\mathbb{R}^3)$; hence $\psi = F_3^{-1}\phi$ belongs to $L^\infty(\mathbb{R}^3)$ and $(2\pi)^{3/2}\|\psi\|_\infty \le \|\phi\|_1$. Furthermore, the Schwarz inequality gives

$$\|\phi\|_\infty \le c\|(I+T_h)\phi\| \le c(\|T_h\phi\| + \|\phi\|),$$

where $c := \|(1+h)^{-1}\|$. Next we define for any $r > 0$ the scaled function ϕ_r : $\phi_r(x) = r^3\phi(rx)$, the relations $\|\phi_r\|^2 = r^3\|\phi\|^2$ and $\|T_h\phi_r\|^2 = r^{-1}\|T_h\phi\|^2$ show that it belongs to D_h. Substituting ϕ_r for ϕ in the last inequality, and using (7.4) together with the unitarity of F_3 we get

$$(2\pi)^{3/2}\|\psi\|_\infty \le \frac{c}{\sqrt{r}}\|T_h\phi\| + cr^{3/2}\|\phi\| = \frac{c}{\sqrt{r}}\|H_0\psi\| + cr^{3/2}\|\psi\|,$$

which yields (7.16) if we choose $r := c^2/a^2(2\pi)^3$. Notice that while the results from the first part of the example extend easily to the n–dimensional case, the inequality (7.16) does not hold for $n \ge 4$ (see the notes).

Let $\mathcal{S}$ be a family of compatible observables. Due to Proposition 7.5.6 it is possible to append to it functions of the observables from $\mathcal{S}$; it may happen that this will exhaust all dynamical variables compatible with $\mathcal{S}$, *i.e.*, that we can find no other "independent" compatible observables. In such a case we call $\mathcal{S}$ a **complete set of compatible observables**. It is clear from Proposition 7.5.1 and Section 6.6 that operators representing the elements of $\mathcal{S}$ then form a complete set of commuting operators; we shall use the shorthand CSCO for both families. It is natural to ask whether a set of compatible observables can always be completed to a CSCO.

7.5.9 Theorem: For any family of compatible observables there is a CSCO $\mathcal{S}_{max}$ which contains it, $\mathcal{S}_{max} \supset \mathcal{S}$.

Proof: The operator set $\mathcal{S}$ is by assumption commutative and symmetric, and according to Problem 6.24 the same is true for $\mathcal{S}_f$; it follows that $\mathcal{S}'' = \mathcal{S}_f''$ is an Abelian $*$–subalgebra in the algebra $\mathcal{A}$ of observables of the given system. The Zorn lemma readily implies the existence of a maximal Abelian $*$–subalgebra $\mathcal{B} \subset \mathcal{A}$ which contains $\mathcal{S}''$; the inclusion $\mathcal{S}'' \subset \mathcal{B}$ then yields $\mathcal{S}' \supset \mathcal{B}'$. Using the notation of Section 6.5, we construct to $\mathcal{B}$ the set $\mathcal{B}_R$ for which $(\mathcal{B}_R)' = \mathcal{B}'$ in view of Problem 6.24. At the same time, we have $\mathcal{B} = \mathcal{B}' = \mathcal{B}''$ by Proposition B.1.2e because $\mathcal{B}$ is maximal Abelian, and therefore $(\mathcal{B}_R)' = (\mathcal{B}_R)''$ so $\mathcal{B}_R$ is a CSCO. The inclusion $\mathcal{S} \subset \mathcal{B}_R$ need not be valid if $\mathcal{S}$ contains unbounded operators. However, one can replace $\mathcal{B}_R$ by the commutative set $\mathcal{S}_{max} := \mathcal{S} \cup \mathcal{B}_R \subset \mathcal{L}_{sa}(\mathcal{H})$ in which $\mathcal{S}$ is contained. We have $\mathcal{S}'_{max} = \mathcal{S}' \cap (\mathcal{B}_R)'$, and since $\mathcal{B}' = \mathcal{B}''$ we get $\mathcal{S}'_{max} = \mathcal{S}''_{max}$; this concludes the proof. ∎

In fact, the set $\mathcal{S}_{max}$ constructed in the proof is unnecessarily large; one can even get a CSCO by adding a single Hermitean operator to $\mathcal{S}$ provided $\mathcal{H}$ is separable (Problem 20). Stated in that way the result has only an abstract meaning; nevertheless in practice we always look for a CSCO consisting of a small number of particular observables.

7.5.10 Example: In view of Example 7.5.4a and Corollary 6.6.6 each of the sets $\{Q_1,\dots,Q_n\}$ and $\{P_1,\dots,P_n\}$ forms a CSCO for a spinless particle (if $n=3$) or a system of such particles. Similarly any of the sets $\mathcal{S}_{qs}$ and $\mathcal{S}_{ps}$ from Example 7.5.4b is a CSCO for a particle of spin s.

Complete sets of compatible observables play an important role in preparation of the state. Measurements included in this procedure will be summarily called a *preparatory measurement.* According to postulate (Q3) its outcome depends on the state before the measurement which we, however, do not know. One can get rid of this dependence provided the observables whose values are determined in the preparatory measurement form a CSCO.

Consider first a CSCO $\mathcal{S} := \{A_1,\dots,A_N\}$ consisting of observables with pure point spectra; for simplicity we assume that the state space $\mathcal{H}$ is separable. We shall use the notation introduced in Section 6.6: the probability of finding the N–tuple of values $\Lambda_{\{k\}} := \{\lambda^{(1)}_{k_1}, \dots, \lambda^{(N)}_{k_N}\}$ for any $\{k\}$ equals $w(\Lambda_{\{k\}}, \mathcal{S}; W) = \mathrm{Tr}\,(P_{\{k\}}W)$, and the state after the measurement is described by the statistical operator $W' := (\mathrm{Tr}\,(P_{\{k\}}W))^{-1} P_{\{k\}} W P_{\{k\}}$. Recall that all the projections $P_{\{k\}}$ are one–dimensional; this enables us to simplify these relations choosing a unit vector $\psi_{\{k\}}$ in each one–dimensional subspace. In particular, the state after the measurement becomes

$$W' = \frac{(\psi_{\{k\}}, \cdot)}{(\psi_{\{k\}}, W\psi_{\{k\}})} P_{\{k\}} W \psi_{\{k\}} = (\psi_{\{k\}}, \cdot)\psi_{\{k\}}\,,$$

so $W' = P_{\{k\}}$. Hence we arrive at the following conclusion.

7.5.11 Proposition: Suppose that $\mathcal{H}$ is separable and $\mathcal{S} := \{A_1, \dots, A_N\}$ is a complete set of compatible observables with pure point spectra. If the measurement yields the values $\lambda^{(1)}_{k_1}, \dots, \lambda^{(N)}_{k_N}$, then the state after the measurement is described by the corresponding common eigenvector $\psi_{\{k\}}$ independently of the state in which the system had been before the measurement.

It is also obvious that we can in this way, at least in principle, obtain all independent states of the system because the vectors $\psi_{\{k\}}$ form an orthonormal basis in $\mathcal{H}$. The situation is more complicated if the CSCO used for the preparatory measurement contains observables with a nonempty continuous spectrum. In such a case it is not possible to suppress the dependence on the original state completely; however, it can be minimalized provided the measurement is exact enough.

7.5.12 Example: Suppose we have a particle in a state Ψ and measure its momentum; if its value is found in a set $\Delta \subset \mathbb{R}^3$, the state after the measurement is described by

$$\psi'_\Delta := N(\Delta)\, F_3^{-1} E(\Delta) F_3 \psi\,,$$

where $E(\Delta) := \chi_\Delta(Q)$ and the normalization factor $N(\Delta) := \|E(\Delta)F_3\psi\|^{-1}$. We take the ball $U_\varepsilon(k_0)$ for Δ and see how the state ψ'_Δ behaves in the limit $\varepsilon \to 0$.

Let $(F_3\psi)(k_0)$ be nonzero; without loss of generality we can assume that it is positive. For the sake of simplicity, assume also the vector ψ belongs to $\mathcal{S}(\mathbb{R}^3)$. Then $F_3\psi \in \mathcal{S}(\mathbb{R}^3)$, and therefore $(F_3\psi)(k) \neq 0$ in $U_\varepsilon(k_0)$ for any ε small enough; this ensures that the normalization factor — which is due to the mean–value theorem equal to $|(F_3\psi)(k_\varepsilon)|^2 V_\varepsilon$, where k_ε is some point of the ball and $V_\varepsilon := \frac{4}{3}\pi\varepsilon^3$ is its volume — is nonzero. Using once more the assumption $\psi \in \mathcal{S}(\mathbb{R}^3)$, which allows us to interchange the integrations, we can express the function ψ'_Δ as

$$\psi'_\Delta(x) = \frac{N(\Delta)}{(2\pi)^3} \int_{\mathbb{R}^3} dy\, \psi(y) \int_{U_\varepsilon(k_0)} e^{ik\cdot(x-y)} dk\,.$$

The inner integral can be computed in an elementary way (see also Problem 19) to be $V_\varepsilon g(\varepsilon|x-y|)\, e^{ik_0\cdot(x-y)}$, where $g(z) := 3z^{-3}(\sin z - z\cos z)$. It is easy to see that

$|g(z)-1| < Cz^2$ for some $C > 0$ and all $z \in \mathbb{R}^+$ (Problem 21); it follows that the limit $\varepsilon \to 0$ may be interchanged with the outer integral. At the same time, the estimate shows that $\lim_{\varepsilon\to 0} g(\varepsilon z) = 1$. Moreover, $k_\varepsilon \to k_0$ as $\varepsilon \to 0$, and since the function $F_3\psi$ is continuous, we finally obtain

$$\lim_{\varepsilon\to 0+} V_\varepsilon^{-1/2}\, \psi'_\Delta(x) = \frac{1}{(2\pi)^{3/2}}\, e^{ik_0\cdot x}\,, \tag{7.17}$$

where the convergence is uniform in any bounded set $M \subset \mathbb{R}^3$. Hence *independently of* ψ we obtain the standard plane–wave expression. It does not belong to $L^2(\mathbb{R}^3)$, of course, but the result tells us that by a sufficiently precise measurement of the momentum we can achieve that the state is in a chosen (bounded) spatial region M approximated arbitrarily closely by the right side of (7.17). Let us remark that the smoothness assumption we have made can be weakened substantially (Problem 21). On the other hand, the argument cannot be used if $(F_3\psi)(k_0) = 0$, this is not surprising, however, because this condition corresponds to the heuristic claim that "the state $|k_0\rangle$ is not contained in the superposition $|\psi\rangle$ ".

7.6 The algebraic approach

Physical theories usually result from unification and generalization of empirical information (like every rule, this one has exceptions; compare with the general theory of relativity). Only later is the effort made to select a few basic ones among the initial facts and to derive the others in a deductive way. Such an activity is, of course, meaningful; we need it to make the theory transparent by finding a suitable mathematical language for it, to check its internal consistency and decide whether some of the starting assumptions are not irrelevant or lacking an empirical foundation.

An axiomatic system on which the theory is based must be free from contradictions; this requirement is common to all theories, which employ mathematical methods. For mathematical theories this is the only requirement; on the other hand, a physical theory should describe a part of the existing world, so its axioms have to be supported by an empirical evidence. It took a long time, in fact, before this difference was fully recognized; it took about twenty two centuries for the mathematical nature of Euclid's axioms to become clear.

Axiomatic systems usually change in the course of time when the development of the theory enables some postulates to be replaced by simpler or more general ones. We restrict ourselves to the example of quantum mechanics: the postulates (Q1)–(Q3) formulated in the preceding sections represent the result of the effort of J. von Neumann to unify the two "pre–quantum" mechanics. However, this was not the end of the quest for an optimal axiomatic system. One reason is that postulate (Q1) has only an indirect empirical justification as we noted in Section 7.1; this flaw can be removed by axiomatizing properties of measurements. We shall discuss this problem in Chapter 13; now we want to mention another approach to quantum theory, which is based on axiomatization of algebraic properties of the observables. Some remarks on the history of this idea are given in the notes.

Let us look in more detail at the structure of the set of observables for a quantum system. Some of them have a well–established meaning and they are measured by particular experimental techniques; sometimes they are called *fundamental observables*. Examples are coordinates, momenta, energies, charges, *etc.*In addition, the set of observables includes other dynamical variables which have no direct experimental meaning but they are functionally related to the fundamental ones. In classical mechanics these observables in a broader sense are all functional expressions of the type $f(q,p)$.

In the quantum case the situation is more complicated because not all fundamental observables are mutually compatible. Nevertheless, some functions can be defined, including in the first place

(i) A real multiple λa of the observable a is measured by the same apparatus which is linearly rescaled.

(ii) The sum of observables a, b, which is understood to be an observable c such that $\langle c\rangle_\phi = \langle a\rangle_\phi + \langle b\rangle_\phi$ holds for any state ϕ.

In particular, if the observables a, b are replaced by the families of *yes–no* experiments in the way described in Section 7.1, then the mean values are nothing else than the probabilities of finding a positive result in such experiments. At the same time, we have to stress that the existence of the sum for a pair of observables is only *assumed*: if a, b are noncompatible it is not clear, in general, how to define the procedure c, *i.e.*, to construct a suitable device so that the mean values $\langle c\rangle_\phi$ would satisfy the above mentioned identity.

Note that the definition (ii) can also cover some fundamental observables. As an example one can take the hydrogen atom, whose Hamiltonian in the center–of–mass frame is $H := (\hbar^2/2m)P^2 - e^2Q^{-1}$ with $Q^{-1} := (Q_1^2+Q_2^2+Q_3^2)^{-1/2}$. The identity $\langle H\rangle_\phi = (\hbar^2/2m)\langle P^2\rangle_\phi - e^2\langle Q^{-1}\rangle_\phi$ is valid, of course, provided the mean values make sense. The eigenvalues of H are determined, however, by measuring the frequencies of the photons coming from transitions between different energy levels; there is no direct correspondence to the measurements of the electron position and momentum.

For simplicity we shall again consider in the following the set $\mathcal{O}_b$ of all bounded observables of the given system. Prescriptions (i) and (ii) define on $\mathcal{O}_b$ the algebraic structure of a real vector space. We have to equip $\mathcal{O}_b$ with a topological structure because its dimension is infinite in all practically interesting cases. A natural way to do this is to define the norm $\|a\| := \sup_\phi |\langle a\rangle_\phi|$ for any observable $a \in \mathcal{O}_b$, where ϕ runs through all states of the system; this makes $\mathcal{O}_b$ a real normed space. The algebraic structure can be further enriched; we use the fact that $\mathcal{O}_b$ also contains powers of the observables which can be easily defined:

(iii) The observable a^n is measured by the appropriately rescaled apparatus.

The operations of summation, multiplication by a real number, and $a \mapsto a^2$ turn $\mathcal{O}_b$ in the way described in the notes into a real commutative, in general nonassociative, algebra on which one can define an involution by $a^* := a$.

The algebra $\mathcal{O}_b$ can be required to satisfy various physically motivated conditions; two such systems of axioms are mentioned in the notes. The classes of algebras they determine are still too wide, and therefore another assumption is added to which no physical foundation has been found up to now: we suppose that

(ao_c) the algebra $\mathcal{O}_b$ can be identified with the set of all Hermitean elements of some C^*–algebra $\mathcal{A}$; the product $a \cdot b$ of the elements of $\mathcal{O}_b$ is related to the multiplication in $\mathcal{A}$ by $a \cdot b = \frac{1}{2}(ab + ba)$.

For simplicity the algebra $\mathcal{A}$ is often also called the *algebra of observables* of the considered system.

If we adopt this postulate we are able to employ the results of Sections 6.1 and 6.2. States of the system are then identified with the positive linear functionals ϕ on $\mathcal{A}$ that satisfy the normalization condition $\phi(e) = 1$ for the unit (trivial) observable e. We denote as $\mathcal{S}_{\mathcal{A}}$ the set of all states on $\mathcal{A}$. Since this notion must be consistent with the physical concept of state introduced in Sections 7.1 and 7.3, one more postulate has to be added:

(ao_s) $\langle a \rangle_\phi = \phi(a)$ holds for any state ϕ on $\mathcal{A}$, where the left side means the mean value of the results of measuring the observable a in the state ϕ.

This also ensures that another definition is consistent: using Theorem 6.2.6 we can check that the norm of a as an element of the algebra $\mathcal{A}$ coincides with the norm introduced above,

$$\|a\| := \sup_{\phi \in S_{\mathcal{A}}} |\phi(a)| \,.$$

We know from Section 6.2 that the set $S_{\mathcal{A}}$ is convex. Its extremal points form the set $P_{\mathcal{A}}$ of pure states, while the other states are called mixed; a necessary and sufficient condition for a state to be pure is given by Theorem 6.2.7.

The most important consequence of the postulate (ao_c) is that it allows us to use Theorem 6.2.6 by which there is an isometric representation of the algebra $\mathcal{A}$ on some Hilbert space $\mathcal{H}$. Then we are able to represent (bounded) observables of the system by Hermitean operators on $\mathcal{H}$. Since $\mathcal{A}$ as an operator algebra supports other topologies we can also strengthen the postulate (ao_c) assuming that

(ao_w) The set $\mathcal{O}_b$ can be identified with the set of all Hermitean elements in some W^*–algebra $\mathcal{A} \subset \mathcal{B}(\mathcal{H})$.

It is appropriate to return now to the definition of the algebra of observables presented in Section 7.4. The arguments presented here show what its meaning is: roughly speaking, we start from the family of fundamental observables for the considered system, extend it algebraically and topologically, and identify observables in a broader sense with all Hermitean elements of the algebra $\mathcal{A}$ obtained in this way. The choice of a complex algebra is motivated by the requirement of simplicity.

The postulate (ao_w) provides finer means to classify the elements of $S_{\mathcal{A}}$. We are interested primarily in normal states since the other states can have rather pathological properties mentioned at the end of Section 6.4. Concerning normal states, we have proven than any such state has a trace representation by means of some operator $W \in \mathcal{J}_1(\mathcal{H})$. For the particular class of type I W^*–algebras in which all minimal projections are one–dimensional we have a stronger result given by Theorem 6.4.7. This result acquires a physical meaning if we suppose that $\mathcal{H}$ decomposes into the orthogonal sum $\sum_{\alpha\in J}^{\oplus} \mathcal{H}_\alpha$ of coherent subspaces. In general, the algebra $\mathcal{A}$ then satisfies the inclusion (7.9). If the latter turns into identity,

$$\mathcal{A} = \sum_{\alpha\in J}^{\oplus} \mathcal{B}(\mathcal{H}_\alpha), \tag{7.18}$$

then to any normal state ϕ on $\mathcal{A}$ we can ascribe just one statistical operator $W \in \mathcal{J}_1(\mathcal{H})$, which is reduced by all the coherent subspaces $\mathcal{H}_\alpha$ in such a way that $\phi(B) = \mathrm{Tr}\,(WB)$ holds for any $B \in \mathcal{A}$, in particular, for any observable of the given system. The state is pure *iff* the operator W is a one–dimensional projection.

If we therefore add the assumption (7.18) to the postulates (ao) we fully recover the standard formalism discussed in the preceding sections. Observable algebras of the type (7.18) are typical for quantum mechanical systems. Recall Example 7.4.5b by which the observable algebra for a system of N spinless particles is $\mathcal{B}(L^2(\mathbb{R}^{3N}))$, but also in more general situations when other degrees of freedom and superselection rules are involved, $\mathcal{A}$ is still of the form (7.18).

We must therefore ask what the algebraic approach is good for. We have to realize first of all that assumption (7.18) plays a crucial role here; once we abolish it we are not able to use the argument which ensures the uniqueness of the trace representation, *i.e.*, of the operator W in Theorem 6.4.7. At the same time, the assumption need not be valid when systems with an infinite number of degrees of freedom are considered; for example, it is known that any algebra of local observables of a free quantum field is of type III (see the notes).

There is a deeper reason, however. The abstract algebra $\mathcal{A}$ can have different nonequivalent representations corresponding to different physical situations. For instance, quantum fields differing by mass or interaction can correspond to nonequivalent representations of *the same* algebra of observables — *cf.* Section 12.3. In such a case the algebraic description of the observables and states has to regarded as primary because it deals with the properties of the system which are representation–independent. Note that in order to decide whether $\mathcal{A}$ can have nonequivalent representations it is not necessary to investigate all its elements; it is sufficient to restrict our attention to the fundamental observables that generate it; an example of fundamental importance will be given in the next chapter.

Notes to Chapter 7

Section 7.1 As mentioned in the preface, the contents of this and the following chapters are not intended as a substitute for a course in quantum theory. There are numerous textbooks which the reader can consult for the physical material discussed here: as a sample let us mention ⟦Bo⟧, ⟦Dav⟧, ⟦Dir⟧, ⟦LL⟧, ⟦Mes⟧ for quantum mechanics; ⟦BD⟧,

⟦ BŠ ⟧, ⟦ IZ ⟧, ⟦ Schwe ⟧ for quantum field theory, in particular, ⟦ IZ ⟧, ⟦ Hua 2 ⟧, ⟦ SF ⟧ for the theory of non–Abelian gauge fields and its applications in physics of elementary particles; ⟦ Fey ⟧, ⟦ Hua 1 ⟧ for quantum statistical physics, and many others. At the same time, applications of quantum theory in different parts of physics, and also chemistry, biology, *etc.*, are nowadays so plentiful and rapidly multiplying that it is meaningless even to attempt to compile a representative list of references.

On the other hand, there is an extensive literature devoted to mathematical aspects of quantum theory. The pioneering role was played by the classical monographs ⟦ vN ⟧ and ⟦ Sto ⟧; following them many authors have analyzed the basic concepts of the theory, their properties, relations, and generalizations — see, *e.g.*, ⟦ BeŠ ⟧; ⟦ Da 1 ⟧, Chaps.2–4; ⟦ Ja ⟧; ⟦ Jor ⟧; ⟦ Ma 1,2 ⟧; ⟦ Pir ⟧; ⟦ Pru ⟧; ⟦ Ri 1 ⟧; ⟦ Var ⟧, and others. As we have also mentioned, the main interest in rigorous quantum theory has shifted gradually from general problems to analysis of particular systems — this is the main topic of monographs ⟦ RS 2–4 ⟧, ⟦ Sche ⟧, ⟦ Si 1 ⟧, ⟦ Thi 3 ⟧, and many others; some of them will be mentioned at the appropriate places below. Rigorous methods of quantum field theory are discussed, *e.g.*, in ⟦ BLOT ⟧, ⟦ Em ⟧, ⟦ Šv ⟧, ⟦ SW ⟧, and also in monographs ⟦ GJ ⟧, ⟦ Sei ⟧, ⟦ Si 2 ⟧ and others which concentrate on the so–called constructive approach. Mathematical aspects of quantum statistical physics are treated, for instance, in ⟦ BR ⟧, ⟦ Em ⟧, ⟦ GJ ⟧, ⟦ Sin ⟧.

The facts from probability theory that we shall need are contained in standard textbooks, *e.g.*, ⟦ Fel ⟧, ⟦ Par ⟧, ⟦ Šir ⟧. The "operational" definitions discussed here can be found, *e.g.*, in ⟦ Ja ⟧, Chap.6. Notice that in reality one has to associate measuring devices only with some important observables; we return to this problem in Section 7.4.

The mentioned consequence of postulate (Q3) for a pair of identical *yes–no* experiments performed immediately one after the other concerns situations when the outcome of the experiment characterizes the state *after* the measurement. An example is the registration of a particle by a Geiger counter: if the apparatus clicks we know that the particle has been found in the sensitive volume of the detector. We usually refer to such situations as *measurements of the first kind.* There are also *second–kind measurements,* where the measured values refer to the state *before* the experiment has been performed; recall, for instance, measuring the excited–level energies of an atom by registration of the photon frequencies coming from their deexcitation. First–kind measurements are simpler and we shall deal mostly with them in the following; their distinctive feature is that they can serve as preparatory measurements — see Section 7.5.

The argument following Example 7.1.3 should not be interpreted as a claim that the *exact* values of physical quantities are never of importance. In the classical theory of dynamical systems and its quantum counterpart, *e.g.*, the systems exhibit a different behavior depending on whether certain parameters (such as ratios of driving frequencies, the sizes of the region in question, *etc.*) are rational or irrational (even the kind of the irrationality is important) — see, *e.g.*, [CG 1], [Com 1], [MŠ 1], [JL 1], or [Šeb 1]. Nevertheless, one cannot decide the value of the parameter experimentally: if we approximate an irrational by a suitable rational number the system will exhibit the behavior characteristic of the irrational value for some time (or in some interval of energy; the longer the interval the better the approximation is), but eventually it comes out that the value is rational after all.

Section 7.2 The "eigenvectors" corresponding to the continuous spectrum used in quantum mechanical textbooks do not, of course, belong to the state space but this does not mean they are useless. In fact, many of the formal considerations involving these eigenfunctions can be made rigorous either in the so-called *rigged Hilbert space* framework — see, *e.g.*, ⟦EG⟧ and references therein — or even within the standard Hilbert-space theory by the *eigenfunction expansion* method, which in a sense represents a generalization of the Fourier transformation — *cf.* ⟦AJS⟧, Chap.10; ⟦RS 3⟧, Sec.XI.6.

Example 7.2.2 should be regarded as a warning against the dangers which may await you if instead of proving a given observable to be (essentially) self-adjoint you merely check formally that it is "Hermitean", *i.e.*, symmetric. At the same time, the employed derivation of a family of Hamiltonians based on the theory of self-adjoint extensions (see also Problems 2–4) gives a glimpse of a powerful method of constructing various solvable models of quantum systems; we shall return to this in the last two chapters. The operator T of Example 4.8.5 also has physically meaningful non-selfadjoint extensions but their interpretation within the standard quantum mechanical formalism requires a longer explanation — *cf.* ⟦Ex⟧, Sec.4.3.

According to Problem 5, $\mathcal{S}(\mathbb{R}^3)$ is a core for the momentum-component operators P_j. Their domains are by definition $D(P_j) := F_3^{-1}D(Q_j)$, but we can also describe them explicitly as the subspaces consisting of those $\psi \in L^2(\mathbb{R}^3)$ for which the right side of (7.5) makes sense as a distribution and belongs to $L^2(\mathbb{R}^3)$. In the same way one can specify the domains of more complicated partial differential operators through so-called *Sobolev spaces.* The proofs are based on properties of the Fourier transformation which go far beyond Example 1.5.6; we refer, *e.g.*, to ⟦Ad⟧ or ⟦RS 2⟧, Chap.IX. The double commutator estimate used to prove the relation (7.6) is due to A. Jaffe — see [Si 2].

Section 7.3 The concept of a mixed state was introduced to quantum theory by L. Landau and J. von Neumann. The state described by the vector $\phi = \sum_k c_k\phi_k$ is often called a *coherent* superposition of the states $\Phi_1, \Phi_2, \dots$ in order to distinguish it from the noncoherent "superposition" given by the density matrix $W = \sum_k w_k E_{\phi_k}$, in the second case we prefer to speak about a *mixture* of the considered states.

Given a probability measure ω_S on $(\mathbb{R}, \mathcal{B})$, we can define its *moments* $m_k(S) := \int_{\mathbb{R}} x^k d\omega_S(x)$, $k = 1, 2, \dots$, provided the integrals exist. The moment $m_1(S)$ is called the *mean value* $\langle S\rangle$ of the random variable S. Furthermore, using the first two moments one defines the *standard* (or *mean-square*) *deviation* as $\Delta S := \sqrt{m_2(S) - m_1(S)^2}$. It is easy to see that $(\Delta S)^2 = \int_{\mathbb{R}}(x - \langle S\rangle)^2 d\omega_S(x)$, this quantity is called *dispersion.*

Section 7.4 The first example of a superselection rule was found by G. Wick, A. Wightman and E. Wigner who deduced from the transformation properties of wave functions that the states with integer and half-integer spins belong to different coherent subspaces — see [WWW 1] and also ⟦Wig⟧, Chap.24. In the same paper they also conjectured that the electric charge and baryon number define superselection rules; the condition under which a mixed state is realizable was formulated in [WWW 2]. There is also an example of a "continuous" superselection rule: in the *nonrelativistic* quantum mechanics the requirement of Galilei covariance implies that states with different masses belong to different coherent subspaces; we shall return to this problem in Remark 10.3.2.

The assumption about finiteness of the energy mean values for realizable states can be found in renown texts such as ⟦SW⟧, Sec.1.1; other authors require even the realizable pure states to belong to the domain of the Hamiltonian — see ⟦BLT⟧, Sec.2.1.3. The fact that such assumptions avoid any experimental verification was discussed in [HE 1], [Ex 1], see also ⟦Ex⟧, Sec.I.6. The argument presented at the end of the section applies to any bounded observables A_k. In the notation of Section 7.6 this can be expressed as $|\phi(A_k) - \phi_0(A_k)| < \varepsilon_k$, where ϕ, ϕ_0 are the states described by the statistical operators W and W_0, respectively, which means that in an actual experiment we do not determine a point in the set $S_{\mathcal{A}} \subset \mathcal{A}^*$ of states but rather some $*$–weak neighborhood.

Section 7.5 The inequality (7.16) of Example 7.5.8 is valid only for $n = 1, 2, 3$, since $(1+h)^{-1}$ does not belong to $L^2(R^n)$ for $n \geq 4$, so we cannot use the factorization trick. On the other hand, $D(H_0) \subset L^q(\mathbb{R}^n)$ in this case holds for any $q \in [2, 2n/(n-4))$, and moreover, for any $a > 0$ there is a b such that $\|\psi\|_q \leq a\|H_0\psi\|_2 + b\|\psi\|_2$ for all $\psi \in D(H_0)$ — see, *e.g.*, [[RS 2]], Sec.IX.7. In the terminology mentioned above, the domain of H_0 is the Sobolev space $H^{2,2}(\mathbb{R}^n)$, which consists of all $\psi \in L^2(\mathbb{R}^n)$ for which $-\Delta\psi$ makes sense as a distribution and belongs to $L^2(\mathbb{R}^n)$, results of the type (7.16) then represent examples of so-called *embedding theorems* telling us which L^p spaces are contained in a given Sobolev space.

The claim contained in Theorem 7.5.9 was formulated for the first time by P. Dirac — see [[Dir]], Sec.III.4. His argument is valid, however, only for the observables with pure point spectra; this is why it is sometimes called *Dirac conjecture.* The proof for the case when $\mathcal{S}$ consists of bounded observables was given by [Mau 1]; *cf.* also [[Mau]], Sec.VIII.5.

It is not the full truth, of course, that we know nothing about the state of the system before the "first" measurement. We certainly have a definite enough idea about the result of the preparatory measurement coming from the theoretical considerations, which guided us during construction of the "source", calibration measurements, *etc.* This all results from a development in which theory mingles with experiment; trying to make a sharp distinction would mean producing a new version of the old "chicken or egg" question.

Section 7.6 The first step towards the axiomatization of properties of quantum systems was made by P. Jordan, J. von Neumann and E. Wigner in 1934 — see [JNW 1]. They started from the observation that the prescriptions (i) and (ii) define on the set $\mathcal{O}_b$ the structure of a real vector space and (iii) can be used to define the *symmetrized product,* $\mathrm{a}\cdot\mathrm{b} := \frac{1}{2}((\mathrm{a}+\mathrm{b})^2 - \mathrm{a}^2 - \mathrm{b}^2)$. In addition, the algebra $\mathcal{O}_b$ is required to satisfy the following conditions *(JNW–axioms):*

(j1) $\mathrm{a}_1^2 + \cdots + \mathrm{a}_n^2 = 0$ implies $\mathrm{a}_1 = \cdots = \mathrm{a}_n = 0$,

(j2) $\mathrm{a}^m \cdot \mathrm{a}^n = \mathrm{a}^{m+n}$,

(j3) $(\mathrm{a}+\mathrm{b}) \cdot \mathrm{c} = \mathrm{a} \cdot \mathrm{c} + \mathrm{b} \cdot \mathrm{c}$.

The structure determined by these postulates is called a *Jordan algebra.* The mentioned authors assumed in addition that $\mathcal{O}_b$ has a finite dimension; this allowed them to prove the existence of a spectral decomposition for any observable. There is also a classification of Jordan algebras; more details and further references can be found in [[Em]], Sec.I.2.3.

The assumption of a finite dimensionality which the authors of [JNW 1] made to avoid introducing a topological structure is, of course, too restrictive. Introducing topology by a norm, I. Segal in 1947 formulated the following set of requirements *(Segal axioms):*

(s1) $\mathcal{O}_b$ is a real Banach space with a norm $\|\cdot\|$,

(s2) $\mathcal{O}_b$ is equipped with the unit element e and the operation $\mathrm{a} \mapsto \mathrm{a}^n$, the polynomials in the variable a obey the standard algebraic rules,

(s3) the map $a \mapsto a^2$ is norm continuous,

(s4) $\|a^2 - b^2\| \le \max\{\|a^2\|, \|b^2\|\}$,

(s5) $\|a^2\| = \|a\|^2$.

The original paper [Seg 2] also contained the requirement $\| \sum_{a\in\mathcal{R}} a^2\| \le \| \sum_{a\in\mathcal{S}} a^2\|$ for any finite subsets $\mathcal{R} \subset \mathcal{S}$ in $\mathcal{O}_b$, only later was it found that this followed from (s1) and (s5). The object satisfying these requirements is called a *Segal algebra.*

One can define on a Segal algebra the symmetrized product by the prescription mentioned above. In general, it is neither distributive nor associative. If we require it to be distributive, then the Segal algebra simultaneously satisfies the JNW–axioms. For associative Segal algebras we have a result analogous to Theorem 6.1.5: any such algebra is isometrically isomorphic to the algebra of continuous functions on some compact Hausdorff space — *cf.* [Seg 2], and also ⟦Em⟧, Thm.I.9. It is clear from this that for quantum theory nonassociative Segal algebras are interesting in the first place.

Let $\mathcal{A}$ be some C^*–algebra; then we can easily check that $\mathcal{A}_R := \{a \in \mathcal{A} : a^* = a\}$ fulfils the axioms (s1)–(s5); the symmetrized product $a \cdot b := \frac{1}{2}(ab + ba)$ is distributive. A Segal algebra is called *special* if it is isometrically isomorphic to the set of Hermitean elements of some C^*–algebra; in a similar way, one defines special real Segal algebras. Other Segal algebras are called *exceptional;* such algebras exist. No "internal" criterion is known, which would allow us to decide whether a given Segal algebra is special or exceptional; for more details and references see ⟦Em⟧, Chap.I.

If we adopt the postulate (ao$_w$), then normal states on $\mathcal{A}$ can be represented by statistical operators $W \in \mathcal{J}_1(\mathcal{H})$. This is why the subspace $\mathcal{J}_1^s \subset \mathcal{J}_1(\mathcal{H})$ consisting of those W which are reduced by all coherent subspaces is sometimes used as a state space — see, *e.g.*, ⟦Da 1⟧, Sec.I.4. It is complete with respect to $\|\cdot\|_1$ and the states on it are represented by the positive elements with the unit trace norm. To distinguish it from the state space introduced in Section 7.1, one usually speak about *state Banach space.*

The proof that any algebra of the so–called local observables of a free quantum field is a type III factor can be found in [Ara 1] (see also the notes to Section 6.3.); the algebraic formulation of quantum field theory will be mentioned again in Section 13.3. C^*–algebras describing systems with an infinite number of degrees of freedom usually have uncountably many nonequivalent irreducible representations of which we actually use a small part only, sometimes a single one, and on other occasions a countable family corresponding to particular superselection rules. This is made possible by the fact mentioned in the notes to Section 7.4 that an experimental determination of a state yields some $*$–weak neighborhood in $\mathcal{A}^* \supset S_{\mathcal{A}}$, at the same time we know from [Fel 1] that pure states corresponding to a single irreducible representation are $*$–weakly dense in $S_{\mathcal{A}}$ — a more detailed discussion of this problem can be found in [Haa 1] or in the appendix to ⟦BLT⟧. A simple example of a problem, which involves nonequivalent representations of the algebra of observables, is represented by the van Hove model — see ⟦Em⟧, Sec.I.5.

Problems

1. Check the spectral decomposition of the operators S_j of Example 7.1.2.

2. Consider a particle on halfline from Example 7.2.2.

   (a) Let $c < 0$, then the operator $H_{0,c} \upharpoonright \{\phi_c\}^\perp$ is positive.

   (b) The reflection amplitude satisfies $|R| = 1$. For which values of c is the phase shift of the reflected wave independent of energy?

3. Consider a particle whose motion is confined to a bounded interval $J = [a, b]$ of the real axis.

   (a) Under which condition can we define on $L^2(J)$ self–adjoint operators of momentum and kinetic energy corresponding to the formal expressions $-i\, d/dx$ and $-d^2/dx^2$, respectively?

   (b) Find the spectra of these operators.

   (c) When does the energy equal the square of some momentum operator?

   Hint: Use Example 4.2.5 and Problem 4.61.

4. Consider the operator H on $L^2(\mathbb{R})$ corresponding to the expression (4.8) with the potential $V(x) = gx^{-2}$.

   (a) Prove that H is *e.s.a.* (in fact, self–adjoint) if $g \geq \frac{3}{4}$, and it decomposes into an orthogonal sum of operators acting on $L^2(\mathbb{R}^\pm)$, respectively.

   (b) Find the self–adjoint extensions of H in the case $g \in \left(-\frac{1}{4}, \frac{3}{4}\right)$.

   Hint: (a) Find solutions to $(H^* \pm i)\psi = 0$. (b) *Cf.* [DE 1].

5. Check that the operators $P_j \upharpoonright \mathcal{S}(\mathbb{R}^3)$ of Example 7.2.3 are *e.s.a.*
   Hint: Use the unitary equivalence (7.4).

6. Let $\mathcal{H}$ be a separable Hilbert space with an orthonormal basis $\{\phi_n\}_{n=0}^\infty$. Consider the operator a defined by $a(\sum_n c_n\phi_n) := \sum_n \sqrt{n}\, c_n\phi_{n-1}$ with the domain $D(a) := \{\, \psi = \sum_n c_n\phi_n \,:\, \sum_n n|c_n|^2 < \infty \,\}$.

   (a) Find the adjoint a^*.

   (b) Check that $D(a^*a) = D(aa^*)$ and $[a, a^*]\psi = \psi$ holds for any $\psi \in D(a^*a)$.

7. Prove that the harmonic–oscillator Hamiltonian of Example 7.2.4 can be expressed as $H = 2a^*a + I$, where $a := 2^{-1/2}(Q + iP)$. Use this result to check that H is self–adjoint and find its spectrum.

8. Let V be a bounded measurable real–valued function on $\mathbb{R}$ such that the limits $V_\pm := \lim_{x\to\pm\infty} V_\pm(x)$ exist; then the operator $H := P^2 + V(Q)$ satisfies $\sigma(H) \supset [v, \infty)$ where $v := \min\{V_+, V_-\}$.

9. Consider the operator P^2 of Example 7.2.1. Prove that its resolvent $(P^2 + \kappa^2)^{-1}$ is for $\mathrm{Re}\,\kappa > 0$ an integral operator with the kernel $G_\kappa(x, y) := \frac{1}{2\kappa}\, e^{-\kappa|x-y|}$.

10. Any statistical operator W satisfies $\mathrm{Tr}\, W^2 \leq 1$, the relation turns into identity *iff* the state W is pure. If $\dim \mathcal{H} = n$, we simultaneously have $\mathrm{Tr}\, W^2 \geq \frac{1}{n}$.

11. Let W be a statistical operator and E a projection such that $\mathrm{Tr}\,(EW) \neq 0$, then

(a) $W' := (\mathrm{Tr}\,(EWE))^{-1}EWE$ is a statistical operator.

(b) If W is a one–dimensional projection determined by a unit vector ϕ, then W' is also a one–dimensional projection and it corresponds to the vector $\phi' := E\phi/\|E\phi\|$.

(c) let $\{E_j : j = 1,\dots,N\}$, $N \le \infty$, be a family of orthogonal projections such that $\mathrm{Tr}\,(EW) \neq 0$ holds for $E := \sum_{j=1}^N E_j$, then $(\mathrm{Tr}\,(EW))^{-1}\sum_{j=1}^N E_jWE_j$ is a statistical operator.

12. Suppose that the system is in a state W and we successively measure the observables $A_1,\dots,A_n$. Prove that $w(\Delta_n, A_n;\dots;\Delta_1, A_1; W) = \mathrm{Tr}\,W_n$, where $W_n := E_n \dots E_1WE_1 \dots E_n$ and $E_j := E_{A_j}(\Delta_j)$, and that the state after such a measurement is described by the statistical operator $W' := (\mathrm{Tr}\,W_n)^{-1}W_n$. In particular, if $A_1 = \dots = A_n =: A$, we have $w(\Delta_n,\dots,\Delta_1, A; W) = \mathrm{Tr}\,\left\{E_A\left(\bigcap_{j=1}^n \Delta_j\right)W\right\}$.

13. The following definition of compatibility is sometimes used for *yes–no* experiments (which we do not regard here as filters): we perform three measurements ordered as $E_1E_2E_1$, if the second measurement E_1 yields *with certainty* the same result as the first one, then E_1 and E_2 are compatible. Show that this occurs *iff* E_1 and E_2 are compatible according to the definition from Section 7.5.

14. The *yes–no* experiments $E_1,\dots,E_n$ are *disjoint* if a positive result in some of them excludes a positive result in another.

(a) Prove that this is true *iff* the projections $E_1,\dots,E_n$ are orthogonal; it implies their compatibility.

(b) Specify the conclusions of Example 7.5.5 to the case when $E_1,\dots,E_n$ are disjoint and extend them to the case of an infinite family of disjoint *yes–no* experiments.

15. Prove the properties of the spin component operators (7.11) mentioned in Example 7.5.4b. Show that the set $\{S_1, S_2, S_3\}$ is irreducible in $\mathbb{C}^{2s+1}$.

16. The operator $-\Delta$ on $L^2(\mathbb{R}^n)$ is *e.s.a.* on $\mathcal{S}(\mathbb{R}^n)$ as well as on $C_0^\infty(\mathbb{R}^n)$.
Hint: Check that $\overline{-\Delta \restriction C_0^\infty(\mathbb{R}^n)} \supset -\Delta \restriction \mathcal{S}(\mathbb{R}^n)$ and use (7.4).

17. Consider the position and momentum operators Q_j, P_k on $L^2(\mathbb{R}^n)$ of Example 7.2.3 and denote $Q := \{Q_1,\dots,Q_n\}$, $P := \{P_1,\dots,P_n\}$. Let $F : \mathbb{R}^n \to \mathbb{C}$ be a Borel function. Prove

(a) $f(P) = F_n^{-1}f(Q)F_n$, in particular, the projection–valued measures corresponding to the operator sets P and Q are unitarily equivalent.

(b) $(f(P)\psi)(x) = (2\pi)^{-n/2}\int_{\mathbb{R}^n}(F_n^{-1}f)(x-y)\psi(y)\,dy$ holds for all $\psi \in D(f(P))$ provided $f \in L^2(\mathbb{R}^n)$.

Hint: Use (7.4) and Example 5.5.1.

18. Extend the result of Problem 9 to higher dimensions: put $m = 1/2$ in Example 7.5.8 and use the previous problem to prove that the resolvent $(H_0+\kappa^2)^{-1}$ is for $\mathrm{Re}\,\kappa > 0$ an integral operator with the kernel

(a) $G_\kappa(x,y) := \frac{i}{4} H_0^{(1)}(i\kappa|x-y|)$ if $n = 2$, where $H_0^{(1)}$ is the Hankel function.

(b) $G_\kappa(x,y) := \frac{e^{-\kappa|x-y|}}{4\pi|x-y|}$ if $n = 3$.

(c) More generally, $G_\kappa(x,y) := (2\pi)^{-n/2} \left(\frac{|x-y|}{\kappa}\right)^{1-n/2} K_{(n/2)-1}(\kappa|x-y|)$ for $n \geq 2$, where K_ν is the modified Bessel function.

19. Let P_1, P_2, P_3 and H_0 be the operators of momentum components and the kinetic energy, respectively. Find the probabilities $w(\Delta, P; \psi)$ and $w(\Delta, H_0; \psi)$.
Hint: Using Problem 18, show that $E_{H_0}[0,\lambda)$ is an integral operator with the kernel $K_\lambda(x,y) := (2\pi^2)^{-1}(2m\lambda)^{3/2}z^{-3}(\sin z - z \cos z)$, where $z := \sqrt{2m\lambda}\,|x-y|$.

20. Let $\mathcal{H}$ be separable. For any commutative set $\mathcal{S} \subset \mathcal{L}_{sa}(\mathcal{H})$ one can find a Hermitean operator $A \in \mathcal{B}(\mathcal{H})$ such that $\mathcal{S} \cup \{A\}$ is a CSCO.
Hint: Use Corollary 6.5.6.

21. Prove the estimate $|g(z)-1| < Cz^2$ used in Example 7.5.12. Show that the relation (7.17) remains valid under the following weakened assumptions:

(i) $\psi \in L^2(\mathbb{R}^3) \cap L^1(\mathbb{R}^3)$ and $\int_{\mathbb{R}^3} |y\,\psi(y)|\,dy < \infty$.

(ii) We replace the balls by a one-parameter family $\{\, U_\varepsilon : \varepsilon > 0\}$ of neighborhoods of the point k_0 such that $\mathrm{diam}\, U_\varepsilon = 2\varepsilon$ and $\limsup_{\varepsilon\to 0} \varepsilon^3(\mathrm{vol}(U_\varepsilon))^{-1} < \infty$, where $\mathrm{vol}(U_\varepsilon) := \int_{U_\varepsilon} dk$ is the Lebesgue measure of U_ε.

Chapter 8

Position and momentum

8.1 Uncertainty relations

We know that the outcome of measuring an observable A is the probability measure $w(\cdot, A; W)$ provided the system is in a state W. This represents a lot of information, and we therefore often use several simpler quantities derived from the measure to characterize the result. Most suitable for this purpose are moments of $w(\cdot, A; W)$ or their combinations; we have to keep in mind that they are not directly measurable, as was pointed out in Section 7.4. The simplest among them is the mean value given by (7.7). It tells us nothing, however, of how much the results of the measurement are spread. This can be done, *e.g.*, by means of the standard deviation; recall that it is defined by

$$(\Delta A)_W := (\langle A^2\rangle_W - \langle A\rangle_W^2)^{1/2} = \left(\int_{\mathbb{R}} (\lambda - \langle A\rangle_W)^2 d\mathrm{Tr}\,(E_\lambda^{(A)} W)\right)^{1/2}.$$

The standard deviation represents a way of gauging how exact the measurement is. Notice first that for a single observable the precision is in principle unrestricted.

8.1.1 Proposition: Let $\lambda \in \sigma(A)$; then

(a) For any $\varepsilon > 0$ there is a pure state represented by a vector ψ_ε such that $|\lambda - \langle A\rangle_{\psi_\varepsilon}| \le \varepsilon$ and $(\Delta A)_{\psi_\varepsilon} \le 2\varepsilon$.

(b) The identity $(\Delta A)_W = 0$ holds *iff* $\lambda_0 := \langle A\rangle_W$ is an eigenvalue of A and $W = E_A(\{\lambda_0\}) W E_A(\{\lambda_0\})$. In particular, for a pure state ψ we have $(\Delta A)_\psi = 0$ *iff* ψ is an eigenvector of A corresponding to the eigenvalue $\langle A\rangle_\psi$.

Proof: By Proposition 5.4.1a, we can choose a unit vector $\psi_\varepsilon \in \mathrm{Ran}\, E_A(\Delta_\varepsilon)$, where $\Delta_\varepsilon := (\lambda-\varepsilon, \lambda+\varepsilon)$. Without loss of generality we may assume that ψ_ε describes a realizable state; otherwise we choose a coherent subspace $\mathcal{H}_\alpha$ such that $E_\alpha \psi_\varepsilon \ne 0$ and set $\psi'_\varepsilon := E_\alpha\psi_\varepsilon / \|E_\alpha\psi_\varepsilon\|$. The projection $E_A(\Delta_\varepsilon)$ as an observable commutes with E_α so $\psi'_\varepsilon \in \mathrm{Ran}\, E_A(\Delta_\varepsilon)$. The mean value can be expressed by (7.7) as $\langle A\rangle_{\psi_\varepsilon} = \int_{\lambda-\varepsilon}^{\lambda+\varepsilon} \xi\, d(\psi_\varepsilon, E_\xi^{(A)}\psi_\varepsilon)$; it easily yields the estimate $\lambda-\varepsilon \le \langle A\rangle_{\psi_\varepsilon} \le \lambda+\varepsilon$, *i.e.*, the first one of the above inequalities. This in turn implies $|\xi - \langle A\rangle_{\psi_\varepsilon}| \le |\xi-\lambda| + \varepsilon$, and therefore

$$(\Delta A)^2_{\psi_\varepsilon} = \int_{\lambda-\varepsilon}^{\lambda+\varepsilon} (\xi - \langle A\rangle_{\psi_\varepsilon})^2 d(\psi_\varepsilon, E_\xi^{(A)}\psi_\varepsilon) \le 4\varepsilon^2 .$$

Concerning part (b), the sufficient condition is easy. On the other hand, suppose that $(\Delta A)_W = 0$; then the definition of the standard deviation gives $w(\mathbb{R}\setminus\{\lambda_0\}, A; W) = 0$, *i.e.*, $w(\{\lambda_0\}, A; W) = 1$. Denoting $E := E_A(\{\lambda_0\})$ and $E' := I - E$ we can rewrite these relations as $\mathrm{Tr}\,(EW) = 1$ and $\mathrm{Tr}\,(E'W) = 0$, respectively. Using the spectral decomposition $W = \sum_j w_j E_j$, $E_j := E_{\phi_j}$, we get $\sum_j w_j \|E\phi_j\|^2 = 1$ which requires $\|E\phi_j\| = 1$, or in other words, $EE_j = E_j$ for those j for which w_j is nonzero in view of the normalization condition $\mathrm{Tr}\,W = 1$. It follows that $EW = WE = EWE$. The operator $E'WE'$ is positive, so the condition $\mathrm{Tr}\,(E'W) = 0$ means $E'WE' = 0$; together we get $EWE = E$. ∎

Moreover, this result extends easily to the case when a family of compatible observables is measured (Problem 1). On the other hand, noncompatibility means not only that the order in which the experiments are performed is important, but also that precision is limited when we measure noncompatible observables on identical copies of the system, *i.e.*, in the same state. We shall use the standard abbreviation for the commutator of two operators, $[A_1, A_2] := A_1A_2 - A_2A_1$; recall that due to Example 5.5.9 the commutator is zero if A_1, A_2 are commuting self–adjoint operators, while the opposite implication is not valid — see Example 8.2.1 below. Then the basic result can be formulated as follows.

8.1.2 Theorem (uncertainty relations): Suppose we measure observables A_1, A_2 on the system in a state W. If the operators A_jW, A_jA_kW belong to the trace class for $j,k = 1,2$, then the standard deviations satisfy the inequality

$$(\Delta A_1)_W (\Delta A_2)_W \ge \frac{1}{2}\,|\mathrm{Tr}\,(i[A_1, A_2]W)|\,.$$

Proof: Without loss of generality we may assume $\langle A_j\rangle_W = 0$; otherwise we take the operators $A_j' := A_j - \langle A_j\rangle_W$, which satisfy the same assumptions and

$$(\Delta A_j)^2_W = (\Delta A_j')^2_W = \langle (A_j')^2\rangle_W\,, \quad i[A_1, A_2]W = i[A_1', A_2']W\,.$$

By assumption, any vector $\phi \in \mathrm{Ran}\,W$ belongs to $D([A_1, A_2]) := D(A_1) \cap D(A_2)$ and $A_j\phi \in D(A_k)$ for $j,k = 1,2$. Then for any real α we have the inequality $0 \le \|(A_1 + i\alpha A_2)\phi\|^2 = (\phi, A_1^2\phi) + \alpha(\phi, i[A_1, A_2]\phi) + \alpha^2(\phi, A_2^2\phi)$, and therefore also

$$\mathrm{Tr}\,(A_1^2W) + \alpha\mathrm{Tr}\,(i[A_1, A_2]W) + \alpha^2\mathrm{Tr}\,(A_2^2W) \ge 0;$$

this yields the result because $\mathrm{Tr}\,(A_j^2W) = (\Delta A_j)^2_W$. ∎

8.1.3 Remarks: (a) Operator $C := i[A_1, A_2]$ need not be densely defined; the theorem requires only $\mathrm{Ran}\, W \subset D(C)$. If it is densely defined, it is symmetric by Proposition 4.1.2; however, it can have no self-adjoint extensions (Problem 3). If there is a self-adjoint $\tilde{C} \supset C$, then the assumed boundedness of CW implies $\tilde{C}W = CW$ and the inequality can be rewritten as

$$(\Delta A_1)_W (\Delta A_2)_W \geq \frac{1}{2} \left| \langle \tilde{C} \rangle_W \right| .$$

(b) If $W := E_\psi$ is a pure state, then operator AE_ψ belongs the trace class *iff* $\psi \in D(A)$. Hence if $\psi \in D(A_j A_k)$ for $j, k = 1, 2$, the standard deviations $(\Delta A_j)_\psi = (\|A_j\psi\|^2 - (\psi, A_j\psi)^2)^{1/2}$ satisfy the relation

$$(\Delta A_1)_\psi (\Delta A_2)_\psi \geq \frac{1}{2} \left| (\psi, i[A_1, A_2]\psi) \right| .$$

As mentioned above, the theorem imposes no restriction if the observables A_1, A_2 are compatible. In other cases the right side may be nontrivial and dependent, in general, on the state W.

8.1.4 Example (spin components): Operators S_j representing the spin components satisfy commutation relations (7.12), so we have

$$(\Delta S_1)_W (\Delta S_2)_W \geq \frac{1}{2} \left| \langle S_3 \rangle_W \right|$$

and the similar relations obtained by cyclic permutations of the indices. If W is a pure state, for instance, described by an eigenvector of S_1 or S_2 we have zero on the left side, and therefore $\langle S_3 \rangle_W = 0$. On the other hand, in the case $W := E_{\chi_m}$ the right side equals $\frac{1}{2}|m|$.

However, the best-known application of Theorem 8.1.2 deals with the position and momentum operators.

8.1.5 Examples: (a) *Heisenberg relations:* Consider operators Q, P of Example 7.2.1. The operator $C := i[P, Q]$ is densely defined because its domain contains, *e.g.*, the set $\mathcal{S}(\mathbb{R})$, and we can easily check that it is a restriction of the unit operator to $D(C) := D(PQ) \cap D(QP)$; hence

$$(\Delta P)_\psi (\Delta Q)_\psi \geq \frac{1}{2}$$

holds for all $\psi \in D(C) \cap D(P^2) \cap D(Q^2)$. The analogous relation is valid for any mixed state which satisfies the assumptions of the theorem.

(b) *n-dimensional Heisenberg relations:* Consider next the operators Q_j, P_k on $L^2(\mathbb{R}^n)$. Due to Proposition 5.7.3, they commute for $j \neq k$. On the other hand, in the case $j = k$ we can reason as in the preceding example; together we get the inequalities

$$(\Delta P_k)_\psi (\Delta Q_j)_\psi \ge \frac{1}{2}\,\delta_{jk}\,,$$

which are valid if $\psi \in D(Q_j^r P_k^s) \cap D(P_k^s Q_j^r)$, where r, s are any non–negative integers fulfilling $r+s \le 2$, and the analogous relation for the mixed states which satisfy the assumptions of the theorem. We can also introduce the global quantities,

$$(\Delta Q)_W^2 := \left\langle \sum_{j=1}^{n} (Q_j - \langle Q_j\rangle_W)^2 \right\rangle_W = \sum_{j=1}^{n} (\Delta Q_j)_W^2$$

and $(\Delta P)_W^2$ defined in the same way, which characterize the uncertainty at the position and momentum measurement, respectively (Problem 4); under the stated assumptions, the Hölder inequality then implies

$$(\Delta P)_W (\Delta Q)_W \ge \frac{n}{2}\,.$$

The position and momentum coordinate operators have purely continuous spectra, so one can never assign an exact value to them in an experiment. Due to Proposition 8.1.1, of course, one can measure them with any desired accuracy, but the results of the above example tell us this cannot be done simultaneously in the same state. It also has consequences for a successive measurement of the two observables on the same copy of the system. If, for instance, we perform a high–precision momentum measurement on a particle, by Example 7.5.12 the resulting state is well approximated by the plane–wave expression in a large spatial region, and therefore the probability density of finding the particle there is approximately uniform. Hence if a position measurement is performed immediately afterwards, the standard deviation of the results is large; the larger the standard deviation, the more precise the original momentum measurement was.

Another conclusion drawn from uncertainty relations is where the borderline between the classical and quantum mechanics should be placed. Recall that in standard units they are of the form

$$(\Delta P_k)_W (\Delta Q_j)_W \ge \frac{\hbar}{2}\,\delta_{jk} \quad \text{and} \quad (\Delta P)_W (\Delta Q)_W \ge \frac{n\hbar}{2}\,,$$

respectively. Quantum effects connected with the nonzero value of the right sides may be disregarded if the left sides, or more generally all quantities of the dimension of action appearing in the description of the given system are large in comparison to the Planck constant; in such a case the classical approach is expected to be adequate.

The reader should be warned that even if the commutator of a pair of observables is a restriction of the unit operator, a formal application of the uncertainty relations can lead to an erroneous conclusion.

8.1.6 Example: When we separate variables in a spherically symmetric problem (see Section 11.5 below) we work with the operators on $L^2(0,2\pi)$ representing the azimuthal angle and the canonically conjugate momentum (the third component of angular momentum). These are defined by $Q_a : (Q_a f)(\varphi) = \varphi f(\varphi)$ and $P_a : (P_a f)(\varphi) = -if'(\varphi)$ with the domain $D(P_a) := \{f \in AC(0,2\pi) : f(0) = f(2\pi)\}$. We know from Examples 4.2.5 and 4.3.3 that both of them are self–adjoint; Q_a is bounded with $\sigma(Q_a) = [0,2\pi]$ while P_a has a pure point spectrum, $\sigma(P_a) = \{0,\pm1,\pm2,\dots\}$. By Theorem 8.1.2,

$$(\Delta P_a)_f(\Delta Q_a)_f \geq \frac{1}{2}$$

holds for $f \in D(P_a^2) \cap D(P_a Q_a) = \{f \in AC^2(0,2\pi) : f(0) = f(2\pi) = 0,\ f'(0) = f'(2\pi)\}$. The danger of formal manipulations is obvious. If we choose $f := f_m$, where $f_m(\varphi) := (2\pi)^{-1/2} e^{im\varphi}$ is an eigenvector of P_a, then the left side is zero. This is no paradox, of course, because $f_m \notin D(P_a Q_a)$.

The states for which the inequality of Theorem 8.1.2 turns into identity are called *minimum–uncertainty states.* For simplicity, we restrict ourselves to the pure states (*cf.* Problem 5) described by vectors $\psi \in D(A_j A_k)$, $j,k = 1,2$; we suppose that $(\Delta A_1)_\psi(\Delta A_2)_\psi = \frac{1}{2}|(\psi, C\psi)|$, where again $C := i[A_1, A_2]$. This means that the quadratic polynomial used in the proof of Theorem 8.1.2 has a double root,

$$\alpha = \frac{|(\psi, C\psi)|}{2(\Delta A_2)^2_\psi} = -\frac{(\Delta A_1)_\psi}{(\Delta A_2)_\psi},$$

which satisfies

$$(A_1 - \langle A_1\rangle_\psi + i\alpha(A_2 - \langle A_2\rangle_\psi))\,\psi = 0\,.$$

Hence if we choose the mean values and the ratio of the standard deviations, ψ can be found as a solution to the last equation.

8.1.7 Example: Suppose that $(\Delta P)_\psi(\Delta Q)_\psi = \frac{1}{2}$ holds for the operators Q, P on $L^2(\mathbb{R})$, and denote $p := \langle P\rangle_\psi$, $q := \langle Q\rangle_\psi$ and $\Delta q := (\Delta Q)_\psi$. The above argument yields a first–order differential equation, which is solved by

$$\psi : \psi(x) = \frac{1}{(2\pi(\Delta q)^2)^{1/4}} \exp\left\{-\frac{(x-q)^2}{4(\Delta q)^2} + ipx - \frac{i}{2}pq\right\} \tag{8.1}$$

(Problem 6a), which is easily seen to belong to $D(P^n Q^m)$ for $n,m = 0,1,2$. The last term in the exponent corresponds to the integration constant and has no meaning for determination of the state. Suppose now for simplicity that $\Delta q = 2^{-1/2}$, *i.e.*, $(\Delta P)_\psi = (\Delta Q)_\psi = 2^{-1/2}$; this can always be achieved by an appropriate choice of units. The vectors (8.1) then become

$$\psi_{q,p} : \psi_{q,p}(x) = \pi^{-1/4} e^{-(x-q)^2/2 + ipx - ipq/2}, \tag{8.2}$$

which is nothing else than ψ_w of Remark 2.2.8 provided we put $w := \frac{1}{2}(q - ip)$ (see Problems 6b and 2.12). Hence we may conclude that the minimum uncertainty conditions define a family of coherent states in $L^2(\mathbb{R})$.

The inequalities of Example 8.1.5, sometimes also called the *uncertainty principle*, are probably the most widely known restriction to the results of position and momentum measurement, but they are by no means the only ones. We shall describe two more, and others will be mentioned in the notes. We again consider the operators Q_j, P_k on $L^2(\mathbb{R}^n)$ and define

$$Q^\alpha := \left(\sum_{j=1}^n Q_j^2\right)^{\alpha/2}, \quad P^\beta := \left(\sum_{k=1}^n P_k^2\right)^{\beta/2}$$

for any real α, β; as real functions of self-adjoint commuting operators these operators are self-adjoint. With this notation, we have the following lower bound.

8.1.8 Theorem (Bargmann): Let $n \geq 3$ and $\mu \geq -2$; then

$$\langle P^2 \rangle_\psi \geq \frac{1}{4}(n+\mu)^2 \frac{\|Q^{\mu/2}\psi\|^4}{\|Q^{\mu+1}\psi\|^2} \tag{8.3}$$

holds for any nonzero $\psi \in D(P^2) \cap D(Q^{\mu+1})$.

Proof: First consider $\psi \in \mathcal{S}(\mathbb{R}^n)$ such that $\psi^{(k)}(0) = 0$ for $k = 0, 1, 2, \ldots$. We have $\langle P^2 \rangle_\psi \geq \| \sum_{j=1}^n Q^{-1} Q_j P_j \psi \|^2$ by (7.5), the relation $\sum_{j=1}^n \left|\frac{x_j}{r}\right|^2 = 1$, and the Schwarz inequality. The right side can be estimated as follows

$$\langle P^2 \rangle_\psi \geq \left| \sum_{j=1}^n \frac{(Q^{-1}Q_j P_j \psi, Q^{\mu+1}\psi)}{\|Q^{\mu+1}\psi\|} \right|^2 \geq \left(\sum_{j=1}^n \frac{\mathrm{Im}\,(Q^{-1}Q_j P_j \psi, Q^{\mu+1}\psi)}{\|Q^{\mu+1}\psi\|} \right)^2$$

provided $Q^{\mu+1}\psi \neq 0$. To express the numerator of the last fraction we use the self-adjointness of the operators contained in it; we have $Q_j Q^{\mu+1}\psi \in \mathcal{S}(\mathbb{R}^n)$ due to the assumption, so (7.5) gives

$$-\mathrm{Im}\,(Q^{-1}Q_j P_j \psi, Q^{\mu+1}\psi) = \frac{1}{2}(\psi, (Q^2 + \mu Q_j^2) Q^{\mu-2}\psi) \,;$$

substituting this into the last inequality and summing over j, we get (8.3).

Next we take any $\psi \in \mathcal{S}(\mathbb{R}^n)$ different from zero. The norm in the denominator then makes sense in view of the inequality

$$\|Q^\alpha \psi\|^2 \geq \frac{2\pi^{n/2}}{\Gamma\left(\frac{n}{2}\right)} \max_{\{x:\, r \leq 1\}} |\psi(x)|^2 \int_0^1 r^{2\alpha+n-1} dr + \int_{\{x:\, r>1\}} r^{2\alpha} |\psi(x)|^2 \, dx \,,$$

where the first integral on the right side converges for $2\alpha + n > 0$; choosing a suitable approximating sequence of functions with vanishing derivatives at the origin, we can check that (8.3) is valid again (Problem 7). Finally, $\mathcal{S}(\mathbb{R}^n)$ is a common core for the closed operators Q^α, $\alpha > -n/2$, and P^2 (see Problem 7.16), so to any $\psi \in D(P^2) \cap D(Q^{\mu+1})$ we can find an approximating sequence $\{\psi_n\} \subset \mathcal{S}(\mathbb{R}^n)$ such that $P^2\psi_n \to P^2\psi$ and $Q^\alpha \psi_n \to Q^\alpha \psi$. ∎

8.1.9 Remarks: (a) Inequality (8.3) also holds for $n = 2$ and $\mu > -2$ (see the notes). Under appropriately stronger assumptions the norms on the right side can be written as mean values (Problem 7d). We can also notice that the commutation relations on which the proof is based do not change when either of the operators Q_j, P_k is "shifted" on a multiple of the unit operator. In this way, *e.g.*, the left side may be replaced by $(\Delta P)^2_\psi$ and the norms on the right side by $(\Delta Q^{\mu/2})^4_\psi$ and $(\Delta Q^{\mu+1})^2_\psi$, respectively — see the following remark. The form (8.3) is, however, more suitable if we use the inequality to get a lower bound on the kinetic energy.

(b) In the case $\mu = 0$ we recover the uncertainty relations. For $\mu = -1, -2$ we get two other frequently used inequalities (see the notes)

$$\langle P^2 \rangle_\psi \geq \frac{(n-1)^2}{4} \langle Q^{-1} \rangle^2_\psi , \quad \langle P^2 \rangle_\psi \geq \frac{(n-2)^2}{4} \|Q^{-1}\psi\|^2 . \tag{8.4}$$

The inequalities (8.3) tell us that the probability measure $w(\cdot, Q; \psi)$ cannot be too concentrated around some point unless the mean value $\langle P^2 \rangle_\psi$ is infinite. This can be expressed in a more illustrative way.

8.1.10 Corollary (a local uncertainty principle): Let $n \geq 3$ and $\psi \in D(P^2)$; then

$$w(U_\delta(a), Q; \psi) \leq C_n \delta^2 (\Delta P)^2_\psi , \tag{8.5}$$

where $C_n := 4(n-2)^{-2}$ and $U_\delta(a) := \{x \in \mathbb{R}^n : |x-a| < \delta\}$, holds for any $\delta > 0$ and $a \in \mathbb{R}^n$.

Proof: As we have remarked, one can consider the case with $\langle P_j \rangle_\psi = 0$ and $a = 0$ only. Choosing $\mu = -2$ in the theorem we get the second of the inequalities (8.4) for any $\psi \in D(P^2)$. Furthermore, we denote $\Delta := U_\delta(0)$; then the simple estimate $|\delta^{-1}\chi_\Delta(x)| \leq r^{-1}$ gives

$$\|\delta^{-1}\chi_\Delta(Q)\psi\|^2 \leq \frac{4}{(n-2)^2} \langle P^2 \rangle_\psi ,$$

i.e., inequality (8.5). ∎

8.2 The canonical commutation relations

The fundamental role in nonrelativistic quantum mechanics is played by the commutation relations between the operators describing position and momentum, which can be formally written as $[P, Q] = -iI$, or more generally as

$$[P_k, Q_j] = -i\delta_{jk} I \tag{8.6}$$

for systems whose configuration space is $\mathbb{R}^n$ (one usually refers to the dimension of the configuration space as the number of degrees of freedom — see also Remark 11.1.1a below). We have already employed the identities, (8.6) which are called **canonical commutation relations** (the shorthand CCR is often used), several times: to check irreducibility of the set $\{Q, P\}$, to prove the uncertainty principle, *etc.* Let us now discuss them more thoroughly.

The basic problem is the existence and uniqueness of their representation; we want to know what the operators satisfying (8.6) may look like. To begin with, notice that the relations (8.6) are indeed formal; there are no bounded operators which would fulfil them (Problem 10), so they make no sense as operator identities. This flaw can be corrected if we replace the original operators by suitable bounded functions of them. Recall that self-adjoint operators A_1, A_2 commute *iff* the same is true for their resolvents or the corresponding unitary groups — see Problem 5.23b and Corollary 5.9.4.

We also know from Example 5.5.9 that if $[A_1, A_2]\psi \neq 0$ holds for some ψ in the domain of the commutator, then operators A_1, A_2 do not commute. On the other hand, caution is needed if we want to use the fact that the commutator vanishes to conclude that the operators commute. To illustrate the hidden danger, suppose that the self-adjoint operators A_1, A_2 are such that

(a) There is a common dense invariant subspace D, *i.e.*, $\overline{D} = \mathcal{H}$ and $A_j D \subset D$ for $j = 1, 2$.

(b) D is a core for A_j.

(c) $A_1 A_2 \psi = A_2 A_1 \psi$ holds for any $\psi \in D$.

Contrary to the natural expectation, these conditions are *not* sufficient for the commutativity of A_1 and A_2.

8.2.1 Example (Nelson): The Riemannian surface of the complex function $z \mapsto \sqrt{z}$ is two-sheeted, *i.e.*, its elements are the pairs $\{z, j\}$ with $z \in \mathbb{C}$ and $j = 1, 2$. The projections $z := x+iy$ of the points of M to $\mathbb{C} \sim \mathbb{R}^2$ can be used to introduce a locally Euclidean metric in M. Globally the topology is more complicated because one can pass through the cut $\{z : y = 0,\, x > 0\}$ from one sheet to the other. Any function $\psi : M \to \mathbb{C}$ can obviously be expressed as the pair $\psi := \{\psi_1, \psi_2\}$ with $\psi_j(z) := \psi(\{z, j\})$; it is continuous *iff* the functions ψ_j are continuous outside the cut and $\lim_{z\to x,\, \pm\mathrm{Im}\, z>0}[\psi_j(z) - \psi_{3-j}(-z)] = 0$. Furthermore, we can equip M with the measure, which identifies locally with the Lebesgue measure on $\mathbb{C} \sim \mathbb{R}^2$, and introduce the corresponding space $L^2(M)$ which consists of (classes of) measurable functions ψ such that $\|\psi\|^2 := \int_{\mathbb{C}}(|\psi_1(x)|^2 + |\psi_2(x)|^2)\, dx\, dy < \infty$. The subspace $D := C_0^\infty(M\backslash\{0\})$ then consists of the functions which have all derivatives continuous in the described sense and a compact support separated from the point $0 := \{0, 1\} = \{0, 2\}$. We define the operators A_1, A_2 on D by

$$A_1\psi := \left\{ -i\,\frac{\partial\psi_1}{\partial x},\, -i\,\frac{\partial\psi_2}{\partial x} \right\}, \quad A_2\psi := \left\{ -i\,\frac{\partial\psi_1}{\partial y},\, -i\,\frac{\partial\psi_2}{\partial y} \right\};$$

it is easy to see that they satisfy conditions (a) and (c). To check that (b) is also valid we shall construct the unitary groups generated by the closures $\overline{A}_j$.

First consider the subset $D_x := \{\psi \in D : \operatorname{supp}\psi$ does not contain the x axes$\}$, which is dense in $L^2(M)$ (Problem 11), and define $U_1(t)\psi := \{\psi_1^t, \psi_2^t\}$ where $\psi_j^t(x,y) := \psi_j(x+t,y)$. These operators are isometric, preserve D_x, and form a one–parameter group. Moreover, the map $U_1(\cdot)\psi$ is continuous for $\psi \in D_x$, so $\{\overline{U_1(t)} : t \in \mathbb{R}\}$ is a strongly continuous unitary group by Problem 5.45. It is not difficult to check that $\lim_{t\to 0}(U_1(t)\psi - \psi) = iA_1\psi$ holds for $\psi \in D_x$, and since D_x is preserved by $U_1(t)$, the operator A_1 is *e.s.a.* on it by Problem 5.46; the same is then true for its symmetric extension A_1. In the same way we define the operators $U_2(s)$ on $D_y := \{\psi \in D : \operatorname{supp}\psi$ does not contain the y axes$\}$, representing shifts in the y–direction,

$$\begin{aligned}(U_2(s)\{\psi_1,\psi_2\})(x,y) \;:=\; &\{(1-\kappa_s(y))\psi_1(x,y+s) + \kappa_s(y)\psi_2(x,y+s),\\ &\kappa_s(y)\psi_1(x,y+s) + (1-\kappa_s(y))\psi_2(x,y+s)\,\}\end{aligned}$$

with $\kappa_s(y) := (\Theta(y+s) - \Theta(y))\operatorname{sgn} s$, where $\Theta := \chi_{[0,\infty)}$ is the Heaviside jump function. In this case, the operators $U(s)$ can move the support of a function from one sheet to the other; repeating the above argument we check that A_2 is *e.s.a.*

We can now use Corollary 5.9.4; if we assume that operators A_1, A_2 commute it would imply that $U_1(t)U_2(s) = U_2(s)U_1(t)$ for all $s,t \in \mathbb{R}$. However, this is not true. Choose, for instance, a function $\psi \in D$ whose support is contained in a sufficiently small neighborhood of the point $\{1+i,\, 1\}$; then

$$U_1(2)U_2(2)\psi \neq U_2(2)U_1(2)\psi\,,$$

because the support of the function on the right side lies in a neighborhood of $\{-1-i,\, 1\}$ while the left side is supported around $\{-1-i,\, 2\}$, *i.e.*, on the other sheet.

8.2.2 Remark: Pathological situations of this type can be prevented if we strengthen the assumptions slightly. We have, *e.g.*, the following result (see the notes): let the operators A_1, A_2 satisfy the conditions (a), (c) together with

(b′) D is a core for $A_1^2 + A_2^2$;

then D is also a core for operators A_j and their closures commute.

Inspired by these considerations we could try to replace the canonical commutation relations stated above by suitable relations between the corresponding unitary groups, *i.e.*, the operators $U(t) := e^{iPt}$ and $V(s) := e^{iQs}$ for all $s,t \in \mathbb{R}$. To cover the n–dimensional case at the same time, we introduce the operators

$$U(t) := \exp\left(i\sum_{k=1}^{n} P_k t_k\right), \quad V(s) := \exp\left(i\sum_{j=1}^{n} Q_j s_j\right)$$

for any $s,t \in \mathbb{R}^n$, where as usual $s\cdot t := \sum_{j=1}^n s_j t_j$. In general, these operators do not commute because the right side of (8.6) is nonzero; instead, by a formal computation, we get the relations

$$U(t)V(s) = e^{is\cdot t}V(s)U(t) \tag{8.7}$$

(Problem 13a), which already make sense as operator identities. We call them **Weyl relations** (or the Weyl form of canonical commutation relations). We expect, of course, that the relations (8.7) will be satisfied for the standard position and momentum coordinate operators.

8.2.3 Example: Consider again the operators Q_j, P_k on $L^2(\mathbb{R}^n)$ of Example 7.2.3. The corresponding unitary groups are of the form

$$(U(t)\psi)(x) = \psi(x+t)\,, \quad (V(s)\psi)(x) = e^{is\cdot x}\psi(x) \tag{8.8}$$

for all $s,t \in \mathbb{R}^n$. The second relation follows from the functional calculus; the first was obtained for $n=1$ in Example 5.5.1b, while for $n \geq 2$ we use the decomposition $U(t) = \prod_{k=1}^n e^{iP_k t_k}$, which is a consequence of Proposition 5.5.7. It is now straightforward to check that the operators (8.8) satisfy the Weyl relations, and moreover, that the operator set $\{U(t) : t \in \mathbb{R}^n\} \cap \{V(s) : s \in \mathbb{R}^n\}$ is irreducible (Problem 14). Notice that the maps $U(\cdot)$ and $V(\cdot)$ are strongly continuous, so we can say that each of the relations (8.8) defines a unitary strongly continuous representation of the group T_n of translations of the space $\mathbb{R}^n$.

We usually refer to these operators Q_j, P_k and the corresponding unitary groups (8.8) as the **Schrödinger representation** of canonical commutation relations. It has a privileged position among all representations of the relations (8.7).

8.2.4 Theorem (Stone–von Neumann): Let $U(\cdot)$, $V(\cdot)$ be unitary strongly continuous representations of the group of translations of the space $\mathbb{R}^n$ on a Hilbert space $\mathcal{H}$ which satisfy the Weyl relations. Then

(a) There is a decomposition $\mathcal{H} = \sum_{\alpha\in I}^{\oplus} \mathcal{H}_\alpha$ such that any of the subspaces $\mathcal{H}_\alpha$ is invariant with respect to $U(t)$ and $V(s)$ for all $s,t \in \mathbb{R}^n$.

(b) A unitary operator $S_\alpha : \mathcal{H}_\alpha \to L^2(\mathbb{R}^n)$ corresponds to any $\alpha \in I$ in such a way that

$$(S_\alpha U(t) S_\alpha^{-1}\psi)(x) = \psi(x+t)\,, \quad (S_\alpha V(s) S_\alpha^{-1}\psi)(x) = e^{is\cdot t}\psi(x)$$

holds for any $L^2(\mathbb{R}^n)$ and all $s,t \in \mathbb{R}^n$.

In particular, any irreducible (unitary, strongly continuous) representation of the relations (8.7) is unitarily equivalent to the Schrödinger representation (8.8).

Proof: Given $U(\cdot)$ and $V(\cdot)$, we define a two-parameter family of unitary operators by

$$R(t,s) := e^{-is\cdot t/2}U(t)V(s) \tag{8.9}$$

for any $s,t \in \mathbb{R}^n$ (Problem 15). First we decompose the set $\{R(t,s) : s,t \in \mathbb{R}^n\}$ to irreducible components. For any $f \in L^1(\mathbb{R}^{2n})$ and vectors $\phi,\psi \in L^2(\mathbb{R}^n)$ we define

$$b_f(\phi,\psi) := \int_{\mathbb{R}^{2n}} f(t,s)\,(\phi, R(t,s)\psi)\,dt\,ds;$$

the right side makes sense because the function $(\phi, R(\cdot,\cdot)\psi)$ is bounded and continuous due to the assumption. It is easy to see that $b_f(\cdot,\cdot)$ is a bounded sesquilinear form; hence there is a unique operator $B_f \in \mathcal{B}(\mathcal{H})$ such that $b_f(\phi,\psi) = (\phi, B_f\psi)$ holds for all $\phi,\psi \in \mathcal{H}$. The map $f \mapsto B_f$ defined in this way is obviously linear; the reader is asked to check its other simple properties in Problem 16. We shall also need its injectivity.

8.2.5 Lemma: $B_f = 0$ *iff* $f(t,s) = 0$ holds a.e. in $\mathbb{R}^{2n}$.

Proof: If $B_f = 0$, then also $R(-v,-u)B_fR(v,u) = 0$ for all $u,v \in \mathbb{R}^n$, *i.e.*,

$$\int_{\mathbb{R}^{2n}} f(t,s)\,(\phi, R(-v,-u)R(t,s)R(v,u)\psi)\,dt\,ds$$
$$= \int_{\mathbb{R}^{2n}} e^{i(t\cdot u - s\cdot v)} f(t,s)\,(\phi, R(t,s)\psi)\,dt\,ds \,=\, 0\,.$$

The Fourier transformation is injective (as we mentioned in the notes to Section 1.5), so $(\phi, f(t,s)R(t,s)\psi) = 0$ for all $\phi,\psi \in \mathcal{H}$ and a.a. $s,t \in \mathbb{R}^n$, and therefore $f(t,s)R(t,s) = 0$ for a.a. $s,t \in \mathbb{R}^n$. Finally, using the fact that operators $R(t,s)$ are unitary, we get $f = 0$. ∎

Proof of Theorem 8.2.4, continued: Now consider the operator $B := B_{f_0}$ corresponding to the function $f_0(t,s) := (2\pi)^{-n}e^{-(t^2+s^2)/4}$ which, in view of the lemma and Problem 16a, is nonzero and Hermitean. Moreover, the relations (8.7) yield two other simple identities (Problem 17); the first of them shows that B is a projection.

Now we are ready for the decomposition. We choose an orthonormal basis $\{\psi_\alpha : \alpha \in I\}$ in $\operatorname{Ran} B$, and denote $M_\alpha := \{R(t,s)\psi_\alpha : s,t \in \mathbb{R}^n\}$ and $\mathcal{H}_\alpha := \overline{(M_\alpha)_{lin}}$. It follows from Problem 17b that $M_\alpha \perp M_\beta$ for $\alpha \neq \beta$, and therefore also $\mathcal{H}_\alpha \perp \mathcal{H}_\beta$; at the same time this gives $\dim(\mathcal{H}_\alpha \cap \operatorname{Ran} B) = 1$, so the projection $B \restriction \mathcal{H}_\alpha$ is one–dimensional. Furthermore, due to Problem 16 $R(t,s)M_\alpha \subset M_\alpha$; hence the subspaces $\mathcal{H}_\alpha$ are invariant with respect to $R(t,s)$ for all $s,t \in \mathbb{R}^n$. The operator set $\mathcal{R} := \{R(t,s) : s,t \in \mathbb{R}^n\}$ is symmetric, which means that it is reduced by all the subspaces $\mathcal{H}_\alpha$. Suppose that there is a nontrivial closed subspace $\mathcal{G}_\alpha \subset \mathcal{H}_\alpha$ which is invariant with respect to $\mathcal{R}$; then its orthogonal complement $\mathcal{G}_\alpha^\perp$ in $\mathcal{H}_\alpha$ is also $\mathcal{R}$–invariant. By definition, operator B is in such a case reduced by the subspaces $\mathcal{G}_\alpha$ and $\mathcal{G}_\alpha^\perp$, so the projections $B\restriction\mathcal{G}_\alpha$ and $B\restriction\mathcal{G}_\alpha^\perp$ should be nonzero due to Lemma 8.2.5. However, this is impossible since their sum is the one–dimensional projection $B\restriction \mathcal{H}_\alpha$.

Next we shall check the decomposition $\mathcal{H} = \sum_{\alpha\in I}^{\oplus} \mathcal{H}_\alpha$. Denote by $\mathcal{G}$ the orthogonal complement to the right side. For a pair of vectors $\phi \in \mathcal{G}$, $\psi \in \mathcal{G}^\perp$ we have $(\psi, R(t,s)\phi) = (R(-t,-s)\psi, \phi) = 0$, *i.e.*, $\mathcal{G}$ is also $\mathcal{R}$–invariant. If $\mathcal{G} \neq \{0\}$, then it follows from Lemma 8.2.5 that the operator $B \restriction \mathcal{G}$ is nonzero; however, it is clear from the construction of the subspaces $\mathcal{H}_\alpha$ that $\operatorname{Ran} B \subset \mathcal{G}^\perp$. Hence we get the sought decomposition and the corresponding decomposition of set $\mathcal{R}$ to irreducible components.

Let us now ask what are the relations between the irreducible representations obtained in this way. Consider a pair of mutually different indices $\alpha,\beta \in I$ and denote $\psi^{\alpha}_{t,s} := R(t,s)\psi_\alpha$, $\psi^{\beta}_{v,u} := R(v,u)\psi_\beta$; then we can define the map $U : M_\alpha \to M_\beta$ by $U\psi^{\alpha}_{t,s} := \psi^{\beta}_{t,s}$. Due to Problem 17b this map is isometric and since the sets M_α, M_β are total in $\mathcal{H}_\alpha$, $\mathcal{H}_\beta$, respectively, it can be extended to a unitary operator $U : \mathcal{H}_\alpha \to \mathcal{H}_\beta$. Using Problem 16 once more we get the relations $R(t,s)\psi^{(j)}_{v,u} = e^{i(t\cdot u - s\cdot v)/2}\psi^{(j)}_{t+v,s+u}$ for $j = \alpha, \beta$, which yield the identity

$$U^{-1}R(t,s)U\psi^{(\alpha)}_{v,u} = e^{i(t\cdot u-s\cdot v)/2}\psi^{(\alpha)}_{t+v,s+u} = R(t,s)\psi^{(\alpha)}_{v,u} ;$$

extending it to $\mathcal{H}_\alpha = \overline{(M_\alpha)_{lin}}$ we get $U^{-1}R_\beta(t,s)U = R_\alpha(t,s)$, where we have denoted $R_j(t,s) := R(t,s) \restriction \mathcal{H}_j$. In particular, we have $U^{-1}U_\beta(t)U = U_\alpha(t)$ and $U^{-1}V_\beta(s)U = V_\alpha(s)$, which means that any two irreducible representations are unitarily equivalent; in combination with the result of Example 8.2.3 this concludes the proof. ∎

8.2.6 Remarks: (a) We have proved the irreducibility of the sets $\mathcal{R} \restriction \mathcal{H}_\alpha$ which are wider than the corresponding restrictions of $\{U(t) : t \in \mathbb{R}^n\} \cap \{V(s) : s \in \mathbb{R}^n\}$; however, it is clear from (8.9) that if $\mathcal{G}$ is an invariant subspace of the operators $U(t)$, $V(s)$ it is also invariant with respect to $R(t,s)$.

(b) In the Schrödinger representation, the relations (8.8) give

$$(R(t,s)\psi)(x) = e^{is\cdot(2x+t)/2}\psi(x+t)$$

for all $s, t \in \mathbb{R}^n$. This allows us to express projection B and to show that it is one–dimensional (Problem 18). It can be seen from the construction performed in the proof that the number of irreducible components of a given representation equals $\dim B$; this again proves that the Schrödinger representation is irreducible.

Operator $R(t,s)$ defined by (8.9) is sometimes called the *Weyl operator*. More often this name is used for

$$W(t,s) := R(-t,s) = e^{i\overline{(s\cdot Q - t\cdot P)}}, \tag{8.10}$$

where we write $s\cdot Q := \sum_{j=1}^{n} s_j Q_j$ *etc.*, which has a rather illustrative meaning (see Problem 19).

8.2.7 Example: Consider the Schrödinger representation with $n = 1$. Given an arbitrary vector $\psi \in L^2(\mathbb{R})$ we denote $\psi_{q,p} := W(q,p)\psi$. Then it follows from Remark 8.2.6a that $\psi_{q,p}(x) = e^{ip(2x-q)/2}\psi(x-q)$; the position and momentum appear symmetrically here as can be seen from the relation

$$(F\psi_{q,p})(k) = e^{-iq(2k-p)/2}(F\psi)(k-p),$$

which can be checked easily by approximating ψ by functions from $\mathcal{S}(\mathbb{R})$. This implies the identities

$$\langle Q\rangle_{\psi_{p,q}} = \langle Q\rangle_\psi + q, \quad \langle P\rangle_{\psi_{p,q}} = \langle P\rangle_\psi + p$$

for $\psi \in D(Q) \cap D(P)$, while the standard deviations, provided they exist, are independent of q, p. In particular, if we take for ψ the harmonic–oscillator ground–state vector $\psi_0 : \psi_0(x) = \pi^{-1/4} e^{-x^2/2}$, then the $\psi_{q,p}$ are nothing else than the coherent states (8.2).

As we remarked in the notes to Section 8.1, the relation $\psi_{q,p} := W(q,p)\psi$ also defines coherent states for other $\psi \in L^2(\mathbb{R})$. The map $(q,p) \mapsto \psi_{q,p}$ is continuous by Problem 14b, so to prove the claim we have to check the relation

$$\phi(x) = \int_{\mathbb{R}^2} (\psi_{q,p}, \phi)\, \psi_{q,p}(x)\, dq\, dp; \tag{8.11}$$

this is not difficult as long as the vectors ψ, ϕ belong to $\mathcal{S}(\mathbb{R}^n)$ and $\|\psi\| = (2\pi)^{-1/2}$ (Problem 20).

8.2.8 Remark: Let us see what Theorem 8.2.4 implies for the original form of canonical commutation relations. If the unitary–operator–valued functions $U(\cdot)$ and $V(\cdot)$ satisfy its assumptions, we have

$$U(t) = \sum_{\alpha\in I}^{\oplus} S_\alpha^{-1} U_S(t) S_\alpha\,, \quad V(s) = \sum_{\alpha\in I}^{\oplus} S_\alpha^{-1} V_S(s) S_\alpha$$

for all $s, t \in \mathbb{R}^n$, where $U_S(\cdot)$ and $V_S(\cdot)$ are defined by (8.8). The properties of the Schrödinger representation mentioned in Example 8.2.3 imply that the operators $P_k := -i\left(\frac{\partial U}{\partial t_k}\right)(0)$ and $Q_j := -i\left(\frac{\partial V}{\partial s_j}\right)(0)$ have the following properties:

(a) There is a common dense invariant subspace, *e.g.*, $D := \sum_{\alpha\in I}^{\oplus} S_\alpha^{-1} \mathcal{S}(\mathbb{R}^n)$.

(b) Operators P_k and Q_j are *e.s.a.* on D.

(c) $[P_k, Q_j]\psi = -i\delta_{jk}\psi$ holds for any $\psi \in D$.

On the other hand, in the same way as in Example 8.2.1 these conditions do *not* imply that the unitary groups associated with P_k, Q_j satisfy the Weyl relations (Problem 21b). As in Remark 8.2.2, a sufficient condition is obtained if we replace (b) by

(b′) the operator $\sum_{j=1}^n (P_j^2 + Q_j^2)$ is *e.s.a.* on D;

notice that in the Schrödinger representation the last condition is valid, *e.g.*, for $D := \mathcal{S}(\mathbb{R}^n)$ (Problem 21a).

We can naturally ask whether Theorem 8.2.4 may be extended to situations where the number of canonical pairs (degrees of freedom) is infinite. A more general form of Weyl relations is obtained if we replace $\mathbb{R}^n$ by a real Hilbert space $\mathcal{S}$ with the inner product $(\cdot,\cdot)$; we are then looking for unitary strongly continuous representations of the group of translations of $\mathcal{S}$ such that

$$U(t)V(s) = e^{i(s,t)} V(s) U(t) \tag{8.12}$$

for all $s, t \in \mathcal{S}$. In the case of a finite–dimensional $\mathcal{S}$, Theorem 8.2.4 can be applied, since $\mathcal{S}$ is then topologically isomorphic to $\mathbb{R}^n$ by Problem 1.21. On the other hand, no analogous result is valid if $\dim \mathcal{S} = \infty$ (see the notes).

8.3 The classical limit and quantization

Since our global picture of the physical world should be free from contradictions, quantum theory has to reproduce the results of classical physics when applied to large objects. It is therefore useful to investigate how the predictions of quantum theory look in such situations, which are usually referred to briefly as the **classical limit**. We mentioned in Section 8.1 that the results of quantum and classical descriptions have to match in cases when quantities of the dimension of action are much larger than the Planck constant. It suggests that a suitable mathematical way to treat this problem is to study how quantum systems behave in the limit $\hbar \to 0$.

8.3.1 Example: The classical harmonic oscillator of mass m and angular frequency ω moves periodically with the amplitude $A := (2E/m\omega^2)^{1/2}$, where E is the total energy; the region outside the interval $[-A, A]$ is classically forbidden. For comparison with the quantum oscillator we can therefore use the probability of finding the particle in a given interval (at a randomly chosen instant), or more generally, the mean values of the quantities $f(x)$ (functions of the particle position) with respect to the corresponding probability measure. The latter is easily found; the probability density of finding the particle at a point x is inversely proportional to its velocity, which yields

$$\langle f(x)\rangle_E = \int_{-A}^{A} \frac{f(x)}{\pi\sqrt{A^2-x^2}}\, dx .$$

In particular, for $f := \chi_J$ the formula yields the classical probability $w_{cl}(J, x : E)$ of finding the particle in the interval J provided it has the total energy E.

Now consider the quantum harmonic oscillator of Example 7.2.4. Suppose, *e.g.*, that the function f is measurable and polynomially bounded; then the mean value of the observable $f(Q)$ in the state ψ_n is by (Q2c) equal to

$$\langle f(Q)\rangle_{\psi_n} = \frac{1}{2^n n!}\sqrt{\frac{m\omega}{\pi\hbar}} \int_{\mathbb{R}} f(x)\, H_n\left(\sqrt{\frac{m\omega}{\hbar}}x\right)^2 e^{-m\omega x^2/\hbar}\, dx .$$

To be able to compare the two cases we have to keep in mind that the energy eigenvalue corresponding to state ψ_n depends on $\hbar$; hence it is necessary to perform simultaneously a suitable limit $n \to \infty$. We choose, for instance, the sequence $\{\hbar_n\}$ in such a way that $\frac{1}{2}\hbar_n\omega(2n+1) = E$ and see how the quantity

$$\langle f(Q)\rangle_{\psi_n,\hbar_n} = \frac{1}{2^n n!\sqrt{\pi}} \int_{\mathbb{R}} f\left(\frac{Ay}{\sqrt{2n+1}}\right) H_n(y)^2 e^{-y^2}\, dy$$

behaves in the limit $n \to \infty$. Choosing $f : f(x) = \cos(kx)$ we get

$$\lim_{n\to\infty} \langle \cos(kQ)\rangle_{\psi_n,\hbar_n} = \lim_{n\to\infty} L_n\left(\frac{k^2A^2}{4n+2}\right) \exp\left(-\frac{k^2A^2}{8n+4}\right) = J_0(kA)\, ,$$

where J_0 is the first-order Bessel function (Problem 23). On the other hand, we have $\langle \cos(kx)\rangle_E = J_0(kA)$, and therefore

$$\lim_{n\to\infty} \langle f(Q)\rangle_{\psi_n,\hbar_n} = \langle f(x)\rangle_E \tag{8.13}$$

for $f(x) := \cos(kx)$. The same is true trivially for the odd function $f(x) := \sin(kx)$, and thus also for $f(x) := e^{ikx}$; we shall show that the relation (8.13) is valid for any f, which is the Fourier transform of some $\hat{f} \in L^1(\mathbb{R})$, in particular, for all $f \in \mathcal{S}(\mathbb{R})$. Under this assumption the Fubini theorem gives $\langle f(x)\rangle_E = (2\pi)^{-1/2}\int_{\mathbb{R}} \hat{f}(k)J_0(kA)\,dk$. On the other hand, the same theorem can be applied to express the quantum mean value,

$$\lim_{n\to\infty} \langle f(Q)\rangle_{\psi_n,\hbar_n} = \lim_{n\to\infty} \frac{1}{\sqrt{2\pi}} \int_{\mathbb{R}} dk \hat{f}(k) \frac{1}{2^n n!\sqrt{\pi}} \int_{\mathbb{R}} \exp\left(\frac{ikAy}{\sqrt{2n+1}}\right) H_n(y)^2 e^{-y^2}\,dy\,.$$

The integrated function in the outer integral is majorized by $|\hat{f}(\cdot)|$ and tends to $\hat{f}(\cdot)J_0(\cdot A)$ as $n\to\infty$; the result then follows from the dominated–convergence theorem.

In a similar way we can deal with the classical limit for other simple systems (Problem 24). A more complicated problem is to describe how quantum systems behave when they approach the classical limit, *i.e.*, how their wave functions, eigenvalues, *etc.*, depend on the value of the Planck constant; we usually speak of the *semiclassical approximation.* A straightforward approach, the formal version of which is well known from quantum mechanical textbooks, is based on expansion of the considered quantities into series in powers of $\hbar$; it can be put on a mathematically sound basis, but it requires a certain effort. We give some references in the notes; the problem will be mentioned again in Section 9.4.

A converse to the classical limit is represented in a sense by **quantization** which is, roughly speaking, a search for the quantum description of a system based on knowledge of its classical counterpart. This problem has an obvious heuristic motivation: if we are looking for operators to represent observables of a system having a classical analogue, it could be useful to have a prescription for constructing them starting from the corresponding classical quantities. On the other hand, the importance of a quantization as a physical method must not be overestimated; the ultimate criterion of adequacy of the quantum model obtained in this way is the correctness of its predictions.

Consider a classical system with phase space $\mathbb{R}^{2n}$, in which the canonical coordinates $q_1,\dots,q_n,p_1,\dots,p_n$ are introduced. The starting object is a suitable family F of functions $f:\mathbb{R}^{2n}\to\mathbb{R}$ which could be required, for example, to obey the following conditions:

(c1) F includes the trivial observable, $1(q_1,\dots,q_n,p_1,\dots,p_n) := 1$ for any $(q_1,\dots,q_n,p_1,\dots,p_n)\in\mathbb{R}^{2n}$, as well as $q_j : q_j(q_1,\dots,q_n,p_1,\dots,p_n) = q_j$ and $p_k : p_k(q_1,\dots,q_n,p_1,\dots,p_n) = p_k$,

(c2) F is a real vector space; in particular, it contains the $(2n+1)$–dimensional subspace $F_1 := \{f = \alpha 1 + \sum_{j=1}^n(\beta_j q_j + \gamma_j p_j) : \alpha,\beta_j,\gamma_j\in\mathbb{R}\}$,

(c3) F is a Lie algebra (usually an infinite–dimensional one) with respect to the product $\{\cdot,\cdot\}_P$ defined by the *Poisson bracket*,

$$\{f,g\}_P := \sum_{j=1}^{n} \left(\frac{\partial f}{\partial q_j}\frac{\partial g}{\partial p_j} - \frac{\partial g}{\partial q_j}\frac{\partial f}{\partial p_j} \right);$$

using the terminology of Section 10.3 we can say that its subalgebra F_1 is just the Heisenberg–Weyl algebra.

The conditions are satisfied for various families of functions, for instance, the set $\mathcal{W}_{2n}$ of all inhomogeneous polynomials in the variables q_j, p_k, the set $C^\infty(\mathbb{R}^{2n})$ of all infinitely differentiable functions, *etc.*

The most common quantization procedure, which is usually referred to as *Dirac quantization*, consists of mapping this algebraic structure to a suitable family of self–adjoint operators: given a set F of classical observables we look for a Hilbert space $\mathcal{H}$ and a linear map $J : F \to \mathcal{L}_{sa}(\mathcal{H})$ such that

(d1) $J(\{f,g\}_P) = (i\hbar)^{-1}[J(f),J(g)]$ holds for all $f,g \in F$,

(d2) $J(1) = I$, *i.e.*, the unit operator corresponds to the trivial observable 1,

(d3) the set $\{J(q_j),\, J(p_k) : j,k = 1,\dots,n\}$ is irreducible;

the inclusion $F \supset F_1$ shows that a part of the problem is to construct a representation of the canonical commutation relations. This means, in particular, that the set $J(F)$ cannot consist of bounded operators only, and condition (d1) has to be understood as being valid on a common dense invariant subspace.

We know from the preceding section that a map J with the properties (d1)–(d3) exists if $F = F_1$, and even that it is not unique unless we add other requirements such as those discussed in Remark 8.2.8. However, the set of classical observables also contains other quantities, such as the kinetic and potential energy, angular momentum, *etc.*, which should be included into F. Unfortunately, it appears that F_1 cannot be extended too much.

8.3.2 Theorem (van Hove–Tilgner): There is no Dirac quantization on the sets $C^\infty(\mathbb{R}^{2n})$ and $\mathcal{W}_{2n}$.

References to the proof are given in the notes. Here we limit ourselves to hinting why such a map J cannot exist; for simplicity we consider the case with a single degree of freedom, $n=1$. Suppose that the self–adjoint operators $Q := J(q)$, $P := J(p)$ correspond to the classical observables q, p, respectively; then the identity $(n+1)q^n = \{q^{n+1},p\}_P$ together with condition (d1) implies $Q^n = J(q^n)$ for all positive integers n and the analogous relations for P^n. Furthermore, the relation $(n+1)q^n p = \{q^{n+1},p^2\}_P$ and (d1) give $J(q^n p) = \frac{1}{2}(Q^n P + PQ^n)$ and similar relations for $J(qp^n)$. Now consider the quantity q^2p^2, which has more than one Poisson–bracket expression, *e.g.*, $q^2p^2 = \frac{1}{9}\{q^3,p^3\}_P = \frac{1}{6}\{q^3p,p^2\}$. Using the first expression, after an elementary computation we get from (d1)

$$J(q^2p^2)\psi = [Q^2P^2 - 2i\hbar QP + \alpha(-i\hbar)^2]\psi$$

for any vector ψ of the common invariant domain where $\alpha = 2/3$, while the second expression yields a similar relation with $\alpha = 1/2$ (Problem 25).

As we shall see in a moment, there are sets $\mathcal{W}_{2n} \supset F \supset F_1$ for which a Dirac quantization can be constructed. However, these sets are too small to meet practical needs; it is therefore more reasonable to change the original question abandoning the requirement (c3) on the family of classical observables.

8.3.3 Example (practical quantization): For simplicity we shall again consider the case $n = 1$; an extension to systems with any finite number of degrees of freedom is easy. The subsets

$$F_2 := \left\{ \sum_{j+k\le 2} a_{jk} q^j p^k \,:\, a_{jk} \in \mathbb{R} \right\}, \qquad F_q := \{ g_1(q)p + g_0(q) \,:\, g_0, g_1 \in C^\infty(\mathbb{R}) \}$$

in $C^\infty(\mathbb{R}^2)$ satisfy conditions (c1)–(c3); we define on them the maps

$$J_2 \,:\, J_2\left(\sum_{j+k\le 2} a_{jk} q^j p^k \right) = \frac{1}{2} \sum_{j+k\le 2} a_{jk} (Q^j P^k + P^k Q^j),$$

$$J_q \,:\, J_q\left(g_1(q)p + g_0(q) \right) = \frac{1}{2}(g_1(Q)P + Pg_1(Q)) + g_0(Q),$$

where Q, P are the position and momentum operators in the Schrödinger representation restricted to a suitable dense invariant subspace $D \subset L^2(\mathbb{R})$. We can choose, *e.g.*, $D = C_0^\infty(\mathbb{R})$; in the first case $\mathcal{S}(\mathbb{R})$ can also be used, while for J_q it is generally not invariant if the functions g_j grow too fast. It is straightforward to check that the two maps satisfy conditions (d1)–(d3), and therefore each of them could be considered as a Dirac quantization — up to the fact that the obtained operators are in general only symmetric. A more elementary and substantial reason, however, why neither of them is satisfactory is the following: they do not contain a typical energy observable of the form $p^2 + v(q)$ with exception of the case when v is a quadratic polynomial.

On the other hand, $J_2(f) = J_q(f)$ holds for any $f \in F_2 \cap F_q$, so we are able to construct a common extension. We take the subset F_{2q} in $C^\infty(\mathbb{R})$ spanned by $F_2 \cap F_q$, *i.e.*, $F_{2q} := \{ g_2 p^2 + g_1(q)p + g_0(p) \,:\, g_2 \in \mathbb{R},\, g_1, g_0 \in C^\infty(\mathbb{R}) \}$, and define the map J_{2q} by

$$J_2(q) := g_2 P^2 + \frac{1}{2}(g_1(Q)P + Pg_1(Q)) + g_0(Q)\,.$$

The set F_{2q} already does not satisfy condition (c3): we have seen, for instance, that $\{q^3 p, p^2\}_P = 6q^2 p^2 \notin F_{2q}$. If the functions $f, g \in F_{2q}$ are such that $\{f, g\}_P \in F_{2q}$, then, of course, condition (d1) is satisfied; for this reason the map J_{2q} may be called a *practical quantization*. Notice also that the operators obtained in this way are symmetric but in general not *e.s.a.* (see the notes).

To illustrate that the quantization procedure described here has a rather heuristic meaning, consider again a typical Hamilton function, $h(q,p) := p^2/2m + v(q)$

with $v \in C^\infty(\mathbb{R})$. We ascribe to it the operator $H := J_{2q}(h) = (2m)^{-1}P^2 + v(Q)$. However, we also postulate the same form of the Hamiltonian for other potentials including those for which function v has discontinuities and other singularities; we care only about the essential self–adjointness of the resulting operator. At the same time, the relations between classical and quantum dynamics include some deep problems — see the notes.

Notes to Chapter 8

Section 8.1 In Example 8.1.7 we again encounter the *coherent states* already mentioned in Section 2.2. This name is used for various subsets $\mathcal{C} := \{\psi_z : z \in I\}$ of the state space $\mathcal{H}$, where I is an uncountable index set equipped with a suitable topology and measure. It is required that

(i) The map $z \mapsto \psi_z$ is continuous.

(ii) A relation of the type (2.2) is valid: $\psi = \int_I (\psi_z, \psi)\psi_z \, dz$ for all $\psi \in \mathcal{H}$.

The family (8.2) of coherent states can be obtained from the harmonic–oscillator ground–state vector ψ_0 by means of the operators of the Heisenberg–Weyl group representation — see Example 8.2.7. If we replace ψ_0 by another vector $\psi \in L^2(\mathbb{R})$ we obtain another family of coherent states which are no longer minimum–uncertainty states and cannot be represented by analytic functions.

The family (8.2) was first noted by E. Schrödinger in [Schr 1]. The notion of a coherent state was formulated at the beginning of the 1960s in the papers [Gla 1, 2], [Kla 1]; the Hilbert space of analytic functions, which is closely related to the states (8.2), was introduced at about the same time — see [Ba 1,2]; ⟦Seg⟧, Chap.VI, and [Seg 3]. The mentioned coherent states are often called *canonical* because they are determined by the HW–group related to the CCR. Coherent states can also be constructed for other groups; a well known example are the *spin* (or Bloch) coherent states in the spaces $\mathbb{C}^{2s+1}$ which were constructed in [Rad 1] using the appropriate representations of the group $SU(2)$; the construction of coherent states for a general Lie group can be found in [Per 1]. As we have already noted (and as we shall see again in Section 9.3) the canonical coherent states are naturally associated with harmonic oscillator potential; similar "coherent" states corresponding to a wider class of potentials have been studied in [NSG 1].

Canonical coherent states were introduced by R. Glauber for purposes of quantum optics; a little earlier they were defined by J. Klauder as an "overcomplete system of states", as a tool to define the Feynman integral — see [Kla 2] and also [KD 1]. As was already noted by E. Schrödinger, coherent states are useful for investigation of the classical limit of quantum mechanics; see also, *e.g.*, [Hep 1], [Lie 1], [Hag 1]. In addition, they have physical applications which are too numerous to be listed here; a representative overview is provided by the book ⟦ KS ⟧, where many important papers are reprinted.

Uncertainty relations are probably discussed in every quantum mechanical textbook; sometimes they are illustrated by an argument adopted from classical optics whose rigorous form can be found in [Wil 1]. The fact that standard deviation is not always the best way to characterize how the state is localized has inspired various modifications. For instance, given numbers $\alpha, \beta \in [1/2, 1)$ we can define $(\delta^\alpha Q)_\psi := \inf\{b - a : w((a,b), Q; \psi) = \alpha\}$

and $(\delta_\beta Q)_\psi := \inf\{|a| : \int_{\mathbb{R}} \overline{\psi}(x)\psi(x-a)\,dx = \beta\}$ (*cf.* Problem 8) and the analogous quantities for operator P. They have an obvious meaning; it is illustrative to inspect what they look like if the support of the function ψ consists of two intervals of a length l at a distance $L \gg l$. Various inequalities between such "uncertainties" have been proven in [HU 1], [UH 1,2], [Pri 2].

Theorem 8.1.8 was proven in the paper [Ba 4] using the partial–wave decomposition $L^2(\mathbb{R}^n) = \bigoplus_{\ell=0}^{\infty} \mathcal{H}_\ell$, where $\mathcal{H}_\ell$ are eigenspaces of the generalized angular momentum (*cf.* Section 11.5). In this way, a stronger result is obtained: if $\psi \in \mathcal{H}_\ell$, then $n+\mu$ in the inequality (8.3) can be replaced by $n+\mu+2\ell$. A generalization to the case when $Q^{\mu+1}$ is replaced by $g(Q)$ with an absolutely continuous $g : \mathbb{R}^+ \to \mathbb{R}$ is given in [Ex 2]. The second of the inequalities (8.4) is the well–known *Hardy inequality* whose origin can be traced to [Har 1]; its concise proof can be found in ⟦RS 2⟧, Sec.X.2, where it is referred to as the *uncertainty principle lemma.* In the case $n = 2$ the result becomes trivial but similar inequalities are valid with P^2 replaced by other operators – see [LW 1], [EkK 1].

Corollary 8.1.10 is adopted from the paper [Far 1], which also contains an alternative proof showing that the bound can be improved for $3 \le n \le 5$, when the constant C_n can be replaced by $4/\pi^2$, and also an example illustrating that the assertion is not valid for $n = 2$. It is elementary to check that for $n = 1$ it is not valid either; in that case we can prove a similar relation with the right side depending linearly on $(\Delta P)_\psi$ (Problem 9). More general inequalities of that type can be found in [Pri 1,2].

Theorem 8.1.8 and Corollary 8.1.10 represent examples of results sometimes called *local uncertainty principles.* They are deeper than the standard Heisenberg inequalities because they show that localizing a particle (or a system of particles) sharply at *any* point of the space is possible only at the expense of raising its kinetic energy. This increase may or not be compensated by potential energy; this is the core of the fundamental problem of *stability of matter* about which we shall say more in the notes to Section 14.3.

Section 8.2 Example 8.2.1 comes from the paper [Nel 1]; see also ⟦RS 1⟧, Sec.VIII.5; ⟦Thi 3⟧, Sec.3.1; basic facts about Riemannian surfaces can be found in most complex–analysis textbooks — see, *e.g.*, ⟦Šab⟧. In the mentioned paper E. Nelson also proved a sufficient condition for integrability of representations of Lie algebras — *cf.* also ⟦BaR⟧, Sec.11.5 — the assertion quoted in Remark 8.2.2 is a particular case of this result for the two–dimensional commutative Lie algebra. The idea is the following: we can expand the involved unitary groups into power series and employ the commutation relations between the generators. This makes sense provided the operators are applied to a common analytic vector; the type (b′) condition ensures the existence of a dense set of such vectors. Another particular case of Nelson's theorem is given in Remark 8.2.8; see also [Di 1]. A simpler example of operators Q, P which satisfy the relations (8.6) on a dense set but (8.7) does not hold for the corresponding $U(t)$, $V(s)$ can be found in Problem 12; however, in that case condition (b) is not valid.

Canonical commutation relations were put in the form (8.7) by H. Weyl in 1927; the result concerning uniqueness of their irreducible representation, *i.e.* Theorem 8.2.4, was first proven in the papers [Sto 1] and [vN 1]. Our proof essentially follows the original von Neumann argument which is reproduced, *e.g.*, in ⟦Pru⟧, Sec.IV.6; other modifications and generalizations can be found in ⟦Hol⟧, Sec.V. ⟦Thi 3⟧, Sec.3.1; [Kas 1], *etc.* A somewhat different proof, which relies on analytic vectors of the operator $P^2 + Q^2 - I$, is sketched in ⟦RS 2⟧, Probl.X.30. Theorem 8.2.4 represents a particular case of a general result proven by G. Mackey within the so–called imprimitivity theory — see, *e.g.*, ⟦BaR⟧, Sec.20.2; ⟦Ja⟧, Sec.12–3; ⟦Var 2⟧, Sec.11.3. Properties of operators of type $i[f(P), g(Q)]$ in the Schrödinger representation are discussed in [Ka 7].

A discussion of the problem of nonequivalent representations of the relation (8.12) for $\dim \mathcal{S} = \infty$ can be found, *e.g.*, in ⟦Seg⟧. Generally speaking, there are plenty of them. If we consider, for instance, operators describing a free scalar field together with the canonically conjugate momenta, then any two families of such operators corresponding to fields with different masses define nonequivalent representations of (8.12), and moreover, representations corresponding to the free and interacting field are also not equivalent — we shall return to this question in Section 12.3.

There we shall also encounter *canonical anticommutation relations* (CAR). In this case the situation is easier because only bounded operators are involved. The *anticommutator* is conventionally defined as $\{B_1, B_2\} := B_1B_2 + B_2B_1$; we say that bounded operators B_j, $j = 1, \dots, N$, satisfy the CAR if $\{B_j, B_k\} = \{B_j^*, B_k^*\} = 0$ and $\{B_j^*, B_k\} = \delta_{jk}I$. If we look for a representation of these relations it is useful to pass to $2N$ unitary operators $U_{2j} := B_j^* + B_j$ and $U_{2j-1} := i(B_j^* - B_j)$, which satisfy $U_n^2 = I$ and $U_nU_m + U_mU_n = 0$ for $n \ne m$. For any positive integer N one can construct an irreducible 2^N–dimensional representation of the CAR which we denote as $U^{(N)}$ (Problem 22). A counterpart to Theorem 8.2.4 is provided by the *Jordan–Wigner theorem:* If $N < \infty$, any irreducible representation of the CAR is unitarily equivalent to $U^{(N)}$ — see, *e.g.*, [Si 1].

Section 8.3 In addition to the correct classical limit, which is required if the quantum theory is to be consistent with classical physics, the former has numerous implications for macroscopic systems which the latter can register but in no way explain. Recall the problem of stability of matter mentioned above, which suggests that the mere existence of our esteemed reader is a quantum effect to which classical physics has little to say, and *a fortiori*, the fact that he or she is living and is reading this book represents a quantum phenomenon so deep that nobody is able to render its full explanation.

The formal scheme of the semiclassical approximation or *WKB–method* in which logarithm of the wave function is supposed to be a meromorphic functions of the "variable" $\hbar$ at the vicinity of the point $\hbar = 0$ is known from textbooks. A detailed discussion together with many applications can be found in [BM 1]; see also ⟦Mas⟧, ⟦MF⟧, and ⟦Vai⟧; one way in which the WKB expansion can be made rigorous is described in [KlS 1]. Semiclassical approximation has been object of intense mathematical interest. We have already touched the problem in the notes to Sections 6.7 and 8.1, see also [GMR 1]; another approach is used in *e.g.*, [CDS 1], [CDKS 1], and yet another one will be mentioned in the notes to Section 9.4. Let us remark that there are situations where semiclassical analysis can be used; however, the expansion parameter has another physical meaning — as an example, the *cascading phenomenon* studied in [GGH 1] can be quoted.

The idea of quantization procedure goes back to P. Dirac — *cf.* ⟦Dir⟧, Sec.IV.21. The proof of Theorem 8.3.2 for $F = C^\infty(\mathbb{R})$ is given in [vH 1] or ⟦vH⟧, Sec.23; for $F = \mathcal{W}_{2n}$ it can be found in [Til 1]. Notice that operators resulting from practical quantization are generally only symmetric; an example is provided by $A := P^2 + Q^2 - Q^4$, which is not *e.s.a.* even on $C_0^\infty(\mathbb{R})$ — see ⟦RS 2⟧, Sec.X.5. As we have mentioned, an extension of the map J_{2q} is possible at the expense of losing the bijective correspondence between the Poisson bracket of the classical observables and the commutator of the corresponding operators. Nevertheless, different extensions have been constructed. They can be distinguished by the rule, which ascribes an operator to the quantity $q^j p^k$, for instance, $\frac{1}{2}(Q^j P^k + P^k Q^j)$ (symmetrization), $(j+1)^{-1}\sum_{m=0}^{j} Q^m P^k Q^{j-m}$ (Born–Jordan), $2^{-j}\sum_{m=0}^{j}\binom{j}{m} Q^m P^k Q^{j-m}$ (Weyl–McCoy), *etc.;* an overview is given in [Wol 1].

More information about various quantization methods can be found, *e.g.*, in [BFF 1], ⟦Hur⟧ or [CdV 1]. Physically this problem is usually motivated as an attempt to find a way to produce a correct quantum description of more complicated classical systems (with constraints, a nonflat configuration space, field theories, *etc.*) but no convincing example has been found up to now. At the same time, quantization has been a rich source of inspiration for mathematics, in particular, for the theory of pseudodifferential operators.

Moreover, there are deep relations between the spectral properties of quantum systems and the phase space behavior of their classical counterparts which are not yet fully understood. For example, in so–called integrable systems the phase–space trajectories consists of families of tori, while in chaotic systems they are rather irregular; in their quantum counterparts the difference is manifested in different distributions of energy–eigenvalue spacings and similar quantities. However, a discussion of this problem goes beyond the scope of the present book; we refer to [Zy 1], [Šeb 1] and also to the monographs ⟦Stock⟧, ⟦Stol⟧ for further reading.

Problems

1. Let $A_1,\dots,A_n$ be compatible observables. Show that if $\lambda_j \in \sigma(A_j)$, $j=1,\dots,n$; then for any $\varepsilon > 0$ there is a pure state ψ_ε such that $|\lambda_j - \langle A_j\rangle_{\psi_\varepsilon}| \le \varepsilon$ and $(\Delta A_j)_{\psi_\varepsilon} \le 2\varepsilon$ holds for $j=1,\dots,n$. In a similar way, generalize assertion (b) of Proposition 8.1.1.

2. Under the assumptions of Theorem 8.1.2, $(\Delta A_1)_W^2 + (\Delta A_2)_W^2 \ge |\mathrm{Tr}\,(i[A_1,A_2]W)|$. In particular, operators Q_j, P_k on $L^2(\mathbb{R}^n)$ satisfy $(\Delta Q_j)_\psi^2 + (\Delta P_k)_\psi^2 \ge \delta_{jk}$ for any $\psi \in \mathcal{S}(\mathbb{R}^n)$.

3. Let A_1, A_2 be unbounded self–adjoint operators; then $C := i[A_1,A_2]$ may have no self–adjoint extensions.
Hint: Consider operators $H_{0,c}$, Q of Example 7.2.2.

4. Show that $(\Delta Q)_W$ of Example 8.1.5b is the standard deviation of the random variable $\left(\sum_{j=1}^{n}(\lambda_j - \langle Q_j\rangle_W)^2\right)^{1/2}$ with respect to the probability measure $w(\cdot, Q; W)$ from Example 7.5.4a, and that the analogous conclusion is valid for $(\Delta P)_W$.

5. Any minimum–uncertainty state is a mixture of pure minimum–uncertainty states.

6. Prove: (a) The relation (8.1) defines a minimum–uncertainty state.
(b) Check that the vectors (8.2) coincide with (2.3) for $w := 2^{-1/2}(q - ip)$.
(c) Find the minimum–uncertainty states for operators Q_j, P_k on $L^2(\mathbb{R}^n)$, *i.e.*, functions ψ, which satisfy the relation $(\Delta P)_\psi(\Delta Q)_\psi = \frac{n}{2}$.

7. Complete the proof of Theorem 8.1.8.

(a) Show that the function $\psi_{\varepsilon,\beta} : \psi_{\varepsilon,\beta}(x) = e^{\varepsilon r^{-\beta}}\psi(x)$ with $\psi \in \mathcal{S}(\mathbb{R}^n)$ and ε, β positive belongs to $\mathcal{S}(\mathbb{R}^n)$ and $\psi^{(k)}(0) = 0$ for $k = 0, 1, 2, \dots$.

(b) Check that $Q^\alpha\psi_{\varepsilon,\beta} \to Q^\alpha\psi$ as $\varepsilon \to 0+$ provided $\alpha > -n/2$.

(c) Check that $|\,\|P_j\psi_{\varepsilon,\beta}\|^2 - \|P_j\psi\|^2| \to 0$ as $\varepsilon \to 0+$ provided $\beta < 1/2$.

(d) For which values of μ can the norms on the right side of (8.3) be written as mean values?

8. Given $\psi \in L^2(\mathbb{R})$ define $f(y) := \int_{\mathbb{R}} \overline{\psi}(x)\psi(x-y)\,dx$. The function f is continuous, satisfies $|f(y)| \le f(0)$, and $\lim_{|y|\to\infty} f(y) = 0$.
Hint: The shift operators form a strongly continuous group.

9. Consider operators Q, P on $L^2(\mathbb{R})$. Prove:

(a) If $n = 1$, the inequality (8.5) holds for no constant C_1.

(b) Let $\psi \in D(P^2)$ and $M \in \mathcal{B}$; then $w(M, Q; \psi) \le \ell(M)(\Delta P)_\psi$ where $\ell(M)$ is the Lebesgue measure of M.

Hint: (a) If $n = 1$ the functions of $D(P^2)$ are continuous. (b) Prove $\frac{1}{2}\langle f'(Q)\rangle_\psi \le \|(P - \langle P\rangle_\psi\|\,\|f(Q)\psi\|$ for an absolutely continuous f and choose the function appropriately with $\operatorname{Ran} f \subset [-1, 1]$.

10. There is no pair of bounded operators Q, P which would fulfil $[P, Q] = -iI$.
Hint: Estimate the norms of the left side in $[P, Q^n] = -inQ^{n-1}$.

11. Prove that sets D_x and D_y of Example 8.2.1 are dense in $L^2(M)$.

12. The operators Q_a, P_a from Example 8.1.6 satisfy $[P_a, Q_a]f = -if$ for all f for which the left side makes sense but the corresponding operators $U_a(t) := e^{iP_a t}$ and $V_a(s) := e^{iQ_a s}$ do not satisfy the Weyl relations.

13. Prove: (a) The Weyl relations follow *formally* from (8.6) together with the expansion of the operators $U(t)$, $V(s)$ into power series.

(b) The argument can be made rigorous in the Schrödinger representation provided the operators are applied to vectors from $C_0^\infty(\mathbb{R}^n)$.

Hint: (a) Use the Hausdorff–Baker–Campbell formula (which is valid for *bounded* operators): $e^{iA}B\,e^{-iA} = \sum_{k=0}^\infty \frac{i^k}{k!}C^k$, where $C_0 := B$ and $C_k := [A, C_{k-1}]$.

14. Consider the operators $U(t)$, $V(s)$ defined by (8.8). Prove that

(a) They satisfy the relations (8.7).

(b) The maps $U(\cdot)$ and $V(\cdot)$ are strongly continuous on $\mathbb{R}^n$.

(c) $\{U(t) : t \in \mathbb{R}^n\}' = \{P_1, \dots, P_n\}'$ and $\{V(s) : s \in \mathbb{R}^n\}' = \{Q_1, \dots, Q_n\}'$.

15. Under the assumptions of Theorem 8.2.4, the operators $R(t, s)$ are unitary, the map $(t, s) \mapsto R(t, s)$ is strongly continuous, and $R(t, s)R(v, u) = e^{i(t\cdot u - s\cdot v)/2}R(t+v, s+u)$ holds for all $s, t, u, v \in \mathbb{R}^n$.

16. Let B_f be the operators defined in the proof of Theorem 8.2.4. Show that

(a) $(B_f)^* = B_{f^*}$, where $f^*(t,s) := \overline{f(-t,-s)}$,

(b) $B_{f_1}B_{f_2} = B_{f_{12}}$, where $f_{12}(t,s) := \int_{\mathbb{R}^{2n}} f_1(t-v, s-u) f_2(v,u)\, e^{i(t\cdot u - s\cdot v)/2} dv\, du$.

17. Consider the operator $B := B_{f_0}$ from the proof of Theorem 8.2.4. Using the result of the previous problem show that

(a) $BR(t,s)B = e^{-(t^2+s^2)/4}B$ for any $s,t \in \mathbb{R}^n$; in particular, $B^2 = B$,

(b) $(R(t,s)\phi, R(v,u)\psi) = e^{-(t-v)^2/4-(s-u)^2+i(s\cdot v - t\cdot u)/2}(\phi,\psi)$ for all $\phi,\psi \in \operatorname{Ran} B$ and any $s,t,u,v \in \mathbb{R}^n$.

Hint: Use the formula $\int_{\mathbb{R}^n} e^{-(\xi^2/2)+i\xi\cdot\eta} d\xi \;=\; (2\pi)^{n/2} e^{-\eta^2/2}$.

18. Prove that the projection B from the proof of Theorem 8.2.4 is expressed in the Schrödinger representation as

$$(B\psi)(x) \;=\; \pi^{-n/2} \int_{\mathbb{R}^n} e^{-(x^2+y^2)/2}\psi(y)\, dy\,.$$

Show that the condition $B\psi = \psi$ is satisfied only by multiples of the vector ψ_0 : $\psi_0(x) = \pi^{-n/4}e^{-x^2/2}$.
Hint: Use the orthonormal basis (2.1).

19. Prove that the Weyl operator satisfies the relation (8.10).
Hint: The operator $A := s\cdot Q - t\cdot P$ is *e.s.a.* on $\mathcal{S}(\mathbb{R}^n)$; check that $W(rt, rs) = e^{i\bar{A}r}$ holds for all $r \in \mathbb{R}$.

20. Consider the vectors $\psi_{q,p} := W(q,p)\psi$ of Example 8.2.7. Let $\phi,\, \psi \in \mathcal{S}(\mathbb{R})$ with $\|\psi\| = (2\pi)^{-1/2}$ and $\chi \in L^2(\mathbb{R})$. Prove:

(a) $\lim_{\varepsilon\to 0+} \int_{\mathbb{R}^2} e^{-\varepsilon^2 x^2} (\psi_{q,p},\phi)\, \psi_{q,p}(x)\, dq\, dp = 2\pi\|\psi\|^2\phi(x)$.

(b) Use this to prove (8.11) under the stated assumptions.

(c) Prove the "weak form" of (2.2): $(\chi,\phi) = \int_{\mathbb{R}^2} (\chi,\psi_{q,p})(\psi_{q,p},\phi)\, dq\, dp$.

21. Prove: (a) Let Q_j, P_k be the operators on $L^2(\mathbb{R}^n)$ from Example 7.2.3. The operator $H := \sum_{j=1}^n (P_j^2 + Q_j^2)$ is self-adjoint and $\mathcal{S}(\mathbb{R}^n)$ is a core for it.

(b) Find an example of operators P, Q, which satisfy conditions (a)–(c) of Remark 8.2.8 but the Weyl relations are not valid for the corresponding unitary groups.

Hint: (b) In the setting of Example 8.2.1, define $P\psi := -i\partial\psi/\partial x$ and $Q\psi := x\psi - i\partial\psi/\partial y$ for any $\psi \in D$.

22. Define the operators $U_n^{(N)}$ with $n = 1,\dots,2N$ on the N–fold tensor product $\mathbb{C}^2 \otimes \cdots \otimes \mathbb{C}^2$ in the following way:

(i) $U_n^{(1)} := \sigma_n$ if $N = 1$,

(ii) $U_n^{(2)} := \sigma_1 \otimes \sigma_n$, $n = 1,2,3$, and $U_4^{(2)} := \sigma_3 \otimes I$ if $N = 2$,

(iii) $U_n^{(N)} := \sigma_1 \otimes U_n^{(N-1)}$, $n = 1,\dots,2N-2$; $U_{2N-1}^{(N)} := \sigma_1 \otimes \sigma_2 \otimes I^{(N-2)}$ and $U_{2N}^{(N)} := \sigma_3 \otimes I^{(N-1)}$ if $N \geq 3$,

where the σ_j are Pauli matrices. Prove that these operators form an irreducible representation of canonical anticommutation relations (*cf.* the notes to Section 8.2).

23. Compute the integrals and the limit in Example 8.3.1.

24. Consider the classical limit for the one–dimensional system described by the square–well Hamiltonian of Example 7.2.5, $H : H\psi = -(\hbar^2/2m)(d^2\psi/dx^2) - V_0\chi_{[-a,a]}\psi$. Take the sequence $\{\hbar_n\}$ with $\hbar_n := \frac{2}{\pi n}\sqrt{2m(E+V_0)a^2}$ and prove that

(a) $\lim_{n\to\infty} w(J,Q;\psi_n,\hbar_n) = w_{cl}(J,x;E) := \frac{1}{2a}\,\ell(J\cap[-a,a])$ holds for any interval $J \subset \mathbb{R}$, where ℓ stands for the Lebesgue measure.

(b) The limit of the probability density $dw((-\infty,x],Q;\psi_n,\hbar_n)/dx$ as $n \to \infty$ does not exist in the classically allowed region, $|x| < a$.

25. Find the general form of operator $J(q^n p^m)$ in a Dirac quantization J.

Chapter 9

Time evolution

9.1 The fundamental postulate

It is a certain idealization to speak about a state at a given instant. We have supposed up to now that the measurements which determine the state are instantaneous processes; however, any real measurement has a finite duration. This fact should be remembered: we expect a quantum–mechanical model to yield correct predictions if the time scale characteristic for the effects under consideration is much longer than the duration of the appropriate measurements.

Other complications may arise if we study relativistic systems since their evolution differs when observed from different reference frames. Recall the well known example of muons in secondary cosmic rays which reach the Earth's surface only due to the fact that their decay — which is a purely quantum process — runs at a much slower pace for us than in their rest system. This does not mean, however, that there is a fundamental difference between relativistic and nonrelativistic systems; the description of time evolution we are going to discuss below is universal, at least as long we use a fixed reference frame. On the other hand, some notions which are useful in nonrelativistic quantum mechanics cannot be transferred simple–mindedly to relativistic quantum field theory; we comment on that at the end of the next section.

To begin with, we introduce the notion of a **unitary propagator**, which is a family $\{U(t,s) \,:\, s,t \in \mathbb{R}\}$ of unitary operators on a Hilbert space $\mathcal{H}$ such that

(i) $U(t,s)U(s,r) = U(t,r)$ holds for any $r,s,t \in \mathbb{R}$, in particular, $U(t,t) = I$ for each $t \in \mathbb{R}$.

(ii) The map $(s,t) \mapsto U(s,t)$ is strongly continuous in $\mathbb{R}^2$.

A statistical operator denoting a state of the system at an instant t will be denoted as W_t; similarly we use the symbol ψ_t for a vector describing a pure state at time t (if necessary we also write $W(t)$, $\psi(t)$, *etc.*).

Suppose that the evolution of the system is not disturbed in a time interval J, *i.e.*, one performs no measurement during this period. Then we postulate

(Q4a) the time evolution of any state of the system is described by a unitary propagator, $W_t = U(t,s)W_sU(t,s)^{-1}$ or $\psi_t = U(t,s)\psi_s$ for any $s,t \in J$.

Let us mention briefly how we can motivate the unitarity requirement for the operator connecting states at different times. For simplicity, we consider pure states only; we assume that at an instant s the state is described by a ray Φ_s, which evolves during $[s,t]$ into Φ_t, provided it is undisturbed by a measurement. We also forget for a moment about superselection rules, in which case any ray can serve as the initial state and the relation $\Phi_t = \tilde{U}(t,s)\Phi_s$ defines a bijective map in the state space. This makes sense also for $t < s$, when Φ_t means the state from which Φ_s has developed during $[t,s]$. Experience tells us that we may suppose that the transition probabilities between states are preserved during an undisturbed time evolution,

$$P(\Phi_s, \Psi_s) = P(\tilde{U}(t,s)\Phi_s, \tilde{U}(t,s)\Psi_s)$$

for any Φ_s, Ψ_s. Then it follows from Wigner's theorem (see the notes) that there is a unitary or antiunitary operator $\hat{U}(t,s)$ such that $\psi_t := \hat{U}(t,s)\psi_s$ belongs to Ψ_t for any $\psi_s \in \Psi_s$; this operator is determined by $\tilde{U}(t,s)$ up to a phase factor. Since $\hat{U}(t,s)\hat{U}(s,r)\psi_r$ and $\hat{U}(t,r)\psi_r$ represent the same state, they differ at most by a phase factor which is, moreover, independent of ψ_r in view of the (anti)linearity of the operators $\hat{U}$, *i.e.*, we have $\hat{U}(t,s)\hat{U}(s,r) = e^{i\alpha(r,s,t)}\hat{U}(t,r)$. Using further the associativity of the operator multiplication we find

$$\alpha(r,s,t) = \beta(r,s) + \beta(s,t) - \beta(r,t),$$

where $\beta(s,t) := \alpha(0,s,t)$, and therefore we can pass to the operators $U(t,s) := e^{-i\beta(s,t)}\hat{U}(t,s)$, which correspond to the same $\tilde{U}(t,s)$ and already obey condition (i). Combining this property with the natural continuity requirement we see that the operators must be unitary (Problem 1).

The family $\{U(t,s) : s,t \in \mathbb{R}\}$ is mostly called briefly a *propagator*; the term *evolution operator* is also sometimes used. It follows from the postulate (Q4a) that

$$\mathrm{Tr}\, W_t^2 = \mathrm{Tr}\, W_s^2 \tag{9.1}$$

for all $s,t \in J$; recall that this quantity tells us "how much" the state W_t is mixed. In particular, a pure state evolves into a pure state again.

A system is called **conservative** if its propagator satisfies $U(t+\tau, s+\tau) = U(t,s)$ for all $s,t,\tau \in \mathbb{R}$. For such systems we define $U(t) := U(t+\tau,\tau)$, where τ is any fixed time instant; the postulate (Q4a) then means that $\{U(t) : t \in \mathbb{R}\}$ is a strongly continuous one–parameter group of unitary operators. Due to the Stone theorem, this group is generated by a self–adjoint operator A. The fundamental dynamical postulate of quantum theory consists of identifying the operator $-A$ with the Hamiltonian of the system (or with $\hbar^{-1}H$ in the standard system of units):

(Q4b) the propagator of a conservative system with the Hamiltonian H is given by $U(t) = e^{-iHt}$ for any $t \in \mathbb{R}$.

An easy consequence is that a quantum system with a time–independent Hamiltonian is conservative. This property is typical for isolated systems; however, those which interact with the environment in a time–independent way also belong to this class. Furthermore, time evolution of conservative systems has a simple differential expression.

9.1.1 Proposition (Schrödinger equation): If $W_sD(H) \subset D(H)$ is valid at some instant s, then the function $t \mapsto W_t\phi$ is differentiable for any $\phi \in D(H)$ and obeys the equation

$$i\frac{d}{dt}W_t\phi = [H, W_t]\phi\,.$$

In particular, if $\psi_s \in D(H)$ for some s, then $t \mapsto \psi_t$ is differentiable and

$$i\frac{d}{dt}\psi_t = H\psi_t\,.$$

Proof: The pure–state part of the task is obvious. For any $t \in \mathbb{R}$ we have $W_tD(H) = U(t-s)W_sU(s-t)D(H) \subset D(H)$ since $U(\tau)$ as a bounded function of H maps $D(H)$ into itself. We have

$$\frac{d}{dt}W_t\phi = \lim_{\delta\to 0}\left\{U(\delta)W_t\,\frac{U(-\delta)-I}{\delta}\,\phi + \frac{U(\delta)-I}{\delta}\,W_t\phi\right\}$$

for any $\phi \in D(H)$; the limit of the second term exists and equals $-iHW_t\varphi$ due to Proposition 5.9.1. As for the first term, we get

$$\left\|U(\delta)W_t\,\frac{U(-\delta)-I}{\delta}\,\phi - i\,W_tH\phi\right\|$$

$$\le \|U(\delta)W_t\|\,\left\|\frac{U(-\delta)-I}{\delta}\,\phi - iH\phi\right\| + \|(U(\delta)-I)W_tH\varphi\|\,,$$

and since $\|U(\delta)W_t\| \le 1$, the right side tends to zero as $\delta \to 0$. ∎

Using the standard argument from the theory of ordinary differential equations, we can check that for a given initial condition $\psi_0 \in \mathcal{H}$, the Schrödinger equation has just one solution, namely $\psi_t = e^{-iHt}\psi_0$; hence in the case of a conservative system it uniquely determines its unitary propagator. This correspondence is useful when we pass to nonconservative systems whose Hamiltonians are time–dependent. Instead of attempting to modify (Q4b), in this case we directly postulate the Schrödinger equation:

(Q4c) the time evolution of a system whose Hamiltonian is generally time–dependent is determined by the equations

$$i\frac{d}{dt}W_t\phi = [H(t), W_t]\phi\,, \quad i\frac{d}{dt}\psi_t = H(t)\psi_t$$

for the mixed and pure states, respectively.

This is clearly consistent with (Q4b); however, we also have to check that the stated equations make sense and define a unitary propagator for the considered operator–valued function $t \mapsto H(t)$. This represents a nontrivial problem; we shall discuss it for some classes of time–dependent Hamiltonians in Section 9.5.

With knowledge of the time evolution of states we can determine how the measurable quantities such as probabilities of the measurement outcomes, mean values of particular observables, *etc.*, change in the course of time. For instance, suppose that $AW_t \in \mathcal{J}_1$ for all t from some interval J; then

$$\langle A\rangle_{W_t} = \mathrm{Tr}\,(A\,e^{-iH(t-s)}W_s e^{iH(t-s)})$$

holds for all $s,t \in J$. Under strengthened assumptions, this relations also admits a differential expression.

9.1.2 Theorem: Let W_t describe a state of a conservative system with the Hamiltonian H for all t of some open interval $J \subset \mathbb{R}$. The mean value of an observable A in state W_t satisfies the equation

$$\frac{d}{dt}\langle A\rangle_{W_t} = \mathrm{Tr}\,(i[H,A]W_t)$$

on the interval J provided

(i) The operators AW_t, HAW_t, and AHW_t belong to the trace class for $t \in J$.

(ii) The map $t \mapsto \|AU(t)\phi\|$ is bounded for $\phi \in \mathrm{Ran}\,W_s$ if $s,t \in J$.

(iii) The sum expressing the right side converges in J uniformly with respect to t.

9.1.3 Remarks: (a) Assumption (iii) is fulfilled, *e.g.*, if $\dim \mathrm{Ran}\,W_t < \infty$ for some $t \in J$ (the dimension obviously does not change with time). This is true, in particular, for any pure state ψ_t; assumption (i) then reads $\psi_t \in D([H,A])$ for all $t \in J$. Assumption (ii) is fulfilled automatically for a bounded observable, in which case the derivative on the left side is easily seen to be continuous. Finally, the operator $i[H,A]$ is symmetric; if it is also *e.s.a.*, we can rewrite the relation as

$$\frac{d}{dt}\langle A\rangle_{W_t} = \left\langle \overline{i[H,A]}\right\rangle_{W_t}.$$

(b) An advantage of the differential expression is again that it extends to non-conservative systems. Formally we get the same relation for them with H replaced by $H(t)$; to give it a rigorous meaning one has to know more about the operator–valued function $H(\cdot)$.

Proof of Theorem 2: Using the orthonormal basis $\{\phi_j^s\}$ which consists of the eigenvectors of W_s corresponding to the eigenvalues w_j we can write

$$\frac{d}{dt}\langle A\rangle_{W_t} = \frac{d}{dt}\sum_j w_j(U(t-s)\phi_j^s, AU(t-s)\phi_j^s),$$

where s is any point of the interval J. For each term of the series we have

$$\frac{d}{dt}\,(U(t-s)\phi_j^s, AU(t-s)\phi_j^s)$$

$$= \lim_{\delta\to 0}\left\{\left(\frac{U(\delta)-I}{\delta}\,\phi_j^t,\, AU(\delta)\phi_j^t\right) + \left(A\phi_j^t,\, \frac{U(\delta)-I}{\delta}\,\phi_j^t\right)\right\},$$

where $\phi_j^t := U(t-s)\phi_j^s$; in the second term we have used the fact that $\phi_j^t \in D(A)$ due to (i). The first term gives

$$\left|\left(\frac{U(\delta)-I}{\delta}\,\phi_j^t,\, AU(\delta)\phi_j^t\right) - i(H\phi_j^t, A\phi_j^t)\right|$$

$$\le \left|\left(\left(\frac{U(\delta)-I}{\delta} + iH\right)\phi_j^t,\, AU(\delta)\phi_j^t\right)\right| + \left|(H\phi_j^t, A(U(\delta)-I)\phi_j^t)\right|$$

$$\le \|AU(\delta)\phi_j^t\|\,\left\|\left(\frac{U(\delta)-I}{\delta} + iH\right)\phi_j^t\right\| + \|AH\phi_j^t\|\,\|(U(\delta)-I)\phi_j^t\| \longrightarrow 0$$

as $\delta \to 0$ in view of (i),(ii), and postulate (Q4b). The sought derivative is therefore $-2\,\mathrm{Im}\,(H\phi_j^t, A\phi_j^t) = (\phi_j^t, i[H,A]\phi_j^t)$ and the assumption (iii) allows us to differentiate the series term by term. ∎

As an illustration, let us mention the known fact that in a quantum mechanical system of N particles interacting through a potential the mean values of position coordinates satisfy the classical equations of motion. Consider a real differentiable function $V : \mathbb{R}^n \to \mathbb{R}$ and denote $F_j := -\partial V/\partial x_j$. Let Q_j, P_k be the operators of Cartesian position and momentum coordinates on $L^2(\mathbb{R}^n)$ introduced in Example 7.2.3; then we define

$$H := \sum_{j=1}^{n} \frac{1}{2m_j}\,P_j^2 + V(Q); \tag{9.2}$$

the operators $V := V(Q)$ and $F_j := F_j(Q)$ are the corresponding functions of the family of compatible observables $Q := \{Q_1, \dots, Q_n\}$.

9.1.4 Corollary (Ehrenfest theorem): Suppose that $\mathcal{S}(\mathbb{R}^n)$ is a core for H and $V \in C^\infty(\mathbb{R}^n)$ leaves it invariant. Assume further that $\psi_t := e^{-i\overline{H}(t-s)}\psi_s$ belongs to $\mathcal{S}(\mathbb{R}^n)$ for t from an open interval $J \subset \mathbb{R}$, and $t \mapsto \max_j \{\|Q_j\psi_t\|, \|P_j\psi_t\|\}$ is bounded in J. Then the functions $t \mapsto \langle Q_j\rangle_{\psi_t}$ are twice differentiable and satisfy the equations

$$m_j\,\frac{d^2}{dt^2}\,\langle Q_j\rangle_{\psi_t} = \langle F_j\rangle_{\psi_t}\,.$$

Proof: Since V maps $\mathcal{S}(\mathbb{R}^n)$ into itself, the identity $F_j\psi = V\partial_j\psi - \partial_j(V\psi)$, where $\partial_j := \partial/\partial x_j$, implies that the same is true for F_j; hence $\mathcal{S}(\mathbb{R}^n) \subset D(F_j)$ follows from the functional calculus. We have $P_j\psi = -i\partial_j\psi$ for any $\psi \in \mathcal{S}(\mathbb{R}^n)$, and therefore $i[H,P_j]\psi = F_j\psi$ and $i[H,Q_j]\psi = m_j^{-1}P_j\psi$. Due to the assumption, $\psi_t \in D([H,Q_j]) \cap D([H,P_j])$ and condition (ii) of Theorem 9.1.2 is valid. It yields the relations

$$\frac{d}{dt}\langle P_j\rangle_{\psi_t} = \langle F_j\rangle_{\psi_t}, \quad \frac{d}{dt}\langle Q_j\rangle_{\psi_t} = m_j^{-1}\langle P_j\rangle_{\psi_t};$$

combining them, we get the result. ∎

The assumptions we have used can be certainly weakened, but since the result serves only an illustrative purpose here, we refrain from doing so. The assertion provides a motivation for postulate (Q4b): it is substantial for the derivation of the "correct" classical relations that we have chosen the operator (9.2) for H which appears in Theorem 9.1.2.

In conclusion, let us mention some quantities which play a distinguished role in time evolution. Consider again a conservative system with the Hamiltonian H. A state W_t is said to be *stationary* if it commutes with the Hamiltonian, $W_tH \subset HW_t$ for all appropriate t. In particular, a pure state is stationary *iff* ψ_t is an eigenvector of H. Due to the Stone theorem, $W_t = e^{-iH(t-s)}W_se^{iH(t-s)} = W_s$ holds for any s,t if W_t is stationary. Similarly a pure stationary state, $H\psi_s = \lambda\psi_s$, satisfies $\psi_t = e^{-i\lambda(t-s)}\psi_s$, so the vectors ψ_t belong to the same ray. With this fact in mind we usually do not indicate the time dependence of stationary states.

The mean value of any observable in a stationary state obviously does not depend on time. On the other hand, if an observable A is such that the mean values $\langle E_A(\Delta)\rangle_{W_t}$ are time–independent *for any realizable state* W_t of the given system and all $\Delta \in \mathcal{B}$ we call it an *integral of motion* (or a *conserved quantity*); the definition applies to nonconservative systems as well. If A is an integral of motion, the probability measure $w(\cdot, A; W_t)$ does not depend on time, so $\langle f(A)\rangle_{W_t}$ is time–independent for any Borel function f if only it makes sense; this applies particularly to $\langle A\rangle_{W_t}$. For conservative systems we have a simple criterion.

9.1.5 Proposition: (a) The observable A is an integral of motion *iff* the operator A commutes with the Hamiltonian.

(b) The total energy of a conservative system is an integral of motion, as is any superselection operator.

Proof: The sufficient condition in (a) verifies easily. Conversely, if A is an integral of motion, the identity

$$(\psi, U(t)E_A(\Delta)U(-t)\psi) = (\psi, E_A(\Delta)\psi)$$

is valid for all $t \in \mathbb{R}$ and $\Delta \in \mathcal{B}$ if ψ belongs to some coherent subspace $\mathcal{H}_\alpha$. The operators H, A describe observables, so they are reduced by all E_α; the same is then true for $U(t)$. Mimicking the argument from the proof of Proposition 7.5.1, we infer that $E_A(\Delta)U(t) = U(t)E_A(\Delta)$ for all $t \in \mathbb{R}$, $\Delta \in \mathcal{B}$, which is equivalent to the commutativity of H and A. Assertion (b) follows easily from (a). ∎

A family $\mathcal{S}$ of observables is called a *complete system of integrals of motion* if any $A \in \mathcal{S}$ is an integral of motion and no independent ones can be added, *i.e.*, any other integral of motion belongs to $\mathcal{S}''_{ex}$. A trivial example of such a system consists of all Hermitean elements in $\{H\}'$; however, we are interested rather in complete systems which consist of a few operators only.

9.1.6 Example: Consider the particle on line of Example 7.2.1, and suppose it is free, *i.e.*, $H = \frac{1}{2m}P^2$. The *parity* observable is represented (up to a sign) by the reflection operator $R: \; (R\psi)(x) = \psi(-x)$; it obviously has eigenvalues ± 1 and the corresponding eigenspaces $\mathcal{H}_\pm$ consist of even and odd functions, respectively. The subsets $D_\pm \subset \mathcal{H}_\pm$ of even and odd functions in $AC^2(\mathbb{R})$ are invariant with respect to P^2, which means that the operator R commutes with H; in other words, the parity is an integral of motion.

Furthermore, the Hamiltonian can be written as $H = H_+ \oplus H_-$, where $H_\pm$ are its parts in the parity eigenspaces. One can check that each of them has a simple spectrum (Problem 3); hence the set $\{H, R\}$ is a CSCO and, at the same time, a complete set of integrals of motion for the free particle on line.

9.2 Pictures of motion

In quantum theory most predictions are expressed in terms of spectral properties of the self–adjoint operators which represent the observables. These properties are unitary invariants, and are therefore insensitive to an operation, which replaces all operators of the observables by the unitary equivalent operators obtained by means of the *same* unitary operator. It is important in the present context that we may use the evolution operator for this purpose.

The way of describing time evolution discussed in the previous section is called the **Schrödinger picture**. States represented by statistical operators or unit vectors depend here on time, while operators of observables are time–independent; the exceptions are those observables whose time dependence is parametrical, *i.e.*, coming from the environment rather than from the system itself.

The **Heisenberg picture** associates with any observable the operator–valued function $A_H : \mathbb{R} \to \mathcal{L}_{sa}(\mathcal{H})$ defined by

$$A_H(t) := U(t,s)^{-1} A\, U(t,s)\,,$$

where s is a fixed instant and $A_H(s) := A$ is the operator representing the observable in the Schrödinger picture. In common parlance, we again do not distinguish between this function and its values, speaking about the observable $A_H(t)$, *etc.* It follows from Proposition 9.1.5b that any observable $A_H(t)$ at an arbitrary instant t is reduced by all the coherent subspaces.

Since the predictions of the theory, such as mean values of observables, must be independent of the chosen picture, the states are represented in the Heisenberg picture by the statistical operators $W_H(t) := U(t,s)^{-1} W_t U(t,s)$ or by the vectors

$\psi_H(t) := U(t,s)^{-1}\psi_t$. Postulate (Q4a) implies that they are time–independent; we write $W_H(t) =: W$ and $\psi_H(t) =: \psi$.

If the system is conservative, we easily get the relation between the operators representing an observable at different time instants,

$$A_H(t) := U(t-\tau)^{-1}A_H(\tau)\,U(t-\tau).$$

A similar relation is *not* in general valid for nonconservative systems because the propagator need not in this case be a commutative family. As in Proposition 9.1.1, the above relation yields a differential form of the equations of motion.

9.2.1 Proposition: Let $A_H(t)$ be a bounded observable preserving the domain of the Hamiltonian, $A_H(s)D(H) \subset D(H)$ for some $s \in \mathbb{R}$. Then the function $A_H(\cdot)\phi$ is differentiable for any $\phi \in D(H)$ and

$$\frac{d}{dt}A_H(t)\phi = i[H, A_H(t)]\phi.$$

It is illustrative to compare this result with Theorem 9.1.2. The norm of the operator representing an observable is preserved during time evolution; hence a bounded observable remains bounded. For unbounded observables we get formally the same equation, but its validity now depends substantially on the relations between the domains of H and $A_H(t)$; we can, however, study the time evolution of unbounded observables using their bounded functions (Problem 4). In the Heisenberg picture the integrals of motion are clearly manifested: they are just those observables for which the operator–valued function $A_H(\cdot)$ is constant, $A_H(t) = A$ for all $t \in \mathbb{R}$. This applies to nonconservative systems as well.

The third frequently used picture is the **interaction** (or **Dirac**) **picture** in which the time dependence is split between the states and observables. It is used for systems with a Hamiltonian of the form $H = \overline{H_0 + V}$, where H_0, V are self–adjoint operators whose sum is *e.s.a.*; if the system is nonconservative the decomposition is usually chosen so that H_0 is time–independent, $H(t) = \overline{H_0 + V(t)}$. The interaction picture is obtained from the Schrödinger picture by the unitary transformation $U_0(t) := e^{-iH_0 t}$, *i.e.*,

$$\begin{aligned} W_D(t) &:= U_0(t-s)^{-1}W_t U_0(t-s) = U_0(s-t)U(t,s)W_s U(s,t)U_0(t-s)\,, \\ \psi_D(t) &:= U_0(t-s)^{-1}\psi_t = U_0(s-t)U(t,s)\psi_s\,, \end{aligned}$$

where s is again a fixed instant. A Schrödinger observable A is represented in the interaction picture by an operator–valued function

$$A_D(t) := U_0(t-s)^{-1}A\,U_0(t-s);$$

hence the observables carry the part of the time dependence connected with the term H_0 in the Hamiltonian. In the same way as above, this last relation can be easily rephrased in a differential form.

9.2.2 Proposition: Let $A_D(t)$ be a bounded observable, $A_D(s)D(H_0) \subset D(H_0)$; then the function $A_D(\cdot)\phi$ is differentiable for any $\phi \in D(H_0)$ and

$$\frac{d}{dt} A_D(t)\phi = i[H_0, A_D(t)]\phi\,.$$

It is slightly more complicated to express the time dependence of states in a differential form. Consider the simplest situation when V is a time–independent Hermitean operator; then if $W_s D(H_0) \subset D(H_0)$ or $\psi_s \in D(H_0)$, we have

$$i\,\frac{d}{dt} W_D(t)\phi = [V_D(t), W_D(t)]\phi$$

for any $\phi \in D(H_0)$, and

$$i\,\frac{d}{dt}\psi_D(t) = V_D(t)\psi_D(t)\,, \tag{9.3}$$

where $V_D(t) := U_0(t{-}s)^{-1} V U_0(t{-}s)$ is the interaction–picture form of V (Problem 5). These equations are also formally satisfied if V is unbounded and time–dependent; however, their actual validity has to be checked again separately in any particular case.

Each of the described pictures has its advantages. In quantum mechanics, where we typically have a few important observables, the Schrödinger picture is mostly preferred. On the other hand, the Heisenberg picture is commonly used in quantum field theory, where we have to treat vast families of observables which include, roughly speaking, field operators at every point of the configuration space.

The interaction picture is particularly useful in situations where we are able to solve the equations of motion exactly for a part of the total Hamiltonian. In many cases the Hamiltonian decomposes naturally into a part H_0, which corresponds to a free motion in some sense, and a part V describing the interaction; a typical example is operator (9.2) consisting of kinetic and potential energy parts. The free problem is often exactly solvable; the interaction picture then allows us to single out the part of the propagator related to the interaction Hamiltonian V (hence the name). Moreover, if the interaction is weak in some sense, this part of the problem can be solved perturbatively; examples will be given in Section 9.5 and Chapter 15.

The interaction picture is used in quantum mechanics as well as in computational methods of quantum field theory. In the last case, however, it cannot be generally justified due to the possible existence of nonequivalent representations of the canonical commutation relations mentioned in Section 8.2. In particular, for any *relativistic* quantum field theory nonexistence of the interaction picture follows from Haag's theorem, which is mentioned in the notes to Section 12.3.

9.3 Two examples

Now we want to discuss in more detail the time evolution of two simple quantum–mechanical systems. First we are going to consider a *system of free particles* described by the Hamiltonian

$$H_0 := \sum_{j=1}^{n} \frac{1}{2m_j} P_j^2$$

on the state space $L^2(\mathbb{R}^n)$ of Example 7.2.3; in the case of N real spinless particles we have $n = 3N$ and $m_{3k+1} = m_{3k+2} = m_{3k+3}$ for $k = 0, 1, \dots, N-1$. In the same way as in Example 7.5.8 we can check that H_0 is self–adjoint and has a purely continuous spectrum, $\sigma(H_0) = \mathbb{R}$; furthermore, C_0^∞ is a core for it. The corresponding propagator $U(t) := e^{-iH_0 t}$ can be written explicitly.

9.3.1 Theorem: Let H_0 be the operator defined above; then the relation

$$(U(t)\psi)(x) = \text{l.i.m.}_{k\to\infty} \prod_{j=1}^{n} \left(\frac{m_j}{2\pi i t}\right)^{1/2} \int_{\mathbb{R}^n} \exp\left(\frac{i}{2t} \sum_{j=1}^{n} m_j |x_j - y_j|^2\right) \psi(y) f_k(y)\, dy$$

holds for all $\psi \in L^2(\mathbb{R}^n)$, where $\{f_k\} \subset L^2(\mathbb{R}^n)$ is an arbitrary sequence such that $|f_k(x)| \le 1$ and $\lim_{k\to\infty} f_k(x) = 1$ for a.a. $x \in \mathbb{R}^n$. Moreover, we have $U(t)\mathcal{S}(\mathbb{R}^n) \subset \mathcal{S}(\mathbb{R}^n)$ for each $t \in \mathbb{R}$.

Proof: Choosing $\varphi_j(x) := \sqrt{2m_j}\, x_j$ in Example 3.3.2 we see that H_0 is unitarily equivalent to P^2; hence it is sufficient to consider the case with $m_j = 1/2$ only. Furthermore, it is enough to prove the formula without regularization because $U(t)$ is a bounded operator and $\psi f_k \to \psi$. Consider the functions

$$u_\varepsilon : u_\varepsilon(k) = e^{-i|k|^2(t-i\varepsilon)}, \quad \varepsilon \ge 0,$$

and $u := u_0$; it follows from the functional–calculus rules that $U(t) = u(P) = \text{s-lim}_{\varepsilon\to 0+} u_\varepsilon(P)$. The operator on the right side can be expressed using Problem 7.17b, and a simple integration yields

$$(u_\varepsilon(P)\psi)(x) = (4\pi i(t-i\varepsilon))^{-n/2} \int_{\mathbb{R}^n} e^{i|x-y|^2/4(t-i\varepsilon)} \psi(y)\, dy$$

for any $\varepsilon > 0$. Moreover, Example 1.2.1 tells us that $(u_{\varepsilon_k}(P)\psi)(x) \to (u(P)\psi)(x)$ holds for any $\psi \in L^2(\mathbb{R}^n)$ a.e. in $\mathbb{R}^n$ as $k \to \infty$; if $\psi \in L^2 \cap L^1$ the dominated–convergence theorem allows us to interchange the limit with the integral, so we obtain the sought result,

$$(U(t)\psi)(x) = (4\pi i t)^{-n/2} \int_{\mathbb{R}^n} e^{i|x-y|^2/4t} \psi(y)\, dy\,.$$

Finally, the last assertion follows from the fact that $\mathcal{S}(\mathbb{R}^n)$ is preserved by the Fourier–Plancherel operator as well as by multiplication by $e^{-it|\cdot|^2}$. ∎

In view of Problem 2a, no state of the free–particle system is stationary. The explicit form of the propagator leads to a stronger conclusion.

9.3.2 Example (spreading of minimum–uncertainty states): Suppose that $n = 1$ and $m_1 = m$, and consider the state (8.2) with some $q, p \in \mathbb{R}$. The vector

$\psi_t := U(t)\psi$ can be obtained by a straightforward computation (Problem 7a); in particular, the probability density of finding the particle at a point x equals

$$|\psi_t(x)|^2 = \frac{1}{\sqrt{2\pi}(\Delta q)_t} \exp\left\{ -\frac{1}{2(\Delta q)_t^2} \left(x - q - \frac{pt}{m} \right)^2 \right\},$$

where

$$(\Delta q)_t^2 := (\Delta q)^2 + \left(\frac{t^2}{2m\Delta q} \right)^2 = (\Delta q)^2 + \left(\frac{\Delta p}{m} \right)^2 ;$$

the last identity holds since ψ is a minimum–uncertainty state, $\Delta q \Delta p = \frac{1}{2}$. The momentum is an integral of motion, so its mean–square deviation is preserved, $(\Delta P)_{\psi_t} = (\Delta P)_\psi := \Delta p$. On the other hand, the mean–square deviation of the position grows quadratically with time; hence ψ_t ceases to be a minimum–uncertainty state immediately, *i.e.*, for any $t > 0$.

The conclusions of the example are valid independently of the initial state and the number of degrees of freedom involved.

9.3.3 Proposition: Let a state of the free–particle system satisfy $\psi_s \in \mathcal{S}(\mathbb{R}^n)$ at some instant s; then

$$\langle Q_j \rangle_{\psi_t} = \langle Q_j \rangle_{\psi_s} + v_j(t-s),$$

$$(\Delta Q_j)^2_{\psi_t} = (\Delta Q_j)^2_{\psi_s} + a_j(t-s) + b_j(t-s)^2$$

holds for all $j = 1, \dots, n$ and $t \geq s$, where

$$v_j := \frac{1}{m_j} \langle P_j \rangle_\psi ,$$

$$a_j := \frac{1}{m_j} \left[\langle P_j Q_j + Q_j P_j \rangle_{\psi_t} - 2\langle Q_j \rangle_{\psi_t} \langle P_j \rangle_\psi \right],$$

$$b_j := \frac{1}{m_j^2} (\Delta P_j)^2_\psi .$$

9.3.4 Remark: By an approximation argument, validity of the relations extends to all states for which the mean values make sense (Problem 7b). The vector $v := (v_1, \dots, v_n)$ is called the *group velocity* of the wave packet. The above relations show, in particular, that the identity $\langle PQ + QP \rangle_\psi = 2\langle Q \rangle_\psi \langle P \rangle_\psi$ holds for minimum–uncertainty states.

Proof of Proposition 3: Put $s = 0$ and $\psi_s =: \psi$. As in Theorem 1 we may consider the case $m_j = 1/2$ only, and furthermore, we know from there that $\psi_t \in \mathcal{S}(\mathbb{R}^n)$ for any $t \geq 0$. We shall again employ the relation $U(t) = F_n^{-1} T_u F_n$ with $u := e^{-it|\cdot|^2}$; denoting for simplicity $u_j := \partial u / \partial x_j$ *etc.*, we get

$$P_j T_u F_n \psi = -iu_j(Q) F_n \psi + u(Q) P_j F_n \psi = u(Q)(P_j - 2tQ_j) F_n \psi .$$

Now $F_n^{-1}P_jF_n = -Q_j$ and $F_n^{-1}u(Q)F_n = U(t)$, so the last relation yields the identity $Q_jU(t)\psi = U(t)(Q_j + 2tP_j)\psi$; from here the first assertion follows easily. To get the standard deviation expression, we similarly use

$$\begin{aligned} P_j^2T_uF_n\psi &= -u_{jj}(Q)F_n\psi - 2iu_j(Q)P_jF_n\psi + u(Q)P_j^2F_n\psi \\ &= u(Q)(4t^2Q_j^2 - 2it - 4tQ_jP_j + P_j^2)\,F_n\psi \\ &= u(Q)(4t^2Q_j^2 - 2t(P_jQ_j + Q_jP_j) + P_j^2)\,F_n\psi\,, \end{aligned}$$

where the canonical commutation relations are involved in the last step (*cf.* Remark 8.2.8). We now have to apply F_n^{-1} to both sides of this equation to get an expression for $\langle Q_j^2\rangle_{\psi_t}$; combining it with the first relation we arrive at the mean–square deviation formula. ∎

Physically the most important consequence of Proposition 9.3.3 is that in a free–particle system *localization is not preserved*: since the momentum operators have purely continuous spectra, and therefore $(\Delta P_j)_\psi$ is nonzero for any state of the system due to Proposition 8.1.1,

$$\lim_{t\to\infty} (\Delta Q_j)_{\psi_t} = \infty$$

regardless of the precision with which we determine the coordinates at the initial instant. Moreover, the speed of the wave packet spreading is not limited (Problem 8); this is connected with the nonrelativistic character of the system under consideration.

Another example that we are going to discuss here concerns the *linear harmonic oscillator* of Example 7.2.4, whose Hamiltonian is

$$H = \frac{1}{2m}\,P^2 + \frac{1}{2}m\omega^2Q^2\,.$$

Since it has a pure point spectrum, the action of the corresponding propagator can be written down easily.

9.3.5 Example t (propagation of minimum–uncertainty states): For simplicity, put again $m^{-1} = \omega = 2$. The vectors ψ_w of Remark 2.2.8 can be expanded in the orthonormal basis $\{\psi_n\}$ of the eigenvectors of H. Using the unitary operator V defined there, together with Problem 8.6b, we easily find $\psi_{q,p} = \psi_w = e^{-|w|^2/2}\sum_{n=0}^{\infty} \bar w^n (n!)^{-1/2}\psi_n$, where $w := 2^{-1/2}(q-ip)$. Then

$$e^{-iHt}\psi_w = e^{-|w|^2/2}\sum_{n=0}^{\infty} \frac{\bar w^n}{\sqrt{n!}}\, e^{-it(2n+1)}\psi_n$$

and the right side is equal to $e^{-it}\psi_{we^{2it}}$. Since the function ψ_w is explicitly known we can compute $e^{-iHt}\psi_w$; in particular, we get

$$\left|(e^{-iHt}\psi_{q,p})(x)\right|^2 = \frac{1}{\sqrt{\pi}}\, e^{-(x\,-\,q\,\cos 2t\,-\,p\,\sin 2t)^2}\,,$$

so the wave packet does not change its form and follows the classical oscillator trajectory with initial position q and momentum p. The simplicity and elegance of this argument is manifested when we compare it with the straightforward computation used in Example 9.3.8 below.

In this case also the propagator has an explicit integral–operator expression.

9.3.6 Theorem: Let $t \neq \frac{n\pi}{\omega}$, where n is an integer; then the propagator of the harmonic oscillator is given by

$$(U(t)\psi)(x) = \text{l.i.m.}_{\,k\to\infty} \int_{\mathbb{R}} K_t(x,y)\,\psi(y) f_k(y)\,dy$$

with

$$K_t(x,y) := \left(\frac{\omega}{2\pi i|\sin\omega t|}\right)^{1/2} \exp\left\{\frac{i\omega}{2\,\sin\omega t}\left[(x^2+y^2)\cos\omega t - 2xy\right] - \frac{\pi i}{2}\left[\frac{\omega t}{\pi}\right]\right\}$$

for all $\psi \in L^2(\mathbb{R})$, where $[\,\cdot\,]$ means the entire part and the regularizing sequence $\{f_k\}$ has the same properties as in Theorem 9.3.1.
Proof is left to the reader (Problem 9).

9.3.7 Remark: For multiples of the oscillator half–period, $t = n\pi/\omega$, the kernel expression is meaningless but the evolution operator still exists and is proportional to a power of the reflection operator, $U(n\pi/\omega) = e^{-in\pi/2}R^n$.

Unlike the free–particle case, the harmonic oscillator exhibits no spreading. The form of the wave packets can change in time; however, the motion is periodic for any initial state because $U(t+2\pi n/\omega) = (-1)^n U(t)$ for any $t \in \mathbb{R}$.

9.3.8 Example (minimum–uncertainty states revisited): Applying the above explicit form of the propagator to the general minimum–uncertainty state ψ of Example 8.1.7, by a straightforward computation (Problem 10) we get

$$\psi_t(x) = \frac{1}{\sqrt[4]{2\pi(\Delta q)_t^2}} \times \exp\left\{-\frac{(\Delta p)_t}{2(\Delta q)_t}\left[x^2 - \frac{i\Delta q}{(\Delta p)_t}\left(2xz - \frac{z^2}{m\omega}\sin\omega t\right)\right] - q^2(\Delta p)^2 - \frac{i}{2}qp\right\},$$

where

$$(\Delta q)_t := \Delta q\,\cos\omega t + \frac{i\Delta p}{m\omega}\sin\omega t\,, \quad (\Delta p)_t := \Delta p\,\cos\omega t + im\omega\,\Delta q\,\sin\omega t\,,$$
$$z := p - \frac{iq}{2(\Delta q)^2} = \frac{p\Delta q - iq\Delta p}{\Delta q}\,;$$

in the last expression we used $\Delta q\Delta p = \frac{1}{2}$. The relations simplify considerably if we put $\Delta q = \Delta p/m\omega = (2m\omega)^{-1/2}$; in this case the vector ψ with $q = p = 0$ describes the harmonic–oscillator ground state and we get, in particular, the expression for probability density,

$$|\psi_t(x)|^2 = \sqrt{\frac{m\omega}{\pi}}\, e^{-m\omega[x - q\,\cos\omega t - (p/m\omega)\sin\omega t]^2}\,.$$

Hence we recover the result obtained in Example 9.3.5 for $m^{-1} = \omega = 2$: the wave packets of coherent states preserve their shape and move along the classical oscillator trajectories. Let us stress, however, that this is true only for states obtained by application of the Weyl operator to the ground state. To make this clear, choose ψ with $q = p = 0$ and $\Delta q \ne (2m\omega)^{-1/2}$; then

$$|\psi_t(x)|^2 = \frac{1}{\sqrt{2\pi(\Delta q)_t^2}}\, e^{-x^2/2(\Delta q)_t^2}$$

with

$$(\Delta q)_t^2 := (\Delta q)^2\cos^2\omega t + \left(\frac{\Delta p}{m\omega}\right)^2 \sin^2\omega t\,,$$

i.e., the wave packet is "breathing", having the Gaussian form of a periodically changing width.

9.4 The Feynman integral

Consider again the system of particles with a potential interaction which is described by the Hamiltonian (9.2). For the present moment we are not going to specify the assumptions concerning V; we suppose that H is *e.s.a.* leaving to Chapter 14 a discussion of the problem for which classes of potentials this is true.

Our aim is to derive a useful expression of the propagator $U(t) = e^{-i\overline{H}t}$. We employ the fact that H is a sum of two operators, $H = H_0 + V$. The first of these is the free Hamiltonian discussed in the previous section, whose propagator is explicitly known; the other is a multiplication operator so that e^{-iVt} is found trivially. With knowledge of the two operator families, we can express the sought propagator by means of Theorem 5.9.7,

$$U(t) = \operatorname*{s-lim}_{N\to\infty} \left(e^{-iH_0t/N} e^{-iVt/N}\right)^N.$$

Hence $U(t)\psi = \lim_{N\to\infty} \psi_t^{(N)}$ for any $\psi \in L^2(\mathbb{R}^n)$, where the approximating sequence is defined recursively by $\psi_t^{(0)} := \psi$ and

$$\psi_t^{(N)} := e^{-iH_0t/N}\, e^{-iVt/N}\, \psi_{t(N-1)/N}^{N-1}\,, \qquad N = 1, 2, \dots\,.$$

We choose, for instance, the characteristic functions of the balls $B_j := \{x \in \mathbb{R}^n : |x| \le j\}$ as the regularization sequence $\{f_j\}$ in Theorem 9.3.1; then by induction we get

$$\psi_t^{(N)}(x) = \prod_{j=1}^{n} \left(\frac{m_j}{2\pi i \delta_N}\right)^{N/2} \tag{9.4}$$

$$\times \,\text{l.i.m.}_{\,j_1,\dots,j_N\to\infty} \int_{B_{j_1}\times\dots\times B_{j_N}} e^{iS_N(y^{(0)},\dots,y^{(N)};t)}\, \psi(y^{(0)})\, dy^{(0)}\dots dy^{(N-1)}$$

for a.a. $x \in \mathbb{R}^n$ (Problem 11), where we have denoted $y^{(N)} := x$, $\delta_N := t/N$ and

$$S_N\left(y^{(0)},\dots,y^{(N)};t\right) = \sum_{k=0}^{N-1}\left(\sum_{j=1}^{n} \frac{m_j}{2\delta_N}\left|y_j^{(k+1)} - y_j^{(k)}\right|^2 - V(y^{(k)})\,\delta_N\right).$$

It is easy to see that we can use any other regularization procedure with the properties specified in Theorem 9.3.1. Hence we arrive at the following conclusion.

9.4.1 Theorem: If the operator (9.2) is *e.s.a.*, then the corresponding propagator is given by (9.4) for any $\psi \in L^2(\mathbb{R}^n)$.

What is the meaning of this result? Consider the system of classical particles whose dynamics is determined by the Hamilton function

$$h(q,p) = \sum_{j=1}^{n} \frac{p_j^2}{2m_j} + V(q),$$

where as usual $q := (q_1,\dots,q_n)$ and $p := (p_1,\dots,p_n)$. For any trajectory $\gamma : [0,t] \to \mathbb{R}^n$ of this system we can define the action

$$S(\gamma) = \int_0^t \left(\frac{1}{2}\sum_{j=1}^{n} m_j \dot\gamma_j(s)^2 - V(\gamma(s))\right) ds.$$

In particular, if $\gamma(\cdot) := \gamma(y^{(0)},\dots,y^{(N)};\cdot)$ is a piecewise linear path whose graph is the polygonal line with vertices at $y^{(k)} = \gamma(kt/N)$, $k = 0,1,\dots,N$, then we have

$$\int_0^t \dot\gamma(y^{(0)},\dots,y^{(N)};s)^2\, ds = \sum_{k=0}^{N=1}\left|y_j^{(k+1)} - y^{(k)}\right|^2 \delta_N^{-1},$$

i.e., the first terms in the expressions of $S_N\left(y^{(0)},\dots,y^{(N)};t\right)$ and $S_N(\gamma)$ coincide.

This is not true for the second terms but their difference is in most cases small for large N (Problem 12). This suggests that the limit in (9.4) might not change if we replace $S_N\left(y^{(0)},\dots,y^{(N)};t\right)$ by $S(\gamma)$; in other words, we conjecture

$$(U(t)\psi)(x) = \lim_{N\to\infty} \prod_{j=1}^{n} \left(\frac{m_j}{2\pi i \delta_N}\right)^{N/2} \tag{9.5}$$

$$\times \,\text{l.i.m.}_{\,j_1,\dots,j_N\to\infty} \int_{B_{j_1}\times\dots\times B_{j_N}} e^{iS(\gamma(y^{(0)},\dots,y^{(N)};t))}\psi(y^{(0)})\, dy^{(0)}\dots dy^{(N-1)}$$

for a.a. $x \in \mathbb{R}^n$. It appears that this relation is indeed valid for wide classes of potentials; however, the corresponding proofs are not easy with exception of the simplest cases (*cf.* Problem 13 and the notes).

Hence $(U(t)\psi)(x)$ is a.e. approximated by expressions, which may be interpreted as the integral of $\gamma \mapsto e^{iS(\gamma)}\psi(\gamma(0))$ over the set of the polygonal paths that are linear in the intervals $(k\delta_N, (k+1)\delta_N)$, $k = 0, 1, \dots, N-1$. Since any continuous trajectory can be approximated by such paths, it is natural to ask whether $(U(t)\psi)(x)$ could be calculated directly, without the limiting procedure, as an integral over the set of all continuous paths ending at point x,

$$(U(t)\psi)(x) = \int_{\gamma(t)=x} e^{iS(\gamma)}\, \psi(\gamma(0))\, D\gamma\,.$$

This idea belongs to R. Feynman, after whom these expressions are usually named. It should be stressed that the object on the right side is only formal, in particular, because the Lebesgue measure has no counterpart in the path space in view of its infinite dimension (Problem 14).

Before we discuss the meaning of the right side in the last formula, we should briefly mention its appealing properties. The Feynman expression of the propagator makes the relations between quantum and classical mechanics very illustrative. In the standard system of units, the formula reads

$$(U(t)\psi)(x) = \int_{\gamma(t)=x} e^{(i/\hbar)S(\gamma)}\psi(\gamma(0))\, D\gamma\,. \tag{9.6}$$

If the Planck constant $\hbar$ is small compared with the action involved, the integral can be treated formally by the stationary–phase method. In this way, we can conjecture that the contributions to the integral cancel with the exception of those coming from the vicinity of stationary points γ_{cl} of the function S. It is known from classical mechanics, however, that the action is stationary just for the paths that solve the equations of motion. Thus only trajectories near to the classical ones contribute significantly to the value of the Feynman integral. Moreover, the formal stationary–phase evaluation of the integral yields the well-known semiclassical approximation to wave function, with the leading term containing the factor $e^{(i/\hbar)S(\gamma_{cl})}$. This heuristic argument admits a rigorous formulation; we comment on it in the notes.

Let us now return to the problem of the meaning of Feynman's integral. It has to be mentioned first that a functional integral similar to (9.6) was studied, long before Feynman's papers, in connection with the mathematical theory of Brownian motion. To describe it briefly, we select a particular class of functions on the path space Γ_x; the latter is chosen as the family of all continuous functions $\gamma : [0,t] \to \mathbb{R}^n$ such that $\gamma(t) = x$. The space Γ_x as a subset of $C([0,t], \mathbb{R}^n)$ is naturally equipped with the norm–induced topology $\|\cdot\|_\infty$, and this topology in turn determines the system $\mathcal{B}_x$ of Borel sets. A function $f : \Gamma_x \to \mathbb{C}$ is called **cylindrical** if $f(\gamma)$ depends only on the values of the function γ at a finite sequence of points $\{\tau_k : 0 = \tau_0 < \tau_1 < \ldots < \tau_n = t\}$; if this is the case we write $f(\gamma) =: f(\gamma(\tau_0), \dots, \gamma(\tau_{N-1}))$.

Denote $\delta_k := \tau_{k+1} - \tau_k$; then for each $\sigma > 0$ there is just one Borel measure w_σ on Γ_x such that the relation

$$\int_{\Gamma_x} f(\gamma(\tau_0), \dots, \gamma(\tau_{N-1}))\, dw_\sigma(\gamma) = \prod_{k=0}^{N-1} (2\pi i \delta_k)^{-n/2}$$
$$\times \int_{\mathbb{R}^{nN}} \exp\left\{ -\frac{1}{2\sigma} \sum_{k=0}^{N-1} |\gamma^{(k+1)} - \gamma^{(k)}|^2 \delta_k^{-1} \right\} f(\gamma^{(0)}, \dots, \gamma^{(N-1)})\, d\gamma^{(0)} \dots d\gamma^{(N-1)} \qquad (9.7)$$

holds for any Borel function which is cylindrical and bounded (see the notes for references to the proof). The measure w_σ is called the *Wiener measure* and the integral with respect to it is the *Wiener integral.*

There are many similarities between Feynman and Wiener integrals. To illustrate this we put for simplicity $m_j = m$, $j = 1, \dots, n$, and denote $f(\gamma) := \exp\left(-i \int_0^t V(\gamma(s))\, ds \right) \psi(\gamma(0))$; then the right side in the expression of $(U(t)\psi)(x)$ can be formally written as

$$\int_{\Gamma_x} f(\gamma) \exp\left(\frac{im}{2} \int_0^t |\dot{\gamma}(s)|^2 ds \right) D\gamma\,.$$

Comparing the Wiener integral (9.7) of a cylindrical function to (9.5) we see that it has the same formal expression with $m = i/\sigma$. In this case, loosely speaking, the singularities of the exponential term and of $D\gamma$ cancel mutually and we can replace $\exp\left(-\frac{1}{2\sigma} \int_0^t |\dot{\gamma}(s)|^2 ds \right) D\gamma$ by $dw_\sigma(\gamma)$.

Unfortunately, this argument depends substantially on the fact that the expression in the exponent is real and nonpositive, as the following result shows (see the notes for references to the proof).

9.4.2 Theorem (Cameron): Let σ be a nonzero complex number, $\mathrm{Re}\,\sigma \geq 0$. A finite complex measure w_σ such that the relation (9.7) holds for any Borel function $f : \Gamma_x \to \mathbb{C}$, which is cylindrical and bounded, exists *iff* $\sigma \in (0, \infty)$.

It follows that *the Feynman integral cannot be interpreted within the standard theory of integration.* Hence if we want to use its advantages suggested by heuristic arguments, which we have indicated briefly above, it must be defined in another way. One possibility is to employ the Trotter formula as we did to derive the relation (9.4); the resulting expression is often called the *product F–integral.* Some other possibilities are reviewed in the notes.

The Feynman idea in turn affected the theory of the Wiener integral: it inspired the so–called **Feynman–Kac formula**, which has found many applications in various parts of mathematical physics. We shall formulate it for $n = 3$ and the potential $V \in L^2 + L^\infty$, *i.e.*, expressible as the sum of a pair of functions which belong to the two classes. The corresponding operator $H = \frac{1}{2m} P^2 + V$ is then self–adjoint by Theorem 14.1.2 and

$$\left(e^{-Ht}\psi\right)(x) = \int_{\Gamma_x} e^{-\int_0^t V(\gamma(s))\, ds}\, \psi(\gamma(0))\, dw_{1/m}(\gamma) \qquad (9.8)$$

holds for all $\psi \in L^2(\mathbb{R}^3)$ and a.a. $x \in \mathbb{R}^3$. Similar assertions are valid under much more general circumstances — see the notes.

9.5 Nonconservative systems

If the Hamiltonian is time–dependent, it is generally not easy to find a solution to equations of motion. We are now going to describe some methods which can be used to this end. The simplest among them is based on expanding the sought solution into a series.

9.5.1 Theorem (Dyson expansion): Let $H : \mathbb{R} \to \mathcal{B}(\mathcal{H})$ be a strongly continuous Hermitean–valued function, and set

$$\psi_t := \phi + \sum_{n=1}^{\infty} U_n(t,s)\phi\,, \tag{9.9}$$

with

$$U_n(t,s)\phi := (-i)^n \int_s^t dt_1 \int_s^{t_1} dt_2 \dots \int_s^{t_{n-1}} dt_n\, H(t_1)\dots H(t_n)\phi$$

for any $\phi \in \mathcal{H}$. Then the series $\sum_n U_n(t,s)$ converges with respect to the operator norm, $U(t,s)\phi := \psi_t$ defines a unitary propagator and the vector–valued function $t \mapsto \psi_t$ solves the Schrödinger equation

$$i\frac{d}{dt}\psi_t = H(t)\psi_t$$

with the initial condition $\psi_s = \phi$.

9.5.2 Remark: If the function $H(\cdot)$ is operator–norm continuous, the existence of the integral

$$U_n(t,s) := (-i)^n \int_s^t dt_1 \int_s^{t_1} dt_2 \dots \int_s^{t_{n-1}} dt_n\, H(t_1)\dots H(t_n)$$

is easily established. In addition, if the family $\{H(t) : t \in \mathbb{R}\}$ is commutative, the multiple integral can be simplified giving the relation $U(t,s) = \exp\left(-i \int_s^t H(\tau)\,d\tau\right)$, which generalizes the expression for propagators of conservative systems. Motivated by this, (9.9) is sometimes written symbolically as

$$\psi_t = \mathrm{T}\, e^{-i\int_s^t H(\tau)\,d\tau}$$

and called the *time–ordered exponential* of the Hamiltonian, the name referring to the ordered arguments in the above integrals.

Proof of Theorem 1: Let us first check the existence and continuity of $U_n(\cdot,s)\phi$ for any $s \in \mathbb{R}$ and $\phi \in \mathcal{H}$. If $n = 1$, this follows from the continuity of $H(\cdot)$ and the absolute continuity of the integral; in the general case the result is obtained by induction from the recursive relation

$$U_{n+1}(t,s)\phi = -i\int_s^t H(t_1)U_n(t_1,s)\phi\,dt_1\,.$$

Denote $K_T := \{[s,t]:\, s^2+t^2 \le T^2\}$; then $|t-s| \le 2T$ holds for each pair $[s,t] \in K_T$ and $J_{st} \subset [-T,T]$ for any closed interval J_{st} with endpoints s,t. By assumption, $\|H(\cdot)\phi\|$ is continuous and therefore bounded on $[-T,T]$, in which case the uniform boundedness principle implies the existence of a positive C_T such that $\|H(t)\| \le C_T$ for all $t \in [-T,T]$. Hence we have

$$\|U_n(t,s)\phi\| \le C_T^n\|\phi\|\,\frac{|t-s|^n}{n!} \le \frac{(2TC_T)^n}{n!}\,\|\phi\|\,, \quad n=1,2,\dots\,,$$

so the operators $U_n(t,s)$ are bounded. Denoting $U^{(N)}(t,s) := I + \sum_{n=1}^N U_n(t,s)$, we deduce from the above estimate that the sequence $\{U^{(N)}(t,s)\}$ converges with respect to the operator norm to some $U(t,s) \in \mathcal{B}(\mathcal{H})$ uniformly on any compact subset of $\mathbb{R}^2$. The relation (9.9) can then be written as $\psi_t = U(t,s)\phi$.

To show that the operator $U(t,s)$ has the needed properties, first note that Hermiticity in combination with a simple transformation of the integration domain yields $(\psi, U_n(t,s)^*\phi) = \overline{(\phi, U_n(t,s)\psi)} = (\psi, U_n(s,t)\phi)$ for any $\phi,\psi \in \mathcal{H}$ (Problem 15), so $U_n(t,s)^* = U_n(s,t)$; then $U(t,s)^* = U(s,t)$ follows for all $s,t \in \mathbb{R}$ from the operator–norm continuity of the adjoint operation. Next we shall check that the functions $U_n(\cdot,\cdot)$ are operator–norm continuous. The just proved property implies

$$\|U_n(t,s) - U_n(t_0,s_0)\| \le \|U_n(t,s) - U_n(t_0,s)\| + \|U_n(s,t_0) - U_n(s_0,t_0)\|\,.$$

The unit ball $B_1([s_0,t_0])$ is contained in K_T for $T := 1+\sqrt{s_0^2+t_0^2}$, so using the recursive expression of $U_n(t,s)$ we get

$$\begin{aligned}\|U_n(t,s) - U_n(t_0,s)\| &\le \sup_{\|\phi\|=1}\left(\operatorname{sgn}(t-t_0)\int_{t_0}^t \|H(t_1)U_{n-1}(t_1,s)\phi\|\,dt_1\right)\\ &\le C_T\,|t-t_0|\,\frac{(2TC_T)^{n-1}}{(n-1)!}\end{aligned}$$

for $(s,t) \in B_1([s_0,t_0])$, and a similar estimate with $|t-t_0|$ replaced by $|s-s_0|$ in the other term; hence continuity follows. Since the sequence $\{U^{(N)}(t,s)\}$ converges uniformly in compact sets, we conclude that $(s,t) \mapsto U(t,s)$ is operator–norm continuous and, *a fortiori*, strongly continuous.

To prove $U(t,s)U(s,r) = U(t,r)$ for all $r,s,t \in \mathbb{R}$, it is sufficient to verify that the sequence of operators

$$U^{(N)}(t,s)U^{(N)}(s,r) = \left(\sum_{n=0}^{N}\sum_{j=0}^{n} + \sum_{n=N+1}^{2N}\,\sum_{j=n-N}^{N}\right) U_j(t,s)U_{n-j}(s,r)\,,$$

where we have set $U_0(t,s) := I$, converges with respect to the operator norm to $U(t,r)$ as $N \to \infty$. The norm of the second part of the sum is bounded above by $\sum_{n=N+1}^{2N} (4TC_T)^n/n!$, which converges to zero, so the sought relation follows from the identity

$$\sum_{j=0}^{n} U_j(t,s)U_{n-j}(s,r) = U_n(t,r)\,,$$

which is easily proved by induction (Problem 15). Combining it with the above results we see that $\{U(t,s) : s,t \in \mathbb{R}\}$ is a unitary propagator.

The last thing to prove is that the vector-valued function $U(\cdot,s)\phi$ satisfies the Schrödinger equation. The recursive expression of $U_n(t,s)$ gives

$$\frac{d}{dt}U^{(N)}(t,s)\phi = -iH(t)U^{(N-1)}(t,s)\phi$$

for any $\phi \in \mathcal{H}$. The right side converges to $-iH(t)\psi_t$ as $N \to \infty$; hence the left side is also convergent and we only have to show that its limit equals $\frac{d}{dt}\psi_t$. In the same way as above we get

$$\begin{aligned}\Bigg\| \frac{1}{h}\sum_{n=0}^{\infty} (\, U_n(t+h,s) - U_n(t,s)\,)\,\phi - \sum_{n=0}^{N} \frac{d}{dt}U_n(t,s)\phi \Bigg\| \\ \le \Bigg\| \sum_{n=0}^{N} \left(\frac{U_n(t+h,s) - U_n(t,s)}{h}\,\phi - \frac{d}{dt}U_n(t,s)\phi \right) \Bigg\| \\ + C_T\,\|\phi\| \sum_{n=N+1}^{\infty} \frac{(2TC_T)^{n-1}}{(n-1)!}\end{aligned}$$

provided $T \ge 1 + \sqrt{s^2+t^2}$ and $|h| < 1$ (Problem 15). For any $\varepsilon > 0$ there are N_0, R such that the last term on the right side is $<\frac{\varepsilon}{2}$ for $N > N_0$, and to such a number N there is $\delta > 0$ such that the first term is $<\frac{\varepsilon}{2}$ if $0 < |h| < \delta$. Hence the norm on the left side is $<\varepsilon$ for all $N < N_0$ and $0 < |h| < \delta$, and the proof is concluded by taking the limit $N \to \infty$. ∎

In most practically interesting situations, however, the Hamiltonians are unbounded, so Theorem 9.5.1 is not directly applicable. The difficulty can often be bypassed by using the interaction picture. Assume, for instance, that $H(t) := H_0 + V(t)$, where H_0 is self-adjoint and $V(\cdot)$ is a strongly continuous Hermitean-valued function, so $H(t)$ is self-adjoint and $D(H(t)) = D(H_0)$ for any $t \in \mathbb{R}$. The function $V_D : V_D(t) = U_0(t-s)^{-1}\,V(t)\,U_0(t-s)$ is again Hermitean-valued and strongly continuous, so the equation (9.3) has the solution $\psi_D(t) := U_D(t,s)\psi_s$ for any $\psi_s \in \mathcal{H}$, which is given by (9.9) with $H(t_j)$ replaced by $V_D(t_j)$. Defining

$$\psi_t := U_0(t-s)U_D(t,s)\psi_s\,,$$

we see that the function $t \mapsto \psi_t$ *formally* obeys the Schrödinger equation with the Hamiltonian $H(t)$. To prove that this is indeed a solution we have to check that

$U_D(t,s)$ maps $D(H_0)$ into itself, which may not be true if the operators H_0, $V(t)$ do not commute.

A more powerful method of establishing the existence of solutions to a time–dependent Schrödinger equation is obtained if the propagator is approximated by a suitable sequence of products of unitary operators. We have

$$U(t,s) = U(t,t-\delta_n)U(t-\delta_n,t-2\delta_n)\dots U(s+\delta_n,s)$$

for any positive integer n where $\delta_n := (t-s)/n$; the approximation consists of replacing the function $H(\cdot)$ in the intervals $(s+(k-1)\delta_n, s+k\delta_n)$ by its value at some point, say, $\tau_k := s+k\delta_n$. This gives

$$U_n(t,s) = e^{-iH(t)\delta_n}e^{-iH(\tau_{n-1})\delta_n}\dots e^{-iH(\tau_1)\delta_n}$$

and one expects these expressions to approximate the propagator,

$$U(t,s) = \operatorname*{s-lim}_{n\to\infty} U_n(t,s)\,, \tag{9.10}$$

provided the function $H(\cdot)$ is smooth enough. We shall present one result of this type without proof; more details can be found in the notes.

9.5.3 Theorem: Suppose the Hamiltonian is of the form $H(t) = H_0 + V(t)$ for t belonging to an open interval $J \subset \mathbb{R}$, with H_0, $V(t)$ self–adjoint, and denote $C(t,s) := (H(t)-i)(H(s)-i)^{-1} - I$. In addition, assume that

(i) $V(t)$ is H_0–bounded with the relative bound <1 for any $t \in J$.

(ii) The function $(s,t) \mapsto (t-s)^{-1}C(t,s)\phi$ is bounded and uniformly continuous for any $\phi \in \mathcal{H}$ if s,t belong to a compact subinterval of J and $s \neq t$.

(iii) The limit $C(t)\phi := \lim_{s\to t}(t-s)^{-1}C(t,s)\phi$ exists for any $\phi \in \mathcal{H}$, uniformly in each compact subinterval, and the function $C(\cdot)$ is strongly continuous and bounded.

Then for any $s,t \in J$, $s \le t$, the convergence in (9.10) is uniform in each compact subinterval and the relation defines $U(t,s)$ such that $\psi_t := U(t,s)\phi$ for any $\phi \in D(H_0)$ solves the Schrödinger equation with the Hamiltonian $H(t)$ and initial condition $\psi_s = \phi$.

One of the consequences is that Dyson's expansion is applicable in the interaction picture provided the function $V(\cdot)$ is strongly differentiable.

9.5.4 Corollary: Let $H(t) := H_0 + V(t)$ with H_0 self–adjoint and $V(t)$ Hermitean for any t from an open interval $J \subset \mathbb{R}$. If $\frac{d}{dt}V(t)\phi =: G(t)\phi$ exists for all $t \in J$, $\phi \in \mathcal{H}$, and the function G is strongly continuous in J; then the conclusions of Theorem 3 are valid.

Proof: The operators $V(t)$ are bounded, so it is sufficient to check the validity of the assumptions (ii), (iii). We define the function F by

$$F(s,t) := \begin{cases} (t-s)^{-1}C(t,s) & \dots \quad s \neq t \\ G(t)(H(t)-i)^{-1} & \dots \quad s = t \end{cases}$$

Since $\|(H(s)-i)^{-1}\| \le 1$, its values are bounded operators; in view of the second resolvent identity, $\|(H(\cdot)-i)^{-1}\|$ is strongly continuous. The sequential continuity of operator multiplication then implies that $F(\cdot,\cdot)$ is continuous for $s \neq t$, and due to the assumption, it is also continuous in the "diagonal" points of $J \times J$. This means, in particular, that $\|F(\cdot,\cdot)\phi\|$ is for any $\phi \in \mathcal{H}$ and a compact interval $J_0 \subset J$ continuous in $J_0 \times J_0$, and is therefore uniformly continuous in this set; the validity of (ii), (iii) follows easily from here. ∎

In fact, Theorem 9.5.3 allows us to establish the existence of the propagator in even more general situations when the interaction term is unbounded — see the notes. Let us now briefly mention an application of Dyson's expansion.

9.5.5 Example: Suppose the system, which at the initial instant $t=0$ is in a state described by an eigenvector ϕ_j of H_0, is exposed to a time-dependent interaction $V(t)$ during $(0,T)$; we are interested in the corresponding transition probability to another eigenstate ϕ_k of the unperturbed Hamiltonian. If the assumptions of Corollary 4 are valid, we get

$$(\phi_k, U(t,0)\phi_j) = \sum_{n=1}^{\infty} (-i)^n \int_0^t dt_1 \dots \int_0^{t_{n-1}} dt_n$$
$$\times\; (\phi_k, U_0(t-t_1)V(t_1)U_0(t_1-t_2)V(t_2)\dots V(t_n)U_0(t_n)\phi_j)$$

since the zeroth term is absent for $k \neq j$. In the first approximation, the transition amplitude equals

$$(\phi_k, U(t,0)\phi_j) = -i\, e^{-i\lambda_k t} \int_0^t e^{i(\lambda_k-\lambda_j)s}\, (\phi_k, V(s)\phi_j)\, ds \,+\, R_2(V,t)\,,$$

where λ_j, λ_k are the respective eigenvalues and the remainder term can be estimated as in the proof of Theorem 9.5.1, $|R_2(V,t)| \le e^{Ct}-1-Ct$ provided $\|V(t)\| \le C$ for $t \in (0,T)$, so it is of order $\mathcal{O}(t^2)$ for small t. In a similar way, we can derive higher approximations. It is natural to expect that the transition probability will be small for $V(t)$ which is only slowly changing. If we replace V by $V_\tau := V(t/\tau)$ then the leading-order contribution to the transition probability approaches that of the time-independent perturbation $V_\infty := V(0)$ as $\tau \to \infty$. This is a particular case of a much more general result — see the notes.

It is usually much more difficult than in the conservative case to derive the propagator of a time-dependent system in a closed form. Sometimes one can find an inspiration in the solution of the corresponding classical problem.

9.5.6 Example: Consider again the harmonic oscillator, this time with time–dependent perturbation,

$$H(t) = \frac{1}{2m}P^2 + \frac{1}{2}m\omega^2 Q^2 + f(t)Q\,.$$

It is known from classical mechanics that the driven–oscillator coordinate and momentum trajectories are shifted by $-(m\omega)^{-1}\beta(t,s)$ and $\alpha(t,s)$, respectively, where

$$\alpha(t,s) := \int_s^t f(\tau)\,\cos\omega(\tau-t)\,d\tau\,, \quad \beta(t,s) := -\int_s^t f(\tau)\,\sin\omega(\tau-t)\,d\tau$$

(Problem 16a). We found in Example 9.3.5 that coherent–state wave packets follow classical trajectories in the unperturbed oscillator. This suggests that we might try to "shift" the corresponding propagator by the Weyl operator $W(-\beta/m\omega,-\alpha)$ (*cf.* Example 8.2.7), *i.e.*, to look for the propagator in the form

$$U(t,s) = e^{-i\gamma(t,s)}e^{-i\alpha(t,s)Q}e^{i\beta(t,s)P/m\omega}U_0(t-s)\,, \tag{9.11}$$

where $U_0(\cdot)$ is the propagator of Theorem 9.3.6 and γ is a phase factor to be found. Using the corresponding Schrödinger equation we find that, choosing

$$\gamma(t,s) := -\frac{1}{2m}\int_s^t \left(\alpha(\tau,s)^2 - \beta(\tau,s)^2\right) d\tau\,,$$

we do indeed get the sought propagator (Problem 16b).

Some classes of nonconservative systems are particularly important in applications. In Example 9.5.5 above we mentioned the case when the Hamiltonian is slowly varying. Another important situation occurs when the time dependence is periodic, $H(t) = H(t+T)$ for some $T > 0$ and all $t \in \mathbb{R}$. It then follows from postulate (Q4c) that the evolution operator satisfies $U(t+T, s+T) = U(t,s)$, and therefore

$$U(t+nT,s) = U(t,s)(U(s+T,s))^n \tag{9.12}$$

for any integer n and arbitrary $s,t \in \mathbb{R}$. Hence the most important contribution to the evolution of states of time–periodic systems over long time intervals comes from powers of the evolution operator over one period, the remaining part being the same within each period. The operator $U(s+T,s)$ for a fixed s is called the *monodromy operator*. The dependence on s is unimportant: the periodicity assumption together with the definition of the propagator implies that all monodromy operators are unitarily equivalent, so they have, in particular, identical spectral properties. Recall that as a unitary operator, $U(s+T,s)$ has a spectral decomposition; we can also write it as $U(s+T,s) =: e^{iAT}$ (*cf.* Problem 5.31) in which case the operator A is usually called the *quasienergy*.

It appears that the evolution of time–periodic systems depends substantially on the spectral properties of $U(s+T,s)$. To illustrate this, consider the situation when the monodromy operator has an eigenvalue. The corresponding state changes, of course, during the period; however, it follows a closed loop in the state space returning to the same ray after each period. In this respect it is similar to a stationary state. On the other hand, if the state belongs to the continuous spectral subspace of $U(s+T,s)$ then it exhibits no recurrent behavior despite the fact that $H(\cdot)$ is periodic. The spectral properties of the quasienergy operator can be rather unstable.

9.5.7 Example: Consider the driven harmonic oscillator of the preceding example and choose $f(t) := \sin \Omega t$ with $\Omega > 0$. The propagator (9.11) is easily found (Problem 16c); in particular, we have

$$\alpha(t,0) = \frac{\Omega(\cos\omega t - \cos\Omega t)}{\Omega^2-\omega^2}, \quad \beta(t,0) = \frac{\Omega\sin\omega t - \omega\sin\Omega t}{\Omega^2-\omega^2} \quad \text{if} \quad \Omega \neq \omega$$

$$\alpha(t,0) = \frac{t\,\sin\omega t}{2}, \qquad \beta(t,0) = -\frac{t\,\cos\omega t}{2} + \frac{\sin\omega t}{2\omega} \quad \text{if} \quad \Omega = \omega$$

We denote by $T := 2\pi/\Omega$ the driving force period; then for $\Omega = \omega$ we obtain $U(T,0) = -e^{-i\gamma(T,0)}e^{-i\pi P/m\omega}$ due to Remark 9.3.7, and therefore the monodromy operator can easily be checked to have a purely continuous spectrum, $\sigma_{ac}(U(T,0)) = \{z \in \mathbb{C} : |z| = 1\}$, in view of the unitary equivalence (4.4). We have mentioned that the spectrum is independent of the initial instant; however, the operators themselves differ: we have, *e.g.*, $U(5T/4, T/4) = -\,e^{-i\gamma(5T/4,T/4)}e^{-i(5T/8)Q}$.

On the other hand, if the driving frequency is not in resonance with the proper frequency of the oscillator, $\Omega \neq \omega$, the character of the spectrum changes. The reader is asked in Problem 16d to check that the point spectrum of $U(T,0)$ is nonempty, but in fact a much stronger result is valid (see the notes).

9.6 Unstable systems

Most microscopic objects in physics are unstable. We know, for instance, that a charged pion will almost surely decay into a muon and a neutrino after 10^{-8} s, or that an excited atom will eventually radiate one or more photons and pass to the ground state as a result of its interaction with the electromagnetic field, *etc.* In spite of the different physical mechanisms which govern decay processes, the latter have some common properties; this is the topic of this section.

The propagator associated with an unstable system should be nonunitary to describe the fact that, roughly speaking, the probability of finding the system in the original undecayed state decreases with time. To avoid a contradiction with postulate (Q4b) the unstable system cannot be regarded as isolated; in other words, we have to treat it as a part of a larger system including its decay products.

Consider an isolated system S with a state space $\mathcal{H}$ on which the evolution operator U acts. If $\mathcal{H}_u$ is a proper subspace in $\mathcal{H}$ which is *noninvariant* with respect to $U(t)$, then $U(t) = E_uU(t)\psi + (I-E_u)U(t)\psi$ holds for $\psi \in \mathcal{H}_u$, where the second term on the right side is generally non-zero; in that case we have $\|E_uU(t)\psi\| < \|\psi\|$.

A state $\psi \in \mathcal{H}_u$ thus evolves into a superposition of vectors from the subspaces $\mathcal{H}_u$ and $\mathcal{H}_u^\perp$; if we identify the rays in these subspaces with the states of the unstable system and its decay products, respectively, we get a natural scheme for description of unstable systems in quantum theory:

(u) the state Hilbert space $\mathcal{H}_u$ of an unstable system is a subspace in the state space $\mathcal{H}$ of a larger isolated system S. Time evolution in $\mathcal{H}$ is described by the unitary propagator $U\,:\,U(t) = e^{-iHt}$, where H is the Hamiltonian of system S. The subspace $\mathcal{H}_u$ fails to be invariant with respect to $U(t)$ for any $t > 0$.

The last part is substantial; it expresses the fact that a state vector can leave the subspace $\mathcal{H}_u$.

As in the previous sections we shall consider pure states only. Time evolution of the unstable system alone is determined by the *reduced propagator*

$$V\,:\,V(t) = E_u U(t) \upharpoonright \mathcal{H}_u\,,$$

where E_u is the projection onto the subspace $\mathcal{H}_u \subset \mathcal{H}$. We assume that the state of the system at the initial time $t=0$ is described by a unit vector $\psi \in \mathcal{H}_u$; then the *decay law* is defined by

$$P_\psi(t) := \|V(t)\psi\|^2 = \|E_u U(t)\psi\|^2\,.$$

Its value is therefore interpreted as the probability of finding the system undecayed in a measurement performed at the instant t, or in other words, of obtaining a positive result in the *yes–no* experiment associated with projection E_u.

9.6.1 Proposition: The function $V(\cdot)$ is strongly continuous, and the relations $V(t)^* = V(-t)$ and $\|V(t)\| \le 1$ hold for any $t \in \mathbb{R}$. The reduced propagator does *not* have the group property: there are $s,t \in \mathbb{R}$ such that $V(s)V(t) \ne V(s+t)$. The decay law $P_\psi(\cdot)$ is for any $\psi \in \mathcal{H}_u$ a continuous function which satisfies the relations $0 \le P_\psi(t) \le P_\psi(0) = 1$.

Proof is left to the reader (Problem 17b).

9.6.2 Example: Let $\mathcal{H}_u$ be a one–dimensional subspace in $\mathcal{H}$ spanned by the unit vector ψ. The reduced propagator acts as a multiplication by $v(t) := (\psi, U(t)\psi)$ and the decay law is $P_\psi(t) = |v(t)|^2$; using the spectral decomposition of the Hamiltonian, we can rewrite it as

$$P_\psi(t) = \left| \int_{\mathbb{R}} e^{-i\lambda t} d(\psi, E_\lambda^{(H)}\psi) \right|^2 .$$

The vector ψ usually belongs to the absolutely continuous subspace $\mathcal{H}_{ac}(H)$; in that case the Riemann–Lebesgue lemma implies $\lim_{t\to\infty} P_\psi(t) = 0$. For instance, if $\psi := \psi_u$ is such that the corresponding measure is given by the Breit–Wigner formula of Example 7.4.6, then the decay law is of the exponential form

$$P_{\psi_u}(t) = \left| \frac{\Gamma}{2\pi} \int_{\mathbb{R}} \frac{e^{-i\lambda t}}{(\lambda-\lambda_0)^2 + \frac14 \Gamma^2}\, d\lambda \right|^2 = e^{-\Gamma t}$$

for all $t \geq 0$ since a simple integration gives $v(t) = e^{-i\lambda_0 t - \Gamma|t|/2}$.

9.6.3 Remark: A typical feature of the exponential decay law is that its initial decay rate, defined as the one-sided derivative $\dot{P}_\psi(0+) = \Gamma$, is positive. This is closely connected to the fact that ψ_u is not a finite-energy state — see Problem 18 and Example 7.4.6. This does not mean, however, that we can use it to decide whether a given unstable state is a finite-energy state or not; like any other measurement of a continuous quantity, time can be determined only with a finite resolution, so in reality we are not able to measure the derivative of $P_\psi(\cdot)$. Recall the result expressed by Proposition 7.4.7; we can again construct the truncated states $\psi_b := E(\Delta_b)\psi$, which are experimentally indistinguishable from ψ for b large enough. For any $\varepsilon > 0$ there is b_0 such that $\|\psi - \psi_b\| < \varepsilon$ for all $b > b_0$, and therefore

$$|P_{\psi_b}(t) - P_\psi(t)| \leq 2\,|\,\|E_u U(t)\psi_b\| - \|E_u U(t)\psi\|\,| \leq 2\,\|E_u U(t)(\psi_b - \psi)\| < 2\varepsilon;$$

at the same time ψ as a finite-energy state has zero initial decay rate (see also Problem 19a).

The reduced evolution operator corresponding to the exponential decay law satisfies the *semigroup condition*,

$$V(s)V(t) = V(s+t), \quad s,t \geq 0,$$

Propagators with this property are also used for the description of more complicated decays.

9.6.4 Example (decay of neutral kaons): If we disregard the configuration-space degrees of freedom we may associate a two-dimensional space $\mathcal{H}_u \approx \mathbb{C}^2$ with a neutral kaon. Its basis is formed either by

$$\phi_{K^0} := \begin{pmatrix} 1 \\ 0 \end{pmatrix}, \quad \phi_{\overline{K}^0} := \begin{pmatrix} 0 \\ 1 \end{pmatrix}$$

or by the pair of nonorthogonal vectors

$$\phi_S := N_\epsilon \begin{pmatrix} 1+\epsilon \\ 1-\epsilon \end{pmatrix}, \quad \phi_L := N_\epsilon \begin{pmatrix} 1+\epsilon \\ -1+\epsilon \end{pmatrix},$$

where $N_\epsilon := (1+|\epsilon|^2)^{-1/2}$ and ϵ is a complex parameter whose experimentally determined value is about $(2.3\times 10^{-3})\, e^{\pi i/4}$. The vectors ϕ_S, ϕ_L represent the short- and long-living component in the weak decay of a kaon-antikaon superposition. The operators P_S, P_L on $\mathcal{H}_u$ are represented in the orthonormal basis by the matrices

$$P_S := \frac{1}{2}\begin{pmatrix} 1 & b_\epsilon \\ b_\epsilon^{-1} & 1 \end{pmatrix}, \quad P_L := \frac{1}{2}\begin{pmatrix} 1 & -b_\epsilon \\ -b_\epsilon^{-1} & 1 \end{pmatrix},$$

where $b_\epsilon := (1+\epsilon)/(1-\epsilon)$; we can check that they are nonorthogonal projections, $P_j P_k = \delta_{jk} P_k$ and $P_j \phi_k = \delta_{jk}\phi_k$ for $j,k = S,L$. Time evolution in $\mathcal{H}_u$ is conventionally described by the reduced propagator

$$V : V(t) = P_S\, e^{-iz_S t} + P_L\, e^{-iz_L t},$$

where $z_j := m_j - \frac{i}{2}\Gamma_j$ with $m_S \approx m_L \approx 498\,\text{MeV}$, $\frac{1}{2}\Gamma_S \approx 3.7 \times 10^{-12}\,\text{MeV}$, and $\frac{1}{2}\Gamma_L \approx 0.64 \times 10^{-14}\,\text{MeV}$. It is easy to see that V satisfies the semigroup condition; however, the decay is exponential only if it starts from some of the states ϕ_S, ϕ_L (Problem 19b).

The trouble with the semigroup condition is that it in fact represents a strong requirement on the energy spectrum. We have seen already in Example 9.6.2 that $\sigma(H) = \mathbb{R}$ must hold for the Hamiltonian corresponding to the exponential decay law; otherwise the integrated function could not be nonzero for all $\lambda \in \mathbb{R}$. It appears that this is also true in a more general situation (see the notes).

9.6.5 Theorem: Suppose that the spaces $\mathcal{H}_u \subset \mathcal{H}$ and the propagator $U(t) := e^{-iHt}$ satisfy assumption (u). If the semigroup condition is valid for the reduced propagator, then $\sigma(H) = \mathbb{R}$.

The Hamiltonian is usually a positive operator; in relativistic quantum field theories one even postulates its positivity. The above result then means that a semigroup reduced evolution cannot satisfy the semigroup condition, in particular, that exponential decay laws are excluded. This is no disaster, however, because it tells us only that the semigroup condition cannot be *exactly* valid; we know from Remark 9.6.3 and Problem 19a that a decay law with a below bounded Hamiltonian can differ so little from the exponential one that the difference is irrelevant from the experimentalist's point of view; the same is also true in a more general context. Hence the semigroup description of decays is necessarily approximate, but the approximation is good enough for every practical purpose.

To be able to grasp better the nature of this approximation, we introduce the *reduced resolvent* of the Hamiltonian with respect to the subspace $\mathcal{H}_u$ by

$$R_H^u(z) := E_u R_H(z) \upharpoonright \mathcal{H}_u$$

and use it to express the reduced propagator. Using the spectral decomposition of the Hamiltonian we can write

$$V(t)\psi = \int_{\mathbb{R}} e^{-i\lambda t} dF_\lambda \psi$$

for any $\psi \in \mathcal{H}_u$, where $F_\lambda \equiv F(-\infty, \lambda] := E_u E_\lambda^{(H)} \upharpoonright \mathcal{H}_u$. Further, using the Stone formula, we can express the vector measure appearing in the last relation,

$$\frac{1}{2}\{F[\lambda_1, \lambda_2] + F(\lambda_1, \lambda_2)\} = \frac{1}{2\pi i} \operatorname*{s-lim}_{\eta \to 0+} \int_{\lambda_1}^{\lambda_2} [R_H^u(\xi + i\eta) - R_H^u(\xi - i\eta)]\, d\xi\,.$$

The support of $F(\cdot)\psi$ is obviously contained in $\sigma(H)$, and the same is true for $\operatorname{supp} F := \bigcup_{\psi \in \mathcal{H}_u} \operatorname{supp} F(\cdot)\psi$. Hence, while the resolvent $R_H(\xi)$ does not exist at a point $\xi \in \operatorname{supp} F$ by definition, it is not excluded at the same time that the limits $\operatorname{s-lim}_{\eta \to 0+} R_H^u(\xi \pm i\eta)$ exist for the *reduced* resolvent, and the subintegral function in the last formula is bounded for $(\xi, \eta) \in [\lambda_1, \lambda_2] \times [-\eta_0, \eta_0]$ with some $\eta_0 > 0$, in which case the limit can be interchanged with the integral by the dominated–convergence theorem. Furthermore, the absolute continuity of the integral implies $F(\{\lambda\}) = 0$ for any $\lambda \in \sigma(H)$, so we get

$$F[\lambda_1, \lambda_2] = \frac{1}{2\pi i} \int_{\lambda_1}^{\lambda_2} \operatorname*{s-lim}_{\eta \to 0+} \left[R_H^u(\xi + i\eta) - R_H^u(\xi - i\eta) \right] d\xi \,.$$

Since the resolvent is analytic in $\rho(H)$, the same is true for $R_H^u(\cdot)$. In general, the reduced resolvent is not continuous when we cross $\sigma(H)$, because otherwise the above relations would imply $V(0) = 0$. It is not excluded, however, that $R_H^u(\cdot)$ has an *analytic continuation* across $\sigma(H)$, *i.e.*, that there is an analytic function in a region $\Omega \subset \mathbb{C}$ containing $\sigma(H)$, which coincides with $R_H^u(\cdot)$ in the upper halfplane. The situation is particularly interesting when this continuation has a meromorphic structure, *i.e.*, isolated poles in the lower halfplane.

9.6.6 Example: Let us return to Example 9.6.2. Since $\dim \mathcal{H}_u = 1$ the reduced resolvent acts as the operator of multiplication by $r_H^u(z)$. Suppose that

$$r_H^u(z) = \frac{\alpha}{z_p - z} + f(z)$$

for $\operatorname{Im} z > 0$, where $\alpha > 0$, f is a holomorphic function, and $z_p := \lambda_p - i\delta_p$ is a point in the lower halfplane, *i.e.*, that $r_H^u(\cdot)$ has the analytic continuation from the upper halfplane to $\mathbb{C} \setminus \{z_p\}$. Using elementary properties of the resolvent we find $r_H^u(\lambda - i\eta) = \overline{r_H^u(\lambda + i\eta)}$, so the measure $F(\cdot)$, which is numerical–valued in the present case, is given by

$$F(\Delta) = \frac{\alpha}{2\pi i} \int_\Delta \left(\frac{1}{\lambda - \bar{z}_p} - \frac{1}{\lambda - z_p} \right) d\lambda + \frac{1}{\pi} \int_\Delta \operatorname{Im} f(\lambda)\, d\lambda \,.$$

This yields further an expression for the reduced propagator,

$$v(t) = \frac{\alpha}{2\pi i} \int_{\mathbb{R}} \frac{2i\delta_p}{(\lambda - \lambda_p)^2 + \delta_p^2}\, e^{-i\lambda t} d\lambda + \frac{1}{\pi} \int_{\mathbb{R}} e^{-i\lambda t} \operatorname{Im} f(\lambda)\, d\lambda \,.$$

The first integral can be evaluated by the residue theorem (Problem 20) giving

$$v(t) = \alpha\, e^{-i\lambda_p t - \delta_p |t|} + \frac{1}{\pi} \int_{\mathbb{R}} e^{-i\lambda t} \operatorname{Im} f(\lambda)\, d\lambda \,.$$

Hence the reduced propagator is expressed as the sum of two terms, the first of which corresponds to the pole term and the other to the remaining analytic part of the resolvent. We often encounter the situation where the modulus of the second term is $\ll 1$ for all $t \geq 0$. Then $\alpha \approx 1$ because v is a continuous function with $v(0) = 1$, and the decay law is approximately exponential with $\Gamma = 2\delta_p$.

The argument described in the example applies in more complicated situations too (see the notes), provided the reduced resolvent has a meromorphic continuation to the lower halfplane. The question of circumstances under which such a continuation exists is nontrivial and the affirmative answer is known only in some cases.

In the rest of this section we are going to illustrate this on one of the simplest examples, which is usually called the **Friedrichs model**. Let us first describe it. The state Hilbert space will be of the form

$$\mathcal{H} := \mathbb{C} \oplus L^2(\mathbb{R}^+),$$

where the one–dimensional subspace is identified with the state space $\mathcal{H}_u$ of the unstable system; the states are thus described by the pairs $\begin{pmatrix} \alpha \\ f \end{pmatrix}$, where α is a complex number and $f \in L^2(\mathbb{R}^+)$. The Hamiltonian is chosen as $H_g := H_0 + gV$, where g is the *coupling constant* and the free Hamiltonian H_0 is defined by

$$H_0 \begin{pmatrix} \alpha \\ f \end{pmatrix} := \begin{pmatrix} \lambda_0 \alpha \\ Qf \end{pmatrix},$$

where λ_0 is a positive parameter and $(Qf)(\xi) = \xi f(\xi)$; hence its continuous spectrum covers the positive real axis, $\sigma_c(H_0) = \mathbb{R}^+$, and the eigenvalue λ_0 is embedded in it. The interaction Hamiltonian is chosen as

$$V \begin{pmatrix} \alpha \\ f \end{pmatrix} := \begin{pmatrix} (v, f) \\ \alpha v \end{pmatrix},$$

where v is a given function from $L^2(\mathbb{R}^+)$; we can easily check that this operator satisfies the *Friedrichs condition* $E_d V E_d = 0$, where E_d is the projection to $\mathcal{H}_d := L^2(\mathbb{R}^+)$. The operator H_g is self–adjoint for any $g \in \mathbb{R}$ and its domain is independent of g (Problem 21a). The above condition makes the Friedrichs model explicitly solvable because it allows us to determine the reduced resolvent.

9.6.7 Proposition: Let $\operatorname{Im} z \neq 0$; then the operator $R^u_{H_g}(z)$ acts on $\mathcal{H}_u$ as the multiplication by

$$r^u_g(z) := \left(-z + \lambda_0 + g^2 \int_0^\infty \frac{|v(\lambda)|^2}{z - \lambda}\, d\lambda \right)^{-1}.$$

Proof: We apply the second resolvent formula to the operators H_g, H_0; it yields

$$E_u R_{H_g}(z) E_u = E_u R_{H_0}(z) E_u$$

$$- g E_u R_{H_0}(z) E_u V E_u R_{H_g}(z) E_u - g E_u R_{H_0}(z) E_u V E_d R_{H_g}(z) E_u,$$

where we have used the commutativity of operators E_u and $R_{H_0}(z)$ together with the identity $E_u + E_d = I$, and furthermore, the relation

$$E_d R_{H_g}(z) E_u = -g E_d R_{H_0}(z) E_d V E_u R_{H_g}(z) E_u,$$

where we have also employed the Friedrichs condition. Next we substitute from the second relation to the first, and apply to the resulting equation $(H_0 - z)E_u$ from the left. Since $(H_0 - z)E_u R_{H_0}(z) = E_u$, after a simple manipulation we get

$$\left[(H_0 - z)E_u + gE_uVE_u - g^2E_uVE_dR_{H_0}(z)E_dVE_u\right]\, E_uR_{H_g}(z)E_u \;=\; E_u\,.$$

Now, using the explicit form of the operators H_0, V, we obtain

$$\left(-z + \lambda_0 + g^2\int_0^\infty \frac{|v(\lambda)|^2}{z-\lambda}\,d\lambda \right)^{-1} r_g^u(z)\psi \;=\; \psi$$

for $\psi \in \mathcal{H}_u$, *i.e.*, the sought result. ∎

To be able to use the conclusions of Example 9.6.6, we have to know whether $r_g^u(\cdot)$ has an analytic continuation. The answer clearly depends on the properties of function v. We shall assume that

(a) there is an entire function $f : \mathbb{C} \to \mathbb{C}$ such that $|v(\lambda)|^2 = f(\lambda)$ for all $\lambda \in (0,\infty)$; for the sake of notational simplicity we write $f(z) = |v(z)|^2$ for nonreal z too.

Then the needed continuation exists.

9.6.8 Proposition: Assume (a); then $\sigma_c(H_g) = \mathbb{R}^+$ for any $g \in \mathbb{R}$, and $r_g^u(\cdot)$ has an analytic continuation from the upper halfplane, $r(z) = [-z + w(z,g)]^{-1}$, where

$$w(\lambda,g) \;:=\; \lambda_0 + g^2 I(\lambda) - \pi i g^2 |v(\lambda)|^2 \qquad \dots \quad \lambda > 0$$

$$w(z,g) \;:=\; \lambda_0 + g^2\int_0^\infty \frac{|v(\xi)|^2}{z-\xi}\,d\xi \;-\; 2\pi i g^2 |v(z)|^2 \qquad \dots \quad \mathrm{Im}\, z < 0$$

and $I(\lambda)$ is defined as the principal value of the integral,

$$I(\lambda) := \mathcal{P}\int_0^\infty \frac{|v(\xi)|^2}{\lambda-\xi}\,d\xi := \lim_{\varepsilon\to 0+}\left(\int_0^{\lambda-\varepsilon} + \int_{\lambda+\varepsilon}^\infty\right)\frac{|v(\xi)|^2}{\lambda-\xi}\,d\xi\,.$$

Proof: As a finite–rank operator, V is compact so $\sigma_{ess}(H_g) = \sigma_{ess}(H_0) = \mathbb{R}^+$ by Theorem 5.4.6. Furthermore, H_g has only simple eigenvalues of which at most one lies in $\mathbb{R}^-$ while the eigenvalues in $(0,\infty)$ can be only at the points where $v(\lambda) = 0$ (Problem 21). Since $|v(\cdot)|^2$ is analytic by assumption, the eigenvalues have no accumulation point, so $\sigma_c(H_g) = \sigma_{ess}(H_g) = \mathbb{R}^+$.

To prove the second assertion we have to check that the function $w(\cdot,g)$ defined as above for $\mathrm{Im}\, z \le 0$, and by

$$w(z,g) \;=\; \lambda_0 \;+\; g^2\int_0^\infty \frac{|v(\xi)|^2}{z-\xi}\,d\xi$$

for $\operatorname{Im} z > 0$, is analytic in $\mathbb{C}\backslash\mathbb{R}^-$. Its analyticity in the upper and lower complex halfplanes verifies easily, and in view of assumption (a), $w(\cdot, g)$ is continuous when crossing $(0,\infty)$ (Problem 22), and therefore uniformly continuous in any compact $M \subset \mathbb{C}\backslash\mathbb{R}^-$. Its analyticity then follows from the edge–of–the–wedge theorem (see the notes). ∎

These properties of the reduced resolvent make it possible to prove the meromorphic structure of its analytic continuation.

9.6.9 Theorem: Suppose that assumption (a) is valid and $v(\lambda_0) \neq 0$. Then there is an $\varepsilon > 0$ such that for all nonzero $g \in (-\varepsilon, \varepsilon)$ the function r of Proposition 9.6.8 has just one simple pole at the point $z_p(g) := \lambda_p(g) - i\delta_p(g)$. The function $z_p(\cdot)$ is infinitely differentiable and the expansion

$$\lambda_P(g) \;=\; \lambda_0 + g^2 I(\lambda_0) + \mathcal{O}(g^4)\,, \quad \delta_p(g) \;=\; \pi g^2 |v(\lambda_0)|^2 + \mathcal{O}(g^4)\,. \tag{9.13}$$

is valid in the vicinity of the point $g = 0$.

Proof: We define the function $F : \mathbb{C}\times(\mathbb{C}\backslash\mathbb{R}^-) \to \mathbb{C}$ by $F(g,z) := z - w(z,g)$; possible poles of $r(\cdot)$ obviously coincide with the zeros of F. In view of Proposition 8, the function F is infinitely differentiable in both variables. It also satisfies $F(0,\lambda_0) = 0$ and $(\partial F/\partial z)(0,\lambda_0) = 1$; thus we may use the implicit–function theorem, which implies the existence of a complex neighborhood U_1 of the point $g = 0$ and an analytic function $z_p : U_1 \to \mathbb{C}$ such that $F(g, z_p(g)) = 0$ for $g \in U_1$. Since F has continuous partial derivatives, $(\partial F/\partial z)(\cdot, z_p(\cdot))$ is also continuous, and therefore nonzero in some neighborhood $U_2 \subset U_1$ of $g = 0$. This means, in particular, that $r(\cdot)$ has at the point $z_p(g)$ a simple pole provided $g \in U_2 \cap \mathbb{R}$. Finally, the Taylor expansion (9.13) is obtained by computing the derivatives of the implicit function $z_p(\cdot)$ at $g = 0$ (Problem 23). ∎

Hence the Friedrichs model fits into the scheme described in Example 9.6.6 provided the coupling is weak enough. Under additional assumptions we can prove that the pole part of reduced resolvent $r(\cdot)$ dominates over the analytic remainder (see the notes); in that case the decay law of the initial state $\psi_u := \begin{pmatrix} 1 \\ 0 \end{pmatrix}$ is approximately exponential and

$$\Gamma(g) \;=\; 2\delta_p(g) \;=\; 2\pi g^2 |v(\lambda_0)|^2 \;+\; \mathcal{O}(g^4)$$

is the corresponding decay rate.

9.6.10 Remark: It is illustrative to compare this expression to the commonly used way of computing the decay rate by the so–called *Fermi golden rule*,

$$\Gamma_F(g) \;=\; 2\pi g^2\, \frac{d}{d\lambda} \left(V\psi_u, E_\lambda^{(0)} P_c(H_0) V\psi_u \right) \Big|_{\lambda=\lambda_0}\,,$$

where $\{E_\lambda^{(0)}\}$ is the spectral decomposition for H_0 and $P_c(H_0)$ the projection to the continuous subspace of this operator. To realize that $\Gamma_F(g)$ is indeed the formula known from quantum–mechanical textbooks, recall that we assume $\hbar = 1$ and formally $\frac{d}{d\lambda}E_\lambda^{(0)} P_c(H_0) = |\lambda\rangle\langle\lambda|$. Using the explicit form of the operators H_0 and V, we find

$$\Gamma_F(g) = 2\pi g^2 \frac{d}{d\lambda} \int_0^\infty |v(\xi)|^2 \, d\xi \bigg|_{\lambda=\lambda_0} = 2\pi g^2 \, |v(\lambda_0)|^2 ,$$

i.e., $\Gamma_F(g)$ is nothing else than the first nonzero term in the Taylor expansion of the function $\Gamma(\cdot)$.

The Friedrichs–model example illustrates a characteristic feature of many decay processes: the free Hamiltonian H_0 has an embedded eigenvalue, which disappears when the interaction is switched on, but leaves a footmark in the form of a pole in the analytically continued resolvent. Another problem of this type will be discussed in Section 15.4, and further references are given in the notes.

Notes to Chapter 9

Section 9.1 The detection of cosmic muons is discussed in the framework of the classical theory of relativity in ⟦ Vot ⟧, Sec.IV.4; for the quantum description of moving–particle decay see [BN 1], [Ex 3]; ⟦Ex⟧, Sec.3.5. The Wigner theorem that we have employed to motivate the postulate (Q4a) is proved, *e.g.*, in ⟦FG⟧, Sec.I.3; ⟦BaR⟧, Sec.13.2. Theorem 9.1.2 is particularly useful when we investigate the classical limit of quantum dynamics, as Corollary 9.1.4 illustrates. However, the Ehrenfest relation itself does not ensure existence of the correct classical limit because in general $\langle F_j(Q)\rangle \neq F_j(\langle Q_1\rangle, \dots, \langle Q_n\rangle)$; more on this subject can be found in [Hep 1]; ⟦Thi 3⟧, Sec.3.3.

Section 9.2 The unitary–valued function $V : V(t,s) = U_0(s-t)U(t,s)$ that appears in the definition of the interaction picture is *not* a unitary propagator; however, we can use it to define a unitary propagator relating the vectors ψ_D at different times (Problem 6). The nonexistence of the interaction picture in relativistic quantum field theories does not mean that the perturbative calculations used there are generally incorrect. Most of them can be justified; however, a presentation of the problem exceeds the scope of this book — we refer, *e.g.*, to ⟦GJ⟧.

Section 9.3 Theorem 1 claims that time evolution preserves smoothness of the initial condition. In fact a stronger result is valid, namely that possible singularities of the function $U(t,s)\psi$ are weaker than those of ψ if $t > s$, and the differentiability properties are improved by the propagator. This is true not only in the free case but also for a wide class of Hamiltonians with potential and magnetic–field interactions including time–dependent ones — see, *e.g.*, [Ya 1].

The treatment of the harmonic oscillator easily generalizes to the n–dimensional situation. In addition to the straightforward proof proposed as Problem 9, the propagator kernel of Theorem 9.3.6 can be derived formally by evaluating the corresponding Feynman integral — see ⟦FH⟧, Sec.3.6; ⟦Schu⟧, Chap.6; with an appropriate definition of the latter, such an argument can be made rigorous. It is also possible to avoid the use of the Feynman integral if the kernel is expressed by means of the Trotter formula — *cf.* [CRRS 1], and also ⟦ Ex ⟧, Sec.6.2. Other proofs of Theorem 9.3.6 are given in [Ito 2], [Tru 1]. In a similar way, the Feynman integral can be employed to derive explicit expressions of the propagator or resolvent kernel for many other simple systems — see, *e.g.*, [Cra 1], [CH 1], [DLSS 1], [CCH 1], [BCGH 1], [KL 1], [Gro 1]; the two last named reviews, as well as ⟦ Schu ⟧, contain many other references.

The last term in the exponential factor of the harmonic–oscillator kernel, which causes a jump in phase every halfperiod, is called the *Maslov correction.* It appears for other potentials too, typically at the instants when the classical particle passes a turning point — see [ETr 1]. The behavior of minimum–uncertainty states discussed in Example 5 was first pointed out by E. Schrödinger in [Schr 1]; the question whether a similar situation can occur for other potentials is discussed in [NSG 1].

Section 9.4 The expression of a propagator in form of the "path sum" was proposed by R. Feynman in 1942; he in turn was inspired by the earlier work of Dirac. He published his results in a series of papers starting from [Fey 1]; they are summarized in ⟦ FH ⟧. A mathematician's view of the Feynman papers can be found in [Jo 2]. Among other books dealing with physical applications of Feynman integrals we may mention, *e.g.*, ⟦ LRSS ⟧, ⟦ Pop ⟧, ⟦ Schu ⟧, and ⟦ SF ⟧ devoted to the quantum theory of gauge fields.

We mentioned in Section 9.3 that the classical limit of quantum mechanics can be interpreted as the limit $\hbar \to 0$. Had the right side of (9.6) been expressed as a standard integral, we might perform this limit by the stationary–phase method. A generalization of this procedure to one of the possible definitions of the Feynman integral can be found in [AH 1]; among other papers studying the classical limit and semiclassical approximation with the help of Feynman integrals let us mention, for instance, [DeW 1], [Tru 2].

The Wiener measure theory is explained, *e.g.*, in the monographs ⟦ Kuo ⟧, ⟦ Si 4 ⟧. A peculiar property of this measure is that smooth paths form only a small subset in the path space Γ_x. A function $\gamma \in \Gamma_x$ is said to be *Hölder continuous of order* α if there is a C such that $|\gamma(r) - \gamma(s)| < C|r-s|^\alpha$ holds for all $r, s \in [0, t]$. Denote by H_α the set of all $\gamma \in \Gamma_x$ with this property; then $w_\sigma(H_\alpha) = 1$ for $0 < \alpha < \frac{1}{2}$ while $w_\sigma(H_\alpha) = 0$ for $\frac{1}{2} < \alpha \le 1$. This means that a "typical" path is rather irregular; in particular, the set of all smooth paths has the Wiener measure zero.

A formula of type (9.8) was obtained for the first time in [Kac 1]; its proof under the stated assumptions can be found in ⟦ RS 2 ⟧, Sec.X.11. A few independent proofs which require stronger hypotheses about the potential but no restriction to dimension n is given in ⟦ Si 4 ⟧, see also ⟦ GJ ⟧, Chap.3; a mathematical generalization can be found in [Lap 1]. The Feynman–Kac formula is a powerful tool for analyzing Schrödinger operators — see, *e.g.*, ⟦ Si 4 ⟧ and [Si 3]; a generalization to systems with relativistic kinetic energy is given in [CMS 1]. The Feynman–Kac formula can also be derived for systems with an infinite number of degrees of freedom. This result plays an important role in the Euclidean approach to constructing models of interacting quantum fields — see, *e.g.*, ⟦ Si 2 ⟧, Chap.V; ⟦ GJ ⟧, Chaps.19, 20.

Theorem 9.5.3, which claims that Feynman integrals are not integrals in the usual sense, is taken from [Cam 1]. The proof of nonexistence of the measure w_σ for $\mathrm{Im}\,\sigma \neq 0$ is based on showing, with the help of a suitably chosen sequence of cylindrical functions, that such a measure should have an infinite variation — see also [Tru 3]; ⟦ Ex ⟧, Sec.5.1. This is due to the oscillatory behavior of the exponential term that formally determines the measure w_σ. On a heuristic level, the following conclusion can be formulated: if we want to give a reasonable meaning to expressions of the type $\int_{\Gamma_x} f(\gamma)\, e^{iS_0(\gamma)} D\gamma$, we have to restrict our attention to sufficiently smooth functions f to suppress the influence of oscillations. It is worth mentioning that the analogous functional integral for relativistic systems described by a Dirac operator has been shown in [Ich 1] to exist for two–dimensional space–time, while in the realistic case of dimension four the corresponding path measure again does not make sense — *cf.* [Zas 1].

There are many ways of defining Feynman integrals rigorously. We will briefly mention the most important of them; a detailed exposition with further references can be found in ⟦ACH⟧ and ⟦Ex⟧, where relations between various approaches are also discussed. The procedure used here to express the propagator by Trotter's formula can be adopted as a definition of the Feynman integral; this idea first appeared in [Nel 2]. However, it is closer to the spirit of Feynman's heuristic approach to use expressions of type (9.5), which contain the *exact* action along a given polygonal path. Such definitions have been formulated and their equivalence with a Trotter–type definition has been proved for various classes of potentials, *e.g.*, in [Cam 1–3], [Tru 1–4]. In [Ito 1,2], the expression $\int_{\Gamma_x} f(\gamma)\, e^{iS_0(\gamma)} D\gamma$ is defined as a limit of a net of integrals with respect to suitably chosen Gaussian measures on the path space.

There are also other definitions which do not require limits; they employ, *e.g.*, objects generalizing the notion of measure [DMN 1]; another approach works with generalizations of Fresnel integrals — see ⟦AH⟧, Chaps. 2 and 4. Finally, there are prescriptions using the Wiener measure theory, in which the sought object is defined by analytic continuation — see [Cam 2], [Nel 2], [JS 1], *etc.* — in mass or in time. It is shown in [Jo 1] that the "Fresnelian" definition can be extended in this way. One of the main aims of the quoted papers was to give meaning to the relation (9.6), and therefore most of them contain expressions of the propagator using the respective definition of the F–integral; pioneering work was done in [Cam 1], [Ito 1], [Nel 2]. Formulas of type (9.6) are also valid for complex potentials — see ⟦Ex⟧, Chap.6.

Section 9.5 The methods of solving the time–dependent Schrödinger equation discussed here are inspired by ways of solving the Cauchy problem for ordinary differential equations. In the case of Theorem 9.5.1 we use the fact that the original equation is formally equivalent to the Volterra equation

$$\psi_t = \phi - i \int_s^t H(\tau)\psi_t \, d\tau \,;$$

we have to check that the iteration series makes sense. The expansion is named after F. Dyson, who used it in his pioneering work on quantum electrodynamics (1949). Theorem 9.5.3 represents a particular case of a general existence result for evolution equations in Banach spaces, which can be found in ⟦RS 2⟧, Sec.X.12. The idea is again inspired by the theory of ordinary differential equations; it was first applied to vector–valued functions in [Ka 1]; see also ⟦Yo⟧, Sec.XIV.4. Similar results can be proved under weaker assumptions on the time–dependent part, so that, *e.g.*, Schrödinger operators with moving Coulomb potentials are included — see [Ya 1], and [KY 1] for the Dirac operator.

If $H(t) := H_0 + V(t)$, where H_0 is the free Hamiltonian (9.2) and $V(t)$ is a time–dependent potential, the relation (9.10) can be rewritten in a form analogous to (9.4), *i.e.*, as a Feynman integral where the action now depends parametrically on time — see, *e.g.*, [Far 2], [EK 1], ⟦Ex⟧, Sec.6.1. This fact has indeed been used to compute explicit expression for propagators of some simple systems — examples can be found, for instance, in [KL 1], Sec.9; [Du 1].

The result mentioned at the end of Example 9.5.5 is related to the *adiabatic theorem*, which can be briefly formulated as follows. An isolated eigenvalue $\lambda(t')$ of the Hamiltonian $H(t')$ is called regular if the corresponding projection $E(t')$ is finite–dimensional and the functions $E(\cdot)$ and $(H(\cdot)-\lambda(\cdot))^{-1}(I-E(\cdot))$ are strongly continuously differentiable. The theorem then says that the probability of transition from a regular state to any other (*i.e.*, orthogonal) state in a fixed time interval $[s,t]$ mediated by the propagator

$$U_\tau(t,s) := \mathrm{T}\exp\left(-i\int_s^t H\left(\frac{t'}{\tau}\right)dt'\right)$$

decays as $\mathcal{O}(\tau^{-1})$ for $\tau\to\infty$. The proof is again based on Dyson's expansion — see, *e.g.*, ⟦Thi 3⟧, Sec.3.3. This is by no means the strongest result which can proved; under not very restrictive assumptions on the spectrum of H_0 and the regularity of the time–dependent part $V(t)$, the transition probability can be shown to decay as a higher power of the adiabatic parameter τ or even exponentially — see, *e.g.*, [Nen 1,2], [JoyP 1,2]. As in the case of the classical limit, we can also ask about expansions of the considered quantities in terms of the parameter τ; then the term *adiabatic approximation* is usually employed. This concerns not only the transition probabilities but other properties of the system as well; for example, the behavior of the eigenvalues and eigenfunctions of H_0 under the influence of the perturbation $V(t)$ — *cf.* [Hag 2], [Joy 1], [HJ 1], [HJ 2]; another application can be found in [ASY 1].

The adiabatic approximation has the following important feature: if the Hamiltonian $H(t)$ depends on time through a time–dependent parameter which makes a loop in the parameter space, *i.e.*, it eventually returns to its original value *slowly enough*, then, during this process, the state acquires a phase factor, the so–called *Berry phase*, which can be observed in an interference experiment, when it is compared to the state which has evolved without this perturbation — *cf.* [Be 1]. Interesting effects occur when adiabatic and semiclassical approximations are combined — see [Be 2].

Propagators of time–dependent systems are often studied using the so–called *Howland method*, which replaces the original problem by investigation of the operator $-i\partial/\partial t + H(t)$ on the enlarged Hilbert space $L^2(\mathbb{R};\mathcal{H}) = L^2(\mathbb{R})\otimes\mathcal{H}$ — see [How 5] and also ⟦RS 2⟧, Sec.X.12. The dynamics of time–periodic systems is discussed, for instance, in [Ya 2], [YK 1], [EV 1]; Examples 9.6.6 and 9.5.7 are taken from the last named paper. In the nonresonant case the spectrum of $U(T,0)$ is known even to be pure point and dense on the unit circle — *cf.* [HLS 1]. Spectra of monodromy operators and their stability with respect to perturbations has been an object of interest recently — see, *e.g.*, [How 6] and [How 7] for the example of a driven anharmonic oscillator, [DuŠ 1] and [DuŠV 1] for an analysis of more general systems, and also the papers on quantum chaos mentioned in the notes to Sections 7.1 and 8.3.

Section 9.6 There are many mathematical problems related to the time evolution of unstable quantum systems. We shall briefly describe some of them; a detailed discussion with more references can be found in ⟦Ex⟧, Chaps.1–4. The investigation of general features of unstable systems based on assumption (u) is sometimes called the *quantum kinematics* of decay processes, because it concerns properties which are independent of the particular form of the total Hamiltonian — see [HM 1], [HLM 1], [Wil 2].

One of the problems formulated in these papers is the *inverse decay problem*: given a continuous operator–valued function $V:\mathbb{R}\to\mathcal{B}(\mathcal{H}_u)$ with $\|V(t)\|\le 1$ for all t, we look for a Hilbert space $\mathcal{H} \supset \mathcal{H}_u$ and a unitary propagator $U(t) = e^{-iHt}$ on $\mathcal{H}$ such that the corresponding reduced propagator coincides with V. A solution exists provided V is of the positive type (see the notes to Section 5.9). This condition is satisfied, in particular, if $\{V(t) : t \geq 0\}$ is an operator semigroup; more information on this subject can be found in ⟦Da 1⟧, Chap.7; ⟦Ex⟧, Sec.1.4. Stone's theorem generalizes to operator semigroups — *cf.* ⟦RS 2⟧, Sec.X.8; ⟦Da 2⟧ — allowing us to characterize a semigroup $\{V(t) : t \geq 0\}$ by means of its generator. In this way we can give a rigorous meaning to phenomenological non–selfadjoint Hamiltonians which appear frequently, *e.g.*, in nuclear physics — see [BEH 1]; ⟦Ex⟧, Chap.4.

Assumption (u) does not exclude the situation when $\mathcal{H}_u^\perp$ is invariant with respect to $U(t)$, or alternatively $\mathcal{H}_u$ is invariant for negative times — see Problem 17a. The non–decay probability $P_\psi(\cdot)$ is continuous by Proposition 1 but need not be smooth as a function of time as one can illustrate on a simple model [EFr 1]. The decay law for a mixed state W is defined by $P_W(t) := \mathrm{Tr}\,(V(t)^*V(t)W)$; its interpretation and properties are analogous to the special case $W = E_\psi$. The proof of Theorem 9.6.5 can be found in [Wil 2]; moreover, the hypothesis can be weakened: $\sigma(H) = \mathbb{R}$ holds provided $V(s)V(t) = V(s+t)$ for all $s \geq 0$ and $t \geq T$, where T is any positive number — see [Sin 1]. The semigroup condition means that the decayed state cannot regenerate; the violation of this property for short times is discussed in [MS 1], [Ex 5]. The approximation of a decaying state by bounded–energy states is investigated in [Ex 1]. The parameter ϵ in Example 4 characterizes the weak CP–violation in neutral–kaon decay.

We mentioned in Example 9.6.6 that the support of the operator–valued measure $F(\cdot)$ is contained in $\sigma(H)$. It is demonstrated in [Ex 4] that under the minimality condition $\mathcal{H} = \overline{\left(\bigcup_{t\in\mathbb{R}} U(t)\mathcal{H}_u\right)_{lin}}$ the opposite inclusion is also valid, so $\mathrm{supp}\, F = \sigma(H)$. The argument of Example 9.6.6 extends to $\dim \mathcal{H}_u > 1$; the approximation obtained by replacing the reduced resolvent by the pole term of the corresponding Laurent expansion is called the *pole approximation.* Justification of this approximation, *i.e.*, estimating the contribution of the neglected terms, represents a difficult problem for particular systems. For the Friedrichs model it was done in [Dem 1]; a related Galilean–invariant model of two–particle decay was analyzed in [DE 2].

The Friedrichs model was first formulated in [Fri 1]. It attracted interest when a similar Hamiltonian was used in the so–called *Lee model* of quantum field theory [Lee 1]; the decay problem in the Lee model has been discussed by [AMKG 1], [Hö 1], and many other authors. Recently the Friedrichs model with a time–periodic coupling constant has been proposed — see [HL 1].

To prove Proposition 9.6.8 we need the *edge–of–the–wedge theorem* which claims the following: suppose the functions F_j, $j = +,-$, are analytic in the regions Ω_j of the upper and lower complex halfplanes, respectively, and their common boundary contains an open interval $J \subset \mathbb{R}$. Assume further that $F_j(\xi) := \lim_{\eta\to 0+} F_j(\xi + ij\eta)$ exist uniformly in J, are continuous with respect to ξ, and $F_+(\xi) = F_-(\xi)$ for all $\xi \in J$. Then there is an analytic function $F : (\Omega_+ \cup \Omega_- \cup J) \to \mathbb{C}$ such that $F(z) = F_j(z)$ for $z \in \Omega_j$. For proof see, *e.g.*, ⟦SW⟧, Sec.2.5. The implicit–function theorem can be found, for example, in ⟦Schw 2⟧, Thms.III.28 and III.31.

In quantum–mechanical textbooks the Fermi golden rule is usually derived formally, using continuous–spectrum "eigenfunctions"; a critical discussion of this procedure is given in ⟦RS 4⟧, notes to Sec.XIII.6. As in the case of the Friedrichs model one can prove that the Fermi rule yields the leading–order contribution to the imaginary part of the pole position for many other models — see, *e.g.*, [Bau 1], [How 3], [Si 4]. The second of these papers presents an example showing that a formal use of the Fermi rule may lead to false conclusions.

The embedded–eigenvalue perturbation theory which applies to most decay models was first studied in [Fri 1]. If the interaction does not fulfil a Friedrichs–type condition, then the algebraic way of obtaining the reduced resolvent used in the proof of Proposition 9.6.7 fails. In that case the factorization technique, the idea of which goes back to [Ka 2], is often used: one assumes $V = B^*A$ for some operators A, B, which allows us to express the resolvent by means of the formula of Problem 1.63. Reduced resolvents for different types of perturbations have been studied in this way — see, *e.g.*, [How 1–4], [Bau 1], [BD 1], [BDW 1]; the difficult part is here usually to check that the results are independent of the chosen factorization.

Another method for the investigation of embedded-eigenvalue perturbations, which is particularly suitable for Schrödinger operators, uses so–called *complex scaling*, *i.e.*, an analytic continuation of unitary scaling transformations (corresponding to linear maps φ in Example 3.3.2) to complex values of the scaling parameter. Adopting certain hypotheses about the analyticity properties of the potentials, one can in this way turn the search for the continued–resolvent poles into the true *eigenvalue* problem of some non–selfadjoint operator — *cf.*, for example, [AC 1], [Si 4].

A typical example discussed in [Si 4], and also in ⟦ RS 4 ⟧, Sec.XII.6, is the *helium autoionization effect.* Its essence is as follows: if we switch off the Coulomb repulsion between the electrons in a helium atom, the Hamiltonian will have a continuous spectrum with embedded eigenvalues. The repulsion represents a perturbation which makes these eigenvalues "dissolve", turning them into resolvent poles. However, the atom remembers the disappeared eigenvalues; they are manifested, for instance, by the cross–section peaks that appear near these energy values if electrons are scattered on He^+ ions — *cf.* the notes to Section 14.4.

In addition to scaling, other families of substitution transformations have been continued to the complex region to enable treatment of other classes of potentials — *cf.* [BB 1], [Cy 1]; these include, in particular, translations — see, *e.g.*, [HH 1], [AF 1]. An extension of the complex scaling to Dirac operators can be found in [Šeb 2]. All these results need some analyticity hypotheses. Without them it no longer makes sense to identify the unstable states with poles of the continued resolvent; one can only prove (for a particular class of Schrödinger operators) that the embedded eigenvalues disappear in the continuous spectrum under the influence of the perturbation — *cf.* [AHS 1].

There are decay processes that are described by perturbation of *isolated* eigenvalues, which disappear in the continuous spectrum once the perturbation is turned on. This concerns, for instance, *decays by tunneling* through a potential barrier; a classical example is represented by the Gamow theory of α–radioactivity [Gam 1]. Unstable states of this type are usually called *shape resonances;* in a mathematical description of such decays one can take for the unperturbed system one with the barrier extended to an infinite width [How 2]; at the same time the embedded–eigenvalue perturbation theory can be used when the height of the barrier is being blown up — see, *e.g.*, [AsH 1], [AS 1], [ES 1]. The semiclassical approximation for shape resonances has been studied, for instance, in [CDS 1], [CDKS 1], and also in [Nak 1], where a complex distortion is used. A perturbation theory for unstable states coming from isolated–eigenvalue perturbations has been discussed in [Hun 1].

Another example of isolated–eigenvalue decay is the *Stark effect*, which represents a linear perturbation (caused typically by an electric field) to Schrödinger operators with a given potential, in particular, a Coulomb–type interaction — see, for instance, ⟦RS 4⟧, Sec.XII.5, and for later results [Her 1], [GG 1], [Nen 1], [Wa 1]. A related and rather difficult problem is the so-called *Wannier ladder*, *i.e.*, a linear perturbation to a periodic potential — see [Avr 1], [HH 1], [AF 1], [CoH 1], [BuD 1], [AEL 1], *etc.*

Problems

1. Let $\{U(t,s)\,:\,s,t\in\mathbb{R}\}$ be a family of operators each of which is either unitary or antiunitary. If conditions (i),(ii) of the definition of the propagator are fulfilled, then all $U(t,s)$ are unitary.

2. Prove: (a) A conservative system whose Hamiltonian has a purely continuous spectrum has no stationary states.

(b) In a system of free particles, functions of the momentum and spin components are integrals of motion.

(c) Integrals of motion need not commute mutually.

(d) Under which condition is any integral of motion a function of the Hamiltonian?

3. Let $\mathcal{H}_\pm$ be the subspaces in $L^2(\mathbb{R})$ consisting of the even and odd functions, respectively. Prove that each of the operators $P^2\upharpoonright\mathcal{H}_\pm$ has a simple spectrum.
Hint: Apply the results of Problem 5.43a to the basis vectors h_0, h_1, respectively.

4. Let $A_H(t)$ be an observable in the Heisenberg picture; then $f(A_H(t))=(f(A))_H(t)$ holds for any Borel function f and all $t\in\mathbb{R}$.

5. Let $H:=H_0+V$ with H_0 self–adjoint and V Hermitean. Prove (9.3) and the corresponding equation for mixed states.

6. Let U_1, U_2 be unitary propagators. If $U_3\,:\,U_3(t,s)=U_1(t,s)U_2(s,t)$ is a unitary propagator, then $U_1(t,s)$ and $U_2(r,s)$ commute for any $r,s,t\in\mathbb{R}$. On the other hand, $U_\tau\,:\,U_\tau(t,s)=U_3(t,\tau)U_3(s,\tau)^{-1}=U_1(\tau,t)U_2(t,s)U_1(s,\tau)$ is a unitary propagator for any fixed $\tau\in\mathbb{R}$.

7. Given a system of free particles,

(a) Verify Example 9.3.2..

(b) Prove that the results of Proposition 9.3.3 remain valid as long as the mean values make sense.

8. Consider again the free–particle system of Section 9.3 and an open bounded set $M\subset\mathbb{R}^n$ in the configuration space. For any $t>0$ and $x\notin M$, there is a state ψ such that $\mathrm{supp}\,\psi\subset M$ and $|\psi_t(x)|^2>0$ for $\psi_t:=U(t)\psi$.

9. Prove Theorem 9.3.6.
Hint: Check that that $U(t)$ maps $\mathcal{S}(\mathbb{R})$ onto itself and its kernel satisfies the equation $\left(i\frac{\partial}{\partial t}+\frac{1}{2m}\frac{\partial^2}{\partial x^2}-\frac{1}{2}m\omega^2x^2\right)K_t(x,y)=0$ for any $x\in\mathbb{R}$.

10. Verify the calculations in Example 9.3.8.

11. Fill in the details in the proof of Theorem 9.4.1.

12. Let potential V be differentiable with $\partial V/\partial x_j$ bounded; then there is a positive C_V such that S_N used in (9.4) differs from the action over the polygonal path $\gamma := \gamma(y^{(0)}, \dots, y^{(N)}; \cdot)$ by $|S_N(y^{(0)}, \dots, y^{(N)}; t) - S(\gamma)| \le C_V \|\gamma\|_\infty N^{-1}$.
Hint: Use the Taylor expansion at the vertices with the Lagrange form of the remainder.

13. Prove (9.5) for a linear potential, $V(x) := \sum_{j=1}^n \alpha_j x_j + \beta$.
Hint: Use the relation $S_N(y^{(0)}, \dots, y^{(N)}; t) = S(\gamma(\tilde{y}^{(0)}, \dots, \tilde{y}^{(N)}; t) + \mathcal{O}(N^{-1})$, where $\tilde{y}_j^{(k)} := y_j^{(k)} + \alpha_j k t^2/2m_j N^2$.

14. Let Γ be a real Hilbert space. A nontrivial measure μ on Γ such that

(i) $\mu(M) < \infty$ for any bounded Borel set $M \subset \Gamma$.

(ii) μ is translation–invariant.

exists *iff* $\dim \Gamma < \infty$.
Hint: The unit ball in an infinite–dimensional Γ contains infinitely many disjoint balls of radius 1/4.

15. Fill in the details in the proof of Theorem 9.5.1.

16. Prove: (a) The trajectories of the classical oscillator driven by a time–dependent force $f(t)$ are given by

$$\begin{aligned} x(t) &= x \cos \omega t + \frac{p}{m\omega} \sin \omega t - \frac{1}{m\omega} \beta(t,s)\,, \\ p(t) &= -m\omega x \sin \omega t + p \cos \omega t + \alpha(t,s)\,, \end{aligned}$$

where α, β are the functions defined in Example 9.5.6, provided the initial conditions are $x(s) = x$ and $p(s) = p$.

(b) In the same way as in Problem 9, check the expression of the propagator (9.11).

(c) Compute the functions α and β for $f(t) := \sin \Omega t$.

(d) Show that if $8\omega/\Omega$ is noninteger, one can choose q, p in such a way that the corresponding state (8.2) is an eigenvector of the monodromy operator corresponding to (9.11).

Hint: The action of $U(T,0)$ on $\psi_{q,p}$ can be written down explicitly using Examples 9.3.5 and 5.5.1b.

17. Using the notation introduced in Section 9.6,

(a) Find an example of Hilbert spaces $\mathcal{H}_u \subset \mathcal{H}$ and a unitary propagator U on $\mathcal{H}$ such that assumption (u) is valid and $\mathcal{H}_u^\perp$ is $U(t)$–invariant for all $t \ge 0$.

(b) Prove Proposition 9.6.1.

Hint: (b) The group property would mean that $V(t)$ is a partial isometry.

18. If ψ is a finite–energy state, the initial decay rate satisfies $\dot{P}_\psi(0+) = 0$.
Hint: Since $|(\psi, U(t)\psi)|^2 \le P_\psi(t) \le 1$ it is enough to differentiate the lower bound.

19. Find the decay laws

(a) For the truncated Breit–Wigner state ψ_b, where

$$d(\psi_b, E_\lambda^{(H)}\psi_b) = N_b\,\chi_{[-b,b]}(\lambda)\left((\lambda-\lambda_0)^2 + \frac{1}{4}\Gamma^2\right)^{-1} d\lambda\,,$$

and N_b is the normalization factor.

(b) For the neutral kaon states of Example 9.6.4.

20. Compute the function $v(t)$ of Example 9.6.6.

21. Let H_g be the Friedrichs–model Hamiltonian. Prove

(a) H_g is self–adjoint on $D(H_g) := \left\{\begin{pmatrix} \alpha \\ f \end{pmatrix} : \alpha \in \mathbb{C},\, f \in D(Q)\right\}$ for any $g \in \mathbb{R}$.

(b) The point spectrum of H_g is simple, and a positive λ can be an eigenvalue only if $v(\lambda) = 0$.

(c) H_g has at most one eigenvalue in the interval $(-\infty, 0]$, and this happens if $g^2 \ge \lambda_0 \left(\int_0^\infty \xi^{-1}|v(\xi)|^2\, d\xi\right)^{-1}$. If $v(0) \ne 0$, then a negative eigenvalue exists for any nonzero g.

Hint: (c) The function $\lambda \mapsto \int_0^\infty (\lambda-\xi)^{-1}|v(\xi)|^2 d\xi$ is monotonic in $(-\infty, 0)$.

22. Fill in the details in the proof of Proposition 9.6.8: check that the function $I(\cdot)$ is continuous in $(0,\infty)$ and $\lim_{[\xi,\eta]\to[\lambda,0+]} w(\xi \pm i\eta, g) = g^2 I(\lambda) \mp \pi i g^2 |v(\lambda)|^2$ is valid for any $\lambda > 0$.
Hint: The function $(|v(\cdot)|^2 - |v(\lambda)|^2)(\lambda - \cdot)^{-1}$ is bounded in the vicinity of $\xi = \lambda$.

23. Verify the Taylor expansion of Theorem 9.6.9.

Chapter 10

Symmetries of quantum systems

The problem we are going to discuss now, namely various symmetry transformations, is studied and used in almost all parts of physics; in some of them, *e.g.*, in the general theory of relativity, it is built into the very foundations of the theory. The importance of symmetry properties for quantum physics stems basically from two facts. First of all, they are related to the different ways in which a system is observed. We already know that quantum systems are generally affected by measurements, and therefore the state after a measurement also depends on the particular observer who performed it. What is even more important, the object on which the transformations are studied is the state Hilbert space, and as a consequence, quantum systems may exhibit additional symmetries which are not related to the underlying space–time structure.

10.1 Basic notions

Any quantum system can, at least in principle, be investigated by different experimentalists, and each of them eventually will convert results of the observation (with some assistance of a theoretician, maybe) to a description of the system along the lines discussed in the preceding chapters: a Hilbert space will be chosen to describe pure states, certain operators will be associated with observables, *etc.* A natural consistency requirement is that all this must allow a translation to the language of another individual or team who has observed the same object in such a way that the results are in mutual agreement. In particular, the probability of transition between any pair of states must be the same for the two observers.

There are numerous ways in which the two observations may differ, and indeed some of them are related rather to the language, *i.e.*, the system of notions used by the theory; recall different experimental setups used to measure the same observable which we mentioned in Section 7.1. Apart from that, there are others which form the matter of our interest here.

10.1.1 Examples: (a) The other apparatus has a different location, it is shifted, rotated, or both. It can also be operated at some other time than the first apparatus was. This amounts to using another coordinate system or a clock adjusted to a different initial instant.

(b) We also can employ different coordinate–axes numbering from that of another observer. If such a renumbering is not cyclic it cannot be achieved by rotating the second apparatus with respect to the first one; instead it corresponds to regarding the object of investigation through a mirror.

(c) Different labels can be used for particle charges, baryon numbers, and other "internal" properties.

(d) We can record a process and later scan its time evolution deliberately choosing the direction in which the tape is run through the projection machine.

These and similar relations between observers have to correspond to some relations between results of their measurements. For definiteness, consider a pair of observers O and O', both performing experiments on the same system S. If they both rely on the standard scheme of quantum theory, they will describe (pure) states of S by rays in a Hilbert space. Strictly speaking, each of them could choose his or her own Hilbert space for this purpose but this would cause no complications. We know from Section 7.1 that the dimension of the state space is in fact the number of independent states in which the system can exist; hence a choice of spaces of a different dimension would eventually lead to a discrepancy. Since all Hilbert spaces of the same dimension are mutually isomorphic, we may assume that the state space is the same for the two observers and call it as usual $\mathcal{H}$.

The ways in which the observers associate rays or statistical operators with states of the system are clearly determined by the sets of observables they use. The simplest possibility is to assume that both observers are endowed with same experimental equipment, *i.e.*, that they use the *same* operator to describe a particular observable (however, there are alternatives — see the notes).

If the system is in a given pure state, the first observer will describe it by a ray Ψ^O, while the second will associate with the same state a ray $\Psi^{O'}$. The above mentioned translation means that there is a bijective map $T_{O'O} : F_O \to F_{O'}$ between the sets of rays they ascribe to admissible states in their formalisms. Now we can repeat the argument used in the previous chapter for time evolution. The transition probabilities determined by the two observers have to be the same,

$$P(\Phi^{O'}, \Psi^{O'}) = P(\Phi^O, \Psi^O),$$

for any $\Phi^O, \Psi^O \in F_O$ and $\Phi^{O'}, \Psi^{O'} \in F_{O'}$. Then it follows from Wigner's theorem that the map $T_{O'O}$ between the rays can be realized by means of a unitary or antiunitary operator on $\mathcal{H}$; with the usual abuse of notation we shall employ the same symbol $T_{O'O}$ for it.

The observers can naturally ask about the validity of the superposition principle. The correspondence between their results means, in particular, that the superselection rules determined by them are the same; if the first observer will rule out the existence of a nontrivial superposition of a proton and neutron, the other must arrive at the same conclusion whatever names he or she uses for these particles. If the system is not coherent, the operators $T_{O'O}$ are not unique in general; however, this nonuniqueness is removed easily provided we make another natural assumption, namely that $T_{O'O}$ is reduced by all coherent subspaces of the system. This discussion may be summarized in the form of another postulate:

(Q5a) the vectors which observers O and O' use to describe the same state of the system are related by a unique unitary or antiunitary operator $T_{O'O}$, which is reduced by all coherent subspaces.

Let us remark that the similarity between these considerations and those of Section 9.1 is not accidental. In view of Example 10.1.1a, time evolution is nothing else than a particular case of the transformations discussed here.

10.1.2 Remarks (on the active approach): (a) The way in which we have treated the problem up to now is sometimes called the *passive approach* to the problem of symmetry transformations. Another possibility is to consider a single observer who performs experiments on a pair of systems S, S', the second of which being obtained by some transformation performed on an identical copy of S and *vice versa*. With this fact in mind, the observer will ascribe to them the same Hilbert space $\mathcal{H}$, but a particular ray will correspond in general to different states for the two systems. Conversely, if a state of S is described by a ray Ψ_S, another ray $\Psi_{S'}$ corresponds to the *same* state of the system S'. A straightforward modification of the above argument allows us to associate with the given pair S, S' a unique operator $T^{S'S}$ which is unitary or antiunitary, reduced by all coherent subspaces, and maps each Ψ_S on the respective $\Psi_{S'}$. Moreover, if S' is obtained from S by the *same* transformation that has related the observers O and O' in the passive approach, then

$$T^{S'S} = T_{O'O}^{-1} \qquad (10.1)$$

follows from the consistency requirement (Problem 1); this makes possible an easy translation between the results obtained in the two approaches.

(b) It is not difficult to imagine a transformed system obtained by changing its position in space or time. On the other hand, in some of the situations mentioned in Examples 10.1.1b–d the active approach cannot be applied because the transformed system does not exist — see Example 10.2.5 below. An attentive reader may object that the same applies to the observers: after all we almost surely have our heart on the left side, to say nothing about the unique orientation of biological time and the fact that we are composed of matter and not antimatter. A partial explanation why the passive approach can nevertheless be used in such situations has already been indicated in Examples 10.1.1, namely that the transformed observation can be realized by technical means like mirrors, recording devices, *etc.* There is a deeper reason, however. What we here briefly call an observation is in fact a complicated cognitive process, in which we are allowed to use any possible mental image of the physical object or process under investigation; the sole criterion for the success of the resulting theory is that its predictions are correct and free from contradictions.

(c) Both approaches are nothing but theoretical idealizations. If the same experiment is performed, say, in *CERN* and *Fermilab*, then two teams equipped with analogous devices each measure on their own proton beam; hence we have, loosely speaking, two observers and two systems. In view of relation (10.1), however, it is no problem to compare the results in such a more realistic setup.

Up to now we have considered pure states only. A generalization to the mixed–state case is straightforward. If the first observer concludes that the state of the system is a mixture of pure states $\Psi_j^{(O)}$ with some weights w_j, $j = 1, 2, \dots$, then the second has to describe it by the mixture of $\Psi_j^{(O')}$ with the *same* weights. Since the corresponding projections E_{ψ_j}, $E_{\psi'_j}$ are related by $E_{\psi'_j} = T_{O'O}E_{\psi_j}T_{O'O}^{-1}$, it is reasonable to extend the postulate (Q5a) as follows:

(Q5a) the statistical operators $W^{(O)}$ and $W^{(O')}$ which observers O and O', respectively, associate with the same state of the system are related by $W^{(O')} = T_{O'O}W^{(O)}T_{O'O}^{-1}$, where $T_{O'O}$ is a unique unitary or antiunitary operator which is reduced by all coherent subspaces.

Notice that since the relation contains the transformation operator together with its inverse, it maps linear statistical operators into linear operators again, even if $T_{O'O}$ is antiunitary.

In some situations only a pair of observers may be involved, for instance, when the transformation in question describes a mirror image or time reversal. However, more often larger families of transformations have to be considered. Then the operators relating different descriptions have to satisfy a natural consistency condition, because the maps $\Psi^{(O)} \mapsto \Psi^{(O')} \mapsto \Psi^{(O'')}$ and $\Psi^{(O)} \mapsto \Psi^{(O'')}$ must yield the same state for any ray $\Psi^{(O)}$ corresponding to a realizable state, *i.e.*, the corresponding operators may differ at most by a phase factor,

$$T_{O''O} = \omega T_{O''O'}T_{O'O}\,,$$

where $|\omega| = 1$. On the other hand, the number ω need not be the same for different triplets $\{O, O', O''\}$.

If we want to proceed further with the analysis of the operators $T_{O'O}$, we have to replace the rather vague notion of a transformation between the observers which we have used up to now by something more specific. It is useful to divide the family of all possible transformations into smaller classes within which different transformations can be compared. We shall assume that a particular class forms a *group*, or equivalently, that there is a group G to which the class can be bijectively mapped, each transformation $O \mapsto O'$ being associated with an element $g \in G$.

10.1.3 Remark: Since obviously any family of transformations is equipped with the identical and the inverse elements, the only problem concerns the composition (group multiplication). Notice that its existence is not automatic. For instance, in time evolution of a *nonconservative* system it generally makes no sense to compose the transformations described by the operators $U(t,r)$ and $U(r',s)$ unless $r=r'$. On the other hand, one can find numerous examples of transformation families which have a natural group structure.

We shall therefore consider a group G of transformations, for the present moment without a more detailed specification, and label the corresponding operators by the respective group elements, $T_{O'O} =: T(g)$. The above consistency requirement then reads

$$T(g')T(g) \;=\; \omega(g',g)T(g'g) \tag{10.2}$$

with $|\omega(g',g)| = 1$; we have now explicitly indicated the dependence of the phase factor on the group elements involved. Since the composition in G is associative, the operator $T(g''g'g)$ can be expressed in two different ways; this yields the condition

$$\omega(g'',g')\omega(g''g',g) = \omega(g',g)\omega(g'',g'g) \tag{10.3}$$

for any $g,g',g'' \in G$. Because of the presence of the factor $\omega(g',g)$ in (10.2), the map $T(\cdot)$ may not be a representation of G. The following more general concept is introduced: a map $T: G \to \mathcal{B}(\mathcal{H})$ is called a *projective representation* of the group G if there is $\omega: G\times G \to \{z\in\mathbb{C}: |z|=1\}$ such that conditions (10.2) and (10.3) are satisfied; the function ω is then the *multiplier* of T. Having introduced this notion, we can summarize the above discussion as follows:

(Q5b) the operators $T(g)$ corresponding to a given group G of transformations form a projective representation of it on the state space.

Let us remark that while in most cases of physical interest the multipliers are trivial, *i.e.*, they can be included into the operators $T(g)$ so that we have a true representation of G, there are situations where the phase factor not only cannot be avoided but also has physical consequences — see Remark 10.3.2 below.

With these prerequisites, we are now prepared to discuss symmetries, *i.e.*, transformations which do not affect observed properties of the system. Suppose that our two observers O and O', whose descriptions are related by the operator $T(g)$, measure an observable A, *i.e.*, they perform the set of *yes–no* experiments corresponding to the projections $E_A(\Delta)$ for all $\Delta\in\mathcal{B}$. They end up with the collections of probabilities $w(\Delta,A;W^{(O)})$ and $w(\Delta,A;W^{(O')})$. The observable is said to be *preserved* by the given group of transformations if the probabilities are the same,

$$w(\Delta,A;W^{(O)}) \;=\; w(\Delta,A;W^{(O')})$$

for any Borel $\Delta\in\mathbb{R}$ and all states of the system. Furthermore, we say that the system is **invariant** with respect to a group of transformations G (alternatively, that it *exhibits a symmetry* with respect to G or that G is a *symmetry group* of the system) if its total energy is preserved by the group.

Consider first the situation when the operator $T(g)$ is unitary. Postulate (Q5a) then yields for any vector ψ representing an admissible pure state the identity $(\psi, E_A(\Delta)\psi) = (\psi, T(g)^{-1}E_A(\Delta)T(g)\psi)$; mimicking the argument of Proposition 7.5.1, we find for any $\Delta \in \mathcal{B}$

$$E_A(\Delta) = T(g)^{-1}E_A(\Delta)T(g), \tag{10.4}$$

so by the spectral theorem the observable A commutes with $T(g)$ (see also Problem 2). In the antiunitary case we have

$$w(\Delta, A; \Psi^{(O')}) = (T(g)\psi, E_A(\Delta)T(g)\psi) = (T(g)^{-1}E_A(\Delta)T(g)\psi, \psi),$$

and since the operators $E_A(\Delta)$ are Hermitean, the right side has to be equal to $(E_A(\Delta)\psi, \psi)$ for any ψ belonging to some coherent subspace; hence we arrive at the same conclusion (Problem 3c). On the other hand, the commutativity of A and $T(g)$ easily implies $w(\Delta, A; W^{(O)}) = w(\Delta, A; W^{(O')})$ for all $\Delta \in \mathcal{B}$ and admissible $W^{(O)}$, so we get the following assertion (see also the notes).

10.1.4 Proposition: Suppose that G is a group of transformations of a quantum system and $T(\cdot)$ is the corresponding representation of G on the state space; then

(a) An observable A is preserved by G *iff* the operator A commutes with $T(g)$ for all $g \in G$. In particular, the system is invariant under G *iff* this is true for its Hamiltonian.

(b) If A is preserved by G, the restriction $T(\cdot) \upharpoonright E_A(\Delta)\mathcal{H}$ is a representation of G for any $\Delta \subset \mathbb{R}$; in particular, $T(\cdot)$ is reduced by each eigenspace of A.

In the present terminology, an observable A of a *conservative* system is therefore an integral of motion if it is preserved by the group of time translations, and any such system is naturally invariant with respect to that group.

10.2 Some examples

The general framework for treatment of transformations of quantum systems and their symmetries constructed up to now covers many different physical situations, as Examples 10.1.1 indicate. On a mathematical level, this diversity is manifested in two ways:

(i) Different types of transformation (symmetry) groups are involved. Some of them are *continuous*, most frequently *Lie groups*; others are *discrete*, often *finite*. With the help of standard group–theory notions, one can classify the transformation (symmetry) groups further, for instance, divide them into *commutative* and *noncommutative*, *etc.*

(ii) The same group G may have different nonequivalent representations, sometimes even a vast family of them, to which different transformation properties of the system correspond. To classify them, one has to find all irreducible representations of G and select those, which fit the description of the system under consideration in view of their dimension, spectra of the involved operators, *etc.*

Rather than attempting a systematic overview, we are going to discuss here several typical and physically important situations. Let us begin with the simplest example of the continuous transformation group, which describes the case when the observers are mutually shifted.

Examples 10.2.1: (a) *Translations on the line.* Consider first the spinless particle of Example 7.2.1 and suppose that the observers parametrize the line by shifted coordinates x and $x' := x+a$ for a fixed $a \in \mathbb{R}$. Since they are investigating the same particle, it is natural to assume that they will describe its state by square integrable functions ψ and ψ', respectively, which are related by $\psi'(x') = \psi(x)$, or equivalently

$$(T(a)\psi)(x) := \psi'(x) = \psi(x-a)$$

for any $x \in \mathbb{R}$. The transformation operators $T(a)$ are unitary and form a continuous group if the parameter a runs through the reals; we know from Example 5.5.1b that $\{T(-a) : a \in \mathbb{R}\}$ is generated by the momentum operator P (in the standard system of units, by $h^{-1}P$).

In view of Stone's theorem and the definition of commutativity, any function of the momentum is preserved at translations; in particular, this concerns the kinetic energy of the particle. On the other hand, mimicking the argument of Example 6.7.2e we find that a function of position Q does not commute with P unless it is trivial, *i.e.*, a multiple of the unit operator. Let the particle be described by the Hamiltonian

$$H = \frac{1}{2m} P^2 + V(Q) ;$$

as earlier we suppose that the right side is self-adjoint (or at least *e.s.a.*, so we can replace it by the closure of the operator; we postpone discussion of the conditions under which this is true to Section 14.1). We see that it is necessary and sufficient for the invariance with respect to translations that the potential V is constant, which means essentially that the particle is free, because the choice of the origin on the energy scale is a matter of convention in *nonrelativistic* quantum mechanics.

Let us stress, however, that this conclusion concerns the whole group T_1 of translations of the real line; it does not exclude the possibility of invariance with respect to a subgroup of it. Consider, *e.g.*, the situation where the function V is periodic, $V(x) = V(x+nb)$ for any integer n, $x \in \mathbb{R}$, and some

$b > 0$. Then $V(Q)$ obviously commutes with the operators $T(nb)$, which form a subgroup in T_1 that is naturally isomorphic to the additive group $\mathbb{Z}$ of integers, and the same is true for the Hamiltonian H. This yields at the same time an example of a symmetry group which is discrete but infinite.

(b) *n–dimensional translations.* The previous considerations easily extend to systems with n configuration–space degrees of freedom. The translation group T_n of $\mathbb{R}^n$ is an n–parameter commutative Lie group, represented on the state space $L^2(\mathbb{R}^n)$ by

$$(T(a)\psi)(x) = \psi(x-a) \tag{10.5}$$

for any $a := (a_1, \dots, a_n) \in \mathbb{R}^n$. Due to Example 8.2.3 the representation is unitary and the one–parameter subgroups $\{T(-ae^{(k)}) : a \in \mathbb{R}\}$ with $(e^{(k)})_j := \delta_{jk}$ are generated by the momentum component operators P_k, $k = 1, \dots, n$. Since $\{P_1, \dots, P_n\}$ is a CSCO, the only observables preserved by the translation group are functions of the momentum; in particular, a system of spinless particles described by the Hamiltonian (9.2) is translation–invariant *iff* it is free, *i.e.*, the potential is a constant function. As above, however, an interacting system can be invariant with respect to a subgroup of T_n provided the potential exhibits such a symmetry; in a multidimensional case the number of different subgroups is much larger. They can be continuous, discrete, or both. For instance, if $n = 2$ and $V(x_1, x_2) = f(x_2)$ where $f : \mathbb{R} \to \mathbb{R}$ is periodic, $f(x) = f(x+n)$, then the corresponding two–dimensional particle interacting with this potential is invariant with respect to the group $T_1 \times \mathbb{Z}$.

Before we proceed further, it is worth mentioning that there are important groups of transformations which can never play the role of a nontrivial symmetry. This happens, for instance, if we extend the group of the previous example to include all *phase–space* translations.

Example 10.2.2 (Heisenberg–Weyl group): The family of Weyl operators introduced in Section 8.2 forms a projective representation of the translation group T_{2n} (Problem 4a). Alternatively, these operators give rise to a *true* representation of the *Heisenberg–Weyl group* defined as the $(2n+1)$–parameter set $G = \{g(s,t,u) : s,t \in \mathbb{R}^n,\ u \in \mathbb{R}\}$ with the binary operation

$$g(s,t,u)g(s',t',u') := g\left(s+s', t+t', u+u' + \frac{1}{2}(t\cdot s' - s\cdot t')\right).$$

It is a $(2n+1)$–dimensional noncommutative Lie group. The corresponding Lie algebra (dubbed the *Heisenberg–Weyl algebra*) is spanned by the elements q_j, p_k, c, which satisfy the relations

$$[p_k, p_j] = [q_k, q_j] = [p_k, c] = [q_k, c] = 0\,, \quad [p_k, q_j] = -i\delta_{jk}c$$

for $j, k = 1, \dots, n$; notice that in distinction to canonical commutation relations it does have a finite–dimensional representation (Problem 4c).

In this way the Schrödinger representation of the CCR yields representations of two closely related transformation groups which are useful for many purposes; just recall the role Weyl operators play in the definition of coherent states. Nevertheless, the set $\{W(s,t) : s,t \in \mathbb{R}^n\}$ is irreducible in $L^2(\mathbb{R}^n)$ by Problem 8.14c, so an observable is preserved by the corresponding transformations only if it is trivial, *i.e.*, a scalar operator.

10.2.3 Examples: (a) *Rotations in the plane.* Suppose that the observers use coordinate frames which are rotated mutually by an angle θ, *i.e.*, they describe a point in the plane by the coordinates $[x_1,x_2]$ and $[x_1',x_2']$, where $x_1' := x_1 \cos\theta + x_2 \sin\theta$, $x_2' := x_1 \sin\theta - x_2 \cos\theta$. The most natural choice for the transformation operator is then

$$(T(\theta)\psi)(x_1,x_2) := \psi(x_1 \cos\theta - x_2 \sin\theta,\, x_1 \sin\theta + x_2 \cos\theta)\,.$$

By Problem 5.51, the family $\{T(\theta) : \theta \in \mathbb{R}\}$ forms a continuous unitary group, which, in addition, satisfies the relation $T(\theta) = T(\theta + 2\pi n)$ for any $n \in \mathbb{Z}$; its generator is the angular momentum operator $L_3 := \overline{Q_1P_2 - Q_2P_1}$, *i.e.*, $T(\theta) = e^{i\theta L_3}$. According to Example B.3.3a, the operators $T(\theta)$ form a representation of the group $SO(2) = T_1/\mathbb{Z}$.

(b) *Spatial rotations.* In contrast to Example 1b, an extension to the case of rotations in $\mathbb{R}^3$ (as well as to higher dimensions) is not straightforward, because the corresponding group $SO(3)$ is not commutative (Problem 5b). On the other hand, any rotation can be composed of rotations around the axes (Problem 5c) to which the results of the previous example apply; the operators representing the rotations can therefore be expressed, *e.g.*, as

$$T_o(\alpha,\beta,\gamma) = e^{i\alpha L_3} e^{i\beta L_2} e^{i\gamma L_3}\,,$$

where

$$L_j := \overline{\sum_{k,l} \epsilon_{jkl} Q_k P_l}\,, \quad j = 1,2,3\,, \tag{10.6}$$

with ϵ_{jkl} being the Levi–Civita symbol, are the generators of the corresponding one–parameter subgroups and α,β,γ are the parameters of the rotation (usually called *Euler's angles*).

The vector–valued observable $L = (L_1,L_2,L_3)$ is called the *angular momentum;* using the relation (7.5) we can write explicitly, for instance, the action of the operators L_j on elements of $\mathcal{S}(\mathbb{R}^3)$. It is also easy to see that the operators L_j are *e.s.a.* on $\mathcal{S}(\mathbb{R}^3)$ and $L_j\mathcal{S}(\mathbb{R}^3) \subset \mathcal{S}(\mathbb{R}^3)$ (*cf.* Problem 5.51); the last mentioned inclusion means that $\mathcal{S}(\mathbb{R}^3)$ belongs to the domain of the operator $L^2 := \sum_{j=1}^3 L_j^2$. In Section 11.5 below we shall demonstrate, using spherical coordinates, that L^2 is unitarily equivalent to $I \otimes \Lambda$, where the operator Λ is given by (11.17); due to Proposition 11.5.2 it has a pure point spectrum, $\sigma(\Lambda) = \{l(l+1) : l = 0,1,\dots\}$.

It is straightforward to check that the angular momentum components satisfy on $\mathcal{S}(\mathbb{R}^3)$ the commutation relations

$$[L_j, L_k] = i\epsilon_{jkl}L_l \tag{10.7}$$

(with the summation convention), which means that they form a representation of the Lie algebra $so(3)$. Hence the relation between the rotation group and its Lie algebra gives rise to a relation between their representations. Furthermore, both of them are reduced by eigenspaces of the operator L^2; this fact can be used to simplify the treatment of systems, which are invariant with respect to rotations — see Section 11.5 below.

Particles with a nonzero spin have multicomponent wave functions, and therefore operators representing rotations in this case have a "matrix" part as well.

Example 10.2.4 (spin rotations): As a kind of angular momentum, spin has again to satisfy the relations (10.7). We know that it is indeed the case; we have constructed the corresponding $(2s+1)$–dimensional irreducible representations (7.11) in Example 7.5.4b. It is therefore natural to associate with rotations operators in the spin state space $\mathbb{C}^{2s+1}$ defined by

$$T_s(\alpha,\beta,\gamma) = e^{i\alpha S_3}e^{i\beta S_2}e^{i\gamma S_3}.$$

They form a unitary representation of $SO(3)$ as in the preceding example. However, there is an important difference. To illustrate this, consider the rotation of the angle 2π around the third axis. In view of (7.11) and functional–calculus rules, $T(2\pi,0,0) = (-1)^{2s}I$, which differs from the unit operator if the spin s is half–integer; this conclusion extends to the rotation of 2π around any axis, because the corresponding operator is unitarily equivalent to $T(2\pi,0,0)$.

Hence if we observe a particle of a half–integer spin, for instance, an electron, then the coordinate system must be rotated *twice*, *i.e.*, by 4π, to arrive at the same spin state. This property manifests that the operators $T_s(\alpha,\beta,\gamma)$ in fact form a representation of the simply connected group $SU(2)$, which is related to the rotation group by $SO(3) = SU(2)/\mathbb{Z}_2$ — *cf.* Example B.3.3b and Problem 6.

If we return to the full state space $L^2(\mathbb{R}^3;\mathbb{C}^{2s+1})$ of a particle with spin s, the spin coordinate operators are, due to Example 7.5.4b, replaced by $\underline{S}_j := \overline{I_c \otimes S_j}$. On the other hand, the "configuration–space part" of the angular momentum (usually called *orbital* in this situation) is $\underline{L}_j := \overline{L_j \otimes I_s}$; it can be expressed in the form (10.6) by means of the corresponding position and momentum operators (Problem 7a). The *total* angular momentum of the particle, whose components are represented by the operators $J_j := \overline{\underline{L}_j + \underline{S}_j}$, then generates a representation $T(\cdot)$ of the rotation group in the same way its orbital and spin parts do; moreover,

$$T(g) = T_o(g) \otimes T_s(g)$$

holds for each element $g = (\alpha,\beta,\gamma)$ of the group. The representation $T(\cdot)$ is, in general, reducible; we return to this problem in Section 11.5.

The rotation group $SO(3)$ considered in the above examples is a subgroup in the group $O(3)$ of all transformations of the configuration space that preserve the vector length. The last named group has two connected components because $\det R = \pm 1$ holds for any orthogonal matrix R; due to this fact, any element of $O(3)$ can be composed of a rotation and a discrete mirror transformation.

Example 10.2.5 (space reflection): The transformations connecting observers, who use the mirrored coordinate systems, form the simplest nontrivial group consisting of two elements. Its representation on the space $L^2(\mathbb{R}^3)$ can be constructed by means of the reflection operator R of Example 3.3.2: the wave functions are related by $P := \eta R$, *i.e.*, $(P\psi)(x) = \eta\psi(-x)$, where η is a phase factor. Since the group property requires $P^2 = I$, there are just two nonequivalent representations of this type corresponding to $\eta = \pm 1$. It is easy to see that the position and momentum operators anticommute with P, *i.e.*,

$$P Q_j P = -Q_j\,, \qquad P P_j P = -P_j\,, \qquad j = 1,2,3\,; \tag{10.8}$$

this in turn implies that the angular momentum is preserved by the space reflection, $PL_jP = L_j$. If the particle has a nonzero spin, in agreement with experience we postulate the same relation for the spin–component operators, so the reflection is represented in this case by $\underline{P} := \overline{P \otimes I_s}$ and $\underline{P}\,\underline{S}_j\underline{P} = \underline{S}_j$; the analogous relation is then also valid for the total angular momentum.

The operator $\underline{P}$, however, is Hermitean at the same time, so it represents an observable. It is called the *parity;* it has a dichotomic character since the eigenvalues of $\underline{P}$ are $\pm\eta$ (*cf.* Problem 8). It should not be confused with the *internal parity,* which is the number η specifying the type of reflection–group representation. The latter is also an observable, but of the type mentioned in Example 7.5.4b: with each of the known elementary particles we can associate just one value of the parameter η. The same is true for composite systems which we shall discuss in the next chapter: the parity operator is then of the form $\eta R \otimes I_{int}$, where R means the space reflection on the configuration–space part of the state space, I_{int} is the unit operator corresponding to the internal degrees of freedom and internal parity η is the product of the internal parities of the constituents.

It is clear from (10.8) that simple systems such as a free particle are invariant under space reflections, so by functional–calculus rules, the parity is an integral of motion. In general, this is not true for operators (9.2) unless the potential V is an even function. This is true, fortunately, for isolated nonrelativistic many–particle systems, where the interaction Hamiltonian typically depends on the distances of particles only. On the other hand, there are isolated systems which are not invariant with respect to the mirror transformation. A famous example is the radioactive nucleus Co^{60}, which emits electrons with a smaller probability in the direction of the nucleus spin than in the opposite direction (see the notes).

10.2.6 Remark: We have mentioned already that elementary particles are grouped into charge multiplets such as a nucleon, a pion, *etc.* The corresponding Hilbert spaces are finite–dimensional and support representations of $SU(2)$ and of the reflection group analogous to those discussed in the preceding two examples. In particular, the operators I_j, $j = 1,2,3$, that generate the representation of $SU(2)$ are called (components of) the *isotopic spin* (or *isospin*). They can again be expressed in the form (7.11); in a similar way we introduce the *isotopic parity*. However, there is an important physical difference between the two cases. The isospin algebra contains the charge operator which equals

$$Q = I_3 + \frac{1}{2}(B+S),$$

where the numbers B, S are the *baryon number* and the *strangeness*, respectively, characterizing the type of particles involved. Since Q defines a superselection rule, only some isospin transformations lead to physically admissible states.

Another important example of a finite group of transformations is obtained if we consider permutations in a system of N particles; this gives rise to a symmetry group if some of the involved particles are identical. We postpone the discussion of this case to Section 11.4.

Example 10.2.7 (time reversal): As in the previous example, transformations connecting observers, who use different directions of time (see Example 10.1.1d) form the two–element group. However, the operator T representing the time inversion is *antiunitary*. To justify this choice, first consider a system which is invariant with respect to switching of the time direction. By Proposition 10.1.4 its Hamiltonian commutes with T, and therefore the corresponding propagator satisfies the relation $TU(t)T = U(-t)$ for all $t \in \mathbb{R}$ (*cf.* Problem 3b) as it should be the case (with a unitary T, the last relation would require the two operators to anticommute, $TH = -HT$, so the sign of the energy will be changed in the transformed system). However, since the operator T should be the same for all systems with the same state space, independently of a particular Hamiltonian, it is also reasonable to assume the antiunitarity in the general case.

In the simplest case of a spinless particle we define the time–reversal operator as $T := K$, where K is the complex conjugation, $(K\psi)(x) = \overline{\psi(x)}$. In general, one may add a phase factor η to the right side (the time parity; in contrast to the space reflections the group condition imposes no restriction on its value due to the antilinearity of T), but in quantum mechanics, where the numbers of elementary particles are preserved, there is no need for it. The basic observables transform as

$$TQ_jT = Q_j\,, \quad TP_jT = -P_j\,, \quad TL_jT = -L_j\,, \quad j = 1,2,3 \tag{10.9}$$

(Problem 9a). For a particle with spin s we put $T := K \otimes V$, where V is a unitary operator on the spin space $\mathbb{C}^{2s+1}$. The relations (10.9) are then obviously valid for the observables $\underline{Q}_j$, $\underline{P}_j$, and $\underline{L}_j$. It is natural to assume that the spin component operators have the same transformation properties, $T\underline{S}_jT = -\underline{S}_j$, $j = 1,2,3$. It appears that this condition determines the operator V uniquely up to an overall phase factor: using the same basis as in (7.11) we have

$$V_{mm'} = \delta_{m,-m'}\, e^{-i\pi m}$$

(Problem 9b). In particular, the time–reversal operator for a particle with spin $\frac{1}{2}$ is $T = K \otimes \sigma_2$, where σ_2 is the Pauli matrix.

To conclude this brief survey, let us add a few comments and observations.

10.2.8 Remarks: (a) The representations of the continuous transformation groups considered above are typically generated by operators associated with some distinguished physical observables: translations are related to momentum components, rotations to the angular momentum, *etc.*These correspondences have the same fundamental importance as the transformation behavior expressed by (Q4b). At the same time, important observables can also be related to some discrete transformation groups as Example 10.2.5 shows; however, there is no general rule which would tell us which groups have this property and what the meaning of the corresponding observables is.

(b) Suppose that an observable A is associated with a transformation group G, either as a generator of its representation or as a representative of a group element itself. If the system is invariant with respect to G, then its Hamiltonian H commutes with the operators $T(g)$, and therefore also with A, so the latter is an integral of motion. This correspondence between symmetries of a system and the existence of conserved quantities is sometimes referred to as *Noether's theorem*; it is valid not only in quantum theory.

(c) The symmetries considered in the above examples have a rather straightforward interpretation. It may happen, however, that the system is invariant under an additional group which is less obvious (see, *e.g.*, Problem 10 and the notes); then we speak about a *hidden symmetry.*

(d) Transformation properties can be expressed in terms of other groups than discrete or Lie groups. As an example, consider a particle in a time–independent electromagnetic field (Problem 11). The Hamiltonians $H(A)$ and $H(A')$ associated with vector potentials A and $A' := A - \nabla\Lambda$ are unitarily equivalent by means of $U_\Lambda := e^{ie\Lambda(Q)}$. The family of these operators forms an Abelian group whose elements are parametrized by functions Λ, and therefore it is infinite–dimensional as a vector space.

(e) The state space can support two representations of the *same* group, both of them physically meaningful — *cf.* Example 10.2.1b and Problem 12. On the other hand, the fact that there is a representation of a group G on the state space, which commutes with the Hamiltonian, does not imply that G is a symmetry of the corresponding system; the point is that G need not be a transformation group. As a simple example, consider a family of linearly independent functions $f_j : \mathbb{R}^3 \to \mathbb{R}$, $j = 1, \dots, n$, and denote $f_j(P) := f_j(P_1, P_2, P_3)$. The operators $\exp\left(\sum_{j=1}^n ia_j f_j(P)\right)$ generate a representation of the Abelian group T_n on $L^2(\mathbb{R}^3)$ which commutes with the free–particle Hamiltonian; however, they can be given a reasonable physical interpretation (invariance with respect to translations) only if f_j are linear functions.

An inspection of the above examples shows that a distinguishing feature of physically relevant space–time transformations is that the corresponding group is represented by substitution operators (*cf.* Example 3.3.2), *i.e.*, that if the group relates points x, x', then the value of the wave function at x is determined only by the value of the transformed wavefunction at x'; this is why such operators are often called *local.*

10.3 General space–time transformations

The translations, rotations, and space reflections considered in the examples of the preceding section together with time translations may be included in a wider group of space–time transformations. However, this group is not unique; its choice depends on whether the system under consideration is nonrelativistic or relativistic. We are going now to review briefly properties of the space–time transformation groups in these two cases; for simplicity we restrict ourselves to continuous transformations, *i.e.*, we shall not speak about space and time reflections.

Consider first the nonrelativistic case, where the general continuous transformation between two inertial frames is given by

$$x' = Rx + vt + a\,, \qquad t' = t + b\,, \tag{10.10}$$

where $R \in SO(3)$ corresponds to rotations, vectors $a,\, v \in \mathbb{R}^3$ to translations and boosts, respectively, and $b \in \mathbb{R}$ is the time shift. Together these transformations form a ten–parameter group whose elements $g := (b, a, v, R)$ satisfy the following composition law

$$g'g = (b + b',\, R'a + a' + v'b,\, R'v + v',\, R'R)\,; \tag{10.11}$$

we call it the **Galilei group** and denote it as $\mathcal{G}$. We also introduce the *Euclidean group* $\mathcal{E}$ as the six–parameter subgroup of $\mathcal{G}$ including the translations and rotations. Using the results of Examples 10.2.1, 10.2.2, and 10.2.4 of the previous section we see that for a particle with spin s the group $\mathcal{E}$ is represented by the operators $T(R, a)$ on $L^2(\mathbb{R}^3; \mathbb{C}^{2s+1})$ defined by

$$(T(R,a)\psi)_m(x) := \sum_{m'=-s}^{s} S_{m,m'}(R)\,\psi(R^{-1}(x-a))\,,$$

where $S(R)$ is the matrix representing the spin–rotation operator $T_s(\alpha,\beta,\gamma)$ for $R \equiv R(\alpha,\beta,\gamma)$. By Example 10.2.1b and Proposition 5.7.3b, the free–particle Hamiltonian $H := \frac{1}{2m}P^2$ is invariant under translations and spin rotations; in combination with the results of Section 11.5 below this shows that it is invariant with respect to the whole group $\mathcal{E}$.

10.3.1 Proposition: Wave functions of a free particle of mass m and spin s transform under the Galilei group by the operators

$$U(g) := e^{imv\cdot(vb-2x-a)/2}\,T(R, a-vb)\,e^{iHb}\,, \tag{10.12}$$

which form a projective (unitary, strongly continuous) representation of $\mathcal{G}$ with the multiplier $\omega(g', g) := \exp\left(\frac{i}{2}m(v'\cdot R'a - a'\cdot R'v - R'v\cdot v'b)\right)$.

Proof: As a product of unitary operators, $U(g)$ is unitary, and in a similar way we obtain the strong continuity of the map $U(\cdot)$. Finally, the relation $U(g')U(g) = \omega(g', g)U(g'g)$ is checked by a straightforward computation (Problem 13). ∎

Note the sign in the exponent of e^{iHb} in (10.12), which is due to the passive approach we use. In the active approach we obtain by (10.1) the standard evolution operator which shifts the system on b in time.

10.3.2 Remark: There is an important consequence of the fact that Galilei–group transformations are realized on the state space by means of a projective representation. The reason is that one can find elements g_a, $g_v \in \mathcal{G}$ such that the composition $g_v^{-1}g_a^{-1}g_vg_a$ equals the unit element but the product of the corresponding operators differs from I by a nontrivial phase factor (Problem 15). The same is true for representations describing transformation properties of many–particle systems. In Section 11.5 below we shall see that these are given by $U(g) := \bigotimes_{j=1}^{N} U^{(j)}(g)$, where $U^{(j)}(\cdot)$ are representations referring to the particles of which the system consists; the phase factor of Problem 15 is in this case replaced by $e^{iMa\cdot v}$, where $M := \sum_{j=1}^{N} m_j$ is the total mass of the system.

Suppose now that a state of such a system is a nontrivial superposition of states Ψ_m, $\Psi_{m'}$ with masses m and m', respectively, *i.e.*, that it is described by $\alpha\psi_m + \beta\psi_{m'}$ for some ψ_m, $\psi_{m'} \in \mathcal{H}$ and nonzero α, β. The argument sketched above then shows that $\alpha\, e^{ima\cdot v}\psi_m + \beta\, e^{im'a\cdot v}\psi_{m'}$ has to describe the same state for all $a, v \in \mathbb{R}^3$; however, this is impossible unless $m = m'$. It follows that in *nonrelativistic* quantum theory states with different masses belong to different coherent subspaces; this fact is usually referred to as the *Bargmann superselection rule*.

Let us now pass to the relativistic case. Following the usual convention, we denote the space–time points as $x = (x_0, \vec{x})$; the components will be numbered by the Greek indices $\mu, \nu = 0, 1, 2, 3$ while the Latin indices, $j, k = 1, 2, 3$, are reserved for space coordinates. Space–time transformations are now described by another ten–parameter group defined by

$$x' = \Lambda x + a\,, \tag{10.13}$$

which is called the **Poincaré group** and is denoted as $\mathcal{P}$. Here the a are four–dimensional vectors representing space–time translations, and the Λ are real matrices, which preserve the indefinite bilinear form $(x, y) \mapsto xy := x_0y_0 - \vec{x}\cdot\vec{y}$. The corresponding six–parameter subgroup $\mathcal{L}$ is called the **Lorentz group**; since the form can be expressed through $xy = x^\mu y_\mu = g_{\mu\nu}x^\mu y^\nu$ with $g_{\mu\nu} := \mathrm{diag}(1, -1, -1, -1)$, the connected component of the unit element in $\mathcal{L}$, which is characterized by the conditions $\det\Lambda = 1$ and $\Lambda_{00} > 0$ (see the notes), is also denoted as $SO(3,1)$. The groups $\mathcal{P}$ and $\mathcal{G}$ are, of course, different but the Euclidean group $\mathcal{E}$ is their common subgroup.

The Poincaré group has many representations. Let us describe a class of them which is physically the most important. The easiest way to do this is to express the four–momentum components as multiplication operators. Given $m > 0$, we take $H_m := \{p \in \mathbb{R}^4 : p_0 > 0,\ p^2 = p_0^2 - \vec p^{\,2} = m^2\}$, *i.e.*, the "mass shell" in the corresponding Minkowski space. It is homeomorphic to $\mathbb{R}^3$ by means of the map $j_m : p \mapsto \vec p$; this makes it possible to equip H_m with a topology and to define on it the Borel measure ω_m by

$$\omega_m(M) := \int_{j_m(M)} \frac{d\vec p}{\sqrt{m^2 + \vec p^{\,2}}} \,.$$

For any $s = 0,\ \frac{1}{2},\ 1,\ \frac{3}{2},\ \dots$ we denote $\mathcal{H}_{ms} := L^2(H_m, d\omega_m; \mathbb{C}^{2s+1})$; then the following assertion is valid (see the notes).

10.3.3 Proposition: For any $m > 0$ and $s = 0,\ \frac{1}{2},\ 1,\ \frac{3}{2},\ \dots$, there is just one irreducible (unitary, strongly continuous) representation of the (proper orthochronous) Poincaré group on $\mathcal{H}_{ms}$ which is of the form

$$(U(\Lambda, a)\psi)(p) = e^{ipa}(S(\Lambda)\psi)(\Lambda^{-1}p)\,, \tag{10.14}$$

where $S(\cdot)$ is an irreducible representation on the Lorentz group on $\mathbb{C}^{2s+1}$.

10.3.4 Remark: The explicit form of the representation S is known but we shall not need it; we refer to the literature quoted in the notes. In particular, for the subgroup of rotations in $\mathcal{L}$ it reduces to the operators of spin rotations discussed in Example 10.2.4; this means again that a fermion wave function changes sign when rotated on the angle 2π.

Now we want to know how the wave functions of relativistic systems transform under the Poincaré group. Due to the postulate (Q5b), the state Hilbert space of such a system supports a representation of $\mathcal{P}$. However, if we have in mind a particular system, for instance, an elementary particle, our situation is substantially different from the nonrelativistic case. There we have first chosen the state Hilbert spaces and operators representing observables for such simple systems, and then we looked for the ways in which they transform. Inspired by the foregoing considerations, we are not going to repeat the reasoning. Instead, we reverse the argument and postulate that

(Q5c) the state Hilbert space $\mathcal{H}$ of a relativistic elementary particle of mass m and spin s supports a representation of the Poincaré group which is unitarily equivalent to (10.14). The operators on $\mathcal{H}$ representing the energy, momentum, and angular momentum coincide with the generators of the subgroups of time translations, space translations, and space rotations, respectively, in $\mathcal{P}$.

Notes to Chapter 10

Section 10.1 In fact, there are theories in which a map analogous to $T_{OO'}$ may not exist; recall again general relativity, where there are space–time manifolds (Schwarzschild horizons), which prevent any possible communication between the observers who had the bad luck to exist on different sides of the barrier. In the existing quantum theory, however, this cannot happen. The argument showing that $T_{OO'}$ is represented by a unitary or antiunitary operator should be regarded as the analogous considerations of Sections 7.1 and 9.1; our aim is to motivate a functioning scheme, which would relate the results of a pair of observers, and not to prove that it is the only possible.

The assumption that the observers use the same operator to describe a particular observable resembles the Schrödinger picture of Section 9.2. An alternative is to keep the states fixed and to suppose that when the first observer describes an observable by an operator A, the other uses the operator $T(g)^{-1}AT(g)$ for the same observable. The conclusions of Section 9.2 extend to the present situation; the Heisenberg–type formulation is usually preferred in situations where a large number of observables is involved, as in quantum field theory. Needless to say, it is also possible to divide the transformation–induced change between the states and observables whenever it is suitable.

Reduction of representations of symmetry groups is an efficient tool for investigation of quantum systems. Suppose that an observable A is preserved by a transformation group G. If the corresponding representation $T(\cdot)$ is reduced by a projection E_1, Proposition 10.2.4 imposes additional restrictions on the operator A. For simplicity, we denote $E_2 := I - E_1$, and furthermore, $T_j(g) := T(g) \upharpoonright \operatorname{Ran} E_j$, $A_{jk} := E_j A E_k$; then in view of the relation $T(g)A \subset AT(g)$ not only $T_j(g)$ commutes with A_{jj}, but also the "off-diagonal" parts have to satisfy the intertwining relations $T_j(g)A_{jk} \subset A_{jk}T_k(g)$ for all $g \in G$.

Section 10.2 Basic facts about Lie groups and algebras are summarized in Appendix B.3; for further reading we recommend, *e.g.*, ⟦BaR⟧, ⟦FG⟧, ⟦Ham⟧, ⟦Kir⟧, ⟦Mac 2⟧, ⟦Nai 3⟧and ⟦Wig⟧. In these books the reader can also find more information about finite groups, which appear here in a few simple examples only, as well as about other applications of the theory of groups and their representations to description of symmetries of quantum as well as classical systems.

More details about the HW–group and HW–algebra can be found, *e.g.*, in [Wol 1]. By Problem 4a, any (unitary, strongly continuous) representation of the Weyl relations generates a projective representation of the translation group T_{2n} with $\omega(s,t;s',t') = \frac{1}{2}(t\cdot s' - s\cdot t')$. Properties of this multiplier can be used to formulate an alternative proof of Theorem 8.2.4 — see [Si 1], Sec.7. Angular momentum, introduced in Example 10.2.3, is one of the fundamental observables. We shall return to it in Section 11.5; more information can be found, *e.g.*, in ⟦ Ja ⟧, Sec.13.3; ⟦ Thi 4 ⟧, Sec.3.2; or in special monographs such as ⟦ Edm ⟧, ⟦ BL ⟧.

The operator R is sometimes also called parity if the value of η is not important. The *parity violation* in weak interactions responsible for β–decays of atomic nuclei mentioned in Example 10.2.5 was predicted by T.D. Lee and C.N. Yang in 1956 and confirmed experimentally by C.S. Wu — see, *e.g.*, ⟦Schwe⟧, Sec.10f; ⟦FG⟧, Chap.3. The essence of the effect is that the momentum component operators change sign in the mirrored system while the spin components do not; hence the assumption of space–reflection invariance would require the momentum distribution of the emitted electron to be symmetric with respect to the plane perpendicular to the nucleus spin.

The isospin group $SU(2)$ extends to more general transformations between internal degrees of freedom of elementary particles. In the early sixties M. Gell-Mann and G. Zweig embedded it into the group $SU(3)$, whose lowest–dimensional representations were connected with a triplet of *quarks* considered as building blocks of elementary particles; this idea led to the currently accepted theory of strong interactions called quantum chromodynamics. A decade later the discovery of J/Ψ and Υ particles demonstrated the existence of two more quarks. In the meantime S. Weinberg and A. Salam constructed a unified theory of weak and electromagnetic interactions by combining the isospin $SU(2)$ group with the Abelian group $U(1)$ of gauge transformations (the latter should not be confused with the infinite–dimensional group of Remark 10.2.8d, which acts in the configuration space). This so–called *standard model* represents one of the most powerful applications of symmetry ideas in quantum physics; the reader can find more details, *e.g.*, in ⟦Hua 2⟧, ⟦CL⟧. However, some ingredients of the standard model have still to be confirmed experimentally, in particular, the existence of the sixth quark.

If the Hamiltonian of a fermion is time–reversal invariant, then by Problem 9c no eigenvalue of it can be simple. The same is true for any system containing an odd number of fermions; we speak about *Kramers' degeneracy*. For most quantum–mechanical systems the symmetry with respect to the time reflection is checked easily — see, *e.g.*, Problem 9d. On the other hand, a violation of T–invariance (the latter is equivalent to the so–called CP–invariance, *i.e.*, the space reflection accompanied by changing signs of the particle charges — *cf.* ⟦SW⟧) have been observed for weakly interacting elementary particles such as neutral kaons mentioned in Example 9.6.4.

A well–known hidden symmetry exists in the hydrogen–atom Hamiltonian — see, *e.g.*, ⟦BaR⟧, Sec.12.1. Hidden symmetries are usually manifested by the fact that eigenspaces of the Hamiltonian which, due to Proposition 10.1.4b, support a representation of the apparent symmetry group, have in fact a larger dimension than corresponds to an appropriate irreducible representation; we also speak about an *accidental degeneracy.*

Section 10.3 A thorough discussion of properties of the Galilei group and its representations can be found in the review [LeL 1]. The reader should pay attention to the fact that in general free–particle observables are *not* preserved by the operators representing elements of $\mathcal{G}$ in the sense of the definition given in Section 10.1 (*cf.* Problem 14). Instead, they transform naturally, *i.e.*, in the same way as the corresponding classical observables. We say that the quantum mechanics of a free particle — and other theories with this property — are *covariant* with respect to $\mathcal{G}$. However, with an abuse of terminology one sometimes speaks about a G–invariant theory having in mind its covariance with respect to the transformation group G.

The fact that the mass represents a superselection rule in nonrelativistic quantum mechanics was first noticed by V. Bargmann — *cf.* [Ba 3], and also [LeL 1]; ⟦BaR⟧, Sec.13.4. We mentioned it in the notes to Section 7.4 as an example of a "continuous" superselection rule; however, one has to take into account that practically every quantum system consists of particles with a finite number of different masses, so the rule is in fact "discrete" after all. While in most quantum mechanical problems mass conservation is self–evident, sometimes this is not the case; for instance, in nonrelativistic decay models; then the Bargmann rule has to kept in mind — see, *e.g.*, [DE 2], Part I.

There are textbooks devoted solely to the special theory of relativity — see, *e.g.*,

⟦ Vot ⟧; on the other hand, information about the Lorentz and Poincaré groups can be derived from some books on group theory such as ⟦ BaR ⟧, Chaps.17 and 21, and almost every book on quantum field theory — *cf.* the notes to Section 7.1.

As we have mentioned above, the "full" Lorentz group also contains the discrete transformations of space and time reflection. The connected component $SO(3,1)$ of the unit element which we consider here consists of $\Lambda \in \mathcal{L}$ with $\det \Lambda = 1$ and $\Lambda_{00} > 0$ (compare to Example B.3.2b). It is called the *proper orthochronous Lorentz group* and is denoted as $\mathcal{L}_+^\uparrow$; the same conventions apply to the Poincaré group. However, if we are not interested in the reflections, we usually drop the adjectives, subscripts, and superscripts.

Some Lie algebras are related by a limiting procedure called *contraction.* In this way, in particular, the Poincaré group passes to Galilei when the velocity of light (which, for simplicity, we have put equal to one here) approaches infinity — see ⟦ BaR ⟧, Sec.1.8.

An explicit expression of the representation $S(\cdot)$ appearing in (10.14) can be found in many places — see, *e.g.*, ⟦ SW ⟧, Secs.1–3; ⟦ BLT ⟧, Sec.2.3. To construct these and other representations of $\mathcal{P}$, one usually replaces $SO(3,1)$ by a simply connected group in analogy with Example 10.2.4; this time it is the group $SL(2,\mathbb{C})$. It appears, however, that the sought representations of $\mathcal{P}$ are determined uniquely by a representation of the subgroup $SU(2) \subset SL(2,\mathbb{C})$; the remaining operators are obtained by means of the group composition law. This is the so–called *induced–representation method* which is described and used for classification of irreducible representations for a wide family of groups, *e.g.*, in ⟦ BaR ⟧, Chap.17. The classification for the Poincaré group appeared for the first time in [Wig 1]. Let us also remark that irreducible representations of the full Poincaré group including reflections are classified by one more discrete index (parity).

Representations of $\mathcal{P}$ associated with relativistic systems other than stable elementary particles are no longer irreducible. For systems consisting of a finite number of elementary particles such representations are easily constructed by tensor product — *cf.* Section 11.5 below — and in a similar way one can proceed for more complicated systems in second–quantization formalism. On the other hand, representations associated with unstable particles are typically direct integrals of the representations (10.14) — see [BN 1], [Ex 3] or ⟦ Ex ⟧, Sec.3.5, and also Problem 16.

Problems

1. Justify the relation (10.1).

2. Using (10.2), we can write the relation (10.4) alternatively in the form $E_A(\Delta) = \omega(g^{-1},g)^{-1}T(g^{-1})E_A(\Delta)T(g)$. Is it possible in general to dispose of the multiplier?

3. Let A be a linear operator, in general unbounded. An antilinear bounded operator T is said to *commute* with A if $TA \subset AT$. Prove:

(a) If A is self–adjoint, an assertion analogous to Theorem 5.3.1b is valid.

(b) The operator T commutes with A *iff* $T\,e^{iAt} = e^{-iAt}T$ for all $t \in \mathbb{R}$, and this is in turn equivalent to the relation $T\,R_A(z) = R_A(\bar{z})T$ for all $z \in \rho(A)$.

(c) If, in addition, T is antiunitary, then A, T commute *iff* $A = T^{-1}AT = TAT^{-1}$.

4. Prove: (a) The operators (8.9) or (8.10) form a projective representation of the translation group T_{2n}.

(b) The Heisenberg–Weyl group of Example 10.2.2 satisfies the group axioms with $e := g(0,0,0)$ and $g(s,t,u)^{-1} := g(-s,-t,-u)$.

(c) The HW–group can be represented by the matrices

$$G(s,t,u) := \begin{pmatrix} 1 & s+it & s+it \\ -s+it & 1+2iu-\frac{1}{2}(s^2+t^2) & 2iu-\frac{1}{2}(s^2+t^2) \\ s-it & -2iu+\frac{1}{2}(s^2+t^2) & 1-2iu+\frac{1}{2}(s^2+t^2) \end{pmatrix}.$$

The matrices Q_j, P_k, C defined by $Q_j := -i(\partial G/\partial s_j)_{g=e}$, *etc.*, form a representation of the HW–algebra.

(d) Why does this fact not violate Theorem 8.2.4?

Hint: (a) Use Problem 8.15.

5. Prove: (a) The relation $x' = Rx$ defines a bijective correspondence between rotations of the Euclidean space $\mathbb{R}^3$ and elements of the group $SO(3)$, *i.e.*, 3×3 orthogonal matrices R, $R^tR = I$, with $\det R = 1$. Extend this result to rotations of $\mathbb{R}^n$.

(b) Find the subgroups of $SO(3)$ corresponding to rotations around the axes. Let $R_j(\theta)$ be the rotation around the j–th axis on the angle θ; compute $R_j(\theta)R_k(\theta')$ for $j \neq k$.

(c) Let $R(\alpha,\beta,\gamma) := R_3(\alpha)R_2(\beta)R_3(\gamma)$. Check that any rotation is of this form for some α,β,γ.

6. Prove: (a) Let $U \in SU(2)$ and $\xi \in \mathbb{R}^3$; then there is a matrix $R \in SO(3)$ such that $U\,\xi{\cdot}\sigma\,U^{-1} = \sum_{j,k} R_{jk}\xi_j\sigma_k$, where σ_k are the Pauli matrices and $\xi{\cdot}\sigma := \sum_k \xi_k\sigma_k$.

(b) The map $SU(2) \to SO(3)$ defined in this way is a homomorphism which is not injective; on the other hand, $\{U,-U\} \mapsto R$ is a bijection.

7. Using the notation of Example 10.2.4b, prove

(a) The orbital–momentum components equal $\underline{L}_j = \overline{\sum_{k,l} \epsilon_{jkl}\underline{Q}_k\underline{P}_l}$, where $\underline{Q}_k$, $\underline{P}_l$ are the operators of Example 7.5.4b.

(b) The total angular–momentum operators $J_j := \overline{\underline{L}_j + \underline{S}_j}$ generate a unitary representation of the rotation group on $L^2(\mathbb{R}^3,\mathbb{C}^{2s+1})$ by

$$T(\alpha,\beta,\gamma) := e^{i\alpha J_3}e^{i\beta J_2}e^{i\gamma J_3}$$

and $T(g) = T_o(g) \otimes T_s(g)$ holds for any $g = (\alpha,\beta,\gamma)$.

8. Let $P := \eta R$ be the parity operator on $L^2(\mathbb{R}^3)$. Prove that

(a) P is Hermitean; write the corresponding spectral decomposition.

(b) P is reduced by the eigenspaces of the operator $\sum_{j=1}^{3} L_j^2$; find its parts in these subspaces.

(c) Extend the conclusions to the parity operator on $L^2(\mathbb{R}^3) \otimes \mathbb{C}^{2s+1}$.

Hint: (b) Use Proposition 11.5.2.

9. Consider the time–reversal operator T of Example 10.2.7.

(a) Prove the relations (10.9).

(b) If the spin component operators (7.11) satisfy the conditions $T\underline{S}_jT = \underline{S}_j$, $j = 1,2,3$, with $T := K \otimes V$, then $V_{mm'} = \delta_{m,-m'}\, e^{i(\alpha-\pi m)}$ for some $\alpha \in \mathbb{R}$.

(c) $T^2 = (-1)^{2s} I$ where s is the spin of the particle. It follows that if s is half–integer and the Hamiltonian is time–reversal invariant, any eigenvalue of it has at least multiplicity two.

(d) If the operator (9.2) with a real–valued potential V is *e.s.a.*, then the corresponding Hamiltonian is time–reversal invariant.

Hint: (b) The assumption implies $V\bar{S}_jV^{-1} = -S_j$, where $\bar{S}_j$ is the complex conjugated matrix of S_j. (c) Apply T^2 to an eigenvector.

10. Let H be the Hamiltonian of the three–dimensional harmonic oscillator on $L^2(\mathbb{R}^3)$, *i.e.*, $H := P^2 + Q^2$ with $P^2 := \sum_{j=1}^{3} P_j^2$ and $Q^2 := \sum_{j=1}^{3} Q_j^2$. Define the operators $T_{jk} := a_j^* a_k + \frac{1}{2}\delta_{jk}$, where $a_j := 2^{-1/2}(Q_j + iP_j)$, and prove

(a) $H = \sum_{j=1}^{3} T_{jj}$ and $[H, T_{jk}]\psi = 0$ for any $\psi \in \mathcal{S}(\mathbb{R}^3)$ and $j,k = 1,2,3$.

(b) The operators T_{jk} form a representation of the Lie algebra $u(3)$, *i.e.*, they satisfy on $\mathcal{S}(\mathbb{R}^3)$ the relations

$$[T_{ij}, T_{kl}] = \delta_{jk}T_{il} - \delta_{il}T_{jk}\,, \quad i,j,k,l = 1,2,3\,,$$

and $\tilde{T}_{jk} := T_{jk} - \frac{1}{3}\delta_{jk}H$ form a representation of $su(3)$.

(c) The Lie algebra of the rotation group is the subalgebra in $su(3)$ whose representation is generated by $L_1 + iL_2 := \tilde{T}_{12}$, its adjoint, and $L_3 := \tilde{T}_{11} - \tilde{T}_{22}$.

11. Suppose that the functions V, A_k satisfy the assumptions of Theorem 14.1.8 and $\Lambda : \mathbb{R}^3 \to \mathbb{R}$ is a function whose first and second partial derivatives are continuous and bounded. Let $H(A) := \frac{1}{2m}(P - eA(Q))^2 + V(Q)$, and in a similar way $H(A')$ corresponds to the gauge–transformed vector potential, $A'_k := A_k - \partial_k\Lambda$. Prove that $H(A') = U_\Lambda H U_\Lambda^{-1}$, where $U_\Lambda := e^{ie\Lambda(Q)}$.

12. Let $\mathcal{G}_d := \{x \mapsto e^s x \,:\, s \in \mathbb{R}\}$ be the dilation group of the space $\mathbb{R}^n$.

(a) Check that the operators $U_d(s) \,:\, (U_d(s)\psi)(x) = e^{ns/2}\psi(e^s x)$ form a continuous unitary representation of $\mathcal{G}_d$ on $L^2(\mathbb{R}^n)$, and find the corresponding representation of the Lie algebra of $\mathcal{G}_d$.

(b) How do the fundamental observables transform with respect to $\mathcal{G}_d$?

(c) Show that in the case $n = 3$ the representation is reduced by the eigenspaces of the operator L^2, and find the parts of $U_d(s)$ in these subspaces. Extend these conclusions to a general $n \geq 2$.

Hint: (a) *Cf.* Example 5.9.5b. (c) Use Remark 11.5.4.

13. With the notation of Theorem 10.3.1, prove

(a) $\left(e^{iHb'}e^{imv\cdot(vb+2x-a)/2}e^{-iHb'}\psi\right)(x) = e^{imv\cdot(v(b+b')+2x-a)/2}\psi(x+vb')$ holds for all $g \in \mathcal{G}$, $b \in \mathbb{R}$, and $\psi \in L^2(\mathbb{R}^3, \mathbb{C}^{2s+1})$.

(b) Use this result to check that the relation $U(g')U(g) = \omega(g',g)U(g'g)$ is valid for any $g, g' \in \mathcal{G}$, where $\omega(g',g) := \exp\left(\frac{i}{2}m(v'\cdot R'a - a'\cdot R'v - R'v\cdot v'b)\right)$.

(c) Check that the condition (10.3) is valid for the multiplier ω.

Hint: (a) Use (7.4) and Example 5.5.1b.

14. Find the transformation properties of the free–particle Hamiltonian $H := \frac{1}{2m}P^2$ under Galilean boost, *i.e.*, the operators $H(v) := U_v H U_v^{-1}$ where $U_v := U(0,0,v,I)$. How do the component operators of position, momentum, angular momentum and spin transform under $\mathcal{G}$?

15. Consider the Galilei–group transformations $g_a := (0,a,0,I)$ and $g_v := (0,0,v,I)$. Check that $g_v^{-1}g_a^{-1}g_v g_a = e$ and $U(g_v^{-1})U(g_a^{-1})U(g_v)U(g_a)\psi = e^{ima\cdot v}\psi$ for any $a, v \in \mathbb{R}^3$ and $\psi \in L^2(\mathbb{R}^3, \mathbb{C}^{2s+1})$.

16. In the setting of Section 9.6, let $U(\cdot)$ be a representation of the Poincaré group on $\mathcal{H}$ and $V : V(\Lambda, a) = U(\Lambda, a) \upharpoonright \mathcal{H}_u$ its restriction to the state space of the unstable system (by Proposition 9.6.1, V is not a representation of $\mathcal{P}$). Show that if the unstable system is preserved by Euclidean transformations (in particular, the projection E_u commutes with the translation operators), then the operators $V(\Lambda, 0)$ associated with Lorentz boosts cannot be unitary.
Hint: Use the identity $U(I,\Lambda a)U(\Lambda^{-1},0)U(I,-a)U(\Lambda,0) = U(I,\Lambda a - \Lambda^{-1}a)$, and choose $\Lambda := \Lambda(\vec{\beta})$ as a boost with a velocity $\vec{\beta}$ and $a := (0,\vec{a})$ where $\vec{a}$ is parallel to $\vec{\beta}$.

Chapter 11

Composite systems

Most information about microscopic physical systems comes from investigation of their mutual interactions and the ways they manifest themselves in bound states, scattering processes, *etc.* A treatment of such situations requires knowledge of how a quantum system composed of a certain number of subsystems can be described. This is the topic of the present chapter.

11.1 States and observables

Suppose that a quantum system S with the state Hilbert space $\mathcal{H}$ consists of a finite number of distinguishable entities $S_1, \dots, S_n$, which we call *subsystems*. Examples are plentiful; recall the hydrogen atom, consisting of a proton and an electron. We could attempt to express in a more formal way what a subsystem should be, but we refrain from doing so limiting ourselves to the assumption that each S_j is by itself a quantum system endowed with its own state Hilbert space $\mathcal{H}_j$. The system S will then be said to be *composed* of $S_1, \dots, S_n$.

Though a practical reduction of a system to its subsystems cannot continue *ad infinitum*, the requirement of finiteness is certainly restrictive, because it is often useful in the theory to consider an infinite number of components. This would lead, however, to additional mathematical problems, so for the present moment we accept this limitation, and postpone the discussion of the infinite case to the next chapter. Moreover, we shall speak mostly about the simplest case of a system consisting of a pair of subsystems, because the extension to any finite number is straightforward.

We have already several times encountered the system of n spinless particles whose state Hilbert space is $\mathcal{H} = L^2(\mathbb{R}^{3n})$; each particle then represents a subsystem with the state space $\mathcal{H}_j = L^2(\mathbb{R}^3)$. Due to Example 2.4.5, $\mathcal{H}$ can be identified in this case with the tensor product of one–particle state spaces; motivated by this fact we adopt the following general assumption:

(Q6a) the state Hilbert space of the composite system is the tensor product of the subsystem state spaces, $\mathcal{H} = \mathcal{H}_1 \otimes \cdots \otimes \mathcal{H}_n$.

It is necessary to make a reservation from the beginning: in the present form the postulate is suitable only if S consists of *mutually different* subsystems. A modification for the situation when some of them are identical will be given in Section 11.4 below.

11.1.1 Remarks: (a) The state Hilbert space can often be written in a tensor product form even if no real physical systems can be associated with the component spaces. The results we shall derive therefore also apply to systems composed of such "fictitious" subsystems; recall, for example, the spinless particle consisting of three "one–dimensional" particles or the electron composed of the "configuration–space" and "spin" electron from Examples 7.2.3 and 7.5.4b. These decompositions are related to the concept of the *number of degrees of freedom;* by this the number of the simplest fictitious systems into which a given system can be split in usually understood.

(b) We mentioned in Section 7.1 that the state Hilbert spaces of quantum mechanical systems are usually supposed to be separable. Now we can formulate an argument in support of this hypothesis. A typical quantum mechanical system consists of a finite number of particles, and each of them can have some internal degrees of freedom such as spin, isospin, *etc.*, whose number is also finite. The total state space is therefore the tensor product of a finite number of Hilbert spaces. Among these, those corresponding to the configuration–space degrees of freedom are isomorphic to $L^2(\mathbb{R})$ by the Stone–von Neumann theorem, and are thus separable. On the other hand, spaces associated with known internal degrees of freedom are even finite–dimensional, *i.e.*, separable again; in view of Proposition 2.4.4b the total state space is then separable.

Let us now look at the relations between the observables of a composite system S and its subsystems S_1, S_2. We shall use the notation introduced in Section 7.4: the symbols $\mathcal{O}$ and $\mathcal{O}_j$ denote the sets of observables of the respective systems, and in a similar way we shall denote the sets of bounded observables and algebras of observables. Consider first the observables which are related to one system only, say, to S_1. With any such observable A we associate a self–adjoint operator A_1 on $\mathcal{H}_1$, and at the same time, an operator A on $\mathcal{H}$ if we regard it as an observable of the whole system S. The values we obtained as results of the measurement are not, at least in principle, influenced by the presence of the other system. It is natural therefore to assume that $A = \underline{A}_1$ in which case $\sigma(A) = \sigma(A_1)$ by Problem 5.38b. Let us stress that, in contrast, the probability of finding the value of A *is* influenced because it depends on the state of the composite system; we shall discuss this in more detail in the next section.

11.1.2 Example: Consider the system of two (different) spinless particles with the state space $L^2(\mathbb{R}^6) = L^2(\mathbb{R}^3) \otimes L^2(\mathbb{R}^3)$ and denote by $Q_k^{(j)}, P_k^{(j)}$ the operators of Cartesian components of the position and momentum of the j–th particle, respectively. Following the above assumption, we have to define $Q_l := \overline{Q_l^{(1)} \otimes I}$ for

$l = 1,2,3$ and $Q_l := \overline{I \otimes Q^{(2)}_{l-3}}$ for $l = 4,5,6$, and the operators P_l in an analogous way. It is clear from Example 7.2.3 that $Q_1 = \overline{Q \otimes I \otimes \cdots \otimes I}$, *etc.*; this means that the choice of the position and momentum operators for a system of particles is consistent.

Observables of the mentioned type do not, of course, exhaust the family $\mathcal{O}$ which also contains $\overline{\underline{A}_1 + \underline{A}_2}$ for $A_j \in \mathcal{O}_j$, other operator polynomials of these operators, *etc.* They nevertheless play an important role.

11.1.3 Proposition: (a) The algebras of observables of the composite system and its subsystems are related by $\mathcal{A} \supset \mathcal{A}_1 \otimes \mathcal{A}_2$.

(b) Suppose that the state spaces $\mathcal{H}_j$ are separable. If $\mathcal{S}_j \subset \mathcal{O}_j$ are complete sets of compatible observables for the subsystems $\mathcal{S}_j$, then $\mathcal{S} := \{\underline{A}_j :\, A_j \in \mathcal{S}_j\}$ is a CSCO for the system S.

Proof: Since the algebras of observables in question are generated by $\underline{\mathcal{O}}_{b1} \cup \underline{\mathcal{O}}_{b2} \subset \mathcal{O}_b$, assertion (a) follows from Theorem 6.6.5; in a similar way, part (b) is implied by Corollary 6.6.6. ∎

Assertion (b) illustrates, in particular, how the state Hilbert space is associated with a given quantum system. If we have a set of compatible observables $A_1, \dots, A_n$ and we believe that it is complete, we choose a space $\mathcal{H}$ and self–adjoint operators $A_1 \dots, A_n$ with the needed spectra, algebraic relations, *etc.*, so that they form a CSCO on $\mathcal{H}$; naturally we do not start from zero because the founding fathers of quantum mechanics have spared us most of this effort. It may happen that in the course of time a new observable B is found, which is not a function of $A_1, \dots, A_n$ but is still compatible with them. We are then forced to find self–adjoint operators $\tilde{A}_1, \dots, \tilde{A}_n, \tilde{B}$ which form a CSCO on a Hilbert space $\tilde{\mathcal{H}}$; at the same time we have to demand that the properties of $A_1 \dots, A_n$ are preserved, in particular, $\sigma(\tilde{A}_j) = \sigma(A_j)$ for $j = 1, \dots, n$. This problem is usually solved by setting $\tilde{\mathcal{H}} := \mathcal{H} \otimes \mathcal{H}_b$ for some Hilbert space $\mathcal{H}_b$ and $\tilde{A}_j := \overline{A_j \otimes I_b}$. The observable B is then represented by $\tilde{B} := \overline{I \otimes B}$, where the operator B on $\mathcal{H}_b$ is chosen in such a way that it has a simple spectrum; then Theorem 5.8.6 in combination with Proposition 11.1.3b ensures that $\tilde{A}_1, \dots, \tilde{A}_n, \tilde{B}$ is a CSCO. A classical example is the addition of the (third component of) spin to the particle coordinate or momenta proposed by H. Uhlenbeck and S. Goudsmith to explain the Zeeman-effect controversy about the number of atomic levels in a magnetic field.

We want to know, of course, under which circumstances the inclusion of Proposition 11.1.3a turns into identity. In some cases, the following simple coherence result provides the answer.

11.1.4 Proposition: Suppose that $\mathcal{S}_j \subset \mathcal{O}_j\,,\, j = 1,2$, are irreducible sets; then the set $\mathcal{S} := \{,\underline{A}_j :\, A_j \in \mathcal{S}_j\}$ is irreducible on $\mathcal{H}_1 \otimes \mathcal{H}_2$ and the system S is coherent.

Proof: The first assertion follows from Theorem 6.7.3; the set $\mathcal{O}$ is then also irreducible, so coherence is implied by Theorem 7.4.4b. ∎

11.1.5 Examples: (a) Since a spinless particle is a coherent system, by Proposition 4 the same is true for any system of (mutually different) spinless particles — *cf.* Example 7.4.5b. The result remains to be valid even if some of the particles are identical, as we shall see in Section 11.4.

(b) By Problem 7.15, the spin component operators $S_j,\ j=1,2,3$, of Example 7.5.4b form an irreducible set on $\mathbb{C}^{2s+1}$; hence $\{\underline{Q}_j, \underline{P}_j, \underline{S}_j : j=1,2,3\}$ is irreducible in the state space $L^2(\mathbb{R}^3)\otimes\mathbb{C}^{2s+1}$, so a particle with a nonzero spin is coherent as well. Combining this with the previous example, we find that a system of particles with arbitrary spins is coherent.

In all these examples we have $\mathcal{A}_j=\mathcal{B}(\mathcal{H}_j)$, and therefore $\mathcal{A}_1\otimes\mathcal{A}_2=\mathcal{B}(\mathcal{H}_1\otimes\mathcal{H}_2)$. This ceases to be true if we add, for example, the isospin; if both subsystems S_1, S_2 are non-coherent the identity in Proposition 11.1.3a may not hold.

11.1.6 Example: Consider the system consisting of a nucleon and a pion; for simplicity we neglect the non-isospin degrees of freedom (Problem 1). The nucleon state space is then $\mathbb{C}^2$ and any observable is of the form $A_N=\sum_{j=0}^{1}\lambda_j E_j^{(N)}$, where $E_1^{(N)}=I-E_0^{(N)}:=\frac{1}{2}(\sigma_3+1)$; similarly the pion space is $\mathbb{C}^3$ and its observables are $A_\pi=\sum_{k=-1}^{1}\mu_k E_k^{(\pi)}$, where $E_k^{(\pi)}$ are again the projections to the eigenspaces of the charge operator $Q=\sum_{k=-1}^{1}kE_k^{(\pi)}$. The state space of the composite system is therefore $\mathcal{H}\approx\mathbb{C}^6$ and it is easy to check that

$$\mathcal{A}_n\otimes\mathcal{A}_\pi = \left\{\sum_{j=0}^{1}\sum_{k=-1}^{1}\beta_{jk}E_{jk} : \beta_{jk}\in\mathbb{C}\right\},$$

where $E_{jk}:=E_j^N\otimes E_k^{(\pi)}$. However, the only superselection rule for the composite system is given by the *total* charge operator $Q_{N\pi}:=\underline{Q}_N+\underline{Q}_\pi$. Its spectral decomposition is $Q_{N\pi}=\sum_{q=-1}^{2}qE_q$, where $E_q:=\sum_{j+k=q}E_{jk}$; this means that its eigenspaces $\mathcal{H}_q:=E_q\mathcal{H}$ are two-dimensional for $q=0,1$ and one-dimensional for $q=-1,2$, and

$$\mathcal{A}_{N\pi}\subset\bigoplus_{q=-1}^{2}\mathcal{B}(\mathcal{H}_q).$$

If the two algebras were identical, any observable of the nucleon–pion system would be reduced by all the projections E_{jk}, and this would in turn mean that any state of the system would be stationary. This contradicts the experimentally established existence of the transitions $p\pi^0\leftrightarrow n\pi^+$ and $p\pi_-\leftrightarrow n\pi^0$; hence $\mathcal{A}_{N\pi}$ must contain observables which do not belong to $\mathcal{A}_n\otimes\mathcal{A}_\pi$.

This conclusion clearly requires that both the involved systems are noncoherent. If we replace, for instance, the pion by an η meson (regarded for this purpose as a stable spinless particle), which is a charge singlet, we get $\mathcal{A}_{N\eta}=\mathcal{A}_N\otimes\mathcal{A}_\eta$.

Thus in general a more detailed analysis of the algebras $\mathcal{A}_j$ is needed to decide whether they span the algebra of observables of the composite system. When dealing with quantum mechanical systems, we usually avoid this problem by *assuming* that

$$\mathcal{A} = \sum_{\alpha\in J}^{\oplus} \mathcal{B}(\mathcal{H}_\alpha), \tag{11.1}$$

where the the coherent subspaces $\mathcal{H}_\alpha$ now correspond to the superselection rules of the composite system.

11.2 Reduced states

Postulate (Q6a) in combination with possible superselection rules tells us what the families of states of the composite system and its subsystems look like. Now we want to consider a more complicated question, namely in which states W_1, W_2 are the subsystems if the composite system is in a state W, and on the other hand, what can be concluded about the state of S from the knowledge of the states of S_1 and S_2. For simplicity, we shall assume throughout this section that we are dealing with the quantum mechanical case, *i.e.*, that the algebra of observables of each of the subsystems is of the form (11.1).

Suppose therefore that system S is in a state W, and let us denote the subsystem states as W_j, $j=1,2$. It is natural to require $\langle \underline{A}_j\rangle_W = \langle A_j\rangle_{W_j}$ for any observable related to one of the subsystems only, where we again use the notation $\underline{A}_1 := A_1 \otimes I_2$, *etc.* It follows then from Theorem 7.3.3b that $\mathrm{Tr}\,(\underline{A}_jW) = \mathrm{Tr}\,(A_jW_j)$, $j=1,2$, must hold for all $A_j \in \mathcal{A}_j$ (in fact, we should index the traces to indicate the space to which they refer, but we refrain from doing so as long as there is no danger of misunderstanding). This requirement is sufficient to determine uniquely the states of the subsystems.

11.2.1 Proposition: Let W be the statistical operator describing a state of S; then there is a unique pair of statistical operators $W_j(W)$ corresponding to realizable states and such that

$$\mathrm{Tr}\,(\underline{A}_jW) = \mathrm{Tr}\,(A_jW_j(W)), \quad j=1,2, \tag{11.2}$$

holds for all $A_j \in \mathcal{A}_j$.

Proof: Consider, for example, the subsystem S_1 and define the functional f_W : $f_W(A_1) = \mathrm{Tr}\,(\underline{A}_1W)$; it is easy to see that f_W is linear, positive, and satisfies $f_W(I_1) = 1$, *i.e.*, it is a state on $\mathcal{A}_1$. We shall check that f_W is normal using the equivalent definition mentioned in Section 6.4. Take a sequence of positive operators $\{A_1^{(k)}\} \subset \mathcal{A}_1$ such that $A_1 := \text{w-lim}_{n\to\infty}\sum_{k=1}^n A_1^{(k)}$ exists. The operator $\underline{A}_1$ is then again positive and equals $\text{w-lim}_{n\to\infty}\sum_{k=1}^n \underline{A}_1^{(k)}$ (Problem 3). Due to Theorem 6.4.1, the operator W determines a normal state on $\mathcal{B}(\mathcal{H}_1) \supset \mathcal{A}_1$, so

$$f_W(A_1) = \mathrm{Tr}\,(\underline{A}_1W) = \sum_{k=1}^{\infty} \mathrm{Tr}\,(\underline{A}_1^{(k)}W) = \sum_{k=1}^{\infty} f_W(A_1^{(k)})\,;$$

the state f_W is therefore normal, and since the algebra $\mathcal{A}_1$ is of the form (11.1), it follows from Theorem 6.4.7 that there is just one statistical operator $W_1(W)$, which is reduced by all $\mathcal{H}_\alpha$ and satisfies the condition $f_W(A_1) = \mathrm{Tr}\,(A_1 W_1(W))$ for all $A_1 \in \mathcal{A}_1$. ∎

The states $W_j(W)$ are called **reduced** (sometimes also *component*) states of the given W. We shall see in a while that we can not only prove their existence, but also find an explicit expression for them. First, however, let us ask the opposite question, namely to what extent is the state of S determined by the states of its subsystems. Given statistical operators W_j on $\mathcal{H}_j$ we easily find W such that the identity (11.2) is valid; in view of Problem 4 it is sufficient to choose $W := W_1 \otimes W_2$. However, the maps $W_j(\cdot)$ are in general not injective.

11.2.2 Example: Consider a pair of spin $\frac{1}{2}$ particles, say, two electrons or a proton and an electron. For simplicity we again neglect the nonspin degrees of freedom and put $\mathcal{H}_j := \mathbb{C}^2$; the eigenvectors of the third component of spin of the j–th particle will be denoted as $\phi_\pm^{(j)}$. A standard orthonormal basis in $\mathcal{H} = \mathcal{H}_1 \otimes \mathcal{H}_2$ consists of the vectors

$$\phi_{10} := \frac{1}{\sqrt{2}}(\phi_+^{(1)} \otimes \phi_-^{(2)} + \phi_-^{(1)} \otimes \phi_+^{(2)}), \quad \phi_{1,\pm 1} := \phi_\pm^{(1)} \otimes \phi_\pm^{(2)},$$
$$\phi_{00} := \frac{1}{\sqrt{2}}(\phi_+^{(1)} \otimes \phi_-^{(2)} - \phi_-^{(1)} \otimes \phi_+^{(2)}),$$

which are eigenvectors of the total spin of the pair and its third component; the first three of them span the *triplet subspace* $\mathcal{H}^{(1)} := E_1\mathcal{H}$, the remaining one spans the *singlet subspace* $\mathcal{H}^{(0)} := E_0\mathcal{H}$ (see Problem 16). The density matrices $W_j := \frac{1}{2} I_j$ describe the unpolarized states of the two particles (*cf.* Example 7.3.2). We have $\mathrm{Tr}\,(A_j W_j) = \frac{1}{2}\mathrm{Tr}\, A_j$ for any $A_j \in \mathcal{B}(\mathbb{C}^2)$; on the other hand a straightforward computation using the above basis shows that $\mathrm{Tr}\,(\underline{A}_j E_1) = 3\,\mathrm{Tr}\,(\underline{A}_j E_0) = \frac{3}{2}\mathrm{Tr}\, A_j$. In this case therefore a statistical operator $W \equiv W(w_0, w_1) := w_0 E_0 + w_1 E_1$, which satisfies (11.2), corresponds to any non–negative w_0, w_1 such that $3w_1 + w_0 = 1$, in particular, $W\left(\frac{1}{4}, \frac{1}{4}\right) = W_1 \otimes W_2$.

The reason for this nonuniqueness is clear: the relation (11.2) defines the state of the composite system on a subset $\underline{\mathcal{A}}_1 \cup \underline{\mathcal{A}}_2 \subset \mathcal{A}$ which is too small to determine the algebra $\mathcal{A}$ completely. In some cases, however, the uniqueness can be guaranteed.

11.2.3 Theorem: (a) If the subsystem S_1 is coherent and the state W_2 is pure, then the relation (11.2) is satisfied by $W = W_1 \otimes W_2$ only.

(b) If the states of both subsystems are pure, $W_j = E_{\psi_j}$, then only the pure state $W = E_\psi$, where $\psi := \psi_1 \otimes \psi_2$, satisfies (11.2).

Proof: We choose orthonormal bases $\{\phi_i\}$, $\{\chi_k\}$ in the spaces $\mathcal{H}_j$ consisting of eigenvectors of the operators W_1 and W_2, respectively; the corresponding sequences of one–dimensional projections are denoted by $\{E^{(i)}\}$ and $\{F^{(k)}\}$. If W_2 is pure, we may assume without loss of generality that $W_2 = F^{(1)}$. Substituting $A_2 := F^{(k)}$ in (11.2), we get $\delta_{k1} = \mathrm{Tr}\,(\underline{F}^{(k)}W) = \sum_i W_{ik,ik}$ where $W_{in,kl} := (\phi_i \otimes \chi_n, W(\phi_k \otimes \chi_l))$ are the matrix elements of the operator W. Its positivity implies $W_{ik,ik} = 0$ unless $k = 1$. Furthermore, by choosing $A_1 := E^{(i)}$, the last result implies

$$W_{i1,i1} = \sum_k W_{ik,ik} = \mathrm{Tr}\,(E^{(i)}W_1) = w_i^{(1)},$$

where $w_i^{(1)}$ is the corresponding eigenvalue of the operator W_1. Let $\{\psi_m\}$ be an orthonormal basis in $\mathcal{H}$ composed of eigenvectors of the operator W, $W\psi_m = w_m\psi_m$. We denote $\alpha_{ij}^{(m)} := (\psi_m, \phi_i \otimes \chi_k)$; using the fact that W is bounded we easily find $W(\phi_k \otimes \chi_l) = \sum_m \alpha_{kl}^{(m)} w_m \psi_m$, and therefore $W_{in,kl} = \sum_m \bar{\alpha}_{in}^{(m)} \alpha_{kl}^{(m)} w_m$. Recall that $W_{ik,ik} = 0$ for $k \neq 1$; the eigenvalues w_m are non–negative, so if $w_m > 0$ and $n \neq 1$ we get $\alpha_{in}^{(m)} = 0$, *i.e.*,

$$W_{in,kl} = 0\,, \qquad n \neq 1 \quad \text{or} \quad l \neq 1\,.$$

If W_1 is pure we may at the same time choose the basis $\{\phi_i\}$ in such a way that $W_1 = E^{(1)}$. Repeating the argument with the roles of W_1, W_2 interchanged we get $W_{in,kl} = 0$ unless $i = k = 1$. Together we have $W_{in,kl} = \delta_{i1}\delta_{n1}\delta_{k1}\delta_{l1} = (W_1 \otimes W_2)_{in,kl}$ for all i, n, k, l, so the operators coincide, $W = W_1 \otimes W_2$. The states W_j are pure by assumption; hence $W^2 = W_1^2 \otimes W_2^2 = W_1 \otimes W_2 = W$ and W is also pure. This proves assertion (b).

Up to now we have not employed the coherence assumption. If S_1 is coherent, then $\mathrm{Tr}\,(B_1W_1) = \mathrm{Tr}\,(\underline{B}_1W)$ holds for any $B_1 \in \mathcal{H}_1$; using the relation $W_{i1,i1} = w_i^{(i)}$ proven above, we get

$$\sum_{i\neq k} (B_1)_{ik} W_{k1,i1} = 0\,.$$

We have $\sum_{i,k} |W_{k1,i1}|^2 = \sum_{i,n,k,l} |W_{kl,il}|^2 = \|W\|_2^2 < \infty$, which means that the matrix $\{\overline{W}_{k1,i1} : i, k = 1, \dots, \dim \mathcal{H}_1\}$ represents some operator $C_1 \in \mathcal{B}(\mathcal{H}_1)$. Substituting $B_1 := C_1$ in the above relation, we obtain $W_{k1,i1} = 0$ for $i \neq k$. In combination with the two displayed relations from the first part of the proof, this implies $W_{ij,kl} = w_i^{(1)} \delta_{ik}\delta_{j1}\delta_{l1} = (W_1 \otimes W_2)_{in,kl}$ for all values of the indices, *i.e.*, $W = W_1 \otimes W_2$. ∎

Let now ask now what the reduced states look like. Without loss of generality we may assume that the composite system is in a pure state $W = E_\psi$ (Problem 5b). We choose orthonormal bases $\mathcal{E} = \{\phi_i\}$ and $\mathcal{F} = \{\chi_k\}$, arbitrary for the moment, in the spaces $\mathcal{H}_1$, $\mathcal{H}_2$, respectively, and express the vector ψ representing the state W in the tensor–product basis, $\psi = \sum_{i,k} \alpha_{ik}\phi_i \otimes \chi_k$. By assumption, ψ is a unit vector, so $\sum_{i,k} |\alpha_{ik}|^2 = 1$. The left side of (11.2) can be expressed formally as $\sum_{i,l,k} \bar{\alpha}_{il}\alpha_{kl}(\phi_i, A_1\phi_k)_1$. We shall check that the series is absolutely convergent, so we can rearrange it obtaining

$$\mathrm{Tr}\,(\underline{A}_1W) = \sum_{i,k} b_{ki}(\phi_i, A_1\phi_k)_1\,,$$

where $b_{ki} := \sum_j \bar{\alpha}_{ij}\alpha_{kj}$. The Hölder inequality together with the normalization condition implies $\sum_{i,k} |b_{ik}|^2 \le 1$, and if A_1 is a Hilbert–Schmidt operator then another application of the Hölder inequality yields absolute convergence. The matrix (b_{ij}) represents in the basis $\mathcal{E}$ an operator of $\mathcal{J}_2(\mathcal{H}_1)$ which we denote as $W^{(1)}$. It is easy to check that $W^{(1)} \ge 0$, and moreover, $\mathrm{Tr}\, W^{(1)} = \sum_i b_{ii} = 1$ so $W^{(1)}$ is a statistical operator. Finally,

$$\mathrm{Tr}\,(A_1 W^{(1)}) = \sum_{i,k} b_{ki}(\phi_i, A_1\phi_k)_1 = \mathrm{Tr}\,(A_1 W_1(W))$$

for all $A_1 \in \mathcal{A}_1 \cap \mathcal{J}_2(\mathcal{H}_1)$. In the same way we can construct the statistical operator $W^{(2)} \in \mathcal{B}(\mathcal{H}_2)$, which is represented in the basis $\mathcal{F}$ by the matrix with the elements $c_{jl} := \sum_i \bar{\alpha}_{ij}\alpha_{il}$ and fulfils the condition $\mathrm{Tr}\,(A_2 W^{(2)}) = \mathrm{Tr}\,(A_2 W_2(W))$ for any observable $A_2 \in \mathcal{A}_2 \cap \mathcal{J}_2(\mathcal{H}_2)$.

If the two subsystems are coherent, we can set $A_j := W^{(j)} - W_j(W)$ in these conditions; it follows that $W_j(W) = W^{(j)}$. This conclusion may not be valid if at least one subsystem is incoherent, since it is then not ensured that $W^{(j)} \in \mathcal{A}_j$. However, we can use the projections $E_\alpha^{(j)}$ referring to the coherent subspaces in $\mathcal{H}_j$ to construct the statistical operators

$$W_j := \sum_{\alpha \in J} E_\alpha^{(j)} W^{(j)} E_\alpha^{(j)} \quad \text{(strong convergence)} \tag{11.3}$$

such that $\mathrm{Tr}\,(A_j W_j) = \mathrm{Tr}\,(A_j W^{(j)})$ holds for all $A_j \in \mathcal{A}_j$ (Problem 6). Now the operators $W_j - W_j(W)$ already belong to $\mathcal{A}_j \cap \mathcal{J}_2(\mathcal{H}_j)$, so the same argument as above implies $W_j(W) = W^{(j)}$.

The relation (11.3) together with the expression of the operators $W^{(j)}$ by means of the Fourier coefficients of the vector ψ,

$$\begin{aligned} W^{(1)}\phi_i &= \sum_{k=1}^{n_1} b_{ik}\phi_k\,, \quad b_{ik} := \sum_{l=1}^{n_2} \bar{\alpha}_{kl}\alpha_{il}\,, \\ W^{(2)}\chi_i &= \sum_{k=1}^{n_2} c_{ik}\chi_k\,, \quad c_{ik} := \sum_{l=1}^{n_1} \bar{\alpha}_{li}\alpha_{lk}\,, \end{aligned} \tag{11.4}$$

where $n_j := \dim \mathcal{H}_j$, fully answers the question of what the reduced states corresponding to a given pure state ψ look like. In the particular case when the subsystems are coherent, the reduced states are given directly by the relations (11.4), which are called **reduction formulas**.

Next we want to show that if the orthonormal bases are suitably chosen the relations (11.4) can be cast into a certain standard form. Suppose therefore that $\mathcal{E} = \tilde{\mathcal{E}}$, where $\tilde{\mathcal{E}} := \{\tilde{\phi}_i\}$ consists of the eigenvectors of $W^{(1)}$; the corresponding eigenvalues will be denoted as $w_i^{(1)}$. The first formula of (11.4) then yields $b_{ik} = w_i^{(1)}\delta_{ik}$ independently of the choice of the basis $\mathcal{F}$. If $w_i^{(1)} > 0$, then at least one of the coefficients α_{ik}, $k = 1, \dots, n_2$, is nonzero, so we can construct the unit vector

$\tilde{\chi}_i := (w_i^{(1)})^{-1/2} \sum_l \alpha_{il}\chi_l$, which satisfies the identity

$$(\chi_k, W^{(2)}\tilde{\chi}_i)_2 = (w_i^{(1)})^{-1/2} \sum_l \alpha_{il} c_{lk} = (w_i^{(1)})^{-1/2} \sum_l \alpha_{il} \sum_n \bar{\alpha}_{nl}\alpha_{nk} .$$

Using the Hölder inequality together with the relations (11.4) and $\sum_l w_l^{(1)} = 1$, we easily check that the series on the right side converges absolutely, $\sum_{n,l} |\alpha_{in}\alpha_{ln}\alpha_{lk}| \le (w_i^{(1)} c_{kk})^{1/2}$. Interchanging the order of summation, we get

$$(\chi_k, W^{(2)}\tilde{\chi}_i)_2 = (w_i^{(1)})^{-1/2} \sum_l \alpha_{lk} b_{li} = \alpha_{ik} (w_i^{(1)})^{1/2} ,$$

and therefore $W^{(2)}\tilde{\chi}_i = w_i^{(1)}\tilde{\chi}_i$. This means that any nonzero eigenvalue of $W^{(1)}$ is at the same time an eigenvalue of the operator $W^{(2)}$. Repeating the argument with the roles of the operators interchanged, we find $\sigma_p(W^{(1)}) \setminus \{0\} = \sigma_p(W^{(2)}) \setminus \{0\}$; these common eigenvalues will be denoted as w_i. The sequence $\{\tilde{\chi}_i\}_{i=1}^n$, where $n := \dim \operatorname{Ran} W^{(1)} = \dim \operatorname{Ran} W^{(2)}$, can be completed to an orthonormal basis $\tilde{\mathcal{F}}$ in $\mathcal{H}_2$. Then $\tilde{\mathcal{E}} \times \tilde{\mathcal{F}}$ is a basis in $\mathcal{H}$ and the Fourier coefficients of the vector ψ with respect to it are

$$(\tilde{\phi}_k \otimes \tilde{\chi}_l, \psi) = \sum_{i,j} \alpha_{ij} w_l^{-1/2} \sum_m \bar{\alpha}_{lm} (\tilde{\phi}_k \otimes \tilde{\chi}_m, \tilde{\phi}_i \otimes \tilde{\chi}_j) = w_l^{-1/2} \sum_m \alpha_{km}\bar{\alpha}_{lm} = \sqrt{w_l}\, \delta_{lk} .$$

Hence we have

$$\begin{aligned} \psi &= \sum_{i=1}^n \sqrt{w_i}\, \tilde{\phi}_i \otimes \tilde{\chi}_i , \\ W^{(1)} &= \sum_{i=1}^n w_i E^{(i)} , \quad W^{(2)} = \sum_{i=1}^n w_i F^{(i)} , \end{aligned} \tag{11.5}$$

where $E^{(i)}$, $F^{(i)}$ are the one–dimensional projections corresponding to the vectors of the bases $\tilde{\mathcal{E}}$ and $\tilde{\mathcal{F}}$, respectively. Let us summarize the results.

11.2.4 Theorem: If the composite system is in a pure state $W = E_\psi$, then the reduced states $W_j(W)$ are given by the relations (11.3) and (11.4). In particular, if the subsystems are coherent, then there are orthonormal bases $\tilde{\mathcal{E}}$ and $\tilde{\mathcal{F}}$ which enable us to express the reduced states in the normal form (11.5).

11.2.5 Example: Consider again the two–particle system of Example 2. The states $E_{1,\pm 1} := E_\pm^{(1)} \otimes E_\pm^{(2)}$ reduce by Problem 5a to $W_j(E_{1,\pm 1}) = E_\pm^{(j)}$. On the other hand, the projection E_0 to the singlet subspace represents an example of a pure state, which cannot be expressed in the form $W_1 \otimes W_2$. The Fourier coefficients of the vector ϕ_{00} with respect to the basis $\{\phi_i^{(1)} \otimes \phi_j^{(2)} : i, j = \pm\}$ are easily found to be $\alpha_{++} = \alpha_{--} = 0$ and $\alpha_{+-} = -\alpha_{-+} = 2^{-1/2}$, so the reduction formulas (11.4) give

$$W_j(E_0) = \frac{1}{2} I_j , \quad j = 1, 2 .$$

Putting further $\tilde{\phi}_\pm := \phi_\pm^{(1)}$ and $\tilde{\chi}_\pm := \pm\phi_\mp^{(2)}$, we are able to write ϕ_{00} in the normal form (11.5),

$$\phi_{00} = \frac{1}{\sqrt{2}}\left(\tilde{\phi}_+ \otimes \tilde{\chi}_+ + \tilde{\phi}_- \otimes \tilde{\chi}_-\right).$$

The pure state described by the vector ϕ_{10} leads again to the reduced states $W_j(E_{10}) = \frac{1}{2}I_j$. The same is true for the mixed triplet state $\frac{1}{3}E_1$, where Problem 5b gives $W_j(\frac{1}{3}E_1) = \frac{1}{6}I_j + \frac{1}{3}(E_+^{(j)} + E_-^{(j)}) = \frac{1}{2}I_j$, and also for any mixture of the states $\frac{1}{3}E_1$ and E_0.

11.3 Time evolution

Let H and H_j be the Hamiltonians of the system S and its subsystems, respectively; in a similar way we denote the corresponding propagators. The operator H is often of the form

$$H = \overline{\underline{H}_1 + \underline{H}_2 + H_{int}}, \tag{11.6}$$

where H_{int} is the *interaction Hamiltonian* of the subsystems. The splitting of the energy observable into this part and the free part, which is the sum of the subsystem Hamiltonians, usually has a direct physical interpretation; it makes sense, of course, if the operator $\underline{H}_1 + \underline{H}_2 + H_{int}$ is *e.s.a.* When the interaction Hamiltonian is Hermitean, this property follows easily from Theorem 5.7.2, while in the general case it is sometimes highly nontrivial to check it.

The subsystems S_1, S_2 are said to be *noninteracting* if $H_{int} = 0$. Since the generator is associated uniquely with a continuous unitary group, Proposition 5.9.6 yields the following simple criterion.

11.3.1 Proposition: The subsystems S_1, S_2 are noninteracting *iff* their propagators are related to the propagator of the composite system by $U(t) = U_1(t) \otimes U_2(t)$ for all $t \in \mathbb{R}$.

11.3.2 Remark: We can speak about noninteracting subsystems even if the subsystems S_j are nonconservative, so the corresponding H_j are replaced by $H_j(t)$ in (11.6), and $H_{int} = 0$. The tensor–product decomposition can in this case be proven again, provided we are able to check the existence of the corresponding unitary propagators — see, *e.g.*, Problem 8.

If the subsystems interact, the relation between their time evolution and that of the composite system is more complicated. We know from the preceding section that for any state W_t of a composite quantum mechanical system there is a unique pair of the reduced states $W_t^{(j)} := W_j(W_t)$ of the subsystems. Their time dependence can exhibit rather pathological properties:

(i) The quantity $\mathrm{Tr}\,((W_t^{(j)})^2)$ may not be preserved; in particular, a pure state may evolve into a mixed one and *vice versa.*

(ii) The operators $U_j(t,s)$ connecting the states of the subsystems at different times, $W_t^{(j)} = U_j(t,s)W_s^{(j)}U_j(t,s)^{-1}$, may not exist, and if they do, they may not have the properties required for a propagator.

11.3.3 Example: Consider again the system of two spin $\frac{1}{2}$ particles of Examples 11.2.2 and 11.2.5 and set

$$H := \frac{\lambda}{2}\,\sigma\otimes\sigma := \frac{\lambda}{2}\sum_{j=1}^{3}\sigma_j\otimes\sigma_j\,,$$

where σ_j are the Pauli matrices and $\lambda\in\mathbb{R}$. If other than spin degrees of freedom are neglected, this operator represents a typical nonrelativistic interaction of the particles. We shall assume that the state is described at the initial time instant $t=0$ by $\phi_+^{(1)}\otimes\phi_-^{(2)}$. The eigenvalues of the operator $\sigma\otimes\sigma$ are 1 and -3; using Example 5.2.3, we find

$$U(t) = e^{-i\lambda t(\sigma\otimes\sigma)/2} = \left[\cos(\lambda t) - \frac{i}{2}(I+\sigma\otimes\sigma)\sin(\lambda t)\right] e^{i\lambda t/2}$$

(*cf.* Problem 9), so

$$\psi(t) := (U(t)(\phi_+^{(0)}\otimes\phi_-^{(2)})) = [\,(\phi_+^{(1)}\otimes\phi_-^{(2)})\cos(\lambda t) - i(\phi_-^{(1)}\otimes\phi_+^{(2)})\sin(\lambda t)\,]\,e^{i\lambda t/2}\,.$$

The reduced states $W_j(t) := W_j(E_{\psi(t)})$ are then

$$W_1(t) = E_+\cos^2(\lambda t) + E_-\sin^2(\lambda t)\,,\quad W_2(t) = E_+\sin^2(\lambda t) + E_-\cos^2(\lambda t)\,,$$

where $E_\pm = \frac{1}{2}(I\pm\sigma_3)$. Since $\mathrm{Tr}\,(W_j(t)^2) = \frac{1}{4}(3+\cos(4\lambda t))$ we see that the reduced states are mixed unless $t=\frac{k\pi}{2\lambda}$ for an integer k; the operators $U_j(t,s)$ exist only if the involved time instants satisfy the condition $t\pm s = \frac{\pi k}{\lambda}$.

From the practical point of view, however, these difficulties are not very important because in most cases we are not interested in the time evolution of reduced states (see the notes). On the other hand, they have an implication for the time–evolution postulates of the previous chapter. We have tacitly assumed that the system is either isolated or it interacts with the environment, which is not influenced by its presence. The example shows that the assumption is substantial; without it the mere existence of the evolution operator is not guaranteed.

11.4 Identical particles

We have stressed in the introduction to this chapter that the postulate (Q6a) applies only to cases when the considered systems are mutually different; now we are going to discuss the situation when the composite system contains two or more identical subsystems. We first have to say a few words about the meaning of identity of microscopic systems. For simplicity we shall speak about elementary particles; however, the conclusions also extend to systems composed of nuclei, atoms and other "nonelementary" constituents.

Properties of a particle can be divided into two groups. Some of them, for instance, its spatial localization or a spin projection, determine its state. The other group includes internal characteristics like the mass, spin, electric charge, *etc.*, which are used to classify the particles. The number of different elementary–particle sorts is known to be finite, and even not too large; just six of them (together with their antiparticles) are stable, among them only the electron and the proton (whose stability has also been challenged) can be claimed with certainty to have a nonzero rest mass. If we add the particles with lifetimes of $\gtrsim 10^{-16}$ s the number will grow to about thirty varieties. The principal assumption of quantum theory is that two particles of the same sort *cannot be distinguished.*

In classical mechanics identical objects, *e.g.* point particles of the same mass, can be distinguished by means of their states: we can ascribe a trajectory to each of them, so it is sufficient to tag them somehow at the initial instant. In quantum theory, however, this idea does not apply because localization is not preserved in general (*cf.* Problem 9.8); once the wave functions of two particles overlap there is no way to determine which one of them has been registered in the intersection of their supports. Hence we have to adopt the mentioned **indistinguishability principle** of identical particles with all its consequences; we have to exclude any observable which would make it possible to tell one particle from the other.

Consider first a pair of particles each of which has the state space $\mathcal{H}$. We are going to show that it is sufficient now to consider a certain subspace in $\mathcal{H}^{(2)} := \mathcal{H} \otimes \mathcal{H}$. We set

$$U_P \left(\sum_{k=1}^{n} \psi_k \otimes \phi_k \right) := \sum_{k=1}^{n} \phi_k \otimes \psi_k \tag{11.7}$$

for any $\phi_k, \psi_k \in \mathcal{H}$ and a positive integer n. This defines an operator on $\mathcal{H} \underline{\times} \mathcal{H}$; its continuous extension to $\mathcal{H}^{(2)}$, denoted again as U_P, is unitary (Problem 10a). Consider now states W and $W_P := U_P W U_P^{-1}$ of the composite system. If they are pure and described by vectors ψ and $\psi_P := U_P \psi$, then the reduction formulas (11.4) yield $W_j(W) = W_{3-j}(W_P)$ for $j = 1, 2$; by Problem 5b this conclusion extends to any admissible state W. The roles of the particles in the states W and W_P are therefore switched, and the principle of indistinguishability is nothing else than the invariance of the system with respect to particle interchange. By Proposition 10.1.4, any observable A of the pair has then to commute with the operator which represents the transformation,

$$U_P A \subset A U_P \,.$$

Define further the operators $S_2 := \frac{1}{2}(I + U_P)$ and $A_2 := \frac{1}{2}(I - U_P)$; since $U_P^2 = I$, they are mutually orthogonal projections such that $S_2 + A_2 = I$. The above relation then says that any observable of a pair of identical particles is reduced by the projections S_2 and A_2. It appears as if the identity would require introducing a new superselection rule. However, we know from experience that admissible states always belong to only one of the subspaces, *i.e.*, that any state of two identical particles satisfies *just one* of the conditions

$$U_P\psi = \pm\psi\,. \tag{11.8}$$

These transformation properties are referred to as *Bose–Einstein statistics* in the case of the upper sign when $\psi \in S_2\mathcal{H}^{(2)}$, and *Fermi–Dirac statistics* for the lower sign with $\psi \in A_2\mathcal{H}^{(2)}$.

It is known that the behavior of the particles is governed by Bose–Einstein statistics if the particles involved have an integer spin, and by Fermi–Dirac statistics in the case of a half-integer spin; for brevity these categories of particles are called the **bosons** and **fermions**, respectively. In the framework of relativistic quantum field theory, this empirical fact can be explained (see the notes).

These conclusions extend to the case of a system consisting of n identical particles. To describe their interchanges, we introduce the group $\mathcal{S}_n$ of permutations of n elements, which we write as

$$p := \begin{pmatrix} 1 & 2 & \dots & n \\ p_1 & p_2 & \dots & p_n \end{pmatrix};$$

the symbol ϵ_p denotes the parity of the permutation, $\epsilon_p = \pm 1$. Since the particles are supposed to be identical, the system is invariant with respect to $\mathcal{S}_n$, so by the postulate (Q5b) its state space has to support a representation of this group. To construct it, we set

$$U(p)(\psi_1 \otimes \cdots \otimes \psi_n) := \psi_{p_1} \otimes \cdots \otimes \psi_{p_n}\,. \tag{11.9}$$

for any $p \in \mathcal{S}_n$ and arbitrary $\psi_j \in \mathcal{H}$; in the same way as above we can check that the map $U(p)$ is well defined and extends continuously to a unitary operator on $\mathcal{H}^{(n)} := \mathcal{H}\otimes\cdots\otimes\mathcal{H}$ denoted by the same symbol. The corresponding transformation leads to the appropriate permutation of the reduced states (Problem 11). Moreover,

$$U(p)U(\tilde{p}) = U(p\tilde{p})$$

holds for any $p, \tilde{p} \in \mathcal{S}_n$; in particular, $U(e) = I$ for the identical permutation $e \in \mathcal{S}_n$ and $U(p^{-1}) = U(p)^*$. This means that $U : \mathcal{S}_n \to \mathcal{B}(\mathcal{H}^{(n)})$ is the sought representation; if $n = 2$ the group consists of two elements represented by the operators I and U_P. Next we define the operators

$$S_n := \frac{1}{n!}\sum_{p\in\mathcal{S}_n} U(p)\,, \qquad A_n := \frac{1}{n!}\sum_{p\in\mathcal{S}_n} \epsilon_p U(p)\,,$$

which are easily seen to be Hermitean; the group property of $U(p)$ yields the relations $U(p)S_n = S_n$ and $U(p)A_n = \epsilon_p A_n$, which in turn imply $S_n^2 = S_n$ and $A_n^2 = A_n$; this means that S_n and A_n are projections.

The indistinguishability principle requires the n–particle system to be invariant with respect to the group $\mathcal{S}_n$, so $U(p)A \subset AU(p)$ must hold for any observable A; it follows that

$$S_n A \subset A S_n\,, \qquad A_n A \subset A A_n\,.$$

In this case the realizable states again correspond only to vectors of one of the subspaces $S_n\mathcal{H}^{(n)}$ and $A_n\mathcal{H}^{(n)}$ depending on the spin of the particles (see also the notes); hence the postulate (Q6a) on the state space can reformulated as follows:

(Q6b) the state space of n identical particles with an integer (half–integer) spin is the subspace $S_n\mathcal{H}^{(n)}$ (respectively, $A_n\mathcal{H}^{(n)}$) in $\mathcal{H}^{(n)} := \mathcal{H} \otimes \cdots \otimes \mathcal{H}$, where $\mathcal{H}$ is the state space of one particle.

Recall the physically most important consequence of this postulate.

10.4.1 Example (Pauli principle): Consider a system of n identical fermions in a state $\psi_1 \otimes \cdots \otimes \psi_n \in A_n\mathcal{H}^{(n)}$. Suppose that $\psi_j = \psi_k$ holds for a pair of different indices j, k, and let p_{jk} denote the transposition of the j–th and k–th elements. Its parity is negative so $U(p_{jk})A_n = -A_n$. We have $A_n\psi = \psi$ by assumption, and at the same time, $U(p_{jk})\psi = \psi$. Combining these relations, we get $\psi = 0$; this means that in a system of identical fermions no two can be simultaneously in the same pure state.

11.5 Separation of variables. Symmetries

The more degrees of freedom a quantum system has, the more difficult it usually is to treat it. We have seen, however, that the task simplifies considerably if the system can be decomposed into noninteracting subsystems, either real or fictitious. It happens often that the Hamiltonian H itself is not of the form $\overline{\underline{H}_1 + \underline{H}_2}$ but we can pass to a unitary equivalent operator $U^{-1}HU$ which already has this property; most frequently U is some "substitution" operator of Example 3.3.2. In this section we are going to illustrate the method by discussing two simple situations of this type.

The first of these concerns the *separation of the center–of–mass motion.* Consider a system of two spinless particles with the state space $\mathcal{H} := L^2(\mathbb{R}^6)$. The configuration–space vectors which appear as the wave function arguments will be denoted as (x_1, x_2), where $x_j = (x_{j1}, x_{j2}, x_{j3})$ refers to the position of the j–th particle, and Q_{jk} are the corresponding coordinate operators; the same double–index notation will be used for the momenta. Suppose that the potential $V : \mathbb{R}^6 \to \mathbb{R}$ by means of which the particles interact depends only on their distance, *i.e.*, it can be expressed as

$$V(x_1, x_2) = v(x_2 - x_1) \tag{11.10}$$

for some Borel function $v : R^3 \to \mathbb{R}$. The Hamiltonian of the two–particle system is then the closure of

$$H : H\psi = \left[\frac{1}{2m_1}P_1^2 + \frac{1}{2m_2}P_2^2 + V\right]\psi\,,$$

provided the operator in the square brackets is *e.s.a.*, where $V := T_V$ with the domain $D_V := \{\psi \in L^2(\mathbb{R}^6) : V\psi \in L^2(\mathbb{R}^6)\}$, $P_j^2 := \sum_{k=1}^3 P_{jk}^2$, and m_j is the mass of the j–th particle. We introduce the center–of–mass and relative position vectors

$$X := \mu_1 x_1 + \mu_2 x_2\,, \qquad x := x_2 - x_1\,,$$

where $\mu_j := m_j/M$ and $M := m_1 + m_2$ is the total mass of the system, and use them to define the substitution operator U on $L^2(\mathbb{R}^6)$ by

$$(U\psi)(x_1, x_2) := \psi(X, x)\,. \tag{11.11}$$

It follows from Example 3.3.2 that U is unitary. Its inverse is $(U^{-1}\psi)(X, x) = \psi(X - \mu_2 x, X + \mu_1 x)$; it is easy to see that both U and U^{-1} preserve $\mathcal{S}(\mathbb{R}^6)$. Next we consider the operator of multiplication by the function

$$W(X, x) := v(x)\,.$$

A simple substitution argument gives $D_W := U^{-1}D_V$ and $W\psi = U^{-1}VU\psi$ for any $\psi \in D_W$, *i.e.*, the operator identity $W = U^{-1}VU$. As for the momentum coordinates, it follows from the relation (7.5) that

$$\begin{aligned} U^{-1}P_{1k}U\phi &= -iU^{-1}[\mu_1 U(\partial_{1k}\phi) - U(\partial_{2k}\phi)] = \mu_1 P_{1k}\phi - P_{2k}\phi\,, \\ U^{-1}P_{2k}U\phi &= -iU^{-1}[\mu_2 U(\partial_{1k}\phi) + U(\partial_{2k}\phi)] = \mu_2 P_{1k}\phi + P_{2k}\phi \end{aligned}$$

for any $\phi \in \mathcal{S}(\mathbb{R}^3)$, so an easy computation yields

$$U^{-1}\left[\frac{1}{2m_1}P_1^2 + \frac{1}{2m_2}P_2^2\right]U\phi = \frac{1}{2M}P_1^2\phi + \frac{1}{2\mu}P_2^2\phi$$

(Problem 12a), where $\mu := m_1 m_2/M$ is the reduced mass of the particles. Together we get the identity

$$U^{-1}HU\psi = \left[\frac{1}{2M}P_1^2 + \frac{1}{2\mu}P_2^2 + W\right]\psi$$

for any $\psi \in \mathcal{S}(\mathbb{R}^6) \cap D_W$. Consider now a dense subspace $L^2(\mathbb{R}^3)$ such that $D \subset \mathcal{S}(\mathbb{R}^3) \cap D_v$. The two–particle state space decomposes as $L^2(\mathbb{R}^6) = L^2(\mathbb{R}^3) \otimes L^2(\mathbb{R}^3)$; using the definition of the function W we find $\mathcal{S}(\mathbb{R}^3) \underline{\times} D \subset \mathcal{S}(\mathbb{R}^6) \cap D_W$, and therefore

$$U^{-1}HU \restriction (\mathcal{S}(\mathbb{R}^3) \underline{\times} D) = T_{cm} \otimes I + I \otimes H_{rel}\,, \tag{11.12}$$

where T_{cm} and H_{rel} are operators with the domains $\mathcal{S}(\mathbb{R}^3)$ and D, respectively, defined by

$$T_{cm}\phi := -\frac{1}{2M}\Delta\phi\,, \qquad H_{rel}\psi := -\frac{1}{2\mu}\Delta\psi + v\psi\,.$$

The operator T_{cm} is *e.s.a.* by Problem 7.16; hence if we are able to choose D in such a way that H_{rel} is also *e.s.a.*, the same will be true for $T_{cm} \otimes I + I \otimes H_{rel}$ due to Theorem 5.7.2; this in turn implies that $\mathcal{S}(\mathbb{R}^3) \underline{\times} D$ is a core for H.

Instead of checking the essential self–adjointness of operator H, we thus have to solve the simpler problem of finding a suitable subspace $D \subset \mathcal{S}(\mathbb{R}^3) \cap D_v$ such that $H_{rel} := (-\frac{1}{2\mu}\Delta + v) \upharpoonright D$ is *e.s.a.* Furthermore, it follows from (11.12) and Problem 4.27d that

$$U^{-1}\overline{H}U = \overline{T_{cm} \otimes I + I \otimes H_{rel}} = \overline{\overline{T}_{cm} \otimes I + I \otimes \overline{H}_{rel}}\,,$$

so other properties of the two–particle Hamiltonian can also be derived from those of the operator H_{rel} (*cf.* Problem 12). In particular, due to Proposition 11.3.1 and Problem 5.12 the corresponding unitary propagator is expressed as

$$U^{-1}e^{-i\overline{H}t}U = \exp(-i\overline{T}_{cm}t) \otimes \exp(-i\overline{H}_{rel}t)\,;$$

in addition, the center–of–mass part is known explicitly because $\overline{T}_{cm}$ is the Hamiltonian of a free particle of mass M, so $\exp(-i\overline{T}_{cm}t)$ is given by Theorem 9.3.1 with $n = 3$ and $m_j = M$.

Concluding this discussion, we see that if the two–particle potential is of the form (11.10), the original problem can be reduced to analysis of the Hamiltonian $\overline{H}_{rel}$ that describes a single particle with the reduced mass μ.

11.5.1 Remark: In a similar way, we can separate the center–of–mass motion in systems of N particles interacting *via* potentials, which depend solely on the differences of the particle position vectors (Problem 13). The new feature here is that while in the two–particle case (11.11) represents the only natural choice of the corresponding operator, for $N \geq 2$ there are different possibilities related to different relative coordinates. The two most frequently used are the *atomic coordinates*

$$y_j := x_j - x_N\,, \quad j = 1, \dots, N-1\,, \tag{11.13}$$

where the positions are related to a chosen particle, and the *Jacobi coordinates*

$$z_j := x_{j+1} - \sum_{i=1}^{j} m_i x_i \left(\sum_{i=1}^{j} m_i\right)^{-1}, \quad j = 1, \dots, N-1\,. \tag{11.14}$$

The former are suitable in situations where some particle is privileged, for example, being much heavier than the rest. On the other hand, z_j in (11.14) describes the relative position of the $(j+1)$–th particle with respect to the center of mass of the preceding j particles, and therefore the ordering is essential in this case.

The second example we are going to discuss concerns particle motion in a *centrally symmetric potential, i.e.*, we consider the operator (9.2) on $L^2(\mathbb{R}^3)$ with $m_1 = m_2 = m_3 =: m$ and the function V such that

$$V(x) = v(r)\,, \quad r := \sqrt{x_1^2 + x_2^2 + x_3^2}\,, \tag{11.15}$$

for some Borel $v : \mathbb{R}^+ \to \mathbb{R}$. Let $S \equiv S^2$ be the unit sphere in $\mathbb{R}^3$. The measure Ω on S is defined using the spherical coordinates,

$$x_1 = r\,\sin\theta\cos\varphi\,, \quad x_2 = r\,\sin\theta\sin\varphi\,, \quad x_3 = r\,\cos\theta\,,$$

by $d\Omega(\theta,\varphi) = \sin\theta\, d\theta\, d\varphi$; it is invariant with respect to rotations of the sphere, because an arbitrary rotation can be expressed as a sequence of rotations around the axes (Problem 14a). Moreover, we can introduce the operator $U : L^2(\mathbb{R}^3) \to L^2(\mathbb{R}^+ \times S,\, dr\, d\Omega)$ by

$$(U\psi)(r,\theta,\varphi) := r\psi(x)\,. \tag{11.16}$$

It is easy to check that U is an isometry of the two spaces; recall that the latter can be identified by Example 2.4.5 with $L^2(\mathbb{R}^+) \otimes L^2(S, d\Omega)$.

Before we use the operator U to transform the Hamiltonian, we need to know something about the space $L^2(S, d\Omega)$. It contains, in particular, the *spherical functions* Y_{lm} defined as

$$Y_{lm}(\theta,\varphi) := (-1)^m \left[\frac{2l+1}{4\pi}\,\frac{(l-|m|)!}{(l+|m|)!}\right]^{1/2} P_l^{|m|}(\cos\theta)\, e^{im\varphi}$$

for any $l = 0,1,\dots$ and $m = -l, -l{+}1, \dots, l$, where P_l^n are the so–called associated Legendre functions on the interval $[-1,1]$ which are given for $n = 0,1,\dots$ and $l = n, n{+}1, \dots$ by

$$P_l^n(z) := (1-z^2)^{n/2}\frac{1}{2^l l!}\,\frac{d^{l+n}}{dz^{l+n}}\,(z^2-1)^l\,.$$

We define the operator Λ on $L^2(S, d\Omega)$ by

$$(\Lambda g)(\theta,\varphi) := \left(\frac{1}{\sin\theta}\frac{\partial}{\partial\theta}\left(\sin\theta\,\frac{\partial g}{\partial\theta}\right) + \frac{1}{\sin^2\theta}\,\frac{\partial^2 g}{\partial\varphi^2}\right)(\theta,\varphi)\,; \tag{11.17}$$

then the following result is valid.

11.5.2 Proposition: Y_{lm} are eigenfunctions of Λ : $\Lambda Y_{lm} = -l(l+1)Y_{lm}$, and $\{Y_{lm} : l = 0,1,\dots,\ m = -l,\dots,l\}$ is an orthonormal basis in $L^2(S, d\Omega)$.

Proof: The first assertion follows from the fact that the functions P_l^n satisfy the equation $(1-z^2)f'' - 2zf' + [l(l{+}1) - n^2/(1-z^2)]f = 0$. To check that the spherical functions form an orthonormal basis in $L^2(S, d\Omega)$, we express the latter in the tensor–product form $L^2(0,2\pi) \otimes L^2((0,\pi), \sin\theta\, d\theta)$ and notice that the last named space is isomorphic to $L^2(-1,1)$. In view of Example 2.2.2 and Problem 2.24, it is sufficient to show that the set

$$\left\{ \left(\frac{2l{+}1}{2}\,\frac{(l{-}n)!}{(l{+}n)!}\right)^{1/2} P_l^n \,:\, l = n, n{+}1, \dots \right\}$$

is for any $n = 0,1,\dots$ an orthonormal basis in $L^2(-1,1)$; this is true, however, because the functions in question are obtained by orthonormalization of the family $\{z \mapsto (1-z^2)^{n/2}z^j : j = 0,1,\dots\}$ which is total due to Example 2.2.2. ∎

Now we are able to use the operator (11.16) to transform the Hamiltonian. Choose $D := \mathcal{S}(\mathbb{R}^3) \cap D_V$ as the domain of H; then a straightforward computation yields

$$UHU^{-1}\Phi = -\frac{1}{2m}\left(\partial_r^2 + \frac{1}{r^2}\Lambda\right)\Phi + v\Phi \tag{11.18}$$

for any $\Phi \in UD$ (*cf.* Problem 14c). By Proposition 2, $L^2(S,d\Omega) = \bigoplus_{l=0}^{\infty} \mathcal{G}_l$, where $\mathcal{G}_l$ is the subspace spanned by Y_{lm}, $m = -l,\dots,l$; it follows that

$$UL^2(\mathbb{R}^3) = L^2(\mathbb{R}^+,dr)\otimes L^2(S,d\Omega) = \bigoplus_{l=0}^{\infty} L^2(\mathbb{R}^+)\otimes \mathcal{G}_l\,.$$

Let D_v be the domain of the operator of multiplication by the function v, *i.e.*, $D_v := \{g \in L^2(\mathbb{R}^+) : vg \in L^2(\mathbb{R}^+)\}$ and suppose for a moment that for any $l = 0,1,\dots$ we can find a subspace $D^{(l)} \subset C_0^\infty(\mathbb{R}^+\setminus\{0\})\cap D_v$ in such way that the operator h_l defined on $D^{(l)}$ by the differential expression

$$h_l := -\frac{1}{2m}\left[\frac{d^2}{dr^2} - \frac{l(l+1)}{r^2}\right] + v(r) \tag{11.19}$$

is *e.s.a.*Then the operator $h_l\otimes I_l$ on $L^2(\mathbb{R}^+)\otimes\mathcal{G}_l$ with the domain $\mathcal{D}_l := \{fY_{lm} : f \in D^{(l)},\, m = -l,\dots,l\}$, where I_l is the unit operator on $\mathcal{G}_l$, is again *e.s.a.* by Theorem 5.7.2. Since $\mathcal{D}_l \subset U\mathcal{S}(\mathbb{R}^3)$ due to Problem 15b and $D^{(l)} \subset D_v$, it follows from Theorem A.3.11 that

$$\int_{\mathbb{R}^3} |(V\psi)(x)|^2 dx = \int_{\mathbb{R}^+\times S} |(UV\psi)(r,\theta,\varphi)|^2 dr\, d\Omega = \|vfY_{lm}\|^2 < \infty$$

holds for $\psi = U^{-1}(fY_{lm}) \in \mathcal{S}(\mathbb{R}^3)$, and therefore $\mathcal{D}_l \subset U(\mathcal{S}(\mathbb{R}^3)\cap D_V) = UD$. Now the relation (11.18) in combination with Proposition 11.5.2 yields the identity

$$UHU^{-1}\restriction \mathcal{D}_l = h_l\otimes I_l\,.$$

Next we construct the operator $\tilde H$ as the algebraic direct sum of $h_l\otimes I_l$, *i.e.*, its domain $\mathcal{D} \subset UD$ consists of all *finite* sums $\Phi = \sum_l \phi_l$ with $\phi_l \in \mathcal{D}_l$, and the operator acts as $\tilde H\Phi := \sum_l (h_l\otimes I_l)\phi_l$ for any $\Phi \in \mathcal{D}$. It is easy to see that $\tilde H \subset \sum_l^{\oplus} h_l\otimes I_l \subset \overline{\tilde H}$ (see the notes to Section 4.4), so the closures of the two operators coincide, and since $h_l\otimes I_l$ are *e.s.a.* the same is true, by Problem 4.29, for $\tilde H$, and *a fortiori*, for its symmetric extension $\sum_l^{\oplus} h_l\otimes I_l$. Moreover, we have $UHU^{-1}\restriction\mathcal{D} = \tilde H$, and therefore

$$U\overline{H}U^{-1} = \overline{\sum_{l=0}^{\infty}{}^{\oplus}\, h_l\otimes I_l} \tag{11.20}$$

due to Problem 4.27d. In this way we have reduced analysis of the operator (9.2) with a centrally symmetric potential to treatment of the sequence of ordinary differential operators (11.19); this procedure is usually referred to as the **partial–wave decomposition**. We have demonstrated that if all the operators h_l are *e.s.a.* the same is true for H. In that case the spectrum of $\overline{H}$ is determined by the spectra of h_l: using Problem 4.24b and a simple induction argument we find

$$\sigma(\overline{H}) = \overline{\bigcup_{l=0}^{\infty} \sigma(\overline{h}_l)}$$

and similar relations for the other types of spectra.

11.5.3 Remark: The presented construction generalizes to operators (9.2) on $L^2(\mathbb{R}^n)$, $n \geq 2$, with a potential V which depends on $r := \left(\sum_{j=1}^n x_j^2\right)^{1/2}$. The generalized spherical coordinates can be introduced as the map $x^{(n)} : \mathbb{R}^+ \times S^{n-1} \to \mathbb{R}^n$, where S^{n-1} is the unit sphere in $\mathbb{R}^n$, defined by the recursive relations

$$x_n^{(n)} := r\cos\theta_{n-2}\,, \quad x_k^{(n)} := x_k^{(n-1)}\sin\theta_{n-2}\,, \quad k = 1,\dots,n-1$$

with $x_1^{(2)} := r\cos\varphi$ and $x_2^{(2)} := r\sin\varphi$; the parameter φ runs through $[0, 2\pi)$ and θ_j, $j = 1,\dots,n-2$, through $[0,\pi)$. The map $U_n : U_n\psi = r^{(n-1)/2}\psi\circ x^{(n)}$ then defines a unitary operator from $L^2(\mathbb{R}^n)$ to $L^2(\mathbb{R}^+ \times S^{n-1}, dr\, d\Omega_n)$, where $d\Omega_n := \left(\prod_{j=1}^{n-2}\sin^j\theta_j\right) d\varphi d\theta_1 \dots d\theta_{n-2}$. The relation (11.18) is now replaced by

$$\begin{aligned} U_n\left(-\frac{1}{2m}\Delta + V\right)U_n^{-1}\phi& \\ = -\frac{1}{2m}\left[\frac{\partial^2}{\partial r^2} - \frac{(n-1)(n-3)}{4r^2} + \frac{1}{r^2}\Lambda_n\right]\phi + v\phi\,,& \end{aligned} \tag{11.21}$$

where Λ_n is the so-called Laplace-Beltrami operator on S^{n-1} (see the notes). We can again show that there are finite-dimensional subspaces $\mathcal{G}_l^{(n)} \subset C^\infty(S^{n-1}) \subset L^2(S^{n-1}, d\Omega_n)$ such that $L^2(S^{n-1}, d\Omega_n) = \bigoplus_{l=0}^\infty \mathcal{G}_l^{(n)}$ and

$$\Lambda_n Y = -l(l+n-2)Y$$

holds for all $Y \in \mathcal{G}_l^{(n)}$, so instead of $-\frac{1}{2m}\Delta + V$ we have to analyze the corresponding countable family of ordinary differential operators on $L^2(\mathbb{R}^+)$.

In both the examples discussed here, symmetry of the problem played an important role. Decomposition into partial waves is based in fact on a reduction of the rotation-group representation generated by the angular momentum operators into irreducible components. On the other hand, if a two-particle potential is of the form (11.10), we are able to decompose the corresponding system into a pair of fictitious subsystems; among these the subsystem associated with the center-of-mass motion is free, and therefore invariant with respect to translations — *cf.* Example 10.2.1b.

It frequently occurs that more than one subsystem of a composite system exhibits a symmetry. Often the subsystems have the same symmetry group. In order to look more closely at transformation properties of composite systems, consider again for simplicity the case of a system S consisting of a pair of subsystems S_1, S_2, and suppose that transformations of a group G can be applied to both of them, and that they are realized on the state spaces $\mathcal{H}_j$ by means of representations $U_j(\cdot)$, $j = 1,2$; the subsystems are invariant with respect to G *iff* their Hamiltonians H_j commute with $U_j(g)$ for all g. In view of the postulate (Q5b), the transformation group G has to be represented on the state space $\mathcal{H} := \mathcal{H}_1 \otimes \mathcal{H}_2$. We assume that

(Q6c) states of S transform with respect to G by means of the representation $U : G \to \mathcal{H}$ defined by $U(g) := U_1(g) \otimes U_2(g)$ for any $g \in G$.

It is evident that $U(\cdot)$ is a representation of G. In most situations, assumption (Q6c) can be justified easily. In particular, if G is a continuous group such that the representations of its one–parameter subgroups are generated by observables having an additive character, we can use Proposition 5.9.6; an illustration is given by the addition of angular momentum and spin in Example 10.2.4 or by Proposition 11.3.1.

It is therefore reasonable to postulate this relation between transformation–group representations for all composite systems. Using it, we obtain the following assertion.

11.5.4 Proposition: If the subsystems are invariant under G and do not interact, the composite system is also invariant with respect to G.

Proof: By the spectral theorem, a system is invariant under G *iff* its propagator commutes with the corresponding representation of G; the result then follows from Proposition 11.3.1 and Theorem 4.5.2a. ∎

We have seen how a reduction of a symmetry–group representation to irreducible components simplifies analysis of a system. It often happens, however, that the representation $U(\cdot)$ obtained as the tensor product of the subsystem representations is reducible, and we have to perform its reduction. A simple illustration is given in Problem 16; more information about these reductions for various groups — which are a frequently used tool in studying symmetries of quantum systems — can be found in the literature quoted in the notes.

Notes to Chapter 11

Section 11.1 A deeper justification of the postulate (Q6a) within the axiomatic approach, which will be discussed in Section 13.1, can be found in [AD 1,2]. We cannot conclude from Example 11.1.6 that any self–adjoint operator reduced by the total charge eigenspaces is an observable of the nucleon–pion system, because the neglection of the non-isospin degrees of freedom is a rather crude simplification. If we should want to justify assumption (11.1) it is worth keeping in mind that in quantum mechanics, where we typically deal with finite systems of particles which preserve their identities, with a few simple superselection rules at most, algebras of observables represent a useful ingredient of the formalism but they are not indispensable. The charge–exchange processes mentioned in the example belong in fact to quantum field theory; we mentioned in Section 7.6 that assumption (11.1) is not valid there even in the coherent case.

Section 11.2 The restriction to the quantum mechanical case with the algebra of observables of the form (11.1) is natural, because in other cases statistical operators may not be an optimal tool for description of states. Notice also that nowhere in this section do we need to know the algebra of observables of the composite system. We repeatedly use countable orthonormal bases without assuming separability of the state spaces involved; this is made possible by the fact that the series in question always have only countably many nonzero terms — *cf.* Remark 6.4.2.

An attentive reader may wonder why we speak about a pair of electrons in Example 2 when we declared above that the choice of the state space based on the postulate (Q6a) does not apply to systems of identical particles. The answer is that the considerations of Section 11.4 concern the whole state space; in this particular situation the two electrons can exist in the triplet as well as in the singlet spin state, but a different symmetry of the configuration–space part of the wave function corresponds to each of these cases.

The relations (11.4) show that the reduction of a pure state leads in general to mixed states; in quantum mechanical textbooks these are usually referred to as *mixed states of the second kind.* It should be stressed, however, that it is irrelevant for the formalism whether a mixed state is obtained in this way or whether it expresses our incomplete knowledge of the state of a particular member of some family of identical systems.

Reduction formulas (11.4) can be written in an elegant form provided we use the particular realization of the tensor product mentioned in Problem 3.39b — see, *e.g.*, ⟦Ja⟧, Sec.11.8. This expression, as well as the normal form (11.5) of the reduction, applies only to the case of a system consisting of two subsystems, while the relations (11.4) generalize easily to any finite number of subsystems (Problem 7).

States of two and more spins such as considered in Example 11.2.2 became important since the first edition of this book because of their vital role in *quantum computing* where a single spin state state represents a basic unit of quantum information called *qubit* — for a review of this field see, *e.g.*, [Ste 1]. Example 11.2.2 comes originally from [BE 1]; the states $W(w_0, w_1)$ appearing there were later used in [We 1] being now called *Werner states* in the quantum–information context with the parameter $F := 1 - 3w_0$ understood as "fidelity". For the reduction of such states see also [Pe 1].

Section 11.3 We have mentioned that we are not always interested in the time evolution of reduced states. Recall two typical situations where we encounter systems consisting of interacting subsystems S_1, S_2. A *bound state* of the subsystems is a pure state described by an eigenvector of the total Hamiltonian H. Such states are stationary and the same is, of course, true for reduced states. On the other hand, if we investigate the *scattering* of S_2 on S_1, we compare the full propagator $U(t) = e^{-iHt}$ of the composite system with $U_0(t) = e^{-iH_1t} \otimes e^{-iH_2t}$ corresponding to the case when the subsystems do not interact; loosely speaking, the time behavior of the reduced states is here of interest in the region where the reduction is trivial. More will be said about bound and scattering states in Section 15.1.

Section 11.4 A more detailed discussion of the concept of identity in quantum theory can be found in ⟦Ja⟧, Sec.15.3. The connection between the type of a particle and the transformation properties (11.8) of its states is usually referred to briefly as the *relation between spin and statistics.* In quantum mechanics it has to be regarded as an empirical fact whose most striking manifestation is the Pauli principle discussed in Example 1; it has deep physical consequences which we shall mention in the notes to Section 14.3.

On the other hand, in quantum field theory the relation between spin and statistics denotes a result closely related to the mentioned transformation property, namely that field operators in points which are causally inaccessible, *i.e.*, separated by a space–like interval cannot commute (anticommute) if the particles associated with this field are fermions (respectively, bosons). In the framework of axiomatic quantum field theory this claim can be demonstrated — see the notes to Section 13.3.

The permutation group S_n and its representations are discussed, *e.g.*, in ⟦Ham⟧,

Chap.7; ⟦BaR⟧, Sec.7.5. If $n \geq 2$ we have $S_n + A_n \neq I$; this is related to the existence of other irreducible representations of S_n than the two we have employed here. We naturally ask whether there are systems with a permutation behavior different from Fermi–Dirac and Bose–Einstein statistics. Various *parastatistics* of this type have been studied (an extensive discussion of the problem is given, for instance, in ⟦OK⟧) but they remain until now a mere mathematical construction.

Section 11.5 The *Laplace–Beltrami operator* Δ_M is defined on a Riemannian manifold M with a metric tensor g by $(\Delta_M\psi)(x) := -g(x)^{-1/2}(\partial_i g(x)^{1/2} g^{ij}(x)\partial_j\psi)(x)$, where $\partial_j := \partial/\partial x^j$ and $g(x) := \det(g^{ij}(x))$. Notice the sign convention: in the case $M = \mathbb{R}^n$ when $g_{ij} = \delta_{ij}$ we have $\Delta_M = -\Delta$. Properties of the operators $\Lambda_n := \Delta_{S^{n-1}}$ for $n = 2, 3$ can be demonstrated directly as we did in Proposition 11.5.2; for the general case we refer, *e.g.*, to the monograph ⟦Mül⟧. Examples of partial wave decomposition for $n > 3$ can be found, for instance, in [Ba 4], [CH 1], [Ex 2].

Problems

1. The conclusions of Example 11.1.6 remain valid even if we also consider the other degrees of freedom of the two particles.

2. A family $\mathcal{S} \subset \mathcal{B}(\mathbb{C}^n)$ is irreducible provided it contains at least n noncommuting Hermitean operators. In the case $n \geq 3$ this condition is sufficient but not necessary; what is the least number of noncommuting Hermitean operators that an irreducible $\mathcal{S}$ must contain?
Hint: Consider the operators S_1, S_3 of Example 7.5.4b.

3. Let $\mathcal{H} = \mathcal{H}_1 \otimes \mathcal{H}_2$. If a sequence $\{B_1^{(k)}\} \subset \mathcal{B}(\mathcal{H}_1)$ converges weakly to an operator B_1, then $\text{w-lim}_{k\to\infty} \underline{B}_1^{(k)} = \underline{B}_1$.

4. Let $\mathcal{H} = \mathcal{H}_1 \otimes \mathcal{H}_2$. Prove:

(a) If $C_j \in \mathcal{J}_p(\mathcal{H}_j)$, $j = 1, 2$, for some $p \geq 1$, then $C_1 \otimes C_2 \in \mathcal{J}_p(\mathcal{H})$ and $\|C_1 \otimes C_2\|_p = \|C_1\|_p \|C_1\|_p$.

(a) If the operators C_j are of the trace class and $B_j \in \mathcal{B}(\mathcal{H}_j)$, $j = 1, 2$, then $\mathrm{Tr}\,(B_1C_1 \otimes B_2C_2) = \mathrm{Tr}\,(B_1C_1)\,\mathrm{Tr}\,(B_2C_2)$.

(c) The tensor product of statistical operators is a statistical operator.

5. Prove: (a) Let the state W of the composite system be pure and $W = W_1 \otimes W_2$, where the W_j are realizable pure states of the subsystems; then the reduced states are also pure and satisfy $W_j(W) = W_j$, $j = 1, 2$.

(b) Suppose that the state of the composite system is described by the statistical operator $W = \sum_j w_j E^{(j)}$, where $\{E^{(j)}\}$ is a sequence of one–dimensional projections; then the reduced states are $W_j(W) = \sum_j w_j W_j(E^{(j)})$ (strong convergence).

6. Suppose that $\{E_\alpha : \alpha \in J\}$ is the projection family corresponding to the coherent subspaces in $\mathcal{H}$ and $\tilde{W} \in \mathcal{J}_1(\mathcal{H})$ is a statistical operator; then the operator

$$W := \sum_{\alpha \in J} E_\alpha \tilde{W} E_\alpha \quad \text{(strong convergence)}$$

is again statistical, it is reduced by all E_α, and $\mathrm{Tr}\,(A\tilde{W}) = \mathrm{Tr}\,(AW)$ holds for any $A \in \sum_{\alpha \in J}^{\oplus} \mathcal{B}(\mathcal{H}_\alpha)$. If W is a projection the same is true for $\tilde{W}$, while the opposite implication is not valid.

7. Derive the reduction formulas for a system consisting of n coherent subsystems.

8. Let $H_j : \mathbb{R} \to \mathcal{B}(\mathcal{H}_j)$, $j = 1, 2$, be strongly continuous Hermitean–valued functions; then the unitary propagator corresponding to $H(t) = \underline{H}_1(t) + \underline{H}_2(t)$ satisfies for all $t, s \in \mathbb{R}$ the relation $U(t,s) = U_1(t,s) \otimes U_2(t,s)$, where the $U_j(t,s)$ are the unitary propagators referring to the Hamiltonians $H_j(t)$.
Hint: It is sufficient to check that the operators coincide on $\mathcal{H}_1 \times \mathcal{H}_2$; use the Dyson expansion.

9. Check the computations of Example 10.3.3.

(a) Find $U(t)$ using Example 5.2.3.

(b) Show that $\{U(t) : t \in \mathbb{R}\}$ is a one–parameter unitary group, and compute its generator.

(c) Find the reduced states.

10. Prove: (a) The operator (11.7) is well defined, *i.e.*, $\psi = 0$ implies $U_P\psi = 0$, its continuous extension is unitary, and $U_P^2 = I$.

(b) Generalize these conclusions to the operators $U(p)$ defined by (11.9).

11. Let W be a state of n identical particles, and denote $W_p := U(p)WU(p)^{-1}$ for any permutation $p \in \mathcal{S}_n$, where the operators $U(p)$ are defined by (11.9); then the reduced states satisfy $W_j(W) = W_{r_j}(W_p)$, $j = 1, \dots, n$, where $r := p^{-1}$.

12. Consider the Hamiltonian of two nonrelativistic spinless particles with the potential (11.10) and suppose that the operator H_{rel} is *e.s.a.*

(a) Prove the relation (11.12).

(b) Find the spectrum of $\overline{H}$.

(c) Given a spectral representation of $\overline{H}_{rel}$, construct that of $\overline{H}$.

Hint: (b) Use Example 5.7.5.

13. Consider the Hamiltonian $H_0 := \sum_{j=1}^{N} (2m_j)^{-1} P_j^2$ describing a system of N noninteracting spinless particles.

(a) Set $U:\,(U\psi)(x_1,\dots,x_n)=\psi(X,y_1,\dots,y_{N-1})$ with $X:=M^{-1}\sum_{j=1}^N m_jx_j$, where $M:=\sum_{j=1}^N m_j$ is the total mass of the system and y_j are the atomic coordinates (11.13); then

$$U^{-1}H_0U=\frac{1}{2M}(P^{cm})^2+\sum_{j=1}^{N-1}\frac{1}{2\mu_j}(P_j^{at})^2-\frac{1}{m_N}\sum_{j=1}^{N-1}\sum_{i=1}^{j-1}P_j^{at}.P_i^{at}\,,$$

where the P_{jk}^{at}, $k=1,2,3$, are the momentum components corresponding to the atomic coordinates of the j–th particle, $P_j^{at}.P_i^{at}:=\sum_{k=1}^3 P_{jk}^{at}P_{ik}^{at}$ and $\mu_j^{-1}:=m_j^{-1}+m_N^{-1}$.

(b) Introducing the analogous substitution operator for the Jacobi coordinates (11.14), we have

$$U^{-1}H_0U=\frac{1}{2M}(P^{cm})^2+\sum_{j=1}^{N-1}\frac{1}{2\mu_j}(P_j^J)^2\,,$$

where the momentum component operators P_{jk}^J correspond to the Jacobi coordinates of the j–th particle and $\mu_j^{-1}:=m_{j+1}^{-1}+(\sum_{i=1}^j m_i)^{-1}$.

14. Let $r,\,\varphi,\,\theta_1,\dots,\theta_{n-2}$ be the generalized spherical coordinates in R^n introduced in Remark 11.5.4. Prove:

(a) The measure Ω on S^{n-1} given by $d\Omega:=\prod_{j=1}^{n-2}\sin^j\theta_j\,d\varphi\,d\theta_1\dots d\theta_{n-2}$ is invariant with respect to the rotations of the sphere around the k–th axis, $k=1,\dots,n$.

(b) The map $U_n:\,L^2(\mathbb{R}^n)\to L^2(\mathbb{R}^+\times S^{n-1},dr\,d\Omega)$ defined by

$$(U_n\psi)(r,\,\varphi,\,\theta_1,\dots,\theta_{n-2})\,:=\,r^{(n-1)/2}(\psi\circ x^{(n)})(r,\,\varphi,\,\theta_1,\dots,\theta_{n-2})$$

is an isometry of the two spaces.

(c) Check the relation (11.21).

15. Prove: (a) Let U be the operator (11.16). Show that for any $f\in C_0^\infty(\mathbb{R}^+\setminus\{0\})$ and Y_{lm} there is a function $\psi\in\mathcal{S}(\mathbb{R}^3)$ such that $(U\psi)(r,\theta,\varphi)=f(r)Y_{lm}(\theta,\varphi)$.

(b) Check that the subspace $C_0^\infty(\mathbb{R}^+\setminus\{0\})$ is dense in $L^2(\mathbb{R}^+,r^\alpha dr)$ for any $\alpha\in\mathbb{R}$, and similarly, $C_0^\infty(\mathbb{R}^n\setminus M)$ is dense in $L^2(\mathbb{R}^n)$ for a positive integer n provided M is an at most countable set without accumulation points.

16. Consider the two–electron system of Example 11.2.2 and denote by $S_k^{(j)}=\frac{1}{2}\sigma_k$ the spin–component operators of the j–th electron.

(a) Let $S_k:=\underline{S}_k^{(1)}+\underline{S}_k^{(2)}$ be the components of the total spin; find the spectral decomposition of the operator $S^2:=\sum_{k=1}^3 S_k^2$.

(b) Construct the corresponding representation of the rotation group.

Hint: Cf. Example 10.2.4.

Chapter 12

The second quantization

12.1 Fock spaces

Now we want to show how some results from the preceding chapter can be extended to situations, where the number of particles is not preserved. The first question is what kind of a state space can be associated with such a system. We start from the one–particle Hilbert space $\mathcal{H}$ and again denote $\mathcal{H}^{(n)} := \mathcal{H} \otimes \cdots \otimes \mathcal{H}$; in addition, we set $\mathcal{H}^{(0)} := \mathbb{C}$. The direct sum of these spaces,

$$\mathcal{F}(\mathcal{H}) := \sum_{n=0}^{\infty}{}^{\oplus} \mathcal{H}^{(n)} ,$$

is then called the **Fock space** over $\mathcal{H}$. Hence if the norm and inner product in $\mathcal{H}^{(n)}$ are indexed by n, the elements of $\mathcal{F}(\mathcal{H})$ are sequences $\Psi = \{\psi_n\}_{n=0}^{\infty}$ with $\psi_n \in \mathcal{H}^{(n)}$ such that $\sum_{n=0}^{\infty} \|\psi_n\|_n^2 < \infty$; the inner product in $\mathcal{F}(\mathcal{H})$ is correspondingly given by

$$(\Phi, \Psi)_F := \sum_{n=0}^{\infty} (\phi_n, \psi_n)_n .$$

Suppose that $\mathcal{H}$ is the state space of a particle; then it is obvious from postulate (Q6a) that pure states of the system of these particles may be described by vectors from $\mathcal{F}(\mathcal{H})$; if the number of particles equals n, they belong to the subspace $\{\Psi = \{\psi_k\} : \psi_k = 0 \text{ for } k \neq n\}$, which we shall for the sake of simplicity denote also as $\mathcal{H}^{(n)}$. The one–dimensional subspace $\mathcal{H}^{(0)}$ then refers to the state with zero number of particles which is called a **vacuum**; we shall use the symbol Ω_0 for the corresponding vector $\{1, 0, 0, \ldots\}$.

Since we are considering systems of identical particles, not every vector of $\mathcal{F}(\mathcal{H})$ can be associated with an admissible state. We know that the subspaces $S_n\mathcal{H}^{(n)}$ and $A_n\mathcal{H}^{(n)}$ in $\mathcal{H}^{(n)}$ are ascribed to systems of n identical bosons and fermions, respectively. This concerns systems with $n \geq 2$, of course, because it makes no sense to speak about the interchange of particles for $n = 0, 1$ and we set $S_n = A_n = I$ in these cases. Now we can define the *symmetric Fock space* as

$$\mathcal{F}_s(\mathcal{H}) := \sum_{n=0}^{\infty}{}^{\oplus} S_n \mathcal{H}^{(n)}$$

and the *antisymmetric Fock space* as

$$\mathcal{F}_a(\mathcal{H}) := \sum_{n=0}^{\infty}{}^{\oplus} A_n \mathcal{H}^{(n)};$$

these play the role of state space if the particles under considerations are bosons and fermions, respectively.

12.1.1 Remark (about notation): The symbols $(\cdot,\cdot)$ and $\|\cdot\|$ will be reserved for the inner product and the norm in $\mathcal{H}$; this space and quantities related to it will be characterized by the adjective *one–particle;* similarly the space $\mathcal{H}^{(n)}$, its vectors, operators on it, *etc.*, will be denoted as *n–particle.* The inner product and the norm in $\mathcal{H}^{(n)}$ and $\mathcal{F}(\mathcal{H})$ will be indexed by n and F, respectively. It is also useful to introduce the symbol P_n as the common notation for the two projections, the symmetrizer S_n and antisymmetrizer A_n; all the formulas in which it appears are valid in both the symmetric and antisymmetric case. In the same way, we write $\mathcal{F}_p(\mathcal{H})$ instead of $\mathcal{F}_s(\mathcal{H})$ and $\mathcal{F}_a(\mathcal{H})$, use the symbol $\mathcal{H}_p^{(n)}$ for the n–particle subspace in $\mathcal{F}_p(\mathcal{H})$, *etc.*

12.1.2 Example: If the one–particle space is $\mathcal{H} := L^2(\mathbb{R}^3)$, then we have $\mathcal{H}^{(n)} = L^2(\mathbb{R}^{3n})$ by Example 2.4.5. The Fock space $\mathcal{F}(L^2(\mathbb{R}^3))$ consists of the sequences $\Psi = \{\psi_n\}_{n=0}^{\infty}$ of (equivalence classes of) functions $\psi_n : \mathbb{R}^{3n} \to \mathbb{C}$ such that

$$\|\Psi\|_F^2 := |\psi_0|^2 + \sum_{n=1}^{\infty} \int_{\mathbb{R}^{3n}} |\psi_n(x_1,\dots,x_n)|^2 \, dx_1 \dots dx_n < \infty .$$

Let us see what the subspace $\mathcal{F}_s(L^2(\mathbb{R}^3))$ looks like. We take any orthonormal basis in $L^2(\mathbb{R}^3)$ and construct an orthonormal basis in $L^2(\mathbb{R}^{3n})$ by Proposition 2.4.4b; using it together with the definition relation (11.9), we readily find that

$$(U(p^{-1})\psi_n)(x_1,\dots,x_n) = \psi_n(x_{p_1},\dots,x_{p_n})$$

holds for any permutation $p \in \mathcal{S}_n$ a.e. in $\mathbb{R}^{3n}$. On the other hand, any $\psi_n \in S_n L^2(\mathbb{R}^{3n})$ satisfies the relations $U(p^{-1})\psi_n = U(p^{-1})S_n\psi_n = S_n\psi_n = \psi_n$, so we have

$$\psi_n(x_{p_1},\dots,x_{p_n}) = \psi_n(x_1,\dots,x_n)$$

a.e. in $\mathbb{R}^{3n}$; hence the subspace $S_n L^2(\mathbb{R}^3)$ consists of functions which, up to a zero–measure set, are symmetric with respect to any permutation of variables. Similarly, $\mathcal{F}_a(L^2(\mathbb{R}^3))$ consists of sequences of functions which are a.e. antisymmetric with respect to permutations of particle positions,

$$\psi_n(x_{p_1},\dots,x_{p_n}) = \epsilon_p \psi_n(x_1,\dots,x_n) .$$

The same results are also valid for other dimensions of the configuration space, *i.e.*, for $\mathcal{H} := L^2(\mathbb{R}^d)$ with any positive integer d.

Let T be a densely defined operator on $\mathcal{H}$; we shall describe a standard procedure by which we can construct to it an operator on $\mathcal{F}_p(\mathcal{H})$. As earlier, the symbol $\underline{T}_j$ will denote the tensor product $I\otimes\cdots\otimes I\otimes T\otimes I\otimes\cdots\otimes I$ in which the operator T stands at the j–th place. For any $n\ge 1$ we define the operators

$$T_n^{\Sigma} := \sum_{j=1}^{n} \underline{T}_j\,, \qquad T_n^{\Pi} := T\otimes\cdots\otimes T$$

whose domain is $D_n(T) := \mathcal{H}^{(n)}$ if T is bounded and $D_n(T) := D(T)\underline{\times}\cdots\underline{\times}D(T)$ otherwise; it is obvious that they are densely defined. In particular, $T_1^{\Sigma} = T_1^{\Pi} = T$, and we also set $T_0^{\Sigma} := 0$ and $T_0^{\Pi} := I$. Then we are able to define the following operators on $\mathcal{F}(\mathcal{H})$,

$$\mathcal{T}^{\Sigma}(T) \;:\; \mathcal{T}^{\Sigma}(T)\{\psi_n\}_{n=0}^{\infty} = \{T_n^{\Sigma}\psi_n\}_{n=0}^{\infty}\,,$$

$$\mathcal{T}^{\Pi}(T) \;:\; \mathcal{T}^{\Pi}(T)\{\psi_n\}_{n=0}^{\infty} = \{T_n^{\Pi}\psi_n\}_{n=0}^{\infty}$$

with the common domain formed by the subspace

$$\mathcal{D}(T) := \{\Psi_N = \{\psi_0,\dots,\psi_N,0,\dots\} :\, \psi_n\in D_n(T),\ N=0,1,\dots\}.$$

12.1.3 Proposition: The operators $\mathcal{T}^{\Sigma}(T)$ and $\mathcal{T}^{\Pi}(T)$ are densely defined. Each of them is reduced by the subspaces $\mathcal{F}_p(\mathcal{H})$, and their parts in these subspaces, which we will be denoted by $\mathcal{T}_p^{\Sigma}(T)$ and $\mathcal{T}_p^{\Pi}(T)$, respectively, are also densely defined.

Proof: It follows from the definition of the norm $\|\cdot\|_F$ that to any $\Psi = \{\psi_n\}_{n=0}^{\infty}$ and $\varepsilon>0$ there is an N such that $\|\Psi-\Psi_N\|_F < \frac{1}{2}\varepsilon$ holds for $\Psi_N := \{\psi_0,\dots,\psi_N,0,\dots\}$. Using the fact that $\overline{D_n(T)} = \mathcal{H}^{(n)}$, we readily find that $\mathcal{D}(T)$ is dense in $\mathcal{F}(\mathcal{H})$.

Next we shall check the assertion about the reduction. Let P the projection onto the (anti)symmetric subspace, $P\Psi = \{P_n\psi_n\}_{n=0}^{\infty}$, and consider an arbitrary n–tuple of vectors $f_j\in D(T)$. We have $f_{p_1}\otimes\cdots\otimes f_{p_n}\in D_n(T)$ for any permutation $p\in\mathcal{S}_n$; due to Problem 11.10b this means $U(p)D_n(T)\subset D_n(T)$, and therefore also $P_nD_n(T)\subset D_n(T)$. The definition of $\mathcal{D}(T)$ then yields $P\Psi\in\mathcal{D}(T)$ for any $\Psi\in\mathcal{D}(T)$, *i.e.*, $P\mathcal{D}(T)\subset\mathcal{D}(T)$. Let $\{f_j\}$, $\{g_j\}$ be n–tuples of vectors from $D(T)$ and $p\in\mathcal{S}_n$. Denoting $r := p^{-1}$, we may write

$$(g_1\otimes\cdots\otimes g_n,\, U(p)T_n^{\Sigma}(f_1\otimes\cdots\otimes f_n))_n$$

$$= \sum_{j=1}^{n} (g_{r_1}\otimes\cdots\otimes g_{r_n},\, f_1\otimes\cdots\otimes Tf_j\otimes\cdots\otimes f_n)_n$$

$$= \sum_{j=1}^{n} (g_{r_j}, Tf_j) \prod_{k\neq j} (g_{r_k}, f_k) = \sum_{l=1}^{n} (g_l, Tf_{p_l}) \prod_{m\neq l} (g_m, f_{p_m})$$

$$= (g_1 \otimes \cdots \otimes g_n, T_n^{\Sigma} U(p)(f_1 \otimes \cdots \otimes f_n))_n .$$

The condition $\overline{D(T)} = \mathcal{H}$ in combination with Proposition 2.4.4a then implies $U(p)T_n(f_1\otimes\cdots\otimes f_n) = T_n U(p)(f_1\otimes\cdots\otimes f_n)$ for any $f_j \in D(T)$. This identity extends linearly to $D_n(T)$, and since it is valid for all $p \in \mathcal{S}_n$, we get $P_n T_n^{\Sigma}\psi_n = T_n^{\Sigma} P_n \psi_n$ for any $\psi_n \in D_n(T)$.

Take now a vector $\Psi_N = \{\psi_0, \dots, \psi_N, 0, \dots\} \in \mathcal{D}(T)$. The inclusion $P\mathcal{D}(T) \subset \mathcal{D}(T)$ means that $P\Psi_N = \{P_0\psi_0, \dots, P_N\psi_N, 0, \dots\}$ belongs to $\mathcal{D}(T)$ and

$$P\mathcal{T}^{\Sigma}(T)\Psi_N = \{P_n T_n^{\Sigma}\psi_n\}_{n=0}^{\infty} = \{T_n^{\Sigma} P_n \psi_n\}_{n=0}^{\infty} = \mathcal{T}^{\Sigma}(T)P\Psi_N ,$$

i.e., the operator $\mathcal{T}^{\Sigma}(T)$ is reduced by the subspaces $\mathcal{F}_p(\mathcal{H})$. The condition $\overline{\mathcal{D}(T)} = \mathcal{F}(\mathcal{H})$ implies that the subspace $P\mathcal{D}(T)$ is dense in $\mathcal{F}_p(\mathcal{H})$, so $\mathcal{T}_p^{\Sigma}(T)$ is densely defined. The proof for the operator $\mathcal{T}_p^{\Pi}(T)$ is similar; we leave it to the reader (Problem 2). ∎

The operator $\mathcal{T}^{\Sigma}(T)$ is called the **second quantization** of the one–particle operator T. The same name is used for its part in the subspace $\mathcal{F}_p(\mathcal{H})$; we are, of course, interested in one of these subspaces only, depending on whether the particles under consideration are bosons or fermions. Such operators are used to describe some observables for systems of identical particles provided the operator $\mathcal{T}_p^{\Sigma}(T)$ is *e.s.a.* (in that case, its closure is often denoted as $d\Gamma(T)$; the origin of this notation is explained in the notes).

12.1.4 Theorem: Let A be a self–adjoint operator on $\mathcal{H}$; then the operators $\mathcal{T}^{\Sigma}(A)$ and $\mathcal{T}_p^{\Sigma}(A)$ are *e.s.a.*

Proof: According to the definition given in the notes to Section 4.4, the operator $\mathcal{T}^{\Sigma}(A)$ is a restriction of the direct sum $\mathcal{T}_{\oplus}(A) := \bigoplus_{n=0}^{\infty} A_n$; both operators are obviously symmetric. Given $\Phi \in D(\mathcal{T}_{\oplus}(A))$, we consider the truncated vectors $\Phi_N := \{\phi_0, \dots, \phi_N, \dots\}$. Each of them belongs to $D(\mathcal{T}^{\Sigma}(A))$ and the sequence $\{\Phi_N\}_{N=1}^{\infty}$ converges; this follows from the fact that $\sum_{n=0}^{\infty} \|A_n\phi_n\|_n^2$ is finite by assumption and

$$\sum_{n=N+1}^{M} \|A_n\phi_n\|_n^2 = \|\mathcal{T}^{\Sigma}(A)(\Phi_M - \Phi_N)\|^2 .$$

In this way, we obtain the inclusion $\mathcal{T}_{\oplus}(A) \subset \overline{\mathcal{T}^{\Sigma}(A)}$, which implies in turn that the closures of the two operators coincide; however, $\mathcal{T}_{\oplus}$ is *e.s.a.* due to Theorem 5.7.2 and Problem 4.29b. The essential self–adjointness of $\mathcal{T}_p^{\Sigma}(A)$ follows from Proposition 12.1.3 and Problem 4.24b. ∎

12.1.5 Example: Consider the simplest nontrivial case: the second quantization of the unit operator I. The operator $\mathcal{T}^{\Sigma}(I)$ is *e.s.a.* by the preceding theorem, and $\sigma(\mathcal{N}) \subset \{0,1,2,\dots\}$ holds for $\mathcal{N} := \overline{\mathcal{T}^\Sigma(I)}$ (*cf.* Problem 3). We shall show that this inclusion is in fact an identity. The operators I_n^Σ are bounded and satisfy $I_n^\Sigma(f_1\otimes\cdots\otimes f_n) = n(f_1\otimes\cdots\otimes f_n)$ for any $f_1,\dots,f_n \in \mathcal{H}^{(n)}$, so we obtain $I_n^\Sigma = nI_n$, where I_n is the unit operator on $\mathcal{H}^{(n)}$. Then the self-adjoint operator $\mathcal{N}$ has a pure point spectrum and the n-particle subspaces $\mathcal{H}^{(n)}$ are the corresponding eigenspaces. Moreover, $D_n(I) = \mathcal{H}^{(n)}$ are obviously invariant with respect to the operators $U(p)$, and therefore also with respect to P_n. The operators $\mathcal{T}_p^\Sigma(I)$ then act on the subspaces $\mathcal{H}_p^{(n)}$ as multiples of the unit operator; their closures $\mathcal{N}_p := \overline{\mathcal{T}_p^\Sigma(I)}$ have pure point spectra, $\sigma(\mathcal{N}_p) = \{0,1,2,\dots\}$, with the eigenspace $\mathcal{H}_p^{(n)}$ corresponding to the eigenvalue n. It is natural to call $\mathcal{N}$ the operator of the *number of particles*; the same name is used for $\mathcal{N}_p$ and also for their *e.s.a.* restrictions.

Theorem 4 allows us to associate with a one-particle observable A the observable $\overline{\mathcal{T}_p^\Sigma(A)}$ of the many-particle system under consideration; the index assumes the value $p = s, a$ according to the type of the particles. Moreover, it is clear from the construction of the second-quantized operator that A and $\overline{\mathcal{T}_p^\Sigma(A)}$ describe the same physical quantity, at least for observables of additive character such as energy, momentum, *etc.* For instance, if H is the one-particle Hamiltonian, then $\overline{\mathcal{T}_p^\Sigma(H)}$ is the Hamiltonian for the noninteracting many-particle system.

This correspondence between the observables induces a simple relation between the unitary groups they generate, in particular, between the evolution operators of one-particle and many-particle systems.

12.1.6 Theorem: Let $\{U(t) : t \in \mathbb{R}\}$ be a continuous unitary group on $\mathcal{H}$ generated by a self-adjoint operator A; then $\{\overline{\mathcal{T}^\Pi(U(t))} : t \in \mathbb{R}\}$ is a continuous unitary group on the Fock space $\mathcal{F}(\mathcal{H})$ generated by $\overline{\mathcal{T}^\Sigma(A)}$, and similarly, $\{\overline{\mathcal{T}_p^\Pi(U(t))} : t \in \mathbb{R}\}$ is a continuous unitary group on $\mathcal{F}_p(\mathcal{H})$ generated by $\overline{\mathcal{T}_p^\Sigma(A)}$.

Proof: The reader is asked to check that $\{\overline{\mathcal{T}^\Pi(U(t))} : t \in \mathbb{R}\}$ is a continuous unitary group (Problem 4a). We denote its generator as $\tilde{A}$. It exists due to the Stone theorem and its domain consists of the vectors $\Psi \in \mathcal{F}(\mathcal{H})$ for which $\lim_{t\to 0}[\overline{\mathcal{T}^\Pi(U(t))}\Psi - \Psi]\,t^{-1}$ exists. It follows from Proposition 5.9.6 that $\{U_n^\Pi(t) : t \in \mathbb{R}\}$ is generated by $A^{(n)} := \overline{A_n^\Sigma}$; in addition, $A^{(1)} = A$ and $A^{(0)} = 0$ obviously holds. We take any vector $\Psi \in \mathcal{D}(A)$, *i.e.*, $\Psi = \{\psi_0,\dots,\psi_N,0,\dots\}$ with $\psi_n \in D_n(A)$; then

$$\lim_{t\to 0}\left\|\frac{1}{t}\left(\overline{\mathcal{T}^\Pi(U(t))}\Psi - \Psi\right) - i\mathcal{T}^\Sigma(A)\Psi\right\|_F = \sum_{n=0}^{N}\lim_{t\to\infty}\left\|\frac{1}{t}\left(U_n^\Pi(t) - I\right)\psi_n - iA_n\psi_n\right\|_n^2 = 0\,,$$

so $\mathcal{T}^\Sigma(A) \subset \tilde{A}$. This implies $\overline{\mathcal{T}^\Sigma(A)} \subset \tilde{A}$, and since both operators are self-adjoint they equal each other. The remaining assertion can be proved in an similar way by restriction to vectors $\Psi \in \mathcal{D}(A) \cap \mathcal{F}_p(\mathcal{H})$. ∎

12.2 Creation and annihilation operators

In the following we shall assume that the one–particle state space $\mathcal{H}$ is separable. Let $\mathcal{E} = \{\phi_j\}_{j=1}^{\infty}$ be an orthonormal basis in it; then the vectors $\phi_{j_1} \otimes \cdots \otimes \phi_{j_n}$ with $j_k = 1, 2, \ldots$ and $k = 1, \ldots, n$ form an orthonormal basis in $\mathcal{H}^{(n)}$ which we denote as $\mathcal{E}_n$. Unfortunately, the latter is not very suitable, since we are interested primarily in the subspaces $P_n\mathcal{H}^{(n)}$ and some vectors of $\mathcal{E}_n$ have nonzero orthogonal components both in $P_n\mathcal{H}^{(n)}$ and in its orthogonal complement. Hence we shall use $\mathcal{E}_n$ to construct another orthonormal basis.

Any vector of $\mathcal{E}_n$ is uniquely determined by an n–tuple of positive integers $j = \{j_1, \ldots, j_n\}$; we shall call j a *variation* and denote the corresponding vector as $\phi_n(j)$. The identity

$$U(p)\phi_n(j) = \phi_{j_{p_1}} \otimes \cdots \otimes \phi_{j_{p_n}} =: \phi_n(j\circ p)$$

obviously holds for any permutation $p \in \mathcal{S}_n$; it is not excluded that the operator $U(p)$ maps some basis vectors onto itself, because in general some of the numbers j_k may coincide. To make clear how the permutations act on the elements of $\mathcal{E}_n$, we introduce the symbol $\hat{\jmath}$ for nondecreasing variations, *i.e.*, such that $j_1 \le \cdots \le j_n$. This ordering makes it possible to define uniquely the numbers of repeated indices,

$$j_1 = \cdots = j_{n_1} < j_{n_1+1} = \cdots = j_{n_1+n_2} < \cdots < j_{n_1+\cdots+n_{m-1}+1} = \cdots = j_n;$$

we have, of course, $\sum_{i=1}^{m} n_i = n$. With the help of permutations, a nondecreasing variation $\hat{\jmath}$ gives rise to the subspace $\mathcal{H}_n(\hat{\jmath}) := \{\phi_n(\hat{\jmath}\circ p) \,:\, p \in \mathcal{S}_n\}_{lin}$; it is not difficult to check that

$$\dim \mathcal{H}_n(\hat{\jmath}) = \frac{n!}{n_1! \ldots n_m!} =: c(\hat{\jmath})\,.$$

Furthermore, the projections P_n preserve this subspace, $P_n\mathcal{H}_n(\hat{\jmath}) \subset \mathcal{H}_n(\hat{\jmath})$; we shall find the part of P_n in $\mathcal{H}_n(\hat{\jmath})$. Consider the vector

$$\phi_n^s(\hat{\jmath}) := \frac{c_s}{\sqrt{n!}} \sum_{p\in\mathcal{S}_n} \phi_n(\hat{\jmath}\circ p)\,,$$

where c_s is a normalization constant which can be found easily: the sum has $n!$ terms, among which just $c(\hat{\jmath})$ vectors are mutually different. They form an orthonormal set and each of them is contained in the sum $n_1! \ldots n_m!$ times; it follows that

$$\|\phi_n^s(\hat{\jmath})\|^2 = \frac{c_s^2}{n!}\,(n_1! \ldots n_m!)^2 c(\hat{\jmath}) = \frac{c_s^2 n!}{c(\hat{\jmath})}$$

so we have to choose $c_s = \sqrt{\frac{c(\hat{\jmath})}{n!}} = (n_1! \ldots n_m!)^{-1/2}$ to make $\phi_n^s(\hat{\jmath})$ a unit vector. Next we express the action of the symmetrizer S_n on $\phi_n(\hat{\jmath}\circ p)$,

$$S_n\phi_n(\hat{\jmath}\circ p) = \frac{1}{n!} \sum_{p'\in\mathcal{S}_n} \phi_n((\hat{\jmath}\circ p)\circ p') = \frac{1}{n!} \sum_{p''\in\mathcal{S}_n} \phi_n(\hat{\jmath}\circ p'') = \frac{1}{\sqrt{n!}\,c_s}\,\phi_n^s(\hat{\jmath})$$

for any $p \in \mathcal{S}_n$; hence by definition of $\mathcal{H}_n(\hat{\jmath})$ we obtain

$$S_n\mathcal{H}_n(\hat{\jmath}) = \{\phi_n^s(\hat{\jmath})\}_{lin},$$

i.e., $S_n \upharpoonright \mathcal{H}_n(\hat{\jmath})$ is the projection onto the subspace spanned by the vector $\phi_n^s(\hat{\jmath})$.

Similar conclusions can be made for the antisymmetrizer A_n. First of all, it is obvious that $A_n\mathcal{H}_n(\hat{\jmath}) = 0$ if a number j_k repeats in the variation $\hat{\jmath}$. If $\hat{\jmath}$ is a variation without repetition, the same argument as above yields

$$A_n\mathcal{H}_n(\hat{\jmath}) = \{\phi_n^a(\hat{\jmath})\}_{lin},$$

where the unit vector $\phi_n^a(\hat{\jmath})$ is defined by

$$\phi_n^a(\hat{\jmath}) := \frac{1}{n!}\sum_{p\in\mathcal{S}_n} \epsilon_p \phi_n(\hat{\jmath}\circ p).$$

The subspaces $\mathcal{H}_n(\hat{\jmath})$ are mutually orthogonal and it follows from the construction that $\{\mathcal{E}_n\}_{lin} = \sum_{\hat{\jmath}}^{\oplus} \mathcal{H}_n(\hat{\jmath}) \subset \mathcal{H}^{(n)}$; however, $\mathcal{E}_n$ is by assumption total in $\mathcal{H}^{(n)}$, and therefore

$$\sum_{\hat{\jmath}}^{\oplus} \mathcal{H}_n(\hat{\jmath}) = \mathcal{H}^{(n)}.$$

Combining this with the above results, we get

$$P_n\mathcal{H}^{(n)} = \sum_{\hat{\jmath}}^{\oplus} P_n\mathcal{H}_n(\hat{\jmath}) = \sum_{\hat{\jmath}}^{\oplus} \{\phi_n^p(\hat{\jmath})\}_{lin},$$

where we put $\phi_n^a(\hat{\jmath}) = 0$ if some number is repeated in $\hat{\jmath}$. This relation shows that the vectors $\phi_n^s(\hat{\jmath})$, with $\hat{\jmath} = \{j_1, \dots, j_n\}$ being an arbitrary nondecreasing variation, form an orthonormal basis in $S_n\mathcal{H}^{(n)}$, and similarly the $\phi_n^a(\hat{\jmath})$, where $\hat{\jmath}$ is any nondecreasing variation without repetition, form a basis in $A_n\mathcal{H}^{(n)}$.

Given a variation $\hat{\jmath}$, we can associate with it the sequence $\{n_i\}_{i=1}^{\infty}$ where n_i denotes the number of elements of $\hat{\jmath}$ equal to i. This fact may be understood so that the state ϕ_i occurs in $\phi_n(\hat{\jmath})$ (is occupied) just n_i times; this is why $\{n_i\}$ is referred to as the *occupation–number sequence* (or briefly, ON sequence). It has m nonzero terms n_{i_k}, $1 \le k \le m$. Clearly, $m \le n$ and

$$\sum_{i=1}^{\infty} n_i = \sum_{k=1}^{m} n_{i_k} = n;$$

the family of ON sequences which satisfy this condition will be denoted as $\mathcal{O}_n^s$. It is easy to check that the map $\hat{\jmath} \mapsto \{n_i\}_{i=1}^{\infty}$ is a bijection between the set of all nondecreasing variations and $\mathcal{O}_n^s$, so we may identify them in the following, $\hat{\jmath} \equiv \{n_i\}_{i=1}^{\infty}$; for the sake of simplicity we shall write $\phi_n(\hat{\jmath}) =: \phi_n\{n_i\}$, *etc.* In a similar way we find that the subset of variations without repetition corresponds to the family $\mathcal{O}_n^a \subset \mathcal{O}_n^s$ of all sequences $\{n_i\}_{i=1}^{\infty}$ composed of the numbers 0 and 1. Let us summarize the results.

12.2.1 Proposition: The set $\mathcal{E}_n^p := \{\phi_n^p\{n_i\} : \{n_i\} \in \mathcal{O}_n^p\}$ is an orthonormal basis in $P_n\mathcal{H}^{(n)}$ for any $n \ge 2$.

The vectors $\phi_n^p\{n_i\}$ can be used to construct an orthonormal basis in $\mathcal{F}_p(\mathcal{H})$ which we denote as $\mathcal{E}_p$. According to Problem 2.15, it consists of the vectors $\Phi_n^p\{n_i\}$ with $n = 0, 1, \dots$ and $\{n_i\} \in \mathcal{O}_n^p$ defined by

$$\Phi_n^p\{n_i\} := \begin{cases} \{1, 0, \dots\} & \dots \quad n = 0 \\ \{0, \phi_1\{n_i\}, 0, \dots\} & \dots \quad n = 1 \\ \{0, \dots, 0, \phi_n^p\{n_i\}, 0, \dots\} & \dots \quad n \ge 2 \end{cases}$$

where $\phi_1\{n_i\} := \phi_j$ for that j for which $n_i = 1$ and the nonzero component stands at the n–th position. Each basis vector is thus determined by a number n and some ON sequence $\{n_i\} \in \mathcal{O}_n^p$; owing to this $\mathcal{E}_p$ is called the **occupation–number basis**.

After this preliminary let us pass to the subject indicated in the title. We associate with arbitrary $f_1, \dots, f_n \in \mathcal{H}$ an element of $\mathcal{F}_p(\mathcal{H})$ defined by $\Psi_n^p := \{0, \dots, 0, P_n(f_1 \otimes \dots \otimes f_n), 0 \dots\}$. The set of such vectors will be denoted as $\mathcal{D}_p^{(n)}$; for $n = 0$ we set $\mathcal{D}_p^{(0)} := \{\Psi_0^p\}$, where $\Psi_0^p := \Omega_0$. We also introduce the set

$$\mathcal{D}_p := \left\{ \bigcup_{n=0}^{\infty} \mathcal{D}_p^{(n)} \right\}_{lin},$$

which is dense in $\mathcal{F}_p(\mathcal{H})$ because it contains the linear envelope of the basis $\mathcal{E}_p$. For any $f \in \mathcal{H}$ we define the **creation operator** $a^*(f)$ and **annihilation operator** $a(f)$ as the maps $\mathcal{D}_p \mapsto \mathcal{D}_p$ obtained by linear extension of the relations

$$\begin{aligned} a^*(f)\Psi_n^p(f_1, \dots, f_n) &:= \sqrt{n+1}\,\Psi_{n+1}^p(f, f_1, \dots, f_n), \\ a(f)\Psi_n^p(f_1, \dots, f_n) &:= \frac{1}{\sqrt{n}} \sum_{j=1}^{n} \delta_{j-1}^{(p)}(f, f_j)\,\Psi_{n-1}^p(f_1, \dots, f_{j-1}, f_{j+1}, \dots, f_n), \end{aligned} \tag{12.1}$$

where $\delta_k^{(p)} = 1$ for $p = s$ and $\delta_k^{(p)} = (-1)^k$ for $p = a$. The second of the relations (12.1) holds for $n \ge 1$, while for $n = 0$ we have

$$a(f)\Omega_0 = 0. \tag{12.2}$$

Notice that the map $f \mapsto a^*(f)$ is by definition *linear* while $f \mapsto a(f)$ is *antilinear*. Using the definition of $a(f)$ we are able to express any vector of $\mathcal{D}_p^{(n)}$ in the form

$$\Psi_n^p(f_1, \dots, f_n) = \frac{1}{\sqrt{n!}} a^*(f_1) \dots a^*(f_n)\Omega_0. \tag{12.3}$$

Correctness of the definition requires that the relations (12.1) define a linear mapping on $\mathcal{D}_p^{(n)}$, *i.e.*, that

$$a^\#(f)\Psi_n^p(f_1, \dots, \alpha f_k + g_k, \dots, f_n) = \alpha\, a^\#(f)\Psi_n^p(\dots, f_k, \dots) + a^\#(f)\Psi_n^p(\dots, g_k, \dots)$$

holds for any $f_1,\dots,f_n,g_k \in \mathcal{H}$ and $\alpha \in \mathbb{C}$; for the sake of brevity, we have introduced here the symbol $a^\#(f)$ as a common notation for $a^*(f)$ and $a(f)$. The linearity can be readily checked, and therefore it is sufficient to know how the operators $a^\#(f)$ act on vectors of a suitable linearly independent subset in $\mathcal{D}_p$, for instance, the occupation–number basis constructed above. To be able to rewrite the relations (12.1) for this particular case, we define for any $\{n_i\} \in \mathcal{O}_n^s$ the sequences $\{n_i\}_k^{(\pm)} := \{n_1,\dots,n_{k-1},n_k \pm 1, n_{k+1},\dots\}$ and the numbers $s_k := \sum_{i=1}^{k-1} n_i$. Naturally, the sequence $\{n_i\}_k^{(-)}$ is defined only if $n_k \ge 1$. If f is some vector of the one–particle basis $\{\phi_j\}_{j=1}^\infty$, then the relations (12.1) in the antisymmetric case yield

$$\begin{aligned} a^*(\phi_k)\,\Phi_n^a\{n_i\} &= (-1)^{s_k}(1-n_k)\,\Phi_{n+1}^a\{n_i\}_k^{(+)}\,, \\ a(\phi_k)\,\Phi_n^a\{n_i\} &= (-1)^{s_k} n_k\,\Phi_{n-1}^a\{n_i\}_k^{(-)} \end{aligned} \tag{12.4}$$

(Problem 6); we take into account only ON sequences $\{n_i\} \in \mathcal{O}_n^a$. In the symmetric case we similarly obtain

$$\begin{aligned} a^*(\phi_k)\,\Phi_n^s\{n_i\} &= \sqrt{n_k+1}\,\Phi_{n+1}^s\{n_i\}_k^{(+)}\,, \\ a(\phi_k)\,\Phi_n^s\{n_i\} &= \sqrt{n_k}\,\Phi_{n-1}^s\{n_i\}_k^{(-)}\,. \end{aligned} \tag{12.5}$$

If f is a general vector in the one–particle space we use the fact that the (anti)linear dependence of the operators $a^\#(f)$ on f may be extended to "infinite linear combinations": we have

$$a^*(f)\Psi = \sum_{k=1}^\infty (\phi_k,f)\,a^*(\phi_k)\Psi\,, \quad a(f)\Psi = \sum_{k=1}^\infty (f,\phi_k)\,a(\phi_k)\Psi \tag{12.6}$$

for any $\Psi \in \mathcal{D}_p$ where the series converge with respect to the norm in $\mathcal{F}_p(\mathcal{H})$ — *cf.* Problem 7. It is obvious from the definition (12.1) that the operators $a^\#(f)$ map the space $\mathcal{D}_p$ onto itself; this means that polynomials in creation and annihilation operators are well defined. An important role among them is played by

$$[a^\#(f),a^\#(g)]_p := \begin{cases} a^\#(f)a^\#(g) - a^\#(g)a^\#(f) & \dots \quad p=s \\ a^\#(f)a^\#(g) + a^\#(g)a^\#(f) & \dots \quad p=a \end{cases}$$

i.e., by the *commutator* and *anticommutator* in the symmetric and antisymmetric cases, respectively; the latter is often alternatively denoted as $[a^\#(f),a^\#(g)]_+$.

12.2.2 Theorem: Creation and annihilation operators satisfy the relations

$$\begin{aligned} [a(f),a(g)]_p\Psi &= [a^*(f),a^*(g)]_p\Psi = 0 \\ [a(f),a^*(g)]_p\Psi &= (f,g)\,\Psi \end{aligned} \tag{12.7}$$

for any $f,g \in \mathcal{H}$ and $\Psi \in \mathcal{D}_p$; moreover, $a^*(f) = a(f)^* \upharpoonright \mathcal{D}_p$ and $a(f) = a^*(f)^* \upharpoonright \mathcal{D}_p$. In the symmetric case the operators $a^{\#}(f)$ are unbounded for any nonzero $f \in \mathcal{H}$, while for $p=a$ they are bounded and $\|a^{\#}(f)\| = \|f\|$.

Proof: The subspace $\mathcal{D}_p$ is dense in $\mathcal{F}_p(\mathcal{H})$, so the adjoint operators $a^{\#}(f)^*$ exist. Let $n \ge 2$; the definition relation (12.1) together with Problem 8 gives

$$\begin{aligned}
&(\Psi_n^p(g_1,\dots,g_n), a^*(f)\Psi_{n-1}^P(f_2,\dots,f_n))_F \\
&= \frac{\sqrt{n}}{n!} \sum_{r\in\mathcal{S}_n} \nu_r^{(p)}\, (g_{r_1},f)(g_{r_2},f_2)\dots(g_{r_n},f_n) \\
&= \frac{1}{\sqrt{n}} \sum_{k=1}^{n} \delta_{k-1}^{(p)}(g_k,f)(\Psi_{n-1}^p(g_1,\dots,g_{k-1},g_{k+1},\dots,g_n),\Psi_{n-1}^p(f_2,\dots,f_n))_F \\
&= (a(f)\Psi_n^p(g_1,\dots,g_n),\Psi_{n-1}^p(f_2,\dots,f_n))_F
\end{aligned}$$

for arbitrary $f_2,\dots,f_n,g_1,\dots,g_n \in \mathcal{H}$; we have introduced here the symbol $\nu_r^{(p)}$, which is $\nu_r^{(p)} = 1$ for $p=s$ and $\nu_r^{(p)} = \epsilon_r$ if $p=a$. It is straightforward to check that this relation holds for $n=1$ as well. Inner–product linearity then implies

$$(\Psi_n^p(g_1,\dots,g_n), a^*(f)\Psi)_F = (a(f)\Psi_n^p(g_1,\dots,g_n),\Psi)_F$$

for any $\Psi \in (\mathcal{D}_p^{(n-1)})_{lin}$. Furthermore, $(\Psi,a^*(f)\Phi)_F = 0$ for $\Psi \in \mathcal{D}_p^{(n)}$, $\Phi \in \mathcal{D}_p^{(m)}$ unless $n-m=1$. We infer that $a^*(f) = a(f)^* \upharpoonright \mathcal{D}_p$; using the identities

$$(\Psi,a(f)\Phi)_F = (\Psi,a^*(f)^*\Phi)_F = (a^*(f)\Psi,\Phi)_F\,,$$

which are valid for any $\Psi,\,\Phi \in \mathcal{D}_p$, we also prove $a(f) = a^*(f)^* \upharpoonright \mathcal{D}_p$.

As for the relations (12.7), it is sufficient to check them for $\Psi \in \mathcal{D}_p^{(n)}$. The identity $[a^*(f),a^*(g)]_p\Psi = 0$ follows from the definition (12.1) and $\Psi_n^p(f_{p_1},\dots,f_{p_n}) = \nu_p^p\Psi_n^p(f_1,\dots,f_n)$; the first line is completed with the help of Problem 9. To prove the second relation, we use (12.1) to express

$$\begin{aligned}
&a(f)a^*(g)\Psi_n^p(f_1,\dots,f_n) \\
&= (f,g)\,\Psi_n^p(f_1,\dots,f_n) + \sum_{j=2}^{n+1} \delta_{j-1}^{(p)}\,(f,f_{j-1})\,\Psi_n^p(g,f_1,\dots,f_{j-2},f,f_j,\dots,f_n) \\
&= (f,g)\,\Psi_n^p(f_1,\dots,f_n) \pm a^*(g)a(f)\,\Psi_n^p(f_1,\dots,f_n)\,,
\end{aligned}$$

where the upper and lower sign correspond to $p=s$ and $p=a$, respectively.

Consider now the symmetric case; in view of the (anti)linearity we need to prove that the operators $a^{\#}(f)$ are unbounded for each unit vector $f \in \mathcal{H}$. This is in fact easy: the vector $\Psi_n^s(f,\dots,f)$ also has a unit norm, so choosing the basis $\{\phi_j\}$ in such a way that $\phi_1 = f$, we obtain from (12.5) $\|a^*(f)\Psi_n^s(f,\dots,f)\|_F = \sqrt{n+1}$ and $\|a(f)\Psi_n^s(f,\dots,f)\|_F = \sqrt{n}$.

In the antisymmetric case we first take the operators $a^*(\phi_k)$ corresponding to some vector of the basis $\{\phi_j\}_{j=1}^{\infty}$. The vectors $\Phi_n^a\{n_i\}$ and $\Phi_m^a\{m_i\}$ corresponding to different ON sequences $\{n_i\}$ and $\{m_i\}$, respectively, are mutually orthogonal; hence it follows from (12.4) that also $a^*(\phi_k)\Phi_n^a\{n_i\} \perp a^*(\phi_k)\Phi_m^a\{m_i\}$. This yields for a linear combination $\Psi := \sum_j \alpha_j \Phi_{n_j}^a\{n_i^{(j)}\}$ the inequality

$$\|a^*(\phi_k)\Psi\|_F^2 = \sum_j |\alpha_j|^2 \left\|a^*(\phi_k)\Phi_{n_j}^a\{n_i^{(j)}\}\right\|_F^2 = \sum_j |\alpha_j|^2(1-n_k^{(j)})^2 \le \|\Psi\|_F^2 ,$$

which turns to identity, *e.g.*, for $\Psi = \Omega_0$. However, vectors of this form are dense in $\mathcal{F}_p(\mathcal{H})$, and therefore $\|a^*(\phi_k)\| = 1$. Finally, if f is an arbitrary nonzero vector we choose the basis $\{\phi_j\}$ so that $\phi_1 = f/\|f\|$; the linearity of $a^*(\cdot)$ together with the already proven property of the adjoint gives $\|a(f)\| = \|a(f)^*\| = \|a^*(f)\|$. ∎

The importance of creation and annihilation operators stems mainly from the fact that we are able to express in terms of them other operators acting on the space $\mathcal{F}_p(\mathcal{H})$ — see the notes. The relations (12.7) are called the **canonical commutation (anticommutation) relations** in the symmetric and antisymmetric situation, respectively. An investigation of them is more difficult in the former case due to the unboundedness of the operators $a^\#(f)$. This difficulty can be removed by replacing (12.7) by suitable relations for bounded functions of $a^\#(f)$ in analogy with Section 8.2; this will be done in the following section.

12.3 Systems of noninteracting particles

Second–quantization formalism has numerous applications but we restrict ourselves to discussion of two simple examples; the reader can find much more information in the literature quoted in the notes.

To describe a particular system, we have to specify the one–particle state space $\mathcal{H}$, the operators representing one–particle observables, among them the Hamiltonian in the first place, and the interaction between the particles. The simplest situation occurs when the particles do not interact. In accordance with what we have said in Section 11.3 about noninteracting systems of a finite number of particles, we assume that the operator $\overline{\mathcal{T}_p^\Sigma(H)}$ corresponding to the one–particle Hamiltonian H plays the role of the total–energy operator. Theorem 12.1.6 then claims that

$$\exp\left(-i\overline{\mathcal{T}_p^\Sigma(H)}t\right) = \overline{\mathcal{T}_p^\Pi(e^{-iHt})} \tag{12.8}$$

holds for any $t \in \mathbb{R}$; in other words, that the propagator of the free–particle system is obtained as the second quantization of the one–particle propagator.

The first system we are going to treat is a *free scalar quantum field.* As a system of spin–zero particles, it is associated with the symmetric Fock space $\mathcal{F}_s(\mathcal{H})$, where the one–particle state space $\mathcal{H}$ will be specified below. Before doing that, however, we have to analyze the relations (12.7) in more detail. In order to draw the line between them and the canonical commutation relations of Section 8.2 we define for any $f \in \mathcal{H}$ the *Segal field operator* as

$$\Phi_S(f) := \frac{1}{\sqrt{2}}\,(a(f) + a^*(f))$$

with the domain $\mathcal{D}_s \subset \mathcal{F}_s(\mathcal{H})$.

12.3.1 Theorem: (a) $\Phi_S(f)$ is *e.s.a.* for any $f \in \mathcal{H}$.

(b) The set $\{\Phi_S(f_1)\dots\Phi_S(f_n)\Omega_0 :\ f_j \in \mathcal{H},\ n = 0,1,\dots\}$ is total in $\mathcal{F}_s(\mathcal{H})$; in other words, the vacuum vector is cyclic for the algebra generated by I_s and the field operators $\Phi_S(f)$ for all $f \in \mathcal{H}$.

(c) $[\Phi_S(f), \Phi_S(g)]\Psi = i\,\mathrm{Im}\,(f,g)\,\Psi$ holds for any $f,g \in \mathcal{H}$ and $\Psi \in \mathcal{D}_S$.

12.3.2 Remark: In addition to field operators, quantum field theories use *canonically conjugate momenta.* In the present context, these are defined by

$$\Pi(f) := \Phi_S(if) := \frac{i}{\sqrt{2}}\,(a^*(f) - a(f))$$

(see the notes; one has to keep in mind that there is no direct relation between $\Pi(f)$ and the momentum observable of a given field). Given an orthonormal basis $\{f_j\}_{j=1}^{\infty}$ in $\mathcal{H}$, assertion (c) of the theorem can rewritten for $\Phi(f_j) := \Phi_S(f_j)$ and $\Psi \in \mathcal{D}_s$ in the form

$$[\Phi_S(f_j), \Phi_S(f_k)]\Psi = [\Pi_S(f_j), \Pi_S(f_k)]\Psi = 0\,, \qquad [\Phi_S(f_j), \Pi_S(f_k)]\Psi = i\delta_{jk}\Psi\,,$$

which is obviously analogous to (8.6).

Proof of Theorem 1: It follows from Theorem 12.2.2 and Proposition 4.1.2d that $\Phi_S(f)$ is symmetric, so it is sufficient to find a dense set of analytic vectors for it. The operator $\Phi_S(f)^m$ is a polynomial in $a^{\#}(f)$ for any positive integer m; hence each $\Psi \in \mathcal{D}_s$ belongs to $D(\Phi_S(f)^m)$. The subspace $\mathcal{D}_s$ contains vectors describing states with a finite number of particles, *i.e.*, for any $\Psi \in \mathcal{D}_s$ there is a positive n such that $\Psi \in \bigoplus_{k=0}^{n} \mathcal{H}_s^{(k)}$. Moreover, it is obvious that $\Phi_S(f)$ maps $\mathcal{H}_s^{(k)}$ to $\mathcal{H}_s^{(k-1)} \oplus \mathcal{H}_s^{(k+1)}$. Problem 10b then yields the estimate

$$\|\Phi_S(f)^m\Psi\|_F \le 2^{m/2}\sqrt{(n+m)(n+m-1)\dots(n+1)}\,\|\Psi\|_F\|f\|^m\,,$$

which implies

$$\sum_{m=0}^{\infty} \frac{t^m}{m!}\,\|\Phi_S(f)^m\Psi\|_F \le \|\Psi\|_F \sum_{m=0}^{\infty} \sqrt{\frac{(n+m)!}{n!}}\,\frac{(\sqrt{2}\,\|f\|\,t)^m}{m!} < \infty$$

for any $t > 0$. The set $\mathcal{D}_s$, which is dense by Proposition 12.1.3, therefore consists of analytic vectors. Assertion (b) follows from the mentioned proposition together with (12.2) and (12.3), and finally, the commutation relations of assertion (c) are obtained by a simple computation from Theorem 12.2.2. ▮

The first assertion of the just proved theorem allows us to cast the canonical commutation relations into an alternative form. For any $f \in \mathcal{H}$ we define the unitary operator

$$W(f) := \exp\left(i\overline{\Phi_S(f)}\right); \tag{12.9}$$

we shall show that it represents a counterpart to the Weyl operator of Section 8.2.

12.3.3 Theorem: The operators (8.10) satisfy for any pair $f, g \in \mathcal{H}$ the relation

$$W(f)W(g) = e^{-i\mathrm{Im}\,(f,g)/2}W(f+g). \tag{12.10}$$

The map $f \mapsto W(f)$ is strongly continuous and the set $\{W(f) : f \in \mathcal{H}\}$ is irreducible.

Proof: As for the strong continuity of $W(\cdot)$ and the relation (12.10) we refer to the literature mentioned in the notes. The irreducibility is a consequence of the following result:

12.3.4 Lemma: Let $\{\phi_j\}$ be an arbitrary orthonormal basis in $\mathcal{H}$; then the set $\{W(t\phi_j),\, W(it\phi_j) : j = 1, 2, \dots,\, t \in \mathbb{R}\}$ is irreducible.

Proof: Since $W(-f) = W^*(f)$ holds by definition, the set under consideration is symmetric, and in view of Schur's lemma it is sufficient to check that any bounded operator B on $\mathcal{F}_s(\mathcal{H})$ such that $[B, W(tf_j)] = 0$ is valid for $f_j = \phi_j$ or $f_j = i\phi_j$ equals a multiple of the unit operator. Differentiating this relation, we find $B\overline{\Phi_S(f_j)} \subset \overline{\Phi_S(f_j)}B$. We introduce

$$N(f) := \frac{1}{2}\left(\overline{\Phi_S(f)}^2 + \overline{\Phi_S(if)}^2 - \|f\|^2\right)$$

for any $f \in \mathcal{H}$; the reader is asked to check simple properties of these operators in Problem 12. Now take a vector $\Psi = \{\psi_0, \psi_1, \dots\} \in \bigcap_j D(N(\phi_j))$ such that a finite limit $n(\Psi) := \lim_{m\to\infty} \sum_{i=1}^m (\Psi, N(\phi_i)\Psi)_F$ exists; since the operators $N(f)$ map each $\mathcal{H}^{(n)}$ onto itself, the inner product on the right side can be written as $\lim_{n\to\infty} \sum_{k=0}^n (\psi_k, N_k(\phi_i)\psi_k)_k$, where $N_k(\phi_i) := N(\phi_i) \restriction \mathcal{H}^{(k)}$, and therefore

$$n(\Psi) = \lim_{m\to\infty} \lim_{n\to\infty} \sum_{i=1}^{m} \sum_{k=0}^{n} (\psi_k, N_k(\phi_i)\psi_k)_k\,.$$

The operators $N(\phi_i)$ are positive by Problem 12b, so the double series may be rearranged; this yields $n(\Psi) = \lim_{n\to\infty} \lim_{m\to\infty} \left(\Psi_n, \sum_{i=1}^m N(\phi_i)\Psi_n\right)_F$ with $\Psi_n := \{\psi_0, \dots, \psi_n, 0, \dots\}$. However, any such vector belongs to $D(\mathcal{N}_s)$ due to Problem 10a and $\mathcal{N}_s\Psi_n = \lim_{m\to\infty} \sum_{i=1}^m N(\phi_i)\Psi_n$; hence

$$n(\Psi) = \lim_{n\to\infty} (\Psi_n, \mathcal{N}_s\Psi_n) = \sum_{k=0} k\|\psi_k\|_k^2\,.$$

Using the fact that the vacuum vector is contained in $\bigcap_j D(N(\phi_j))$ together with Problem 12a, we find that $B\Omega_0 \in \bigcap_j D(N(\phi_j))$. Furthermore, Problem 12b gives $N(\phi_j)B\Omega_0 = BN(\phi_j)B\Omega_0 = 0$, so $n(B\Omega_0)$ exists and equals zero; the identity displayed above then implies $B\Omega_0 = \alpha\Omega_0$ for some $\alpha \in \mathbb{C}$. Finally, $a^*(f) = 2^{-1/2}(\Phi_S(f) - i\Phi(if))$, and therefore

$$B\Psi_n^s(\phi_{i_1},\dots,\phi_{i_n}) = \alpha\,\Psi_n^s(\phi_{i_1},\dots,\phi_{i_n})$$

follows from the relations $B\overline{\Phi_S(f_j)} \subset \overline{\Phi_S(f_j)}B$ and (12.3). This identity extends linearly to the subspace $\mathcal{D}_s$, and since the latter is dense in $\mathcal{F}_s(\mathcal{H})$ and B is bounded, we obtain $B = \alpha I_s$. ∎

The relations (12.10) imply, in particular, that the operators (12.9) satisfy for any pair $f, g \in \mathcal{H}$ the relations

$$W(f)W(g) = e^{-i\mathrm{Im}\,(f,g)}W(g)W(f)\,. \tag{12.11}$$

Given an orthonormal basis $\{\phi_j\} \subset \mathcal{H}$, we may define the continuous one–parameter unitary groups $U_j(t) := W(it\phi_j)$ and $V_k(s) := W(s\phi_j)$; then

$$U_j(t)V_k(s) = e^{i\delta_{jk}ts}V_k(s)U_j(t)$$

holds for all $s, t \in \mathbb{R}$ and $j, k = 1, 2, \dots$. Alternatively, we may introduce the real Hilbert space S as the closure of the *real* linear envelope of $\{\phi_j\}$ and define unitary–operator valued maps $S \to \mathcal{F}_s(\mathcal{H})$ by $U(f) := W(if)$ and $V(g) := W(g)$. The relations (12.11) may then be rewritten in the form

$$U(f)V(g) = e^{i(f,g)}V(g)U(f)\,,$$

which is nothing else than (8.12). In this way, we have associated with any one–particle Hilbert space an irreducible (unitary, strongly continuous) representation of the canonical commutation relations which is usually referred to as the **Fock representation**.

The properties of this representation depend, of course, on the choice of $\mathcal{H}$. In the case of a quantum field the one–particle state space is infinite–dimensional and the Stone–von Neumann theorem is no longer valid. Our next goal is to illustrate this claim on the physically important case of a free relativistic field, which we shall introduce below.

It is convenient for our present purpose to choose $\mathcal{H} := L^2(\mathbb{R}^3)$. For any $m > 0$, we define on $\mathcal{H}$ the operator $\mu_m : (\mu_m f)(p) = \sqrt[4]{m^2+p^2}\,f(p)$ with the usual domain, and the antilinear operator $C : (Cf) = \overline{f(-p)}$. We shall also need the inverse μ_m^{-1}, which belongs to $\mathcal{B}(L^2(\mathbb{R}^3))$. It is clear that $\mathcal{S}(\mathbb{R}^3)$ is a common invariant subspace of these operators; thus we shall not discriminate between them and their restrictions to $\mathcal{S}(\mathbb{R}^3)$. We also denote $\mathcal{S}_C(\mathbb{R}^3) := \{f \in \mathcal{S}(\mathbb{R}^3) : Cf = f\}$ and define

$$\Phi_m(f) := \Phi_S(\mu_m^{-1}f)\,, \quad \Phi_m(if) := \Phi_S(i\mu_m f) \tag{12.12}$$

for an arbitrary $f \in \mathcal{S}_C(\mathbb{R}^3)$, where $\Phi_S(\cdot)$ is the Segal field on $L^2(\mathbb{R}^3)$, and furthermore,

$$W_m(g) := \exp\left(i\overline{\Phi_m(g)}\right)$$

for $g = f,\, if$. Now we are able to formulate the following important assertion.

12.3.5 Theorem: Let an operator $U \in \mathcal{B}(\mathcal{F}_s(L^2(\mathbb{R}^3)))$ satisfy the condition $UW_m(g) = W_{m'}(g)U$ for some $m \neq m'$ and all $g = f,\, if$ with $f \in \mathcal{S}_C(\mathbb{R}^3)$; then $U = 0$.

Proof: Suppose on the contrary that $U \neq 0$; without loss of generality we may assume that U is unitary (Problem 13). We shall first show that the vacuum is an eigenvector of U. For a fixed $a \in \mathbb{R}^3$ we define the unitary operator V on $L^2(\mathbb{R}^3)$ by $(Vf)(p) := e^{ia\cdot p} f(p)$ (with the choice of observables given below, it represents the translation on a). It is easy to see that $\mathcal{S}_C(\mathbb{R}^3)$ is invariant under V and the unitary operator $\overline{\mathcal{T}_s^{\Pi}(V)}$ acts as

$$(\mathcal{T}_s^{\Pi}(V)F_n)(p_1,\dots,p_n) = e^{ia\cdot(p_1+\dots+p_n)} F_n(p_1,\dots,p_n)$$

for any $F_n \in S_n L^2(\mathbb{R}^{3n})$. Using the transformation properties of the Segal field given in Problem 11b we find $\Phi_{\tilde m}(Vg) = \overline{\mathcal{T}_s^{\Pi}(V)}\Phi_{\tilde m}(g)\overline{\mathcal{T}_s^{\Pi}(V)}^{-1}$, where $\tilde m := m$ or m'. It follows from the functional–calculus rules that

$$W_{\tilde m}(Vg) = \overline{\mathcal{T}_s^{\Pi}(V)}W_{\tilde m}(g)\overline{\mathcal{T}_s^{\Pi}(V)}^{-1}$$

holds for any $g \in L^2(\mathbb{R}^3)$, in particular for $g = f,\, if$ with $f \in \mathcal{S}_C(\mathbb{R}^3)$. Using this relation together with $UW_m(g) = W_{m'}(g)U$ we readily check that $W_m(g)$ commutes with the operator $U^{-1}\,\overline{\mathcal{T}_s^{\Pi}(V)}^{-1} U\,\overline{\mathcal{T}_s^{\Pi}(V)}$, and since the assumptions of Lemma 4 are satisfied, we have

$$U^{-1}\,\overline{\mathcal{T}_s^{\Pi}(V)}^{-1} U\,\overline{\mathcal{T}_s^{\Pi}(V)} = \gamma I_s$$

for some nonzero $\gamma \in \mathbb{C}$. Let F_n be the n–particle component of the vector $U\Omega_0$; then the last relation together with the explicit expression of $\overline{\mathcal{T}_s^{\Pi}(V)}$ given above implies

$$e^{-ia\cdot(p_1+\dots+p_n)} F_n(p_1,\dots,p_n) = \gamma F_n(p_1,\dots,p_n)\,,$$

which is possible with a nonzero γ only if $F_n = 0$ for $n = 0, 1, \dots$, *i.e.*, $U\Omega_0 = \gamma I_s$.

To complete the proof, we use the just obtained result to compute the norm $\|\Phi_m(f)\Omega_0\|_F^2$ for $f \in \mathcal{S}_C(\mathbb{R}^3)$ in two different ways. Evaluating it directly, we get

$$\|\Phi_m(f)\Omega_0\|_F^2 = \left\| \frac{1}{\sqrt{2}}\left[a(\mu_m^{-1}f) + a^*(\mu_m f)\right]\Omega_0 \right\|_F^2 = \frac{1}{2}\int_{\mathbb{R}^3} \frac{|f(p)|^2}{\sqrt{m^2+p^2}}\, dp\,.$$

On the other hand, using the unitarity of the operator U together with the inclusion $U\overline{\Phi_m(f)} \subset \overline{\Phi_{m'}(f)}U$, which is obtained by differentiating the condition $UW_m(g) = W_{m'}(g)U$, we find

$$\|\Phi_m(f)\Omega_0\|_F^2 = \|U\Phi_m(f)\Omega_0\|_F^2 = |\gamma|^2 \|\Phi_{m'}(f)\Omega_0\|_F^2 = \frac{|\gamma|^2}{2}\int_{\mathbb{R}^3} \frac{|f(p)|^2}{\sqrt{(m')^2+p^2}}\, dp;$$

it is obvious that the two expressions cannot coincide for all $f \in \mathcal{S}_C(\mathbb{R}^3)$. ∎

Choose now an orthonormal basis $\{\phi_j\} \subset L^2(\mathbb{R}^3)$ which is contained in $\mathcal{S}_C(\mathbb{R}^3)$; such a basis can be constructed easily, *e.g.*, using the Hermite functions (2.1). Let S be the real Hilbert space spanned by $\{\phi_j\}$; then the relations

$$U_m(f) := W_m(if)\,, \quad V_m(g) := W_m(g)\,, \quad f,g \in S$$

define for any $m > 0$ an irreducible (unitary, strongly continuous) representation of the canonical commutation relations.

12.3.6 Corollary: The representations $\{U_m(\cdot),\, V_m(\cdot)\}$ corresponding to different values of m are nonequivalent, *i.e.*, there is no bounded invertible operator U such that

$$U\,U_m(f)\,U^{-1} = U_{m'}(f)\,, \quad U\,V_m(g)\,U^{-1} = V_{m'}(g)$$

holds for all $f,g \in S$ unless $m = m'$.

Proof: Suppose that the operator U exists; then it follows from the definition of the representations and from (12.10) that $UW_m(h) = W_{m'}(h)U$ holds for all $h := g+if$ with $f,g \in S$. However, such a set is dense in $L^2(\mathbb{R}^3)$, and we therefore find to any $h \in \mathcal{S}_C(\mathbb{R}^3)$ a sequence $\{e_n\}$ that converges to $\mu_{m'}h \in \mathcal{S}_C(\mathbb{R}^3)$. Then the sequence $\{h_n\}$, $h_n := \mu_{m'}^{-1}e_n$, converges to the function h, and similarly, $\{\mu_m h_n\}$ converges to $\mu_m h$. In addition, the limits of the sequences $\{\mu_{\tilde m}^{-1}h_n\}$ are $\mu_{\tilde m}^{-1}h$ as a consequence of the boundedness of the operators $\mu_{\tilde m}^{-1}$; here again we denote $\tilde m := m$ or m'. Since the map $W(\cdot)$ is continuous by Theorem 3, the definition of $W_m(\cdot)$ implies

$$W_{\tilde m}(h_n) \longrightarrow W_{\tilde m}(h)\,, \quad W_{\tilde m}(ih_n) \longrightarrow W_{\tilde m}(ih)$$

for any $h \in \mathcal{S}_C(\mathbb{R}^3)$. Moreover, the operator U is bounded by assumption, so the relation $UW_m(h) = W_{m'}(h)U$ should hold for all $h = f,\, if$ with $f \in \mathcal{S}_C(\mathbb{R}^3)$; however, this contradicts the preceding theorem. ∎

Let us now return to the construction of a free quantum field. We have first to select the one–particle state space. The most straightforward way to do that is to use momentum representation. Since we have in mind relativistic particles of spin zero, we choose $\mathcal{H}_m := \mathcal{H}_{m,0} = L^2(H_m, d\omega_m)$, which we introduced in Section 10.3. The momentum components are then represented by multiplication operators, $(P_j f)(p) = p_j f(p)$, and the free–particle Hamiltonian $H := P_0$ acts as

$$(Hf)(p) = \sqrt{m^2 + \vec p^{\,2}}\, f(p)$$

for all f of the natural domain; the generators of the Lorentz–group representation are obtained by differentiating (10.14) for $s = 0$.

It is more complicated to express the wave functions in terms of the space–time variables $x = (x_0, \vec x)$. By analogy with nonrelativistic quantum mechanics we expect that this can be achieved by the Fourier transformation. However, in the relativistic case we may no longer separate the time from the spatial coordinates. Given $f \in \mathcal{S}(\mathbb{R}^4)$, we define

$$\tilde{f}(p) := (2\pi)^{-3/2} \int_{\mathbb{R}^4} e^{i(p_0 x_0 - \vec{p}\cdot\vec{x})} f(x)\, dx\,.$$

The function $\tilde{f}$ differs from the Fourier transform of f on $\mathcal{S}(\mathbb{R}^4)$ by the factor $\sqrt{2\pi}$ and the sign in the exponent; this choice is motivated by the requirement that the extension of the transformation $f \mapsto \tilde{f}$ to generalized functions of the type $f(x) := g(\vec{x})\delta(t)$ (see the notes) should preserve the standard correspondence (7.4) between the coordinate and momentum representation, *i.e.*, $\tilde{f}(p) = \hat{g}(\vec{p})$.

If a function has to describe a state of a particle of mass m in the momentum representation, its support must be contained in H_m. We introduce the map $E_m : \mathcal{S}(\mathbb{R}^4) \to \mathcal{H}_m$ by $E_m f := \tilde{f} \restriction H_m$; then we are able to define

$$\Phi_m(f) := \Phi_S(E_m \mathrm{Re}\, f) + i\Phi_S(E_m \mathrm{Im}\, f) \tag{12.13}$$

for any $f \in \mathcal{S}(\mathbb{R}^4)$, where Φ_S is the Segal field on $\mathcal{F}_s(\mathcal{H}_m)$. The map $f \mapsto \Phi_m(f)$ is called the **free scalar Hermitean field** of mass m. Its time evolution is given by (12.8), and in a similar way we can obtain its behavior with respect to the other Poincaré group transformations from Proposition 10.3.3 and Problem 11b.

12.3.7 Remark: The restriction to functions of $\mathcal{S}(\mathbb{R}^4)$ is not just a matter of convenience. The field operators are often written in a formal analogy with (12.6) as

$$\Phi(x) = \frac{1}{\sqrt{2}} \sum_k \left(\overline{\phi_k(x)} a^*(\phi_k) + \phi_k(x) a(\phi_k) \right)$$

or $\Phi(x) = 2^{-1/2} \left(\Phi^{(+)}(x) + \Phi^{(-)}(x) \right)$, where $\Phi^{(\pm)}(x)$ denotes the "creation" and "annihilation" part, respectively; $\{\phi_k\} \subset \mathcal{S}(\mathbb{R}^4)$ is an orthonormal basis with respect to an orthonormal product which need not coincide with $(\cdot,\cdot)$. The expansion on the right side does not converge to an operator. However, on a formal level it yields for $a^*(f)$ the expression

$$a^*(f) = \sum_k a^*(\phi_k) \int_{\mathbb{R}^4} \overline{\phi_k(x)} f(x)\, dx = \int_{\mathbb{R}^4} \Phi^{(+)}(x) f(x)\, dx$$

and a similar relation for $a(f)$, where the left side is well defined. Motivated by this, we regard quantum fields as *operator–valued distributions* which associate the operators $\Phi^{(+)}(f) := a^*(f)$ and $\Phi^{(-)}(f) := a(f)$ with each test function $f \in \mathcal{S}(\mathbb{R}^4)$. An alternative formulation uses "matrix elements" of the field operators: the map $(\Psi_1, \Phi_m(\cdot)\Psi_2)$ is a tempered distribution for all $\Psi_1, \Psi_2 \in \mathcal{D}_s$ (see the notes).

The second–quantization formalism is not restricted to description of quantum fields. As another illustration, let us mention an application in statistical physics.

12.3.8 Example: The system of a large number of nonrelativistic particles (fermions or bosons), which are confined to a certain region $M \subset \mathbb{R}^3$ and do not interact mutually, is called an **ideal** (respectively, Fermi or Bose) **gas**. The one– article state space is then $L^2(M;\mathbb{C}^{2s+1})$, with s being the spin of the particles, and the one–particle Hamiltonian is usually chosen as $H := \tilde{H}\otimes I_s$, where $\tilde{H}$ is a suitable self–adjoint extension of the symmetric operator $H_0 : H_0\psi = -\Delta\psi$ with the domain $D(H_0) = C_0^\infty(\mathbb{R}^3)$ and I_s means the unit operator on $\mathbb{C}^{2s+1}$. The number of particles is preserved in this case, so it would be sufficient to work with the subspace $\mathcal{H}_p^{(n)}$ in $\mathcal{F}_p(\mathcal{H})$ for some fixed large number n (a typical value is $n \approx 10^{23}$), but it is more convenient to use the Fock space.

In the antisymmetric case it is customary to choose for the algebra of observables the C^*–algebra $\mathcal{A}\subset\mathcal{B}(\mathcal{F}_a(\mathcal{H}))$ generated by the creation and annihilation operators for all $f\in\mathcal{H}$. In the symmetric case, where the $a^\#(f)$ are unbounded, we define $\mathcal{A}$ as the C^*–algebra generated by the corresponding Weyl operators (12.9). An important role among states on the algebra $\mathcal{A}$ is played by the *Gibbs states*

$$\omega : \omega(A) = \frac{\mathrm{Tr}\,\left(A\,e^{-\beta K_\mu}\right)}{\mathrm{Tr}\,\left(e^{-\beta K_\mu}\right)}\,, \tag{12.14}$$

where $K_\mu := \overline{\mathcal{T}_p^\Sigma(H-\mu I)} = \overline{\mathcal{T}_p^\Sigma(H)-\mu\mathcal{N}}$, which describe the gas at thermodynamical equilibrium for an inverse temperature β and a chemical potential μ.

There is an important physical difference between the symmetric and antisymmetric cases. To understand it, suppose that the one–particle Hamiltonian has a purely discrete spectrum consisting of eigenvalues $E_1\le E_2\le\cdots$ (with repetition according to multiplicity) which correspond to eigenvectors $\phi_1,\phi_2,\ldots$ (this is a natural assumption as long as the region M is bounded — see the notes). It is easy to see that the elements of the corresponding occupation–number basis are eigenvectors of the second–quantized Hamiltonian,

$$\overline{\mathcal{T}_p^\Sigma(H)}\,\Phi_n^p\{n_i\} = E\,\Phi_n^p\{n_i\}\,,\qquad E\equiv E\{n_i\} := \sum_{i=1}^\infty n_iE_i \tag{12.15}$$

(Problem 14). The energetically lowest state, in which an ideal Fermi gas consisting of n particles may exist, therefore corresponds to the vector $\Phi_n^a\{1,\ldots,1,0,\ldots\}$, *i.e.*, each of the n lowest states of the Hamiltonian H is occupied by one particle. In contrast, in an ideal Bose gas the one–particle ground state ϕ_1 may be occupied by a larger number of particles, in the extreme case even by all of them; the energetically lowest state here is $\Phi_n^s\{n,0,\ldots\}$. This effect is called the *Bose–Einstein condensation.*

Notes to Chapter 12

Section 12.1 There are two motivations for the introduction of Fock spaces. In some physical systems the number of particles is not preserved in the course of time evolution; this is typical for processes studied in quantum field theories. On the other hand, in statistical physics we usually deal with large systems where the number of particles is preserved, but it is technically impossible to determine it; in that case it also is more convenient to work in the Fock–space framework. Examples will be given in Section 12.3. The spaces $\mathcal{F}_s(\mathcal{H})$ and $\mathcal{F}_a(\mathcal{H})$ are sometimes called *Bose–Fock* and *Fermi–Fock*, respectively. Of the two possibilities discussed in Example 12.1.2, only $\mathcal{F}_s(L^2(\mathbb{R}^3))$ has a direct physical meaning, because the state space $L^2(\mathbb{R}^3)$ is associated with no real fermion.

We have mentioned that the operators $\overline{\mathcal{T}_p^{\Sigma}(A)}$ corresponding to a self–adjoint A are often denoted as $d\Gamma(A)$; similarly we put $\Gamma(A) := \overline{\mathcal{T}_p^{\Pi}(A)}$. The origin of this notation is illustrated by Theorem 12.1.6. We have $\Gamma(e^{iAt}) = e^{id\Gamma(A)}$; hence Γ maps $e^{iAt} \mapsto \{I,\, e^{iAt}, e^{iAt}\otimes e^{iAt}, \dots\}$ and $d\Gamma$ may be formally identified with the linear part of this map at the point $t=0$ which associates $A \mapsto \{0,\, A,\, \underline{A}_1+\underline{A}_2, \dots\}$.

The idea of the second–quantization method was formulated in [Fo 1]; on a formal level it is discussed in most quantum field theory textbooks — see, *e.g.*, ⟦Schwe⟧, Part II; ⟦BŠ⟧, Chap.II, *etc.* A pioneering role in the mathematically correct treatment was played by the papers [Coo 1] and [Seg 4]; in addition to them, there are nowadays numerous sources from which information about the rigorous formulation of second quantization can be derived, for instance, ⟦ Ber ⟧; ⟦ BR 2 ⟧, Sec.5.2; ⟦ GJ ⟧, Chap.6; ⟦ RS 2 ⟧, Sec.X.7; ⟦ Šv ⟧, Chaps.3 and 6; ⟦ Thi 4 ⟧, Sec.1.3.

Section 12.2 Creation and annihilation operators, and also the occupation number basis, are standard tools of the second–quantization method which were already introduced in [Fo 1]. One can employ $a^{\#}(f)$ to express other operators on $\mathcal{F}_p(\mathcal{H})$. A simple example is the operator of the number of particles (Problem 10a); in a similar way for the second quantization $\mathcal{T}_p^{\Sigma}(T)$ of a one–particle operator T and an orthonormal basis $\{\phi_j\}_{j=1}^{\infty}$ of $\mathcal{H}$ we obtain the formal expression

$$\mathcal{T}_p^{\Sigma}(T) = \sum_{j,k} (\phi_j, T\phi_k)\, a^*(\phi_j)a(\phi_k)\,.$$

The meaning of the series depends on the operator T; however, if we use the formula to compute matrix elements of the operator $\mathcal{T}_p^{\Sigma}(T)$ in the basis $\mathcal{E}_p$ corresponding to $\{\phi_j\}$, the series reduces to a finite sum and no problems arise. The expression is particularly illustrative if T has a pure point spectrum and $\{\phi_j\}$ is the basis of its eigenvectors.

The same is true for the second quantization of a k–particle operator $T^{(k)}$, which can be constructed by a straightforward generalization of the procedure described in the previous section. We formally obtain the expansion

$$\mathcal{T}_p^{(\Sigma)}(T^{(k)}) = \frac{1}{k!} \sum_{j_1,\dots,j_k;l_1\dots,l_k} (\phi_{j_1}\otimes\cdots\otimes\phi_{j_k}, T^{(k)}(\phi_{l_1}\otimes\cdots\otimes\phi_{l_k})_k \\ \times\, a^*(\phi_{j_1})\dots a^*(\phi_{j_k})a(\phi_{l_k})\dots a(\phi_{l_1})\,,$$

where the meaning of the series again depends on the operator $T^{(k)}$. Expansions of this type with the creation operators placed to the left of the annihilation operators are usually referred to as the *normal form* of the operator under consideration. More details on these problems can be found, for instance, in ⟦ Ber ⟧, Sec.I.1; ⟦ Šv ⟧, Chap.6.

Section 12.3 For a complex function f the operator (12.9) can be written in the form $W(f) = \exp\left(i(\overline{\Phi(\mathrm{Re}\, f) + \Pi(\mathrm{Im}\, f)})\right)$, which illustrates that it does indeed generalize the Weyl operator of Section 8.2. As in the case of a finite number of degrees of freedom, one can use $W(f)$ to define coherent states on $\mathcal{F}_s(\mathcal{H})$ — see ⟦ Da 1 ⟧, Chap.8. The relation

(12.10) can be checked, *e.g.*, by a direct computation using power–series expansions on the set $\mathcal{D}_s$ of analytic vectors; this method is used in ⟦RS 2⟧, Sec.X.7. An alternative proof is given in ⟦BR 2⟧, Sec.5.2. At the same places the reader can find a proof of the strong continuity of the map $W(\cdot)$, which is based on the result of Problem 11b and the fact that the domain $\mathcal{D}_s$ is a core for $\Phi_S(f)$.

The Segal field operator was introduced in [Seg 4]. It corresponds to the formal expression of Remark 12.3.7; its advantage is that it allows an easier formulation of the CCR. On the other hand, from the distribution–theory point of view it is more convenient to have the field $\Phi(\cdot)$ *complex* linear; this motivates the definition (12.13). For properties of operator–valued distributions, the reader may consult, *e.g.*, ⟦ BLT ⟧, Sec.3.1; ⟦ SW ⟧, Sec.3.1; ⟦ Ber ⟧, Sec.I.1; ⟦ Si 2 ⟧, Sec.II.1. The definition (12.13) is usually extended to distributions of the type $f_{g,\tau}\,:\; f_{g,\tau}(x) = g(\vec{x})\delta(t)$ with $g \in \mathcal{S}(\mathbb{R}^3)$; the corresponding modification of the map $\Phi_m(\cdot)$ is called the *field at an instant* τ. Of course, we lose the relativistic covariance, but on the other hand some properties formulate more easily for fields at a fixed instant.

The representations of Corollary 12.3.6 do not exhaust the list of nonequivalent representations of canonical commutation (or Weyl) relations. There are other representations which are not of the Fock type. Such representations correspond, in particular, to all interacting fields. This is a consequence of the *Haag theorem* which, loosely speaking, claims the following: suppose that $\Phi^{(1)}(\cdot)$ and $\Phi^{(2)}(\cdot)$ are quantum fields which satisfy the Gårding–Wightman axioms (see Section 13.3 below) and $\Phi^{(1)}(\cdot)$ is a free field of mass m. If the fields are unitarily equivalent, then $\Phi^{(2)}(\cdot)$ is also a free field of mass m.

Haag's theorem appeared for the first time in [Haa 2]; its proof and extension can be found in ⟦SW⟧, Sec.4.5; ⟦BLT⟧, Sec.5.4. The exact formulation of the unitary equivalence uses fields at a fixed instant mentioned above; what is essential is that the equivalence is assumed for complex–valued test functions, *i.e.*, for both fields and their canonically conjugated momenta. The main importance of the Haag theorem is that perturbative methods based on the interaction picture cannot be applied to interacting fields in a straightforward way such as is common in formal perturbative computations. Let us remark, on the other hand, that our ability to distinguish nonequivalent representations of the CCR in an actual physical experiment is limited. This problem is analogous to the one discussed in Section 7.4; a detailed discussion can be found in [Haa 1].

While the central place undoubtedly belongs to relativistic theories, quantum field theory is not *a priori* relativistic. An illustration of this claim is provided by the Lee model mentioned in the notes to Section 9.6, which describes a field of light nonrelativistic "θ–particles" interacting with a heavy nucleon having two internal states. In spite of the nonrelativistic character of the model, the number of particles in it may not be preserved, provided the parameters are chosen in such a way that emission of a θ–particle is possible; this process is closely related to the Friedrichs model discussed in Section 9.6. A more general discussion of nonrelativistic field theories can be found in [LeL 1,2].

Furthermore, since the first edition of this book there were numerous rigorous studies of the non–relativistic quantum electrodynamics and other models of non–relativistic quantum field theory — we refer to the papers [AKa 1], [BFS 1], [CEH 1], [CVV 1], [GLL 1], [HS 1], [HSp 1] and the bibliography there.

We did not specify the region M in Example 12.3.8. The standard statistical–physics procedure is to treat the system first in a bounded M, and then to study the behavior of the results when M blows out; this is referred to as the *thermodynamic limit*. The boundedness of M means that the one–particle Hamiltonians considered in the example have purely discrete spectra; this can be checked easily for parallelepipeds, balls, *etc.*, but the result is valid for much more general regions — see Theorem 14.5.2. Notice that the discrete character of the spectrum is essential for the existence of the states (12.14). More information about ideal gases can be found, for instance, in ⟦BR⟧, Sec.5.3; ⟦Thi 4⟧ and ⟦Ja⟧, Chap.15.

Problems

1. The Fock space $\mathcal{F}(\mathcal{H})$ is separable *iff* the same is true for $\mathcal{H}$.

2. Prove Proposition 12.1.3 for the operator $\mathcal{T}_p^{\Pi}(T)$.

3. Let A be a self–adjoint operator. Its spectrum is related to that of $\overline{\mathcal{T}^{\Sigma}(A)}$ by
$$\sigma\left(\overline{\mathcal{T}^{\Sigma}(A)}\right) \supset \bigcup_{n=0}^{\infty} \sigma\left(\overline{A_n^{\Sigma}}\right) \supset \{0\} \cup \bigcup_{n=1}^{\infty} P_{\Sigma}^{(n)}(\sigma(A) \times \cdots \times \sigma(A)),$$
where $P_{\Sigma}^{(n)}(t_1, \dots, t_n) := t_1 + \cdots + t_n$.

4. With the notation of Theorem 12.1.6, prove

   (a) $\{\overline{\mathcal{T}^{\Pi}(U(t))} : t \in \mathbb{R}\}$ is a strongly continuous group of unitary operators which leaves the subspaces $\mathcal{H}^{(n)}$ invariant.

   (b) The group $\{\overline{\mathcal{T}_p^{\Pi}(U(t))} : t \in \mathbb{R}\}$ is generated by $\overline{\mathcal{T}_p^{\Sigma}(A)}$.

   Hint: (a) Use Problem 5.45.

5. Let $\hat{\jmath}$ be a nondecreasing variation. Given $p \in \mathcal{S}_n$, show that the number of permutations $p' \in \mathcal{S}_n$ such that $\hat{\jmath}\circ p = \hat{\jmath}\circ p'$ is independent of p and equals $n_1! \dots n_m!$, where n_i are the numbers of repeated values in $\hat{\jmath}$.

6. Prove the relations (12.4) and (12.5).
   Hint: Check the relations for $a^*(\phi_k)$; then use Theorem 12.2.2.

7. Prove the relation (12.6).
   Hint: Check it first for $\Psi \in \mathcal{D}_p^{(n)}$.

8. Given a map $F_n : \mathbb{Z}_+ \times \cdots \times \mathbb{Z}_+ \to \mathbb{C}$, where $\mathbb{Z}_+$ is the set of positive integers, prove
$$\sum_{p \in \mathcal{S}_n} \nu_p^{(p)} F_n(j_{p_1}, \dots, j_{p_n}) = \sum_{k=1}^{n} \delta_{k-1}^{(p)} \sum_{r \in \mathcal{S}_{n-1}^{(k)}} \nu_r^{(p)} F_n(j_k, j_{r_1}, \dots, j_{r_{k-1}}, j_{r_{k+1}}, \dots, j_{r_n}),$$
where $\delta_j^{(p)}$, $\nu_p^{(p)}$ are as in the proof of Theorem 12.2.2 and $\mathcal{S}_{n-1}^{(k)}$ is the set of permutations of the numbers $(1, \dots, k-1, k+1, \dots, n)$.

9. Prove $(a^*(f_1)\dots a^*(f_n)a(g_1)\dots a(g_m))^* \upharpoonright \mathcal{D}_p = a^*(g_m)\dots a^*(g_1)a(f_n)\dots a(f_1)$ and the analogous relation with the interchange of the creation and annihilation operators.

10. Let $\mathcal{N}_p$ be the operator of the number of particles of Example 12.1.5. Prove:

(a) If $\{\phi_j\}_{j=1}^{\infty}$ is an orthonormal basis in $\mathcal{H}$, then $\mathcal{N}_p\Psi = \sum_{j=1}^{\infty} a^*(\phi_j)a(\phi_j)\Psi$ holds for an arbitrary vector $\Psi \in \mathcal{D}_p$, where the series converges with respect to the norm of $\mathcal{F}_p(\mathcal{H})$.

(b) In the symmetric case, the inequalities $\|a^*(f)\Psi\|_F \le \|f\|\,\|(\mathcal{N}_s+I)^{1/2}\Psi\|_F^2$ and $\|a(f)\Psi\|_F \le \|f\|\,\|\mathcal{N}_s^{1/2}\Psi\|_F^2$ hold for any $\Psi \in \mathcal{D}(\mathcal{N}_s^{1/2})$.

Hint: (b) Use the occupation–number basis.

11. Check the following properties of Segal field operators:

(a) Let a sequence $\{f_j\}_{j=1}^{\infty} \subset \mathcal{H}$ converge to some $f \in \mathcal{H}$; then $\Phi_S(f_j)\Psi \to \Phi_S(f)\Psi$ holds for any $\Psi \in \mathcal{D}_s$.

(b) $\overline{\mathcal{T}_s^{\Pi}(U)}\,\Phi_S(f)\,\overline{\mathcal{T}_s^{\Pi}(U)}^{-1} = \Phi_S(Uf)$ holds for any unitary operator U on the one–particle space $\mathcal{H}$.

(c) The preceding assertion extends to unitary operators $V : \mathcal{H} \to \mathcal{H}'$ between different Hilbert spaces. We define the map $\overline{\mathcal{T}_s^{\Pi}(V)} : \mathcal{F}_s(\mathcal{H}) \to \mathcal{F}_s(\mathcal{H}')$ by $\overline{\mathcal{T}_s^{\Pi}(V)}\psi_n(f_1,\dots,f_n) := S_n\psi_n(Vf_1,\dots,Vf_n)$; then the relation

$$\overline{\mathcal{T}_s^{\Pi}(V)}\,\Phi_S(f)\,\overline{\mathcal{T}_s^{\Pi}(V)}^{-1} = \Phi'_S(Vf)$$

is valid between the corresponding Segal field operators on the spaces $\mathcal{F}_s(\mathcal{H})$ and $\mathcal{F}_s(\mathcal{H}')$, respectively.

Hint: (a) Use the result of the previous problem.

12. Consider the operators $N(f)$ from the proof of Lemma 12.3.4. Show that

(a) If a bounded B commutes with $W(tf_j)$, then $BN(\phi_j) \subset N(\phi_j)B$.

(b) $N(f)$ is positive for any $f \in \mathcal{H}$ and $N(f)\upharpoonright \mathcal{D}_s = \tilde{N}(f)\upharpoonright \mathcal{D}_s = a^*(f)a(f)$, where $\tilde{N}(f) := \overline{\mathcal{T}_s^{\Sigma}((f,\cdot)f)}$.

13. Show that if U satisfies the assumptions of Theorem 12.3.5 and $U \ne 0$, then there is $\alpha \in \mathbb{R}$ such that αU is unitary.
Hint: The operators U^*U, UU^* commute with W_m and $W_{m'}$, respectively; use Lemma 12.3.4.

14. Prove the relation (12.15).

Chapter 13

Axiomatization of quantum theory

13.1 Lattices of propositions

In Section 7.6 we mentioned the motives to axiomatize a physical theory. Now we are going to discuss two axiomatic approaches to quantum theory. The first of them is concerned mostly with a justification of the postulates formulated above; it is based on a deeper analysis of the properties of *yes–no* experiments.

For brevity we replace the expression "outcome of a given *yes–no* experiment" by the term **proposition**; the dichotomic character means that a proposition is either *valid* (*true*) or *not valid* (*false*). Various relations can exist between the propositions concerning a particular system. In general, they depend on the state of the system. Some of the relations are state–independent, however, and we shall be interested in them in the following. For instance, the inclusion $\Delta_a \subset \Delta_b \subset \mathbb{R}^3$ means that validity of the proposition "the particle has been found within Δ_b" follows from the validity of "the particle has been found within Δ_a" irrespective of the particle state. In such a case we say that a *implies* b and write symbolically $a \subset b$. In accordance with experience we adopt the following assumption:

(pl1) the relation $\subset$ defines a partial ordering on the set $\mathcal{L}$ of propositions corresponding to a physical system.

It implies, in particular, that the relations $a \subset b$ and $a \supset b$ mean *identity* of a and b, *i.e.*, that the two propositions are simultaneously either true or false.

Furthermore, we introduce the proposition $a \cap b$, which has the meaning of the *simultaneous validity* of the propositions a and b. Mathematically it can be defined as the *infimum* (or *greatest lower bound*) of the set $\{a, b\}$, by which we understand the element $a \cap b \in \mathcal{L}$ such that $c \subset a \cap b$ holds *iff* $c \subset a$ and $c \subset b$; in the same way we introduce the infimum $\bigcap_{a\in\mathcal{S}} a$ for any subset $\mathcal{S} \subset \mathcal{L}$. Physically the meaning of the proposition $a \cap b$ is clear provided the order in which the corresponding *yes–no* experiments are performed is unimportant; its validity expresses the fact that the positive result has been found in both of them. In general, the situation is more complicated since the observables may be noncompatible, but even then the proposition $a \cap b$ makes sense: we can identify it with the positive outcome of the idealized measurement consisting of the sequence $\{a, b, a, b, \dots\}$ — see Example 13.13 below and Remark 3.2.10. We shall not go into details and shall adopt the following postulate:

(pl2) the infimum $\bigcap_{a\in\mathcal{S}} a$ exists for any subset $\mathcal{S} \subset \mathcal{L}$.

This implies the existence of the *absurd proposition* $\emptyset$, which has the property that $\emptyset \subset a$ for any $a \in \mathcal{L}$; by definition $\emptyset = \bigcap_{a\in\mathcal{L}} a$.

The third basic postulate equips the proposition set $\mathcal{L}$ with the operation of *orthocomplementation* defined in the following way:

(pl3) for any $a \in \mathcal{L}$, there is a proposition $a' \in \mathcal{L}$ such that

(i) $(a')' = a$.

(ii) $a \cap a' = \emptyset$.

(iii) The relation $a \subset b$ implies $b' \subset a'$.

In this case the interpretation is easy; the proposition a' refers to the same *yes–no* experiment, with the results switched: if a is true, a' is false and *vice versa.* The postulates (pl1)–(pl3) may be expressed in a more concise form if we use the terminology of the algebraic lattice theory (see the notes):

(pl) the family $\mathcal{L}$ of propositions corresponding to any physical system forms a **complete orthocomplemented lattice**;

for brevity we shall drop the adjectives and speak about $\mathcal{L}$ as a *lattice*. The postulates have some simple consequences. Any subset $\mathcal{S} \subset \mathcal{L}$ has the *supremum* (or *least upper bound*) $\bigcup_{a\in\mathcal{S}} a$, which is by definition a proposition such that $b \supset \bigcup_{a\in\mathcal{S}} a$ *iff* $b \supset a$ for all $a \in \mathcal{S}$. The supremum and infimum are related by a de Morgan–type formula (Problem 1). The proposition $a \cup b$ has the meaning "a or b is valid". The complement to the absurd proposition is the *trivial proposition* $I := \bigcap_{a\in\mathcal{L}} a$ which is always valid.

The requirements on the family of propositions we have formulated until now are of a general nature and contain nothing peculiar to quantum theory.

13.1.1 Example (proposition lattices in classical mechanics): The phase space $\mathbb{R}^{2n}$ is associated with an unconstraint system with n degrees of freedom; its points are described generalized coordinates $q_1, \dots, q_n$ and the canonically conjugated momenta $p_1, \dots, p_n$. A typical proposition about such a system reads "the value of a Borel function $F : \mathbb{R}^{2n} \to \mathbb{R}$ lies in a Borel set Δ " (for instance, the energy of the harmonic oscillator, $H(q,p) = q^2 + p^2$, assumes a value E or is contained in an interval (E_1, E_2)). The admissible values of the coordinates and momenta then have to satisfy the relation $(q_1, \dots, p_n) \in F^{(-1)}(\Delta)$; this allows us to ascribe to any proposition $a \in \mathcal{L}_{cl}$ a Borel set $M \subset \mathbb{R}^{2n}$ in such a way that

$$M(a \cap b) = M(a) \cap M(b) , \quad M(a') = \mathbb{R}^{2n} \backslash M(a) .$$

The propositions to which the same subset of the phase space corresponds are physically equivalent, and therefore they can be regarded as identical and the map $M(\cdot)$ can be regarded as a bijection.

The validity of the postulates (pl) for $\mathcal{L}_{cl}$ then follows from properties of the set operations, with the exception of (pl2): the family $\mathcal{B}^{2n}$ is closed with respect to at most countable intersections, so $\mathcal{L}_{cl}$ is an *countably* complete orthocomplemented lattice. On the other hand, $\mathcal{L}_{cl}$ has two additional properties. First of all,

(pl$_b$) any $a, b, c \in \mathcal{L}_{cl}$ satisfy the *distributive law, i.e.*,

$$a \cup (b \cap c) = (a \cup b) \cap (a \cup c) , \qquad a \cap (b \cup c) = (a \cap b) \cup (a \cap c) ;$$

a lattice with this property is called *Boolean.*

Furthermore, a *point* in a lattice is a minimal proposition different from $\emptyset$; if b is such a proposition, then the inclusions $a \subset c \subset (a \cup b)$ imply $c = a$ or $c = a \cup b$. We can easily check that

(pl$_a$) for any proposition $a \in \mathcal{L}_{cl}$ different from $\emptyset$ one can find a point $b \subset a$; a lattice with this property is called *atomic.*

The proposition lattices of quantum mechanical systems have a different structure; the main difference is that they are not Boolean.

13.1.2 Example: Let a_j be the proposition "the j–th component of spin has the value $\frac{1}{2}$ ". The proposition $a_j \cap a_k$ is absurd if $j \neq k$, and similarly $a'_j \cap a'_k = \emptyset$, so $a_j \cup a_k = I$. It implies, *e.g.*, $(a_1 \cup a_2) \cap (a_1 \cup a_3) = I$ while $a_1 \cup (a_2 \cap a_3) = a_1 \neq I$.

In order to be able to weaken the requirement (pl$_b$), we introduce a few notions. A *sublattice* is a subset in $\mathcal{L}$, which is itself a (complete, orthocomplemented) lattice with respect to the same operations. For any set $M \subset \mathcal{L}$ there is a minimal lattice which contains it (*cf.* Problem 3); we call it the lattice *generated* by M. Propositions a, b are said to be *compatible* if the lattice generated by the set $\{a, b\}$ is Boolean; we use the symbol $a \leftrightarrow b$. The family of propositions in $\mathcal{L}$ that are compatible with all $a \in \mathcal{L}$ is called the *center* of $\mathcal{L}$; if the center contains only the propositions $\emptyset$, I we say that the lattice is *irreducible.*

The requirement which replaces distributivity for quantum systems can now be formulated as follows:

(pl$_{wr}$) the relation $a \subset b$ between propositions implies their compatibility, $a \leftrightarrow b$.

It is usually called *weak modularity;* we shall check that it is satisfied in standard formalism of quantum theory.

13.1.3 Example: Consider a coherent quantum system with the state space $\mathcal{H}$; the family of *yes–no* experiments is in this case identified with the set $\mathcal{B}(\mathcal{H})^E$ of all projections on $\mathcal{H}$. The partial ordering is given by the inequalities between the projections. It has the needed interpretation: $a \subset b$ is equivalent to $E(a) \leq E(b)$. To any two propositions a, b we define $E(a \cap b)$ as the projection onto the subspace $E(a)\mathcal{H} \cap E(b)\mathcal{H}$; in a similar way we define the infimum for any family of propositions. The orthogonal complement is described by the projection $E(a') := I - E(a)$, and the absurd and trivial propositions are represented by the zero and unit operator, respectively. It is easy to see that the conditions (pl1)–(pl3) are fulfilled, *i.e.*, that $\mathcal{B}(\mathcal{H})^E$ with the mentioned operations is a complete orthocomplemented lattice; in addition, it is atomic, the role of points being played by one–dimensional projections (Problem 4).

We shall prove that $\mathcal{B}(\mathcal{H})^E$ is also weakly modular. This requires that we first check that the notion of compatibility introduced above is consistent with the definition of Section 7.5. If $a \leftrightarrow b$, then the propositions a, b as well as a', b' belong to a Boolean sublattice in $\mathcal{B}(\mathcal{H})^E$, *i.e.*, we have $(a \cap b) \cup (a \cap b') = a$ and $(a \cap b) \cup (a' \cap b) = b$. The inclusion $a \cap b' \subset (a \cap b)'$ follows from $a \cap b' \subset b' \subset a' \cup b'$ and Problem 1b for $\mathcal{S} = \{a', b'\}$; in a similar way we get $a' \cap b \subset (a \cap b)'$ and $a \cap b' \subset (a' \cap b)'$. Using Problem 4a, we then find that the projections $E(a \cap b)$, $E(a' \cap b)$ and $E(a \cap b')$ are mutually orthogonal and satisfy the relations

$$E(a) = E(a \cap b) + E(a \cap b')\,, \qquad E(b) = E(a \cap b) + E(a' \cap b)\,;$$

it yields the commutativity of $E(a)$, $E(b)$. On the other hand, suppose that $E := E(a)$ and $F := E(b)$ commute. We can check directly that

$$\begin{aligned} \mathcal{S} = \{&0,\, I,\, E,\, F,\, I-E,\, I-F,\, EF,\, I-EF,\, E-EF,\, F-EF,\, I-E+EF, \\ &I-F+EF,\, E+F-EF,\, I-E-F+EF,\, E+F-2EF,\, I-E-F+2EF\,\} \end{aligned}$$

this the lattice generated by $\{E, F\}$ since it is closed with respect to the multiplication, which represents the operation $\cap$, and $P \in \mathcal{S}$ implies $I - P \in \mathcal{S}$. The set $\mathcal{S}$ is obviously commutative and $P \cup Q = I - (I-P)(I-Q) = P+Q-PQ$ holds for any $P, Q \in \mathcal{S}$; it follows that the projections $E(a)$ and $E(b)$ are compatible in the lattice sense. Hence the two definitions are equivalent and the weak modularity of $\mathcal{B}(\mathcal{H})^E$ is a consequence of Proposition 3.2.9b.

13.1.4 Remarks: (a) The proved equivalence shows at the same time that the irreducibility of the lattice $\mathcal{B}(\mathcal{H})^E$ is equivalent to its irreducibility as an operator set. This correspondence can be extended to more general situations. Consider, for instance, a quantum system whose algebra of observables is of the form (7.18). The set $\mathcal{A}^E$ of all projections in $\mathcal{A}$ forms an (atomic, weakly modular) lattice, which is reducible unless the index set J consists of a single point; its center is generated by the projections onto the coherent subspaces (Problem 5).

(b) Propositions a, b are said to be *disjoint* if $a \subset b'$; the requirement of weak modularity means that disjoint propositions are mutually compatible.

These considerations show that in the quantum case the postulate (pl) can be appended as follows:

(pl_q) the set of propositions for a quantum system is an atomic, weakly modular (complete, orthocomplemented) lattice.

In order not to repeat the adjectives all the time, we shall refer to a family $\mathcal{L}$ which satisfies the condition (pl_q) as a *proposition system.* Two such systems are *isomorphic* if there is a bijection between them which preserves the ordering and the orthogonal complement, and maps the infimum of any subset $M \subset \mathcal{L}$ to the infimum of its image.

The basic question that the discussed axiomatic approach should answer is whether there are proposition systems, which are not isomorphic to the standard system of Example 13.1.3, or whether on the contrary the quantum mechanical formalism treated in the previous chapters is the only possible one. We return in this way to the question considered on a heuristic level in Section 7.1.

This problem is highly nontrivial and we restrict ourselves to a brief description of the result; references to the proofs are given in the notes. First we have to mention how a proposition system can be reduced. If the center of the lattice $\mathcal{L}$ contains a finite number of propositions different from $\emptyset$, I, it is easy to decompose $\mathcal{L}$ to irreducible components (Problem 6). The same procedure can be applied in the general case; it is just more difficult to check the existence and uniqueness of the decomposition. With this fact in mind, we shall suppose in the following that the proposition system $\mathcal{L}$ is irreducible.

A few more notions are needed. The *dimension* of an atomic lattice is the minimal number of compatible points $b_1, b_2, \dots$ such that $\bigcup_j b_j = I$. Let F be a field; an *involutive automorphism* is a map $* : F \to F$ such that

$$(\alpha+\beta)^* = \alpha^* + \beta^* , \quad (\alpha\beta)^* = \beta^*\alpha^* , \quad (\alpha^*)^* = \alpha$$

holds for any $\alpha, \beta \in F$; examples are the identical mapping in $\mathbb{R}$ or $\mathbb{C}$, the complex conjugation in $\mathbb{C}$, *etc.* Given a vector space V over F, we define a symmetric sesquilinear form $f : V \times V \to F$ in the same way as in Section 1.1 replacing only the complex conjugation by the corresponding involutive automorphism. A particular case of the sesquilinear form is an inner product, which is again defined by the requirement that $f(x,x) = 0$ *iff* $x = 0$. In addition to real and complex Hilbert spaces, one can in this way define, for instance, Hilbert spaces over the field $\mathbb{Q}$ of quaternions (see the notes). The mentioned result now reads as follows.

13.1.5 Theorem (Piron): (a) Let $\mathcal{L}$ be an irreducible proposition system of a dimension ≥ 4. Then there is a vector space V over F, an involutive automorphism of the field F, and an inner product $\langle\cdot,\cdot\rangle : V \times V \to F$ such that $\mathcal{L}$ is isomorphic to the lattice consisting of those subspaces in V, which are orthogonal complements of some subset $M \subset V$ with respect to $\langle\cdot,\cdot\rangle$.

(b) In addition, if the field F is required to contain the reals, then the proposition system $\mathcal{L}$ is isomorphic to the lattice of closed subspaces in a Hilbert space over some of the fields $\mathbb{R}$, $\mathbb{C}$, or $\mathbb{Q}$.

Loosely speaking, the existence of partial ordering and the operation $\cap$ together with the assumptions of atomicity and weak modularity determine the vector structure of the state space, leaving the choice of the field open. The fact that $\mathcal{L}$ is equipped with orthocomplementation induces the existence of an inner product.

If we adopt the hypothesis of part (b), for which we have no empirical support, the theorem leaves three possibilities open; in addition to the standard quantum theory the postulate (pl_q) also allows theories in which the state space is a Hilbert space over the field $\mathbb{R}$ or $\mathbb{Q}$. Both the alternative schemes, *real* and *quaternionic* quantum theories, have indeed been constructed. In some cases, the equivalence of their predictions with those of "complex" quantum theory has been demonstrated (see the notes), but in general this problem remains open.

13.2 States on proposition systems

We know that states of a quantum system are determined experimentally by measuring a family of transition probabilities by repeating a certain collection of *yes–no* experiments on a large number of copies of the system. It is therefore natural in the axiomatic approach under consideration to identify a *state* with a function $p:\,\mathcal{L}\to[0,1]$ whose values are the probabilities that a given proposition is true. We require it to satisfy the following conditions:

(s1) $p(\emptyset)=0$ and $p(I)=1$.

(s2) $p(\bigcup_j a_j)=\sum_j p(a_j)$ for any sequence $\{a_j\}$ of mutually disjoint propositions.

(s3) If $\{\,a_\alpha:\,\alpha\in J\,\}$ is a subset in $\mathcal{L}$ whose elements are valid with certainty, $p(a_\alpha)=1$ for all $\alpha\in J$, then $p(\bigcap_{\alpha\in J}a_\alpha)=1$.

(s4) For any pair $a,b\in\mathcal{L}$, $a\ne b$, there is a state p such that $p(a)\ne p(b)$.

Conditions (s2) and (s3) can be justified for a finite number of propositions; for the sake of simplicity we make the usual idealization and extend their validity to infinite subsystems as well. Requirement (s2) means, in particular, that the inclusion $a\subset b$ implies $p(a)\le p(b)$. The purpose of condition (s4) is to eliminate "superfluous" elements of system $\mathcal{L}$; if two propositions can be separated by no state, *i.e.*, $p(a)=p(b)$ for all p, it is natural to regard them as identical. It is easy to see that states in the standard quantum mechanical formalism fulfil these requirements (Problem 7).

Suppose that p_1,p_2 are states on $\mathcal{L}$ and α,β are non–negative numbers such that $\alpha+\beta=1$; then $\alpha p_1+\beta p_2$ is again a state. Hence the set of all states on $\mathcal{L}$ is convex; its extremal points will be called *pure states* as before while the other states are *mixed*. For any state p and a proposition a we define its *dispersion* by

$$\sigma_p(a)\;:=\;p(a)-p(a)^2\;;$$

this is consistent with the definition mentioned in the notes to Section 7.3 due to the fact that propositions are represented in the standard formalism by idempotent operators, $E(a)^2 = E(a)$. Furthermore, we define the *total dispersion* of a state p,

$$\sigma_p := \sup_{a\in\mathcal{L}} \sigma_p(a);$$

its value lies in the interval $[0, \frac{1}{4}]$. If $\sigma_p = 0$, the state p is called *dispersionless.* Any such state is pure (Problem 8) while the opposite implication is not valid as we shall see in a moment.

The existence of dispersionless states represents an important distinction between classical and quantum systems. In classical mechanics such states do exist. Consider, *e.g.*, the lattice $\mathcal{L}_{cl}$ of Example 13.1.1 and the state p determined by a point $(q_1^{(p)}, \dots, p_n^{(p)})$ of the phase space. We know that $\mathcal{L}_{cl}$ may be identified with the σ–algebra of all Borel sets $M \subset \mathbb{R}^{2n}$; the function $M \mapsto p(M)$ then assumes only the values 1 or 0 depending on whether the point $(q_1^{(p)}, \dots, p_n^{(p)})$ is contained in M or not. It follows that p is dispersionless, and therefore pure. Such states are typical of classical mechanics; we usually take no other states into account.

The situation in quantum mechanics is different; owing to the existence of noncompatible observables we can find pure states with a nonzero dispersion.

13.2.1 Example: Consider again the propositions corresponding to the electron spin of Example 13.1.2. The state p in which the spin is oriented along the third axis, *i.e.*, $p(a_3) = 1$ and $p(a_3') = 0$, is obviously pure; however, $p(a_1) = \frac{1}{2}$, so its dispersion is $\sigma_p = \sigma_p(a_1) = \frac{1}{4}$.

Experience tells us that dispersionless states do not exist in quantum systems. This claim can easily be illustrated in the standard formalism. However, we can also prove it using only the properties of system $\mathcal{L}$. For simplicity we restrict ourselves to the situation without superselection rules.

13.2.2 Proposition: Suppose that the system is coherent, *i.e.*, the corresponding proposition system $\mathcal{L}$ is irreducible; then there are no dispersionless states on it unless $\mathcal{L}$ is trivial.

Proof: Let p be a dispersionless state; then $p(a) = 1$ or $p(a) = 0$ holds for all $a \in \mathcal{L}$. We denote by $\mathcal{L}_1$ the set of those propositions for which the first named possibility takes place, and set $e := \inf \mathcal{L}_1$; it follows from condition (s3) that $p(e) = 1$. If a proposition $b \subset e$ is different from e, then $p(b) = 0$. The propositions b, b' are obviously disjoint, so (s2) gives $p(b) + p(b') = p(b \cup b') = 1$, *i.e.*, $b' \in \mathcal{L}_1$. We obtain $b = b \cap e \subset b \cap b' = \emptyset$, which means that the inclusion $b \subset e$ is satisfied only if $b = e$ or $b = \emptyset$; hence e is a point.

We shall check that e is contained in the center of the lattice $\mathcal{L}$. If $a \in \mathcal{L}_1$, then $e \subset a$ and the weak modularity of $\mathcal{L}$ yields $a \leftrightarrow e$. Conversely, $p(a) = 0$ implies $a' \in \mathcal{L}_1$ as we have demonstrated above, and therefore $e \leftrightarrow a'$; using Problem 3e we again get $e \leftrightarrow a$, *i.e.*, together $e \leftrightarrow a$ for all $a \in \mathcal{L}$. However, the center of $\mathcal{L}$ is trivial by assumption, so $e = I$. Since we know that e is a point, it follows that the lattice is trivial, $\mathcal{L} = \{\emptyset, I\}$. ∎

This fact has been regarded by some physicists as a serious defect showing that quantum description of reality is incomplete. A possible explanation was that microscopic systems could exist in dispersionless states but for some reason we were able to prepare only their mixtures. The quantities that should be appended to complete sets of observables to enable preparation of dispersionless states has therefore been named *hidden parameters.* If such a "complete" theory could be constructed, of course, it must reproduce all experimentally verifiable predictions of standard quantum theory.

It appears, however, that this is impossible. To be able to formulate this result more exactly, we need the appropriate definition. We say that a quantum system with the proposition lattice $\mathcal{L}$ admits a description by means of hidden parameters if there is a measure space $(X, \mathcal{A}, \varrho)$ such that $\varrho(X) = 1$, and to any state p on $\mathcal{L}$, there is a map $p : \mathcal{L} \times X \to [0,1]$ with the following properties:

(i) $p(a, \cdot)$ is measurable on $(X, \mathcal{A})$ for any $a \in \mathcal{L}$.

(ii) $p(\cdot, \xi)$ is a dispersionless state on $\mathcal{L}$ for ϱ–a.a. $\xi \in X$.

(iii) The state p expresses as a mixture of $p(\cdot, \xi)$, *i.e.*,

$$p(a) = \int_X p(a,\xi)\, d\varrho(\xi)$$

holds for all $a \in \mathcal{L}$.

In particular, if the measure ϱ has a discrete support $\{\xi_1, \dots, \xi_n\} \subset X$, then the last relation may be written as $p(a) = \sum_k w_k p_k(a)$, where $w_k := \varrho(\xi_k)$ and $p_k(\cdot) := p(\cdot, \xi_k)$, $k = 1, 2, \dots$, are dispersionless states; we have, of course, $\sum_k w_k = \varrho(X) = 1$. We shall need the following auxiliary results.

13.2.3 Proposition: (a) Let $\mathcal{L}$ admit description by means of hidden parameters; then $p(a) + p(b) = p(a \cap b) + p(a \cup b)$ holds for each state p and any $a, b \in \mathcal{L}$.

(b) Propositions a, b in a weakly modular $\mathcal{L}$ are compatible *iff* $(a \cap b') \cup b \supset a$.

Proof: In order to prove (a), it is sufficient according to condition (iii) and linearity of the integral to check the identity for a dispersionless p. In that case, $p(a)$ and $p(b)$ may assume only the values 0, 1. Let us review the four possibilities. If $p(a) = p(b) = 0$; then the inequality $p(a \cap b) \le p(a)$ implies $p(a \cap b) = 0$. Moreover, we have $p(a') = p(b') = 1$, *i.e.*, $p(a' \cap b') = 1$ by the postulate (s3), which yields

$$p(a \cup b) = 1 - p((a \cup b)') = 1 - p(a' \cap b') = 0;$$

hence the identity is valid. Consider next the case with $p(a) = 1$ and $p(b) = 0$. Using the inequalities $p(a \cup b) \ge p(a) = 1$ and $p(a \cap b) \le p(b) = 0$, we see that $p(a \cup b) = 1$ and $p(a \cap b) = 0$, so the relation again holds true; the same conclusion can be made in the case $p(a) = 0$, $p(b) = 1$, which differs from the previous one by the interchange of the propositions. Finally, if $p(a) = p(b) = 1$ we use the identities $p(a \cup b) = 1 - p(a' \cap b')$ and $p(a \cap b) = 1 - p(a' \cup b')$, which give

$$p(a) + p(b) - p(a \cap b) - p(a \cup b) \;=\; p(a' \cup b') + p(a' \cap b') - p(a') - p(b')\,,$$

and since we have already proven the assertion for the propositions a', b' fulfilling $p(a') = p(b') = 0$, part (a) is valid. A reference to the proof of (b) is given in the notes. ∎

With these preliminaries, we are now able to prove the mentioned result.

13.2.4 Theorem (Jauch–Piron): If a proposition system $\mathcal{L}$ admits description by means of hidden parameters, then any two propositions $a, b \in \mathcal{L}$ are compatible.

Proof: Using Proposition 13.2.3a repeatedly, we get

$$\begin{aligned} p((a \cap b') \cup b) &= p(a \cap b') + p(b) \;=\; p(a) + p(b') - p(a \cup b') + p(b) \\ &= p(a) + 1 + p(a \cup b') \;=\; p(a) + p(a' \cap b) \;=\; p(a \cup (a' \cap b))\,. \end{aligned}$$

By assumption, this relation is valid for all p, so $(a \cap b') \cup b = a \cup (a' \cap b)$ holds due to postulate (s4). However, the proposition on the right side is implied by a, and therefore the assertion follows from Proposition 13.2.3b. ∎

After this digression, let us return to the main question, to what extent the formulated postulates determine the structure of quantum theory. In the preceding section we dealt with the state space; now we want to know whether the postulates (s1)–(s4) lead to the standard description of states. For simplicity we shall suppose that the assumptions of Theorem 13.1.5b are satisfied, *i.e.*, that the state space is a Hilbert space $\mathcal{H}$ over some of the fields $\mathbb{R}$, $\mathbb{C}$, or $\mathbb{Q}$.

Let W be a statistical operator W on $\mathcal{H}$; its definition and properties for $F = \mathbb{R}, \mathbb{Q}$ are analogous to the complex case. The map $p_W : \mathcal{B}(\mathcal{H})^E \to [0,1]$ defined by

$$p_W(E) := \mathrm{Tr}\,(EW)$$

satisfies the mentioned requirements (Problem 9); in other words, p_W is a state on $\mathcal{L} = \mathcal{B}(\mathcal{H})^E$. Conversely, we may ask whether for any p, which satisfies the conditions (s1) and (s2), *i.e.*,

$$p(0) \;=\; 1 - p(I) \;=\; 0 \quad \text{and} \quad p\left(\sum_k E_k\right) \;=\; \sum_k p(E_k)\,, \qquad (13.1)$$

where $\{E_k\}$ is an arbitrary sequence of mutually orthogonal projections, we can find a statistical operator W such that $p = p_W$. It is not difficult to see, for the question stated like that, that the answer is negative.

13.2.5 Example: Any one–dimensional projection on $\mathcal{H} := \mathbb{C}^2$ may be characterized by a pair of complex numbers, $E = E(\alpha, \beta)$ with $|\alpha|^2 + |\beta|^2 = 1$, so that the corresponding subspace is spanned by the unit vector $(\alpha, \beta) \in \mathbb{C}^2$. It is obvious that $E(\lambda\alpha, \lambda\beta) = E(\alpha, \beta)$ if $|\lambda| = 1$. Any sequence of mutually orthogonal projections consists in this case either of a single operator $E(\alpha,\beta)$ or of a pair $E(\alpha,\beta)$, $E(\bar\beta,-\bar\alpha)$. Let S be the unit sphere in $\mathbb{C}^2$ and suppose that a function $f : S \to \mathbb{R}$ satisfies the conditions

$$f(\lambda\alpha,\lambda\beta) = f(\alpha,\beta) \quad \text{for} \quad |\lambda| = 1\,, \qquad f(\alpha,\beta) = 1 - f(\bar\beta,-\bar\alpha)\,. \tag{13.2}$$

Then we may define the function $p_f : \mathcal{B}(\mathbb{C}^2)^E \to [0,1]$ by

$$p_f(0) = 1 - p_f(I) := 0\,, \quad p_f(E(\alpha,\beta)) := f(\alpha,\beta)\,.$$

It obeys the conditions (13.1), and we readily check that $p_f = p_W$ may hold for some $W \in \mathcal{B}(\mathbb{C}^2)$ only if the function f is continuous. However, the conditions (13.2) are satisfied by other functions, for instance, by

$$f \,:\, f(\alpha,\beta) = \begin{cases} |\alpha|^2 & \dots \quad 0 < |\alpha| < 1 \\ 1 & \dots \quad \alpha = 0 \\ 0 & \dots \quad |\alpha| = 1 \end{cases}$$

and therefore the lattice $\mathcal{B}(\mathbb{C}^2)^E$ supports states without a trace representation.

Fortunately, it appears to be substantial that we have considered a two-dimensional space, which contains "too few" projections. In spaces of a higher dimension we are able to describe states by statistical operators. This result is highly nontrivial and we present it without proof (references are given in the notes).

13.2.6 Theorem (Gleason): Let $\mathcal{H}$ be a Hilbert space over one of the fields $\mathbb{R}$, $\mathbb{C}$, or $\mathbb{Q}$ and $\dim\mathcal{H} \ge 3$. Then for any function $p : \mathcal{B}(\mathcal{H})^E \to [0,1]$ which satisfies the conditions (13.1) there is just one statistical operator W such that $p(E) = \mathrm{Tr}\,(EW)$ holds for any $E \in \mathcal{B}(\mathcal{H})^E$.

Hence the postulates (s1) and (s2) together with the additional restriction to the state space dimension lead to the standard description of states, provided the system is coherent and its algebra of observables is $\mathcal{A} = \mathcal{B}(\mathcal{H})$. Moreover, these assumptions may be weakened: the Gleason theorem extends to systems with superselection rules and more general algebras of observables — see the notes.

13.3 Axioms for quantum field theory

The considerations of the two preceding sections apply to quantum mechanics as well as to quantized fields. However, in the latter case the axiomatic approach usually means something else; the reason is that the situation here is entirely different. In quantum mechanics it is not difficult to associate a well-defined mathematical structure with the systems under consideration, *i.e.*, to define the state Hilbert space and operators on it representing observables. Although this task is sometimes performed for particular physical problems in a mathematically superficial way, there is no doubt that with some effort these flaws can be corrected.

Quantum field theory, which was born almost simultaneously with quantum mechanics, represents a considerably more complicated object. Its first three decades brought a host of results and some remarkably exact predictions, in particular, in quantum electrodynamics, but the methods used abounded with mathematically dubious arguments. Beginning from the early fifties, this motivated the formulation of several axiomatic systems, which a correct quantum field theory was expected to satisfy; the question was whether the existing formal theories could be put into accord with these postulates. The intricacy of the problem is clear from the fact that, in spite of great progress, forty years later there is still no exhaustive answer. Here we limit ourselves to a description of two systems of axioms; some more will be mentioned in the notes.

The oldest and best known are the *Gårding–Wightman axioms.* We shall formulate them for the simplest case of a scalar field.

(GW1) *Relativistic transformations:* States are represented by rays in a complex Hilbert space $\mathcal{H}$, and there exists a unitary strongly continuous representation U of the proper orthochronous Poincaré group $\mathcal{P}_+^\uparrow$ on $\mathcal{H}$, which describes transformations of states when we pass from one inertial frame to another.

The subgroup $\{(I,a) \,:\, a \in \mathbb{R}^4\} \subset \mathcal{P}$ representing space–time translations is Abelian, so by Problem 5.48 there are four commuting self–adjoint operators P_μ, $\mu = 0,1,2,3$, such that $U(I,a) = e^{i(P_0a_0 - \vec P\cdot\vec a)}$ with $\vec P\cdot\vec a := \sum_{j=1}^3 P_ja_j$ holds for $a\in\mathbb{R}^4$. The operator $H := P_0$ is the *Hamiltonian* of the system while P_j, $j=1,2,3$, has the meaning of momentum components. Moreover, we know from Chapter 5 that a unique projection $E(\cdot)$ on $\mathbb{R}^4$ is associated with the operators P_μ and

$$(\phi, U(I,a)\psi) = \int_{\mathbb{R}^4} e^{i(a_0p_0 - \vec a\cdot\vec p)}\, d(\phi, E(p)\psi)$$

holds for all $\phi,\psi\in\mathcal{H}$. Recall also that the *closed future light cone* is the set $\overline{V}_+ := \{\, p = (p_0,\vec p) \,:\, p_0 \ge 0,\ p^2 := p_0^2 - \vec p^{\,2} \ge 0 \,\}$.

(GW2) *Spectral condition:* The support of the measure $E(\cdot)$ associated with the operators P_μ is contained in $\overline{V}_+$; in other words, the set $\mathbb{R}^4 \setminus \overline{V}_+$ is E–zero.

Note that the postulate (GW2) requires the Hamiltonian P_0 to be positive. The same has to be true for $M^2 := P_0^2 - \sum_{j=1}^3 P_j^2$, so it makes sense to define $M := (M^2)^{1/2}$, which is called the *mass operator.*

(GW3) *The existence and uniqueness of the vacuum:* The space $\mathcal{H}$ contains just one vector ψ_0 such that $U(\Lambda,a)\psi_0 = \psi_0$ holds for all $(\Lambda,a)\in\mathcal{P}_+^\uparrow$; we call it the *vacuum.*

This implies, in particular, that the E–measure of the point $(0,\vec 0)\in\overline{V}_+$ is nonzero and $\dim\operatorname{Ran} E(0,\vec 0) = 1$. Now we can pass to the requirements imposed on the field itself. We mentioned in Remark 12.3.7 that it is reasonable to regard it as an operator–valued distribution. In addition, field operators are generally unbounded, and therefore they are not defined on the whole $\mathcal{H}$.

(GW4) *The invariant domain:* There is a dense subspace $D \subset \mathcal{H}$ containing the vector ψ_0, and the map $\varphi : \mathcal{S}(\mathbb{R}^4) \to \mathcal{L}(\mathcal{H})$ with the following properties:

(i) D belongs for any $f \in \mathcal{S}(\mathbb{R}^4)$ to the domains of the operators $\varphi(f), \varphi(f)^*$ and $\varphi(f)^* \upharpoonright D = \varphi(\bar{f}) \upharpoonright D$.

(ii) $\varphi(f)D \subset D$ holds for all $f \in \mathcal{S}(\mathbb{R}^4)$.

(iii) The map $f \mapsto \varphi(f)\psi$ is linear for each $\psi \in D$.

Condition (i) implies that the operators $\varphi(f)$ associated with real-valued test functions $f \in \mathcal{S}(\mathbb{R}^4)$ are symmetric, and (ii) ensures that polynomials in the operators $\varphi(f)$, $\varphi(f)^*$ are well-defined. This fact in combination with (GW3) means that the expressions $(\psi_0, \varphi(f_1)\dots\varphi(f_n)\psi_0)$, called *vacuum means*, make sense for any $f_1, \dots, f_n \in \mathcal{S}(\mathbb{R}^4)$. The distribution character of the field is expressed as follows.

(GW5) *Field regularity:* The map $f \mapsto (\psi_1, \varphi(f)\psi_2)$ is a tempered distribution for any $\psi_1, \psi_2 \in D$.

Next we have to postulate the transformation properties of field operators.

(GW6) *Relativistic covariance of the field:* The relation $U(\Lambda, a)D \subset D$ holds for all $(\Lambda, a) \in \mathcal{P}_+^\uparrow$, and moreover

$$U(\Lambda, a)\,\varphi(f)\,U(\Lambda, a)^{-1}\psi = \varphi(f_{\Lambda,a})\psi$$

for all $f \in \mathcal{S}(\mathbb{R}^4)$ and $\psi \in D$, where $f_{\Lambda,a}(x) := f(\Lambda^{-1}(x-a))$.

If we follow the standard convention by which distributions may be written as functions, *i.e.*, we use formal "fields at a point x", the last condition acquires the illustrative form $U(\Lambda, a)\varphi(x)U(\Lambda, a)^{-1}\psi = \varphi(\Lambda x + a)\psi$. Recall further that points x, y are *space-like separated* if $(x_0 - y_0)^2 - (\vec{x} - \vec{y})^2 < 0$; sets M_1, $M_2 \in \mathbb{R}^4$ are space-like separated if this is true for any $x \in M_1$ and $y \in M_2$.

(GW7) *Microcausality* (or local commutativity): If the supports of $f, g \in \mathcal{S}(\mathbb{R}^4)$ are space-like separated, then $[\varphi(f), \varphi(g)]\psi = 0$ holds for any $\psi \in D$.

Finally we have to add a requirement which will ensure that the state space is not too large.

(GW8) *Vacuum cyclicity:* $\{\varphi(f_1)\dots\varphi(f_n)\psi_0 : f_1, \dots, f_n \in \mathcal{S}(\mathbb{R}^4),\, n \in \mathbb{Z}_+\}$ is total in the state space $\mathcal{H}$.

We have to check, of course, that the system of axioms is free of contradictions. It is sufficient to find an example of the quadruplet $\{\mathcal{H}, U, \varphi, \psi_0\}$ which would satisfy all the stated requirements. The simplest system that comes to mind is naturally the free scalar field (12.13). In this case we already know the state space and the vacuum vector, as well as the appropriate representation of the Poincaré group which is obtained by second quantization of the one-particle representation of

Proposition 10.3.3, $(U_m(\Lambda, a)\psi)(p) := e^{ipa}\psi(\Lambda^{-1}p)$, for any $\psi \in \mathcal{H}_m$. In a similar way we can check the other postulates, *i.e.*, to prove the following assertion (see the notes).

13.3.1 Theorem: The quadruplet $\left\{\mathcal{F}_s(L^2(H_m, d\omega_m)),\, \overline{\mathcal{T}_s^{\Pi}(U_m(\cdot))},\, \Phi_m(\cdot),\, \Omega_0\right\}$ satisfies the postulates (GW1)–(GW8). In addition,

$$\Phi_m((\Box + m^2)f) = 0 \tag{13.3}$$

holds for any $f \in \mathcal{S}(\mathbb{R}^4)$, where $\Box f := \frac{\partial^2 f}{\partial t^2} - \Delta f$.

The result is not surprising, of course, because the axioms have been formulated with the free field on mind. Much less trivial is the question whether one can construct examples of fields with a nontrivial interaction which will obey the GW–axioms. This question stimulated a search for rigorous quantum–field model which has produced some remarkable results, though a satisfactory positive answer is still missing; we present a few comments on this problem in the notes.

The second set of postulates that we are going to mention here are the *Haag–Kastler axioms;* in contrast to the preceding case they concern rather the algebraic structure of quantum field theory. We can formulate them, for instance, as follows:

(HK1) *Algebras of local observables:* A C^*–algebra $\mathcal{A}(M)$ with the unit element is associated with any bounded open region $M \subset \mathbb{R}^4$.

(HK2) *Isotony:* $M_1 \subset M_2$ implies $\mathcal{A}(M_1) \subset \mathcal{A}(M_2)$.

The existence of a partial ordering between the algebras of local observables is important because it allows us (by a standard procedure called inductive limit) to associate the C^*–algebra $\mathcal{A}(M)$ corresponding to $M := \bigcup_{j=1}^{\infty} M_j$ with a monotonic sequence $\{M_j\}$ of sets, $M_{j+1} \supset M_j$. In particular, one can in this way construct the algebra $\mathcal{A} := \mathcal{A}(\mathbb{R}^4)$, whose Hermitean elements are sometimes called *quasilocal observables.*

(HK3) *Microcausality:* If the regions $M_1, M_2 \subset \mathbb{R}^4$ are space–like separated, then $\mathcal{A}(M_1) \subset \mathcal{A}(M_2)'$.

(HK4) *Relativistic covariance:* For any $(\Lambda, a) \in \mathcal{P}_+^{\uparrow}$ there is a $*$–automorphism $\alpha_{\Lambda,a}$ of the algebra $\mathcal{A}$ such that

$$\alpha_{\Lambda,a}(\mathcal{A}(M)) = \mathcal{A}(\Lambda^{-1}(M - a))$$

holds for any bounded open region $M \subset \mathbb{R}^4$. The map $(\Lambda, a) \mapsto \alpha_{\Lambda,a}$ is a representation of $\mathcal{P}_+^{\uparrow}$.

Although the two described axiomatic systems lean on the same physical principles, they are not directly related. The Haag–Kastler axioms are more abstract and better suited for analysis of general properties of fields such as the structure of superselection rules, *etc.* We expect, of course, that physically reasonable quantum–field models will obey both sets of postulates, as well as other sets which we mention in the notes.

Notes to Chapter 13

Section 13.1 As we saw in Section 7.6, a stumbling point in the algebraic approach to axiomatization of quantum mechanics is the definition of the sum for noncompatible observables. This led G. Birkhoff and J. von Neumann to an attempt at formulating a system of axioms which would avoid summation of observables — *cf.* [BvN 1]. Their idea has been further developed by G. Mackey, G. Ludwig, J.M. Jauch, C. Piron and other authors; a fairly complete overview can be found in ⟦Ja⟧, ⟦Var 1,2⟧, ⟦Lud 1–3⟧, ⟦Pir⟧; see also the review paper [Gud 1].

The requirement (pl2) for any subset in $\mathcal{L}$ (which is usually referred to as completeness) again represents an idealization, because we are in fact able to compare only finite proposition systems. It has sometimes to be weakened for mathematical reasons — *cf.* Example 13.1.1. A more complete justification of the relations between propositions based on analysis of the structure of experiments can be found, *e.g.*, in [Herb 1]. A lattice $\mathcal{L}$ of propositions about a quantum system is often called a *quantum logic* since, like formal logic, it contains the relation of implication and the operations of negation, logical product "a and b", and logical sum "a or b"; the name appeared for the first time in the mentioned paper [BvN 1]. This analogy should not obscure the fact that $\mathcal{L}$ represents a collection of empirical information about a particular physical system; at the same time, we have to keep in mind that formal logic has the Boolean property which, in general, is not true for $\mathcal{L}$ — see Example 13.1.2. A more detailed discussion of this problem is given in ⟦Ja⟧, Sec.5.3. The non–Boolean nature of quantum proposition systems is manifested in various ways; for instance, the uncertainty relations in this formalism are treated in [Pul 1]. The structure of proposition lattices for classical systems is analyzed in ⟦Ja⟧, Sec.5.4; ⟦Var 1⟧, Chap.I.

The name "weak modularity" comes from the fact that this condition weakens the requirement of *modularity* of the lattice $\mathcal{L}$, which means $a \cup (b \cap c) = (a \cup b) \cap c$ for $a \subset c$. This property was proposed by Birkhoff and von Neumann as a replacement of the distributive law. However, it appears that the lattice $\mathcal{B}(\mathcal{H})^E$ is modular only if $\dim \mathcal{H} < \infty$, and therefore modularity cannot be required for the proposition lattice of a general quantum system — *cf.* ⟦Ja⟧, Sec.5.7.

The reduction of a proposition system to irreducible components is discussed, for instance, in ⟦Ja⟧, Sec.8.2. Some ideas of the proof of Theorem 13.1.5 already appeared in the paper [BvN 1], which treated proposition lattices of a finite dimension; the general result was obtained by C.Piron — see [Pir 1] or ⟦Var 1⟧, Sec.VIII.5. The hypothesis of Theorem 13.1.5b lacks a convincing justification. Nevertheless, one can at least prove that the field F cannot be finite unless it is noncommutative — *cf.* [EZ 1], [IS 1]. It is also worth recalling that the field of *quaternions* or hypercomplex numbers consists of all real linear combinations $a = a_1 + a_2 i + a_3 j + a_4 k$, where the quantities i, j, k satisfy the same algebraic relations as the multiples of Pauli matrices $-i\sigma_l$, $l = 1, 2, 3$.

Many results concerning quantum theory in real Hilbert spaces can be found in the papers [Stu 1–4]; interest in it vanished when it became clear that predictions for simple systems are the same as in the "complex" case. The "quaternionic" quantum theory has attracted more attention. It was first formulated in [FJSS 1,2]. Although for a single relativistic elementary particle it reproduces the "complex" results — see [Em 1]— in general this theory is expected to have a richer structure. More recent results in "quaternionic" quantum mechanics and field theory can be found, *e.g.*, in [HB 1], [Adl 1].

Section 13.2 The definition of state requires, to interpret the function $p : \mathcal{L} \to [0,1]$, a modified way of understanding the probability. As we have mentioned in Section 7.1, the conventional way to introduce this notion leans on a non–negative σ–additive measure whose domain is a certain σ–algebra, *i.e.*, a Boolean lattice; however, proposition lattices corresponding to quantum systems are non–Boolean.

The first result on the nonexistence of hidden parameters was formulated by J. von Neumann within the standard quantum–mechanical formalism — *cf.* ⟦ vN ⟧, Sec.4.2. Theorem 13.2.4 was proved for the first time in [JP 1], see also ⟦ Ja ⟧, Sec.7.3; the proof of Proposition 13.2.3b which is used here can be found in [Pir 1]. It is worth mentioning that Theorem 13.2.4 holds irrespective of the assumption of atomicity of the lattice $\mathcal{L}$, and that the proof employs postulate (s3) for finite proposition subsystems only. There are various more recent results concerning the problem of existence of hidden variables in quantum theory; we refer to the book ⟦ BLM ⟧ for more information.

The proof of Gleason's theorem is a nice problem, which can be managed with the tools elaborated in this book; unfortunately it is too long and complicated to be included. The original paper [Gle 1] contains the demonstration for $F = \mathbb{R}, \mathbb{C}$; the extension to the quaternionic case is given in ⟦ Var 1 ⟧, Sec.VII.2. These proofs also used the assumption that $\mathcal{H}$ is separable, but this could be eliminated — see [EH 1]. As we have mentioned, a state represents a probability "measure" on the proposition system $\mathcal{L}$; Theorem 13.2.6 extends to a class of signed "measures" on $\mathcal{L}$ — *cf.* [Dri 1].

The Gleason theorem readily generalizes to the situation with superselection rules, where $\mathcal{L} := \mathcal{A}^E$ with the algebra of observables (7.18) in which the dimension of each coherent subspace is at least three — see ⟦ Jor ⟧, Thm.28.2. In ⟦ Mac 1 ⟧ a more complicated question has been posed, namely whether the analogous result is also valid for the lattices of projections in W^*–algebras which are not of type I. The answer is positive provided the decomposition of $\mathcal{A}$ does not contain type I_2 factors, and the same is also true for a certain class of Jordan algebras — see, *e.g.*, [BW 1], [Chri 1], [Yea 1].

Section 13.3 The Gårding–Wightman axioms were formulated at the beginning of the fifties and published a decade later in [WG 1]. We group them following ⟦ RS 2 ⟧, Sec.IX.8; usually they are condensed into a lesser number of postulates — see ⟦ SW ⟧, Sec.3.1; ⟦ Si 2 ⟧, Sec.II.1; ⟦ GJ ⟧, Sec.6.1. Different formulations contain slight modifications; for instance, postulate (GW5) is sometimes replaced by the weaker assumption that the test–function space is $C_0^\infty(\mathbb{R}^4)$, the state space $\mathcal{H}$ is in addition supposed to be separable, or one assumes from the beginning the existence of a representation of the full Poincaré group including the reflections, *etc.*

It follows from the physical interpretation of the generators of the representation U that $U(I,a)$ with $a := (-t, \vec{0})$ is the evolution operator of the field φ; postulate (GW6) then tells us that the field operators $\varphi(f)$ with a real $f \in \mathcal{S}(\mathbb{R}^4)$ are observables in the Heisenberg picture — provided they are not only symmetric but self–adjoint as well. Postulate (GW3) can be weakened; it is sufficient to require vacuum invariance with respect to space–time translations — *cf.* ⟦ RS 2 ⟧, Sec.IX.8. As for distribution character of the field, we are *forced* to work with "smeared" field operators: one can show that a scalar field obeying the GW–axioms cannot be expressed in the form $\varphi(f) = \int_{\mathbb{R}^4} \varphi(x) f(x)\, dx$ for some (nondistributional) function $\varphi : \mathbb{R}^4 \to \mathcal{L}(\mathcal{H})$ — see ⟦ RS 2 ⟧, Problem IX.53.

We have presented the GW–axioms for the simplest case of a scalar field, which is called *Hermitean* in connection with requirement (i) in (GW4). We may consider more general scalar fields such that the operators $\varphi(f)$ corresponding to real-valued $f \in \mathcal{S}(\mathbb{R}^4)$ are not symmetric. More important, however, are modifications to vector and spinor fields describing particles with nonzero spins:

(a) In this case the fields are multicomponent operator-valued functions, *i.e.*, $\varphi(f) = (\varphi_1(f), \dots, \varphi_{2s+1}(f))$ with $\varphi_j : \mathcal{S}(\mathbb{R}^4) \to \mathcal{L}(\mathcal{H})$, and the transformation property of (GW6) is replaced by $U(\Lambda,a)\,\varphi(f)\,U(\Lambda,a)^{-1}\psi = (S(\Lambda^{-1})\varphi)(f_{\Lambda,a})\psi$ for any $\psi \in D$, where $S(\Lambda^{-1})$ corresponds to the matrix operator of Proposition 10.3.3. Fields with a half-integer spin change sign when rotated by 2π.
(b) The microcausality postulate differs with respect to the type of statistics: if the supports of f, g are space-like separated, then we have $[\varphi_j(f), \varphi(g)]_{\mp}\psi = 0$ for all the field components and any $\psi \in D$, where the upper and lower sign correspond to the boson and fermion cases, respectively.

In distinction to quantum mechanics, one can demonstrate here the connection between spin and statistics; namely, one can prove that a quantum field, which satisfies the GW-axioms with the listed modifications, is governed by Bose-Einstein or Fermi-Dirac statistics *iff* the spin of the particles involved is integer or half-integer, respectively. This result was obtained by M. Fierz and W. Pauli for free fields; the general proof was given by G. Lüders, B. Zumino and N. Burgoyne — see, *e.g.*, ⟦SW⟧, Sec.4.4; ⟦BLT⟧, Sec.5.3.

Proof of Theorem 13.3.1 can be found, *e.g.*, in ⟦ SW ⟧, Sec.3.2; ⟦ RS 2 ⟧, Sec.X.7. Relation (13.3) expresses the fact that the formal field operators satisfy the Klein-Gordon equation. Another example of systems which obey the GW-axioms is provided by the so-called *generalized free fields* — see, *e.g.*, ⟦ BLT ⟧, Sec.3.4. They are superpositions of free fields of different masses, *i.e.*, the one-particle space equals the direct integral $\int_{\mathbb{R}^+}^{\oplus} L^2(H_m, d\omega_m)\, d\varrho(m)$ where ϱ is a positive measure; the generalized free field is then defined as in Section 12.3. If ϱ is supported by a single point we get a free field.

Axiomatic quantum field theory splits in two main directions. One is concerned with general properties of fields, relations between different postulate systems, *etc.* On the other hand, *constructive field theory* aims at finding explicitly formulated examples of quantum fields satisfying some system of axioms. Since the problem is very complicated; the first attempts were concerned with models in a space-time of dimension $d < 4$. The first nontrivial model for $d = 2$ was found at the beginning of the seventies; a few years later examples were also obtained in the case $d = 3$ — see [GJ 1], ⟦Si 2⟧, ⟦GJ⟧, and more recently, *e.g.*, [BFS 1], [Ost 1]. The drawback of the methods used was that they worked only for superrenormalizable models, *i.e.*, such that the number of singularities in the formal perturbation-theory expansion was finite. In the middle of the eighties new models appeared (for $d = 2, 3$) which were not superrenormalizable — see, *e.g.*, [FMRS 1], [GK 1] and the review [Ost 1] — this gave a new impetus for study of physically most interesting case $d = 4$. Let us remark that constructive methods are not the only way to establish the existence of interacting quantum fields; for instance, there is an abstract result derived in [Hof 1] due to which a quantum field must exist, which satisfies the GW-axioms and differs from a generalized free field.

Using the vacuum means we can define for any positive integer n a distribution $W_n : W_n(f_1, \dots, f_n) = (\psi_0, \varphi(f_1) \dots \varphi(f_n)\psi_0)$, which is called the ($n$-point) *Wightman function*. Starting from the GW-axioms, one can derive what they imply for the sequence $\{W_n\}_{n=1}^{\infty}$: the behavior at relativistic transformations, the consequences of the spectral condition and microcausality, *etc.* However, it is also possible to reverse the order and to postulate these properties of the functions W_n. This collection of requirements is called the *Wightman axioms;* their equivalence with the GW–axioms follows from the *reconstruction theorem,* due to which a quantum field satisfying the GW–axioms corresponds to a sequence $\{W_n\}$ which obeys the W–axioms, and this field is determined uniquely up to a unitary equivalence — see, *e.g.*, ⟦SW⟧, Sec.3.4; ⟦Si 2⟧, Sec.II.1.

Another axiomatic method is based on the fact that Wightman functions have certain analytical properties which allow us to study their continuation to regions of complexified configuration space where time is purely imaginary. These analytic continuations, so–called *Schwinger functions,* are then invariant with respect to the Euclidean group of nonhomogeneous linear transformations of the space $\mathbb{R}^4$ rather than to $\mathcal{P}_+^\uparrow$. Properties of Schwinger functions can be postulated as *Osterwalder–Schrader axioms;* these were formulated for the first time in [OS 1,2]; see also ⟦Si 2⟧, Chap.II; ⟦GJ⟧, Sec.6.1. In this case, too, a reconstruction theorem is valid, which enables us to find the Wightman functions corresponding to the Schwinger functions which satisfy the OS–axioms.

An alternative formulation of the Euclidean approach to quantum field theory employs the notion of a Euclidean field, *i.e.*, an operator object whose vacuum means are the Schwinger functions. This idea was proposed by J. Schwinger, T. Nakano and K. Szymanczik; later, the required properties of such fields were formulated as the so–called *Nelson axioms* — see [Nel 3] and ⟦Si 2⟧, Sec.IV.1. If these axioms are satisfied the same is true for the OS–axioms, while the opposite implication is not valid.

To illustrate the advantages of this approach we use an analogy with quantum mechanics. Let H be a Schrödinger operator on $L^2(\mathbb{R}^d)$. The treatment of the propagator e^{-iHt} can be simplified if we are able to reconstruct it from knowledge of the operators e^{-Ht} for $t \ge 0$; this is possible, *e.g.*, when operator H is below bounded so the function $z \mapsto e^{-iHz}$ is strongly analytic in the halfplane $\{z : \operatorname{Im} z < 0\}$ (see the notes to Section 1.7). The operators e^{-Ht} can be expressed by means of the Feynman–Kac formula (9.8), whose right side is a well-defined mathematical object which may be treated by methods of integration theory. In a similar way the Feynman–Kac formula is used to express Euclidean fields and their vacuum means — *cf.* the notes to Section 9.4.

Haag–Kastler axioms were formulated in [HK 1]; the construction of the algebra $\mathcal{A}$, which is usually called the *quasilocal algebra,* can be found in ⟦BR 1⟧, def.2.6.3. There is a modification of this system of postulates in which one associates a W^*–algebra $\mathcal{A}(M)$ with any bounded open $M \subset \mathbb{R}^4$. In this case we speak about the *Haag–Araki axioms;* the postulate of relativistic covariance then requires existence of a unitary strongly continuous representation U of $\mathcal{P}_+^\uparrow$ such that $U(\Lambda,a)\mathcal{A}(M)U(\Lambda,a)^{-1} = \mathcal{A}(\Lambda^{-1}(M-a))$ holds for any (Λ,a) and M. In this approach quasilocal observables are operator–norm limits of local observables. An extensive overview of these axioms, their modifications, and consequences can be found in the monographs ⟦Haa⟧ and ⟦Hor⟧; see also the reviews [Haa 1,3], [Ara 2]. Relations between HK–axioms and GW/OS–axioms are discussed in ⟦GJ⟧, Chap.19; however, some problems remain open — see, *e.g.*, [HV 1].

Problems

1. Let a be any element of a proposition lattice $\mathcal{L}$ and $\mathcal{S} \subset \mathcal{L}$. Prove:

(a) $a \subset I$ and $a \cup a' = I$

(b) $\bigcup_{a\in\mathcal{S}} a = \left(\bigcap_{a\in\mathcal{S}} a'\right)'$

2. Let $E_j^{(+)} := \frac{1}{2}(I+\sigma_j)$ be the eigenprojection of the j–th component of spin corresponding to the value $\frac{1}{2}$. Show that $\lim_{n\to\infty}(E_j^{(+)}E_k^{(+)})^n = 0$ holds for $j \neq k$.

3. Let $\mathcal{L}$ be a complete orthocomplemented lattice; then

   (a) The intersection of any family of sublattices in $\mathcal{L}$ is a sublattice in $\mathcal{L}$.

   (b) If $\mathcal{L}$ is Boolean, so is any sublattice of it.

   (c) The propositions a, a' are compatible for any $a \in \mathcal{L}$.

   (d) The propositions $\emptyset$, I are compatible with any $a \in \mathcal{L}$.

   (e) propositions $a, b \in \mathcal{L}$ are compatible *iff* the same is true for a, b'.

4. Let $\mathcal{B}(\mathcal{H})^E$ be the lattice of Example 13.1.3. Prove that

   (a) $E(a \cup b)$ is for any a, b the projection onto the subspace $(E(a)\mathcal{H} \cup E(b)\mathcal{H})^{\perp\perp}$. In particular, if $a \subset b'$, then the projections $E(a)$, $E(b)$ are orthogonal and $E(a \cup b) = E(a) + E(b)$.

   (b) The lattice $\mathcal{B}(\mathcal{H})^E$ is atomic.

5. Let $\mathcal{A}$ be the algebra (7.18); then $\mathcal{A}^E$ is an atomic weakly modular lattice and its center consists of the projections $P := \sum_{\alpha\in J} c_\alpha E_\alpha$, where the E_α are the projections onto the subspaces $\mathcal{H}_\alpha$ and the coefficients c_α assume the values 0 or 1.

6. Suppose that $\mathcal{L}$ is a proposition system and its center contains an element c different from $\emptyset$, I; then any $a \in \mathcal{L}$ can be expressed as $a = a_1 \cup a_2$, where $a_1 := a \cap c$ and $a_2 := a \cap c'$. Prove that the lattices $\mathcal{L}_j := \{a_j : a \in \mathcal{L}\}$ satisfy the condition (pl$_q$) and extend this result to the case of a center generated by a finite number of propositions.

7. Assume that $\mathcal{A}$ is an algebra of observables of the form (7.18) and W is a statistical operator, which is reduced by all the subspaces $\mathcal{H}_\alpha$. The function $p_W : \mathcal{A}^E \to [0,1]$ defined by $p_W(E) := \mathrm{Tr}\,(EW)$ satisfies the conditions (s1)–(s4) of Section 13.2.

8. Any dispersionless state is pure.
   Hint: Such states satisfy $p(a) = 0$ or $p(a) = 1$ for any $a \in \mathcal{L}$.

9. Prove that the result of Problem 7 remains valid if we replace the complex Hilbert space by a real or quaternionic space.

Chapter 14

Schrödinger operators

The most important operator class in nonrelativistic quantum mechanics is represented by **Schrödinger operators** on $L^2(\mathbb{R}^n)$, *i.e.*, operators of the form (9.2),

$$H := \sum_{j=1}^{n} \frac{1}{2m_j} P_j^2 + V(Q) \tag{14.1}$$

for a given measurable function $V : \mathbb{R}^n \to \mathbb{R}$ called the **potential**, which we have repeatedly encountered in earlier chapters. Now we want to discuss their properties in a more systematic manner. Even so, however, we are able provide only an introduction to the subject; a guide for further reading is given in the notes.

14.1 Self–adjointness

Since operator (14.1) typically represents the total–energy observable, it has to be self–adjoint, or at least essentially self–adjoint to have a unique self–adjoint extension. This is obviously true provided the potential is essentially bounded, *i.e.*, $V \in L^\infty(\mathbb{R}^n)$. Unfortunately, such a class is too narrow for practical purposes; it is sufficient to recall the electrostatic interaction of a pair of charged particles. Hence we must look for more general self–adjointness criteria; in this section we are going to discuss some widely used conditions.

14.1.1 Remarks: (a) Though the physically most important case corresponds to $n = 3$, and operators with $n = 1, 2$ also play a distinguished role, it is reasonable to consider operators (14.1) for any n. Apart from mathematical interest, some results obtained in this way can be useful for treatment of the physically interesting case of N–particle systems.

(b) On the other hand, without loss of generality we may assume that $2m_1 = \cdots = 2m_n = 1$, since otherwise we can repeat the argument we have already used in the proof of Theorem 9.3.1, namely we can employ the substitution operator $U : (U\psi)(x_1, \dots, x_n) = \prod_{j=1}^{n}(2m_j)^{1/2}\psi(\sqrt{2m_1}x_1, \dots, \sqrt{2m_n}x_n)$, which transforms (14.1) to $UHU^{-1} = \sum_{j=1}^n P_j^2 + V(\sqrt{2m_1}Q_1, \dots, \sqrt{2m_n}Q_n)$; all the conditions on the potential that we shall discuss below are invariant with respect to this scaling transformation.

(c) For the sake of brevity, we shall again use $P^2 := \sum_{j=1}^n P_j^2$ and similar shortened notation for functions of the momentum and position coordinate operators. The free Schrödinger operator corresponding to $V = 0$ will be denoted alternatively as H_0.

First we are going to derive a useful sufficient condition based on a perturbation–theory argument. We shall say that a function V belongs to $(L^p + L^\infty)(\mathbb{R}^n)$ if it can be expressed in the form of a sum, $V = V_p + V_\infty$, with $V_p \in L^p$ and $V_\infty \in L^\infty$; speaking about functions we have in mind, as usual, classes of functions which coincide a.e. in $\mathbb{R}^n$.

14.1.2 Theorem: Suppose that $V \in (L^p + L^\infty)(\mathbb{R}^n)$, where

$$p = 2 \quad \text{if} \quad n \le 3, \qquad p > \frac{n}{2} \quad \text{if} \quad n \ge 4; \tag{14.2}$$

then $H := H_0 + V(Q)$ is self–adjoint and $C_0^\infty(\mathbb{R}^n)$ is a core for it.

Proof: In view of the Kato–Rellich theorem and Problem 7.16, it is sufficient to check that $V(Q)$ is H_0–bounded with the relative bound less than one. If $n \le 3$ and $\psi \in D(H_0)$, we have

$$\|V\psi\| \le \|V_2\psi\| + \|V_\infty\psi\| \le \|V_2\|\,\|\psi\|_\infty + \|V_\infty\|_\infty \|\psi\|,$$

where the first term on the right side is finite due to (7.16). On the other hand, if $n \ge 4$ we obtain in a similar way using the Hölder inequality

$$\|V\psi\| \le \|V_p\|_p\,\|\psi\|_q + \|V_\infty\|_\infty \|\psi\|$$

with $p^{-1} + q^{-1} = \frac{1}{2}$. This last named condition is valid for $q > 2$ if $n = 4$ and $2 \le q < 2n/(n-4)$ for $n \ge 5$, so the right side is again finite — see the notes to Section 7.5. Hence for any positive integer n and $a > 0$, there is a $b > 0$ such that

$$\|V(Q)\psi\| \le a\,\|V_p\|_p\,\|H_0\psi\| + (b\|V_p\|_p + \|V_\infty\|_\infty)\,\|\psi\|$$

holds for all $\psi \in D(H_0)$; if we choose a in such a way that $a\|V_p\|_p < 1$, the assumptions of Theorem 4.3.12 are satisfied. ∎

14.1.3 Example: Suppose that the potential V is essentially bounded with the exception of a finite number of point singularities $x_1, \dots, x_N$ in the vicinity of which it obeys the restriction $|V(x)| \le c_j|x - x_j|^{-\alpha_j}$ for some positive c_j, α_j. If the powers satisfy the condition $\alpha_j < \min(2, n/2)$, the theorem may be applied.

However, in case of Schrödinger operators in $L^2(\mathbb{R}^n)$ with $n > 3$ we are often interested in potentials, which have singularities with respect to a part of the coordinates only. The most important consequence of Theorem 14.1.2 concerns systems of N particles, which interact

(i) With an external field

(ii) With the other particles by means of translation–invariant two–particle forces

i.e., interactions involving more than two particles are excluded. A common property of the corresponding potentials V is that there is a three-dimensional projection E on $\mathbb{R}^{3N}$ such that

$$V(x) = V(Ex)\,. \tag{14.3}$$

In other words, $V(x)$ depends on one three–dimensional vector argument only; given V, we can choose a basis in $\mathbb{R}^{3N}$ and a function $\tilde{V} : \mathbb{R}^3 \to \mathbb{R}$ in such a way that $V(x_1,\dots,x_N) = \tilde{V}(x_1)$; in case (ii) the vector x_1 is associated with the relative position of the two particles.

14.1.4 Theorem (Kato): Suppose that $V := \sum_{k=1}^m V_k$, where each of the potentials V_k satisfies condition (14.3) and the corresponding $\tilde{V}_k \in (L^2 + L^\infty)(\mathbb{R}^3)$ for any $k = 1,\dots,m$. Then the operator $H := H_0 + V(Q)$ is self–adjoint and $C_0^\infty(\mathbb{R}^{3N})$ is a core for it.

Proof: We use the fact that the assumption of the Kato–Rellich theorem may be equivalently expressed in a quadratic form — *cf.* Section 4.4. Given V_k we choose a basis in $\mathbb{R}^{3N}$ in the way described above, and use $\tilde{H}_0^{(k)}$ to denote the self–adjoint extension of $-\sum_{j=1}^3 \partial^2/\partial x_{1j}^2$ with the domain $\mathcal{S}(\mathbb{R}^3)$. Mimicking the argument from the proof of Theorem 2, we find that for any $\alpha > 0$ there is a $\beta_k > 0$ such that

$$\|\tilde{V}_k(Q_1)\psi\|^2 \le \alpha^2 \|\tilde{H}_0^{(k)}\psi\|^2 + \beta_k^2\|\psi\|^2$$

holds for any $\psi \in D(\tilde{H}_0^{(k)})$, where the norms are related to $L^2(\mathbb{R}^3)$ and $\tilde{V}_k(Q_1)$ is shorthand for $\tilde{V}_k(Q_{11},Q_{12},Q_{13})$. Next we employ the decomposition $L^2(\mathbb{R}^{3N}) = L^2(\mathbb{R}^3)\otimes L^2(\mathbb{R}^{3N-3})$. Due to Example 5.5.8 and Theorem 5.7.2, we have $V_k(Q) = \overline{\tilde{V}_k(Q_1)\otimes I}$, and similarly, $H_0^{(k)} = \overline{\tilde{H}_0^{(k)}\otimes I}$. Since the last named operator is self–adjoint, and therefore closed, we readily see that

$$\|V_k(Q)\psi\|^2 \le \alpha^2 \|H_0^{(k)}\psi\|^2 + \beta_k^2\|\psi\|^2$$

holds for all $\psi \in D(H_0^{(k)})$. The identity $F_{3N} = F_3 \otimes F_{3N-3}$ in combination with (7.4) implies that $H_0^{(k)}$ is unitarily equivalent to the operator of multiplication by h_k : $h_k(x_1,\dots,x_N) = |x_1|^2$; it follows that $D(H_0^{(k)}) \subset D(H)$ and $\|H_0^{(k)}\psi\|^2 \le \|H_0\psi\|^2$ for any $\psi \in D(H_0)$, and therefore

$$\|V_k(Q)\psi\|^2 \le \alpha^2 \|H_0\psi\|^2 + \beta_k^2\|\psi\|^2\,.$$

What is important is that this estimate is independent of the choice of the coordinate frame in $\mathbb{R}^{3N}$ that we used to derive it. Now we denote $\beta := \max_{1\le k\le m}\beta_k$ and use the Schwarz inequality; we obtain

$$\|V(Q)\psi\|^2 \le \sum_{k,l=1}^m \|V_k(Q)\psi\|\,\|V_l(Q)\psi\| \le m^2\left(\alpha^2\|H_0\psi\|^2 + \beta^2\|\psi\|^2\right)$$

for any $\psi \in D(H_0)$; hence, to conclude the proof it is sufficient to choose α in such a way that $m^2\alpha^2 < 1$. ∎

The proved theorem shows, in particular, that the standard quantum mechanical description of atoms and molecules as particle systems with a pair electrostatic interaction is consistent.

14.1.5 Example: Let Δ_j be the Laplace operator corresponding to the coordinates $x_j = (x_{j1}, x_{j2}, x_{j3})$ of the j–th particle. Then the Hamiltonian

$$H : H\psi = -\frac{1}{2M}\Delta_0\psi - \frac{1}{2m}\sum_{j=1}^{Z}\Delta_j\psi - \sum_{j=1}^{Z}\frac{Ze^2}{|x_j - x_0|}\psi + \frac{1}{2}\sum_{j,k=1}^{Z}\frac{e^2}{|x_j - x_k|}\psi\,, \tag{14.4}$$

which describes a neutral atom with Z electrons, is *e.s.a.* on $C_0^\infty(\mathbb{R}^{3Z+3})$ (Problem 1a). The same is true for other systems consisting of a finite numbers of particles with Coulomb interaction: molecules, atomic and molecular ions, *etc.*

Another class of potentials, for which one can prove the essential self–adjointness of operator (14.1), consists of functions which are *below bounded;* the idea is that in such a case, sufficiently large negative numbers will be regular values of $\overline{H}$ and Proposition 4.7.4b will apply. Of course, the potential has to obey some regularity requirements; we shall assume that it belongs to $L^2_{loc}(\mathbb{R}^n)$, *i.e.*, it is square integrable on any compact subset of $\mathbb{R}^n$. We shall not discuss technical aspects of the problem (see the notes) and shall restrict ourselves to quoting the result. Before doing so, however, we remark that a stronger result can be obtained when the mentioned idea is combined with the perturbation–theory argument of Theorem 14.1.2; this allows us to treat potentials which have some negative singularities as well. We define

$$V_\pm : V_\pm(x) := \pm\max(\pm V(x), 0);$$

then the following assertion is valid.

14.1.6 Theorem: Suppose that potential V is such that $V_+ \in L^2_{loc}(\mathbb{R}^n)$ and $V_- \in (L^p + L^\infty)(\mathbb{R}^n)$ for some p which satisfies condition (14.2); then the operator $H := H_0 + V(Q)$ is *e.s.a.* on $C_0^\infty(\mathbb{R}^n)$.

This result applies to some physically interesting potentials which are not covered by Theorems 14.1.2 and 14.1.4.

14.1.7 Example: Heavy mesons, such as J/Ψ and Υ, are unstable but long–living (in the appropriate time scale), so they are successfully modelled as nonrelativistic quark–antiquark systems interacting by means of a confining potential, for instance, $V : V(r) = \lambda r + \beta - \gamma/r$, $\lambda > 0$, where $r := |x_2 - x_1|$ is the quark distance, γ is the squared quark charge, and λ, β are phenomenological constants. Due to Theorem 14.1.6 and Problem 1a, such a description is consistent.

There are other methods for proving the self–adjointness of Schrödinger operators. Some of them have been mentioned above; recall the application of the Nelson theorem to the anharmonic–oscillator Hamiltonian in Example 5.6.4, or the commutator estimate used in Example 7.2.5 — the applicability of these ideas is not restricted to the particular situations, in which they have been discussed.

In addition to (14.1), we often encounter various types of generalized Schrödinger operators. One of these involves interactions which depend not only on particle positions but also on their momenta. A typical example is the Hamiltonian of spinless particles of charges $q_1, \dots, q_N$ placed into a magnetic field described by a vector potential A,

$$H := \sum_{j=1}^{N} \frac{1}{2m_j} \left(P^{(j)} + \frac{q_j}{c} A^{(j)}(Q) \right)^2 + V(Q) , \tag{14.5}$$

where $A^{(j)}(Q)$ is a shorthand for the operator–valued vector with the components $(A_k^{(j)}\psi)(x_1, \dots, x_n) := A_k(x_j)\psi(x_1, \dots, x_n)$; a possible electrostatic interaction is included into the potential V.

For the sake of simplicity, we limit ourselves to the case of a single particle; in addition, we put $2m = 1$ in accordance with Remark 14.1.1b and $q/c = -1$. Using canonical commutation relations we find that the Hamiltonian may then be rewritten in the form

$$H \,:\, H\psi \,=\, -\Delta\psi \,+\, 2iA\cdot\nabla\psi \,+\, i(\nabla\cdot A)\psi \,+\, A^2\psi \,+\, V\psi \tag{14.6}$$

provided we choose a suitable domain, say, $C_0^\infty(\mathbb{R}^3)$.

14.1.8 Theorem: Suppose that $V \in (L^2 + L^\infty)(\mathbb{R}^3)$ and the vector–potential components are continuously differentiable functions such that $A_k,\ \nabla\cdot A \in L^\infty(\mathbb{R}^3)$. Then operator (14.6) is *e.s.a.* on $C_0^\infty(\mathbb{R}^3)$.

Proof: The interaction term in (14.6) is a symmetric operator. Furthermore, it is easy to check that the relative bound of a sum of perturbations is less than or equal to the sum of the relative bounds; since the relative bound of $V(Q)$ is zero and the functions $\nabla\cdot A$, A^2 are bounded, it is sufficient to check the operator $A\cdot P$. We have

$$\|A\cdot P\| \,=\, \left(\sum_{k=1}^{3} \int_{\mathbb{R}^3} \left| A_k(x)\, \frac{\partial\psi(x)}{\partial x_k} \right|^2 dx \right)^{1/2} \,\le\, a \left(\sum_{k=1}^{3} \|P_k\psi\|^2 \right)^{1/2} \,=\, a\|H_0^{1/2}\psi\|$$

for any $\psi \in C_0^\infty(\mathbb{R}^3)$, where $a^2 := \max_{1\le k\le 3} \|A_k^2\|_\infty$. Estimating the right side we obtain

$$\|A\cdot P\| \,\le\, a\beta\|H_0\psi\| \,+\, \frac{a}{4\beta}\,\|\psi\|$$

(Problem 2); hence we need only to choose β so that $a\beta < 1$. ∎

14.1.9 Remarks: (a) The fact that vector potential itself is not an observable quantity extends the applicability of this result. It is sufficient to find a suitable gauge–transformed potential $A' = A - \nabla\Lambda$ which satisfies the assumption of the theorem; then the Hamiltonian corresponding to the original A is also *e.s.a.* and its core is readily obtained from the result of Problem 10.11.

(b) The theorem admits a substantial generalization. We shall comment on it briefly in the notes; it is important that essential self–adjointness can be proven for physically interesting situations such as an N–electron atom in a homogeneous magnetic field described by $A(x) := \frac{1}{2}x \wedge B_0$ for a fixed vector B_0; this is the situation in which the *Zeeman effect* is observed.

14.2 The minimax principle. Analytic perturbations

Before proceeding, we want to discuss two methods which make it possible to obtain information about spectra by comparing different operators. Their applicability is not restricted to Schrödinger operators. For instance, the first method may be used for any below bounded self–adjoint operator; we have already encountered a modification of it in Corollary 3.5.10.

14.2.1 Theorem (minimax principle): Let H be a self–adjoint below bounded operator on an infinite–dimensional Hilbert space $\mathcal{H}$. We denote by $\mathcal{D}_n$ an arbitrary n–dimensional subspace in $\mathcal{H}$ and set

$$\begin{aligned} \mu_1(H) &= \inf_{\substack{\psi \in D(H) \\ \|\psi\| = 1}} (\psi, H\psi)\,, \\ \mu_{n+1}(H) &= \sup_{\mathcal{D}_n \subset \mathcal{H}} \inf_{\substack{\psi \in \mathcal{D}_n^\perp \cap D(H) \\ \|\psi\| = 1}} (\psi, H\psi)\,, \qquad n = 1, 2, \dots\,. \end{aligned} \tag{14.7}$$

The sequence $\{\mu_n(H)\}_{n=1}^\infty$ is nondecreasing, and for any positive integer n the following alternative is valid:

(i) If there is $k > n$ such that $\mu_k(H) > \mu_n(H)$, then H has at least n eigenvalues (taking their multiplicity into account) and $\mu_n(H)$ is an eigenvalue of H.

(ii) If $\mu_k(H) = \mu_n(H)$ holds for all $k > n$, then H has at most $n-1$ eigenvalues (counting the multiplicity) and $\mu_n(H) = \inf \sigma_{ess}(H)$.

Proof: For brevity, we write $\mu_n := \mu_n(H)$. The inequality $\mu_2 \ge \mu_1$ obviously holds. If $n \ge 2$, then $\mathcal{D}_n$ is uniquely determined by a subspace $\mathcal{D}_{n-1} \subset \mathcal{D}_n$ and a vector $\phi \in \mathcal{D}_n \backslash \mathcal{D}_{n-1}$. Using the identity $(\mathcal{D}_{n-1} \cup \{\phi\})^\perp = \mathcal{D}_{n-1}^\perp \cap \{\phi\}^\perp$, we obtain

$$\mu_{n+1} = \sup_{\mathcal{D}_{n-1} \subset \mathcal{H}} \sup_{\phi \notin \mathcal{D}_{n-1}} \inf_{\substack{\psi \in D(H) \cap (\mathcal{D}_{n-1} \cup \{\phi\})^\perp \\ \|\psi\| = 1}} (\psi, H\psi)$$

$$\geq \sup_{\mathcal{D}_{n-1}\subset\mathcal{H}} \sup_{\phi\notin\mathcal{D}_{n-1}} \inf_{\substack{\psi\,\in\, D(H)\,\cap\,\mathcal{D}_{n-1}^{\perp}\\ \|\psi\|\,=\,1}} (\psi, H\psi) = \mu_n\,,$$

i.e., the sequence $\{\mu_n\}_{n=1}^{\infty}$ is nondecreasing. Next we as usual denote $E_\lambda := E_H(-\infty,\lambda]$ and verify the inequalities

$$\dim E_\lambda < n \quad \text{if} \quad \lambda < \mu_n\,, \qquad \dim E_\lambda \geq n \quad \text{if} \quad \lambda > \mu_n\,. \tag{14.8}$$

By Corollary 4.3.5b and Proposition 5.4.1, they are valid for $n = 1$; hence it is sufficient to assume $n \geq 2$. Since operator H is supposed to be below bounded, it follows from the spectral theorem that $\psi \in \operatorname{Ran} E_\lambda$ belongs to $D(H)$ for any $\lambda \in \mathbb{R}$; this fact is decisive for the rest of the proof.

Suppose that the first of the inequalities (14.8) is not valid for some $\lambda < \mu_n$; then there is an n–dimensional subspace $\mathcal{L} \subset \operatorname{Ran} E_\lambda \subset D(H)$ and $(\psi, H\psi) \leq \lambda\|\psi\|^2$ holds for all $\psi \in \mathcal{L}$. Due to Problem 3, to any $(n-1)$–dimensional subspace $\mathcal{D}_{n-1}$ we can find a nonzero vector $\psi \in \mathcal{D}_{n-1}^{\perp} \cap \mathcal{L}$. We infer that

$$\inf_{\substack{\psi\,\in\, D(H)\,\cap\,\mathcal{D}_{n-1}^{\perp}\\ \|\psi\|\,=\,1}} (\psi, H\psi) \leq \lambda\,,$$

i.e., $\mu_n \leq \lambda$; this proves the first inequality. Assume next that the second of the relations (14.8) does not hold, *i.e.*, $\dim E_\lambda \leq n{-}1$ for some $\lambda > \mu_n$. In that case we may choose an $(n{-}1)$–dimensional subspace $\mathcal{D}_{n-1} \supset \operatorname{Ran} E_\lambda$; this inclusion together with properties of the spectral measure implies

$$D(H) \cap \mathcal{D}_{n-1}^{\perp} \subset \mathcal{D}_{n-1}^{\perp} \subset (\operatorname{Ran} E_\lambda)^{\perp} = \operatorname{Ran} E_H(\lambda,\infty)\,,$$

so $(\psi, H\psi) \geq \lambda\|\psi\|^2$ holds for any $\psi \in D(H) \cap \mathcal{D}_{n-1}^{\perp}$, and therefore $\mu_n \geq \lambda$ which contradicts the assumption; it means that the second inequality is also valid.

Relations (14.8) and Proposition 5.4.1a show that each μ_n belongs to the spectrum of H, because $\dim E_H(\mu_n - \varepsilon, \mu_n + \varepsilon) \geq \dim E_H[\mu_n - \varepsilon/2, \mu_n + \varepsilon/2] \geq 1$ holds for any $\varepsilon > 0$. Furthermore, the function $\lambda \mapsto \dim E_\lambda$ is nondecreasing and its values are non–negative integers or $+\infty$. At each $\lambda := \mu_n$, the following mutually exclusive possibilities arise:

(i) $\dim E_{\mu_n+\eta} < \infty$ for some $\eta > 0$. Then $\dim E_H(\mu_n{-}\varepsilon, \mu_n{+}\varepsilon) \leq \dim E_{\mu_n+\varepsilon} < \infty$ for any $\varepsilon \in (0,\eta)$, and therefore $\mu_n \in \sigma_d(H)$ in view of Theorem 5.4.3a; in other words, μ_n is an isolated eigenvalue of a finite multiplicity and we have $\sigma(H) \cap (\mu_n - \delta, \mu_n + \delta) = \{\mu_n\}$ for all sufficiently small $\delta > 0$. The second of the inequalities (14.8) gives $\dim E_{\mu_n} = E_{\mu_n+\delta} \geq n$; hence the operator H has at least n eigenvalues in $(-\infty, \mu_n)$. We arrange them in ascending order, $\lambda_1 \leq \lambda_2 \leq \cdots$. The inequality $\lambda_n < \mu_n$ would mean $\dim E_{\lambda_n} \geq n$ but this contradicts the first of the inequalities (14.8), and therefore $\mu_n = \lambda_n$. Since the multiplicity of μ_n is finite and the points $\mu_{n+1}, \mu_{n+2}, \ldots$ belong to $\sigma(H)$, there is a k such that $\mu_k > \mu_n$.

(ii) $\dim E_{\mu_n+\varepsilon} = \infty$ for all $\varepsilon > 0$. Then (14.8) implies $\dim E_{\mu_n-\varepsilon/2} \le n-1$, so $\dim E_H(\mu_n - \varepsilon, \mu_n + \varepsilon) \ge \dim E_{\mu_n+\varepsilon/2} - \dim E_{\mu_n-\varepsilon/2} = \infty$ and $\mu_n \in \sigma_{ess}(H)$ by Theorem 5.4.3a. Assume further that $\lambda < \mu_n$ belongs to $\sigma_{ess}(H)$; then $\dim E_H(\lambda-\eta, \lambda+\eta) \le \dim E_{\lambda+\eta} < n$ holds for all $\eta \in (0, \mu_n-\lambda)$ which contradicts the assumption; this means that $\mu_n = \inf \sigma_{ess}(H)$. Next we suppose that $\mu_k > \mu_n$ for some $k > n$; then (14.8) gives $\dim E_{(\mu_k+\mu_n)/2} < k$ which is again impossible. Finally, assume that there are eigenvalues $\lambda_1 \le \cdots \le \lambda_n < \mu_n$; then $\dim E_{(\mu_n+\lambda_n)/2} \ge n$, which is once more in contradiction with the first of the inequalities (14.8). This concludes the proof. ∎

14.2.2 Remarks: (a) The theorem also holds, of course, in the case $\dim \mathcal{H} < \infty$, where we may drop the domain $D(H)$ in the definition of $\mu_n(H)$. The same is true for any bounded H on an infinite–dimensional $\mathcal{H}$. However, there is an important difference. The theorem describes only that part of the spectrum which is below $\inf \sigma_{ess}(H)$ (or above $\sup \sigma_{ess}(H)$; if H is bounded from above, we can apply the result to the operator $-H$). In the first case $\sigma_{ess}(H)$ is necessarily empty, so the theorem describes the whole spectrum; this is not true in general if $\dim \mathcal{H} = \infty$ (*cf.* Problem 4).

(b) The relations (14.7) can be written in a unified manner if we set $\mathcal{D}_0 := \{0\}$. Suppose further that $\lambda_1 \le \cdots \le \lambda_{n+1} \le \cdots$ are the eigenvalues of H situated below the bottom of the essential spectrum (with the multiplicity taken into account), and $\psi_1, \dots, \psi_{n+1}, \dots$ are the corresponding normalized eigenvectors. Substituting $\mathcal{D}_n := \{\psi_1, \dots, \psi_n\}_{lin}$ and $\psi := \psi_{n+1}$ into the relations (14.7), we see that the supremum and infimum may be replaced by the maximum and minimum, respectively; hence the name of the theorem.

(c) The relations (14.7) allow various modifications. For instance, we have

$$\mu_n(H) = \sup_{\phi_1,\dots,\phi_{n-1}\in\mathcal{H}} \; \inf_{\substack{\psi \in \{\phi_1,\dots,\phi_{n-1}\}^\perp \\ \psi \in D(H),\ \|\psi\|=1}} (\psi, H\psi)\,,$$

where the vectors $\phi_1, \dots, \phi_{n_1}$ are not supposed to be linearly independent, or

$$\mu_n(H) = \sup_{\mathcal{D}_{n-1}\subset\mathcal{H}} \; \inf_{\substack{\psi \in \mathcal{D}_{n-1}^\perp \cap Q(H) \\ \|\psi\|=1}} (\psi, H\psi)\,,$$

where $Q(H)$ is the form domain of operator H (Problem 5).

As a simple consequence of the minimax principle, let us mention the coupling–constant dependence of eigenvalues.

14.2.3 Proposition: Let H_0, V be self–adjoint operators and $H_0 \ge 0$. Suppose further that V is H_0–bounded with zero relative bound, and that the operator $H_g := H_0 + gV$ satisfies $\sigma_{ess}(H_g) = \mathbb{R}^+$ for all $g \ge 0$. Then for any n, the function

$g \mapsto \mu_n(H_g)$ is continuous and nonincreasing on $[0,\infty)$; moreover, it is strictly monotonic at the points where $\mu_n(H_g) < 0$.

Proof: In view of the Kato–Rellich theorem, H_g is self–adjoint and $D(H_g) = D(H_0)$ for all $g \ge 0$. The numbers $\mu_n(H_g)$ are nonpositive by assumption; hence

$$\mu_n(H_g) = \sup_{\mathcal{D}_{n-1}\subset\mathcal{H}} \inf_{\substack{\psi\,\in\,\mathcal{D}_{n-1}^{\perp}\cap D(H_0)\\ \|\psi\|=1}} \min\{\,0,\ (\psi,H_0\psi)+g(\psi,V\psi)\,\}\,.$$

Since $(\psi,H_0\psi) \ge 0$, we readily check that $g \mapsto \min\{0,\,(\psi,H_0\psi)+g(\psi,V\psi)\}$ is nonincreasing in $\mathbb{R}^+$ for any $\psi \in D(H_0)$; the same is then also true for the function $g \mapsto \mu_n(H_g)$. The continuity follows from the analytic perturbation theory for isolated eigenvalues — see Theorem 14.2.7 below.

Assume further that $\mu_n(H_g) < 0$ and denote by $\psi_1,\dots,\psi_n$ the normalized eigenvectors of H_g corresponding to the lowest n eigenvalues (counting the multiplicity). We have $(\psi,V\psi) < 0$ for any $\psi \in L := \{\psi_1,\dots,\psi_n\}_{lin}$, because $(\psi,H_0\psi) \ge 0$ holds by assumption. The subspace L is finite–dimensional; hence the unit sphere $S := \{\,\psi \in L :\ \|\psi\| = 1\,\}$ is compact, and since $\psi \mapsto (\psi,V\psi)$ is continuous on S, Corollary 1.3.7c implies existence of a positive c such that $(\psi,V\psi) < -c$ holds for any $\psi \in S$. In view of Problem 3, we can find to any $\mathcal{D}_{n-1} \subset \mathcal{H}$ a unit $\psi \in \mathcal{D}_{n-1}^{\perp} \cap L$; this vector satisfies

$$(\psi,H_{g+\delta}\psi) = (\psi,H_g\psi)+\delta(\psi,V\psi) \le (\psi,H_g\psi)-c\delta$$

for any $\delta > 0$. At the same time, it is not difficult to check that $(\psi,H_g\psi) \le (\psi_n,H_g\psi_n) = \mu_n(H_g)$ holds for each $\psi \in L$, and therefore

$$\inf_{\substack{\psi\,\in\,\mathcal{D}_{n-1}^{\perp}\cap D(H_0)\\ \|\psi\|=1}} (\psi,H_{g+\delta}\psi) \le \inf_{\substack{\psi\,\in\,\mathcal{D}_{n-1}^{\perp}\cap L\\ \|\psi\|=1}} (\psi,H_{g+\delta}\psi) \le \mu_n(H_g)-c\delta$$

for any $\mathcal{D}_{n-1} \subset \mathcal{H}$, *i.e.*, $\mu_n(H_{g+\delta}) < \mu_n(H_g)$. ∎

14.2.4 Example: Consider the operator $H_g := H_0 + gV(Q)$ on $L^2(\mathbb{R}^n)$ with a potential $V \in L^p(\mathbb{R}^n)$, where p satisfies condition (14.2). By Theorem 14.1.2, $V(Q)$ is H_0–bounded with zero relative bound; in Section 14.4 we shall show that $\sigma_{ess}(H_g) = \sigma_{ess}(H_0) = \mathbb{R}^+$ holds for any $g \in \mathbb{R}$. Proposition 14.2.3 may then be applied either to the operators H_g with $g \in [0,\infty)$, or to $H_g = H_0 - g(-V)$ with $g \in (-\infty,0]$. We are interested, of course, in the nontrivial situation when H_g has isolated (*i.e.*, negative) eigenvalues of a finite multiplicity — in dependence on the potential this may occur in both cases, in one of them, or not at all. Proposition 14.2.3 implies that the number of eigenvalues (counting the multiplicity) is nondecreasing as a function of $|g|$, and the bottom of the spectrum $\mu_1(H_g) = \inf\sigma(H_g)$ is continuously decreasing with respect to $|g|$ provided there is at least one negative eigenvalue. This conclusion extends to potentials $V \in (L^p + L^\infty_\varepsilon)(\mathbb{R}^n)$ which we shall discuss below (*cf.* Problem 1).

We know from Section 5.4 that if H is a below bounded self–adjoint operator and ψ is a unit vector from $D(H)$, then $(\psi,H\psi) \ge \inf\sigma(H)$. The minimax principle allows us to generalize this result substantially.

14.2.5 Proposition: Assume that H is self–adjoint and L is an N–dimensional subspace in $D(H)$; we denote the corresponding projection as P and set $H_L := PHP$. Let $\lambda_1^P \le \cdots \le \lambda_N^P$ be the eigenvalues of the operator $H_L \upharpoonright L$; then

$$\mu_n(H) \le \lambda_n^P\,, \quad n = 1,\dots,N\,.$$

In particular, if $\sigma_d(H)$ consists of the eigenvalues $\lambda_1 \le \cdots \le \lambda_m$ (counting the multiplicity), then $\lambda_n \le \lambda_n^P$ holds for $n = 1,\dots,\min(m,N)$.

Proof: The operator $H_L \upharpoonright L$ is clearly Hermitean and satisfies $(\psi, H_L\psi) = (\psi, H\psi)$ for $\psi \in L$; then Theorem 14.2.1 gives

$$\lambda_n^P = \sup_{\mathcal{D}_{n-1}\subset L} \inf_{\substack{\psi\in\mathcal{D}_{n-1}^\perp\cap L\\ \|\psi\|=1}} (\psi, H\psi) = \sup_{\mathcal{D}_{n-1}\subset\mathcal{H}} \inf_{\substack{\psi\in(P\mathcal{D}_{n-1})^\perp\cap L\\ \|\psi\|=1}} (\psi, H\psi)$$

(*cf.* Remark 14.2.2c). Furthermore, using the fact that $(\psi, P\phi) = (\psi, \phi)$ holds for $\psi \in L$, we obtain

$$\lambda_n^P = \sup_{\mathcal{D}_{n-1}\subset\mathcal{H}} \inf_{\substack{\psi\in\mathcal{D}_{n-1}^\perp\cap L\\ \|\psi\|=1}} (\psi, H\psi) \ge \sup_{\mathcal{D}_{n-1}\subset\mathcal{H}} \inf_{\substack{\psi\in\mathcal{D}_{n-1}^\perp\cap D(H)\\ \|\psi\|=1}} (\psi, H\psi) = \mu_n(H)\,. \quad \blacksquare$$

The obtained result makes it possible to estimate the eigenvalues of H from above by finite–dimensional approximations; this is the basic idea of the so–called *Rayleigh–Ritz method*, which is widely used in practical calculations.

The second topic we want to speak about here is the perturbation theory. This term covers an extremely wide range of problems. A few of them have been discussed above; let us recall

(a) The stability of self–adjointness with respect to relatively bounded perturbations which forms the content of Theorems 4.3.12 and 4.6.14

(b) The stability of σ_{ess} under relatively compact perturbations — Theorem 5.4.6

(c) Time–dependent perturbations — *cf.* Example 9.5.5 and the notes to Section 9.5

(d) Perturbations of embedded eigenvalues — see Section 9.6 and the notes to it

However, for a quantum physicist the term is associated primarily with perturbations of isolated eigenvalues. It is a vast subject which has not been mentioned up to now; we limit ourselves to an illustration of the ideas on a particular class of operators. We shall need the following result, which is not difficult to prove for self–adjoint operators (Problem 6); for the general case see the notes.

14.2.6 Theorem: Let T be a closed operator and z_0 an isolated point of $\sigma(T)$. Suppose that $\Gamma \subset \rho(T)$ is a closed curve enclosing the point z_0 clockwise, and define

$$P := -\frac{1}{2\pi i}\int_\Gamma R_T(z)\,dz\,.$$

Then operator P is a projection, in general nonorthogonal, and its range is an invariant subspace of T. If $\operatorname{Ran} P$ is one–dimensional, it is an eigenspace of T corresponding to the eigenvalue z_0; more generally, if $\dim \operatorname{Ran} P < \infty$, then the identity $(T - z_0)^n \psi = 0$ holds for each $\psi \in \operatorname{Ran} P$ and a positive integer n.

Any perturbation problem concerns a class of operators depending on a parameter (or parameters); the simplest situation corresponds to the case when this dependence is linear. We shall again consider relatively bounded perturbations.

14.2.7 Theorem: Suppose that H_0, V are self–adjoint and V is H_0–bounded; we set $H_g := H_0 + gV$ for all $g \in \mathbb{C}$.

(a) Let $z \in \rho(H_0)$; then there is a complex neighborhood U of the point $g = 0$ such that the operator–valued function $g \mapsto (H_g - z)^{-1}$ is analytic in U.

(b) Let λ_0 be a simple isolated eigenvalue of H_0 with an eigenvector ψ_0. Then for all sufficiently small $|g|$, the operator H_g has a simple eigenvalue $\lambda(g)$; the function $\lambda(\cdot)$ is analytic in the vicinity of $g = 0$ and $\lambda(0) = \lambda_0$.

Proof: Suppose that the operators H_0, V satisfy the inequality analogous to (4.3). Mimicking the argument from the proof of Theorem 4.3.12, we find

$$\|V(H_0 - z)^{-1}\| \le a\|H_0(H_0 - z)^{-1}\| + b\|(H_0 - z)^{-1}\|$$

for any $z \in \rho(H_0)$, so the operator on the left side is bounded; now we do not care about the value of a because the perturbation V enters H_g with the coupling constant g. By Lemma 1.7.3 and Theorem 1.7.5, the operators $(I + gV(H_0 - z)^{-1})^{-1}$ then exist for small enough $|g|$ and $g \mapsto (I + gV(H_0 - z)^{-1})^{-1}$ is analytic in the vicinity of $g = 0$. The last named property is preserved when the function is multiplied by a bounded g–independent operator; furthermore, by Problem 1.63 we have $z \in \rho(H_g)$ and

$$(H_g - z)^{-1} = (H_0 - z)^{-1} (I + gV(H_0 - z)^{-1})^{-1} \qquad (14.9)$$

for sufficiently small $|g|$; this proves assertion (a).

Next we use Theorem 14.2.6. We set $\Gamma_\varepsilon(\lambda_0) := \{z = z_\varphi := \lambda_0 + \varepsilon\, e^{-i\varphi} : \varphi \in [0, 2\pi)\}$. Since λ_0 is isolated due to the assumption, we can achieve $\Gamma_\varepsilon(\lambda_0) \in \rho(H_0)$ by choosing ε small enough. Then

$$P(g) := -\frac{1}{2\pi i} \int_{\Gamma_\varepsilon(\lambda_0)} (H_0 + gV - z)^{-1} dz \qquad (14.10)$$

makes sense if $g = 0$; we want to check that this remains true in some neighborhood of this point. By the already proven first part, $g \mapsto (H_g - z_\varphi)^{-1}$ is analytic in an open set U_φ containing $g = 0$; hence it is sufficient to check that $U := \bigcap_{0 \le \varphi < 2\pi} U_\varphi$ is open. The resolvent $(H_g - z)^{-1}$ is expressed by a power series whose radius of convergence is $g_0 := \|V(H_0 - z)^{-1}\|^{-1}$. Using Example 4.3.6 we readily find that

$\|(H_0 - z)^{-1}\| \le \varepsilon^{-1}$ and $\|H_0(H_0 - z)^{-1}\| \le 2 + |\lambda_0|\varepsilon^{-1}$ provided ε is chosen small enough; this gives the inequality

$$g_0 \ge \left[a\|H_0(H_0 - z)^{-1}\| + b\|(H_0 - z)^{-1}\|\right]^{-1} \ge \frac{\varepsilon}{a(|\lambda_0| + 2\varepsilon) + b},$$

which shows that the radius of U is positive. It now readily follows from Theorem A.5.3 that the operator–valued function $P(\cdot)$ is analytic around $g = 0$.

This means, in particular, that it is continuous; then $\|P(\cdot)\|$ is also continuous and since it assumes integer values, it is constant in the vicinity of $g = 0$. Hence $\dim \operatorname{Ran} P(g) = 1$ so, by Theorem 14.2.6, for $|g|$ sufficiently small, H_g has a simple eigenvalue $\lambda(g)$ and $P(g)\psi_0$ is the corresponding eigenvector. The latter is not normalized, in general, but its norm is a continuous function of g; this means that it is nonzero around $g = 0$ and we may introduce $\psi(g) := \|P(g)\psi_0\|^{-1}P(g)\psi_0$. Finally, we have $(H_g - z)^{-1}\psi(g) = (\lambda(g) - z)^{-1}\psi(g)$, and therefore

$$\frac{(\psi_0, (H_g - z)^{-1}P(g)\psi_0}{(\psi_0, P(g)\psi_0)} = (\lambda(g) - z)^{-1}$$

for $z \in \Gamma_\varepsilon(\lambda_0)$; the left side is analytic with respect to g so the same is true for the right side; then the function $\lambda(\cdot)$ is also analytic in the vicinity of the point $g = 0$. ∎

In this way one is able not only to prove the existence of the perturbed eigenvalues and their analyticity with respect to the parameter (coupling constant), but also to compute the coefficients of the appropriate Taylor expansion, which is usually named the *Rayleigh–Schrödinger series*. The starting point is the operator $P(g)$; note that we are usually interested in the *real* coupling constant, in which case H_g is self–adjoint and by Problem 6, $P(g)$ is an orthogonal projection (a spectral projection of the perturbed operator). Writing the last operator on the right side of (14.9) as a geometric series and substituting into (14.10), we obtain

$$P(g) = -\frac{1}{2\pi i}\sum_{n=0}^{\infty}(-g)^n \int_{\Gamma_\varepsilon(\lambda_0)} (H_0 - z)^{-1}\left[V(H_0 - z)^{-1}\right]^n dz\,.$$

This formula determines the perturbed eigenvalue by $\lambda(g) = (\psi(g), (H_0 + gV)\psi(g))$; after the necessary algebra we not only recover the first few terms of the Rayleigh–Schrödinger series known from quantum mechanics textbooks, but can also obtain an expression for the general term — see the references mentioned in the notes.

14.3 The discrete spectrum

In quantum mechanics we associate eigenvectors of the operators (14.1) with bound states of the corresponding system of particles; a motivation for this interpretation will be mentioned in Section 15.1 below. The most important are isolated eigenvalues of a finite multiplicity, *i.e.*, points of the discrete spectrum. In this section we are going to discuss some of their properties.

The first question to be asked concerns the *number of bound states*, which is understood as the dimension of the spectral projection corresponding to $\sigma_d(H)$. We denote this quantity for operator (14.1) with a potential V as $N(V)$, and present some results which allow us to estimate $N(V)$ from above.

Let us begin with the case of a centrally symmetric potential which we discussed in Section 11.5, $V(x) = v(r)$ for some $V : \mathbb{R}^+ \to \mathbb{R}$. To check the essential self–adjointness of $H := H_0 + V(Q)$, we have to prove this property for the partial–wave operators (11.19). In fact, in most cases only the s–wave operator h_0 is important (*cf.* Problem 7a); on the other hand, one has to choose properly the boundary condition at the origin which specifies the self–adjoint extension h_0 (Problem 7b).

Let $n_\ell(V)$ be the number of bound states of the operator $\overline{h}_\ell$. Since $\dim \mathcal{G}_\ell = 2\ell+1$, the decomposition $UL^2(\mathbb{R}^3) = \bigoplus_{\ell=0}^{\infty} L^2(\mathbb{R}^+) \otimes \mathcal{G}_\ell$ of Section 11.5 gives

$$N(V) = \sum_{\ell=0}^{\infty} (2\ell+1)\, n_\ell(V)\,. \tag{14.11}$$

Hence it is sufficient to find the numbers $n_\ell(V)$ of "partial–wave" bound states; in this way the problem is reduced to analysis of ordinary differential operators. We shall quote a typical result for potentials of the class $L^2 + L^\infty_\varepsilon$ introduced in Problem 1.

14.3.1 Theorem (GMGT bound): Let $V \in (L^2 + L^\infty_\varepsilon)(\mathbb{R}^3)$ be a centrally symmetric potential, $V(x) = v(r)$; then the number of "partial–wave" bound states of operator (14.1) with $2m = 1$ satisfies the inequality

$$n_\ell(V) \le \frac{(p-1)^{p-1}\Gamma(2p)}{p^p\,\Gamma(p)^2}\,(2\ell+1)^{-(2p-1)} \int_0^\infty r^{2p-1}|v(r)|^p\,dr \tag{14.12}$$

for any $p \ge 1$.

14.3.2 Remarks: (a) The bound for a general mass is obtained easily using the scaling transformation of Remark 14.1.1b. In the particular case of $p = 1$ the inequality (14.12) yields the classical *Bargmann bound*

$$n_\ell(V) \le \frac{1}{2\ell+1} \int_0^\infty r|v(r)|\,dr\,.$$

(b) By Theorem 14.1.2 and Problem 4.22, operator H is below bounded under the assumptions of the theorem. The minimax principle then allows us to replace the potential by its negative part. We have, for instance,

$$n_\ell(V) \le n_\ell(V_-) \le \frac{1}{2\ell+1} \int_0^\infty r|v_-(r)|\,dr\,,$$

and and analogous relations for $p > 1$; similar upper bounds exist for the number of bound states with energies smaller than a given value (Problem 8).

(c) The bound shows, in particular, that operator (14.1) on $L^2(\mathbb{R}^3)$ with a potential which is *purely attractive*, *i.e.*, assuming nonpositive values only, may have no bound state at all if the interaction is weak enough. For example, the Bargmann bound shows that this is the case if

$$\int_0^\infty r|v(r)|\,dr < 1\,.$$

A similar conclusion can be made for Schrödinger operators on $L^2(\mathbb{R}^n)$ with $n > 3$; in contrast, we shall see below that this cannot happen for $n = 1, 2$.

Let us turn to potentials without a rotational symmetry. Since our aim is primarily to illustrate the methods, in the proofs we shall impose rather strong regularity assumptions on the potentials; in the notes we indicate how they can be weakened. We begin with the following argument: if the Schrödinger operator $H_g := H_0 + gV(Q)$ with a potential $V \in L^\infty(\mathbb{R}^n)$ has a negative eigenvalue λ, then the corresponding eigenvector satisfies the identity $\psi = -g(H_0 - \lambda)^{-1}V(Q)\psi$. We denote $V(Q)^{1/2} := |V(Q)|^{1/2}\operatorname{sgn} V(Q)$ and $\phi := |V(Q)|^{1/2}\psi$; then the last relation can be rewritten as

$$\phi = -g|V(Q)|^{1/2}(H_0 - \lambda)^{-1}V(Q)^{1/2}\phi\,.$$

On the other hand, suppose that a vector $\phi \in L^2(\mathbb{R}^n)$ solves the last equation. The operator $V(Q)^{1/2}$ is bounded by assumption, so $V(Q)^{1/2}\phi \in L^2(\mathbb{R}^n)$ and the vector $\psi := -g(H_0 - \lambda)^{-1}V(Q)^{1/2}\psi$ belongs to $D(H_0)$ and $\phi = |V(Q)|^{1/2}\psi$, *i.e.*,

$$\psi = -g(H_0 - \lambda)^{-1}V(Q)^{1/2}|V(Q)|^{1/2}\psi = -g(H_0 - \lambda)^{-1}V(Q)\psi\,.$$

Furthermore, Problem 1.63 gives $I + g(H_0 - \lambda)^{-1}V(Q) = (H_0 - \lambda)^{-1}(H_g - \lambda)$, and therefore $(H_0 - \lambda)^{-1}(H_g - \lambda)\psi = 0$; however, $\lambda \notin \sigma(H_0)$ in view of (7.14), which means that ψ is an eigenvector of H_g with the eigenvalue λ. In this way, we have obtained the following useful result.

14.3.3 Lemma (Birman–Schwinger principle): Let $V \in L^\infty(\mathbb{R}^n)$; then $-\kappa^2$ is an eigenvalue of $H_g := H_0 + gV(Q)$ for some $\kappa > 0$ *iff* the operator

$$K_\kappa := |V(Q)|^{1/2}(H_0 + \kappa^2)^{-1}V(Q)^{1/2} \tag{14.13}$$

has an eigenvalue $-g^{-1}$.

14.3.4 Remark: Once we drop the assumption $V \in L^\infty(\mathbb{R}^n)$ in the lemma, the vector $|V(Q)|^{1/2}\psi$ may not belong to L^2, in which case the right side of (14.13) has no meaning. However, the resolvent $(H_0 + \kappa^2)^{-1}$ is explicitly known from Problem 7.18; this allows us to define K_κ as an integral operator and extend the validity of the lemma to a much wider class of potentials (see the notes). A similar integral–operator representation of $|V(Q)|^{1/2}(H_0 - z)^{-1}V(Q)^{1/2}$ exists for any $z \in \mathbb{C} \setminus \mathbb{R}^+$. In a more general context, we shall encounter this problem in Proposition 15.1.3 below.

Using Lemma 14.3.3, one can prove a bound on $N(V)$ for potentials of the Rollnik class R introduced in the notes to Section 14.1. The Schrödinger operator must then in general be defined as the form sum $H_0 \dot{+} V(Q)$; however, if the corresponding operator sum is *e.s.a.*, then the two definitions coincide — see Remark 4.6.12.

14.3.5 Theorem (Birman–Schwinger bound): Consider the Schrödinger operator $H := H_0 \dot{+} V(Q)$ on $L^2(\mathbb{R}^3)$ with a potential $V \in R$. The number $N(V)$ of bound states is finite and satisfies the inequality

$$N(V) \le \frac{1}{16\pi^2} \int_{\mathbb{R}^6} \frac{|V(x)|\,|V(y)|}{|x-y|^2}\, dx\, dy\,. \tag{14.14}$$

Proof will be given for $V \in C_0^\infty(\mathbb{R}^3)$; for the general case see the notes. As in Remark 14.3.2b, in view of the minimax principle we may restrict ourselves to the case when $V(x) \le 0$ holds for all $x \in \mathbb{R}^3$. Let $H_g := H_0 \dot{+} gV(Q)$, *i.e.*, $H = H_1$; then another consequence of this principle contained in Proposition 14.2.3 tells us that the values $\mu_n(H_g)$ are continuous and nondecreasing functions of the coupling constant g. We consider $N_\lambda(V)$ defined in Problem 8 for $\lambda := -\kappa^2 < 0$. This quantity is obviously equal to the maximal n for which $\mu_n(H_1) < \lambda$; it follows from the continuity of $g \mapsto \mu_n(H_g)$ that for any such n there is a number $g_n \in (0,1)$ such that $\mu_n(H_{g_n}) = \lambda$. These numbers satisfy the inequalities $0 < g_1 \le g_2 \le \cdots$; for the other n with $\mu_n(H_1) \ge \lambda$ the numbers g_n may exist but $g_n \ge 1$. This means that the quantity $N_\lambda(V)$ is at the same time equal to the maximal n for which $g_n < 1$; this yields the inequality

$$N_\lambda(V) \le \sum_{n=1}^{N_\lambda(V)} g_n^{-2} \le \sum_{n=1}^{\infty} g_n^{-2}\,.$$

Next we use Lemma 14.3.3 according to which H_{g_n} has an eigenvalue λ *iff* $-g_n^{-1}$ is an eigenvalue of the operator (14.13); this implies

$$N_\lambda(V) \le \sum_{n=1}^{\infty} \kappa_n^2\,,$$

where κ_n, $n = 1, 2, \ldots$, are the eigenvalues of K_κ. By Remark 14.3.4, the latter is an integral operator with the kernel

$$K_\kappa(x,y) := -\frac{|V(x)|^{1/2} e^{-\kappa|x-y|} |V(y)|^{1/2}}{4\pi|x-y|}\,,$$

which is real and symmetric, *i.e.*, K_κ is Hermitean. Moreover, it follows from the assumption $V \in C_0^\infty \subset R$ that $K_\kappa(\cdot,\cdot) \in L^2(\mathbb{R}^6)$, which means that the operator K_κ is Hilbert–Schmidt. Hence we may use Proposition 3.6.6 and Remark 3.6.3 to estimate

$$N_\lambda(V) \le \mathrm{Tr}\,(K_\kappa^* K_\kappa) = \frac{1}{16\pi^2} \int_{\mathbb{R}^6} \frac{|V(x)|\,|V(y)|}{|x-y|^2}\, e^{-2\kappa|x-y|}\, dx\, dy\,. \tag{14.15}$$

To conclude the proof; it is sufficient to express $N(V) = \lim_{\lambda\to 0-} N_\lambda(V)$ by means of the dominated–convergence theorem. ∎

As another illustration of the Birman–Schwinger principle, let us prove the claim made in Remark 2c.

14.3.6 Theorem: Let $H_g := H_0 + gV(Q)$ be the Schrödinger operator on $L^2(\mathbb{R}^n)$ with a potential $V \in C_0^\infty(\mathbb{R}^n)$, which is nonzero and purely attractive. If $n = 1, 2$, then H_g has at least one bound state for $g > 0$.

Proof: In view of Proposition 14.2.3 we have to check that $\mu_1(H_g) < 0$ holds for all $g > 0$. As in the preceding proof, a number $g_1 = g_1(\kappa)$ such that $\mu_1(H_{g_1}) = \lambda$ corresponds to any $\lambda = -\kappa^2 < 0$, and by Lemma 3 this is equivalent to the fact that $-g_1^{-1}$ is an eigenvalue of K_κ. The mentioned requirement will then be satisfied if for any $g > 0$ there is a $\kappa > 0$ such that $g_1(\kappa) < g$, *i.e.*, that the corresponding $-K_\kappa$ has an eigenvalue $g_1(\kappa)^{-1} > g^{-1}$. Since the potential is of C_0^∞ and purely attractive, $-K_\kappa$ is for any $\kappa > 0$ a positive Hilbert–Schmidt operator; hence it has a purely discrete spectrum and its norm equals its maximal eigenvalue. It is then sufficient to show that $\lim_{\kappa\to 0+} \|K_\kappa\| = \infty$, *i.e.*, to find a vector $\phi \in L^2(\mathbb{R}^n)$ such that the relation

$$\lim_{\kappa\to 0+} (\phi, |V(Q)|^{1/2}(H_0 + \kappa^2)^{-1}|V(Q)|^{1/2}\phi) = \infty \tag{14.16}$$

is satisfied; this can easily be achieved if $n = 1, 2$ (Problem 9a). ∎

We shall mention one more upper bound which, in distinction to most of the above ones, represents at a good estimate in the semiclassical situation when there are many bound states. It is useful to do it in a more general context. Consider the operator H given by (14.1) on $L^2(\mathbb{R}^n)$ with $2m_j = 1$, $j = 1, \dots, n$, and the potential $V \in L^{\delta+\frac{n}{2}}(\mathbb{R}^n)$, where the parameter $\delta \ge 0$ will be specified below; we suppose that H is *e.s.a.* Denote by $\{\lambda_j\}$, the family its negative eigenvalues and consider its moments $S_\delta(V) := \mathrm{Tr}\,\left((H)_-^\delta\right) = \sum_j (-\lambda_j)^\delta$. Then we have the following important result (see the notes).

14.3.7 Theorem (Lieb–Thirring inequalities): Let $\delta \ge 0$ for $n \ge 3$, $\delta > 0$ for $n = 2$, and $\delta \ge \frac{1}{2}$ if $n = 1$; then the moments $S_\delta(V)$ are estimated by

$$S_\delta(V) \le R(\delta, n) S_{\delta,n}^{\mathrm{cl}}(V) \tag{14.17}$$

for some $R(\delta, n) \ge 1$, where

$$S_{\delta,n}^{\mathrm{cl}}(V) := \int_{\mathbb{R}^{2n}} (|\xi|^2 - V(x))_-^\delta \,\frac{d\xi dx}{(2\pi)^n} = L_{\delta,n}^{\mathrm{cl}} \int_{\mathbb{R}^n} V_-(x)^{\delta+\frac{n}{2}}\, dx$$

with $L_{\delta,n}^{\mathrm{cl}} := \Gamma(\delta+1)\left[2^n \pi^{n/2} \Gamma\left(\delta + \frac{n}{2} + 1\right)\right]^{-1}$.

14.3.8 Remarks: (a) In particular, for $\delta = 0$ and $n \ge 3$ we get from here the mentioned bound, known as the *Cwikel–Lieb–Rozenblium theorem*,

$$N(V) \le c_n \int_{\mathbb{R}^n} |V_-(x)|^{n/2}\, dx\,,$$

with the constant $c_n \geq c_n^{\text{cl}} := \left((4\pi)^{n/2}\,\Gamma\left(1+\frac{n}{2}\right)\right)^{-1}$.
(b) If the potential V is replaced by gV the factor $g^{\delta+\frac{n}{2}}$ arises at the right side of (14.17), which is the expected behavior. In the semiclassical regime one expects the spectrum to behave as if $R(\delta,n)=1$. This need not be true, however, over the whole spectrum. The present knowledge about the constants involved is as follows,

$R(\delta,n)=1$ if $\delta\geq 3/2,\ n\in\mathbb{N}$,
$R(\delta,n)\leq 2$ if $1\leq\delta<3/2,\ n\in\mathbb{N}$ or $1/2\leq\delta<1,\ n=1$,
$R(\delta,n)\leq 4$ if $1/2\leq\delta<1,\ n=1$.

For $\delta=1/2$ and $n=1$ an opposite inequality is known to be valid which shows that the above bound is saturated with $R(1/2,1)=2$ as $g\to 0$ (see the notes).

Let us now ask more generally under which circumstances the discrete spectrum is finite or infinite. The above discussed bounds suggest that the answer may depend on the behavior of the function V at large distances. In the three–dimensional case, we have the following result.

14.3.9 Theorem: Let $H := H_0 + V(Q)$ be the operator (14.1) on $L^2(\mathbb{R}^3)$ with a potential $V\in(L^2+L^\infty_\varepsilon)(\mathbb{R}^3)$.

(a) If there are $c\in\left[0,\frac14\right)$ and $r_0>0$ such that $V(x)\geq -c\,r^{-2}$ holds for any $r:=|x|\geq r_0$, then $\sigma_d(H)$ is finite.

(b) If there are positive r_0, d, ε such that $V(x)\leq -d\,r^{-2+\varepsilon}$ holds for all $r\geq r_0$, then $\sigma_d(H)$ is infinite.

Proof: In the next section we shall prove that $\sigma_{ess}(H)=[0,\infty)$; hence we have to find the number of negative eigenvalues (including their multiplicities). Had the potential decayed slightly faster at infinity, say $V(x)\geq -c\,e^{-2-\varepsilon}$ for some $\varepsilon>0$, assertion (a) would follow from Theorem 14.3.7. Under the present assumption, the argument is more complicated. We define $\tilde V(x) := V(x)+c\,r^{-2}$. Since any $\psi\in C_0^\infty(\mathbb{R}^3)$ belongs to $D(Q^{-1})$, we may use the second of the inequalities (8.4), which gives

$$\int_{\mathbb{R}^3}\overline{\psi(x)}\left(-\Delta\psi-\frac{1}{4r^2}\,\psi\right)(x)\,dx \;=\; (\psi,H_0\psi)-\frac14\,\|Q^{-1}\psi\|^2\,\geq\,0\,,$$

and therefore

$$\begin{aligned}(\psi,H\psi) &= \int_{\mathbb{R}^3}\overline{\psi(x)}\left(-(1-4c)\,\Delta\psi+\tilde V\psi+4c\left(-\Delta-\frac{1}{4r^2}\right)\psi\right)(x)\,dx\\ &\geq (1-4c)\int_{\mathbb{R}^3}\overline{\psi(x)}\left(-\Delta\psi+(1-4c)^{-1}\tilde V\psi\right)(x)\,dx\,\geq\,(1-4c)\,(\psi,H_c\psi)\,,\end{aligned}$$

where we have denoted $H_c := H_0+(1-4c)^{-1}\tilde V_-(Q)$. Since $|\tilde V_-|\leq|V_-|\leq|V|$, the function $\tilde V_-$ belongs to L^2+L^∞, and the same is true for $(1-4c)^{-1}\tilde V_-$. By

Theorem 14.1.2, the operators H and H_c are then self–adjoint with the common domain $D(H_0)$ and $C_0^\infty(\mathbb{R}^3)$ is a core for both of them — *cf.* Problem 7.16. Using the fact that $V(Q)$ and $(1-4c)^{-1}\tilde{V}_-(Q)$ are H_0–bounded, we readily check that the above inequality can be extended to any $\psi \in D(H_0)$ (Problem 10a). The minimax principle then implies

$$\mu_n(H) \ge (1-4c)\,\mu_n(H_c)\,, \qquad n = 1,2,\dots\,,$$

and since $\sigma_{ess}(H_c) = [0,\infty)$, it is sufficient to check that the operator H_c has a finite number of bound states. By assumption, the function $(1-4c)^{-1}\tilde{V}_-$ has a compact support, so it belongs to $L^p(\mathbb{R}^3)\,,\; p \in [1,2]$ (Problem 10b). Assertion (a) now follows from Theorem 14.3.7, or alternatively, from Theorem 14.3.5 (see the notes to Section 14.1).

To prove (b) we again employ the fact that $\sigma_{ess}(H) = [0,\infty)$; in view of the minimax principle, it is sufficient to show that $\mu_n(H) < 0$ for any positive integer n. We shall employ Proposition 14.2.5, due to which we have to find to any N an N–dimensional subspace $L \subset D(H)$ such that the projected operator $PH \restriction L$ has negative eigenvalues. We shall emulate the scaling argument of Example 7.5.8, taking $\phi_\alpha :\, \phi_\alpha(x) = \alpha^{-3/2}\phi(x/\alpha)$ for a unit vector $\phi_\alpha \in C_0^\infty(\mathbb{R}^3)$, which we choose in such a way that $\mathrm{supp}\,\phi \subset \{\, x :\, 1 < r < 2\,\}$. If $\alpha > r$, the hypothesis of part (b) implies

$$(\phi_\alpha, H\phi_\alpha) \le \alpha^{-2}(\phi, H_0\phi) \;-\; d\,\alpha^{-2+\varepsilon}(\phi, Q^{-2+\varepsilon}\phi)\;;$$

hence there is an $r_1 > 0$ such that $(\phi_\alpha, H\phi_\alpha) < 0$ holds for $\alpha > r_1$. The sought subspace may be now constructed by means of the vectors ϕ_α. Putting $\alpha_n := 2^n r_1$ and $\psi_n := \phi_{\alpha_n}$, we have $(\psi_n, H\psi_n) < 0$ and $(\psi_n, H\psi_k) = 0$ for $k \ne n$ since $\mathrm{supp}\,\psi_k \cap \mathrm{supp}\,\psi_n = \emptyset$. Thus we choose $L := \{\psi_1,\dots,\psi_N\}_{lin}$ for which the mentioned proposition yields

$$\mu_n(H) \le \sup\left\{ \sum_{n=1}^N |\beta_n|^2(\psi_n, H\psi_n) \,:\, \sum_{n=1}^N |\beta_n|^2 = 1 \right\} \;=\; \sup_{1\le n\le N}\,(\psi_n, H\psi_n) \;<\; 0$$

for all $n = 1,\dots,N$; this concludes the proof. ∎

14.3.10 Example: The Coulomb potential $V :\, V(x) = -\gamma/r\,,\; \gamma := Ze^2$, belongs to $(L^2\!+L^\infty_\varepsilon)(\mathbb{R}^3)$ and satisfies the assumption of part (b) for all r large enough; hence hydrogen–type atoms have an infinite discrete spectrum. This is not surprising, of course, because the spectrum is known explicitly in this case: the eigenvalues are $\lambda_n = -m\gamma^2/2n^2$ with the multiplicities n^2, and we also know the corresponding eigenfunctions $\psi_{n\ell m}$.

The situation is less simple in higher dimensions, where we are interested primarily in the Schrödinger operators which play the role of Hamiltonians of atoms and molecules. We shall discuss the example of helium–type atoms; more complicated systems will be mentioned in the notes. If we neglect the spins and regard the nucleus as infinitely heavy, then we associate with a two-electron atom the Hamiltonian

$$H := \frac{1}{2m}\left(P_1^2+P_2^2\right) - Ze^2\left(|Q_1|^{-1}+|Q_2|^{-1}\right) + e^2|Q_1-Q_2|^{-1}$$

on $L^2(\mathbb{R}^6)$, where Ze is the nucleus charge; we again use the notation $Q_j := (Q_{j1}, Q_{j2}, Q_{j3})$, *etc.* To simplify the problem, we use the unitary "substitution" operator U : $(U\psi)(x_1,x_2) = (mZe^2)^{-6}\psi\left(\frac{x_1}{mZe^2}, \frac{x_2}{mZe^2}\right)$, which transforms H into $UHU^{-1} = \frac{1}{2}m(Ze^2)^2 H_\gamma$, where

$$H_\gamma := P_1^2 + P_2^2 - 2|Q_1|^{-1} - 2|Q_2|^{-1} + \gamma\,|Q_1-Q_2|^{-1} \tag{14.18}$$

and $\gamma := 2/Z$ is a positive parameter. The operators H_γ are self-adjoint by Theorem 14.1.4; the physically interesting cases are $\gamma = 2, 1, 2/3, 1/2, \dots$ corresponding to H^-, He, Li^+, Be^{++}, *etc.*, respectively.

14.3.11 Theorem (Kato): $\sigma_d(H_\gamma)$ is infinite if $\gamma < 2$.

Proof: By Examples 14.3.10 and 5.7.5, $\sigma_{ess}(H_\gamma) = [-1,\infty)$ for $\gamma = 0$; the same is also true for a nonzero γ due to the HVZ-theorem, which we shall discuss below. By analogy with the proof of Theorem 14.3.9b, it is then sufficient to find to any positive integer N an N-dimensional subspace $L \subset D(H_\gamma)$ such that the eigenvalues of the "cut-off" operator are less than -1, *i.e.*,

$$\sup\{(\psi, H_\gamma\psi) \,:\, \psi\in L,\ \|\psi\|=1\} < -1\,. \tag{14.19}$$

Let $\psi_{100} : \psi_{100}(x) = \pi^{-1/2}e^{-r}$ be the wave function of the hydrogen-atom ground state in the used system of units, $(P_1^2 - 2|Q_1|^{-1})\psi_{100} = -\psi_{100}$; then we have

$$(\psi_{100}\otimes\phi,\, H_\gamma(\psi_{100}\otimes\phi)) = -1 + \left(\phi, \left(P_2^2-(2-\gamma)|Q_2|^{-1}\right)\phi\right)$$

$$+\,\gamma\left(\psi_{100}\otimes\phi, \left(|Q_1-Q_2|^{-1}-|Q_2|^{-1}\right)(\psi_{100}\otimes\phi)\right)$$

for any unit vector $\phi\in\mathcal{S}(\mathbb{R}^3)$. Using the fact that the function ψ_{100} is rotationally invariant, we obtain

$$\begin{aligned}&(\psi_{100}\otimes\phi,\, |Q_1-Q_2|^{-1}(\psi_{100}\otimes\phi))\\ &\quad= \int_{\mathbb{R}^3} dx_2\,|\phi(x_2)|^2 \int_{\mathbb{R}^3} dx_1\,\min(r_1^{-1},r_2^{-1})\,|\psi_{100}(x_1)|^2\\ &\quad\le \int_{\mathbb{R}^3} r_2^{-1}|\phi(x_2)|^2\,dx_2 \le \left(\psi_{100}\otimes\phi,\, |Q_2|^{-1}(\psi_{100}\otimes\phi)\right)\end{aligned}$$

(Problem 11), so the last term in the above identity is nonpositive. In view of the preceding example the operator appearing in the second term has the eigenvalues $\lambda_n = -(2-\gamma)^2/4n^2$, $n = 1,2,\dots$, with the multiplicities n^2. Let $\{E_k\}_{k=1}^\infty$ be the sequence obtained by ordering these eigenvalues (with repetition according to the multiplicity); the corresponding sequence of mutually orthogonal eigenfunctions will be denoted as $\{\phi_k\}_{k=1}^{\infty}$ (they differ from hydrogen–atom eigenfunctions by the above mentioned scaling transformation only, and therefore they belong to $\mathcal{S}(\mathbb{R}^3)$). We define $L := \{\psi_{100}\otimes\phi_1,\dots,\psi_{100}\otimes\phi_N\}_{lin}$; then

$$(\psi, H_\gamma\psi) \le -1+\sum_{k=1}^{N}|\alpha_k|^2E_k < -1$$

holds for any $\psi = \sum_{k=1}^N \alpha_k\psi_{100}\otimes\phi_k$, which satisfies the normalization condition $\sum_{k=1}^N|\alpha_k|^2 = 1$, because all the eigenvalues E_k are negative. Using further the compactness of the unit sphere in L in the same way as in the proof of Proposition 14.2.3, we arrive at the sought inequality (14.19). ∎

Of course, finding the cardinality of the discrete spectrum is just a part of its analysis; one also has to study the distribution of eigenvalues, the behavior of the eigenfunctions, *etc.*Such a task goes beyond the scope and purpose of the present book; the reader can find a lot of information on these subjects in the literature quoted in the notes. In conclusion, let us mention one often used result.

14.3.12 Theorem: Let $H := H_0+V(Q)$ be the operator (14.1) on $L^2(\mathbb{R}^3)$ with a potential $V \in (L^2+L^\infty)(\mathbb{R}^3)$. Suppose that $\lambda_0 := \inf\sigma(H)$ is an eigenvalue; then its multiplicity is one and the corresponding eigenvector ψ_0 can be (with a suitable choice of phase factor) represented by a function which is everywhere positive.

The lowest eigenvalue of a Hamiltonian is conventionally called the *ground state;* the theorem says that it is nondegenerate and the corresponding wave function has no nodes. This is true also many other Schrödinger operators (see the notes).

14.4 The essential spectrum

The investigation of the essential spectrum of Schrödinger operators again involves many problems; we shall mention just a few of them. Let us begin with the question of stability. The classical Weyl theorem of Problem 5.26 does not apply to Schrödinger operators with the exception of the trivial case, because the operators of multiplication by a function are not compact. Fortunately, for a wide class of potentials they are H_0–compact.

14.4.1 Theorem: Let $H := H_0+V(Q)$ be the operator (14.1) with a potential $V \in (L^2+L^\infty_\varepsilon)(\mathbb{R}^n)$, $n\le 3$; then $\sigma_{ess}(H) = [0,\infty)$.

Proof: In view of Theorem 5.4.6 and Example 7.5.8, it is sufficient to check that the operator $V(Q)(H_0+I)^{-1}$ is compact. If $V\in L^2$, this follows from Proposition 15.1.3a and the fact that $g: \; g(p) = (p^2+1)^{-1}$ is square integrable. Furthermore, for any positive integer n there is a $V_n \in L^2$ such that $\|V-V_n\|_\infty < n^{-1}$; hence

$$\begin{aligned} \|(V-V_n)(H_0+I)^{-1}\psi\|^2 &= \int_{\mathbb{R}^3} |V(x)-V_n(x)|^2|((H_0+I)\psi)(x)|^2 dx \\ &\le \|V-V_n\|_\infty^2 \|(H_0+I)^{-1}\psi\|^2 \le n^{-2}\|\psi\|^2 \end{aligned}$$

holds for any $\psi \in L^2(\mathbb{R}^n)$ due to Example 4.3.6. This means that the sequence $\{V_n(Q)(H_0+I)^{-1}\}$ converges to $V(Q)(H_0+I)^{-1}$ with respect to the operator norm, so the result follows from Theorem 3.5.2. ∎

The result extends to Schrödinger operators on $L^2(\mathbb{R}^n)$, $n \ge 4$, with potentials $V \in (L^p + L^\infty_\varepsilon)(\mathbb{R}^n)$, where p satisfies the condition (14.2), and other situations (see the notes). This assumption is valid, *e.g.*, if $V \in L^p_{loc}$ and $\lim_{|x|\to\infty} V(x) = 0$ (Problem 1b). In the case that the limit exists and assumes a nonzero value c we may apply the result to the operator with the shifted potential $V-c$, obtaining $\sigma_{ess}(H) = [c,\infty)$. This suggests that the spectrum of Schrödinger operators with potentials, which grow at large distances, is purely discrete.

14.4.2 Theorem: Let H be the operator (14.1) on $L^2(\mathbb{R}^n)$ with a potential $V \in L^2_{loc}$, which is below bounded and satisfies the condition $\lim_{|x|\to\infty} V(x) = \infty$; then $\sigma_{ess}(\overline{H}) = \emptyset$.

Proof: The operator $H = H_0 + V(Q)$ is symmetric and below bounded, and since $D_V = D(V(Q)) \supset C_0^\infty(\mathbb{R}^n)$, it is also *e.s.a.* due to Theorem 14.1.6. Without loss of generality we may assume that $V \ge 0$; then we define the form $s : s[\psi] = \|H_0^{1/2}\psi\|^2 + \|V^{1/2}\psi\|^2$ with the domain $D(H_0) \cap D_V$. It is closable by Proposition 4.6.7, and the self-adjoint operator associated with $\overline{s}$, *i.e.*, the Friedrichs extension of H, coincides with $\overline{H}$, because the latter is essentially self-adjoint. Due to Theorem 4.6.5b, for any $\psi \in D(\overline{s}) := Q(\overline{H})$ there is a sequence $\{\psi_n\} \subset D(H)$ such that $\psi_n \to \psi$ and $s[\psi_n - \psi_m] \to 0$ as $n, m \to \infty$. Since s is a sum of two positive forms, it follows that $\|H_0^{1/2}(\psi_n - \psi_m)\|^2 \to 0$ as $n, m \to \infty$, *i.e.*, $\psi \in D(H_0^{1/2}) = Q(H_0)$. In this way, we have obtained the inclusion $Q(\overline{H}) \subset Q(H_0)$.

Now we may use the minimax principle; it is sufficient to check that $\mu_n(\overline{H}) \to \infty$ as $n \to \infty$. By assumption, for any $b > 0$ there is an $r_b > 0$ such that $V(x) \ge b$ holds for $|x| > r_b$. The function

$$\tilde{V}_b : \tilde{V}_b(x) = \begin{cases} -b & \dots \quad |x| > r_b \\ 0 & \dots \quad |x| \le r_b \end{cases}$$

is bounded; hence the corresponding Schrödinger operator $H_b := H_0 + \tilde{V}_b(Q)$ has the domain $D(H_0)$, and similarly $Q(H_b) = Q(H_0)$. It now follows from the inequality $V(x) \ge b + \tilde{V}_b(x)$ and Theorem 14.2.1 that

$$\mu_n(\overline{H}) = \sup_{\mathcal{D}_{n-1}\subset\mathcal{H}} \inf_{\substack{\psi \in \mathcal{D}_{n-1}^\perp \cap Q(\overline{H}) \\ \|\psi\|=1}} s[\psi] \ge \sup_{\mathcal{D}_{n-1}\subset\mathcal{H}} \inf_{\substack{\psi \in \mathcal{D}_{n-1}^\perp \cap Q(\overline{H}) \\ \|\psi\|=1}} \|(H_b+b)^{1/2}\psi\|^2$$

$$\geq \sup_{\mathcal{D}_{n-1}\subset\mathcal{H}} \inf_{\substack{\psi\,\in\,\mathcal{D}_{n-1}^{\perp}\cap\, Q(H_0)\\ \|\psi\|=1}} \|(H_b+b)^{1/2}\psi\|^2 = \mu_n(H_b+b) = b+\mu_n(H_b)\,.$$

The potential $\tilde{V}_b$ obviously belongs to $L^{n/2}(\mathbb{R}^n)$, so it has a finite number of bound states due to Theorem 14.3.7. Hence there is an n_b such that $\mu_n(H_b) \geq 0$ holds for all $n > n_b$, and therefore $\mu_n(\overline{H}) \geq b$ for all $n > n_b$; this concludes the proof. ∎

The condition given by the last theorem is sufficient but not necessary. In the one–dimensional case, we have the following simple criterion (see the notes).

14.4.3 Theorem (Molčanov): Let H be the operator (14.1) on $L^2(\mathbb{R})$ with a potential $V \in L^2_{loc}(\mathbb{R})$ which is below bounded. Then $\sigma_{ess}(\overline{H}) = \emptyset$ holds *iff* the conditions

$$\lim_{a\to\pm\infty} \int_a^{a+\delta} V(x)\,dx = \infty$$

are valid for any $\delta > 0$.

Let us now return to operators with a nonempty essential spectrum. Consider the Hamiltonian of a system of N particles interacting via two–particle forces,

$$H := \sum_{j=1}^{N} \frac{1}{2m_j} P_j^2 + \sum_{j=1}^{N}\sum_{k=1}^{j-1} V_{jk}(Q_j - Q_k)\,. \tag{14.20}$$

In this case the $3N$–dimensional version of Theorem 14.4.1 is of little use, because even if the full potential approaches a limit at large distances, the latter generally depends on the chosen direction; the bundle of lines $x_j - x_k = const$, along which this term is constant, corresponds to each two–particle potential V_{jk}. The way to determine the essential spectrum of H is thus more complicated; however, it has a transparent physical meaning. At the same time, it is technically complicated, so we limit ourselves to the formulation of the result.

First we shall ask how the system under consideration, which we denote as S, can be decomposed into subsystems. We assume $N \geq 2$. However, for $N = 2$ the following argument reduces to the comparison of the center–of–mass Hamiltonian from the decomposition (11.12) to the Hamiltonian of the free relative motion. If $N \geq 3$, there are different ways to decompose the system (four possibilities exist for $N = 3$, fourteen for $N = 4$, *etc.*). Let $D := \{C_1, \dots, C_{n(D)}\}$ be such a *partition* of S into *clusters* of particles; we exclude the trivial case with $n(D) = 1$. We denote $I := \{1, \dots, N\}$ and define the function $\epsilon^D : I \times I \to \{0, 1\}$ in such a way that $\epsilon^D_{jk} = 1$ if the j–th and k–th particle belong to the same cluster and $\epsilon^D_{jk} = 0$ otherwise. Then we can define the operator

$$H_D := \sum_{j=1}^{N} \frac{1}{2m_j} P_j^2 + \sum_{j=1}^{N}\sum_{k=1}^{j-1} \epsilon^D_{jk} V_{jk}(Q_j - Q_k)\,,$$

with a clear physical meaning: it describes the system in which we have "switched off" the interactions between the clusters. In a similar way, we define the Hamiltonian of the cluster C_i as

$$\underline{H}_{C_i} := \sum_{j\in C_i} \frac{1}{2m_j} P_j^2 + \sum_{\substack{j,k\,\in\, C_i \\ j<k}} V_{jk}(Q_j - Q_k)\,;$$

we know from Section 11.1 that $\underline{H}_{C_i} = \overline{H_{C_i} \otimes I}$, where H_{C_i} is the operator on $L^2(\mathbb{R}^{3n_i})$ defined by the right side of the last relation, where P_{jl}, Q_{nl}, $l = 1,2,3$, are now regarded as operators on $L^2(\mathbb{R}^{3n_i})$; we denote here by n_i the number of particles in the cluster C_i.

Let us stress that rather than in the spectrum of the Hamiltonian (14.20) itself, we are interested in the spectrum of the operator obtained by separating the center–of–mass motion, because only the latter contains nontrivial dynamical information about the system. Therefore let H^{rel} and H_D^{rel} be the operators obtained after separation of the center–of–mass motion from H and H_D, respectively, and let $H_{C_i}^{rel}$ be the operator originating from the elimination of the center of mass of the cluster C_i from H_{C_i} (*cf.* Remark 11.5.1). The operators H^{rel}, H_D^{rel} act on the space $\mathcal{H}_{rel} := L^2(\mathbb{R}^{3N-3})$, in which suitable relative coordinates are introduced, and $H_{C_i}^{rel}$ acts on $\mathcal{H}_{C_i}^{rel} := L^2(\mathbb{R}^{3n_i-3})$ determined by the internal coordinates of the cluster; if $n_i = 1$ the corresponding relative state space is one–dimensional and $H_{C_i}^{rel}$ is a zero operator on it. As above, we denote $\underline{H}_{C_i}^{rel} := \overline{H_{C_i}^{rel} \otimes I}$.

Concluding this argument, we can say that $\sum_{i=1}^{n(D)}(3n_i - 3) = 3N - 3n(D)$ degrees of freedom are associated with the motion within the clusters. The remaining $3n(D) - 3$ degrees of freedom correspond to the relative motion of the clusters in the center–of–mass frame (since we assume $n(D) \ge 2$, this quantity is positive); the space $L^2(\mathbb{R}^{3n(D)-3})$ determined by these coordinates will be denoted as $\mathcal{H}_D$. By analogy with Section 11.5 and Problem 11.13, one can prove the following result (see the notes).

14.4.4 Proposition: The decomposition $\mathcal{H}_{rel} = \mathcal{H}_D \otimes \mathcal{H}_{C_1}^{rel} \otimes \cdots \otimes \mathcal{H}_{C_{n(D)}}^{rel}$ holds for any partition D, and the corresponding operator H_D^{rel} is expressed as

$$H_D^{rel} = \overline{\underline{T}_D + \underline{H}_{C_1}^{rel} + \cdots + \underline{H}_{C_{n(D)}}^{rel}}\,,$$

where $\underline{T}_D := \overline{T_D \otimes I}$ and T_D is the operator on $\mathcal{H}_D$ describing the kinetic energy of the relative motion of the clusters $C_1, \dots, C_{n(D)}$ in the center–of–mass frame.

The spectrum of H_D^{rel} is given by Theorem 5.7.4. Moreover, choosing the coordinates suitably (see the notes) one can check that $\sigma(T_D) = \mathbb{R}^+$; it follows that $\sigma(H_D^{rel}) = [t_D, \infty)$ where

$$t_D := \sum_{i=1}^{n(D)} \inf \sigma(H_{C_i}^{rel})\,.$$

Now we are able to state the mentioned result.

14.4.5 Theorem (HVZ–theorem): Let H be the Schrödinger operator (14.20) with two–particle potentials $V_{jk} \in (L^2 + L^\infty_\varepsilon)(\mathbb{R}^3)$. Then $\sigma_{ess}(H) = [t, \infty)$, where

$$t := \min_{n(D)\geq 2} t_D = \min_{n(D)=2} t_D .$$

After we have characterized the essential spectrum of Schrödinger operators, we have to say something about its subsets, such as the absolutely and singularly continuous spectrum or possible embedded eigenvalues. Some results may be derived by tools adopted from scattering theory. As an illustration, let us mention an assertion which we shall prove in the next chapter (*cf.* Example 15.3.2 and Remark 15.3.3).

14.4.6 Proposition: Let $H := H_0 + V(Q)$ be the operator (14.1) on $L^2(\mathbb{R}^3)$ with a potential $V \in L^2_{loc}(\mathbb{R}^3)$ which has the following property: there are positive C, R, ε such that $|V(x)| \geq C|x|^{-1-\varepsilon}$ holds for $|x| \geq R$. Then $\sigma_{ac}(H) = [0, \infty)$.

Let us now turn to the singularly continuous spectrum. In Section 15.1 we shall see that vectors from the corresponding spectral subspace give rise to states with pathological scattering properties; this suggests that we should seek conditions under which $\sigma_{sc}(H) = \emptyset$. Let us first mention the following general result.

14.4.7 Theorem: Let H be a self–adjoint operator on $\mathcal{H}$. If the condition

$$\sup_{0<\varepsilon<1} \int_\alpha^\beta |\operatorname{Im}(\psi, (H - \lambda - i\varepsilon)^{-1}\psi)|^p d\lambda < \infty \tag{14.21}$$

holds for a bounded interval $(\alpha, \beta) \subset \mathbb{R}$, some $p > 1$, and a vector $\psi \in \mathcal{H}$, then $E_H(\alpha, \beta)\psi \in \mathcal{H}_{ac}(H)$.

Proof employs the Stone formula of Example 5.5.2. We denote $J := (\alpha, \beta)$ and use the fact that any open set $G \subset J$ can be expressed as an at most countable disjoint union $\bigcup_{k\in I} J_k$ of open intervals. We have $E_H(J_k) \leq E_H(\bar{J}_k)$ for any $k \in I$, and therefore

$$(\psi, E_H(J_k)\psi) \leq \frac{1}{\pi} \lim_{\varepsilon\to 0+} \int_{J_k} \operatorname{Im}(\psi, (H - \lambda - i\varepsilon)^{-1}\psi)\, d\lambda .$$

Next we denote $J^{(n)} = \bigcup_{k=1}^n J_k$ and use the Hölder inequality with $p^{-1} + q^{-1} = 1$ obtaining

$$\begin{aligned}(\psi, E_H(J^{(n)})\psi) &\leq \frac{1}{\pi} \lim_{\varepsilon\to 0+} \int_{J^{(n)}} \operatorname{Im}(\psi, (H - \lambda - i\varepsilon)^{-1}\psi)\, d\lambda \\ &\leq \frac{1}{\pi} \lim_{\varepsilon\to 0+} \left(\int_{J^{(n)}} |\operatorname{Im}(\psi, (H - \lambda - i\varepsilon)^{-1}\psi)|^p\, d\lambda \right)^{1/p} \ell(J^{(n)})^{1/q} ,\end{aligned}$$

where $\ell(\cdot)$ is the Lebesgue measure. If the index set I is finite, this estimate together with the inclusion $G \subset J$ gives the inequality

$$(\psi, E_H(G)\psi) \leq C(J, p)\, \ell(G)^{1/q} ,$$

where the number

$$C(J,p) := \frac{1}{\pi} \limsup_{\varepsilon\to 0+} \left(\int_J |\,\mathrm{Im}\,(\psi,(H-\lambda-i\varepsilon)^{-1}\psi)|^p \, d\lambda \right)^{1/p}$$

is finite by assumption. The same is true for a countable I in view of Proposition 5.1.1, the inequality $C(J^{(n)},p) \le C(J,p)$, and the σ–additivity of $\ell(\cdot)$. To prove that the measure $(\psi, E_H(\cdot)\psi)$ is absolutely continuous on J, we have to check that $(\psi, E_H(M)\psi) = 0$ holds for any ℓ–zero set $M \subset J$. Since the Lebesgue measure is regular, for any positive integer j there is an open set $G_j \supset M$ such that $\ell(G_j) < j^{-1}$; hence

$$(\psi, E_H(M)\psi) \le \inf_j\, (\psi, E_H(G_j)\psi) \le C(J,p) \inf_j\, \ell(G_j)^{1/q} = 0$$

follows from the above inequality. ∎

Let us present an illustration of how the proved theorem can be applied to single–particle Schrödinger operators.

14.4.8 Theorem: Let $H := H_0 + V(Q)$ be the operator (14.1) on $L^2(\mathbb{R}^3)$ with a potential $V \in (L^1 \cap L^2)(\mathbb{R}^3)$ and $2m = 1$; then the condition

$$\|V\|_R := \left(\int_{\mathbb{R}^6} \frac{|V(x)|\,|V(y)|}{|x-y|^2}\, dx\, dy \right)^{1/2} < 4\pi \tag{14.22}$$

implies $\sigma_{sc}(H) = \emptyset$.

Proof: The operator H is self–adjoint by Theorem 14.1.2 and $D(H) = D(H_0)$. The set $L^1 \cap L^2 \subset L^{3/2}$ is contained in R (*cf.* the notes to Section 14.1), so the integral on the right side of (14.22) is finite. We need to estimate the resolvent of H. To this end we use the integral–operator representation of $|V(Q)|^{1/2}(H_0 - z)^{-1}V(Q)^{1/2}$ mentioned in Remark 4.3.4; with the help of Problem 7.18 we easily estimate its Hilbert–Schmidt norm,

$$\|\, |V(Q)|^{1/2}(H_0 - z)^{-1}V(Q)^{1/2}\|_2 \le \frac{1}{4\pi}\, \|V\|_R < 1$$

for any $z \in \mathbb{C}\backslash\mathbb{R}^+$. It follows then from Theorem 3.6.2 and Lemma 1.7.2 that the operator $I + |V(Q)|^{1/2}(H_0 - z)^{-1}V(Q)^{1/2}$ is invertible and its inverse has a bound independent of z, namely $1 - (4\pi)^{-1}\|V\|_R$. Next we use the modification of the second resolvent identity from Problem 1.63, which gives

$$(H - z)^{-1} = (H_0 - z)^{-1} - (H_0 - z)^{-1}V(Q)^{1/2}$$

$$\times\ \left[I + |V(Q)|^{1/2}(H_0 - z)^{-1}V(Q)^{1/2}\right]^{-1} |V(Q)|^{1/2}(H_0 - z)^{-1}.$$

Consider an arbitrary $\phi \in C_0^\infty(\mathbb{R}^3)$ and denote $f := |\phi|^{1/2}$. Since it has a compact support, the function f belongs to R, and we can check in the same way as above that the inequality $\|T_f(H_0-z)^{-1}T_f\| \le C_1 := (4\pi)^{-1}\|f\|_R$ is valid for the corresponding multiplication operator T_f and any $z \in \mathbb{C}\backslash\mathbb{R}^+$. It is a little more complicated to estimate the mixed products: we have

$$\|T_f(H_0-z)^{-1}V(Q)^{1/2}\|_2^2$$

$$\le \int_{\mathbb{R}^6} \frac{f(x)\,|V(y)|}{|x-y|^2}\,dx\,dy \;=\; \frac{1}{2}\int_{\mathbb{R}^6} \frac{f(x)|V(y)|+f(y)|V(x)|}{|x-y|^2}\,dx\,dy$$

$$\le \frac{1}{2}\int_{\mathbb{R}^6} \frac{f(x)f(y)+|V(x)|\,|V(y)|}{|x-y|^2}\,dx\,dy \;=\; \frac{1}{2}\,\|f\|_R^2 + \frac{1}{2}\,\|V\|_R^2$$

because the function $|V(\cdot)|-f(\cdot)$ is real-valued (*cf.* Problem 9b). If we denote the right side as C_2^2, we obtain $\|T_f(H_0-z)^{-1}V(Q)^{1/2}\| \le C_2$; in a similar way, we can find $\|\,|V(Q)|^{1/2}(H_0-z)^{-1}T_f\| \le C_2$. Putting the estimates together, we obtain

$$\|T_f(H_0-z)^{-1}T_f\| \;\le\; C_1 + C_2^2\left(1-(4\pi)^{-1}\|V\|_R\right)$$

for any $z \in \mathbb{C}\backslash\mathbb{R}^+$. Since $\phi = T_f\phi^{1/2}$, where $\phi^{1/2} := |\phi|^{1/2}\mathrm{sgn}\,\phi$, it follows that

$$(\phi,(H_0-z)^{-1}\phi) \;=\; (\phi^{1/2},T_f(H_0-z)^{-1}T_f\phi^{1/2}) \;\le\; C_1 + C_2^2\left(1-(4\pi)^{-1}\|V\|_R\right)\,\|\phi^{1/2}\|^2.$$

This means that the function $(\lambda,\varepsilon) \mapsto (\phi,(H_0-\lambda-i\varepsilon)^{-1}\phi)$ is uniformly bounded in $[\alpha,\beta]\times(0,1]$ for any $\alpha,\beta\in\mathbb{R}$; Theorem 14.4.7 then implies $E_H(\alpha,\beta)\phi \in \mathcal{H}_{ac}(H)$ for all $\alpha,\beta\in\mathbb{R}$, *i.e.*, $\phi\in\mathcal{H}_{ac}(H)$. In this way, we have checked the inclusion $C_0^\infty(\mathbb{R}^3)\subset\mathcal{H}_{ac}(H)$, and since the set on the left side is dense in $L^2(\mathbb{R}^3)$, we conclude that $\mathcal{H}_{ac}(H) = L^2(\mathbb{R}^3)$. ∎

14.4.9 Remarks: (a) In fact, we have proven a stronger result, namely that the spectrum of H is purely absolutely continuous. Recall that condition (14.22) ensures, due to Theorem 14.3.5, that H has no bound state, *i.e.*, $\sigma_p(H)=\emptyset$. Then the operators H, H_0 are unitarily equivalent by means of the wave operators $\Omega(H,H_0)$, which we shall introduce in the next chapter.

(b) The singularly continuous spectrum is difficult to investigate because it is highly unstable. While σ_{ess} and σ_{ac} are stable with respect to wide classes of perturbations (including unbounded operators — see Theorem 5.4.6, Theorems 15.2.6–15.2.10 and Remark 15.3.3), σ_{sc} can be changed by adding a one–dimensional projection (see the notes). Nevertheless, rather strong sufficient conditions for the absence of the singularly continuous spectrum have been proven; in Section 15.3 we shall mention two of them without proof.

Another question to which we should pay attention is whether $\sigma_{ess}(H)$ may contain embedded eigenvalues. Note first that an eigenvalue may occur at the boundary of the essential spectrum in a rather natural way, even if the potential is purely attractive (Problem 13). It is less trivial that an interior point of $\sigma_{ess}(H)$ may be an eigenvalue.

14.4.10 Example (Wigner — von Neumann): Consider the Schrödinger operator H on $L^2(\mathbb{R}^3)$ with the centrally symmetric potential

$$V : V(x) = -\frac{32 \sin r}{(1+f(r)^2)^2}\,[\,f(r)(1+f(r)^2)\cos r + (1-3f(r)^2)\sin^3 r\,],$$

where we set $f(r) := 2r - \sin(2r)$. The right side is a smooth bounded function of r; hence H is self-adjoint. Moreover, the potential behaves at large distances as $V(x) = \frac{8}{r}\sin(2r) + \mathcal{O}(r^{-2})$, so $\sigma_{ess}(H) = [0,\infty)$ due to Theorem 1. By a straightforward computation, we can check that H has a positive eigenvalue,

$$H\psi = \psi, \quad \psi(x) := \frac{r^{-1}\sin r}{1+f(r)^2}.$$

The existence of a positive eigenvalue in the example is due to the clever choice of the family of potential barriers which prevent the particle escaping to infinity. It is essential that the oscillating potential is slowly decaying, so the following result does not apply (see the notes).

14.4.11 Theorem (Kato): Let $H := H_0 + V(Q)$ be the operator (14.1) on $L^2(\mathbb{R}^n)$ with a potential, which satisfies the following conditions:

(i) $V \in L^p + L^\infty$, where p satisfies condition (14.2).

(ii) There is zero–measure set $M \subset \mathbb{R}^n$ such that $\mathbb{R}^n \setminus M$ is connected and V is bounded on each compact subset of $\mathbb{R}^n \setminus M$.

(iii) $\lim_{|x|\to\infty} |x|V(x) = 0$.

Then H has no positive eigenvalue.

We shall describe one more method which allows us to exclude the existence of positive eigenvalues. It is based on the scaling transformation which we have already employed in Theorem 14.3.9. Using the unitary operator $U_\alpha : (U_\alpha\psi)(x) = \alpha^{n/2}\psi(\alpha x)$, we define $V_\alpha(Q) := U_\alpha V(Q) U_\alpha^{-1}$ for a potential V; it is the operator of multiplication by the function $x \mapsto V(\alpha x)$.

14.4.12 Theorem (the virial theorem): Let $H := H_0 + V(Q)$ be the operator (14.1) on $L^2(\mathbb{R}^n)$ with a potential which satisfies the assumptions of Theorem 14.1.2 or Theorem 14.1.4. Suppose further that there is a measurable function $f : \mathbb{R}^n \to \mathbb{R}$ such that $D(T_f) \supset D(H_0)$, and

$$\lim_{\alpha\to 1} \frac{V_\alpha(Q) - V(Q)}{\alpha - 1}\,\psi = T_f\psi$$

holds for any $\psi \in D(H_0)$. Let λ be an eigenvalue of H corresponding to an eigenvector ψ; then

$$2(\psi, H_0\psi) = 2(\psi, (\lambda - V(Q))\psi) = (\psi, T_f\psi). \tag{14.23}$$

Proof: H is self–adjoint with $D(H) = D(H_0)$ by Theorem 14.1.2, and it is easy to check that $U_\alpha H_0 U_\alpha^{-1} = \alpha^{-2} H_0$. The last identity together with the assumption $H\psi = \lambda\psi$ implies $(H_0 + \alpha^2 V_\alpha) U_\alpha \psi = \lambda \alpha^2 U_\alpha \psi$. Since the operators under consideration are symmetric, we infer that

$$\begin{aligned}\lambda(\alpha^2 - 1)(U_\alpha\psi, \psi) &= ((H_0 + \alpha^2 V_\alpha(Q))U_\alpha\psi, \psi) - (U_\alpha\psi, (H_0 + V(Q))\psi) \\ &= (\alpha^2 - 1)(U_\alpha\psi, V(Q)\psi) - \alpha^2(U_\alpha\psi, (V(Q) - V_\alpha(Q))\psi),\end{aligned}$$

i.e.,

$$\lambda(\alpha+1)\,(U_\alpha\psi, \psi) = (\alpha+1)\,(U_\alpha\psi, V(Q)\psi) + \alpha^2\left(U_\alpha\psi, \frac{V_\alpha(Q) - V(Q)}{\alpha - 1}\,\psi\right).$$

By Problem 10.12, the dilations form a continuous group on $L^2(\mathbb{R}^n)$, so $U_\alpha\psi \to \psi$ as $\alpha \to 1$. The limit of the second term on the right side exists by assumption; together we obtain $2\lambda\|\psi\|^2 = 2(\psi, V(Q)\psi) + (\psi, T_f\psi)$, *i.e.*, identity (14.23). ∎

14.4.13 Remark: In view of Example 5.9.5b and Problem 10.12, the dilation group $\{U_\alpha : \alpha \in (0,\infty)\}$ is generated by the (closure of the) operator

$$D := \sum_{j=1}^{n} Q_j P_j - \frac{in}{2}.$$

On a formal level therefore, the definition of $V_\alpha(Q)$ gives $T_f = i[D, V(Q)]$, and the right side can be expressed by means of the canonical commutation relations as $i[D, V(Q)] = \sum_{j=1}^{n} Q_j(\nabla_j V)(Q)$. Another formal manipulation yields the identity $i[D, H_0] = -2H_0$, *i.e.*,

$$2(\psi, H_0\psi) = -i(\psi, [D, H]\psi) + i(\psi, [D, V(Q)]\psi) = \left(\psi, \sum_{j=1}^{n} Q_j(\nabla_j V)(Q)\psi\right).$$

As usual, the danger is hidden in the fact that the set of vectors, on which this reasoning makes sense, is too small. Nevertheless, the argument shows that

$$f : f(x) = \sum_{j=1}^{n} x_j(\nabla_j V)(x)$$

is a natural candidate for the role of the function f appearing in the theorem. To check that it satisfies the assumption, it is sufficient to find a function g which satisfies the condition $D(T_g) \subset D(H_0)$ and majorizes $(V_\alpha - V)(\alpha - 1)^{-1}$ in the vicinity of the point $\alpha = 1$, and to prove the existence of the limit by means of the dominated–convergence theorem. This is particularly easy if V is a *homogeneous function of order* $-\beta$, *i.e.*, $V(\alpha x) = \alpha^{-\beta} V(x)$ for all $x \in \mathbb{R}^n$; then $f = -\beta V$.

14.4.14 Corollary: Let the potential V satisfy the assumptions of Theorem 13 and at least one of the following conditions:

(i) V is *repulsive, i.e.*, $V(\alpha x) \le V(x)$ holds for all $x \in \mathbb{R}^n$ and $\alpha \ge 1$.

(ii) V is homogeneous of order $-\beta$ for some $\beta \in (0,2)$.

Then H has no positive eigenvalue.

Proof: If ψ is an eigenvector, $H\psi = \lambda\psi$, relation (14.23) is valid. In case (i) we readily check that $f \le 0$, so $(\psi, T_f\psi) \le 0$; and since $(\psi, H_0\psi) \ge 0$, it follows that $(\psi, H_0\psi) = 0$. However, the spectrum of H_0 is purely continuous; hence $\operatorname{Ker} H_0 = \{0\}$, *i.e.*, $\psi = 0$. In case (ii) we have $f = -\beta V$, and therefore

$$2\lambda\|\psi\|^2 \;=\; 2(\psi, H_0\psi) + 2(\psi, V(Q)\psi) \;=\; (2-\beta)(\psi, V\psi)$$

$$=\; \frac{\beta-2}{\beta}\,(\psi, T_f\psi) \;=\; \frac{2(\beta-2)}{\beta}\,(\psi, H_0\psi)\,,$$

where we have repeatedly used the relation (14.23). The coefficient on the right side is negative by assumption and $(\psi, H_0\psi) > 0$ holds for a nonzero ψ; hence the last identity cannot be valid if $\lambda \ge 0$. ∎

14.4.15 Example: Let H be the atomic Hamiltonian (14.4). Since Coulomb potentials are homogeneous functions of order -1, the corollary implies that H has no positive eigenvalues; the same is true for Hamiltonians of molecules, atomic and molecular ions, *etc.* Let us stress that in spite of this result, $\sigma_{ess}(H)$ may contain embedded eigenvalues, because its threshold is negative for multiparticle systems with attractive potentials, if only they have at least one bound subsystem (see the notes).

14.5 Constrained motion

Up to now we have considered systems whose configuration space was $\mathbb{R}^n$ with the dimension given by the number of degrees of freedom; the probability of finding the particle(s) in a particular region was determined by the interaction encoded in the potential. Now we want to mention the situation when the motion is constrained to a subset of the configuration space.

To motivate the question, recall that if the potential forms a sharply distinguished well and the particle energy is well below its bank, the eigenfunctions decay rapidly in the classically forbidden region — *cf.* Example 7.2.5. It is therefore useful in some situations to simplify the problem and to assume that certain regions are not accessible at all, *i.e.*, that the value of the potential is "infinite" there. This is an idealization, of course, because any potential is in fact a sum of elementary interactions (Coulomb, spin–orbit, *etc.*), so even finite–depth rectangular wells do not exist in nature. On the other hand, in some cases "sharp" constraints represent a reasonable approximation, which is, moreover, often easier to solve than a more realistic model. In statistical physics, restriction to a bounded region represents a basic ingredient of the theory — see Example 12.3.8.

The subject again includes numerous problems, and we limit ourselves to the presentation of several basic results concerning free motion within an open connected region $\Omega \subset \mathbb{R}^n$. First we have to ask about the self–adjointness of the corresponding Schrödinger operator. This question is now more complicated, because there are different candidates for the role of the free Hamiltonian associated, roughly speaking, with the boundary conditions we impose on $\partial G := \mathrm{bd}(G)$. One class is important for our purpose; we shall describe it and mention also another one.

Recall that $C_0^\infty(\Omega)$ is a dense subset in $L^2(\Omega)$, which consists of infinitely differentiable functions $\psi : \Omega \to \mathbb{C}$ whose supports are contained in Ω; moreover, since Ω is open by assumption, the distance of $\mathrm{supp}\,\psi$ to $\partial\Omega$ is positive. The form t defined by

$$t(\phi,\psi) := \sum_{j=1}^{n} \int_G \overline{\frac{\partial\phi(x)}{\partial x_j}}\, \frac{\partial\psi(x)}{\partial x_j}\, dx\,, \qquad \phi,\psi \in C_0^\infty(\Omega)\,, \tag{14.24}$$

is thus densely defined; it is obviously symmetric and below bounded. We also introduce the operator H_0 with the domain $C_0^\infty(\Omega)$ by $H_0\psi := -\Delta\psi$. Since the functions $\phi,\psi \in C_0^\infty(\Omega)$ and their partial derivatives belong to $L^2(\Omega)$, and the same is true for their extensions to a cube in $\mathbb{R}^n$ containing $\mathrm{supp}\,\phi \cup \mathrm{supp}\,\psi$ which are zero outside the supports, the Fubini theorem may be applied; integrating by parts, we obtain

$$(\phi, H_0\psi) = t(\phi,\psi) = (H_0\phi,\psi)$$

for any $\phi,\psi \in C_0^\infty(\Omega)$. This means that the operator H_0 is symmetric and the form t generated by it is closable due to Proposition 4.6.7. Then there is a unique self–adjoint operator associated with the closure $\bar{t}$, *i.e.*, the Friedrichs extension of H_0; we call this operator the **Dirichlet Laplacian** referring to the region Ω, and denote it as $-\Delta_D^\Omega$ or H_D^Ω. Similarly we introduce the **Neumann Laplacian** $-\Delta_N^\Omega$ or H_N^Ω as the self-adjoint operator associated with the form given by again (14.24), however, with the domain consisting of all $\psi \in L^2(\Omega)$ such that $\nabla_j\psi \in L^2(\Omega)\,,\ j = 1,\dots,n$, holds for components of the gradient, understood in the sense of distributions, *i.e.*, $\int_\Omega \phi(x)(\nabla\psi)(x)\,dx = -\int_\Omega (\nabla\phi)(x)\psi(x)\,dx$ for any $\phi \in C_0^\infty(\mathbb{R}^n)$. To illustrate where these names come from, consider a few simple cases.

14.5.1 Examples: (a) Let $n = 1$ and $\Omega := (a,b) \subset \mathbb{R}$; then H_D^Ω is by Problem 4.40b the operator corresponding to the differential expression $-d^2/dx^2$ with *Dirichlet* boundary conditions at the endpoints of the interval, *i.e.*, T_D in the notation of Problem 4.61. Similarly, H_N^Ω corresponds to *Neumann* boundary conditions being identical with T_N of Problem 4.61.

(b) Consider next a parallelepiped $\Omega := \mathsf{X}_{i=1}^n J_i$ in $\mathbb{R}^n$, where $J_i := (a_i,b_i)$. It is sufficient to discuss the case $n = 2$; the extension to higher dimensions is straightforward. We denote $H_i := H_D^{J_i}$ and define $H = \overline{\underline{H_1} + \underline{H_2}}$. By Theorem 5.7.2, this operator is self–adjoint. Moreover, Example 5.7.5 and Problem 4.61c show that the spectrum of H is purely discrete, consisting of the eigenvalues $\lambda_{jk} := \left(\frac{\pi j}{\ell_1}\right)^2 + \left(\frac{\pi k}{\ell_2}\right)^2$, $j,k = 1,2,\dots$, where $\ell_i := b_i - a_i$. The corresponding eigenfuctions are $\chi_{jk} := \chi_j^{(1)} \otimes \chi_k^{(2)}$, where $\chi_j^{(i)}$ are the (trigonometric) eigenfunctions of H_i. By Example 5.3.11a, the operator H is *e.s.a.* on $\dot{D} := \{\chi_{jk}\}_{lin}$.

The set $C^\infty(\Omega)$ consists of functions ψ, which are infinitely differentiable in Ω and have a finite limit at each point of $\partial\Omega$; the latter defines the function $\partial\Omega \mapsto \mathbb{C}$, which we denote as $\psi \restriction \partial\Omega$. Then we can introduce the set

$$D := \{\psi \in C^\infty(\Omega) :\, \psi \restriction \partial\Omega = 0\}; \tag{14.25}$$

in the same way as above, we can check that the operator $H^{(D)} :\, H^{(D)}\psi := -\Delta\psi$ with the domain $D(H^{(D)}) := D$ is symmetric, and as a symmetric extension of $H \restriction \dot{D}$ it is *e.s.a.* Its closure coincides with the corresponding Dirichlet Laplacian, *i.e.*, *D is a core for H_D^Ω*; since $H^{(D)}$ extends H_0, by Theorem 4.6.11 it is sufficient to check that D is contained in $Q(H_D^\Omega) := D(\bar{t})$ (Problem 14a). In a similar way one can treat $H^{(N)}$ and to show, in particular, that its core is $C^\infty(\Omega)$ (Problem 14d).

(c) The explicit form of the operators also allows us to say something about the distribution of the eigenvalues. Let us denote

$$N_\lambda^D(\Omega) := \dim E_H(-\infty,\lambda)$$

for $H = H_D^\Omega$. In the one–dimensional case, the eigenvalues of H_D^Ω are $(\pi j/\ell)^2$, $j = 1,2,\dots$, where $\ell := b-a$; hence $N_\lambda^D(\Omega) = N$ holds provided $N < \frac{\ell\sqrt{\lambda}}{\pi} \le N+1$. This means, in particular, that $\lim_{\lambda\to\infty} N_\lambda^D(\Omega)\,\lambda^{-1/2} = \ell/\pi$. In view of Problem 4.61c we have at the same time $\lim_{\lambda\to\infty} N_\lambda^N(\Omega)\,\lambda^{-1/2} = \ell/\pi$, and similar conclusions can be made for $n \ge 2$ (Problems 14b,d).

In this way, we have in this particular case obtained an expression for the Dirichlet Laplacian and its Neumann counterpart through explicitly known operators with a purely discrete spectra; the same can be done for some other simple bounded sets Ω (Problems 14c,d). This may not be true for a general Ω, but the essential conclusions of the example remain valid provided the boundary is not too wild.

14.5.2 Theorem: Let Ω be an open bounded and connected set with a smooth boundary or a polyhedron in R^n.

(a) The subset $D_\infty(\Omega) := \{\psi \in D :\, -\Delta\psi \in L^2(\Omega)\}$ of (14.25) is a core for the Dirichlet Laplacian H_D^Ω.

(b) In addition, if Ω is bounded, then $D_\infty(\Omega) = D$, the spectrum of H_D^Ω is purely discrete and

$$\lim_{\lambda\to\infty} \frac{N_\lambda^D(\Omega)}{\lambda^{n/2}} = \frac{v_n \operatorname{vol}(\Omega)}{(2\pi)^n}\,, \tag{14.26}$$

where $v_n := 2\pi^{n/2}\left(n\Gamma\left(\frac{n}{2}\right)\right)^{-1}$ and $\operatorname{vol}(\Omega)$ is the Lebesgue measure of Ω.

We will not give a full proof of the theorem referring to the notes; we limit ourselves with mentioning the main ingredients needed to obtain (14.26). They are two. The first is the eigenvalue behavior of Dirichlet and Neumann Laplacians on rectangular "blocks" described in Example 1c, the other is a trick which allows us to get estimates from a decomposition of Ω into such blocks.

14.5.3 Proposition: $H_D^\Omega \ge H_N^\Omega \ge 0$ holds for any open connected Ω. Moreover, suppose that Ω decomposes into a pair of subsets with a common boundary, *i.e.*, there are disjoint open Ω_1, Ω_2 such that $\overline{\Omega_1 \cup \Omega_2} = \overline{\Omega}$ and the common boundary $\Omega\backslash(\Omega_1 \cup \Omega_2)$ has zero measure, then the following inequalities hold,

$$0 \le H_D^\Omega \le H_D^{\Omega_1} \oplus H_D^{\Omega_2}\,, \qquad 0 \le H_N^{\Omega_1} \oplus H_N^{\Omega_2} \le H_N^\Omega\,. \tag{14.27}$$

Proof: Both claims follow easily from the minimax principle and comparison of the form domains of the operators involved (see Remark 14.2.2c); notice that the common boundary has zero measure by assumption so the L^2 integrability of a function in Ω and $\Omega_1 \cup \Omega_2$ are equivalent. ∎

The second part is a useful tool in many spectral estimates. One usually refers to it as to **Dirichlet-Neumann bracketing**; it holds also for other operators. To understand it one has to realize that adding a Dirichlet boundary, we restrict the form domain of the operator by imposing a new condition. On the other hand, the domain is extended when a Neumann boundary is added because in this case we relax a condition requiring no longer the smoothness at this boundary.

Speaking about asymptotic properties of the discrete spectrum we have assumed that the region Ω is bounded. For an unbounded region the spectrum usually has also a continuous component, although this may not be true in general (Problem 15). Spectral properties of H_D^Ω depend on the geometry of Ω; we quote some results in the notes. The cases when Ω is unbounded in the form of a tube or a layer are of a particular interest; we will discuss them in detail in Chapter 16.

14.6 Point and contact interactions

It is sometimes useful to assume that the interaction is strongly localized in the sense that the particle behaves as free outside some small region of the configuration space. If we shrink this region to a zero measure set, then the considerations of the previous sections yield trivial results only, because such a potential belongs to the same L^p class as $V = 0$. In this section we are going to show that one can nevertheless construct a wide class of generalized Schrödinger operators, which correspond to such situations, using the theory of self-adjoint extensions.

Let us start with the simplest example. Consider the free-particle Hamiltonian on line, $H_0 := P^2$ of Example 7.2.1, and choose a point $y \in \mathbb{R}$. Let A_y be the restriction of H_0 to the subspace $\{\psi \in AC^2(\mathbb{R}) : \psi(y) = 0\}$. In the same way as in Section 4.8, we can check that the adjoint operator is again given by the differential expression $-d^2/dx^2$ and its domain is $\{\psi : \psi_\pm \in AC^2(J_y^\pm),\ \psi'' \in L^2(\mathbb{R})\}$, where we have denoted $J_y^+ := (y,\infty)$, $J_y^- := (-\infty,y)$, and $\psi_\pm := \psi \upharpoonright J_y^\pm$. The equation $(A_y^* - z)\psi = 0$ then has for any $z \in \mathbb{C}\backslash\mathbb{R}$ a unique solution up to a multiplicative constant, namely $\psi(x) := e^{i\sqrt{z}|x-y|}$ with the cut in the square root taken along the positive real axis, *i.e.*, the deficiency indices are $(1,1)$. The corresponding self–adjoint extensions and their properties can be found easily (Problems 16a–c).

14.6.1 Proposition: For any α, $-\infty < \alpha \le \infty$, there is one self–adjoint extension $H_{\alpha,y}$ of A_y which is the restriction of A_y^* to

$$D(H_{\alpha,y}) := \{\psi \in D(A_y^*) : \psi'(y+) - \psi'(y-) = \alpha\psi(y)\}. \tag{14.28}$$

In particular, $\alpha = 0$ corresponds to the free Hamiltonian H_0, while $\alpha = \infty$ refers to the Dirichlet boundary condition $\psi(y) = 0$; in this case the operator decomposes into an orthogonal sum, $H_{\infty,y} = H_D^{(-)} \oplus H_D^{(+)}$, where $H_D^{(\pm)}$ are the Dirichlet Laplacians on $J_y^\pm$.

We shall call $H_{\alpha,y}$ the **δ–interaction** Hamiltonian supported by the point y and with the interaction strength α. To find a correspondence between it and Schrödinger operators discussed in the previous sections, consider a one–parameter family $H_\varepsilon(V,y)$ defined for a given V and $\varepsilon \in (0,\infty)$ by

$$H_\varepsilon(V,y) := P^2 + V_\varepsilon(Q), \quad V_\varepsilon(x) := \frac{1}{\varepsilon} V\left(\frac{x-y}{\varepsilon}\right). \tag{14.29}$$

The family of the scaled potentials V_ε "shrinks" as $\varepsilon \to 0$ preserving the integral; in the sense of distributions it converges to a multiple of $\delta(\cdot - y)$. It is therefore natural to ask whether $H_{\alpha,y}$ is in a sense a limit of the operators $H_\varepsilon(V,y)$, so that it can give a rigorous meaning to the formal Schrödinger operator $-\frac{d^2}{dx^2} + \alpha\delta(x-y)$. However, since we have not introduced convergence of families of unbounded operators, it is necessary to replace them first by suitable bounded functions.

14.6.2 Theorem: Let $V \in C_0^\infty(\mathbb{R})$. A number $z \in \rho(H_{\alpha,y})$ belongs to $\rho(H_\varepsilon(V,y))$ for all ε small enough and the corresponding resolvent converges in the operator–norm sense,

$$\operatorname*{u-lim}_{\varepsilon \to 0+} (H_\varepsilon(V,y) - z)^{-1} = (H_{\alpha,y} - z)^{-1}, \tag{14.30}$$

where the interaction strength is $\alpha := \int_{\mathbb{R}} V(x)\, dx$.

Proof: The resolvent of $H_{\alpha,y}$ can be expressed by Krein's formula (Problem 16b). On the other hand, the resolvent of $H_\varepsilon(V,y)$ is given by Problem 1.63, *cf.* the proof of Theorem 14.4.8; with the notation introduced in Section 14.3 we can write it as

$$(H_\varepsilon(V,y) - k^2)^{-1} = (H_0 - k^2)^{-1} - (H_0 - k^2)^{-1} V_\varepsilon(Q)^{1/2}$$

$$\times\, [I + |V_\varepsilon(Q)|^{1/2}(H_0 - k^2)^{-1} V_\varepsilon(Q)^{1/2}]^{-1} |V_\varepsilon(Q)|^{1/2}(H_0 - k^2)^{-1},$$

where $k := \sqrt{z}$. It is thus sufficient to check that the second term on the right side, which we denote as $D_{k,\varepsilon}$, converges in the operator norm to the rank–one operator

$\frac{2\alpha k}{2k+i\alpha}\left(\overline{G_k(\cdot,y)},\cdot\right) G_k(\cdot,y)$. The operator in question is a composition of integral operators with known kernels; changing the integration variable x' to $(x'-y)/\varepsilon$, we can rewrite it as $D_{k,\varepsilon} = B_{k,\varepsilon}(I+C_{k,\varepsilon})^{-1}\tilde{B}_{k,\varepsilon}$, where the involved operators are determined by their kernels,

$$\begin{aligned} B_{k,\varepsilon}(x,x') &= G_k(x,y+\varepsilon x')\,V(x')^{1/2}\,, \quad \tilde{B}_{k,\varepsilon}(x,x') = |V(x)|^{1/2}G_k(y+\varepsilon x,x')\,,\\ C_{k,\varepsilon}(x,x') &= |V(x)|^{1/2}G_k(y+\varepsilon x,y+\varepsilon x')\,V(x')^{1/2} \end{aligned}$$

(Problem 16d). We want to show that they converge to the integral operators with the kernels

$$\begin{aligned} B_k(x,x') &:= G_k(x,y)\,V(x')^{1/2}\,, \quad \tilde{B}_k(x,x') := |V(x)|^{1/2}G_k(y,x')\,,\\ C_k(x,x') &:= |V(x)|^{1/2}G_k(y,y)\,V(x')^{1/2} \end{aligned}$$

in the Hilbert–Schmidt norm, and thus *a fortiori*, with respect to the operator norm. Since the function $\Omega_k(\cdot,\cdot)$ is continuous, the kernels converge pointwise; then the assertion will follow from Theorems 3.6.5 and A.3.7 if we are able to majorize the kernel difference by a square integrable function independent of ε. By assumption, there is N such that $\operatorname{supp} V \subset (-N,N)$; using the explicit form of Ω_k we readily check that

$$|B_{k,\varepsilon}(x,x') - B_k(x,x')|^2 \le \|V\|_\infty^2\, e^{-2(\operatorname{Im} k)(|x|-N)}$$

holds for any $x,x' \in \mathbb{R}$ and $\varepsilon > 0$. The convergence for the other two operators can be proven in a similar way.

Hence $D_{k,\varepsilon}$ converges to the operator $D_k := B_k(I+C_k)^{-1}\tilde{B}_k$ provided -1 is not an eigenvalue of C_k. Since we know the last named operator explicitly, we can express $(I+C_k)^{-1}$; using Problem 1.63 once more we obtain

$$(I+C_k)^{-1} = I - \frac{i}{2k+i(V^{1/2},|V|^{1,2})}\,|V|^{1/2}(V^{1/2},\cdot)\,.$$

This shows that the condition is satisfied if the denominator is nonzero, *i.e.*, if $\alpha := (V^{1/2},|V|^{1,2}) \neq 2ik$. Finally, the above identity together with the explicit form of B_k, $\tilde{B}_k$ yields the sought result, $D_k = \frac{2\alpha k}{2k+i\alpha}\left(\overline{G_k(\cdot,y)},\cdot\right) G_k(\cdot,y)$. ∎

14.6.3 Remarks: (a) Since the analytical structure of the resolvent determines the spectrum, the proved result suggests that for ε small enough the approximating operator has just one eigenvalue if $\alpha := \int_{\mathbb{R}} V(x)\,dx < 0$, and that it approaches $-\alpha^2/4$ as $\varepsilon \to 0$; this can actually be proven (see the notes). To grasp the meaning of this result, note that $H_\varepsilon(V,0)$ is by Problem 16d unitarily equivalent to the ε^{-2} multiple of the operator $P^2 + \varepsilon V(Q)$, which has for small ε a single eigenvalue whose asymptotic behavior is $-\frac{1}{4}(\varepsilon\alpha)^2 + \mathcal{O}(\varepsilon^3)$ as mentioned in the notes to Section 14.3. On the other hand, there is no eigenvalue for $\alpha > 0$, while in the case $\alpha = 0$ the operator $H_\varepsilon(V,y)$ has a negative eigenvalue for small ε but it disappears in the essential spectrum in the limit.

(b) The δ–interactions constructed above do not represent the most general class of point interaction for a particle on line. If we restrict H_0 to the subspace of $AC^2(\mathbb{R})$, which consists of functions ψ such that $\psi(0) = \psi'(0) = 0$, then the deficiency indices are $(2,2)$ due to Example 4.8.5 and Problem 4.56, and there is a four–parameter family of self–adjoint extensions. It includes, of course, δ–interactions. Another important one–parameter subclass are the so–called **δ'–interactions** (supported by a point y and with an interaction strength β), which are specified by the boundary conditions

$$\psi'(y+) = \psi'(y-) =: \psi'(y)\,, \qquad \psi(y+) - \psi(y-) = \beta\psi'(y)\,; \tag{14.31}$$

the corresponding self–adjoint operator will be denoted as $H'_{\beta,y}$ (*cf.* Problem 17). In contrast to the preceding case, δ'–interactions are *not* approximated by $P^2{+}V_\varepsilon$ but rather by several more complicated operator families which we will describe briefly in the notes.

The proved result thus provides a rigorous background for the heuristic idea of the Schrödinger operator with a δ–function potential. Recall that such Hamiltonians are typically introduced in order to the describe the behavior of particles which are very slow, and therefore the region to which their wave functions are localized in the configuration space is large in view of the uncertainty relations. If such an "extended" particle interacts with a potential supported by a much smaller set, it "sees" it as a point interaction with an averaged interaction strength — this is the physical content of Theorem 14.6.2.

It is natural to ask whether similar point–interaction Hamiltonians can be constructed in higher dimensions. Since the method used above was based on constructing self–adjoint extensions to a symmetric operator, obtained by restricting the free Hamiltonian to functions, which are zero together with their derivatives at the points, which are supposed to support the interaction, it is necessary that such a restriction is not *e.s.a.*; this shows that the construction does not work if $n \ge 4$ (Problem 18a). On the other hand, in the cases $n = 2, 3$ the deficiency indices of a one–point restriction are $(1,1)$ (Problem 18b), so there exist one–parameter families of point interaction Hamiltonians.

We shall describe briefly the three–dimensional situation; the case $n = 2$ which can be treated analogously is left to the reader (Problem 19). Without loss of generality, we may assume that the point interaction is supported by the origin of the coordinates. The starting point is then the partial–wave decomposition (11.20) of the free Hamiltonian H_0; for simplicity, we put again $2m = 1$. By Problem 18b, the operators h_ℓ, $\ell \ge 1$, are *e.s.a.* on $C_0^\infty(\mathbb{R} \setminus \{0\})$ while the deficiency indices of h_0 are $(1,1)$. Self–adjoint extensions $h_{0,\alpha} := T(4\pi\alpha)$ of this operator are known from Examples 4.9.6 and 7.2.2; they are specified by the boundary condition

$$f'(0+) - 4\pi\alpha f(0+) = 0\,, \tag{14.32}$$

where $-\infty < \alpha \le \infty$. In the case $\alpha = \infty$, (14.32) is replaced by the Dirichlet condition $f(0+) = 0$, which corresponds to the free Hamiltonian — *cf.* Problem 7b.

If we look at (14.32) from the point of the original wavefunctions in $L^2(\mathbb{R}^3)$ it can be understood as a relation between generalized boundary values, $f(0+)$ being the coefficient at the pole – *cf.* (11.16). This is useful in situations when the partial–wave decomposition cannot be applied; in Section 16.2 below we shall see an example (in two dimensions, where the pole singularity is replaced by a logarithmic one).

14.6.4 Theorem: The operator $H_0 \upharpoonright C_0^\infty(\mathbb{R}^3\backslash\{y\})$ has for any $y \in \mathbb{R}^3$ a one–parameter family of self–adjoint extensions given by

$$H_{\alpha,y} := e^{-iy\cdot P} U^{-1} \left(\overline{\overline{h_{0,\alpha} \otimes I_0} \,\oplus\, \bigoplus_{\ell=1}^{\infty} h_\ell \otimes I_\ell} \right) U\, e^{iy\cdot P} ,$$

where U is the operator (11.16) and $y \cdot P := \sum_{j=1}^3 y_j P_j$. Any self–adjoint extension of $H_0 \upharpoonright C_0^\infty(\mathbb{R}^3\backslash\{y\})$ coincides with $H_{\alpha,y}$ for some $\alpha \in (-\infty,\infty]$. The essential spectrum is the same as in the free case, $\sigma_{ess}(H_{\alpha,y}) = \sigma_{ac}(H_{\alpha,y}) = \mathbb{R}^+$ while $\sigma_{sc}(H_{\alpha,y}) = \emptyset$. If $\alpha < 0$, there is one simple eigenvalue, namely $-(4\pi\alpha)^2$, corresponding to the normalized eigenfunction

$$\psi_\alpha :\ \psi_\alpha(x) = \sqrt{-\alpha}\, \frac{e^{4\pi\alpha|x-y|}}{|x-y|} ;$$

otherwise the point spectrum of $H_{\alpha,y}$ is empty.

Proof: The above argument together with (11.20) proves the first part for $y = 0$; the general result follows from the fact that due to (10.1) and Example 10.2.1b, $e^{iy\cdot P}$ is the operator of (active) shift on y. Theorem 4.7.14 implies the relation $\sigma_{ess}(H_{\alpha,y}) = \mathbb{R}^+$, and by Proposition 4.7.12, $H_{\alpha,y}$ has at most one eigenvalue; it is found by solving the radial Schrödinger equation with the boundary condition (14.32). Finally, $\sigma_{sc}(H_{\alpha,y}) = \emptyset$ follows from the explicit form of the resolvent in combination with Theorem 14.4.7 (Problem 20b). ∎

The operator $H_{\alpha,y}$ is again called the Hamiltonian with **δ–interaction** of the interaction strength α supported by the point y. In a similar way, one can introduce point interaction supported by a finite subset of $\mathbb{R}^3$, or even an infinite subset provided it has no accumulation points; these Hamiltonians are again approximated by suitably chosen families of Schrödinger operators (see the notes). Hence they have the same physical meaning: they model an interaction of slow particles with potentials supported by regions much smaller than the wave function spread. This model simplification brings a significant computational advantage. We have seen, for instance, that the spectral problem for a one–center point interaction is explicitly solvable: the singularities of the resolvent kernels in Problems 16b and 20b are given by simple linear equations. If the point interaction is supported by finitely many points, the situation is more complicated, but we still have to solve a single transcendental equation instead of a differential one.

Singular interactions of the type constructed above can be supported not only by discrete point sets. For instance, we can easily define Hamiltonians with an interaction concentrated at a line or plane in $\mathbb{R}^3$; the difference is that now the restriction of

H_0 to functions vanishing with their derivatives at the interaction support leads to a symmetric operator with infinite deficiency indices, and the operators in question represent only a "small part" of the vast family of self-adjoint extensions. Similar singular interactions can be constructed, however, even for manifolds which exhibit no symmetry that would allow us to use a separation of variables. One usually speaks in these cases about Hamiltonians with **contact interactions** to emphasize the fact that the particle is influenced only if it "hits" a set of lower dimensionality which supports the interaction. They can be again regarded as a low-energy limit to a more realistic quantum mechanical description in situations, where the interaction part of the Hamiltonian is negligible except in the vicinity of a curve, surface, or other manifold in the configuration space. We will discuss a class of such systems in Section 17.5 below.

Notes to Chapter 14

Section 14.1 Theorem 14.1.2 and its extension to N-body systems with Coulomb interactions contained in Theorem 4 were proved for the first time in [Ka 3]; see also ⟦Ka⟧, Sec.5.5; ⟦RS 2⟧, Sec.X.2. An alternative approach to Schrödinger operators is based on quadratic forms: we may define the operator in question as the form sum of H_0 and $V(Q)$ (*cf.* Remark 4.6.12). The counterpart of the perturbative argument of Theorem 14.1.2 is then provided by the KLMN-theorem; the advantage is that this allows us to define Schrödinger operators for potentials with stronger local singularities — see, *e.g.*, ⟦Si 1⟧.

To illustrate this claim, recall that by Example 14.1.3, the result of Theorem 14.1.2 applies to potentials with local singularities $|x-x_0|^{-\alpha}$, where $\alpha < 3/2$. Form sums allow us to define Schrödinger operators for potentials $V \in R+L^\infty$, where R denotes the set of potentials which satisfy the condition

$$\int_{\mathbb{R}^6} \frac{|V(x)|\,|V(y)|}{|x-y|^2}\,dx\,dy < \infty;$$

these are called the *Rollnik class.* Since $R \supset L^{3/2}(\mathbb{R}^3)$ — *cf.*, for example, [Ka 2], Sec.6 — we see that this approach admits singularities with $\alpha < 2$. The form-sum method also works for potentials of the so-called *Kato classes* — *cf.* ⟦CFKS⟧, Sec.I.2. Moreover, it is also applicable in some cases when the form referring to the potential is represented by no operator at all; an example is given by point interactions which we discuss in Section 14.6.

As we have mentioned, the proof of Theorem 14.1.6 requires to check that a sufficiently large negative λ is a regular value of $\overline{H}$; this will be ensured, *e.g.*, if $(H_0+V+\lambda)^*\psi = 0$ implies $\psi = 0$. If $V_- \neq 0$, we also have to use the result of Problem 4.22, due to which H_0+V_- is below bounded. Technically the argument is based on the so-called *Kato inequality* — see, *e.g.*, ⟦RS 2⟧, Sec.X.4; ⟦CFKS⟧, Sec.1.3.

The self-adjointness problem for Schrödinger operators with a magnetic field was discussed for the first time in [IK 1]; an overview of the known results can be found in ⟦We⟧, Sec.10.4; ⟦RS 2⟧, Secs.X.2,4, and also ⟦CFKS⟧, Chap.6. For instance, one need not assume the boundedness of A_k and $\nabla\cdot A$ in Theorem 14.1.8 (then the result applies to the important case of a homogeneous magnetic field) or the smoothness of A_k.

Section 14.2 The history of the minimax principle is described in ⟦ RS 4 ⟧, notes to Sec.XIII.1. The finite–dimensional case is discussed in detail in ⟦ Ka ⟧, Sec.I.6. Various modifications of Theorem 14.2.1 can be found in ⟦ DS 2 ⟧, Sec.XIII.9, or ⟦ LiL ⟧, Chap.12; a generalization to certain sums of eigenvalues is given in ⟦ Thi 3 ⟧, Sec.3.5.

An extensive and detailed exposition of perturbation theory can be found in the classical monograph ⟦ Ka ⟧; the reader may also consult ⟦ RS 4 ⟧, Chap.XII; ⟦ Thi 3 ⟧, Sec.3.5; ⟦ We ⟧, Chap.9, *etc.* The core of analytic perturbation theory is the treatment of operator families $\mathcal{T} := \{T(g) \in \mathcal{L}_c(\mathcal{H}) : g \in G\}$, where Ω is a domain in the complex plane, with suitable analytic properties. $\mathcal{T}$ is called the *Kato-type analytic family* if $\rho(T(g)) \neq \emptyset$, and for any $g_0 \in G$ there is a $z_0 \in \rho(T(g_0))$ which belongs to $\rho(T(g))$ for g of some neighborhood U_{g_0} of g_0, and $(T(\cdot)-z_0)^{-1}$ is an analytic operator–valued function in U_{g_0}. The assertion of Theorem 14.2.7 extends to analytic families of this type. Proof of Theorem 14.2.6 can be found in ⟦ Ka ⟧, Thm.III.6.17; see also ⟦ RS 4 ⟧, Thm.XII.5.

One of the classes which satisfies the above requirements consists of the families $\mathcal{T}$, such that the operators $T(g)$ have a common domain D independent of g and the vector–valued function $g \mapsto T(g)\psi$ is analytic in G for any $\psi \in D$; such $\mathcal{T}$ is called the *analytic family of type (A)*. An example is provided by operators $T(g) := H_0 + gV$, where V is H_0–bounded, considered in Theorem 14.2.7. Another often used class of analytic perturbations is the form counterpart of the type (A) families.

A result analogous to Theorem 14.2.7 can also be proved if the unperturbed eigenvalue is *degenerate, i.e.*, its multiplicity is greater than one. In general, however, the perturbation removes the degeneracy — the eigenvalue splits and the dependence on the perturbation parameter is more complicated; a simple example of such a situation will be given in Theorem 15.4.4 below. The algebraic derivation of the Rayleigh–Schrödinger series coefficients is based on expanding the resolvent (or its projection onto a chosen subspace) in a Laurent series; an exhaustive discussion is performed in ⟦ Ka ⟧, Chap.II. Finally, let us mention that in many cases of practical interest the hypotheses of analytic perturbation theory are not satisfied; nevertheless, the perturbation series can be written down and summed in some generalized sense — see ⟦ Ka ⟧, Chap.VIII; ⟦ RS 4 ⟧, Secs.XII.2,3.

Section 14.3 If the operator (14.1) is understood in the form–sum sense, Theorem 14.3.1 may be extended to potentials belonging to the family $R + L^\infty_\varepsilon$ corresponding to the Rollnick class, which is defined as the set of (equivalence classes of) functions with the property that for any $\varepsilon > 0$ there is a decomposition $f = f_R + f_\infty$ and the L^∞–component satisfies the condition $\|f_\infty\|_\infty < \varepsilon$. In addition to (14.12), there are other estimates. For instance, the *Calogero bound*

$$n_\ell(V) \leq \frac{2}{\pi} \int_0^\infty |v(r)|^{1/2} dr$$

holds under the assumptions of Theorem 14.3.1 provided the function v is nonpositive and nondecreasing. These estimates were derived in [Ba 5], [Cal 1], [GMGT 1]; see also ⟦ RS 4 ⟧, Sec.XIII.3; the condition for the absence of bound states of Remark 14.3.2c appeared for the first time in [JoP 1].

The Birman–Schwinger principle was formulated independently in the papers [Bir 1] and [Schw 1]. Lemma 14.3.3 can in the sense of Remark 14.3.4 be extended to other potentials, for instance, to the class $R + L^\infty_\varepsilon$ in the case $n = 3$ — see ⟦ Si 1 ⟧, Chap.3. To prove Theorem 14.3.5 in the general case, we have to approximate V by potentials from $C_0^\infty(\mathbb{R}^3)$ — *cf.* ⟦ RS 4 ⟧, Sec.XIII.3. The relation (14.15) is called the *Ghirardi–Rimini bound.* The

Birman–Schwinger principle can also be used to derive the Bargmann bound — see [Si 5]; ⟦Thi 3⟧, Sec.3.5. Various sufficient conditions for existence of bound states can be found in [ChD 1], [CMF 1], [BChK 1].

The conclusion of Theorem 14.3.6 is valid for a considerably wider class of potentials — see, *e.g.*, [Si 6], [BGS 1]. What is important is that the potential need not be purely attractive: apart from some regularity conditions, it is sufficient that $\int V(x)\,dx \le 0$. It is clear from the proof that the existence of the bound state follows from the fact that the resolvent in the p–representation, *i.e.*, the function $p \mapsto (p^2+\kappa^2)^{-1}$ ceases to be integrable in the limit $\kappa \to 0$ if $n = 1,2$. A more detailed analysis of this singularity yields the weak–coupling behavior of the lowest eigenvalue,

$$\mu_1(g) = -\frac{g^2}{4}\left(\int_{\mathbb{R}} V(x)\,dx\right)^2 + \mathcal{O}(g^3) \quad \text{and} \quad \mu_1(g) \approx -\exp\left(\left(\frac{g}{4\pi}\int_{\mathbb{R}^2} V(x)\,dx\right)^{-1}\right)$$

for $n = 1$ and $n = 2$, respectively, provided the integrals are negative. A similar idea can be used to modify the Birman–Schwinger bound to obtain bounds on the number of bound states in these cases — *cf.* [Set 1], [Kl 1], [New 1].

The Lieb–Thirring inequalities were proved in [LiT 1], *cf.* also [GGM 1], [Lie 3]; for newer results, values of the constants given in Remark 14.3.8b, and a bibliography see [HLT 1], [HLW 1], and [LW 2]. The Cwikel–Lieb–Rozenblioum bound given in Remark 8a was derived in different ways in the papers [Ros 1], [Lie 2] and [Cwi 1]; see also ⟦Si 4⟧, Sec.III.9; ⟦RS 4⟧, Sec.XIII.3. As we have said, the distinguishing property of these results is their correct behavior for large values of the coupling constant. For the number of bound states, *e.g.*, the semiclassical relation

$$\lim_{g\to\infty} g^{-n/2}N(gV) = c_n^{\mathrm{cl}} \int_{\mathbb{R}^n} |V_-(x)|^{n/2}\,dx$$

is valid — see [Lie 3] and ⟦RS 4⟧, Thm.XIII.80, and compare this to Theorem 14.5.2b below. A more detailed analysis of the number of bound states for potentials having the critical decay at infinity can be found in [St 1].

The discrete spectrum of hydrogen–type atoms is probably discussed in every quantum mechanics textbook; a detailed treatment of Schrödinger operators describing atoms with one or two electrons can be found in ⟦Thi 3⟧, Secs.4.1–4.3. Theorem 14.3.11 was proved in [Ka 4], where the helium atom with a finite nucleus mass M was also considered. In that case the center–of–mass Hamiltonian contains the *Hughes–Eckart term* $-M^{-1}P_1\cdot P_2$ (*cf.* Problem 11.13a); T. Kato demonstrated that the number of bound states is not less than $\frac{1}{6}n_0(n_0+1)(2n_0+1)$, where n_0 is the greatest integer satisfying the inequality $n_0^2 \le (1-Z^{-1})^2M/m$. It follows that the real helium atom has at least 25,585 bound states; a more sophisticated estimate due to [Ži 1] shows that the discrete spectrum of the neutral–atom Hamiltonian is again infinite. Let us remark that a more realistic treatment of atomic and molecular Hamiltonians requires electron spins to be taken into account even if the interaction is purely Coulomb; the reason is that the electrons obey the Pauli principle, and therefore we are in fact interested in the spectrum of the appropriate Schrödinger operator restricted to the subspace with a prescribed permutation symmetry. A review of this problem is given in ⟦AŽŠ⟧; see also ⟦RS 4⟧, notes to Sec.XIII.3.

The mentioned conclusion about the infiniteness of $\sigma_d(H)$ for atomic and molecular systems may not be valid for negative ions. The simplest example is the H^- ion which was shown in [Hil 1] to have just one bound state; this illustrates at the same time that Theorem 14.3.11 is not valid for $\gamma = 2$. More generally, there are upper estimates on the maximal number $N(Z)$ of electrons which a nucleus of charge Ze can bind; in particular, it has been shown that large atoms are asymptotically neutral, *i.e.*, $N(Z)/Z \to 1$ as $Z \to \infty$ — see, *e.g.*, [Rus 1], [Lie 4], [LSST 1], [ELS 1]; ⟦CFKS⟧, Chap.3. Similar properties have been demonstrated for diatomic molecules — [Rus 2], [Sol 1]; for other properties of ions see, *e.g.*, [Rus 3], [SSS 1].

Theorem 14.3.12 can be proven for a substantially wider class of Schrödinger operators including Hamiltonians of N particles interacting via two–particle potentials $V_{jk} \in R+L^\infty_\varepsilon$, Schrödinger operators in $L^2(\mathbb{R}^n)$ with potentials $V \in L^2_{loc}(\mathbb{R}^n)$ satisfying the condition $\lim_{|x|\to\infty} V(x) = \infty$, *etc.* — see ⟦RS 4⟧, Sec.XIII.12; ⟦We⟧, Sec.10.5.

One of the important questions for Schrödinger operators is finding lower bounds for the ground–state eigenvalue. In some cases such estimates are of fundamental physical importance. Consider, for instance, a system of atoms consisting of N electrons and K nuclei described by the appropriate Hamiltonian H which involves their Coulomb interactions, and denote $E_{N,K} := \inf\sigma(H)$. The problem of ***stability of matter*** means to determine whether the binding energy of such a system is an extensive quantity, *i.e.*, whether $E_{N,K} \ge -C(N+K)$ holds for some constant C. This inequality was proved by A. Lenard and F.J. Dyson; using the inequalities (14.17) E.H. Lieb and W. Thirring managed to obtain a realistic estimate of the constant C — *cf.* the reviews [Lie 5,6] and ⟦Thi 3⟧, Sec.4.3. It is crucial for the proof that electrons are *fermions*, because the bound holds only on the "antisymmetrized" subspace in the domain of the corresponding Schrödinger operator. The ground state energy E_N in a system of N charged bosons fulfils, on the contrary, the inequality $E_N \le -D\,N^{7/5}$ — see [Co 1], [CLY 1]. The stability of atoms and molecules interacting with a magnetic field, *i.e.*, the spectral boundedness from below in dependence on the field intensity has been studied in [FLL 1], [LL 1].

Section 14.4 In addition to the mentioned extension of Theorem 14.4.1 to $V \in L^p+L^\infty$ in the n–dimensional case, the same result holds for $p = n/2$ if $n \ge 5$, as well as for $V \in R+L^\infty$ in the case $n = 3$ — see ⟦RS 4⟧, Thm.XIII.15. The assumption about the boundedness of the potential from below in Theorem 14.4.2 can also be weakened; it is sufficient that the negative part $V_- \in L^p+L^\infty$, where p satisfies condition (14.2). Stronger negative singularities of the potential are admissible provided operator H is defined as a form sum — *cf.* ⟦RS 4⟧, Thm.XIII.69.

Theorem 14.4.3 appeared for the first time in [Mol 1]; see also ⟦Gl⟧, Sec.28; if the operator H is defined as a form sum, it is sufficient to assume $V \in L^1_{loc}(\mathbb{R})$. In the original paper A. Molčanov proved a similar criterion for the n–dimensional case; however, the latter is of little practical importance because the interval $(a, a+\delta)$ is not replaced by a cube in $\mathbb{R}^n$ but by a more complicated set.

The described transformation of an N–particle Hamiltonian for a given clustering of the particles is usually referred to as *cluster decomposition;* to obtain a better idea about it we recommend the reader to work out in detail a simple nontrivial case (Problem 12). The proof of Proposition 14.4.4 and the positivity of T_D uses the so–called *cluster Jacobi coordinates*, *i.e.*, the Jacobi coordinates for the $3n(D)$–3 degrees of freedom corresponding to the motion of cluster barycenters, completed by arbitrary $3N-3n(D)$ coordinates describing the motion within the clusters. The operator T_D is then unitarily equivalent to the Hamiltonian of $n(D)-1$ free particles — a more detailed discussion can be found in ⟦RS 3⟧, Sec.XI.5.

Theorem 14.4.5 is named after G.M. Žislin, C. van Winter and W. Hunziker, who proved it for different classes of potentials; for Schrödinger operators defined as form sums its validity can be extended to potentials $V_{jk} \in R+L^\infty_\varepsilon$ — see ⟦ Si 1 ⟧, Sec.VII.3 or ⟦RS 4⟧, Sec.XIII.5. Since in physically interesting cases some of the particles involved are usually identical, one has to study the restriction of H to the appropriate permutation–symmetry subspace; a detailed discussion for atomic systems including the interaction with an external potential is given in ⟦JW⟧.

Proposition 14.4.6 follows from scattering theory which we shall discuss in the next chapter. It is amusing that the same method can also be used to prove the *absence* of an absolutely continuous spectrum for a class of one–dimensional Schrödinger operators with a family of growing or broadening potential barriers to both sides of a given point — *cf.* [SSp 1].

Theorem 14.4.8 and the unitary equivalence mentioned in Remark 14.4.9a appeared for the first time in [Ka 2]; see also ⟦RS 4⟧, Thm.XIII.21; their validity extends to an arbitrary potential $V \in R$ provided H is defined as a form sum. Examples showing that the singularly continuous spectrum may be unstable with respect to finite–dimensional perturbations can be found in ⟦RS 4⟧, Sec.XIII.6.

The idea of Example 14.4.10 was formulated by E. Wigner and J. von Neumann in 1930; it is also possible to construct examples of potentials which produce any prescribed finite set of positive eigenvalues — *cf.* [Al 1]. Theorem 14.4.11 was proved in [Ka 5]; it represents a particular case of the Kato–Agmon–Simon theorem, which also admits potentials with a slow decay at infinity provided they do not oscillate — see, *e.g.*, ⟦RS 4⟧, Thm.XIII.58. The formal virial theorem mentioned in Remark 14 was introduced in quantum mechanics by B. Finkelstein in 1928; the proof of Theorem 12 was given for the first time in [We 1]; see also [Al 1], [Si 4].

The existence of embedded eigenvalues for multiparticle systems with bound subsystems is quite natural. Consider, for instance, operator (14.18), whose point spectrum for $\gamma=0$ is $\sigma_p(H_\gamma)=\{-n_1^{-2}-n_2^{-2}\,:\,n_1,n_2=1,2,\dots\}$; since $\sigma_{ess}(H_\gamma)=[-1,\infty)$ by Theorem 14.4.5, the eigenvalues with $n_1,n_2\ge 2$ are embedded in the essential spectrum. If we switch in the electron repulsion, *i.e.*, take $\gamma\ne 0$, a part of these eigenvalues turn into resonances as we mentioned in the notes to Section 9.6.

In addition to the interactions considered here, there are other important classes such as *periodic potentials.* Consider the simplest situation, where $H:=H_0+V(Q)$ is the Schrödinger operator on $L^2(\mathbb{R})$ corresponding to a bounded measurable function V which satisfies the condition $V(x+b)=V(x)$ for some $b>0$. The operator H can then be shown to be unitarily equivalent to the direct integral $\frac{b}{2\pi}\int^\oplus_{[-\pi/b,\pi/b)}(P^2_{\theta b}+V(Q))\,d\theta$, where P_θ is defined in Example 4.7.11 and V is the restriction of the potential to the interval $(0,b)$ denoted by the same symbol; this decomposition is usually named *Floquet* by mathematicians and *Bloch* by solid–state physicists, and θ is the *quasimomentum.*

As in the case of a direct sum, the spectrum of the direct integral is a union of the component spectra, $\sigma(H)=\bigcup_{\theta b\in[-\pi,\pi)}\sigma(P^2_{\theta b}+V(Q))$. If $V=0$, the eigenvalue $\lambda_k(\theta)$ runs by Example 4.9.1 through the interval $[(2\pi k/b)^2,(2\pi(k+1)/b)^2)$, so we obtain $\sigma(H)=\mathbb{R}^+$ as expected. On the other hand, if $V\ne 0$ the intervals covered by the eigenvalues may open *gaps* ; the spectrum consists in this case of a finite or infinite number of *bands.* A discussion of periodic potentials with a literature guide can be found in ⟦RS 4⟧, Sec.XIII.16. If a periodic Schrödinger operator is perturbed by a localized potential, eigenvalues may appear in the gaps — see, *e.g.*, [ADH 1]. The importance of such operators stems from the fact that they can be used to model crystal impurities.

Section 14.5 The definitions of the Dirichlet and Neumann Laplacians again conventionally include the minus sign. As we have stressed, a selection of these or other "boundary conditions" means different choices of the dynamics of the problem. Theorem 14.5.2 holds for a considerably wider class of regions Ω than those considered here. The relation (14.26) was derived by H. Weyl in 1911 using the idea hinted here; its detailed exposition can be found in various textbooks — see, *e.g.*, ⟦RS 4⟧, Secs.XIII.14 and XIII.15. Its meaning in the quantum–mechanical context is similar to that of the Cwikel–Lieb–Rozenblioum theorem: it expresses the well–known principle that in a semiclassical regime when there are many eigenvalues, their number equals roughly the phase space volume occupied by the corresponding classical system, divided by $(2\pi)^n$ (or $(2\pi\hbar)^n$ in the standard system of units). Using this, we can derive the asymptotic relation for $N(gV)$ mentioned above; notice that c_n^{cl} of Remark 14.3.8 is nothing else than $(2\pi)^{-n}v_n$. Estimates of the asymptotic behavior for sets with a less regular boundary are given in [vdB 1]. In general, the spectrum of Neumann Laplacians is more sensitive to geometric properties of the region Ω and its boundary — see, *e.g.*, examples in [Si 8], [HSS 1].

In addition to the asymptotics (14.26), the distribution of eigenvalues of H_D^{Ω} for regions of different shapes offers many interesting problems. For instance, various bounds can be derived for ratios and distances of the eigenvalues — see, *e.g.*, [AsHS 1], [AB 1,2], [AE 1]. An inverse problem for Dirichlet and Neumann Laplacians is to reconstruct the shape of Ω from the knowledge of the spectrum; this is content of the M. Kac's famous question whether one can *hear the shape of the drum* — for a recent negative result see [GWW 1]. Still another problem is represented by the distribution of the eigenvalue *spacings* in billiard–type regions; a type of this distribution determines whether such a system is integrable or chaotic — *cf.* the notes to Section 8.3.

Section 14.6 Point interactions regarded as formal δ–function potentials were introduced into quantum mechanics by R. Kronig, W.G. Penney, E. Fermi, and others in the thirties; rigorous interpretation using self–adjoint extensions is due to [BF 1]. Later they became an object of intensive study, in the first place by the group of S. Albeverio and R. Høegh–Krohn; the results are summarized in the monograph ⟦AGHH⟧, where a rich bibliography can also be found; see also the proceedings ⟦EŠ 1,2⟧, ⟦DE⟧, ⟦EN⟧, and ⟦AFHL⟧. Physical aspects of point interactions are discussed in ⟦DO⟧.

Theorem 14.6.2 is valid for any $V \in L^1(\mathbb{R})$ provided the approximating operators are understood in the quadratic form sense — *cf.* ⟦AGHH⟧, Chap.I.3, where one can also find the proof of the claim contained in Remark 14.6.3a. δ–interactions can also be approximated by Schrödinger operators with scaled short–range potentials in dimension two and three, even in the situation when they are supported by an infinite set without accumulation points; the difference is that the coupling constant requires in these cases a *renormalization*, *i.e.*, other than natural scaling as $\varepsilon \to 0$ — see ⟦AGHH⟧.

The approximation of the δ'–interactions introduced in Remark 14.6.3b is more complicated. One way is to use the family $P^2 + g(\varepsilon)E_{\psi(\varepsilon)}$, where the $E_{\psi(\varepsilon)}$ are projections to one–dimensional subspaces spanned by vectors $\psi(\varepsilon)$ that "shrink" as $\varepsilon \to 0$ and the coupling constant $g(\varepsilon)$ diverges in the limit, as discussed in [Šeb 3]. In a fixed interval of energy, the δ'–interactions can alternatively approximated by Schrödinger operators on suitable graphs — see [AEL 1]. A "potential–type" approximation exists too: T. Cheon and T. Shigehara noticed that the family of operators with three δ–interactions,

$$-\frac{d^2}{dx^2}+\frac{\beta}{a^2}\delta(x)+\sum_{\varsigma=\pm}\left(\frac{2}{\beta}-\frac{1}{a}\right)\delta(\varsigma a),$$

converges to $H'_{\beta,0}$ formally as $a\to 0$ [CS 1], and the convergence was shown to be norm–resolvent in [ENZ 1], [AN 1]; in combination with Theorem 14.6.2 it yields an approximation by regular potentials [ENZ 1]. Some of the "remaining" point interactions in one dimension have been treated, *e.g.*, in [Šeb 4], [ChH 1], and [SMMC 1].

The δ'–interaction properties differ in many respects from that of the δ case. One example is a different high–energy behavior of the corresponding Kronig–Penney models, namely that in the δ' case the spectrum is dominated by the gaps there (Problem 21). This has interesting consequences for the so–called *Wannier–Stark model* in which a linear potential is added to the periodic system, *i.e.*, the δ' WS system is characterized by the Hamiltonian $H'_{\beta,Y}(f):=H'_{\beta,Y}-fQ$ for some $\beta\in\mathbb{R}$, $f>0$ and $Y:=\{nL:\,n\in\mathbb{Z}\}$. It appears that the absolutely continuous spectrum of $H'_{\beta,Y}$ is empty for any $f>0$ [AEL 1], [Ex 8], and that for a "large" set of parameter values the spectrum is pure point and depends on the rationality of $fL^3\pi^{-2}$ [ADE 1]. The question about the spectrum of the analogous Wannier–Stark system with an array of δ–interactions, which represents a simple model of a crystal in an electric field, is still open.

There are other examples of how we can use a symmetry of a submanifold in the configuration space to construct a contact–interaction Hamiltonian; let us mention interactions supported by a sphere [AGS 1], a family of concentric spheres [Sha 1], *etc.* Other methods make it possible to define and analyze contact interactions on sets with no symmetry, and also to obtain wider classes of Hamiltonians in symmetric cases; loosely speaking, the coupling constant is allowed to vary along the interaction support — see, *e.g.*, [Te 1], [BT 1], [BEKŠ 1]. Let us remark that such constructions also work for "wild" sets such as curves which are continuous but nowhere differentiable — *cf.* [Che 1]. We will return to this question in Section 17.5 where we will analyze a class of systems with a contact interaction supported by a graph.

* * *

In addition to Schrödinger operators, there are other important classes of quantum mechanical Hamiltonians. One of them consists of *Dirac operators*

$$H:=\alpha\cdot P+\beta m+V(Q)$$

on $L^2(\mathbb{R}^3;\mathbb{C}^4)$ describing a relativistic particle of spin $\frac{1}{2}$. The matrix coefficients are $\alpha_j:=\begin{pmatrix}0&\sigma_j\\ \sigma_j&0\end{pmatrix}$, where σ_j are the Pauli matrices, and $\beta:=\begin{pmatrix}I&0\\0&-I\end{pmatrix}$. In general, the potential V is a matrix–valued function, but usually it of the form of a function V multiplied either by the 4×4 unit matrix (in which case we abuse the terminology and speak about a function) or by β. Among physically most interesting are naturally potentials such that V is a sum of Coulomb potentials. Dirac operators can be defined also in $\mathbb{R}^n$, $n\neq 3$; the dimension of the matrices, *i.e.*, the number of wave function components depends on the dimension n of the configuration space.

The present status of the theory of Dirac operators is described in detail in the recent monograph ⟦ Tha ⟧; here we limit ourselves to a few basic results. In distinction to the

Schrödinger case, the self–adjointness may here depend on the value of a coupling constant. For instance, a Dirac operator with the Coulomb potential $v(x) = \gamma|x|^{-1}$ is *e.s.a.* on $C_0^\infty(\mathbb{R}^3 \setminus \{0\}; \mathbb{C}^4)$ provided $|\gamma| < \sqrt{3}/2$; if we express γ in the standard system of units, this corresponds to the condition $Z \le 118$, which is satisfied for all really existing nuclei. This fact was noted already in [Rel 1]; a simple proof of the essential self–adjointness for $|\gamma| < 1/2$ can be found in ⟦We⟧, Sec.10.6. On the other hand, if the condition is violated there is a distinguished self–adjoint extension of such a Dirac operator — the only one which exhibits finite mean values of the kinetic and potential energy; these conclusions extend to potentials with several Coulomb–type singularities — *cf.* [Nen 3], [Kla 3], [Kar 1]. It is interesting that if the Dirac particle has an *anomalous magnetic moment*, *i.e.*, the potential V is replaced by $V - \frac{\mu}{2m}\tau \cdot \nabla V$, where $\tau_j := \begin{pmatrix} 0 & i\sigma_j \\ -i\sigma_j & 0 \end{pmatrix}$, then the Dirac operator with $V(x) = \gamma|x|^{-1}$ is *e.s.a.* on $C_0^\infty(\mathbb{R}^3; \mathbb{C}^4)$ for any $|\gamma|$ — see [Beh 1], [GST 1].

Under the above mentioned conditions, too, the essential spectrum of the Dirac operator is $(-\infty, -m] \cup [m, \infty)$ as in the free case; in the standard system of units m is replaced by the rest energy mc^2. Since H is below unbounded, we lose some efficient tools of the Schrödinger theory such as the quadratic form method, the minimax theorem, *etc.* The discrete spectrum is contained in the gap; some results concerning this can be found, *e.g.*, in [Kl 2], [Gr 1]. In comparison to Schrödinger operators, Dirac theory has very few exactly solvable models. One of them, which is known from quantum mechanical textbooks, concerns a single–center Coulomb potential (relativistic hydrogen atom); a generalization to an arbitrary dimension is given in [Wo 1].

We have mentioned that the mass parameter m in the Dirac operator is in fact $mc/\hbar$. By a natural consistency requirement, the operator $H - (mc/\hbar)I$ has, as $c \to \infty$, in some sense to approach the corresponding Schrödinger (or Pauli) operator describing the nonrelativistic spin $\frac{1}{2}$ particle interacting with the considered potential. There are numerous papers treating the nonrelativistic limit of the Dirac theory; as an example, let us quote [GGT 1,2].

Since Dirac operators are first–order differential operators, they are *e.s.a.* when restricted to a set of functions vanishing at a set of codimension greater than one — *cf.* [Sve 1] — and therefore point interactions analogous to Theorem 14.6.4 and Problem 19 cannot be constructed. On the other hand, room is open for contact interactions supported by subsets of codimension one, say, by a sphere in $\mathbb{R}^3$; such Dirac operators have been constructed in [DEŠ 1,2]. Point interactions exist in dimension one — see [GŠ 1]; Dirac operators on graphs have been discussed in [BT 1].

Still another class of Hamiltonians consists of operators of $H := \sqrt{P^2 + m^2} + V(Q)$, which may be used to describe relativistic spinless particles in a fixed reference frame. In distinction to the Dirac case, these operators are below bounded and many spectral properties of Schrödinger theory can be extended to them — see, *e.g.*, [DL 1], [Lie 7], [LY 1], [Nar 1].

Problems

1. Let $(L^p + L^\infty_\varepsilon)(\mathbb{R}^n)$ be the set of (classes of) functions $f : \mathbb{R}^n \to \mathbb{C}$ with the following property: for any $\varepsilon > 0$ there is a decomposition $f = f_{p,\varepsilon} + f_{\infty,\varepsilon}$ such that $f_{p,\varepsilon} \in L^p$, $f_{\infty,\varepsilon} \in L^\infty$, and $\|f_{\infty,\varepsilon}\|_\infty < \varepsilon$.

(a) The Coulomb potential $V: V(x) = \gamma|x|^{-1}$, belongs to $(L^2 + L^\infty_\varepsilon)(\mathbb{R}^3)$.

(b) More generally, what local singularities and decay at infinity are allowed if a function belongs to $(L^p + L^\infty_\varepsilon)(\mathbb{R}^n)$?

2. Let $H_0 := P^2$ be the free Hamiltonian on $L^2(\mathbb{R}^n)$. Show that the inequality $\|H_0^{1/2}\psi\| \le \|(\beta H_0 + (4\beta)^{-1})\psi\| \le \beta\|H_0\psi\| + (4\beta)^{-1}\|\psi\|$ is valid for any $\beta > 0$ and an arbitrary $\psi \in D(H_0)$.

3. Let $\mathcal{D}_n$ and $\mathcal{D}_{n-1}$ be subspaces of dimension n and $n-1$, respectively, in a Hilbert space; then there is a nonzero vector $\psi \in \mathcal{D}_{n-1}^\perp \cap \mathcal{D}_n$.
Hint: $\mathcal{D}_n \cap \mathcal{D}_{n-1}^\perp = \{0\}$ would mean that the projection of a nonzero $\phi \in \mathcal{D}_n$ to $\mathcal{D}_{n-1}$ is nonzero; choose linearly independent $\phi_1, \dots, \phi_n \in \mathcal{D}_n$.

4. Find examples of Hermitean operators B on an infinite–dimensional $\mathcal{H}$ such that

(a) $\sigma_{ess}(B)$ consists of a single point λ and the parts of the discrete spectrum, which we denote as $\sigma_d(B)$, in both the intervals $(-\infty,\lambda)$ and (λ,∞) are nonempty.

(b) $\sigma_{ess}(B)$ is a two–point set $\{\lambda_1,\lambda_2\}$ and $\sigma_d(B)$ is contained in (λ_1,λ_2).

Hint: (a) Consider T_s of Example 4.1.4 with a suitable alternating sequence.

5. Check the relations of Remark 14.2.2c.

6. Using functional calculus, prove Theorem 14.2.6 for a self–adjoint T; show that the projection P is orthogonal in this case.

7. Let $v \in L^\infty(\mathbb{R}^+)$. Prove:

(a) The operator $h: (hf)(r) = -f''(r) + (\beta r^{-2} + \gamma r^{-1} + v(r))f(r)$ on $L^2(\mathbb{R}^+)$ is *e.s.a.* on $C_0^\infty(0,\infty)$ for any $\gamma \in \mathbb{R}$ *iff* $\beta \ge \frac{3}{4}$; in the opposite case its deficiency indices are $(1,1)$.

(b) Let $H := H_0 + V(Q)$ be the Schrödinger operator on $L^2(\mathbb{R}^3)$ with the centrally symmetric potential, $V(x) = v(r)$. Check that the s–wave operator $h_0: h_0 f = -f'' + vf$ satisfies the Dirichlet boundary condition at the origin, *i.e.*, $D(h_0) = \{ f \in AC^2(0,\infty) : f(0) = 0 \}$.

Hint: (a) Use Theorem 4.8.7. (b) If $f \in D(h_0)$, then $U^{-1}(fY_{00})$ given by (11.16) must belong to the domain of H.

8. Using Theorem 14.3.1, find upper bounds to the number $N_\lambda(V) := \dim E_H(-\infty,\lambda)$ of bound states with energies smaller than λ.
Hint: Put $V_-(\lambda,x) := \min\{V(x)-\lambda, 0\}$ and use the minimax principle.

9. Prove: (a) If $n = 1,2$, we can choose $\phi \in L^2(\mathbb{R}^n)$ so that (14.16) is valid.

(b) $\int_{\mathbb{R}^6} \frac{f(x)f(y)}{|x-y|^2}\, dx\, dy \le 0$ holds for a real function $f \in (L^1 \cap L^2)(\mathbb{R}^3)$.

Hint: (a) The expression in question equals $\int_{\mathbb{R}^n} |\chi(p)|^2(p^2+\kappa^2)^{-1}dp$, where $\chi := F_n|V(Q)|^{1/2}\phi$; choose $\phi \in C_0^\infty$ with $\operatorname{supp}\phi \supset \operatorname{supp} V$. As for (b), replace the denominator by $|x-y|^2+\varepsilon^2$ and use Problem 7.17b.

10. Fill the details into the proof of Theorem 14.3.9.

(a) Show that $(\psi, H\psi) \geq (1-4c)\,(\psi, H_c\psi)$ holds for any $\psi \in D(H_0)$.

(b) If $f \in L^2+L^\infty$ has a compact support, it belongs to L^p for $p \in [1,2]$.

11. Let $\psi \in L^2(\mathbb{R}^3)$ be rotationally invariant, $\psi(x) = \tilde\psi(r)$ for some $\tilde\psi \in L^2(\mathbb{R}^+, r^2dr)$; then

$$\int_{\mathbb{R}^3} |x-a|^{-1}|\psi(x)|^2dx = \int_{\mathbb{R}^3} \min\{r^{-1}, |a|^{-1}\}\, |\psi(x)|^2dx$$

holds for any vector $a \in \mathbb{R}^3$.

12. Perform the cluster decomposition for a system of three particles. Prove Proposition 14.4.4 in this case, and check that if D consists of n clusters, the operator T_D is unitarily equivalent to P^2 on $L^2(\mathbb{R}^{3(n-1)})$.

13. Let $H := H_0 + V(Q)$ be the operator (14.1) on $L^2(\mathbb{R}^3)$ with a centrally symmetric potential, $V(x) = v(r)$ for some $v : \mathbb{R}^+ \to \mathbb{R}$.

(a) Find an example of a purely attractive V such that H has a *zero-energy bound state, i.e.*, there is $\psi \in L^2(\mathbb{R}^3)$ such that $H\psi = 0$.

(b) Check the result discussed in Example 14.4.10.

Hint: (a) Consider an eigenspace of L^2 with $\ell \geq 2$ for a rectangular well.

14. Using the notation of Section 14.5, prove

(a) The inclusion $D \subset Q(H_D^\Omega)$ holds in Example 14.5.1b, *i.e.*, for any $\psi \in D$ there is a sequence $\{\psi_m\} \subset C_0^\infty(\Omega)$ such that $\psi_m \to \psi$ and $\partial_i\psi_m \to \partial_i\psi$ in $L^2(\Omega)$.

(b) The relation (14.26) for a parallelepiped.

(c) Using the partial-wave decomposition, check that the conclusions of Example 14.5.1 extend to the ball $\Omega := \{x : |x| < R\}$ in $\mathbb{R}^n$.

(d) The analogous properties for the Neumann Laplacian on $L^2(\Omega)$.

Hint: (a) Let x_0 be the center of Ω, and $\{\alpha_m\}$ an increasing sequence with $\alpha_m \to 1$. A given $\psi \in D$ can be approximated by functions, which are convolutions of $\psi(x_0+\alpha_m(\cdot - x_0))$ with $\beta_m^n j(\beta_m\cdot)$ for a suitably chosen sequence $\{\beta_m\}$, where $j \in C_0^\infty(\mathbb{R}^n)$ with $\int j(x)\,dx = 1$. (b) The volume of an n-dimensional ellipsoid with semi-axes $s_1, \dots, s_n$ is $v_n \prod_{j=1}^n s_j$.

15. Let $\Omega := \{(x,y) \in \mathbb{R}^2 : |xy| < \frac{1}{2}\pi\}$. Show that the spectrum of the corresponding operator H_D^Ω is purely discrete and the formula (14.26) does not apply.
Hint: Use a bracketing argument placing additional Neumann conditions at the cuts of Ω with the lines '$|x|, |y| = b$ for an arbitrary $b > 0$ and show that the "tails" do not contribute to the spectrum below b^{-2}. Check the area of Ω.

16. Consider the δ–interaction for a particle on line of Section 14.6.

(a) Prove Proposition 14.6.1.

(b) Given k with $\operatorname{Im} k > 0$, denote by G_k the free resolvent kernel of Problem 7.9, $G_k(x,x') = K_{-ik}(x,x') = \frac{i}{2k}\, e^{ik|x-x'|}$. If $k \neq -i\alpha/2$, the resolvent $(H_{\alpha,y} - k^2)$ is an integral operator with the kernel

$$G_k(x,x') - \frac{2\alpha k}{2k+i\alpha}\, G_k(x,y) G_k(x',y)\,.$$

(c) $\sigma_{ess}(H_{\alpha,y}) = \sigma_{ac}(H_{\alpha,y}) = \mathbb{R}^+$ and $\sigma_{sc}(H_{\alpha,y}) = \emptyset$. The point spectrum of $H_{\alpha,y}$ is empty if $\alpha \geq 0$ or $\alpha = \infty$, while for $\alpha \in (-\infty, 0)$ there is exactly one eigenvalue $-\frac{1}{4}\alpha^2$ corresponding to the eigenfunction $\psi_\alpha :\ \psi_\alpha(x) = \sqrt{-2\alpha}\, e^{\alpha|x-y|}$.

(d) Fill the details into the proof of Theorem 14.6.2. Check that up to an overall scaling factor, the approximating operator is transformed by the dilation group of Example 5.9.5b into

$$\varepsilon^2 U_d(\varepsilon) H_\varepsilon(V,y) U_d(-\varepsilon) \;=\; P^2 + \varepsilon\, e^{-iPy/\varepsilon} V(Q)\,.$$

(e) Let $Y = \{ y_j \,:\, j = 1,\dots,N \}$ be a finite subset of the real axis. To any $\alpha_1,\dots,\alpha_N \in (-\infty,\infty]$ construct the point–interaction Hamiltonian supported by Y with the δ–interaction of strength α_j at the point y_j. Generalize to this case assertions (b), (c), and Theorem 14.6.2.

Hint: (b) Use Theorem 4.7.15. (c) Use Theorem 14.4.7.

17. Consider the δ'–interaction specified by the boundary conditions (14.31).

(a) Find the spectrum of $H'_{\beta,y}$ and an integral–operator expression for its resolvent.

(b) Extend the results to a δ'–interaction supported by a finite subset of $\mathbb{R}$.

(c) Generalize the conclusions to the point interaction specified by the boundary conditions

$$\begin{aligned}
\psi'(y+) - \psi'(y-) &= \frac{\alpha}{2}\,(\psi(y+) + \psi(y-)) + \frac{\gamma}{2}\,(\psi'(y+) + \psi'(y-))\,,\\
\psi(y+) - \psi(y-) &= -\frac{\bar\gamma}{2}\,(\psi(y+) + \psi(y-)) + \frac{\beta}{2}\,(\psi'(y+) + \psi'(y-))
\end{aligned}$$

with $\alpha, \beta \in \mathbb{R}$ and $\gamma \in \mathbb{C}$.

18. Let H_0 be the free Schrödinger operator on $L^2(\mathbb{R}^n)$ and denote by $\dot H_0$ its restriction to $C_0^\infty(\mathbb{R}^n \setminus \{0\})$. Using the partial–wave decomposition, prove

(a) $\dot H_0$ is *e.s.a.* if $n \geq 4$; this conclusion remains valid if the origin of the coordinates is replaced by any point $y \in \mathbb{R}^n$.

(b) In the cases $n = 2,3$ the partial–wave components of $\dot H_0$ with a nonzero ℓ are *e.s.a.* while the s–wave part of the operator has the deficiency indices $(1,1)$.

Hint: Use Problem 7a and Remark 11.5.3.

19. Construct single–center point–interaction Hamiltonians in $L^2(\mathbb{R}^2)$. Show that in contrast to the one– and three–dimensional cases, such an operator has always a negative eigenvalue unless it is trivial.

20. Consider the point–interaction Hamiltonian $H_{\alpha,y}$ of Theorem 14.6.4.

(a) Prove that the resolvent $(H_{\alpha,y} - k^2)^{-1}$ is for any $k \neq -4\pi i\alpha$ with $\operatorname{Im} k > 0$ an integral operator with the kernel

$$G_k(x,x') + \frac{4\pi}{4\pi\alpha - ik}\, G_k(x,y)G_k(x',y)\,,$$

where G_k denotes the free resolvent kernel of Problem 7.19b, *i.e.*, $G_k(x,x') = K_{-ik}(x,x') = \frac{e^{ik|x-y|}}{4\pi|x-y|}$.

(b) Use this result to show that $\sigma_{sc}(H_{\alpha,y})$ is empty.

21. The *Kronig–Penney model* describes equidistantly spaced δ–interactions of the same strength α on line. In other words, there is $L > 0$ such that the corresponding Hamiltonian $H_{\alpha,Y}$ acts as $H_{\alpha,Y}\psi = -\psi''$ in each interval $J_n := (nL, nL{+}L)$, and its domain consists of all functions ψ such that $\psi \upharpoonright J_n \in AC^2(J_n)$ and the boundary conditions

$$\psi(nL+) = \psi(nL-) =: \psi(nL)\,, \quad \psi'(nL+) - \psi'(nL-) = \alpha\psi(nL)$$

are valid at each point of $Y := \{\, nL :\; n \in \mathbb{Z}\,\}$.

(a) Find the spectrum of $H_{\alpha,Y}$ and show how the band and gap widths behave for large energies.

(b) How does the spectrum change if we replace δ by δ', or more generally, by the point interaction of Problem 17c?

Hint: Use the Floquet–Bloch decomposition described in the notes to Section 14.4.

Chapter 15

Scattering theory

15.1 Basic notions

The problem of scattering of certain objects (particles, waves) on an obstacle (target) can be found in classical mechanics and field theory as well as in quantum theory. In the last case it is particularly important because intentionally prepared collisions of particles (nuclei, atoms, *etc.*) represent one of the very few efficient ways of studying their structure.

Recall first some basic concepts. Scattering may be a natural process such as penetration of cosmic rays through the Earth's atmosphere, part of an artificially prepared system like the cooling of neutrons in an atomic reactor, or an experiment in which we collide a beam of certain objects (for brevity we shall speak mostly about particles) produced by an accelerator with other particles which belong either to a fixed target or to another particle beam. Many scattering processes involve three or more particles some of which may be clustered (*e.g.*, into atomic nuclei). The scattering is said to be *elastic* if the number of particles, their sorts, and clustering are the same before the collision as after; otherwise we speak about an *inelastic process.* These possibilities are not mutually exclusive; in most cases the same projectile and target can collide both elastically and inelastically; a simultaneous description of all scattering processes is called *multichannel scattering.* Another classification concerns the number of collisions involved: if it is only one we speak about a *simple* scattering; otherwise we have a *multiple scattering* process.

A common feature of all scattering situations is that at the initial and final stage, when the projectile and target are far apart, their interaction can be neglected and the time evolution is governed by a "free" Hamiltonian H_0 rather than the full Hamiltonian of the system. For instance, if two spinless particles which interact via a potential are scattered, then the full Hamiltonian is the corresponding Schrödinger operator, while the free Hamiltonian is the Schrödinger operator with zero potential; one can easily find other examples. Hence a scattering system is characterized by a pair of self–adjoint operators $H,\, H_0$ to which the propagators

$$U(t) := e^{-iHt}, \quad U_0(t) := e^{-iH_0t}$$

correspond; we again call them the full and free propagator, respectively.

Now we want to compare the two dynamics. First of all, we have to give a precise meaning to the heuristic claim that for very large positive or negative times the system behaves as it would be free. It is obvious that this is not true generally: if the projectile and the target are in a bound state described by an eigenvector of H, then they stay in this state forever. Motivated by this observation we associate with H a set $M_s(H) \subset \mathcal{H}$ whose elements describe *scattering states*. We have to specify, of course, the properties that such states are supposed to have. This will be done below; for the present moment we assume only that

(s1) $M_s(H)$ is $U(t)$–invariant for all $t \in \mathbb{R}$.

This requirement is physically natural, expressing the *time homogeneity*: the set of scattering states certainly should not depend on the instant when we decide to perform the experiment. Denoting the corresponding projection by $E_s(H)$, we can write the assumption (s1) in the form $[E_s(H), U(t)] = 0$ for all $t \in \mathbb{R}$.

Using scattering states we can express the basic idea of scattering theory as the *asymptotic conditions*: we assume that for any state $\psi \in M_s(H)$ there are vectors $\psi_\pm \in M_s(H_0)$ such that

$$\lim_{t\to\pm\infty} \|U(t)\psi - U_0(t)\psi_\pm\| = 0; \tag{15.1}$$

the states associated with them are called *asymptotic states*; these allow us to express the "initial" and "final" state as $U_0(\pm t)\psi_\mp$, respectively, for large negative t. Using the unitarity of the propagators, we may rewrite (15.1) as

$$\psi_\pm = \lim_{t\to\pm\infty} U_0(t)^*U(t)\psi\,, \qquad \psi = \lim_{t\to\pm\infty} U(t)^*U_0(t)\psi_\pm;$$

the first of these relations shows that the maps $\psi \mapsto \psi_\pm$ are injective.

On the other hand, the ranges of these maps should be the whole set $M_s(H_0)$; in particular, an "initial" state which belongs to the set of scattering states of the free Hamiltonian should evolve into a scattering state of the system. In other words, we suppose that for any $\psi_- \in M_s(H_0)$ there is a vector $\psi \in M_s(H)$ such that (15.1) holds, and the same for $\psi_+ \in M_s(H_0)$. In view of the above relations, this is equivalent to the assumption of existence of the operators

$$\Omega_\pm := \operatorname*{s-lim}_{t\to\pm\infty} U(t)^*U_0(t)E_s(H_0)\,. \tag{15.2}$$

These are called the **wave operators**; we also denote them as $\Omega_\pm(H, H_0)$ when we want to stress the pair of operators to which they correspond.

By definition, the wave operators map into the set of scattering states of the full Hamiltonian, $\operatorname{Ran}\Omega_\pm \subset M_s(H)$. It may happen, however, that one or both of the asymptotic conditions are not fulfilled for some $\psi \in M_s(H)$, *i.e.*, $\operatorname{Ran}\Omega_\pm \neq M_s(H)$. An example of the situation, where the "positive" asymptotic condition is violated, is provided by a process in which the scattered particle is eventually captured by the target and never leaves the interaction region.

We are particularly interested in the case when the wave operators satisfy the asymmetric relation

$$\operatorname{Ran}\Omega_- \subset \operatorname{Ran}\Omega_+; \tag{15.3}$$

then we have the following scheme of a scattering event: for any $\psi_- \in M_s(H_0)$ the vector–valued function $U_0(\cdot)\psi_-$ describes a state of the system in the distant past. The map $\psi_- \mapsto \psi := \Omega_-\psi_- \in M_s(H)$ defines a vector describing the state at $t = 0$, to which the initial state has evolved by means of the full propagator $U(\cdot)$. Since ψ belongs by assumption to $\operatorname{Ran}\Omega_+$, there is a vector $\psi_+ \in M_s(H_0)$ such that $\psi = \Omega\psi_+$; in view of (15.1) we have $U(t)\psi \approx U_0(t)\psi_+$ as $t \to +\infty$. Speaking figuratively, we may say that the vectors $\psi_\pm$ determine "free asymptotes" to the trajectory $U(\cdot)\psi$ of the state vector.

The scattering experiment now consists of comparing in the distant future the state $U_0(t)\psi_+$ with $U_0(t)\phi$ for different $\phi \in M_s(H_0)$, *i.e.*, with states whose evolution was governed all the time by the free propagator. The corresponding transition–probability amplitude $(U_0(t)\phi, U_0(t)\psi_+) = (\phi, \psi_+)$ can be expressed through the original vector ψ_- because assumption (15.3) implies the existence of the operator

$$S := \Omega_+^*\Omega_-, \tag{15.4}$$

which connects the asymptotic states, $\psi_+ = S\psi_-$; it is called the **scattering operator** (or S–operator, S–matrix).

To make use of these definitions we have to know more about scattering states, in particular, to give meaning to the above described intuitive distinction between bound and scattering states. Intuition leans on properties of simple quantum mechanical systems consisting of one or several particles, where the probability of finding the system within bounded spatial regions is clearly essential. Consider therefore a family $\{M_r : r \geq 0\}$ of subsets in the configuration space C which characterize the localization; we suppose that they are fully ordered by inclusion and such that $\bigcup_{r\geq 0} M_r = C$. For instance, if the scattering system consists of just two particles, we can choose $M_r := \mathbb{R}^3 \times B_r$, where the B_r are concentric balls of radius r in the center–of–mass frame. Furthermore, denote by F_r the projection to the subspace of functions with supports in M_r; then we have

$$\operatorname*{s-lim}_{r\to\infty} F_r = I\,.$$

A natural requirement on *scattering states* is that the probability of finding the system within a fixed bounded region vanishes as $t \to \pm\infty$. Since the latter equals $\|F_r e^{-iHt}\psi\|^2$, we define

$$M_s(H) := \left\{ \psi \in \mathcal{H} : \lim_{|t|\to\infty} F_r e^{-iHt}\psi = 0 \quad \text{for all } r > 0 \right\}. \tag{15.5}$$

On the other hand, *bound states* have just the opposite property: the probability of finding the system *outside* a sufficiently large region remains small for all times. Hence we put

$$M_b(H) := \left\{ \psi \in \mathcal{H} : \lim_{r\to\infty} \sup_{t\in\mathbb{R}} \|(I - F_r)e^{-iHt}\psi\| = 0 \right\}. \tag{15.6}$$

Using the definitions we can derive the elementary properties of these sets.

15.1.1 Proposition: The sets of scattering and bound states are mutually orthogonal subspaces in $\mathcal{H}$ and

$$M_b(H) \supset \mathcal{H}_p(H)\,, \qquad M_s(H) \subset \mathcal{H}_c(H),$$

where $\mathcal{H}_p(H)$ is as usual the closed subspace spanned by the eigenvectors of H and $\mathcal{H}_c(H) = \mathcal{H}_p(H)^\perp = \mathcal{H}_{ac}(H) \oplus \mathcal{H}_{sc}(H)$.

Proof: Using the inequality $\|\psi+\phi\|^2 \le 2\|\psi\|^2 + 2\|\phi\|^2$ we check that $M_s(H)$ and $M_b(H)$ are subspaces in $\mathcal{H}$. Let $\{\psi_n\} \subset M_s(H)$ be a sequence converging to some $\psi \in \mathcal{H}$. The operators F_r and $U(t)$ have unit norm, so

$$\|F_rU(t)\psi\|^2 \le 2\|\psi-\psi_n\|^2 + 2\|F_rU(t)\psi_n\|^2 .$$

By assumption, for any $\varepsilon > 0$ there is an n such that $\|\psi-\psi_n\|^2 < \frac{\varepsilon}{4}$, and since $\psi_n \in M_s(H)$, we can find a t_0 so that $\|F_rU(t)\psi_n\|^2 < \frac{\varepsilon}{4}$ holds for all t with $|t| > t_0$. Then $\|F_rU(t)\psi\|^2 < \varepsilon$, so $\psi \in M_s(H)$ and the subspace $M_s(H)$ is closed. On the other hand, choosing $\{\psi_n\} \subset M_b(H)$, we deduce in a similar way from the inequality

$$\sup_{t\in\mathbb{R}} \|(I-F_r)U(t)\psi\|^2 \le 2\|\psi-\psi_n\|^2 + 2\sup_{t\in\mathbb{R}} \|(I-F_r)U(t)\psi_n\|^2$$

that $\psi \in M_b(H)$, *i.e.*, the subspace $M_b(H)$ is closed too. Consider now arbitrary vectors $\psi \in M_s(H)$ and $\phi \in M_b(H)$; then we have

$$\begin{aligned} |(\psi,\phi)|^2 &= |(U(t)\psi, U(t)\phi)|^2 \\ &= |(F_rU(t)\psi, U(t)\phi) + (U(t)\psi, (I-F_r)U(t)\phi)|^2 \\ &\le 2\,\|\phi\|^2\|F_rU(t)\psi\|^2 + 2\,\|\psi\|^2\|(I-F_r)U(t)\phi\|^2 \\ &\le 2\,\|\phi\|^2\|F_rU(t)\psi\|^2 + 2\,\|\psi\|^2 \left(\sup_{t\in\mathbb{R}} \|(I-F_r)U(t)\phi\|\right)^2 . \end{aligned}$$

After the limit $t \to \pm\infty$, the first term on the right side disappears and the other limit $r \to \infty$ also annuls the second term, so $(\phi,\psi) = 0$, *i.e.*, $M_s(H) \perp M_b(H)$. Finally, assume that ψ is an eigenvector of H, $H\psi = \lambda\psi$; then we have

$$\sup_{t\in\mathbb{R}} \|(I-F_r)e^{-iHt}\psi\|^2 = \|(I-F_r)\psi\|^2 \to 0$$

as $r \to \infty$, so ψ belongs to $M_b(H)$, and the same is true for the subspace $\mathcal{H}_p(H)$ spanned by all the eigenvectors. Combining this result with the orthogonality of the subspaces, we get

$$M_s(H) \subset M_b(H)^\perp \subset \mathcal{H}_p(H)^\perp = \mathcal{H}_c(H)\,,$$

so the second inclusion is also valid. ∎

Next we are going to give an abstract sufficient condition under which the above inclusions turn to identities.

15.1.2 Proposition: Let $\sigma_{sc}(H) = \emptyset$ and denote $P_\alpha := E_H(-\infty, \alpha]$ for any $\alpha > 0$. If the operators $F_r P_\alpha$ are compact for all positive r, α, then

$$M_b(H) = \mathcal{H}_p(H), \quad M_s(H) = \mathcal{H}_c(H) = \mathcal{H}_{ac}(H). \tag{15.7}$$

Proof: By assumption, $\mathcal{H}_c(H) = \mathcal{H}_{ac}(H)$. Choosing a vector ψ from this subspace and an arbitrary $\varepsilon > 0$, in the same way as above we obtain the estimate

$$\|F_r U(t)\psi\|^2 \le 2\|(I - P_\alpha)\psi\|^2 + 2\|F_r P_\alpha U(t)\psi\|^2 .$$

Due to properties of the spectral measure, the first term on the right side is $< \frac{\varepsilon}{2}$ for α large enough. Next we use Problem 1a, which gives $\lim_{|t|\to\infty} F_r P_\alpha U(t)\psi = 0$; hence there is a t_0 such that the second term is also $< \frac{\varepsilon}{2}$ if $|t| > t_0$. It follows that $\psi \in M_s(H)$, so the second of the relations (15.7) is proved; combining it with the preceding proposition, we find $\mathcal{H}_p(H) \subset M_b(H) \subset M_s(H)^\perp = \mathcal{H}_p(H)$. ∎

We want to know, of course, whether the identities of Proposition 2 are valid for concrete scattering systems. We restrict ourselves here to the simplest case of two spinless particles interacting via a potential; references to more general results are given in the notes. After separating the center-of-mass motion, our problem reduces to the scattering of a particle with the reduced mass on a potential $V : \mathbb{R}^3 \to \mathbb{R}$; the free Hamiltonian is H_0 of Example 7.5.8 while H is (the closure of) $H_0 + V(Q)$.

We begin with an auxiliary result concerning operators of the form $f(Q)g(P)$ and $g(P)f(Q)$. If $g \in L^2(\mathbb{R}^3)$, their action, in view of Problem 7.17, can be written down explicitly as a composition of an integral operator and multiplication by a function. The operators $f(Q)$, $g(P)$ are generally unbounded; however, it may happen that the product is densely defined and bounded on its domain; following the usual convention, in such a case we employ the symbols $f(Q)g(P)$ and $g(P)f(Q)$ for the corresponding continuous extension to $L^2(\mathbb{R}^3)$.

15.1.3 Proposition: (a) Suppose that $f, g \in L^2(\mathbb{R}^3)$ and the operators $f(Q)g(P)$, $g(P)f(Q)$ are densely defined; then they are Hilbert–Schmidt and

$$\|f(Q)g(P)\| \le \|f(Q)g(P)\|_2 = (2\pi)^{-3/2}\, \|f\|\, \|g\|;$$

the analogous relation is valid for $g(P)f(Q)$.

(b) The operators $F_r(H_0 - z)^{-1}$ and $(H_0 - z)^{-1}F_r$ are compact (even Hilbert–Schmidt) for any $z \in \rho(H_0)$, $r > 0$, and

$$M_s(H_0) = \mathcal{H}_{ac}(H_0) = L^2(\mathbb{R}^3) .$$

Proof: By Problem 7.17, $f(Q)g(P)$ acts on its domain as an integral operator with the kernel $(x,y) \mapsto K(x,y) := (2\pi)^{-3/2} f(x)(F_3^{-1}g)(y-x)$. It is straightforward to compute its Hilbert–Schmidt norm: using a simple substitution and unitarity of the FP–operator, we obtain the relation

$$\int_{\mathbb{R}^3} |K(x,y)|^2 dx\, dy = (2\pi)^{-3} \|f\|^2 \|g\|^2 ,$$

which implies the above inequality. In a similar way, we can check that $g(P)f(Q)$ with the kernel $(x,y) \mapsto (2\pi)^{-3/2} f(y)(F_3^{-1}g)(y-x)$ is Hilbert–Schmidt; this proves assertion (a). The operators F_r and $(H_0-z)^{-1}$ are bounded, and the function χ_{F_r} belongs to $L^2(\mathbb{R}^3)$ as well as $g : g(k) = |k^2 - z|^{-1}$ (Problem 2); hence (b) follows from (a) and Proposition 2 in combination with Example 7.5.8. ∎

By part (b) of the just proved result, any state is a scattering state of the free Hamiltonian. On the other hand, the set $M_s(H)$ of the full Hamiltonian depends, of course, on the chosen potential. We have, for instance, the following result.

15.1.4 Theorem: Let $H = H_0 + V(Q)$ with the potential $V \in (L^2 + L^\infty)(\mathbb{R}^3)$ and $\sigma_{sc}(H) = \emptyset$; then the relations (15.7) are valid.

Proof: We know from Section 14.1 that $D(H) = D(H_0) \subset D(V)$; hence the second resolvent identity gives

$$(H-z)^{-1}F_r = (H_0-z)^{-1}F_r - (H-z)^{-1}V_2(Q)(H_0-z)^{-1}F_r$$

$$- (H-z)^{-1}V_\infty(Q)(H_0-z)^{-1}F_r$$

for any $z \in \rho(H)\cap\rho(H_0)$. In view of Proposition 15.1.3b, the operators $V_2(Q)(H_0 - z)^{-1}$ and $(H_0-z)^{-1}F_r$ with an arbitrary $r > 0$ are Hilbert–Schmidt. Furthermore, $V_\infty \in L^\infty$, so the corresponding operator $V_\infty(Q)$ is bounded, and the operators $(H-z)^{-1}$ and F_r are also bounded. Hence $(H-z)^{-1}F_r$ is Hilbert–Schmidt, and the same is true for the adjoint operator $F_r(H-\bar{z})^{-1}$; the result then follows from Problem 1b. ∎

15.1.5 Remark: It is essential in the above proof that the singularly continuous spectrum of H was supposed to be empty; this allowed us to use the Riemann–Lebesgue lemma in the proof of Proposition 15.1.2. If $\sigma_{sc}(H) \neq \emptyset$ there may exist states orthogonal to both the sets $M_b(H)$ and $M_s(H)$. In a sense, they are "between" the bound and scattering states: the probability of finding the system within a fixed bound region may not have a zero limit as $|t| \to \pm\infty$, but its mean value is arbitrarily small if taken over a sufficiently long time interval (see the notes).

The identification of the bound states and scattering states with the pure point and absolutely continuous subspaces, respectively, of the full Hamiltonian, which we have made here for the system of two Schrödinger particles, can be also proved for other scattering systems. With this fact in mind, in scattering theory we usually replace the physically justified *definition* (15.5) of scattering states by the *assumption* (s2) $M_s(H) = \mathcal{H}_{ac}(H)$.

This is clearly consistent with (s1); the advantage is that now the set $M_s(H)$ depends automatically on the spectral properties of operator H only.

Now we are able to formulate two basic problems of scattering theory. The first of them concerns *existence of the wave operators* for a given pair H, H_0; if we adopt assumption (s2), definition (15.2) is reformulated as

$$\Omega_\pm(H, H_0) := \operatorname*{s-lim}_{t\to\pm\infty} U(t)^* U_0(t) P_{ac}(H_0) \,. \tag{15.8}$$

We have mentioned that the scattering process may be asymmetric with respect to the direction of time, in particular, that the S–operator exists under condition (15.3). If we require at the same time that an "initial" asymptotic state ψ_- corresponds to each "final" ψ_+, then the opposite inclusion must hold too (*cf.* Problem 4a). Finally, if we add the assumption that the validity of the asymptotic condition is not restricted, *i.e.*, a pair ψ_-, ψ_+ corresponds to any scattering state ψ (recall that $\psi_\pm$ are always mapped into $\mathcal{H}_{ac}(H)$ — see Problem 3b) we obtain

$$\operatorname{Ran} \Omega_+(H, H_0) = \operatorname{Ran} \Omega_-(H, H_0) = \mathcal{H}_{ac}(H) \,. \tag{15.9}$$

If this condition is satisfied the wave operators are said to be **complete**. This is one of two important properties we try to check for any scattering system; the other is the absence of the pathological states mentioned in Remark 15.1.5. The wave operators are **asymptotically complete** if they are complete and $\sigma_{sc}(H) = \emptyset$. Proof of asymptotic completeness represents the second basic problem of scattering theory, which is usually more complicated than the first.

Of course, a rigorous scattering theory is not restricted to proving existence and asymptotic completeness of the wave operators. One also has to deduce the relations between the S–matrix and the cross section, which represents the true observable quantity, to give meaning to the stationary scattering theory, *e.g.*, by proving the direct integral decomposition $S = \int^\oplus S(\lambda)\, d\lambda$ where $S(\lambda)$ is the on–shell S–matrix of the stationary theory; furthermore, one has to prove the dispersion relations, to introduce the Born series and check its convergence, to investigate scattering in a centrally symmetric potential (phase analysis in partial waves, Jost solutions, analytical properties of the scattering amplitude, *etc.*); we suppose the reader has already encountered a formal version of these results in quantum mechanical textbooks.

One should also introduce multichannel scattering formalism, allowing us to describe scattering of three or more particles (or two particles with an internal structure), to modify the definition of wave operators and other results for long–range potentials and for scattering on a time–dependent target, and also to investigate specific features of scattering in quantum field theory. However, such a program would require a separate book and we can only refer to the sources mentioned in the notes (see also an example in Section 15.4).

15.2 Existence of wave operators

Our aim is now to discuss some of the problems formulated at the end of the previous section, *i.e.*, to find some sufficient conditions for the existence and completeness of wave operators. We start with several simple assertions which are valid in any scattering system if only the operators (15.8) exist.

15.2.1 Proposition: (a) The wave operators $\Omega_\pm(H,H_0)$ are partial isometries with the initial subspace $\mathcal{H}_{ac}(H_0)$.

(b) The *intertwining relations* are valid:

$$\begin{aligned} U(t)\Omega_\pm(H,H_0) &= \Omega_\pm(H,H_0)U_0(t)\,, \qquad t\in\mathbb{R}\,,\\ \Omega_\pm(H,H_0)H_0 &\subset H\Omega_\pm(H,H_0)\,. \end{aligned}$$

Proof: The wave operators are bounded by definition. We have $\mathcal{H}_{ac}(H_0)^\perp\subset \operatorname{Ker}\Omega_\pm$, and on the other hand, $\|U(t)^*U_0(t)P_{ac}(H_0)\psi\| = \|\psi\|$ holds for any $\psi\in\mathcal{H}_{ac}(H_0)$ and $t\in\mathbb{R}$. Since $\Omega_\pm\equiv\Omega_\pm(H,H_0)$ are supposed to exist, the left side tends to $\|\Omega_\pm\psi\|$ as $t\to\pm\infty$; this proves (a). Consider next a fixed $t\in\mathbb{R}$; then

$$\Omega_\pm = \operatorname*{s-lim}_{s\to\pm\infty} U(s+t)^*U_0(s+t)P_{ac}(H_0) = U(t)^*\left[\operatorname*{s-lim}_{s\to\pm\infty} U(s)^*U_0(s)P_{ac}(H_0)\right]U_0(t)\,,$$

because $U(t)$, $U_0(t)$ are bounded, the group $\{U(t)\}$ is commutative and $U_0(t)$ commutes with $P_{ac}(H_0)$; the first intertwining relation then follows from the unitarity of operator $U(t)$. Now let ψ be an arbitrary vector from $D(H_0)$; then the boundedness of $\Omega_\pm$ implies

$$\left\|\Omega_\pm\,\frac{U_0(t)-I}{t}\psi + i\,\Omega_\pm H_0\psi\right\| \le \left\|\frac{U_0(t)-I}{t}\psi + iH_0\psi\right\|.$$

Due to the Stone theorem, the right side tends to zero as $t\to 0$, so the function $\Omega_\pm U_0(\cdot)\psi$ is differentiable at $t=0$ and its derivative equals $-i\,\Omega_\pm H_0\psi$. It follows from the already proven identity that

$$\operatorname*{s-lim}_{s\to 0}\frac{U(s)-I}{s}\Omega_\pm\psi = -i\Omega_\pm H_0\psi,$$

and therefore using the Stone theorem once again, we obtain $\Omega_\pm\psi\in D(H)$ and $H\Omega_\pm\psi = \Omega_\pm H_0\psi$. ∎

Similar intertwining relations hold for the adjoint operators $\Omega_\pm^*$ (Problem 5).

15.2.2 Proposition (chain rule): Let H, H_1, and H_0 be self–adjoint. If $\Omega_\pm(H,H_1)$ and $\Omega_\pm(H_1,H_0)$ exist; then the same is true for the wave operators $\Omega_\pm(H,H_0)$ and

$$\Omega_\pm(H,H_0) = \Omega_\pm(H,H_1)\,\Omega_\pm(H_1,H_0)\,.$$

Proof: Denote $U_1(t) := e^{-iH_1t}$. We have $\operatorname{Ran}\Omega_\pm(H_1,H_0) \subset \mathcal{H}_{ac}(H_1)$, and therefore

$$\operatorname*{s-lim}_{t\to\pm\infty} (I-P_{ac}(H_1))U_1^*(t)U_0(t)P_{ac}(H_0) = (I-P_{ac}(H_1))\,\Omega_\pm(H_1,H_0) = 0\,.$$

Next we use the decomposition

$$\begin{aligned} U^*(t)U_0(t)P_{ac}(H_0) &= U^*(t)U_1(t)P_{ac}(H_1)U_1(t)^*U_0(t)P_{ac}(H_0) \\ &\quad + U^*(t)U_1(t)(I-P_{ac}(H_1))U_1(t)^*U_0(t)P_{ac}(H_0)\,, \end{aligned}$$

where the second term vanishes in the limit $t\to\pm\infty$ in view of the above relation. By the sequential continuity of operator multiplication, the first term gives $\Omega_\pm(H,H_1)\,\Omega_\pm(H_1,H_0)$. ∎

We shall also prove a simple completeness criterion.

15.2.3 Proposition: The wave operators $\Omega_\pm(H,H_0)$ are complete *iff* $\Omega_\pm(H_0,H)$ exist.

Proof: If $\Omega_\pm(H,H_0)$ are complete, $P_\pm = P_{ac}(H)$ holds for the projections $P_\pm$ onto the subspaces $\operatorname{Ran}\Omega_\pm$, and Problem 5 implies $\Omega_\pm(H_0,H) = \Omega_\pm(H,H_0)^*$. Conversely, if $\Omega_\pm(H_0,H)$ exist, then the chain rule gives $\Omega_\pm(H,H_0)\,\Omega_\pm(H_0,H) = \Omega_\pm(H,H) = P_{ac}(H)$; hence $\operatorname{Ran}\Omega_\pm(H,H_0) \supset \mathcal{H}_{ac}(H)$, and the opposite inclusion follows from Problem 3b. ∎

After this introduction we want to derive some conditions which ensure the existence of wave operators.

15.2.4 Theorem (Cook criterion): Let H, H_0 be self-adjoint, and assume that there is a set $D\subset D(H_0)\cap\mathcal{H}_{ac}(H_0)$, which is dense in $\mathcal{H}_{ac}(H_0)$ and such that for any $\psi\in D$ there is $T_\psi>0$ with the following properties:

(i) $U_0(t)\psi\in D(H)$ for $|t|>T_\psi$.

(ii) The vector-valued functions $(H-H_0)U_0(\cdot)\psi$ are continuous on any compact subinterval of $\mathbb{R}\backslash(-T_\psi,T_\psi)$.

(iii) $\int_{T_\psi}^\infty \|(H-H_0)U_0(\pm t)\psi\|\,dt<\infty$.

Then the wave operators $\Omega_\pm(H,H_0)$ exist.

Proof: Denote $\psi(t) := U(t)^*U_0(t)\psi$ for $\psi\in D$. In view of Problem 6a, the function $\psi(\cdot)$ is continuously differentiable in (T_ψ,∞) with $\psi'(t) = iU(t)^*(H-H_0)U_0(t)\psi$. By Proposition A.5.1, we have

$$\|\,\psi(s)-\psi(r)\,\| \le \int_r^s \|\psi'(t)\|\,dt = \int_r^s \|\,(H-H_0)U_0(t)\psi\,\|\,dt\,.$$

for any $s>t>T_\psi$. Furthermore, using the absolute continuity of the integral in combination with assumption (iii), we see that $\|\psi(s)-\psi(r)\|\to 0$ as $r,s\to\infty$, *i.e.*, the limit

$$\lim_{t\to\infty} U(t)^*U_0(t)P_{ac}(H_0)\psi$$

exists for all $\psi \in D$. It also exists for $\psi \in \mathcal{H}_{ac}(H_0)^\perp$ (being equal to zero), and therefore for each $\psi \in (D \cup \mathcal{H}_{ac}(H_0)^\perp)_{lin}$. Due to the assumption, we can find ψ from this set for any $\phi \in \mathcal{H}$, $\varepsilon > 0$ in such a way that $\|\psi - \phi\| < \frac{\varepsilon}{3}$; using the standard trick we check that

$$\|(U(s)^* U_0(s) - U(r)^* U_0(r))\, P_{ac}(H_0)\phi\,\| < \varepsilon$$

for all r, s large enough, *i.e.*, that Ω_+ exists. The existence of Ω_- is proved in the same way. ∎

Let us remark that it is sometimes not difficult to check assumption (ii) of the theorem (Problem 7). It is often useful to introduce the *generalized wave operators* corresponding to a given $B \in \mathcal{B}(\mathcal{H})$; they are defined by

$$\Omega_\pm(H, H_0; B) := \operatorname*{s-lim}_{t\to\pm\infty} U(t)^* B U_0(t) P_{ac}(H_0)\,.$$

As an illustration, we prove the following modification of the previous theorem.

15.2.5 Theorem (Kupsch–Sandhas): Let H, H_0 be self–adjoint and $F \in \mathcal{B}(\mathcal{H})$. Suppose that the assumptions of Theorem 15.2.4 are valid with the exception of (iii), which is replaced by

$$\int_{T_\psi}^{\infty} \|(HB - BH_0)U_0(\pm t)\psi\|\, dt < \infty\,,$$

where $B := I - F$. If $F(H_0 - z)^{-\beta}$ is compact for some $z \in \rho(H)$, $\beta > 0$, and $D \subset D((H_0 - z)^\beta)$, the wave operators $\Omega_\pm(H, H_0)$ exist.

Proof: As in the previous theorem, we can prove the existence of $\Omega_\pm(H, H_0; B)$; hence it is sufficient to check that $\Omega_\pm(H, H_0; F) = 0$, which is equivalent to

$$\lim_{t\to\pm\infty} \|FU_0(t)\psi\| = 0$$

for any $\psi \in \mathcal{H}_{ac}(H_0)$. We choose first $\psi \in D$; then we have $\|FU_0(t)P_{ac}(H_0)\psi\| = \|F(H_0 - z)^{-\beta}U_0(t)(H_0 - z)^\beta\psi\|$, where we have used the fact that functions of H_0 commute mutually, together with the inclusion $D \subset \mathcal{H}_{ac}(H_0)$. By assumption, $F(H_0 - z)^{-\beta}$ is compact, so the condition is satisfied for any $\psi \in D$ in view of Problem 1a; the proof is completed by the density argument. ∎

Next we are going to formulate several existence conditions based, roughly speaking, on trace–class properties of the interaction Hamiltonian. They can be derived from the following fundamental theorem.

15.2.6 Theorem (Pearson): Let H, H_0 be self–adjoint and $B \in \mathcal{B}(\mathcal{H})$. Suppose that there is a trace–class operator C such that $(H\phi, B\psi) - (\phi, BH_0\psi) = (\phi, C\psi)$ for all $\phi \in D(H)$, $\psi \in D(H_0)$; then the generalized wave operators $\Omega_\pm(H, H_0; B)$ exist.

First we shall derive some general properties of the absolutely continuous spectral subspace $\mathcal{H}_{ac}(A)$ for an arbitrary $A \in \mathcal{L}_{sa}$. Recall that the Radon–Nikodým theorem associates with each $\psi \in \mathcal{H}_{ac}(A)$ a unique $f_\psi \in L^1(\mathbb{R})$, which is positive a.e. in $\mathbb{R}$ and satisfies the relation

$$\mu_\psi(M) := \|E_A(M)\psi\|^2 = \int_M f_\psi(t)\,dt$$

for any Borel $M \subset \mathbb{R}$, in particular, $\|\psi\|^2 = \|f_\psi\|_1$.

15.2.7 Lemma: The set $\mathcal{M}(A) := \{\psi \in \mathcal{H}_{ac}(A) : \|f_\psi\|_\infty < \infty\}$ has the following properties:

(a) $\mathcal{M}(A)$ is a dense subspace in $\mathcal{H}_{ac}(A)$,

(b) $\int_{\mathbb{R}} |(\phi, e^{-iAt}\psi)|^2 dt \le 2\pi\, \|\phi\|^2 \|f_\psi\|_\infty$ holds for any $\psi \in \mathcal{M}(A)$ and $\phi \in \mathcal{H}$.

Proof: The fact that $\mathcal{M}(A)$ is a subspace in $\mathcal{H}_{ac}(A)$ follows from the inequality $\mu_{\alpha\psi+\phi} \le 2|\alpha|^2\mu_\psi + 2\mu_\phi$, which implies $\|f_{\alpha\psi+\phi}\|_\infty \le 2|\alpha|^2\|f_\psi\|_\infty + 2\|f_\phi\|_\infty$ for any $\psi, \phi \in \mathcal{M}(A)$ and $\alpha \in \mathbb{C}$. Next we shall construct to an arbitrary $\eta \in \mathcal{H}_{ac}(A)$ a sequence $\{\psi_n\} \subset \mathcal{M}(A)$ such that $\psi_n \to \eta$. The corresponding vector $f_\eta \in L^1(\mathbb{R})$ can be represented by a Borel function which is *everywhere* positive and finite; with an abuse of notation we shall employ the symbol f_η again. The Borel sets $S_n := \{t \in \mathbb{R} : f_\eta(t) > n\}$ obviously form a nonincreasing family and $\bigcap_{n=1}^\infty S_n = \emptyset$. We set $\psi_n := (I - E_A(S_n))\eta$; then

$$\mu_{\psi_n}(M) = \int_M (1-\chi_{S_n}(t))\,d\mu_\eta(t) = \int_M (1-\chi_{S_n}(t))\,f_\eta(t)\,dt\,,$$

which shows that $\psi_n \in \mathcal{H}_{ac}(A)$ and the corresponding function is $f_{\psi_n} := (1-\chi_{S_n})f_\eta$; it follows from the construction that $\|f_{\psi_n}\|_\infty \le n$, so $\psi_n \in \mathcal{M}(A)$. Finally, the relation $\text{s-lim}_{n\to\infty} E_A(S_n) = E_A(\emptyset) = 0$ implies $\psi_n \to \eta$, and consequently, $\mathcal{M}(A)$ is a dense subspace in $\mathcal{H}_{ac}(A)$.

To prove (b) we introduce the measure $\nu_{\psi\phi}(\cdot) := (\psi, E_A(\cdot)\phi)$. Using the Schwarz inequality together with the fact that the projection $P_{ac}(A)$ onto $\mathcal{H}_{ac}(A)$ commutes with A, we find

$$|\nu_{\psi\phi}(M)|^2 \le \mu_\psi(M)\mu_\eta(M)\,,$$

where $\eta := P_{ac}(A)\phi$. This means that $\nu_{\psi\phi}(\cdot)$ is absolutely continuous with respect to the Lebesgue measure, so there is a unique $g_{\psi\phi} \in L^1(\mathbb{R})$ such that $\nu_{\psi\phi}(M) = \int_M g_{\psi\phi}(t)\,dt$ holds for all $M \in \mathcal{B}$. It follows then from Proposition A.4.5 that

$$(\psi, e^{-iAt}\phi) = \int_{\mathbb{R}} e^{-it\lambda} d\nu_{\psi\phi}(\lambda) = \int_{\mathbb{R}} e^{-it\lambda} g_{\psi\phi}(\lambda)\,d\lambda\,.$$

Hence we need to check that $g_{\psi\phi} \in L^2(\mathbb{R})$, in which case the last relation may be rewritten as $(\psi, e^{-iAt}\phi) = \sqrt{2\pi}\,(Fg_{\psi\phi})(t)$ and the unitarity implies

$$\int_{\mathbb{R}} |(\psi, e^{-iAt}\phi)|^2 dt = 2\pi\, \|g_{\psi\phi}\|_2^2\,.$$

The function $\nu_{\psi\phi}[\cdot] := \nu_{\psi\phi}((-\infty,\cdot))$ is absolutely continuous in $\mathbb{R}$, so $\frac{d}{dt}\nu_{\psi\phi}[t] = g_{\psi\phi}(t)$ with the possible exception of a set $N_{\psi\phi}$ of Lebesgue measure zero. The same is true for $\mu_\psi = \nu_{\psi\psi}$; then using the above mentioned consequence of the Schwarz inequality for the interval M with the endpoints t and $t+h$, we obtain

$$\left|\frac{\nu_{\psi\phi}[t+h]-\nu_{\psi\phi}[t]}{h}\right|^2 \leq \frac{1}{h}\left(\mu_\psi[t+h]-\mu_\psi[t]\right)\frac{1}{h}\left(\mu_\eta[t+h]-\mu_\eta[t]\right)$$

for all $t \in \mathbb{R}\backslash(N_{\psi\phi}\cup N_{\psi\psi}\cup N_{\eta\eta})$ and $h \neq 0$. Since $\psi \in \mathcal{M}(A)$, the limit $h\to 0$ yields $|g_{\psi\phi}(t)|^2 \leq f_\psi(t)f_\eta(t) \leq \|f_\psi\|_\infty f_\eta(t)$. We have $f_\eta \in L^1(\mathbb{R})$ by assumption, and therefore $g_{\psi\phi} \in L^2(\mathbb{R})$ with

$$\|g_{\psi\phi}\|^2 \leq \|f_\psi\|_\infty \|f_\eta\|_1 \leq \|f_\psi\|_\infty \|\eta\|^2 \leq \|f_\psi\|_\infty \|\phi\|^2,$$

where we have used $\eta = P_{ac}(A)\phi$; this concludes the proof. ∎

Proof of Theorem 6: Denote $\Omega(t) := U(t)^*BU_0(t)$ and $\Omega_{t,s} := \Omega(t)-\Omega(s)$. We shall prove the existence of $\Omega_+(H,H_0;B)$; the argument for $\Omega_-(H,H_0;B)$ is analogous. As in the proof of Theorem 4, it is sufficient to show that

$$\lim_{s,t\to\infty} \|\Omega_{t,s}\eta\| = 0$$

holds for all η of some dense subspace; we choose for it the set $\mathcal{M}(H_0)$ introduced above. First we shall find to any $\phi \in \mathcal{H}$ an integral representation of the vector $\Omega_{t,s}\phi$. Using Problem 6b, we readily check that the function $(\psi,\Omega(\cdot)\tilde{\psi})$ is differentiable for all $\psi \in D(H)$, $\tilde{\psi} \in D(H_0)$ and

$$\frac{d}{dt}(\psi,\Omega(t)\tilde{\psi}) = \frac{d}{dt}(B^*U(t)\psi, U_0(t)\tilde{\psi}) = (\psi, U(t)^*CU_0(t)\tilde{\psi}).$$

The derivative is clearly continuous, so the function $(\psi,\Omega(\cdot)\tilde{\psi})$ is absolutely continuous; then we have

$$(\psi,\Omega_{t,s}\tilde{\psi}) = (\psi,(\Omega(t)-\Omega(s))\tilde{\psi}) = i\int_s^t (\psi, U(\tau)^*CU_0(\tau)\tilde{\psi})\,d\tau$$

for all $s,t \in \mathbb{R}$. Since the vector-valued function $t \mapsto U(t)^*CU_0(t)\phi$ is continuous for any $\phi \in \mathcal{H}$, the Bochner integral $J_{ts}(\phi) := \int_s^t U(\tau)^*CU_0(\tau)\phi\,d\tau$ exists; the map $J_{ts}(\cdot)$ is linear and $\|J_{ts}(\phi)\| \leq |t-s|\,\|C\|\,\|\phi\|$ follows from Proposition A.5.1, so $J_{ts} \in \mathcal{B}(\mathcal{H})$. Moreover, Proposition A.5.2 gives $(\psi,\Omega_{t,s}\tilde{\psi}) = i(\psi, J_{ts}(\tilde{\psi}))$ for all $\psi \in D(H)$ and $\tilde{\psi} \in D(H_0)$. Thus $\Omega_{t,s} = iJ_{ts}$ and the sought integral expression is of the form

$$\Omega_{t,s}\phi = i\int_s^t C(\tau)\phi\,d\tau, \quad C(\tau) := U(\tau)^*CU_0(\tau).$$

Due to Problems 6b and 8b, the function $\tau \mapsto \|\Omega_{t+\tau,s+\tau}\phi\|^2$ is continuously differentiable for any $\phi \in \mathcal{H}$ and

$$\begin{aligned}\omega_{ts}^{(\phi)}(\tau) := \frac{d}{d\tau}\|\Omega_{t+\tau,s+\tau}\phi\|^2 &= 2\,\mathrm{Re}\,(\Omega_{t+\tau,s+\tau}\phi, (C(t+\tau)-C(s+\tau))\phi) \\ &= 2\,\mathrm{Re}\,(\Omega_{t,s}U_0(\tau)\phi, (C(t)-C(s))U_0(\tau)\phi).\end{aligned}$$

Since the function on the left side is continuous, we have

$$\|\Omega_{t+r,s+r}\phi\|^2 - \|\Omega_{t,s}\phi\|^2 = \int_0^r \omega_{ts}^{(\phi)}(\tau)\,d\tau\,.$$

By Problem 8a, the operator Ω_{ts} is compact for any $s,t \in \mathbb{R}$, and therefore $\lim_{r\to\infty}\Omega_{t+r,s+r}\eta = \lim_{r\to\infty} U(r)^*\Omega_{t,s}U_0(r)\eta = 0$ holds for $\eta \in \mathcal{M}(H_0) \subset \mathcal{H}_{ac}(H_0)$; it follows that

$$\|\Omega_{t,s}\eta\|^2 = -\lim_{r\to\infty}\int_0^r \omega_{ts}^{(\eta)}(\tau)\,d\tau\,.$$

It remains to check that $\omega_{ts}^{(\eta)}$ is integrable and $\lim_{s,t\to\infty}\int_0^\infty \omega_{t,s}^{(\eta)}(\tau)\,d\tau = 0$. It is sufficient to show that the continuous function $v_{ts}^{(\eta)} : \mathbb{R}^+ \to \mathbb{R}^+$ defined by

$$v_{ts}^{(\eta)}(\tau) := |(\Omega_{t,s}U_0(\tau)\eta, C(t)U_0(\tau)\eta)| = |(U(t)\Omega_{t,s}U_0(\tau)\eta, CU_0(t+\tau)\eta)|\,,$$

because it satisfies the inequality $|\omega_{ts}^{(\eta)}(\tau)| \le 2(v_{ts}^{(\eta)}(\tau) + v_{st}^{(\eta)}(\tau))$. To this end, we write the operator C in the canonical form (3.10), $C = \sum_j \mu(j)\,(\psi_j,\cdot)\phi_j$, where $\{\psi_j\}$, $\{\phi_j\}$ are orthonormal bases in $\mathcal{H}$ and $\sum_j \mu(j) = \operatorname{Tr}|C| < \infty$. Using the monotone–convergence theorem together with the Hölder inequality, we find

$$\int_0^\infty v_{ts}^{(\eta)}(\tau)\,d\tau \le \sum_j \mu(j)$$

$$\times\left(\int_0^\infty |(\psi_j, U_0(t+\tau)\eta)|^2 d\tau \int_0^\infty |(U(t)\Omega_{t,s}U_0(\tau)\eta, \phi_j)|^2 d\tau\right)^{1/2}.$$

Since $\eta \in \mathcal{M}(H_0)$, both integrals can be estimated by Lemma 15.2.7b,

$$\int_0^\infty |(\psi_j, U_0(t+\tau)\eta)|^2 d\tau = \int_t^\infty |(\psi_j, U_0(z)\eta)|^2 dz \le 2\pi\|f_\eta\|_\infty$$

$$\int_0^\infty |(U(t)\Omega_{t,s}U_0(\tau)\eta, \phi_j)|^2 d\tau \le 2\pi\|f_\eta\|_\infty\|\Omega_{t,s}^*U(t)^*\phi_j\|^2 \le 8\pi\|f_\eta\|_\infty\|B\|^2\,;$$

the last inequality follows from $\|\Omega_{t,s}^*\| = \|\Omega_{t,s}\| \le \|\Omega(t)\|+\|\Omega(s)\| \le 2\|B\|$. Denoting $\gamma_j(t) := \left(\int_t^\infty |(\psi_j, U_0(z)\eta)|^2 dz\right)^{1/2}$, we obtain in this way the inequality

$$0 \le \int_0^\infty v_{ts}^{(\eta)}(\tau)\,d\tau \le 2\|B\|(2\pi\|f_\eta\|_\infty)^{1/2}\sum_j \mu(j)\gamma_j(t) \le 8\pi\|B\|\,\|f_\eta\|\operatorname{Tr}|C|\,.$$

The functions $\gamma_j(\cdot)$ are uniformly bounded, so the series $\sum_j \mu(j)\gamma_j(t)$ converges uniformly with respect to t. On the other hand, $\lim_{t\to\infty}\gamma_j(t) = 0$ holds by Lemma 15.2.7b; then the last estimate gives the relation $\lim_{t\to\infty}\int_0^\infty v_{t,s}^{(\eta)}(\tau)\,d\tau = 0$, and therefore also $\lim_{s,t\to\infty}\int_0^\infty v_{t,s}^{(\eta)}(\tau)\,d\tau = 0$. ∎

Choosing $B = I$ in the just proved theorem, in combination with Proposition 15.2.3, we obtain the following result.

15.2.8 Corollary (Kato–Rosenblum theorem): Let $H = H_0 + V$, where H_0 is self–adjoint and V is a Hermitean trace–class operator; then the wave operators $\Omega_\pm(H, H_0)$ exist and are complete.

A simple application is given in Problem 9a. However, Theorem 6 allows us to derive other sufficient conditions covering much wider classes of interactions.

15.2.9 Theorem (Birman–Kuroda): Suppose that H, H_0 are self–adjoint and $(H-z)^{-1} - (H_0 - z)^{-1} \in \mathcal{J}_1(\mathcal{H})$ for some $z \in \rho(H) \cap \rho(H_0)$; then the wave operators $\Omega_\pm(H, H_0)$ exist and are complete.

Proof: We set $B := (H-z)^{-1}(H_0 - z)^{-1}$ and $C := (H_0 - z)^{-1} - (H-z)^{-1}$; then

$$(H\phi, B\psi) - (\phi, BH_0\psi) = ((H-\bar{z})\phi, B\psi) - (\phi, B(H_0 - z)\psi) = (\phi, C\psi)$$

holds for any $\phi \in D(H)$, $\psi \in D(H_0)$. Theorem 12.2.6 then implies existence of the limits $\lim_{t\to\pm\infty} U(t)^*(H-z)^{-1}(H_0-z)^{-1}U_0(t)P_{ac}(H_0)\eta$ for any $\eta \in \mathcal{H}$. Choosing $\eta := (H_0-z)\psi$ we see that $\lim_{t\to\pm\infty} U(t)^*(H-z)^{-1}U_0(t)P_{ac}(H_0)\psi$ exist for all $\psi \in D(H_0)$. However, this domain is dense and $\|U(t)^*(H-z)^{-1}U_0(t)P_{ac}(H_0)\| \le \|(H-z)^{-1}\|$ for all $t \in \mathbb{R}$, so repeating the standard density trick once more, we establish the existence of

$$\operatorname*{s-lim}_{t\to\pm\infty} U(t)^*(H-z)^{-1}U_0(t)P_{ac}(H_0)\,.$$

Next we use Problem 1a; since operator C is compact by assumption, we obtain $\text{s-lim}_{t\to\pm\infty} U(t)^*CU_0(t)P_{ac}(H_0) = 0$, which means in combination with the above result that the limits $\lim_{t\to\pm\infty} U(t)^*(H_0 - z)^{-1}U_0(t)P_{ac}(H_0)\eta$ exist for any $\eta \in \mathcal{H}$. Choosing $\eta = (H_0 - z)\psi$, we conclude that

$$\lim_{t\to\pm\infty} U(t)^*U_0(t)P_{ac}(H_0)\psi$$

exists for all $\psi \in D(H_0)$, and by the density argument, $\Omega_\pm(H, H_0)$ exist. Finally, the roles of H and H_0 in the above argument may be interchanged; hence completeness follows from Proposition 15.2.3. ∎

To derive one more consequence of Theorem 15.2.6 we introduce the following relation between a pair of self–adjoint operators H, H_0 : we say that H is *dominated* by H_0 if there are functions f, $f_0 : \mathbb{R} \to [1,\infty)$ such that $\lim_{|x|\to\infty} f(x) = \infty$, $D(f(H)) \supset D(f_0(H_0))$ and the operator $f(H)f_0(H_0)^{-1}$ is bounded. If H_0 is at the same time dominated by H, the operators are called *mutually dominated;* this is true, *e.g.*, if $D(H) = D(H_0)$ (Problem 10).

15.2.10 Theorem (Birman): Let the self–adjoint operators H, H_0 satisfy the following conditions:

(i) $E_H(J)(H-H_0)E_{H_0}(J)$ belongs to the trace class for any bounded interval J.

(ii) H and H_0 are mutually dominated.

Then the wave operators $\Omega_\pm(H,H_0)$ exist and are complete.

Proof: To any $\lambda\in\mathbb{R}^+$ we define $B_\lambda := E_H(J_\lambda)E_{H_0}(J_\lambda)$, where $J_\lambda := (-\lambda,\lambda)$. The identity

$$(H\phi, B_\lambda\psi) - (\phi, B_\lambda H_0\psi) = (\phi, E_H(J_\lambda)(H-H_0)E_{H_0}(J_\lambda)\psi)$$

holds for any $\phi\in D(H)$ and $\psi\in D(H_0)$, so assumption (i) together with Theorem 6 ensures existence of the generalized wave operators $\Omega_\pm(H,H_0;B_\lambda)$. The spectral–measure properties imply for a fixed $\lambda_0\in\mathbb{R}$ and $\psi\in\operatorname{Ran}E_{H_0}(\lambda_0)$ that the limits

$$\lim_{t\to\pm\infty} U(t)^*E_H(J_\lambda)U_0(t)P_{ac}(H_0)\psi$$

exist provided $\lambda\ge\lambda_0$. We shall prove that

$$\lim_{\lambda\to\infty}\sup_{t\in\mathbb{R}} \|(I-E_H(J_\lambda))U_0(t)P_{ac}(H_0)\psi\| = 0$$

holds at the same time for $\psi\in\operatorname{Ran}E_{H_0}(J_{\lambda_0})$. In view of (ii), H is dominated by H_0; hence there are functions f, f_0 with the above stated properties. We denote $F(\lambda) := \inf\{f(\xi) : |\xi|\ge\lambda\}$; by definition we have $\lim_{\lambda\to\infty}F(\lambda)=\infty$ and

$$\begin{aligned}
&\|\,(I-E_H(J_\lambda))\,U_0(t)P_{ac}(H_0)\psi\,\|\\
&\quad\le \|\,(I-E_H(J_\lambda))\,f(H)^{-1}\|\;\|f(H)f_0(H_0)^{-1}\|\;\|f_0(H_0)U_0(t)P_{ac}(H_0)E_{H_0}(J_{\lambda_0})\psi\,\|\\
&\quad\le F(\lambda)^{-1}\,\|f(H)f_0(H_0)^{-1}\|\;\|f_0(H_0)U_0(t)E_{H_0}(J_{\lambda_0})\|\;\|\psi\|
\end{aligned}$$

for any $\lambda\ge\lambda_0$. The first of the norms on the right side is finite due to (ii), and since J_{λ_0} is a bounded interval, the same follows for the second from functional–calculus rules; this yields the above limiting relation. Now we have

$$\begin{aligned}
&\|\,(U(t)^*U_0(t)-U(s)^*U_0(s))\,P_{ac}(H_0)\psi\,\|\\
&\quad\le \|\,(U(t)^*E_H(J_\lambda)U_0(t)-U(s)^*E_H(J_\lambda)U_0(s))\,P_{ac}(H_0)\psi\,\|\\
&\qquad+\|\,(I-E_H(J_\lambda))\,U_0(t)P_{ac}(H_0)\psi\,\| + \|\,(I-E_H(J_\lambda))\,U_0(s)P_{ac}(H_0)\psi\,\|\,.
\end{aligned}$$

Given $\varepsilon>0$ we can find a $\lambda\ge\lambda_0$ such that each of the last two terms is $<\frac{\varepsilon}{3}$. For this λ, there is a t_0 such that the first term is also $<\frac{\varepsilon}{3}$ for $s,t>t_0$; together we have shown that the limits

$$\lim_{t\to\pm\infty} U(t)^*U_0(t)P_{ac}(H_0)\psi$$

exist for $\psi\in\operatorname{Ran}E_{H_0}(J_{\lambda_0})$. Since λ_0 is an arbitrary number, the set of these vectors is dense in $\mathcal{H}$ and $\Omega_\pm(H,H_0)$ exist. Finally, conditions (i),(ii) are symmetric with respect to operators H, H_0, so completeness follows again from Proposition 15.2.3. ∎

In conclusion, we shall mention one result which allows us to extend substantially the domain of applicability of the above derived existence and completeness conditions; references to the proof are given in the notes.

15.2.11 Theorem (invariance principle): Let the function $\varphi : J \to \mathbb{R}$ on an open interval $J \subset \mathbb{R}$ be strictly monotonic and piecewise differentiable with φ' absolutely continuous. Assume that A_0, A_1 are self-adjoint, $\sigma(A_j) \subset \overline{J}$; and at each of the endpoints of J either φ has a finite limit or the point is not an eigenvalue of any of the operators A_j. Moreover, let one of the following conditions be valid:

(i) $A_1 - A_0 \in \mathcal{J}_1(\mathcal{H})$,

(ii) $(A_1 - z)^{-1} - (A_0 - z)^{-1} \in \mathcal{J}_1(\mathcal{H})$ for some $z \in \rho(A_0) \cap \rho(A_1)$.

Then the wave operators $\Omega_\pm(\varphi(A_1), \varphi(A_0))$ exist, are complete, and

$$\Omega_\pm(\varphi(A_1), \varphi(A_0)) = \begin{cases} \Omega_\pm(A_1, A_0) & \ldots \quad \varphi \text{ increasing} \\ \Omega_\mp(A_1, A_0) & \ldots \quad \varphi \text{ decreasing} \end{cases}$$

15.2.12 Corollary: Let self-adjoint H, H_0 satisfy one of the conditions:

(a) both the operators H, H_0 are positive, and either $H^2 - H_0^2 \in \mathcal{J}_1(\mathcal{H})$ or $(H^2 + \alpha^2)^{-1} - (H_0^2 + \alpha^2)^{-1} \in \mathcal{J}_1(\mathcal{H})$ for some $\alpha > 0$,

(b) $e^{-\beta H} - e^{-\beta H_0} \in \mathcal{J}_1(\mathcal{H})$ for some positive β.

Then the wave operators $\Omega_\pm(H, H_0)$ exist and are complete.

Proof: In the first case, it is sufficient to choose $J := \mathbb{R}$ with $\varphi(x) := \sqrt{x}$ for $x \ge 0$ and $\varphi(x) := x$ for $x < 0$. If assumption (b) is valid, we use $\varphi(x) := -\frac{1}{\beta} \log x$ on $J := (0, \infty)$. ∎

15.3 Potential scattering

Now we want to show what follows from the existence results of the preceding section in the case when H is a Schrödinger operator with a potential V and H_0 is the corresponding free Hamiltonian. Let us first consider two-particle scattering, where after separating the center-of-mass motion we have to compare Schrödinger operators H, H_0 on $L^2(\mathbb{R}^3)$.

15.3.1 Theorem (Hack–Cook): Let $H = H_0 + V(Q)$, where $V \in (L^2 + L^s)(\mathbb{R}^3)$ for some $s \in [2, 3)$; then the wave operators exist.

Proof: We shall use the Cook criterion. Operator H is self-adjoint due to Problem 11; we choose for D the subspace in $L^2(\mathbb{R}^3)$ spanned by the vectors

$$\psi_q : \psi_q(x) = \pi^{-3/4} e^{-|x-q|^2/2}$$

with $q \in \mathbb{R}^3$; then we obviously have $D \subset \mathcal{S}(\mathbb{R}^3) \subset D(H_0)$ and it is necessary to check that D is dense in $\mathcal{H}_{ac}(H_0) = L^2(\mathbb{R}^3)$. However, the ψ_q can be expressed as tensor products of the vectors $\psi_{q,p}$ from Example 8.1.7 with $p=0$ and $q \in \mathbb{R}$, which are dense in $L^2(\mathbb{R})$ in view of Theorem 2.2.7 (because an entire function vanishes everywhere if it vanishes on $\mathbb{R}$), so the required property follows from Proposition 2.4.4a. Since $D(H) = D(H_0)$ follows from Problem 11, the condition $U_0(t)D \subset D(H)$ is fulfilled automatically and assumption (ii) of Theorem 15.2.4 holds due to Problem 7; it remains for us to check assumption (iii). The free propagator acts on vectors of the subspace D in the way given by Theorem 9.3.1,

$$(U_0(t)\psi_q)(x) = \pi^{-3/4}\, a(t)^{3/4}\, e^{-(a(t)+ib(t))|x-q|^2/2},$$

where $a(t) := (1+4t^2)^{-1}$ and $b(t) := -2ta(t) + \frac{3}{2}\arg\left(\frac{1}{2}+it\right)$ (Problem 12a). Next we define $\psi_q^{(\beta,t)} := (I+|Q|)^\beta U_0(t)\psi_q$ for any positive β and $t \in \mathbb{R}$, where $|Q| := \left(\sum_{j=1}^3 Q_j^2\right)^{1/2}$. This function satisfies the inequality

$$\sup_{x\in\mathbb{R}^3} |\psi_q^{(\beta,t)}(x)| \le \left(\pi(1+4t^2)\right)^{-3/4} \max_{y\ge 0}\,(1+|q|+y)^\beta\, e^{-y^2/2(1+4t^2)}$$

and the maximum of the right side is not difficult to find; using the inequality $\sqrt{1+z} \le 1+\sqrt{z}$, we get the estimate

$$\sup_{x\in\mathbb{R}^3} |\psi_q^{(\beta,t)}(x)| = \left(\pi(1+4t^2)\right)^{-3/4}\left(1+|q|+\sqrt{\beta(1+4t^2)}\right)^\beta$$

so there is a positive K_q such that

$$\|(I+|Q|)^\beta U_0(t)\psi_q\|_\infty \le K_q\,(1+|t|)^{-3/2+\beta}$$

for all $t \in \mathbb{R}$; this allows us to estimate the expression, which appears in assumption (iii), for $H - H_0 = V(Q)$ as follows:

$$\|V(Q)U_0(t)\psi_q\| \le \|V(Q)(I+|Q|)^{-\beta}\|\, K_q\,(1+|t|)^{-3/2+\beta}$$

(Problem 12b). Due to the assumption, $V = V_2 + V_s$ with $V_r \in L^r$, so we obtain

$$\|V(Q)U_0(t)\psi_q\| \le \left\{\|V_2\| + \|V_s(Q)(I+|Q|)^{-\beta}\|\right\} K_q\,(1+|t|)^{-3/2+\beta},$$

because $\|(I+|Q|)^{-\beta}\|_\infty \le 1$. The first norm in the curly brackets is finite, and the second can be estimated by the Hölder inequality,

$$\|V_s(Q)(I+|Q|)^{-\beta}\|^2 \le \|V_s\|_s^2 \left(4\pi\int_0^\infty (1+r)^{-\frac{2\beta s}{s-2}}\, r^2\, dr\right)^{1-\frac{2}{s}};$$

the right side is finite provided $\beta > \frac{3(s-2)}{2s}$. The integral of assumption (iii) exists if $\beta < \frac{1}{2}$ and we readily verify that for any $s \in [2,3)$ there is a β which obeys these restrictions. ∎

The existence of wave operators in some cases provides us with information about the absolutely continuous spectrum of the full Hamiltonian.

15.3.2 Example: Due to Problem 3a, the operators $H_0 \upharpoonright \mathcal{H}_{ac}(H_0)$ and $H \upharpoonright \mathrm{Ran}\, \Omega_\pm$ are unitarily equivalent; combining this result with Proposition 4.2.6, we obtain

$$\sigma_{ac}(H_0) \subset \sigma_{ac}(H) \subset \sigma_{ess}(H)\,.$$

Consider, *e.g.*, Schrödinger operator $H = H_0 + V(Q)$ on $L^2(\mathbb{R}^3)$ with a potential V, which satisfies the following conditions:

(i) $V \in L^2_{loc}(\mathbb{R}^3)$.

(ii) There are positive C, R, ε such that $|V(x)| < Cr^{-1-\varepsilon}$ if $r := |x| > R$.

Such a potential belongs to $L^2 + L^s$ with $s \in (3(1+\varepsilon)^{-1}, 3)$, so the wave operators exist. At the same time, $V \in L^2 + L^\infty_\varepsilon$, so $\sigma_{ess}(H) = \mathbb{R}^+$ by Theorem 14.4.1. Finally, we know from Example 7.5.8 that $\sigma_{ac}(H_0) = \mathbb{R}^+$. Hence we get $\sigma_{ac}(H) = \mathbb{R}^+$, *i.e.*, Proposition 14.4.6.

15.3.3 Remark: If the wave operators are complete, then we have the identity $\sigma_{ac}(H_0) = \sigma_{ac}(H)$ instead of the inclusion, and we need not know the essential spectrum. The trouble is that it is usually easier to find $\sigma_{ess}(H)$ than to check the completeness of wave operators.

One of the ways to prove the completeness is based on the Kato–Birman theory discussed in the previous section. Let us mention a typical result.

15.3.4 Theorem: Let $H = H_0 + V(Q)$ be a Schrödinger operator on $L^2(\mathbb{R}^3)$ with $V \in (L^1 \cap L^2)(\mathbb{R}^3)$; then the wave operators $\Omega_\pm(H, H_0)$ exist and are complete.

Proof: In view of Theorem 14.1.2, the domains of the two operators coincide, $D(H) = D(H_0)$, so they are mutually dominated and the Birman theorem may be used. It is sufficient to check that $E_H(J)|V(Q)|^{1/2}$ and $|V(Q)|^{1/2} E_{H_0}(J)$ are Hilbert–Schmidt for any bounded $J \subset \mathbb{R}$, since then their product multiplied by $\mathrm{sgn}\, V(Q) \in \mathcal{B}(\mathcal{H})$ will belong to $\mathcal{J}_1(\mathcal{H})$. As for the second operator, this follows from Proposition 15.1.3 because $|V(Q)|^{1/2} \in L^2$ due to the assumption. In the first case it is enough to check that $(H-\lambda)^{-1}|V(Q)|^{1/2}$ is Hilbert–Schmidt, because $E_H(J)(H-\lambda)$ is bounded. The second resolvent formula gives

$$(H-\lambda)^{-1}|V(Q)|^{1/2} = \left[I + (H_0-\lambda)^{-1}V(Q)\right]^{-1} (H_0-\lambda)^{-1}|V(Q)|^{1/2}$$

provided the inverse operator exists. Proposition 15.1.3 and Problem 2 yield for $\lambda < 0$ the inequality

$$\left\|(H_0-\lambda)^{-1}V(Q)\right\| \le \frac{\|V\|}{\sqrt{8\pi}}\, |\lambda|^{-1/4};$$

hence by choosing a sufficiently large negative λ, the norm can be made smaller than one, in which case the operator $C := I + (H_0-\lambda)^{-1}V(Q)$ is invertible. By another application of Proposition 15.1.3, $(H_0-\lambda)^{-1}|V(Q)|^{1/2}$ is Hilbert–Schmidt and its product with the bounded operator C^{-1} is also Hilbert–Schmidt. ∎

Compared with Theorem 15.3.1, the just proved result requires a much faster decay of the potential at large distances. Roughly speaking, we need $|V(x)| \le Cr^{-3-\varepsilon}$ for large values of r. This restriction can be weakened, *e.g.*, for central potentials where the Birman–Kuroda theorem applies (see the notes).

15.3.5 Theorem: Let V be a central potential on $\mathbb{R}^3$, $V(x)=v(r)$ for a measurable function $v:\mathbb{R}^+\to\mathbb{R}$, which satisfies the condition

$$\int_0^1 r\,|v(r)|\,dr + \int_1^\infty |v(r)|\,dr < \infty .$$

Then $V(Q)$ is H_0–bounded with zero relative bound, and the corresponding wave operators $\Omega_\pm(H_0+V(Q),H_0)$ exist and are complete.

Let us finally say a few words about asymptotic completeness. The most difficult part is usually to check that the singularly continuous spectrum is void. In Section 14.4 we have derived one sufficient condition; in combination with Theorem 4 it provides the following result.

15.3.6 Theorem: Let $H=H_0+V(Q)$ be a Schrödinger operator on $L^2(\mathbb{R}^3)$ with a potential $V\in(L^1\cap L^2)(\mathbb{R}^3)$ which satisfies the condition of Theorem 14.4.8; then the wave operators $\Omega_\pm(H,H_0)$ are asymptotically complete. In addition, the spectrum of H is purely absolutely continuous, so $\Omega_\pm(H,H_0)$ are unitary.

One can prove asymptotic completeness for much wider classes of potentials; however, the methods used to this aim are not simple. We limit ourselves to quoting without proof two important results (see the notes). A potential $V:\mathbb{R}^n\to\mathbb{R}$ is said to satisfy the *Agmon condition* if there is an $\varepsilon>0$ such that the operator T_f of multiplication by $f: f(x)=(1+|x|^2)^{1/2+\varepsilon}V(x)$ is H_0–compact. An example is any potential of the form $V(x):=(1+|x|^2)^{-1/2-\varepsilon}f(x)$ with $f\in(L^p+L^\infty_\varepsilon)(\mathbb{R}^n)$, where p fulfils the conditions (14.2).

In order to introduce the second mentioned class of potentials we employ the "localization" projections F_r onto the subspaces $L^2(B_r)$ which we described loosely in Section 15.1; recall that in the simplest case of a single particle scattered by a potential the natural choice for B_r is the family of concentric balls B_r parametrized by the radius. Assume that the potential $V:\mathbb{R}^n\to\mathbb{R}$ gives rise to a H_0–bounded operator with a relative bound <1, and furthermore, that $V(Q)(H_0+i)^{-1}$ is bounded and one may define

$$h: h(r) = \left\|(I-F_r)V(Q)(H_0+i)^{-1}\right\| .$$

It is easy to check that $h:\mathbb{R}^+\to\mathbb{R}^+$ is nonincreasing with $\lim_{r\to\infty}h(r)=0$; if it belongs to $L^2(\mathbb{R}^+)$ the potential is said to satisfy the *Enss condition.*

15.3.7 Theorem: Let $H=H_0+V(Q)$ be a Schrödinger operator on $L^2(\mathbb{R}^n)$ with a potential V, which satisfies one of the two above stated conditions; then the wave operators exist and are asymptotically complete.

15.3.8 Remarks: (a) The classes of potentials specified by the Agmon and Enss conditions are similar. To illustrate this, take $V(x) := (1+|x|^2)^{-1/2-\varepsilon} f(x)$, for which we find

$$h(r) \leq (1+r^2)^{-1/2-\varepsilon} \| T_f(H_0+i)^{-1} \| ;$$

hence V satisfies the Enss condition if $V(Q)$ is H_0–bounded with a relative bound <1; recall also that a H_0–compact operator is H_0–bounded with zero relative bound. On the other hand, the methods by which one proves Theorem 15.3.7 are substantially different for the two conditions; the Enss method is technically simpler and admits an extension to scattering in N–particle systems (see the notes).

(b) It T follows from the theorem that $\sigma_{ac}(H) = \sigma_{ac}(H_0) = \mathbb{R}^+$; if the potential satisfies the Agmon condition, we have also $\sigma_{ess}(H) = \mathbb{R}^+$ because the operator $V(Q)$ is H_0–compact. This is not true for the Enss condition; however, one can prove that H has in this case no nonzero eigenvalues of infinite multiplicity, and that $\lambda = 0$ is the only possible accumulation point of $\sigma_p(H)$. Since $\sigma_{sc}(H) = \emptyset$ and $\sigma_{ac}(H) = \mathbb{R}^+$, we see that any negative $\lambda \in \sigma(H)$ is an isolated point of the spectrum with $\dim E_H(\{\lambda\}) < \infty$, so $\sigma_{ess}(H) \subset \mathbb{R}^+$. Since the opposite implication is also valid due to Example 2, we find in case of the Enss condition again $\sigma_{ess}(H) = \mathbb{R}^+$.

15.4 A model of two–channel scattering

As mentioned in Section 15.1, the scope of the present book does not allow us to discuss the rigorous scattering theory to its full extent. To give the reader just a flavor of what was left out, we are going to discuss in the final section a model which is explicitly solvable, and at the same time, it exhibits some features of realistic scattering systems, namely simultaneous occurrence of the elastic and inelastic scattering, and a resonance scattering in the elastic channel.

The model will describe a pair of nonrelativistic particles, which may exist in two states; for definiteness we can imagine a neutron plus a nucleus having two internal states: the ground state and an excited state. After separating the center–of–mass motion, the state Hilbert space is therefore $\mathcal{G} = \mathcal{G}_1 \oplus \mathcal{G}_2$ with $\mathcal{G}_j := L^2(\mathbb{R}^3)$. Since the model is supposed to be Galilei–invariant, the Bargmann superselection rule requires the reduced masses in the two channels to be the same (see Remark 10.3.2); for simplicity we put $\mu_1 = \mu_2 = \frac{1}{2}$. Our basic assumption is that the full interaction between the particles, including the part which is responsible for the transitions between the two channels, has a *contact nature*, *i.e.*, it is supported by the origin of coordinates in the relative–coordinate space only.

Such an interaction can be constructed by the procedure described in Section 14.6; we start from the Hamiltonian (or its *e.s.a.* restriction) which describes the free motion in the two channels, and restrict it to the set of functions whose supports are separated from the origin. In this way we get a non–selfadjoint operator $A_0 := A_{0,1} \oplus A_{0,2}$ with

$$A_{0,1} = -\Delta\,, \qquad D(A_{0,1}) = C_0^\infty(\mathbb{R}^3 \setminus \{0\})\,,$$

$$A_{0,2} = -\Delta + E\,, \quad D(A_{0,2}) = C_0^\infty(\mathbb{R}^3 \setminus \{0\})\,,$$

where E is a positive number representing the *threshold energy* of the inelastic channel. To specify the dynamics, we have to say what happens when the particles "hit each other"; this will again be achieved by choosing the Hamiltonian among self–adjoint extensions of A_0.

These extensions can easily be constructed since the deficiency indices of A_0 are $(2,2)$. This is seen from the partial–wave decomposition in the same way as in Theorem 14.6.4; due to the presence of the centrifugal barrier, the component of $A_{0,j}$ in the ℓ–th partial wave is *e.s.a.* if $\ell \geq 1$, so only the s–wave parts can be coupled. In this way the problem is substantially simplified: using the unitary operator (11.16) we pass to the reduced radial wave functions $f\,:\, f(r) := r\psi(r)$; then we may consider $\mathcal{H} = \mathcal{H}_1 \oplus \mathcal{H}_2$ with $\mathcal{H}_j := L^2(\mathbb{R}^+)$ as the state space of the problem, and the "starting operator" will be $H_0 := H_{0,1} \oplus H_{0,2}$, where

$$H_{0,1} = -\frac{d^2}{dr^2}\,, \qquad D(H_{0,1}) = C_0^\infty(0,\infty)\,,$$
$$H_{0,2} = -\frac{d^2}{dr^2} + E\,, \quad D(H_{0,2}) = C_0^\infty(0,\infty)\,. \tag{15.10}$$

The operator H_0 has the deficiency indices $(2,2)$, and therefore a four–parameter family of self–adjoint extensions. The most convenient way to characterize them is again through suitable boundary conditions.

15.4.1 Proposition: For any $u := \{a,b,c\}$ with $a,b \in \mathbb{R}$ and $c \in \mathbb{C}$ denote by H_u the operator defined by the same differential expression (15.10) as H_0, with the domain $D(H_u) \subset D(H_0^*)$ specified by the boundary conditions

$$f_1'(0) = af_1(0) + cf_2(0)\,, \quad f_2'(0) = \bar{c}f_1(0) + bf_2(0)\,. \tag{15.11}$$

Then H_u is a self–adjoint extension of H_0.

Proof: Using the results of Section 4.8, we obtain $D(H_0^*) = AC^2(R^+) \oplus AC^2(\mathbb{R}^+)$, and the Lagrange formula of Problem 4.53 yields

$$(H_0^* f, g) = \bar{f}_1'(0)g_1(0) + \bar{f}_2'(0)g_2(0) - \bar{f}_1(0)g_1'(0) - \bar{f}_2(0)g_2'(0) + (f, H_0 g)$$

for any $f,g \in D(H_0^*)$; it is simple to check that the boundary term vanishes under the conditions (15.11). ∎

In addition to the extensions specified here there are some exceptional ones (for instance, those obtained formally by inserting $a=\infty$ or $b=\infty$ into the boundary conditions). These can be described by the standard methods discussed in Chapter 4 (*cf.* Problem 13a), but for the purposes of our model it is sufficient to know the "most part" of the admissible Hamiltonians.

It is obvious that c is the complex *coupling constant* between the channels; if $c=0$ we have $H_u = H_a \oplus H_b$, where the last named operators correspond to the s–wave parts of the point–interaction Hamiltonians $H_{\alpha,0}$ and $H_{\beta,0}$ in the two channels (*cf.* Theorem 14.6.4) with the interaction strengths $\alpha := a/4\pi$ and $\beta := b/4\pi$, respectively. Hence the spectrum of H_u in the case of *decoupled* channels is easily obtained from the spectra of H_α and H_β; below we shall discuss $\sigma(H_u)$ in the interacting case.

15.4.2 Remark: The above specified family of Hamiltonians is restricted further if we demand the interaction to be time–reversal invariant — see Problem 13b. In fact, we can easily check that H_u and $H_{u'}$ with $a'=a$, $b'=b$ and $c'=ce^{i\alpha}$ are unitarily equivalent by the operator $U\,:\, U(f_1\oplus f_2) = e^{i\alpha/2}f_1 \oplus e^{-i\alpha/2}f_2$.

To find the spectrum of H_u we first have to determine its resolvent. The form of the Hamiltonian allows us to do it explicitly. For simplicity, we shall write elements of $\mathcal{H}$ as columns, $f := \begin{pmatrix} f_1 \\ f_2 \end{pmatrix}$.

15.4.3 Proposition: For any $z\in\rho(H_u)$, we denote $k:=\sqrt{z}$. The resolvent $(H_u - z)^{-1}$ is an integral operator on $\mathcal{H}$ with the kernel

$$G_u(r,r';k) = \begin{pmatrix} \frac{e^{ik|r+r'|}-e^{ik|r-r'|}}{2ik} & 0 \\ 0 & \frac{e^{i\kappa|r+r'|}-e^{i\kappa|r-r'|}}{2i\kappa} \end{pmatrix} + D^{-1}\begin{pmatrix} (b-i\kappa)\,e^{ik(r+r')} & -c\,e^{(ikr+\kappa r')} \\ -\bar c\,e^{i(\kappa r+kr')} & (a-ik)\,e^{i\kappa(r+r')} \end{pmatrix},$$

where $D := (a-ik)(b-i\kappa) - |c|^2$ and $\kappa := \sqrt{k^2-E}$.

Proof: If $u_0 := \{a,b,0\}$ the operator H_{u_0} decouples into an orthogonal sum, so its resolvent kernel expresses through the known resolvent kernels of the point–interaction Hamiltonians H_a and H_b as

$$G_{u_0}(r,r';k) = \begin{pmatrix} g_a(r,r';k) & 0 \\ 0 & g_b(r,r';\kappa) \end{pmatrix},$$

where

$$g_a(r,r';k) = \frac{e^{ik|r+r'|}-e^{ik|r-r'|}}{2ik} + \frac{e^{ikr}e^{ikr'}}{a-ik}$$

and g_b is obtained by replacing a,k with b,κ, respectively. Since H_u and H_{u_0} are self–adjoint extensions of the same symmetric operator H_0 with finite deficiency indices, the resolvent of H_u can be expressed by means of Theorem 4.7.15: its kernel equals

$$G_u(r,r';k) = G_{u_0}(r,r';k) + \sum_{j,l=1}^{2} \lambda_{jl}(k^2)\, F_j^k(r) F_l^k(r')^t ,$$

where t means transposition and the 2×1 columns F_j^k can be chosen as

$$F_1^k(r) := \begin{pmatrix} g_a(r,0;k) \\ 0 \end{pmatrix}, \quad F_2^k(r) := \begin{pmatrix} 0 \\ g_b(r,0;\kappa) \end{pmatrix};$$

this ensures that the F_j^k solve the equation $(H_0^* - k^2)F_j^k = 0$, *i.e.*, they belong to the deficiency subspaces of operator H_0. It remains to find the functions $\lambda_{jl}(\cdot)$. We use the fact that the resolvent $(H_u - z)^{-1}$ maps $\mathcal{H}$ into $D(H_u)$, which means that for any $f \in \mathcal{H}$, the component functions of $(H_u - z)^{-1} f$ must satisfy the boundary conditions (15.11). This requirement yields a system of four linear equations,

$$-\lambda_{11}(k^2) = \frac{c}{b - i\kappa}\lambda_{21}(k^2), \qquad -\lambda_{21}(k^2) = \bar{c}\left(1 + \frac{\lambda_{11}(k^2)}{a - ik}\right),$$

$$-\lambda_{12}(k^2) = c\left(1 + \frac{\lambda_{22}(k^2)}{b - i\kappa}\right), \quad -\lambda_{22}(k^2) = \frac{\bar{c}}{a - ik}\lambda_{12}(k^2),$$

which is readily solved (Problem 13c); substituting into the above expression of $G_{u_0}(r,r';k)$, we arrive at the result. ∎

The explicit form of the resolvent makes it possible to study its singularities. In fact, it is sufficient to find zeros of the "discriminant" D appearing in the resolvent kernel. To see this, notice that for $z \in \mathbb{C}\backslash\mathbb{R}^+$, the resolvent can be written as $(H_u - z)^{-1} = A(z) + D^{-1}(z)B(z)$, where the two operators corresponding to the matrices in Proposition 15.4.3 are bounded, and therefore they cannot give rise to a pole. On the other hand, we can check directly from Proposition 15.4.1 that H_u has no embedded eigenvalue unless $c = 0$. Of course, the resolvent also has cuts associated with the continuous spectrum, which come from the square roots in definitions of k and κ, but they are independent of the interaction.

Consider first the decoupled case, $c = 0$. The discriminant then factorizes, so $D = 0$ holds *iff* $k = -ia$ or $\kappa = -ib$. By Problem 4.24b, the spectrum is the union of spectra of the two point–interaction operators, the second of which is shifted by the constant E. We have to distinguish several cases:

$a < 0$ ... H_u has the eigenvalue $-a^2$ corresponding to the normalized eigenfunction $f(r) = \sqrt{-2a}\begin{pmatrix} e^{ar} \\ 0 \end{pmatrix}$.

$a \ge 0$ ... the pole now corresponds to a "zero–energy resonance" ($a = 0$) or to an "antibound state" (*i.e.*, a resonance on the negative imaginary axis of k — deeply "hidden" on the second sheet of energy).

$b < 0$ ... H_u has the eigenvalue $E - b^2$ corresponding to $\sqrt{-2b}\begin{pmatrix} 0 \\ e^{br} \end{pmatrix}$.

$b \geq 0 \quad \ldots \quad H_u$ has a zero–energy resonance or an antibound state.

The continuous spectrum of the decoupled operator is again the union of the continuous spectra of H_a and H_b, *i.e.*, it covers the interval $[0,\infty)$ being simple in $[0,E)$ and of multiplicity two in $[E,\infty)$ (*cf.* Problem 13d; we say that the spectrum of a self–adjoint A has *multiplicity* n in an interval J if $E_A(J)A$ is unitarily equivalent to $\bigoplus_{j=1}^n A_j^{(J)}$, where each $A_j^{(J)}$ has a simple spectrum). Our main interest concerns the cases when both a,b are negative; among these, we shall pay particular attention to the situation when $b^2 < E$, *i.e.*, the eigenvalue of H_b is embedded into the continuous spectrum of H_a.

After this preliminary, let us turn to the interacting case, $c \neq 0$. Since the deficiency indices of H_0 have been finite, the essential spectrum is not affected by the interaction due to Theorem 4.7.14. To find the eigenvalues (or resonances) of H_u, we have to solve the equation

$$(a-ik)\left(b-i\sqrt{k^2-E}\right) = |c|^2. \tag{15.12}$$

It reduces to a quartic equation, and can therefore be solved in terms of radicals; however, the formulas expressing the roots for general a,b, and c are too complicated to be of practical use. We shall therefore concentrate our attention on the *weak–coupling case* in which the equation (15.12) can be solved perturbatively; at the same time it represents one of the physically most interesting situations.

15.4.4 Theorem: (a) Let the point spectrum of the decoupled operator H_{u0} be nondegenerate, $-a^2 \neq E-b^2$. then the perturbed first–channel eigenvalue (resonance) behaves for small $|c|$ as

$$e_1(c) = -a^2 + \frac{2a|c|^2}{b+\sqrt{a^2+E}} + \frac{a^2-E-b\sqrt{a^2+E}}{\sqrt{a^2+E}\left(b+\sqrt{a^2+E}\right)^3}|c|^4 + \mathcal{O}(|c|^6)\,.$$

In particular, the zero–energy resonance corresponding to $a=0$ turns into an antibound state if H_{u0} has an isolated eigenvalue in the second channel, $b<-\sqrt{E}$, and into a bound state otherwise.

(b) Under the same assumption, let H_{u0} have an isolated eigenvalue in the second channel, $b<-\sqrt{E}$; then the perturbation changes it into

$$e_2(c) = E - b^2 + \frac{2b|c|^2}{a+\sqrt{b^2-E}} + \frac{b^2+E-a\sqrt{b^2-E}}{\sqrt{b^2-E}\left(a+\sqrt{b^2-E}\right)^3}|c|^4 + \mathcal{O}(|c|^6)\,.$$

On the other hand, if H_{u0} has an embedded eigenvalue, $-\sqrt{E} < b < 0$, then the latter under the perturbation gives rise to a pole of the analytically continued resolvent with

$$\mathrm{Re}\, e_2(c) = E - b^2 + \frac{2ab|c|^2}{a^2-b^2+E} + \mathcal{O}(|c|^4)\,,$$

$$\mathrm{Im}\, e_2(c) = \frac{2b|c|^2\sqrt{E-b^2}}{a^2-b^2+E} + \mathcal{O}(|c|^4)\,.$$

(c) Finally, suppose that H_{u_0} has a nonsimple eigenvalue, $b = -\sqrt{a^2+E}$. Under the perturbation, it splits intov

$$e_{1,2}(c) \;=\; -a^2 \mp 2\sqrt{-a}\sqrt[4]{a^2+E}\,|c| \;+\; \frac{2a^4+4a^2E+E^2}{2a(a^2+E)^{3/2}}\,|c|^2 \;+\; \mathcal{O}(|c|^3)\,.$$

Proof: We rewrite the spectral condition (15.12) as $(\alpha - iz)\left(\beta - i\sqrt{z^2-1}\right) = |\gamma|^2$, where $\alpha := a/\sqrt{E}$ and β,γ,z are similarly renormalized a,b,k. To prove part (a), we put $z = -i\alpha + \sum_{n=1}^{\infty} c_n|\gamma|^{2n}$. Substituting this into the above condition and expanding the left side in powers of $|\gamma|^2$, we get an infinite system of equations,

$$\begin{aligned} -ic_1(\beta+A) \;-\; 1 \;&=\; 0\,,\\ -ic_2(\beta+A) \;+\; \alpha c_1^2A^{-1} \;&=\; 0\,,\\ -ic_3(\beta+A) \;+\; 2\alpha c_1c_2A^{-1} + \frac{i}{2}\,c_1^3A^{-3} \;&=\; 0\,, \end{aligned}$$

etc., where $A := \sqrt{\alpha^2+1}$, which can be solved recursively (Problem 13e). Returning to the original quantities, we have

$$\begin{aligned} k_1 \;=\; -ia \;&+\; \frac{i|c|^2}{b+\sqrt{a^2+E}} \;+\; \frac{ia|c|^4}{\sqrt{a^2+E}\left(b+\sqrt{a^2+E}\right)^3}\\ &-\,i\,\frac{bE+(E-4a^2)\sqrt{a^2+E}}{2(a^2+E)^{3/2}\left(b+\sqrt{a^2+E}\right)^5}\,|c|^6 \;+\; \mathcal{O}(|c|^8)\,, \end{aligned}$$

so taking the square we arrive at the expression for $e_1(c)$. Notice that k_1 is purely imaginary; if $a=0$ its sign for small $|c|$ is given by the denominator of the second term on the right side. Assertion (b) is proved in a similar manner; we obtain

$$k_2 = i\sqrt{b^2-E} - \frac{ib|c|^2}{\sqrt{b^2-E}\left(a+\sqrt{b^2-E}\right)} - i\,\frac{aE+(2b^2+E)\sqrt{b^2-E}}{2(b^2-E)^{3/2}\left(a+\sqrt{b^2-E}\right)^3}\,|c|^4 + \mathcal{O}(|c|^6)\,,$$

which leads to the expression for the eigenvalue $e_2(c)$, and

$$k_2 = \sqrt{E-b^2} + \frac{b|c|^2}{\sqrt{E-b^2}\left(a-i\sqrt{E-b^2}\right)} + \frac{i(2b^2+E)\sqrt{E-b^2}-aE}{2(E-b^2)^{3/2}\left(a-i\sqrt{E-b^2}\right)^3}\,|c|^4 + \mathcal{O}(|c|^6)\,,$$

respectively. In the last named case k_2 is no longer purely imaginary and its square equals

$$e_2(c) \;=\; E \;-\; b^2 \;+\; \frac{2b|c|^2}{a-i\sqrt{E-b^2}} \;+\; \frac{i(b^2+E)-a\sqrt{E-b^2}}{\sqrt{E-b^2}\left(a-i\sqrt{E-b^2}\right)^3}\,|c|^4 \;+\; \mathcal{O}(|c|^6)\,;$$

taking the real and imaginary parts we obtain the remaining relations of part (b). As in Section 9.6, the resolvent of H_u is an analytic function out of the real axis.

The corresponding pole is not on the first sheet but can be reached by analytic continuation; note that the imaginary part of the pole position is negative.

Finally, in case (c) the above perturbation expansions lose meaning; we replace them by $z = -i\alpha + i\sum_{k=1}^{\infty} b_k|\gamma|^k$. Substitution into the spectral condition (15.12) yields a system of equations which can again be solved recursively; however, for the odd coefficients we now get two solutions differing in sign,

$$k_{1,2} = -ia \pm i\,\frac{\sqrt[4]{a^2+E}}{\sqrt{-a}}\,|c| - \frac{iE^2|c|^2}{4a^2(a^2+E)^{3/2}} \pm i\,\frac{\sqrt{-a}E(5E-8a^2)}{32a^4(a^2+E)^{5/4}}\,|c|^3 + \mathcal{O}(|c|^4)\,.$$

For both signs and $|c|$ small enough, this value is negative imaginary; the proof is concluded by taking its square. ∎

15.4.5 Remarks: (a) It follows from the first two assertions that the interaction always *repels* the eigenvalues; in the case (c) it removes the degeneracy. The appearance of the powers of $|c|$ in the last expansion is related to the fact that the original eigenvalue had multiplicity two; we have obtained an example of the so–called *Puiseaux series* — see the notes.

(b) We have mentioned that the continuous spectrum of H_u is the same as in the decoupled case; we can check, moreover, that it is also absolutely continuous (Problem 14a).

After this preliminary, let us now consider the model in question as a scattering problem. By Proposition 15.4.3, the interaction represents a finite–rank perturbation to the free resolvent; hence the existence and completeness of the wave operators follow from the Birman–Kuroda theorem. Furthermore, Remark 15.4.5b shows that $\Omega_\pm(H_u, H_{u0})$ are also asymptotically complete. More interesting, however, is the explicit form of the S–matrix.

We can find it easily using a procedure, which is common in the stationary scattering theory, and which we used already in Example 7.2.2, namely by looking for generalized eigenfunctions of the corresponding Hamiltonian with a suitable asymptotic behavior. In the present case we take the function

$$f(r) := \begin{pmatrix} e^{-ikr} - A\,e^{ikr} \\ B\,e^{i\kappa r} \end{pmatrix},$$

which is *locally* square integrable, and require it to belong — again locally — to the domain of H_u, in other words, to obey the boundary conditions (15.11). In this way, we get a system of linear equations which is readily solved by

$$A := S_0(k) := \frac{(a+ik)(b-i\kappa) - |c|^2}{D}\,,$$

where D is the same as in Proposition 15.4.3, and

$$B := \frac{2ik\bar{c}}{D}$$

Consider first the case when $k^2 \le E$. A simple calculation (Problem 14b) shows then that $|A| = 1$, and since the reflection amplitude A is (with the choice of sign that we have made) just the s–wave part of the first–channel on–shell scattering matrix (see the notes), we have checked that the *scattering is elastic for* $k^2 \le E$. In other words, there is no solution with a "plane–wave" component in the excited channel, because the incoming particle does not have enough energy to excite the target permanently; we usually say that the inelastic channel is closed in this case.

Furthermore, recall that the on–shell S–matrix (in a given partial wave, in our case $\ell = 0$) is expressed conventionally through the phase shift as $S_0(k) = e^{2i\delta_0(k)}$. Another simple calculation then yields

$$\delta_0(k) = \arctan \frac{k(b-i\kappa)}{a(b-i\kappa) - |c|^2} \quad (\mathrm{mod}\ \pi)\,.$$

The most interesting situation occurs when H_{u_0} has an embedded eigenvalue, which turns under the perturbation into a resonant state whose lifetime, according to Example 9.6.6 and Theorem 15.4.4b, is expected to be

$$T(c) := -\frac{a^2-b^2+E}{4b|c|^2\sqrt{E-b^2}}\left(1+\mathcal{O}(|c|^2)\right);$$

we shall return to this problem a little later. At the same time, the phase shift should exhibit a sheer change peculiar to a resonance. Since κ is purely imaginary in this case, we may rewrite the expression of the phase shift as

$$\delta_0(k) = \arctan \frac{k(b+\sqrt{E-k^2})}{a(b+\sqrt{E-k^2}) - |c|^2}\,.$$

If there is a bound state in the first channel, $a < 0$, we have $\delta_0(k) < 0$ in the interval $(0, \sqrt{E-b^2})$ with zero value at its endpoints. Furthermore, $\delta_0(k)$ reaches the value $\frac{\pi}{2}$ at $k_0 := \sqrt{E - (|c|^2a^{-1} - b)^2}$ provided $|c|^2 < ab$; in the opposite case the argument of the arctan in the above expression has no singularity at all in $(0, \sqrt{E})$. For a, b fixed and $|c|$ small enough, the phase–shift plot is composed of the background $\arctan(\frac{k}{a})$ and the jump to the next branch of this function around the singularity. The energies at which $\delta_0(k) = \pm\frac{\pi}{4}$, *e.g.*, are

$$\lambda_\pm = E - b^2 + \frac{2b|c|^2}{a^2-b^2+E}(a \pm \sqrt{E-b^2}) + \mathcal{O}(|c|^4)\,,$$

i.e., $\mathrm{Re}\, e_2(c) \pm \mathrm{Im}\, e_2(c)$. However, it has to be pointed out that the jump may be substantially deformed, even for small $|c|$, if $|a|$ is small enough; to see this it is sufficient to take an arbitrarily but fixed $|c|$ and $a := |c|^2/b$.

We can also observe the resonance behavior in the cross section. Since the scattering is obviously isotropic, we have $\sigma_{tot}(k^2) = \sigma_{\ell=0}(k^2) = \frac{4\pi}{k^2}\sin^2\delta_0(k)$. In the embedded–eigenvalue case therefore the above expression of the phase shift yields for $k^2 \le E$ the following relation,

$$\sigma_{tot}(k^2) = \frac{4\pi\left(b+\sqrt{E-k^2}\right)^2}{\left[a\left(b+\sqrt{E-k^2}\right) - |c|^2\right]^2 + k^2\left(b+\sqrt{E-k^2}\right)^2}\,.$$

For a, b fixed and $|c|$ small enough, the cross section varies rapidly (has a dip followed by a peak) in the region of the width $\approx T(c)^{-1}$ around the value $k = k_0$.

Let us now turn to the case when $k^2 > E$, *i.e.*, the incident particle energy is greater than the excitation energy of the target. The reflection and transmission amplitudes then satisfy

$$|A|^2 + \frac{\kappa}{k}|B|^2 = 1.$$

This relation illustrates that the requirement of Hamiltonian self–adjointness (which is contained in the boundary conditions determining the amplitudes A, B) means physically the conservation of probability flow (*cf.* Problem 14.24d): it is easy to see that its radial components $J_j := 2\,\mathrm{Im}\,\bar{f}_j f_j'$ are

$$J_1 = -2k\left(1 - |A|^2\right), \quad J_2 = 2\kappa|B|^2$$

in the first and the second channel, respectively. The elastic scattering is now nonunitary since $B \neq 0$ in the interacting case, and therefore $|A|^2 < 1$; this is due to the fact that the "nucleus" may now leave the interaction region in the excited state. The above relation can also be expressed as

$$|S_{0,1\to 1}(k)|^2 + |S_{0,1\to 2}(k)|^2 = 1,$$

i.e., as a part of the full two-channel S-matrix unitarity condition (Problem 14b).

Let us return to the resonance–scattering situation. We have already mentioned the generally accepted concept, according to which scattering resonances are quasi–stable objects of a sort. To illustrate this in the present example, we are going in the rest of this section to examine in more detail what the resonant state arising from the embedded eigenvalue of H_{u_0} looks like and how it decays. The most natural choice for the "compound nucleus" wave function is the eigenstate of the unperturbed Hamiltonian (see the notes),

$$f\,:\, f(r) = \sqrt{-2b}\begin{pmatrix} 0 \\ e^{br} \end{pmatrix}; \tag{15.13}$$

recall that $-\sqrt{E} < b < 0$. To find the decay law of this state, we have to compute according to Section 9.6 the reduced resolvent $(f, (H_u - k^2)^{-1} f)$ and investigate its analytic structure. Using Proposition 15.4.3, we obtain after a simple integration

$$(f, (H_u - k^2)^{-1} f) = \frac{|c|^2 + (a - ik)(b + i\kappa)}{(b + i\kappa)^2[|c|^2 - (a - ik)(b - i\kappa)]}$$

(Problem 15a). Notice first that the function on the right side corresponds to a four-sheeted Riemann surface with respect to the complex parameter $z = k^2$ (one usually speaks about the complex energy plane, or briefly, z–plane): it has cuts along the intervals $[0, \infty)$ and $[E, \infty)$ on the positive real axis corresponding to the square roots in the definitions of k and κ. It is customary in stationary scattering theory to pass to the variable k (we speak about the momentum plane, or k–plane);

then the above expression corresponds to a two–sheeted surface with the cuts along $(-\infty, -\sqrt{E}]$ and $[\sqrt{E}, \infty)$.

The reduced evolution operator of the resonant state, which acts in the present case as multiplication by $v_u(t) := (f, e^{-iH_u t} f)$ on the one–dimensional subspace of $\mathcal{H}$ spanned by the vector f can be calculated as in Section 9.6. Using the fact that the reduced resolvent is (in the z–plane) analytic in $\mathbb{C} \setminus \mathbb{R}^+$, we may replace the integral appearing in the Stone formula by integration over a curve encircling the spectrum of H_u in a clockwise way. In the k–plane this gives

$$v_u(t) = \frac{1}{\pi i} \int_\Gamma e^{-ik^2 t} (f, (H_u - k^2)^{-1} f)\, k\, dk\,,$$

where Γ is a curve in the upper halfplane (corresponding to the "physical sheet" in the z–plane) which runs from $-\infty + i\varepsilon$ to $\infty + i\varepsilon$ passing *above* the set $\{k : k^2 \in \sigma(H_u)\}$ (draw a picture). Notice that we have to consider only $t \geq 0$ since $v_u(-t) = \overline{v_u(t)}$.

Evaluating the integral (Problem 15a) we have to distinguish two cases. If there is an antibound state in the first channel, the corresponding pole is not on the first sheet and the line $\{k_\varepsilon = k + i\varepsilon : k \in \mathbb{R},\, \varepsilon > 0\}$ can be used as Γ. On the other hand, if the pole *is* placed on the first sheet, then moving Γ towards the real axis in the k–plane, we get in addition the integral over a clockwise circle around the pole. The last named contribution to the integral can be calculated by the residue theorem; we get

$$\begin{aligned} v_u(t) &= -\lim_{k \to k_1} (k - k_1)\, 2k_1 \frac{|c|^2 + (a - ik_1)(b + i\kappa_1)}{(b + i\kappa_1)^2 \left[|c|^2 - (a - ik)(b - i\kappa)\right]}\, e^{-ik_1^2 t} \\ &+ \frac{1}{\pi i} \lim_{\varepsilon \to 0+} \int_{\mathbb{R}} e^{-i(k + i\varepsilon)^2 t} \frac{|c|^2 + (a - ik + \varepsilon)(b + i\kappa_\varepsilon)}{(b + i\kappa_\varepsilon)^2 \left[|c|^2 - (a - ik + \varepsilon)(b - i\kappa_\varepsilon)\right]} (k + i\varepsilon)\, dk\,, \end{aligned}$$

where κ_1 corresponds to the first–channel bound–state pole position k_1 (*cf.* the proof of Theorem 15.4.4a) and $\kappa_\varepsilon := \sqrt{(k + i\varepsilon)^2 - E}$. If $a > 0$ we get the same expression without the first term; in the case of zero–energy resonance, which we shall not discuss in detail, the first term enters with the factor $\frac{1}{2}$.

By a straightforward computation using the identity $|c|^2 + (a - ik_1)(b + i\kappa_1) = 2b(a - ik_1)$, we find that for $a < 0$ the first term equals

$$\frac{4ab|c|^2}{(a^2 - b^2 + E)^2}\, e^{-ik_1^2 t} \left(1 + \mathcal{O}(|c|^2\right)\,.$$

On the other hand, the limit $\varepsilon \to 0+$ in the second term cannot be interchanged with the integral directly because the limiting function is not integrable. Fortunately, we can rewrite the integral using a simple substitution as

$$\int_0^\infty e^{-ik_\varepsilon^2 t} \frac{|c|^2 + (a - ik_\varepsilon)(b + i\kappa_\varepsilon)}{(b + i\kappa_\varepsilon)^2 \left[|c|^2 - (a - ik_\varepsilon)(b - i\kappa_\varepsilon)\right]}\, k_\varepsilon\, dk$$

$$- \int_0^\infty e^{-i\bar{k}_\varepsilon^2 t} \frac{|c|^2 + (a+i\bar{k}_\varepsilon)(b+i\tilde{\kappa}_\varepsilon)}{(b+i\tilde{\kappa}_\varepsilon)^2 \Big[|c|^2 - (a+i\bar{k}_\varepsilon)(b-i\tilde{\kappa}_\varepsilon)\Big]} \bar{k}_\varepsilon \, dk \,,$$

where $k_\varepsilon := k+i\varepsilon$ and

$$\tilde{\kappa}_\varepsilon := \begin{cases} \kappa_\varepsilon & \dots \quad 0 \le k \le \sqrt{E} \\ -\kappa_\varepsilon & \dots \quad \sqrt{E} \le k \end{cases}$$

Here we have used the fact that the limit of κ when k approaches the real axis from the upper halfplane (on the first sheet) equals $\pm\sqrt{k^2 - E}$ on the right and left cut, respectively, and $i\sqrt{E-k^2}$ in between. When we now write the expression in question as a single integral over $(0,\infty)$, the dominated–convergence theorem may already be applied giving

$$\frac{1}{\pi i} \int_0^{\sqrt{E}} \frac{e^{-ik^2 t}}{(b+i\kappa)^2} \frac{-4ik^2 b|c|^2}{|c|^4 - 2a|c|^2(b-i\kappa) + (a^2+k^2)(b-i\kappa)^2} \, dk$$
$$+ \frac{1}{\pi i} \int_{\sqrt{E}}^\infty \frac{e^{-ik^2 t}}{(b^2+\kappa^2)^2} \frac{-4ib\kappa|c|^4 - 4ibk|c|^2(b^2+\kappa^2)}{|c|^4 - 2|c|^2(ab-k\kappa) + (a^2+k^2)(b^2+\kappa^2)} \, k \, dk \,.$$

The first integral is for small $|c|$ expected to be dominated by the contribution from the second–sheet resonance pole; recall that the "discriminant" D is contained in the denominator. The corresponding residue equals

$$\frac{1}{\pi i} \lim_{k \to k_2} (k-k_2)k \frac{|c|^2 + (a-ik)(b+i\kappa)}{(b+i\kappa)^2 \Big[|c|^2 - (a-ik)(b-i\kappa)\Big]} e^{-ik^2 t}$$
$$= \frac{1}{\pi i} \frac{2bk_2\kappa_2(a-ik_2)}{(b+i\kappa_2)^2 \Big[E + i(ak_2 - b\kappa_2)\Big]} e^{-ik_2^2 t} \,,$$

where $\kappa_2 := \sqrt{E-k_2^2}$ and k_2 is given in the proof of Theorem 15.4.4b. Using its expansion in powers of $|c|^2$, after a tedious but straightforward calculation we find that the preexponential factor equals $-1/2\pi i$ up to higher-order terms, *i.e.*, that the resonance–pole contribution to $v_u(t)$ is $e^{-ik_2^2 t}[1 + \mathcal{O}(|c|^2)]$ as expected.

It remains for us to determine the background term. In the first of the above two integrals, we close the integration curve by extending it to a point R; then continuing over a circle segment of radius R, and returning to the origin on the fourth–quadrant axis. After the limit $R \to \infty$, the contribution to the background consists of the integrals over $(\sqrt{E},\infty)$ and the fourth–quadrant axis. The first of them, however, can be shown easily to cancel, up to higher order terms in $|c|^2$, with the second integral; hence we finally obtain the following result for the reduced propagator and the corresponding decay law $P_u(t) := |v_u(t)|^2$.

15.4.6 Theorem: Assume $a \neq 0$ and $-\sqrt{E} < b < 0$. The reduced propagator and the decay law of the resonant state (15.13) corresponding to the Hamiltonian H_u are given by

$$v_u(t) = \Bigg\{ e^{-ik_2^2 t} - |c|^2 \Bigg[\frac{2(|a|-a)b}{(a^2-b^2+E)^2}\, e^{-ik_1^2 t} + \frac{ib}{\sqrt{E-b^2}(a-i\sqrt{E-b^2})^2}\, e^{-ik_2^2 t} + \frac{4b}{\pi}\, e^{-\pi i/4} \int_0^\infty \frac{z^2 e^{-z^2 t}\, dz}{(z^2+ia^2)(z^2-i(E-b^2))^2} \Bigg] \Bigg\} \left(1+\mathcal{O}(|c|^2)\right)$$

and

$$P_u(t) = \Bigg\{ e^{2(\mathrm{Im}\, e_2)t} - 2|c|^2 \mathrm{Re} \Bigg[\frac{2(|a|-a)b}{(a^2-b^2+E)^2}\, e^{-i(k_1^2-\bar{k}_2^2)t} + \frac{ib}{\sqrt{E-b^2}(a-i\sqrt{E-b^2})^2}\, e^{2(\mathrm{Im}\, e_2)t} + \frac{4b}{\pi}\, e^{i(\bar{k}_2^2 t-\pi/4)} \int_0^\infty \frac{z^2 e^{-z^2 t}\, dz}{(z^2+ia^2)(z^2-i(E-b^2))^2} \Bigg] \Bigg\} \left(1+\mathcal{O}(|c|^2)\right), \qquad (15.14)$$

where $e_j := e_j(c) := k_j^2$ are specified in Theorem 15.4.4.

Hence we have obtained the expected result, namely the exponential decay law with a lifetime proportional to the inverse distance of the resonance pole from the real axis, up to higher order terms in $|c|^2$.

These corrections are also interesting. First of all, the model illustrates that the initial decay rate can be zero even if the decaying state does not belong to the domain of the Hamiltonian (*cf.* Problems 15b and 9.18). On the other hand, the long–time behavior of the decay law depends substantially on the spectrum of the unperturbed Hamiltonian: if it has an eigenvalue in the first channel (so H_u also has an eigenvalue for a sufficiently weak coupling), then the decay law contains a term of order of $|c|^4$, which does not vanish as $t \to \infty$, and therefore dominates the expression for large time values; it is clear that this term comes from the component of the first–channel bound state contained in the resonant state (15.13).

Notes to Chapter 15

Section 15.1 One often speaks about particle beams of a given energy when describing scattering experiments. This is not precise, of course, because scattering states belong to the continuous spectrum of the Hamiltonian. The correct meaning is the following: using spectral representation for the free Hamiltonian, we describe the incoming–particle state by a wave packet with energy support in the vicinity of a given energy value, assuming that the support size is small enough in the energy scale characteristic for the processes under consideration. If we study resonance scattering, for instance, we have to choose the "beam monochromaticity" small compared to the widths of the resonances in question.

The primary, or *direct*, problem of scattering theory is to find measurable quantities such as cross sections, phase shifts, *etc.*, from knowledge of the interaction between the projectile and the target. The *inverse scattering problem* poses the opposite question: to find the interaction, most frequently the potential in a Schrödinger operator, from scattering data; more information can be found, *e.g.*, in the monograph ⟦ChS⟧.

The assumption that the projectile and target behave as free at large distances, *i.e.*, that there are asymptotic states $\psi_\pm$ to a given scattering state ψ, which we have used to introduce the wave operators, is not valid for long-range interactions like the Coulomb force. In this case the definition of the wave operators has to be modified — see, *e.g.*, ⟦AJS⟧, Chap.13; ⟦RS 3⟧, Sec.XI.9. Notice also that in physical literature for historical reasons one mostly uses the "reversed" convention in which the wave operators $\Omega_\pm$ correspond to the limits $t \to \mp\infty$.

Proposition 3 generalizes in various ways — see ⟦RS 3⟧, App. to Sec.XI.3. The result we have derived for scattering states modifies to the case $\sigma_{sc}(H) \neq \emptyset$; the set $M_s(H)$ has to replaced by the subspace

$$\tilde{M}_s(H) := \left\{ \psi \in \mathcal{H} : \lim_{T\to\infty} \frac{1}{2T} \int_{-T}^{T} \|F_r e^{-iHt}\psi\|^2 \, dt = 0 \quad \text{for all } r > 0 \right\},$$

which satisfies $\tilde{M}_s(H) = \mathcal{H}_c(H)$ — *cf.* [AG 1], and also ⟦AJS⟧, Sec.7.6; ⟦RS 3⟧, App. to Sec.XI.7. Hence the pathological states mentioned in Remark 15.4.5 belong to the singularly continuous spectral subspace of the Hamiltonian.

Basic information about scattering theory can be found in most quantum mechanical textbooks; an extensive exposition is given, *e.g.*, in ⟦Ta⟧, ⟦New⟧. Rigorous formulation of scattering theory is the content of the monographs ⟦Am⟧, ⟦AJS⟧, ⟦BW⟧, ⟦Pea⟧, ⟦Pe⟧, and ⟦RS 3⟧. We should keep in mind that the terminology is sometimes not unified: the (asymptotic) completeness is often called (strong) asymptotic completeness, *etc.*Some authors use the term *Møller operators* for $\Omega_\pm$, after the author who introduced them formally for the first time in 1945. Finally, let us mention that Hilbert space methods are also used in acoustic and electromagnetic–wave scattering — see ⟦LP⟧.

Section 15.2 Proposition 15.2.3 looks like a simple tool for proving the completeness; however, in order to use it, one has to check the existence of $\Omega_\pm(H_0, H)$, which is often difficult. The Cook criterion was proved in the paper [Coo 2], which in 1957–1958 together with [Ja 1,2], [Ka 6], [LF 1] formulated the foundations of rigorous scattering theory. Theorems 15.2.6 and 15.2.8–15.2.10 form the backbone of the so–called *Kato–Birman theory*, named after the authors who formulated it together with S. Kuroda, R. Putnam, M.G. Krein, D.B. Pearson and others. Theorem 15.2.6, which allows us to simplify earlier proofs considerably, comes from [Pea 2]. The Kato–Rosenblum theorem is the oldest result of the theory; it can be found in [Ka 6], [Ro 1]. Theorem 15.2.9 was proved in [Kur 1,2], [Bir 2,3]; note that there are examples showing that $\mathcal{J}_1(\mathcal{H})$ cannot be replaced in the hypothesis by $\mathcal{J}_p(\mathcal{H})$ for some $p > 1$. One possible application of Theorem 15.2.9 is to prove existence and completeness of wave operators for point–interaction Hamiltonians with a finite–number of centers introduced in the same way as in Section 14.6. Theorem 15.2.10 was proved in [Bir 4]. Further references to Kato–Birman theory can be found in ⟦RS 3⟧, notes to Sec.XI.3.

The invariance principle was discussed for the first time in [Bir 3]; later it was repeatedly generalized. Proof of Theorem 15.2.11 and further details can be found in ⟦Ka⟧, Sec.X.4; ⟦RS 3⟧, Sec.XI.3 and App.3; ⟦We⟧, Thm.11.13. A thorough discussion of general principles of the scattering theory can be found in the monograph ⟦Ya⟧.

Section 15.3 Theorem 15.3.1 was proved in the paper [Coo 2] mentioned above for potentials from $L^2(\mathbb{R}^3)$ and generalized in [Ha 1] to potentials $V \in L^2_{loc}(\mathbb{R}^3)$, which behave as $\mathcal{O}(r^{-1-\varepsilon})$ at large distances. This result was further extended in [KuS 1] with the help of Theorem 15.2.5 to potentials of the same decay but stronger local singularities. Theorem 4 extends to Schrödinger operators on $L^2(\mathbb{R}^n)$ — see ⟦RS 3⟧, Thm.XI.30; its disadvantage is that the requirement $V \in L^1 \cap L^2$ represents a rather strong decay restriction. Theorem 6 follows from the Birman–Kuroda theorem combined with the partial–wave decomposition — *cf.* ⟦RS 3⟧, Thm.XI.31; an alternative proof can be found in [DF 1].

To avoid the impression that the completeness of wave operators is a natural property, which is just difficult to prove, recall the example constructed in [Pea 1]. It represents a short–range central potential composed of rectangular barriers and wells chosen so that the corresponding wave operators are not complete; physically this result means that the scattered particle may be captured, so it is not able to leave the interaction region.

The Agmon and Enss conditions formulate in the same way for Schrödinger operators on $L^2(\mathbb{R}^n)$, and both Theorem 15.3.7 and Remarks 15.3.8 extend to this case. The first part of Theorem 15.3.7 represents a particular case of the Agmon–Kato–Simon theorem — see ⟦RS 3⟧, Thm.XIII.33. The remaining part is proved by the "geometric" method proposed by V. Enss, which is based on estimates of the asymptotic behavior of spreading wave packets — see [En 1,2] and also [Si 7]; ⟦RS 3⟧, Sec.XI.17; ⟦Am⟧; ⟦Pe⟧, Sec.2.7; ⟦CFKS⟧, Chap.5; [En 3]. Similar methods have been used in recent proofs of asymptotic completeness for N–particle scattering for some classes of short–range and long–range interactions — see, *e.g.*, [SiS 1,2], [Sig 1], and also [Der 1,2], [Gra 1], [Kit 1]. An exhaustive overview of the N–particle scattering can be found in the monograph ⟦DG⟧.

Section 15.4 The model discussed in this section was introduced in [Ex 6]. It is known from perturbation theory that if the unperturbed eigenvalue has a multiplicity $m > 1$, then the perturbed eigenvalue is given by the so-called Puiseaux series, *i.e.*, rather than in powers of the perturbation parameter g itself it expands in powers of $g^{1/m}$ — see ⟦Ka⟧, Chap.2; in the present case we have $g = |c|^2$ and $m = 2$.

As we have mentioned, the relation between stationary scattering theory and the concepts discussed in Section 15.1 is given by the direct–integral decomposition $S = \int^{\oplus} S(\lambda)\, d\lambda$, where $S(\lambda)$ is the on–shell scattering matrix for energy λ — see ⟦AJS⟧, Section 5.7, for more details, and also the notes to Section 7.2. The generalized–eigenfunction expansion can also be written for the model in question, but we shall not discuss this problem.

In the decoupled case, $c = 0$, the expression obtained for the phase shift reduces to the known formula for the point–interaction phase shift — *cf.* ⟦AGHH⟧, Sec.I.1.4. In the same way as the phase shift and cross section, other scattering quantities can be computed for the present model, *e.g.*, the *time delay* caused by the scattering which is related to the S–matrix by the so–called *Eisenbud–Wigner formula* — see ⟦AJS⟧, Sec.7.2; and also [Mar 1], [AmC 1]. The appearance of the factor κ/k in the relation between the reflection and transmission amplitudes expresses the fact that the particle has different velocities in the two channels. Of course, true scattering states which are wave packets composed of generalized eigenfunctions do not have a sharp value of velocity, but even then the mean values of the velocities differ because a part of the incident kinetic energy has been absorbed to excite the target.

The ways to choose the resonant–state wave function have been discussed in [Hun 1]; it appears that for an embedded–eigenvalue perturbation problem like that treated here, (15.13) is essentially the only possibility — see [Ex 6] for more details. The resonances in the present model come from zeros of the function D, so they are simultaneously poles of the (analytically continued) reduced resolvent and scattering matrix. In general, one expects that the same should be true in other cases too, *i.e.*, that different possible ways to define a resonance will lead to the same result; however, it is not always easy to prove this — see, *e.g.*, [How 3], [BDW 1], and also [Si 4], [Hag 1] for the equivalence between the scattering resonances and the dilation–analytic resonances mentioned in the notes to Section 9.6.

Problems

1. Prove: (a) Suppose that A is self–adjoint, C is compact, and $\psi \in \mathcal{H}_{ac}(H)$; then $\lim_{|t|\to\infty} C\, e^{-iAt}\psi = 0$.

(b) The conclusion of Proposition 15.1.2 remains valid if $\sigma_{sc}(H) = \emptyset$ and the operators $F_r(H-z)^{-\beta}$ are compact for some $z \in \rho(H)$, $\beta > 0$, and all $r > 0$.

Hint: (a) The Riemann-Lebesgue lemma implies $\text{w-lim}_{|t|\to\infty} e^{-iAt}\psi = 0$.
(b) $(H-z)^{\beta} P_\alpha$ with $P_\alpha := E_H(-\alpha,\alpha)$ is bounded for any $\alpha, \beta > 0$.

2. The function $g:\, g(k) = |k^2 - z|^{-1}$ belongs for $\text{Im}\, z \neq 0$ to $L^2(\mathbb{R}^3)$ and its norm is equal to $\|g\| = \pi|\text{Im}\,\sqrt{z}|^{-1/2}$ for $\text{Re}\, z > 0$.

3. The wave operators $\Omega_\pm(H, H_0)$ have the following properties:

(a) The subspaces $\text{Ran}\,\Omega_\pm$ reduce the group $\{U(t)\}$, and $H \upharpoonright \text{Ran}\,\Omega_\pm$ is unitarily equivalent to $H_0 \upharpoonright \mathcal{H}_{ac}(H_0)$.

(b) $\text{Ran}\,\Omega_\pm \subset \mathcal{H}_{ac}(H)$.

Hint: (a) Use the intertwining relations.

4. Suppose that the wave operators $\Omega_\pm$ exist and satisfy relation (15.3); then

(a) The scattering operator is a partial isometry with the initial subspace $\mathcal{H}_{ac}(H_0)$ and $S \upharpoonright \mathcal{H}_{ac}(H_0)$ is unitary *iff* $\text{Ran}\,\Omega_+ = \text{Ran}\,\Omega_-$.

(b) The operator S commutes with the free Hamiltonian, *i.e.*, $U_0(t)S = SU_0(t)$ for all $t \in \mathbb{R}$.

Hint: (b) Use the intertwining relations.

5. Prove: (a) $\Omega_\pm^* = \text{s-lim}_{t\to\pm\infty} U_0(t)^* U(t) P_\pm$, where the $P_\pm$ are the projections onto the subspaces $\text{Ran}\,\Omega_\pm$.

(b) $U_0(t)\Omega_\pm^* = \Omega_\pm^* U(t)$ for all $t \in \mathbb{R}$, and furthermore, $\Omega_\pm^* H \subset H_0 \Omega_\pm^*$.

6. Prove: (a) Let $\psi \in D(H_0)$ and $U_0(t)\psi \in D(H)$ for some $t \in \mathbb{R}$. Then the function $\psi:\, \psi(s) = U_0^*(s)U(s)\psi$ is differentiable at the point t and $\psi'(t) = iU(t)^*(H - H_0)U_0(t)\psi$.

(b) Suppose that $\psi, \phi : \mathbb{R} \to \mathcal{H}$ are differentiable functions; then $f : f(t) = (\psi(t), \phi(t))$ is also differentiable and $f'(t) = (\psi'(t), \phi(t)) + (\psi(t), \phi'(t))$.

(c) If a function $\psi : \mathbb{R} \to \mathcal{H}$ is continuously differentiable on an interval $[a, b]$, then $\psi(d) - \psi(c) = \int_c^d \psi'(t)\, dt$ holds for $a < c < d < b$.

7. Let H_0, V be self–adjoint operators and $U_0(t) := e^{-iH_0 t}$. If V is H_0–bounded; then the vector–valued function $VU_0(\cdot)\psi$ is continuous for any $\psi \in D(H_0)$.

8. Let $\Omega_{t,s}$ be the operator defined in the proof of Theorem 15.2.6. Show that

(a) $\Omega_{t,s}$ is compact.

(b) the function $\tau \mapsto \Omega_{t+\tau,s+\tau}\phi$ is continuously differentiable for any $\phi \in \mathcal{H}$ and

$$\frac{d}{d\tau}\Omega_{t+\tau,s+\tau}\phi = (C(t+\tau) - C(s+\tau))\phi\,.$$

Hint: (a) Using the continuity of $C(\cdot)\phi$, check that $\phi_n \xrightarrow{w} \phi$ implies $\Omega_{t,s}\phi_n \xrightarrow{w} \Omega_{t,s}\phi$.

9. Consider the Friedrichs model of Section 9.6 as a scattering system. Prove that

(a) The wave operators $\Omega_\pm(H_g, H_0)$ exist and are complete.

(b) If the function v is continuously differentiable, then the $\Omega_\pm(H_g, H_0)$ are also asymptotically complete.

Hint: (b) In analogy with the proof of Proposition 9.6.7 express the full resolvent and use Theorem 14.4.7 — *cf.* [DE 2], Part IV.

10. If self-adjoint H, H_0 have the same domains, they are mutually dominated.
Hint: Use the closed–graph theorem for $I + |H|$, $I + |H_0|$.

11. The Schrödinger operator $H_0 + V(Q)$ on $L^2(\mathbb{R}^n)$, $n \geq 3$, with $V \in L^p + L^s$ is self–adjoint on $D(H_0)$ provided p satisfies condition (14.2) and $s \in [p, n)$.
Hint: Check that $L^s \subset L^p + L^\infty$.

12. Using the notation from the proof of Theorem 15.3.1,

(a) Compute $U_0(t)\psi_q$.

(b) Verify the estimate of the expression $\|V(Q)U_0(t)\psi_q\|$ used there.

(c) Generalize the Haak-Cook theorem to Schrödinger operators $H = H_0 + V(Q)$ on $L^2(\mathbb{R}^n)$, $n \geq 3$, with a potential obeying the assumptions of Problem 11. In the same way, extend the conclusions of Example 15.3.2.

13. Using the notation of Section 15.4,

(a) Find all self–adjoint extensions of the operator H_0.

(b) Show that the system with the Hamiltonian H_u is invariant with respect to the time reversal *iff* the coupling constant c is real.

(c) Compute the functions $\lambda_{jk}(\cdot)$ from the proof of Proposition 15.4.3.

(d) Find the multiplicity of $\sigma(H_{u_0})$ corresponding to $u_0 := \{a, b, 0\}$.

(e) Fill the details into the proof of Theorem 15.4.4.

Hint: (d) *Cf.* Example 9.1.6.

14. Consider the model of Section 15.4 as a scattering system.

(a) Prove that $\sigma_{sc}(H_u) = \emptyset$.

(b) Find the on–shell scattering matrix and prove its unitarity. Derive the phase–shift expressions for elastic–channel scattering.

Hint: (a) Use Proposition 15.4.3 and Theorem 14.4.7.

15. Consider the operator H_u discussed in Section 15.4 as the full Hamiltonian of the unstable system whose state space is the one–dimensional subspace in $\mathcal{H}$ spanned by the vector (15.13).

(a) Fill the details into the sketched proof of Theorem 15.4.6.

(b) Show that the decay law (15.14) has zero initial decay rate, $\dot{P}_u(0+) = 0$.

Hint: $\mathrm{Im}\,(f, (H_u - \lambda)^{-1} f)$ is $\mathcal{O}(\lambda^{-5/2})$ for large positive λ; use Problem 9.18.

Chapter 16

Quantum waveguides

Let us return now to the subject we discussed in Section 14.5 and analyze in more detail the situation when a particle is confined to an unbounded region Ω of the particular shape of a tube or layer; in view of the natural analogy we will speak about such systems as *quantum waveguides.* There are two reasons why this topic deserves a separate chapter. First of all, such systems can be useful as models of various systems studied by the experimentalists, such as semiconductor "quantum wires", carbon nanotubes, hollow-fibre atomic waveguides, etc. The second, and no less important, reason is that one encounters here effects which show again that an intuition based on our everyday "macroscopic" experience can be a false guide when dealing with objects governed by the laws of quantum theory.

16.1 Geometric effects in Dirichlet stripes

From the physical point of view, the object of our interest will be a spinless particle confined to a region $\Omega \subset \mathbb{R}^d$. Typically we assume that the boundary acts as a hard wall corresponding to Dirichlet boundary conditions, *i.e.*, the Hamiltonian is

$$H = -\frac{\hbar^2}{2m^*}\Delta_D^\Omega , \qquad (16.1)$$

or a perturbation of this operator, where m^* is the effective mass of the particle. Usually the latter will not be important and we will use rationalized units putting $\hbar^2/2m^* = 1$, identifying H with the Dirichlet Laplacian as we defined it in Section 14.5.

For simplicity we restrict ourselves in this section to the two–dimensional case when Ω is a strip in the plane of width $d = 2a > 0$ leaving extensions to higher dimensions to the notes and problems. As long as the strip is straight, $\Omega_0 = \mathbb{R} \times (-a, a)$, the energy spectrum is easily found.

16.1.1 Proposition: $\sigma(H_D^{\Omega_0}) = [\kappa_1^2, \infty)$, where $\kappa_1 := \pi/d$, being purely absolutely continuous.

Proof is left to the reader (Problem 1).

Let us see what will happen if Ω_0 is replaced by a smoothly bent strip Ω of the same width. The geometry of Ω can be described by means of its axis, which is a smooth curve Γ of infinite length in $\mathbb{R}^2$; we assume that it does not intersect itself and identify Ω with the set $\{x \in \mathbb{R}^2 : \operatorname{dist}(x,\Gamma) < a\}$. Its points can be expressed using the natural curvilinear coordinates. Let s be the arc length of Γ and u the normal distance of a strip point from the curve, then its Cartesian coordinates are

$$x(s,u) = \xi(s) - u\eta'(s)\,, \quad y(s,u) = \eta(s) + u\xi'(s)\,, \tag{16.2}$$

where the functions ξ, η represent a parametric expression of Γ; they must satisfy the condition $\xi'(s)^2+\eta'(s)^2 = 1$. This allows us to define further the *signed curvature* of Γ by

$$\gamma(s) := \eta'(s)\xi''(s) - \xi'(s)\eta''(s);$$

this function in turn determines the curve Γ uniquely up to Euclidean transformations of the plane (Problem 2). Up to the sign, γ coincides with the inverse radius of the osculation circle to Γ at the point $(\xi(s),\eta(s))$.

The curvilinear coordinates s, u are locally orthogonal so the metric properties of Ω express through a diagonal metric tensor, $dx^2+dy^2 = g_{ss}ds^2+g_{uu}du^2$, where the transverse component $g_{uu} = 1$ and the longitudinal one is $g_{ss} \equiv g = (1+u\gamma)^2$. They define a map $f : \mathbb{R}\times(-a,a) \to \Omega$; it is easy to compute the Jacobi matrix of the transformation and its determinant, which is equal to $\sqrt{g} = 1+u\gamma$. Thus the map is regular provided the right side is positive, which has an obvious meaning: if the boundary of Ω is supposed to be smooth, the halfwidth if the strip must not exceed the curvature radius, $a\|\gamma\|_\infty < 1$. We will also assume that the map f is globally injective, *i.e.*, Ω does not intersect itself, and that $\gamma \in C^2$ (see the notes).

With these preliminaries, we are ready to discuss the operator H_D^Ω for the curved strip. First we define the map $U : L^2(\Omega) \to L^2(\mathbb{R}) \otimes L^2(-a,a)$ by

$$(U\psi)(s,u) := \sqrt{1+u\gamma(s)}\,(\psi\circ f)(s,u)\,. \tag{16.3}$$

It is a unitary operator by Example 3.3.2, and a straightforward computation (Problem 3a) yields the explicit form of the transformed Hamiltonian.

16.1.2 Proposition: Assume that the map f is injective, $\gamma \in C^2$ with $a\|\gamma\|_\infty < 1$, and the functions $\gamma^{(k)}$, $k = 0,1,2$, are bounded in $\mathbb{R}$. Then the operator $H := UH_D^\Omega U^{-1}$ is *e.s.a.* on $D_\infty(\Omega_0)$ and its action there is given by

$$\begin{aligned} H\psi &= -\frac{\partial}{\partial s}(1+u\gamma)^{-2}\frac{\partial\psi}{\partial s} - \frac{\partial^2\psi}{\partial u^2} + V\psi\,, \\ V(s,u) &= -\frac{\gamma(s)^2}{4(1+u\gamma(s))^2} + \frac{u\gamma''(s)}{2(1+u\gamma(s))^3} - \frac{5}{4}\,\frac{u^2\gamma'(s)^2}{(1+u\gamma(s))^4}\,. \end{aligned} \tag{16.4}$$

Hence we have transformed the problem to a simpler region, a straight strip. The price we pay for this is the more complicated structure of the operator itself;

apart from the weight factor in the "longitudinal" part of the kinetic energy, the transformed operator has acquired the curvature–induced *effective potential* V. It is important that the geometrical properties are now expressed through the coefficients so we can investigate the operator by standard functional–analytic methods.

Let us start from a heuristic discussion of the situation when the strip is thin, $a\|\gamma\|_\infty \ll 1$. Then the factor $g^{1/2}$ does not differ much from one and the effective potential is dominated by its first term, in other words

$$H = -\frac{\partial^2}{\partial s^2} - \frac{1}{4}\gamma(s)^2 - \frac{\partial^2}{\partial u^2} + \mathcal{O}(a) \tag{16.5}$$

as $a \to 0+$. The transverse–part has discrete and simple spectrum, with eigenvalues $\{\kappa_n^2\}_{n=1}^\infty$, where $\kappa_n := \pi n/d$, corresponding to the eigenfunctions

$$\chi_{2j+1}(u) = \sqrt{\frac{2}{d}}\,\cos\kappa_{2j+1}u\,, \quad \chi_{2j}(u) = \sqrt{\frac{2}{d}}\,\sin\kappa_{2j}u\,. \tag{16.6}$$

The longitudinal part is a one–dimensional Schrödinger operator with the potential $-\frac{1}{4}\gamma^2$ which is attractive whenever Ω is not straight. If the curvature vanishes at large distances, the essential spectrum of this part covers the positive halfline and there is at least one isolated eigenvalue – *cf.* Theorem 14.3.6. Without the mode–coupling terms therefore the operator would have, by Theorem 5.7.4, an eigenvalue below the essential spectrum whose threshold is κ_1^2. In a thin strip the perturbation coming from mode coupling is weak, however, which leads us to the expectation that the discrete spectrum will survive (Problem 4a).

Such results are indeed valid, and moreover, they do not require the strip to be thin. First of all, the essential spectrum of $H_D^{\Omega_0}$ is preserved if Ω is asymptotically straight in a suitable sense.

16.1.3 Proposition: In addition to the assumptions of Proposition 16.1.2, suppose that γ and its two first derivatives belong to $C_\infty(\mathbb{R})$; then $\sigma_{ess}(H_D^\Omega) = [\kappa_1^2, \infty)$.

Proof is left to the reader (Problem 5a).

It is maybe a lot more surprising that the bound states indicated above exist in *any* bent and asymptotically straight strip.

16.1.4 Theorem: Suppose that f is injective, $\gamma \in C^1$ with $a\|\gamma\|_\infty < 1$; then $\inf\sigma(H_D^\Omega) < \kappa_1^2$ holds unless $\gamma = 0$ identically. In particular, let $\inf\sigma_{ess}(H_D^\Omega) \geq [\kappa_1^2, \infty)$, for instance under the conditions of Proposition 16.1.3; in such a case H_D^Ω has at least one isolated eigenvalue of finite multiplicity.

Proof: In view of the unitary equivalence of Proposition 2 we have to check the desired property for the operator H or the associated quadratic form described in Problem 3a. To this aim it is sufficient to find a trial function ψ, not necessarily of unit norm, from the form domain $Q(H)$ such that

$$q[\psi] = \left\|g^{-1/4}\frac{\partial\psi}{\partial s}\right\|^2 + \left\|g^{1/4}\frac{\partial\psi}{\partial u}\right\|^2 - \kappa_1^2\|g^{1/4}\psi\|^2 < 0\,.$$

We will seek it in the form $\psi = \phi_\lambda\chi_1 - \varepsilon h$, where χ_1 is the lowest transverse eigenfunction (16.6) and the functions ϕ_λ and h have to be properly chosen. The former should have a slow decay. We denote $K := [-s_0, s_0]$ for some $s_0 > 0$ and choose a function $\phi \in C_0^\infty(\mathbb{R})$ such that $\phi(s) = 1$ on K; a suitable family of ϕ_λ is then obtained by a scaling exterior to the compact set K,

$$\phi_\lambda(s) := \begin{cases} \phi(s) & \dots \quad |s| \le s_0 \\ \phi(\pm s_0 + \lambda(s \mp s_0)) & \dots \quad \pm s > s_0 \end{cases} \tag{16.7}$$

This allows us to make the contribution from the trial function tails small, since

$$q[\phi_\lambda\chi_1] := \int_{\mathbb{R}} \left(\int_{-a}^{a} g(s,u)^{-1/2}\chi_1(u)^2 du \right) |\phi_\lambda'(s)|^2 \, ds \le \frac{\lambda}{1 - a\|\gamma\|_\infty} \|\phi'\|^2 .$$

Next we modify the trial function in the central part of the strip Ω_0 choosing $h := \jmath^2(H - \kappa_1^2)\phi_\lambda\chi_1$, where $\jmath$ is a function from $C_0^\infty(K \times (-a,a))$, or more explicitly

$$h(s,u) = -\left(\jmath^2 \frac{\gamma}{1+u\gamma} \chi_1' \right)(s,u) = \jmath^2(s,u) \sqrt{\frac{2}{d}} \frac{\kappa_1\gamma(s)}{1+u\gamma(s)} \sin \kappa_1 u \, .$$

Under our assumptions both the functions h and $\phi_\lambda\chi_1$ belong to $Q(H)$, and a straightforward computation yields the value of the form,

$$q[\phi_\lambda\chi_1 - \varepsilon h] = q[\phi_\lambda\chi_1] - 2\varepsilon \|g^{1/4}\jmath(H - \kappa_1^2)\phi_\lambda\chi_1\|^2 + \varepsilon^2 (h, g^{1/2}(H - \kappa_1^2)h) \, .$$

Notice that the coefficient of the linear term at the right side is independent of λ, because the scaling acts only out of the support of $\jmath$. If γ is nonzero in K, we can always choose $\jmath$ in such a way that this coefficient is nonzero. Then the sum of the last two terms is negative for all sufficiently small $\varepsilon > 0$. Fixing such an ε we see that for λ small enough we have $q[\phi_\lambda\chi_1 - \varepsilon h] < 0$ what we wanted to prove. ∎

Geometric binding has a truly quantum nature. To appreciate this fact one has to realize that in a bent strip, as considered here, there are no classically closed trajectories, apart from the trivial ones (Problem 6). Moreover, validity of this result is not restricted to smoothly bent two–dimensional strips: other examples and properties of such bound states are mentioned in Problems 7–12 and the notes.

Let us mention one more geometric mechanism which can produce a discrete spectrum, namely a *local deformation* of a quantum waveguide. Again we restrict ourself to the simplest example of a planar strip, *i.e.*, we consider the region

$$\Omega := \{ (x,y) \in \mathbb{R}^2 : \, 0 < y < d(1 + f(x)) \} \tag{16.8}$$

corresponding to a bounded, piecewise smooth function $f : \mathbb{R} \to (-1, \infty)$ with a compact support. Spectral properties of H_D^Ω depend on the sign of f.

16.1.5 Proposition: $\sigma_{ess}(H_D^\Omega) = [\kappa_1^2, \infty)$ where $\kappa_1 := \pi/d$ as above. The discrete spectrum is empty if $f(x) \le 0$ for all $x \in \mathbb{R}$. On the other hand, if $f(x) \ge 0$

everywhere and there is an interval where the inequality is sharp, the operator H_D^Ω has at least one eigenvalue in $(0,\kappa_1^2)$.

Proof: The essential spectrum is found as in Proposition 16.1.3 (*cf.* Problem 5a), similarly the absence of a discrete spectrum in a "squeezed" tube is verified by means of Dirichlet bracketing. For a "protruded" Ω we use a variational argument again: one has to find $\psi \in Q(H_D^\Omega)$ such that the shifted quadratic form $q[\psi] := \|\nabla\psi\|^2 - \kappa_1^2\|\psi\|^2 < 0$. The transverse spectrum is purely discrete for any $x \in \mathbb{R}$; the eigenvalues are $\kappa_j(x)^2$ where $\kappa_j(x) := \kappa_j(1+f(x))^{-1}$, corresponding to the eigenfunctions $\chi_{j,x}$ analogous to (16.6), with κ_j replaced by $\kappa_j(x)$.

We choose the trial function as $\psi(x,y) := \phi_\lambda(x)\chi_{1,x}(y)$, where ϕ_λ is again given by (16.7) with the parameter s_0 picked in such a way that $f(x)=0$ for $|x|>s_0$. A straightforward calculation gives

$$q[\psi] = \lambda\|\phi'\|^2_{L^2(\mathbb{R})} + \int_{-s_0}^{s_0} (\kappa_j(x)^2 - \kappa_j^2)\,dx\,.$$

Under the assumption on f the second term on the right side is negative, hence it is sufficient to choose the parameter λ small enough. ∎

On the other hand, no general claim can be made if f is sign–changing, except for the case when the deformation is weak in a suitable sense; we will return to this question in Section 16.4. In a similar way one can treat strips with both boundaries modified as well as locally deformed tubes in higher dimensions.

After looking into the bound states let us turn to the continuous spectrum of geometrically perturbed waveguides. We will again restrict ourselves to bent planar strips and show that in addition to the discrete spectrum, a nontrivial curvature can yield also *resonances* in the vicinity of the higher spectral thresholds, *i.e.*, the values κ_j^2, $j=2,3,\dots$, where the spectral multiplicity changes.

Let us start with a heuristic discussion of the problem. The operator H_D^Ω corresponding to a bent strip can be decomposed into transverse modes (16.6), *i.e.*, to be mapped unitarily to the space $\bigoplus_{j=1}^\infty L^2(\mathbb{R})$ of Problem 1, however, it will no longer be diagonal in this representation. Instead it has "matrix elements"

$$H_{jk} := -\frac{\partial}{\partial s}\,b_{jk}(s)\,\frac{\partial}{\partial s} + V_{jk}(s) + \kappa_j^2\delta_{jk}\,, \qquad (16.9)$$

where $b_{jk}(s) := \int_{-a}^{a}(1+u\gamma(s))^{-2}\chi_j(u)\chi_k(u)\,du$ and $V_{jk}(s)$ is given similarly through the effective potential appearing in (16.4). Now we notice that for a thin strip, $a\|\gamma\|_\infty \ll 1$, we have $b_{jk}(s) = \delta_{jk} + \mathcal{O}(a)$ where, of course, the small term is a differential operator of the second order, and $V_{jk}(s) = -\frac{1}{4}\gamma(s)^2 + \mathcal{O}(a)$. Consequently, we can write the "matrix elements" (16.9) in this case as

$$H_{jk} = H_{jk}^{(0)} + \mathcal{O}(a) = H_\gamma\delta_{jk} + \kappa_j^2\delta_{jk} + \mathcal{O}(a)\,,$$

where we have denoted by H_γ the one–dimensional Schrödinger operator appearing in (16.5) and in Problem 4b which involves the "longitudinal" variable s. In the absence of the last term on the right side the operator would have the form of an orthogonal sum with spectrum which is, by Problem 4.24b, a union of the spectra of its parts. Each of these is just a translation by κ_j^2 of the same set, the spectrum of H_γ, and as long as $\gamma(s) \to 0$ for $|s| \to \infty$ the latter equals $\{\mu_n\}_{n=1}^N \cup [0,\infty)$, where the family of negative eigenvalues is nonempty provided $\gamma \neq 0$ by Theorem 4.3.6. In particular, the eigenvalues of the operator $H^{(0)}$ are of the form $\mu_n + \kappa_j^2$. Since $\kappa_j^2 - \kappa_1^2 = (\pi/2a)^2(j^2-1) \to \infty$ as $a \to 0$ unless $j=1$, those with $j \geq 2$ become embedded in the continuous spectrum of the lower modes for a small enough.

The difference $H - H^{(0)}$ can be regarded as a perturbation. However, it is not reduced by projections of subspaces associated with a fixed transverse mode χ_j, and therefore it can turn the embedded eigenvalues into resonances. As in the analogous situations earlier, *cf.* Sections 9.6 and 15.4, we will look for lower–halfplane poles $E_{jn}(a)$ of the analytically continued resolvent coming from perturbation of $\mu_n + \kappa_j^2$. We will need suitable analyticity assumptions, for instance

(i) γ extends to an analytic function, denoted by the same symbol, in the region $\mathcal{M}_{\alpha_0,\eta_0} := \{\, z \in \mathbb{C} :\ |\arg z| \leq \alpha_0,\ |\mathrm{Im}\, z| < \eta_0 \}$ for some positive α_0, η_0.

(ii) To each $\alpha \in (0,\alpha_0)$ and $\eta \in (0,\eta_0)$ one can find positive $c_{\alpha,\eta}$ and δ such that the inequality $|\gamma(z)| < c_{\alpha,\eta}(1+|z|)^{-1-\delta}$ holds in $\mathcal{M}_{\alpha_0,\eta_0}$.

The perturbation analysis is not simple and we will just quote the result referring to the notes for the proof and further information.

16.1.6 Theorem: Suppose that the strip Ω does not intersect itself and the corresponding nonzero curvature satisfies assumptions (i) and (ii). Then for fixed $n = 1,\dots,N$ and $j \geq 2$ and any a small enough the analytically continued resolvent has a unique pole $E_{jn}(a)$ such that $E_{jn}(a) \to \mu_n + \kappa_j^2$ as $a \to 0$. Moreover, for any $\eta \in (0,\eta_0)$ there is a $C_j(\eta) > 0$ such that the inequalities

$$0 \leq -\mathrm{Im}\, E_{jn}(a) \leq C_j(\eta)\, e^{-\pi\eta a^{-1}\sqrt{2j-1}}$$

hold for all sufficiently small strip halfwidth a.

16.2 Point perturbations

Now we turn to the situation where the perturbation of a straight waveguide is due to singular pointlike potentials. This is a good model, *e.g.* for a non–ideal quantum wire which allows us to take material impurities, or alien atoms in the crystalline lattice into account. Equally important, such a model is solvable.

We will again discuss the simplest case of a two–dimensional strip only, mentioning extensions to higher dimensions in the notes. We will write the strip in the form $\Omega := \mathbb{R} \times (0,d)$ and denote its points as $\vec{x} = (x,y)$. Neglecting the constants as usual we suppose that the Dirichlet Laplacian $-\Delta_D^\Omega =: H_0$ describes the free dynamics. Moreover, using the scaling properties we put for simplicity $d = \pi$ everywhere in this section (*cf.* Problem 14). The perturbation will consist of two–dimensional point interactions which we have introduced in Section 14.6. What is important is that they are characterized through a *local* change of the operator domain; this allows us to employ the same boundary conditions as in the case $\Omega = \mathbb{R}^2$. They are written in terms of generalized boundary values at a point $\vec{a} \in \Omega$ defined as

$$L_0(\psi,\vec{a}) := -\lim_{\vec{x}\to\vec{a}} \frac{2\pi\psi(\vec{x})}{\ln|\vec{x}-\vec{a}|}\,, \quad L_1(\psi,\vec{a}) := \lim_{\vec{x}\to\vec{a}} \left[\psi(\vec{x}) + L_0(\psi,\vec{a})\,\frac{\ln|\vec{x}-\vec{a}|}{2\pi}\right],$$

which relate ψ to the corresponding fundamental solution of the Laplace equation.

Given now N–tuples $\alpha := \{\alpha_1,\cdots,\alpha_N\} \subset \mathbb{R}$ and $\vec{a} := \{\vec{a}_1,\ldots,\vec{a}_N\} \subset \Omega$ we define the Hamiltonian $H_{\alpha,\vec{a}}$ describing a straight strip with point perturbations by

$$L_1(\psi,\vec{a}_j) - \alpha_j L_0(\psi,\vec{a}_j) = 0\,, \quad j = 1,\ldots,N; \tag{16.10}$$

in other words, the operator acts as $(H_{\alpha,\vec{a}}\psi)(\vec{x}) = -(\Delta\psi)(\vec{x})$ for $\vec{x} \neq \vec{a}$ in the sense of distributions and its core consists of all $\psi \in C_0^\infty(\Omega\setminus\vec{a})$ such that $-\Delta\psi \in L^2$ and the boundary conditions (16.10) are satisfied. The parameters α_j play the role of coupling constants; the absence of a point interaction at some $\vec{a}_j$ means $L_0(\psi,\vec{a}_j) = 0$ for all ψ from the domain which can be formally achieved by putting $\alpha_j = \infty$.

Consider first a single perturbation, $N = 1$, supported by a point $\vec{a} := (a,b)$ with $b \in (0,\pi)$. To find the resolvent of $H_{\alpha,\vec{a}}$ we start with that of H_0. We use a unitary equivalence analogous to that of Problem 1, with (16.6) replaced by a basis modified for the "one-sided" strip, $\chi_n(y) = \sqrt{\frac{2}{\pi}}\,\sin(ny)$. Using Problem 7.9 we find that the free resolvent is an integral operator with the kernel

$$G_0(\vec{x}_1,\vec{x}_2;z) \equiv (H_0-z)^{-1}(\vec{x}_1,\vec{x}_2) = \frac{i}{\pi}\sum_{n=1}^{\infty} \frac{e^{ik_n(z)|x_1-x_2|}}{k_n(z)}\,\sin(ny_1)\,\sin(ny_2)\,, \tag{16.11}$$

where $\vec{x}_j = (x_j,y_j)$ and $k_n(z) := \sqrt{z-n^2}$ for $z \in \rho(H_0) = \mathbb{C}\setminus[1,\infty)$. The function $G_0(\cdot,\cdot;z)$ is defined and smooth except at $\vec{x}_1 = \vec{x}_2$, but the sum at the right side may not converge absolutely if $x_1 = x_2$. Moreover, the latter makes sense also for all non-integer $z > 1$, where it gives the boundary value of the kernel at the cut; one has to properly choose the branch of the square root in $k_n(z)$. For $z < 1$ we have $G_0(\vec{x}_1,\vec{x}_2;z) > 0$ for all mutually different $\vec{x}_1,\vec{x}_2 \in \Omega$ (see the notes).

Following the general theory of self-adjoint extensions the resolvent of $H_{\alpha,\vec{a}}$ differs from $(H_0-z)^{-1}$ by a rank one operator given by

$$(H_{\alpha,\vec{a}}-z)^{-1}(\vec{x}_1,\vec{x}_2) \;=\; G_0(\vec{x}_1,\vec{x}_2;z) + \frac{G_0(\vec{x}_1,\vec{a};z)\,G_0(\vec{a},\vec{x}_2;z)}{\alpha-\xi(\vec{a};z)}\,, \tag{16.12}$$

where

$$\xi(\vec{a};z) := \frac{i}{\pi}\sum_{n=1}^{\infty}\left(\frac{\sin^2(nb)}{k_n(z)} - \frac{1}{2in}\right); \tag{16.13}$$

it follows from Theorem 4.7.15 and a direct computation (Problem 15). It is clear from the formula (16.12) that the function ξ is important for determining the spectrum of $H_{\alpha,\vec{a}}$. Some of its properties are summarized in Problem 16. The strip with a single impurity has then the following spectral properties.

16.2.1 Proposition: For any $\vec{a} \in \Omega$ and $\alpha \in \mathbb{R}$ we have $\sigma_{ess}(H_{\alpha,\vec{a}}) = \sigma_{ac}(H_{\alpha,\vec{a}}) = [1,\infty)$ and $\sigma_{sc}(H_{\alpha,\vec{a}}) = \emptyset$. Moreover, the operator $H_{\alpha,\vec{a}}$ has exactly one eigenvalue $\lambda_{\alpha,\vec{a}} \in (-\infty, 1)$. The function $\alpha \mapsto \lambda_{\alpha,\vec{a}}$ is real-analytic and increasing and has the following asymptotic behavior,

$$\begin{aligned} \lambda_{\alpha,\vec{a}} &= 1 - (\pi\alpha)^{-2} \sin^4 b + \mathcal{O}(\alpha^{-3}), \\ \lambda_{\alpha,\vec{a}} &= -4\, e^{-4\pi\alpha - 2\gamma_E} \left(1 - \mathcal{O}\left(\exp\left(-2\varrho\, e^{2\pi\alpha}\right)\right)\right) \end{aligned}$$

with any $\varrho < \mathrm{dist}(\vec{a}, \partial\Omega) = \frac{\pi}{2} - \left|b - \frac{\pi}{2}\right|$, for $\alpha \to \pm\infty$, respectively. Furthermore,

$$\lambda_{\alpha,\vec{a}} < \lambda_{\alpha,\vec{a}'} \qquad \text{if} \quad \left|b - \frac{\pi}{2}\right| < \left|b' - \frac{\pi}{2}\right| .$$

Finally, there are no eigenvalues embedded in the continuous spectrum.

Proof: Both the essential and absolutely continuous spectrum are preserved in view of Theorems 5.4.6 and 15.2.9, respectively. Let $\{E(t)\}$ be the spectral measure of $H_{\alpha,\vec{a}}$; to prove the absence of the embedded eigenvalues away from the thresholds, it is sufficient to check that $t \mapsto (\psi, E(t)\psi)$ is, for all $\psi \in L^2(\Sigma)$, a continuous function in any open interval $I \subset (n^2, (n+1)^2)$, $n \in \mathbb{N}$. This follows from Stone's formula and the fact that the boundary values of the full Green's function are smooth by (16.12) and Problem 16; notice that the denominator in the expression for the resolvent kernel cannot be zero in $[1,\infty)$ – *cf.* Problem 17. One can check directly that no resolvent poles can exist at thresholds, $z = n^2$. In a similar way we get

$$(\psi, (E(t) - E(t_0))\psi) = \frac{1}{\pi} \int_{t_0}^{t} \mathrm{Im}\, \left(\psi, (H_{\alpha,\vec{a}} - u)^{-1}\psi\right) du\,.$$

The left side is thus a smooth function of t away from the thresholds, so the singularly continuous spectrum is absent by Theorem 15.4.7. For the point spectrum, one has to solve the equation

$$\xi(\vec{a}; z) = \alpha\,; \tag{16.14}$$

the existence, uniqueness, and properties of the solution follow easily from the properties of the function ξ listed in Problem 16. ∎

Notice also that the residue at the pole given by (16.14) is a projection determining the eigenfunction of the bound state, $\phi_{\alpha,\vec{a}} = G_0(\cdot, \vec{a}; \epsilon_{\alpha,\vec{a}})$, or more explicitly

$$\phi_{\alpha,\vec{a}}(x) = \frac{1}{\pi} \sum_{n=1}^{\infty} \frac{e^{-\kappa_n(\lambda_{\alpha,\vec{a}})|x-a|}}{\kappa_n(\lambda_{\alpha,\vec{a}})} \sin(ny)\, \sin(nb)$$

up to a normalization, where we have denoted $\kappa_n(z) := -ik_n(z) = \sqrt{n^2 - z}$.

In distinction to the case of geometrically perturbed strips considered in the previous section the problem of scattering by the point impurity is easily solvable.

16.2.2 Proposition: The wave operators for the pair $(H_0, H_{\alpha,\vec{a}})$ exist and are asymptotically complete. Moreover, the on-shell scattering operator $S(k)$ at the energy $z = k^2 \in (1,\infty)\setminus\{n^2\}_{n\in\mathbb{N}}$ is a unitary $2[\sqrt{z}\,]\times 2[\sqrt{z}\,]$ matrix with the blocks

$$S_{nm}(k) = \sqrt{\frac{k_m(z)}{k_n(z)}}\begin{pmatrix} t_{nm}(k) & r_{nm}(k) \\ \tilde{r}_{nm}(k) & \tilde{t}_{nm}(k) \end{pmatrix},$$

where the indices n,m run from one to the integer part $[\sqrt{z}]$ of $\sqrt{z}$ and the transmission and reflection amplitudes are given by

$$(t_{nm}(k)-\delta_{nm})\,e^{i(k_m-k_n)a} = r_{nm}(k)\,e^{-i(k_n+k_m)a} = \frac{i}{\pi}\,\frac{\sin(nb)\sin(mb)}{k_m(z)(\alpha-\xi(\vec{a};z))},$$

with the tilded (right–to–left) amplitudes obtained by changing the sign of a, *i.e.*, as $\tilde{r}_{nm}(k) := r_{nm}(k)\,e^{-2i(k_n+k_m)a}$ and $\tilde{t}_{nm}(k) := \delta_{nm} + (t_{nm}(k)-\delta_{nm})\,e^{2i(k_m-k_n)a}$.
Proof: Since the two resolvents differ by a rank–one operator, the first claim follows from Theorem 15.2.9. To find the scattering matrix, we employ the formula (16.12) by which, for any $\psi\in D(H_{\alpha,\vec{a}})$ and a nonreal z, there is a unique decomposition

$$\psi(\vec{x}) = \psi_z(\vec{x}) + \frac{G_0(\vec{x},\vec{a};z)}{\alpha-\xi(\vec{a};z)}\,\psi_z(\vec{a})$$

with $\psi_z\in D(H_0)$ and $(H_{\alpha,\vec{a}}-z)\psi = (H_0-z)\psi_z$. If we choose for ψ_z, for instance, $\psi_z^\varepsilon(\vec{x}) := e^{ik_n(z)x-\varepsilon x^2}\chi_n(y)$, then the corresponding $\psi^\varepsilon\in D(H_{\alpha,\vec{a}})$ for all $\varepsilon>0$ and

$$((H_{\alpha,\vec{a}}-z)\psi^\varepsilon)(\vec{x}) = 2\varepsilon\left(2\varepsilon x^2-1-2ik_n(z)x\right)\psi_z^\varepsilon(\vec{x}).$$

The right side makes sense on the real line, hence $\psi^\varepsilon\in D(H_{\alpha,\vec{a}})$ holds for $z\in[1,\infty)$ and the last relation is valid there also. Furthermore, the pointwise limit $\psi(\vec{x}) = \lim_{\varepsilon\to 0}\psi^\varepsilon(\vec{x})$ exists and

$$\psi(\vec{x}) = e^{ik_n(z)x}\chi_n(y) + \frac{e^{ik_n(z)a}}{\alpha-\xi(\vec{a};z)}\,G_0(\vec{x},\vec{a};z)\chi_n(b).$$

This function belongs to L^2_{loc}, satisfies the appropriate boundary conditions and solves $(H(\alpha,\vec{a})-z)\psi = 0$ as a differential equation, *i.e.*, it is a generalized eigenvector. Substituting for G_0 and comparing the coefficients of the plane waves at different transverse modes as $x\to\pm\infty$, we get $r_{nm}(k)$ and $t_{nm}(k)$ for the incident wave corresponding to the n–th transverse mode; for the right–to–left amplitudes one has to change $k_n(z)$ to $-k_n(z)$ in the above Ansatz. It is clear that the solution which does not vanish asymptotically is obtained only if both the involved channels are open, *i.e.* $z>\max\{n^2,m^2\}$. Finally, the S–matrix must be normalized with respect to the relative velocities; its unitarity follows from the completeness of the wave operators but it can also be checked directly (Problem 17). ∎

The situation when there are $N>1$ point impurities in the waveguide is similar but has some different features. The resolvent of the Hamiltonian $H_{\alpha,\vec{a}}$ is again expressed by Krein's formula, this time its kernel equals

$$(H_{\alpha,\vec a}-z)^{-1}(\vec x_1,\vec x_2) \,=\, G_0(\vec x_1,\vec x_2;z)+\sum_{j,m=1}^{N}[\Lambda_{\alpha,\vec a}(z)]^{-1}_{jm}\,G_0(\vec x_1,\vec a_j;z)\,G_0(\vec a_m,\vec x_2;z)\,, \tag{16.15}$$

the coefficients being entries of the inverse to the $N\times N$ matrix $\Lambda\equiv\Lambda_{\alpha,\vec a}(z)$ with the elements $\Lambda_{jm}:=(\alpha_j-\xi(\vec a;z))\,\delta_{jm}-G_0(\vec a_j,\vec a_m;z)(1-\delta_{jm})$, where $\xi(\vec a;z)$ is given by (16.13) (Problem 15). This determines the spectral properties of $H_{\alpha,\vec a}$.

16.2.3 Theorem: For any $\vec a=\{\vec a_j\}$ with $\vec a_j\in\Omega$, $j=1,\dots,N$, and $\alpha\in\mathbb{R}^N$ we have $\sigma_{ess}(H_{\alpha,\vec a})=\sigma_{ac}(H_{\alpha,\vec a})=[1,\infty)$ and $\sigma_{sc}(H_{\alpha,\vec a})=\emptyset$. The discrete spectrum consists of k eigenvalues $\lambda_i^{\alpha,\vec a}\in(-\infty,1)$, $i=1,\dots,k$, counting multiplicity, arranged in ascending order, with $1\le k\le N$. They are real–analytic functions of the parameters α_j. The ground–state eigenvalue $\lambda_1^{\alpha,\vec a}$ is simple while the other ones may be degenerate. Eigenfunctions corresponding to $\lambda_i^{\alpha,\vec a}$ are of the form

$$\phi_i^{\alpha,\vec a}(\vec x) \,=\, \sum_{j=1}^{N} d_j\,G_0(\vec x,\vec a_j;\lambda_i^{\alpha,\vec a})\,,$$

where the coefficients solve the system $\sum_{m=1}^N \Lambda(\lambda_i^{\alpha,\vec a})_{jm}d_m=0$. In particular, for the eigenfunction $\phi_1^{\alpha,\vec a}$ all of them can be chosen positive. Finally, $z>1$ cannot be an eigenvalue with an eigenvector from the subspace $\bigoplus_{n=1}^{[\sqrt z]}L^2(\mathbb{R})\otimes\{\chi_n\}$.

Proof: The continuous–spectrum argument is analogous to that of Proposition 1. The discrete spectrum is again determined by poles of the resolvent coming from the coefficients in the Krein formula; they are given by the condition

$$\det\Lambda_{\alpha,\vec a}(z) \,=\, 0\,. \tag{16.16}$$

Let us next determine the eigenfunctions. Denote $H\equiv H_{\alpha,\vec a}$ and suppose that $\phi\in L^2(\Omega)$ is such that $H\phi=z\phi$ for some $z\in\mathbb{R}$. We pick an arbitrary $z'\in\rho(H)$, then by (16.15) there is a $\psi_0\in D(H_0)$ such that the eigenvector ϕ is expressed as

$$\phi \,=\, \psi_0+\sum_{j=1}^{N} d_jG_0(\cdot,\vec a_j;z') \tag{16.17}$$

with the coefficients given explicitly by $d_j:=\sum_{k=1}^N[\Lambda(z')]^{-1}_{jk}\,\psi_0(\vec a_k)$. Moreover, the relations $(H_0-z')\psi_0=(H-z')\phi=(z-z')\phi$ hold; applying $(H_0-z')^{-1}$ to this identity we obtain

$$\psi_0 \,=\, (z-z')\left[(H_0-z')^{-1}\psi_0+\sum_{j=1}^{N} d_j(H_0-z')^{-1}G_0(\cdot,\vec a_j;z')\right]\,,$$

which in turn yields $(H_0-z)\psi_0=(z-z')\sum_{j=1}^N d_jG_0(\cdot,\vec a_j;z')$. If $z<1$, the resolvent $(H_0-z)^{-1}$ exists and may be applied to both sides of the last relation giving

$$\psi_0 \,=\, \sum_{j=1}^{N} d_j\,(G_0(\cdot,\vec a_j;z)-G_0(\cdot,\vec a_j;z'))$$

with the help of the first resolvent identity. Substituting ψ_0 into (16.17) and setting $z = \lambda_i^{\alpha,\vec{a}}$ we arrive at the expression for the (non–normalized) eigenfunction given in the theorem. To determine the coefficients, we use the explicit form of the matrix elements $\Lambda(z)_{jm}$ and the above expression for ψ_0 to infer that $\psi_0(\vec{a}_j) = \sum_{m=1}^N (\Lambda(z')_{jm} - \Lambda(z)_{jm})\, d_m$. On the other hand, inverting the formula for the coefficients in (16.17) we get $\psi_0(\vec{a}_j) = \sum_{m=1}^N \Lambda(z')_{jm} d_m$; comparing these two expressions at $z = \lambda_i^{\alpha,\vec{a}}$ we arrive at the sought claim.

It is straightforward to check with the help of Theorem 4.7.15b that $\frac{d}{dz}\Lambda(z)$ for fixed $\alpha, \vec{a}$ is a strictly positive matrix, so all the eigenvalues of $\Lambda(z)$ are strictly monotonous as functions of z. The matrix elements are real-analytic functions of z, hence the eigenvalues are continuous; this shows that $k \le N$. The next question is about the existence of solutions to the equation (16.16). We observe that the matrix $\Lambda(z)$ has the following asymptotic behavior,

$$\Lambda_{\alpha,\vec{a}}(z) = \frac{1}{4\pi} \ln\left(-\frac{z}{4}\right) I + \mathcal{O}(1)\,, \quad \Lambda_{\alpha,\vec{a}}(z) = -\frac{1}{\pi\sqrt{1-z}}\, M_1 + \mathcal{O}(1)$$

as $z \to -\infty$ and $z \to 1-$, respectively, where $M_1 := (\sin b_j \, \sin b_m)_{j,m=1}^N$. The last named matrix has, in particular, an eigenvector $(\sin b_1, \dots, \sin b_N)$ corresponding to the *positive* eigenvalue $\sum_{j=1}^N \sin^2 b_j$, and therefore one of the eigenvalues of $\Lambda_{\alpha,\vec{a}}(z)$, tends to $-\infty$ as $z \to 1-$. Using again the eigenvalue continuity we see that at least one of them must cross zero in $(-\infty, 1)$ giving rise to an eigenvalue. Furthermore, $\det \Lambda_{\alpha,\vec{a}}(z)$ is a real-analytic function of α_j and z, so the analytic dependence of $\lambda_i^{\alpha,\vec{a}}$ on the coupling constants follows by the implicit-function theorem. For the remaining part of the proof concerning nondegeneracy of the ground state, positivity of the coefficients d_j involved, and embedded eigenvalues we refer to the notes. ∎

In distinction to the case $N = 1$, embedded eigenvalues are not excluded.

16.2.4 Example: Consider a pair of impurities with the *same* coupling constant α placed at $\vec{a}_1 := (0, b)$ and $\vec{a}_2 := (0, \pi - b)$. The eigenvalue problem can be divided into the part symmetric and antisymmetric with respect to the strip axis. In view of Problem 14 the antisymmetric part is obtained by scaling the single–interaction problem with $\vec{a} := (0, 2b)$ and coupling constant $\alpha - \frac{1}{2\pi}\ln 2$; the scaled eigenvalue tends to 4 as $\alpha \to \infty$, hence it becomes embedded in $\sigma_c(H_{\alpha,\vec{a}}) = [1, \infty)$ for all α large enough. Similarly one can construct other examples of embedded eigenvalues based on existence of a symmetry which prevents the decay of such an eigenstate; a violation of this symmetry turns these eigenvalues naturally into resonances.

One can also ask about possible degeneracy of the discrete spectrum. Theorem 3 says that the maximum multiplicity is $N-1$, in particular, the discrete spectrum is always simple for $N = 2$. This is not true in general, however.

16.2.5 Example: For brevity we put $g_{jm}(z) := -G_0(\vec{a}_j, \vec{a}_m; z)$. Let $N = 3$ with $\vec{a}_{1,3} := \left(\pm a, \frac{\pi}{2}\right)$ and $\vec{a}_2 := (0, b)$, and fix an energy $z < 1$. We clearly have $g_{12}(z) = g_{23}(z)$ for any $b \in (0, \pi)$. If $b = \frac{\pi}{2}$ this value is obviously strictly greater than $g_{13}(z)$; on the other hand, $\lim_{b\to 0} g_{12}(z) = 0$, so there is a $b \in \left(0, \frac{\pi}{2}\right)$ for which all the three $g_{jm}(z)$ have the same value. Choosing now the coupling constants α_j in such a way that $\alpha_j - \xi(\vec{a}_j; z) = g_{jm}(z)$, $j = 1, 2, 3$, we find that z must be an eigenvalue of multiplicity two.

The number k of eigenvalues depends on the coupling constants. In the *strong-coupling case* when all the α_j's are large negative, the matrix $\Lambda_{\alpha,\vec{a}}(z)$ is dominated by its diagonal and $H_{\alpha,\vec{a}}$ has N eigenvalues with strongly localized eigenfunctions; in the opposite case there is just one bound state (Problem 18).

Let us finally turn to the scattering problem with $N > 1$ impurities.

16.2.6 Theorem: The wave operators for the pair $(H_0, H_{\alpha,\vec{a}})$ exist and are asymptotically complete. The on-shell operator $S(k)$ at energy $k^2 > 1$ with $k \notin \mathbb{N}$ is a unitary $2[k] \times 2[k]$ matrix with block structure as in Proposition 2 and the transmission and reflection coefficients given by

$$\begin{aligned} r_{nm}(k) &= \frac{i}{\pi} \sum_{j,l=1}^{N} [\Lambda(z)]^{-1}_{jl} \, \frac{\sin(mb_j)\sin(nb_l)}{k_m(z)} \, e^{i(k_m a_j + k_n a_l)} \,, \\ t_{nm}(k) &= \delta_{nm} + \frac{i}{\pi} \sum_{j,l=1}^{N} [\Lambda(z)]^{-1}_{jl} \, \frac{\sin(mb_j)\sin(nb_l)}{k_m(z)} \, e^{-i(k_m a_j - k_n a_l)} \,, \end{aligned}$$

with the right-to-left amplitudes obtained by the mirror transformation which replaces all the a_j by $-a_j$.

Proof will be just sketched because it is analogous to that of Proposition 16.1.2. The generalized eigenfunction at a non-threshold $z > 1$ is now replaced by

$$\psi(\vec{x}) = e^{ik_n(z)x}\chi_n(y) + \sum_{j,k=1}^{N} [\Lambda(z)]^{-1}_{jk} G_0(\vec{x}, \vec{a}_j; z)\, e^{ik_n(z)a_k}\chi_n(b_k)$$

with the incident wave in the n-th channel. Its asymptotic behavior as $x \to \pm\infty$ gives the above reflection and transmission amplitudes. ∎

Let us finally mention that in application to solid-state physics where such waveguides model "quantum wires" it is not the S-matrix which is of primary interest but rather the *conductance* given by the *Landauer formula*

$$G(z) = \frac{2e^2}{h} \sum_{n,m=1}^{[\sqrt{z}]} \frac{k_m(z)}{k_n(z)} \, |t_{nm}(z)|^2 \,, \tag{16.18}$$

where z is determined by the Fermi energy and chemical potential of the reservoirs. We write it in proper units; notice that the summed quantities are dimensionless. The presence of impurities (or other perturbations) deforms the steplike shape of this function corresponding to an ideal straight channel, in particular, resonances are typically manifested by dips near the thresholds.

16.3 Curved quantum layers

Another wide class of constrained systems concerns the situation where the configuration space is a layer Ω, in particular, a deformation of $\Omega_0 = \mathbb{R}^2 \times (-a,a)$ and similar infinite regions. As before we will consider a nonrelativistic quantum particle and, neglecting unimportant constants, we will identify the corresponding Dirichlet Laplacian $H_D^\Omega := -\Delta_D^\Omega$ with the Hamiltonian of the problem.

We will concentrate on the cases when the layer has a finite width $d = 2a$ being generated by a smooth surface Σ. Some preliminary considerations are needed, because the geometry of surfaces is more complicated than that of curves. The first task is to parametrize Σ by means of a map from $\mathbb{R}^2$ to $\mathbb{R}^3$, or more generally, through a suitable atlas — recall that in contrast to Section 16.1 there is no natural system of coordinates (see the notes). For simplicity we will first suppose that Σ is a C^2–smooth surface equipped with *geodesic polar coordinates*. This means that Σ is diffeomorphic to $\mathbb{R}^2$ and as such is simply connected and non-compact; the coordinate lines are the geodesics emanating from a *pole* and the geodesic circles which connect points at the same geodesic distance from the pole.

In such a case the surface Σ is characterized by the map $p:\, \Sigma_0 \to \mathbb{R}^3$, where $\Sigma_0 := (0,\infty) \times S^1$ is the plane equipped with polar coordinates, $q = (s,\theta)$. The tangent vectors $p_{,\mu} := \partial p/\partial q^\mu$ are linearly independent and their cross–product defines the unit normal n at a given point. We will consider the *layer* $\Omega := \mathcal{L}(\Omega_0)$ of the indicated width generated by the surface Σ which is the image of the flat layer $\Omega_0 := \Sigma_0 \times (-a,a)$ by the map $\mathcal{L}:\, \Omega_0 \to \mathbb{R}^3$ defined as

$$\mathcal{L}(q,u) := p(q) + un(q)\,. \tag{16.19}$$

To classify such layers by their metric properties, let us inspect first the geometry of Σ. The surface metric tensor, $g_{\mu\nu} := p_{,\mu}\cdot p_{,\nu}$, has in geodesic polar coordinates a diagonal form, $(g_{\mu\nu}) = \mathrm{diag}(1,r^2)$, where $r^2 \equiv g := \det(g_{\mu\nu})$ is the square of the appropriate Jacobian, analogous to that of the tube case.

The *Gauss curvature* K and *mean curvature* M are determined in the usual way: the second fundamental form $h_{\mu\nu} := -n_{,\mu}\cdot p_{,\nu}$ gives rise to the Weingarten tensor $h_\mu^{\;\nu} := h_{\mu\rho}g^{\rho\nu}$, which in turn defines the two curvatures by $K := \det(h_\mu^{\;\nu})$ and $M := \frac{1}{2}\mathrm{Tr}(h_\mu^{\;\nu})$. The corresponding global quantities obtained by integration with respect to the invariant surface element, $d\sigma := g^{1/2}dq$, namely the total Gauss curvature $\mathcal{K}$ and the quantity $\mathcal{M}$ are

$$\mathcal{K} := \int_{\Sigma_0} K(q)\,d\sigma\,, \quad \mathcal{M} := \left(\int_{\Sigma_0} M(q)^2\,d\sigma\right)^{1/2}.$$

Let us next turn to the layer geometry. It is convenient to distinguish the tensor component indices using Greek letters for the surface variables and Latin for those of Ω including $q^3 := u$. It follows from the definition (16.19) that the metric tensor of the layer, as of a submanifold in $\mathbb{R}^3$, has the block form

$$(G_{ij}) = \begin{pmatrix} (G_{\mu\nu}) & 0 \\ 0 & 1 \end{pmatrix}, \quad G_{\nu\mu} = (\delta_\nu^\sigma - uh_\nu^{\;\sigma})(\delta_\sigma^\rho - uh_\sigma^{\;\rho})g_{\rho\mu}\,.$$

From here the determinant $G := \det(G_{ij})$ is easily computed. Recall that the eigenvalues of the Weingarten map matrix are the *principal curvatures* k_1, k_2 through which the Gauss and mean curvatures defined above are expressed as $K = k_1 k_2$ and $M = \frac{1}{2}(k_1 + k_2)$, respectively; it follows that

$$G = g[(1-uk_1)(1-uk_2)]^2 = g(1-2Mu+Ku^2)^2.$$

In particular, this expression defines the volume element by $d\omega := G^{1/2} dq\, du$.

In analogy with Section 16.1 we will suppose that Ω is not self-intersecting, and moreover, that it is thin enough to make the boundary smooth. The last requirement is in the present case ensured if $a < \rho_m := (\max\{\|k_1\|_\infty, \|k_2\|_\infty\})^{-1}$ where the principal curvatures are assumed to be uniformly bounded, $\|k_j\|_\infty < \infty$ for $j = 1, 2$. The number ρ_m is the minimal normal curvature radius of Σ; the condition implies that the factor $1-2Mu+Ku^2$ is bounded from below on Ω_0 by a positive number. Moreover, $C_- \le 1-2Mu+Ku^2 \le C_+$ with $C_\pm := (1 \pm a\rho_m^{-1})^2$. Together with the C^2–smoothness of Σ it means that $\mathcal{L}$ is a global diffeomeorphism. Another consequence is that $G_{\mu\nu}$ can be estimated by the surface metric,

$$C_- g_{\mu\nu} \le G_{\mu\nu} \le C_+ g_{\mu\nu}\,. \tag{16.20}$$

This bound is important, because in distinction to the tube case the "straightening" transformation does not allow us here to get rid of the geometry of the generating manifold fully — roughly speaking, one cannot unfold Σ into a plane.

Let us look further at how the parametrization of Ω will be manifested in the Hamiltonian H_D^Ω. In the coordinates q, u it acquires the Laplace-Beltrami form

$$\tilde{H} := -G^{-1/2}\partial_i G^{1/2} G^{ij} \partial_j$$

acting on $L^2(\Omega_0, G^{1/2} dq\, du)$, where the usual relation between covariant and contravariant tensor components, $G_{ij}G^{jk} := \delta_i^k$, is employed. The right side of the last relation can be made more explicit using the block structure of $G_{\nu\mu}$ (Problem 19a).

As we have said in the layer case one cannot remove the Jacobian appearing in $d\omega := G^{1/2} dq\, du$ fully, however, it is possible at least to get rid of the "transverse" factor $1-2Mu+Ku^2$ in the inner-product weight by means of the unitary operator $\hat{U} : L^2(\Omega_0, d\omega) \to L^2(\Omega_0, d\sigma\, du)$ which acts as $\hat{U}\psi := (1-2Mu+Ku^2)^{1/2}\psi$. We employ the symbol H for the transformed operator on $L^2(\Omega_0, d\sigma\, du)$; the norm and inner product in this Hilbert space will be indicated by the subscript "g". Evaluating the operator explicitly (Problem 19b) we find that it contains an effective potential which can be conveniently written using $J := \frac{1}{2}\ln(1-2Mu+Ku^2)$; we have

$$H = -g^{-1/2}\partial_i g^{1/2} G^{ij} \partial_j + V\,, \quad V := g^{-1/2}(g^{1/2}G^{ij}J_{,j})_{,i} + J_{,i}G^{ij}J_{,j}\,, \tag{16.21}$$

which is well defined as a self-adjoint operator as long as Σ is C^4–smooth. Using again the block form of the metric tensor, we can write $H = H_1 + H_2$, where H_1 contains the part of (16.21) with the summation over Greek indices, and

$$H_2 := -\partial_3^2 + V_2\,, \quad V_2 = \frac{K-M^2}{(1-2Mu+Ku^2)^2}\,. \tag{16.22}$$

An advantage of this form is that it allows us to see how the transverse and longitudinal variable become asymptotically decoupled in the thin layer situation, $a \ll \rho_m$, in analogy with (16.5); using the bounds (16.20) one finds

$$H = -g^{-1/2}\partial_\mu g^{1/2} g^{\mu\nu}\partial_\nu - \partial_3^2 + K - M^2 + \mathcal{O}(a)\,. \tag{16.23}$$

In distinction to the tube case the geometry of the surface is not suppressed, the first term being the Laplace-Beltrami operator Δ_g on Σ. At the same time, the mode–coupling terms vanish again in the limit $a \to 0$. To assess the leading term of the effective potential, $K - M^2$, let us rewrite it in terms of the principal curvatures as $-\frac{1}{4}(k_1 - k_2)^2$. This expression can vanish not only when Σ is planar, but also if it is locally spherical, $k_1 = k_2 \neq 0$. Nevertheless, a non–compact and non–planar surface cannot be spherical everywhere, thus at some parts of it the curvature–induced interaction is attractive and the discrete spectrum may be nonempty.

After this preliminary we can formulate some results about $\sigma(H_D^\Omega)$. The first question concerns again the essential spectrum, which for a planar layer Ω_0 is the interval $[\kappa_1^2, \infty)$. We will say that Ω is *asymptotically planar* if the curvatures vanish at large geodesic distance from the pole, *i.e.*, $K(s,\theta)$, $M(s,\theta) \to 0$ as $s \to \infty$.

16.3.1 Proposition: $\inf \sigma_{ess}(-\Delta_D^\Omega) \geq \kappa_1^2$ holds provided $\mathcal{L}$ is injective, $a < \rho_m$, and the layer Ω is asymptotically planar.

Proof is left to the reader (Problem 20).

In the curved–layer case no universal result is known, however, we have various sufficient conditions under which the curvature pushes the spectral threshold down. The trick used in Theorem 16.1.4 works for layers of a *non–positive curvature.*

16.3.2 Theorem: Let $\mathcal{L}$ be injective, $a < \rho_m$, with K, M uniformly bounded and $K \in L^1(\Sigma_0, d\sigma)$. If $\mathcal{K} \leq 0$, then $\inf \sigma(H_D^\Omega) < \kappa_1^2$. In particular, the discrete spectrum is nonempty if Ω is asymptotically planar.

Proof: We seek a function ψ from the form domain of the transformed Hamiltonian $\tilde{H}$ such that the shifted form $q[\psi] := t_\Omega[\psi] - \kappa_1^2 \|\psi\|_G^2 < 0$, with the t_Ω specified in Problem 19a. We begin with $\psi :\ \psi(s,\theta,u) = \phi(s)\chi_1(u)$, where ϕ is a function radially symmetric in the geodesic polar coordinates and χ_1 is the lowest transverse mode (16.6). It is convenient to split t_Ω into two parts, $t_\Omega = t_1 + t_2$ associated with the decomposition of $\tilde{H}$ described in Problem 19a. Using the explicit form of $\tilde{H}_2$ we get by a straightforward computation

$$t_2[\psi] - \kappa_1^2 \|\psi\|_G^2 = (\phi, K\phi)_g\,.$$

On the other hand, the "longitudinal part" $t_1[\psi]$ can be estimated by means of (16.20) and Problem 21 as

$$t_1[\psi] \leq C_1 \int_0^\infty |\phi'(s)|^2\, s\, ds\,,$$

where the right side depends on the surface geometry through the constant $C_1 := (C_+/C_-)^2(2\pi + \|K\|_{g,1})$ only, which is finite by assumption. This positive contribution from the trial function tails has to be made small by the choice of a mollifier.

In two dimensions we can not simply scale as in (16.7); instead we take a family $\{\phi_\lambda\}$ with

$$\phi_\lambda(s) := \min\left\{1, \frac{K_0(\lambda s)}{K_0(\lambda s_0)}\right\} \tag{16.24}$$

for some $s_0 > 0$ and $\lambda > 0$, where K_0 is the Macdonald function. These functions are not smooth, of course, but the corresponding $\psi_\lambda := \phi_\lambda \chi_1$ is an admissible trial function belonging to $Q(\tilde{H})$. The integral can be calculated; we find

$$\int_0^\infty |\phi'_\lambda(s)|^2 s\, ds < \frac{C_2}{|\ln \lambda s_0|} \tag{16.25}$$

for all λs_0 small enough and some $C_2 > 0$; hence $t_1[\psi_\lambda]$ can be made arbitrarily small if λ tends to zero. On the other hand, by dominated convergence we get $t_2[\psi_\lambda] - \kappa_1^2 \|\psi_\lambda\|_G^2 \to \mathcal{K}$ in the same limit; hence in the case $\mathcal{K} < 0$ it is sufficient to choose λ sufficiently small to conclude to proof.

To deal with the remaining case $\mathcal{K} = 0$ we again add a small deformation to the central part of the layer setting $\psi_{\lambda,\varepsilon} := \psi_\lambda + \varepsilon f$, where $f(q,u) := j(q) u \chi_1(u)$ with $j \in C_0^\infty((0, s_0) \times S^1)$. Since f belongs obviously to $Q(\tilde{H})$, we can write

$$q[\psi_{\lambda,\varepsilon}] = q[\psi_\lambda] + 2\varepsilon q(f, \psi_\lambda) + \varepsilon^2 q[f]$$

using the fact that the scaling acts out of the support of j. The coefficient of the linear term, $2q(f, \psi_\lambda) = -2(j, M)_g$, can be made nonzero by choosing j supported on a compact where the mean curvature M does not change sign; such a set certainly exists because Σ is not a plane and the parameter s_0 can be chosen arbitrarily large. This concludes the proof in the same way as in Theorem 16.1.4. ∎

The fact that negative Gauss curvature leads to a non–void discrete spectrum is understandable from the relation (16.23): the leading term of the effective potential is more attractive in the parts of Σ where the two principal curvatures have different signs, which means $K < 0$. The proved theorem covers a wide class of layers.

16.3.3 Example: Suppose that Ω is *locally curved*, *i.e.*, it is built over a surface Σ which is planar outside a compact set. By Gauss–Bonnet theorem we have $\mathcal{K} = 0$ so Theorem 16.3.2 applies. Since the polar coordinate system in the planar region exists trivially, the above proof can be extended to such surfaces without a pole (see the notes), or even those which are not diffeomorphic to $\mathbb{R}^2$.

Another family to which the result can be applied are layers built over *Cartan–Hadamard surfaces*, which are by definition complete simply connected non–compact surfaces with non–positive Gauss curvature. Recall that each point of such a Σ is a pole so there are infinitely many geodesic polar coordinate systems. Excluding the trivial planar case, $\mathcal{K}$ is always strictly negative for them, so such layers possess at least one bound state if the other assumptions of Theorem 16.3.2 are satisfied.

16.3.4 Examples: Two surfaces belonging to this class is the *hyperbolic paraboloid*, the simple quadric in $\mathbb{R}^3$ given by the equation $z = x^2 - y^2$, and the *monkey saddle* defined by $z = x^3 - 3xy^2$. Both are asymptotically planar having $\mathcal{K} = -2\pi$ and $\mathcal{K} = -4\pi$, respectively. Hence Theorem 16.3.2 applies as long as $a < \rho_m$, specifically in the first case for $a < \frac{1}{2}$ (Problem 22).

The geometry of surfaces is much richer, however, than that of curves and it is natural to ask whether the above assumptions about parametrization and simple connectedness of the generating surface are indeed needed. In the following we will suppose only that Σ is C^2–smooth, connected and orientable, embedded in $\mathbb{R}^3$, and moreover, noncompact and *complete*, *i.e.*, no geodetic on Σ is terminated.

The atlas of such a surface need no longer consist of a single chart, so it is useful to employ the surface itself as a basis for the parametrization, *i.e.*, to replace the set Σ_0 by Σ. The notions appearing in Theorem 16.3.2 can be used again in this context. In particular, a layer will be called *asymptotically planar* if the corresponding curvatures $K(x)$, $M(x)$ vanish at infinity; recall that a function f on a noncompact manifold Σ has this property if for any $\varepsilon > 0$ there are $R_\varepsilon > 0$ and a point $x_\varepsilon \in \Sigma$ such that $|f(x)| < \varepsilon$ holds for any $x \in \Sigma$ with geodesic distance from x_ε not smaller that R_ε. Proposition 16.3.1 holds again in this setting (Problem 20).

We employ the shorthand ∇_g for the two–dimensional covariant gradient on Σ so one has, *e.g.*, $M_{,\mu}g^{\mu\nu}M_{,\nu} = |\nabla_g M|_g^2$. The only place in the proof of Theorem 16.3.2 where the geodesic polar coordinates were needed was the construction of the mollifier (16.24). Instead one can use the following abstract result (see the notes).

16.3.5 Lemma: Let $K \in L^1(\Sigma)$, then there is a sequence $\{\phi_n\}_{n=1}^\infty$ of smooth functions with compact supports in Σ such that

(a) $0 \le \phi_n \le 1$ holds for all $n \in \mathbb{N}$ and $x \in \Sigma$.

(b) $\|\nabla_g \phi_n\|_g \to 0$ holds as $n \to \infty$.

(c) $\phi_n(x) \to 1$ as $n \to \infty$ uniformly on compact subsets of Σ.

This allows us to prove the claim of Theorem 16.3.2 under weaker assumptions. At the same time we will derive two other sufficient conditions which apply to surfaces of a positive total Gauss curvature as well.

16.3.6 Theorem: Let Σ be a C^2–smooth and non-planar surface with the properties described above and $K \in L^1(\Sigma)$. Suppose that the layer Ω built over Σ does not intersect itself and $a < \rho_m$ with K, M uniformly bounded; then the inequality $\inf \sigma(H_D^\Omega) < \kappa_1^2$ is valid if any of the following conditions holds true:

(a) $\mathcal{K} \le 0$.

(b) The layer is sufficiently thin, *i.e.* a is small enough, and $\nabla_g M \in L^2_{loc}(\Sigma)$.

(c) $\mathcal{M} = \infty$ and $\nabla_g M \in L^2(\Sigma)$.

Consequently, the discrete spectrum is nonempty if Ω is asymptotically planar.

Proof: The condition (a) is proved in a way similar to Theorem 16.3.2. We take the sequence $\{\phi_n\}$ of Lemma 16.3.5 and construct trial functions $\psi_n := \phi_n\chi_1$ for which

$$q[\psi_n] = \| |\nabla\phi_n|\, \chi_1\|^2 + (\phi_n, K\phi_n)_g \,.$$

Since $|\nabla\varphi_n|$ can be estimated by $|\nabla_g\varphi_n|_g$ using (16.20) the first term at the right side tends to zero as $n \to \infty$, while the second one gives $\mathcal{K}$ in the limit by Lemma 16.3.5c and dominated convergence; this concludes the proof if $\mathcal{K} < 0$.

In the critical case, $\mathcal{K} = 0$, we again add a small deformation, $\psi_{n,\varepsilon} := \psi_n + \varepsilon\theta$, where $\theta(x,u) := j(x)u\chi_1(u)$ with j being a C_0^∞ function supported in a region where M is nonzero and does not change sign. This yields

$$q[\psi_{n,\varepsilon}] = q[\psi_n] + 2\varepsilon\, q(\theta,\psi_n) + \varepsilon^2 q[\theta]\,.$$

Since $\mathcal{K} = 0$, the first term at the right side tends to zero as $n \to \infty$. The form in the second term needs now more attention, because the supports of θ and $\nabla\psi_n$ may not be disjoint in general. We express it as

$$q(\theta,\psi_n) = (\theta, 2M_u\,\phi_n\chi_1') + (\nabla\psi_n, \nabla\phi_n) - 2\,(\theta\nabla\chi_1, \nabla\phi_n)\,,$$

where the last two terms tend to zero as $n \to \infty$ by Schwarz inequality, the estimates (16.20) and Lemma 16.3.5, while the first integral is calculated to be $-(j, M\phi_n)_g$ as before, with a nonzero limit $-(j, M)_g$. This means that one can conclude the argument in the same way as in Theorem 16.3.2.

To prove the conditions (b) and (c) it is no longer possible to begin the trial function construction from the threshold–resonance function determined by the lowest mode, $(x,u) \mapsto \chi_1(u)$. We will modify the above ψ_n by a multiplicative variable–mixing factor taking $\tilde\psi_\lambda(x,u) := (1+M(s,\theta)u)\,\phi_n(x)\chi_1(u)$. Since $\nabla\psi_n(\cdot,u)$ equals

$$(1+Mu)(\nabla\varphi_n)\chi_1(u) + (\nabla M)u\,\varphi_n\chi_1(u) + \big((1+Mu)\kappa_1\,\varphi_n\chi_1'(u) + M\varphi_n\chi_1(u)\big)\nabla u\,,$$

it is easy to see, using again (16.20), that $\tilde\psi_n \in Q(H_D^\Omega)$ provided $\nabla_g M \in L^2_{loc}(\Sigma)$; recall that K and M are uniformly bounded by assumption. We have

$$\begin{aligned} q[\tilde\psi_n] \;\le\; & 2\left((1+a\|M\|_\infty)^2\, \||\nabla\phi_n|\,\chi_1\|^2 + a^2\||\nabla M|\,\phi_n\chi_1\|^2\right) \\ & + \left(\phi_n, (K-M^2)\phi_n\right)_g + \frac{\pi^2-6}{12\kappa_1^2}\left(\phi_n, KM^2\phi_n\right)_g\,, \end{aligned}$$

where we have used (16.20) and the Minkovski inequality, the second line following from a direct computation using the other two terms of the gradient.

Consider first the condition (c). If $\nabla_g M \in L^2(\Sigma)$ and $K \in L^1(\Sigma)$, all terms at the right side of the above estimate have finite limits as $n \to \infty$, except for the one containing $K - M^2$ which tends to $-\infty$, hence there is an n_0 such that $q[\tilde\psi_{n_0}] < 0$. The validity of condition (b) follows from two observations. First of all, the integral containing $K - M^2$ is always negative for any non–planar and non–compact surface in view of the observation made above. Furthermore, the first term in the estimate tends to zero as $n \to \infty$ because of (16.20) and Lemma 16.3.5, and the remaining ones vanish for any fixed n as $a \to 0$; recall that $\kappa_1^{-2} = 4a^2/\pi^2$. Hence we can find n_0 large enough so that the sum of the first and the third term is negative, and then choose the layer halfwidth a so small that $q[\tilde\psi_{n_0}] < 0$. ∎

Let us mention consequences of this result for layers of nontrivial topology. By that we mean that the generating surface Σ has *handles* or ends; recall that an open set $\mathcal{E} \subset \Sigma$ is called an *end* of Σ if it is connected, unbounded and its boundary $\mathrm{bd}\,\mathcal{E}$ is compact (in particular, it may be empty).

16.3.7 Corollary: Under the assumptions of Theorem 16.3.6 we have $\inf \sigma(H_D^\Omega) < \kappa_1^2$ whenever Σ is not conformally equivalent to the plane, in particular, for any surface Σ which is not simply connected.

Proof: Follows from the *Cohn–Vossen inequality* which says that $\mathcal{K} \le 2\pi\,(2-2h-e)$, where h is the genus of Σ, *i.e.* the number of handles, and e is the number of ends. Consequently, $\mathcal{K} < 0$ whenever h is nonzero. ∎

Condition (b) in Theorem 16.3.6 confirms the expectation coming from (16.23), namely that curved layers do have bound states if they are asymptotically planar and thin enough. It gives no quantitative estimate, however, on the critical thickness. For layers of positive curvature sometimes condition (c) can be used.

16.3.8 Example: An *elliptic paraboloid* is another simple quadric in $\mathbb{R}^3$, this time given by the equation $z = \frac{x^2}{b^2} + \frac{y^2}{c^2}$ with $b, c > 0$. One can check easily that $K = 4(gbc)^{-2}$ with $g = 1 + 4(x^2 b^{-2} + y^2 c^{-2})$ and $M = \mathcal{O}(r^{-1})$ as $r \to \infty$. Hence Σ is an asymptotically planar surface with $\mathcal{K} = 2\pi$ and $\mathcal{M} = \infty$, and Theorem 6 applies to the respective layers for any $a < \min\left(\frac{1}{2}b^2, \frac{1}{2}c^2\right)$.

There is another class of layers with $\mathcal{K} \ge 0$ for which one can establish existence of the discrete spectrum with no restriction on the width other than $a < \rho_m$, namely those invariant with respect to rotations around a fixed axis. Suppose that Σ is a surface of revolution parametrized by a function $p : \Sigma_0 \to \mathbb{R}^3$ of the form

$$p(s,\theta) := (r(s)\cos\theta, r(s)\sin\theta, z(s)),$$

where $r, z \in C^2((0,\infty))$ and $r > 0$. These functions define a geodesic polar coordinate chart if we impose the condition $(r')^2 + (z')^2 = 1$, which means that s is the arc length of the curves which are radial cuts of Σ; in such a case we have also $r'r'' + z'z'' = 0$. The Weingarten tensor is diagonal, $(h_\mu^{\ \nu}) = \mathrm{diag}(k_s, k_\theta)$, with the principal curvatures $k_s = r'z'' - r''z'$ and $k_\theta v = z'r^{-1}$, in fact, it is sufficient to know the function $s \mapsto k_s(s)$ only, because r, z can then be reconstructed as in Problem 2. Recall that the total Gauss curvature of a cylindrically symmetric Σ cannot be negative in view of the Gauss–Bonnet theorem,

$$\mathcal{K} + 2\pi r'(\infty) = 2\pi\,, \quad r'(\infty) := \lim_{s\to\infty} r'(s)\,, \tag{16.26}$$

since $r'(\infty)$ cannot exceed one in the chosen parametrization. On the other hand, we assume that $\mathcal{K}$ exists which means that the limit value $r'(\infty)$ makes sense; in addition, positivity of r requires $\mathcal{K} \le 2\pi$. Since the case $\mathcal{K} = 0$ is covered by Theorem 16.3.6 we will suppose that $\mathcal{K} \in (0, 2\pi]$, or equivalently, $0 \le r'(\infty) < 1$. In that case one checks easily that there are $\delta, s_0 > 0$ such that $\delta r(s)^{-1} \le |k_\theta(s)| \le r(s)^{-1}$

holds for all $s > s_0$ and the function k_θ does not change sign there. In combination with Problem 21 this implies that $k_\theta \notin L^1(\mathbb{R}^+)$. On the other hand, the meridian curvature k_s is integrable provided $\mathcal{K} \in L^1(\Sigma)$ in view of the following estimate

$$\int_{s_0}^\infty |k_s(s)|\, ds \le \delta^{-1} \int_{s_0}^\infty |k_s(s) k_\theta(s)|\, r(s)\, ds \le \delta^{-1} \int_0^\infty |K(s)|\, r(s)\, ds\,.$$

These observations give us a hint how to demonstrate the existence of bound states in the present situation. Although the mean curvature may decay at infinity, it is not negligible there in the integral sense, hence choosing a family of trial functions supported far from the pole of Σ one can attempt to eliminate the positive term coming from the Gauss curvature.

16.3.9 Theorem: Let Σ be a C^2–smooth non–planar surface of revolution with $\mathcal{K} \in L^1(\Sigma)$ satisfying the assumption of Proposition 16.3.1. Then $\inf \sigma(H_D^\Omega) < \kappa_1^2$, in particular, the discrete spectrum is nonempty if Ω is asymptotically planar.

Proof: We have noted that $\mathcal{K} > 0$ can be supposed without loss of generality. We employ trial functions $\psi_{n,\varepsilon}(s,u) := (\varphi_n(s) + \varepsilon\phi_n(s)u)\chi_1(u)$, where ε will be specified later and φ_n, ϕ_n are defined as follows: we take three integer sequences $\{b_n\}$, $\{c_n\}$, $\{d_n\}$, such that $0 < b_n < c_n < d_n$ and $b_n \to \infty$ as $n \to \infty$, and set

$$\varphi_n(s) := \frac{\ln(s/b_n)}{\ln(c_n/b_n)}\, \chi_{[b_n,c_n]}(s) + \frac{\ln(s/d_n)}{\ln(c_n/d_n)}\, \chi_{[c_n,d_n]}(s)$$

and $\phi_n(s) := s^{-1}\varphi_n(s)$. These functions are obviously positive and uniformly bounded, and as before, the corresponding $\psi_{n,\varepsilon}$ are not smooth but belong to the form domain $Q(\tilde H)$. Using inequalities (16.20) and Problem 21 we can estimate the longitudinal kinetic parts of $q[\psi_{n,\varepsilon}]$ by means of the integrals

$$t_1[\varphi_n\chi_1] \le C_1 \int_0^\infty \varphi_n'(s)^2 s\, ds\,, \quad t_1[\phi_n u\chi_1] \le 2a^2 C_1 \int_0^\infty \phi_n'(s)^2 s\, ds\,,$$

which both tend to zero as $n \to \infty$ provided the three sequences diverge at different rates, *i.e.* c_n/b_n and d_n/c_n tend to infinity as $n \to \infty$, and the same is true for the mixed term $t_1(\varphi_n\chi_1, \phi_n u\chi_1)$ by Schwarz inequality. On the other hand, by an explicit integration in the variable u we find that the remaining part of the form,

$$t_2[\psi_{n,\varepsilon}] - \kappa_1^2 \|\psi_{n,\varepsilon}\|_G = (\varphi_n, K\varphi_n)_g - 2\varepsilon(\varphi_n, M\phi_n)_g + \varepsilon^2 \left[\|\phi_n\|_g^2 + \frac{\pi^2-6}{3\kappa_1^2}\, (\phi_n, K\phi_n)_g \right].$$

In the limit $n \to \infty$ the terms containing Gauss curvature vanish in view of $\mathcal{K} \in L^1(\Sigma)$ and the fact that φ_n and ϕ_n are uniformly bounded and their supports move towards infinity as n increases. In this way we arrive at

$$\lim_{n\to\infty} q[\psi_{n,\varepsilon}] = \lim_{n\to\infty} \left[\varepsilon^2 \|\phi_n\|_g^2 - 2\varepsilon(\varphi_n, M\phi_n)_g\right] \tag{16.27}$$

provided the limit on the right side exists. Next we make the parameter ε also dependent on n putting $\varepsilon_n := (\varphi_n, M\phi_n)_g^{-1}$ which is a reasonable choice as long as the scalar product tends to infinity with $n \to \infty$; we are trying to show that

$$\lim_{n\to\infty} \frac{(\phi_n, \phi_n)_g}{(\varphi_n, M\phi_n)_g^2} < 2\,.$$

Now the cylindrical symmetry comes into play allowing us to demonstrate that the limit on the left side is zero. Indeed, since k_s is integrable and ϕ_n is chosen in a way to eliminate the weight r with the help of Problem 21, the contribution from the meridian curvature in the denominator can be neglected as long as the other part diverges. Furthermore, we have mentioned that the factor $|k_\theta r|$ is bounded from below by a positive constant at large distances. Using finally the result of Problem 21 in the numerator, we find that it is sufficient to check that the expression

$$\frac{\int_0^\infty \phi_n(s)^2 s\, ds}{\left(\int_0^\infty \varphi_n(s)\phi_n(s)\, ds\right)^2} = \frac{3}{\ln(d_n/b_n)}$$

vanishes as $n \to \infty$, which happens, *e.g.* if we choose $b_n = n$, $c_n = n^2$, $d_n = n^3$ with $n \geq 2$, the other requirements being also satisfied. ∎

16.3.10 Example: Consider a *hyperboloid* sheet given by $x^2+y^2-z^2\tan^2\vartheta = 1$. It is an asymptotically planar and cylindrically symmetric surface with total Gauss curvature $\mathcal{K} = 2\pi\cos^2\vartheta(1+\sin\vartheta)^{-1}$ which varies over the interval $(0, 2\pi)$ as a function of ϑ, hence Theorem 16.3.9 applies as long as $a < \rho_m = \tan\vartheta$.

16.4 Weak coupling

One of the natural questions is how the bound states in tubes and layers behave when the coupling is weak. We have already encountered an example of such a situation in Problem 18, now we will discuss the geometrically induced discrete spectra in the case when the deformations which cause them are mild in a suitable sense.

A powerful tool for this purpose can be obtained by adapting to the present problem Lemma 14.3.3. We will first reformulate it in an abstract way for a perturbation which allows a factorization, $H_g = H_0 + g\tilde{U}^*U$, by means of a pair of operators $U, \tilde{U}$; we assume naturally that H_g is self-adjoint. For any $z \in \rho(H_0)$ the resolvent $(H_0 - z)^{-1}$ is bounded, hence one can define the operator

$$K_g^z := gU(H_0 - z)^{-1}\tilde{U}^* \tag{16.28}$$

provided $D(H_0) \subset D(U)$. Then we can make the following claim.

16.4.1 Proposition (Birman–Schwinger principle): Suppose that $D(H_g) \subset D(U)$; then $z \in \sigma_d(H_\lambda)$ holds *iff* $-1 \in \sigma_d(K_g^z)$.

Proof: If $K_g^z\psi = -\psi$, the vector $\phi := -g(H_0 - z)^{-1}\tilde{U}^*\psi$ is easily seen to satisfy

$H_g\phi = z\phi$. Conversely, if $H_g\phi = z\phi$ we have $\varphi \in D(H_g) \subset D(U)$ by assumption, so $\psi := U\phi$ exists and $K_g^z\psi = -\psi$. ∎

Let us apply this result first to a Schrödinger operator in a straight tube, $H_g = H_D^{\Omega_0} + gV$ with a bounded potential to be specified later. An important role will be played again by its matrix representation similar to that appearing in (16.9), namely

$$V_{mn} : V_{mn}(x) = \int_{-a}^{a} V(x,y)\,\chi_m(y)\chi_n(y)\,dy\,; \tag{16.29}$$

if V is bounded, these functions are well defined. One can derive a weak–coupling expansion which modifies the result mentioned in the notes to Section 14.3.

16.4.2 Theorem: Suppose that $V \in L^\infty(\Omega_0)$ is such that $\lim_{|x|\to\infty} V(x,y) = 0$ for a.a. $y \in (-a,a)$ and $|V|_{11} \in L^1(\mathbb{R}, |x|\,dx)$. Then H_g has for small $|g|$ at most one simple eigenvalue $\lambda(g) < \kappa_1^2$, and this happens *iff* $\int_{\mathbb{R}} gV_{11}(x)\,dx \le 0$. Moreover, if this condition holds the following expansion is valid,

$$\sqrt{\kappa_1^2 - \lambda(g)} = -\frac{g}{2}\int_{\mathbb{R}} V_{11}(x)\,dx - \frac{g^2}{4}\left\{ \int_{\mathbb{R}^2} V_{11}(x)|x-x'|\,V_{11}(x')\,dx\,dx' \right. \tag{16.30}$$
$$\left. - \sum_{n=2}^{\infty} \int_{\mathbb{R}^2} V_{1n}(x)\,\frac{e^{-\sqrt{\kappa_n^2-\kappa_1^2}\,|x-x'|}}{\sqrt{\kappa_n^2-\kappa_1^2}}\,V_{n1}(x')\,dx\,dx' \right\} + \mathcal{O}(g^3)\,.$$

Proof: It is easy to check that $\inf\sigma_{ess}(H_g) = \kappa_1^2$. By Proposition 16.4.1 we have to find the spectrum of $K_g^z = g|V|^{1/2}(H_D^{\Omega_0} - z)^{-1}V^{1/2}$. The free resolvent is given by (16.11); its part coming from the lowest transverse mode has a singularity as $z \to \kappa_1^2-$ which we will single out introducing the decomposition $K_g^z = gQ_z + \lambda P_z$ with

$$Q_z(\vec{x},\vec{x}') = \frac{e^{-\kappa_1(z)\,|x|}}{2\kappa_1(z)}\,|V(x,y)|^{1/2}\chi_1(y)\,e^{-\kappa_1(z)\,|x'|}\,\chi_1(y')V(x',y')^{1/2},$$

where we again use $\kappa_n(z) := \sqrt{\kappa_1^2 - z}$, and $P_z := A_z + |V|^{1/2}B_zV^{1/2}$ with

$$A_z(\vec{x},\vec{x}') := |V(x,y)|^{1/2}\chi_1(y)\,\frac{e^{-\kappa_1(z)\,|x|_>}\sinh\kappa_1(z)\,|x|_<}{\kappa_1(z)}\,\chi_1(y')V(x',y')^{1/2}$$

and $B_z(\vec{x},\vec{x}')$ representing the sum of the higher–mode contributions to the Green function $G_0(\vec{x},\vec{x}';z)$. In the last formula we use the notation $|x|_> :=, \max\{|x|,|x'|\}$ and $|x|_< := \max\{0, \min(|x|,|x'|)\,\mathrm{sgn}\,(xx')\}$. It is obvious that $\|B_z\|$ has for any $z \le z_0 < 4\kappa_2^2$ a bound independent of z, and the same is true, under the boundedness hypothesis on the potential, for $\||V|^{1/2}B_zV^{1/2}\|$. In the first term $\|A_z\|$ can be estimated by its Hilbert–Schmidt norm,

$$\|A_z\|_2^2 \le \int_{\Omega_0\times\Omega_0} |V(x,y)\chi_1(y)|^2|x|_<^2|V(x',y')\chi_1(y')|^2\,d\vec{x}\,d\vec{x}' \le \left(\int_{\mathbb{R}} |x|\,|V|_{11}(x)\,dx\right)^2$$

with the right side finite by assumption, where we have used $e^{-z'}\sinh z \le z$ for $z' \ge z$ and $|x|_<^2 \le |xx'|$. Moreover, the same estimate shows by dominated convergence that A_z tends to a bounded operator in the limit $z \to \kappa_1^2-$; hence $\|P_z\|$ has in $(-\infty, \kappa_1^2]$ a bound independent of z and $\|gP_z\| < 1$ for g small enough.

In such a case $I + gP_z$ is invertible and we may rewrite the operator, the singularities of which we are interested in, using the identity

$$(I + K_g^z)^{-1} = \left[I + g(I + gP_z)^{-1}Q_z\right]^{-1}(I + gP_z)^{-1}. \tag{16.31}$$

It follows that K_g^z has eigenvalue -1 *iff* the same is true for $\lambda(I+\lambda P_z)^{-1}Q_z$. This is a rank–one operator, however, which acts as $(\psi, \cdot)\phi$ with $\psi := \frac{g}{2\kappa_1(z)}\, e^{-\kappa_1(z)|\cdot|}V^{1/2}\chi_1$ and $\phi := (I + gP_z)^{-1}e^{-\kappa_1(z)|\cdot|}|V|^{1/2}\chi_1$, so it has just one eigenvalue which is (ψ, ϕ). Putting it equal to -1 we get for $\kappa_1(z) =: \zeta$ the equation $\zeta = \mathcal{G}(\lambda, \zeta)$, where

$$\mathcal{G}(\lambda, \zeta) := -\frac{g}{2}\int_{\Omega_0} e^{-z|x|}V(x,y)^{1/2}\chi_1(y)\left[(I + gP_{z(\zeta)})^{-1}e^{-z|\cdot|}|V|^{1/2}\chi_1\right](x,y)\, dx\, dy\,,$$

which has for small $|g|$ exactly one solution with the given expansion (see the notes).

It remains to check that the second–order term is positive for $\int_{\mathbb{R}} V_{11}(x)\, dx = 0$, so the bound state exists in this case too. If the corresponding functions V_{mn} belongs to $L^2(\mathbb{R})$, the corresponding coefficient in the Taylor expansion is proportional to

$$\begin{aligned}
&\lim_{\varepsilon\to 0+} 2\int_{\mathbb{R}^2} V_{11}(x)\,\frac{e^{-\varepsilon|x-x'|}-1}{2\varepsilon}\, V_{11}(x')\, dx\, dx' \\
&\quad + \sum_{n=2}^{\infty}\int_{\mathbb{R}^2} V_{1n}(x)\,\frac{e^{-\sqrt{\kappa_n^2-\kappa_1^2}\,|x-x'|}}{\sqrt{\kappa_n^2-\kappa_1^2}}\, V_{n1}(x')\, dx\, dx' \\
&= \lim_{\varepsilon\to 0+} 2\int_{\mathbb{R}} |\hat V_{11}(k)|^2 \frac{dk}{k^2+\varepsilon^2} + \sum_{n=2}^{\infty}\int_{\mathbb{R}} |\hat V_{1n}(k)|^2\,\frac{dk}{k^2+\kappa_n^2-\kappa_1^2} > 0\,,
\end{aligned}$$

otherwise we use cut-off approximations and check that the sign is preserved in the limit when the regularization is removed. This concludes the proof. ∎

Now we will apply this result to bound states in bent strips. To compare different shapes we need to introduce a parametrization which would allow us to say what is a weak curvature. We will consider families of strips with the generating curves Γ_β characterized by the curvature

$$\gamma_\beta(s) := \beta\gamma(s) \tag{16.32}$$

for a fixed function γ and $\beta > 0$. If $\int_{\mathbb{R}} \gamma(s)\, ds \neq 0$ we may put it equal to one without loss of generality; then β is just the total bending – *cf.* Problem 2. If the integral is zero, decreasing β means straightening the curve. This is seen from the example of a compactly supported γ: the length of the curved part remains preserved and the curvature radius at each point is proportional to β^{-1}.

16.4.3 Theorem: Let $\{\Omega_\beta\} \subset \mathbb{R}^2$ be a family of strips of width $2a$ built over the curves Γ_β corresponding to (16.32). If the assumptions of Proposition 16.1.3 are satisfied and Γ_1 is not straight, the operator $H_D^{\Omega_\beta}$ has, for small enough nonzero β, exactly one isolated eigenvalue $\lambda(\beta)$ in $(0, \kappa_1^2)$, and moreover the expansion

$$\sqrt{\kappa_1^2 - \lambda(\beta)} = \frac{\beta^2}{8} \Big\{ \|\gamma\|^2 - \frac{1}{2} \sum_{n=2}^{\infty} (\chi_n, u\chi_1)^2 \times \varrho_n \int_{\mathbb{R}^2} \gamma'(s)\, e^{-\varrho_n |s-s'|} \gamma'(s')\, ds\, ds' \Big\} + \mathcal{O}(\beta^3)$$

holds, where $\varrho_n := \kappa_1 \sqrt{n^2 - 1}$ and the sum runs, in fact, over even n only.

Proof: The argument is based on two-sided estimates of the operator (16.4) by

$$H_\pm := -(1 \mp a\beta\|\gamma\|_\infty)^{-2} \frac{\partial^2}{\partial s^2} - \frac{\partial^2}{\partial u^2} + V_\beta(s,u)\,,$$

where V_β is the effective potential associated with γ_β. By Theorem 16.4.2 and the minimax principle there is a single eigenvalue for small nonzero β which is squeezed between the eigenvalues $\lambda_\pm(\beta)$ of the estimating operators; rescaling the longitudinal variable to $s_\pm := (1 \mp a\beta\|\gamma\|_\infty)s$ and changing appropriately the integrations, we get

$$\begin{aligned} \sqrt{\kappa_1^2 - \lambda_\pm(\beta)^2} = &-\frac{1}{2}(1 \mp a\beta\|\gamma\|_\infty) \int_{\mathbb{R}} (V_\beta)_{11}(s)\, ds \\ &- \frac{1}{4}(1 \mp a\beta\|\gamma\|_\infty)^2 \Big\{ \int_{\mathbb{R}^2} (V_\beta)_{11}(s)\, |s-s'| (V_\beta)_{11}(s')\, ds\, ds' \\ &- \sum_{n=2}^{\infty} \varrho_n^{-1} \int_{\mathbb{R}^2} (V_\beta)_{1n}(s)\, e^{-\varrho_n |s-s'|}\, (V_\beta)_{n1}(s')\, ds\, ds' \Big\} + \mathcal{O}(\beta^3)\,. \end{aligned}$$

Since V_β may be expanded in terms of β and we are interested in the leading term only, the scaling factors $1 \mp a\beta\|\gamma\|_\infty$ play no role. Let us denote the explicit part of the right side as $I_1 + \sum_{n=1}^\infty I_{2,n}$. The first term can be calculated by integration by parts,

$$\begin{aligned} I_1 &:= \frac{\beta^2}{8} \int_{-a}^{a} du \int_{\mathbb{R}} ds\, \chi_1(u)^2 \left[\frac{\gamma(s)^2}{(1+u\beta\gamma(s))^2} - \frac{u^2\gamma'(s)^2}{(1+u\beta\gamma(s))^4} \right] \\ &= \frac{\beta^2}{8} \Big\{ \|\gamma\|^2 - \|u\chi_1\|^2 \|\gamma'\|^2 \Big\} + \mathcal{O}(\beta^3)\,, \end{aligned}$$

where we have used the fact that $\gamma'(s) \to 0$ as $|s| \to \infty$ by assumption; the important thing is that in the integration by parts the term linear in β is canceled. Hence we need to compute $\sum_{n=1}^\infty I_{2,n}$ too because it contains terms of order β^2 coming from the second term of the effective potential. In particular, $I_{2,1}$ equals

$$-\frac{\beta^2}{16} \int_{-a}^{a} \int_{-a}^{a} du\, du'\, uu'\, \chi_1(u)^2 \chi_1(u')^2 \int_{\mathbb{R}^2} ds\, ds'\, \gamma''(s)\, |s-s'|\, \gamma''(s') + \mathcal{O}(\beta^3)\,,$$

where the inner integral is $-2\int_{\mathbb{R}}\gamma'(s)^2ds$ by a double integration by parts, so

$$I_{2,1} = \frac{\beta^2}{8}(\chi_1, u\chi_1)^2\|\gamma'\|^2 + \mathcal{O}(\beta^3).$$

Splitting off the leading terms in the other expressions we get

$$I_{2,n} = \frac{\beta^2}{16}\int_{-a}^{a}\int_{-a}^{a} du\,du'\,uu'\,\chi_1(u)\chi_N(u)\chi_1(u')\chi_N(u') \times \int_{\mathbb{R}^2} ds\,ds'\,\gamma''(s)\,\frac{e^{-\varrho_n|s-s'|}}{\varrho_n}\,\gamma''(s') + \mathcal{O}(\beta^3).$$

The inner integral can be rewritten by a repeated integration by parts as

$$2\int_{\mathbb{R}}\gamma'(s)^2ds - \varrho_n\int_{\mathbb{R}^2} ds\,ds'\,\gamma'(s)\,e^{-\varrho_n|s-s'|}\,\gamma'(s'),$$

hence putting all the leading–term contributions together and using the Parseval relation, $\|u\chi_1\|^2 = \sum_{n=1}^{\infty}(\chi_n, u\chi_1)^2$, we find that the terms containing $\|\gamma'\|^2$ cancel and we arrive finally at the sought asymptotic expansion. ∎

In a similar way one can can analyze bound states in weakly curved three-dimensional tubes (Problem 27). One can also derive an expansion analogous to Theorem 16.4.2 for Schrödinger operators in layers and apply it to systems discussed in the preceding section. We will restrict ourselves to a description of the result referring to the notes for the proof. We will consider a class of layers built over surfaces which are graphs of a function $\mathbb{R}^2 \to \mathbb{R}^3$ containing a scaling parameter,

$$\Sigma_\beta := p_\beta(\mathbb{R}^2), \quad p_\beta(x^1, x^2) = \left(x^1, x^2, \beta f(x^1, x^2)\right),$$

where f is a C^4–smooth function of compact support and $\beta > 0$; the layer $\Omega_\beta := \mathcal{L}(\Omega)$ is defined by (16.19). Then we can make the following claim:

16.4.4 Theorem: If Σ_1 is not planar, then for all β small enough the operator $H_D^{\Omega_\beta}$ has exactly one eigenvalue $\lambda(\beta)$ below the threshold of the essential spectrum. The latter can be expressed as $\lambda(\beta) = \kappa_1^2 - e^{2w(\beta)^{-1}}$, where the function w has the following asymptotic expansion,

$$w(\beta) = -\beta^2\sum_{n=2}^{\infty}(\chi_1, u\chi_n)^2\left(\kappa_n^2 - \kappa_1^2\right)^2\int_{\mathbb{R}^2}\frac{|\hat{m}_0(\omega)|^2}{|\omega|^2 + \kappa_n^2 - \kappa_1^2}\,d\omega + \mathcal{O}(\beta^3).$$

The sum runs over even n only and $\hat{m}_0$ is the Fourier image of $m_0 = \frac{1}{2}\Delta f$.

The described derivation of the asymptotic expansion follows the scheme used in the Schrödinger operator theory. It is sometimes more simple, however, to directly apply the abstract Birman–Schwinger principle as given by Proposition 16.4.1. As an example consider a strip with a weak one–sided deformation, with (16.8) replaced by

$$\Omega_\beta := \{(x, y) \in \mathbb{R}^2 : 0 < y < d(1 + \beta f(x))\};$$

for simplicity we assume that $f \in C_0^\infty(\mathbb{R})$. We need not assume a definite sign, however, in the weak–coupling case it is only the total area change which matters.

16.4.5 Theorem: Let $\{\Omega_\beta\}$ be the described family of strips with a nonzero function $f \in C_0^\infty(\mathbb{R})$. Then $H_D^{\Omega_\beta}$ has, for small enough $\beta > 0$, at most one simple eigenvalue in $(0, \kappa_1^2)$. This is the case if $\langle f \rangle := \int_{\mathbb{R}} f(x)\, dx > 0$, and then the eigenvalue $\lambda(\beta)$ is a real–analytic function at $\lambda = 0$ with

$$\lambda(\beta) = \kappa_1^2 - \beta^2 \kappa_1^4 \langle f \rangle^2 + \mathcal{O}(\beta^3)\,.$$

Proof is left to the reader (Problem 28).

The Birman–Schwinger method is not the only tool for weak–coupling analysis. Sometimes one can use variational estimates. In this way, *e.g.*, one can treat the critical case left out in Theorem 16.4.5. We will just quote the result referring to notes for the proof and further information.

16.4.6 Proposition: Let $\{\Omega_\beta\}$ be the family from Theorem 16.4.5 with a nonzero $f \in C_0^\infty(\mathbb{R})$ such that $\operatorname{supp} f \subset [-b, b]$ and $\int_{-b}^{b} f(x)\, dx = 0$. Then $\sigma_d(H_D^{\Omega_\beta}) = \emptyset$ holds for small enough $|\beta|$ provided $8b < d\sqrt{3}$. On the other hand, there is $C > 0$ such that the same operator has, for small nonzero $|\beta|$, an isolated eigenvalue $\lambda(\beta)$ provided $\|f'\| < C\kappa_1 \|f\|$, and in that case there are positive c_1, c_2 such that

$$c_1 \beta^4 \le \kappa_1^2 - \lambda(\beta) \le c_2 \beta^4\,.$$

Notes to Chapter 16

Section 16.1 It is not self-evident that the systems discussed in this chapter are suitable models for semiconductor "quantum wires", that is, that the idealization we make is not too large; recall that systems in the micrometer or even nanometer range still consist of a large number of atoms. Several factors play a decisive role. First of all, the material is usually very pure so the electron mean free path significantly exceeds the size of the system. At the same time, the crystalline structure allows us to reduce the characteristics of the material to a single constant, the effective mass of the electron. In addition, the electron gas density is often low in which case the one-particle effects dominate, and the system boundary is usually an interface between two semiconductor materials where the wave functions are suppressed, which justifies the use of Dirichlet boundary conditions. For more information see ⟦ LCM ⟧; similarly one can justify the use of quantum waveguide models for other meso- and nanoscopic systems. Atomic waveguides based on hollow fibers were proposed in [SMZ 1], their use to produce Bose-Einstein condensate geometrically in [EV 3], [EZ 1]. Moreover, the results on two–dimensional Dirichlet strips discussed here apply also to flat electromagnetic waveguides as observed in [EŠ 7], [SM 1], and experimentally confirmed in [CLM 1,2], [CLMTY 1,2].

The condition $a\|\gamma\|_\infty < 1$ ensures local injectivity of the map f, the global property represents an additional assumption. The regularity assumptions we use are not optimal.

For instance the first part of Theorem 4 is valid for $\gamma \in L^\infty(\mathbb{R})$ provided $\gamma(s) \geq c > 0$ on a set of nonzero measure. Also the sharp inequality $a\|\gamma\|_\infty < 1$ is needed only outside a compact, so $\partial\Omega$ may have angles in the central part. The main idea of the proof of Theorem 16.1.4 came from [GoJ 1] and was elaborated in [DuE 1], [RB 1]. It works also for higher-dimensional waveguides, in particular, for tubes in $\mathbb{R}^3$, see [GoJ 1], [DuE 1] and Problem 9; a generalization to tubes in $\mathbb{R}^n$, $n \geq 3$, is worked out in [CDFK 1].

The existence of a bound state in a channel with a rectangular bend considered in Problem 7 was observed in [LLMRSY 1], [EŠŠ 1]. Strips of other angles were discussed in [ABGM 1], many other examples of such waveguides are analyzed in ⟦LCM⟧. The cross-shaped region of Problem 8 was discussed first in [SRW 1], for similar systems with an arbitrary angle see [BEPS 1]. The results of Problem 9 come from [EŠTV 1], a system of two parallel guides separated by a semitransparent wall was analyzed in [EKr 1].

Various results can be derived about properties of these geometrically induced bound states. If the strip is thin, the leading term in the binding energy is determined by the appropriate one-dimensional Schrödinger operator – *cf.* Problem 4b. One can also write the complete perturbation series for planar strips and spatial tubes [DuE 1]. Another question concerns the location of eigenvalues in the interval $(0, \kappa_1^2)$. Various lower bounds can be found — see [AE 1], [EFK 1] and Problem 11 — and more generally, a version of Lieb–Thirring inequalities is valid for such discrete spectra – *cf.* [ELW 1]. The fact that the effective attraction comes for curvature poses also an isoperimetric problem for closed annular-shaped strips and tubes – *cf.* [EHL 1] and Problem 12. Most known results about bent-tube discrete spectra concern a single particle only; a system of N particles in a guide with a Coulomb repulsion is discussed in [EV 3].

For curved Neumann waveguides the essential spectrum typically has no gaps in $\mathbb{R}^+$. However, if one boundary is Dirichlet and the other Neumann, an interesting effect occurs: roughly speaking, the discrete spectrum is nonempty *iff* the Neumann condition is at the outer part of the bend [DK 1]. Also combinations of Dirichlet and Neumann conditions in a straight guide lead to nontrivial spectral properties [DK 2].

Scattering in bent strips was discussed first in [DuEŠ 1], where the asymptotic completeness was proved and a perturbation series was derived for the pole positions $E_{jn}(a)$. However, its coefficients depend on a so, in distinction to the situation in Sections 9.6 and 15.4, it is not sufficient to consider the first nonzero contribution to the imaginary part. Theorem 16.1.6 was proved in [Ned 1] and [DEM 1], where also an upper bound for $\eta = \eta_0$ can be found. It is expected that bending gives rise to true resonances, *i.e.*, that the first inequality in Theorem 16.1.6 is sharp but the proof of this fact is at present missing.

There are situations in which waveguide resonances are easier to analyze. For instance, if a straight channel is perturbed by a potential or a local deformation with a mirror symmetry, the spectrum may exhibit embedded eigenvalues which turn into resonances if the symmetry is violated, geometrically or by applying a magnetic field. This effect was predicted by Nöckel [Nö 1] and analyzed rigorously in [DEM 2].

Solving the scattering problem in waveguides means, in the first place, finding the S-matrix (and other physically important quantities derived from it — see (16.18) and notes to Section 2). This is not easy in general if the perturbation is of a geometrical character as considered here. A class of systems where the on-shell S-matrix can be found numerically consists of those which are composed of rectangular components to which the mode-matching technique mentioned in the hint to Problem 7 can be applied. An example taken from [EŠTV 1] is proposed as Problem 13, many others are discussed in [[LCM]]. Recall also that even the bound states discussed in this section give rise to resonances which are manifested in conductivity properties — see, *e.g.*, [TBV 1] — because in practice a "quantum wire" is always coupled to macroscopic leads; a model analysis of this effect can be found in [Ex 7].

Section 16.2 The model discussed in this section comes from [EGŠT 1], where the reader can also find a discussion of the situations when the point interactions are periodically or randomly arranged, combined with a regular potential, *etc.*An extension to point interactions in tubes can be found in [Ex 17], for the analogous problem in layers see [EN 1] and [EN 2] where a periodic lattice in a homogeneous magnetic field is considered.

Positivity of the resolvent kernel below the spectral threshold is proved, *e.g.*, in Sec. XIII.12 of [[RS 4]]. In combination with the concept of positivity improving semigroups introduced there it yields simplicity of the ground state and positivity of the corresponding eigenfunction; we refer to [EGŠT 1] for this claim and for the embedded–eigenvalue part of Theorem 16.2.3, the proof of which is just too long. As for Example 16.2.5, it should be noted that degenerated eigenvalues occur rather exceptionally. The reason is that the matrix $\Lambda_{\alpha,\vec{a}}(z)$ has $3N-1$ real parameters, because one of the coordinates a_j may be chosen arbitrarily, while the number of different $(N-1)\times(N-1)$ minors is N^2, and the number of conditions required for a multiplicity larger than two grows even faster with N. Finally, one can also study resonances in this model which are given by complex solutions of the condition (16.16) on the other Riemann sheets specified by the choice of square–root branch of the transverse–mode momentum $k_n(z)$, see [EGŠT 1] for more details.

The formula (16.18) was proposed in [La 1], its generalization to systems with a larger number of input and output channels, called the *Landauer–Büttiker formula*, in [Bü 2]. A rigorous proof of these relations can be found in [CJM 1].

Section 16.3 The material of this section is taken from [DEK 1] and [CEK 1]. There are many textbooks which may be consulted for the differential–geometric notions used in the section, *e.g.*, [[Kli]]. The geodesic polar coordinates are used for simplicity; recall that there are surfaces diffeomorphic to $\mathbb{R}^2$ without such a coordinate system [GMa 1]. While the lower bound of Proposition 1 is sufficient for our purpose, the opposite inequality can also be demonstrated [CEK 1]. Moreover, under stronger assumptions one can also construct a family of Weyl sequences showing that $\sigma_{ess}(H_D^\Omega)=[\kappa_1^2,\infty)$. Sometimes, however, the last relation is easy to prove, *e.g.*, for layers which are curved in a compact part only.

Proof of Lemma 16.3.5 uses the fact that (Σ, g) is by Huber's lemma conformally equivalent to a closed surface with a finite number of points removed – *cf.* [CEK1]. Theorem 6 and Corollary 16.3.7 come from this paper, Theorem 16.3.9 was demonstrated in [DEK 1]. In fact, a stronger claim can be made (Problem 23); notice the effective potential in (16.23) behaves as $\mathcal{O}(s^{-2})$ at large distances and compare to Theorem 14.3.9. The surface must be asymptotically planar, though, since otherwise the discrete spectrum may be empty due to the shift of the essential spectrum threshold (Problem 24). Moreover, the fact that any compact neighborhood of the origin can be left out in the construction of the trial functions in the proof of Theorem 16.3.9 makes it possible to extend the results to a wider class of layers, namely those which have at least one cylindrically symmetric end with a positive total Gauss curvature, and also to compact deformations of such layers — *cf.* [CEK 1]. Recall finally that interesting effects of a different type can be seen in curved layers which are not asymptotically planar, for instance, if such a layer has a straight edge, particles with energies below κ_1^2 move along it — *cf.* [EŠŠ 2].

Naturally there are other geometric perturbations of a straight layer which give rise to a discrete spectrum — we mention two examples in Problems 25 and 26. Scattering in layers due to point perturbations was analyzed in [EN 1], apart from that few results are known despite the fact that the problem is of obvious importance.

Section 16.4 The Birman–Schwinger analysis of weakly curved strips and tubes comes from the paper [DE 1]. Concerning the implicit equation following from the factorization (16.31), if $V(x,y)$ decays sufficiently fast as $|x| \to \infty$ the function $\mathcal{G}$ is analytic around $(0,0)$ and the assertion follows by the implicit–function theorem; in the general case one can use instead an argument adapted from the theory of one–dimensional Schrödinger operators [Si 6]. Theorem 16.4.4 comes from [EKr 2]. The assumption about compact support of f we have made can be abandoned, however, then the error term may depend on the decay rate of this function and its derivatives. The function m_0 in the expansion comes from the mean curvature, the Gauss curvature influences only the error term.

Theorem 16.4.5 was proved in [BGRS 1], an analogous result for weakly deformed layers can be found in [BEGK 1]. Proposition 6 shows that the analogy with Theorem 14.3.6 is not complete. It was demonstrated in [EV 1] with $C < 0.99$, in [EV 2] a similar variational argument was applied to window–coupled strips of Problem 9 where the binding energy of a weakly coupled state can be squeezed between multiples of a^4. Resolvent expansions in combination with more involved PDE techniques make it possible to find the leading terms in these situations exactly — see [BEGK 1], [BEG 1] — and also to find the spectral behavior around the critical values a_m of Problem 9d — *cf.* [BE 1].

Problems

1. Prove Proposition 16.1.1 by showing that $H_D^{\Omega_0}$ is unitarily equivalent to the operator $\bigoplus_{j=1}^{\infty}(P^2+\kappa_j^2)$ on $\bigoplus_{j=1}^{\infty} L^2(\mathbb{R})$, where $\kappa_j := \pi j/d$, $j = 1,2,\dots$. What is the multiplicity of the spectrum in the interval $(\kappa_j^2, \kappa_{j+1}^2)$? Using the same unitary equivalence, prove the formula (16.9).

2. Suppose that $\Gamma : \mathbb{R} \to \mathbb{R}^2$ is a smooth curve of curvature γ. Show that the quantity $\beta(s_0,s_1) := \int_{s_0}^{s_1} \gamma(s)\,ds$ represents the bending of the curve, *i.e.*, the angle between the tangent vectors taken at s_1 and s_0. Points of Γ are then expressed by

$$\xi(s) = \xi(s_0) + \int_{s_0}^{s} \cos\beta(s_0,s_1)\,ds_1\,, \qquad \eta(s) = \eta(s_0) + \int_{s_0}^{s} \sin\beta(s_0,s_1)\,ds_1$$

for a fixed s_0; a change of the latter is equivalent to a Euclidean transformation of the coordinates.

3. (a) Prove Proposition 16.1.2.

(b) The corresponding quadratic form acts as $t_\Omega[\psi] = \|g^{-1/4}\partial_s\psi\|^2 + \|g^{1/4}\partial_u\psi\|^2$ with the core $D_\infty(\Omega_0)$, the result being valid whenever $\gamma \in C^1$.

4. Suppose that $\gamma(s), \gamma'(s), |\gamma(s)''|^{1/2} = \mathcal{O}(|s|^{-1-\varepsilon})$ as $|s| \to \infty$, in addition to the assumption of Proposition 16.1.2. Show that

(a) There is $a_0 > 0$ such that H_D^Ω has an isolated eigenvalue below the essential–spectrum threshold κ_1^2 for all $a \in (0, a_0)$.

(b) In the limit $a \to 0$ the n–th eigenvalue of H_D^Ω behaves as

$$\lambda_n(a) = \left(\frac{\pi}{2a}\right)^2 + \mu_n + \mathcal{O}(a),$$

where μ_n is the n–th eigenvalue of the operator $P^2 - \frac{1}{4}\gamma(Q)^2$ on $L^2(\mathbb{R})$.

Hint: Use the minimax principle. Estimate the effective potential for a fixed a by a suitable function of the variable s and find the error dependence on a.

5. (a) Prove Proposition 16.1.3 and the first claim of Proposition 16.1.5.

(b) Show that an infinite planar curve has asymptotes if $\gamma(s) = \mathcal{O}(|s|^{-2-\varepsilon})$ as $s \to \pm\infty$. What is the curvature decay for a parabola?

Hint: (a) To get a lower bound, use a bracketing argument placing Neumann conditions at $s = \pm s_0$, further employ Theorem 5.4.4. (b) $|\sin\beta(s,\infty)| \le |\beta(s,\infty)|$.

6. Let Ω be a bent strip with a piecewise continuous curvature γ satisfying $a\|\gamma\|_\infty < 1$. Show that the only closed geometric–optic trajectories in Ω are those perpendicular to the strip axis.
Hint: Map such a trajectory on a straight strip Ω_0 and look for turning points.

7. Consider the operator H_D^Ω associated with a planar L–shaped strip $\Omega := \{(x,y) \in \mathbb{R}^2 : 0 < \min\{x,y\} < \pi\}$. Show that $\sigma_{ess}(H_D^\Omega) = [1,\infty)$ and there is a single isolated eigenvalue $\lambda \approx 0.9291...$.
Hint: Use Dirichlet–Neumann bracketing and the symmetry of Ω. To find the eigenvalue, write the sought eigenfunction as series through trigonometric bases and match these expansions smoothly at $x, y = \pi$.

8. Let H_D^Ω correspond to a cross–shaped region in the plane, $\Omega := \{(x,y) \in \mathbb{R}^2 : 0 < \min\{|x|,|y|\} < \frac{1}{2}\pi\}$. The essential spectrum is again $[1,\infty)$ and there is a single isolated eigenvalue $\lambda \approx 0.66...$. Moreover, H_D^Ω has an eigenvalue $\lambda \approx 3.72...$ embedded in the continuous spectrum.
Hint: To prove the last claim use the result of the previous problem.

9. Consider the operator H_D^Ω on the region consisting of a pair of adjacent strips coupled by a window in the common boundary, *i.e.*, $\Omega := \{\, (x,y) \in \mathbb{R}^2 : \, y \in (-d_2, d_1)\,,\, x \in \mathbb{R}\,\} \setminus B_a$, where $B_a := ((-\infty,-a] \cup [a,\infty)) \times \{0\}$. Denote $d := \max\{d_1,d_2\}$, $D := d_1 + d_2$, and $\varrho := d^{-1}\min\{d_1,d_2\}$. Furthermore, set $\epsilon_d := \left(\frac{\pi}{d}\right)^2$ and $\epsilon_D = \epsilon_d(1+\varrho)^{-2}$. Prove

(a) $\sigma_{ess}(H_D^\Omega) = [\epsilon_d,\infty)$ and $\sigma_d(H_D^\Omega) \neq \emptyset$ for any $a > 0$.

(b) The eigenvalues $\lambda_m = \lambda_m(a)$, $m = 1,\dots,N$, are contained in (ϵ_D,ϵ_d) and decrease continuously as functions of a; their number satisfies the bounds $N_a \le N \le N_a + 1$, where $N_a := \min\left\{1, \left[-\frac{2a}{d}\sqrt{1-(1+\varrho)^{-2}}\right]\right\}$ and $[\cdot]$ denotes the integer part.

(c) The spectrum is simple and the eigenvalues are bound by

$$\left(\frac{d}{2a}\right)^2 (m-1)^2 \le \frac{\lambda_m}{\epsilon_d} - (1+\varrho)^{-2} < \left(\frac{d}{2a}\right)^2 m^2.$$

(d) Similarly, for the critical values a_m, $m = 2, \ldots$, at which the m–th eigenvalue emerges from the continuum we have

$$\frac{d(m-1)}{\sqrt{1-(1+\varrho)^{-2}}} \le a_m < \frac{dm}{\sqrt{1-(1+\varrho)^{-2}}}.$$

(e) Finally, show how the eigenvalues and eigenfunctions can be found numerically.

Hint: Use Dirichlet–Neumann bracketing; the existence of an eigenvalue for small a can be demonstrated in a way similar to the proof of Theorem 16.1.4. In the last part use the mode–matching technique mentioned in the hint to Problem 7.

10. Let $\Gamma : \mathbb{R} \to \mathbb{R}^3$ be a C^4–smooth curve for which the global Frenet triad frame (t, n, b) exists. Take an open precompact set $M \subset \mathbb{R}^2$ which contains zero and put $a := \sup_{x\in M} |x|$. Given a function $\alpha \in C^1(\mathbb{R})$ we define a map $f : \mathbb{R} \times M \to \mathbb{R}^3$ by $f(s, r, \theta) := \Gamma(s) - r\,[\,n \cos(\theta-\alpha(s)) + b \sin(\theta-\alpha(s))\,]$. Consider the tube in $\mathbb{R}^3$ defined as $\Omega := f(\mathbb{R}\times M)$ and prove

(a) The Jacobian of the transformation is $g^{1/2} = r(1 + r\gamma \cos(\theta-\alpha))$, where γ, τ are the curvature and torsion of Γ, respectively. The coordinates s, r, θ are locally orthogonal if the *Tang condition*, $\alpha'(s) = \tau(s)$ holds for all $s \in \mathbb{R}$.

(b) If this is true and the map f is injective, the quadratic form of H_D^Ω can be written as $t[\psi] = \|rg^{-1/4}\partial_s\psi\|^2 + \|g^{1/4}\partial_r\psi\|^2 + \|g^{1/4}r^{-1}\partial_\theta\psi\|^2$ with the core $D_\infty(\mathbb{R}\times M)$, corresponding to the operator on the straight tube $\Omega_0 := \mathbb{R}\times M$ acting as $H = -\partial_s h^{-2}\partial_s - \Delta_D^M + V(s, r, \theta)$ where $-\Delta_D^M$ is the Dirichlet Laplacian on the tube cross section and the effective potential is given by

$$V = -\frac{\gamma^2}{4h^2} + \frac{1}{2}\frac{h_{ss}}{h^3} - \frac{5}{4}\frac{h_s^2}{h^4}\,;$$

here we have denoted $h := g^{1/2}r^{-1}$ and $h_s = \partial_s h$, *etc.*

(c) The above assumptions together with $a\|\gamma\|_\infty < 1$ imply $\inf\sigma(H_D^\Omega) < \nu_1$, where ν_1 is the lowest eigenvalue of $-\Delta_D^M$, unless $\gamma = 0$ identically.

(d) If, in addition, $\gamma, \gamma', \gamma'' \in C_\infty(\mathbb{R})$ holds, we have $\sigma_{ess}(H_D^\Omega) = [\nu_1, \infty)$, so the discrete spectrum of H_D^Ω is nonempty.

Hint: Part (c) is proved similarly as in Theorem 16.1.4.

11. Let $\mathcal{A} \equiv \mathcal{A}(r, a)$ be a circular annulus with radii $r \pm a$ and denote by $k(r, a)^2$ the ground–state eigenvalue of $-\Delta_D^{\mathcal{A}}$. Given a function $\gamma \in L^\infty(\mathbb{R})$ with $a\|\gamma\|_\infty < 1$ we define $\kappa_1[\gamma, a] := \min\{\kappa_1, \inf\{k(|\gamma(s)|^{-1}, a) : \gamma(s) \ne 0\}\}$. Let Ω be a strip without self–intersections of width $2a$ and the axis curvature γ; then the corresponding operator H_D^Ω satisfies the bounds

$$\inf\sigma(-\Delta_D^\Omega) \ge \kappa_1[\gamma, a]^2 > \left(\frac{j_{0,1}}{2a}\right)^2 \approx 0.586\,\kappa_1^2\,,$$

where $j_{0,1} \approx 2.401$ is the first zero of the Bessel function J_0.
Hint: Neglect the first term in the expression for the quadratic form in Problem 3b and estimate the second one using the ground state of the Dirichlet Laplacian in $\mathcal{A}$.

12. Let Γ be a closed C^2-smooth curve without self-intersections of a fixed perimeter $L > 0$. Fix $a > 0$ and consider all strips Ω centered at Γ such that the condition $a\|\gamma\|_\infty < 1$ is satisfied. Prove that within this class the ground-state eigenvalue of the corresponding operator H_D^Ω is uniquely maximized when Γ is a circle.
Hint: Use a family of trial functions independent of s.

13. Analyze the scattering problem for a pair of two parallel planar waveguides coupled by a lateral window. Using the notation of Problem 9

(a) Show that the on-shell S-matrix $S(k)$ at energy k^2 is a $2(j_1+j_2)\times 2(j_1+j_2)$ matrix if $j_1 < k\,\epsilon_d^{-1/2} < j_1+1$ and $j_2 < k\,\rho\epsilon_d^{-1/2} < j_2+1$, its entries being the transmission and reflection amplitudes in the particular transverse modes.

(b) Derive the system of equations which allows one to find $S(k)$ numerically.

Hint: For the second part use the mode-matching technique.

14. Let $H_{\alpha,\vec a}$ be the Hamiltonian of the drip with point interactions from Sec. 2 and denote $\kappa_1 := \pi/d$ as usual. Show that changing the strip width from π to d is equivalent to scaling the spectrum of $H_{\alpha,\vec a}$ by κ_1^2 provided the coupling constants are simultaneously changed from α to $\alpha-(2\pi)^{-1}\ln\kappa_1$.

15. Prove the relations (16.12) and (16.15).
Hint: To find the coefficients $\lambda_{jk}(z)$ in Krein's formula use the fact that $(H_{\alpha,\vec a}-z)^{-1}$ maps $L^2(\Omega)$ into $D(H_{\alpha,\vec a})$. To compute $L_j(\psi,\vec a)$ for $\psi = (H_{\alpha,\vec a}-z)^{-1}\phi$ recall that $k_n^{-1}-(in)^{-1}$ and $e^{ik_n+n-z/2n}-1$ are $\mathcal{O}(n^{-3})$ as $n\to\infty$.

16. The function $\xi(\vec a;\cdot)$ defined by (16.13) is, for a fixed vector $\vec a\in\Omega$, analytic in $\rho(H_0)$, its value depending on the transverse component b of $\vec a$ only. In the interval $(-\infty,1)$ it is increasing with $\xi(\vec a;z) = \pi^{-1}(1-z)^{-1/2}\sin^2 b + \mathcal{O}(1)$ as $z\to 1-$ and

$$\xi(\vec a;z) = -\frac{1}{4\pi}\ln\left(-\frac{z}{4}\right) - \frac{1}{2\pi}\gamma_E + \mathcal{O}\left(e^{-c\sqrt{-z}}\right)$$

as $z\to-\infty$ for any $c<1$, where $\gamma_E = -\psi(1)\approx 0.577$ is the Euler constant. We also have

$$\xi(\vec a;z) > \xi(\vec a\,';z) \qquad \text{if} \quad \left|b-\frac{\pi}{2}\right| < \left|b'-\frac{\pi}{2}\right|.$$

Finally, the defining series converges also in $[1,\infty)\setminus\{n^2\}_{n\in\mathbb{N}}$ giving boundary values of $\xi(\vec a;\cdot)$ at the cut which are smooth in the intervals $(n^2,(n+1)^2)$; the choice of the branch of the square roots in $\kappa_n(z) := -ik_n(z) = \sqrt{n^2-z}$ determines the corresponding sheet of the Riemann surface.
Hint: $\xi(\vec a;z)-\xi(\vec a;z')$ converges absolutely for any $z,z'\in\mathbb{C}\setminus[1,\infty)$ and the existence of $\xi(\vec a;0)$ can be checked directly. To prove the monotonicity across a halfstrip, use the identity $\sin^2(nb)-\sin^2(nb') = \sin(n(b+b'))\sin(n(b-b'))$ for $0 < b' < b \le \frac{\pi}{2}$ together with positivity of the free-resolvent kernel.

17. Prove that the single–impurity on–shell S–matrix in Proposition 16.2.2 is unitary, in particular, the probability current is preserved, $\sum_{m=1}^{[\sqrt{z}]} k_m(|t_{nm}|^2+|r_{nm}|^2)=k_n$.
Hint: Use $t_{nn}=1+r_{nn}\,e^{-2ik_na}$ and $\mathrm{Im}\,\xi(\vec a;z)=\pi^{-1}\sum_{m=1}^{[\sqrt{z}]}\frac{\sin^2(mb)}{k_m(z)}$, $z\neq n^2$.

18. Analyze the *weak–coupling regime* in a straight waveguide with point perturbations: if $\alpha_- := \min_{1\le j\le N}\alpha_j$ is large enough, $H_{\alpha,\vec a}$ has a single eigenvalue which behaves as

$$\lambda_1^{\alpha,\vec a} = 1-\left(\sum_{j=1}^N \frac{\sin^2 b_j}{\pi\alpha_j}\right)^2 + \mathcal{O}(\alpha_-^{-3})$$

in the limit $\alpha_-\to\infty$. What does the corresponding eigenfunction look like?
Hint: Denote $\mathcal{A} := \mathrm{diag}(\alpha_1,\dots,\alpha_N)$ and decompose $\Lambda_{\alpha,\vec a}(z)$ into the sum of $A(z) := \mathcal{A}-\pi^{-1}(1-z)^{-1/2}M_1$ and a remainder term.

19. (a) Show that the substitution $\psi\mapsto\psi\circ\mathcal{L}$ allows us to rewrite the quadratic form associated with H_D^Ω as $t_\Omega[\psi]=(\psi_{,i},G^{ij}\psi_{,j})_G$ with the core $C_0^\infty(\Omega_0)$, where the subscript indicates the inner product in $L^2(\Omega_0,d\omega)$. Moreover, if the surface Σ is C^3–smooth the substitution allows us to map H_D^Ω unitarily to $\tilde H=\tilde H_1+\tilde H_2$ with

$$\begin{aligned}\tilde H_1 &:= -G^{-1/2}\partial_\mu G^{1/2}G^{\mu\nu}\partial_\nu = -\partial_\mu G^{\mu\nu}\partial_\nu - 2F_{,\mu}G^{\mu\nu}\partial_\nu\,,\\ \tilde H_2 &:= -G^{-1/2}\partial_3 G^{1/2}\partial_3 = -\partial_3^2+2\,\frac{M-Ku}{1-2Mu+Ku^2}\,\partial_3\,,\end{aligned}$$

where $F:=\ln G^{1/4}$ and $F_{,3}$ is expressed explicitly in $\tilde H_2$.
(b) Prove the relations (16.21) and (16.22).

20. Prove Proposition 16.3.1 and show that it remains valid for the more general class of asymptotically planar layers considered in Theorem 16.3.6.
Hint: Employ Neumann bracketing at a geodesic distance s_0 from the pole which can be made arbitrarily large, and estimate the spectral threshold of the outer part.

21. Let $\|K\|_{g,1}$ be the norm of the Gauss curvature $K\in L^1(\Sigma_0,d\sigma)$; then the inequality $\int_0^{2\pi} r(s,\theta)\,d\theta\le(2\pi+\|K\|_{g,1})s$ holds for any $s>0$.
Hint: Use the Jacobi equation, $r''(s,\theta)+K(s,\theta)\,r(s,\theta)=0$, with the initial conditions $r(0,\theta)=1-r'(0,\vartheta)=0$, where the primes mean derivatives w.r.t. s . Integrating over s we get the estimate $r'(s,\theta)\le 1+\int_0^\infty |K(\zeta,\theta)|r(\zeta,\theta)\,d\theta$ and further integrations over θ and s yield the result.

22. Check the claims made in Examples 16.3.4.
Hint: $k_{1,2}=2g^{-1}\left[2(y^2-x^2)\pm\sqrt{g+4(y^2-x^2)^2}\right]$ with $g=1+4(x^2+y^2)$ holds in the first case and $K=-36g^{-2}(x^2+y^2)$ with $g=1+9(x^2+y^2)^2$ in the second one.

23. $\mathrm{card}\,\sigma_d(H_D^\Omega)=\infty$ holds under the assumptions of Theorem 16.3.9 provided the generating surface Σ is asymptotically planar.
Hint: Choosing b_n,c_n,d_n properly the trial functions can be made orthogonal.

24. Construct an example of a layer over a surface Σ diffeomorphic to $\mathbb{R}^2$ and equipped with geodesic polar coordinates, which has no bound states.
Hint: Consider a semiinfinite cylinder with a hemispherical "cap".

25. Let $\Omega := \{ (x,y) \in \mathbb{R}^3 : 0 < y < d(1+f(x)) \}$ be a locally deformed layer characterized by a bounded compactly supported smooth function $f : \mathbb{R}^2 \to (-1,\infty)$. Show that $\sigma_{ess}(H_D^\Omega) = [\kappa_1^2, \infty)$ holds for the corresponding Dirichlet Laplacian. Moreover, the discrete spectrum is empty if $f(x) \le 0$ for all $x \in \mathbb{R}^2$ and, on the other hand, there is at least one eigenvalue in $(0,\kappa_1^2)$ if $f(x) \ge 0$ everywhere and there is a set of nonzero Lebesgue measure where the inequality is sharp.

26. Consider the operator H_D^Ω associated with a pair of adjacent layers coupled by a window M in the common boundary, *i.e.*, $\Omega := \{ (x,y) \in \mathbb{R}^3 : y \in (-d_2, d_1)\,,\ x \in \mathbb{R}^2 \} \backslash B_M$, where $B_M := (\mathbb{R}^2 \setminus M) \times \{0\}$. Show that $\sigma_d(H_D^\Omega) \ne \emptyset$ holds whenever the Lebesgue measure of M is nonzero, and estimate the number and location of the eigenvalues in way similar to Problem 9.

27. (a) Generalize Theorem 16.4.2 to Schrödinger operators in a d–dimensional tube, $\Omega := \mathbb{R} \times M$, where M is an open connected and precompact set in $\mathbb{R}^{d-1}$, with κ_n^2 and χ_n replaced by the n–th eigenfunction of $-\Delta_D^M$ and the corresponding eigenfunction, respectively.

(b) Using the above result, show that the bound–state energy in a weakly curved three-dimensional tube satisfying the assumptions of Problem 10 has the asymptotic expansion

$$\sqrt{\nu_1 - \epsilon(\beta)} = \frac{\beta^2}{8}\,\|\gamma\|^2 - \frac{\beta^2}{16}\sum_{n=2}^{\infty}\int_{M\times M} dy\,dy'\,\chi_1(y)\chi_1(y')\chi_n(y)\chi_n(y')$$
$$\times\, \varrho_n \int_{\mathbb{R}^2} ds\,ds'\,h_s(s,y)\,e^{-\varrho_n|s-s'|}\,h_s(s',y') + \mathcal{O}(\beta^3)\,,$$

where h_s is the expression defined in Problem 10b and $\varrho_n := \sqrt{\nu_n - \nu_1}$.

28. Prove Theorem 16.4.5.
Hint: Pass to the straight strip using the unitary operator associated with the substitution $U_\beta :\ (U_\beta\psi)(x,y) = \sqrt{1+\beta f(x)}\,\psi(x,(1+\beta f(x))y)$ rewriting the operator $H_D^{\Omega_\beta} \equiv H_\beta$ as $H_\beta = H_0 + \beta C_\beta^* D$ with suitable $C_\beta,\ D : L^2(\Omega_0) \to L^2(\Omega_0) \otimes \mathbb{C}^7$. In analogy with Theorem 16.4.2, analyze $K_\beta^z := \beta D(H_0 - z)^{-1} C^*$ splitting off the singularity in the lowest mode.

Chapter 17

Quantum graphs

In experiments the quantum wires considered in the previous chapter are often combined into more complicated structures. This opens the way to numerous interesting physical effects, however, from the theoretical point of view such systems are not easy to describe. In many cases a simplified model in which the transverse size of the wires is neglected and one investigates only the graph-shaped skeleton of such a structure has proved to be useful. This naturally motivates the question how quantum mechanics on a graph looks, which is the subject of this chapter.

17.1 Admissible Hamiltonians

The configuration space will be in the following sections a **graph** Γ which we identify with the family of *vertices* $\mathcal{V} = \{v_i : i \in I\}$ and *edges (links)* $\mathcal{E} = \{e_{ij} : (i,j) \in I_\mathcal{E} \subset I \times I\}$. The index set I is supposed to be at most countable and $I_\mathcal{E}$ determines adjacency of the graph, *i.e.* it tells us which vertices are connected by a link. We suppose here that each pair of vertices is connected by not more than one link, which can be done without loss of generality; in the opposite case we divide the superfluous links by additional vertices. The *degree of a vertex* is defined as the number of edges connected in this vertex; we restrict our attention to graphs where all the vertex degrees are finite.

Let us add that it is often useful to employ other ways of labelling graph elements, *e.g.*, one may number instead of the vertices the graph edges writing $\mathcal{E} = \{e_j\}_{j\in J}$ with an appropriate index set J. We shall suppose that Γ is a **metric graph**, *i.e.* it has a *local* metric structure in the sense that every edge e_n is isometric with a line segment $I_j := [0, \ell_j]$. Sometimes the graph can be also equipped with a *global* metric, for instance, if we identify it with a subset of a Euclidean space; for simplicity we will assume that in such a case the two metrics coincide. Using the local metric, we are able to introduce the Hilbert space $L^2(\Gamma) := \bigoplus_{j\in J} L^2(I_j)$ with elements written as $\psi = \{\psi_j\}_{j\in J}$ or simply $\{\psi_j\}$; this will be the state space of the system we are going to study.

As we have said our intention is to study a quantum mechanical particle confined to the described metric graph Γ. We shall suppose that the particle is non–relativistic and spinless so that its motion at each edge is described by a one–dimensional Schrödinger operator. Given then a family of potentials $V := \{V_j\}_{j\in J}$ with properties to be specified later, we look for operators acting on the state Hilbert space $L^2(\Gamma)$ as

$$H\{\psi_j\}_{j\in J} := \{-\psi_j'' + V_j\psi_j\}_{j\in J} \tag{17.1}$$

with the domain being a subset of $AC^2(\Gamma) := \bigoplus_{j\in J} AC^2(I_j)$, which we could use as Hamiltonians of such a particle. The first question, to be addressed in this section, is what one needs to obtain in this way a self–adjoint operator.

The considerations of Section 4.9 suggest that the behavior at the edge endpoints will be important, *i.e.*, the way in which the wave functions are coupled at the graphs vertices. To single out the substantial features, let us consider a star–shaped Γ with a single vertex and n halfline edges joined there, $\mathcal{H} = \bigoplus_{j=1}^n L^2(\mathbb{R}^+)$. Furthermore, since the coupling will not depend on the potentials V_j, at least if the latter are not too singular, we shall assume in the following $V_j = 0$, $j = 1,\dots,n$, and discuss operators corresponding to the formal expression $H\{\psi_j\} = \{-\psi_j''\}$. To describe the family of admissible couplings we introduce the boundary values at the vertex,

$$\psi_j(0+) := \lim_{x\to 0+} \psi_j(x)\,, \qquad \psi_j'(0+) := \lim_{x\to 0+} \psi_j'(x)\,, \tag{17.2}$$

which make sense for any $\psi \in AC^2(\Gamma)$. We know, of course, ways to define a self–adjoint operator, *e.g.*, by imposing the Dirichlet boundary condition, $\psi_j(0+) = 0$, at the endpoint of each halfline; we call this operator H_D. In the present situation it represents a trivial case because the edges are then independent of each other.

Restriction of this fully decoupled operator to the domain $\{\psi \in AC^2(\Gamma) : \psi_j(0+) = \psi_j'(0+) = 0\}$ gives by Example 4.6.9 a symmetric operator H_0 with deficiency indices (n,n). Then we may employ Theorem 4.7.10 by which this operator has a family of self–adjoint extensions that can be parametrized by $n\times n$ unitary matrices, *i.e.* by n^2 real parameters; we identify them with the admissible Hamiltonians of the star–graph system.

However, the von Neumann formula is not the optimal tool to characterize these extensions. Since we deal with second–order differential operators it is more convenient to use suitable boundary conditions in analogy with Theorem 4.9.2. In analogy with Example 4.9.6 one can check that the adjoint of H_0 is defined on $AC^2(\Gamma)$ acting as $H_0^*\{\psi_j\} = \{-\psi_j''\}$. Then we have to find restrictions of H_0^* which annulate the boundary form

$$\mathcal{B}(\phi,\psi) := (\phi, H_0^*\psi) - (H_0^*\phi,\psi) = \sum_{j=1}^n \Big(\overline{\phi}_j(0+)\psi_j'(0+) - \overline{\phi}_j'(0+)\psi_j(0+)\Big)\,, \tag{17.3}$$

which can also be rewritten as a symplectic form $([\phi], J[\psi])_{\mathbb{C}^{2n}}$ on $\mathbb{C}^{2n}$, where $\left(\begin{smallmatrix} 0 & I \\ -I & 0 \end{smallmatrix}\right)$ and the symbol $[\psi] := \left(\begin{smallmatrix} \psi(0+) \\ \psi'(0+) \end{smallmatrix}\right)$ denotes the $2n$–dimensional vector of boundary values. The task is then to find the *Lagrangean subspaces*, *i.e.*, maximal subspaces in $\mathbb{C}^{2n}$ on which the form (17.3) vanishes.

17.1.1 Proposition: Any self–adjoint extension of H_0 is characterized by the conditions

$$A\psi(0+) + B\psi'(0+) = 0, \tag{17.4}$$

where A, B are $n \times n$ matrices such that the $n \times 2n$ matrix (A, B) has maximal rank (*i.e.*, rank n) and AB^* is Hermitean. Conversely, any pair of matrices with these properties determines through (17.4) a self–adjoint extension of H_0.

Proof is left to the reader (Problem 1).

It is clear that, for the most part, the boundary conditions can be written in a simpler form, $\psi'(0+) = C\psi(0+)$ or $\psi(0+) = D\psi(0+)$ with Hermitean matrices C, D, however, the condition (17.4) is universal, *i.e.*, independent of the invertibility of B or A. On the other hand, this characterization of the extensions is certainly not unique since the matrices can be replaced by VA, VB, respectively, for any regular $n\times n$ matrix V. This introduces an equivalence relation into the set of all admissible matrix pairs, and there is a simple way how to characterize the equivalence classes.

17.1.2 Proposition: Any self–adjoint extensions of H_0 is uniquely characterized by (17.4) with $A = (U - I)$ and $B = i(U + I)$, where U is an $n \times n$ unitary matrix.

Proof: The claim can be verified by inspecting properties of the matrices A, B (Problem 2). Alternatively, there is a direct argument which starts from the squared norms $\|\psi(0+) \pm i\ell\psi'(0+)\|^2_{\mathbb{C}^n}$ with a fixed $\ell > 0$. It is straightforward to check that their difference is just the $2i\ell$ multiple of the boundary form (17.3) with $\phi = \psi$. Consequently, for any ψ from the domain of a self–adjoint extension the two vectors have the same norm, hence there is a unitary U such that $\psi(0+) - i\ell\psi'(0+) = U(\psi(0+) + i\ell\psi'(0+))$ which gives the boundary conditions (17.4) with $A = (U - I)$ and $B = i\ell(U + I)$. To conclude the proof, we observe that the parameter ℓ fixes the length scale of the problem and can be set to one without loss of generality (Problem 3). ∎

We can thus label the extensions by the corresponding matrices U. It is clear from the construction that the family of the operators H_U is rich, being characterized by n^2 real parameters. Several interesting vertex couplings can be singled out.

17.1.3 Examples: *(a) The δ–coupling:* If we require the wave functions to be continuous at the vertex we obtain, independently of n, a one–parameter subfamily of extensions described by the boundary conditions

$$\psi_j(0+) = \psi_k(0+) =: \psi(0+)\,,\; j,k = 1,\dots,n\,, \quad \sum_{j=1}^{n} \psi_j'(0+) = \alpha\psi(0+)\,, \tag{17.5}$$

for a fixed $\alpha \in \mathbb{R} \cup \{\infty\}$. We refer to this as the δ–coupling because in the particular case $n = 2$ it is nothing else that the δ–interaction of Proposition 14.6.1.

The conditions (17.5) can be naturally cast into the form (17.4). For the sake of brevity we denote by $\mathcal{J}$ the $n\times n$ matrix whose all entries are equal to one (equivalently, $\mathcal{J} = nP$ where P is the projection onto the subspace spanned by the vector $(1,\dots,1)^t$). Then it is straightforward to check that the δ–coupling with the parameter α corresponds to the matrix $U := \frac{2}{n+i\alpha}\mathcal{J} - I$.

The extension corresponding formally to $\alpha = \infty$ is identical to the fully decoupled operator with Dirichlet boundary conditions, $\psi_j(0+) = 0$, $j = 1,\dots,n$, from which the construction has started. On the other hand, $\alpha = 0$ corresponds to **free boundary conditions**; the name reflects the fact that in the particular case $n = 2$ the corresponding operator is just the free Hamiltonian of Example 7.2.1 (see also the notes).

(b) The δ'_s–coupling: This is a direct counterpart to the previous one with the roles of the function values and derivatives reversed, *i.e.*, it is given by the boundary conditions

$$\psi'_j(0+) = \psi'_k(0+) =: \psi'(0+)\,,\; j,k = 1,\dots,n\,,\quad \sum_{j=1}^{n}\psi_j(0+) = \beta\psi'(0+) \tag{17.6}$$

with $\beta \in \mathbb{R}\cup\{\infty\}$, which corresponds to $U = I - \frac{2}{n-i\beta}\mathcal{J}$. The extension corresponding to $\beta = \infty$ gives again an operator fully decoupled into the orthogonal sum of the edge Hamiltonians, this time with Neumann boundary conditions, $\psi'_j(0+) = 0$, $j = 1,\dots,n$.

(c) The δ'–coupling: It is natural to ask about the coupling which generalizes the δ'–interaction of Remark 14.6.3b. It is easy to see that it is characterized by the conditions

$$\sum_{j=1}^{n}\psi'_j(0+) = 0\,,\quad \psi_j(0+)-\psi_k(0+) = \frac{\beta}{n}(\psi'_j(0+)-\psi'_k(0+))\,,\; j,k = 1,\dots,n\,, \tag{17.7}$$

with $\beta \in \mathbb{R}\cup\{\infty\}$, corresponding to $U = -\frac{n+i\beta}{n-i\beta}I + \frac{2}{n-i\beta}\mathcal{J}$. The infinite value of the parameter refers again to the Neumann decoupling of the graph edges, and furthermore, some other properties of this coupling are similar to these of the previous one (Problem 4).

All the operators H_U listed in these examples exhibit a particular symmetry: they are invariant with respect to interchanges of the edge indices. It is manifested by the fact that the corresponding U are linear combinations of the matrices I and $\mathcal{J}$. One can describe such *permutation–invariant couplings* generally.

17.1.4 Proposition: The boundary conditions of Proposition 17.1.2 are permutation invariant *iff* the matrix U equals $U = aI + b\mathcal{J}$ with complex coefficients a,b satisfying the relations

$$|a| = 1 \quad\text{and}\quad |a+nb| = 1\,.$$

Proof: The claim is equivalent to the fact that the boundary conditions are satisfied at the same time by the vectors $P\psi(0+)$ and $P\psi'(0+)$ for any $P \in \mathcal{S}_n$. Multiplying them by P^{-1} from the left we get (17.4) with $A = P^{-1}UP - I$ and $B = i(P^{-1}UP + I)$. The matrix $P^{-1}UP$ is obviously unitary; since by Proposition 17.1.2 the map $U \mapsto H_U$ is a bijection we get $P^{-1}UP = U$ for any $P \in \mathcal{S}_n$. Next we notice that a simultaneous permutation of the rows and columns leaves the diagonal elements on the diagonal, and the off-diagonal ones off the diagonal; it follows that $U = aI + b\mathcal{J}$ for some $a, b \in \mathbb{C}$. The conditions restricting the values of a, b follow from the unitarity of U,

$$(UU^*)_{ij} = |a|^2\delta_{ij} + 2\mathrm{Re}\,(a\bar{b}) + n|b|^2 \;=\; \delta_{ij}\,,$$

which yields the relations $|a|^2 + 2\mathrm{Re}\,(a\bar{b}) + n|b|^2 = 1$ and $2\mathrm{Re}\,(a\bar{b}) + n|b|^2 = 0$ for $i = j$ and $i \neq j$, respectively. Substituting from the second to the first one we get $|a|^2 = 1$. Finally, using $|a+nb|^2 = |a|^2 + 2n\mathrm{Re}\,(a\bar{b}) + n^2|b|^2$ we see that the left side of the second relation is a multiple of $|a+nb|^2 - |a|^2$. ∎

These boundary conditions can also be written explicitly as the following system,

$$(a-1)\psi_j(0+) + b\sum_{k=1}^{n}\psi_k(0+) + i(a+1)\psi_j'(0+) + ib\sum_{k=1}^{n}\psi_k'(0+) \;=\; 0\,, \quad j = 1,\dots,n\,;$$

this expression shows, in particular, that the linear combination $(a-1)\psi_j(0+) + i(a+1)\psi_j'(0+)$ is independent of j. To get a useful equivalent formulation we subtract the k–th one of these conditions from the j–th one obtaining

$$(a-1)\left(\psi_j(0+) - \psi_k(0+)\right) + i(a+1)\left(\psi_j'(0+) - \psi_k'(0+)\right) \;=\; 0\,, \quad j,k = 1,\dots,n\,, \tag{17.8}$$

while summing all of them gives

$$(a-1+nb)\sum_{k=1}^{n}\psi_k(0+) + i(a+1+nb)\sum_{k=1}^{n}\psi_k'(0+) \;=\; 0\,. \tag{17.9}$$

On the other hand, the zoology of self–adjoint extensions for star–shaped graphs is extensive and the permutation–invariant class is still a small part of the whole family. Sometimes it is useful to consider extension classes which allow us to "switch off" the coupling of an edge by changing an appropriate parameter. Let us describe one simple case.

17.1.5 Example: Let $n = 3$ and consider the boundary conditions

$$\begin{aligned} \psi_1(0+) &= \psi_2(0+)\,, \\ \psi_3(0+) &= b\psi_1(0+) + c\psi_3'(0+)\,, \\ \psi_1'(0+) + \psi_2'(0+) &= d\psi_1(0+) - b\psi_3'(0+) \end{aligned}$$

with real parameters b, c, d. It is easy to check that they annulate the boundary form (17.3), see also Problem 5. This coupling is permutation–invariant *iff* $b = 1$ and $c = 0$ when it is reduced to the δ–coupling with the parameter $\alpha = d$. On the other hand, if $b = 0$ the graph decouples into two parts. The first is the line consisting of the first two halflines; using the wave function $\psi(x) = \chi_{\mathbb{R}^+}(x)\psi_1(x)+\chi_{\mathbb{R}^-}(x)\psi_2(-x)$ the first and third of the above boundary conditions give a δ–interaction of strength d. The other part is the halfline with the Hamiltonian of Example 7.2.2. A nonzero b means that the two systems are coupled in a way which can be tuned by changing this parameter.

The construction described above can be used not only for a star–shaped Γ but for other graphs as well. Restricting the domain of a fully decoupled operator to functions which vanish together with their derivatives at the vertices we get the symmetric operator with deficiency indices $\left(\sum_j n_j, \sum_j n_j\right)$, where n_j is the degree of the j–th vertex. What is important, however, is that we are not interested in all the self–adjoint extensions of such an H_0 but only in those described by *local* boundary conditions, *i.e.* those coupling the boundary values at each vertex separately. This is given by the physical meaning of the model we are discussing; we disregard the dynamics allowing the particle to "jump" from one vertex to another. With this restriction Proposition 17.1.2 gives us a full description in which the j–th vertex is described by an $n_j \times n_j$ unitary matrix U_j.

17.2 Meaning of the vertex coupling

We already know how to construct graph Hamiltonians with different wave function coupling at the vertices. If we want to use this knowledge to construct a realistic model, *e.g.*, of a graph–shaped semiconductor nanostructure we have to understand the physical meaning of the boundary conditions discussed in the previous section. At first glance this appears easy. One has to take a "fat graph", *i.e.* a waveguide system of thin tubes centered at the edges of the given Γ, and to analyze the behavior of the waveguide Hamiltonian – which contains no free parameters – as the tube width tends to zero.

Unfortunately this problem is harder than it seems. Let us describe one result of this type. For simplicity we shall suppose that Γ is a planar graph, *i.e.*, that it can be identified with a subset of $\mathbb{R}^2$ from which it inherits the metric structure. We assume that it consists from a finite number of finite–length edges, each of which is a C^1–smooth curve. Let H_f be the Hamiltonian of the previous section, $H_f\{\psi_j\} = \{-\psi_j''\}$, with the *free* boundary conditions of Example 17.1.3a at each vertex. Construct now the family of "fat graphs" as neighborhoods of the graph, $\Gamma_\varepsilon := \{x \in \mathbb{R}^2 :\ \varrho(x,\Gamma) < \varepsilon\}$, and denote by $H_N^{\Gamma_\varepsilon}$ the corresponding Neumann Laplacian. Let $\lambda_j(A)$ be the j–th eigenvalue of an operator A, numbered in ascending order as in Theorem 14.2.1.

17.2.1 Theorem: $\lambda_j(H_N^{\Gamma_\varepsilon}) \to \lambda_j(H_f)$ as $\varepsilon \to 0$ holds for each $j \in \mathbb{N}$.

For the proof and various generalizations of this result we refer to the notes. In general, "fat graphs" with Neumann boundary lead to the most simple coupling represented by free boundary conditions. On the other hand, in the previous chapter we have seen that waveguides with *Dirichlet* boundary conditions are important often giving a physically plausible model. The "squeezing" limit is very different in this case and some understanding of it was reached only recently; we again refer to the notes for references.

An easier way to approach the mentioned problem is to look for an approximation by means of operators on the graph itself in a way similar to that used in Section 14.6. For the sake of simplicity we shall again consider a star-shaped graph of n halflines connected at a single vertex to which the state Hilbert space $\bigoplus_{j=1}^{n} L^2(\mathbb{R}^+)$ corresponds. This time the motion at the edges need not be free; we suppose that the j-th edge supports a real-valued potential $V_j \in L^1_{loc}(\mathbb{R}^+)$ such that the differential expression $-\frac{d^2}{dx^2} + V_j(x)$ is limit-point at infinity (*cf.* Remark 4.8.8 and Proposition 4.8.9).

Let us begin with the *δ-coupling.* For a fixed $\alpha \in \mathbb{R} \cup \{\infty\}$, we define the operator $H_\alpha(V)$ which acts according to (17.1) on the domain of all$\psi \in AC^2(\Gamma)$ satisfying the boundary conditions (17.5); it is straightforward to check that $H_\alpha(V)$ is self-adjoint (Problem 6a). To find an approximation to the δ-coupling we first need an expression for the resolvent of $H_\alpha(V)$. Given k with $\operatorname{Im} k \geq 0$ such that k^2 belongs to the resolvent set of the decoupled operator $H_\infty(V)$, we denote by $u_j \equiv u_j(\cdot;k)$ and $v_j \equiv v_j(\cdot;k)$ solutions to the equation $-\psi_j'' + V_j\psi_j = k^2\psi_j$ with the appropriate behavior at 0 and infinity, respectively, *i.e.*, $u_j(0;k) = 0$ and $v_j \in L^2$ in (c,∞) for $c > 0$. The decoupled edge parts of $H_\infty(V)$ are by Theorem 4.9.7 characterized by the resolvent kernels

$$g_j(x,y;k) := -\frac{u_j(x_<;k)v_j(x_>;k)}{W(u_j,v_j)}, \tag{17.10}$$

where $W(u_j,v_j)$ is the Wronskian of the two solutions, $x_< := \min\{x,y\}$ and $x_> := \max\{x,y\}$. This allows us to express the resolvent of the nontrivial graph Hamiltonian.

17.2.2 Lemma: Fix k with $\operatorname{Im} k \geq 0$, $k^2 \in \rho(H_\infty(V))$. Under the stated assumption about $\{V_j\}$ the resolvent $(H_\alpha(V) - k^2)^{-1}$ is the matrix integral operator with the kernel

$$G_{ij}^\alpha(x,y;k) = \delta_{ij}g_i(x,y;k) + \frac{v_i(x;k)v_j(y;k)}{v_i(0;k)v_j(0;k)(\alpha - M(k))}, \quad i,j = 1,\dots,n,$$

where we have denoted $M(k) := \sum_{j=1}^{N} \frac{v_j'(0;k)}{v_j(0;k)}$.

Proof: By Theorem 4.7.15 the matrix elements of the resolvent kernel are of the form $\delta_{ij}g_j(x,y;k) + \Lambda_{ij}v_i(x;k)v_j(y;k)$. To find the unknown coefficients, we notice that $\psi := (H_\alpha(V) - k^2)^{-1}\phi$ has to satisfy the boundary conditions (17.5) for any $\phi \in \mathcal{H}$; in this way we arrive at a system of linear equations for Λ_{ij} which yields the result (Problem 6b). ▮

We notice in passing that the resolvent formula shows that the discrete spectrum can be determined from the equation $M(k) = \alpha$. We shall discuss this problem in more generality in the next section. Another application is the weak–coupling asymptotic expansion for star–graph Schrödinger operators with free boundary conditions (Problem 7) which generalizes the one–dimensional result mentioned in the notes to Section 14.3.

After this preliminary, let us pass to the approximation of the operators $H_\alpha(V)$. Given functions $W_j\,:\,\mathbb{R}^+\to\mathbb{R}$ we define the scaled potentials, as in (14.29), by

$$W_{\epsilon,j} := \frac{1}{\varepsilon}\, W_j\left(\frac{x}{\varepsilon}\right), \quad j=1,\dots,n\,. \tag{17.11}$$

As in the case of the δ–interaction on the line one can get the sought boundary conditions by "shrinking" this potential family with the mean preserved.

17.2.3 Theorem: Suppose that $V_j\in L^1_{loc}(\mathbb{R}^+)$ are bounded from below and $W_j\in L^1(\mathbb{R}^+)$ for $j=1,\dots,n$. Then

$$H_0(V+W_\varepsilon) \longrightarrow H_\alpha(V) \qquad \text{as} \quad \varepsilon\to 0+ \tag{17.12}$$

in the norm resolvent sense, where $\alpha=\langle W\rangle := \sum_{j=1}^n \int_0^\infty W_j(x)\,dx$.

Proof: The argument is analogous to the proof of Theorem 14.6.2, so we sketch it only briefly leaving the details to the reader (Problem 6d). Let $G^{W_\epsilon}_{ij}(x,y;k)$ be the resolvent kernel of $H_0(V+W_\epsilon)$. Using Lemma 17.2.2 we may rewrite it explicitly as

$$\begin{aligned} G^{W_\epsilon}_{j\ell}(x,y;k) &= G^0_{ij}(x,y;k) - \sum_{l,m} \int_0^\infty\!\!\int_0^\infty G^0_{il}(x,y';k)\, W_{\varepsilon,l}(x')^{1/2} \\ &\quad\times \left(I+|W_\varepsilon|^{1/2}\left(H_0(V)-k^2\right)^{-1} W_\varepsilon^{1/2}\right)^{-1}_{lm} (x',x'')\, |W_{\varepsilon,r}(x'')|^{1/2} \\ &\quad\times G^0_{mj}(x'',y;k)\, dx'\, dx''\,. \end{aligned}$$

Changing the integration variables to x'/ϵ and x''/ϵ, the second term of the resolvent becomes $-B_{k,\varepsilon}\,(I+C_{k,\varepsilon})^{-1}\,\tilde B_{k,\varepsilon}$, with the factors determined by their kernels,

$$\begin{aligned} (B_{k,\varepsilon})_{ij}(x,y) &= G^0_{ij}(x,\varepsilon y;k) W_j(y)^{1/2}\,, \\ (\tilde B_{k,\varepsilon})_{ij}(x,y) &= |W_i(x)|^{1/2} G^0_{ij}(\varepsilon x,y;k)\,, \\ (C_{k,\varepsilon})_{ij}(x,y) &= |W_i(x)|^{1/2} G^0_{ij}(\varepsilon x,\varepsilon y;k)\, W_j(y)^{1/2}\,, \end{aligned}$$

which converge pointwise to

$$\begin{aligned} (B_k)_{ij}(x,y) &= G^0_{ij}(x,0;k) W_j(y)^{1/2}\,, \\ (\tilde B_k)_{ij}(x,y) &= |W_i(x)|^{1/2} G^0_{ij}(0,y;k)\,, \\ (C_k)_{ij}(x,y) &= |W_i(x)|^{1/2} G^0_{ij}(0,0;k)\, W_j(y)^{1/2}\,, \end{aligned}$$

respectively, as $\varepsilon \to 0+$. The explicit form of the last operator makes it possible to find the inverse of the middle term in the factorization. It has the kernel

$$(I + C_k)^{-1}_{ij}(x,y) = \delta(x-y)\delta_{ij} - \frac{|W_i(x)|^{1/2}W_j(y)^{1/2}}{\langle W \rangle - M(k)}.$$

Hence the limiting resolvent kernel differs from the free one by

$$- \sum_r \int_0^\infty dx'\, W_l(x')\, G^0_{il}(x,0;k)G^0_{lj}(0,y;k)$$

$$+ \sum_{l,m} \int_0^\infty \int_0^\infty dx'\, dx''\, W_l(x')W_m(x'')\, \frac{G^0_{il}(x,0;k)G^0_{mj}(0,y;k)}{\langle W \rangle - M(k)}$$

$$= \frac{v_i(x;k)v_j(y;k)}{v_i(0;k)v_j(0;k)}\, \frac{\langle W \rangle}{M(k)(\langle W \rangle - M(k))},$$

where we have used (17.10); this expression coincides with the analogous term in Lemma 17.2.2 if we set $\alpha := \langle W \rangle$. The norm–resolvent convergence has to be checked for a particular k. Since all the V_j are bounded from below by assumption, the functions $v_j(\cdot;k)$ are exponentially decaying for κ large enough; using this fact together with the dominated–convergence theorem one can check that the Hilbert–Schmidt norm of the resolvent difference tends to zero as $\varepsilon \to 0$, and the same holds by Theorem 3.6.2 for the operator norm. ∎

Approximation of other vertex couplings with wave functions discontinuous at the vertex is more complicated. We shall describe the case of the δ'_s–coupling of Example 17.1.3b where an approximation can be constructed by extending the idea mentioned in the notes to Section 14.6. For simplicity we again put $V_j = 0$ and denote the corresponding free Hamiltonian of the star–shaped Γ with the boundary conditions (17.6) by H_β. The approximating operators will act as $\psi_j \mapsto -\psi''_j$ at each edge, however, with modified boundary conditions. First of all, we suppose that at the vertex of Γ the wave functions will satisfy the δ conditions (17.5) with coupling parameter $\alpha = b$ to be specified later. We also add a δ–interaction of a coupling strength c at each edge on the distance a from the vertex; alternatively one can say that we pass to a graph with n additional vertices of degree two, each with the same δ-coupling. We call such operators $H^{b,c}(a)$.

Let us inspect how b, c should depend on the distance a. Notice that the problem has a symmetry which allows us to simplify it: each of the operators $H^{b,c}(a)$ and H_β decomposes into a nontrivial part which acts on the "one–dimensional" subspace of $\mathcal{H} = \bigoplus_{j=1}^n L^2(\mathbb{R}^+)$ consisting of functions symmetric with respect to permutations, $\psi_i(x) = \psi_j(x)$ for all i, j, and the "$(n-1)$–dimensional" part corresponding to Dirichlet and Neumann conditions at the central vertex for the δ and δ'_s–coupling, respectively (this reflects the fact the matrices U corresponding to these couplings by Proposition 17.1.2 each have one simple eigenvalue and another eigenvalue equal to ∓ 1, respectively, of multiplicity $n-1$).

Consider first the nontrivial part. As we have said, in this case all the functions are the same so we may drop the edge index. For small a the boundary values of ψ at $x=0$ and $x=a$ are then related by

$$\psi(a) = \psi(0) + a\psi'(0) + \mathcal{O}(a^2)\,, \quad \psi'(a-) = \psi'(0+) + \mathcal{O}(a)\,,$$
$$\psi'(a+) = \psi'(a-) + c\psi(a)\,, \quad n\psi'(0+) = b\psi(0)\,.$$

Eliminating $\psi(0)$ and $\psi'(0+)$ from these relations we get to leading order the relation $B(a)\psi(a) = \psi'(a+)$ with $B(a) := c + b(n+ab)^{-1}$, hence the required limit, $\beta\psi'(0+) = n\psi(0)$, is satisfied as $a \to 0+$ provided we choose

$$b(a) := -\frac{\beta}{a^2}\,, \quad c(a) := -\frac{1}{a}\,. \tag{17.13}$$

In the orthogonal complement to the permutation–symmetric subspace the edge index can again be dropped, because the operators act in the same way on all the linear combinations $\sum_{j=1}^n d_j\psi_j(x)$ with $\sum_{j=1}^n d_j = 0$. The last of the above boundary–value conditions is now replaced by $\psi(0) = 0$. Eliminating then the values at $x=0$ we get to leading order the relation $\psi'(a+) = (c+a^{-1})\psi(a) + \mathcal{O}(a)$; the right side vanishes with the parameter choice (17.13) giving the Neumann condition, $\psi'(0+) = 0$, in the limit.

What is important is that the family of approximation operators $H^{b,c}(a)$, found in this heuristic way, gives the correct answer.

17.2.4 Theorem: $H^{b,c}(a) \to H_\beta$ holds as $a \to 0+$ in the norm–resolvent sense provided the dependence of the coupling parameters b, c on a is given by (17.13).

Proof: The resolvent difference can be written explicitly by means of Krein's formula so that the result follows by a straightforward computation (Problem 8). ∎

17.3 Spectral and scattering properties

After explaining what quantum graphs are, let us turn to the discussion of their properties. An attractive aspect of these models is that the analysis of the spectra and scattering behavior is equivalent to the solution of a family of ordinary differential equations and the proper coupling of their solutions. Moreover, the last part can be reduced to an algebraic problem which involves boundary values of certain standard solutions only. We shall start by presenting a general method of calculating the spectrum of a quantum–graph Hamiltonian with a δ–coupling at the vertices.

We consider a connected graph Γ with the properties which will be specified a little later. It is useful for the present purpose to use the index set I and to number the edges by elements of $I_\mathcal{E}$, *i.e.* by the pairs of the vertex indices which they connect. Given a vertex v_j we introduce the set $\mathcal{N}(v_j) \equiv \{v_n : n \in \nu(j) \subset I\setminus\{j\}\}$ consisting of its *neighbors*, *i.e.* the vertices connected with v_j by a single link; by assumption $\mathcal{N}(v_j)$ is nonempty for any $j \in I$. The graph *boundary* $\mathcal{B}$ consists of vertices having a single neighbor, in general it may be empty. We denote by $I_\mathcal{B}$ and

$I_{\mathcal{I}}$ the index subsets in I corresponding to $\mathcal{B}$ and to the graph *interior* $\mathcal{I} := \mathcal{V}\setminus\mathcal{B}$, respectively.

The Schrödinger operator on Γ we are going to study corresponds to a family of potentials $V := \{V_{jn}\}$ where we, for simplicity, assume that $V_{jn} \in L^\infty(I_{jn})$; their action is given by (17.1) with the appropriate change of numbering. The domain consists of all functions from $AC^2(\Gamma)$ which satisfy the appropriate boundary conditions: given the families $\alpha = \{\alpha_j\}_{j\in I_{\mathcal{I}}}$ and $\omega = \{\omega_j\}_{j\in I_{\mathcal{B}}}$ we impose the δ–coupling (17.5) at graph interior vertices and $\psi_j \cos\omega_j + \psi'_j \sin\omega_j = 0$ at the boundary (identifying the point $x=0$ with v_j we denote the boundary values as $\psi_{jn}[j] := \lim_{x\to 0+}\psi_{jn}(x)$ and $\psi'_{jn}[j] := \lim_{x\to 0+}\psi'_{jn}(x)$, or simply ψ_j and ψ'_j). The operator obtained in this way will be denoted by $H_{\alpha,\omega}(V)$; for bounded potentials it is obviously self–adjoint.

It is useful to simultaneously look for generalized eigenfunctions corresponding to the continuous spectrum if the latter is nonempty; this happens if some links of Γ are infinite. Let us denote by $D_{loc}(H_{\alpha,\omega}(V))$ the set of functions $\{\psi_{jn}\}$ which satisfy all the requirements except for global square integrability, *i.e.*, ψ_{jn}, ψ'_{jn} are absolutely continuous, $\psi''_{jn} \in L^2_{loc}(\Gamma)$, and the above specified boundary conditions at the graph vertices are satisfied.

To formulate the result we need to introduce some notation. First of all, the family described above contains operators with Dirichlet decoupled edges, *i.e.* such that $\alpha_j = \infty$ for all $j \in I_{\mathcal{I}}$. The spectrum of such an operator, denoted by $H_{D,\omega}(V)$, will be of importance; we introduce $\mathcal{K}_D := \{k :\, k^2 \in \sigma(H_{D,\omega}(V)),\, \mathrm{Im}\, k \geq 0\}$. Our aim is to express the eigenfunctions for the given problem in terms of standard elements of $\mathcal{H}$. On $e_{nj} \equiv (0,\ell_{jn})$, where the *right* endpoint is identified with v_j, we shall denote by u_{jn}, the solution to $-\psi'' + V_{jn}\psi = k^2\psi$ which satisfies the boundary conditions $u_{jn}(\ell_{jn}) = 1 - u'_{jn}(\ell_{jn}) = 0$. Similarly v_{jn} is the solution to the same equation determined by the behavior at the left endpoint, $v_{jn}(0) = 1 - v'_{jn}(0) = 0$ if $n \in I_{\mathcal{I}}$, and $v_{jn}(0) = \sin\omega_n,\ v'_{jn}(0) = -\cos\omega_n$ if $n \in I_{\mathcal{B}}$. The Wronskian $W_{jn} \equiv W(u_{jn}, v_{jn})$ of these solutions equals $W_{jn} = -v_{jn}(\ell_{jn})$, or equivalently $W_{jn} = u_{jn}(0)$ if $n \in I_{\mathcal{I}}$ and $W_{jn} = -u_{jn}(0)\cos\omega_n - u'_{jn}(0)\sin\omega_n$ if $n \in I_{\mathcal{B}}$. Now we are able the state the desired result.

17.3.1 Theorem: (a) Suppose that $\psi = \{\psi_{jn}\} \in D_{loc}(H_{\alpha,\omega}(V))$ solves the equation $H_{\alpha,\omega}(V)\psi = k^2\psi$ with $V_{jn} \in L^\infty(I_{jn}),\ (j,n) \in I_{\mathcal{E}}$, for some $k \neq \mathcal{K}_D$ with $k^2 \in \mathbb{R}$, $\mathrm{Im}\, k \geq 0$; then the corresponding boundary values satisfy the equations

$$\sum_{n\in\nu(j)\cap I_{\mathcal{I}}} \frac{\psi_n}{W_{jn}} - \left(\sum_{n\in\nu(j)} \frac{v'_{jn}(\ell_{jn})}{W_{jn}} - \alpha_j \right) \psi_j = 0\,, \quad j \in I_{\mathcal{I}}\,. \tag{17.14}$$

Conversely, any solution $\{\psi_j :\, j \in I_{\mathcal{I}}\}$ of the system to (17.14) determines a solution of the Schrödinger equation on Γ by

$$\psi_{jn}(x) = \frac{\psi_n}{W_{jn}}\, u_{jn}(x) - \frac{\psi_j}{W_{jn}}\, v_{jn}(x) \qquad \text{if} \quad n \in \nu(j)\cap I_{\mathcal{I}}\,,$$

$$\psi_{jn}(x) = -\frac{\psi_j}{W_{jn}}\, v_{jn}(x) \qquad \text{if} \quad n \in \nu(j)\cap I_{\mathcal{B}}.$$

(b) If $\ell_0 := \inf\{\ell_{jn} : (j,n)\in I_{\mathcal{L}}\} > 0$, then $\psi \in L^2(\Gamma)$ implies that the solution $\{\psi_j\}$ of the system (17.14) belongs to $\ell^2(I_{\mathcal{I}})$.

(c) The opposite implication, $\{\psi_j\}\in \ell^2(I_{\mathcal{I}}) \Rightarrow \psi \in L^2(\Gamma)$, is also valid provided the quantities $L_0 := \sup\{\ell_{jn} : (j,n)\in I_{\mathcal{L}}\}$ and $N_0 := \max\{\mathrm{card}\nu(j) : j\in I\}$ are finite, and k has a positive distance from the set $\mathcal{K}_D$.

17.3.2 Remarks: (a) The assumptions about the graph in the parts (b) and (c) are satisfied automatically if Γ is finite and all its edges are of a finite length, respectively. If two vertices are joined by more than a single edge, the theorem can be still used; it is sufficient to add a vertex with the free coupling, $\alpha_j = 0$, at each extra link. In a similar way one can consider graphs with singular δ–type potentials.

(b) An analogous result is valid for graphs with the δ'_s–coupling (Problem 9).

(c) The difference equation (17.14) is in a sense reminiscent of discrete Schrödinger operators, often called *Jacobi matrices*. In particular, the first term contains a summation over the neighbors of a site as in the case of a discrete Laplacian, however, in distinction to it here the coefficients depend on the spectral parameter.

Proof of Theorem 17.3.1: For simplicity we use the symbol H for $H_{\alpha,\omega}(V)$. If $n\in I_{\mathcal{I}}$, the solution at $x\in I_{nj}$ and its derivative can be expressed in terms of the boundary values $\psi_{jn}[n] =: \psi_n$ and $\psi'_{jn}[n]$ by means of the transfer matrix

$$T_{nj}(x,0) \;=\; W_{jn}^{-1}\begin{pmatrix} u_{jn}(x) - u'_{jn}(0)v_{jn}(x) & u_{jn}(0)v_{jn}(x) \\ u'_{jn}(x) - u'_{jn}(0)v'_{jn}(x) & u_{jn}(0)v'_{jn}(x) \end{pmatrix};$$

the Wronskian is nonzero for $k\notin \mathcal{K}_D$. In particular, we get

$$\psi_j \;:=\; \psi_{jn}[j] = u'_{jn}(0)\psi_n + v_{jn}(\ell_{jn})\psi'_{jn}[n]\,,$$

$$-\psi'_{jn}[j] \;=\; \frac{1-u'_{jn}(0)v'_{jn}(\ell_{jn})}{W_{jn}}\psi_n + v'_{jn}(\ell_{jn})\psi'_{jn}[n];$$

the sign change at the left side of the last condition reflects the fact that in (17.2) we consider the outward derivative at v_j. We express $\psi'_{jn}[n]$ using the first relation and substitute into the second one; this yields

$$\psi'_{jn}[j] \;=\; -\frac{\psi_n}{W_{jn}} + \frac{v'_{jn}(\ell_{jn})}{W_{jn}}\,\psi_j\,. \tag{17.15}$$

If $n\in I_{\mathcal{B}}$ we have instead $\psi_j = u'_{jn}(0)\psi_n - u_{jn}(0)\psi'_{jn}[n]$ and

$$-\psi'_{jn}[j] = \frac{u'_{jn}(0)v'_{jn}(\ell_{jn}) - \cos\omega_n}{W_{jn}}\,\psi_n + \frac{u_{jn}(0)v'_{jn}(\ell_{jn}) - \sin\omega_n}{W_{jn}}\,\psi'_{jn}[n]\,;$$

we may write, of course, $\psi_n' := \psi_{jn}'[n]$. Using the relations $\psi_n \cos\omega_n + \psi_n' \sin\omega_n = 0$, we find that in this case $\psi_{jn}'[j]$ is given by the second term on the left side of (17.15) only. The equations (17.14) then follow from (17.5) with the boundary contribution absent in the first sum. The remaining part of claim (a) follows from the transfer–matrix expression for $\psi_{jn}(x)$ together with the mentioned formula for $\psi_{jn}'[n]$. For the proof of the claims (b) and (c) we refer to the notes. ∎

Theorem 1 provides us with a tool to determine spectra of graph Hamiltonians which is not limited to eigenvalues of finite graphs. Let us mention one important particular case.

17.3.3 Example (rectangular lattices): Let Γ be an infinite periodic planar graph whose basic cell is a rectangle of sides $\ell_j,\ j=1,2$, and suppose for simplicity that $V_{jn}=0$. Using the natural vertex labelling, we can write the equation (17.14) in the form

$$\begin{aligned}(\psi_{n,m+1}+\psi_{n,m-1})\sin k\ell_1 + (\psi_{n+1,m}+\psi_{n-1,m})\sin k\ell_2 \\ -\left(\frac{\alpha_{nm}}{k}\sin k\ell_1\sin k\ell_2 + 2\sin k(\ell_1+\ell_2)\right)\psi_{nm} = 0\,.\end{aligned} \tag{17.16}$$

The case $\alpha_{mn}=\alpha$ for all $m,n\in\mathbb{Z}$ corresponding to a periodic Hamiltonian H_α on Γ is of particular interest. In analogy with Problem 14.21 one has to employ Floquet–Bloch decomposition, now with respect to shifts in both directions. Correspondingly, the values at neighboring vertices differ by the factors $e^{i\theta_j\ell_j},\ j=1,2$, in the "horizontal" and "vertical" directions, respectively, where $\theta_j\in[-\pi\ell_j^{-1},\pi\ell_j^{-1})$ are quasimomentum components. The spectral condition resulting from (17.16) is, for $k^2>0$, of the following form

$$\frac{\alpha}{2k} = F(k;w_1,w_2) := \sum_{j=1}^{2}\frac{w_j-\cos k\ell_j}{\sin k\ell_j}\,, \tag{17.17}$$

where $w_j := \cos\theta_j\ell_j$. Since these parameters run through the interval $[-1,1]$, it is easy to find the extremum values of $F(k;\cdot,\cdot)$ for a fixed k,

$$F_+(k) := \max_{w_j\in[-1,1]} F(k;w_1,w_2) = \sum_{j=1}^{2}\tan\left(\frac{k\ell_j}{2}-\frac{\pi}{2}\left[\frac{k\ell_j}{\pi}\right]\right)$$

and $F_-(k) := \min_{w_j\in[-1,1]} F(k;v_1,v_2)$ is similar with tan replaced by $-\cot$; the symbol $[\cdot]$ again denotes the integer part. For the negative part of the spectrum we take $\kappa := -ik$ and replace the above functions by $F_\pm(\kappa) := -\sum_{j=1}^2 \binom{\tanh}{\coth}\left(\frac{\kappa\ell_j}{2}\right)$, respectively. As in the notes to Section 14.4, the spectrum of H_α consists of the points k^2 for which the condition (17.17) is satisfied. This means that the *gap* condition can, in this case, be expressed as

$$\pm\frac{\alpha}{2k} > \pm F_\pm(k) \tag{17.18}$$

for $\pm\alpha > 0$, respectively. It is useful to introduce the following notation,

$$\ell := \sqrt{\ell_1 \ell_2}\,, \qquad \theta := \frac{\ell_2}{\ell_1} \qquad \text{and} \qquad L := \max(\ell_1, \ell_2)\,. \tag{17.19}$$

Inspecting the behavior of the functions $F_\pm$ one can find simple properties of the spectrum (Problem 10). In comparison with the one-dimensional analogue of Example 14.21, however, $\sigma(H_\alpha)$ is more complicated (with exception of the square-lattice case, $\ell_1 = \ell_2$) due to the irregular dependence of $F_\pm$ on θ coming from the existence of competing periods, in particular, if θ is irrational.

Recall that irrational numbers are classified with respect to how well they can be approximated by rationals; a number θ is *badly approximable* if there is a $\delta > 0$ such that

$$\left| \theta - \frac{p}{q} \right| > \frac{\delta}{q^2}\,. \tag{17.20}$$

Such θ form a non-empty subset in the family of all Diophantine numbers which contains, for instance, all algebraic numbers of degree 2, *i.e.*, irrational solutions of a quadratic equation with rational coefficients. On the other hand, the Lebesgue measure of this set is zero. An alternative characterization uses the (unique) representation of θ by a continued fraction $[a_0, a_1, \dots]$ with nonnegative integer coefficients. For badly approximable numbers the sequence $\{a_n\}$ is bounded; the worst irrational from this point of view is the golden mean $\frac{1}{2}(1+\sqrt{5}) = [1,1,1,\dots]$. On the other hand, for *well approximable* irrationals the sequence is unbounded, $\limsup_{n\to\infty} a_n = \infty$, or equivalently, there are sequences $\{m_r\}_{r=1}^\infty$, $\{n_r\}_{r=1}^\infty$ of pairwise relatively prime integers such that

$$\lim_{r\to\infty} n_r^2 \left| \theta - \frac{m_r}{n_r} \right| = 0\,. \tag{17.21}$$

Using these notions one can show that the spectrum has the following properties:

(a) If θ is rational or well approximable, $\sigma(H_\alpha)$ has infinitely many gaps for any nonzero coupling constant α.
(b) For a badly approximable θ there is $\alpha_0 > 0$ such that for $|\alpha| < \alpha_0$ the spectrum has no gaps above the threshold.
(c) And finally, $\sigma(H_\alpha)$ has infinitely many gaps for any θ provided $|\alpha| L > 5^{-1/2}\pi^2$.

Indeed, for $\alpha > 0$ it is sufficient to investigate the minima of F_+; we employ (for this particular purpose) the number-theory symbols for the fractional part, $\{x\} := x - [x]$, and $\|x\| := \min(\{x\}, 1-\{x\})$. If $\theta = \frac{p}{q}$ the function F_+ has infinitely many zeros in $\mathbb{R}^+$ without accumulation points which by (17.18) gives rise to open gaps. On the other hand, in the irrational case all the minima are positive; it is easy to see that they occur at $k_n := \frac{\pi n}{\ell_1}$ and $\tilde{k}_m := \frac{\pi m}{\ell_2}$ with $n, m \in \mathbb{Z}_+$, and the corresponding values are

$$F_+(k_n) = \tan\left(\frac{\pi}{2}\{n\theta\}\right), \qquad F_+(\tilde{k}_m) = \tan\left(\frac{\pi}{2}\{m\theta^{-1}\}\right).$$

If θ is badly approximable, the condition (17.20) yields

$$F_+(k_n) > \frac{\pi}{2}\{n\theta\} \geq \frac{\pi}{2}\|n\theta\| > \frac{\pi\delta}{2n}.$$

Furthermore, θ^{-1} is also badly approximable in this case, so we get a similar lower bound for $F_+(\tilde{k}_m)$, possibly with another δ. This means that for α small enough we have $F_+(k_n) > \frac{c}{2k_n}$ and an analogous bound for the tilded quantities, *i.e.* the claim (b).

It is known that δ in (17.20) cannot be too big: by Hurwitz's extension of Dirichlet's theorem one can find for any irrational θ sequences $\{n_r\}$, $\{m_r\}$ such that $|n_r\theta - m_r| < 5^{-1/2}n_r^{-1}$. Moreover, these approximations can be constructed explicitly in terms of truncated continued fractions, and choosing the truncations of even lengths (without the integer part) we obtain a sequence $\{m_r n_r^{-1}\}$ approaching θ from below, so

$$F_+(k_{n_r}) < \frac{\pi}{2}(1+\varepsilon)\{n_r\theta\} < \frac{\pi(1+\varepsilon)}{2\sqrt{5}\,n_r}$$

holds for any $\varepsilon > 0$ and r large enough. In the same way one can approximate θ^{-1}, and by (17.18) we find that there are infinitely many gaps if $\alpha > 5^{-1/2}\pi^2 L^{-1}(1+\varepsilon)$ for any $\varepsilon > 0$, in other words, we obtain the claim (c). Notice that the critical value $5^{-1/2}\pi^2$ cannot decreased as it is saturated for the golden mean, $\theta = \frac{1}{2}(1+\sqrt{5})$, and moreover, there are *no* gaps in this case if $|\alpha|L$ is below this value (see the notes for references giving more details).

Finally let θ be a well approximable irrational. Since the inverseof$\theta = [a_0, a_1, \dots]$ equals $[0, a_0, a_1, \dots]$ if $a_0 \neq 0$ and $[a_1, a_2, a_3, \dots]$ otherwise we may without loss of generality suppose that $\text{limsup}_{n\to\infty} a_{2n} = \infty$. Then the difference in (17.21) is positive, *i.e.*, there is a sequence $\{n_r\}$ such that $n_r\{n_r\theta\} \to 0$ which implies

$$n_r F_+(k_{n_r}) = n_r \tan\left(\frac{\pi}{2}\{n_r\theta\}\right) < \pi n_r \{n_r\theta\} \to 0$$

as $r \to \infty$. Hence $k_{n_r}F_+(k_{n_r}) \to 0$ also holds, and thus there are infinitely many values of k accumulating at infinity for which $kF_+(k) < \frac{1}{2}\alpha$. This concludes the proof for $\alpha > 0$, the argument for the opposite sign is analogous.

Let us end the example with a caution. While the solutions of the condition (17.17) are nonconstant as functions of θ_1 and θ_2 the spectrum of H_α nevertheless may not be absolutely continuous if there are eigenvalues at band edges (Problem 11); this feature distinguishes graph Hamiltonians from the Schrödinger operators of Chapter 14.

Other examples of graph Hamiltonians are mentioned in the notes. It is useful to observe that operators of the type $H_{\alpha,\omega}(V)$ are not the only ones for which the spectral problem can be rephrased as a difference equation. A prominent example are magnetic graph Hamiltonians describing a charged particle on Γ under the influence of a magnetic field characterized by a smooth vector potential A. In analogy with (14.5) such an operator acts on e_{jn} as $\psi_{jn} \mapsto (-i\psi'_{jn} + A_{jn})^2$, where

$A_{jn}(x)$ is the tangent component of A to the edge e_{jn} at the point labelled by x and the particle charge is conventionally assumed to be $e=-1$. The boundary conditions (17.5) have to be modified by replacing $\psi'_{jn}[j]$ by $\psi'_{jn}[j]-iA_{jn}(0)$. One can then check self-adjointness of the operator under similar hypotheses, *e.g.* to those of Theorem 14.1.8 (see the notes).

Theorem 17.3.1 can now be modified without repeating the proof, because the magnetic field can be handled using the unitary operator $U: L^2(\Gamma)\to L^2(\Gamma)$ defined by

$$(U\psi)_{jn}(x) := \exp\left(-ia_{jn}-i\int_0^x A_{jn}(y)\,dy\right)\psi_{jn}(x),$$

where the numbers a_{jn} are chosen to make the functions $(U\psi)_{jn}$ continuous at the vertices. In such a case the latter satisfy (17.5) and the equations (17.14) are replaced by

$$\sum_{n\in\nu(j)\cap I_{\mathcal{I}}}\frac{e^{-iA_n}}{W_{jn}^\alpha}\psi_n-\left(\sum_{n\in\nu(j)}\frac{(v_{jn}^\alpha)'(\ell_{jn})}{W_{jn}^\alpha}-\alpha_j\right)e^{-iA_j}\psi_j=0 \tag{17.22}$$

with the phase factors A_j which have to satisfy the consistency conditions, $A_n-A_j=\int_0^{\ell_{jn}}A_{jn}(y)\,dy$ for any pair of neighboring vertices. In view of the gauge freedom the only important quantities are then integrals of the vector potential over loops of Γ, which are naturally gauge–independent representing magnetic fluxes through these loops.

17.3.4 Example (magnetic lattices): Let Γ be lattice graph of Example 17.3.3, to which we now add a homogeneous magnetic field of intensity B perpendicular to the graph plane. Using the circular gauge, we associate with it the vector potential $A(x,y)=\frac{1}{2}B(-x,y)$ which allows us to choose $A_{nm}:=\Phi(n-m)/2$ where $\Phi:=B\ell_1\ell_2$ is the flux through a graph cell. If no other potential is present, the equation (17.16) is replaced by

$$\begin{aligned}(e^{-i\Phi m/2}\psi_{n,m+1}+e^{i\Phi m/2}\psi_{n,m-1})\sin k\ell_1+(e^{i\Phi m/2}\psi_{n+1,m}+e^{-i\Phi m/2}\psi_{n-1,m})\sin k\ell_2\\ -\left(\frac{\alpha_{nm}}{k}\sin k\ell_1\sin k\ell_2+2\sin k(\ell_1+\ell_2)\right)\psi_{nm}=0\end{aligned} \tag{17.23}$$

The presence of the magnetic field makes the spectral problem more complicated and interesting. In particular, if $\ell_1=\ell_2$ and the coupling constants vanish, $\alpha_{nm}=0$, the spectrum is the same as for the so-called *almost Mathieu equation* (see the notes).

The scattering problem for graphs typically concerns a situation where some number of halflines are attached to a finite compact graph. Since the coupling of these "leads" is realized by boundary conditions at a finite number of points, it is usually easy to establish the existence of the scattering operator. What is more difficult and important is to find the elements of the on–shell scattering matrix, *i.e.*, the reflection and transmission amplitudes at an energy k^2. There are general methods to solve this problem which we mention in the notes; here we limit ourselves to presenting a simple example.

17.3.5 Example (scattering on a line with a stub): Consider Γ consisting of a line to which a segment $J := (0,\ell)$ is attached at $x=0$. Elements of the state Hilbert space $\mathcal{H} := L^2(\mathbb{R})\oplus L^2(J)$ will be written as columns $\psi = \binom{f}{u}$. The particle motion is governed by a Hamiltonian (17.1); we suppose that the motion is free on the line and denote by V the potential on the stub assuming that it belongs to L^∞. Then coupling of the edges is given by the "partially continuous" boundary conditions of Example 17.1.5 with the natural identification $f(x) = \chi_{\mathbb{R}^+}(x)\psi_1(x)+\chi_{\mathbb{R}^-}(x)\psi_2(-x)$ and $u=\psi_3$; since the stub is finite we have to add a boundary condition at its other end which is assumed to be Dirichlet, $u(\ell)=0$. We use the shorthand $\mathcal{K} = \left(\begin{smallmatrix} b & c \\ d & -b \end{smallmatrix}\right)$ for the real-valued matrix of parameters and denote the corresponding self-adjoint extension by $H_{\mathcal{K}}(V)$.

The parameter b plays the role of a coupling constant. If $b=0$ the two parts of Γ are disjoint, on one hand we have motion on the line with δ-interaction of coupling d, while the stub dynamics is described by the Schrödinger operator $h_c := -\frac{d^2}{dx^2}+V$ with boundary condition $u(0)-cu'(0)=0$ at $x=0$ and the Dirichlet one at the other end of J. To find the on-shell scattering matrix in the coupled case we have to solve the Schrödinger equation on Γ using the Ansatz

$$u(x) = \beta u_\ell(x)\,, \qquad f(x) := \begin{cases} e^{ikx}+r(k)\,e^{-ikx} & \dots \quad x<0 \\ t(k)\,e^{ikx} & \dots \quad x>0 \end{cases}$$

where u_ℓ solves $-u''+Vu=k^2u$ with the boundary condition $u_\ell(\ell)=0$, uniquely up to a multiplicative constant. We require that these function satisfies the boundary conditions so the corresponding ψ belongs *locally* to the domain of $H_{\mathcal{K}}(V)$; this yields

$$r(k) = -\,\frac{b^2u_\ell'(0)+d(cu_\ell'-u_\ell)(0)}{b^2u_\ell'(0)+(d-2ik)(cu_\ell'-u_\ell)(0)}\,, \quad t(k) = \frac{-2ik(cu_\ell'-u_\ell)(0)}{b^2u_\ell'(0)+(d-2ik)(cu_\ell'-u_\ell)(0)} \tag{17.24}$$

and one can check that these amplitudes satisfy $|t(k)|^2+|r(k)|^2=1$ (Problem 13b).

The present example also allows us to illustrate resonance effects in quantum graphs. If $b=0$ the stub part has a simple purely discrete spectrum; the eigenvalues are positive and therefore embedded into the continuous spectrum of the line Hamiltonian. When the coupling is switched on, we expect these eigenvalues to turn into resonances.

To show that this is the case we need to find the resolvent of $H_{\mathcal{K}}(V)$. Since all of these operators are self-adjoint extension of the same symmetric operator with deficiency indices $(2,2)$ we can use Theorem 4.7.15, starting from the resolvent of the "free" operator corresponding to the zero matrix which we denote as $\mathcal{K}_0$. The corresponding resolvent decomposes into the line part which is an integral operator with the kernel $R_1(x,y;z) = \frac{i}{2k}\,e^{ik|x-y|}$, where $k := \sqrt{z}$. On the other hand, by Theorem 4.9.7 the stub resolvent kernel is $R_2(x,y;z) = u_0(x_<)u_\ell(x_>)\,W(u_0,u_\ell)^{-1}$, where $x_>$, $x_<$ means respectively the greater (smaller) of x, y and u_0 is a solution satisfying the condition $u_0(0) = 0$. By Krein's formula, the resolvent kernel of $H(\mathcal{K})$ is in our notation a 2×2 matrix,

$$(H(\mathcal{K}) - z)^{-1}(x,y) = (H(\mathcal{K}_0) - z)^{-1}(x,y) + \sum_{j=1,2} \lambda_{jk} F_j(x) F_k(y), \tag{17.25}$$

with F_j chosen, *e.g.*, as $F_1(x) := \binom{R_1(x,y)}{0}$ and $F_2(x) := \binom{0}{u_\ell(x)}$ (we do not indicate the dependence on z if not necessary); the coefficients λ_{jk} are easily found to be

$$\lambda_{11} = \frac{b^2 u_\ell'(0) + d(cu_\ell' - u_\ell)(0)}{D(k)}, \quad \lambda_{12} = \lambda_{21} = \frac{b}{D(k)}, \quad \lambda_{22} = \frac{u_\ell(0)^{-1}}{D(k)} \left(c + i\frac{cd+b^2}{2k} \right),$$

where

$$D(k) := -\frac{ib^2}{2k} u_\ell'(0) - \left(1 + \frac{id}{2k}\right)(cu_\ell'(0) - u_\ell(0))$$

(Problem 13c). Notice that the last expression, which determines the resonances of $H_\mathcal{K}(V)$ through the condition $D(k) = 0$, coincides up to a factor with the denominator of the amplitudes (17.24). This means that in this case the resonances, defined as poles of the continued resolvent, again coincide with the poles of the scattering matrix.

The embedded eigenvalues for $b = 0$ are easily found. Let k_n refer to the n-th eigenvalue of h_c and denote by χ_n the corresponding *normalized* eigenfunction. If the coupling between the two parts of the graph is weak one can check that the condition $D(k) = 0$ has a unique solution in the vicinity of k_n and write an expansion for it,

$$k_n(b) = k_n - \frac{ib^2 \chi_n'(0)^2}{2k_n(2k_n + id)} + \mathcal{O}(b^4), \tag{17.26}$$

which yields an expansion for the imaginary part of the pole position $k_n(b)^2$ (Problem 13d).

If the coupling is not weak, the pole condition can be solved analytically in particular cases only, *e.g.* when the stub motion is free and the decoupled operator is specified by the Dirichlet boundary condition, $V = 0$ and $c = d = 0$. In that case we have $k_n := \frac{\pi n}{\ell}$ and $D(k) = 0$ is reduced to $\tan(k\ell) = -\frac{ib^2}{2}$ which is easily seen to be solved by

$$k_n(b) = \begin{cases} \frac{\pi n}{\ell} + \frac{i}{2\ell} \ln \frac{2-b^2}{2+b^2} & \text{if } \; |b| < \sqrt{2} \\ \frac{\pi(2n-1)}{2\ell} + \frac{i}{2\ell} \ln \frac{b^2-2}{b^2+2} & \text{if } \; |b| > \sqrt{2} \end{cases}$$

Hence the resonance moves down in the momentum plane as $|b|$ increases until they disappear at $|b| = \sqrt{2}$ when the system has *no* resonances. For a stronger coupling constant the poles reappear and move towards the real axis, however, their trajectories are now shifted and end up, in the limit $|b| \to \infty$, at the points corresponding to the decoupled system again, now with Neumann condition at the connection point.

More examples of scattering on graphs are mentioned in the notes.

17.4 Generalized graphs

The quantum graph concept discussed above can be generalized in various ways. A natural possibility is to consider "edges" which are not line segments but manifolds of higher dimension. This looks like an exercise in pure mathematics but in fact such *generalized graphs* can be used to model motion of a quantum particle in situations of practical interest. To mention just two examples, recall scanning microscopy in which a needle almost touches a surface and a tunneling current is measured, or structures composed of spherically shaped fulleren molecules to which carbon nanotubes are attached.

To define Schrödinger operators describing particles in such configuration spaces one has to find conditions replacing (17.4) which would couple wave functions at the "vertices" of the generalized graph. We restrict ourselves here to a brief description of the simplest nontrivial situation, and mention some extensions in the notes.

17.4.1 Example (a halfline attached to a plane): Suppose that the configuration space has two components, a plane and a halfline perpendicular to it. We assume that they have one common point, the endpoint of the halfline, and equip the plane with polar coordinates centered at the junction. The state Hilbert is thus $L^2(\mathbb{R}^-)\oplus L^2(\mathbb{R}^2)$, *i.e.* wave functions are pairs $\phi := \binom{\phi_1}{\Phi_2}$ of square integrable functions. We suppose further that the motion on the two parts of the configuration space is free so the Hamiltonian acts as

$$H\phi = \begin{pmatrix} -\phi_1'' \\ -\Delta\Phi_2 \end{pmatrix}. \tag{17.27}$$

One has to couple the components in such a way that the resulting operator is self-adjoint. We can proceed as in Section 14.6 taking first $C_0^\infty(\mathbb{R}^-\setminus\{0\})\oplus C_0^\infty(\mathbb{R}^2\setminus\{0\})$ as the domain and to look for self-adjoint extensions of such an operator. In the plane part we can perform the partial-wave decomposition; in view of Problem 14.18 only the s-wave part contributes nontrivially, and consequently, the operator (17.27) with the above mentioned domain is symmetric with deficiency indices $(2,2)$.

To construct the self-adjoint extensions one has to properly match the wave functions at the connection point. As we have said one can restrict to the s-wave component in the plane considering functions $\Phi_2(r,\varphi) = (2\pi)^{-1/2}\phi_2(r)$ independent of the azimuthal angle. The appropriate boundary conditions in the plane part will involve generalized boundary values, the same as those appearing in the relations (16.10). The boundary conditions can be written in a form analogous to (17.4), with a pair of 2×2 matrices relating vectors of boundary values. We restrict ourselves to the generic case when one of them can be taken as the unit matrix: for any $a,d\in\mathbb{R}$ and $c\in\mathbb{C}$ the relations

$$\begin{aligned} \phi_1'(0-) &= a\phi_1(0-) + \bar{c}L_0(\Phi_2)\,, \\ L_1(\Phi_2) &= c\phi_1(0-) + dL_0(\Phi_2)\,; \end{aligned} \tag{17.28}$$

determine a self-adjoint operator acting away from the connection point according to (17.27); we denote it by $H_{\mathcal{A}}$ using the abbreviation $\mathcal{A} := \begin{pmatrix} a & \bar{c} \\ c & d \end{pmatrix}$. This claim can be checked by the method of Section 4.7. However, it is not difficult to verify it directly using the boundary form of H_0^*, the adjoint of H_0 obtained by restriction of a "decoupled" operator, say, with $c = 0$, to functions which vanish in the vicinity of the junction. Since the action of H_0^* is given by the differential expression (17.27) we find, after an integration by parts, that the difference $(\phi, H_0^*\psi) - (H_0^*\phi, \psi)$ equals

$$\bar{\phi}_1'(0)\psi_1(0) - \bar{\phi}_1(0)\psi_1'(0) + \lim_{\varepsilon \to 0+} \varepsilon \left(\bar{\phi}_2(\varepsilon)\psi_1'(\varepsilon) - \bar{\phi}_2'(\varepsilon)\psi_2(\varepsilon)\right) .$$

Using the asymptotics $\phi_2(\varepsilon) = \frac{1}{2\pi} L_0(\Phi_2) \ln \varepsilon + L_1(\Phi_2) + \mathcal{O}(\varepsilon)$ as $\varepsilon \to 0$ we compute the limit on the right side to be $L_1(\bar{\Phi}_2)L_0(\Psi_2) - L_0(\bar{\Phi}_2)L_1(\Psi_2)$; it follows easily that $(\phi, H_0^*\psi) = (H_0^*\phi, \psi)$ holds whenever the boundary values satisfy the conditions (17.28) with a Hermitean matrix $\mathcal{A}$.

Properties of the operator $H_{\mathcal{A}}$ depend on the coupling at the contact point. This can be seen in various ways, *e.g.* by investigating scattering for the pair $(H_{\mathcal{A}}, H_0)$. Existence and completeness are easily verified with the on–shell scattering matrix found as in Example 17.3.5. In particular, the reflection and transmission amplitudes for a plane wave e^{ikx} approaching the junction on the halfline are given by

$$r(k) = -\frac{\mathcal{D}_-}{\mathcal{D}_+}, \qquad t(k) = \frac{2ick}{\mathcal{D}_+}, \tag{17.29}$$

where

$$\mathcal{D}_\pm := (a \pm ik)\left[1 + \frac{2i}{\pi}\left(\gamma_E - d + \ln\frac{k}{2}\right)\right] + \frac{i}{\pi^2}|c|^2$$

and $\gamma_E = 0.5772\ldots$ is, as before, Euler's constant (Problem 15). Notice that reflection dominates at high energies, since $\lim_{k\to\infty} t(k) = 0$ holds for any values of the parameters; if $c = 0$ the two parts of the configuration space are decoupled and we return to the situation of Example 7.2.2.

17.5 Leaky graphs

In view of their simplicity and versatility, quantum graph models have become a rather useful tool in solid–state physics and other areas; we mention some references in the notes to Section 17.1. They also have, however, some drawbacks. One is the presence of *ad hoc* parameters in the boundary conditions (17.4) which describe how the wave functions are matched at the vertices; this problem can be addressed in the way described in Section 17.2. On the other hand, graph models also contain an idealization hidden in the assumption that particles are strictly confined to graph edges. While this is often a reasonable approximation when dealing, say, with electrons in semiconductor quantum wires, such a model neglects tunneling effects which can play a role once the wires modeled by graphs edges are placed close to each other.

This motivates us to describe in this section another type of graph models. In contrast to the above considerations we regard the graphs now as subsets of a Euclidean space. For simplicity we restrict ourselves to the simplest situation when

$\Gamma \subset \mathbb{R}^2$ is a planar graph; for higher dimensions we refer to the notes. We are going to consider operators in $L^2(\mathbb{R}^2)$ which can be formally written as

$$-\Delta - \alpha\delta(x-\Gamma) \tag{17.30}$$

with an attractive interaction, $\alpha(x) > 0$; for obvious reasons we speak about *leaky quantum graphs*. It is clear that such a model does not suffer from the deficiencies mentioned above. On the other hand, it is more difficult to analyze because it involves a partial differential operator.

The first question is how to define the operator (17.30) properly. One way is to employ the associated quadratic form, using the following result (see the notes).

17.5.1 Proposition: Suppose that each edge $e_j \in \Gamma$ is the graph of a C^1 function $\gamma_j : I_j \to \mathbb{R}^2$ where I_j is an interval, such that $|\gamma_j'(s)| = 1$. Assume further that at the vertices of Γ the edges meet at nonzero angles and that each compact subset of $\mathbb{R}^2$ contains at most a finite number of edges and vertices of Γ. Then the form

$$q[\psi] := \int_{\mathbb{R}^2} |(\nabla\psi)(x)|^2\, dx - \alpha \sum_{e_j\in\Gamma} \int_{I_j} |\psi(\gamma_j(s))|^2\, ds \tag{17.31}$$

defined on $\mathcal{S}(R^2)$ is closable and bounded from below.

In view of Corollary 4.6.9 there is thus a unique self-adjoint operator, denoted $H_{\alpha,\Gamma}$, associated with the form (17.31); we regard it as the Hamiltonian corresponding to the formal expression (17.30). There are alternative ways to define $H_{\alpha,\Gamma}$, *e.g.* using the appropriate boundary conditions (Problem 26); this approach is illustrative because it shows the relation with the definition (14.28) of a one-dimensional δ interaction making thus clear the meaning of the interaction term in (17.30).

A convenient way to analyze the spectrum of leaky-graph Hamiltonians is to use the Birman–Schwinger method. For a singular interaction one cannot use Lemma 14.3.3 directly, of course. To modify it we use the resolvent $R_0^k = (H_0 - k^2)^{-1}$ of the two-dimensional Laplacian which is, according to Problem 7.18, an integral operator with kernel $G_k(x,y) = \frac{i}{4} H_0^{(1)}(k|x-y|)$. We employ it to define the needed embedding operators. Denoting by $m \equiv m_\Gamma$ the Dirac measure on $\mathbb{R}^2$ associated with Γ we define the map $R_{m,dx}^k : L^2(\mathbb{R}^2) \to L^2(\Gamma)$ as the integral operator with the kernel $G_k(x,y)$, the first variable running over Γ; in a similar way we define the maps $R_{dx,m}^k : L^2(\Gamma) \to L^2(\mathbb{R}^2)$ and $R_{m,m}^k : L^2(\Gamma) \to L^2(\Gamma)$. With this notation we have the following result (see the notes):

17.5.2 Proposition: Suppose that $I - \alpha R_{m,m}^k$ is invertible on $L^2(\Gamma)$ with $\mathrm{Im}\, k > 0$ and the operator

$$R^k := R_0^k + \alpha R_{dx,m}^k [I - \alpha R_{m,m}^k]^{-1} R_{m,dx}^k \tag{17.32}$$

is defined everywhere, which is true, *e.g.*, for $k = i\kappa$ with a large enough positive κ. Then k^2 belongs to the resolvent set $\rho(H_{\alpha,\Gamma})$ and $(H_{\alpha,\Gamma} - k^2)^{-1} = R^k$. Moreover, $\dim\mathrm{Ker}\,(H_{\alpha,\Gamma} - k^2) = \dim\mathrm{Ker}\,(I - \alpha R_{m,m}^k)$ holds for any k with $\mathrm{Im}\, k > 0$. Finally, an eigenfunction of $H_{\alpha,\Gamma}$ associated with such an eigenvalue k^2 is expressed as

$$\psi(x) = \int_0^L R^k_{dx,m}(x,s)\phi(s)\,ds\,, \tag{17.33}$$

where ϕ is the corresponding eigenfunction of $\alpha R^k_{m,m}$ with the eigenvalue one.

The formula (17.32) allows us to find the spectrum of the leaky–graph Hamiltonian $H_{\alpha,\Gamma}$, in analogy with Lemma 14.3.3, by looking for eigenvalues of the corresponding integral operator $R^k_{m,m}$ on $L^2(\Gamma)$. To illustrate how the geometry of the graph influences its spectral properties, let us present a result which is a leaky–graph analogy of Theorem 16.1.4; recall that for the quantum graphs discussed in the previous sections the *shapes* of the edges were of no importance.

Consider first a straight line $\Gamma = \Gamma_0$ being the graph of $\gamma_0(s) = as + b$ for some $a, b \in \mathbb{R}^2$ with $|a| = 1$. In that case the spectral problem can be solved directly by separation of variables; it follows from Theorem 5.7.4 and Problem 14.16c that

$$\sigma(H_{\alpha,\Gamma_0}) = [\,-\tfrac{1}{4}\alpha^2, \infty)\,, \tag{17.34}$$

and the spectrum is purely absolutely continuous. Suppose now that Γ is not straight but merely asymptotically straight. Let us state this hypotheses more precisely. Since it is parametrized by arc length the corresponding piecewise C^1 function γ satisfies $|\gamma(s) - \gamma(s')| \le |s - s'|$; we suppose that the inequality is sharp for some $s, s' \in \mathbb{R}$. On the other hand we assume that there is $c \in (0,1)$ such that $|\gamma(s) - \gamma(s')| \ge c|s - s'|$ excluding, in particular, self-intersections and parallel asymptotes. Finally, the asymptotic straightness means that there are positive d, $\mu > \frac{1}{2}$, and $\omega \in (0,1)$ such that the inequality

$$1 - \frac{|\gamma(s) - \gamma(s')|}{|s - s'|} \le d\left[1 + |s + s'|^{2\mu}\right]^{-1/2} \tag{17.35}$$

holds true in the sector $S_\omega := \left\{(s,s') : \omega < \frac{s}{s'} < \omega^{-1}\right\}$ (see also Problem 17). Under these assumptions we have the following result:

17.5.3 Theorem: For any $\alpha > 0$ the essential spectrum is preserved under the deformation, $\sigma_{\mathrm{ess}}(H_{\alpha,\Gamma}) = \left[-\frac{1}{4}\alpha^2, \infty\right)$. On the other hand, the operator $H_{\alpha,\Gamma}$ has in addition at least one isolated eigenvalue below $-\frac{1}{4}\alpha^2$.

Proof: We will describe the main steps of the argument leaving the details to the reader (Problem 17). Since for any $r > 0$ there is a circle of radius r which does not intersect Γ it is not difficult to construct Weyl sequences for $H_{\alpha,\Gamma}$ showing that any non–negative number belongs to $\sigma_{\mathrm{ess}}(H_{\alpha,\Gamma})$. To deal with the negative part we employ Proposition 17.5.2: the idea is to treat the difference between the operator $\mathcal{R}^\kappa_{\alpha,\Gamma} := \alpha R^{i\kappa}_{m,m}$ on $L^2(\mathbb{R})$ and its counterpart corresponding to Γ_0 as a perturbation. The integral kernel of said operator is

$$\mathcal{R}^\kappa_{\alpha,\Gamma}(s,s') := \frac{\alpha}{2\pi}\, K_0\left(\kappa|\gamma(s) - \gamma(s')|\right)\,,$$

where K_0 is the Macdonald (*i.e.*, modified Bessel) function; for $\Gamma = \Gamma_0$ one has to replace $|\gamma(s)-\gamma(s')|$ by $|s-s'|$. In the last named case the operator is of convolution type and using Problem 7.17 it is easy to check that its spectrum is absolutely continuous covering the interval $[0, \alpha/2\kappa]$ in correspondence with (17.34).

The key observation is that the kernel of $\mathcal{D}_\kappa := \mathcal{R}^\kappa_{\alpha,\Gamma} - \mathcal{R}^\kappa_{\alpha,\Gamma_0}$ satisfies

$$\mathcal{D}_\kappa(s,s') := \frac{\alpha}{2\pi}\Big(K_0\left(\kappa|\gamma(s)-\gamma(s')|\right) - K_0\left(\kappa|s-s'|\right)\Big) \ge 0 \tag{17.36}$$

in view of $|\gamma(s)-\gamma(s')| \le |s-s'|$ and the monotonicity of K_0, and the inequality is sharp for at least some values of the variables s, s'. The argument then proceeds in three steps:

(i) A variational estimate in combination with (17.36) shows that the spectrum is "pushed up" by the perturbation, $\sup\sigma\left(\mathcal{R}^\kappa_{\alpha,\Gamma}\right) > \frac{\alpha}{2\kappa}$ if Γ is not straight; it is sufficient to take a trial function which is sufficiently smeared.

(ii) The inequality (17.35) allows us to estimate the Hilbert–Schmidt norm of $\mathcal{D}_\kappa$ to check that the operator is compact. Consequently, the spectrum of $\mathcal{R}^\kappa_{\alpha,\Gamma}$ above $\frac{\alpha}{2\kappa}$, nonempty in view of (i), consists of isolated eigenvalues of finite multiplicity.

(iii) Finally, the map $\kappa \mapsto \mathcal{R}^\kappa_{\alpha,\Gamma}$ is operator–norm continuous and $\mathcal{R}^\kappa_{\alpha,\gamma} \to 0$ as $\kappa \to \infty$; hence any eigenvalue of $\mathcal{R}^\kappa_{\alpha,\Gamma}$ regarded as a function of κ must cross the value one at a point to the right of $\frac{\alpha}{2}$.

This gives, in view of Proposition 17.5.2, the sought result. ∎

Using Theorem 14.2.1 we arrive at the following easy consequence:

17.5.4 Corollary: Suppose that a graph Γ has a subgraph in the form of an infinite curve satisfying the assumptions of the preceding theorem, and $\sigma_{\mathrm{ess}}(H_{\alpha,\Gamma}) = \left[-\frac{1}{4}\alpha^2, \infty\right)$, then the discrete spectrum of $H_{\alpha,\Gamma}$ is non-empty.

It is important that the assumption about preservation of the essential spectrum can be often verified easily, for instance, in the situation when the graph Γ has outside a compact set a finite number of straight edges which are not mutually parallel; an example can be found in Problem 18.

The fact that the geometry of Γ has implications for the spectrum of $H_{\alpha,\Gamma}$ is manifested also in other ways. Recall that the relation between the shape of a region and the corresponding spectral properties is a problem frequently occurring in mathematical physics. The well–known *inequality of Faber and Krahn* says, for instance, that among all regions Ω of a fixed volume, the ground–state eigenvalue of the Dirichlet Laplacian $-\Delta^\Omega_D$ is sharply minimized by a ball.

We will now show that the situation here is in a sense opposite. If Γ is a finite curve it is not difficult to see that $\sigma_{\mathrm{ess}}(H_{\alpha,\Gamma}) = \mathbb{R}^+$ and $\lambda_1(\alpha,\Gamma) := \inf\sigma\left(H_{\alpha,\Gamma}\right) < 0$. Suppose now that Γ is a loop of a fixed length, *i.e.*, a closed curve, and ask which shape of it makes $\lambda_1(\alpha,\Gamma)$ *maximal.* To state the assumptions, let $\gamma : [0,L] \to \mathbb{R}^2$ be a C^2 smooth function with $\gamma(0) = \gamma(L)$ which generates the loop Γ. We introduce the equivalence relation: $\Gamma \sim \Gamma'$ if one can be obtained from the other by a Euclidean transformation of the plane; spectral properties of the corresponding operators $H_{\alpha,\Gamma}$ and $H_{\alpha,\Gamma'}$ are obviously the same, hence we will speak about a curve Γ having in mind the corresponding equivalence class. The assumptions are satisfied, in particular, by the circle, say $\mathcal{C} := \{(\frac{L}{2\pi}\cos\frac{2\pi s}{L}, \frac{L}{2\pi}\sin\frac{2\pi s}{L}) : s\in[0,L]\}$, and its equivalence class. Then we have the following result:

17.5.5 Theorem: Within the described class of loops, the eigenvalue $\lambda_1(\alpha,\Gamma)$ is, for any fixed $\alpha>0$ and $L>0$, sharply maximized by the circle.

The proof is based on a geometric lemma about means of chords (Problem 19).

17.5.6 Lemma: Let Γ be as above, then for $p\in(0,2]$ and $u\in(0,\frac12 L]$ we have

$$\int_0^L |\gamma(s+u)-\gamma(s)|^p\,ds \le \frac{L^{1+p}}{\pi^p}\sin^p\frac{\pi u}{L},$$

where the right side is the value of the integral for the circle of radius $L/2\pi$.

Proof of Theorem 17.5.5: We shall rely again on Proposition 17.5.2 which relates our eigenvalue problem to the integral equation $\mathcal{R}^\kappa_{\alpha,\Gamma}\phi=\phi$ on $L^2([0,L])$ with $\mathcal{R}^\kappa_{\alpha,\Gamma}$ defined as in the proof of Theorem 17.5.3. We note that the operator-valued function $\kappa\mapsto\mathcal{R}^\kappa_{\alpha,\Gamma}$ is strictly decreasing in $(0,\infty)$ and $\|\mathcal{R}^\kappa_{\alpha,\Gamma}\|\to 0$ as $\kappa\to\infty$; the same is then, by the minimax principle, true for any eigenvalues of $\mathcal{R}^\kappa_{\alpha,\Gamma}$.

Next we need information about the ground-state eigenfunction. Despite the fact that the potential is singular we still have a result analogous to Theorem 14.3.12, namely that the ground state of $H_{\alpha,\Gamma}$ is simple (see the notes) and by Proposition 17.5.2 the same is true for the maximum eigenvalue of $\mathcal{R}^\kappa_{\alpha,\Gamma}$. Furthermore, if Γ is a circle then the ground-state eigenfunction of $H_{\alpha,\Gamma}$ exhibits rotational symmetry with respect to its center, and by (17.33) the respective eigenfunction of $\mathcal{R}^{\tilde\kappa_1}_{\alpha,\mathcal{C}}$ corresponding to the unit eigenvalue is constant, $\tilde\phi_1(s)=L^{-1/2}$. Then we have

$$\max\sigma(\mathcal{R}^{\tilde\kappa_1}_{\alpha,\mathcal{C}}) = (\tilde\phi_1, \mathcal{R}^{\tilde\kappa_1}_{\alpha,\mathcal{C}}\tilde\phi_1) = \frac1L\int_0^L\int_0^L \mathcal{R}^{\tilde\kappa_1}_{\alpha,\mathcal{C}}(s,s')\,dsds',$$

while for a general loop Γ a simple variational estimate gives

$$\max\sigma(\mathcal{R}^{\tilde\kappa_1}_{\alpha,\Gamma}) \ge (\tilde\phi_1, \mathcal{R}^{\tilde\kappa_1}_{\alpha,\Gamma}\tilde\phi_1) = \frac1L\int_0^L\int_0^L \mathcal{R}^{\tilde\kappa_1}_{\alpha,\Gamma}(s,s')\,dsds';$$

hence to check that the circle is a maximizer it sufficient to show that

$$\int_0^L\int_0^L K_0(\kappa|\Gamma(s)-\Gamma(s')|)\,dsds' \ge \int_0^L\int_0^L K_0(\kappa|\mathcal{C}(s)-\mathcal{C}(s')|)\,dsds'$$

holds for *any* $\kappa>0$ and Γ of the considered curve class. By a simple change of variables we find that this is equivalent to positivity of the functional

$$F_\kappa(\Gamma) := \int_0^{L/2} du\int_0^L ds\Big[K_0(\kappa|\Gamma(s+u)-\Gamma(s)|) - K_0(\kappa|\mathcal{C}(s+u)-\mathcal{C}(s)|)\Big],$$

where the second term is easily seen to be equal to $K_0(\frac{\kappa L}{\pi}\sin\frac{\pi u}{L})$. Next we employ the fact that the Macdonald function K_0 is strictly convex in $(0,\infty)$ which means that the integral can be estimated from below as follows,

$$\frac{1}{L}F_\kappa(\Gamma) \geq \int_0^{L/2}\left[K_0\left(\frac{\kappa}{L}\int_0^L |\Gamma(s+u)-\Gamma(s)|ds\right) - K_0\left(\frac{\kappa L}{\pi}\sin\frac{\pi u}{L}\right)\right] du\,,$$

and the inequality is sharp unless $|\Gamma(s+u)-\Gamma(s)|\,ds$ is independent of s. Finally, we note that K_0 is decreasing in $(0,\infty)$, hence the positivity of the expression in the square bracket follows from Lemma 17.5.6 with $p=1$. ∎

So far it is not clear whether there is a relation between the leaky graphs discussed here and the quantum graphs of the previous sections. To illuminate this point, consider first the case when Γ is a straight line; using separation of variables and Problem 14.16c we see that the states in the negative part of the spectrum are exponentially localized around the line, their transverse spread being of order $\mathcal{O}(\alpha^{-1})$. This motivates us to look at the behavior of leaky-graph Hamiltonians as the coupling constant $\alpha\to\infty$. Consider a single finite edge which is sufficiently smooth; then we have the following result.

17.5.7 Theorem: Suppose that $\gamma:\,[0,L]\to\mathbb{R}^2$ is a C^4 smooth function, $|\gamma'|=1$, which defines a curve Γ; then $\operatorname{card}\sigma_{\mathrm{disc}}(H_{\alpha,\Gamma}) = \frac{\alpha L}{2\pi} + \mathcal{O}(\ln\alpha)$ holds as $\alpha\to\infty$. In addition, if Γ is a closed curve without self-intersections, then the j-th eigenvalue of the operator $H_{\alpha,\Gamma}$ behaves asymptotically as

$$\lambda_j(\alpha) = -\frac{1}{4}\alpha^2 + \mu_j + \mathcal{O}(\alpha^{-1}\ln\alpha)\,,$$

where μ_j is the j-th eigenvalue of the operator $S_\Gamma := -\frac{d}{ds^2} - \frac{1}{4}k(s)^2$ on $L^2(0,L)$ with periodic b.c., counted with multiplicity, and $k(\cdot)$ is the signed curvature of Γ.

Proof: Suppose first that Γ is closed and without self-intersections, and consider its strip neighborhood analogous to the waveguides of Section 16.1, in other words, the set Σ_a onto which the function $\Phi_a:\,[0,L)\times(-a,a)\to\mathbb{R}^2$ defined in analogy with (16.2) maps diffeomorphically for all $a>0$ small enough. The main idea is to apply to $H_{\alpha,\Gamma}$ the bracketing of Proposition 14.5.3 at the boundary of Σ_a,

$$(-\Delta^{\mathrm{N}}_{\Lambda_a})\oplus L^-_{a,\alpha} \leq H_{\alpha,\Gamma} \leq (-\Delta^{\mathrm{D}}_{\Lambda_a})\oplus L^+_{a,\alpha}\,,$$

where $\Lambda_a = \Lambda_a^{\mathrm{in}}\cup\Lambda_a^{\mathrm{out}}$ is the region exterior to the strip, and $L^\pm_{a,\alpha}$ are self-adjoint operators associated with the closures of the forms

$$q^\pm_{a,\alpha}[\psi] = \|\nabla\psi\|^2_{L^2(\Sigma_a)} - \alpha\int_\Gamma |\psi(\gamma(s))|^2\,ds\,,$$

where $f\in C_0^\infty(\Sigma_a)$ and $(C^\infty\cap L^2)(\Sigma_a)$ for the $\pm$ sign, respectively. It is important that the exterior region $\mathbb{R}^2\setminus\overline{\Sigma}_a$ does not contribute to the negative part of the spectrum, so it is sufficient to consider the operators $L^\pm_{a,\alpha}$ only.

We use the curvilinear coordinates (s,u) analogous to those of (16.2) and pass from $L^{\pm}_{a,\alpha}$ to unitarily equivalent operators given by closures of the quadratic forms

$$b^{+}_{a,\alpha}[\psi] = \int_0^L \int_{-a}^{a} (1+uk(s))^{-2} \left|\frac{\partial\psi}{\partial s}\right|^2 (s,u)\,du\,ds + \int_0^L \int_{-a}^{a} \left|\frac{\partial\psi}{\partial u}\right|^2 (s,u)\,du\,ds$$
$$+ \int_0^L \int_{-a}^{a} V(s,u)|\psi(s,u)|^2\,ds\,du - \alpha \int_0^L |\psi(s,0)|^2\,ds$$

with $\psi \in C^\infty((0,L)\times(-a,a))$ satisfying periodic boundary conditions in the variable s and Dirichlet b.c. at $u=\pm a$, and its counterpart

$$b^{-}_{a,\alpha}[\psi] = b^{+}_{a,\alpha}[f] - \frac{1}{2}\sum_{j=0}^{1}(-1)^j \int_0^L \frac{k(s)}{1+(-1)^j ak(s)}\,|\psi(s,(-1)^j a)|^2\,ds\,,$$

where V is the curvature–induced potential of (16.4) with $\gamma(s)$ replaced by $k(s)$. The variables s and u are still coupled here, hence it is useful to pass to rougher bounds squeezing $H_{\alpha,\Gamma}$ between the closures of $\tilde H^{\pm}_{a,\alpha} = A^{\pm}_a \otimes 1 + 1 \otimes T^{\pm}_{a,\alpha}$, where $A^{\pm}_a$ are self-adjoint operators on $L^2(0,L)$ given by

$$A^{\pm}_a = -(1\mp a\|k\|_\infty)^{-2}\frac{d^2}{ds^2} + V_{\pm}(s)$$

with periodic b.c., where $V_-(s) \le V(s,u) \le V_+(s)$ with an $\mathcal{O}(a)$ error as $a\to 0$, and the transverse operators are associated with the forms

$$\begin{aligned} t^{+}_{a,\alpha}[f] &= \int_{-a}^{a} |f'(u)|^2\,\mathrm{d}u - \alpha|f(0)|^2 \\ t^{-}_{a,\alpha}[f] &= t^{-}_{a,\alpha}[f] - \|k\|_\infty(|f(a)|^2+|f(-a)|^2)\,, \end{aligned}$$

where $f \in C_0^\infty(-a,a)$ and $(C^\infty \cap L^2)(-a,a)$ for the $\pm$ sign, respectively. Their negative spectrum can be localized with an exponential precision: there is a $c>0$ such that $T^{\pm}_{\alpha,a}$ has for α large enough a single negative eigenvalue $\kappa^{\pm}_{\alpha,a}$ satisfying

$$-\frac{\alpha^2}{4}\left(1+c\,\mathrm{e}^{-\alpha a/2}\right) < \kappa^{-}_{\alpha,a} < -\frac{\alpha^2}{4} < \kappa^{+}_{\alpha,a} < -\frac{\alpha^2}{4}\left(1-8\mathrm{e}^{-\alpha a/2}\right) \qquad (17.37)$$

(Problem 20). To complete the proof one has to check that the eigenvalues of $A^{\pm}_a$ differ by $\mathcal{O}(a)$ from those of the comparison operator, then we choose $a = 6\alpha^{-1}\ln\alpha$ as the width of the neighborhood Σ_a. Putting the estimates together we get the eigenvalue asymptotic formula; we leave the details to the reader (Problem 20). If Γ is not closed, a similar argument can be used with the comparison operators $S^{\mathrm{D,N}}_\Gamma$ having the appropriate b.c., Dirichlet or Neumann, at the endpoints of Γ. This gives the estimate on card $\sigma_{\mathrm{disc}}(H_{\alpha,\Gamma})$, concluding the proof. ∎

In a similar way one can treat the strong–coupling asymptotics for other curve classes and analogous problems in higher dimensions. At the same time, various questions remain open here; we give some references in the notes.

Notes to Chapter 17

Section 17.1 The idea to study quantum mechanics on graphs is not new. A proposal to model organic molecules by considering electrons living on the molecule "skeleton" was formulated by L. Pauling in 1936 and worked out in [RuS 1]; for an early review see [EŠ 5]. Then it was mostly forgotten, however, and reappeared only in the 1980's when experiments with microscopic structures became possible and various interesting properties of them, in particular, concerning the transport in presence of electric or magnetic fields, were revealed — see, *e.g.*, [Bü 1], [WWU 1], [ARZ 1], [EŠŠ 3], [AAAS 1], [GP 1]. Since then the subject has been intensively investigated — *cf.* [Kuch 1], ⟦ EKKST ⟧ for recent reviews and an extensive bibliography.

The construction of admissible Hamiltonians for quantum graphs was proposed in [EŠ 6], where also some examples of boundary conditions were given. The general form (17.4) of the boundary conditions was formulated in [KoS 1], and the unique characterization of Proposition 17.1.2 was proposed independently in [Harm 1] and [KoS 2], as well as in [FT 1] in the particular case of the generalized point interaction, $n = 2$; the trick used in the proof of Proposition 17.1.2 comes from the last named paper.

The most frequently used vertex couplings are the free boundary conditions, which are also historically the first having been introduced in [RuS 1]. They are often called *Kirchhoff conditions* (sometimes also *Neumann*). The first name alludes to classical electrical circuits, however, it is misleading in a sense, because the physical meaning of the condition $\mathcal{B}(\psi, \psi) = 0$ is that the probability current is preserved at the vertex which is true for any extension, not just for $U := \frac{2}{n}\mathcal{J} - I$ associated with the "Kirhhoff case". Among the other boundary conditions discussed here the δ–coupling, the permutation–invariant one, and the coupling with partially continuous wave functions of Example 17.1.5 were introduced in [EŠ 6], δ'_s in [Ex 10] and δ_p of Problem 4b in [CE 1]; for further examples of vertex couplings see [KoS 1].

Section 17.2 Theorem 17.2.1 is a particular case of a result derived in [KuZ 1], similar approximations were worked out in [RSch 1] and [Sai 1]. For further extensions to "fat graphs" as manifolds of any dimension see [EPo 1]; in this paper the situation when the fat edges and vertices have different scaling rates is also analyzed, however, it again basically leads to free boundary conditions. A stronger convergence which applies not only to the discrete spectrum is derived in [Pos 2] and convergence of resonances in [EPo 2]. A common feature of all these results is that the approximating operators have Neumann boundary, or none at all, so the lowest transverse eigenfunction is constant.

Physically more interesting is the case of Dirichlet boundary where, naturally, the energy has to be renormalized because the whole spectrum blows up as the tube diameters tend to zero. A natural remedy is to subtract the diverging lowest transverse eigenvalue. The situation differs substantially from the Neumann case and for most values of the parameters the limit leads to the trivial situation of the graph with decoupled edges see [Pos 1] and also [DAT 1]. A nontrivial outcome, however, is possible if the fat graph system has a resonance at the spectral threshold [ACF 1], [CaE 1], [Gri 1]. Moreover, a generically nontrivial limit is obtained if the renormalization is done not with respect to the first transverse eigenvalue but rather at a point higher in the spectrum [MV 1].

Lemma 17.2.2 and Theorem 17.2.3, which shows how a δ–coupling can be obtained using a squeezed potential on the graph, come from [Ex 11]. The resolvent formula also yields a weak–coupling expansion for the free boundary conditions (Problem 7) and a modified Birman–Schwinger bound similar to that mentioned in the notes to Section 14.3. Theorem 17.2.4 is taken from [CE 1], an extension of this result to the permutation–symmetric b.c. of Proposition 17.1.4 and other vertex couplings can be found in [ETu 1].

Section 17.3 Theorem 1 and its analogue for δ'_s interactions as in Problem 9 come from [Ex 12] where the reader can also find the proof of parts (b) and (c) based on properties of Sturm–Liouville operators. An introduction to the theory of Jacobi matrices can be found in ⟦Te⟧. The rectangular–lattice graph model of Example 17.3.3 was proposed in [Ex 9], [Ex 10] together with its δ'_s–counterpart (Problem 12); further spectral properties of these Hamiltonians were discussed in [EG 1]. Basic notions and results from the number theory used in the example can be found, *e.g.*, in ⟦HW⟧. For eigenvalues of Problem 11 see [Kuch 2]; their existence demonstrates that graph Hamiltonians need not satisfy the *unique continuation principle* which says that if a solution to a Schrödinger equation is zero in an open set then it vanishes everywhere — *cf.* ⟦RS 4⟧, App. to Sec.XIII.13.

Another interesting class are the *tree graphs*, *i.e.*, those which have no loops; they are called *rooted* if they sprout from a single vertex and *regular* if all the vertex degrees are the same and the edges of the same generation (distance from the root) have the same length. The spectrum of the corresponding graph Hamiltonians can have different forms and natures depending on the geometry of such a Γ even if we suppose free boundary conditions at the vertices — see [Solo 1] and references therein.

Magnetic graph Hamiltonians were introduced in [ARZ 1], self–adjointness for general boundary conditions is discussed in [KoS 4]. The equations (17.23) for a square lattice and $\alpha_{nm} = 0$ correspond to the discrete magnetic Laplacian [Shu 1], also called the Harper operator [Harp 1]. Its spectrum coincides, for any irrational $\phi = \frac{1}{2\pi}\Phi$, with that of the almost Mathieu operator with $\lambda = 1$; the latter acts on $\ell^2(\mathbb{Z})$ as

$$(H_{\phi,\theta,\lambda}\psi)_n := \psi_{n+1} + \psi_{n-1} + \lambda\cos(2\pi\phi n + \theta)\psi_n\,,$$

see [Shu 1]. The dependence of the spectrum on ϕ represents a complicated pattern known as the *Hofstadter butterfly* [Hof 1]. It was proved recently that the spectrum for irrational ϕ is a Cantor set — see [AJi 1] where references to related results can also be found. On the other hand, spectra of square lattice graphs with a magnetic field also contain a point part — see [BGP 1] for a thorough spectral analysis of such systems.

There are various general methods to treat scattering on metric graphs, *e.g.*, a difference–equation reformulation similar to Theorem 17.3.1 — see [Ex 13]. Another idea is based on the observation that, in view of their one–dimensional character, solutions of the Schrödinger equation have a known form on the whole leads, not only asymptotically. This makes it possible to decompose Γ into suitable subgraphs and construct the S–matrix in a factorized form through the scattering quantities of the subgraphs [KoS 3]. In the particular case of *finitely periodic* graphs (also called serial structures) there is a simple explicit formula for the reflection and transmission amplitudes [ETV 1].

Numerous simple cases of scattering on graphs, interpreted as models of nanostructures, are discussed in the physical literature. Example 17.3.5 comes from [EŠe 1], where pole trajectories for other potentials and couplings are also investigated. Extension to serial stub structures can be found in [ETV 1]. There one can also find a discussion of scattering on graph loops under the influence of external fields which is one of the most intensively studied systems, see [EŠŠ 3] and recall that micron rings were among the first systems in which the *Aharonov–Bohm effect*, *i.e.*, dependence of quantum mechanical observable quantities on loop integrals of the vector potential, was demonstrated [CRWP 1].

Section 17.4 Example 17.4.1, coming from [EŠ 4], gives the simplest nontrivial example of such a generalized graph. The method of self–adjoint extensions allows us to construct many other Schrödinger–type operators on configuration spaces composed of sets of a different dimensionality provided that the dimensions of the adjacent "edges" do not differ by more than two. For such models with applications to so–called point–contact spectroscopy experiments see [EŠ 3–5]. One may also describe a plane with a "lead" for an electron with spin-orbit interaction [EŠ 8]. In a similar way, one can model the Helmholz resonator, *i.e.*, a cavity with a small opening [Pop 1], a point interaction with an internal structure [Pav 1] , *etc.* Still another model of this type was discussed in Section 15.4.

As in the previous section one can ask how to choose parameters in the boundary conditions (17.28). The problem is open in general, a heuristic comparison of the scattering derived in Example 1 with that for a thin cylinder of radius ρ attached to the plane suggests that a suitable choice could be $a = (2\rho)^{-1}$, $c = (2\pi\rho)^{-1}$, and $d = -\ln\rho$; using these values one can successfully model microwave resonators with a thin antenna [EŠ 9].

One can consider more complicated scattering problems. The case of a sphere with two leads attached was investigated in [Kis 1], [ETV 1] and [BGMP 1]. The picture is now more complicated because of resonances resulting from the discrete spectrum of the sphere, however, at high energies reflection again dominates for most parameter values. A general theory of scattering on such "hedgehog" manifolds was formulated in [BG 1]. One can also study periodic systems of spheres connected through line segments or contact points. The fact that the passage through such contacts becomes more difficult at high energies is manifested by an asymptotic dominance of gaps in the spectrum, cf. [BEG 1], in analogy with the δ' Kronig–Penney model mentioned in the notes to Section 14.6.

Section 17.5 Singular interactions supported by manifolds of codimension one were studied in particular cases in the papers mentioned in the notes to Section 14.6. The application to the investigation of leaky graphs is a new topic studied in recent years. A summary of this work is contained in the review [Ex 15] where the reader can also find many other results such as geometric perturbations, scattering theory, *etc.*, as well as extensions to higher dimensions. A way to prove Propositions 17.5.1 and 17.5.2 is shown in [BEKŠ 1], the last claim of Proposition 17.5.2, the relation (17.33), comes from [Posi 1]. Theorem 17.5.3 is taken from the paper [EI 1]. Similar results about curvature–induced binding can be derived for singular Schrödinger operators with interactions supported by curves and surfaces in $\mathbb{R}^3$ — *cf.* [Ex 15] for references.

Theorem 17.5.5 comes from the paper [EHL 2]; it is illustrative to compare it with the situation considered in Problem 16.12. The geometric result employed in the proof, Lemma 6, was originally proved in a different way in [Lü 1], see also [ACFGH 1]. For negative powers, $p \in [-2, 0)$, an analogous reverse inequality is valid; the particular case $p = -1$ has applications in electrostatics [ACFGH 1], [EHL 2]. An alternative argument showing that the circle maximizes the ground–state energy *locally* can be found in [Ex 16]. This isoperimetric problem has a point–interaction counterpart: if we have N point interactions in $\mathbb{R}^d$, $d = 2, 3$, cyclically numbered, of the same coupling constant, and such that the neighbor distances are $\leq \ell$ for some $\ell > 0$, the corresponding N–center Hamiltonian (see Section 14.6) achieves the maximum ground–state energy if the point interactions are placed at the vertices of a regular planar polygon [Ex 17].

Theorem 17.5.7 was proved in [EY 1]. The asymptotic formula also holds for bounds associated with infinite curves according to Theorem 17.5.3 provided they are sufficiently smooth, and similar strong–coupling asymptotic expansions are valid in higher dimensions, see again [Ex 15]. The situation becomes more complicated if the curve Γ is not smooth or it has endpoints. For a finite curve in $\mathbb{R}^2$ which is not closed the argument from the proof of Theorem 17.5.7 can be used again, with S replaced by the operators having Dirichlet and Neumann b.c. at $s = 0, L$, giving the same asymptotics for the number of bound states. The estimates are too rough, however, to yield the eigenvalue behavior; a conjecture is that the appropriate comparison operator is the one with Dirichlet boundary conditions but the proof is missing. Furthermore, the asymptotics of Theorem 17.5.7 is not valid for a non–smooth curve, for instance, one having a point where the tangents at the left and right meet at a nonzero angle, and the same is true for the case when Γ is a nontrivial graph with branching points. To find these asymptotic expansions is an open and difficult problem, analogous to limits of "fat" Dirichlet graphs mentioned in the notes to Section 17.2.

Problems

1. Prove Proposition 17.1.1 by showing that a pair of $n \times n$ matrices A, B such that the $n \times 2n$ matrix (A, B) has maximal rank defines a Lagrangean subspace through (17.4) *iff* the product AB^* is Hermitean.

2. Check directly that the matrices $A = (U - I)$ and $B = i(U + I)$ with a unitary U have the properties stated in Proposition 17.1.1, and conversely, that any such matrix pair can be expressed as $A = V(U - I)$, $B = iV(U + I)$ where V is a regular $n \times n$ matrix.

3. Suppose that a unitary matrix U characterizes a self–adjoint extension by the boundary conditions (17.4) with $A = (U - I)$ and $B = i\ell(U + I)$ with $\ell > 0$. Then for any $\ell' > 0$ the same extension is described by $A' = (U' - I)$ and $B = i\ell'(U' + I)$ with $U' := [(\ell + \ell')U + \ell - \ell'][(\ell - \ell')U + \ell + \ell']^{-1}$.

4. Analyze the star–graph Hamiltonians H_U discussed in Section 17.1.

(a) Check the claims made in Examples 17.1.3.

(b) Show that the δ_p*–coupling* given by the boundary conditions

$$\sum_{j=1}^{n} \psi_j(0+) = 0\,, \quad \psi_j'(0+) - \psi_k'(0+) = \frac{\alpha}{n}(\psi_j(0+) - \psi_k(0+))\,,\; j,k = 1,\dots,n\,,$$

with $\alpha \in \mathbb{R} \cup \{\infty\}$ corresponds to $U = \frac{n-i\alpha}{n+i\alpha} I - \frac{2}{n+i\alpha}\mathcal{J}$.

(c) Find the spectra of these operators, in particular, the possible eigenvalues and their multiplicities.

(d) Show that the on–shell scattering matrix at energy k^2 for the pair (H_U, H_D) is given by $S_U(k) = [k - 1 + (k + 1)U][k + 1 + (k - 1)U]^{-1}$. Calculate this explicitly for the listed operators H_U and analyze the behavior of the reflection and transmission amplitudes at high energies, $k \to \infty$.

5. Find the matrix U corresponding to the boundary conditions of Example 17.1.5.
Hint: Writing $\psi(0+) = M\psi'(0+)$ we have $U = (M-i)(M+i)^{-1}$.

6. Consider the operators $H_\alpha(V)$ of Section 17.2.

   (a) Show that $H_\alpha(V)$ is self–adjoint for the class of potentials $\{V_j\}$ specified there.

   (b) Fill in the details in the proof of Lemma 17.2.2.

   (c) Analyze the spectrum of $H_\alpha(V)$ and of the analogous operator on a star graph with edges of finite lengths ℓ_j at the endpoints of which the boundary conditions $\psi(\ell_j-)\cos\omega_j + \psi'(\ell_j-)\sin\omega_j = 0$ are imposed.

   (d) Fill in the details in the proof of Theorem 17.2.3.

7. Consider the operator $H_0(gV)$ on a star–shaped Γ of n halfline edges with the potentials $V_j \in L^1(\mathbb{R}^+, (1+|x|)\,dx)$, $j = 1,\dots,n$. Show that for all sufficiently small $g>0$ this operator has a single negative eigenvalue $\lambda(g) = -\kappa(g)^2$ *iff* the condition $\sum_{j=1}^n \int_0^\infty V_j(x)\,dx \le 0$ is satisfied, and in that case its weak–coupling asymptotic behavior is given by

$$\begin{aligned}\kappa(g) \;&=\; -\frac{g}{n}\sum_{j=1}^n \int_0^\infty V_j(x)\,dx \;-\; \frac{g^2}{2n}\Bigg\{\sum_{j=1}^n \int_0^\infty\!\!\int_0^\infty V_j(x)|x-y|V_j(y)\,dx\,dy \\ &+\; \sum_{i,j=1}^n \left(\frac{2}{n}-\delta_{ij}\right)\int_0^\infty\!\!\int_0^\infty V_i(x)(x+y)V_j(y)\,dx\,dy\Bigg\} + \mathcal{O}(g^3)\,.\end{aligned}$$

Hint: By the Birman–Schwinger principle an eigenvalue of $H_0(gV)$ exists *iff* $K := |V|^{1/2}(H_0(0)-k^2)^{-1}V^{1/2}$ has eigenvalue $-g^{-1}$. Write $K = P_\kappa + Q_\kappa$ with $\kappa := -ik$ where the first term contains the pole singularity, and use the identity $(I+gK)^{-1} = [I + g(I+gP_\kappa)^{-1}Q_\kappa]^{-1}(I+gP_\kappa)^{-1}$ to cast the spectral condition in the form $(\psi_g, \phi_g) = -1$, where $\psi_g := \frac{g}{\kappa n} e^{-\kappa\cdot}V^{1/2}$ and $\phi_g := (I+gP_\kappa)^{-1}e^{-\kappa\cdot}|V|^{1/2}$.

8. Prove Theorem 17.2.4.
Hint: Express the resolvents in question by Theorems 4.7.15 and 4.9.7, prove the pointwise convergence of the kernel difference, and estimate the Hilbert–Schmidt norm of the corresponding operator.

9. Prove the analogue of Theorem 17.3.1 for graphs with a δ_s'–coupling.

10. Consider the periodic–lattice graph Hamiltonian H_α of Example 17.3.3; using the notation (17.19) describe its band spectrum.

   (a) Show that $\sigma(H_\alpha) = \bigcup_{r=1}^N [a_r, b_r]$ for some $N \ge 1$, where $a_r < b_r < a_{r+1}$. In particular, $\pm a_1 > 0$ holds *iff* $\pm\alpha > 0$, respectively, and $\sigma(H_0) = [0,\infty)$.

   (b) If $\alpha > 0$, each b_r is of the form $\left(\frac{\pi n}{\lambda}\right)^2$ for $\lambda = \ell_1$ or ℓ_2, and some $n \in \mathbb{Z}$. Similarly, $a_r = \left(\frac{\pi m}{\lambda}\right)^2$ with $m \in \mathbb{Z}$ for $\alpha < 0$ and $r \ge 2$.

   (c) If $\alpha < -4\ell^{-1}(\theta^{1/2} + \theta^{-1/2})$ one has $b_1 < 0$ and $a_2 = \left(\frac{\pi}{L}\right)^2$.

   (d) $\sigma(H_{\alpha'}) \cap \mathbb{R}^+ \subset \sigma(H_\alpha) \cap \mathbb{R}_+$ holds for $|\alpha'| > |\alpha|$.

(e) All the gaps above the threshold are finite. If there is an infinite number of them, their widths are asymptotically bounded, $a_{r+1} - b_r < 2|\alpha|(\ell_1+\ell_2)^{-1} + \mathcal{O}(r^{-1})$ as $r \to \infty$.

(f) Compare these results with Example 14.21. Show that each gap is contained in the intersection of a pair of gaps of the Kronig–Penney model with the coupling constant α and spacings ℓ_1 and ℓ_2.

11. Let H_α be the lattice-graph Hamiltonian considered above with a fixed coupling $\alpha \in \mathbb{R}$. Show that its spectrum is not purely absolutely continuous if θ is rational. *Hint:* Construct eigenfunctions which vanish at the vertices of Γ.

12. Consider a modification of the periodic lattice graph Hamiltonian of Problem 10 in which the δ–coupling at the vertices is replaced by a δ'_s–coupling with a fixed $\beta \in \mathbb{R}$. Show that the spectral condition in the notation (17.17) changes to $F(k; -w_1, -w_2) + \frac{1}{2}\beta k = 0$ and use it to analyze the spectrum of this operator.

13. Consider the Hamiltonians $H_{\mathcal{K}}(V)$ of Example 17.3.5 for the line with a stub.

(a) Find the spectrum of $H_{\mathcal{K}}(V)$. Show that it can have a negative eigenvalue for some values of the parameters.

(b) Find the on–shell scattering matrix $S(k)$: prove (17.24) and the analogous formulæ for the left–to–right scattering. Check that $S(k)$ is unitary.

(c) Prove the formula (17.25). Calculate it explicitly for the cases when the particle on the stub (i) is free, $V = 0$, (ii) is under influence of an electric field, $V(x) = Fx$.

(d) Derive (17.26) and $\mathrm{Im}\, z_n(b) = -\frac{ib^2 \chi'_n(0)^2}{2k_n + id} + \mathcal{O}(b^4)$ for the resonance width.

Hint: (a) The condition is $b^2 u'_\ell(0) + (d+2\kappa)(cu'_\ell - u_\ell)(0) = 0$ for $\kappa > 0$; a solution exists if $b = 0$ and $d < 0$ and the eigenvalue persists for $|b|$ small enough. (c) The resolvent of $H_{\mathcal{K}}(V)$ maps $L^2(\Gamma)$ onto the domain of the operator. (d) Employ the implicit function theorem and express $\frac{d}{dk}(cu'_\ell(0) - u_\ell(0))_{k=k_n}$.

14. Prove the self–adjointness of the operators determined by the relations (17.28) using the von Neumann theory of Section 4.7. Find the remaining self–adjoint extensions. *Hint:* The deficiency subspaces can be constructed explicitly — *cf.* [EŠ 4].

15. Consider the Hamiltonians $H_{\mathcal{A}}$ of Example 17.4.1.

(a) Prove the relations (17.29) and find the whole on–shell scattering matrix; check that it is unitary.

(b) For which values of the parameters is the discrete spectrum of $H_{\mathcal{A}}$ non-empty?

(c) Solve the scattering problem if the plane is replaced by an open precompact set $\Omega \subset \mathbb{R}^2$, for instance, a rectangle, and the Hamiltonian is subject to Dirichlet boundary conditions at the boundary of Ω. Find the reflection amplitude and scattering resonances.

Hint: (a) [EŠ 4]. (c) Look for an Ansatz for the scattering solutions employing the Green function of $-\Delta_D^\Omega$ — *cf.* [EŠ 9].

16. The operator associated with the quadratic form (17.31) can be alternatively defined using boundary conditions. Consider an operator acting as $\psi(x) \mapsto -(\Delta\psi)(x)$ for $x \notin \Gamma$ on any ψ with compact support which is continuous, C^∞ outside Γ, and at the interior points of each edge $e_j \in \Gamma$ has a jump in its normal derivative,

$$\frac{\partial\psi}{\partial n_+}(x) - \frac{\partial\psi}{\partial n_-}(x) = -\alpha\psi(x)\,.$$

Show that this operator is *e.s.a.* and its closure coincides with $H_{\alpha,\Gamma}$.
Hint: Use Green's formula — *cf.* [BEKŠ 1], Remark 4.1.

17. Fill in the details in the proof of Theorem 17.5.3. Furthermore, show that if the curve is smooth, $\gamma \in C^2$, then the condition (17.35) required in the theorem is satisfied if the curvature of Γ behaves as $\mathcal{O}(|s|^{-5/4-\varepsilon})$ for some $\varepsilon > 0$ when $|s| \to \infty$.

18. Analyze leaky *star graphs* for which Γ consists of N halflines emanating from the origin, characterized by an N-tuple of positive angles $\beta = \{\beta_1, \dots, \beta_N\}$ satisfying the condition $\sum_{j=1}^N \beta_j = 2\pi$. Put $H_N(\beta) := H_{\alpha,\Gamma_\beta}$.

(a) Prove that $\sigma_{\rm ess}(H_N(\beta)) = [-\frac14\alpha^2, \infty)$.

(b) The eigenvalues of $H_2(\pi-\phi, \pi+\phi)$ are decreasing functions of ϕ in $(0,\pi)$.

(c) Analyze $H_4(\beta_s)$ with $\beta_s = \left\{\frac{\pi}{2}, \frac{\pi}{2}, \frac{\pi}{2}, \frac{\pi}{2}\right\}$ using separation of variables.

(d) Fix N and a positive integer n. If at least one of the angles β_j is small enough, $\operatorname{card}\sigma_{\rm disc}(H_N(\beta)) \geq n$, in particular, the number of bound states can exceed any fixed integer for N large enough.

Hint: (a) For the negative part use a Neumann bracketing. (d) It is sufficient to consider $N=2$. Take a trial function which in the direction transverse to the angle axis is constant within the angle and decreases as $e^{-\alpha|y|}$ outside.

19. Fill in the details in the proof of Theorem 17.5.5, in particular, prove Lemma 17.5.6.
Hint: Use first convexity of $x \mapsto x^\alpha$, $\alpha > 1$, in $(0,\infty)$ to show that it is sufficient to prove the claim for $p=2$. After that, write $\gamma(s) = \sum_{0\neq n\in\mathbb{Z}} c_n e^{ins}$, use the fact that $\int_0^L |\gamma'(s)|^2\, ds = L$ and the inequality $|\sin nx| \leq n\sin x$ for $x \in (0, \frac12\pi)$.

20. Check the estimate (17.37) and complete the proof of Theorem 17.5.7. Derive the analogous eigenvalue asymptotics for bound states corresponding to a non–straight C^4–smooth curve Γ satisfying the assumptions of Theorem 17.5.3.

Appendix A. Measure and integration

We suppose the reader is familiar with the basic facts concerning set theory and integration as they are presented in the introductory course of analysis. In this appendix, we review them briefly, and add some more which we shall need in the text. Basic references for proofs and a detailed exposition are, *e.g.*, ⟦Hal 1⟧, ⟦Jar 1,2⟧, ⟦KF 1,2⟧, ⟦LiL⟧, ⟦Ru 1⟧, or any other textbook on analysis you might prefer.

A.1 Sets, mappings, relations

A set is a collection of objects called elements. The symbol $\operatorname{card} X$ denotes the cardinality of the set X. The subset M consisting of the elements of X which satisfy the conditions $P_1(x), \dots, P_n(x)$ is usually written as $M = \{ x \in X : P_1(x), \dots, P_n(x) \}$. A set whose elements are certain sets is called a *system* or *family* of these sets; the family of all subsystems of a given X is denoted as 2^X.

The operations of union, intersection, and set difference are introduced in the standard way; the first two of these are commutative, associative, and mutually distributive. In a system $\{M_\alpha\}$ of any cardinality, the *de Morgan relations*,

$$X \setminus \bigcup_\alpha M_\alpha = \bigcap_\alpha (X \setminus M_\alpha) \quad \text{and} \quad X \setminus \bigcap_\alpha M_\alpha = \bigcup_\alpha (X \setminus M_\alpha),$$

are valid. Another elementary property is the following: for any family $\{M_n\}$, which is at most countable, there is a disjoint family $\{N_n\}$ of the same cardinality such that $N_n \subset M_n$ and $\bigcup_n N_n = \bigcup_n M_n$. The set $(M \setminus N) \cup (N \setminus M)$ is called the *symmetric difference* of the sets M, N and denoted as $M \triangle N$. It is commutative, $M \triangle N = N \triangle M$, and furthermore, we have $M \triangle N = (M \cup N) \setminus (M \cap N)$ and $M \triangle N = (X \setminus M) \triangle (X \setminus N)$ for any $X \supset M \cup N$. The symmetric difference is also associative, $M \triangle (N \triangle P) = (M \triangle N) \triangle P$, and distributive with respect to the intersection, $(M \triangle N) \cap P = (M \cap P) \triangle (N \cap P)$.

A family $\mathcal{R}$ is called a *set ring* if $M \triangle N \in \mathcal{R}$ and $M \cap N \in \mathcal{R}$ holds for any pair $M, N \in \mathcal{R}$. The relation $M \setminus N = (M \triangle N) \cap M$ also gives $M \setminus N \in \mathcal{R}$, and this in turn implies $\emptyset \in \mathcal{R}$ and $M \cup N \in \mathcal{R}$. If the symmetric difference and intersection are understood as a sum and product, respectively, then a set ring is a ring in the sense of the general algebraic definition of Appendix B.1.

A.1.1 Example: Let $\mathcal{J}^d$ be the family of all bounded intervals in $\mathbb{R}^d$, $d \geq 1$. The family $\mathcal{R}^d$, which consists of all finite unions of intervals $J \subset \mathcal{J}^d$ together with the empty set, is a set ring, and moreover, it is the smallest set ring containing $\mathcal{J}^d$. As mentioned above, any $R \in \mathcal{R}^d$ can be expressed as a finite union of *disjoint* bounded intervals.

A set ring $\mathcal{R} \subset 2^X$ is called a *set field* if it contains the set X (notice that in the terminology of Appendix B.1 it is a ring with a unit element but *not* an algebra). A set field $\mathcal{A} \subset 2^X$ is called a *σ–field* if $\bigcup_{n=1}^{\infty} M_n \in \mathcal{A}$ holds for any countable system $\{M_n\}_{n=1}^{\infty} \subset \mathcal{A}$. De Morgan relations show that also $\bigcap_{n=1}^{\infty} M_n \in \mathcal{A}$, and furthermore that a family $\mathcal{A} \subset 2^X$ containing the set X is a σ–field *iff* $X \setminus M \in \mathcal{A}$ for all $M \in \mathcal{A}$ and $\bigcup_n M_n \in \mathcal{A}$ for any at most countable subsystem $\{M_n\} \subset \mathcal{A}$. Given a family $\mathcal{S} \subset 2^X$ we consider all σ–fields $\mathcal{A} \subset 2^X$ containing $\mathcal{S}$ (there is at least one, $\mathcal{A} = 2^X$). Their intersection is again a σ–field containing $\mathcal{S}$; we call it the *σ–field generated by* $\mathcal{S}$ and denote it as $\mathcal{A}(\mathcal{S})$.

A.1.2 Example: The elements of $\mathcal{B}^d := \mathcal{A}(\mathcal{J}^d)$ are called *Borel sets* in $\mathbb{R}^d$. In particular, all the open and closed sets, and thus also the compact sets, are Borel. The σ–field $\mathcal{B}^d$ is also generated by other systems, *e.g.*, by the system of all open sets in $\mathbb{R}^d$. In general, Borel sets in a topological space (X, τ) are defined as the elements of the σ–field $\mathcal{A}(\tau)$.

A sequence $\{M_n\}_{n=1}^{\infty}$ is *nondecreasing* or *nonincreasing* if $M_n \subset M_{n+1}$ or $M_n \supset M_{n+1}$, respectively, holds for $n = 1, 2, \dots$. A set family $\mathcal{M}$ is *monotonic* if it contains the set $\bigcup_n M_n$ together with any nondecreasing sequence $\{M_n\}$, and $\bigcap_n M_n$ together with any nonincreasing sequence $\{M_n\}$. Any σ–field represents an example of a monotonic system. To any $\mathcal{S}$ there is the smallest monotonic system $\mathcal{M}(\mathcal{S})$ containing $\mathcal{S}$ and we have $\mathcal{M}(\mathcal{S}) \subset \mathcal{A}(\mathcal{S})$. If $\mathcal{R}$ is a ring, the same is true for $\mathcal{M}(\mathcal{R})$; in addition, if $\bigcup_{M \in \mathcal{R}} M \in \mathcal{M}(\mathcal{R})$, then $\mathcal{M}(\mathcal{R})$ is a σ–field and $\mathcal{M}(\mathcal{R}) = \mathcal{A}(\mathcal{R})$.

A *mapping* (or *map*) f from a set X to Y is a rule, which associates with any $x \in X$ a unique element $y \equiv f(x)$ of the set Y; we write $f : X \to Y$ and also $x \mapsto f(x)$. If $Y = \mathbb{R}$ or $Y = \mathbb{C}$ the map f is usually called a real or a complex *function*, respectively. It is also often useful to consider maps which are defined on a subset $D_f \subset X$ only. The symbol $f : X \to Y$ must then be completed by specifying the set D_f which is called the *domain* of f; we denote it also as $D(f)$. If D_f is not specified, it is supposed to coincide with X. The sets $\operatorname{Ran} f := \{\, y \in Y : \, y = f(x),\, x \in D_f \,\}$ and $\operatorname{Ker} f := \{\, x \in D_f : \, f(x) = 0 \,\}$ are the *range* and *kernel* of the map f, respectively.

A map $f : X \to Y$ is *injective* if $f(x) = f(x')$ holds for any $x, x' \in X$ only if $x = x'$; it is *surjective* if $\operatorname{Ran} f = Y$. A map which is simultaneously injective and surjective is called *bijective* or a *bijection*. The sets X and Y have the same cardinality if there is a bijection $f : X \to Y$ with $D_f = X$. The relation $f = g$ between $f : X \to Y$ and $g : X \to Y$ means by definition $D_f = D_g$ and $f(x) = g(x)$ for all $x \in D_f$. If $D_f \supset D_g$ and $f(x) = g(x)$ holds for all $x \in D_g$ we say that f is an *extension* of g while g is a *restriction* of f to the set D_g; we write $f \supset g$ and $g = f \restriction D_g$.

A.1.3 Example: For any $X \subset M$ we define a real function $\chi_M : \chi_M(x) = 1$ if $x \in M$, $\chi_M(x) = 0$ if $x \in X \setminus M$; it is called the *characteristic* (or *indicator*) function of the set M. The map $M \mapsto \chi_M$ is a bijection of the system 2^X to the set of all functions $f : X \to \mathbb{R}$ such that $\operatorname{Ran} f = \{0, 1\}$.

The set $f^{(-1)}(N) := \{\, x \in D_f : \, f(x) \in N \,\}$ for given $f : X \to Y$ and $N \subset Y$ is called the *pull–back* of the set N by the map f. One has $f^{(-1)}\left(\bigcup_{\alpha \in I} N_\alpha\right) = \bigcup_{\alpha \in I} f^{(-1)}(N_\alpha)$ for any family $\{N_\alpha\} \subset Y$, and the analogous relation is valid for intersections. Furthermore, $f^{(-1)}(N_1 \setminus N_2) = f^{(-1)}(N_1) \setminus f^{(-1)}(N_2)$ and $f(f^{(-1)}(N) = \operatorname{Ran} f \cap N$. On the other hand, $f^{(-1)}(f(M)) \supset M$; the inclusion turns to identity if f is injective.

A.1.4 Example: Let $f : X \to Y$ with $D_f = X$. To a σ–field $\mathcal{B} \subset 2^Y$ we can construct the family $f^{(-1)} := \{\, f^{(-1)}(N) : \, N \in \mathcal{B} \,\}$, which is obviously again a σ–field. Similarly, if

$\mathcal{A} \subset 2^X$ is a σ–field, then the same is true for $\{ N \subset Y : f^{(-1)}(N) \in \mathcal{A} \}$. Hence for any family $\mathcal{S} \subset 2^Y$ we can construct the σ–field $f^{(-1)}(\mathcal{A}(\mathcal{S})) \subset 2^X$ and the latter coincides with the σ–field generated by $f^{(-1)}(\mathcal{S})$, *i.e.*, $f^{(-1)}(\mathcal{A}(\mathcal{S})) = \mathcal{A}(f^{(-1)}(\mathcal{S}))$.

Given $f : X \to Y$ and $g : Y \to Z$ we can define the *composite map* $g \circ f : X \to Z$ with the domain $D(g \circ f) := f^{(-1)}(D_g) = f^{(-1)}(D_g \cap \operatorname{Ran} f)$ by $(g \circ f)(x) := g(f(x))$. We have $(g \circ f)(P)^{(-1)} = f^{(-1)}(g^{(-1)}(P))$ for any $P \subset Z$.

If $f : X \to Y$ is injective, then for any $y \in \operatorname{Ran} f$ there is just one $x_y \in D_f$ such that $y = f(x_y)$; the prescription $y \mapsto g(y) := x_y$ defines a map $g : Y \to X$, which is called the *inverse* of f and denoted as f^{-1}. We have $D(f^{-1}) = \operatorname{Ran} f$, $\operatorname{Ran} f^{-1} = D_f$ and $f^{-1}(f(x)) = x$, $f(f^{-1}(y)) = y$ for any $x \in D_f$ and $y \in \operatorname{Ran} f$, respectively. These relations further imply $f^{(-1)}(N) = f^{-1}(N)$ for any $N \subset \operatorname{Ran} f$. Often we have a pair of mappings $f : X \to Y$ and $g : Y \to X$ and we want to know whether f is invertible and $f^{-1} = g$; this is true if one of the following conditions is valid:

(i) $D_g = \operatorname{Ran} f$ and $g(f(x)) = x$ for all $x \in D_f$
(ii) $\operatorname{Ran} f \subset D_g$, $g(f(x)) = x$ for all $x \in D_f$ and $\operatorname{Ran} g \subset D_f$, $f(g(y)) = y$ for all $y \in D_g$

If $f : X \to Y$ is injective, then f^{-1} is also injective and $(f^{-1})^{-1} = f$. If $g : Y \to Z$ is also injective, then the composite map $g \circ f$ is invertible and $(g \circ f)^{-1} = f^{-1} \circ g^{-1}$.

The *Cartesian product* $M \times N$ is the set of ordered pairs $[x, y]$ with $x \in M$ and $y \in N$; the Cartesian product of the families $\mathcal{S}$ and $\mathcal{S}'$ is defined by $\mathcal{S} \times \mathcal{S}' := \{ M \times N : M \in \mathcal{S},\, N \in \mathcal{S}' \}$. For instance, the systems of bounded intervals of Example 1 satisfy $\mathcal{J}^{m+n} = \mathcal{J}^m \times \mathcal{J}^n$. If $M \times N$ is empty, then either $M = \emptyset$ or $N = \emptyset$. On the other hand, if $M \times N$ is nonempty, then the inclusion $M \times N \subset P \times R$ implies $M \subset P$ and $N \subset R$. We have $(M \cup P) \times N = (M \times N) \cup (P \times N)$ and similar simple relations for the intersection and set difference. Notice, however, that $(M \times N) \cup (P \times R)$ can be expressed in the form $S \times T$ only if $M = P$ or $N = R$.

The definition of the Cartesian product extends easily to any finite family of sets. Alternatively, we can interpret $M_1 \times \cdots \times M_n$ as the set of maps $f : \{1, \dots, n\} \to \bigcup_{j=1}^n M_j$ such that $f(j) \in M_j$. This allows us to define $\mathsf{X}_{\alpha \in I} M_\alpha$ for a system $\{ M_\alpha : \alpha \in I \}$ of any cardinality as the set of maps $f : I \to \bigcup_{\alpha \in I} M_\alpha$ which fulfil $f(\alpha) \in M_\alpha$ for any $\alpha \in I$. The existence of such maps is related to the axiom of choice (see below).

Given $f : X \to \mathbb{C}$ and $g : Y \to \mathbb{C}$, we define the function $f \times g$ on $X \times Y$ by $(f \times g)(x, y) := f(x)g(y)$. Let $M \subset X \times Y$; then to any $x \in X$ we define the x*–cut* of the set M by $M_x := \{ y \in Y : [x, y] \in M \}$; we define the y–cuts analogously.

Let $\mathcal{A} \subset 2^X$, $\mathcal{B} \subset 2^Y$ be σ–fields; then the σ–field $\mathcal{A}(\mathcal{A} \times \mathcal{B})$ is called the *direct product* of the fields $\mathcal{A}$ and $\mathcal{B}$ and is denoted as $\mathcal{A} \otimes \mathcal{B}$.

A.1.5 Example: The Borel sets in $\mathbb{R}^m$ and $\mathbb{R}^n$ in this way generate all Borel sets in $\mathbb{R}^{m+n}$, *i.e.*, we have $\mathcal{B}^m \otimes \mathcal{B}^n = \mathcal{B}^{m+n}$.

On the other hand, the cuts of a set $M \in \mathcal{A} \otimes \mathcal{B}$ belong to the original fields: we have $M_x \in \mathcal{B}$ and $M_y \in \mathcal{A}$ for any $x \in X$ and $y \in Y$, respectively.

A subset $R_\varphi \subset X \times X$ defines a *relation* φ on X: if $[x, y] \in R_\varphi$ we say the element x is in relation with y and write $x \varphi y$. A common example is an *equivalence*, which is a relation $\sim$ on X that is *reflexive* ($x \sim x$ for any $x \in X$), *symmetric* ($x \sim y$ implies $y \sim x$), and *transitive* ($x \sim y$ and $y \sim z$ imply $x \sim z$). For any $x \in X$ we define the *equivalence class* of x as the set $T_x := \{ y \in X : y \sim x \}$. We have $T_x = T_y$ *iff* $x \sim y$, so the set X decomposes into a disjoint union of the equivalence classes.

Another important example is a *partial ordering* on X, which means any relation $\prec$ that is reflexive, transitive, and *antisymmetric*, *i.e.*, such that the conditions $x \prec y$ and $y \prec x$ imply $x = y$. If X is partially ordered, then a subset $M \subset X$ is said to be *(fully) ordered* if any elements $x, y \in M$ satisfy either $x \prec y$ or $x \succ y$. An element $x \in X$ is an *upper bound* of a set $M \subset X$ if $y \prec x$ holds for all $y \in M$; it is a *maximal element* of M if for any $y \in M$ the condition $y \succ x$ implies $y = x$.

A.1.6 Theorem (Zorn's lemma): Let any ordered subset of a partially ordered set X have an upper bound; then X contains a maximal element.

Zorn's lemma is equivalent to the so–called *axiom of choice*, which postulates for a system $\{ M_\alpha : \alpha \in I \}$ of any cardinality the existence of a map $\alpha \mapsto x_\alpha$ such that $x_\alpha \in M_\alpha$ for all $\alpha \in I$ — see, *e.g.*, ⟦ DS 1 ⟧, Sec.I.2, ⟦ Ku ⟧, Sec.I.6. Notice that the maximal element in a partially ordered set is generally far from unique.

A.2 Measures and measurable functions

Let us have a pair $(X, \mathcal{A})$, where X is a set and $\mathcal{A} \subset 2^X$ a σ–field. A function $f : X \to \mathbb{R}$ is called *measurable* (with respect to $\mathcal{A}$) if $f^{(-1)}(J) \in \mathcal{A}$ holds for any bounded interval $J \subset \mathbb{R}$, *i.e.*, $f^{(-1)}(\mathcal{J}) \subset \mathcal{A}$. This is equivalent to any of the following statements: (i) $f^{(-1)}((c, \infty)) \in \mathcal{A}$ for all $c \in \mathbb{R}$, (ii) $f^{(-1)}(G) \in \mathcal{A}$ for any open $G \subset \mathbb{R}$, (iii) $f^{(-1)}(\mathcal{B}) \subset \mathcal{A}$. If X is a topological space, a function $f : X \to \mathbb{R}$ is called *Borel* if it is measurable w.r.t. the σ–field $\mathcal{B}$ of Borel sets in X.

A.2.1 Example: Any continuous function $f : \mathbb{R}^d \to \mathbb{R}$ is Borel. Furthermore, let $f : X \to \mathbb{R}$ be measurable (w.r.t. some $\mathcal{A}$) and $g : \mathbb{R} \to \mathbb{R}$ be Borel; then the composite function $g \circ f$ is measurable w.r.t. $\mathcal{A}$.

If functions $f, g : X \to \mathbb{R}$ are measurable, then the same is true for their linear combinations $af + bg$ and product fg as well as for the function $x \mapsto (f(x))^{-1}$ provided $f(x) \neq 0$ for all $x \in X$. Even if the last condition is not valid, the function h, defined by $h(x) := (f(x))^{-1}$ if $f(x) \neq 0$ and $h(x) := 0$ otherwise, is measurable. Furthermore, if a sequence $\{f_n\}$ converges pointwise, then the function $x \mapsto \lim_{n \to \infty} f_n(x)$ is again measurable.

The notion of measurability extends to complex functions: a function $\varphi : X \to \mathbb{C}$ is *measurable* (w.r.t. $\mathcal{A}$) if the functions $\mathrm{Re}\,\varphi(\cdot)$ and $\mathrm{Im}\,\varphi(\cdot)$ are measurable; this is true *iff* $\varphi^{(-1)}(G) \in \mathcal{A}$ holds for any open set $G \subset \mathbb{C}$. A complex linear combination of measurable functions is again measurable. Furthermore, if φ is measurable, then $|\varphi(\cdot)|$ is also measurable. In particular, the modulus of a measurable $f : X \to \mathbb{R}$ is measurable, as are the functions $f^\pm := \frac{1}{2}(|f| \pm f)$.

A function $\varphi : X \to \mathbb{C}$ is *simple* (*σ–simple*) if $\varphi = \sum_n y_n \chi_{M_n}$, where $y_n \in \mathbb{C}$ and the sets $M_n \in \mathcal{A}$ form a finite (respectively, at most countable) disjoint system with $\bigcup_n M_n = X$. By definition, any such function is measurable; the sets of (σ–)simple functions are closed with respect to the pointwise defined operations of summation, multiplication, and scalar multiplication. The expression $\varphi = \sum_n y_n \chi_{M_n}$ is not unique, however, unless the numbers y_n are mutually different.

A.2.2 Proposition: A function $f : X \to \mathbb{R}$ is measurable *iff* there is a sequence $\{f_n\}$ of σ–simple functions, which converges to f uniformly on X. If f is bounded, there is a sequence of simple functions with the stated property.

In fact, the approximating sequence $\{f_n\}$ can be chosen even to be *nondecreasing.* If f is not bounded it can still be approximated pointwise by a sequence of simple functions, but not uniformly.

Given $(X, \mathcal{A})$ and $(Y, \mathcal{B})$ we can construct the pair $(X \times Y, \mathcal{A} \otimes \mathcal{B})$. Let $\varphi : M \to \mathbb{C}$ be a function on $M \in \mathcal{A} \otimes \mathcal{B}$; then its x–cut is the function φ_x defined on M_x by $\varphi_x(y) := \varphi(x, y)$; we define the y–cut similarly. Cuts of a measurable functions may not be measurable in general, however, it is usually important to ensure measurability a.e. – *cf.* Theorem A.3.13 below.

A mapping λ defined on a set family $\mathcal{S}$ and such that $\lambda(M)$ is either non–negative or $\lambda(M) = +\infty$ for any $M \in \mathcal{S}$ is called (a non–negative) *set function.* It is *monotonic* if $M \subset N$ implies $\lambda(M) \le \lambda(N)$, *additive* if $\lambda(M \cup N) = \lambda(M) + \lambda(N)$ for any pair of sets such that $M \cup N \in \mathcal{S}$ and $M \cap N = \emptyset$, and *σ–additive* if the last property generalizes, $\lambda\left(\bigcup_n M_n\right) = \sum_n \lambda(M_n)$, to any disjoint at most countable system $\{M_n\} \subset \mathcal{S}$ such that $\bigcup_n M_n \in \mathcal{S}$.

A set function μ, which is defined on a certain $\mathcal{A} \subset 2^X$, is σ–additive, and satisfies $\mu(\emptyset) = 0$ is called a (non–negative) *measure* on X. If at least one $M \in \mathcal{A}$ has $\mu(M) < \infty$, then $\mu(\emptyset) = 0$ is a consequence of the σ–additivity. The triplet $(X, \mathcal{A}, \mu)$ is called a *measure space;* the sets and functions measurable w.r.t. $\mathcal{A}$ are in this case often specified as *μ–measurable.* A set $M \in \mathcal{A}$ is said to be *μ–zero* if $\mu(M) = 0$, a proposition–valued function defined on $M \in \mathcal{A}$ is valid *μ–almost everywhere* if the set $N \subset M$, on which it is not valid, is μ–zero. A measure μ is *complete* if $N \subset M$ implies $N \in \mathcal{A}$ for any μ–zero set M; below we shall show that any measure can be extended in a standard way to a complete one.

Additivity implies that any measure is monotonic, and $\mu(M \cup N) = \mu(M) + \mu(N) - \mu(M \cap N) \le \mu(M) + \mu(N)$ for any sets $M, N \in \mathcal{A}$ which satisfy $\mu(M \cap N) < \infty$. Using the σ–additivity, one can check that $\lim_{k \to \infty} \mu(M_k) = \mu\left(\bigcup_{n=1}^{\infty} M_n\right)$ holds for any *nondecreasing* sequence $\{M_n\} \subset \mathcal{A}$, and a similar relation with the union replaced by intersection is valid for *nonincreasing* sequences. A measure μ is said be *finite* if $\mu(X) < \infty$ and *σ–finite* if $X = \bigcup_{n=1}^{\infty} M_n$, where $M_n \in \mathcal{A}$ and $\mu(M_n) < \infty$ for $n = 1, 2, \ldots$.

Let (X, τ) be a topological space, in which any open set can be expressed as a countable union of compact sets (as, for instance, the space $\mathbb{R}^d$; recall that an open ball there is a countable union of closed balls). Suppose that μ is a measure on X with the domain $\mathcal{A} \supset \tau$; then the following is true: if any point of an open set G has a μ–zero neighborhood, then $\mu(G) = 0$.

Given a measure μ we can define the function $\varrho_\mu : \mathcal{A} \times \mathcal{A} \to [0, \infty)$ by $\varrho_\mu(M \times N) := \mu(M \bigtriangleup N)$. The condition $\mu(M \bigtriangleup N) = 0$ defines an equivalence relation on $\mathcal{A}$ and ϱ_μ is a metric on the corresponding set of equivalence classes.

A point $x \in X$ such that the one–point set $\{x\}$ belongs to $\mathcal{A}$ and $\mu(\{x\}) \ne 0$ is called a *discrete point* of μ; the set of all such points is denoted as P_μ. If μ is σ–finite the set P_μ is at most countable. A measure μ is *discrete* if $P_\mu \in \mathcal{A}$ and $\mu(M) = \mu(M \cap P_\mu)$ for any $M \in \mathcal{A}$.

A measure μ is said to be *concentrated on a set* $S \in \mathcal{A}$ if $\mu(M) = \mu(M \cap S)$ for any $M \in \mathcal{A}$. For instance, a discrete measure is concentrated on the set of its discrete points. If (X, τ) is a topological space and $\tau \subset \mathcal{A}$, then the *support* of μ denoted as $\operatorname{supp} \mu$ is the smallest closed set on which μ is concentrated.

Next we are going to discuss some ways in which measures can be constructed. First we shall describe a construction, which starts from a given non–negative σ–additive set function $\dot{\mu}$ defined on a ring $\mathcal{R} \subset 2^X$; we assume that there exists an at most countable disjoint system $\{B_n\} \subset \mathcal{R}$ such that $\bigcup_n B_n = X$ and $\dot{\mu}(B_n) < \infty$ for $n = 1, 2, \dots$.

Let $\mathcal{S}$ be the system of all at most countable unions of the elements of $\mathcal{R}$; it is closed w.r.t. countable unions and finite intersections, and $M \backslash R \in \mathcal{S}$ holds for all $M \in \mathcal{S}$ and $R \in \mathcal{R}$. Any $M \in \mathcal{S}$ can be expressed as $M = \bigcup_j R_j$, where $\{R_j\} \subset \mathcal{R}$ is an at most countable disjoint system; using it we can define the set function $\ddot{\mu}$ on $\mathcal{S}$ by $\ddot{\mu}(M) := \sum_j \dot{\mu}(R_j)$. It is monotonic and σ–additive. Furthermore, we have $\ddot{\mu}\left(\bigcup_n M_n\right) \le \sum_n \ddot{\mu}(M_n)$; this property is called *countable semiadditivity*. Together with the monotonicity, it is equivalent to the fact that $M \subset \bigcup_{k=1}^{\infty} M_k$ implies $\ddot{\mu}(M) \le \sum_{k=1}^{\infty} \ddot{\mu}(M_k)$.

The next step is to extend the function $\ddot{\mu}$ to the whole system 2^X by defining the *outer measure* by $\mu^*(A) := \inf\{\, \ddot{\mu}(M) : M \in \mathcal{S},\, M \supset A \,\}$. The outer measure is again monotonic and countably semiadditive; however, it is not additive so it is *not* a measure. Its importance lies in the fact that the system

$$\mathcal{A}_\mu := \{\, A \subset X : \inf_{M \in \mathcal{S}} \mu^*(A \triangle M) = 0 \,\}$$

is a σ–field. This finally allows us to define $\mu := \mu^* \restriction \mathcal{A}_\mu$; it is a complete σ–additive measure on the σ–field $\mathcal{A}_\mu \supset \mathcal{A}(\mathcal{R})$, which is determined uniquely by the set function $\dot{\mu}$ in the sense that any measure ν on $\mathcal{A}(\mathcal{R})$, which is an extension of $\dot{\mu}$, satisfies $\nu = \mu \restriction \mathcal{A}(\mathcal{R})$. The measure μ is called the *Lebesgue extension* of $\dot{\mu}$.

A measure μ on a topological space (X, τ) is called *Borel* if it is defined on $\mathcal{B} \equiv \mathcal{A}(\tau)$ and $\mu(C) < \infty$ holds for any compact set C. We are particularly interested in Borel measures on $\mathbb{R}^d$, where the last condition is equivalent to the requirement $\mu(K) < \infty$ for any compact interval $K \subset \mathbb{R}^d$.

Any Borel measure on $\mathbb{R}^d$ is therefore σ–additive and corresponds to a unique σ–additive set function $\dot{\mu}$ on $\mathcal{R}^d$. The space $\mathbb{R}^d$, however, has the special property that for any bounded interval $J \in \mathcal{J}^d$ we can find a nonincreasing sequence of open intervals $I_n \supset J$ and a nondecreasing sequence of compact intervals $K_n \subset J$ such that $\bigcap_n I_n = \bigcup_n K_n = J$. This allows us to replace the requirement of σ–additivity by the condition $\dot{\mu}(J) = \inf\{\, \dot{\mu}(I) : I \in \mathcal{G}_J \,\} = \sup\{\, \dot{\mu}(K) : K \in \mathcal{F}_J \,\}$ for any $J \in \mathcal{J}^d$, where $\mathcal{G}_J \subset \mathcal{J}^d$ is the system of all open intervals containing J, and $\mathcal{F}_J \subset \mathcal{J}^d$ is the system of all compact intervals contained in J. A set function $\tilde{\mu}$ on $\mathcal{J}^d$ which is finite, additive, and fulfils the last condition is called *regular.*

A.2.3 Theorem: There is a one–to–one correspondence between regular set functions $\tilde{\mu}$ and $\mu := \mu^* \restriction \mathcal{B}^d$ on $\mathbb{R}^d$. In particular, Borel measures μ and ν coincide if $\mu(J) = \nu(J)$ holds for any $J \in \mathcal{J}^d$.

A.2.4 Example: Let $f : \mathbb{R} \to \mathbb{R}$ be a nondecreasing right–continuous function. For any $a, b \in \mathbb{R}$, $a < b$, we set $\tilde{\mu}_f(a, b) := f(b-0) - f(a)$, $\tilde{\mu}_f(a, b] := f(b) - f(a)$, and analogous expressions for the intervals $[a, b]$ and $[a, b]$. This defines the regular set function $\tilde{\mu}_f$ on

$\mathcal{J}^1$; the corresponding Borel measure μ_f is called the *Lebesgue–Stieltjes measure* generated by the function f. In particular, if f is the identical function, $f(x) = x$, we speak about the *Lebesgue measure* on $\mathbb{R}$. Let us remark that the Lebesgue–Stieltjes measure is sometimes understood as a Lebesgue extension with a domain which is generally dependent on f; however, it contains $\mathcal{B}$ in any case.

A.2.5 Example: Let $\tilde{\mu}$ and $\tilde{\nu}$ be regular set functions on $\mathcal{J}^m \subset \mathbb{R}^m$ and $\mathcal{J}^n \subset \mathbb{R}^n$, respectively; then the function $\tilde{\varrho}$ on $\mathcal{J}^{m+n}$ defined by $\tilde{\varrho}(J \times L) := \tilde{\mu}(J)\tilde{\nu}(L)$ is again regular; the corresponding Borel measure is called the *direct product* of the measures μ and ν which correspond to $\tilde{\mu}$ and $\tilde{\nu}$, respectively, and is denoted as $\mu \otimes \nu$. In particular, repeating the procedure d times, we can in this way construct the *Lebesgue measure* on $\mathbb{R}^d$ which associates its volume with every parallelepiped.

A.2.6 Proposition: Any Borel measure on $\mathbb{R}^d$ is *regular, i.e.*, $\mu(B) = \inf\{\mu(G) : G \supset B,\, G \text{ open}\} = \sup\{\mu(C) : C \subset B,\, C \text{ compact}\}$.

As a consequence of this result, we can find to any $B \in \mathcal{B}^d$ a nonincreasing sequence of open sets $G_n \supset B$ and a nondecreasing sequence of compact sets $C_n \subset B$ (both dependent generally on the measure μ) such that $\mu(B) = \lim_{n\to\infty} \mu(G_n) = \mu\left(\bigcap_n G_n\right) = \lim_{n\to\infty} \mu(C_n) = \mu\left(\bigcup_n C_n\right)$. Proposition A.2.6 generalizes to Borel measures on a locally compact Hausdorff space, in which any open set is a countable union of compact sets – see ⟦Ru 1⟧, Sec.2.18.

Let us finally remark that there are alternative ways to construct Borel measures. One can use, *e.g.*, the *Riesz representation theorem*, according to which Borel measures correspond bijectively to positive linear functionals on the vector space of continuous functions with a compact support — *cf.* ⟦Ru 1⟧, Sec.2.14; ⟦RS 1⟧, Sec.IV.4.

A.3 Integration

Now we shall briefly review the Lebesgue integral theory on a measure space $(X, \mathcal{A}, \mu)$. It is useful from the beginning to consider functions which may assume infinite values; this requires to define the algebraic operations $a + \infty := \infty$, $a \cdot \infty := \infty$ for $a > 0$ and $a \cdot \infty := 0$ for $a = 0$, *etc.*, to add the requirement $f^{(-1)}(\infty) \in \mathcal{A}$ to the definition of measurability, and several other simple modifications.

Given a simple non–negative function $s := \sum_n y_n \chi_{M_n}$ on X, we define its integral by $\int_X s\, d\mu := \sum_n y_n \mu(M_n)$; correctness of the definition follows from the additivity of μ. In the next step, we extend it to all measurable functions $f : X \to [0, \infty]$ putting

$$\int_X f\, d\mu \equiv \int_X f(x)\, d\mu(x) := \sup\left\{ \int_X s\, d\mu \,:\, s \in S_f \right\},$$

where S_f is the set of all simple functions $s : X \to [0, \infty)$ such that $s \le f$. We also define $\int_M f\, d\mu := \int_X f\chi_M\, d\mu$ for any $M \in \mathcal{A}$; in this way we associate with the function f and the set M a number from $[0, \infty]$, which is called the (*Lebesgue*) *integral* of f over M w.r.t. the measure μ.

A.3.1 Proposition: Let f, g be measurable functions $X \to [0, \infty]$ and $M \in \mathcal{A}$; then $\int_M (kf)\, d\mu = k \int_M f\, d\mu$ holds for any $k \in [0, \infty)$, and moreover, the inequality $f \le g$ implies $\int_M f\, d\mu \le \int_M g\, d\mu$.

Notice that the integral of $f = 0$ is zero even if $\mu(X) = \infty$. On the other hand, the relation $\int_X |\varphi|\, d\mu = 0$ for any measurable function $\varphi : X \to \mathbb{C}$ implies that $\mu(\{ x \in X : \varphi(x) \neq 0 \}) = 0$, *i.e.*, that the function φ is zero μ–a.e. Let us turn to limits which play the central role in the theory of integration.

A.3.2 Theorem (monotone convergence): Let $\{f_n\}$ be a nondecreasing sequence of non-negative measurable functions; then $\lim_{n\to\infty} \int_X f_n\, d\mu = \int_X (\lim_{n\to\infty} f_n)\, d\mu$.

The right side of the last relation makes sense since the limit function is measurable. However, we often need some conditions under which both sides are finite. The corresponding modification is also called the *monotone-convergence* (or *Levi's*) *theorem:* if $\{f_n\}$ is a nondecreasing sequence of non–negative measurable functions and there is a $k > 0$ such that $\int_X f_n\, d\mu \le k$ for $n = 1, 2, \dots$, then the function $x \mapsto f(x) := \lim_{n\to\infty} f_n(x)$ is μ–a.e. finite and $\lim_{n\to\infty} \int_X f_n\, d\mu = \int_X f\, d\mu \le k$. The monotone–convergence theorem implies, in particular, that the integral of a measurable function can be approximated by a nondecreasing sequence of integrals of simple functions.

A.3.3 Corollary (Fatou's lemma): $\int_X (\liminf_{n\to\infty} f_n)\, d\mu \le \liminf_{n\to\infty} \int_X f_n\, d\mu$ holds for any sequence of measurable functions $f_n : X \to [0, \infty]$.

This result has the following easy consequence: let a sequence $\{f_n\}$ of non–negative measurable functions have a limit everywhere, $\lim_{n\to\infty} f_n(x) = f(x)$, and $\int_X f_n\, d\mu \le k$ for $n = 1, 2, \dots$; then $\int_X f\, d\mu \le k$. Applying the monotone–convergence theorem to a sequence $\{f_n\}$ of non–negative measurable functions we get the relation $\int_X (\sum_{n=1}^\infty f_n)\, d\mu = \sum_{n=1}^\infty \int_X f_n\, d\mu$. In particular, if $f : X \to [0, \infty]$ is measurable and $\{M_n\}_{n=1}^\infty \subset \mathcal{A}$ is a disjoint family with $\bigcup_{n=1}^\infty M_n = M$, then putting $f_n := f\chi_{M_n}$ we get $\int_M f\, d\mu = \sum_{n=1}^\infty \int_{M_n} f\, d\mu$. This relation is called *σ–additivity of the integral*; it expresses the fact that the function f together with the measure μ *generates another measure.*

A.3.4 Proposition: Let $f : X \to [0, \infty]$ be a measurable function; then the map $M \mapsto \nu(M) := \int_M f\, d\mu$ is a measure with the domain $\mathcal{A}$, and $\int_X g\, d\nu = \int_X gf\, d\mu$ holds for any measurable $g : X \to [0, \infty]$.

Let us pass to integration of complex functions. A measurable function $\varphi : X \to \mathbb{C}$ is *integrable* (over X w.r.t. μ) if $\int_X |\varphi|\, d\mu < \infty$ (recall that if φ is measurable so is $|\varphi|$). The set of all integrable functions is denoted as $\mathcal{L}(X, d\mu)$; in the same way we define $\mathcal{L}(M, d\mu)$ for any $M \in \mathcal{A}$. Given $\varphi \in \mathcal{L}(X, d\mu)$ we denote $f := \mathrm{Re}\, \varphi$ and $g := \mathrm{Im}\, \varphi$; then $f^\pm$ and $g^\pm$ are non–negative measurable functions belonging to $\mathcal{L}(X, d\mu)$. This allows us to define the *integral* of complex functions through the positive and negative parts of the functions f, g as the mapping

$$\varphi \longmapsto \int_X \varphi\, d\mu := \int_X f^+\, d\mu - \int_X f^-\, d\mu + i \int_X g^+\, d\mu - i \int_X g^-\, d\mu$$

of $\mathcal{L}(X, d\mu)$ to $\mathbb{C}$. If, in particular, μ is the Lebesgue measure on $\mathbb{R}^d$ we often use the symbol $\mathcal{L}(\mathbb{R}^d)$ or $\mathcal{L}(\mathbb{R}^d, dx)$ instead of $\mathcal{L}(\mathbb{R}^d, d\mu)$, and the integral is written as $\int \varphi(x)\, dx$, or occasionally as $\int \varphi(\vec{x})\, d\vec{x}$.

The above definition has the following easy consequence: if $\int_M \varphi\, d\mu = 0$ holds for all $M \in \mathcal{A}$, then $\varphi(x) = 0$ μ–a.e. in X. Similarly $\int_M \varphi\, d\mu \ge 0$ for all $M \in \mathcal{A}$ implies $\varphi(x) \ge 0$ μ–a.e. in X; further generalizations can be found in ⟦ Ru 1 ⟧, Sec.1.40. The integral has the following basic properties:

(a) linearity: $\mathcal{L}(X,d\mu)$ is a complex vector space and $\int_X(\alpha\varphi+\psi)\,d\mu = \alpha\int_X\varphi\,d\mu + \int_X\psi\,d\mu$ for all $\varphi,\psi\in\mathcal{L}(X,d\mu)$ and $\alpha\in\mathbb{C}$,

(b) $\left|\int_X\varphi\,d\mu\right| \le \int_X|\varphi|\,d\mu$ holds for any $\varphi\in\mathcal{L}(X,d\mu)$.

A.3.5 Examples: A simple complex function $\sigma=\sum_n\eta_n\chi_{M_n}$ on X is integrable *iff* $\sum_n|\eta_n|\mu(M_n)<\infty$, and in this case $\int_X\sigma\,d\mu=\sum_n\eta_n\mu(M_n)$. The same is true for σ–simple functions. Further, let functions $f:X\to[0,\infty]$ and $\varphi:X\to\mathbb{C}$ be measurable and $d\nu:=f\,d\mu$; then $\varphi\in\mathcal{L}(X,d\nu)$ *iff* $\varphi f\in\mathcal{L}(X,d\mu)$ and Proposition A.3.4 holds again with g replaced by φ.

For a *finite* measure, we have an equivalent definition based on approximation of integrable functions by sequences of σ–simple functions.

A.3.6 Proposition: If $\mu(X)<\infty$, then a measurable function $\varphi:X\to\mathbb{C}$ belongs to $\mathcal{L}(X,d\mu)$ *iff* there is a sequence $\{\tau_n\}$ of σ–simple integrable functions such that $\lim_{n\to\infty}(\sup_{x\in X}|\varphi(x)-\tau_n(x)|)=0$; if this is the case, then $\int_X\varphi\,d\mu=\lim_{n\to\infty}\int_X\tau_n\,d\mu$. Moreover, if φ is bounded the assertion is valid with simple functions τ_n.

One of the most useful tools in the theory of integral is the following theorem.

A.3.7 Theorem (dominated convergence, or Lebesgue): Let $M\in\mathcal{A}$ and $\{\varphi_n\}$ be a sequence of complex measurable functions with the following properties: $\varphi(x):=\lim_{n\to\infty}\varphi_n(x)$ exists for μ–almost all $x\in M$ and there is a function $\psi\in\mathcal{L}(X,d\mu)$ such that $|\varphi_n(x)|\le\psi(x)$ holds μ–a.e. in M for $n=1,2,\dots$. Then $\varphi\in\mathcal{L}(X,d\mu)$ and

$$\lim_{n\to\infty}\int_M|\varphi-\varphi_n|\,d\mu=0,\qquad \lim_{n\to\infty}\int_M\varphi_n\,d\mu=\int_M\varphi\,d\mu.$$

Suppose that we have non–negative measures μ and ν on X (without loss of generality, we may assume that they have the same domain) and $k>0$; then we can define the non–negative measure $\lambda:=k\mu+\nu$. We obviously have $\mathcal{L}(X,d\lambda)=\mathcal{L}(X,d\mu)\cap\mathcal{L}(X,d\nu)$ and $\int_X\psi\,d\lambda=k\int_X\psi\,d\mu+\int_X\psi\,d\nu$ for any $\psi\in\mathcal{L}(X,d\lambda)$. In particular, if non–negative measures μ and λ with the same domain $\mathcal{A}$ satisfy $\mu(M)\le\lambda(M)$ for any $M\in\mathcal{A}$, then $\mathcal{L}(X,d\lambda)\subset\mathcal{L}(X,d\mu)$ and $\int_X f\,d\mu\le\int_X f\,d\lambda$ holds for each non–negative $f\in\mathcal{L}(X,d\lambda)$.

Let μ, ν again be measures on X with the same domain $\mathcal{A}$. We say that ν is *absolutely continuous* w.r.t. μ and write $\nu\ll\mu$ if $\mu(M)=0$ implies $\nu(M)=0$ for any $M\in\mathcal{A}$. On the other hand, if there are disjoint sets S_μ, $S_\nu\in\mathcal{A}$ such that μ is concentrated on S_μ and ν on S_ν we say that the measures are *mutually singular* and write $\mu\perp\nu$.

A.3.8 Theorem: Let λ and μ be non–negative measures on $\mathcal{A}$, the former being finite and the latter σ–finite; then there is a unique decomposition $\lambda=\lambda_{ac}+\lambda_s$ into the sum of non–negative mutually singular measures such that $\lambda_{ac}\ll\mu$ and $\lambda_s\perp\mu$. Moreover, there is a non–negative function $f\in\mathcal{L}(X,d\mu)$, unique up to a μ–zero measure subset of X, such that $d\lambda_{ac}=f\,d\mu$, *i.e.*, $\lambda_{ac}(M)=\int_M f\,d\mu$ for any $M\in\mathcal{A}$.

The relation $\lambda=\lambda_{ac}+\lambda_s$ is called the *Lebesgue decomposition* of the measure λ. The second assertion implies the *Radon–Nikodým theorem:* let μ be σ–finite and λ finite; then $\lambda\ll\mu$ holds *iff* there is $f\in\mathcal{L}(X,d\mu)$ such that $d\lambda=f\,d\mu$.

A.3.9 Remark: There is a close connection between these results (and their extensions to complex measures mentioned in the next section) and the theory of the indefinite Lebesgue integral, properties of absolutely continuous functions, *etc.* We refer to the literature mentioned at the beginning; in this book, in fact, we need only the following facts: a function $\varphi : \mathbb{R} \to \mathbb{C}$ is *absolutely continuous* on a compact interval $[a,b]$ if for any $\varepsilon > 0$ there is $\delta > 0$ such that $\sum_j |\varphi(\beta_j) - \varphi(\alpha_j)| < \varepsilon$ holds for a finite disjoint system of intervals $(\alpha_j, \beta_j) \subset [a,b]$ fulfilling $\sum_j (\beta_j - \alpha_j) < \delta$. The function φ is absolutely continuous in a (noncompact) interval J if it is absolutely continuous in any compact $[a,b] \subset J$. A function $\varphi : \mathbb{R} \to \mathbb{C}$ is absolutely continuous in $\mathbb{R}$ *iff* its derivative φ' exists almost everywhere w.r.t. the Lebesgue measure and belongs to $\mathcal{L}(J, dx)$ for any bounded interval $J \subset \mathbb{R}$ with the endpoints $a \le b$; in such a case we have $\varphi(b) - \varphi(a) = \int_a^b \varphi'(x)\, dx$.

Next we shall mention integration of composite functions. Let $w : X \to \mathbb{R}^d$ be a map such that $w^{(-1)}(\mathcal{B}^d) \subset \mathcal{A}$; this requirement is equivalent to measurability of the "component" functions $w_j : X \to \mathbb{R}$, $1 \le j \le d$. Suppose that $\mu(w^{(-1)}(J)) < \infty$ holds for any $J \in \mathcal{J}^d$; then the relation $B \mapsto \mu^{(w)}(B) := \mu(w^{(-1)}(B))$ defines a Borel measure $\mu^{(w)}$ on $\mathbb{R}^d$.

A.3.10 Theorem: Adopt the above assumptions, and let a Borel function $\varphi : \mathbb{R}^d \to \mathbb{C}$ belong to $\mathcal{L}(\mathbb{R}^d, d\mu^{(w)})$; then $\varphi \circ w \in \mathcal{L}(X, d\mu)$ and

$$\int_B \varphi \, d\mu^{(w)} = \int_{w^{(-1)}(B)} (\varphi \circ w)\, d\mu$$

holds for all $b \in \mathcal{B}^d$.

In particular, if $X = \mathbb{R}^d$, $\mathcal{A} = \mathcal{B}^d$, and $\mu^{(w)}$ is the Lebesgue measure on R^d, the latter formula can under additional assumptions be brought into a convenient form. Suppose that $w : \mathbb{R}^d \to \mathbb{R}^d$ is injective, its domain is an open set $\mathcal{D} \subset \mathbb{R}^d$, the component functions $w_j : \mathcal{D} \to \mathbb{R}$ have continuous partial derivatives, $(\partial_k w_j)(\cdot)$ for $j,k = 1, \dots, d$, and finally, the Jacobian determinant, $D_w := \det(\partial_k w_j)$, is nonzero a.e. in $\mathcal{D}$. Such a map is called *regular;* its range $\mathcal{R} := \operatorname{Ran} w$ is an open set in $\mathbb{R}^d$, and the inverse w^{-1} is again regular.

A.3.11 Theorem (change of variables): Let w be a regular map on $\mathbb{R}^d$ with the domain $\mathcal{D}$ and range $\mathcal{R}$; then a Borel function $\varphi : \mathcal{R} \to \mathbb{C}$ belongs to $\mathcal{L}(\mathcal{R}, dx)$ *iff* $(\varphi \circ w) D_w \in \mathcal{L}(\mathcal{D}, dx)$; in that case we have

$$\int_B \varphi(x)\, dx = \int_{w^{(-1)}(B)} ((\varphi \circ w)|D_w|)(x)\, dx$$

for any Borel $B \subset \mathcal{R}$.

In Example A.2.5 we have mentioned how the measure $\mu \otimes \nu$ can be associated with a pair of Borel μ, ν on $\mathbb{R}^d$. An analogous result is valid under much more general circumstances.

A.3.12 Theorem: Let $(X, \mathcal{A}, \mu)$ and $(Y, \mathcal{B}, \nu)$ be measure spaces with σ–finite measures; then there is just one measure λ on $X \times Y$ with the domain $\mathcal{A} \otimes \mathcal{B}$ such that $\lambda(A \times B) = \mu(A)\nu(B)$ holds for all $A \in \mathcal{A}$, $B \in \mathcal{B}$; this measure is σ–finite and satisfies $\lambda(M) = \int_X \nu(M_x)\, d\mu(x) = \int_Y \mu(M^y)\, d\nu(y)$ for any $M \in \mathcal{A} \otimes \mathcal{B}$.

The measure λ is again called the *product measure* of μ and ν and is denoted as $\mu \otimes \nu$. Using it we can formulate the following important result.

A.3.13 Theorem (Fubini): Suppose the assumptions of the previous theorem are valid and $\varphi : X \times Y \to \mathbb{C}$ belongs to $\mathcal{L}(X \times Y, d(\mu \otimes \nu))$. Then the cut $\varphi_x \in \mathcal{L}(Y, d\nu)$ for μ–a.a. $x \in M$ and the function $\Phi : \Phi(x) = \int_Y \varphi_x \, d\nu$ belongs to $\mathcal{L}(X, d\mu)$; similarly $\varphi^y \in \mathcal{L}(X, d\mu)$ ν–a.e. in N and $\Psi : \Psi(y) = \int_X \varphi^y \, d\mu$ belongs to $\mathcal{L}(Y, d\nu)$. Finally, we have $\int_X \Phi \, d\mu = \int_Y \Psi \, d\nu = \int_{X \times Y} \varphi \, d(\mu \otimes \nu)$, or in a more explicit form,

$$\int_X \left(\int_Y \varphi(x,y) \, d\nu(y) \right) d\mu(x) = \int_Y \left(\int_X \varphi(x,y) \, d\mu(x) \right) d\nu(y) = \int_{X \times Y} \varphi(x,y) \, d(\mu \otimes \nu)(x,y).$$

We should keep in mind that the latter identity may not be valid if at least one of the measures μ and ν is not σ–finite. It is also not sufficient that both double integrals exists — counterexamples can be found, *e.g.*, in ⟦Ru 1⟧, Sec.7.9 or ⟦KF⟧, Sec.V.6.3. However, if at least one double integral of the *modulus* $|\varphi|$ is finite, then all the conclusions of the theorem are valid.

A.4 Complex measures

A σ–additive map $\nu : \mathcal{A} \to \mathbb{C}$ corresponding to a given $(X, \mathcal{A})$ is called *complex measure* on X; if $\nu(M) \in \mathbb{R}$ for all $m \in \mathcal{A}$ we speak about a *real* (or *signed*) measure. Any pair of non–negative measures μ_1, μ_2 with a common domain $\mathcal{A}$ determines a signed measure by $\varrho := \mu_1 - \mu_2$; similarly a pair of real measures ϱ_1, ϱ_2 defines a complex measure by $\nu := \varrho_1 + i\varrho_2$.

Any at most countable system $\{M_j\}$, which is disjoint and satisfies $M = \bigcup_j M_j$, will be called *decomposition* of the set M; the family of all decompositions of M will be denoted by $\mathcal{S}_M$. To a given complex measure ν and $M \in \mathcal{A}$, we define $|\nu|(M) := \sup \left\{ \sum_j |\nu(M_j)| : \{M_j\} \in \mathcal{S}_M \right\}$. One has $|\nu|(M) \ge |\nu(M)|$; the set function $|\nu|(\cdot)$ is called the (*total*) *variation* of the measure ν.

A.4.1 Proposition: The variation of a complex measure is a non–negative measure; it is the smallest non–negative measure such that $\mu(M) \ge |\nu(M)|$ holds for all $M \in \mathcal{A}$.

Using the total variation, we can decompose in particular any signed measure ϱ in the form $\varrho = \mu^+ - \mu^-$, where $\mu^\pm : \mu^\pm(M) = \frac{1}{2}[|\varrho|(M) \pm \varrho(M)]$. Since in general the decomposition of a signed measure into a difference of non–negative measures is not unique, one is interested in the minimal decomposition $\varrho = \mu_\varrho^+ - \mu_\varrho^-$ such that any pair of non–negative measures μ_1, μ_2 on $\mathcal{A}$ with the property $\varrho = \mu_1 - \mu_2$ satisfies $\mu_1(M) \ge \mu_\varrho^+(M)$ and $\mu_2(M) \ge \mu_\varrho^-(M)$ for each $M \in \mathcal{A}$.

The minimality is ensured if there is a disjoint decomposition $Q^+ \cup Q^- = X$ such that $\mu_\varrho^\pm := \pm\varrho(M \cap Q^\pm) \ge 0$; the pair $\{Q^+, Q^-\}$ is called *Hahn decomposition* of X w.r.t. the measure ϱ. The Hahn decomposition always exists but it is not unique. Nevertheless, if $\{\tilde{Q}^+, \tilde{Q}^-\}$ is another Hahn decomposition, one has $\varrho(M \cap Q^\pm) = \varrho(M \cap \tilde{Q}^\pm)$ for any $M \in \mathcal{A}$, so the measures $\mu_\varrho^\pm$ depend on ϱ only; we call them the *positive* and *negative variation* of the measure ϱ. The formula $\varrho = \mu_\varrho^+ - \mu_\varrho^-$ is named the *Jordan decomposition* of the measure ϱ.

One has $\mu_\varrho^+(M)+\mu_\varrho^-(M) = |\varrho|(M)$ and $\mu_\varrho^\pm(M) = \sup\{\pm\varrho(A) : A \subset M,\, A \in \mathcal{A}\}$ for any $M \in \mathcal{A}$. As a consequence, the positive and negative variations of a signed measure as well as the total variation of a complex measure are finite. One can introduce also infinite signed measures; however, we shall not need them in this book.

Complex Borel measures on $\mathbb{R}^d$ have $\mathcal{A} = \mathcal{B}^d$ for the domain. Variation of a complex Borel measure is a non–negative Borel measure. As in the non–negative case, a complex Borel measure can be approximated using monotonic sequences of compact sets from inside and open sets from outside of a given $M \in \mathcal{B}^d$. Also the second part of Theorem A.2.3 can be generalized.

A.4.2 Proposition: Let complex Borel measure ν and $\tilde{\nu}$ on $\mathbb{R}^d$ satisfy $\nu(J) = \tilde{\nu}(J)$ for all $J \in \mathcal{J}^d$; then $\nu = \tilde{\nu}$.

Before proceeding further, let us mention how the notion of absolute continuity extends to complex measures. The definition is the same: a complex measure ν is *absolutely continuous* w.r.t. a non–negative μ if $\mu(M) = 0$ implies $\nu(M) = 0$ for all $M \in \mathcal{A}$. There is an alternative definition.

A.4.3 Proposition: A complex measure ν satisfies $\nu \ll \mu$ *iff* for any $\varepsilon > 0$ there is a $\delta > 0$ such that $\mu(M) < \delta$ implies $|\nu(M)| < \varepsilon$.

In particular, if $\varphi \in \mathcal{L}(X, d\mu)$ and ν is the measure generated by this function, $\nu(M) := \int_M \varphi\, d\mu$, then $\nu \ll \mu$, so for any $\varepsilon > 0$ there is a $\delta > 0$ such that $\mu(M) < \delta$ implies $\left|\int_M \varphi\, d\mu\right| < \varepsilon$; this property is called *absolute continuity of the integral.*

Theorem A.3.8 holds for a complex measure λ as well. The measure can be even σ–finite; however, then the function f belongs no longer to $\mathcal{L}(X, d\mu)$, it is only measurable and integrable over any set $M \in \mathcal{A}$ with $\lambda(M) < \infty$. The Radon–Nikodým theorem yields the *polar decomposition of a complex measure*:

A.4.4 Proposition: For any complex measure ν there is a measurable function h such that $|h(x)| = 1$ for all $x \in X$ and $d\nu = h\, d|\nu|$.

Let us pass now to integration with respect to complex measures. We start with a signed measure $\varrho = \mu_\varrho^+ - \mu_\varrho^-$: a function $\varphi : X \to \mathbb{C}$ is *integrable* w.r.t. ϱ if it belongs to $\mathcal{L}(X, d\mu_\varrho^+) \cap \mathcal{L}(X, d\mu_\varrho^-) =: \mathcal{L}(X, d\varrho)$. Its integral is then defined by

$$\int_X \varphi\, d\varrho := \int_X \varphi\, d\mu_\varrho^+ - \int_X \varphi\, d\mu_\varrho^- \,;$$

the correctness follows from the uniqueness of the Jordan decomposition. The integral w.r.t. a complex measure ν represents then a natural extension of the present definition: for any function $\varphi : X \to \mathbb{C}$ belonging to $\mathcal{L}(X, d\nu) := \mathcal{L}(X, d\,\mathrm{Re}\,\nu) \cap \mathcal{L}(X, d\,\mathrm{Im}\,\nu)$ we set

$$\int_X \varphi\, d\nu := \int_X \varphi\, d\,\mathrm{Re}\,\nu + \int_X \varphi\, d\,\mathrm{Im}\,\nu.$$

The set of integrable functions can be expressed alternatively as $\mathcal{L}(X, d\nu) = \mathcal{L}(X, d|\nu|)$; it is a complex vector space and the map $\varphi \mapsto \int_X \varphi\, d\nu$ is again linear. Also other properties of the integral discussed in the previous section extend to the complex–measure case. For instance, the inequality $\left|\int_X \varphi\, d\nu\right| \le \int_X |\varphi|\, d|\nu|$ holds for any $\varphi \in \mathcal{L}(X, d\mu)$. We shall not continue the list, restricting ourselves by quoting the appropriate generalization of Proposition A.3.4.

A.4.5 Proposition: Let $\varphi \in \mathcal{L}(X, d\nu)$ for a complex measure ν; then the map $M \mapsto \gamma(M) := \int_M \varphi \, d\nu$ defines a complex measure γ on $\mathcal{A}$. The conditions $\psi \in \mathcal{L}(X, d\gamma)$ and $\psi\varphi \in \mathcal{L}(X, d\nu)$ are equivalent for any measurable $\gamma : X \to \mathbb{C}$; if they are satisfied one has $\int_X \psi \, d\gamma = \int_X \psi\varphi \, d\nu$.

A.5 The Bochner integral

The theory of integration recalled above can be extended to vector–valued functions $F : Z \to \mathcal{X}$, where $\mathcal{X}$ is a Banach space; they form a vector space denoted as $\mathcal{V}(Z, \mathcal{X})$ when equipped with pointwise defined algebraic operations. Let $(Z, \mathcal{A}, \mu)$ be a measure space with a positive measure μ. A function $S \in \mathcal{V}(Z, \mathcal{X})$ is *simple* if there is a disjoint decomposition $\{M_j\}_{j=1}^n \subset \mathcal{A}$ of the set Z and vectors $y_1, \dots, y_n \in \mathcal{X}$ such that $S = \sum_{j=1}^n y_j \chi_{M_j}$. The integral of such a function is defined by $\int_Z S(t) \, d\mu(t) := \sum_{j=1}^n y_j \mu(M_j)$; as above, it does not depend on the used representation of the function S.

To any $F \in \mathcal{V}(Z, d\mu)$ we define the non–negative function $\|F\| := \|F(\cdot)\|$. A vector–valued function F is integrable w.r.t. μ if there is a sequence $\{S_n\}$ of simple functions such that $S_n(t) \to F(t)$ holds for μ–a.a. $t \in Z$ and $\int_Z \|F - S_n\| \, d\mu \to 0$. The set of all integrable functions $F : Z \to \mathcal{X}$ is denoted by $\mathcal{B}(Z, d\mu; \mathcal{X})$. If F is integrable, the limit

$$\int_Z F(t) \, d\mu(t) := \lim_{n\to\infty} \int_Z S_n(t) \, d\mu(t)$$

exists and it is independent of the choice of the approximating sequence; we call it the *Bochner integral* of the function F. The function $\chi_M F$ is integrable for any set $M \in \mathcal{A}$ and $F \in \mathcal{B}(Z, d\mu; \mathcal{X})$, so we can also define $\int_M F(t) \, d\mu(t) := \int_Z \chi_M(t) F(t) \, d\mu(t)$. If $\{M_k\}$ is a finite disjoint decomposition of M, we have

$$\int_M F(t) \, d\mu(t) = \sum_k \int_{M_k} F(t) \, d\mu(t),$$

which means that the Bochner integral is *additive.*

A.5.1 Proposition: The map $F \mapsto \int_Z F(t) \, d\mu(t)$ from the subspace $\mathcal{B}(Z, d\mu; \mathcal{X}) \subset \mathcal{V}(Z, \mathcal{X})$ to $\mathcal{X}$ is linear. Suppose that for a vector–valued function F there is a sequence $\{S_n\}$ of simple functions that converges to F μ–a.e.; then F belongs to $\mathcal{B}(Z, d\mu; \mathcal{X})$ *iff* $\|F\| \in \mathcal{L}(Z, d\mu)$, and in that case

$$\left\| \int_Z F(t) \, d\mu(t) \right\| \le \int_Z \|F(t)\| \, d\mu(t).$$

The existence of an approximating sequence of simple functions has to be checked for each particular case; it is easy in some situations, *e.g.*, if Z is a compact subinterval in $\mathbb{R}$ and F is continuous, or if Z is any interval, F is continuous and its one–sided limits at the endpoints exist. The continuity of $F : \mathbb{R} \to \mathcal{X}$ in an interval $[a, b]$ also implies the relation $\frac{d}{dt} \int_a^t F(u) \, du = F(t)$ for any $t \in (a, b)$. Proposition A.5.1 shows that the Bochner integral is *absolutely continuous*: for any $\varepsilon > 0$ there is a $\delta > 0$ such that $\|\int_N F(t) \, d\mu(t)\| < \varepsilon$ holds for any $N \in \mathcal{A}$ with $\mu(N) < \delta$. Another useful result is the following.

A.5.2 Proposition: If $B : \mathcal{X} \to \mathcal{Y}$ is a bounded linear map to a Banach space $\mathcal{Y}$; then the condition $F \in \mathcal{B}(Z, d\mu; \mathcal{X})$ implies $BF \in \mathcal{B}(Z, d\mu; \mathcal{Y})$, and

$$\int_Z (BF)(t)\, d\mu(t) = B\left(\int_Z F(t)\, d\mu(t) \right).$$

Many properties of the Lebesgue integral can be extended to the Bochner integral. Probably the most important among them is the dominated–convergence theorem.

A.5.3 Theorem: Let $\{F_n\} \subset \mathcal{B}(Z, d\mu; \mathcal{X})$ be a sequence such that $\{F_n(t)\}$ converges for μ–a.a. $t \in Z$ and $\|F_n(t)\| \le g(t)$, $n = 1, 2, \dots$, for some $g \in \mathcal{L}(Z, d\mu)$. Assume further that there is a sequence $\{S_n\}$ of simple functions, which converges to the limiting function $F : F(t) = \lim_{n\to\infty} F_n(t)$ μ–a.e. in Z; then $F \in \mathcal{B}(Z, d\mu; \mathcal{X})$ and

$$\lim_{n\to\infty} \int_Z F_n(t)\, d\mu(t) = \int_Z F(t)\, d\mu(t).$$

An analogue to Theorem A.3.10 can be proven for some classes of functions, *e.g.*, for a monotonic $w : \mathbb{R} \to \mathbb{R}$.

Since $\mathcal{B}(\mathcal{X})$ is a Banach space, the Bochner integral is also used for operator–valued functions. For instance, suppose that a map $B : \mathbb{R} \to \mathcal{B}(\mathcal{X})$ is such that the vector–valued function $t \mapsto B(t)x$ is continuous for any $x \in \mathcal{X}$. Further, let $K \subset \mathbb{R}$ be a compact interval and μ a Borel measure on $\mathbb{R}$; then $\lim_{n\to\infty} \int_K B(t)x_n\, d\mu(t) = \int_K B(t)x\, d\mu(t)$ holds for any sequence $\{x_n\} \subset \mathcal{X}$ converging to a point x. Moreover, if an operator $T \in \mathcal{C}$ commutes with $B(t)$ for all $t \in K$, then $\int_K B(t)y\, d\mu(t)$ belongs to $D(T)$ for any $y \in D(T)$ and $T \int_K B(t)y\, d\mu(t) = \int_K B(t)Ty\, d\mu(t)$.

Appendix B. Some algebraic notions

In this appendix we collect some algebraic definitions and results needed in the text. There are again many textbooks and monographs in which this material is set out extensively; let us name, *e.g.*, ⟦BR 1⟧, ⟦Nai 1⟧, ⟦Ru 2⟧, or ⟦Ti⟧ for associative algebras, and ⟦BaR⟧, ⟦Kir⟧, ⟦Pon⟧, or ⟦Žel⟧ for Lie groups and algebras.

B.1 Involutive algebras

A *binary operation* in a set M is a map $\varphi : M \times M \to M$; it is *associative* or *commutative* if $\varphi(\varphi(a,b),c) = \varphi(a,\varphi(b,c))$ or $\varphi(a,b) = \varphi(b,a)$, respectively, holds for all $a,b,c \in M$. A set G equipped with an associative binary operation is called a *group* if there exist the *unit element* $e \in G$, $\varphi(g,e) = \varphi(e,g) = g$ for any $g \in G$, and the *inverse element* $g^{-1} \in G$ to any $g \in G$, $\varphi(g,g^{-1}) = \varphi(g^{-1},g) = e$.

Consider next a set R equipped with two binary operations, which we call *summation*, $\varphi_a(a,b) := a+b$, and *multiplication*, $\varphi_m(a,b) := ab$. The triplet (R,φ_a,φ_m) is a *ring* if (R,φ_a) is a commutative group and the two operations are distributive, $a(b+c) = ab+ac$ and $(a+b)c = ac + bc$ for all $a,b,c \in R$. If there is an $e \in R$ such that $ae = ea = a$ holds for all $a \in R$, we call it the *unit element* of R.

Let $\mathcal{A}$ be a vector space over a field F. The vector summation gives it the structure of a commutative group; if we define a multiplication which is distributive with the summation and satisfies $\alpha(ab) = (\alpha a)b = a(\alpha b)$ for any $a,b \in \mathcal{A}$, $\alpha \in \mathbb{C}$, then $\mathcal{A}$ becomes a ring, which we call a *linear algebra* over the field F, in particular, a *real* or *complex* algebra if $F = \mathbb{R}$ or $F = \mathbb{C}$, respectively. An algebra is said to be *associative* if its multiplication is associative. The term "algebra" without a further specification always means a complex associative algebra in what follows; we should stress, however, that many important algebras are nonassociative, *e.g.*, the Lie algebras discussed in Sec.B.3 below. An algebra is *Abelian* or *commutative* if its multiplication is commutative.

A *subalgebra* of an algebra $\mathcal{A}$ is a subset $\mathcal{B}$, which is itself an algebra with respect to the same operations. If $\mathcal{A}$ has the unit element, which is not contained in $\mathcal{B}$, then we can extend the subalgebra to $\tilde{\mathcal{B}} := \{\alpha e + b \,:\, \alpha \in \mathbb{C},\, b \in \mathcal{B}\}$; in a similar way, any algebra can be completed with the unit element by extending it to the set of pairs $[\alpha,a]$, $\alpha \in \mathbb{C}$, $a \in \mathcal{A}$, with the appropriately defined operations. A proper subalgebra $\mathcal{B} \subset \mathcal{A}$ is called a *(two–sided) ideal* in $\mathcal{A}$ if the products ab and ba belong to $\mathcal{B}$ for all $a \in \mathcal{A}$, $b \in \mathcal{B}$; we define the *left* and *right* ideal analogously. A trivial example of an ideal is the zero subalgebra $\{0\} \subset \mathcal{A}$. The algebra $\mathcal{A}$ itself is not regarded as an ideal;

thus no ideal can contain the unit element. A *maximal* ideal in $\mathcal{A}$ is such that it is not a proper subalgebra of another ideal in $\mathcal{A}$; any ideal in an algebra with the unit element is a subalgebra of some maximal ideal. An algebra is called *simple* if it contains no nontrivial two–sided ideal. The intersection of any family of subalgebras (ideals, one–sided ideals) in $\mathcal{A}$ is respectively a subalgebra (ideal, one–sided ideal), while the analogous assertion for the unions is *not* valid.

Let $\mathcal{A}$ be an algebra with the unit element. We say that an element $a \in \mathcal{A}$ is *invertible* if there exists an *inverse element* $a^{-1} \in \mathcal{A}$ such that $a^{-1}a = aa^{-1} = e$; we define the *left* and *right* inverse in the same way. For any $a \in \mathcal{A}$ there is at most one inverse; an element is invertible *iff* it belongs to no one–sided ideal of the algebra $\mathcal{A}$, which means, in particular, that in an algebra without one–sided ideals any nonzero element is invertible. Recall that a *field* is a ring with the unit element which has the last named property; the examples are $\mathbb{R}$, $\mathbb{C}$ or the noncommutative field $\mathbb{Q}$ of quaternions.

We define the *spectrum* of $a \in \mathcal{A}$ as the set $\sigma_{\mathcal{A}}(a) := \{\lambda : (a-\lambda e)^{-1} \text{ does not exist}\}$. The complement $\rho_{\mathcal{A}}(a) := \mathbb{C} \setminus \sigma_{\mathcal{A}}(a)$ is called the *resolvent set;* its elements are *regular values* for which the the *resolvent* $r_a(\lambda) := (a-\lambda e)^{-1}$ exists.

B.1.1 Proposition: Let $\mathcal{A}$ be an algebra with the unit element; then

(a) If a, ab are invertible, b is also invertible. If ab, ba are invertible, so are a and b.
(b) If $ab = e$, the element ba is idempotent but it need not be equal to the unit element unless $\dim \mathcal{A} < \infty$.
(c) If $e-ab$ is invertible, the same is true for $e-ba$.
(d) $\sigma_{\mathcal{A}}(ab) \setminus \{0\} = \sigma_{\mathcal{A}}(ba) \setminus \{0\}$, and moreover, $\sigma_{\mathcal{A}}(ab) = \sigma_{\mathcal{A}}(ba)$ provided one of the elements a, b is invertible.
(e) $\sigma_{\mathcal{A}}(a^{-1}) = \{\lambda^{-1} : \lambda \in \sigma_{\mathcal{A}}(a)\}$ holds for any invertible $a \in \mathcal{A}$.

For any set $\mathcal{S} \subset \mathcal{A}$ we define the algebra $\mathcal{A}_0(\mathcal{S})$ *generated by* $\mathcal{S}$ as the smallest subalgebra in $\mathcal{A}$ containing $\mathcal{S}$; it is easy to see that it consists just of all polynomials composed of the elements of $\mathcal{S}$ without an absolute term. We say that $\mathcal{S}$ is commutative if $ab = ba$ holds for any $a, b \in \mathcal{S}$; the algebra $\mathcal{A}_0(\mathcal{S})$ is then Abelian. A *maximal Abelian* algebra is such that it is not a proper subalgebra of an Abelian subalgebra; any Abelian subalgebra in $\mathcal{A}$ can be extended to a maximal Abelian subalgebra. We also define the *commutant* of a set $\mathcal{S} \subset \mathcal{A}$ as $\mathcal{S}' := \{a \in \mathcal{A} : ab = ba,\, b \in \mathcal{S}\}$; in particular, the *center* is the set $\mathcal{Z} := \mathcal{A}'$. We define the *bicommutant* $\mathcal{S}'' := (\mathcal{S}')'$ and higher–order commutants in the same way.

B.1.2 Proposition: Let $\mathcal{S}, \mathcal{T}$ be subsets in an algebra $\mathcal{S}$; then

(a) $\mathcal{S}'$ and $\mathcal{S}''$ are subalgebras containing the center $\mathcal{Z}$, and also the unit element if $\mathcal{A}$ has one. Moreover, $\mathcal{S}' = \mathcal{S}''' = \cdots$ and $\mathcal{S}'' = \mathcal{S}^{IV} = \cdots$.
(b) The inclusion $\mathcal{S} \subset \mathcal{T}$ implies $\mathcal{S}' \supset \mathcal{T}'$.
(c) $\mathcal{S} \subset \mathcal{S}''$, and $\mathcal{S}$ is commutative *iff* $\mathcal{S} \subset \mathcal{S}'$, which is further equivalent to the condition that $\mathcal{S}''$ is Abelian.
(d) $\mathcal{A}_0(\mathcal{S})' = \mathcal{S}'$ and $\mathcal{A}_0(\mathcal{S})'' = \mathcal{S}''$.
(e) A subalgebra $\mathcal{B} \subset \mathcal{A}$ is maximal Abelian *iff* $\mathcal{B} = \mathcal{B}'$; in that case also $\mathcal{B} = \mathcal{B}''$.

Let us turn to algebras with an additional unary operation. Recall that an involution $a \mapsto a^*$ on a vector space $\mathcal{A}$ is an antilinear map $\mathcal{A} \to \mathcal{A}$ such that $(a^*)^* = a$ holds for all $a \in \mathcal{A}$; an *involution on an algebra* is also required to satisfy the condition $(ab)^* = b^*a^*$ for any $a, b \in \mathcal{A}$. An algebra equipped with an involution is called an *involutive algebra* or briefly a $*$–*algebra.* A subalgebra in $\mathcal{A}$, which is itself a $*$–algebra w.r.t. the same involution, is called a $*$–*subalgebra;* we define the $*$–*ideal* in the same way. The element a^* is said to be *adjoint* to a. Given a subset $\mathcal{S} \subset \mathcal{A}$ we denote $\mathcal{S}^* := \{a^* : a \in \mathcal{S}\}$; the set $\mathcal{S}$ is *symmetric* if $\mathcal{S}^* = \mathcal{S}$; in particular, an element a fulfilling $a^* = a$ is called *Hermitean.* By $\mathcal{A}_0^*(\mathcal{S})$ we denote the smallest $*$–subalgebra in $\mathcal{A}$ containing the set $\mathcal{S}$.

B.1.3 Proposition: Let $\mathcal{A}$ be a $*$–algebra; then

(a) Any element is a linear combination of two Hermitean elements, and $e^* = e$ provided $\mathcal{A}$ has the unit element.

(b) a^* is invertible *iff* a is invertible, and $(a^*)^{-1} = (a^{-1})^*$.

(c) $\sigma_{\mathcal{A}}(a^*) = \overline{\sigma_{\mathcal{A}}(a)}$ holds for any $a \in \mathcal{A}$.

(d) A subalgebra $\mathcal{B} \subset \mathcal{A}$ is a $*$–subalgebra *iff* it is symmetric; the intersection of any family of $*$–subalgebras ($*$–ideals) is a $*$–subalgebra ($*$–ideal).

(e) Any $*$–ideal in $\mathcal{A}$ is two–sided.

(f) $\mathcal{A}_0^*(\mathcal{S}) = \mathcal{A}_0(\mathcal{S} \cup \mathcal{S}^*)$ holds for any subset $\mathcal{S} \subset \mathcal{A}$; if $\mathcal{S}$ is symmetric, then $\mathcal{S}'$ and $\mathcal{S}''$ are $*$–subalgebras in $\mathcal{A}$.

B.1.4 Example (bounded–operator algebras): The set $\mathcal{B}(\mathcal{H})$ with the natural algebraic operations and the involution $B \mapsto B^*$ is a $*$–algebra whose unit element is the operator I. Let us mention a few of its subalgebras:

(a) If E is a nontrivial projection, then $\{EB : B \in \mathcal{B}(\mathcal{H})\}$ is a right ideal but not a $*$–subalgebra; on the other hand, $\{EBE : B \in \mathcal{B}(\mathcal{H})\}$ is a $*$–subalgebra but not an ideal.

(b) If $\dim \mathcal{H} = \infty$, the sets $\mathcal{K}(\mathcal{H}) \supset \mathcal{J}_2(\mathcal{H}) \supset \mathcal{J}_1(\mathcal{H})$ of compact, Hilbert–Schmidt, and trace–class operators, respectively, are ideals in $\mathcal{B}(\mathcal{H})$; similarly $\mathcal{J}_p(\mathcal{H})$ is an ideal in any $\mathcal{J}_q(\mathcal{H})$, $q > p$, *etc.*

(c) The algebra $\mathcal{A}_0(B)$ generated by an operator $B \in \mathcal{B}(\mathcal{H})$ consists of all polynomials in B without an absolute term. It is a $*$–algebra if B is Hermitean, while the opposite implication is not valid; for instance, the Fourier–Plancherel operator F is non–Hermitean but $\mathcal{A}_0(F)$ is a $*$–algebra because $F^3 = F^{-1} = F^*$.

The algebras of bounded operators, which represent our main topic of interest, inspire some definitions. We have already introduced the notions of spectrum and hermiticity; similarly an element $a \in \mathcal{A}$ is said to be *normal* if $aa^* = a^*a$, a *projection* if $a^* = a = a^2$, and *unitary* if $a^* = a^{-1}$, *etc.* Of course, we also employ other algebras than $\mathcal{B}(\mathcal{H})$ and its subalgebras, *e.g.*, the Abelian $*$–algebra $C(M)$ of continuous complex functions on a compact space M with natural summation and multiplication, and the involution given by complex conjugation, $(f^*)(x) := \overline{f(x)}$.

An ideal $\mathcal{J}$ in an algebra $\mathcal{A}$ is a subspace, so we can construct the factor space $\mathcal{A}/\mathcal{J}$. It becomes an algebra if we define on it a multiplication by $\tilde{a}\tilde{b} := \widetilde{ab}$, where a, b are any elements representing the equivalence classes $\tilde{a}$ and $\tilde{b}$; it is called the *factor algebra* (of $\mathcal{A}$ w.r.t. the ideal $\mathcal{J}$). If $\mathcal{A}$ has the unit element, then the class $\tilde{e} := \{e+c : c \in \mathcal{J}\}$ is the unit element of $\mathcal{A}/\mathcal{J}$.

A *morphism* of algebras $\mathcal{A}, \mathcal{B}$ is a map $\varphi : \mathcal{A} \to \mathcal{B}$ which preserves the algebraic structure, $\varphi(\alpha a + b) = \alpha\varphi(a) + \varphi(b)$ and $\varphi(ab) = \varphi(a)\varphi(b)$ for all $a, b \in \mathcal{A}$, $\alpha \in \mathbb{C}$. In particular, if φ is surjective, then the image of the unit element (an ideal, maximal ideal, maximal Abelian subalgebra) in $\mathcal{A}$ is respectively the unit element (an ideal, ...) in $\mathcal{B}$. If φ is bijective, we call it an *isomorphism*; in the case $\mathcal{A} = \mathcal{B}$ one uses the terms *endomorphism* and *automorphism* of $\mathcal{A}$, respectively. The *null–space* of a morphism φ is the pull–back $\varphi^{-1}(0_{\mathcal{B}})$ of the zero element of the algebra $\mathcal{B}$; it is an ideal in $\mathcal{A}$. If $\mathcal{A}$, $\mathcal{B}$ are $*$–algebras and φ preserves the involution, $\varphi(a^*) = \varphi(a)^*$, it is called $*$–morphism.

B.1.5 Example: Let $\mathcal{J}$ be a ($*$–)ideal in a ($*$–)algebra $\mathcal{A}$; then the map $\varphi_c : \varphi_c(a) = \tilde{a}$ is a ($*$–)morphism of $\mathcal{A}$ to $\mathcal{A}/\mathcal{J}$. It is called a *canonical morphism*; its null–space is just the ideal $\mathcal{J}$. The factor algebra $\mathcal{A}/\mathcal{J}$ is simple *iff* the ideal $\mathcal{J}$ is maximal. Moreover, any ($*$–)morphism $\varphi : \mathcal{A} \to \mathcal{B}$ can be expressed as a composite mapping, $\varphi = \pi\circ\varphi_c$, where φ_c is the canonical morphism corresponding to the ($*$–)ideal $\mathcal{J} := \varphi^{-1}(0_{\mathcal{B}})$ and $\pi : \mathcal{A}/\mathcal{J} \to \varphi(\mathcal{A})$ is the ($*$–)isomorphism defined by $\pi(\tilde{a}) := \varphi(a)$ for any $a \in \mathcal{A}$.

Let us finally recall a few notions concerning representations. This term usually means the mapping of an algebraic object onto a suitable set of operators, which preserves the algebraic structure. We shall most often (but not exclusively) use representations by bounded operators: by a *representation* of a ($*$–)algebra $\mathcal{A}$ we understand in this case a ($*$–)morphism $\pi : \mathcal{A} \to \mathcal{B}(\mathcal{H})$, the space $\mathcal{H}$ is called the *representation space* and $\dim\mathcal{H}$ the *dimension* of the representation π. If the morphism π is injective, the representation is said to be *faithful*. Representations $\pi_j : \mathcal{A} \to \mathcal{B}(\mathcal{H}_j)$, $j = 1, 2$, are *equivalent* if there is a unitary operator $U : \mathcal{H}_1 \to \mathcal{H}_2$ such that $\pi_2(a)U = U\pi_1(a)$ holds for any $a \in \mathcal{A}$. A representation $\pi : \mathcal{A} \to \mathcal{B}(\mathcal{H})$ is called *irreducible* if the operator family $\pi(\mathcal{A})$ has no nontrivial closed invariant subspace. A vector $x \in \mathcal{H}$ is *cyclic* for the representation π if the set $\pi(\mathcal{A})x := \{\pi(a)x : a \in \mathcal{A}\}$ is dense in $\mathcal{H}$. Representations of groups, Lie algebras, *etc.*, are defined in the same way.

B.2 Banach algebras

Algebras can be equipped with a topological structure. Suppose that an algebra $\mathcal{A}$ is at the same time a locally convex topological space with a topology τ; then we call it a *topological algebra* if the multiplication is *separately* continuous, *i.e.*, the maps $a \mapsto ab$ and $a \mapsto ba$ are continuous w.r.t. the topology τ for any fixed $b \in \mathcal{A}$. A subalgebra $\mathcal{B} \subset \mathcal{A}$ is *closed* if it is closed as a subset in $\mathcal{A}$. The closed subalgebra $\mathcal{A}(\mathcal{S})$ *generated by a set* $\mathcal{S} \subset \mathcal{A}$ is the smallest closed subalgebra in $\mathcal{A}$ containing $\mathcal{S}$. Isomorphisms $\varphi : \mathcal{A} \to \mathcal{B}$ of topological algebras are classified by their continuity: the algebras $\mathcal{A}$, $\mathcal{B}$ are *topologically isomorphic* if there is a continuous isomorphism φ such that φ^{-1} is also continuous.

B.2.1 Proposition: Let $\mathcal{A}$ be a topological algebra and $\mathcal{S}$, $\mathcal{B}$ its subset and subalgebra, respectively; then

(a) $\overline{\mathcal{B}}$ is a closed subalgebra in $\mathcal{A}$. If $\mathcal{B}$ is Abelian, the same is true for $\overline{\mathcal{B}}$, and any maximal Abelian subalgebra is closed.
(b) $\mathcal{A}(\mathcal{S}) = \overline{\mathcal{A}_0(\mathcal{S})}$.
(c) The subalgebras $\mathcal{S}'$, $\mathcal{S}''$ are closed and $(\overline{\mathcal{S}})' = \mathcal{S}'$.
(d) If $\mathcal{B}$ is an ideal, $\overline{\mathcal{B}} \neq \mathcal{A}$, then $\overline{\mathcal{B}}$ is also an ideal in $\mathcal{A}$. Any maximal ideal is closed.
(e) the null–space of a continuous morphism $\varphi : \mathcal{A} \to \mathcal{C}$ is a closed ideal in $\mathcal{A}$.

There are various ways how of defining a topology on an algebra.

B.2.2 Example: The strong and weak operator topologies on $\mathcal{B}(\mathcal{H})$ are both locally convex, and the operator multiplication is separately continuous with respect to them (compare with Theorem 3.1.9 and Problem 3.9); thus $\mathcal{B}_s(\mathcal{H})$ and $\mathcal{B}_w(\mathcal{H})$ are topological algebras.

One of the most natural ways is to introduce a topology by means of a norm. An algebra $\mathcal{A}$ is called a *normed algebra* provided

(i) $\mathcal{A}$ is a normed space with a norm $\|\cdot\|$.
(ii) $\|ab\| \le \|a\|\,\|b\|$ for any $a, b \in \mathcal{A}$.
(iii) If $\mathcal{A}$ has the unit element, then $\|e\| = 1$.

The last condition may be replaced by $\|e\| \le 1$ because the opposite inequality follows from (ii). The multiplication in a normed algebra is jointly continuous. If $\mathcal{A}$ is complete w.r.t. the norm $\|\cdot\|$, it is called a *Banach algebra.* We can again assume without loss of generality that a normed algebra $\mathcal{A}$ has the unit element; otherwise we extend it in the above described way, defining the norm by $\|[\alpha, a]\| := |\alpha| + \|a\|_{\mathcal{A}}$. If $\mathcal{J}$ is a closed ideal in a Banach algebra $\mathcal{A}$, then $\mathcal{A}/\mathcal{J}$ is a Banach algebra w.r.t. the norm $\|a\| := \inf_{b\in\mathcal{J}} \|a-b\|_{\mathcal{A}}$.

A *complete envelope* of a normed algebra $\mathcal{A}$ is a Banach algebra $\mathcal{B}$ such that it contains $\mathcal{A}$ as a dense subalgebra and $\|a\|_{\mathcal{A}} = \|a\|_{\mathcal{B}}$ holds for any $a \in \mathcal{A}$.

B.2.3 Theorem: Any normed algebra $\mathcal{A}$ has a complete envelope, which is unique up to an isometric isomorphism preserving the elements of $\mathcal{A}$.

The space $\mathcal{B}(\mathcal{H})$ equipped with the operator norm provides an example of a Banach algebra; the complete envelope of any subalgebra $\mathcal{B} \subset \mathcal{B}(\mathcal{H})$ is its closure $\overline{\mathcal{B}}$. By a direct generalization of the methods of Section 1.7, we can prove the following claims.

B.2.4 Theorem: Let $\mathcal{A}$ be a Banach algebra with the unit element; then

(a) Any element $a \in \mathcal{A}$ fulfilling $\|a-e\| < 1$ is invertible. The set $\mathcal{R}$ of all invertible elements in $\mathcal{A}$ is open and the map $a \mapsto a^{-1}$ is continuous in it.
(b) The resolvent set $\rho_{\mathcal{A}}(a)$ of any element $a \in \mathcal{A}$ is open in $\mathbb{C}$ and the resolvent $r_a : \rho_{\mathcal{A}} \to \mathcal{A}$ is analytic.
(c) The spectrum $\sigma_{\mathcal{A}}(a)$ of any element $a \in \mathcal{A}$ is a nonempty compact set.
(d) The spectral radius $r(a) := \sup\{\, |\lambda| : \lambda \in \sigma_{\mathcal{A}}(a) \,\}$ is independent of $\mathcal{A}$ and equals

$$r(a) = \lim_{n\to\infty} \|a^n\|^{1/n} = \inf_n \|a^n\|^{1/n} ;$$

it does not exceed the norm, $r(a) \le \|a\|$.

The independence feature of part (d) is not apparent in the case of bounded operators, where the spectral quantities are related to a single algebra $\mathcal{B}(\mathcal{H})$. To appreciate this result, notice that the spectrum, and in particular its radius, is a purely algebraic property, while the right side of the formula depends on the metric properties of the algebra $\mathcal{A}$.

B.2.5 Proposition: Let $\mathcal{A}, \mathcal{B}$ be Banach algebras; then

(a) A morphism $\varphi : \mathcal{A} \to \mathcal{B}$ is continuous *iff* there is C such that $\|\varphi(a)\|_{\mathcal{B}} \le C\|a\|_{\mathcal{A}}$ for any $a \in \mathcal{A}$; if φ is a continuous isomorphism, then the algebras $\mathcal{A}, \mathcal{B}$ are topologically isomorphic.
(b) If $\mathcal{A}, \mathcal{B}$ are complete envelopes of normed algebras $\mathcal{A}_0, \mathcal{B}_0$, then any continuous morphism $\varphi_0 : \mathcal{A}_0 \to \mathcal{B}_0$ has just one continuous extension $\varphi : \mathcal{A} \to \mathcal{B}$.
(c) If $\mathcal{J}$ is the null–space of a continuous surjective morphism $\varphi : \mathcal{A} \to \mathcal{B}$, then $\mathcal{A}/\mathcal{J}$ and $\mathcal{B}$ are topologically isomorphic.

An isomorphism $\varphi : \mathcal{A} \to \mathcal{B}$ is called *isometric* if $\|\varphi(a)\|_{\mathcal{B}} = \|a\|_{\mathcal{A}}$ holds for all $a \in \mathcal{A}$.

B.2.6 Theorem (Gel'fand–Mazur): A Banach algebra with the unit element, in which any nonzero element is invertible, is isometrically isomorphic to the field $\mathbb{C}$ of complex numbers.

B.3 Lie algebras and Lie groups

A *Lie algebra* (real, complex, or more generally, over a field F) is a finite–dimensional (nonassociative) linear algebra L with the multiplication which we conventionally denote as $(a,b) \mapsto [a,b]$. The latter is antisymmetric, $[a,b] = -[b,a]$, and satisfies the *Jacobi identity* $[a,[b,c]] + [b,[c,a]] + [c,[a,b]] = 0$ for any $a,b,c \in L$. The *dimension* of L is its vector–space dimension; if $\{e_j\}_{j=1}^n$ is a basis in L, the product is fully determined by the relations $[e_j, e_k] = c^i_{jk} e_i$, where the coefficients c^i_{jk} are called the *structure constants* (one uses the *summation convention*, according to which the sum is taken over any repeated index). A *complex extension* $L_{\mathbb{C}}$ is a complex extension of L as a vector space with the product $[a_1 + ib_1, a_2 + ib_2] := [a_1,a_2] - [b_1,b_2] + i[a_1,b_2] + i[b_1,a_2]$.

A Lie algebra is *commutative* if $[a,b] = 0$ for all $a,b \in L$, *i.e.*, all the structure constants are zero; this definition differs from the associative case. In contrast, other definitions like those of a subalgebra, ideal, and also morphisms, representations, *etc.*, modify easily for Lie algebras.

B.3.1 Examples: Consider the following matrix algebras:

(a) The set of all $n \times n$ real matrices forms an n^2–dimensional real Lie algebra called the *general linear algebra* and denoted as $gl(n,\mathbb{R})$; its subalgebra $sl(n,\mathbb{R})$ consists of all traceless $g \in gl(n,\mathbb{R})$. Its complex extension is the algebra $gl(n,\mathbb{C})$ of $n \times n$ complex matrices; similarly traceless complex matrices form the algebra $sl(n,\mathbb{C})$, which is often also denoted as A_{n-1}.

(b) The algebra $gl(m,\mathbb{C})$ has other Lie subalgebras. A matrix g is said to be *skew–symmetric* if $g^t = -g$, where g^t denotes the transposed matrix of g (this property should not be confused with antihermiticity). The set of skew–symmetric $m \times m$ matrices forms the *orthogonal* Lie algebra, which is denoted as $o(m,\mathbb{C})$; we alternatively speak about the algebras B_n and D_n for $o(2n+1,\mathbb{C})$ and $o(2n,\mathbb{C})$, respectively. On the other hand, consider the subset in $gl(2n,\mathbb{C})$ consisting of matrices such that $g^t j_{2n} + j_{2n} g = 0$, where $j_{2n} := \begin{pmatrix} 0 & e_n \\ e_n & 0 \end{pmatrix}$ and e_n is the $n \times n$ unit matrix. The corresponding Lie algebra is called *symplectic* and denoted as $sp(n,\mathbb{C})$ or C_n.

Given a Lie algebra L, we define the subalgebras $L^{(n)}$ by the recursive relations $L^{(0)} := L$ and $L^{(n+1)} := [L^{(n)}, L^{(n)}]$; similarly we define $L_{(n)}$ by $L_{(0)} := L$ and $L_{(n+1)} := [L_{(n)}, L]$. The algebra is *solvable* if $L^{(n)} = \{0\}$ for some n; it is *nilpotent* if $L_{(n)}$ for some n. Any nilpotent algebra is solvable. A commutative Lie algebra is, of course, nilpotent; a less trivial example is the Heisenberg–Weyl algebra, which is nilpotent of order two.

On the other hand, a Lie algebra L is *semisimple* if it has no commutative ideal; it is *simple* if it has no (nontrivial) ideal at all. An equivalent characterization leans on the notion of the *Cartan tensor* g : $g_{rs} = c^i_{rk} c^k_{si}$, through which one defines the *Killing form* $L \times L \to F$ by $(a,b) := g_{rs} a^r b^s$. The algebra L is semisimple *iff* its Killing form is nondegenerate, *i.e.*, $\det g \neq 0$. For *real* Lie algebras, the Killing form may be used to introduce the following notion: L is *compact* if the form is positive, and noncompact otherwise. A compact L can be expressed as $L = Z \oplus S = Z \oplus S_1 \oplus \cdots \oplus S_n$, where Z is its center, S is semisimple, and $S_1, \dots, S_n$ are simple algebras.

Simple Lie algebras allow a full classification. It appears that, up to an isomorphism, *complex* simple algebras are almost exhausted by the types A_n, B_n, C_n, and D_n listed in Example 1; there are just five more simple Lie algebras called *exceptional*. For small values of n, some of these algebras are isomorphic, namely $A_1 \sim B_1 \sim C_1$, $B_2 \sim D_2$, and $A_3 \sim D_3$, while D_2 is semisimple and isomorphic to $A_1 \oplus A_1$. One can classify real forms of simple complex algebras in a similar way.

The notion of a group was introduced above; for simplicity we shall here denote the group operation as a multiplication. A *subgroup* of a group G is a subset $H \subset G$ which is itself a group w.r.t. the same operation. H is a *left invariant subgroup* if $hg \in H$ holds for any $h \in H$ and $g \in G$. We define the right invariant subgroup in a similar way; a subgroup is *invariant* provided it is left and right invariant at the same time (these notions play a role analogous to ideals in algebras). The notions of the *direct product* of groups and a *factor group* are easy modifications of the above discussed algebraic definitions.

A *topological group* is a group G, which is simultaneously a T_1 topological space such that the map $g \mapsto g^{-1}$ is continuous and the group multiplication is *jointly* continuous. An *isomorphism* of topological groups is a map which is a group isomorphism and, at the same time, a homeomorphism of the corresponding topological spaces. One introduces various classes of topological groups according to the properties of G as a topological space, *e.g.*, *compact* groups or *locally compact* groups. In a similar way, one defines a *connected* group. If G is not connected, it can be decomposed into connected components; the component containing the unit element is a closed invariant subgroup.

B.3.2 Examples: (a) The group T_n of translations of the Euclidean space $\mathbb{R}^n$ is a commutative topological group, which is locally but not globally compact.

(b) The orthogonal group $O(n)$ consists of real orthogonal $n \times n$ matrices, *i.e.*, such that $g^t g = e$. It is compact and has two connected components specified by the conditions $\det g = \pm 1$; the connected component of the unit element is the rotation group denoted as $SO(n)$.

(c) The group $U(n)$ of unitary complex $n \times n$ matrices is locally compact and connected; the same is true for its subgroup $SU(n)$ of matrices with $\det g = 1$.

(d) The substitution operators U_φ of Example 3.3.2 form a group. If we equip the set of mappings φ with the metric $\varrho(\varphi, \tilde\varphi) := \sup_{x \in \mathbb{R}^n} |\varphi(x) - \tilde\varphi(x)|$, it becomes a topological group which is not locally compact.

A topological group can also be equipped with a measure. The easiest way to introduce it is through linear functionals — *cf.* the concluding remark in Section A.2. Consider such a functional μ on the space $C_0(G)$ of continuous functions with compact supports which is positive, *i.e.*, $\mu(f) \geq 0$ holds for all $f \geq 0$. If it satisfies $\mu(f(g^{-1}\cdot)) = \mu(f)$ for any $f \in C_0(G)$ and $g \in G$, it is called a *left Haar* measure on G. On a locally compact group G, there is always a left Haar measure and it is unique up to a multiplicative constant. We introduce the right Haar measure in a similar way; a measure on G is said to be *invariant* if it combines the two properties.

An important class of topological groups consists of those which allow a locally Euclidean parametrization. To be more precise, the notion of an *analytic manifold* is needed. This is a Hausdorff space M together with a family of pairs $(U_\alpha, \varphi_\alpha)$, $\alpha \in I$, where U_α is an open set in M and φ_α is a homeomorphism $U_\alpha \to \mathbb{R}^n$ for a fixed n with the following properties: $\bigcup_{\alpha \in I} U_\alpha = M$ and for any $\alpha, \beta \in I$ the component functions of the map $\varphi_\beta \circ \varphi_\alpha^{-1}$ are real analytic on $\varphi_\alpha(U_\alpha \cap U_\beta)$. The number n is called the *dimension* of the manifold; replacing $\mathbb{R}^n$ by $\mathbb{C}^n$ in the definition we introduce complex analytic manifolds in the same way.

Given analytic manifolds M, N, one can associate with their topological product $(M \times N, \tau_{M \times N})$ the family of pairs $(U_\alpha \times V_\beta,\, \varphi_\alpha \times \psi_\beta)$. The obtained structure again satisfies the above conditions; we call it the *product manifold* of M and N. The dimension of the product manifold is $m = m_M + m_N$. A map $M \to N$ is *analytic* if the component functions of all the maps $\psi_\beta \circ \varphi_\alpha^{-1}$ are analytic on their domains.

A group G is called a (real or complex) *Lie group* if it is an analytic manifold (real or complex, respectively) and its multiplication and inversion as maps $G \times G \to G$ and $G \to G$, respectively, are analytic. For instance, the groups of Examples B.3.2a–c belong to this class; this is not true for the group of Example B.3.2d, where the dimension of the group manifold is infinite. Any Lie group is locally compact. A subgroup of G which is itself a Lie group with the same multiplication is called a *Lie subgroup*.

As an analytic manifold, at the vicinity of any point a Lie group admits a description through local coordinates defined by the corresponding map φ_α. This concerns, in particular, the unit element e: there is a neighborhood U of the point $0 \in \mathbb{R}^n$ where we can parametrize the group elements by $g \equiv (g_1, \dots, g_n)$. The composition law of G is then *locally* expressed by real analytic functions f_j, $j = 1, \dots, n$, from $U \times U$ to $\mathbb{R}^n$ so that $(gh)_j = f_j(g_1, \dots, g_n, h_1, \dots, h_n)$; the consistency requires $f_j(g, 0) = g_j$, $f_j(0, h) = h_j$ and $(\partial f_j / \partial g_k)_{(0,0)} = (\partial f_j / \partial h_k)_{(0,0)} = \delta_{jk}$. The *structure constants* of the group G are defined by $c^i_{jk} := (\partial^2 f^i / \partial g_j \partial h_k - \partial^2 f^i / \partial h_j \partial g_k)_{(0,0)}$; they satisfy the same conditions as the structure constants of Lie algebras.

This is not a coincidence; there is a close connection between Lie groups and Lie algebras. Let U be the neighborhood of the parameter–space origin used above and consider the space $C^\infty(\overline{U})$. The operators $T_j : (T_j\phi)(g) = (\partial\phi/\partial g_j)(gg_j^{-1})|_{g=0}$ are then well–defined and span a Lie algebra L which is said to be *associated* with G. The correspondence extends to subalgebras: if H is a Lie subgroup of G, then its Lie algebra M is a subalgebra of L. Moreover, if H is an invariant subgroup, then M is an ideal in L, *etc.*On the other hand, the association $G \mapsto L$ is not injective.

B.3.3 Examples: (a) The group of rotations (translations on a circle) $SO(2)$ can be expressed as $SO(2) = T_1/\mathbb{Z}$, where T_1 means translations on a line and $\mathbb{Z}$ is the additive group of integers. Both $SO(2)$ and T_1 have the same (one–dimensional) Lie algebra $so(2)$.

(b) Let $\mathbb{Z}_2$ be the two–point group $\{0, 1\}$ with the addition modulo 2. The Lie groups $SU(2)$ and $SO(3) = SU(2)/\mathbb{Z}_2$ have the same Lie algebra $so(3)$ as was discussed in Example 10.2.3e.

Basic notions concerning the representation theory of Lie groups and algebras can be readily adapted from the preceding sections. The correspondence discussed above induces a natural relation between some representations of a Lie group and those of its Lie algebra; in the simplest case of a one–dimensional G this is the content of Stone's theorem. In general, however, the representation theory of Lie groups and algebras is a complicated subject which goes beyond the scope of the present book; we refer the reader to the literature quoted at the beginning of the appendix.

References

a) Monographs, textbooks, proceedings:

⟦Ad⟧ R.A. Adams: *Sobolev Spaces*, Academic, New York 1975.

⟦AG⟧ N.I. Akhiezer, I.M. Glazman: *Theory of Linear Operators in Hilbert space*, 3rd edition, Viša Škola, Kharkov 1978 *(in Russian; English translation of the 1st edition: F. Ungar Co., New York 1961, 1963).*

⟦ACH⟧ S. Albeverio *et al.*, eds.: *Feynman Path Integrals*, Lecture Notes in Physics, vol. 106, Springer, Berlin 1979.

⟦AFHL⟧ S. Albeverio *et al.*, eds.: *Ideas and Methods in Quantum and Statistical Physics. R. Høegh–Krohn Memorial Volume*, Cambridge University Press, Cambridge 1992.

⟦AGHH⟧ S. Albeverio, F. Gesztesy, R. Høegh–Krohn, H. Holden: *Solvable Models in Quantum Mechanics*, 2nd edition, with an appendix by P. Exner, AMS Chelsea Publishing, Providence, RI, 2005

⟦AH⟧ S. Albeverio, R. Høegh–Krohn: *Mathematical Theory of Feynman Path Integrals*, Lecture Notes in Mathematics, vol. 523, Springer, Berlin 1976.

⟦Al⟧ P.S. Alexandrov: *Introduction to Set Theory and General Topology*, Nauka, Moscow 1977 *(in Russian).*

⟦Am⟧ W.O. Amrein: *Non–Relativistic Quantum Dynamics*, Reidel, Dordrecht 1981.

⟦AJS⟧ W.O. Amrein, J.M. Jauch, K.B. Sinha: *Scattering Theory in Quantum Mechanics. Physical Principles and Mathematical Methods*, Benjamin, Reading, MA 1977.

⟦AŽŠ⟧ M.A. Antonec, G.M. Žislin, I.A. Šereševskii: *On the discrete spectrum of N–body Hamiltonians*, an appendix to the Russian translation of the monograph ⟦JW⟧, Mir, Moscow 1976 *(in Russian).*

⟦BaR⟧ A.O. Barut, R. Raczka: *Theory of Group Representations and Applications*, 2nd edition, World Scientific, Singapore 1986.

⟦BW⟧ H. Baumgärtel, M. Wollenberg: *Mathematical Scattering Theory*, Akademie Verlag, Berlin 1983.

⟦Ber⟧ F.A. Berezin: *The Second Quantization Method*, 2nd edition, Nauka, Moscow 1986 *(in Russian; English transl. of the 1st edition: Academic, New York 1966).*

⟦BeŠ⟧ F.A. Berezin, M.A. Šubin: *Schrödinger Equation*, Moscow State University Publ., Moscow 1983 *(in Russian; English translation: Kluwer, Dordrecht 1996).*

⟦BL⟧ L.C. Biedenharn, J.D. Louck: *Angular Momentum in Quantum Theory. Theory and Applications*, Addison–Wesley, Reading, MA 1981.

⟦BS⟧ M.Š. Birman, M.Z. Solomyak: *Spectral Theory of Self-Adjoint Operators in Hilbert Space*, Leningrad State University Lenninguad. 1980 *(in Russian; English translation: Kluwer, Dordrecht 1987).*

⟦BD 1,2⟧ J.D. Bjorken, S.D. Drell: *Relativistic Quantum Theory, I. Relativistic Quantum Mechanics, II. Relativistic Quantum Fields*, McGraw–Hill, New York 1965.

⟦Boe⟧ H. Boerner: *Darstellungen von Gruppen*, 2.Ausgabe, Springer, Berlin 1967 *(English translation: North-Holland, Amsterdam 1970).*

⟦BLOT⟧ N.N. Bogolyubov, A.A. Logunov, A.I. Oksak, I.T. Todorov: *General Principles of Quantum Field Theory*, Nauka, Moscow 1987 (*in Russian;* a revised edition of *Foundations of the Axiomatic Approach to Quantum Field Theory* by the first two and the last author, Nauka, Moscow 1969; *English translation: W.A. Benjamin, Reading, MA 1975*, referred to as ⟦BLT⟧).

⟦BŠ⟧ N.N. Bogolyubov, D.V. Širkov: *An Introduction to the Theory of Quantized Fields*, 4th edition, Nauka, Moscow 1984 *(in Russian; English translation of the 3rd edition: Wiley–Interscience, New York 1980).*

⟦Bo⟧ D. Bohm: *Quantum Theory*, Prentice-Hall, New York 1952.

⟦BR 1,2⟧ O. Bratelli, D.W. Robinson: *Operator Algebras and Quantum Statistical Mechanics I, II*, Springer, New York 1979, 1981.

⟦BLM⟧ P. Busch, P.J. Lahti, P. Mittelstaedt: *The Quantum Theory of Measurement*, 2nd revised edition, Springer LNP m2, Berlin 1996.

⟦ChS⟧ K. Chadan, P. Sabatier: *Inverse Problems in Quantum Scattering Theory*, 2nd edition, Springer, New York 1989.

⟦CL⟧ Tai–Pei Cheng, Ling–Fong Li: *Gauge Theory of Elementary Particle Physics*, Clarendon Press, Oxford 1984.

⟦Cher⟧ P.R. Chernoff: *Product Formulas, Nonlinear Semigroups and Addition of Unbounded Operators*, Mem. Amer. Math. Soc., Providence, RI 1974.

⟦CFKS⟧ H.L. Cycon, R.G. Froese, W. Kirsch, B. Simon: *Schrödinger operators, with Application to Quantum Mechanics and Global Geometry*, Springer, Berlin 1987; corrected and extended 2nd printing, Springer, Berlin 2007.

⟦Da 1⟧ E.B. Davies: *Quantum Theory of Open Systems*, Academic, London 1976.

⟦Da 2⟧ E.B. Davies: *One–Parameter Semigroups*, Academic, London 1980.

⟦Dav⟧ A.S. Davydov: *Quantum Mechanics*, 2nd edition, Nauka, Moscow 1973 *(in Russian; English translation of the 1st edition: Pergamon Press, Oxford 1965).*

⟦DO⟧ Yu.N. Demkov, V.N. Ostrovskii: *The Zero–Range Potential Method in Atomic Physics*, Leningrad University Press, Leningrad 1975 *(in Russian).*

⟦DENZ⟧ M. Demuth, P. Exner, H. Neidhardt, V.A. Zagrebnov, eds.: *Mathematical Results in Quantum Mechanics*, Operator Theory: Advances and Applications, vol. 70; Birkhäuser, Basel 1994.

⟦DG⟧ J. Derezinski, Ch. Gerard: *Scattering theory of classical and quantum N–particle systems*, Texts and Monographs in Physics, Springer, Berlin 1997.

⟦Dir⟧ P.A.M. Dirac: *The Principles of Quantum Mechanics*, 4th edition, Clarendon Press, Oxford 1969.

⟦DE⟧ J. Dittrich, P. Exner, eds.: *Rigorous Results in Quantum Dynamics*, World Scientific, Singapore 1991.

⟦DET⟧ J. Dittrich, P. Exner, M. Tater, eds.: *Mathematical Results in Quantum Mechanics*, Operator Theory: Advances and Appl., vol. 108; Birkhäuser, Basel 1999.

⟦Di 1⟧ J. Dixmier: *Les algèbres des opérateurs dans l'espace hilbertien (algèbres de von Neumann)*, 2me edition, Gauthier–Villars, Paris 1969.

⟦Di 2⟧ J. Dixmier: *Les C^*–algèbras and leur représentations*, 2me edition, Gauthier-Vilars, Paris 1969.

⟦DS 1–3⟧ N. Dunford, J.T. Schwartz: *Linear Operators, I. General Theory, II. Spectral Theory, III. Spectral Operators*, Interscience Publications, New York 1958, 1962, 1971.

⟦Edm⟧ A.R. Edmonds: *Angular Momentum in Quantum Mechanics*, Princeton University Press, Princeton, NJ 1957; a revised reprint Princeton 1996.

⟦EG⟧ S.J.L. van Eijndhoven, J. de Graaf: *A Mathematical Introduction to Dirac Formalism*, North–Holland, Amsterdam 1986.

⟦Em⟧ G.G. Emch: *Algebraic Methods in Statistical Mechanics and Quantum Field Theory*, Wiley–Interscience, New York 1972.

⟦Ex⟧ P. Exner: *Open Quantum Systems and Feynman Integrals*, D. Reidel, Dordrecht 1985.

⟦EKKST⟧ P. Exner, J. Keating, P. Kuchment, T. Sunada, A. Teplyaev, eds.: *Analysis on Graphs and Applications*, Proceedings of the Isaac Newton Institute programme "Analysis on Graphs and Applications", AMS "Contemporary Mathematics" Series, 2008

⟦EN⟧ P. Exner, J. Neidhardt, eds.: *Order, Disorder and Chaos in Quantum Systems*, Operator Theory: Advances and Applications, vol. 46; Birkhäuser, Basel 1990.

⟦EŠ 1⟧ P. Exner, P. Šeba, eds.: *Applications of Self–Adjoint Extensions in Quantum Physics*, Lecture Notes in Physics, vol. 324; Springer, Berlin 1989.

⟦EŠ 2⟧ P. Exner, P. Šeba, eds.: *Schrödinger Operators, Standard and Non–standard*, World Scientific, Singapore 1989.

⟦Fel 1,2⟧ W. Feller: *An Introduction to Probability Theory and Its Applications I, II*, 3rd and 2nd edition, resp., Wiley, New York 1968, 1971.

⟦Fey⟧ R.P. Feynman: *Statistical Mechanics. A Set of Lectures*, W.A. Benjamin, Reading, MA. 1972.

⟦FH⟧ R.P. Feynman, A.R. Hibbs: *Quantum Mechanics and Path Integrals*, McGraw–Hill, New York 1965.

⟦FG⟧ L. Fonda, G.C. Ghirardi: *Symmetry Principles in Quantum Physics*, Marcel Dekker, New York 1970.

⟦GŠ⟧ I.M. Gel'fand, G.M. Šilov: *Generalized Functions and Operations upon Them, vol.I*, 2nd edition, Fizmatgiz, Moscow 1959 *(in Russian; English translation: Academic, New York 1969).*

⟦Gl⟧ I.M. Glazman: *Direct Methods of Qualitative Analysis of Singular Differential Operators*, Fizmatgiz, Moscow 1963 *(in Russian).*

⟦GJ⟧ J. Glimm, A. Jaffe: *Quantum Physics: A Functional Integral Point of View*, 2nd edition, Springer, New York 1987.

⟦Gr⟧ A. Grothendieck: *Produits tensoriels topologiques et espaces nucléaires*, Mem. Am. Math. Soc., vol. 16, Providence, RI 1955.

⟦Haa⟧ R. Haag: *Local Quantum Physics. Fields, Particles, Algebras*, 2nd revised and enlarged edition Springer, Berlin 1996.

⟦Hal 1⟧ P. Halmos: *Measure Theory*, 2nd edition, Van Nostrand, New York 1973.

⟦Hal 2⟧ P. Halmos: *A Hilbert Space Problem Book*, Van Nostrand, Princeton, NJ 1967.

⟦Ham⟧ M. Hamermesh: *Group Theory and Its Applications to Physical Problems*, Addison-Wesley, Reading, MA. 1964.

⟦HW⟧ G.H. Hardy, E.M. Wright: *An Introduction to the Theory of Numbers*, 5th edition, Oxford University Press, oxford 1979.

[[Hel]] B. Helffer: *Semi–Classical Analysis for the Schrödinger Operator and Applications*, Lecture Notes in Mathematics, vol. 1366, Springer, Berlin 1988.

[[HP]] E. Hille, R.S. Phillips: *Functional Analysis and Semigroups*, Am. Math. Soc. Colloq. Publ., vol. 31, Providence, Rhode Island 1957; 3rd printing 1974.

[[Hol]] A.S. Holevo: *Probabilistic and Statistical Aspects of the Quantum Theory*, Nauka, Moscow 1980 *(in Russian; English translation: North-Holland, Amsterdam 1982)*.

[[Hör]] L. Hörmander: *The Analysis of Linear Partial Differential Operators III*, Springer, Berlin 1985; corrected reprint 1994.

[[Hor]] S.S. Horužii: *An Introduction to Algebraic Quantum Field Theory*, Nauka, Moscow 1986 *(in Russian; English translation: Kluwer, Dordrecht 1990)*.

[[vH]] L. van Hove: *Sur certaines représentations unitaires d'un group infini des transformations*, Memoires Acad. Royale de Belgique XXVI/6, Bruxelles 1951.

[[Hua 1]] K. Huang: *Statistical Mechanics*, Wiley, New York 1963.

[[Hua 2]] K. Huang: *Quarks, Leptons and Gauge Fields*, World Scientific, Singapore 1982.

[[Hur]] N.E. Hurt: *Geometric Quantization in Action*, D. Reidel, Dordrecht 1983.

[[IZ]] C. Itzykson, J.–B. Zuber: *Quantum Field Theory*, McGraw–Hill, New York 1980.

[[Jar 1]] V. Jarník: *Differential Calculus II*, 3rd edition, Academia, Prague 1976 *(in Czech)*.

[[Jar 2]] V. Jarník: *Integral Calculus II*, 2nd edition, Academia, Prague 1976 *(in Czech)*.

[[Ja]] J.M. Jauch: *Foundations of Quantum Mechanics*, Addison-Wesley, Reading, MA 1968.

[[Jor]] T.F. Jordan: *Linear Operators for Quantum Mechanics*, Wiley, New York 1969.

[[JW]] K. Jörgens, J. Weidmann: *Spectral Properties of Hamiltonian Operators*, Lecture Notes in Mathematics, vol. 313, Springer, Berlin 1973.

[[Kam]] E. Kamke: *Differentialgleichungen realer Funktionen*, Akademische Verlagsgesselschaft, Leipzig 1956.

[[Kas]] D. Kastler, ed.: *C^*–algebras and Their Applications to Statistical Mechanics and Quantum Field Theory*, North–Holland, Amsterdam 1976.

[[Ka]] T. Kato: *Perturbation Theory for Linear Operators*, Springer, 2nd edition, Berlin 1976; reprinted in 1995.

[[Kel]] J.L. Kelley: *General Topology*, Van Nostrand, Toronto 1957; reprinted by Springer, Graduate Texts in Mathematics, No. 27, New York 1975.

[[Kir]] A.A. Kirillov: *Elements of the Representation Theory*, 2nd edition, Nauka, Moscow 1978 *(in Russian; French translation: Mir, Moscow 1974)*.

[[KGv]] A.A. Kirillov, A.D. Gvišiani: *Theorems and Problems of Functional Analysis*, Nauka, Moscow 1979 *(in Russian; French translation: Mir, Moscow 1982)*.

[[KS]] J.R. Klauder, B.–S. Skagerstam, eds.: *Coherent States. Applications in Physics and Mathematical Physics*, World Scientific, Singapore 1985.

[[Kli]] W. Klingenberg: *A Course in Differential Geometry*, Springer, New York 1978.

[[KF]] A.N. Kolmogorov, S.V. Fomin: *Elements of Function Theory and Functional Analysis*, 4th edition, Nauka, Moscow 1976 *(in Russian; English translation of the 2nd ed.: Graylock 1961; French translation of the 3rd edition: Mir, Moscow 1974)*.

[[Kuo]] H.-H. Kuo: *Gaussian Measures in Banach Spaces*, Lecture Notes in Mathematics, vol. 463, Springer, Berlin 1975.

[[Ku]] A.G. Kuroš: *Lectures on General Algebra*, 2nd edition, Nauka, Moscow 1973 *(in Russian)*.

[[LL]] L.D. Landau, E.M. Lifšic: *Quantum Mechanics. Nonrelativistic Theory*, 3rd ed., Nauka, Moscow 1974 *(in Russian; English translation: Pergamon, New York 1974)*.

[[LP]] P.D. Lax, R.S. Phillips: *Scattering Theory*, Academic, New York 1967; 2nd edition, with appendices by C.S. Morawetz and G. Schmidt, 1989.

[[LiL]] E.H. Lieb, M. Loss: *Analysis*, 2nd edition, Graduate Studies in Mathematics, vol. 14, AMS, Providence, RI, 2001

[[LSW]] E.H. Lieb, B. Simon, A.S. Wightman, eds.: *Studies in Mathematical Physics. Essays in Honor of V. Bargmann*, Princeton University Press, Pinceton, NJ 1976.

[[Loe 1–3]] E.M. Loebl, ed.: *Group Theory and Its Applications I–III*, Academic, New York 1967.

[[LCM]] J.T. Londergan, J.P. Carini, D.P. Murdock: *Binding and Scattering in Two-Dimensional Systems. Applications to Quantum Wires, Waveguides and Photonic Crystals*, Springer LNP m60, Berlin 1999.

[[Lud 1,2]] G. Ludwig: *Foundations of Quantum Mechanics I, II*, Springer, Berlin 1983, 1985.

[[Lud 3]] G. Ludwig: *An Axiomatic Basis for Quantum Mechanics, I. Derivation of Hilbert Space Structure*, Springer, Berlin 1985.

[[LRSS]] S. Lundquist, A. Ranfagni, Y. Sa–yakanit, L.S. Schulman: *Path Summation: Achievements and Goals*, World Scientific, Singapore 1988.

[[Mac 1]] G. Mackey: *Mathematical Foundations of Quantum Mechanics*, Benjamin, New York 1963; reprinted by Dover Publications INC., Mineola, NY, 2004.

[[Mac 2]] G. Mackey: *Induced Representations of Groups and Quantum Mechanics*, W.A. Benjamin, New York 1968.

[[MB]] S. MacLane, G. Birkhoff: *Algebra*, 2nd edition, Macmillan, New York 1979.

[[Mar]] A.I. Markuševič: *A Short Course on the Theory of Entire Functions*, 2nd edition, Fizmatgiz, Moscow 1961 *(in Russian; English translation: Amer. Elsevier 1966)*.

[[Mas]] V.P. Maslov: *Perturbation Theory and Asymptotic Methods*, Moscow State University Publ., Moscow 1965 *(in Russian; French translation: Dunod, Paris 1972)*.

[[MF]] V.P. Maslov, M.V. Fedoryuk: *Semiclassical Approximation to Quantum Mechanical Equations*, Nauka, Moscow 1976 *(in Russian)*.

[[Mau]] K. Maurin: *Hilbert Space Methods*, PWN, Warsaw 1959 *(in Polish; English translation: Polish Sci. Publ., Warsaw 1972)*.

[[Mes]] A. Messiah: *Mécanique quantique I, II*, Dunod, Paris 1959 *(English translation: North–Holland, Amsterdam 1961, 1963)*.

[[Mül]] C. Müller: *Spectral Harmonics*, Lecture Notes In Mathematics, vol. 17, Springer, Berlin 1966.

[[Nai 1]] M.A. Naimark: *Normed Rings*, 2nd edition, Nauka, Moscow 1968 *(in Russian; English translation: Wolters–Noordhoff, Groningen 1972)*.

[[Nai 2]] M.A. Naimark: *Linear Differential Operators*, 2nd edition, Nauka, Moscow 1969 *(in Russian; English translation of the 1st edition: Harrap & Co., London 1967, 1968)*.

[[Nai 3]] M.A. Naimark: *Group Representation Theory*, Nauka, Moscow 1976 *(in Russian; French translation: Mir, Moscow 1979)*.

[[vN]] J. von Neumann: *Mathematische Grundlagen der Quantenmechanik*, Springer Verlag, Berlin 1932 *(English translation: Princeton University Press, Princeton, NJ 1955, reprinted 1996)*.

[[New]] R.G. Newton: *Scattering Theory of Waves and Particles*, 2nd edition, Springer Verlag, New York 1982; reprinted by Dover Publications Inc., Mineola, NY 2002.

⟦OK⟧ Y. Ohnuki, S. Kamefuchi: *Quantum Field Theory and Parastatistics*, Springer, Heidelberg 1982.

⟦Par⟧ K.R. Parthasarathy: *Introduction to Probability and Measure*, New Delhi 1980.

⟦Pea⟧ D.B. Pearson: *Quantum Scattering and Spectral Theory*, Techniques of Physics, vol. 9, Academic, London 1988.

⟦Pe⟧ P. Perry; *Scattering Theory by the Enss Method*, Harwood, London 1983.

⟦Pir⟧ C. Piron: *Foundations of Quantum Physics*, W.A. Benjamin, Reading, MA 1976.

⟦Pon⟧ L.S. Pontryagin: *Continuous Groups*, 3rd edition, Nauka, Moscow 1973 *(in Russian; English translation of the 2nd edition: Gordon and Breach, New York 1966).*

⟦Pop⟧ V.S. Popov: *Path Integrals in Quantum Theory and Statistical Physics*, Atomizdat, Moscow 1976 *(in Russian; English translation: D. Reidel, Dordrecht 1983).*

⟦Pru⟧ E. Prugovečki: *Quantum Mechanics in Hilbert Space*, 2nd edition, Academic, New York 1981.

⟦RS 1–4⟧ M. Reed, B. Simon: *Methods of Modern Mathematical Physics, I. Functional Analysis, II. Fourier Analysis. Self–Adjointness, III. Scattering Theory, IV. Analysis of Operators*, Academic, New York 1972–79.

⟦Ri 1⟧ D.R. Richtmyer: *Principles of Advanced Mathematical Physics I*, Springer, New York 1978.

⟦RN⟧ F. Riesz, B. Sz.–Nagy: *Lecons d'analyse fonctionelle*, 6me edition, Akademic Kiadó, Budapest 1972.

⟦Ru 1⟧ W. Rudin: *Real and Complex Analysis*, 3rd edition, McGraw–Hill, New York 1987.

⟦Ru 2⟧ W. Rudin: *Functional Analysis*, 2nd edition, McGraw–Hill, New York 1991.

⟦Šab⟧ B.V. Šabat: *Introduction to Complex Analysis*, Nauka, Moscow 1969 *(in Russ.)*.

⟦Sa⟧ S. Sakai: *C^*–Algebras and W^*–Algebras*, Springer, Berlin 1971.

⟦Sch⟧ R. Schatten: *A Theory of Cross Spaces*, Princeton University Press, Princeton, 1950.

⟦Sche⟧ M. Schechter: *Operator Methods in Quantum Mechanics*, North–Holland, New York 1981; reprinted by Dover Publications Inc., Mineola, NY 2002.

⟦Schm⟧ K. Schmüdgen: *Unbounded Operator Algebras and Representation Theory*, Akademie–Verlag, Berlin 1990.

⟦Schu⟧ L.S. Schulman: *Techniques and Applications of Path Integration*, Wiley–Interscience, New York 1981.

⟦Schw 1⟧ L. Schwartz: *Théorie des distributions I, II*, Hermann, Paris 1957, 1959.

⟦Schw 2⟧ L. Schwartz: *Analyse Mathématique I, II*, Hermann, Paris 1967.

⟦Schwe⟧ S.S. Schweber: *An Introduction to Relativistic Quantum Field Theory*, Row, Peterson & Co., Evanston, IL 1961.

⟦Seg⟧ I.E. Segal: *Mathematical Problems of Relativistic Physics, with an appendix by G.W. Mackey*, Lectures in Appl. Math., vol. 2, American Math. Society, Providence, RI 1963.

⟦Sei⟧ E. Seiler: *Gauge Theories as a Problem of Constructive Quantum Field Theory and Statistical Mechanics*, Lecture Notes in Physics, vol. 159, Springer, Berlin 1982.

⟦Si 1⟧ B. Simon: *Quantum Mechanics for Hamiltonians Defined as Quadratic Forms*, Princeton University Press, Princeton, NJ 1971.

⟦Si 2⟧ B. Simon: *The $P(\Phi)_2$ Euclidian (Quantum) Field Theory*, Princeton University Press, Princeton, NJ 1974.

⟦Si 3⟧ B. Simon: *Trace Ideals and Their Applications*, Cambridge University Press, Cambridge 1979.

⟦Si 4⟧ B. Simon: *Functional Integration and Quantum Physics*, Academic, New York 1979; 2nd edition, AMS Chelsea Publishing, Providence, RI 2005.

⟦Sin⟧ Ya.G. Sinai: *Theory of Phase Transitions. Rigorous Results*, Nauka, Moscow 1980 *(in Russian).*

⟦Šir⟧ A.N. Širyaev: *Probability*, Nauka, Moscow 1980 *(in Russian).*

⟦SF⟧ A.A. Slavnov, L.D. Faddeev: *Introduction to the Quantum Theory of Gauge Fields*, Nauka, Moscow 1978 *(in Russian).*

⟦Stock⟧ H.-J. Stöckmann: *Quantum chaos. An introduction*, Cambridge University Press, Cambridge, 1999.

⟦Stol⟧ P. Stollmann: *Caught by Disorder: Bound States in Random Media*, Birkhäuser, Basel 2001.

⟦Sto⟧ M.H. Stone: *Linear Transformations in Hilbert Space and Their Applications to Analysis*, Amer. Math. Colloq. Publ., vol. 15, New York 1932.

⟦Str⟧ R.F. Streater, ed.: *Mathematics of Contemporary Physics*, Academic, London 1972.

⟦SW⟧ R.F. Streater, A. Wightman: *PCT, Spin, Statistics and All That*, W.A. Benjamin, New York 1964.

⟦Šv⟧ A.S. Švarc: *Mathematical Foundations of the Quantum Theory*, Atomizdat, Moscow 1975 *(in Russian).*

⟦Tay⟧ A.E. Taylor: *Introduction to Functional Analysis*, 6th edition, Wiley, New York 1967.

⟦Ta⟧ J.R. Taylor: *Scattering Theory. The Quantum Theory of Nonrelativistic Collisions*, Wiley, New York 1972.

⟦Te⟧ G. Teschl: *Jacobi operators and completely integrable nonlinear lattices*, Math. Surveys and Monographs, vol. 72; AMS, Providence, RI 2000.

⟦Tha⟧ B. Thaller: *The Dirac Equation*, Springer, Berlin 1992.

⟦Thi 3,4⟧ W. Thirring: *A Course in Mathematical Physics, 3. Quantum Mechanics of Atoms and Molecules, 4. Quantum Mechanics of Large Systems*, Springer, New York 1981, 1983.

⟦Ti⟧ V.M. Tikhomirov: *Banach Algebras*, an appendix to the monograph ⟦KF⟧, pp. 513–528 *(in Russian).*

⟦Vai⟧ B.R. Vainberg: *Asymptotic methods in the Equations of Mathematical Physics*, Moscow State University Publishing, Moscow 1982 *(in Russian).*

⟦Var 1,2⟧ V.S. Varadarajan: *Geometry of Quantum Theory I, II*, Van Nostrand Reinhold, New York 1968, 1970.

⟦Vot⟧ V. Votruba: *Foundations of the Special Theory of Relativity*, Academia, Prague 1969 *(in Czech).*

⟦WEG⟧ R. Weder, P. Exner, B. Grebert, eds.: *Mathematical Results in Quantum Mechanics*, Contemporary Mathematics, vol. 307, AMS, Providence, RI 2002.

⟦We⟧ J. Weidmann: *Linear Operators in Hilbert Space*, Springer, Heidelberg 1980 *(2nd edition in German: Lineare Operatoren in Hilberträumen, I. Grundlagen, II. Anwendungen, B.G. Teubner, Stuttgart 2000, 2003).*

⟦Wig⟧ E.P. Wigner: *Symmetries and Reflections*, Indiana University Press, Bloomington, IN 1971.

⟦Ya⟧ D.R. Yafaev: *Mathematical Scattering Theory: General Theory*, Transl. Math. Monographs, vol. 105; AMS, Providence, RI, 1992

⟦Yo⟧ K. Yosida: *Functional Analysis*, 3rd edition, Springer, Berlin 1971.

⟦Žel⟧ D.P. Želobenko: *Compact Lie Groups and Their Representations*, Nauka, Moscow 1970 *(in Russian).*

b) Research and review papers:

[ACFGH 1] A. Abrams, J. Cantarella, J.G. Fu, M. Ghomi, R. Howard: Circles minimize most knot energies, *Topology* **42** (2003), 381–394.

[Adl 1] S.L. Adler: Quaternionic quantum field theory, *Commun. Math. Phys.* **104** (1986), 611–656.

[AD 1] D. Aerts, I. Daubechies: Physical justification for using the tensor product to describe two quantum systems as one joint system, *Helv. Phys. Acta* **51** (1978), 661–675.

[AD 2] D. Aerts, I. Daubechies: A mathematical condition for a sublattice of a propositional system to represent a physical subsystem, with a physical interpretation, *Lett. Math. Phys.* **3** (1979), 19–27.

[AF 1] J. Agler, J. Froese: Existence of Stark ladder resonances, *Commun. Math. Phys.* **100** (1985), 161–172.

[AHS 1] S. Agmon, I. Herbst, E. Skibsted: Perturbation of embedded eigenvalues in the generalized N–body problem, *Commun. Math. Phys.* **122** (1989), 411–438.

[AC 1] J. Aguilar, J.–M. Combes: A class of analytic perturbations for one–body Schrödinger Hamiltonians, *Commun. Math. Phys.* **22** (1971), 269–279.

[AAAS 1] E. Akkermans, A. Auerbach, J. Avron, B. Shapiro: Relation between persistent currents and the scattering matrix, *Phys. Rev. Lett.* **66** (1991), 76–79.

[ADH 1] S. Alama, P. Deift, R. Hempel: Eigenvalue branches of Schrödinger operator $H-\lambda W$ in a gap of $\sigma(H)$, *Commun. Math. Phys.* **121** (1989), 291–321.

[Al 1] S. Albeverio: On bound states in the continuum of N–body systems and the virial theorem, *Ann. Phys.* **71** (1972), 167–276.

[ACF 1] S. Albeverio, C. Cacciapuoti, D. Finco: Coupling in the singular limit of thin quantum waveguides, *J. Math. Phys.* **48** (2007), 032103.

[AH 1] S. Albeverio, R. Høegh–Krohn: Oscillatory integrals and the method of stationary phase, *Inventiones Math.* **40** (1977), 59–106.

[AN 1] S. Albeverio, L. Nizhnik: Approximation of general zero-range potentials, *Ukrainian Math. J.* **52** (2000), 582–589.

[AmC 1] W.O. Amrein, M.B. Cibils: Global and Eisenbud–Wigner time delay in scattering theory, *Helv. Phys. Acta* **60** (1987), 481–500.

[AG 1] W.O. Amrein, V. Georgescu: Bound states and scattering states in quantum mechanics, *Helv. Phys. Acta* **46** (1973), 635–658.

[And 1] J. Anderson: Extensions, restrictions and representations of C^*–algebras, *Trans. Am. Math. Soc.* **249** (1979), 303–329.

[AGS 1] J.–P. Antoine, F. Gesztesy, J. Shabani: Exactly solvable models of sphere interactions in quantum mechanics, *J. Phys.* **A20** (1987), 3687–3712.

[AIT 1] J.–P. Antoine, A. Inoue, C. Trapani: Partial $*$–algebras of closed operators, I. Basic theory and the Abelian case, *Publ. RIMS* **26** (1990), 359–395.

[AK 1] J.–P. Antoine, W. Karwowski: Partial $*$–algebras of Hilbert space operators, in *Proceedings of the 2nd Conference on Operator Algebras, Ideals and Their Applications in Theoretical Physics* (H. Baumgärtel *et al.*, eds.), Teubner, Leipzig 1984; pp. 29–39.

[AKa 1] A. Arai, H. Kawano: Enhanced binding in a general class of quantum field models, *Rev. Math. Phys.* **15** (2003), 387–423.

[Ara 1] H. Araki: Type of von Neumann algebra associated with free field, *Progr. Theor. Phys.* **32** (1964), 956–965.

[Ara 2] H. Araki: C^*–approach in quantum field theory, *Physica Scripta* **24** (1981), 981–985.

[AJ 1] H. Araki, J.–P. Jurzak: On a certain class of $*$–algebras of unbounded operators, *Publ. RIMS* **18** (1982), 1013–1044.

[AMKG 1] H. Araki, Y. Munakata, M. Kawaguchi, T. Goto: Quantum Field Theory of Unstable Particles, *Progr. Theor. Phys.* **17** (1957), 419–442.

[ArJ 1] D. Arnal, J.–P. Jurzak: Topological aspects of algebras of unbounded operators, *J. Funct. Anal.* **24** (1977), 397–425.

[ADE 1] J. Asch, P. Duclos, P. Exner: Stability of driven systems with growing gaps. Quantum rings and Wannier ladders, *J. Stat. Phys.* **92** (1998), 1053–1069.

[AB 1] M.S. Ashbaugh, R.D. Benguria: Optimal bounds of ratios of eigenvalues of one–dimensional Schrödinger operators with Dirichlet boundary conditions and positive potentials, *Commun. Math. Phys.* **124** (1989), 403–415.

[AB 2] M.S. Ashbaugh, R.D. Benguria: A sharp bound for the ratio of the first two eigenvalues of Dirichlet Laplacians and extensions, *Ann. Math.* **135** (1992), 601–628.

[AE 1] M.S. Ashbaugh, P. Exner: Lower bounds to bound state energies in bent tubes, *Phys. Lett.* **A150** (1990), 183–186.

[AsH 1] M.S. Ashbaugh, E.M. Harrell II : Perturbation theory for resonances and large barrier potentials, *Commun. Math. Phys.* **83** (1982), 151–170.

[AsHS 1] M.S. Ashbaugh, E.M. Harrell II, R. Svirsky: On the minimal and maximal eigenvalue gaps and their causes, *Pacific J. Math.* **147** (1991), 1–24.

[AS 1] M.S. Ashbaugh, C. Sundberg: An improved stability result for resonances, *Trans. Am. Math. Soc.* **281** (1984), 347–360.

[AJI 1] A. Avila, S. Jitomirskaya: Solving the ten Martini problem, *Lecture Notes in Physics* **690** (2006), 5–16.

[ABGM 1] Y. Avishai, D. Bessis, B.M. Giraud, G. Mantica: Quantum bound states in open geometries, *Phys. Rev.* **B44** (1991), 8028–8034.

[Avr 1] J. Avron: The lifetime of Wannier ladder states, *Ann. Phys.* **143** (1982), 33–53.

[AEL 1] J. Avron, P. Exner, Y. Last: Periodic Schrödinger operators with large gaps and Wannier–Stark ladders, *Phys. Rev. Lett.* **72** (1994), 896–899.

[ARZ 1] J. Avron, A. Raveh, B. Zur: Adiabatic quantum transport in multiply connected systems, *Rev. Mod. Phys.* **60** (1988), 873–915.

[ASY 1] J. Avron, R. Seiler, L.G. Yaffe: Adiabatic theorem and application to the quantum Hall effect, *Commun. Math. Phys.* **110** (1987), 33–49.

[BB 1] D. Babbitt, E. Balslev: Local distortion techniques and unitarity of the S–matrix for the 2–body problem, *J. Math. Anal. Appl.* **54** (1976), 316–349.

[BFS 1] V. Bach, J. Fröhlich, I.M. Sigal: Spectral analysis for systems of atoms and molecules coupled to the quantized radiation field, *Commun. Math. Phys.* **207** (1999), 249–290.

[Ba 1] V. Bargmann: On a Hilbert space of analytical functions and an associate integral transform I, II, *Commun. Pure Appl. Math.* **14** (1961), 187–214; **20** (1967), 1–101.

[Ba 2] V. Bargmann: Remarks on Hilbert space of analytical functions, *Proc. Natl. Acad. Sci. USA* **48** (1962), 199–204, 2204.

[Ba 3] V. Bargmann: On unitary ray representations of continuous groups, *Ann. Math.* **59** (1954), 1–46.

[Ba 4] V. Bargmann: Note on some integral inequalities, *Helv. Phys. Acta* **45** (1972), 249–257.

[Ba 5] V. Bargmann: On the number of bound states in a central field of forces, *Proc. Natl. Acad. Sci. USA* **38** (1952), 961–966.

[BFF1] F. Bayen, M. Flato, C. Fronsdal, A. Lichnerowicz, D. Sternheimer: Deformation theory and quantization I, II, *Ann. Phys.* **111** (1978), 61–110, 111–151.

[Bau 1] H. Baumgärtel: Partial resolvent and spectral concentration, *Math. Nachr.* **69** (1975), 107–121.

[BD 1] H. Baumgärtel, M. Demuth: Perturbation of unstable eigenvalues of finite multiplicity, *J. Funct. Anal.* **22** (1976), 187–203.

[BDW 1] H. Baumgärtel, M. Demuth, M. Wollenberg: On the equality of resonances (poles of the scattering amplitude) and virtual poles, *M. Nachr.* **86** (1978), 167–174.

[Beh 1] H. Behncke: The Dirac equation with an anomalous magnetic moment, *Math. Z.* **174** (1980), 213–225.

[BF 1] F.A. Berezin, L.D. Faddeev: A remark on Schrödinger's equation with a singular potential, *Sov. Acad. Sci. Doklady* **137** (1961), 1011–1014 *(in Russian)*.

[vdB 1] M. van den Berg: On the spectral counting function for the Dirichlet Laplacian, *J. Funct. Anal.* **107** (1992), 352–361.

[Be 1] M.V. Berry: Quantal phase factor accompanying adiabatic changes, *Proc. Roy. Soc. London* **A392** (1984), 45–57.

[Be 2] M.V. Berry: The adiabatic limit and the semiclassical limit, *J. Phys.* **A17** (1984), 1225–1233.

[BM 1] M.V. Berry, K.E. Mount: Semiclassical approximations in wave mechanics, *Rep. Progr. Phys.* **35** (1972), 315–389.

[BN 1] A. Beskow, J. Nilsson: The concept of wave function and the irreducible representations of the Poincaré group, II. Unstable systems and the exponential decay law, *Arkiv för Physik* **34** (1967), 561–569.

[BvN 1] G. Birkhoff, J. von Neumann: The logic of quantum mechanics, *Ann. Math.* **37** (1936), 823–843.

[Bir 1] M.Š. Birman: On the spectrum of singular boundary problems, *Mat. Sbornik* **55** (1961), 125–174 *(in Russian)*.

[Bir 2] M.Š. Birman: Existence conditions for the wave operators, *Doklady Acad. Sci. USSR* **143** (1962), 506–509 *(in Russian)*.

[Bir 3] M.Š. Birman: An existence criterion for the wave operators. *Izvestiya Acad. Sci. USSR, ser. mat.* **27** (1963), 883–906 *(in Russian)*.

[Bir 4] M.Š. Birman: A local existence criterion for the wave operators, *Izvestiya Acad. Sci. USSR, ser. mat.* **32** (1968), 914–942 *(in Russian)*.

[BE 1] J. Blank, P. Exner: Remarks on tensor products and their applications in quantum theory, I. General considerations, II. Spectral properties, *Acta Univ. Carolinae, Math. Phys.* **17** (1976), 75–89; **18** (1977), 3–35.

[BEH 1] J. Blank, P. Exner, M. Havlíček: Quantum–mechanical pseudo–Hamiltonians, *Czech. J. Phys.* **B29** (1979), 1325–1341.

[BGS 1] R. Blankenbecler, M.L. Goldberger, B. Simon: The bound states of weakly coupled long–range one–dimensional Hamiltonians, *Ann. Phys.* **108** (1977), 69–78.

[BChK 1] D. Bollé, K. Chadan, G. Karner: On a sufficient condition for the existence of N–particle bound states, *J. Phys.* **A19** (1986), 2337–2343.

[Bor 1] H.J. Borchers: Algebras of unbounded operators in quantum field theory, *Physica* **124A** (1984), 127–144.

[BoE 1] D. Borisov, P. Exner: Exponential splitting of bound states in a waveguide with a pair of distant windows, *J. Phys.* **A37** (2004), 3411–3428.

[BEG 1] D. Borisov, P. Exner, R. Gadyl'shin: Geometric coupling thresholds in a two-dimensional strip, *J. Math. Phys.* **43** (2002), 6265–6278.

[BEGK 1] D. Borisov, P. Exner, R. Gadyl'shin, D. Krejčiřík: Bound states in weakly deformed strips and layers, *Ann. H. Poincaré* **2** (2001), 553–572.

[BCGH 1] F. Boudjeedaa, L. Chetouani, L. Guechi, T.F. Hamman: Path integral treatment for a screened potential, *J. Math. Phys.* **32** (1991), 441–446.

[BEKŠ 1] J. Brasche, P. Exner, Yu. Kuperin, P. Šeba: Schrödinger operators with singular interactions, *J. Math. Anal. Appl.* **184** (1994), 112–139.

[BrT 1] J. Brasche, A. Teta: Spectral analysis and scattering theory for Schrödinger operators with an interaction supported by regular curve, in the proceedings volume ⟦AFHL⟧, pp. 197–211.

[EBG 1] J. Brüning, P. Exner, V. Geyler: Large gaps in point-coupled periodic systems of manifolds, *J. Phys. A: Math. Gen.* **36** (2003), 4875–4890.

[BG 1] J. Brüning, V.A. Geyler: Scattering on compact manifolds with infinitely thin horns, *J. Math. Phys.* **44** (2003), 371–405.

[BGMP 1] J. Brüning, V.A. Geyler, V.A. Margulis, M.A. Pyataev: Ballistic conductance of a quantum sphere, *J. Phys.* **A35** (2002), 4239–4247.

[BGP 1] J. Brüning, V. Geyler, K. Pankrashkin: Cantor and band spectra for periodic quantum graphs with magnetic fields, *Commun. Math. Phys.* **269** (2007), 87–105.

[BFS 1] D.C. Brydges, J. Fröhlich, A. Sokal: A new proof of existence and nontriviality of the continuum Φ_2^4 and Φ_3^4 quantum field theories, *Commun. Math. Phys.* **91** (1983), 141–186.

[BEPS 1] E.N. Bulgakov, P. Exner, K.N. Pichugin, A.F. Sadreev: Multiple bound states in scissor-shaped waveguides, *Phys. Rev.* **B66** (2002), 155109

[BGRS 1] W. Bulla, F. Gesztesy, W. Renger, B. Simon: Weakly coupled bound states in quantum waveguides, *Proc. Amer. Math. Soc.* **127** (1997), 1487–1495.

[BT 1] W. Bulla, T. Trenkler: The free Dirac operator on compact and non–compact graphs, *J. Math. Phys.* **31** (1990), 1157–1163.

[BW 1] L.J. Bunce, J.D.M. Wright: Quantum measures and states on Jordan algebras, *Commun. Math. Phys.* **98** (1985), 187–202.

[BuD 1] V.S. Buslaev, L. Dmitrieva: Bloch electrons in an external electric field, *Leningrad Math. J.* **1** (1991), 287–320.

[Bü 1] M. Büttiker: Small normal–metal loop coupled to an electron reservoir, *Phys. Rev.* **B32** (1985), 1846–1849.

[Bü 2] M. Büttiker: Absence of backscattering in the quantum Hall effect in multiprobe conductors, *Phys. Rev.* **B38** (1988), 9375–9389.

[CaE 1] C. Cacciapuoti, P. Exner: Nontrivial edge coupling from a Dirichlet network squeezing: the case of a bent waveguide, *J. Phys.* **A40** (2007), F511–F523.

[Ca 1] J.W. Calkin: Two–sided ideals and congruence in the ring of bounded operators in Hilbert space, *Ann. Math.* **42** (1941), 839–873.

[Cal 1] F. Calogero: Upper and lower limits for the number of bound states in a given central potential, *Commun. Math. Phys.* **1** (1965), 80–88.

[Cam 1] R.H. Cameron: A family of integrals serving to connect the Wiener Feynman integrals, *J. Math. Phys.* **39** (1961), 126–141.

[Cam 2] R.H. Cameron: The Ilstow and Feynman integrals, *J d'Analyse Math.* **10** (1962–63), 287–361.

[Cam 3] R.H. Cameron: Approximation to certain Feynman integrals, *J. d'Analyse Math.* **21** (1968), 337–371.

[CLM 1] J.P. Carini, J.T. Londergan, K. Mullen, D.P. Murdock: Bound states and resonances in quantum wires, *Phys. Rev.* **B46** (1992), 15538–15541.

[CLM 2] J.P. Carini, J.T. Londergan, K. Mullen, D.P. Murdock: Multiple bound states in sharply bent waveguides, *Phys. Rev.* **B48** (1993), 4503–4514.

[CLMTY 1,2] J.P. Carini, J.T. Londergan, D.P. Murdock, D. Trinkle, C.S. Yung: Bound states in waveguides and bent quantum wires, I. Applications to waveguide systems, II. Electrons in quantum wires, *Phys. Rev.* **B55** (1997), 9842–9851, 9852–9859.

[CMS 1] R. Carmona, W.Ch. Masters, B. Simon: Relativistic Schrödinger operators, *J. Funct. Anal.* **91** (1990), 117–142.

[CEK 1] G. Carron, P. Exner, D. Krejčiřík: Topologically non-trivial quantum layers, *J. Math. Phys.* **45** (2004), 774–784.

[CG 1] G. Casati, I. Guarneri: Non–recurrent behaviour in quantum dynamics, *Commun. Math. Phys.* **95** (1984), 121–127.

[CM 1] D. Castrigiano, U. Mutze: On the commutant of an irreducible set of operators in a real Hilbert space, *J. Math. Phys.* **26** (1985), 1107–1110.

[CEH 1] I. Catto, P. Exner, Ch. Hainzl: Enhanced binding revisited for a spinless particle in non-relativistic QED, *J. Math. Phys.* **45** (2004), 4174–4185.

[ChD 1] K. Chadan, Ch. DeMol: Sufficient conditions for the existence of bound states in a potential without a spherical symmetry, *Ann. Phys.* **129** (1980), 466–478.

[CRWP 1] V. Chandrasekhar, M.J. Rooks, S. Wind, D.E. Prober: Observation of Aharonov-Bohm electron interference effects with periods h/e and $h/2e$ in individual micron-size, normal-metal rings, *Phys. Rev. Lett.* **55** (1985), 1610–1613.

[CVV 1] T. Chen, V. Vougalter, S. Vugalter: The increase of binding energy and enhanced binding in nonrelativistic QED, *J. Math. Phys.* **44** (2003), 1961–1970.

[CDFK 1] P. Chenaud, P. Duclos, P. Freitas, D. Krejčiřík: Geometrically induced discrete spectrum in curved tubes, *Diff. Geom. Appl.* **23** (2005), 95–105.

[CE 1] T. Cheon, P. Exner: An approximation to δ' couplings on graphs, *J. Phys.* **A37** (2004), L329–335.

[CS 1] T. Cheon, T. Shigehara: Realizing discontinuous wave functions with renormalized short-range potentials, *Phys. Lett.* **A243** (1998), 111–116.

[Che 1] S. Cheremshantsev: Hamiltonians with zero–range interactions supported by a Brownian path, *Ann. Inst. H. Poincaré: Phys. Théor.* **56** (1992), 1–25.

[Cher 1] P.R. Chernoff: A note on product formulas for operators, *J. Funct. Anal.* **2** (1968), 238–242.

[ChH 1] P.R. Chernoff, R.Hughes: A new class of point interactions in one dimension, *J. Funct. Anal.* **111** (1993), 97–117.

[CH 1] L. Chetouani, F.F. Hamman: Coulomb's Green function in an n–dimensional Euclidean space, *J. Math. Phys.* **27** (1986), 2944–2948.

[CCH 1] L. Chetouani, A. Chouchaoui, T.F. Hamman: Path integral solution for the Coulomb plus sector potential, *Phys. Lett.* **A161** (1991), 87–97.

[CdV 1] Y. Colin de Verdière: Bohr–Sommerfeld rules to all orders, *Ann. H. Poincaré* **6** (2005), 925–936.

[CRRS 1] Ph. Combe, G. Rideau, R. Rodriguez, M. Sirugue–Collin: On the cylindrical approximation to certain Feynman integrals, *Rep. Math. Phys.* **13** (1978), 279–294.

[CDS 1] J.–M. Combes, P. Duclos, R. Seiler: Krein's formula and one–dimensional multiple well, *J. Funct. Anal.* **52** (1983), 257–301.

[CDKS 1] J.–M. Combes, P. Duclos, M. Klein, R. Seiler: The shape resonance, *Commun. Math. Phys.* **110** (1987), 215–236.

[CoH 1] J.–M. Combes, P.D. Hislop: Stark ladder resonances for small electric fields, *Commun. Math. Phys.* **140** (1991), 291–320.

[Com 1] M. Combescure: Spectral properties of periodically kicked quantum Hamiltonians, *J. Stat. Phys.* **59** (1990), 679–690.

[Co 1] J. Conlon: The ground state of a Bose gas with Coulomb interaction I, II, *Commun. Math. Phys.* **100** (1985), 355–379; **108** (1987), 363–374.

[CLY 1] J.G. Conlon, E.H. Lieb, H.–T. Yau: The $N^{7/5}$ law for charged bosons, *Commun. Math. Phys.* **116** (1988), 417–488.

[Con 1] A. Connes: The Tomita–Takesaki theory and classification of type III factors, in the proceedings volume ⟦Kas⟧, pp. 29–46.

[Coo 1] J.M. Cook: The mathematics of second quantization, *Trans. Am. Math. Soc.* **74** (1953), 222–245.

[Coo 2] J.M. Cook: Convergence of the Møller wave matrix, *J. Math. Phys.* **36** (1957), 82–87.

[CJM 1] H. Cornean, A. Jensen, V. Moldoveanu: A rigorous proof for the Landauer-Büttiker formula, *J. Math. Phys.* **46** (2005), 042106.

[CMF 1] F.A.B. Coutinho, C.P. Malta, J. Fernando Perez: Sufficient conditions for the existence of bound states of N particles with attractive potentials, *Phys. Lett.* **100A** (1984), 460–462.

[Chri 1] E. Christensen: Measures on projections and physical states, *Commun. Math. Phys.* **86** (1982), 529–538.

[Cra 1] R.E. Crandall: Exact propagator for reflectionless potentials, *J. Phys.* **A16** (1983), 3005–3011.

[Cwi 1] M. Cwikel: Weak type estimates for singular values and the number of bound states of Schrödinger operators, *Ann. Math.* **106** (1977), 93–100.

[Cy 1] H.L. Cycon: Resonances defined by modified dilations, *Helv. Phys. Acta* **58** (1985), 969–981.

[DL 1] I. Daubechies, E.H. Lieb: One electron relativistic molecule with Coulomb interaction, *Commun. Math. Phys.* **90** (1983), 497–510.

[DAT 1] G. Dell'Antonio, L. Tenuta: Quantum graphs as holonomic constraints, *J. Math. Phys.* **47** (2006), 072102.

[Dem 1] M. Demuth: Pole approximation and spectral concentration, *Math. Nachr.* **73** (1976), 65–72.

[Der 1] J. Derezyński: A new proof of the propagation theorem for N-body quantum systems, *Commun. Math. Phys.* **122** (1989), 203–231.

[Der 2] J. Derezyński: Algebraic approach to the N–body long–range scattering, *Rep. Math. Phys.* **3** (1991), 1–62.

[DLSS 1] C. DeWitt, S.G. Low, L.S. Schulman, A.Y. Shiekh: Wedges I, *Found. Phys.* **16** (1986), 311–349.

[DeW 1] C. DeWitt–Morette: The semiclassical expansion, *Ann. Phys.* **97** (1976), 367–399; **101**, 682–683.

[DMN 1] C. DeWitt–Morette, A. Maheswari, B. Nelson: Path–integration in nonrelativistic quantum mechanics, *Phys. Rep.* **50** (1979), 255–372.

[DE 1] J. Dittrich, P. Exner: Tunneling through a singular potential barrier, *J. Math. Phys.* **26** (1985), 2000–2008.

[DE 2] J. Dittrich, P. Exner: A non–relativistic model of two–particle decay I–IV, *Czech. J. Phys.* **B37** (1987), 503–515, 1028–1034; **B38** (1988), 591–610; **B39** (1989), 121–138.

[DEŠ 1] J. Dittrich, P. Exner, P. Šeba: Dirac operators with a spherically symmetric δ–shell interaction, *J. Math. Phys.* **30** (1989), 2975–2982.

[DEŠ 2] J. Dittrich, P. Exner, P. Šeba: Dirac Hamiltonians with Coulombic potential and spherically symmetric shell contact interaction, *J. Math. Phys.* **33** (1992), 2207–2214.

[DK 1] J. Dittrich, J. Kříž: Curved planar quantum wires with Dirichlet and Neumann boundary conditions, *J. Phys.* **A35** (2002), L269–275.

[DK 2] J. Dittrich, J. Kříž: Bound states in straight quantum waveguides with combined boundary conditions, *J. Math. Phys.* **43** (2002), 3892–3915.

[Di 1] J. Dixmier: Sur la relation $i(PQ-QP)=I$, *Compos. Math.* **13** (1956), 263–269.

[DF 1] J.D. Dollard, C.N. Friedmann: Existence and completeness of the Møller wave operators for radial potentials satisfying $\int_0^1 r|v(r)|\,dr+\int_1^\infty |v(r)|\,dr<\infty$, *J. Math. Phys.* **21** (1980), 1336–1339.

[DS 1] E. Doron, U. Smilansky: Chaotic spectroscopy, *Phys. Rev. Lett.* **68** (1992), 1255–1258.

[Dri 1] T. Drisch: Generalization of Gleason's theorem, *Int. J. Theor. Phys.* **18** (1978), 239–243.

[DuE 1] P. Duclos, P. Exner: Curvature–induced bound states in quantum waveguides in two and three dimensions, *Rev. Math. Phys.* **7** (1995), 73–102.

[DEK 1] P. Duclos, P. Exner, D. Krejčiřík: Bound states in curved quantum layers, *Commun. Math. Phys.* **223** (2001), 13–28.

[DEM 1] P. Duclos, P. Exner, B. Meller: Exponential bounds on curvature–induced resonances in a two–dimensional Dirichlet tube, *Helv. Phys. Acta* **71** (1998), 133–162.

[DEM 2] P. Duclos, P. Exner, B. Meller: Open quantum dots: resonances from perturbed symmetry and bound states in strong magnetic fields, *Rep. Math. Phys.* **47** (2001), 253–267.

[DuEŠ 1] P. Duclos, P. Exner, P. Šťovíček: Curvature–induced resonances in a two–dimensional Dirichlet tube, *Ann.Inst. H.Poincaré: Phys. Théor.* **62** (1995), 81–101.

[DuŠ 1] P. Duclos, P. Šťovíček: Floquet Hamiltonians with pure spectrum, *Commun. Math. Phys.* **177** (1996), 327–347.

[DuŠV 1] P. Duclos, P. Šťovíček, M. Vittot: Perturbations of an eigen-value from a dense point spectrum: a general Floquet Hamiltonian, *Ann. Inst. H. Poincaré* **71** (1999), 241–301.

[Du 1] I.H. Duru: Quantum treatment of a class of time–dependent potentials, *J. Phys.* **A22** (1989), 4827–4833.

[EZ 1] J.–P. Eckmann, Ph.C. Zabey: Impossibility of quantum mechanics in a Hilbert space over a finite field, *Helv. Phys. Acta* **42** (1969), 420–424.

[EH 1] M. Eilers, M. Horst: The theorem of Gleason for nonseparable Hilbert spaces, *Int. J. Theor. Phys.* **13** (1975), 419–424.

[EkK 1] T. Ekholm, H. Kovařík: Stability of the magnetic Schrödinger operator in a waveguide, *Comm. PDE* **30** (2005), 539–565.

[ETr 1] K.D. Elworthy, A. Truman: A Cameron–Martin formula for Feynman integrals (the origin of Maslov indices), in *Mathematical Problems in Theoretical Physics*, Lecture Notes in Physics, vol. 153, Springer, Berlin 1982; pp. 288–294.

[Em 1] G.G. Emch: Mécanique quantique quaternionique et relativité restreinte I, II, *Helv. Phys. Acta* **36** (1963), 739–769, 770–788.

[ES 1] G.G. Emch, K.B. Sinha: Weak quantization in a non–perturbative model, *J. Math. Phys.* **20** (1979), 1336–1340.

[En 1,2] V. Enss: Asymptotic completeness for quantum–mechanical potential scattering, I. Short–range potentials, II. Singular and long–range potentials, *Commun. Math. Phys.* **61** (1978), 285–291; *Ann. Phys.* **119** (1979), 117–132.

[En 3] V. Enss: Topics in scattering theory for multiparticle systems: a progress report, *Physica* **A124** (1984), 269–292.

[EV 1] V. Enss, K. Veselič: Bound states and propagating states for time dependent Hamiltonians, *Ann. Inst. H. Poincaré: Phys. Théor.* **39** (1983), 159–191.

[Epi 1] G. Epifanio: On the matrix representation of unbounded operators, *J. Math. Phys.* **17** (1976), 1688–1691.

[ET 1] G. Epifanio, C. Trapani: Remarks on a theorem by G. Epifanio, *J. Math. Phys.* **20** (1979), 1673–1675.

[ET 2] G. Epifanio, C. Trapani: Some spectral properties in algebras of unbounded operators, *J. Math. Phys.* **22** (1981), 974–978.

[ET 3] G. Epifanio, C. Trapani: V^*–algebras: a particular class of unbounded operator algebras, *J. Math. Phys.* **25** (1985), 2633–2637.

[ELS 1] W.D. Evans, R.T. Lewis, Y. Saito: Some geometric spectral properties of N–body Schrödinger operators, *Arch. Rat. Mech. Anal.* **113** (1991), 377–400.

[Ex 1] P. Exner: Bounded–energy approximation to an unstable quantum system, *Rep. Math. Phys.* **17** (1980), 275–285.

[Ex 2] P. Exner: Generalized Bargmann inequalities, *Rep. Math. Phys.* **19** (1984), 249–255.

[Ex 3] P. Exner: Representations of the Poincaré group associated with unstable particles, *Phys. Rev.* **D28** (1983), 3621–2627.

[Ex 4] P. Exner: Remark on the energy spectrum of a decaying system, *Commun. Math. Phys.* **50** (1976), 1–10.

[Ex 5] P. Exner: One more theorem on the short–time regeneration rate, *J. Math. Phys.* **30** (1989), 2563–2564.

[Ex 6] P. Exner: A solvable model of two–channel scattering, *Helv. Phys. Acta* **64** (1991), 592–609.

[Ex 7] P. Exner: A model of resonance scattering on curved quantum wires, *Ann. Physik* **47** (1990), 123–138.

[Ex 8] P. Exner: The absence of the absolutely continuous spectrum for δ' Wannier–Stark ladders, *J. Math. Phys.* **36** (1995), 4561–4570.

[Ex 9] P. Exner: Lattice Kronig–Penney models, *Phys. Rev. Lett.* **74** (1995), 3503–3506.

[Ex 10] P. Exner: Contact interactions on graph superlattices, *J. Phys.* **A29** (1996), 87–102.

[Ex 11] P. Exner: Weakly coupled states on branching graphs, *Lett. Math. Phys.* **38** (1996), 313–320.

[Ex 12] P. Exner: A duality between Schrödinger operators on graphs and certain Jacobi matrices, *Ann. Inst. H. Poincaré: Phys. Théor.* **66** (1997), 359–371.

[Ex 13] P. Exner: Magnetoresonances on a lasso graph, *Found. Phys.* **27** (1997), 171–190.

[Ex 14] P. Exner: Point interactions in a tube, in *"Stochastic Processes: Physics and Geometry: New Interplayes II"* (A volume in honor of S. Albeverio; F. Gesztesy et al., eds.); CMS Conference Proceedings, vol. 29, Providence, R.I. 2000; pp. 165–174.

[Ex 15] P. Exner: *Leaky quantum graphs: a review*, in the proceedings volume ⟦EKKST⟧; `arXiv: 0710.5903 [math-ph]`

[Ex 16] P. Exner: An isoperimetric problem for leaky loops and related mean-chord inequalities, *J. Math. Phys.* **46** (2005), 062105

[Ex 17] P. Exner: Necklaces with interacting beads: isoperimetric problems, AMS "Contemporary Math" Series, vol. 412, Providence, RI, 2003; pp. 141–149.

[EFr 1] P. Exner, M. Fraas: The decay law can have an irregular character, *J. Phys.* **A40** (2007), 1333–1340.

[EFK 1] P. Exner, P. Freitas, D. Krejčiřík: A lower bound to the spectral threshold in curved tubes, *Proc. Roy. Soc.* **A460** (2004), 3457–3467.

[EG 1] P. Exner, R. Gawlista: Band spectra of rectangular graph superlattices, *Phys. Rev.* **B53** (1996), 7275–7286.

[EGŠT 1] P. Exner, R. Gawlista, P. Šeba, M. Tater: Point interactions in a strip, *Ann. Phys.* **252** (1996), 133–179.

[EHL 1] P. Exner, E.M. Harrell, M. Loss: Optimal eigenvalues for some Laplacians and Schrödinger operators depending on curvature, *Proceedings of the Conference "Mathematical Results in Quantum Mechanics"* (QMath7, Prague 1998); Operator Theory: Advances and Applications, Birkhäuser, Basel; pp. 47–53.

[EHL 2] P. Exner, E.M. Harrell, M. Loss: *Inequalities for means of chords, with application to isoperimetric problems, Lett. Math. Phys.* **75** (2006), 225–233; addendum **77** (2006), 219.

[EI 1] P. Exner, T. Ichinose: Geometrically induced spectrum in curved leaky wires, *J. Phys.* **A34** (2001), 1439–1450.

[EK 1] P. Exner, G.I. Kolerov: Uniform product formulae, with application to the Feynman–Nelson integral for open systems, *Lett. Math. Phys.* **6** (1982), 151–159.

[EKr 1] P. Exner, D. Krejčiřík: Quantum waveguide with a lateral semitransparent barrier: spectral and scattering properties, *J. Phys.* **A32** (1999), 4475–4494.

[EKr 2] P. Exner, D. Krejčiřík: Bound states in mildly curved layers, *J. Phys.* **A34** (2001), 5969–5985.

[ELW 1] P. Exner, H. Linde, T. Weidl: Lieb-Thirring inequalities for geometrically induced bound states, *Lett. Math. Phys.* **70** (2004), 83–95.

[ENZ 1] P. Exner, H. Neidhardt, V.A. Zagrebnov: Potential approximations to δ': an inverse Klauder phenomenon with norm-resolvent convergence, *Commun. Math. Phys.* **224** (2001), 593–612.

[EN 1] P. Exner, K. Němcová: Quantum mechanics of layers with a finite number of point perturbations, *J. Math. Phys.* **43** (2002), 1152–1184.

[EN 2] P. Exner, K. Němcová: Magnetic layers with periodic point perturbations, *Rep. Math. Phys.* **52** (2003), 255–280.

[EPo 1] P. Exner, O. Post: Convergence of spectra of graph-like thin manifolds, *J. Geom. Phys.* **54** (2005), 77–115.

[EPo 2] P. Exner, O. Post: Convergence of resonances on thin branched quantum wave guides, *J. Math. Phys.* **48** (2007), 092104

[EŠ 1] P. Exner, P. Šeba: Bound states in curved quantum waveguides, *J. Math. Phys.* **30** (1989), 2574–2580.

[EŠ 2] P. Exner, P. Šeba: Electrons in semiconductor microstructures: a challenge to operator theorists, in the proceedings volume ⟦EŠ 2⟧, pp. 79–100.

[EŠ 3] P. Exner, P. Šeba: Quantum motion in two planes connected at one point, *Lett. Math. Phys.* **12** (1986), 193–198.

[EŠ 4] P. Exner, P. Šeba: Quantum motion on a halfline connected to a plane, *J. Math. Phys.* **28** (1987), 386–391, 2254.

[EŠ 5] P. Exner, P. Šeba: Schrödinger operators on unusual manifolds, in the proceedings volume ⟦AFHL⟧, pp. 227–253.

[EŠ 6] P. Exner, P. Šeba: Free quantum motion on a branching graph, *Rep. Math. Phys.* **28** (1989), 7–26.

[EŠ 7] P. Exner, P. Šeba: Trapping modes in a curved electromagnetic waveguide with perfectly conducting walls, *Phys. Lett.* **A144** (1990), 347–350.

[EŠ 8] P. Exner, P. Šeba: A "hybrid plane" with spin-orbit interaction, *Russ. J. Math. Phys.* **14** (2007), 401–405.

[EŠ 9] P. Exner, P. Šeba: Resonance statistics in a microwave cavity with a thin antenna, *Phys. Lett.* **A228** (1997), 146–150.

[EŠŠ 1] P. Exner, P. Šeba, P. Šťovíček: On existence of a bound state in an L-shaped waveguide, *Czech. J. Phys.* **B39** (1989), 1181–1191.

[EŠŠ 2] P. Exner, P. Šeba, P. Šťovíček: Semiconductor edges can bind electrons, *Phys. Lett.* **A150** (1990), 179–182.

[EŠŠ 3] P. Exner, P. Šeba, P. Šťovíček: Quantum interference on graphs controlled by an external electric field, *J. Phys.* **A21** (1988), 4009–4019.

[EŠTV 1] P. Exner, P. Šeba, M. Tater, D. Vaněk: Bound states and scattering in quantum waveguides coupled laterally through a boundary window, *J. Math. Phys.* **37** (1996), 4867–4887.

[EŠe 1] P. Exner, E. Šerešová: Appendix resonances on a simple graph, *J. Phys.* **A27** (1994), 8269–8278.

[ETV 1] P. Exner, M. Tater, D. Vaněk: A single-mode quantum transport in serial-structure geometric scatterers, *J. Math. Phys.* **42** (2001), 4050–4078.

[ETu 1] P. Exner, O. Turek: Approximations of singular vertex couplings in quantum graphs, *Rev. Math. Phys.* **19** (2007), 571–606.

[EV 1] P. Exner, S.A. Vugalter: Bounds states in a locally deformed waveguide: the critical case, *Lett. Math. Phys.* **39** (1997), 59–68.

[EV 2] P. Exner, S.A. Vugalter: Asymptotic estimates for bound states in quantum waveguides coupled laterally through a narrow window, *Ann. Inst. H. Poincaré: Phys. Théor.* **65** (1996), 109–123.

[EV 3] P. Exner, S.A. Vugalter: On the number of particles that a curved quantum waveguide can bind, *J. Math. Phys.* **40** (1999), 4630–4638.

[EY 1] P. Exner, K. Yoshitomi: Asymptotics of eigenvalues of the Schrödinger operator with a strong δ-interaction on a loop, *J. Geom. Phys.* **41** (2002), 344–358.

[EZ 1] P. Exner, V.A. Zagrebnov: Bose-Einstein condensation in geometrically deformed tubes, *J. Phys.* **A38** (2005), L463–470.

[Far 1] W.G. Faris: Inequalities and uncertainty principles, *J. Math. Phys.* **19** (1978), 461–466.

[Far 2] W.G. Faris: Product formulas for perturbation of linear operators, *J. Funct. Anal.* **1** (1967), 93–107.

[FMRS 1] J. Feldmann, J. Magnen, V. Rivasseau, R. Senéor: A renormalizable field theory: the massive Gross–Neveu model in two dimensions, *Commun. Math. Phys.* **103** (1986), 67–103.

[Fel 1] J.G.M. Fell: The dual spaces of C^*–algebras, *Trans. Am. Math. Soc.* **94** (1960), 365–403.

[Fey 1] R.P. Feynman: Space–time approach to nonrelativistic quantum mechanics, *Rev. Mod. Phys.* **20** (1948), 367–387.

[FJSS 1] D. Finkelstein, J.M. Jauch, S. Schminovich, D. Speiser: Foundations of quaternionic quantum mechanics, *J. Math. Phys.* **3** (1962), 207–220.

[FJSS 2] D. Finkelstein, J.M. Jauch, S. Schminovich, D. Speiser: Principle of general Q covariance, *J. Math. Phys.* **4** (1963), 788-796.

[Fo 1] V.A. Fock: Konfigurationsraum und zweite Quantelung, *Z. Phys.* **75** (1932), 622–647.

[Fri 1] K.O. Friedrichs: On the perturbation of continuous spectra, *Commun. Appl. Math.* **1** (1948), 361–406.

[FLL 1] J. Fröhlich, E.H. Lieb, M. Loss: Stability of Coulomb systems with magnetic fields, I. The one–electron atom, *Commun. Math. Phys.* **104** (1986), 251–270.

[FT 1] T. Fülöp, I. Tsutsui: A free particle on a circle with point interaction, *Phys. Lett.* **A264** (2000), 366–374.

[Gam 1] G. Gamow: Zur Quantetheorie des Atomkernes, *Z. Phys.* **51** (1928), 204–212.

[GK 1] K. Gawedzki, A. Kupiainen: Gross–Neveu model through convergent expansions, *Commun. Math. Phys.* **102** (1985), 1–30.

[GN 1] I.M. Gel'fand, M.A. Naimark: On embedding of a normed ring to the ring of operators in Hilbert space, *Mat. Sbornik* **12** (1943), 197–213 *(in Russian)*.

[GMR 1] C. Gerard, A. Martinez, D. Robert: Breit–Wigner formulas for the scattering phase and the total cross section in the semiclassical limit, *Commun. Math. Phys.* **121** (1989), 323–336.

[GP 1] N.I. Gerasimenko, B.S. Pavlov: Scattering problem on non–compact graphs, *Teor. mat. fiz.* **74** (1988), 345–359 *(in Russian)*.

[GGT 1] F. Gesztesy, H. Grosse, B. Thaller: Efficient method for calculating relativistic corrections to spin 1/2 particles, *Phys. Rev. Lett.* **50** (1983), 625–628.

[GGT 2] F. Gesztesy, H. Grosse, B. Thaller: First order relativistic corrections and the spectral concentration, *Adv. Appl. Math.* **6** (1985), 159–176.

[GGH 1] F. Gesztesy, D. Gurarie, H. Holden, M. Klaus, L. Sadun, B. Simon, P. Vogel: Trapping and cascading in the large coupling limit, *Commun. Math. Phys.* **118** (1988), 597–634.

[GŠ 1] F. Gesztesy, P. Šeba: New analytically solvable models of relativistic point interactions, *Lett. Math. Phys.* **13** (1987), 213–225.

[GST 1] F. Gesztesy, B. Simon, B. Thaller: On the self–adjointness of Dirac operator with anomalous magnetic moment, *Proc. Am. Math. Soc.* **94** (1985), 115–118.

[GGM 1] V. Glaser, H. Grosse, A. Martin: Bounds on the number of eigenvalues of the Schrödinger operator, *Commun. Math. Phys.* **59** (1978), 197–212.

[GMGT 1] V. Glaser, A. Martin, H. Grosse, W. Thirring: A family of optimal conditions for the absence of bound states in a potential, in ⟦LSW⟧, pp. 169–194.

[Gla 1] R.J. Glauber: Photon correlations, *Phys. Rev. Lett.* **10** (1963), 84–86.

[Gla 2] R.J. Glauber: Coherent and incoherent states of the radiation fields, *Phys. Rev.* **131** (1963), 2766–2788.

[Gle 1] A.M. Gleason: Measures on the closed subspaces of a Hilbert space, *J. Math. Mech.* **6** (1957), 91–110.

[GJ 1] J. Glimm, A. Jaffe: Boson quantum field models, in ⟦Str⟧, pp. 77–143.

[GoJ 1] J. Goldstone, R.L. Jaffe: Bound states in twisting tubes, *Phys. Rev.* **B45** (1992), 14100–14107.

[GWW 1] C. Gordon, D.L. Webb, S. Wolpert: One cannot hear the shape of a drum, *Bull. Am. Math. Soc.* **27** (1992), 134–138.

[Gra 1] G.M. Graf: Asymptotic completeness for N–body short range quantum systems: a new proof, *Commun. Math. Phys.* **132** (1990), 73–101.

[GG 1] S. Graffi, V. Grecchi: Resonances in Stark effect of atomic systems, *Commun. Math. Phys.* **79** (1981), 91–110.

[GLL 1] M. Griesemer, E.H. Lieb, M. Loss: Ground states in non-relativistic quantum electrodynamics, *Invent. Math.* **145** (2001), 557–595.

[Gri 1] D. Grieser: Spectra of graph neighborhoods and scattering, `arXiv:0710.3405v1 [math.SP]`

[GMa 1] D. Gromoll, W. Meyer, On complete open manifolds of positive curvature, *Ann. Math.* **90** (1969), 75–90.

[Gro 1] C. Grosche: Coulomb potentials by path integration, *Fortschr. Phys.* **40** (1992), 695–737.

[Gr 1] H. Grosse: On the level order for Dirac operators, *Phys. Lett.* **B197** (1987), 413–417.

[Gud 1] S.P. Gudder: The Hilbert space axiom in quantum mechanics, in *Old and New Questions in Physics, Cosmology, Philosophy and Theoretical Biology. Essays in Honor of W. Yourgrau*, Plenum Press, New York 1983; pp. 109–127.

[Haa 1] R. Haag: Quantum field theory, in the proceedings volume ⟦Str⟧, pp. 1–16.

[Haa 2] R. Haag: On quantum field theories, *Danske Vid. Selsk. Mat.-Fys. Medd.* **29** (1955), No. 12.

[Haa 3] R. Haag: Local relativistic quantum physics, *Physica* **A124** (1984), 357–364.

[HK 1] R. Haag, D. Kastler: An algebraic approach to quantum field theory, *J. Math. Phys.* **5** (1964), 848–861.

[Ha 1] M. Hack: On the convergence to the Møller wave operators, *Nuovo Cimento* **9** (1958), 731–733.

[Hag 1] G. Hagedorn: Semiclassical quantum mechanics I–IV, *Commun. Math. Phys.* **71** (1980), 77–93; *Ann. Phys.* **135** (1981), 58–70; *Ann. Inst. H. Poincaré* **A42** (1985), 363–374.

[Hag 2] G. Hagedorn: Adiabatic expansions near adiabatic crossings, *Ann. Phys.* **196** (1989), 278–295.

[HJ 1] G. Hagedorn, A. Joye: Molecular propagation through small avoided crossings of electron energy levels, *Rev. Math. Phys.* **11** (1999), 41–101.

[HJ 2] G. Hagedorn, A. Joye: Time development of exponentially small non-adiabatic transitions, *Commun. Math. Phys.* **250** (2004), 393–413.

[HLS 1] G. Hagedorn, M. Loss, J. Slawny: Non–stochasticity of time–dependent quadratic Hamiltonians and the spectra of transformation, *J. Phys.* **A19** (1986), 521–531.

[HS 1] Ch. Hainzl, R. Seiringer: Mass renormalization and energy level shift in non-relativistic QED, *Adv. Theor. Math. Phys.* **6** (2002), 847–871.

[Har 1] G.H. Hardy: A note on a Theorem by Hilbert, *Math. Z.* **6** (1920), 314–317.

[Harm 1] M. Harmer: Hermitian symplectic geometry and extension theory, *J. Phys.* **A33** (2000), 9193–9203.

[Harp 1] P.G. Harper: Single band motion of conduction electrons in a uniform magnetic field, *Proc. Roy. Soc. London* **A68** (1955), 874–878.

[HE 1] M. Havlíček, P. Exner: Note on the description of an unstable system, *Czech. J. Phys.* **B19** (1973), 594–600.

[HSS 1] R. Hempel, L.A. Seco, B. Simon: The essential spectrum of Neumann Laplacians on some bounded singular domains, *J. Funct. Anal.* **102** (1991), 448–483.

[Hep 1] K. Hepp: The classical limit for quantum correlation function, *Commun. Math. Phys.* **35** (1974), 265–277.

[Her 1] I.W. Herbst: Dilation analycity in constant electric field, I. The two–body problem, *Commun. Math. Phys.* **64** (1979), 279–298.

[HH 1] I.W. Herbst, J.S. Howland: The Stark ladder resonances and other one–dimensional external field problems, *Commun. Math. Phys.* **80** (1981), 23–42.

[Herb 1] F. Herbut: Characterization of compatibility, comparability and orthogonality of quantum propositions in terms of chains of filters, *J. Phys.* **A18** (1985), 2901–2907.

[HU 1] J. Hilgevoord, J.B.M. Uffink: Overall width, mean peak width and uncertainty principle, *Phys. Lett.* **95A** (1983), 474–476.

[Hil 1] R.N. Hill: Proof that the H^- ion has only one bound state, *Phys. Rev. Lett.* **38** (1977), 643–646.

[HSp 1] F. Hiroshima, H. Spohn: Enhanced binding through coupling to a quantum field, *Ann. Henri Poincaré* **2** (2001), 1159–1187.

[Hof 1] G. Hofmann: On the existence of quantum fields in space–time dimension 4, *Rep. Math. Phys.* **18** (1980), 231–242.

[Hof 1] R. Hofstadter: Energy levels and wave functions of Bloch electrons in rational and irrational magnetic fields, *Phys. Rev.* **B14** (1976), 2239–2249.

[Hö 1] G. Höhler: Über die Exponentialnäherung beim Teilchenzerfall, *Z. Phys.* **152** (1958), 546–565.

[HV 1] S.S. Horužii, A.V. Voronin: Field algebras do not leave field domains invariant, *Commun. Math. Phys.* **102** (1986), 687–692.

[HB 1] L.P. Horwitz, L.C. Biedenharn: Quaternion quantum mechanics: second quantization and gauge fields, *Ann. Phys.* **157** (1984), 432–488.

[HLM 1] L.P. Horwitz, J.A. LaVita, J.–P. Marchand: The inverse decay problem, *J. Math. Phys.* **12** (1971), 2537–2543.

[HL 1] L.P. Horwitz, J. Levitan: A soluble model for time dependent perturbation of an unstable quantum system, *Phys. Lett.* **A153** (1991), 413–419.

[HM 1] L.P. Horwitz. J.–P. Marchand: The decay–scattering system, *Rocky Mts. J. Math.* **1** (1971), 225–253.

[vH 1] L. van Hove: Sur le probleme des relations entre les transformations unitaires de la Mécanique quantique et les transformations canoniques de la Mécanique classique, *Bull. Acad. Roy. de Belgique, Classe des Sciences* **37** (1951), 610–620.

[How 1] J.S. Howland: Perturbations of embedded eigenvalues by operators of finite rank, *J. Math. Anal. Appl.* **23** (1968), 575–584.

[How 2] J.S. Howland: Spectral concentration and virtual poles I, II, *Am. J. Math.* **91** (1969), 1106–1126; *Trans. Am. Math. Soc.* **162** (1971), 141–156.

[How 3] J.S. Howland: Puiseaux series for resonances at an embedded eigenvalue, *Pacific J. Math.* **55** (1974), 157–176.

[How 4] J.S. Howland: The Livsic matrix in perturbation theory, *J. Math. Anal. Appl.* **50** (1975), 415–437.

[How 5] J.S. Howland: Stationary theory for time–dependent Hamiltonians, *Math. Ann.* **207** (1974), 315–333.

[How 6] J.S. Howland: Floquet operator with singular spectrum I, II, *Ann. Inst. H. Poincaré: Phys.Théor.* **49** (1989), 309–323, 325–335.

[How 7] J.S. Howland: Stability of quantum oscillators, *J. Phys.* **A25** (1992), 5177–81.

[HLT 1] D. Hundertmark, E.H. Lieb, L.E. Thomas: A sharp bound for an eigenvalue moment of the one-dimensional Schrödinger operator, *Adv. Theor. Math. Phys.* **2** (1998), 719–731.

[HLW 1] D. Hundertmark, A. Laptev, T. Weidl: New bounds on the Lieb-Thirring constants, *Invent. Math.* **140** (2000), 693–704.

[Hun 1] W. Hunziker: Resonances, metastable states and exponential decay laws in perturbation theory, *Commun. Math. Phys.* **132** (1990), 177–188.

[Ich 1] T. Ichinose: Path integral for a hyperbolic system of the first order, *Duke Math. J.* **51** (1984), 1–36.

[IK 1] T. Ikebe, T. Kato: Uniqueness of the self–adjoint extension of singular elliptic differential operators, *Arch. Rat. Mech. Anal.* **9** (1962), 77–92.

[Ito 1] K. Ito: Wiener integral and Feynman integral, in *Proceedings of the 4th Berkeley Symposium on Mathematical Statistics and Probability*, vol. 2, University of California Press, Berkeley, CA 1961; pp. 227–238.

[Ito 2] K. Ito: Generalized uniform complex measures in the Hilbertian metric space with their applications to the Feynman integral, in *Proceedings of the 5th Berkeley Symposium on Mathematical Statistics and Probability*, vol.2/1, University of California Press, Berkeley, CA 1967; pp. 145–167.

[IS 1] P.A. Ivert, T. Sjödin: On the impossibility of a finite proposition lattice for quantum mechanics, *Helv. Phys. Acta* **51** (1978), 635–636.

[Ja 1,2] J.M. Jauch: Theory of the scattering operator I, II, *Helv. Phys. Acta* **31** (1958), 127–158, 661–684.

[JP 1] J.M. Jauch, C. Piron: Can hidden variables be excluded in quantum mechanics?, *Helv. Phys. Acta* **36** (1963), 827–837.

[JL 1] H. Jauslin, J.L. Lebowitz: Spectral and stability aspects of quantum chaos, *Chaos* **1** (1991), 114–121.

[Jo 1] G.W. Johnson: The equivalence of two approaches to the Feynman integral, *J. Math. Phys.* **23** (1982), 2090–2096.

[Jo 2] G.W. Johnson: Feynman's paper revisited, *Suppl. Rend. Circ. Mat. Palermo*, Ser.II, **17** (1987), 249–270.

[JS 1] G.W. Johnson, D.L. Skoug: A Banach algebra of Feynman integrable functions with applications to an integral equation which is formally equivalent to Schrödinger equation, *J. Funct. Anal.* **12** (1973), 129–152.

[JNW 1] P. Jordan, J. von Neumann, E. Wigner: On an algebraic generalization of the quantum mechanical formalism, *Ann. Math.* **35** (1934), 29–64.

[JoP 1] R. Jost, A. Pais: On the scattering of a particle by a static potential, *Phys. Rev.* **82** (1951), 840–850.

[Joy 1] A. Joye: Proof of the Landau–Zener formula, *Asympt. Anal.* **9** (1994), 209–258.

[JoyP 1] A. Joye, Ch.–Ed. Pfister: Exponentially small adiabatic invariant for the Schrödinger equation, *Commun. Math. Phys.* **140** (1991), 15–41.

[JoyP 2] A. Joye, Ch.–Ed. Pfister: Superadiabatic evolution and adiabatic transition probability between two non–degenerate levels isolated in the spectrum, *J. Math. Phys.* **34** (1993), 454–479.

[Kac 1] M. Kac: On distributions of certain Wiener functionals, *Trans. Am. Math. Soc.* **65** (1949), 1–13.

[Kad 1] R.V. Kadison: Normal states and unitary equivalence of von Neumann algebras, in the proceedings volume ⟦Kas⟧, pp. 1–18.

[Kar 1] B. Karnarski: Generalized Dirac operators with several singularities, *J. Operator Theory* **13** (1985), 171–188.

[Kas 1] D. Kastler: The C^*–algebra of a free boson field, I. Discussion of basic facts, *Commun. Math. Phys.* **1** (1965), 14–48.

[Ka 1] T. Kato: Integration of the equations of evolution in a Banach space, *J. Math. Soc. Jpn* **5** (1953), 208–234.

[Ka 2] T. Kato: Wave operators and similarity for some non–selfadjoint operators, *Math. Ann.* **162** (1966), 258–279.

[Ka 3] T. Kato: Fundamental properties of Hamiltonian of the Schrödinger type, *Trans. Am. Math. Soc.* **70** (1951), 195–211.

[Ka 4] T. Kato: On the existence of solutions of the helium wave equations, *Trans. Am. Math. Soc.* **70** (1951), 212–218.

[Ka 5] T. Kato: Growth properties of solutions of the reduced wave equation with variable coefficients, *Commun. Pure Appl. Math.* **12** (1959), 403–425.

[Ka 6] T. Kato: Perturbations of continuous spectra by trace class operators, *Proc. Jpn. Acad.* **33** (1057), 260–264.

[Ka 7] T. Kato: Positive commutators $i[f(P), g(Q)]$, *J. Funct. Anal.* **96** (1991), 117–129.

[KY 1] T. Kato, K. Yajima: Dirac equations with moving nuclei, *Ann. Inst. H. Poincaré: Phys. Théor.* **54** (1991), 209–221.

[KL 1] J.C. Khandekar, S.V. Lawande: Feynman path integrals: some exact results and applications, *Phys. Rep.* **137** (1986), 115–229.

[Kis 1] A. Kiselev: Some examples in one–dimensional "geometric" scattering on manifolds, *J. Math. Anal. Appl.* **212** (1997), 263–280.

[Kit 1] H. Kitada: Asymptotic completeness of N–body operators, I. Short–range systems, *Rep. Math. Phys.* **3** (1991), 101–124.

[Kla 1] J.R. Klauder: Continuous–representation theory I, II, *J. Math. Phys.* **4** (1963), 1055–1058, 1058–1073.

[Kla 2] J.R. Klauder: The action option and Feynman quantization of spinor fields in terms of ordinary C-numbers, *Ann. Phys.* **11** (1960), 123–164.

[KD 1] J.R. Klauder, I. Daubechies: Quantum mechanical path integrals with Wiener measure for all polynomial Hamiltonians, *Phys. Rev. Lett.* **52** (1984), 1161–1164.

[Kl 1] M. Klaus: On the bound states of Schrödinger operators in one dimension, *Ann. Phys.* **108** (1977), 288–300.

[Kl 2] M. Klaus: On the point spectrum of Dirac operators, *Helv. Phys. Acta* **53** (1980), 453–462.

[Kl 3] M. Klaus: Dirac operators with several Coulomb singularities, *Helv. Phys. Acta* **53** (1980), 463–482.

[KlS 1] M. Klein, E. Schwarz: An elementary proof to formal WKB expansions in $\mathbb{R}^n$, *Rep. Math. Phys.* **2** (1990), 441–456.

[KoS 1] V. Kostrykin, R. Schrader: Kirhhoff's rule for quantum wires, *J. Phys.* **A32** (1999), 595–630.

[KoS 2] V. Kostrykin, R. Schrader: Kirhhoff's rule for quantum wires. II: The inverse problem with possible applications to quantum computers, *Fortschr. Phys.* **48** (2000), 703–716.

[KoS 3] V. Kostrykin, R. Schrader: The generalized star product and the factorization of scattering matrices on graphs, *J. Math. Phys.* **42** (2001), 1563–1598.

[KoS 4] V. Kostrykin, R. Schrader: Quantum wires with magnetic fluxes, *Commun. Math. Phys.* **237** (2003), 161–179.

[Kuch 1] P. Kuchment: Quantum graphs, I. Some basic structures, *Waves Rand. Media* **14** (2004), S107–128.

[Kuch 2] P. Kuchment: Quantum graphs, II. Some spectral properties of quantum and combinatorial graphs, *J. Phys.* **A38** (2005), 4887–4900.

[KuZ 1] P. Kuchment, H. Zeng: Convergence of spectra of mesoscopic systems collapsing onto a grap, *J. Math. Anal. Appl.* **258** (2001), 671–700.

[KuS 1] J. Kupsch, W. Sandhas: Møller operators for scattering on singular potentials, *Commun. Math. Phys.* **2** (1966), 147–154.

[Kur 1,2] S. Kuroda: Perturbations of continuous spectra by unbounded operators I, II, *J. Math. Soc. Jpn* **11** (1959), 247–262; **12** (1960), 243–257.

[LF 1] O.A. Ladyženskaya, L.D. Faddeev: On perturbations of the continuous spectrum, *Doklady Acad. Sci. USSR* **120** (1958), 1187–1190 *(in Russian)*.

[Lan 1] C. Lance: Tensor products of C^*–algebras, in ⟦Kas⟧, pp. 154–166.

[La 1] R. Landauer: Electrical resistance of disordered one-dimensional lattices, *Phil. Mag.* **21** (2000), 863–867.

[Lap 1] M.L. Lapidus: The Feynman–Kac formula with a Lebesgue–Stieltjes measure and Feynman operational calculus, *Studies Appl. Math.* **76** (1987), 93–132.

[LW 1] A. Laptev, T. Weidl: Hardy inequalities for magnetic Dirichlet forms, *Oper. Theory Adv. Appl.* **108** (1999), 299–305.

[LW 2] A. Laptev, T. Weidl: Sharp Lieb-Thirring inequalities in high dimensions, *Acta Math.* **184** (2000), 87–111.

[Las 1] G. Lassner: Topological algebras of operators, *Rep. Math. Phys.* **3** (1972), 279–293.

[Las 2] G. Lassner: Topologien auf Op^*–Algebren, *Wiss. Z. KMU Leipzig, Math.-Naturwiss.* **24** (1975), 465–471.

[Las 3] G. Lassner: Algebras of unbounded operators and quantum dynamics, *Physica* **A124** (1984), 471–480.

[LT 1] G. Lassner, W. Timmermann: Normal states on algebras of unbounded operators, *Rep. Math. Phys.* **3** (1972), 295–305.

[LT 2] G. Lassner, W. Timmermann: Classification of domains of operator algebras, *Rep. Math. Phys.* **9** (1976), 205–217.

[Lee 1] T.D. Lee: Some special examples in renormalizable field theory, *Phys. Rev.* **95** (1954), 1329–1334.

[LLMRSY 1] F. Lenz, J.T. Londergan, R.J. Moniz, R. Rosenfelder, M. Stingl, K. Yazaki: Quark confinement and hadronic interactions, *Ann. Phys.* **170** (1986), 65–254.

[LeL 1] J.–M. Lévy–Leblond: Galilei group and Galilean invariance, in the proceedings volume ⟦Loe 2⟧, pp. 221–299.

[LeL 2] J.–M. Lévy–Leblond: Galilean quantum field theories and a ghostless Lee model, *Commun. Math. Phys.* **4** (1967), 157–176.

[Lie 1] E.H. Lieb: The classical limit of quantum spin systems, *Commun. Math. Phys.* **31** (1973), 327–340.

[Lie 2] E.H. Lieb: Bounds on the eigenvalues of the Laplace and Schrödinger operators, *Bull. Am. Math. Soc.* **82** (1976), 751–753.

[Lie 3] E.H. Lieb: The number of bound states of one–body Schrödinger operators and the Weyl problem, *Proc. Symp. Pure Math.* **36** (1980), 241–252.

[Lie 4] E.H. Lieb: A bound on the maximal ionization of atoms and molecules, *Phys. Rev.* **A29** (1984), 3018–3028.

[Lie 5] E.H. Lieb: The stability of matter, *Rev. Mod. Phys.* **48** (1976), 553–569.

[Lie 6] E.H. Lieb: The stability of matter: from atoms to stars, *Bull. Am. Math. Soc.* **22** (1990), 1–49.

[Lie 7] E.H. Lieb: A bound on maximum ionization of atoms and molecules, *Phys. Rev.* **A29** (1984), 3018–3028.

[LL 1] E.H. Lieb, M. Loos: Stability of Coulomb systems with magnetic fields, II. The many–electron atom and the one–electron molecule, *Commun. Math. Phys.* **104** (1986), 271–282.

[LiT 1] E.H. Lieb, W. Thirring: Inequalities for the momenta of the eigenvalues of the Schrödinger Hamiltonian and their relations to Sobolev inequalities, in the proceedings volume ⟦LSW⟧, pp. 269–304.

[LSST 1] E.H. Lieb, I.M. Sigal, B. Simon, W. Thirring: Asymptotic neutrality of large Z ions, *Phys. Rev. Lett.* **52** (1984), 994–996.

[LY 1] E.H. Lieb, H.–T. Yau: The stability and instability of the relativistic matter, *Commun. Math. Phys.* **118** (1988), 177–213.

[LöT 1] F. Löffler, W. Timmermann: The Calkin representation for a certain class of algebras of unbounded operators, *Rev. Roum. Math. Pures Appl.* **31** (1986), 891–903.

[Lü 1] G. Lükő: On the mean lengths of the chords of a closed curve, *Israel J. Math.* **4** (1966), 23–32.

[Mar 1] Ph. Martin: Time delay in quantum scattering processes, *Acta Phys. Austriaca Suppl.* **XXIII** (1981), 157–208.

[Mau 1] K. Maurin: Elementare Bemerkungen über komutative C^*–Algebren. Beweis einer Vermutung von Dirac, *Stud. Math.* **16** (1957), 74–79.

[MS 1] B. Misra, K.B. Sinha: A remark on the rate of regeneration in decay processes, *Helv. Phys. Acta* **50** (1977), 99–104.

[MŠ 1] B. Milek, P. Šeba: Quantum instability in the kicked rotator with rank–one perturbation, *Phys. Lett.* **A151** (1990), 289–294.

[Mol 1] A.M. Molčanov: On conditions of the spectrum discreteness of self–adjoint second–order differential equations, *Trudy Mosk. mat. obščestva* **2** (1953), 169–200 *(in Russian)*.

[MV 1] S. Molchanov, B. Vainberg: Scattering solutions in a network of thin fibers: small diameter asymptotics, *Commun. Math. Phys.* **273** (2007), 533–559.

[Nak 1] S. Nakamura: Shape resonances for distortion analytic Schrödinger operators, *Commun. PDE* **14** (1989), 1385–1419.

[Nar 1] F. Nardini: Exponential decay for the eigenfunctions of the two–body relativistic Hamiltonian, *J. d'Analyse Math.* **47** (1986), 87–109.

[Nas 1] A.H. Nasr: The commutant of a multiplicative operator, *J. Math. Phys.* **23** (1982), 2268–2270.

[Ned 1] L. Nedelec: Sur les résonances de l'opérateur de Dirichlet dans un tube, *Comm. PDE* **22** (1997), 143–163.

[Nel 1] E. Nelson: Analytic vectors, *Ann. Math.* **70** (1959), 572–614.

[Nel 2] E. Nelson: Feynman integrals and the Schrödinger equation, *J. Math. Phys.* **5** (1964), 332–343.

[Nel 3] E. Nelson: Construction of quantum fields from Markoff fields, *J. Funct. Anal.* **12** (1973), 97–112.

[Nen 1] G. Nenciu: Adiabatic theorem and spectral concentration I, *Commun. Math. Phys.* **82** (1981), 121–135.

[Nen 2] G. Nenciu: Linear adiabatic theory. Exponential estimates, *Commun. Math. Phys.* **152** (1993), 479–496.

[Nen 3] G. Nenciu: Distinguished self–adjoint extension for Dirac operators dominated by multicenter Coulomb potentials, *Helv. Phys. Acta* **50** (1977), 1–3.

[vN 1] J. von Neumann: Mathematische Begründung der Quantenmechanik, *Nachr. Gessel. Wiss. Göttingen, Math. Phys.* (1927), 1–57.

[vN 2] J. von Neumann: Allgemeine Eigenwerttheorie Hermitescher Funktionaloperatoren, *Math. Ann.* **102** (1930), 49–131.

[vN 3] J. von Neumann: On infinite direct products, *Compos. Math.* **6** (1938), 1–77.

[New 1] R.G. Newton: Bounds on the number of bound states for the Schrödinger equations in one and two dimensions, *J. Operator Theory* **10** (1983), 119–125.

[NSG 1] M.M. Nieto, L.M. Simmons, V.P. Gutschik: Coherent states for general potentials I–VI, *Phys. Rev.* **D20** (1979), 1321–1331, 1332–1341, 1342–1350; **D22** (1980), 391–402, 403–418; **D23** (1981), 927–933.

[Nö 1] J.U. Nöckel: Resonances in quantum-dot transport, *Phys. Rev.* **B46** (1992), 15348–15356.

[Ost 1] K. Osterwalder: Constructive quantum field theory: goals, methods, results, *Helv. Phys. Acta* **59** (1986), 220–228.

[OS 1,2] K. Osterwalder, R. Schrader: Axioms for Euclidean Green's functions, *Commun. Math. Phys.* **31** (1973), 83–112; **42** (1975), 281–305.

[Pav 1] B.S. Pavlov: A model of a zero–range potential with an internal structure, *Teor. mat. fiz.* **59** (1984), 345–353 *(in Russian).*

[Pea 1] D.B. Pearson: An example in potential scattering illustrating the breakdown of asymptotic completeness, *Commun. Math. Phys.* **40** (1975), 125–146.

[Pea 2] D.B. Pearson: A generalization of Birman's trace theorem, *J. Funct. Anal.* **28** (1978), 182–186.

[Per 1] A.M. Perelomov: Coherent states for arbitrary Lie group, *Commun. Math. Phys.* **26** (1972), 222–236.

[Pe 1] A. Peres: Separability criterion for density matrices, *Phys. Rev. Lett.* **77** (1996), 1413–1415.

[Pir 1] C. Piron: Axiomatique quantique, *Helv. Phys. Acta* **37** (1964), 439–468.

[Pop 1] I.Yu. Popov: The extension and the opening in semitransparent surface, *J. Math. Phys.* **33** (1982), 1685–1689.

[Posi 1] A. Posilicano: Boundary triples and Weyl functions for singular perturbations of self-adjoint operator, *Meth. Funct. Anal. Topol.* **10** (2004), 57–63.

[Pos 1] O. Post: Branched quantum wave guides with Dirichlet boundary conditions: the decoupling case, *J. Phys.* **A38** (2005), 4917–4931.

[Pos 2] O. Post: Spectral convergence of non-compact quasi-one-dimensional spaces, *Ann. H. Poincaré* **7** (2006), 933–973.

[Pow 1] R. Powers: Self–adjoint algebras of unbounded operators, *Commun. Math. Phys.* **21** (1971), 85–124.

[Pri 1] J.F. Price: Inequalities and local uncertainty principles, *J. Math. Phys.* **24** (1983), 1711–1714.

[Pri 2] J.F. Price: Position versus momentum, *Phys. Lett.* **105A** (1984), 343–345.

[Pul 1] S. Pulmannová: Uncertainty relations and state spaces, *Ann. Inst. H. Poincaré: Phys. Théor.* **48** (1988), 325–332.

[Rad 1] J.M. Radcliffe: Some properties of coherent spin states, *J. Phys.* **A4** (1971), 314–323.

[Rel 1] F. Rellich: Die zulässigen Randbedingungen bei den singulären Eigenwertproblemen der Mathematischen Physik, *Math. Z.* **49** (1943–44), 702–723.

[RB 1] W. Renger, W. Bulla: Existence of bound states in quantum waveguides under weak conditions, *Lett. Math. Phys.* **35** (1995), 1–12.

[Ro 1] M. Rosenblum: Perturbations of continuous spectra and unitary equivalence, *Pacific J. Math.* **7** (1957), 997–1010.

[Ros 1] G.V. Rozenblium: Discrete spectrum distribution of singular differential operators, *Doklady Acad. Sci. USSR* **202** (1972), 1012–1015.

[RSch 1] J. Rubinstein, M. Schatzmann: Variational problems on multiply connected thin strips, I. Basic estimates and convergence of the Laplacian spectrum, *Arch. Rat. Mech. Anal.* **160** (2001), 271–308.

[RuS 1] K. Ruedenberg, C.W. Scherr: Free–electron network model for conjugated systems, I. Theory, *J. Chem. Phys.* **21** (1953), 1565–1581.

[Rus 1] M.B. Ruskai: Absence of discrete spectrum in highly negative ions I, II, *Commun. Math. Phys.* **82** (1982), 457–469; **85** (1982), 325–327.

[Rus 2] M.B. Ruskai: Limits of excess negative charge of a dynamic diatomic molecule, *Ann. Inst. H. Poincaré: Phys. Théor.* **52** (1990), 397–414.

[Rus 3] M.B. Ruskai: Limits on stability of positive molecular ions, *Lett. Math. Phys.* **18** (1989), 121–132.

[Sai 1] T. Saito: Convergence of the Neumann Laplacian on shrinking domains, *Analysis* **21** (2001), 171–204.

[SMZ 1] C.M. Savage, S. Marksteiner, and P. Zoller. Atomic Waveguides and Cavities from Hollow Optical Fibres. In: *Fundamentals of Quantum Optics III* (Ed.- F. Ehlotzky), Springer, Lecture Notes in Physics, vol. 420, 1993; p. 60

[Schm 1] K. Schmüdgen: On trace representation of linear functionals on unbounded operator algebras, *Commun. Math. Phys.* **63** (1978), 113–130.

[Schm 2] K. Schmüdgen: On topologization of unbounded operator algebras, *Rep. Math. Phys.* **17** (1980), 359–371.

[Schr 1] E. Schrödinger: Der stetige Übergang von der Mikro–zur Makromechanik, *Naturwissenschaften* **14** (1926), 664–666.

[SRW 1] R.L. Schult, D.G. Ravenhall, H.W. Wyld: Quantum bound states in a classically unbounded system of crossed wires, *Phys. Rev.* **B39** (1989), 5476–5479.

[Schw 1] J. Schwinger: On the bound states of a given potential, *Proc. Natl. Acad. Sci. USA* **47** (1961), 122–129.

[Šeb 1] P. Šeba: Wave chaos in singular quantum billiard, *Phys. Rev. Lett.* **64** (1990), 1855–1858.

[Šeb 2] P. Šeba: Complex scaling method for Dirac resonances, *Lett. Math. Phys.* **16** (1988), 51–59.

[Šeb 3] P. Šeba: Some remarks on the δ'–interaction in one dimension, *Rep. Math. Phys.* **24** (1986), 111–120.

[Šeb 4] P. Šeba: The generalized point interaction in one dimension, *Czech. J. Phys.* **B36** (1986), 667–673.

[SSS 1] L.A. Seco, I.M. Sigal, J.P. Solovej: Bounds on a ionization energy of large atoms, *Commun. Math. Phys.* **131** (1990), 307–315.

[Seg 1] I.E. Segal: Irreducible representations of operator algebras, *Bull. Am. Math. Soc.* **53** (1947), 73–88.

[Seg 2] I.E. Segal: Postulates for general quantum mechanics, *Ann. Math.* **48** (1947), 930–948.

[Seg 3] I.E. Segal: Mathematical characterization of the physical vacuum for a linear Bose–Einstein field (Foundations of the dynamics of infinite systems III), *Illinois Math. J.* **6** (1962), 500–523.

[Seg 4] I.E. Segal: Tensor algebras over Hilbert spaces I, *Trans. Am. Math. Soc.* **81** (1956), 106–134.

[Set 1] N. Seto: Bargmann inequalities in spaces of arbitrary dimension, *Publ. RIMS* **9** (1974), 429–461.

[Sha 1] J. Shabani: Finitely many δ–interactions with supports on concentric spheres, *J. Math. Phys.* **29** (1988), 660–664.

[SMMC 1] T. Shigehara, H. Mizoguchi, T. Mishima, T. Cheon: Realization of a four parameter family of generalized one-dimensional contact interactions by three nearby delta potentials with renormalized strengths, *IEICE Trans. Fund. Elec. Comm. Comp. Sci.* **E82-A** (1999), 1708–1713.

[Shu 1] M.A. Shubin: Discrete magnetic Laplacian, *Commun. Math. Phys.* **164** (1994), 259–275.

[Sig 1] I.M. Sigal: On long–range scattering, *Duke Math. J.* **60** (1990), 473–496.

[SiS 1] I.M. Sigal, A. Soffer: The N–particle scattering problem: asymptotic completeness for short–range potentials, *Ann. Math.* **126** (1987), 35–108.

[SiS 2] I.M. Sigal, A. Soffer: Long–range many–body scattering. Asymptotic clustering for Coulomb–type potentials, *Invent. Math.* **99** (1990), 1155–143.

[Si 1] B. Simon: Topics in functional analysis, in the proceedings ⟦Str⟧, pp. 17–76.

[Si 2] B. Simon: Coupling constant analyticity for the anharmonic oscillator, *Ann. Phys.* **58** (1970), 76–136.

[Si 3] B. Simon: Schrödinger semigroups, *Bull. Am. Math. Soc.* **7** (1982), 447–526.

[Si 4] B. Simon: Resonances in N–body quantum systems with dilation analytic potentials and the foundations of time–dependent perturbation theory, *Ann. Math.* **97** (1973), 247–272.

[Si 5] B. Simon: On the number of bound states of two–body Schrödinger operators – a review, in the proceedings volume ⟦LSW⟧, pp. 305–326.

[Si 6] B. Simon: The bound state of weakly coupled Schrödinger operators in one and two dimensions, *Ann. Phys.* **97** (1976), 279–288.

[Si 7] B. Simon: Phase space analysis of some simple scattering systems: extensions of some work of Enss, *Duke Math. J.* **46** (1979), 119–168.

[Si 8] B. Simon: The Neumann Laplacians of a jelly roll, *Proc. AMS* **114** (1992), 783–785.

[SSp 1] B. Simon, T.Spencer: Trace class perturbations and the absence of absolutely continuous spectra, *Commun. Math. Phys.* **125** (1989), 111–125.

[Sin 1] K.B. Sinha: On the decay of an unstable particle, *Helv. Phys. Acta* **45** (1972), 619–628.

[Solo 1] M. Solomyak: On the spectrum of the Laplacian on regular metric trees, *Wave Random Media* **14** (2004), S155–S171.

[Sol 1] J.P. Solovej: Asymptotic neutrality of diatomic molecules, *Commun. Math. Phys.* **130** (1990), 185–204.

[SM 1] F. Sols, M. Macucci: Circular bends in electron waveguides, *Phys. Rev.* **B41** (1990), 11887–11891.

[Ste 1] A. Steane: Quantum computing, *Rep. Progr. Phys.* **61** (1998), 117–173.

[St 1] J. Stubbe: Bounds on the number of bound states for potentials with critical decay at infinity, *J. Math. Phys.* **31** (1990), 1177–1180.

[Stu 1–4] E.C.G. Stueckelberg, M. Guenin, C. Piron, H. Ruegg: Quantum theory in real Hilbert space I–IV, *Helv. Phys. Acta* **33** (1960), 727–752; **34** (1961), 621–628, 675–698; **35** (1962), 673–695.

[Sto 1] M.H. Stone: Linear transformations in Hilbert space, III. Operational methods and group theory, *Proc. Nat. Acad. Sci. USA* **16** (1930), 172–175.

[Sve 1] E.C. Svendsen: The effect of submanifolds upon essential self–adjointness and deficiency indices, *J. Math. Anal. Appl.* **80** (1980), 551–565.

[Te 1] A. Teta: Quadratic forms for singular perturbations of the Laplacian, *Publ. RIMS* **26** (1990), 803–817.

[Til 1] H. Tilgner: Algebraical comparison of classical and quantum polynomial observables, *Int. J. Theor. Phys.* **7** (1973), 67–75.

[Tim 1] W. Timmermann: Simple properties of some ideals of compact operators in algebras of unbounded operators, *Math. Nachr.* **90** (1979), 189–196.

[Tim 2] W. Timmermann: Ideals in algebras of unbounded operators, *Math. Nachr.* **92** (1979), 99–110.

[TBV 1] G. Timp *et al.*: Propagation around a bend in a multichannel electron waveguide, *Phys. Rev. Lett.* **60** (1988), 2081–2084.

[Tro 1] H. Trotter: On the product of semigroups of operators, *Proc. Am. Math. Soc.* **10** (1959), 545–551.

[Tru 1] A. Truman: The classical action in nonrelativistic quantum mechanics, *J. Math. Phys.* **18** (1977), 1499–1509.

[Tru 2] A. Truman: Feynman path integrals and quantum mechanics as $\hbar \to 0$, *J. Math. Phys.* **17** (1976), 1852–1862.

[Tru 3] A. Truman: The Feynman maps and the Wiener integral, *J. Math. Phys.* **19** (1978), 1742–1750; **20** (1979), 1832–1833.

[Tru 4] A. Truman: The polynomial path formulation of the Feynman path integrals, in the proceedings volume ⟦ACH⟧, pp. 73–102.

[UH 1] J.B.M. Uffink, J. Hilgevoord: Uncertainty principle and uncertainty relations, *Found. Phys.* **15** (1985), 925–944.

[UH 2] J.B.M. Uffink, J. Hilgevoord: New bounds for the uncertainty principle, *Phys. Lett.* **105A** (1984), 176–178.

[Vas 1] A.N. Vasiliev: Algebraic aspects of Wightman axiomatics, *Teor. mat. fiz.* **3** (1970), 24–56 *(in Russian).*

[VSH 1] A.V. Voronin, V.N. Suško, S.S. Horužii: The algebra of unbounded operators and vacuum superselection in quantum field theory, 1. Some properties of Op^*–algebras and vector states on them, *Teor. mat. fiz.* **59** (1984), 28–48.

[Wa 1] X.–P. Wang: Resonances of N–body Schrödinger operators with Stark effect, *Ann. Inst. H. Poincaré: Phys. Théor.* **52** (1990), 1–30.

[WWU 1] R.A. Webb *et al.*: The Aharonov–Bohm effect in normal–metal non–ensemble averaged quantum transport, *Physica* **A140** (1986), 175–182.

[We 1] J. Weidmann: The virial theorem and its application to the spectral theory of Schrödinger operators, *Bull. Am. Math. Soc.* **73** (1967), 452–456.

[We 1] R.F. Werner: Quantum states with Einstein–Podolsky–Rosen correlation admitting a hidden–variable model, *Phys. Rev.* **A40** (1989), 4277–4281.

[WWW 1] G.C. Wick, A.S. Wightman, E.P. Wigner: The intrinsic parity of elementary particles, *Phys. Rev.* **88** (1952), 101–105.

[WWW 2] G.C. Wick, A.S. Wightman, E.P. Wigner: Superselection rule for charge, *Phys. Rev.* **D1** (1970), 3267–3269.

[WG 1] A.S. Wightman, L. Gårding: Fields as operator–valued distributions in relativistic quantum theory, *Arkiv för Fysik* **28** (1964), 129–184.

[Wig 1] E.P. Wigner: On unitary representations of the inhomogeneous Lorentz group, *Ann. Math.* **40** (1939), 149–204.

[Wil 1] D.N. Williams: New mathematical proof of the uncertainty relations, *Amer. J. Phys.* **47** (1979), 606–607.

[Wil 2] D.N. Williams: Difficulty with a kinematic concept of unstable particles: the Sz.–Nagy extension and Matthews-Salam-Zwanziger representation, *Commun. Math. Phys.* **21** (1971), 314–333.

[Wol 1] K.B. Wolf: The Heisenberg–Weyl ring in quantum mechanics, in the proceedings volume ⟦Loe 3⟧, pp. 189–247.

[Wo 1] M.F.K. Wong: Exact solutions of the n–dimensional Dirac–Coulomb problem, *J. Math. Phys.* **31** (1991), 1677–1680.

[Ya 1] K. Yajima: Existence of solutions for Schrödinger equations, *Commun. Math. Phys.* **110** (1987), 415–426.

[Ya 2] K. Yajima: Quantum dynamics of time periodic systems, *Physica* **A124** (1984), 613–620.

[YK 1] K. Yajima, H. Kitada: Bound states and scattering states for time periodic Hamiltonians, *Ann. Inst. H. Poincaré: Phys. Théor.* **39** (1983), 145–157.

[Yea 1] F.J. Yeadon: Measures on projections in W^*–algebras of type II_1, *Bull. London Math. Soc.* **15** (1983), 139–145.

[Zas 1] T. Zastawniak: The non–existence of the path measure for the Dirac equation in four space–time dimensions, *J. Math. Phys.* **30** (1989), 1354–1358.

[Ži 1] G.M. Žislin: Discussion of the spectrum of the Schrödinger operator for systems of many particles, *Trudy Mosk. mat. obščestva* **9** (1960), 81–128.

[Zy 1] K. Życzkowski: Classical and quantum billiards, nonintegrable and pseudointegrable, *Acta Phys. Polon.* **B23** (1992), 245–269.

List of symbols

$\dot{+}$	form sum	*116*
$\oplus$	direct, orthogonal sum	
$\sum^{\oplus}$	direct, orthogonal sum	
$\int^{\oplus}$	direct integral	*54, 140*
$\otimes$	tensor product, product measure	
$\breve{\times}$	algebraic tensor product	*56, 108*
$\times$	Cartesian product	*597*
$/$	factorization	*2*
^{o}	interior of a set	*23*
$'$	commutant, dual space	
$''$	bicommutant	*214, 610*
*	adjoint, involution, dual space	
$\hat{}\ \check{}$	Fourier transform	*18*
$^{\perp}$	orthogonal complement	*4*
^{t}	transposition	*614*
$\overline{}$	closure	
$\circ$	composite mapping	*597*
$^{-1}$	inverse	*597*
$^{(-1)}$	pull-back	*596*
$\upharpoonright$	restriction	*596*
$\supset$	extension, inclusion	
$\cap$	intersection	
$\cup$	union	
$\setminus$	set difference	
Δ	symmetric difference, Laplacian	
$\to$	limit, map	
$\mapsto$	map (point to point)	
$\xrightarrow{s}$	strong operator limit	*66*
$\xrightarrow{w}$	weak limit, weak operator limit	*22, 67*
$\Rightarrow$	implication	
$\lvert\cdot\rvert$	modulus, norm in $\mathbb{R}^n$	*3*
$[\cdot]$	integer part	
$\lVert\cdot\rVert$	norm	*3*
$\lVert\cdot\rVert_p$	norm	*3, 88*
$\lVert\cdot\rVert_\infty$	norm	*3, 4*
$(\cdot,\cdot)$	inner product	*4*
$[\cdot,\cdot]$	commutator, boundary form	

Index

MIX
Papier aus verantwortungsvollen Quellen
Paper from responsible sources
FSC® C105338

FSC
www.fsc.org

If you have any concerns about our products,
you can contact us on
ProductSafety@springernature.com

In case Publisher is established outside the EU,
the EU authorized representative is:
Springer Nature Customer Service Center GmbH
Europaplatz 3, 69115 Heidelberg, Germany

Printed by Libri Plureos GmbH
in Hamburg, Germany